MARRIAGE AND THE FAMILY

Marriage and the Family: Mirror of a Diverse Global Society is a comprehensive text about marriage and the family in sociology, family science, and diversity studies. The book is divided into four parts: studying marriage patterns and understanding family diversity; developing and maintaining intimate relationships; tackling family issues and managing household crises; and appreciating contemporary living arrangements in a diverse American society and across the global community. *Marriage and the Family* is unique in its focus on diversity as well as its global perspective. Diversity Overview boxes feature vignettes of family diversity in America. Global Overview boxes invite students to experience family life in different areas of the world. Indeed, families become a mirror that helps students see a diversifying American society and a globalizing world.

Dr. Julie Xuemei Hu is Senior Professor of Sociology at Union County College. She holds a doctorate in Sociology, a master's degree in Sociology, and a bachelor's degree in English and American literature. Her interests are in social stratification, marriage and the family, racial and ethnic relations, social problems, student learning outcome assessment, and distance learning. She is a QM-Certified Peer Reviewer. She has enjoyed teaching Marriage and the Family since 1998. She has published the text *Gender and Work*, co-authored the text *Introduction to Sociology: Collaborative Approach*, and created supplementary teaching materials for this book.

Dr. Shondrah Tarrezz Nash is Professor of Sociology at Morehead State University. She holds a doctorate in Sociology from the University of Kentucky and conducted postdoctoral research at the University of Illinois at Chicago. Her research interests include violence against women and spousal violence and religious coping. Dr. Nash co-authored the textbook *Collective Sociology: An Introduction to Sociology*. She has published in *Violence against Women*, *Journal of Interpersonal Violence*, *Qualitative Sociology*, *Journal of Transformative Education*, and *Counseling and Values*, in addition to book chapters in *Sourcebook on Violence against Women* and *Battleground: The Family*.

MARRIAGE AND THE FAMILY

MIRROR OF A DIVERSE GLOBAL SOCIETY

JULIE XUEMEI HU AND SHONDRAH TARREZZ NASH

NEW YORK AND LONDON

First published 2019
by Routledge
52 Vanderbilt Avenue, New York, NY 10017

and by Routledge
2 Park Square, Milton Park, Abingdon, Oxon, OX14 4RN

Routledge is an imprint of the Taylor & Francis Group, an informa business

Library of Congress Cataloging-in-Publication Data
A catalog record for this book has been requested

ISBN: 978-1-138-18575-3 (hbk)
ISBN: 978-1-138-18576-0 (pbk)
ISBN: 978-1-315-64138-6 (ebk)

Typeset in Amasis
by Apex CoVantage, LLC

To my son, Wen

CONTENTS

PART IV
DYNAMICS IN FAMILY LIVING ARRANGEMENTS

PREFACE

Marriage has been declining since the 1960s. The decrease in marriage and family size and increase in marriage age reflect a change in American culture. Today children are raised by two married parents, stepparents, cohabiting parents, single parents, grandparents, LGBT parents, and others. American families and classrooms are becoming increasingly diverse. Understanding family diversity requires exploring the changing marriage patterns and feeling the pulse of contemporary families. This text aims to keep track of increasingly *diverse families* that mirror a diverse global society.

Marriage and the Family: Mirror of a Diverse Global Society introduces students to a diverse American society and global community through the lens of the family. The text has several unique features. The book presents the material with an eye toward diversity and from a global perspective. The text uses a diversity framework to explore family diversity and family inequality. One person's family experience does not represent that of others in terms of race, gender, class, education, religion, etc. It is equally important to navigate family life and marriage patterns around the world we all live in. The book is deeply diverse and global. Diversity Overview boxes discuss family diversity in America. Global Overview boxes invite students to experience family life in different areas of the world. Indeed, families become a mirror that helps students see a diversifying American society and a globalizing world.

Each chapter opens with chapter learning outcomes and thought-provoking questions. Student accounts of their family lives help students develop a better understanding of the material not only through the eyes of the authors but also through the experiences of their peers.

The book is a comprehensive text about marriage and the family in sociology, family science, and diversity studies. The book has sixteen chapters that are arranged to provide a comprehensive resource for introducing students to the field of marriage and the family, developing intimate relationships, dealing with the challenges of maintaining family relationships, and appreciating contemporary living arrangements. Upon completion of this course, students are expected to achieve the following course learning outcomes:

1. Develop critical thinking skills using sociological concepts and theories to analyze the changing family forms and marriage patterns.
2. Use research methods to conduct research on marriage and the family and incorporate research findings into written and oral presentations.

3. Explore the ways in which factors such as gender, race, social class affect family life.
4. Explain relationship formation processes and factors that affect the development and maintenance of intimate relationships.
5. Evaluate family issues that individuals encounter and possible methods of resolving them throughout the life cycle.
6. Explain strengths and challenges of contemporary living arrangements.

Organization of the Book

The textbook is laid out sequentially and thematically. The sixteen chapters are organized into four parts. Each part includes multiple chapters that correspond to a subject area. **Part I** covers Chapters 1 through 5. The first five chapters begin with an introduction to the field of marriage and the family, research on marriage and the family, and family diversity along the lines of social class, race and ethnicity, and gender in history.

Chapter 1 (The Meaning of the Family and Marriage) examines how family performs functions as a social group and social institution. It explores the definitions of family and forms of family on the basis of structure, kinship, descent, authority, residence, and trends. The chapter also focuses on forms of marriage, a universal social institution that forms a family. From a diversity and global perspective, the chapter discusses single life and marriage patterns across countries.

Chapter 2 (Studying Marriage and the Family) explores macro and micro sociological theories on marriage and the family that guide research. The chapter examines research design, research process, and research methods used to collect data and conduct research on marriage and the family. From a diversity and global perspective, the chapter applies sociological theory and explanatory research to the analysis of ethnic social mobility as well as refugee families.

Chapter 3 (Family Diversity) examines Native American, early European, and slave families in Colonial America. It explores changing marriage patterns and family life from the nineteenth to the twenty-first century onward. It further examines family life along the lines of social class. From a diversity and global perspective, the chapter discusses immigrant families and world families.

Chapter 4 (Race and Ethnicity in Families) explores race and ethnicity as a marker of family diversity. Recognizing race and ethnicity remains a central framework through which to analyze an arena of racial and ethnic families under minority influence. The chapter examines the strengths and challenges of racial and ethnic families. It explores the characteristics of White American, Native American, African American, Hispanic, and Asian American families. From a diversity and global perspective, the chapter discusses interracial families and transnational families.

Chapter 5 (Gender Roles and Socialization) examines how gender roles are socialized within and beyond family relationships. The chapter discusses agents of socialization and different treatment experienced by women, men, girls, and boys. From a diversity and global perspective, the chapter explores the division of household labor in relationships with transgender men and a global depiction of gender in children's textbooks.

Part II covers Chapters 6 through 10. The five chapters introduce students to processes that help develop intimate relationships. The conceptual material provides critical guidance for students who wish to develop a strong foundation for a happy relationship and marriage.

Chapter 6 (Sexuality and Sexual Relationships throughout Life) explores sexual identity, sexual attraction, sexual desire, sexual orientation, sexual behavior, and sexual response. It delves into sexual relationships throughout a life cycle. Sexual health and healthy sexual relationships are also discussed. From a diversity and global perspective, the chapter examines LGBT families and same-sex marriage.

Chapter 7 (Love) examines eight types of love, elements of love, and theoretical explanations of love. It explores the functions of love and the challenges of experiencing love. From a diversity and global perspective, the chapter delves into gender differences in expressions of love and provides a view of love across countries.

Chapter 8 (Courtship, Dating, and Mate Selection) examines two types of courtship systems. It explores traditional and modern dating behaviors as well as contemporary dating services. Problems in dating are explored and strategies for avoiding dating issues are provided. The chapter further discusses rules and theories of mate selection. From a diversity and global perspective, the chapter discusses gender differences in the mate selection process and provides a view of mate selection across countries.

Chapter 9 (Marriage and Marital Relationships) focuses on marriage and marital relationships throughout the life cycle. Marital adjustment and marital satisfaction are contributors to marital quality in young adulthood, parenthood, middle adulthood, and late adulthood. From a diversity and global perspective, the chapter discusses different wedding rituals and collective marriages across countries.

Chapter 10 (Preparing for Children and Child Adoption) examines fertility patterns in the United States and discusses why couples decide to have a child, or not. It examines how couples prevent pregnancy and deal with pregnancy if they have decided not to have children. The chapter explores how couples deal with infertility and prepare for a child when they are pregnant. From a diversity and global perspective, the chapter examines childfree families and provides a view of sex selection across countries.

Part III includes Chapters 11 through 14. The four chapters address the challenges of maintaining family relationships and tackle family issues concerning parenthood, paid work and family work, family issues, and communication.

Chapter 11 (Parenting) focuses on the transition to successful parenthood, parental roles, parenting over the life course, and parenting styles for raising successful children. From a diversity and global perspective, the chapter delves into the parenting experiences of children by race and provides a view of Third-Culture Kids across countries.

Chapter 12 (Balancing Paid Work, Family Work, and School Work) focuses on work as a means of earning income and supporting the family. This chapter also explores the ways through which work affects family life. It explores paid work, and unpaid work, and examines the strategies for balancing paid work, housework, and school work. From a diversity and global perspective, the chapter discusses military families and the future of work.

Chapter 13 (Family Crises, Intimate Partner Violence, and Alcohol and Drug Abuse) explores the multiple types, effects, and risks associated with intimate partner violence, as well as theories and strategies addressing their occurrence and prevention. The chapter further discusses mental illness, suicide, and eating disorders in families. From a diversity and global perspective, the chapter explores the roles of gender and cultural norms in the promotion and prevention of family violence.

Chapter 14 (Communication, Power, Conflict, and Resiliency) explores communication and power in marriage and family relationships. The chapter discusses common couple conflicts and factors of maintaining resilient relationships. From a diversity and global perspective, the chapter discusses power in same-sex committed relationships and global attitudes on gender, marriage, and work.

Part IV covers Chapters 15 and 16. These two chapters explore marital dissolution and new family formation. The chapters discuss contemporary living arrangements in a diverse American society and across the global community.

Chapter 15 (Breaking up, Marital Separation, and Divorce) examines factors associated with the disintegration of dating relationships and marriages. The chapter focuses on marital separation, factors of divorce, the influence of divorce on children and adults, post-divorce adjustment, and the positive effects of divorce. From a diversity and global perspective, the chapter discusses divorce pressures unique to same-sex families and associations between divorce and homicide in the Caribbean.

Chapter 16 (Remarriage and Trends in Living Arrangements) examines remarriage and trends in living arrangements. The chapter delves into a variety of living arrangements, including remarried families, stepfamilies, single-parent families, cohabiting families, and aging families. From a diversity and global perspective, the chapter analyzes retirement inequality and the "Living Apart Together" (LAT) living arrangement.

Dr. Julie Xuemei Hu

ACKNOWLEDGMENTS

I was excited about the idea of writing *Marriage and the Family: Mirror of a Diverse Global Society* with a diverse and global approach, so I wrote a prospectus and created a detailed table of contents. I thank Dr. Victor Shaw for introducing me to Routledge.

At Routledge, Editor Samantha Barbaro strongly encouraged me to submit the prospectus. I deeply appreciate her consistent support and generous help throughout the review and editing processes. She is a role model of professional editors. I thank Assistant Editor Erik Zimmerman for meticulously handling my manuscript pre-production. Equally, I thank former Assistant Editor Athena Bryan for providing invaluable assistance with manuscript preparation. I am grateful to Production Editor Olivia Hatt for skillfully taking my book through the production process to publication.

I would like to thank Dr. Victor Shaw and Dr. Cheryl Boudreaux for reviewing the prospectus, detailed table of contents, and sample chapter that I wrote. I also want to acknowledge the contributions of the reviewers of the prospectus and detailed table of contents. In this respect, I am thankful to the reviewers from Central Piedmont Community College, Fresno Pacific University, Judson University, New Mexico State University, North Virginia Community College, and Pennsylvania State University for their positive comments and suggestions.

I thank Dr. Shondrah Nash for having written Chapters 5, 13, 14, and 15. I would like to acknowledge the students whose accounts of their family lives helped motivate me to write this book.

Writing a book means cutting back on housework. I finally want to thank my husband, Eric, for having doubled time spent doing household chores and encouraging me to devote my time and effort to writing the text *Marriage and the Family: Mirror of a Diverse Global Society.*

Dr. Julie Xuemei Hu

PART I

INTRODUCTION TO MARRIAGE AND THE FAMILY

1
THE MEANING OF THE FAMILY AND MARRIAGE

Learning Outcomes

Upon completion of the chapter, students should be able to:

1. explain how the family is a primary group and a social institution
2. analyze the functions of the family in traditional and modern society
3. describe various definitions of family and household
4. compare nuclear families and extended families
5. explain the formation of conjugal and consanguineous families
6. apply unilineal and bilateral descent to the analysis of traditional family and modern family
7. describe the characteristics of the patriarchal family and the matriarchal family
8. describe forms of family based on residence
9. explain trends in forms of family
10. compare monogamy and polygamy
11. analyze the impact of open marriage and polyamory on the family
12. explore family life from a diversity and global approach.

Brief Chapter Outline

Pre-test

Family as a Mirror of a Diverse Global Society

Family as the Center of Culture in Society

Family as Primary Group; Family as Social Institution

Functions of the Family

Provision of Intimate and Sexual Relationships; Reproduction; Economic Cooperation; Socialization of Children

Pre-test

Engaged or active learning is a powerful strategy that leads to better learning outcomes. One way to become an active learner is to begin the chapter and try to answer the following true/false statements from the material as you read. You will find that you have ready answers to these questions upon the completion of the chapter.

The family is a primary group.	T	F
Family members include the householder and related people by birth, marriage, or adoption.	T	F
Family household members include the householder, family members, and all unrelated people.	T	F
Unilateral descent is divided into patrilineal and matrilineal descent.	T	F
Another term for extended family is *multigenerational household*.	T	F
A family of neolocal residence is a family in which people live in their own home.	T	F
Trends in family forms include stepfamilies, cohabiting families, childfree families, single-parent families, and grandparent families.	T	F
Polygamy is practiced on a traditional and religious basis in most countries.	T	F
Polyamory is a form of open marriage or relationship that is practiced in modern society.	T	F
Family diversity is a way of embracing such dimensions as race, ethnicity, class, gender, sexual orientation, and religion.	T	F

Upon completion of this section, students should be able to:
LO1: Explain how the family is a primary group and a social institution.
LO2: Analyze the functions of the family in traditional and modern society.

Angela was 21 and home on winter break from her third year at the university. After seeing the Broadway show *Cabaret*, Angela and her parents walked slowly toward Rockefeller Center to admire the Christmas tree there, which had become a symbol of that holiday for many people in New York and beyond. Angela and her parents gathered with a sea of other visitors to look up at the tree, to listen to Christmas carols, and to celebrate the spirit of Christmas. Later, as she walked away, Angela felt filled with special warmth from being around her family at Christmastime.

All American families have their own celebrations. For some, like Angela's family, Christmas is an important celebration. Other Americans celebrate Kwanzaa—a weeklong celebration honoring African American culture. Jewish American families celebrate Hanukkah or the Jewish Festival of Lights. Indian American families celebrate the Hindu festival Diwali. And when Muslim Americans celebrate Eid al-Fitr and Eid al-Adha, New York City closes schools for these two holy days. New York City schools also close for the Chinese Lunar New Year. What traditional holidays do you celebrate in your family?

Family as a Mirror of a Diverse Global Society

As these examples show, at the center of any celebration is the family. The activity of traditional celebrations takes place at home. This makes the holiday season a perfect time for families to come together and celebrate. This section explores how families are the center of culture and share a unique heritage. In fact, we can say that families are a mirror that helps us understand our diverse global society.

Family as the Center of Culture in Society

The metaphor of the family as a mirror of society can be traced back to the Greek philosophers Socrates (ca. 470–399 BCE), Plato (ca. 428–348 BCE), and Aristotle (ca. 384–322 BCE). They considered the family a model for the ways in which societies are organized. For example, a *monarchy* is a political system in which power is embodied in one person or family through lines of inheritance and is reflective of one way of organizing society. Monarchal societies mirror the *patriarchal family*, with people obeying the king as children obey their father, which in turn helps maintain social order. Healthy families are a mirror of a healthy and ordered society.

Human societies include hunting and gathering, horticultural and pastoral, agricultural, industrial, and post-industrial societies according to technological and economic development. The United States is a post-industrial society—post-industrialism (Bell, 1973) is based on computer technology that produces information and supports a service industry. **Society** refers to groups of people who share a common culture in a geographic location. This definition can be broken down into three components.

1. First, *individual people* are the basic component of society.
2. When individuals interact and develop relationships with each other, they form the second component: *social groups*. People living and working in social groups share a common culture and maintain their own heritage.
3. *Social institutions* are the third component of society. Social institutions fulfill a society's basic needs. For example, families are one important social institution.

Why do families embrace the culture to which they belong? Culture and society are closely connected. Society cannot exist without culture because culture helps hold society together. **Culture**—a way of life within a society or social group—is central to family life. Culture includes material culture and non-material culture. **Material culture** refers to tangible objects that the people of a society make and use. For example, on her way to Rockefeller Center, Angela bumped into other visitors bundled in warm clothes and enjoyed the street food. She saw vehicles yielding to pedestrians in the crosswalk. The food, clothes, and cars that Angela saw are examples of material culture. Food is a universal item of material culture, but the way people eat—a non-material item—varies widely from one culture to the next. For example, people in Japan and China eat with chopsticks, while people in Western countries eat with forks and knives. **Non-material culture** refers to intangible creations that influence the behavior of the members of a society. Non-material culture includes beliefs, values, norms, symbols, and sanctions. The languages people speak, the holidays they celebrate, traffic rules they obey, and American values (e.g., freedom) they accept are examples of non-material culture.

Subculture is a culture shared by a category of people within a society. This concept has been applied to cultural differences based on gender, age, ethnicity, religion, occupation, social class, etc. If Angela is a 21-year-old pharmacist with an Italian background, she falls within the subcultures of Generation Y, the middle class, Italian Americans, and health-care professionals. Subcultures are not opposed to the principles of the mainstream culture. For example, most American people celebrate Independence Day and believe in the power of the Constitution. At the same time, ethnic families may observe family traditions that help keep them grounded in their subcultures and help children retain ethnic identity. Indeed, the family functions as transmitter of ethnic subculture. For example, research has shown the persistence of ethnic identity among second- and fourth-generation Italian American adults (Alessandria *et al.*, 2016).

Family as Primary Group

Families maintain their cultural heritage within a social group. Formally defined, **social groups** are characterized by more than two people, frequent interaction, a sense of belongingness, and a feeling of interdependence. Sociologists distinguish primary groups from secondary groups. A **primary group** is a small group whose relationships are face to face and personal. The family is an example of a primary group. The primary group is what we generally refer to as family members related by blood, marriage, adoption, and relatives. Families share lasting relationships and support and buffer each other against stress. A **secondary group** is a large group whose relationships are impersonal and goal-oriented. People engage in educational and career pursuits in secondary groups on a daily basis. Social groups such as workplaces and universities are secondary groups.

Family as Social Institution

Sociologists viewed the family as a social institution (Komarovsky & Waller, 1945). As noted earlier, a **social institution** is an organized system of social relationships which embodies values and procedures and meets basic needs of the society (Horton *et al.*, 1983). Basic social institutions include the family, marriage, religion, politics, the economy, and education. Families are related to other social institutions, and reflect the different religious, educational, political, and economic features of a society. For example, a study of four time periods (up to 1500, from 1500 to 1700, from 1700 to 1900, and after 1900) in England reveals that the family and education are structurally interrelated (Musgrave, 1971). Human beings develop different social institutions depending upon the basic needs of the society. These social institutions will be changed to suit the changing requirements of the society. For example, the educational institution of schools underwent a series of changes from elite boarding schools, charity school, to compulsory education. In the management of a diverse society, religious institutions strive harder to create an environment of respect for people of all faiths (Ramakrishnan & Balgopal, 1995).

The family is among the oldest social institutions. To describe the family as a social institution, one should first reveal its universal character because there is no society in which family is missing (Ionuţ, 2012). The sociologists were concerned with the family's forms, functions, and relations to other institutions (Komarovsky & Waller, 1945). Social change has an impact on the forms of family. Today children live in an array of arrangements, including married stepfamilies, cohabiting families, single-parent families, and grandparent families. No matter how families have changed in structure, the basic functions that families perform remain the same.

Functions of the Family

The family has been performing a variety of functions since the existence of human beings. Sociologist MacIver (1931) divides the functions of the family into essential and non-essential functions. *Essential functions* include satisfaction of sex needs, production, sustenance, provision of home, and socialization. *Non-essential functions* include the economic function, religious function, educative function, recreational function, wish fulfillment, property transformation, etc. During periods of social change, some functions that families performed were taken over by other institutions. Despite these structural changes, the main functions of the family have remained essentially unaltered (Ionuţ, 2012). American anthropologist Murdock (1949) has examined 250 societies in different cultures and has classified functions of family into (1) the provision of intimate relationships and regulation of sexual behavior, (2) reproduction, (3) economic cooperation, and (4) the socialization of children.

Provision of Intimate and Sexual Relationships

Modern views of the family emphasize the function of providing stable relationships and setting rules for sexual behavior in society. Families are where people are attached to each other and offer a source of intimate relationship. **Intimate relationships** are interpersonal relationships that involve physical or emotional intimacy. Intimacy plays a central role in marital relationships. When a couple shares an intimate and sexual relationship, they are likely to be healthy and happy, providing stability for a family. Also, the family is a social unit for teaching cultural rules for sexual behavior in society. For example, in many societies, the practice of sexuality is not permitted during periods of mourning, in the periods of various religious ceremonies, while the woman is pregnant, or during menstruation (Ionuţ, 2012). An **incest taboo** is a cultural rule that prohibits sexual relations between closely related persons. In many cultures, certain types of cousin relations are preferred as sexual and marital partners, whereas in others these are taboo. Parent–child and sibling–sibling unions are almost universally taboo (Rosman *et al.*, 2009).

Reproduction

The reproductive function relates to childbearing and childrearing. In the absence of biological reproduction, any society is sentenced to disappearance (Ionuţ, 2012). Situations of this type were found mainly in religious communities, such as the *self-entitled Shakers*—the society of believers in the second appearance of Jesus Christ, in which social equality was essential but sexual relations were rejected under the notion of human dignity, which resulted in a strong decrease in the number of members of this community (DiRenzo,1990). In order to develop, each society needs new generations of young people to replace the old people (Ionuţ, 2012). Indeed, traditional views of the family emphasized the reproductive function of the family. Today marriage rates have declined, the age of first marriage has risen, and family size has decreased. Yet families still provide society with new members.

Economic Cooperation

Another important function of the family is that it provides economic benefits to its members. The family provides its members with necessities such as food. In pre-industrial societies, families acted as productive economic units that pooled their resources together to provide food and shelter and a stable environment within which to raise children. In such societies, labor was often divided along gender lines, with the male members working outside the home and the female members taking care of the home and the children. The "social impact of technology" (Schroeder, 2007) has altered the family economic system. In modern societies, the productive economic units traditionally belonging to the family have been taken over by economic institutions. Many economic functions of the family are being performed by manufacturing or service industries. From this point of view, a major change has occurred at this level which has the function to transform the family's producer function into a consumer function (Ionuţ, 2012). In modern societies, both spouses often work outside the home and share family responsibilities.

Socialization of Children

Another function of the family is the socialization of children. In pre-industrial American society, part of the socialization role of the family was to educate children. However, this changed, starting in the seventeenth century, when children began to attend school in certain parts of the United States. By the mid-nineteenth century, the role of schools in New England had expanded to such an extent that they took over many of the educational tasks traditionally handled by parents (Cremin, 1970). Although it is mandatory for children across the United States to receive formal compulsory education in schools today, families are still regarded as the primary agent of socialization of children. **Primary socialization** occurs during early childhood within the family. For example, parents teach children how to speak their home language and transmit societal and cultural norms to their children. **Intergenerational cultural transmission** refers to the transmission of cultural ideas (e.g., values, beliefs, knowledge, practices) from one generation to the next generation (Tam, 2015). At Christmastime, for example, Polish families enjoy a traditional wafer called *oplatek* that parents and grandparents pass on to their children. Before partaking of the Christmas Eve meal, Polish families gather around the table. The eldest member holds a large wafer and breaks off a piece to begin the ritual. In this way, the family socializes its youngest members. **Secondary socialization** occurs outside the family. As children enter school, they are introduced to new people and experiences and continue to learn how to become functioning members of society.

Sociologists have suggested further family functions. Another function that the family performs is provision of social status. In this sense, our ascribed statuses (e.g., race, ethnicity, social class, gender) are directly related to and granted by the family to which we belong. Also, within the family, the parents are not limited to ensuring their children are fed, but equally importantly, they are to provide protection, self-confidence, and emotional and medical support in relation to difficult situations (Ionuţ, 2012).

Meanings of Family and Household

Out of marriage, a family is formed. Families are the unit into which children are born, become attached to each other, are nurtured, and are socialized to become functioning members of society. This section examines the definitions of family and household. Table 1.1 provides a summary of our definitions of family and household.

Definitions of Family

The definition of family has changed over time. Sociologists believe that families are socially and culturally constructed. Murdock (1949) defines *family* as a "social group characterized by common residence, economic cooperation, and reproduction." The defining characteristics of the family are the fact that its members are living together to undertake various joint activities, to contribute resources necessary to life, and to have children (Ionuţ, 2012). According to the U.S. Census Bureau (2015), a **family** consists of a householder and one or more other people living in the same household who are related to the householder by birth, marriage, or adoption. A **householder** refers to the person in whose name the housing unit is owned, rented, or maintained. If a house is owned or rented jointly by a **married couple**—a husband and wife enumerated as members of the same household—the householder may be either the husband or the wife. Figure 1.1 shows more householders age 65 and over than those under 30 from 1960 to 2017.

Definitions of Household

The household is the basic unit of analysis in social, economic, and government research models. Traditionally, households refer to groupings for the domestic purposes of food gathering, processing,

Upon completion of this section, students should be able to:
LO3: Describe various definitions of family and household.

Table 1.1 Summary of Family and Household Definitions

Family	*Definitions*	*Examples*
Family	It consists of a householder and one or more other people living in the same household who are related to the householder by birth, marriage, or adoption.	Eric as householder, his wife, and one biological son and one adopted daughter
Family Household	It consists of (1) a householder and family members related by blood, marriage, or adoption and (2) may include unrelated people who live in that household.	Eric as householder, related family members such as wife and children, and unrelated people such as tenants
Non-family Household	It consists of a householder living alone or a householder who shares the housing unit with unrelated people.	Eric as householder lives alone (single person household) Or Eric as householder shares the house with a roommate, a tenant, or foster child Or cohabiting or unmarried couples, or POSSLQs (persons of the opposite sex sharing living quarters)

Source: U.S. Census Bureau. https://www.census.gov/programs-surveys/cps/technical-documentation/subject-definitions.html#-familyhousehold

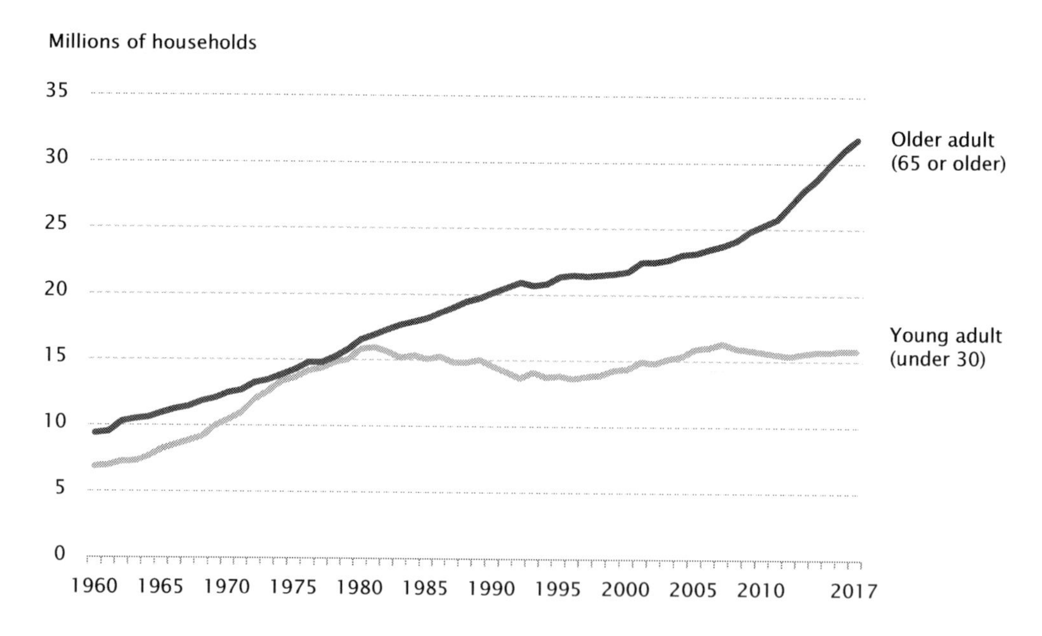

Figure 1.1 Households by Age of the Householder

Source: U.S. Census Bureau, Current Population Survey, Annual Social and Economic Supplements, 1960–2017. https://www.census.gov/content/dam/Census/library/visualizations/time-series/demo/families-and-households/hh-3.pdf

distribution and consumption to support members of both sexes and all ages (Leonetti & Chabot-Hanowell, 2011). According to the U.S. Census Bureau (n.d.), a **household** consists of all people who occupy a housing unit. There are two types of households: family households and non-family households. A **family household** consists of (1) a householder and family members related by blood, marriage, or adoption and (2) unrelated people who live in that household. In the United States, family households include married households and may include other unrelated people. There were 83 million family households in 2017 (U.S. Census Bureau, 2018b). A **non-family household** consists of a householder living alone or a householder who shares the housing unit with unrelated people. Households that consist of unmarried couples living together would be counted as non-family households. A household in which a person who owns the house lives alone or cohabits with an unmarried partner is also considered a non-family household. The U.S. Census Bureau has been exploring new ways to describe Americans' living arrangements with terms such as unmarried couples, shared households, gay and lesbian couples, and POSSLQs (persons of the opposite sex sharing living quarters) (Bianchi & Casper, 2003). In the United States, there were 43 million non-family households in 2017 (U.S. Census Bureau, 2018b). Figure 1.2 shows that married households decreased from about 80 percent to 50 percent while non-family households increased from 10 percent to 30 percent from 1947 to 2017.

Forms of Family

Every year, a traditional Thanksgiving meal with touches of Italian heritage takes place at Angela's home. Those who attend enjoy a variety of courses—antipasto, lasagna, salads, turkey, desserts, and wines. Other families come to celebrate with Angela's. Some fit the category of nuclear families while others are extended families. Regardless of their nature, all the families who attend enjoy being with each other on the holiday. Forms of family are based on several parameters. This section discusses forms of the family based on structure, kinship, ancestry, authority, and residence.

Forms of Family Based on Structure

When it comes to family structure, there are two main types: nuclear families and extended families. Traditionally, **nuclear family** is a form of family consisting of a married couple and their biological children. The nuclear family arose among the English elites in the late seventeenth century (Ariès, 1962; Trumbach, 1978) and it was universally present in all societies either as such or as a basic building block of more complex family forms (Murdock, 1949).

Types of Nuclear Families

Nuclear families are classified into three types depending on the employment status of their members. As defined above, the traditional nuclear family is *male breadwinner and female homemaker family model,*

Upon completion of this section, students should be able to:
LO4: Compare nuclear families and extended families.
LO5: Explain the formation of conjugal and consanguineous families.
LO6: Apply unilineal and bilateral descent to the analysis of traditional and modern families.
LO7: Describe the characteristics of the patriarchal family and the matriarchal family.
LO8: Describe forms of family based on residence.

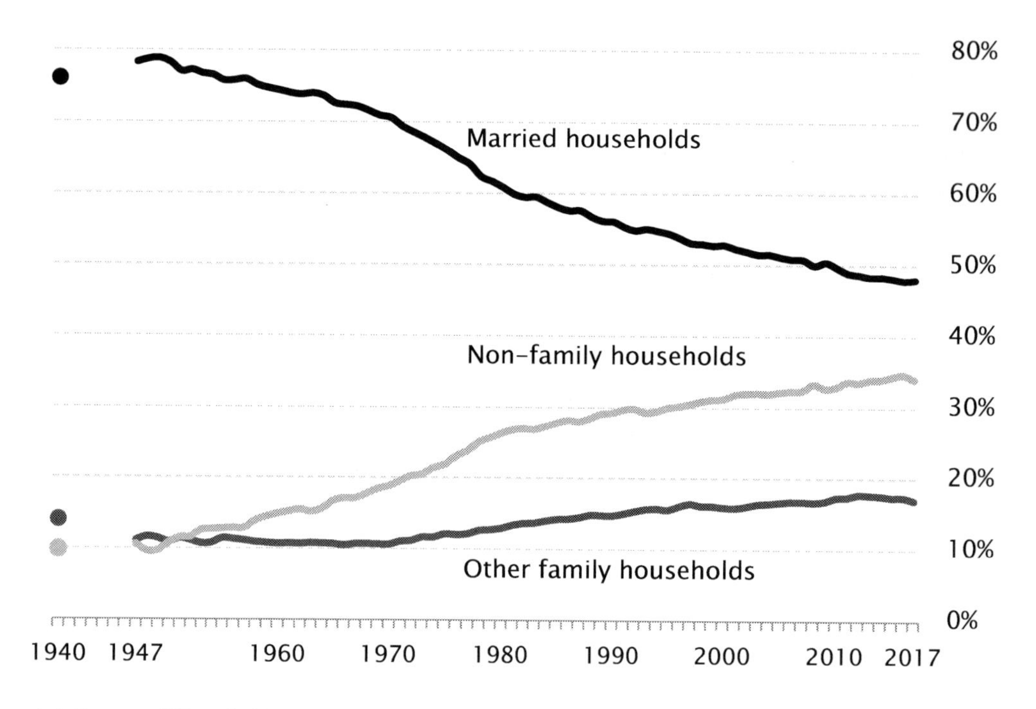

Figure 1.2 Percent of Households by Type
Source: U.S. Census Bureau, Decennial Census, 1940, and Current Population Survey, Annual Social and Economic Supplements, 1968–2017. https://www.census.gov/content/dam/Census/library/visualizations/time-series/demo/families-and-households/hh-1.pdf

involving two married persons providing care for their biological children. The dominant familial image for the nineteenth and twentieth centuries in the United States was one in which married women were viewed as caretakers of the home and family, and married men were responsible for securing the family's financial well-being (Cunningham, 2008). In this structure, the husband serves as the **breadwinner**—a person who earns money to support others and usually works outside the home while the wife serves as the homemaker and stays at home to care for children. This so-called "Ozzie and Harriet" model is derived from *The Adventures of Ozzie and Harriet,* a television program that ran in the United States in the 1950s. Men and women assumed traditional gender roles, which were represented on many American television programs. The rise of the male breadwinner family (MBWF) produced a "compact" covering the sexual division of labor, the economic support of family members, and the regulation of marriage and parenthood (Creighton, 1999). This traditional type of nuclear family (a mother and a father—usually married—and their biological children) has long been assumed to be the standard North American family (Smith, 1993) and continues to be the standard form to which all others are compared (Powell *et al.*, 2010).

The second type of nuclear family is the **dual-earner family**—both husband and wife providing income for the family, usually by working outside the home. The traditional male breadwinner model, where men are responsible for economic provision while women are responsible for the home, is in decline across the Western world as women are increasingly taking up paid employment (Nadim,

2015). For example, whereas the labor market participation of West German women increased over time, the employment pattern of East German women adjusted to the West German pattern after unification, resulting in an increase of part-time employment among mothers (Trappe *et al.*, 2015). In the United States, between 1968 and 1978, the number of families in which only the husband worked declined by 4.1 million, while dual-earner families rose by 4.5 million (Hayghe, 1981). In 2017, more than 60 percent of husband and wife pairs in the U.S. labor force are living in dual-earner families, compared with about 20 percent in husband-only and 10 percent in wife-only (see Figure 1.3). As increasing proportions of women entered the paid labor market during the latter decades of the twentieth century, the male breadwinner and female homemaker family model came under significant challenge both as a practice (Sayer *et al.*, 2004) and as an ideology (Brewster & Padavic, 2000).

A female breadwinner and a full-time male homemaker becomes the third type of nuclear family. *Female breadwinner family* (FBWF) takes place when the female provides the main source of income for the family. The women's movement pushed for women to engage in traditionally masculine pursuits in society. Women's support for gender specialization in marriage declined rapidly from the late 1970s through the mid-1980s (Cunningham, 2008). Women chose to sac-

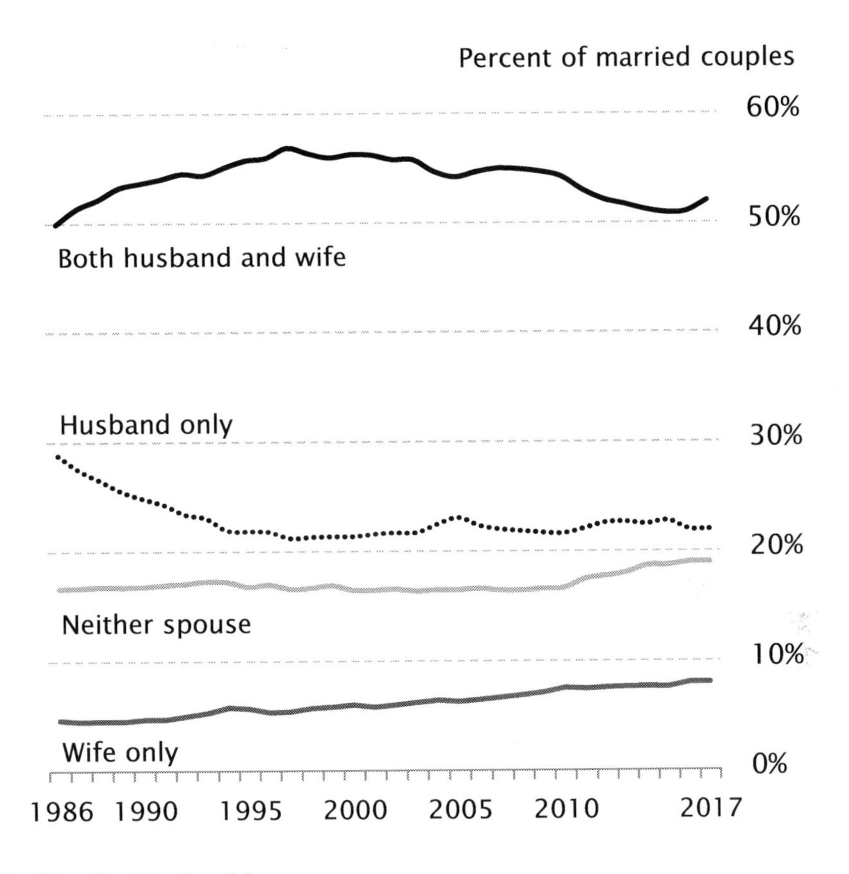

Figure 1.3 Labor Force Participation of Spouses

Source: U.S. Census Bureau, Current Population Survey, Annual Social and Economic Supplements, 1986–2017. https://www.census.gov/content/dam/Census/library/visualizations/time-series/demo/families-and-households/mc-1a.pdf

rifice their childbearing years to establish their careers, and men felt increasing pressure to be involved with tending to children (Fine, 1992). In the United States, "breadwinner moms" were the sole or primary breadwinners in 40 percent of heterosexual relationships with children (Wang *et al.*, 2013) and 37 percent of the breadwinner moms were married mothers who had a higher income than their husbands (Pew Research Center, 2010). The stay-at-home fathers or *househusbands* had been out of the labor force for the past year and stayed at home to raise their children and care for family. Although there have been an increasing number of families in which wives are in the workforce while husbands stay at home to care for children (see Figure 1.3), stay-at-home females still outnumber males, with 5 million stay-at-home mothers nationwide in 2017, compared with 267,000 stay-at-home fathers (U.S. Census Bureau, 2018a). In sum, these three types of nuclear family—male breadwinner and female homemaker family model, dual-earner family, female breadwinner model—are characterized by their small size and simple structure.

Extended Families

A family that extends beyond the nuclear family is an extended family. An **extended family** (or multigenerational household) is a form of family in which the nuclear family and relatives live in the same household or nearby. The extended family includes the nuclear family (the householder, spouse, children) and relatives. An extended family with four generations may consist of a household with a parent, child, grandchild, and grandparent living in the same household. Living in an extended family is nothing new. Globally, extended families are common in almost every society and usually occur in places in which economic conditions make it difficult for the nuclear family to achieve self-sufficiency. Extended families play a multifunctional role. In pre-industrial societies, extended families were popular because families were engaged in agricultural pursuits. Family members worked on family-based farms, so grandmothers often took care of the grandchildren while the parents worked on the farm. Families also served as nursing homes to grandparents who were too old to work. In addition, families played a role in the religious upbringing of children and provided a form of entertainment and education for their members. Extended families were the norm before the industrial society. However, social change has had an impact on the family structure. For example, the arrival of industrialization led to a shift from extended families to nuclear families as the nuclear unit fits the needs of an industrial society (Parsons, 1955). The decline of dependency on agricultural subsistence, which results in a weakening of extended family ties, is another cause of nuclear family creation (Ruggles, 2009). After that time, the nuclear family became the most common form of family as young adults migrated elsewhere for job and education opportunities and became economically independent in modern society.

Extended families are making a comeback today, with a record 64 million Americans living in multigenerational households (Cohn & Passel, 2018). It means that one in five Americans lived in multigenerational households in 2016. Reasons for extended families include sharing space and saving money. Divorced adult children may return home and live in their parents' home. Young adults may return home temporarily after graduation as they look for a job. And at times, grandparents may move in to help care for grandchildren, while adult children at times care for and support their aging parents by having them move in with them. Finally, immigrants in the United States are likely to live in extended families. Adolescents are often close to and exchange support with extended family members, including grandparents, aunts and uncles, or cousins (Sterrett *et al.*, 2011).

Forms of Family Based on Kinship

Another term that people use to describe family relationship is kin. **Kinship** is the relationship that links family members through blood, marriage, adoption, or remarriage. Kinship relationships form a network of bonds of varying intensity across time and across members (Reis & Sprecher, 2009). Research shows that women's pivotal contributions in food processing and distribution in hunting and gathering societies promoted kinship and appeared to be the foundation from which households evolved (Leonetti & Chabot-Hanowell, 2011).

Consanguineous Family

A **consanguineous family** is a form of family that is created by blood ties and is the equivalent of the extended family—a household that includes other kin in addition to the members of the nuclear family. The family is created at birth and establishes ties across generations (Beutler *et al.*, 1989). There are three types of consanguineous kins: *lineal kins* who are the descendants of the same ancestor (e.g., children, parents, grandparents, and great grandparents), *siblings* who are the brothers and sisters (e.g., children of the same parent), and *collateral kins* who are related through a relative (e.g., father's brother). Cross-cultural research has explored variations in the prioritization of kin relationships. An emphasis on lineal bonds is more typical of Caucasian and Asian families, whereas an emphasis on collateral bonds is more typical of Black families (Reis & Sprecher, 2009).

Conjugal Family

The nuclear family is known as a **conjugal family**—a form of family that is created by marriage, adoption, or remarriage. As a principle of kinship, marriage differs from blood in that it can be terminated (Reis & Sprecher, 2009). Kinship structure is either a victim of or a barrier to modernization (Inkeles & Smith, 1974). Indeed, changes over the last 30 years in patterns of family formation and dissolution have given rise to questions about the definition of kin relationships (Reis & Sprecher, 2009). The family has changed over time (e.g., cohabiting family, same-sex family, stepfamily, and adoptive family). *The Brady Bunch*, a television show that aired in the late 1960s and early 1970s in America and consisted of two remarried parents and their respective children, is an example of a conjugal family. Given the potential for marital break-up, blood is recognized as the more important principle of kinship, but births resulting from infertility treatments such as gestational surrogacy and in vitro fertilization with ovum donation challenge the biogenetic basis for kinship (Reis & Sprecher, 2009).

Types of Kinship

American anthropologist Morgan (1871) argues that all human societies share a basic set of principles for social organization along kinship lines, based on the principles of consanguinity (kinship by blood) and affinity (kinship by marriage). Social scientists distinguish between primary kin, secondary kin, and tertiary kin. **Primary kinship**—members of the families of origin (immediate family)—is divided into *primary consanguineous kinship* (i.e., related to each other by blood) and *primary affinal kinship* (i.e., related to each other by marriage). The eight primary kins are husband–wife, father–son, mother–son, father–daughter, mother–daughter, brother–sister, younger brother–elder brother, younger sister–elder sister. Kinship bonds including parent–child relations are the core networks of such shared relationships (Carol, 2014). **Secondary kinship**—the primary kin of a primary kin (extended family)—is divided into *secondary consanguineous kinship* (e.g., grandparent and grandchild) and *secondary affinal kinship* (e.g., sibling's spouse). Therefore, people who are directly related to the primary kin of their primary kin become their secondary kin. **Tertiary kinship**—the primary kin of a secondary kin—is divided into *tertiary consanguineous kinship* (e.g., great grandchildren and great grandparents) and *tertiary affinal kinship* (e.g., spouse's grandparents). Let's consider Angela's family as an example. Angela's family consists of eight primary kins. Outside her nuclear family, Angela has 33 secondary kins and 151 types of tertiary kins. The relationship between two persons is based on the degree of closeness or distance of that relationship (Table 1.2). While primary kin (parents, children) remain significant throughout life, extended kin fulfill supplementary functions—as family historians, mediators, mentors, and buffers in conflict (Reis & Sprecher, 2009).

Kinship Norms

Kinship norms are culturally defined and specify how kin-related persons are expected to behave towards each other (Rossi & Rossi, 1990). The reasons proposed to explain the motives for assisting older kin include: a *norm of family obligation*—culturally prescribed duties based on kinship (Seelbach, 1978), a *norm of reciprocity*—the belief that children owe a debt to parents that should be repaid as parents age (Albert, 1990), a *norm of gratitude*—the belief that offspring want to help parents because they are grateful for past parental help and sacrifices (Brakman, 1995), and a norm of *filial obligations*—a moral duty (Finch, 1989). Globally, the negotiation of kin responsibilities

Table 1.2 Table of Consanguinity: Showing Degrees of Relationship

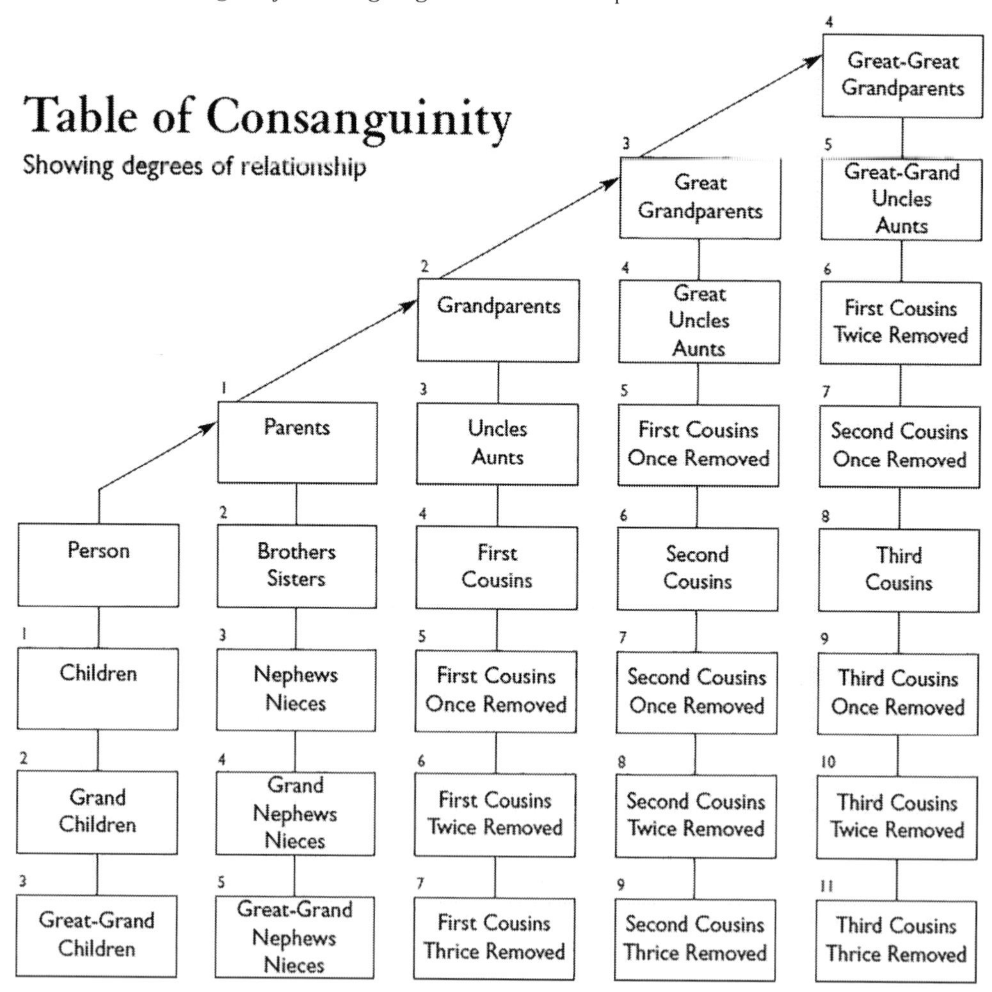

Source: https://upload.wikimedia.org/wikipedia/commons/2/28/Table_of_Consanguinity_showing_degrees_of_relationship.png

is more likely in individualistic (European and American) cultures, which stress independence and the pursuit of personal goals, than in collectivistic (Islamic or Confucian) cultures, which stress kin group membership and the submission of individual goals to the needs and wishes of the family (Schneider, 1980).

The *hierarchical model of family obligations* suggests that relationships that are supported both by genetic and/or legal ties tend to be higher and those that lack the genetic connection are lower (Rossi & Rossi, 1990). In most Western societies, the strongest kinship norm is the obligation toward children, followed by that toward parents; kinship norms (e.g., obligation to provide support) are weaker for distant family members and are also weaker for ascendant (up lineal lines) than descendent (down lineal lines) kinship (Reis & Sprecher, 2009). Sociologically based research shows that members of larger families or extended families are much more likely to adhere to cultural norms and traditions than members of smaller families (De Vries *et al.*, 2009).

Kinkeeping—consisting of efforts expended on behalf of keeping family members in touch with one another—captures how kinship operates as a network. Research shows that having a family kinkeeper is related to greater extended family interaction and greater emphasis on family ritual at both extended family and lineage levels. Usually, kinkeeping is primarily a female activity and is related to the importance sibling relationships hold for people (Rosenthal, 1985). Women often fulfill the role of keeping others informed about what is happening in the family, organizing get-togethers, and encouraging direct interactions (Reis & Sprecher, 2009).

Kinship norms help maintain unity and regulate the behavior of different kin. **Iroquois kinship**—the Iroquoian (a Native American group) kinship system—was identified by American anthropologist Morgan (1871) and is used to define family. Interestingly, this kinship system used the same kin terms for all male blood relatives on the father's side (i.e., father's brother is mentioned with the same term as father), and all female blood relatives on the mother's side (i.e., mother's sisters are mentioned with the same term as mother). The children of an aunt or an uncle are cousins (cross cousins) (see image below). Preferential cross-cousin marriage can be useful in reaffirming alliances between unilineal lineages or clans. Thus, the Iroquois kinship system encourages wedding ties between cross cousins but discourages parallel cousin marriage because parallel cousins are considered siblings. Around the globe, people in South India and Sri Lanka also employ the Iroquois kinship system (Haviland, 2002).

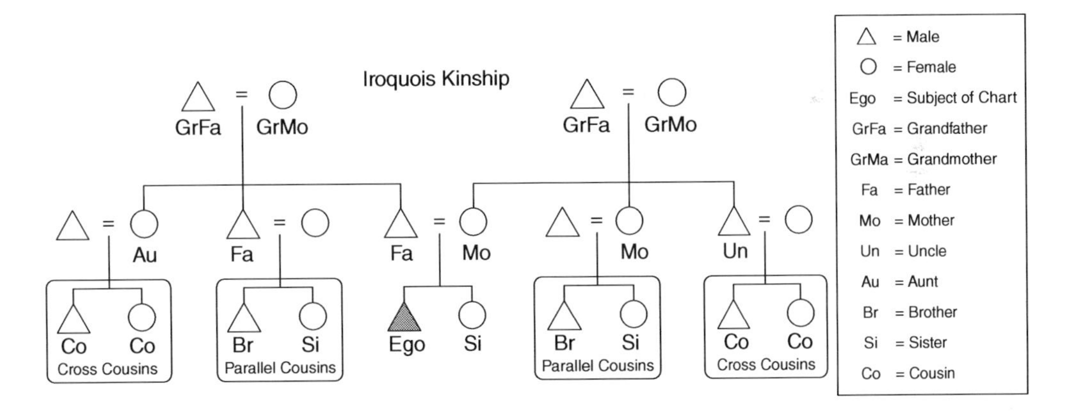

Source: https://commons.wikimedia.org/wiki/File:Iriquois-kinship-chart.svg

Forms of Family Based on Descent

Have you ever tried to complete a family tree assignment for a class? If so, you were looking into your family based on your descent. **Descent** refers to the biological relationship between individuals in society. **Lineage**—the line through which descent is traced—is used to describe everyone who descends from an ancestor. Two forms of family lineage are unilateral descent and bilateral descent.

Unilineal Descent

The extended family is usually built around a unilineal descent group. **Unilineal descent** (or one-sided descent) is a form of family lineage that traces a person's descent through either the maternal or the paternal line. In a unilineal descent, membership may rest either on patrilineal descent (patrilineage) or on matrilineal descent (matrilineage). The ethnic Mosuos (Na) of southwest China are a small-scale society in which patrilineal descent and matrilineal descent coexist and distinguish two sub-groups with many otherwise shared cultural characteristics (Mattison *et al.*, 2016).

Patrilineal Family

A **patrilineal family** is characterized by the inheritance of property rights, surnames, or titles through the male line. The link between kinship and gender inequality has been a focus of study (Collier, 1988). The patrilineal family is reinforced by male dominance or a strict gender-based division of labor. In hunting and gathering societies, males usually hunted while females gathered wild berries and vegetables for food. Horticultural (growing crops with simple tools) and pastoral (raising livestock) societies were less egalitarian than hunter-gatherer societies. In agricultural societies, as men began to be engaged in the labor-intensive food production sphere, they became provider, property and land owner, and breadwinner. Men became dominant while women became dependent. The gender-based division of labor has affected family and society in many ways. One example is the rise of patrilineal descent—the relations that come from the father's blood line. Male and female children belong to their father's kin group. A family tree is traced through males from a founding male ancestor.

Another aspect of kinship is the link between descent and inheritance (Radcliffe-Brown, 1952). In ancient Rome, males were the sole owners of family property. Various European aristocracies adopted a patrilineal ideology around the year 1000 (Trumbach, 1978). In an analysis of 20,000 patrilineages and multigenerational families from eighteenth- and nineteenth-century China, Song *et al.* (2015) found that patrilineal families with high-status males had higher growth rates for the next 150 years. Today, the patrilineal Mosuos in southwest China still trace their descent through the fraternal line. Lineage membership and transmission of resources occur via the male line (Mattison *et al.*, 2016). In Indonesia, the ethnic groups of North Sumatra and Bali follow patrilineal norms (Rammohan & Robertson, 2012). Kinship norms are male focused, with only sons being allowed to inherit their father's property (Frankenberg & Thomas, 2001). Son preference is more common as sons continue the line of descent. Daughters, when married to members of other lineages from different clans, must drop their own clans and face limited inheritance rights. Research shows that patrilineal descent systems are associated with poor education and health outcomes for women (Rammohan & Robertson, 2012).

Matrilineal Family

The opposite of the patrilineal family is the **matrilineal family**—a form of family in which descent is traced through females from a female ancestor to her descendants. In a matrilineal family, the mother is the basis of descent. Daughters pass on the family line to their offspring and inherit their status from their mothers. Matrilineal surnames are names transmitted from mother to daughter, in contrast to patrilineal surnames transmitted from father to son (Sykes, 2001). Property is transferred through the mother. Women pass hereditary leadership through the maternal line.

Morgan (1851) examined the Iroquois, a Native American group in the northeastern United States, and discovered their matrilineal system of kinship reckoning. The six Iroquois tribes had matrilineal systems, in which children were born into the mother's clan and gained status through it. Operated by the Great Binding Law of Peace before the Americas became independent, women retained matrilineal rights and participated in political decision-making, including deciding whether to proceed to war (Jacobs, 1991). In Indonesia, the Minangkabau of West Sumatra is one of the world's largest matrilineal ethnic groups (Rammohan & Robertson, 2012) and classifies property into "ancestral property" and "earned property" (Kahn, 1980). A woman has lifelong rights to specific pieces of land, and rights of use are inherited from mothers by daughters, not by sons (Dube, 1997). Since ancestral property is always inherited by women, and almost always passes from mother to daughter, females are crucial to the continuity of the matriline (Rammohan & Robertson, 2012). Thus, matrilineal practices form the basis of social, labor, and land relations in these communities (Blackwood, 2008). In southwest China, the Mosuos residing in the basins of the Hengduan Mountains practice matrilineal descent and inheritance (Mattison *et al.*, 2016) and engage in a non-marital reproductive union known as "walking marriage" (Shih, 2010). Reproductive partners remain in their natal matrilineal households (Cai, 2001).

Bilateral Descent

Bilateral descent (or two-sided descent) is a form of family lineage that traces a person's descent through both the mother's and the father's sides. Descent and inheritance are passed bilaterally through both males and females. Traditionally, bilateral descent is found among a few groups in West Africa, Indonesia, Polynesia, etc. For example, the Urapmin people, a small tribe in Papua New Guinea, have a system of kinship classes known as tanum miit. The classes are inherited bilaterally from both parents (Robbins, 2004). A tribal structure based on bilateral descent helps members live in extreme environments because it allows individuals to rely on two sets of families dispersed over a wide area (Ezzell, 2001). Indeed, a woman's dowry and her husband's inheritance are pooled to form a conjugal fund or estate, which has the effect of minimizing lineage and emphasizing the conjugal unit (Johnson, 1988).

Bilateral descent is increasingly the norm and is used by most societies today. Angela tried to dig up her ancestry and complete a family tree assignment for her family class. She asked her parents for help. Her mother listed her grandparents and great-grandparents and shared stories from her maternal side. Her father did the same from his paternal side. All the relatives on the mother's and father's sides were considered equally important as descent was traced bilaterally.

Egalitarian Family

The notion of the patrilineal family—and its implicit female subordination—began being questioned in seventeenth-century England. The eighteenth century saw the rise of the egalitarian family in England as there was a shift from a patriarchal system to a more egalitarian system (Trumbach, 1978). The **egalitarian family** is a form of family in which power and descent is shared equally by both wife and husband. For example, romantic marriages were practiced and property rights of married women were expected. In the United States, the Civil Rights movement in the 1950s and the women's movement in the 1960s, as well as female entry into the labor force, paved the way for an egalitarian family.

Forms of Family Based on Authority

A patrilineal family takes place within a patriarchal society, in which males are dominant. *Patriarchy* ("the rule of the father") has manifested itself in the social, legal, political, religious, and economic organization of a range of different cultures (Malti-Douglas, 2007). Throughout most of American history and world history, families were patriarchal. Historically, the term patriarchy was used to refer to autocratic rule by the male head of a family but in modern times, it refers to social systems in which power is primarily held by adult men (Cannell & Green, 1996). Even if not explicitly defined as patriarchal in their own constitutions and laws, most contemporary societies are, in practice, patriarchal (Lockard, 2007).

Patriarchal Family

A **patriarchal family** is a form of family in which the father holds authority over his wife and children. The father is the head of the family and the owner of the family property. The patriarchal family is of ancient provenance. In hunting and gathering societies, the role of meat sharing by men has dominated the discourse on food sharing (Hawkes & Bliege Bird, 2002). The agricultural societies that give power over women and sons to fathers should be called *patriarchal* and in these societies brides are likely to be brought into the patriarchal household (Johnson, 1988). Domination of women by men is found in the ancient Near East as far back as 3100 BCE (Strozier, 2002). The works of Greek philosopher Aristotle portrayed women as morally, intellectually, and physically inferior to men; saw women as the property of men; claimed that women's role in society was to reproduce and to serve men in the household; and saw male domination of women as natural and virtuous (Fishbein, 2002). The basic presumption was that at the head of each household stood a man who, in his roles as master, father, and husband, owned his wife, his children, his slaves, his animals, and his land (Trumbach, 1978).

Matriarchal Family

The opposite of a patriarchy is a matriarchy in which women hold positions of power. A **matriarchal family** is a form of family in which the mother is the head of the family with authority over other family members. In some hunting and gathering societies, the earliest type of the family was matriarchal. Research shows how men's hunting led to dependence on women's contributions, bonded men to women and bonded generations together (Leonetti & Chabot-Hanowell, 2011). In fact, men were able to become hunters, apparently dependent on women to provision them during periods of low hunting success with foods reliably obtained and differing from those mainly pursued by men (Marlowe, 2003). Females focus on plant foods more than do males (Marlowe, 2007). For example, among African foragers (hunters and gatherers), females tend to collect most of the kilocalories eaten with reliable daily productivity (Marlowe, 2010). Furthermore, the cooking, chopping, grinding, pounding, and mixing entailed are usually done by women. In many cultures, women keep the fire burning (Wrangham, 2009). Apportionment usually falls to the mother or grandmother after processing and this can represent a key role for women in the power dynamics of the group (Leonetti & Chabot-Hanowell, 2011).

In matriarchal families, women's roles are critical to household formation, pair bonding, and intergenerational bonds (Leonetti & Chabot-Hanowell, 2011). For example, among the Iroquois, a Native American group in the northeastern United States, unrelated men lived in a long-house with wives who were each other's sisters. Brother men have less chance to exercise authority over their sisters' sons, and husbands have less authority over their own wives and children (Johnson, 1988).

Forms of Family Based on Residence

Deciding where to reside as newlyweds can be challenging. A family can be classified into three forms based on where a newlywed couple lives.

Patrilocal Family

A **patrilocal family** is a form of family in which a newlywed couple resides with or near the husband's parents. That is, sons will stay and daughters will move in with their husbands' families. Children will follow the same pattern. After marriage, the wife comes to live in the home of her husband. The patrilocal family is also patriarchal and patrilineal because it is maintained by the senior man, who holds authority over other family members. Patrilocal residence existed during the Early Neolithic (the New Stone Age) in southwestern Central Europe (Schiesberg, 2016). Today, in southwest China, the patrilineal Mosuos reside with the husbands' parents and are typically patrilocal (Mattison *et al.*, 2016).

Matrilocal Family

A **matrilocal family** is a form of family in which a newlywed couple lives with the wife's side. After marriage, the daughter does not leave her maternal home and still lives with her mother. For example, the Native American families in the pre-Columbia era were matrilocal. When a man married a woman, he moved to her tribe and became a member of her tribe. The matrilocal family is also matriarchal and matrilineal. When descent is matrilineal and the residence of the married couple is matrilocal, there is the potential both for breaking up the solidarity of brothers and for weakening the authority of husbands, since brothers leave their own territory when they marry to become husbands whose wives own the land (Johnson, 1988).

Circumstantial evidence shows that matrilocal residence has existed since the beginning of the Early Iron Age (Schiesberg, 2016). Korotayev (2003) suggests that the female contribution to subsistence (e.g., necessary resources for survival) does correlate significantly with matrilocal (as opposed to patrilocal) residence. In matrilocal societies, women are more likely to connect both women and men as equals, a connection related ultimately to the early dependency of both female and male infants on women (Johnson, 1988). For example, matrilocal daughters in Minangkabau society, Indonesia, bring in male labor resources at marriage and produce more labor by having children (Rammohan & Robertson, 2012). As women contribute to subsistence and connect women and men as equals, matrilocal societies among foraging (i.e., searching for wild food resources) groups are less likely to commit female infanticide than are patrilocal societies (Pinker, 2011).

Neolocal Family

Most societies were patrilineal, patriarchal, and patrilocal. In the United States today, it is common for newlyweds to form a **neolocal family** in which the newlywed couple resides in their own home. In fact, neolocal residence was brought to the British colonies in the Americas and as American colonists expanded westward, this form of residence remained

Upon the completion of the next section, students should be able to:
LO9: Explain trends in forms of family.

Student Accounts of Forms of Family

My parents made choices to bring me and my sister into the world. It is their responsibility to raise the kids to become functioning members of society. My mom became our first educator and taught us the language and the norms of society. Then, my mother got divorced when I was five. She got remarried and had a baby. The family that I grew up with was my mother, my grandmother, my sister, and my stepfather. In a couple of years, I plan to live with my current boyfriend and create a family of cohabitation. I will carry the things that I have learned from my family growing up to my future family (Personal communication, July 20, 2017).

(Fawver, 2012). Countries that experience economic development tend to experience increases in neolocal forms of residence (Ruggles, 2009) because of the higher mobility of nuclear families in modern economies (Gordon, 1970). Also, upon marriage, adults are expected to move out of their parents' home and establish a new residence, thus forming the nuclear family.

Trends in Family Forms

The above student account of her family life might help you develop a better understanding of the material not only through the eyes of the authors but also through the experience of your peers. While the old "ideal" involved couples marrying young, then starting a family, and staying married till "death do they part," the family has become more complex and less "traditional" (Livingston, 2014). The perspective on the family has suffered substantial changes in the second half of the twentieth century (Ionuţ, 2012). Advances in society have created new family forms. The traditional nuclear families with breadwinning fathers and stay-at-home childrearing mothers during the 1950s have become a dated form of family in relation to diverse living arrangements. Since 1960, fewer Americans are married, and more are living at their parental homes or cohabiting before marriage (Fry, 2016). Figure 1.6 (see p. 25) shows the growing complexity of living arrangements in the United States. Today parental forms have become more heterogeneous and fluid (see Table 1.3). These changes are due to increases in never-married single parents, divorce, cohabitation, same-sex parenting, multipartnered fertility, and coresidence with grandparents (Pearce *et al.*, 2018). Let's look at trends in family.

Table 1.3 Traditional Nuclear Family and New Alternatives

Traditional Nuclear Family	*New Alternatives*
Legally married	Never-married singlehood, non-marital cohabitation
With children	Voluntary childlessness
Two-parent	Single-parent
Permanent	Divorce, remarriage (including binuclear family involving joint custody, stepfamily or "blended" family)
Male primary provider, ultimate authority	Egalitarian marriage (including dual-career and commuter marriage)
Sexually exclusive	Extramarital relationships (including sexually open marriage, swinging, and intimate friendships)
Heterosexual	Same-sex intimate relationships or households
Two-adult household	Multi-adult households (including multiple spouses, communal living, affiliated families, and multigenerational families)

Source: Brym, R. J. & Lie, J. (2007). *Sociology: Your Compass for a New World*, 3rd edn. Thomson Wadsworth.

Demographic Trends in Families

American society moves at such a fast pace that people are adapting to new trends in a changing society. Many new family forms exist and meet the challenges of the post-industrial society and globalizing world regardless of marital status, gender role, or sexual orientation. One trend in families in the United States is that they are becoming more racially and ethnically diverse. The changing racial makeup of the United States is most visible among children. By 2020, a majority of children are projected to be a race other than White (Vespa & Armstrong, 2018). Figure 1.4 shows the growth in American households headed by racial and ethnic families from 1970 to 2017.

Many Americans put off starting a family until they are older. Trends show up in historical data that illustrate a retreat from marriage and childbearing at younger ages. The declining marriage rate, rising age at first marriage, and available alternative living arrangements are key factors involved in an increase

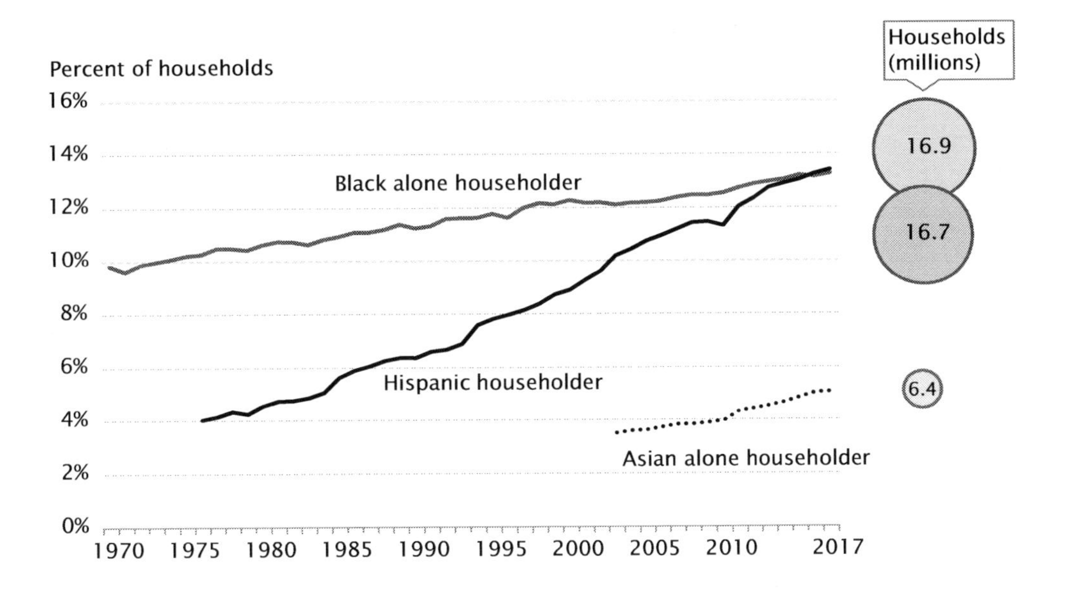

Figure 1.4 Growth in Households Headed by Racial or Ethnic Minorities
Source: U.S. Census Bureau, Current Population Survey, Annual Social and Economic Supplements, 1970–2017. https://www.census.gov/content/dam/Census/library/visualizations/time-series/demo/families-and-households/hh-2.pdf

in singlehood in the United States. Singlehood is a state of being unmarried. It includes young adults who have not yet been married as well as divorced and widowed adults. The U.S. Census Bureau (2017a) found that there were 63.5 percent of never-married U.S. residents age eighteen and older in 2016; another 23.1 percent were divorced and 13.4 percent were widowed. Figure 1.5 illustrates that the gap between married and unmarried Americans age fifteen and older has narrowed since 1950.

Forms of Family through Marriage and Remarriage

People who divorce may remarry and form a **stepfamily**—a form of family in which at least one marital or cohabiting partner has children who are not biologically related to the other spouse or partner. In this form of family, either partner has a child from a prior relationship regardless of whether the children live in the household. Another family form is **lesbian, gay, bisexual, or transgender (LGBT) families.** These lesbian, gay, bisexual, and transgender people marry and may raise children. One in ten LGBT Americans (10.2 percent) are married to a same-sex partner, up from the months before the High Court decision (7.9 percent); a majority (61 percent) of same-sex cohabiting couples are now married, up from 38 percent before the ruling (Masci *et al.*, 2017). Today, a married or cohabiting couple or single parent in a household could be biological, adoptive, grandparent, or stepparents to the child or children in the household.

Even when couples settle down and marry, many couples decide not to have children. In 2006, 26.2 percent of women ages 30 to 34 had never given birth to a child and by 2016 that number had risen to 30.8 percent (U.S. Census Bureau, 2017b). **Childfree families** are a form of family in which married partners choose not to have children. More women age 15 to 44 voluntarily choose not to have children: 6 percent of currently married middle-aged adults had never had a child in 1992; this figure rose to 11 percent in 2012 (Wu *et al.*, 2016).

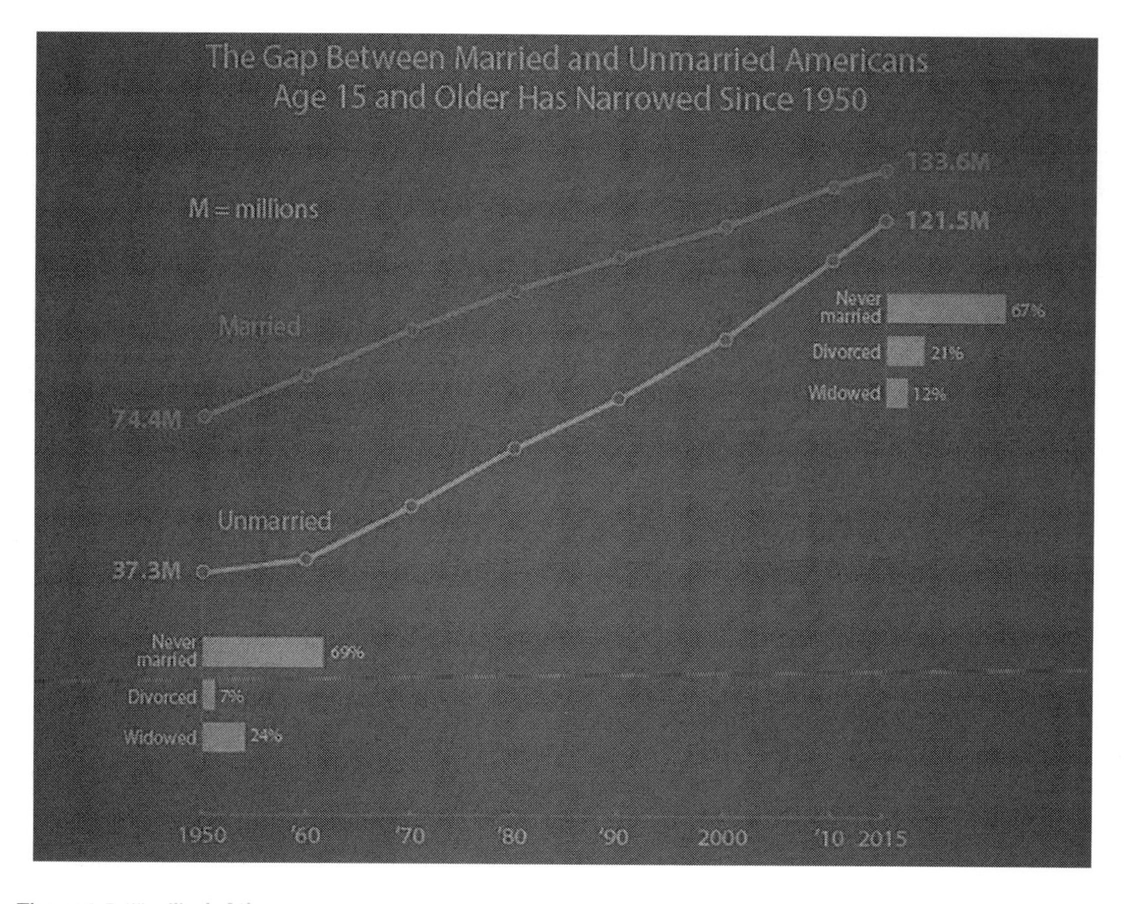

Figure 1.5 The Single Life

Source: 1950 and 1960 Censuses and the Current Population Survey Annual Social and Economic Supplement, U.S. Census Bureau 2016. https://www.census.gov/library/visualizations/2016/comm/cb16-ff18_single_americans.html

Forms of Family Outside Marriage

Due to the declining marriage rate, the rising marriage age, shrinking family size, and high divorce rates in the United States, almost half of U.S. adults today do not live with a married partner. Although young people are delaying marriage, they are not putting off romantic relationships. Over the last 40 years, the number of young people living with a boyfriend or girlfriend has increased more than twelve times, making it the fastest growing living arrangement among young adults (Vespa, 2017). From 1967 to 2016, the number of unmarried people has increased; nationwide 53.2 percent of unmarried residents age eighteen and older were women in 2016 and 46.8 percent were men (U.S. Census Bureau, 2017a). This has led to alternative living arrangements.

Young adults are still starting relationships at the same age that their parents did, but they are trading marriage for cohabitation (Vespa, 2017). A **cohabiting family** is a form of family in which a couple, with or without children, lives together without being legally married. As marriage rates have fallen, the number of U.S. adults in cohabiting relationships has continued to climb, reaching about 18 million in 2016, and the number of cohabiting adults age 50 and older has risen 75 percent since 2007 (Stepler, 2017). Figure

1.6 shows that almost half of adults today do not live with a spouse.

There has also been a significant increase in the number of grandparent-headed families. A **grandparent-headed family** is a form of family in which grandparents take care of a grandchild or grandchildren with or without a parent of the child being present. Increasing gains in longevity translate to a higher likelihood that adolescents know their grandparents longer than in previous generations (Kemp, 2007). The closer grandparents live to their grandchildren, the more emotionally close they are; frequent phone or email conversations build closeness (Harwood, 2000). Kinds of support that grandparents provide include emotional support, peace-keeping, "straight talking," and sharing family history (Soliz, 2008). Today, grandparenting occurs when a grandparent assumes responsibility for a grandchild because the grandchild's parents are not able to care for the child. In 2015, 5.9 million children under age eighteen were living with a grandparent householder (U.S. Census Bureau, 2017c). Figure 1.7 shows that a total of 3,484,140 grandparents are in the labor force and at the same time are responsible for care of coresident grandchildren.

Although marriage rates have decreased in the United States, non-marital births have increased. In 2015, 35.7 percent of women age 15 to 50 with a birth in the last twelve months were widowed, divorced, or never married (U.S. Census Bureau, 2017a). Children born to unmarried mothers are likely to grow up in a **single-parent family**—a form of family

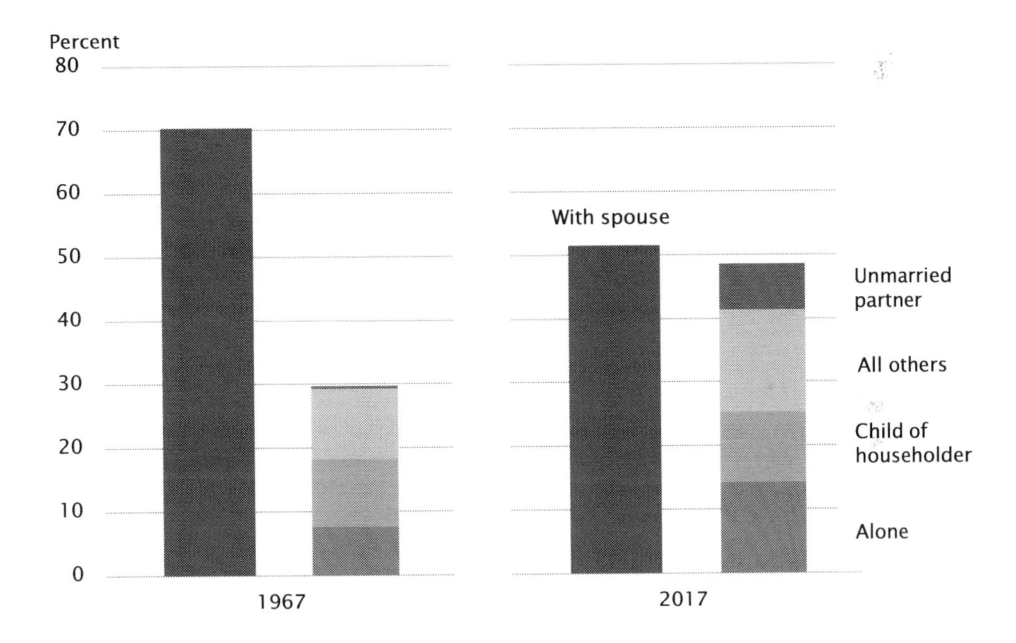

Figure 1.6 Living Arrangements of Adults 18 and Over, 1967–2017

Source: Current Population Survey, Annual Social and Economic Supplement, 1967–2017. https://www.census.gov/content/dam/Census/library/visualizations/time-series/demo/families-and-households/ad-3a.pdf

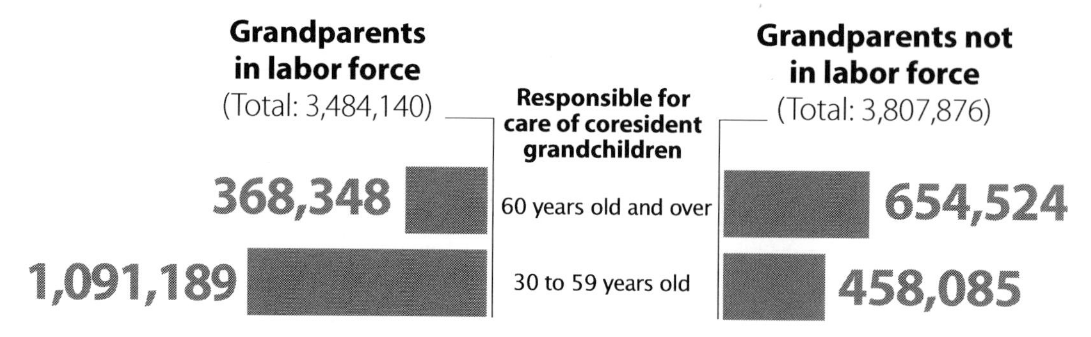

Figure 1.7 Grandparents Still Work to Support Grandchildren
Source: 2015 American Community Survey. https://www.census.gov/library/visualizations/2017/comm/grandparents-support-grandchildren.html

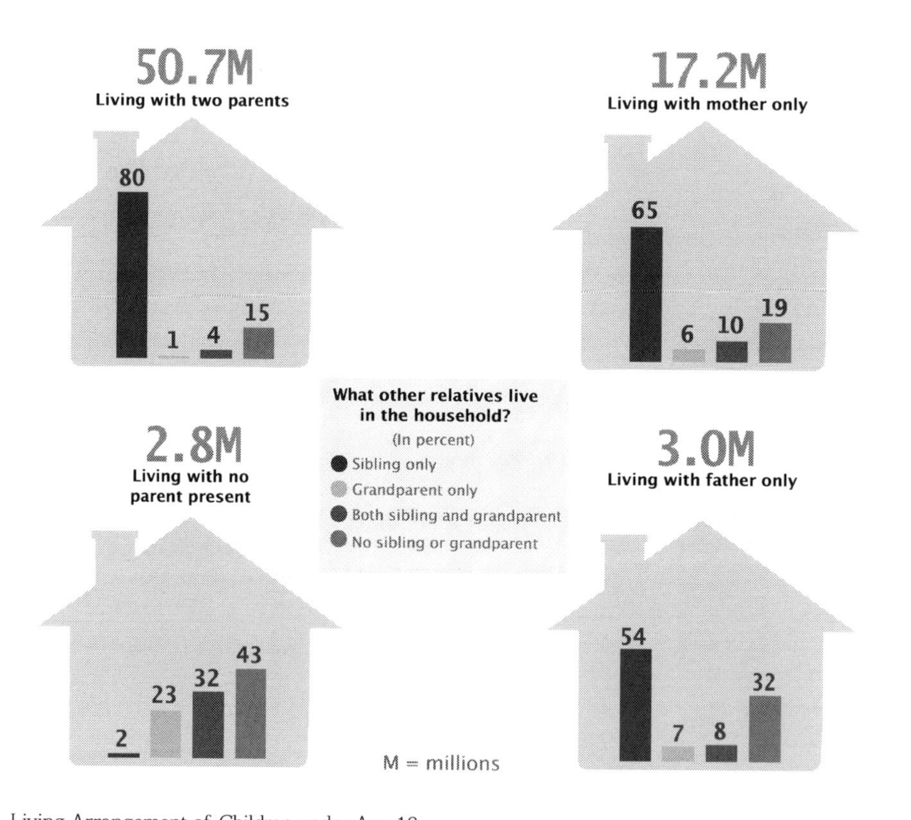

Figure 1.8 Living Arrangement of Children under Age 18
Source: Current Population Survey, Annual Social and Economic Supplement, 2007–2015. https://www.census.gov/library/visualizations/2016/comm/cb16-192_living_arrangements.html

that includes children under age eighteen and that is headed by a parent who is widowed or divorced and not remarried, or never married. Living in a single-parent family can have profound effects on family members. Single-parent family households run by women had the lowest median income in 2017 ($41,703), followed by households maintained by men with no wife present ($60,843). In contrast, married-couple family households had the highest median income ($90,386) in 2017 (Fontenot *et al.*, 2018). Figure 1.8 shows that 17.2 million children under age eighteen live with mothers only and 3.0 million live with fathers.

Diversity Overview on Single Life

There were 110.6 million unmarried people age eighteen and older living in the United States in 2016, making up 45.2 percent of all U.S. residents eighteen and older (U.S. Census Bureau, 2017a). "Unmarried and Single Americans Week" takes place in September every year to celebrate single life and to recognize singles and their contributions to society. What are the living arrangements of singles in the United States today?

Unmarried Young Adults Living Alone

Traditionally, as *young adults* (those age 18 to 39) finished schooling and established a career, they would marry and have children. Today, however, many young adults choose to delay getting married and having children. The share of singles is at a historic high, with one in five, or 42 million adults, age 25 and older never married in 2012, compared with one in ten adults or 9 million in 1960 (Wang & Parker, 2014). In 2016, 35.4 million people lived alone and comprised 28.1 percent of all households, up from 17.1 percent in 1970 (U.S. Census Bureau, 2017a). Figure 1.9 shows the growth of people living alone from 1960 to 2017.

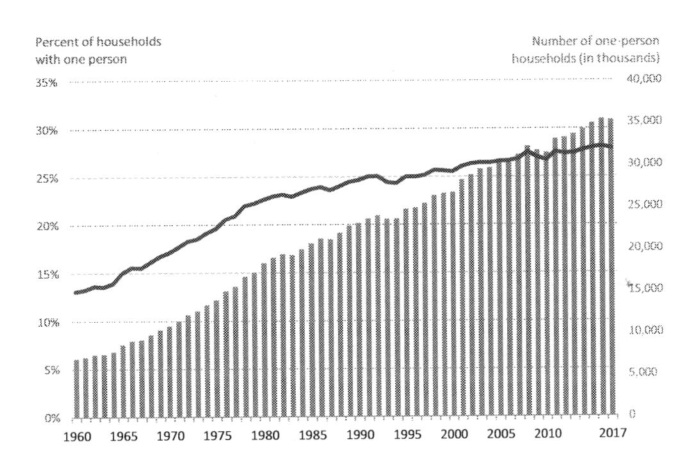

Figure 1.9 The Rise of Living Alone

Source: U.S. Census Bureau, Current Population Survey, Annual Social and Economic Supplements, 1960–2017. https://www.census.gov/content/dam/Census/library/visualizations/time-series/demo/families-and-households/hh-4.pdf

Unmarried Young Adults Living with Their Parents

The act of leaving the parental home and establishing an independent residence is considered an important marker of the transition to adulthood (Koc, 2007). Today, however, alongside the rise of living together without being married, there are more young adults who are choosing to live alone, move in with roommates, stay in their parents' home, or live with other family members such as siblings (Vespa, 2017). Figure 1.10 shows more than one in three, or about 22.9 million (31 percent), of 18- to 34-year-olds lived in their parents' home in 2016, up from 14.7 million (26 percent) in 1975.

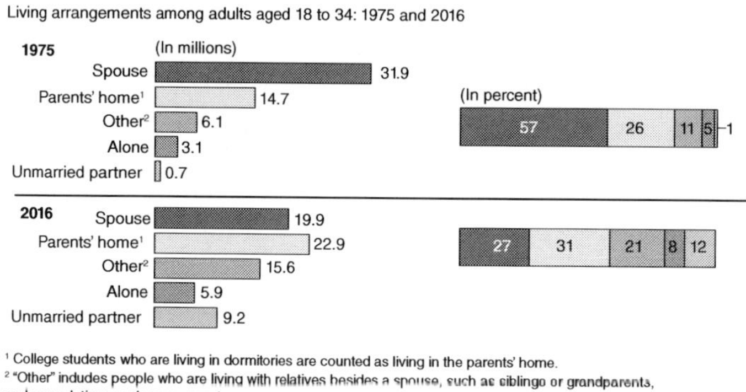

Figure 1.10 More Young Adults Lived with Parents than a Spouse in 2016
Source: U.S. Census Bureau, 1975 and 2016 Current Population Survey Annual Social and Economic Supplement. https://www.census.gov/content/dam/Census/newsroom/press-kits/2017/figure3.png

The number of single adults varies by state and gender. At the state level, New Jersey had the highest percentage of 18- to 34-year-olds living in their parents' home (46.9 percent), followed by Connecticut (41.6 percent) and New York (40.6 percent) (U.S. Census Bureau, 2016). Figure 1.11 shows that men age 18–34 were more likely to live in the parental home than women.

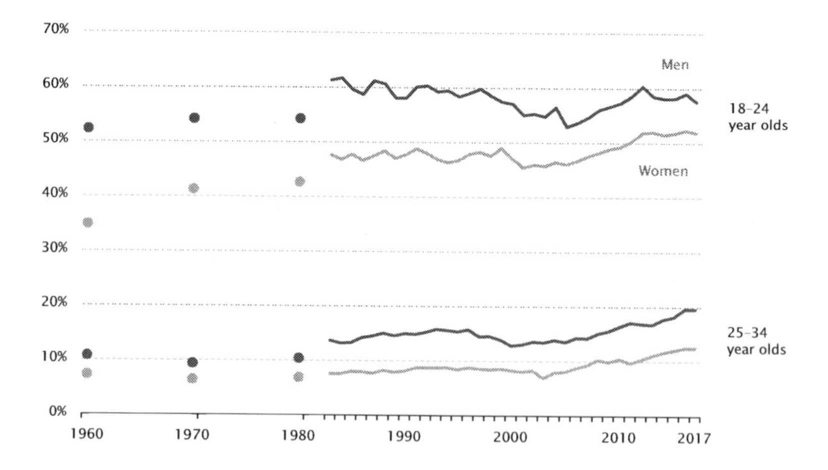

Figure 1.11 Young Adults Living in Parental Home
Source: U.S. Census Bureau, Decennial Censuses, 1960–1980, and Current Population Survey, Annual Social and Economic Supplements, 1983–2017. https://www.census.gov/content/dam/Census/library/visualizations/time-series/demo/families-and-households/ad-1.pdf

Young adults choose to live with their parents for many reasons, including low level of education, low income, limited employment opportunities, falling wages, student debt, and the high cost of living on their own. Younger male adults are falling to the bottom of the income ladder. In 1975, only 25 percent of men age 25 to 34 had incomes of less than $30,000 per year but by 2016 that share rose to 41 percent of young men (Vespa, 2017). Young adults without higher education are more likely to live in their parental home (Fry, 2016). Figure 1.12 illustrates changes in rates of marriage, parenthood, education, employment, and homeowning status among 30-year-old adults between 1975 and 2015. In 1975, three out of four 30-year-olds had married, had a child, were not enrolled in school, and lived somewhere other than their parents' home. In 2015, this number had reduced to just one in three. Figure 1.12 also shows that more 30-year-olds (81 percent) were in the labor force in 2015, compared with 71 percent in 1975, but that fewer (55 percent) earned a moderate income in 2015 compared with 71 percent in 1975. In addition, fewer 30-year-olds (33 percent) were homeowners in 2015, compared with 56 percent in 1975.

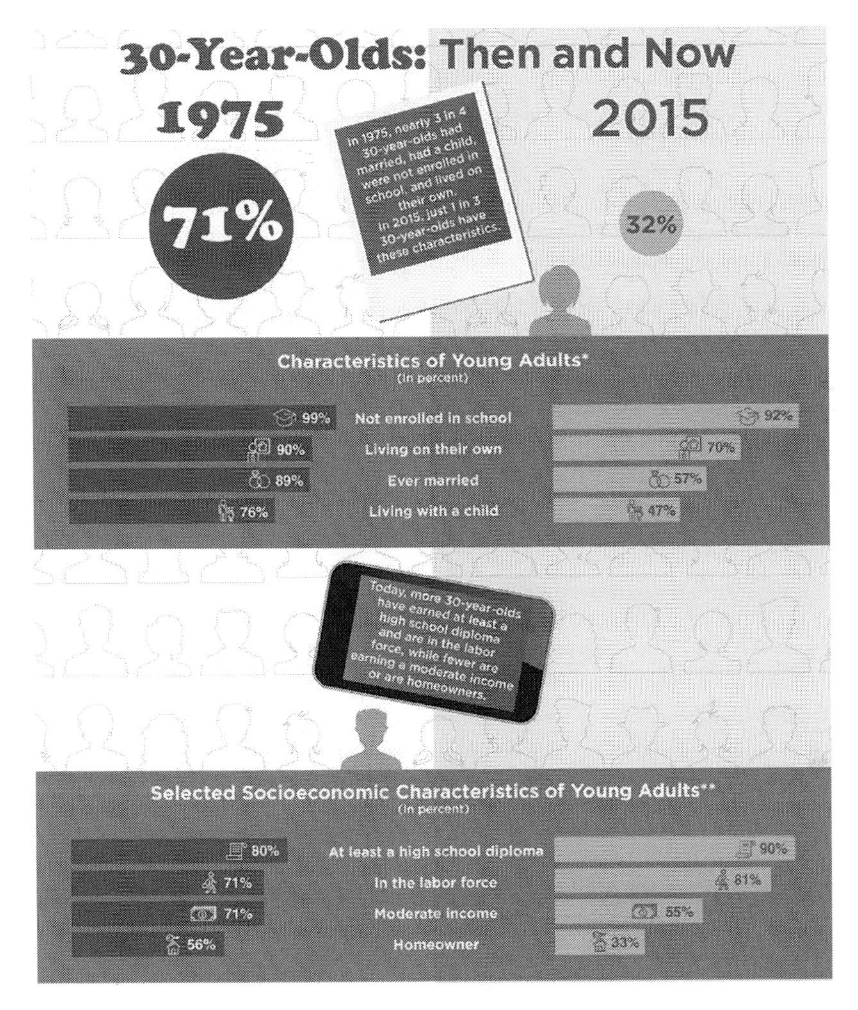

Figure 1.12 Thirty-year-olds Then and Now

Source: U.S. Census Bureau, Current Population Survey, Annual Social and Economic Supplement, 1975 and 2015. https://census.gov/library/visualizations/2016/comm/30-year-olds.html

Single Older Adults

There were 19.5 million unmarried adults age 65 and older in 2016 who made up 17.7 percent of all unmarried people age eighteen and older; 790,103 unmarried grandparents were responsible for most of the basic care of a coresident grandchild and 30.7 percent of coresident grandparents responsible for their grandchildren were unmarried (U.S. Census Bureau, 2017a).

Upon completion of this section, students should be able to:
LO10: Compare monogamy and polygamy.
LO11: Analyze the impact of open marriage and polyamory on the family.

Forms of Marriage

Throughout all cultures across the globe, marriage is a universal institution that is viewed as a passage into family formation. **Marriage** is a socially approved or legally recognized union between two people that establishes sexual and reproductive rights and obligation. Murdock (1981) recorded the marital composition of 1,231 societies from 1960 to 1980. Of these societies, 186 were found to be monogamous; 453 had occasional polygyny; 588 had more frequent polygyny; and 4 had polyandry. Forms of marriage have changed historically across the United States and around the globe. This section explores the forms of marriage and trends in marriage forms.

Polygamy

Polygamy is a form of marriage in which an individual has more than one spouse at a given time. Polygamy, also called plural marriage, was taught by leaders of the Church of Jesus Christ of Latter-day Saints (LDS) for more than half of the nineteenth century in the United States and was practiced publicly from 1852 to 1890 by a minority of LDS families (Flake, 2004). Polygamy is a cultural norm in other parts of the world, but the practice of polygamy violates marriage laws in the United States.

Polygyny

One type of polygamy is polygyny. **Polygyny** is a form of marriage in which a husband has multiple wives at a time. Polygyny is practiced in many countries, such as some Middle Eastern nations and indigenous cultures in Southeast Asia and Africa. In rural parts of Africa, where many families engage in agricultural pursuits, polygynous marriages are prominent because men who have multiple wives can produce more children to become agricultural workers for the family. Since 80 percent of patrilocal societies are polygynous, the presumption about male patrilocal residence seems consistent with presumptions about one-male polygynous family systems, and led to the conviction that early human mating systems could safely be assumed to be patrilocal and polygynous (Hrdy, 2000).

The polygyny threshold model suggests that polygyny should be positively associated with wealth inequality. Polygyny is thought to be more prevalent where male wealth inequality is greater (Oh *et al.*, 2017). In societies that practice polygyny, high-status men are viewed as dominant and can hoard and procreate with multiple marriage partners. This usually leaves lower-status men out. However, research shows that polygyny has been found to be more common among relatively egalitarian horticultural societies (growing crops with simple tools) than in unequal agricultural societies. In fact, the decline of polygyny in most of the world occurred with the emergence of capital-intensive farming with unprecedented wealth disparities (Oh *et al.*, 2017). In modern agricultural societies, where wealth is highly concentrated, few men are wealthy enough to afford more wives, and the wealthy do not take on wives in proportion to their wealth (Santa Fe Institute, 2018).

Polyandry

In marriages that practice polygyny, men may have more wives, but in **polyandry** a woman has more than one husband at a given time. Around the world, polygynous marriage (one man, several women) is vastly more common than polyandrous marriage (one woman, several men); formal polyandrous marriages are exceedingly rare (fewer than 2 percent of human cultures are so classified) (Hrdy, 2000). In the Masai communities in Kenya, polyandry is the usual form to regulate relations between spouses. It means that a married woman may have permitted sexual relations not only with her husband but also with all the men from a similar age group (Ionuţ, 2012). Informally, polyandrous arrangements are far more common due to extramarital affairs, a shortage of women (Peters & Hunt, 1975), the inability of one man to provide security (Smedley, 1980), to a husband's "sharing" of his wife with kin, age-mates, or allies (which by estimates is found in one-third of all human cultures) (Broude, 1994), or due to women's taking up with sequential mates over a lifetime (Sangree, 1980; Guyer, 1994). Once we broaden our definition of polyandrous behavior to include situations where one mother is linked to several men husbands, polyandry is not so rare after all for both pre- and postcolonial West African societies (Hrdy, 2000; Guyer, 1994), for densely populated areas of East Africa (Hakansson, 1988), for tribal South America (Crocker & Crocker, 1994), for aboriginal Australia (Berndt & Berndt, 1951), for traditional societies in Central Japan and southwestern China (Befu, 1968; Shih, 2010), and for additional examples from contemporary mother-centered families in rural and urban Africa, the Americas, and the Caribbean (Hrdy, 1999). Under a range of circumstances, females enhance their reproductive success by mating with multiple partners and use polyandrous mating (soliciting copulations from several or more males) to circumvent male-imposed costs on their free choice of mates (Hrdy, 2000).

Monogamy

Monogamy is a form of marriage in which an individual has only one spouse at any one time. Monogamy is the most preferred form of marriage in most countries today and it is the only marriage pattern sanctioned by law in the United States. This marriage pattern may be rooted in religion. For example, in the Christian Bible, Adam and Eve (who the Bible claims are the world's first couple) are monogamous (Glaeser, 2014). Our modern shift from polygamy to monogamy has also been caused by social, economic, and cultural changes over time. A study of 29 human populations reveals that in stratified agricultural economies, (1) the population frequency of relatively poor individuals increased and that (2) diminishing marginal fitness returns to additional wives prevents extremely wealthy men from obtaining as many wives as their relative wealth would otherwise predict (Ross *et al.*, 2018). For example, a wealthy man with four wives will have fewer kids than two men with two wives and with the same wealth divided between them. These two conditions lead to a high population-level frequency of monogamy (Ross *et al.*, 2018). Monogamy provides many benefits to a married couple. For one, it provides a more stable home in which to raise children. In modern agricultural societies, taking on additional wives usually reduces the amount of a male's material wealth (e.g., land, cattle, equipment available to each wife) (Santa Fe Institute, 2018). The incidence of unwanted pregnancies and the spreading of sexually transmitted diseases might be also reduced in monogamous relationships.

Today, due to high divorce and remarriage rates, serial monogamy is a common marriage pattern. With **serial monogamy**, an individual has several spouses over a lifetime but only one partner at any given time. When men marry women sequentially, or women marry men sequentially, we call it serial monogamy (Borgerhoff Mulder, 2009). For example, British-American film star Elizabeth Taylor was married eight times in a pattern of marriage, divorce, and remarriage, but she had only one husband at a time. Serial monogamy is viewed as favorable to male fitness and unfavorable to women's fitness (Forsberg & Tullberg, 1995) and is viewed as a form of *polygyny*—a strategy whereby some men monopolize more than a single female reproductive life span through repeated divorce and remarriage (Starks & Blackie, 2000).

Marriage has undergone a process of **deinstitutionalization**—a weakening of the social norms that define partners' behavior—over the past few decades. The transitions in the meaning of marriage that occurred in the United States during the twentieth century have created the social context for deinstitutionalization (Cherlin, 2004). Ever since the 1960s, there has been a trend toward individualized marriage in the United States. **Individualized marriage** is a form of marriage in which spouses maintain independence in their relationship. The individualized marriage occurred when the emphasis on personal choice and self-development expanded (Cherlin, 2004). Couples are expected to facilitate such growth and become sources of support for each other. In individualized marriage, for example, both partners earn an income but do not have a joint bank account. After examining changing gender roles in 31 country contexts, Lauer and Yodanis (2011) concluded that when couples practiced individualized marriage, they were more likely to keep their money separate. When it comes to parental roles, both partners can divide up the work and complete a range of tasks.

Consanguineous Marriage

Consanguineous marriage is defined as a union of second cousins or closer relatives (Saadat, 2015). Globally, one billion populations live in communities with a preference for consanguineous marriage (Modell & Darr, 2002). Consanguineous marriage is traditional and respected in most communities of North Africa, the Middle East, and West Asia, where intra-familial unions collectively account for 20–50+ percent of all marriages (Hamamy *et al.*, 2011). Relevant rules specify who may marry each other. For *matrilateral cross-cousin marriage*, a man marries his mother's brother's daughter. In *patrilateral cross-cousin marriage*, a man marries his father's sister's daughter. For example, beginning around 250 CE, there were two different marriage systems in Maya society: a bilateral cross-cousin marriage system for commoners and a matrilateral cross-cousin marriage for royalty and nobility (Hage, 2003). It is likely that 80 percent of all marriages in history may have been between second cousins or closer (Fox, 1983).

The reasons given for the preference of consanguineous marriages are primarily social and economic. Historically, marriage often served to create and maintain alliances and exchanges between social groups through *political marriages* between the children of important leaders or cross-cousin marriages. Also, these kinds of cousin marriages were traditionally encouraged because they strengthened the clan. For example, the Rothschild family is a highly wealthy family descending from Mayer Amschel Rothschild, a court Jew to the German Landgraves of Hesse-Kassel, in the Free City of Frankfurt, who established his banking business in the 1760s (Elon, 1996). An essential part of their family's strategy for success was to keep control of their business in family hands through cousin marriages. Research shows that consanguinity has some protective roles against divorce and also survival of marriages increased among consanguineous marriages compared to unrelated marriages (Saadat, 2015). For example, consanguineous marriage could enforce the couples' stability due to higher compatibility between husband and wife who share the same social relationships after marriage as before marriage, as well as the compatibility between the couple and other family members (Hamamy, 2011). Some countries around the world legally prohibited cousin marriage as incest following a debate about an increased risk of genetic disorder. Consanguineous marriages are associated with an increased risk for congenital malformations and autosomal recessive diseases, with some resultant increased postnatal mortality in the offspring of first cousin couples (Hamamy, 2011). In the United States, each state determines whether cousins can marry and sets legal ages for marriage.

Forms of Consensual Non-monogamy (CNM)

The marriage rate has fallen from 80 percent in 1947 to 50 percent in 2017 (see Figure 1.2). Marriage has been the subject of extensive research across academic disciplines. While monogamy remains the most common romantic relationship arrangement in North America, *consensual non-monogamy* (CNM) is prominent (Rubin *et al.*, 2014). **Consensual non-monogamy (CNM) relationships** are those in which partners explicitly agree that they or their partners can enter romantic and/or sexual relationships with other people (Conley *et al.*, 2013a). Estimates derived from internet samples suggest that approximately 4–5 percent of individuals are currently involved in some form of consensually non-monogamous relationship (Rubin *et al.*, 2014). Haupert *et al.* (2017) suggest that approximately one in five people have previously been a part of a CNM relationship at some point during their lifetime. CNM relationships can take many forms: *open relationships* and *polyamory*. The "open" in open relationship usually refers to the sexual aspect of a non-closed relationship, whereas "polyamory" refers to the extension of a relationship by allowing bonds to form (which may be sexual or otherwise) as additional long-term relationships (Taormino, 2008).

Open Relationships

With the rising marriage age and declining marriage rate today, marriage is often seen as optional. In addition, married couples often expect different rewards from marriage than people did in the past. They often do not marry primarily for financialreasons. Childbearing might not be a motivating factor for marrying either. In consensually non-monogamous relationships there is an open agreement that one, both, or all individuals involved in a romantic relationship may also have other sexual and/or romantic partners (Balzarini *et al.*, 2017). **Open relationships** are relationships in which partners agree on sexual relations with others, either as a couple or independently, but operate with minimal emotional and romantic capacity (Conley *et al.*, 2013b). In open relationships, we would not expect substantial commitment or investment to occur with partners beyond the initial dyad, because these relationships are typically premised around sex (Balzarini *et al.*, 2017).

Open relationships are a generalization of the concept of a relationship beyond monogamous relationships and a form of open relationship is the open marriage, in which the participants in a marriage have an open relationship (Taormino, 2008). A growing number of married couples in the United States today expect an open relationship. An **open marriage** is a form of marriage in which two partners agree that each partner has room for personal growth and to develop a relationship outside marriage. In an open marriage, for example, spouses may agree that each can engage in extramarital sexual relationships. Back in the 1960s, the term open marriage was used to describe individual freedom in choosing marriage partners. O'Neill & O'Neill (1972) defined an *open marriage* as one having eight characteristics: here-and-now living, respect for personal privacy, open and honest communication, role flexibility, open companionship, equality, pursuit of identity, and mutual trust. Wachowiak and Bragg (1980) find that for women, marital adjustment increases as the amount of consensus between husbands' and wives' views on marital openness increases. Couples involved in open marriages adopt ground rules to guide their activities. For example, extramarital sexual relationships may not be allowed or may not be regarded as infidelity.

Polyamorous Relationships

Another form of consensual non-monogamy (CNM) is polyamory. **Polyamory** refers to an identity in which people philosophically agree with and/or practice multi-partner relationships, with the consent of everyone involved (Conley *et al.*, 2013b). The relationship may be sexual, emotional, spiritual, or any combination thereof, depending on the desires and agreements of all partners involved. Although the term polyamory indicates permission to engage in sexual or romantic relationships with more than one partner, the nature of these relationships and how individuals approach them can vary from one person partnering with multiple people, to members of a couple dating a third (triad), to two couples in a relationship with each other (quad), to networks of people involved with each other in various configurations (Sheff, 2013). **Open polyamorous relationships** occur when the committed partners agree to permit their partner to have romantic or sexual relationships with others. **Closed polyamorous relationships** occur when the committed partners do not engage in sexual relationships outside the committed partnership.

Polyamory includes many different styles of intimate involvements and has two concurrent partners (Wosick-Correa, 2010). People who practice polyamory use the terms *primary* and *secondary relationships* to distinguish between degrees of relationships and to indicate the intimate experience of all partners involved. **Primary relationships** refer to relationships that include a close degree of involvement and intimacy. In this configuration, a primary relationship is between two partners who typically share a household (live together) and finances, who are married (if marriage is desired), and/or who have or are raising children together (if children are desired) (Klesse, 2006). A polyamorous person's spouse or live-in partner can be referred to as his or her primary relationship based on the breadth and depth of knowledge the couple has about each other. The type of involvement includes emotional involvement and logistical involvement such as shared childrearing. Research shows that individuals in CNM relationships are equally satisfied with and committed to their relationships as individuals in monogamous relationships (Conley *et al.*, 2013b). Mogilski *et al.* (2015) found no significant differences between relationship satisfaction ratings of monogamous partners and CNM primary partners. Partners beyond the primary relationship are often referred to as non-primary partners or "secondary" partners (Balzarini *et al.*, 2017). **Secondary relationships** refer to relationships that include lesser degrees of involvement and intimacy. A secondary relationship often consists of partners who live in separate households and do not share finances (Klesse, 2006). Secondary partners are afforded relatively less time, energy, and priority in a person's life than are primary partners (Balzarini *et al.*, 2017). Lovers may be referred to as a secondary relationship when they do not live together and see each other only once a week. Furthermore, a secondary relationship often consists of fewer ongoing commitments, such as plans for the future (Veaux *et al.*, 2014). Based on characteristics of involvement, some polyamorists consider themselves to have more than one primary relationship but others prefer only secondary relationships. Self-identified polyamorous individuals (N=1,308) reported less stigma as well as more investment, satisfaction, commitment, and greater communication about the relationship with primary compared to secondary relationships (Balzarini *et al.*, 2017).

Many people across the United States and around the world are supporters of polyamory. Bennett (2009) finds that polyamory is a thriving phenomenon in the United States, with over half a million families openly living in relationships that are between multiple consenting partners. The study of polyamorous relationships has been the focus of social sciences since the evolution of relationship science in the 1990s. Proponents of polyamory claim that it embraces such cultural values as freedom and individualism. Some scholars argue that monogamy is not innate in humans. Ferrer (2008) finds that social monogamy frequently masks biological polyamory in an increasingly significant number of couples. As polyamory extends beyond sexual connection, individuals may report that commitment does exist with partners beyond the initial dyad (Balzarini *et al.*, 2017). **Interdependence theory** posits that individuals initiate and maintain relationships because of the benefits of interactions in a relationship (Rusbult, 2000). As relationships develop, the interaction among partners yields outcomes in the forms of rewards (e.g., sexual pleasure, relationship satisfaction, security), and costs (e.g., increased responsibility, distress or anxiety, despair, fear) (Rusbult & Buunk, 1993). For example, in polyamorous relationships, primary partners may afford certain rewards because primary partners can share in major life decisions and can help to promote greater levels of interdependence (e.g., joint finances, cohabitate, etc.) (Sheff, 2013). It should be more difficult to develop interdependence in secondary relationships compared to primary relationships (Balzarini *et al.*, 2017). The **Investment Model** proposes that motivation to maintain a relationship is the product of four variables that represent the ways one is bound to the relationship: (1) *investment size* (e.g., time invested, plans for the future); (2) *satisfaction* (or how rewarding the relationship is); (3) *quality of alternatives* (or the degree to which one believes that one's needs could be fulfilled in another relationship); and (4) *commitment* (a feeling of attachment to the partner and desire to maintain the relationship) (Rusbult & Buunk, 1993). For example, if the primary partner is the recipient of many of the investments typical in traditional relationship trajectories (moving in together, getting married, having children, etc.), there are simply fewer resources left to invest into relationships with secondary partners, and thus, fewer opportunities to become truly interdependent (Balzarini *et al.*, 2017).

Group Marriage

Group marriage is a form of a non-monogamous marriage arrangement consisting of three to six people living together, sharing finances, children, and household and family responsibilities. Group marriage was practiced in communal societies in the nineteenth and twentieth centuries. For example, the Oneida community (a Christian religious commune) in New York in the 1800s believed strongly in a system of free love known as *complex marriage* (Foster, 2010), where any member was free to have sex with any other who consented (though it was usually the males who did the asking) (Stoehr, 1979). Oneida commune members lived together as a social group and shared parental responsibilities.

Constantine and Constantine (1973) explored the benefits of group marriage or contemporary multilateral marriage. For example, particularly useful is the idea of assigning to each responsibility for his or her domain of adeptness, without competition or dominance. Also from the most involved relationships, the complicated details of separation are worked out without legal advice. However, group marriage has

Upon completion of the chapter, students are about to:
LO12: Apply diversity and global approach to the analysis of family life.

problems because of its inherent complexity (Miller & Miller, 1974). Scholars wonder whether it is possible for all partners in a group marriage to love one another and form attachments equally in the same ways. Childrearing might be another problem. In multilateral as in nuclear families, "good marriages are good for children, bad marriages are not" (Constantine & Constantine, 1973). It was difficult to estimate the number of people who practice group marriage because this form of marriage is not officially permitted in any jurisdiction in the United States.

Framework of the Text: Diversity and Global Approach

The family is a mirror of the changing society in the United States. As forms of families change, society also changes. This text takes a diversity approach, examining diverse families emerging in the United States. From a global approach, this text helps students see how families are changing around the globe.

Diversity Approach: Dimensions of Diverse Family Life

Cultural diversity refers to cultural differences within a society and across societies. Cultural diversity occurs within an increasingly diverse American society and in the process of globalization where countries become interconnected. It is important to be aware of other cultures and appreciate their cultural differences. Family diversity is a way of embracing such dimensions as race, ethnicity, class, gender, sexual orientation, and religion. Indeed, family life differs widely based on the family members' race and ethnicity, social class, gender, sexual orientation, age, and religion, among other factors. Multiple factors influence family life and determine what job and how much income families receive that shape family life. One person's family experience does not represent that of others in terms of race, gender, class, education, religion, etc. These diversity dimensions make every family special.

Race and Ethnicity

Before we discuss the effect of race and ethnicity on families, we must first define some terms. Sociology distinguishes race from ethnicity. **Race** refers to a group of people who share similar biological characteristics such as skin color, hair texture, and eye shape and color. **Ethnicity** refers to a group of people who identify with each other based on a shared cultural heritage or national origin. Race is examined in terms of biological differences and ethnicity is examined in terms of cultural differences. One's ethnicity is often reflected in how one speaks, in what one eats, and in what one wears. Learning about race and ethnicity in families is a great way to understand individual family life. Figure 1.4 illustrates the growth in households headed by racial or ethnic families.

Social Class

Social class refers to a group of people with similar shares of resources in hierarchical social divisions such as upper, middle, and lower classes. Experiences for upper-class and lower-class families are a complete opposite. Family experience differs by social class. People are born into different social classes, which affects their family lives. For example, wealthy families can enjoy the luxury of eating and buying what they want, paying for expensive schools and tutors for their children, and expanding their children's lives through travel. Poor families, on the other hand, must work hard to survive and to meet their basic needs. Sadly, there were 39.7 million people living in poverty in the United States in 2017 (Fontenot *et al.*, 2018).

Gender

Gender is another aspect of diversity that affects family life. **Gender** refers to the social and cultural meanings and practices associated with masculinity and femininity in a culture. Gender is examined in terms of gender identity, gender role, and gender inequality. Gender inequality between males and females can create issues both inside and outside the home. In dual-earner families, females are income providers but are often still expected to take on primary responsibility for the family. Although men sometimes share family responsibilities, working mothers often struggle to handle their dual roles. One gender issue is the pay gap between men and women, which has a large effect on family life. The median annual earnings of women who worked full time and year-round in 2017 were $41,977 while the median annual earnings of men were $52,146 (Fontenot *et al.*, 2018).

Sexual Orientation

Sexual orientation is an enduring pattern of emotional, romantic, or sexual attraction to persons of the opposite sex (heterosexuality), the same sex (homosexuality), or both sexes (bisexuality), or multiple sexes, or lack of sexual attraction to others. Those who are attracted to persons of the opposite sex are heterosexual, those who are attracted to persons of the same sex are homosexual, and those who are attracted to persons of both sexes are bisexual. Same-sex marriage became legal in the United States in June 2015, enabling same-sex couples to be married no matter where they lived in the United States. Americans are becoming more accepting in their views of LGBT people and homosexuality: 63 percent of Americans said in 2016 that homosexuality should be accepted by society, compared with 51 percent in 2006 (Brown, 2017). LGBT families are likely to experience a different life. In the United States, one in four LGBT adults (27 percent), or 2.2 million people, experienced a time in 2015 when they did not have enough money for the food that they or their families needed, compared to 17 percent of non-LGBT adults (Brown *et al.*, 2016).

Religion

One's religion also affects one's family life. This is in part because family life is affected by the rituals, holidays, and ceremonies that make up the religion. Religion lays down guidelines for remaining monogamous within relationships. Adhering to such guidelines may allow a family to live in harmony.

Our global society is becoming more religiously and spiritually diverse, and the religious makeup of the world is changing. For example, it is estimated that Christians will decline from more than three-quarters of the population in 2010 to two-thirds in 2050 while the Muslim population is expected to increase from 1.6 billion people (23 percent of the world's population as of 2010) to 2.76 billion people in 2050 (Lipka, 2015a). The United States has a long history as a majority Protestant nation. More than half of U.S. adults (51.3 percent) identified as Protestants in 2007 (Lipka, 2015b). Protestantism is therefore the dominant religion in the United States and the rest of the religions are considered minority religions. People practicing certain minority religions may experience prejudice and be discriminated against, having an impact on family life. However, Protestants no longer make up a majority of U.S. adults as that figure has fallen in 2014 to 46.5 percent (Lipka, 2015b). Most U.S. adults describe themselves as Catholic, Baptist, Methodist, Jewish, Mormon or Muslim, etc. and can be sorted into highly religious (39 percent), somewhat religious (32 percent), and non-religious groups (29 percent) based on the religious and spiritual beliefs they share (Pew Research Center, 2018).

The Global Approach

The world is becoming more global and connected, which also influences families. This text will examine the impact of American families on other countries and the impact of the world's families on American families. Let's start by exploring how where we come from affects family structure and marriage patterns (see Global Overview on Family Structure and Marriage Patterns).

Global Overview on Family Structure and Marriage Patterns

Globally, marriage patterns may be influenced by culture, religion, parental choice, and individual desire. As we will see in the next sections, in some areas of the world, fraternal polyandry, ghost marriage, and child marriage are practiced.

Fraternal Polyandry

Fraternal polyandry is a form of marriage in which several brothers share the same wife. The eldest brother is responsible for finding a wife, but all the brothers share the wife equally. Children consider their uncles as their fathers. Fraternal polyandry can help reduce population growth when the region has limited resources. In feudal times, it also reduced tax obligations to feudal lords as all the brothers shared the same household. Proponents of feudal polyandry claim that the practice preserves the productive resources of the family units across generations (Goldstein, 1987). Fraternal polyandry was traditionally practiced in Nepal, parts of China, and part of northern India; even though it is currently illegal in China, polyandry in Tibet is *de facto* the norm in rural areas (Gielen, 1998). The stem family in mid-nineteenth-century England served the same purpose. In *stem families*, the father selected one child to remain near the parental homestead to work on the farm and eventually inherit it, thus continuing the family line (Le Play, 1855; Ruggles, 2010). The stem family prevented brothers from dividing family resources among male heirs and kept the home estate's resources intact by allowing the eldest son to inherit the family estate.

Ghost Marriage

Ghost marriage is a form of marriage in which the two partners are deceased. Such marriages were traditionally practiced in Asia. Similar ghost marriages are practiced in Sudan as the brother serves as a replacement to the bride once her husband dies. Their offspring are considered children of the deceased husband. Similar marriages were also practiced in France, especially as many died during war times. **Posthumous marriage** occurs when one of the participating members is deceased. Since World War I, there have been hundreds of people requesting such marriages. This practice is legal in France. For example, a French man recently posthumously married his partner who died in a terror attack (Gonzalez-Ramirez, 2017).

Child Marriage

Child marriage occurs when an individual marries before reaching the age of eighteen. About 15 million girls a year marry before age eighteen, one every two seconds (UNESCO, 2014). If current trends continue, the number of girls who marry as children will reach nearly 1 billion by 2030 (Gray, 2016). At least 117 countries around the world allow it (Sandstrom & Theodorou, 2016). Some countries practice child marriage due to cultural traditions or poor socioeconomic situations. For example, people suffering from poverty may marry their daughter off for money (often called a bride price). Girls who marry as children are less likely to achieve their full potential. They are more likely to leave education early, suffer domestic violence, contract HIV/AIDS, and die due to complications during pregnancy and childbirth (Gray, 2016). Such practices are outlawed out of concerns for women's rights in other areas of the world. In the United States, children under the age of eighteen can marry in many jurisdictions with parental consent or in special circumstances such as teenage pregnancy. The United Nations marked the first International Day of the Girl Child by calling for an end to child marriage, and stressing education as one of the best strategies for protecting girls against this harmful practice (UN, 2012).

Source: © Stephanie Sinclair/VII/Tooyoungtowed.org. https://news.un.org/en/story/2012/10/423262-worlds-first-international-day-girl-child-un-calls-end-child-marriage

A Final Note

Before we conclude this chapter, consider the following lyrics from the 2008 Beijing Summer Olympics theme song called 'You and Me,' which express our global connectedness:

> You and me
> From one world
> We are family.

With its global approach, this text will allow you to explore marriage patterns and family life in the United States and beyond. As you will see, you will learn how the world and its diverse families are indeed connected. We can also say that families are a mirror that helps us understand our diverse global society.

Summary

Chapter 1 has discussed the fundamentals of the family and marriage. The family is a primary group and social institution. Family is formed by marriage, remarriage, or adoption, and the affiliated kin. Based on structure, there are nuclear families and extended families. Consanguineous families are the equivalent of extended families. Conjugal families are formed through marriage and are the equivalent of nuclear families.

Patrilineal families trace descent from the father's side while matrilineal families trace descent from the mother's side. In the patriarchal family, men hold authority over family members and in a matriarchal family women hold authority over family members. An egalitarian family is a form of family in which both the father and mother share power equally. Trends in forms of family include stepfamily, cohabiting family, single-parent family, LGBT, childfree, and grandparent families. Marriage forms include monogamy and polygamy. Polygamy is practiced on a traditional and religious basis. Consensual non-monogamy (CNM) relationships such as open relationships and polyamory are practiced in modern society.

The textbook has been written from a diversity approach and global approach. Family life is examined in terms of race and ethnicity, social class, gender, sexual orientation, religion, etc. Monogamy is the only form of marriage sanctioned by law in the United States but fraternal polyandry and child marriage are practiced in other parts of the world.

Families serve as a mirror, helping us understand a diversifying American society and globalizing world.

Key Terms

Bilateral descent (or two-sided descent) is a form of family lineage that traces a person's descent through both the mother's and the father's sides.

Breadwinner is a person who earns money to support others and usually works outside the home.

Childfree families are a form of family in which married partners choose not to have children.

Child marriage occurs when an individual marries before reaching the age of eighteen.

Closed polyamorous relationships occur when the committed partners do not engage in sexual relationships outside the committed partnership.

Cohabiting family is a form of family in which a couple, with or without children, lives together without being legally married.

Conjugal family is a form of family that is created by marriage, adoption, or remarriage.

Consanguineous family is a form of family that is created by blood ties.

Consanguineous marriage is defined as a union of second cousins or closer relatives.

Consensual non-monogamy (CNM) relationships are those in which partners explicitly agree that they or their partners can enter romantic and/or sexual relationships with other people.

Cultural diversity refers to cultural differences within a society and across societies.

Culture is a way of life within a society or social group and is central to family life. Culture includes material culture and non-material culture.

Deinstitutionalization is weakening of the social norms that define partners' behavior.

Descent refers to the biological relationship between individuals in society.

Dual-earner family refers to both husband and wife providing income for the family, usually by working outside the home.

Egalitarian family is a form of family in which power and descent is shared equally by both wife and husband.

Ethnicity refers to a group of people who identify with each other based on a shared cultural heritage or national origin.

Extended family (or multigenerational household) is a form of family in which the nuclear family and relatives live in the same household or nearby.

Family consists of a householder and one or more other people living in the same household who are related to the householder by birth, marriage, or adoption.

Family household consists of a householder and family members related by blood, marriage, or adoption and unrelated people who live in that household.

Fraternal polyandry is a form of marriage in which several brothers share the same wife.

Gender refers to the social and cultural meanings and practices associated with masculinity and femininity in a culture.

Ghost marriage is a form of marriage in which the two partners are deceased.

Grandparent-headed family is a form of family in which grandparents take care of a grandchild or grandchildren with or without a parent of the child being present.

Group marriage is a form of non-monogamous marriage arrangement consisting of three to six people living together, sharing finances, children, and household and family responsibilities.

Household consists of all people who occupy a housing unit.

Householder refers to the person in whose name the housing unit is owned, rented, or maintained.

Incest taboo is a cultural rule that prohibits sexual relations between closely related persons.

Individualized marriage is a form of marriage in which spouses maintain independence in their relationship.

Interdependence theory posits that individuals initiate and maintain relationships because of the benefits of interactions in a relationship.

Intergenerational cultural transmission refers to the transmission of cultural ideas (e.g., values, beliefs, knowledge, practices) from one generation to the next generation.

Intimate relationships are interpersonal relationships that involve physical or emotional intimacy.

Investment Model proposes that motivation to maintain a relationship is the product of four variables (investment size, satisfaction, quality of alternatives, and commitment).

Iroquois kinship is the Iroquoian (a Native American group) kinship system.

Kinkeeping consists of efforts expended on behalf of keeping family members in touch with one another.

Kinship is the relationship that links family members through blood, marriage, adoption, or remarriage.

Kinship norms are culturally defined and specify how kin-related persons are expected to behave towards each other.

LGBT families are lesbian, gay, bisexual, or transgender families.

Lineage is the line through which descent is traced.

Marriage is a socially approved or legally recognized union between two people that establishes sexual and reproductive rights and obligations.

Married couple is a husband and wife enumerated as members of the same household.

Material culture refers to tangible objects that the people of a society make and use.

Matriarchal family is a form of family in which the mother is the head of the family with authority over other family members.

Matrilineal family is a form of family in which descent is traced through females from a female ancestor to her descendants.

Matrilocal family is a form of family in which a newlywed couple lives with the wife's side.

Monogamy is a form of marriage in which an individual has only one spouse at any one time.

Neolocal family is a form of family in which the newlywed couple resides in their own home.

Non-family household consists of a householder living alone or a householder who shares the housing unit with unrelated people.

Non-material culture refers to intangible creations that influence the behavior of the members of a society.

Nuclear family is a form of family consisting of a married couple and their biological children.

Open marriage is a form of marriage in which two partners agree that each partner has room for personal growth and to develop a relationship outside marriage.

Open polyamorous relationships occur when the committed partners agree to permit their partner to have romantic or sexual relationships with others.

Open relationships are relationships in which partners agree on sexual relations with others, either as a couple or independently, but operate with minimal emotional and romantic capacity.

Patriarchal family is a form of family in which the father holds authority over his wife and children.

Patrilineal family is characterized by the inheritance of property, rights, surnames, or titles through the male line.

Patrilocal family is a form of family in which a newlywed couple resides with or near the husband's parents.

Polyamory refers to an identity in which people philosophically agree with and/or practice multi-partner relationships, with the consent of everyone involved.

Polyandry is a form of marriage in which a woman has more than one husband at a given time.

Polygamy is a form of marriage in which an individual has more than one spouse at a given time.

Polygyny is a form of marriage in which a husband has multiple wives at a time.

Posthumous marriage occurs when one of the participating members is deceased.

Primary group is a small group whose relationships are face to face and personal. The family is an example of a primary group.

Primary kinship refers to members of the families of origin (immediate family).

Primary relationships refer to relationships that include a close degree of involvement and intimacy.

Primary socialization occurs during early childhood within the family.

Race refers to a group of people who share similar biological characteristics.

Secondary group is a large group whose relationships are impersonal and goal-oriented.

Secondary kinship is the primary kin of a primary kin (extended family).

Secondary relationships refer to relationships that include lesser degrees of involvement and intimacy.

Secondary socialization occurs outside the family.

Serial monogamy is where an individual has several spouses over a lifetime but only one partner at any given time.

Sexual orientation is an enduring pattern of emotional, romantic, or sexual attraction to persons of the opposite sex (heterosexuality), the same sex (homosexuality), or both sexes (bisexuality), or multiple sexes, or lack of sexual attraction to others.

Single-parent family is a form of family that includes children under age eighteen and that is headed by a parent who is widowed or divorced and not remarried, or never married.

Social class refers to a group of people with similar shares of resources in hierarchical social divisions such as upper, middle, and lower classes.

Social groups are characterized by more than two people, frequent interaction, a sense of belongingness, and a feeling of interdependence.

Social institution is an organized system of social relationships which embodies values and procedures and meets basic needs of the society.

Society refers to groups of people who share a common culture in a geographic location.

Stepfamily is a form of family in which at least one marital or cohabiting partner has children who are not biologically related to the other spouse or partner.

Subculture is a culture shared by a category of people within a society.

Tertiary kinship is the primary kin of a secondary kin.

Unilineal descent (or one-sided descent) is a form of family lineage that traces a person's descent through either the maternal or the paternal line.

Discussion Questions

- The family is a social institution in every society. Do you think that there could be a society without the family institution? Which social institution would replace the family institution? How is the family institution related to other social institutions?
- What is the ideal traditional family? What positive and negative effects do this kind of family have on each spouse and child?
- Which marriage pattern do you think fits modern society? Why?

Suggested Film and Video

ABC (Jul 27, 2015). ABC Pushes Polyamory with 'Trailblazing Triad' [Video file] Video posted to https://www.youtube.com/watch?v=P0SOCcRWrrE

Feingold, L. (June 11, 2015). TED Talks. Feingold L: Polyamory [Video file] Video posted to http://tedxtalks.ted.com/video/Polyamory-Leon-Feingold-TEDxBus

MacDonald, K. (2009). Layers of Gender Identity. [Video file] Video posted to http://fod.infobase.com/p_Search.aspx?rd=a&q=Layers%20of%20Gender%20identity

Internet Sources

Family Life, Help for Today and Hope for Tomorrow: http://www.familylife.com/

More than Two. (2011). *Dos and don'ts for happy polyamorous relationships:* https://www.morethantwo.com/polytips.html

National Resource Center for Marriage and Families: https://www.healthymarriageandfamilies.org/

The United Nations, Division for Social Policy and Development Family: https://www.un.org/development/desa/family/international-day-of-families/2016idf.html

References

Albert, S. M. (1990). Caregiving as a Cultural System: Conceptions of Filial Obligation and Parental Dependency in Urban America. *American Anthropologist*, 92, 319–331.

Alessandria, K. P., Kopacz, M. A., Goodkin, G., Valerio, C. & Lappi, H. (2016). Italian American Ethnic Identity Persistence: A Qualitative Study. *Identity: an International Journal of Theory and Research,* 16(4), 1–17.

Ariès, P. (1962). *Centuries of Childhood.* New York: Vintage Books.

Balzarini, R. N., Campbell, L., Kohut, T., Holmes, B. M., Lehmiller, J. J., Harman, J. J. *et al.* (2017). Perceptions of Primary and Secondary Relationships in Polyamory. *PLoS ONE,* 12(5), e0177841.

Befu, H. (1968). Origins of Large Households and Duo-local Residence in Central Japan. *Am. Anthropol.*, 70, 309–319.

Bell, D. (1973). *The Coming of Post-Industrial Society: A Venture in Social Forecasting.* New York: Basic Books.

Bennett, J. (June 29, 2009). Polyamory—Relationships with Multiple, Mutually Consenting Partners—Has a Coming-out Party. *Newsweek Magazine Online.* Retrieved on August 19, 2016 from http://www.newsweek.com/polyamory-next-sexual-revolution-82053

Berndt, R. M. & Berndt, C. H. (1951). *Sexual Behavior in Western Arnhem Land.* Viking Fund Publications in Anthropology, No. 16. New York: The Viking Fund.

Beutler, I., Burr, W., Bahr, K. & Herrin, D. (1989). The Family Realm: Theoretical Contributions for Understanding Its Uniqueness. *Journal of Marriage and Family,* 51(3), 805–816.

Bianchi, S. M. & Casper, L. M. (2003). American Families, Population Bulletin 55, no. 4 (2000), 3–6 and 14; and Jason Fields, Children's Living Arrangements and Characteristics: March 2002, *Current Population Reports* P20–547.

Blackwood, E. (2008). Not Your Average Housewife: Minangkabau Women Rice Farmers in West Sumatra. In M. Ford & L. Parker (Eds.), *Women and Work in Indonesia*. New York: Routledge.

Borgerhoff Mulder, M. (2009). Serial Monogamy as Polygyny or Polyandry?: Marriage in the Tanzanian Pimbwe. *Human Nature*, 20(2), 130–150.

Brakman, S. V. (1995). Filial Responsibility and Decision-making. In L. B. McCullough & N. L. Wilson (Eds.), *Long-term Care Decisions* (pp. 181–196). Baltimore: Johns Hopkins.

Brewster, K. L. & Padavic, I. (2000). Change in Gender-Ideology, 1977–1996: The Contributions of Intracohort Change and Population Turnover. *Journal of Marriage and the Family*, 62(2), 477–487.

Broude, G. J. (1994). *Marriage, Family and Relationships: A Cross-Cultural Encyclopedia*. Santa Barbara, CA: ABC-CLIO.

Brown, A. (June 13, 2017). *5 Key Findings about LGBT Americans*. Retrieved on August 11, 2017 from http://www.pewresearch.org/fact-tank/2017/06/13/5-key-findings-about-lgbt-americans/

Brown, T. N. T., Romero, A. P. & Gates, G. J. (July 2016). *Food Insecurity and SNAP Participation in the LGBT Community*. The Willman Institute. Retrieved on December 4, 2016 from http://williamsinstitute.law.ucla.edu/research/lgbt-food-insecurity-2016/

Cai, H. (2001). *A Society without Fathers or Husbands: the Na of China*. New York: Zone Books.

Cannell, F. & Green, S. (1996). Patriarchy. In A. Kuper & S. Kuper, *The Social Science Encyclopedia*. London: Taylor & Francis.

Carol, S. (2014). The Intergenerational Transmission of Intermarriage Attitudes and Intergroup Friendships: The Role of Turkish Migrant Parents. *Journal of Ethnic and Migration Studies*, doi: 10.1080/1369183X.2013.872557

Cherlin, A. (2004). The Deinstitutionalization of American Marriage. *Journal of Marriage and Family*, 66(4), 848–861. Retrieved on September 30, 2018 from http://www.jstor.org/stable/3600162

Cohn, D'V. & Passel, J. S. (April 5, 2018). *A Record 64 Million Americans Live in Multigenerational Households*. Washington, DC: Pew Research Center. Retrieved on September 17, 2018 from http://www.pewresearch.org/fact-tank/2018/04/05/a-record-64-million-americans-live-in-multigenerational-households/

Collier, J. (1988). *Marriage and Inequality in Classless Societies*. Stanford, CA: Stanford University Press.

Conley, T. D., Moors, A. C., Matsick, J. L. & Ziegler, A. (2013a). The Fewer the Merrier: Assessing Stigma Surrounding Non-Normative Romantic Relationships. *Analyses of Social Issues and Public Policy*, 13, 1–30.

Conley, T. D., Ziegler, A., Moors, A. C., Matsick, J. L. & Valentine, B. (2013b). A Critical Examination of Popular Assumptions about the Benefits and Outcomes of Monogamous Relationships. *Personality and Social Psychology Review*, 17, 124–141.

Constantine, L. L. & Constantine, J. M. (1973). *Group Marriage: A Study of Contemporary Multilateral Marriage*. Virginská univerzita: Macmillan.

Creighton, C. (1999). The Rise and Decline of the 'Male Breadwinner Family' in Britain. *Cambridge Journal of Economics*, 23(5), 519–541. Retrieved on September 22, 2018 from https://doi.org/10.1093/cje/23.5.519

Cremin, L. (1970). *American Education: The Colonial Experience, 1607–1783*. New York: Harper & Row.

Crocker, W. & Crocker, J. (1994). *The Canela: Bonding through Kinship, Ritual and Sex*. Fort Worth, TX: Harcourt Brace.

Cunningham, M. (2008). Changing Attitudes toward the Male Breadwinner, Female Homemaker Family Model: Influences of Women's Employment and Education over the Lifecourse. *Social Forces*, 87(1), 299–323. Retrieved on September 22, 2018 from ttp://www.jstor.org/stable/20430858

De Vries, J., Kalmijn, M. & Liefbroer, A. C. (2009). Intergenerational Transmission of Kinship Norms? Evidence from Siblings in a Multi-Actor Survey. *Social Science Research*, 38(1), 188–200.

DiRenzo, G. J. (1990). *Human Social Behavior: Concepts and Principles of Sociology*. USA: Holt, Rinehart and Winston, Inc.

Dube, L. (1997). *Women and Kinship: Comparative Perspectives on Gender in South and South-East Asia*. Tokyo: United Nations University Press.

Elon, A. (1996). *Founder: Meyer Amschel Rothschild and His Time*. New York: HarperCollins.

Ezzell, C. (2001). The Himba and the Dam. *Scientific American*, 284(6), 80–90.

Fawver, K. (2012). Neolocality and Household Structure in Early America. *The History of the Family*, 17, 407–433.

Ferrer, J. N. (2008). Beyond Monogamy and Polyamory: A New Vision of Intimate Relationships for the Twenty-first Century. *Revision*, 30(1/2), 53–58.

Finch, J. (1989). *Family Obligations and Social Change*. Oxford. Polity Press.

Fine, M. A. (1992). Families in the United States: Their Current Status and Future Prospects. *Family Relations*, 41.

Fishbein, H. D. (2002). *Peer Prejudice and Discrimination: The Origins of Prejudice*. Psychology Press.

Flake, K. (2004). *The Politics of American Religious Identity*. Chapel Hill, NC: University of North Carolina Press.

Fontenot, K., Semega, J. & Kollar, M. (2018). *Income and Poverty in the United States: 2017 Current Population Reports*.

U.S. Census Bureau. Retrieved on September 29, 2018 from https://www.census.gov/content/dam/Census/library/publications/2018/demo/p60-263.pdf

Forsberg, A. J. L. & Tullberg, B. S. (1995). The Relationship between Cumulative Number of Cohabiting Partners and Number of Children from Men and Women in Modern Sweden. *Ethology and Sociobiology*, 16, 221–232.

Foster, L. (2010). Free Love and Community: John Humphrey Noyes and the Oneida Perfectionists. In D. E. Pitzer (Ed.), *America's Communal Utopias* (pp. 253–278). Chapel Hill, NC: University of North Carolina Press.

Fox, R. (1983). *Kinship and Marriage: An Anthropological Perspective.* Cambridge Studies in Social Anthropology. Cambridge, UK: Cambridge University Press.

Frankenberg, E. & Thomas, D. (2001). *Measuring Power*, FCND Discussion Paper No. 113. International Food Policy Research Institute.

Fry, R. (May 24, 2016). *For First Time in Modern Era, Living With Parents Edges Out Other Living Arrangements for 18- to 34-Year-Olds.* Washington, DC: Pew Research Center. Retrieved on July 26, 2016 from http://www.pewsocialtrends.org/2016/05/24/for-first-time-in-modern-era-living-with-parents-edges-out-other-living-arrangements-for-18-to-34-year-olds/

Gielen, U. (1998). Gender Roles in Traditional Tibetan Cultures. In L. L. Adler (Ed.), *International Handbook on Gender Roles* (pp. 413–437). Westport, CT: Greenwood.

Glaeser, K. (2014). Till Death Us Do Part: The Evolution of Monogamy. *Oglethorpe Journal of Undergraduate Research, 4*(2), 1–21. Retrieved February 5, 2016 from http://digitalcommons.kennesaw.edu/cgi/viewcontent.cgi?article=1043&context=ojur

Goldstein, M. C. (1987). When Brothers Share a Wife. *Natural History*, 96(3), 109–112.

Gonzalez-Ramirez, A. (2017). French Man Posthumously Marries His Partner Who Died in a Terror Attack. Retrieved on August 10, 2017 from http://www.refinery29.com/2017/06/157269/france-policeman-terror-attack-posthumous-wedding

Gordon, D. N. (1970). Societal Complexity and Kinship: Family Organization or Rules of Residence? *The Pacific Sociological Review,* 13, 252–262.

Gray, A. (2016). These Are the Countries Where Child Marriage Is Legal. *World Economic Forum.* Retrieved on September 28, 2018 from https://www.weforum.org/agenda/2016/09/these-are-the-countries-where-child-marriage-is-legal/

Guyer, J. (1994). Lineal Identities and Lateral Networks: The Logic of Polyandrous Motherhood. In C. Bledsoe & G. Pison (Eds.), *Nuptiality in Sub-Saharan Africa—Contemporary Anthropological and Demographic Perspectives* (pp. 231–252). Oxford, UK: Clarendon Press.

Hage, P. (2003). The Ancient Maya Kinship System. *Journal of Anthropological Research,* 59(1), 5–21. Retrieved September 20, 2018 from http://www.jstor.org/stable/3631442

Hakansson, T. (1988). Bridewealth, Women and Land: Social Change among the Gusi of Kenya. *Uppsala Studies in Cultural Anthropology*, 10.

Hamamy, H. (2011). Consanguineous Marriages: Preconception Consultation in Primary Health Care Settings. *Journal of Community Genetics*, 3(3), 185–92. Retrieved on November 11, 2018 from https://www.ncbi.nlm.nih.gov/pmc/articles/PMC3419292/#CR11

Hamamy, H., Antonarakis, S. E., Cavalli-Sforza, L. L., Temtamy, S., Romeo, G. *et al.* (2011). Consanguineous Marriages, Pearls and Perils: Geneva International Consanguinity Workshop Report. *Genet Med*, 13, 841–847.

Harwood, J. (2000). Communication Media Use in the Grandparent–Grandchild Relationship. *Journal of Communication,* 50(4), 56–78.

Haupert, M, Gesselman, A. N. & Moors, A. C., Fisher H. & Garcia J.R. (2017). Prevalence of Experiences with Consensual Nonmonogamous Relationships: Findings from Two National Samples of Single Americans. *Journal of Sex & Marital Therapy*, 43(5), 424–440.

Haviland, W. (2002). *Cultural Anthropology.* Belmont, CA: Wadsworth Publishing.

Hawkes, K. & Bliege Bird, R. (2002). Showing Off, Handicap Signaling, and the Evolution of Men's Work. *Evolutionary Anthropology,* 11, 58–67.

Hayghe, H. (1981). Husbands and Wives as Earners: An Analysis of Family Data. *Monthly Labor Review,* 104(2), 46–53. Retrieved September 22, 2018 from http://www.jstor.org/stable/41841417

Horton, P. B., Cohen, B. J. & Hunt, C. L. (1983). *Sociology.* Columbus, OH: McGraw-Hill Higher Education.

Hrdy, S. B. (1999). *Mother Nature: A History of Mothers, Infants, and Natural Selection.* New York: Pantheon.

Hrdy, S. B. (2000). The Optimal Number of Fathers: Evolution, Demography, and History in the Shaping of Female Mate Preferences. *Annals of the New York Academy of Sciences,* 907. 10.1111/j.1749–6632.2000.tb06617.x

Inkeles, A. & Smith, D. H. (1974). *Becoming Modern: Individual Change in Six Developing Countries.* Cambridge, MA: Harvard University Press.

Ionuţ, A. (2012). The Social Functions of the Family. *Euromentor Journal—Studies about Education.* http://www.ceeol.com/aspx/getdocument.aspx?logid=5&id=296DFE8B-AADB-42AA-9647-299D9A63AD03

Jacobs, R. E. (1991). Iroquois Great Law of Peace and the United States Constitution: How the Founding Fathers

Ignored the Clan Mothers. *American Indian Law Review,* 16(2), 497–531, esp. pp. 498–509.
Johnson, M. M. (1988). *Strong Mothers, Weak Wives: The Search for Gender Equality.* Berkeley, CA: University of California Press.
Kahn, J. S. (1980). *Minangkabau Social Formations: Indonesian Peasants and the World Economy.* Cambridge, UK: Cambridge University Press.
Kemp, C. L. (2007). Grandparent–Grandchild Ties: Reflections on Continuity and Change across Three Generations. *Journal of Family Issues,* 28, 855–881.
Klesse, C. (2006). Polyamory and Its 'Others': Contesting the Terms of Non-Monogamy. *Sexualities,* 9, 565–583.
Koc, I. (2007). The Timing of Leaving Parental Home and Its Relationship with Other Life Course Events in Turkey. *Marriage and Family Review,* 42(1), 15–22. https://www.ncbi.nlm.nih.gov/pmc/articles/PMC3785225/
Komarovsky, M. & Waller, W. (1945). Studies of the Family. *American Journal of Sociology,* 50(6), 443–451. Retrieved from http://www.jstor.org/stable/2771388
Korotayev, A. (2003). Form of Marriage, Sexual Division of Labor, and Postmarital Residence in Cross-Cultural Perspective: A Reconsideration. *Journal of Anthropological Research,* 59(1), 69–89.
Lauer, S. R. & Yodanis, C. (2011). Individualized Marriage and the Integration of Resources. *Journal of Marriage and Family,* 73, 669–683.
Leonetti, D. L. & Chabot-Hanowell, B. (2011). The Foundation of Kinship: Households. *Human Nature,* 22(1–2), 16–40.
Le Play, F. (1855). *Les ouvriers européens: Etudes sur les travaux, la vie domestique et la condition morale des populations ouvrières de l'Europe, précédées d'un exposé de la méthode d'observation.* Paris: Imprimerie impériale.
Lipka, M. (April 2, 2015a). *7 Key Changes in the Global Religious Landscape.* Washington, DC: Pew Research Center. Retrieved on October 31, 2016 from http://www.pewresearch.org/fact-tank/2015/04/02/7-key-changes-in-the-global-religious-landscape/
Lipka, M. (August 27, 2015b). *10 Facts about Religion in America.* Washington, DC: Pew Research Center. Retrieved on October 31, 2016 from http://www.pewresearch.org/fact-tank/2015/08/27/10-facts-about-religion-in-america/
Livingston, G. (December 22, 2014). *Fewer than Half of U.S. Kids Today Live in a 'Traditional' Family.* Retrieved on August 10, 2017 from http://www.pewresearch.org/fact-tank/2014/12/22/less-than-half-of-u-s-kids-today-live-in-a-traditional-family/
Lockard, C. (2007). *Societies, Networks, and Transitions.* Boston, MA: Cengage Learning.
MacIver, R.M. (1931). *Society: Its Structure and Changes.* New York: R. Long & R. R. Smith, Inc.
Malti-Douglas, F. (2007). *Encyclopedia of Sex and Gender.* Detroit, MI: Macmillan.
Marlowe, F. (2003). A Critical Period for Provisioning by Hadza Men. Implication for Pair Bonding. *Evolution and Human Behavior,* 24, 217–229.
Marlowe, F. (2007). Hunting and Gathering. The Human Sexual Division of Foraging Labor. *Cross-Cultural Research,* 41, 170–195.
Marlowe, F. (2010). *The Hadza: Hunter-gatherers of Tanzania.* Berkeley, CA: University of California Press.
Masci, D., Brown, A. & Kiley, J. (June 26, 2017). *5 Facts about Same-Sex Marriage.* Washington, DC: Pew Research Center. Retrieved on August 10, 2017 from http://www.pewresearch.org/fact-tank/2017/06/26/same-sex-marriage/
Mattison, S. M., Beheim, B., Chak, B. & Buston, P. (2016). Offspring Sex Preferences among Patrilineal and Matrilineal Mosuo in Southwest China Revealed by Differences in Parity Progression. *Royal Society Open Science,* 3(9), 160526.
Miller, D. & Miller, J. (1974). Reviewed Work: Group Marriage: A Study of Contemporary Multilateral Marriage by Larry L. Constantine, Joan M. Constantine. *Journal of Marriage and Family,* 36(2), 416–419.
Modell, B. & Darr, A. (2002). Science and Society: Genetic Counselling and Customary Consanguineous Marriage. *Nat Rev Genet.* 3, 225–229.
Mogilski, J. K., Memering, S. L., Welling, L. L. & Shackelford, T. K. (2015). Monogamy versus Consensual Non-Monogamy: Alternative Approaches to Pursuing a Strategically Pluralistic Mating Strategy. *Archives of Sexual Behavior,* 1–11.
Morgan, L. H. (1851). *The League of the Ho-dé-no-sau-nee or Iroquois.* Rochester, NY: Sage and Brothers.
Morgan, L. H. (1871). *Systems of Consanguinity and Affinity of the Human Family.* Washington, DC: Smithsonian Institution.
Murdock, G. P. (1949). *Social Structure.* New York: Macmillan.
Murdock, G. P. (1981). *Atlas of World Cultures.* Pittsburgh: University of Pittsburgh Press.
Musgrave, P. W. (1971). The Relationship between the Family and Education in England: A Sociological Account. *British Journal of Educational Studies,* 19(1), 17–31.
Nadim, M. (2015). Undermining the Male Breadwinner Ideal? Understandings of Women's Paid Work among Second-Generation Immigrants in Norway. *Sociology,* 50(1), 109–124.
Oh, S.-Y., Ross, C. T., Borgerhoff, M. & Bowles, S. (2017). *The Decline of Polygyny: An Interpretation.* Santa Fe Institute. Retrieved on September 29, 2018 from https://sfi-edu.s3.amazonaws.com/sfi-edu/production/uploads/working_paper/pdf/2017-12-037-rev_fedae8.pdf
O'Neill, N. & O'Neill, G. (1972). *Open Marriage: A New Life Style for Couples.* New York: M. Evans & Company.

Parsons, T. (1955). The American Family: Its Relations to Personality and the Social Structure. In T. Parsons and R. F. Bales (Eds.), *Family Socialization and Interaction Process*. New York: Free Press.

Pearce, L. D., Hayward, G. M., Chassin, L. & Curran, P. J. (2018). The Increasing Diversity and Complexity of Family Structures for Adolescents. *Journal of Research on Adolescence*, 28(3), 591–608. Retrieved on September 17, 2018 from http://curran.web.unc.edu/files/2018/08/PearceHaywardChassinCurran2018.pdf

Peters, J. F. & Hunt, C. L. (1975). Polyandry among the Yanomana Shirishana. *J. Comp. Fam. Stud.*, 6, 197–207.

Pew Research Center (2010). *The Decline of Marriage and Rise of New Families*. Retrieved on September 23, 2018 from http://www.pewsocialtrends.org/2010/11/18/the-decline-of-marriage-and-rise-of-new-families/

Pew Research Center (2018). *The Religious Typology: A New Way to Categorize Americans by Religion*. Retrieved on September 29, 2018 from http://www.pewforum.org/2018/08/29/the-religious-typology/

Pinker, S. (2011). *The Better Angels of Our Nature: Why Violence Has Declined*. New York: Viking.

Powell, B., Bolzendahl, C., Geist, C. & Steelman, L. C. (2010). *Counted Out: Same-sex Relations and Americans' Definitions of Family*. New York: Russell Sage Foundation.

Radcliffe-Brown, A. R. (1952). *Structure and Function in Primitive Society, Essays and Addresses*. London: Routledge.

Ramakrishnan, K. R. & Balgopal, P. R. (1995). Role of Social Institutions in a Multicultural Society. *Journal of Sociology & Social Welfare*, 22(1), Article 3. Available at: https://scholarworks.wmich.edu/jssw/vol22/iss1/3

Rammohan, A. & Robertson, P. (2012). Do Kinship Norms Influence Female Education? Evidence from Indonesia. *Oxford Development Studies*, 40(3), 283–304.

Reis, H. T. & Sprecher, S. (Eds.) (2009). *Encyclopedia of Human Relationships*. Thousand Oaks, CA: Sage.

Robbins, J. (2004). *Becoming Sinners: Christianity and Moral Torment in a Papua New Guinea Society*. Berkeley, CA: University of California Press. Retrieved from http://www.jstor.org/stable/10.1525/j.ctt1pp8f0

Rosenthal, C. (1985). Kinkeeping in the Familial Division of Labor. *Journal of Marriage and Family*, 47(4), 965–974.

Rosman, A., Rubel, P. G. & Weisgrau, M. (2009). *The Tapestry of Culture: An Introduction to Cultural Anthropology*. Plymouth, UK: AltaMira Press.

Ross, C., Borgerhoff Mulder, M., Oh, S.-Y., Bowles, S., Beheim, B., Bunce, J., Caudell, M., Clark, G., Colleran, H., Cortez, C., Draper, P., Greaves, R., Gurven, M., Headland, T., Headland, J., Hill, K., Hewlett, B., Kaplan, H., Koster, J. & Ziker, J. (2018). Greater Wealth Inequality, Less Polygyny: Rethinking the Polygyny Threshold Model. *Journal of the Royal Society Interface*, 15(144), 20180035.

Rossi, A. & Rossi, P. (1990). *Of Human Bonding*. New York: Aldine deGruyter.

Rubin, J. D., Moors, A. C., Matsick, J. L., Ziegler, A. & Conley, T. D. (2014). On the Margins: Considering Diversity among Consensually Non-Monogamous Relationships. *Journal fur Psychologie*, 22, 1–23.

Ruggles, S. (2009). Reconsidering the Northwest European Family System: Living Arrangements of the Aged in Comparative Historical Perspective. *Population and Development Review*, 35, 249–273.

Ruggles, S. (2010). Stem Families and Joint Families in Comparative Historical Perspective. *Population and Development Review*, 36(3), 563–577.

Rusbult, C. E. (2000). Understanding Responses to Dissatisfaction in Close Relationships: The Exit, Voice, Loyalty, and Neglect Model. In S. Worchel, & J. A. Simpson (Eds.), *Conflict between People and Groups: Causes, Processes, and Resolutions*. Chicago: Nelson-Hall.

Rusbult, C. E. & Buunk, B. P. (1993). Commitment Processes in Close Relationships: An Interdependence Analysis. *Journal of Social and Personal Relationships*, 10, 175–204.

Saadat, M. (2015). Association between Consanguinity and Survival of Marriages. *Egyptian Journal of Medical Human Genetics*, 16(1), 67–70.

Sandstrom, A. & Theodorou, A. (2016). Many Countries Allow Child Marriage. Retrieved on September 28, 2018 from http://www.pewresearch.org/fact-tank/2016/09/12/many-countries-allow-child-marriage/

Sangree, W. H. (1980). The Persistence of Polyandry in Irigwe, Nigeria. *J. Comp. Fam. Stud.* XI(3), 335–343.

Santa Fe Institute (July 17, 2018). Concentrated wealth in agricultural populations may account for the decline of polygyny. *ScienceDaily*. Retrieved on September 29, 2018 from www.sciencedaily.com/releases/2018/07/180717194550.htm

Sayer, L. C., Bianchi, S. M. & Robinson, J. P. (2004). Are Parents Investing Less in Children? Trends in Mothers' and Fathers' Time with Children. *American Journal of Sociology*, 110(1), 1–43.

Schiesberg, S. (2016). Post-marital Residence Patterns and Descent Rules in Prehistory. In T. Kerig, K. Nowak & G. Roth (Eds.), *Alles was zählt ... Festschrift für Andreas Zimmermann*. Habelt.

Schneider, D. M. (1980). *American Kinship: A Cultural Account*, 2nd edn. Chicago, IL: University of Chicago Press.

Schroeder, R. (2007). *Rethinking Science, Technology, and Social Change*. Stanford, CA: Stanford University Press.

Seelbach, W. C. (1978). Correlates of Aged Parents' Filial Responsibility Expectations and Realizations. *Family Coordinator*, 27, 341–350.

Sheff, E. (2013). *The Polyamorists Next Door: Inside Multiple Partner Relationships and Families.* Lanham, MD: Rowman and Littlefield.

Shih, C.-K. (2010). *Quest for Harmony: The Moso Traditions of Sexual Union & Family Life.* Stanford, CA: Stanford University Press.

Smedley, A. (1980). The Implications of Birom Ciscisbeism. *J. Comp. Fam. Stud.*, XI(3), 345–357.

Smith, D. E. (1993). The Standard North American Family: SNAF as an Ideological Code. *Journal of Family Issues*, 14, 50–65.

Soliz, J. (2008). Intergenerational Support and the Role of Grandparents in Post-divorce Families: Retrospective Accounts of Young Adult Grandchildren. *Qualitative Research Reports in Communication*, 9, 72–80.

Song, X., Campbell, C. D. & Lee, J. Z. (2015). Ancestry Matters: Patrilineage Growth and Extinction. *American Sociological Review*, 80(3), 574–602.

Starks, P. T. & Blackie, C. A. (2000). The Relationship between Serial Monogamy and Rape in the United States (1960–1995). *Proceedings of the Royal Society (London), B: Biological Sciences*, 267, 1259–1263.

Stepler, R. (April 6, 2017). *Number of U.S. Adults Cohabiting with a Partner Continues to Rise, Especially among Those 50 and Older.* Retrieved on August 10, 2017 from http://www.pewresearch.org/fact-tank/2017/04/06/number-of-u-s-adults-cohabiting-with-a-partner-continues-to-rise-especially-among-those-50-and-older/

Sterrett, E. M., Jones, D. J., McKee, L. G. & Kincaid, C. (2011). Supportive Non-Parental Adults and Adolescent Psychosocial Functioning: Using Social Support as a Theoretical Framework. *American Journal of Community Psychology*, 48(3–4), 284–295.

Stoehr, T. (1979). *Free Love in America: A Documentary History.* New York: AMS Press, Inc.

Strozier, R. M. (2002). *Foucault, Subjectivity, and Identity: Historical Constructions of Subject and Self.* Detroit, MI: Wayne State University Press.

Sykes, B. (2001). *The Seven Daughters of Eve.* New York: W.W. Norton.

Tam, K. P. (2015). Cultural Transmission through the Role of Perceived Norms. *Journal of Cross-Cultural Psychology*, 1–7.

Taormino, T. (2008). *Opening Up: A Guide to Creating and Sustaining Open Relationships.* California: Cleis Press.

Trappe, H., Pollmann-Schult, M. & Schmitt, C. (2015). The Rise and Decline of the Male Breadwinner Model: Institutional Underpinnings and Future Expectations. *European Sociological Review*, 31(2), 230–242.

Trumbach, R. (1978). *The Rise of the Egalitarian Family: Aristocratic Kinship and Domestic Relations in Eighteenth-Century England.* New York: Academic Press.

UN (United Nations) (2012). On World's First International Day of the Girl Child, UN Calls for End to Child Marriage. Retrieved on September 29, 2018 from https://news.un.org/en/story/2012/10/423262-worlds-first-international-day-girl-child-un-calls-end-child-marriage

UNESCO (United Nations Educational, Scientific and Cultural Organization) (2014). *Education for All: Global Monitoring Report 2014: Teaching and Learning: Achieving Quality for All.* Paris: UNESCO.

U.S. Census Bureau (Jan 28, 2015). *Statistical Definition of 'Family' Unchanged Since 1930.* Retrieved on September 22, 2018 from https://www.census.gov/newsroom/blogs/random-samplings/2015/01/statistical-definition-of-family-unchanged-since-1930.html

U.S. Census Bureau (August 25, 2016). *FFF: Unmarried and Single Americans Week: Sept. 18–24, 2016.* Retrieved on July 22, 2016 from https://www.census.gov/newsroom/facts-for-features/2016/cb16-ff18.html

U.S. Census Bureau (August 16, 2017a). *FFF: Unmarried and Single Americans Week: Sept. 17–23, 2017.* Retrieved on August 15, 2017, 2016 from https://www.census.gov/newsroom/facts-for-features/2017/single-americans-week.html

U.S. Census Bureau (May 3, 2017b). *Childlessness Rises for Women in Their Early 30s.* Retrieved on August 2, 2017 from https://www.census.gov/newsroom/blogs/random-samplings/2017/05/childlessness_rises.html

U.S. Census Bureau (July 12, 2017c). *FFF: National Grandparents Day 2017: Sept. 10.* Retrieved on July 28, 2017 from https://www.census.gov/newsroom/facts-for-features/2017/grandparents-day.html

U.S. Census Bureau (2018a). *Stats for Stories: Father's Day: June 17, 2018.* Retrieved on September 15, 2018 from https://www.census.gov/newsroom/stories/2018/fathers-day.html

U.S. Census Bureau (September 16, 2018b). *Unmarried and Single Americans Week: Sept. 16–22, 2018.* Table H1. Households by Type and Tenure of Householder for Selected Characteristics: 2017. Retrieved on September 17, 2018 from https://www.census.gov/newsroom/stories/2018/unmarried-single-americans-week.html

U.S. Census Bureau (n.d.). *Subject Definitions.* Retrieved on August 23, 2017 from https://www.census.gov/programs-surveys/cps/technical-documentation/subject-definitions.html#family

Veaux, F., Hardy, J. & Gill, T. (2014). *More than Two: A Practical Guide to Ethical Polyamory.* Portland, OR: Thorntree Press.

Vespa, J. (April 2017). *The Changing Economics and Demographics of Young Adulthood: 1975–2016.* U.S. Census Bureau. Retrieved on August 14, 2017 from https://www.census.gov/content/dam/Census/library/publications/2017/demo/p20-579.pdf

Vespa, J. & Armstrong, D. M. (2018). *Demographic Turning Points for the United States: Population Projections for 2020 to 2060.* U.S Census Bureau. Retrieved on September 30, 2018 from https://www.census.gov/content/dam/Census/library/publications/2018/demo/P25_1144.pdf

Wachowiak, D. & Bragg, H. (1980). Open Marriage and Marital Adjustment. *Journal of Marriage and Family,* 42(1), 57–62.

Wang, W. & Parker, K. (September 14, 2014). *Record Share of Americans Have Never Married As Values, Economics and Gender Patterns Change.* Washington, DC: Pew Research Center. Retrieved on July 27, 2016 from http://www.pewsocialtrends.org/2014/09/24/record-share-of-americans-have-never-married/

Wang, W., Parker, K. & Taylor, P. (2013). *Breadwinner Moms.* Washington, DC: Pew Research Center Social & Demographic Trends. Retrieved on September 22, 2018 from http://www.pewsocialtrends.org/2013/05/29/breadwinner-moms/

Wosick-Correa, K. (2010). Agreements, Rules, and Agentic Fidelity in Polyamorous Relationships. *Psychology & Sexuality,* 1, 44–61.

Wrangham, R. (2009). *Catching Fire: How Cooking Made Us Human.* New York: Basic Books.

Wu, H., Brown, S. L. & Payne, K. K. (2016). *Marital Status and Childlessness among Middle-aged U.S. Adults, 1992–2012 (FP-16–02).* National Center for Family & Marriage Research. Retrieved from: http://www.bgsu.edu/ncfmr/resources/data/family-profiles/wu-brown-lin-childlessness-marital-status-middle-age-fp-16-02

2
STUDYING MARRIAGE AND THE FAMILY

Learning Outcomes

Upon completion of the chapter, students should be able to:

1. describe how assimilation and pluralism contribute to intergroup relationships according to structural functionalist theory
2. apply Karl Marx's class theory, Max Weber's stratification theory, and socioeconomic status to the analysis of family life
3. explain how labels affect family life according to symbolic interactionalist theory
4. apply social exchange theory to the analysis of spousal relationships
5. analyze the impact of children on spousal relationships according to family development theory
6. explain how individuals meet role expectations and play a spousal and parental role according to role theory
7. distinguish between descriptive research and explanatory research
8. use an example to explain each step of the seven-step research process
9. explain the advantages and disadvantages of surveys, experiments, observation, case studies, and secondary data analysis research methods
10. explain reliability and validity with regards to quantitative and qualitative research
11. describe how refugee families adapt to a new country.

Brief Chapter Outline

Pre-test
Sociological Theories on Family Research
Structural Functionalist Theory
Structural Functionalist Theory: Family Research
Assimilation; Pluralism
Ethnic Pluralism

Pre-test

Engaged or active learning is a powerful strategy that leads to better learning outcomes. One way to become an active learner is to begin the chapter and try to answer the following true/false statements from the material as you read. You will find that you have ready answers to these questions upon the completion of the chapter.

Structural functionalist theory and conflict theory operate on a macro-level of analysis.	T	F
Symbolic interactionist theory operates on a micro-level of analysis.	T	F
Pluralism can be described by the equation A + B + C = A + B + C.	T	F
Max Weber saw wealth, prestige, and power as sources of social stratification.	T	F
Family development theory explores how couples and family members deal with roles and tasks as they move through each stage of life.	T	F
Social exchange theory explores how people interact within reciprocal relationships on a benefit and cost basis.	T	F
The deductive approach can be referred to as a "top-down" approach because researchers work from general to specific.	T	F
The inductive approach can be referred to as a "bottom-up" approach because researchers work from specific to general.	T	F
Independent variable is assumed to be the cause of the relationship.	T	F
APA style describes rules for the preparation of manuscripts for researchers.	T	F

Upon completion of this section, students are expected to:

LO1: Describe how assimilation and pluralism contribute to intergroup relationships according to structural functionalist theory.

LO2: Apply Karl Marx's class theory, Max Weber's stratification theory, and socioeconomic status to the analysis of family life.

Sociology of the family refers to the scientific study of changing marriage patterns and family forms. The first period in sociological studies and the first twenty years beginning with 1895 were devoted to the evolution of the family and an appraisal of institutional trends (Komarovsky & Waller, 1945). **Sociological theories** are at the heart of the sociological study of the family as they guide researchers in their studies and try to explain how families work and fit into society. A **theory** consists of statements that (1) illustrate proposed relationships between variables and (2) try to explain social phenomenon. Sociological theories must be supported by research, and they also provide a framework for conducting this research. The second period in sociological studies is characterized by more scientific methods (Komarovsky & Waller, 1945). **Scientific research methods** such as surveys and experiments are used to collect data. This chapter introduces sociological theories on marriage and the family and scientific research methods used to conduct research on the family.

Sociological Theories on Family Research

There are two main levels of sociological theory. **Macro-level theories** study a large-scale social structure at the societal and global level. For example, *structural functionalist theory*, *conflict theory*, *world systems theory*, and *feminist theory* all focus on macro-oriented analysis. **Micro-level theories** study everyday behavior and interaction patterns in order to research what happens within families. *Symbolic interactionist theory*, *social exchange theory*, *family development theory*, and *role theory* all focus on micro-level analysis. Let's look at each of these theories in a bit of detail.

Structural Functionalist Theory

Structural functionalist theory operates at a macro-level of analysis and explains how a society is organized to perform functions. Structural functionalist theory views society as a social structure, with many different parts all working for social stability. The key question posed by structural functionalist theory is whether each part of society performs functions and contributes to the smooth functioning of society.

Structural functionalist theory is based on three major propositions. First, the theory posits that society functions like the human body. According to sociologist Herbert Spencer (1820–1903), just as the human body needs the brain, heart, and stomach internally and the arms, legs, and skin externally in order to function properly, society needs certain building blocks to function properly. These building blocks include social institutions such as the family, politics, education, the economy, and religion. Structural functionalists contend that all of these building blocks are necessary for society to maintain social stability. Each block provides a function that keeps society working like a healthy body. Therefore, structural functionalist theory argues that society is understood as a complex system striving for equilibrium.

Second, structural functionalist theory examines *functions*. According to sociologist Robert Merton (1910–2003), functions can be divided into two categories: manifest and latent (Merton, 1949). **Manifest functions** are consequences or outcomes that are intended and expected. For example, students are expected to achieve a course's learning outcomes—that is the result of the function that was designed and intended. **Latent functions** are unintended or unexpected consequences of functions. The manifest function of higher education is to get a degree. A latent function might be that students interact with peers and make friends.

Finally, structural functionalist theory emphasizes the way that social institutions are structured to maintain social stability. Researchers analyze what specific systems are working or not working, and then diagnose problems and devise solutions to restore balance.

French sociologist Émile Durkheim (1858–1917) is one of the founding fathers of structural functionalist theory. During his lifetime, he explored topics such as religion, suicide, and crime, and developed sociology as an academic field. During the transition to an industrial society in Europe, many social problems appeared. The main concern was social instability and the main question was how to bring back social order. Durkheim believed that society was held together by a shared social solidarity in which people worked together to achieve social cohesion.

Durkheim distinguished between two types of social solidarity that correlate with two types of society: mechanical solidarity and organic solidarity. **Mechanical solidarity** is a form of social solidarity in traditional societies based on kinship ties of familial networks with minimal division of labor. For example, in rural areas in pre-industrial societies, people work and live in family-based farms and engage in agricultural pursuits. Social cohesion results from personal relationships, as people feel connected through family ties and bonds. **Organic solidarity** is a form of social solidarity that takes place in industrial societies based on interdependence from specialization of work and a high level of division of labor. With organic solidarity, a diverse group of people work in specialized occupations and achieve social cohesion from their need to rely on each other for survival. In urban areas in industrial societies, for example, families depend on others to provide them with jobs, food, clothes, shelter, education, public safety, health care, and so on.

Structural Functionalist Theory: Family Research

Structural functionalist theory views the family as a social system that provides new members through reproduction and in turn socializes those new members. When it comes to socialization, children are taught to play the roles and become functioning members of society. Structural functionalist theory emphasizes supportive interactions, with people treating each other as means of support.

Assimilation

Structural functionalist theory initially advocated assimilation in an effort to maintain social order and avoid potential conflict between families of different races and ethnicities. **Assimilation** refers to a process in which distinct groups merge to become part of the dominant group. Assimilation can be described by the equation A + B + C = A. After the United States became an independent nation in 1776, President George Washington and U.S. military officer Henry Knox conceived of the idea of "civilizing" Native Americans in preparation for their assimilation as American citizens. Assimilation thus became an official policy of the federal government. In the United States, this type of assimilation results in Anglo-conformity, with racial and ethnic groups adapting to the Anglo-American culture as a precondition to access to education and job opportunities. This assimilation may be forced or voluntary. An example of forced assimilation is the General Allotment Act of 1887, in which Native American tribal reservations were broken up and an allotment of land was given to each Native American family. The purpose of this Act was to assimilate Native Americans into White society by making them land owners. Voluntary assimilation allows minority families the opportunity to adapt to the dominant culture or adopt parts of the American culture as they wish.

Sociologist Milton Gordon (1918–) contends that full assimilation includes cultural assimilation or acculturation, structural assimilation or integration, and marital assimilation or intermarriage (Gordon, 1964). *Cultural assimilation* involves learning the language and basic values of the dominant group. *Structural assimilation* refers to integration and participation in the dominant group's social institutions such as schools, jobs, clubs, and information networks. *Marital assimilation* involves intermarriage with the dominant group.

One problem with assimilation is that it has been a one-sided process that consists of ethnic and racial minorities conforming to the Anglo majority. Sociologist Horace Kallen (1882–1974) questioned the Anglo-conformity and assimilation model and argued that people of diverse cultures can retain their heritage and still live peacefully in American society (1915). Indeed, since the 1960s, assimilation has been viewed in a negative light in the United States. For example, the Civil Rights movement criticized assimilation, claiming that minority families were being judged for not conforming to Anglo-American standards.

Pluralism

In response to the issues posed by assimilation, structural functionalist theorists advocated pluralism in an effort to maintain social order. **Pluralism** refers to the process in which racial and ethnic groups maintain their own cultural distinctiveness and participate in society without prejudice. Pluralism can be described by the equation A + B + C = A + B + C. Pluralism implies mutual respect in a diverse society. Immigrant families are quick to integrate with the mainstream culture while still embracing their unique heritage.

Ethnic Pluralism Before 1820, the majority of immigrant families to the United States were British. However, between the 1820s and the 1920s, the United States witnessed waves of immigration from southern and eastern Europe and East Asia. By 1905, three-fourths of the newcomers were from southern and eastern Europe, and their religions and cultures were notably different from those of northern and western European immigrants. From the late 1960s, immigrant families from Latin America, Asia, and the Caribbean Basin further increased family diversity in the United States. These immigrant families assimilated to the mainstream culture but struggled to retain their own heritages. They therefore created communities in order to stick together for mutual support and to fight discrimination. Today, many cities and metropolitan areas still have ethnic enclaves such as Little

Italy, Little Warsaw, Little Ireland, Chinatown, and Little Tokyo that are inhabited by families of the same background and that enable them to build businesses and support one another. For example, the Amish, sometimes called the Pennsylvania Dutch, are a cultural group committed to a simple way of living on farms who maintain an institutional life separate from the mainstream American culture.

A number of terms have been used to describe pluralism. Whereas the United States was once seen as a **mixing bowl**, in which people of different cultures all mixed together to assimilate to a single homogeneous culture, with pluralism, diverse family cultures create a **salad bowl** of ingredients that co-exist yet keep their separate identities. The pluralistic culture of America is also described as a **cultural mosaic**, reflecting the mixture of ethnic groups with different cultures who coexist within society. The term **kaleidoscope** can also be used to refer to the different cultural patterns and colors one sees in American society, similar to those one sees when looking through a kaleidoscope. Finally, **multiculturalism** refers to the co-existence of multiple cultures within a society.

Conflict Theory

Let's now turn our attention to conflict theory, another macro-level theory of society which explains how society is made up of people competing for limited resources (e.g., wealth, money, property, power, status). **Conflict theory** views society as being in a constant struggle over control and allocation of scarce resources. The key question conflict theory poses is who benefits, and why? Conflict theory is useful in understanding wealth and poverty, discrimination and prejudice, and other conflict-related social phenomena.

Conflict theory is based on three major propositions. First, as noted above, because society has limited resources to allocate among social groups, people struggle to compete for these resources. **Social inequality** refers to unequal distribution of resources among members of a society. Second, conflict theorists believe that unequal distribution of resources is harmful to society as it benefits the rich at the expense of the poor. There is always conflict between the "haves" and the "have-nots." Finally, conflict theorists contend that conflict between social groups is beneficial to social change. Therefore, in contrast to the structural functionalist view of social stability, conflict theorists seek social change and a redistribution of resources. This change may occur only as a result of a revolution when the "have-nots" unite and overthrow the "haves." Conflict theorists view society as being stratified by social class, gender, race, etc. The indicators and measurement of what determines social class have varied over time.

Social Class

German sociologist Karl Marx (1818–1863) is the founding father of conflict theory. **Social class** refers to a group of people with similar shares of resources in hierarchical social divisions such as upper, middle, and lower classes. At the early stage of capitalist development, Marx identified two social classes according to an individual's relationship to the means of production. Social class is thus determined by ownership or non-ownership of means of production such as factories. The **working class** (or the **proletariat)** do not own the means of production and must sell their muscle to earn a minimum amount of wages. The **capitalist class** (or the **bourgeoisie**) lives off the surplus generated by the working class. According to Marx, the capitalist class exploits the working class by having them work more and paying them less. Marx considered such exploitation to be a social problem. *The Communist Manifesto,* which Marx wrote in 1848, encouraged all the working class in the world to unite and overthrow the capitalist society and build a socialist society, with the goal of socialism being to equalize.

Social Stratification

German sociologist Max Weber (1864–1920) broadened the class theory of Marx and developed the theory of social stratification. **Social stratification** is the categorizing of people into various socioeconomic strata with different shares of resources arranged in the social hierarchy. Weber identified three dimensions of social stratification and saw wealth, prestige, and power as sources of social stratification. Wealth, power, and prestige are unequally distributed among members of a society. **Wealth** refers to the total value of household assets, including income, personal property, and income-producing property. **Prestige** refers to the respect or status conferred on a person by others. **Power** is the ability of people to achieve their goals despite opposition from others. Weber believed that people having wealth, power, or prestige are likely to be advantaged in society.

Conflict Theory: Family Research

Based on Weber's approach, researchers use **socioeconomic status (SES)** as a combined measure to classify people who have the same income, occupation, and education and to determine their position in the social hierarchy. Conflict theorists view socioeconomic status as a major influence on family life. Social stratification in employment, house ownership, and health-care coverage further affects family life. According to conflict theorists, upper-class families have superior resources that translate to a better family life as compared to lower-class families. Let's look at some family research issues examined by conflict theorists, including income inequality, health inequality, and poverty.

Wealth Inequality

Wealth inequality or wealth gap is the unequal distribution of assets among residents of the United States. Wealth is an important measure of economic health. Wealth allows families to build up savings and meet spending demands in the future, buffer economic security against family crisis, finance a comfortable retirement, or provide an inheritance to children. For example, racial wealth gap persists. More than half of White families end up with more wealth than their parents, while only 23 percent of Blacks are able to do the same. White families are twice as likely to receive an inheritence as Black families, and that inheritance is nearly three times as much (Jones, 2017a), as shown in Figure 2.1.

Income Inequality

Income refers to the revenue streams individuals receive from salaries, wages, interests on savings accounts, dividends from shares of stock, rent, and profits from elsewhere. **Income inequality** refers to the extent to which income is distributed in an uneven manner among a population. U.S. Bureau of Labor Statistics views inequality as the differences between people with the highest levels of wealth, income, or earnings and those with the lowest levels. Those whose earnings are at or above the 90th percentile increased from a minimum of $1,422 per week in 1979 to $1,898 per week in 2014, whereas those whose earnings are at or below the 10th percentile decreased from $383 per week in 1979 to less than $379 per week in 2014 (U.S. Bureau of Labor Statistics, 2015). Income inequality continues to grow in the United States. The top 5 percent of households in the income distribution had incomes of $237,035 or more while households in the lowest quintile had incomes of $24,638 or less in 2017 (U.S. Census Bureau, 2018a). Conflict theory encourages the "have nots" to close this income gap. For example, a labor protest movement of fast-food workers in New York City led to higher wages for workers all across the country and legislation passed in Los Angeles, Seattle, and San Francisco to raise their minimum wage (McGeehan, 2015). While the federal minimum wage sits at $7.25, many

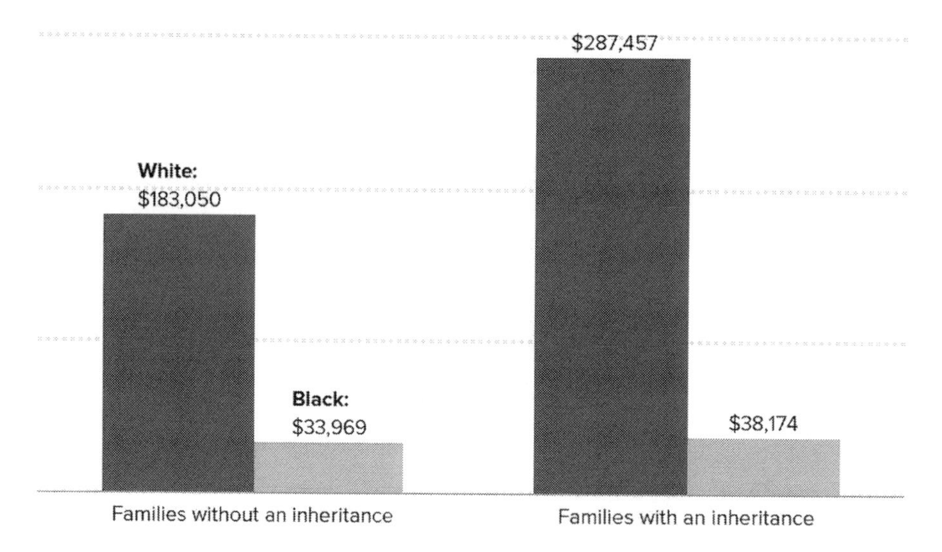

Figure 2.1 Receiving an Inheritance Helps White Families More than Black Families
Source: Thompson, J. P. and Suarez, G. A. (2015). *Exploring the Racial Wealth Gap Using the Survey of Consumer Finances, Finance and Economics Discussion Series 2015–076.* Washington: Board of Governors of the Federal Reserve System, Economic Policy Institute, http://www.epi.org/publication/receiving-an-inheritance-helps-white-families-more-than-black-families/

Source: Fibonacci Blue/flickr

states and localities have increased their minimum wages, which helps lift working families out of poverty (Jones, 2017b).

Health Inequality

Income inequality leads to health inequality. **Health inequality** refers to differences in health status and in the distribution of health care between different population groups. Scholars identify three social problems involved with health inequality. The first problem involves disparities in health issues such as heart disease and cancer, with poor people having a greater risk for many illnesses. For example, children from the poorest 20 percent of households are nearly twice as likely to die before their fifth birthday as are children in the richest 20 percent (WHO, 2017). The second problem involves disparities in the care available to cope with these diseases, with poor people having less access to care. Considerable evidence indicates that the racial and ethnic health disparities observed in the United States arise mostly through the effects of discrimination, differences in treatment, poverty, lack of access to health care, health-related behaviors, racism, stress, and other socially mediated forces (Berg *et al.*, 2005).

The third problem involves disparities in health insurance and the ability to pay for health care. Health insurance coverage is classified into private coverage, government coverage, and the uninsured. **Private health insurance** is defined as a plan provided through an employer or a union and coverage purchased directly by an individual from an insurance company or through an exchange. **Exchanges** include coverage purchased through the federal Health Insurance Marketplace, as well as other state-based marketplaces, and include both subsidized and unsubsidized plans. **Government insurance coverage** includes federal programs, such as Medicare, Medicaid, the Children's Health Insurance Program (CHIP), individual state health plans, TRICARE, CHAMPVA (Civilian Health and Medical Program of the Department of Veterans Affairs), as well as care provided by the Department of Veterans Affairs and the military (Berchick *et al.*, 2018). Individuals are considered to be uninsured if they do not have health insurance coverage for the entire calendar year. In 2017, private health insurance coverage continued to be more prevalent than government coverage, at 67.2 percent and 37.7 percent, respectively. Employer-based insurance covered 56.0 percent of the population for some or all of the calendar year, followed by Medicaid (19.3 percent), Medicare (17.2 percent), direct-purchase (16.0 percent), and military coverage (4.8 percent) (Berchick *et al.*, 2018). Figure 2.2 shows that in 2017, the percentage of people without health insurance coverage at any point during the year was 8.8 percent, or 28.5 million people.

Household income leads to health inequality. Private coverage rates increased as income increased. In 2017, people with household income between $25,000 and $49,999 had a higher private health insurance coverage rate (51.1 percent) than people with household income below $25,000 (30.1 percent) (Berchick *et al.*, 2018). Government coverage rates decreased as income increased. In 2017, people with household income between $25,000 and $49,999 had a lower government coverage rate (53.0 percent) than people with household income of less than $25,000 (68.4 percent). Figure 2.3 shows that for children under age nineteen and adults ages 19 to 64, uninsured rates were higher for people with household income below $25,000, compared with people with household income of $125,000 or more in 2016. The uninsured rate for adults in poverty (24.7 percent) was over twice that for adults not in poverty (10.4 percent). Hispanic children had the highest uninsured rate, at 7.9 percent.

Health-care inequality differs in part by state. In 2017, the state with the lowest percentage of people without health insurance was Massachusetts (2.8 percent), while the state with the highest percentage was Texas (17.3 percent) (Berchick *et al.*, 2018). People

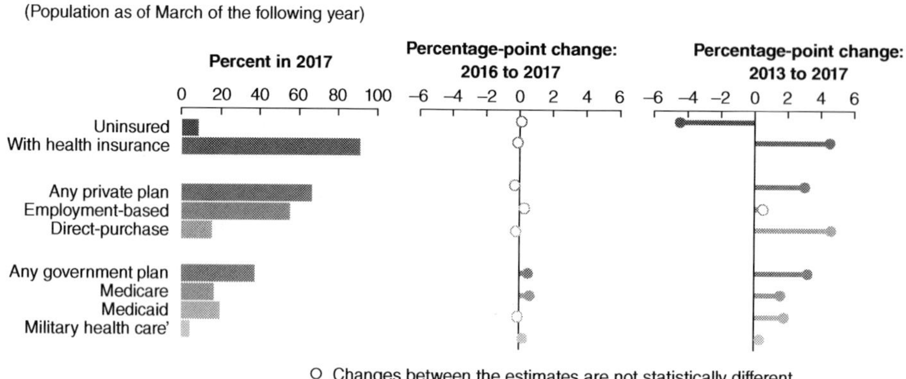

Figure 2.2 Percentage of People by Type of Health Insurance Coverage and Change, 2013–2017
Source: U.S. Census Bureau, Current Population Survey, 2014, 2017, 2018 Annual Social and Economic Supplements. https://www.census.gov/library/visualizations/2018/demo/p60-264.html

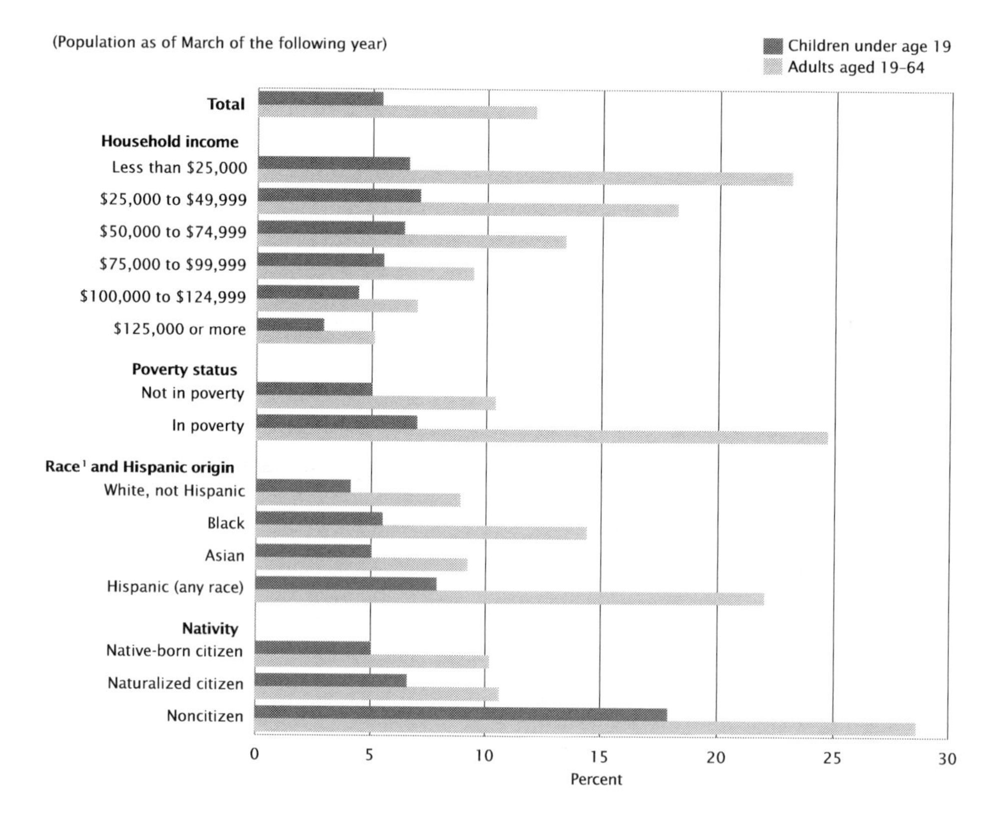

Figure 2.3 Percentage of Children Under Age 19 and Adults Aged 19–64 Without Health Insurance Coverage by Selected Characteristics, 2016

Source: U.S. Census Bureau, Current Population Survey, 2017 Annual Social and Economic Supplement. https://www.census.gov/content/dam/Census/library/visualizations/2017/demo/p60-260/figure6.pdf

without health insurance suffer. In low-resource settings, health-care costs for noncommunicable diseases can quickly drain household resources, driving families into poverty (WHO, 2017). For example, many new American immigrants struggle with health inequality. Often, their employers classify them as part-time workers or independent contractors, therefore avoiding any obligation to provide health care (O'Donnell & Ungar, 2016). Many Medicaid expansion states had uninsured rates lower than the national average, while many non-expansion states had uninsured rates above the national average (see Figure 2.4).

Poverty

Poverty refers to a state of deprivation and a lack of material possessions. **Absolute poverty** refers to a condition where a person or the family does not have the minimum amount of income needed to meet basic human needs. A family that suffers from absolute poverty may be homeless, for example, and unable to afford enough food to support themselves. **Relative poverty** refers to a condition where a person or the family can afford the basic daily necessities but can't maintain an average standard of living. For example, although some immigrant families who settle in the United States enjoy a better standard of living than where they came from, they may experience relative poverty as compared to those around them. In 2017 there were 39.7 million (12.3 percent) people in poverty, down 0.4 percentage points from 12.7 percent in 2016 (Fontenot *et al.*, 2018). Figure 2.5 illustrates poverty and poverty rates in the United States from 1959 to 2017. The chances of experiencing poverty vary by family structure, employment status of parents, and the presence of children.

Culture of Poverty The **culture of poverty theory** suggests that living in conditions of poverty leads to the development of a culture adapted to those conditions. In 1959, American anthropologist Oscar Lewis (1914–70) described the lives of slum dwellers and explained that a global culture of poverty existed in which cultural similarities occurred across nations because they were "common adaptations to common problems." For example, according to the culture of poverty theory, children are socialized into behaviors and attitudes that perpetuate their inability to escape the underclass. This theory offers one explanation of why poverty exists despite anti-poverty programs.

The Importance of Policy The issue of **working poverty** or the **working poor**—working people whose incomes fall beneath the poverty line—came to the attention of the public in the Progressive Era (1890s–1920s). American sociologist W. E. B. Du Bois (1868–1963) viewed social inequality as the cause of working poverty. In his study of Philadelphia's African American neighborhoods, Du Bois drew a distinction between "hardworking" poor people who fail to escape poverty due to racial discrimination (i.e., the working poor) and those who are poor due to the fact that they are not working (what we now term the non-working poor) (1899).

Research has shown that the non-working poor have a difficult time overcoming basic barriers to entry into the labor market, such as arranging for affordable childcare, finding housing near potential jobs, or arranging for transportation to and from work (Edin & Lein, 1997). In order to help the non-working poor gain entry into the labor market, liberal scholars and policymakers argue that the government should provide more housing assistance, childcare, and other kinds of aid to poor families (de Souza Briggs *et al.*, 2010).

The solution to the problem of the non-working poor has thus far been to provide the poor with welfare benefits. Government programs including Social Security, refundable tax credits, and Supplemental Nutrition Assistance Program (SNAP) are directly responsible for keeping tens of millions out of poverty across the country (Jones, 2017a). Social Security kept 26.6 million people out of poverty in the United States; food stamps (SNAP) kept 4.6 million people out of poverty; and unemployment insurance

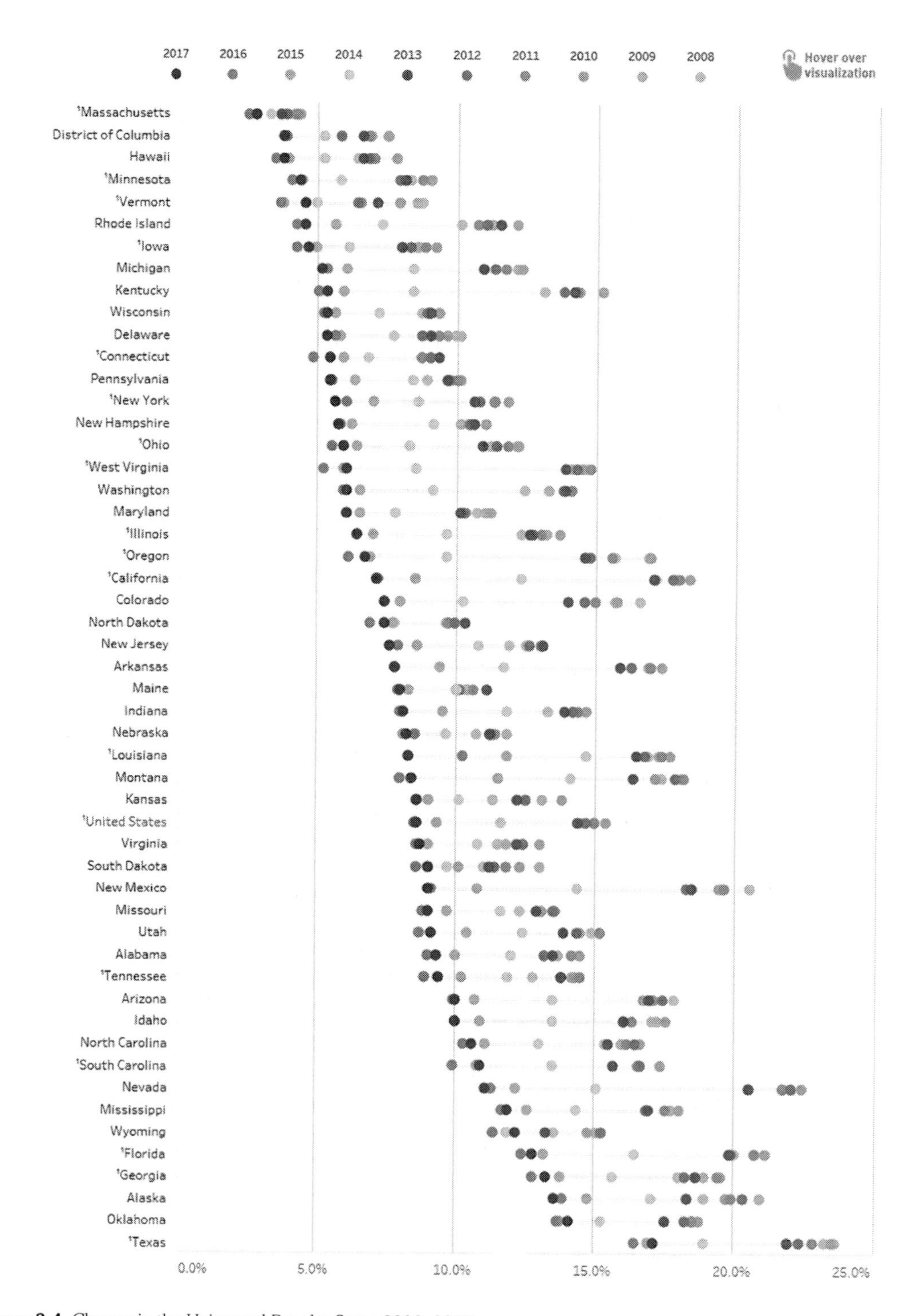

Figure 2.4 Change in the Uninsured Rate by State, 2008–2017

Source: U.S. Census Bureau, 2008–2017 1-Year American Community. https://www.census.gov/library/visualizations/interactive/health-insurance-coverage.html

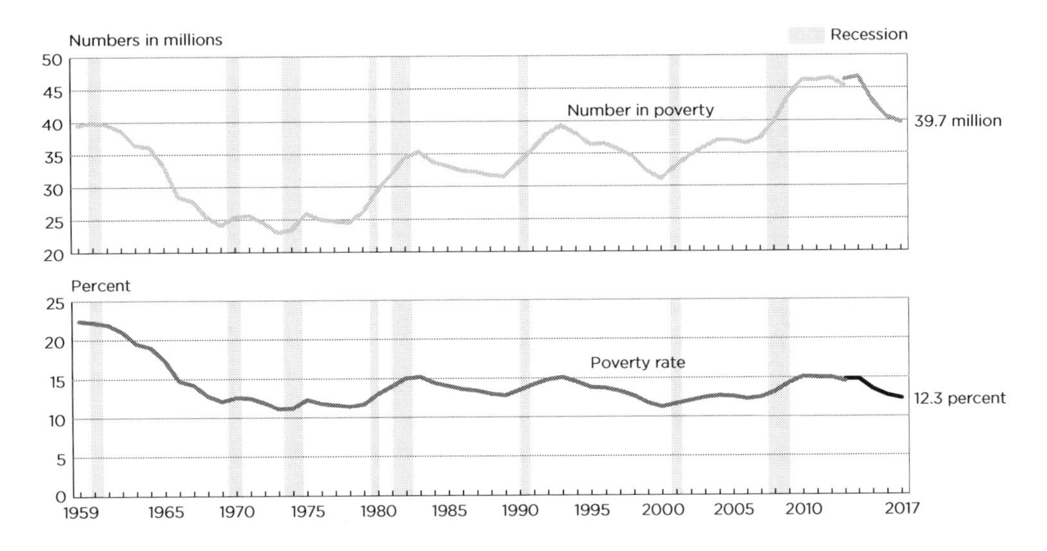

Figure 2.5 Number in Poverty and Poverty Rates, 1959–2017
Source: U.S. Census Bureau, Current Population Survey, 1960–2017 Annual Social and Economic Supplements. https://www.census.gov/content/dam/Census/library/visualizations/2018/demo/p60-263/figure4.pdf

kept nearly 650,000 people out of poverty in 2015 (Gould & Schieder, 2016).

Family Structure and Poverty Family structure has a great effect on poverty. For example, the poverty rate in the United States in 2017 for married-couple families was 8.4 percent (4.2 million), compared with 40.8 percent (7.2 million) for a female-headed family and 19.1 percent (1.0 million) for a male-headed family (Fontenot *et al.*, 2018). Researchers suggest that any serious effort to address poverty must recognize that children need a stable and secure home environment with two caring and committed parents (Corrigan, 2015). Broad-based wage growth is the best way to fight poverty (Jones, 2017b). The Working Poor Families Project (WPFP) is a national initiative that seeks to strengthen state policies on behalf of low-income working families.

World Systems Theory

Another macro-level theory is world systems theory. **World systems theory** suggests that the world is divided into core, semi-periphery, and periphery nations. **Core nations** are developed and industrialized countries such as the United Kingdom and the United States. Core nations receive the greatest share of resources. **Semi-periphery nations** such as Brazil are midway between core and periphery nations and are moving toward industrialization. Based on gross national product per capita between 1975 and 2002, core, periphery, and semi-periphery countries consistently fell into a particular class (Babones, 2015). According to American sociologist Immanuel Wallerstein (1930–), a world systems analysis can be used to trace the rise of the capitalist world economy from the sixteenth century, which has resulted in **global stratification**—unequal distribution of resources on a global basis (Wallerstein, 1974).

Periphery nations are developing and poor countries such as Zambia. Periphery nations receive the least amount of resources. People living below a poverty line do not have enough to meet their basic needs. Every country typically defines national poverty lines. As differences in the cost of living across the world evolve, the global poverty line has to be

Division of the world into Core, Semi-periphery, and Periphery Nations
Source: Salvatore J. Babones 2015

periodically updated to reflect these changes. The new global poverty line is set at $1.90 using 2011 prices and over 900 million people globally lived under this line in 2012 (World Bank, 2015). We use the lines of a group of the poorest countries to define the international extreme poverty line of $1.90 per day (World Bank, 2017). Indeed, it is clear that global stratification and disparity among nations is profound. As noted, periphery countries are the poorest or the least developed countries. The United Nations' so-called "Least Developed Countries" such as Afghanistan and Haiti exhibit the lowest indicators of socioeconomic development. In fact, families of the children in these countries live in extreme poverty, defined by the World Bank as living on less than $1.90 per day.

Feminist Theory

Feminist theory is another macro-level theory. **Feminist theory** focuses on gender equality and equal rights for women and men. The theory suggests that men maintain power over women in family and society due to socially constructed gender roles.

One issue that feminist theorists are concerned about is equal pay. Women make up a growing number of breadwinners in families in the United States. In fact, 76.8 million (47.4 percent) of females age sixteen and older participated in the civilian labor force in 2016 (U.S. Census Bureau, 2018b). However, a **gender wage gap** exists, in which working women are paid less than working men. The average American female worker loses more than $530,000 over the course of her lifetime because of the gender wage gap, and the average college-educated woman loses even more—$800,000 (Gould *et al.*, 2016). Equal pay affects family life in many ways. For example, when females are paid less than they deserve, couples may find themselves with less money for tuition or childcare and less to retire on (White House, 2009).

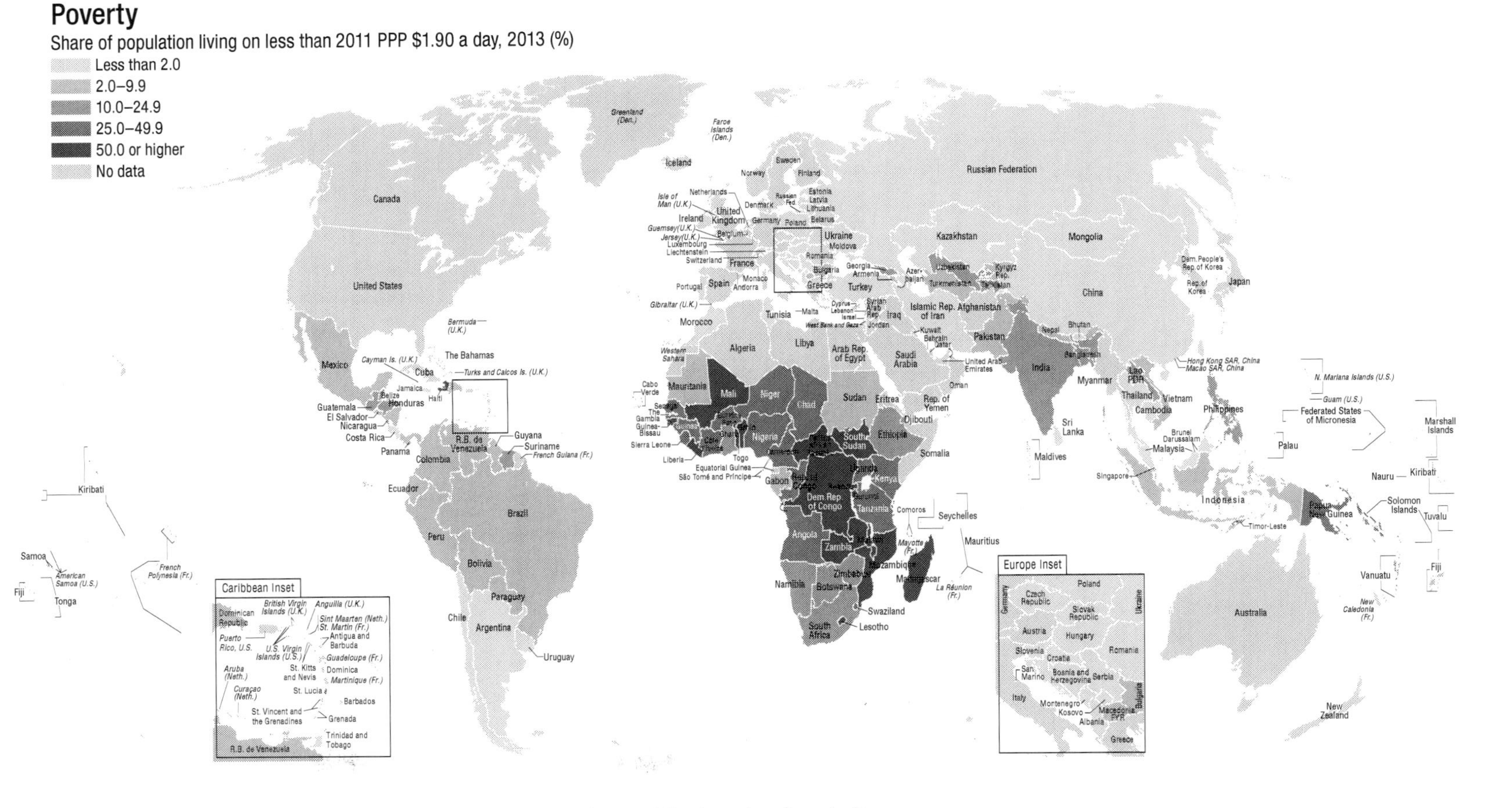

Source: The World Bank: The World Development Indicators. http://data.worldbank.org/products/wdi-maps

Upon completion of this section, students are expected to:
LO3: Explain how labels affect family according to symbolic interactionalist theory.
LO4: Apply social exchange theory to the analysis of spousal relationships.
LO5: Analyze the impact of children on spousal relationships according to family development theory.
LO6: Explain how individuals meet role expectations and play a spousal and parental role according to role theory.

Symbolic Interactionalist Theory

Symbolic interactionalist theory focuses on the micro-level of analysis and attempts to understand the relationship between humans and society. **Symbolic interactionalist theory** views society as the sum of social interactions of people. The key question of symbolic interactionist theory is how people make sense of their everyday social interactions. Symbolic interactionist theory is based on two basic propositions. First, society creates symbols from which meanings are derived and shared by people. **Symbols** refer to anything that represents something else. Symbols take the form of language or words, gestures, signs, and visual images, and are used to convey other ideas. People interact with each other through the use of symbols. Social behaviors and social interactions are understandable only through the exchange of meaningful symbols. For example, a thumbs-up or thumbs-down sign is a common symbol of approval or disapproval in the United States, whereas the V-sign symbolizes victory or peace. Symbolic interaction takes place verbally and non-verbally, and some symbols have multiple meanings depending on the context in which they are used.

Second, symbolic interactionalist theory posits that people create their own social world through their interactions with people around them. **Social interaction** refers to reciprocal communication between two or more people through symbols, verbal language and non-verbal language.

Every time people engage in an interaction, they have to define and interpret the situation. For example, imagine an older woman is struggling with a heavy object. A young man walking by wishes to be helpful and so he offers his assistance. The woman could interpret this action in a number of ways. For example, she may be happy that the young man has offered to help her and in exchange offers to buy him a cup of coffee. Alternatively, she may be offended that the young man assumed that because she was an older female that she was weak, and she may therefore rebuff his offer of help. Or, perhaps the woman may fear that the man is actually a thief and that his offer of help is a ruse; she may therefore yell at him to leave her alone. In the future, the young man may therefore be reluctant to offer his help to others. As you can see, how both parties define a situation affects the social interaction. Symbolic interactionists believe that social order is maintained when people share a common understanding of everyday behaviors.

Symbolic interactionists also seek to examine human behavior from a cultural perspective. How people define a situation varies according to their cultural context and becomes the foundation for how they behave. For example, when it comes to greeting, people from different cultural backgrounds might shake hands, or kiss, or bow, or hug.

Symbolic Interactionalist Theory: Family Research

Symbolic interactionists are interested in the labels that people attach to the world, the meanings and consequences of the labels, and how they change over time. For example, consider the meanings and consequences attached to different names. Ethnically diverse companies are 35 percent more likely to outperform their non-diverse counterparts (WEF, 2017), yet people with ethnic names are less likely to be hired. Based on 13,000 fake résumés sent out to over 3,000 job postings, people with Chinese, Indian, or Pakistani-sounding names were 28 percent less likely to get invited to an interview than the fictitious candidates with English-sounding names, even when their qualifications were the same (WEF, 2017).

Labeling Theory

Related to symbolic interactionalist theory is labeling theory. Sociologist Howard Becker (1928–) developed **labeling theory**, which studies labels put on individuals and the consequences that the labels produce (Becker, 1963). For example, people are likely to label minority groups negatively. Negative labels are identified as **stereotypes** (i.e., oversimplified generalizations about a group of people) that affect an individual's self-identity. One aspect related to labeling theory is deviance. Sociologist Edwin Lemert (1912–96) distinguishes primary deviance from secondary deviance (1967). **Primary deviance** is the initial act of deviant behavior. For example, children who do not have a great deal of parental supervision may start committing deviant behavior such as smoking marijuana or taking illicit drugs. These children may then be labeled as drug addicts by society. **Secondary deviation** is deviant behavior that results from being labeled as a deviant by society. When people commit primary acts of deviance, they hardly consider themselves deviants or criminals. However, when they are labeled as such and internalize the label, they seem to accept the label and continue to act out accordingly and receive unequal treatment in society.

Scapegoat Theory

Another theory related to symbolic interactionalist theory is scapegoat theory. **Scapegoat theory** refers to the tendency to blame a social group for one's own problems. For example, Adolf Hitler used the Jews as scapegoats for the social and economic problems in Germany at the time. Hitler's solution to these problems was to wipe out all Jews. In the United States, immigrants have often been the scapegoat for the nation's woes, with some people placing labels on ethnic groups and blaming immigrant families for social problems. Laws have even been enacted that let the dominant group scapegoat minority families. For example, the Chinese Exclusion Act of 1882 excluded the Chinese from citizenship and prevented them from bringing over spouses and children to the United States for over 80 years. In the twenty-first century, some people are still judged and mistreated because of their religion or the Muslim faith. For example, Lewiston, Maine, became a thriving town when the Somalis, an ethnic group inhabiting the Horn of Africa (Somali Peninsula), moved in and established a small group. When the 9/11 event happened, people started to blame them for the tragedy and wanted them out.

Social Exchange Theory

Borrowing from the symbolic interactionalist perspective, **social exchange theory**, another micro-level theory, explores how people interact within reciprocal relationships and interpret their experiences of self and others to determine the benefits and costs of the relationship. Benefits are rewarding because they meet a perceived need such as financial, physical, or emotional security. Costs are actions that meet the needs of another or partner, such as providing financial, physical, or emotional support. The basic tenets of exchange theory suggest that people choose to participate in a particular relationship because of the relationship's ability to provide a satisfactory level of outcomes—defined as the *rewards* derived from the relationship minus the *costs* of participating in the relationship—and that these outcomes are better than those available in other competing relationships (Nye, 1979).

The fundamental principle of social exchange theory is that people seek to maximize their rewards and minimize their costs through their interactions with others. When a relationship bears more costs than benefits for a partner, that person is more likely to end the relationship. Sabatelli (1988) focuses on the role that relationship expectations play in the evaluations that people make of their relationships. Social exchange theorists also have explored the effects of children on marital satisfaction. Spousal relationships are likely to be stable when benefits that each individual receives are in balance with the costs of the relationship.

Family Development Theory

Family development theory, another micro-level theory, focuses on the systematic and patterned changes experienced by families and explores how couples and family members deal with roles and tasks as they move through each stage of life. Family developmental theorist Evelyn Duvall (1957) highlights the following eight stages and family developmental roles or tasks:

- Stage 1: Partners are newly married with no children. The newlywed couple assumes the spousal role.
- Stage 2: The parental role is added as the first infant is born.
- Stage 3: Parents adapt to the needs of preschool children when the child is between two and six.
- Stage 4: When children are school-aged, parents try to encourage children's educational achievements.
- Stage 5: Parents try to balance freedom with responsibility and establish post-parental interests.
- Stage 6: Adult children leave home and the parental role involves helping children become independent.
- Stage 7: Parents who now live without their children try to refocus on their marriage.
- Stage 8: The work role ends as spouses retire and may cope with death and living alone.

Family development literature has discussed whether having a baby causes declines in marital relationships. Across countries, the feeling of well-being in parenthood is significantly related to macroeconomic conditions: The negative relationship between parenthood and life satisfaction is stronger in countries with higher GDP per capita or a higher unemployment rate (Stanca, 2016). The process of transitioning to parenthood can bring up difficulties in spousal relationships. About two-thirds of couples see the quality of their relationship drop within three years of the birth of a child (Petersen, 2011). Parents often feel that at the birth of a baby, any previous routine and romantic time that the couple had disappears. The couple spends all their time with the baby, attending to its constant needs. Female satisfaction in the marriage often decreases because of hormonal changes and the physical demands of childbirth and nursing. Marital conflict may occur as emotional and physical distance develops with little time for conversation and sex between couples. If the couple does not recognize and deal with these changes, marital problems might occur.

However, children might get an unfair share of the blame for parental distress. Cowan and Cowan (1992) find that the seeds of new parents' individual and marital problems are sown long before their first baby arrives. Marital problems are less likely to occur if couples strengthen their families and cope with changes after babies are born.

Role Theory

Role theory, another micro-level theory, suggests that most of our everyday activities consist of acting out socially defined roles associated with a given status. How people behave in a society is determined by status and role. A **status** is a socially defined position characterized by rights, responsibilities, and obligations. Examples are student or professor. A **role**

is a set of behavioral expectations associated with a status.

In terms of families, role theory can help explain how working parents juggle multiple roles. Everyone has different socially defined roles such as professor and student, marital roles such as husband and wife, and parental roles such as mother and father, etc. Sociologists distinguish role expectation from role performance. **Role expectation** relates to the expectations required of a parental role. Human behavior is guided by expectations held both by the individual and by others. The role expectation attached to the status of mother is that she cares and provides for her child. **Role performance** refers to the actual delivery of those expectations. The role performance would be her actual day-to-day mothering activities. If expectations are met, the mother plays a good parental role.

Diversity Overview: Ethnic Social Mobility

In the nineteenth century, immigrant families from southern and eastern Europe had a different cultural heritage from previous English settlers. These new immigrants often experienced discrimination in employment. However, researchers were curious why Jewish immigrants had been upwardly mobile sooner than had Italian immigrants. Did culture or social class pave the way for certain ethnic groups to achieve success?

Researchers started to look to cultural causes to explain the economic progress of Jewish immigrants as compared to Italians. They found that Jewish immigrants tended to value educational achievement, which paved their road to success. By contrast, Italian immigrants did not value learning as much as they did family loyalty. Over the decades, many sociologists have attributed the rapid upward mobility of Jewish immigrants to cultural advantages they brought with them from eastern Europe (Steinberg, 1974).

Perlmann (1988) used class theory to explain the economic progress of Jewish immigrants as compared to Italian immigrants. He found that Jewish success was based on their superior starting position; for example, Jewish men in 1915 were skillful and often self-employed and engaged in business at some level. Steinberg (2001) further formulated that social class theory explained the rapid upward mobility of Jewish immigrants as compared to Italian immigrants as the two groups did not start off on equal footing. Jewish immigrants were more likely to be engineers, whereas Italians were more likely to be ditch diggers in terms of their social class origins. Being an engineer meant earning higher wages than those of a ditch digger. Upon arriving in new surroundings, people with higher-status backgrounds are likely to be able to regain higher-status positions (Steinberg, 2001). People of lower-class origins must acquire new skills to move upward in a stratified system. Steinberg's theory suggests that the Jewish immigrants' status attainment in the United States reflected their higher status back in eastern Europe. They arrived in North America with experience and technical skills qualifying them for highly skilled jobs and enabling them to successfully pursue business and commercial opportunities.

This research illustrates that researchers must look not only for evidence that supports their theory but also for evidence that has the potential to falsify their preferred explanations. The above example explains that researchers started by using culture to explain why Jewish immigrants had been upwardly mobile sooner than had Italian immigrants. However, they later refuted this notion and used social class theory to explain the social mobility of Jewish immigrants as compared to Italian immigrants.

Upon completion of this section, students are expected to:
LO7: Distinguish between descriptive research and explanatory research.
LO8: Use an example to explain each step of the seven-step research process.
LO9: Explain the advantages and disadvantages of surveys, experiments, observation, case studies, and secondary data analysis research methods.
LO10: Explain reliability and validity with regards to qualitative and quantitative research.

Research on Marriage and the Family

Research is the process of collecting data for the purpose of testing an existing theory or generating a new theory. The important period in sociological studies of the family is marked by the emergence of the social psychology of the family and the use of case studies in addition to quantitative methods (Komarovsky & Waller, 1945). Today, scholars have used a variety of methods to conduct research on marriage and the family. This section focuses on the research process and research methods.

Research Design: Description and Explanation

Research design refers to the strategy that researchers choose to integrate the components of the study in a logical way and effectively address the *research* problem. When it comes to research design, researchers determine whether the research is descriptive or explanatory as it affects what information is collected. Researchers ask two basic types of research questions: what is going on (descriptive research) and why is it going on (explanatory research)?

Descriptive Research

Descriptive research is used by sociologists to describe characteristics of social situations, circumstances, or groups of people. For example, a sociologist may use descriptive research to determine the number of single mothers residing in a certain state. Descriptive research can be conducted to determine the degree of a social problem, but it does not provide conclusive evidence for what causes a problem. For example, policymakers and researchers are concerned about income inequality and want to know the nature and degree of the social problem before they find out the main causes of the problem and solution to the problem. They have found that there has been rising income inequality since the 1970s in the United States. Figure 2.6 shows that incomes for the bottom 99 percent of families grew by 3.3 percent over 2013 levels but incomes for those families in the top 1 percent of earners grew by 10.8 percent over the same period. Descriptive research can establish the facts of the income inequality situation in the United States but it cannot be used to determine a causal relationship for why such income inequality exists. Descriptive research may be followed up with examinations of why a social situation exists and what the implications of the findings are.

Descriptive research can be concrete or abstract (De Vaus, 2001). Researchers might use a concrete description to discuss contemporary families along social class lines. For example, researchers may provide a concrete description of income inequality such as the following: In the most unequal states such as New York, Connecticut, and Wyoming, the top 1 percent earned average incomes more than 40 times those of the bottom 99 percent (Sommeiller *et al.*, 2016). High-income families are more likely to accumulate assets and retirement accounts than low-income families. Figure 2.7 shows that high-income families are ten times as likely to have retirement accounts as low-income families.

Abstract questions explore relationships between facts. For example, during the nineteenth century, German sociologist Karl Marx detected and described

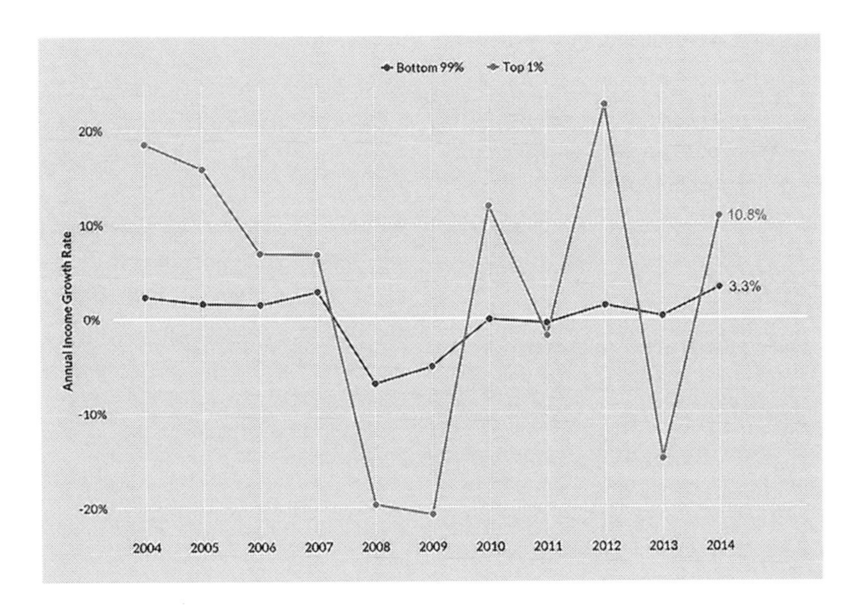

Figure 2.6 The Wealthiest American Families Growing Ever Richer
Source: Emmanuel Saez and Thomas Pikkety's analysis of IRS data.

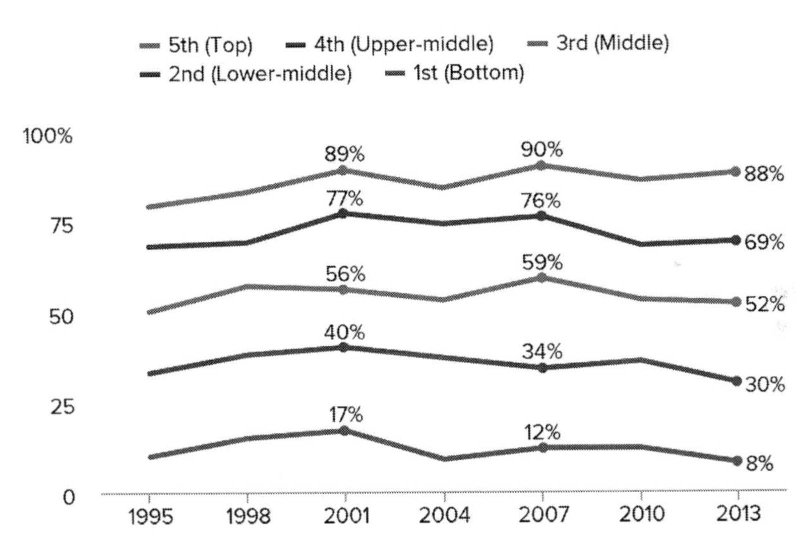

Figure 2.7 High-income Families Are Ten Times as Likely to Have Retirement Accounts as Low-income Families
Source: Economic Policy Institute Analysis of Survey of Consumer Finance data, 2013. http://www.epi.org/publication/retirement-in-america/

class polarization between the capitalist class and working class. He believed the main cause of income inequality was due to class polarization between the capitalist class as owners of the means of production such as factories and the working class as non-owners or "have-nots."

Explanatory Research

The purpose of **explanatory research** is to develop and evaluate causal explanations. Explanatory research focuses on "why" questions (De Vaus, 2001). It explores a research topic in depth and forms the basis of conclusive research. When it comes to income inequality, researchers try to find the main causes of pay gap between men and women. Let's consider an example: a sociological researcher may use explanatory research to argue that phenomenon Y (family income level) is affected by factor X (gender).

Gender ⟶ Family income

Explanatory research indicates that gender might indeed affect family income, as men are likely to earn more than women. As of 2015, women's hourly median wages were 82.7 percent of men's hourly median wages (Gould *et al*., 2016). Figure 2.8 shows that the median hourly wage of women was $15.67 compared with $18.94 for men.

Causal relationships may involve a number of interrelated causal chains. Because education, occupation, and income are correlated, researchers further argue that gender affects the level of education, which in turn affects occupational options, which in turn affects income level.

Gender ⟶ Education ⟶ Occupation ⟶ Family income

When exploring the main causes of income inequality in terms of gender, researchers wonder whether level of education is related to the gender wage gap. Women are more likely to graduate from college than men and are more likely to receive a graduate degree than men (Gould & Schieder, 2016). Yet, at every education level, women are paid less than men (see Figure 2.9).

Women are almost half of the workforce and the sole or co-breadwinner in half of American families with children, receiving more college and graduate degrees than men but continuing to make only 80

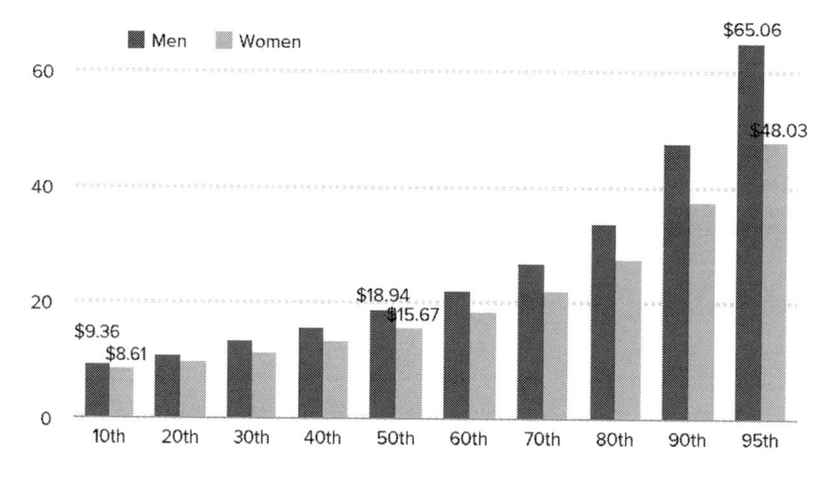

Figure 2.8 Women Earn Less than Men at Every Wage Level, Hourly Wages by Gender and Wage Percentile, 2015
Source: Economic Policy Institute Analysis of Current Population Survey Outgoing Rotation Group microdata. http://www.epi.org/publication/what-is-the-gender-pay-gap-and-is-it-real/

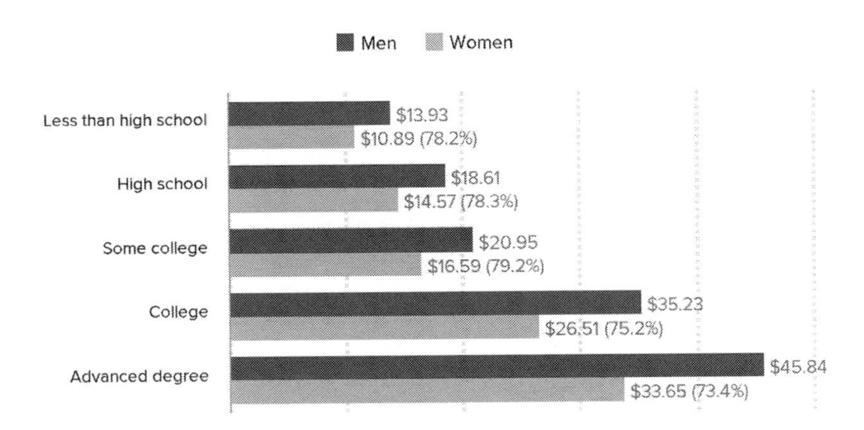

Figure 2.9 Women Earn Less than Men at Every Education Level, Average Hourly Wages by Gender and Education, 2015

Source: EPI Analysis of Current Population Survey Outgoing Rotation Group microdata. http://www.epi.org/publication/what-is-the-gender-pay-gap-and-is-it-real/

cents for every dollar earned by men in 2015 (Institute for Women's Policy Research, 2017). Gould *et al.* (2016) wonder what causes these gaps and use the data to shed light on "why" questions. They have examined pay gaps in terms of education, race, jobs, state, and union membership. They have found that reasons for the gender wage gap are multifaceted. When factors such as occupation and parental or marital status are used as control variables in statistical models aiming to explain what "causes" the wage gap, the size of that gap is reduced, and what is left unexplained is generally thought to possibly be the result of discrimination (Institute for Women's Policy Research, 2016).

Sociologists distinguish correlation from causation. A **correlation** is a single number that describes the degree of relationship between two variables. Two factors may be correlated—that is, they may be related in some way—but that does not mean that one factor causes the other. The link between two factors may be coincidental rather than causal. For example, marriage rates have been declining. American divorce rates rose from the 1950s to the 1970s, peaked around 1980, and have fallen ever since (Rotz, 2011). Is divorce causing the declining marriage rate? Rather than divorce causing the declining marriage rate, the declining marriage rate might be due to other factors such as changing views toward marriage or increasing poverty. Rotz (2011) suggested that the increase in women's age at marriage was the main proximate cause of the fall in divorce.

Research Approaches

There are two main approaches to research: deductive and inductive. Most research involves both deductive and inductive approaches. Let's explore each in a bit of detail.

Deductive Approach

When researchers use a **deductive approach,** they develop a *hypothesis*—a tentative statement about the relationship between two or more variables—then test the hypothesis with quantitative research. **Quantitative research** is the systematic empirical study of observable phenomena via statistical, mathematical, or computational techniques. The deductive approach can be referred to as a "top-down" approach because researchers work from general to specific. Researchers begin with a *theory* about their topic of interest and then narrow it down into specific *hypotheses* that they can test. For example, a researcher may use a deductive approach to test the hypothesized effects of a treatment on some outcome. The researcher starts with a theory and formulates a new hypothesis that proposes a relationship between two variables. After testing a hypothesis with research methods and examining the results, researchers confirm or reject the theory and modify the theory if the hypothesis is falsified. As we discussed earlier, researchers started by using culture to explain why Jewish immigrants moved up the ladder sooner than Italian immigrants but refuted this hypothesis and used social class theory to explain the social mobility of Jewish immigrants as compared to Italian immigrants. The deductive approach can be thought of like this:

Theory ⟶ Hypothesis ⟶ Observation ⟶ Confirmation (or not of the original theory)

Inductive Approach

The **inductive approach** is concerned with generating a new theory based on data and is associated with qualitative research. **Qualitative research** is the examination, analysis, and interpretation of observable phenomena for the purpose of discovering underlying patterns of relationships with no mathematical models. The inductive approach starts with observations and leads to theories. The inductive approach is open-ended and exploratory at the beginning. This approach is referred to as a "bottom-up" approach because researchers work from specific to general. Researchers begin with specific observations, detect patterns, develop empirical generalizations, and identify relationships through their research. Researchers use observations to describe a picture of the phenomenon that is being studied and to build a theory. The inductive approach can be thought of like this:

Observation ⟶ Pattern ⟶ Hypothesis ⟶ Theory

The Research Process

Researchers aim to identify a social problem or social phenomenon, find the main causes or effects of the problem, and provide possible solutions to the problem. This research process involves seven steps. Let's look at each step in the research process.

Step 1: Select and Define a Problem Statement

The first step in the research process is to select and define a situation or problem. The **problem statement** is a description of the social issue that needs to be addressed before trying to solve the problem. For example, a researcher may want to look into the increasing suicide rate among a certain group of people. In 1897, French sociologist Émile Durkheim did just that: he explored the social problem of suicide after observing higher suicide rates. Durkheim compared suicide rates in Catholic and Protestant populations and pioneered quantitative methods in criminology during his suicide case study. Since then, suicide has been a social problem that has been the subject of inquiry by many researchers.

Step 2: Perform a Literature Review

When researchers perform a **literature review**, they look at how previous research findings fit into the proposed research. Before beginning the research, it is important to analyze what others have written about the topic in order to determine gaps and avoid errors. In doing so, researchers retrieve the relevant literature such as books and articles related to the topic and review the research findings and theoretical contributions to a topic of study. For example, when Durkheim began his research on suicide, very little sociological literature existed to review, but he studied the works of several moral philosophers. Researchers should understand the literature they have gathered, make connections between the literature and their topic of study, and evaluate the strengths and weaknesses of the literature and research findings they have collected before they formulate a hypothesis.

Step 3: Formulate a Hypothesis

Based on a literature review, a researcher is able to formulate a hypothesis. A **hypothesis** is a tentative statement about the relationship between variables. Therefore, part of formulating a hypothesis is determining the variables. A **variable** is something with measurable traits that can change. The most fundamental relationship in a hypothesis is between an independent variable and a dependent variable. The **independent variable** is a variable that can affect a dependent variable. The **dependent variable** is a variable that is affected by a change in the independent variable. The dependent variable is the effect that is assumed to be caused by independent variables. When a researcher tests relationships between variables, it is testing whether an independent variable affects a dependent variable. For example, Durkheim hypothesized that the rate of suicide varies inversely with the degree of social integration. He believed that a low degree of social integration (the independent variable) is likely to cause a high rate of suicide (the dependent variable).

A research study may have many variables depending on the topic being studied. For example, researchers may consider divorce to be related to lack of communication, age, and education. The dependent variable (or the effect) is divorce. Lack of communication, age, and education are independent variables or the causes. For example, the least (no high school diploma or GED) and the highest (college degree) educated women share the lowest rate of first divorce; Asian women have the lowest first divorce rate at 10 divorces per 1,000 women in a first marriage, compared with White (16.3), Hispanic (18.1), and African American women (30.4) (Bowling Green State University, 2011). The research findings show that the divorce rate varies depending on race and level of education.

Note that if researchers use an explanatory research design and explore the causes of a problem, they are likely to formulate one or more hypotheses to test their theories. If they plan to use a descriptive research design, researchers are less likely to use hypotheses as they desire only to describe social reality and facts.

Step 4: Develop the Research Design

Researchers must determine the unit of analysis to be used in the study. A **unit of analysis** is what is being studied. The unit of analysis is the entity that a researcher is analyzing in the study. In social science research, units of analysis include individuals, groups, families, social organizations, and social artifacts.

Then researchers must decide how to collect data and support the hypothesis that is formulated. The function of a research design is to ensure that the evidence obtained enables researchers to answer the initial hypothesis as unambiguously as possible (De Vaus, 2001). Research designs are equated with qualitative and quantitative research. For example, a researcher can use quantitative research and generate quantitative data through surveys to support their hypothesis. Also, researchers must choose and use a research method to collect data (see research methods section) and support the stated hypothesis. De Vaus (2001) emphasizes that when designing research, it is essential that we identify the type of evidence required to answer the research question in a convincing way.

Step 5: Collect and Analyze Data

The next step is to collect and analyze data. Researchers decide what population will be observed or questioned and then carefully select a sample. A **population** is the entire group of people about whom researchers are able to draw conclusions. In the research, the first item that researchers need to decide is which subjects should be studied and to whom the findings can be applied. For example, a population can be all the people in the United States or only females in New York. However, gathering data from each member of a population is time consuming and costly (and sometimes impossible). Therefore, researchers gather data from a select number of people. A **sample** is a subset of the population to be studied. A **representative sample** is a selection from a population that has the essential characteristics of the population. The research findings can represent the entire population if each person from the population has an equal opportunity to be selected for participation in a study, known as a **random sample**. A random sample is chosen by chance as every member of an entire population being studied has the same chance of being selected. A **stratified sample** goes beyond a random sample and ensures that the random sample matches the proportions of the population being studied. For example, if a population was 70 percent male, researchers would make sure that their sample was 70 percent male.

Researchers collect data in a number of ways, which we will discuss a bit later in the chapter. After collecting the data, researchers must analyze it. **Data analysis** refers to the process of applying theoretical and statistical techniques to gather, evaluate, condense, and describe raw data with the purpose of extracting useful information. When examining the causes of higher suicide rates, for example, French sociologist Durkheim used quantitative data to support his hypothesis that single or widowed or divorced people were more likely to commit suicide than married people. Durkheim analyzed these results to show that individuals who were less integrated into society had higher rates of suicide.

Step 6 and Step 7: Draw Conclusions and Report the Research Findings

The last steps in the research process are to draw conclusions and report the research findings. The report covers a review of each step taken in the research process for the purpose of further study. Hypotheses come about as a research finding or product of research. The limitations of the study and problems with the sample are also discussed. Researchers present their research findings at professional meetings and publish them in journals and books. For example, Durkheim (1897 [1951]) reported his findings on suicide in *Suicide: A Study in Sociology*, arguing that the suicide rate of a group is a social fact that cannot be explained in terms of the personality traits of individuals.

Referencing is one of the most important aspects of any academic research. On a reference page, researchers list resources that they referred to within the body of the work. A **bibliography** is a list of resources that researchers have read during the research process in order to widen their knowledge about the research area. The most popular referencing system used in social science academic works is APA (American Psychological Association).

Data Collection and Data Analysis

Data collection is the process of collecting and measuring data that are used to support hypotheses and evaluate outcomes. The goal of data collection is to enable researchers to use the data that have been collected to answer sociological questions and test the hypotheses that have been posed. Earlier we defined the two types of data that can be collected: quantitative and qualitative. Let's look at these in a bit more detail.

Quantitative Research Data

Quantitative data is data that can be measured in numerical form, such as statistics and percentages. **Measurement** is the process of recording the observations that are collected and investigating causal associations. Researchers can use surveys to gather and analyze quantitative data. Quantitative data is associated with quantitative research. The purpose of quantitative research is to quantify data and make generalizations from a sample to the population of interest.

In quantitative data analysis, researchers are expected to turn raw numbers into meaningful data and critically analyze the primary data through comparing the collected data to the previous research findings within the framework of the same research. For example, if we argue that there is strong positive correlation between social class theory and ethnic social mobility, we have to provide the data and provide explanation about how the Jewish immigrants' social position contributed to their ethnic mobility in the early twentieth century. Analytical software such as Microsoft Excel and SPSS can be used to assist with analysis of quantitative data.

Qualitative Research Data

Qualitative data is descriptive data that can be qualified and explained with words. Qualitative data is associated with qualitative research, which gathers data that is not in numerical form. The purpose of qualitative research is to uncover a social situation and explore the nature of the problem. Researchers often use open-ended questions to which respondents are free to give any answer. Although qualitative study cannot make generalizations, researchers can use techniques to identify relationships within responses of a sample group. For example, researchers can analyze the words and phrases repeatedly used by respondents when they are analyzing the data obtained from an open-ended question interview. Researchers then use quotations from the respondents to illustrate the points within the research findings. Another technique of qualitative data analysis is to compare the primary data that they gathered through the interview with the secondary data or existing data they collect through literature review, discuss differences between them, and link their research findings to research questions.

Qualitative research is used to interpret the research findings and understand the meaning of the conclusions produced by quantitative methods. In the social sciences, to combine quantitative and qualitative data in the research process is referred to as *mixed-methods research* and is recommended.

Research Methods

Research methods are techniques for collecting data and conducting research. After a research question and research design have been established and a hypothesis has been formulated, the next step is to figure out how to collect data. There are several research methods that researchers can use to collect data based on factors like time, money, and topic.

Surveys

A **survey** is a research method that allows researchers to collect and analyze data from a group of people and determine the relationships between variables. Surveys are the most widely used research method in the social sciences because they make it possible to study things that are not directly observable, such as people's attitudes and beliefs, and to describe a population too large to observe directly (Babbie, 1999).

One reason for their popularity is that surveys can be conducted over the phone, in person, over the internet, or via mail questionnaire. **Telephone surveys** are conducted by telephone in which the questions are read to respondents. **Interviews** are face-to-face data-collection encounters in which a researcher asks questions and records the answers. A **structured interview** or standardized interview is a quantitative research method commonly used in survey research. In a structured interview, each interviewee is presented with the same questions in the same order and the interviewer records the answers that are scored on a standard grid. **Unstructured interviews** or non-directive interviews are interviews in which questions are not prearranged. This allows the interviewer to focus on interacting with the respondent and follow the discussion. Because unstructured interviews often contain open-ended questions, researchers might tape-record interviews for analysis. **Questionnaires** are a research instrument containing a series of questions for the purpose of collecting data from respondents. Questionnaires may be self-administered by respondents or administered in face-to-face-encounters. They can be both quantitative and qualitative in nature.

With surveys, researchers select a representative sample from a population to provide responses. Examples of large-scale surveys are the Gallup, the Harris, and the U. S. Census Bureau. Responses obtained through closed-ended questions with multiple-choice answer options are analyzed using quantitative methods such as pie charts, bar charts, and percentages. Responses obtained to open-ended questions are analyzed using qualitative methods.

Advantages of surveys include gathering a large amount of cost-effective data in a short period of time and allowing participants to remain anonymous. Survey research is useful in describing the characteristics of a large population. Computer technology helps do multivariate analysis and examine more than two independent variables. The major disadvantages include difficulties of ensuring depth for the research. Researchers might not get truthful answers through interviews as respondents may provide the answers they believe the researchers expect.

Experiments

Experimental research is used to test hypotheses. An **experiment** is a research method that involves creating two groups and manipulating an independent variable to assess its impact on the dependent variable in a controlled environment. Subjects are randomly classified into either an experimental group or a control group. The **experimental group** contains the subjects who are exposed to an independent variable or treatment. The **control group** contains the subjects who are not exposed to the independent variable or treatment. The results from the two groups are compared. This form of research is generally used for causal studies. If an experiment is conducted in a laboratory, it is difficult to know whether its results apply to the real world, as the real world is different from an artificial laboratory.

Observation

Observation is a research method of collecting qualitative data through observing people in a natural setting. **Detached observation** (or **non-participant observation**) is a type of observation in which researchers observe without getting involved and are neither seen nor noticed by participants. In non-participant observation, the researcher observes a group of people but does not interact with them. For example, a researcher watches fathers and their children in a park to find out how fathers interact with their children. Through observing their actual behaviors, researchers take notes and apply a theory to the analysis of the observations.

Participant observation is a type of observation in which researchers observe a group's behavior from within the group. A researcher joins the group and participates in the group activity. The people who are being observed may not know that they are being studied. Participant researchers can get information with accuracy if they try not to let their presence influence the attitudes and behaviors of the people they are observing. Participant observation is widely used among social scientists who want to observe the social behaviors and cultural patterns of a group in the short time available to them.

One advantage of observation is that researchers can see how people interact in real-life situations and collect data in the natural environment. One disadvantage is that people who are being studied may behave differently if they know they are being observed. This phenomenon is known as the **Hawthorne effect** and occurs when individuals improve their behavior in response to their awareness of being observed (McCarney *et al.*, 2007). This first came to light in a series of studies on factory workers between 1924 and 1932. Although the change in behavior observed in the factory worker case was positive, the results of a research study in which the Hawthorne effect occurred would not be valid (Mayo, 1924).

Case Studies

Case study is a research method that is used to collect data and analyze issues within a particular environment. *Exploratory case studies* focus on the questions of "what?" or "who?" using survey questionnaires to collect and analyze quantitative data. *Descriptive case studies* focus on describing what is going on and discovering social phenomena. Researchers, for example, want to study the impact of increasing levels of cultural diversity on learning outcomes. *Explanatory case studies* focus on a social phenomenon and explain the reasons for the social phenomenon. For example, the Institute for Women's Policy Research consistently conducts exploratory case study and provides comprehensive data on women of color in the United States. Eight out of ten (80.6 percent) Black mothers are breadwinners and earn 40 percent of household income; in Louisiana, Black women earned less than half of White men's earnings (46.3 percent) (DuMonthier *et al.*, 2017). Sociologists use **intersectionality theory**—examining experiences of minority people who are subjected to multiple forms of subordination in society to explain how intersections of race, ethnicity, and class affect Black women's outcomes.

Advantages of the case study method include explaining social phenomenon in greater levels of depth by using qualitative and quantitative data. However, case studies lack the basis for applying the research findings to a large social group because a researcher cannot make generalizations on the *basis* of a particular case.

Secondary Data Analysis

Primary data is collected by researchers for the purpose of their research. **Secondary data analysis** is the practice of analyzing and using existing data that has been collected by others for the purpose of prior study. Researchers are able to tap into databases and public records and collect existing data, such as population statistics. In 1897, French sociologist Durkheim studied existing death records to examine the relationship among variables such as age, marital status, and the circumstances surrounding a person's suicide.

Secondary data analysis includes content analysis. **Content analysis** is a research method through which researchers look at a variety of content and draw conclusions about social life. It allows researchers to consider all aspects of content, such as written records (books, public records), visual texts such as movies, TV shows, music, art, even garbage, and examine patterns of behavior that occur over time.

Secondary data analysis is the cheapest way to collect data. Researchers do not gather their own data but instead analyze existing data that someone else has gathered. Researchers simply use the data and move to the data analysis. One major disadvantage of secondary data analysis is the lack of variables in which researchers are interested. Researchers can only choose from the available variables out of a database.

Criteria for Conducting Research

Reliability and validity are the two criteria used to evaluate a measurement and assess research findings.

Reliability

Reliability refers to the degree to which a measure produces consistent results. It explains the extent to which the same research findings are obtained using the same instruments. For example, the most reliable research and the most consistent findings that the National Research Council conducted are that paternal imprisonment results in both behavioral problems and delinquency (National Research Council, 2014). Hagan (2013) finds that the U.S. college graduation rate of 40 percent among youth drops to 2 percent among children of incarcerated mothers and 15 percent for imprisoned fathers. Moody and Warcholik

(2017) further find that 5 million children with at least one parent in prison at some point in their life have to deal with a number of additional challenges including a higher number of other major, potentially traumatic life events. Therefore, a measure is considered to have high reliability if it produces similar research findings under consistent conditions.

Validity

Validity refers to how well a test measures what it is purported to measure. **Internal validity** refers to the extent to which the effect of the independent variable on the dependent variable is real and not caused by extraneous factors. If parental incarceration has an independent effect on a child's behavior as noted above, researchers should be more confident that changes in a child's behavior (dependent variable) are caused by parental incarceration (the independent variable). It means that researchers are able to conclude that they have strong evidence of causality because no other variables except the one they are studying caused the result. **External validity** refers to the extent to which the research findings are able to be generalized to other contexts. External validity is important also because it is the extent to which results of a study can be generalized to the world at large.

Reliability and validity are two cornerstones of the scientific method. When designing a questionnaire during the data collection process, researchers may ask whether the questionnaire is understandable and answerable and whether it is both valid and reliable.

APA Guidelines

The American Sociological Association (ASA) provides guidelines to which researchers must adhere. **APA (American Psychological Association) style** describes guidelines and rules for the preparation of manuscripts for researchers and students. APA style is commonly used to cite sources within the social sciences and establishes standards of written communication concerning the organization of content, writing style, and citing references. Researchers are required to follow ethical rules that govern their research and protect the privacy of the respondents and maintain confidentiality.

Global Overview: Refugee Families

On World Refugee Day, held every year on June 20, we commemorate the strength, courage, and perseverance of millions of refugees. A **refugee** is a person who has been forced to leave their country in order to escape war, persecution, or natural disaster. The United States admitted 84,995 refugees in 2016 and almost half (46 percent) of these were Muslim, with the largest numbers going to California and Texas (Krogstad & Radford, 2017).

Refugee Families in the 1970s

Refugee families have often escaped from their home countries due to wars and persecution. **Resettlement** involves the relocation of refugees from an asylum country to a third country (UN Refugee Agency, 2017). Resettlement is a life-changing experience. Refugee families are resettled to a country where the society, language, and culture are completely new to them. In 1975, after the Vietnam War and the communist victory in Indochina, the Refugee Act of 1980 allowed Vietnamese, Cambodian, and Laotian refugee families to the United States for political asylum. The second wave of refugees in 1981 was predominantly family-reunification cases sponsored by relatives (Ng, 1998). A third wave came in the 1990s and 2000s. The various groups differ dramatically in language, culture, and history. In a study of Lao living in Alabama and middle Tennessee, researchers found 80 percent of the refugees experience stress through problems understanding the behavior of Americans and 82 percent reported difficulties because America did not understand Lao cultural ways (Nicassio *et al.*, 1986).

Census Data Collection through Ethnographic Observations

Language, cultural differences, and the particular circumstances of resettlement of any new Americans pose barriers to obtain complete and accurate census data. Since 1975, almost 850,000 Southeast Asian refugees have been admitted to the United States and in 1980 there was a 28 percent undercount of Southeast Asian refugees in the United States as cultural differences posed barriers to obtain complete census data (Rynearson & Gosebrink, 1989).

In order to assist the 1990 Census achieve an accurate count of Southeast Asian refugee families, researchers drew on ethnographic observations and helped obtain comprehensive coverage of Southeast Asian refugees in the 1990 Census. **Ethnographic observation** is a fieldwork method that allows participant researchers to observe a cultural group and write detailed accounts of their social behaviors and cultural patterns.

A team of researchers conducted ethnographic fieldwork from January 1, 1989 and focused primarily on Lowland Lao refugees in St. Louis. Both participant observation and guided discussion with the Lao concerning their lives reveal that there is very little interaction with outsiders, i.e., non-Lao, in the neighborhood setting. Barriers to interacting with neighbors are reinforced because Lao and other refugees are initially settled in the city's most dangerous neighborhoods, primarily because housing is cheapest there (Rynearson & Gosebrink, 1989). These refugee families were traditionally patriarchal, with elders and men holding authority over wives and children. Caplan *et al.* (1989) find that the children of Southeast Asian boat people excel in the American school system. Today Vietnamese Americans (2.1 million) are the fourth largest population of Asian Americans in the United States, concentrated in the West, including Orange County, California, San Jose, California, and Houston, Texas (U. S. Census Bureau, 2018c/2017). "Techniques now being developed to assure an accurate count of refugees for the 1990 Census will continue to be valuable well into the twenty-first century" (Rynearson & Gosebrink, 1989).

Summary

This chapter discussed sociological theories and scientific research on marriage and the family. It looked at a number of macro-level theories, including structural functionalist theory, conflict theory, and feminist theory. It also looked at a number of micro-level theories, including symbolic interactionist theory, social exchange theory, family developmental theory, and role theory.

Research is the process of collecting data for the purpose of testing an existing theory or generating a new theory. Descriptive research examines the nature of the problem and generates qualitative data. Explanatory research focuses on "why" questions and explores the degree of the problem. The research process involves selecting and defining a problem statement, performing a literature review, formulating a hypothesis, developing the research design, collecting and analyzing data, and drawing conclusions and reporting research findings. Research methods that are used to collect data include surveys, experiment, observation, and secondary data analysis. Reliability and validity are research criteria and explain whether research findings are reliable and valid. From a diversity and global approach, the chapter applies sociological theory and explanatory research to the analysis of ethnic social mobility and refugee families.

Key Terms

Absolute poverty refers to a condition where a person or the family does not have the minimum amount of income needed to meet basic human needs.

APA (American Psychological Association) style describes guidelines and rules for the preparation of manuscripts for researchers and students.

Assimilation refers to a process in which distinct groups merge together to become part of the dominant group.

Bibliography is a list of resources that researchers have read during the research process in order to widen their knowledge about the research area.

Bourgeoisie comprises the people who live off the surplus generated by the working class.

Capitalist class comprises the people who live off the surplus generated by the working class.

Case study is utilized to collect data and analyze issues within an environment.

Conflict theory views society as being in a constant struggle over control and allocation of scarce resources.

Content analysis is a research method through which researchers look at a variety of content and draw conclusions about social life.

Control group contains the subjects who are not exposed to the independent variable or treatment.

Core nations are developed and industrialized countries such as the United Kingdom and the United States.

Correlation is a single number that describes the degree of relationship between two variables.

Cultural mosaic is the mixture of ethnic groups with different cultures that coexist within society.

Culture of poverty theory suggests that living in conditions of poverty leads to the development of a culture adapted to those conditions.

Data analysis refers to the process of applying theoretical and statistical techniques to gather, evaluate, process, condense, and describe raw data with the purpose of extracting useful information.

Data collection is the process of collecting and measuring data that are used to support hypotheses and evaluate outcomes.

Deductive approach is to develop a hypothesis based on existing theory and design a research to test the hypothesis with quantitative research.

Dependent variable is a variable that is affected by a change in the independent variable.

Descriptive research is used by sociologists to describe characteristics of social situations, circumstances, or groups of people.

Detached observation (or non-participant observation) is a type of observation in which researchers observe without getting involved and are neither seen nor noticed by participants.

Ethnographic observation is a fieldwork method that allows participant researchers to observe a cultural group and write detailed accounts of their social behaviors and cultural patterns.

Exchanges include coverage purchased through the federal Health Insurance Marketplace, as well as other state-based marketplaces, and include both subsidized and unsubsidized plans.

Experiment is a research method that involves creating two groups and manipulating an independent variable to assess its impact on the dependent variable in a controlled environment.

Experimental group contains the subjects who are exposed to an independent variable or treatment.

Explanatory research focuses on "why" questions and evaluates causal explanations.

External validity refers to the extent to which the research findings are warranted to be generalized to other contexts.

Family development theory focuses on the systematic and patterned changes experienced by families and explores how couples and family members deal with roles and tasks as they move through each stage of life.

Feminist theory focuses on gender inequality and equal rights for women and men.

Gender wage gap is the situation in which working women are paid less than working men.

Global stratification is unequal distribution of resources on a global basis.

Government insurance coverage includes federal programs, such as Medicare and Medicaid.

Hawthorne effect is a type of reactivity in which individuals improve their behavior in response to their awareness of being observed.

Health inequality refers to differences in health status and in the distribution of health care between different population groups.

Hypothesis is a tentative statement of relationship between variables.

Income refers to the revenue streams from salaries, wages, and interest on savings accounts, dividends from shares of stock, rent, and profits from elsewhere.

Income inequality means the extent to which income is distributed in an uneven manner among a population.

Independent variable is a variable that the research can manipulate in an experiment or that can affect a dependent variable.

Inductive approach is concerned with generating a new theory based on data and is associated with qualitative research.

Internal validity refers to the extent to which the effect of the independent variable on the dependent variable is real and not caused by extraneous factors.

Intersectionality theory examines the experiences of minority people who are subjected to multiple forms of subordination in society.

Interviews are face-to-face data-collection encounters in which a researcher asks questions and records the answers.

Kaleidoscope is used to refer to the different cultural patterns and colors one sees in American society, similar to those one sees when looking through a kaleidoscope.

Labeling theory studies labels put on individuals and the consequences that the labels produce.

Latent functions are unintended or unexpected consequences of functions.

Literature review is to look at how previous research findings fit into the proposed research.

Macro-level theories study a large-scale social structure at the societal and global level.

Manifest functions are consequences of functions that are intended and expected.

Measurement is the process of recording the observations that are collected and investigating causal associations

Mechanical solidarity is a form of social solidarity based on kinship ties of familial networks with minimal division of labor in traditional societies.

Micro-level theories study everyday behavior and interaction patterns in order to research what happens within families.

Mixing bowl refers to the place in which people of different cultures all mix together to assimilate to a single homogeneous culture.

Multiculturalism refers to the coexistence of multiple cultures within a society.

Observation is a research method of collecting qualitative data through observing people in a natural setting.

Organic solidarity is a form of social solidarity based on interdependence from the specialization of work and with a high level of division of labor in industrial society.

Participant observation is a type of observation that allows a researcher to observe a group's behavior from within the group.

Periphery nations are developing and poor countries such as Zambia.

Pluralism refers to the process in which racial and ethnic groups maintain their own cultural distinctiveness and participate in society without prejudice.

Population is the entire group of people about whom researchers are able to draw conclusions.

Poverty is a state of deprivation and a lack of material possessions.

Power is the ability of people to achieve their goals despite opposition from others.

Prestige refers to the respect or status position regarded by others.

Primary data is collected by researchers for the purpose of their research.

Primary deviance is the initial act of deviant behavior that receives little social reaction.

Private health insurance is defined as a plan provided through an employer or a union and coverage purchased directly by an individual from an insurance company or through an exchange.

Problem statement is a description of the social issue that needs to be addressed before trying to solve the problem.

Proletariat comprises the people who do not own the means of production and must sell their muscle to earn a minimum amount of wages.

Qualitative data is descriptive data that can be qualified and explained with words.

Qualitative research is the examination, analysis, and interpretation of observable phenomena for the purpose of discovering underlying patterns of relationships with no mathematical models.

Quantitative data is data that can be measured in numerical form such as statistics and percentages.

Quantitative research is the systematic empirical study of observable phenomena via statistical, mathematical, or computational techniques.

Questionnaires are a research instrument containing a series of questions for the purpose of collecting data from respondents.

Random sample means that the research findings can represent the entire population if each person from the population has an equal opportunity to be selected for participation.

A **refugee** is a person who has been forced to leave their country in order to escape war, persecution, or natural disaster.

Relative poverty refers to a condition where a person or the family can afford the basic daily necessities but can't maintain an average standard of living.

Reliability refers to the degree to which a measure produces consistent results.

Representative sample is a selection from a population that has the essential characteristics of the population.

Research design refers to the strategy that researchers choose to integrate the components of the study in a logical way and effectively address the *research* problem.

Research methods are techniques for collecting data and conducting research.

Research is the process of collecting data for the purpose of testing the existing theory or generating a new theory.

Resettlement is the only durable solution that involves the relocation of refugees from an asylum country to a third country.

A **role** is a set of behavioral expectations associated with a status.

Role expectation relates to the expectations required of a parental role.

Role performance refers to the actual delivery of those expectations.

Role theory suggests that most of our everyday activities consist of acting out socially defined roles associated with a given status.

Salad bowl concept encourages discrete *cultures* to coexist yet keep their separate identities.

Sample is a subset of a population from the population to be studied and represents that population.

Scapegoat theory refers to the tendency to blame a social group for one's own problems.

Scientific research methods, such as surveys and experiments, are used to collect data.

Secondary data analysis is the practice of analyzing and using existing data that has been collected by others for the purpose of prior study.

Secondary deviation is a deviant behavior that results from being labeled as a deviant by society.

Semi-periphery nations such as Brazil are midway between core and periphery nations and are moving toward industrialization.

Social class refers to a group of people with similar shares of resources.

Social exchange theory explores how people interact within reciprocal relationships and interpret their experiences of self and others to determine the benefits and costs of the relationship.

Social inequality refers to unequal distribution of resources among members of a society.

Social interaction refers to reciprocal communication between two or more people through symbols, verbal language and non-verbal language.

Social stratification is the categorizing of people into various socioeconomic strata with different shares of resources arranged in the social hierarchy.

Socioeconomic status (SES) is a combined measure to classify people who have the same income, occupation, and education and to determine their position in the social hierarchy.

Sociological theories are at the heart of the sociological study of the family as they guide researchers in their studies and try to explain how families work and fit into society.

Sociology of the family refers to the scientific study of changing marriage patterns and family forms.

Status is a socially defined position characterized by rights, responsibilities, and obligations.

Stereotypes are oversimplified generalizations about a group of people.

Stratified sample goes beyond a random sample and ensures that the random sample matches the proportions of the population being studied.

Structural functionalist theory views society as a social structure, with many different parts all working for social stability.

Structured interview is a quantitative research method in survey research through which each interviewee is presented with the same questions in the same order and the interviewer records the answers that are scored on a standard grid.

Survey is a research method that allows researchers to collect and analyze data from a group of people and determine the relationships between variables.

Symbolic interactionalist theory views society as the sum of people's everyday interactions.

Symbols refer to anything that represents something else.

Telephone survey is a survey conducted by telephone in which the questions are read to the respondents.

Theory consists of statements that illustrate proposed relationships between variables and try to explain social phenomenon.

Unit of analysis is what is being studied.

Unstructured interviews or non-directive interviews are interviews in which questions are not prearranged.

Validity refers to how well a test measures what it is purported to measure.

Variable is something with measurable traits that can change.

Wealth refers to the total value of household assets that include income, personal property, and income-producing property.

Wealth inequality or wealth gap is the unequal distribution of assets among residents of the United States.

Working class comprises the people who do not own the means of production and must sell their muscle to earn a minimum amount of wages.

Working poor are the working people whose incomes fall beneath the poverty line.

Working poverty is the situation of those working people whose incomes fall beneath the poverty line.

World systems theory suggests that the world is divided into core, semi-periphery, and periphery nations.

Discussion Questions

- Sociologist Émile Durkheim was curious about high suicide rates and was interested in employing data and theory to offer an explanation. He hypothesized that weakening social solidarity (independent variable) is likely to cause higher suicide rates (dependent variable). French sociologist Durkheim's study is a classic example of the use of theory to explain the relationship between two variables. Proposing cause-and-effect relationships is the major component of sociological theory. Based on this example, choose a topic of interest (e.g., divorce, cohabitation), formulate a hypothesis, and use a theory to explain the relationship between variables.
- If you decide to use a questionnaire to collect primary data to support your hypothesis, what kinds of questions would you ask and what kind of detailed information would you want to gather through survey research?

Suggested Film and Video

Economic Policy Institute: http://www.epi.org/publications/

Inequality in the UK: http://voxeu.org/videovox

Internet Sources

American Psychological Association APA Style: http://www.apastyle.org/

Minimum Wage Tracker: http://www.epi.org/minimum-wage-tracker/#/min_wage/Montana

Research and Analysis: http://equitablegrowth.org/research-analysis/

Status of Women in the States: https://statusofwomendata.org/

The Working Poor Families Project (WPFP): http://www.workingpoorfamilies.org/

What are Poverty Lines?: http://www.worldbank.org/en/news/video/2017/04/14/what-are-poverty-lines

References

Babbie, E. (1999). *The Basics of Social Research*. Belmont, CA: Wadsworth Publishing Company.

Babones, S. J. (2015). The Country-Level Income Structure of the World-Economy. *Journal of World-Systems Research*, 11(1), 29–55.

Becker, H. (1963). *Outsiders*. New York: Free Press.

Berchick, E. R., Hood, E. & Barnett, J. C. (September 2018). *Health Insurance Coverage in the United States: 2017*. U.S. Census Bureau. Retrieved on October 1, 2018 from https://www.census.gov/content/dam/Census/library/publications/2018/demo/p60-264.pdf

Berg, K., Bonham, V., Boyer, J., Brody, L., Brooks, L. *et al.* (2005). The Use of Racial, Ethnic, and Ancestral Categories in Human Genetics Research. *American Journal of Human Genetics*, 77(4), 519–532. See https://www.ncbi.nlm.nih.gov/pmc/articles/PMC1275602/pdf/AJHGv77p519.pdf

Bowling Green State University (2011, November 7). First-time Divorce Rate Tied to Education, Race. Retrieved on October 2, 2018 from https://www.bgsu.edu/news/2011/11/first-time-divorce-rate-tied-to-education-race.html

Caplan, N., Choy, M. H. & Whitmore, J. K. (1989). Indochinese refugee families and academic achievement. In D. W. Halnes (Ed.), *Refugees as Immigrants: Cambodians, Laotians, and Vietnamese in America.* New York: Rowman & Littlefield.

Corrigan, M. (September 22, 2015). *We Can't Stop Poverty without Addressing Families.* Institute for Family Studies. Retrieved on November 18, 2016 from https://www.aei.org/publication/we-cant-solve-poverty-without-addressing-families/

Cowan, C. P. & Cowan, P. A. (1992). *When Parents Become Partners: The Big Life Change for Couples.* New York: Basic Books.

de Souza Briggs, X., Popkin, S. J. & Goering, J. (2010). *Moving to Opportunity.* Oxford: Oxford University Press.

De Vaus, D. A. (2001). *Research Design in Social Research.* London: Sage.

Du Bois, W.E.B. (1899). *The Philadelphia Negro.* Philadelphia, PA: University of Pennsylvania Press.

DuMonthier, A., Childers, C. & Milli, J. (2017). *The Status of Black Women in the United States.* IWPR. Retrieved on August 25, 2017 from http://statusofwomendata.org/wp-content/uploads/2017/06/SOBW_report2017_compressed.pdf

Durkheim, E. (1897) [1951]. *Suicide: A Study in Sociology.* New York: The Free Press.

Duvall, E. M. (1957). *Family Development.* Philadelphia, PA: Lippincott.

Edin, K. & Lein, L. (1997). Work, Welfare, and Single Mothers' Economic Survival Strategies. *American Journal of Sociology,* 62(2), 253–266.

Fontenot, K., Semega, J. & Kollar, M. (2018). *Income and Poverty in the United States: 2017 Current Population Reports.* U.S. Census Bureau. Retrieved on September 29, 2018 from https://www.census.gov/content/dam/Census/library/publications/2018/demo/p60-263.pdf

Gordon, M. M. (1964). *Assimilation in American Life: The Role of Race, Religion, and National Origins.* New York: Oxford University Press.

Gould, E. & Schieder, J. (September 13, 2016). *By the Numbers: Income and Poverty, 2015.* Economic Policy Institute. Retrieved on November 15, 2016 from http://www.epi.org/blog/by-the-numbers-income-and-poverty-2015/

Gould, E., Schieder, J. & Geier, K. (October 20, 2016). *What is the Gender Pay Gap and is it Real?* Economic Policy Institute. Retrieved on November 13, 2016 from http://www.epi.org/publication/what-is-the-gender-pay-gap-and-is-it-real/

Hagan, J. (2013). *Parental Incarceration in the United States: Bringing Together Research and Policy to Reduce Collateral Costs to Children, A Report for the American Bar Foundation (2013).* Retrieved on August 25, 2017 from http://www.americanbarfoundation.org/uploads/cms/documents/white_house_conference_summa

Institute for Women's Policy Research (IWPR) (September 16, 2016). *Five Ways to Win an Argument about the Gender Wage Gap.* Retrieved on August 25, 2017 from https://iwpr.org/publications/five-ways-to-win-an-argument-about-the-gender-wage-gap/

Institute for Women's Policy Research (IWPR) (March 22, 2017). *Pay Equity & Discrimination.* Retrieved on August 25, 2017 from https://iwpr.org/issue/employment-education-economic-change/pay-equity-discrimination/

Jones, J. (February 17, 2017a). *Receiving an Inheritance Helps White Families More Than Black Families.* Economic Policy Institute. Retrieved on September 21, 2017 from http://www.epi.org/publication/receiving-an-inheritance-helps-white-families-more-than-black-families/

Jones, J. (September 20, 2017b). *One-third of Native American and African American Children Are (Still) in Poverty.* Economic Policy Institute. Retrieved on September 21, 2017 from http://www.epi.org/publication/one-third-of-native-american-and-african-american-children-are-still-in-poverty/

Kallen, H. M. (1915). Democracy versus the Melting Pot. *Nation 100,* (February 18 & 25), 190–194, 217–220.

Komarovsky, M. & Waller, W. (1945). Studies of the Family. *American Journal of Sociology,* 50(6), 443–451. Retrieved on October 1, 2018 from http://www.jstor.org/stable/2771388

Krogstad, J. M. & Radford, J. (Jan 30, 2017). *Key Facts about Refugees to the U.S.* Washington, DC: Pew Research Center. Retrieved on November 23, 2018 from http://www.pewresearch.org/fact-tank/2017/01/30/key-facts-about-refugees-to-the-u-s/

Lemert, E. (1967). *Human Deviance, Social Problems and Social Control.* Englewood Cliffs, NJ: Prentice-Hall.

Lewis, O. (1959). *Five Families: Mexican Case Studies in the Culture of Poverty.* New York: Basic Books.

Mayo, E. (1924). Recovery and Industrial Fatigue. *Journal of Personnel Research,* 3, 273–281.

McCarney, R., Warner, J., Iliffe, S., van Haselen, R., Griffin, M. & Fisher, P. (2007). The Hawthorne Effect: A Randomised, Controlled trial. *BMC Medical Research Methodology,* 7, 30.

McGeehan, P. (July 22, 2015). New York Plans $15-an-Hour Minimum Wage for Fast Food Workers. *The New York Times.*

Merton, R. (1949). *Social Theory and Social Structure.* New York: Free Press.

Moody, J. S. & Warcholik, W. (2017). *The Family Prosperity Index: Family Self Sufficiency.* American Conservative Union Foundation. Retrieved on August 25, 2017 from http://familyprosperity.org/media/2017-family-prosperity-index/2017-family-prosperity-index-family-self-sufficiency.

National Research Council (NRC) (2014). *The Growth of Incarceration in the United States: Exploring Causes and Consequences.* Washington, DC: The National Academies Press.

Ng, F. (1998). *The History and Immigration of Asian Americans.* New York: Garland.

Nicassio, P., Solomon, G. S., Guest, S. S. & McCullough, J. E. (1986). Emigration Stress and Language Proficiency as Correlates of Depression in a Sample of Southeast Asian Refugees. *The International Journal of Social Psychiatry,* 32, 22–28.

Nye, F. (1979). Choice, Exchange and the Family. In W. Burr, R. Hill, F. Nye & I. Reiss (Eds.), *Contemporary Theories about the Family* (Vol. 2, pp. 1–41). New York: Free Press.

O'Donnell, J. & Ungar, L. (January 8, 2016). Most of those without Medicaid are the Working Poor. *USA Today.*

Perlmann, J. (1988). *Ethnic Differences: Schooling and Social Structure among the Irish, Italians, Jews, and Blacks in an American City, 1880–1935.* Interdisciplinary Perspectives on Modern History. New York: Cambridge University Press.

Petersen, A. (April 28, 2011). After Baby, Men and Women Are Unhappy in Different Ways; Pushing Pre-Emptive Steps. *The Wall Street Journal.* Retrieved on November 20, 2016 from http://www.wsj.com/articles/SB100014240527487040997045762889540116759OO

Rotz, D. (2011). *Why Have Divorce Rates Fallen? The Role of Women's Age at Marriage.* Retrieved on October 2, 2018 from SSRN: https://ssrn.com/abstract=1960017 or http://dx.doi.org/10.2139/ssrn.1960017

Rynearson, A. M. & Gosebrink, T. A. (1989). *Ethnographic Exploratory Research Report # 2 Census-related Behavior of Southeast Asian Refugees in U.S.* U.S. Census Bureau. Retrieved on Nov. 13, 2016 from http://www.census.gov/srd/papers/pdf/ex89-02.pdf

Sabatelli, R. (1988). Exploring Relationship Satisfaction: A Social Exchange Perspective on the Interdependence between Theory, Research, and Practice. *Family Relations,* 37(2), 217–222.

Saez, E. (June 29, 2015). *U.S. Income Inequality Persists Amid Overall Growth in 2014.* Washington Center for Equitable Growth. Retrieved on November 11, 2016 from http://equitablegrowth.org/research-analysis/u-s-income-inequality-persists-amid-overall-growth-2014/

Sommeiller, E., Price, M. and Wazeter, E. (June 16, 2016). *Income Inequality in the U.S. by State, Metropolitan Area, and County.* Economic Policy Institute. Retrieved on August 8, 2016 from http://www.epi.org/publication/income-inequality-in-the-us/

Stanca, L. (2016). The Geography of Parenthood and Well-being: Do Children Make Us Happy, Where and Why? *World Happiness Report 2016.* Retrieved on November 20, 2016 from http://worldhappiness.report/wp-content/uploads/sites/2/2016/03/HR-V2Ch4_web.pdf

Steinberg, S. (1974). *The Academic Melting Pot.* New York: McGraw-Hill.

Steinberg, S. (2001). *The Ethnic Myth. Race, Ethnicity, and Class in America.* Boston, MA: Beason Press.

UN Refugee Agency (UNHCR). (2017). *Resettlement.* Retrieved on August 25, 2017 from http://www.unhcr.org/en-us/resettlement.html

U.S. Bureau of Labor Statistics (May 2015). *A Look at Pay at the Top, the Bottom, and in between.* Retrieved on November 13, 2016 from http://www.bls.gov/spotlight/2015/a-look-at-pay-at-the-top-the-bottom-and-in-between/home.htm

U.S. Census Bureau (March 14, 2017). *FFF: Asian/Pacific American Heritage Month: May 17.* Retrieved on June 25, 2017 from https://www.census.gov/newsroom/facts-for-features/2017/cb17-ff07.html

U.S. Census Bureau (September 2018a). *Income and Poverty in the United States: 2017.* Retrieved on October 1, 2018 from https://www.census.gov/content/dam/Census/library/publications/2018/demo/p60-263.pdf

U.S. Census Bureau (February 15, 2018b). *FFF: Women's History Month: March 2018.* Retrieved on October 1, 2018 from https://www.census.gov/newsroom/facts-for-features/2018/womens-history.html

U.S. Census Bureau (March 1, 2018c). *FFF: Asian-American and Pacific Islander Heritage Month: May 2018.* Retrieved on October 2, 2018 from https://www.census.gov/newsroom/facts-for-features/2018/asian-american.html

Wallerstein, I. (1974). *The Modern World-System I: Capitalist Agriculture and the Origins of the European World-Economy in the Sixteenth Century.* New York: Academic Press.

WEF (World Economic Forum) (2017). *Here's Why You Didn't Get That Job: Your Name.* Retrieved on August 24, 2017 from https://www.weforum.org/agenda/2017/05/job-applications-resume-cv-name-descrimination/

White House (January 29, 2009). Remarks of President Barack Obama on the Lilly Ledbetter Fair Pay Restoration Act bill signing. Retrieved on June 28, 2017 from https://obamawhitehouse.archives.gov/realitycheck/the-press-office/remarks-president-barack-obama-lilly-ledbetter-fair-pay-restoration-act-bill-signin

WHO (World Health Organization) (April 2017). 10 Facts on Health Inequities and Their Causes. Retrieved on August 24, 2017 from http://www.who.int/features/factfiles/health_inequities/en/

World Bank (September 30, 2015). *FAQs: Global Poverty Line Update.* Retrieved on August 31, 2017 from http://www.worldbank.org/en/topic/poverty/brief/global-poverty-line-faq

World Bank. (April 14, 2017). *What are Poverty Lines?* Retrieved on August 31, 2017 from http://www.worldbank.org/en/news/video/2017/04/14/what-are-poverty-lines

3
FAMILY DIVERSITY

Learning Outcomes

Upon completion of the chapter, students should be able to:

1. describe the diversity of the first Americans during the earliest years
2. describe Colonial family life during European colonization
3. describe slave family life in Colonial America
4. describe the multifunctional role of the family in pre-industrial society
5. analyze the impact of the Industrial Revolution on the American family
6. explain the changing of American families in post-industrial society
7. compare upper-upper-class families with lower-upper-class families
8. explain how middle-class household income affects family life
9. explain the effect of post-industrial society on working-class families
10. analyze the causes and consequences of working poor and underclass families
11. explore family life of immigrant families and world families from a diversity and global approach.

Brief Chapter Outline

Pre-test

American Families from the Earliest Years to Colonial Era

Early Americans

Hunting and Gathering; Matriarchal Family; Assimilation

Colonial Families (1607–1776)

Thanksgiving

Slave Families

Slave Family Life

Pre-test

Engaged or active learning is a powerful strategy that leads to better learning outcomes. One way to become an active learner is to begin the chapter and try to answer the following true/false statements from the material as you read. You will find that you have ready answers to these questions upon the completion of the chapter.

Before 1492, America was inhabited by Native Americans, who were the first Americans.	T	F
Early English immigrants settled into thirteen colonies such as New York.	T	F
Slavery is a closed stratification system in which slaves are classified as property.	T	F
In the 1950s, the ideal family consisted of a breadwinner father and a homemaker mother.	T	F
The dual-earner family became widely accepted in the twentieth century.	T	F
Upper-upper-class families are the wealthiest families with old money.	T	F
The lower-upper-class consists of self-made billionaires with new money from investments.	T	F
Physicians, lawyers, scientists, professors, IT professionals, and CEOs are largely considered to be upper middle class.	T	F
Baby boomers are people born between 1946 and 1964 during the demographic post-World War II baby boom.	T	F
Working poor families live paycheck to paycheck.	T	F

Upon the completion of this section, students should be able to:
LO1: Describe the diversity of the first Americans during the earliest years.
LO2: Describe Colonial family life during European colonization.
LO3: Describe slave family life in Colonial America.

The history of American families starts with Native Americans, who lived in what is now the United States before European settlers arrived. A number of diverse groups of immigrants have come to the United States since the early European settlers, in a number of large waves, adding a diverse mix of families to the country. Today, the United States continues to be home to many immigrant families. In fact, about 1 million green cards—a permit that allows foreign nationals to live and work in the United States permanently—are issued every year (Cohn & Ruiz, 2017). Let's look at the history of the diversity of families in the United States.

American Families from the Earliest Years to the Colonial Era

The United States is currently home to roughly 328.7 million residents (U.S. Census Bureau, 2018a), and as we will see, the nation has been a diverse one from the earliest years to today.

Early Americans

Williams *et al.* (2018) have dated a significant assemblage of stone artifacts to 16–20,000 years of age, pushing back the timeline of the first human inhabitants of North America before Clovis by at least 2,500 years. Before the arrival of Christopher Columbus in 1492, what is now the United States was inhabited by indigenous people who can claim to be the first Americans. These Native Americans themselves came from Asia and populated North and South America, forming various tribes with distinct cultures, customs, and languages.

Hunting and Gathering

Native Americans historically lived in hunting and gathering societies. In North America, nomadic tribes roamed over the Great Plains in search of buffalo and other game. Many tribes came into conflict with each other over the use of hunting grounds. Due to over-hunting or scarce game, many Native

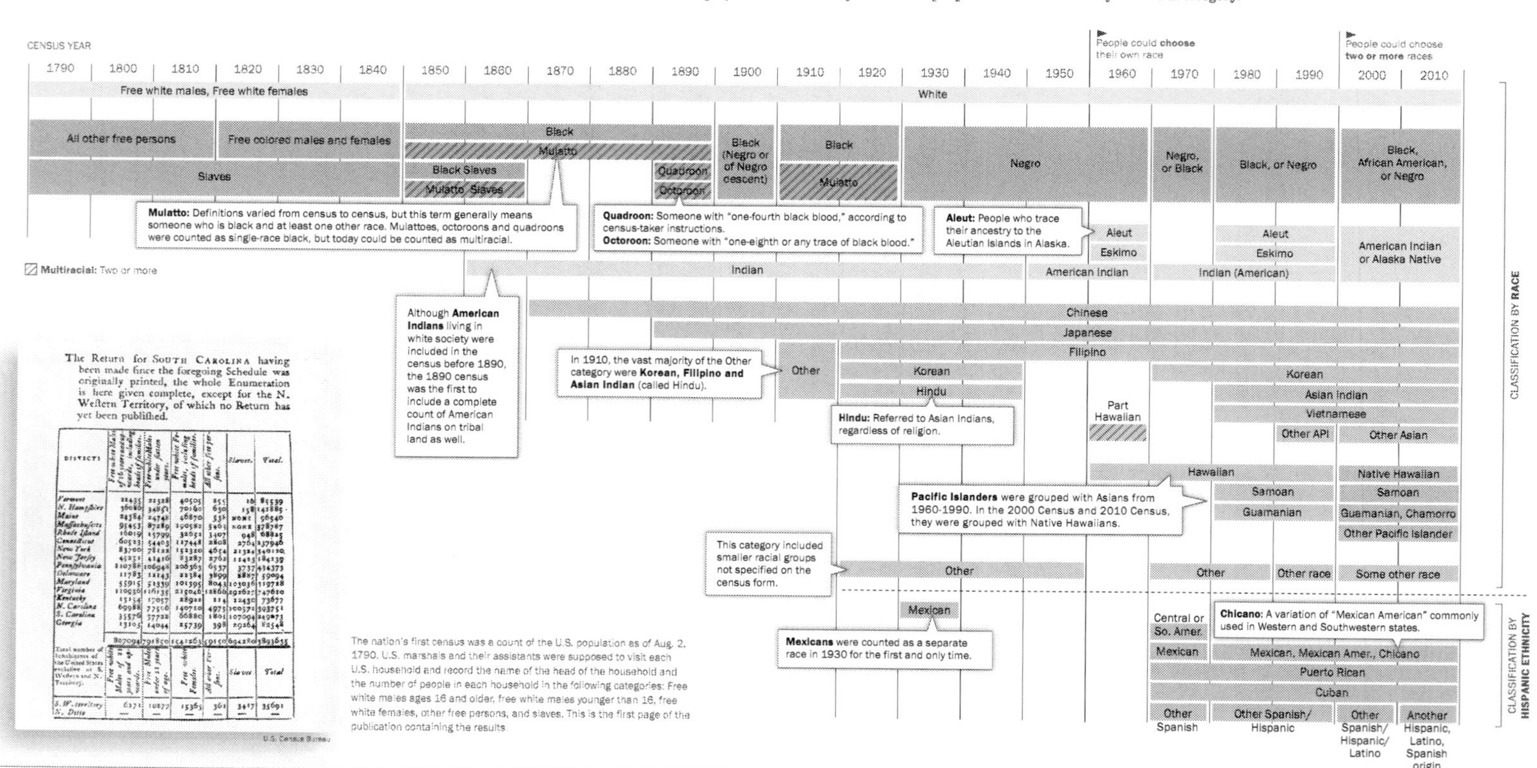

What Census Calls Us: A Historical Timeline 1960 onward: Americans could choose their own race and 2000 onward could be recorded in more than one race category on the census form.

Source: U.S. Census Bureau; Pew Research Center 2015. http://www.pewsocialtrends.org/interactives/multiracial-timeline/

American tribes turned to agriculture as a means of subsistence. For example, colonists in the eastern part of the continent observed that Natives cleared large areas for cropland. Their fields in New England sometimes covered hundreds of acres. Colonists in Virginia noted thousands of acres under cultivation by Native Americans (Krech III, 1999). Native American families maintained their hunting grounds and agricultural lands for use by the entire tribe. They used the natural resources available to them to build plank houses and make clothing. They refined farming methods and developed new crops such as corn, potatoes, and beans—the same foodstuffs that American families still consume today. They also gathered herbs used in remedies from their surrounding environment.

Matriarchal Families

Native American families were matriarchal, with wives having control over the children. Usually, the mother's brother mentored the child in a male child's life within the mother's clan. The Iroquois, a powerful Northeast Native American confederacy, were known during the colonial years to the French as the Iroquois League, and later as the Iroquois Confederacy, and then to the English they became known as the Six Nations (Beauchamp, 1905). Many tribes had matrilineal systems, in which property and hereditary leadership were controlled by and passed through the maternal lines. Married males resided in their wives' tribes as matriarchal families are matrilocal. Some English settlers who were fur traders married Native American women and assimilated into their tribes as well.

Native American families performed multiple functions as they lived, worked, played, learned, and worshiped at home. When it comes to gender roles, wives sustained family affairs, growing crops and erecting house frames, while husbands went hunting and fishing. Indeed, the women had primary responsibility for the survival of the families. For example, they gathered and cultivated plants, used plants and herbs to treat illnesses, cared for the young and the elderly, and tanned hides to make clothing (Waldman, 2006). Because hunting and fishing were recreational in Europe, male hunters were not regarded as providers of the families by early settlers. When demand for furs grew, however, male hunters became more important providers than female farmers. Girls married early and children began working at an early age. Native American families preserved their histories and religions through oral traditions.

Religion was an important element of their tribal way of family life. Traditional Native American religions exhibit a great deal of diversity as tribes were spread out across North and South America, allowing for the evolution of different beliefs and practices between tribes. Native American religion is connected to the land in which Native Americans reside. For example, the Ghost Dance was meant to serve as a connection with traditional ways of life and to honor the dead while predicting their resurrection (Waldman, 2009).

Assimilation

The history of the interaction between Native Americans and early Europeans began in 1492. When Italian explorer Christopher Columbus reached America, he mistook it for India in Asia, where he had intended to travel. Columbus called the natives of the land he visited *indios* (Phillips & Phillips, 1992). European impact was profound. For example, many tribes had patrilineal systems in which hereditary leadership passed through the male line, and children were considered to belong to the father and his clan.

As domestic "foreigners," Native Americans could not seek naturalized citizenship, for they were not "white" (Takaki, 2008). After the United States became a newly independent nation in 1776, European settlers tried to weaken tribal institutions in an effort to assimilate Native Americans into Anglo culture and White society. Native American families experienced forced migration and forced assimilation through relocation, allotments, and termination. In 1830, the U.S. Congress passed the Indian Removal

Act and called for the forcible relocation of all eastern tribes across the Mississippi River in order to expand the United States. Native American families were forced to remove from their ancestral homelands to federal territory west of the Mississippi River in exchange for their lands. This resulted in the ethnic cleansing of Native American tribes, with the brutal and forced marches to be known as "The Trail of Tears."

In 1887, the U.S. Congress passed the General Allotment Act of 1887, whose purpose was to divide land between European settlers and Native American families. Each family was given 160 acres of land and became a land owner, but families were not allowed to sell the land for 25 years. The Allotment Act intended to impose upon Native Americans the concept of private property. Indeed, the biggest conflict between European settlers and Native Americans was the concept of land ownership. In the eyes of the Native Americans, the land was communal and sacred to their tribes, having been used by them for hundreds of years for hunting or fishing. To European settlers, land in the new country was something to be owned by individuals.

In 1924, Native Americans who were not already citizens were granted citizenship by Congress. The Indian Reorganization Act of 1934 forced the tribal people to adopt the European way of voting and election of leaders. House concurrent resolution 108 (HCR-108) of 1953 established that Congress would pass termination acts on a tribe-by-tribe basis, abolishing federal supervision over American Indian tribes and subjecting the Indians to the same laws, privileges, and responsibilities as other US citizens (U.S. Government Publishing Office, 1953). The Indian Termination Act of 1953 ended the tribal rights as sovereign nations and terminated federal support of health-care and education programs, utility services, and police and fire departments available to Indians on reservations. In this way, Native American families were subordinated and European settlers became dominant.

Colonial Families (1607–1776)

Following Christopher Columbus in 1492, many Europeans set their sights on the Americas and became the main group of immigrants. **Colonial America** refers to the history of European settlements from the colonization of America until their incorporation into the United States. The Colonial Era (1607–1776) began when the English landed at Jamestown in 1607 and ended with the Declaration of Independence on July 4, 1776.

On April 10, 1606, King James I of England issued the First Virginia Charter, which created the Virginia Company, authorizing the English to colonize Virginia. In this way, Jamestown became the first English colony in North America, established by Protestants from England on May 13, 1607. The **Pilgrims** were members of a Puritan separatist sect who set sail in the *Mayflower* bound for the Americas and landed at Plymouth, Massachusetts in 1621 to establish another colony where they could enjoy religious freedom. The colonialists settled into thirteen colonies (Virginia, New York, Massachusetts, Maryland, Rhode Island, Connecticut, New Hampshire, Delaware, North Carolina, South Carolina, New Jersey, Pennsylvania, and Georgia). The early families generally married within their social class for many generations, and as a result, most surnames of so-called "First Families" date back to the Colonial Era (Tyler, 1915). Protestants gradually emerged from the Colonial period as the dominant group politically and economically.

Colonial life centered on the *patriarchal family*. When the first colonists landed in Jamestown, Virginia in 1607, they brought the English common law with them. The English common law was characterized by male dominance and male inheritance of familial property. Husband and wife were considered one person under the law (National Women's History Project, 2016). The notion of conjugal unity has biblical origin and Genesis, II, 24, is explicit that husband and wife "shall be one flesh," and this is repeated in the New Testament (Williams, 1947). The father had full legal authority over his family, made all decisions, and guided the family in daily activities. Colonial families traced their descent through the male side in a patrilineal system. *Households* at the time often included a husband, wife, children, servants, and apprentices. Family members of all social classes were engaged in agricultural pursuits in the fields, hunting and fishing, or working as craftsmen or merchants. Women worked inside the home, doing everything from taking care of children to producing goods such as cheese, candles, and clothes.

Thanksgiving

When the early Europeans landed in the New World, Native Americans helped them survive. In the fall of 1621, the Massachusetts Pilgrims held a three-day Thanksgiving feast to celebrate a bountiful harvest. The Wampanoag Indians in attendance at the first Thanksgiving were essential to the survival of the colonists during the newcomers' first year (U.S. Census Bureau, 2016a). However, as noted earlier, their initial friendship deteriorated after the thirteen colonies revolted against Great Britain, gained strength, and formed the United States on July 4, 1776 when the Continental Congress approved the Declaration of Independence, setting the thirteen colonies on the road to freedom as a sovereign nation. The estimated number of people living in the newly independent nation was 2.5 million (U.S. Census Bureau, 2017a).

Slave Families

Following the early European settlers, beginning in 1641, Africans were forcibly brought to Colonial America as slaves. At slave markets in cities such as Charleston, South Carolina, those from Africa were thrust before a crowd of jeering men and thrown into a cell until the next slave auction, where they were sold off to the highest bidder.

Before slavery started, a system of indentured servitude was in place in Colonial America to meet

Source: Jean Louis Gerome Ferris, 1863–1930

the growing demand for cheap labor in the agricultural economy of the South. An **indentured servant** was a person under contract to work for another person for a period of time without pay but in exchange for free passage to the new country. In this way indentured servants joined the settlers and became the labor pool for the Colonial economy. However, European settlers soon realized that they wanted cheaper labor to produce profitable tobacco. Slavery was thus introduced to provide cheap labor on Southern tobacco plantations. The first recorded slaves, twenty Africans, were brought to the colony of Jamestown in August 1619 by a Dutch-flagged ship. Starving and in need of supplies, a Dutch trader negotiated with Jamestown settlers and traded these Africans who were captured from a Spanish ship by the Dutch sailors.

In 1641, Massachusetts became the first colony to authorize slavery through enacted law (Higginbotham, 1978). **Slavery** is a closed stratification system in which slaves are classified as property. In 1662, Virginia passed a law that children of enslaved women took the status of the mother under English common law. In 1790, when the first U.S. Census was taken, Africans numbered about 760,000 and made up about 19.3 percent of the population (Brunner, 2004), the second largest ethnic group after the European settlers.

Slave Family Life

The home was viewed as a cultural and political force in the middle of the nineteenth century. It consisted of a large physical space for the more fortunate, a simple cabin for the less fortunate, and slave quarters for many (Riedy, 1997). Harriet Beecher Stowe's *Uncle Tom's Cabin* (or *Life Among the Lowly*) was the best-selling novel of the nineteenth century and told much about the lives of slaves. Published in 1852, it is credited with helping to spur the cause of the abolitionists, who wished to put an end to slavery.

Slave codes defined the social position of slaves. Marriages between Whites and slaves were prohibited. Slaves were not allowed to legally marry and own property because they themselves were regarded as property. In most cases, slave families were owned by one slave owner. Female and male slaves were treated as breeders. Such **slave breeding** was used to reproduce slaves and increase the wealth of slave owners. Female slaves also worried about being raped by their owners. Slave children were assigned tasks like running errands and working in the fields with their mother. In some slave families, the father had a different owner and lived miles away from his wife and children. Both slaves and slave owners referred to these relationships between male and female slaves who did not live together as "abroad marriages" (Thompson & Hickey, 2005). Regardless of where they lived, slaves constantly feared that their families would be separated through the sale of one of their members.

Blacks placed a very high priority on their families both during and after slavery despite the overwhelming difficulties they faced (Tolman, 2011). Slaves frequently named their children for parents, grandparents, aunts, uncles, and sometimes even great-aunts, great-uncles, and deceased siblings (Berlin & Rowland, 1998).

Student Accounts of Slave Family

Slavery was a terrible way of family life. I have heard tons of stories from my own family members. My great-grandmother was a midwife in Abilene, Texas. She was able to deliver babies of all races. Yet there were certain "rules" she had to follow due to the fact she was Black. I can recall my grandmother telling stories to myself where Black people felt they were less than a White person's family pet (personal communication, September 3, 2017).

Upon the completion of this section, students are expected to:
LO4: Describe the multifunctional role of the family in pre-industrial society.
LO5: Analyze the impact of the Industrial Revolution on the American family.
LO6: Explain the changing of American families in post-industrial society.

Religion was also a major part of the slave family. Although Christianity was initially forced on slaves, it helped get many slave families through the most difficult periods of their lives.

The Thirteenth Amendment to the U.S. Constitution was ratified in 1865, putting an end to slavery in the United States. However, the legacy of slavery lingers in our cities' ghettos (Loury, 1998). The legacy of slavery and efforts to fight for equality remained an issue in the United States for decades.

American Families from the Nineteenth to the Twenty-first Century

Following the European settlers and African slaves, the next waves of immigrants came from Europe and Asia between the 1820s and the 1920s. During this time, 38 million immigrants came to the United States, the majority from Germany, Ireland, and Italy (Simkin, 2015). From that time until today, families in the United States have become more and more diverse.

Changing Family Structure and Increasing Family Diversity in the Nineteenth Century

Rapid **industrialization** first began in Britain, starting with mechanized spinning in the 1780s, with high rates of growth in steam power and iron production occurring after 1800 (Broadberry *et al.*, 2005). The British colonization of America soon ushered in the **Industrial Revolution**—the transition from home-based hand production methods to factory manufacturing production. In the early nineteenth century, while Britain experienced an economic recession and decline, the United States became a growing industrial nation. The new innovations included new steel-making processes, the large-scale manufacture of machine tools, and the use of increasingly advanced machinery in steam-powered factories (Roe, 1916). For example, steam-powered manufacturing overtook water-powered manufacturing, leading to increased mechanization of production. The steam engine underwent great increases in power due to the use of higher pressure steam (Hunter, 1985), allowing the manufacturing industry to spread across the United States. Inventions such as electric lights in 1880 allowed American people to work around the clock.

Growing Immigrant Families from the 1820s to the 1880s

Immigrants and families from western Europe brought money with which to secure a new life and shared the same White Anglo-Saxon Protestant background as the early settlers. The Homestead Act of 1862 offered to sell public lands to citizens and to immigrants. After six months of residency, homesteaders had the option of purchasing the land from the government for $1.25 per acre. The promise of land at a low price also attracted hundreds of thousands of people from Europe.

Between 1820 and 1860, a wave of immigrants came from northern and western Europe. For example, the potato famine in Ireland (1845–52) brought large numbers to the United States. The unsuccessful 1848 revolution in Germany created considerable emigration. The 1880s saw a huge immigration explosion. A wave of immigrants from southern and eastern Europe, such as eastern European Jews fleeing religious persecution, arrived in the United States. These immigrants did not come from the Anglo-American culture of many of the earlier immigrants, and some struggled to make a living in the New World. Immigrants from Asia arrived to work on Hawaiian sugar plantations, to search for gold in the California Gold Rush from 1848 to 1855, and to work on the transcontinental railroads in the 1860s.

Immigrant families became the labor pool that fueled the growing economy in the United States. Most workers labored ten hours a day, often six days a week, which left them with little time or energy for family life (Coontz, 2016). Many of these immigrant families assimilated into American culture and embraced their heritage, resulting in growing family diversity in the United States.

Impact of Industrialization on Family Life

Industrialization led to changes in family life in the United States. First, family production was replaced by market production in which capitalists paid workers wages to produce goods in factories (Padavic & Reskin, 2002). The wage-labor system was seen in the mid-nineteenth century. Many previous farming families left for towns, seeking employment in factories that processed raw materials into finished goods. Meanwhile, many male farmers left their villages in the south and flocked to cities in the north to become wage earners there. The functions of the family began to change as well. As a result of workplaces being separated from the home, the family became a consumption unit rather than a production unit. As factories and businesses replaced family-based farms, young adults became wage earners and therefore economically independent. In this way, some traditional functions that the family performed were taken over by other social institutions.

As many farmers became industrial workers in cities, it changed the way Americans perceived the home itself. In preindustrial society, the home was seen as a productive and economic center. In the mid-nineteenth century, the home was seen as a safe refuge from the chaotic society beyond its walls—a quiet paradise for the exhausted father at the end of his working day. Lyrics from the song *Home Sweet Home*—"Home! Home! Sweet, sweet home! There's no place like home"—reflect the domestic ideal to which families aspired. Domestic labor provided a means of agency, economic independence, and social influence in a way other forms of labor could not (Tompkins, 2012). For example, this domestic ideal led to housing reform. Andrew Jackson Downing led the movement that created the suburban home. The mid-nineteenth century saw an architectural shift from classic and simple revival homes to gothic revival and Italianate villa styles (Goffigon, 2014). Popular architectural plan books associated house reform and decoration with the cultivation of middle-class standards of refined taste and the "moral uplift" of a nation in transition from "rudeness to civility" (West, 2013). Articles in *Godey's Lady's Book* in the mid-nineteenth century encouraged women to decorate their homes to appear "luxurious" and soft (Goffigon, 2014).

The home also became a mirror of the family's status. Wealthy families hired domestic workers to do housework, and wealthy wives planned social gatherings that were crucial to their husbands' political and business success. The middle class also turned to their homes as a display of their wealth. From a financial standpoint, middle-class wives could contribute to their family's earnings by using their needlework skills to sew or repair clothing and linens for pay (Boydston, 1994). Working-class wives did their own baking, gardening, and other chores or sewed on a piece-work basis to balance the family budget. Widowed or divorced women often struggled to support themselves and their children.

Nuclear Family

The domestic ideal was also applied to the notion of the ideal family. A **nuclear family**, consisting of a father, a mother, and their children, became the male breadwinner model. Research has explored gender in the wage-labor system, claiming, "Gender shapes our perception of what constitutes work, of who is working, and of the value of that labor" (Boydston, 1994). From roughly the mid to late nineteenth century, the division of labor transformed men into workers and put women in the domestic sphere. Women became associated with homemaking, while men were breadwinners. The term **breadwinner** was coined to describe the male role as provider. Men left the home to work in offices or factories and assumed sole responsibility for the financial support of the family. The term **homemaker** was coined to describe the female role as keeper of the home. Married women were expected to devote themselves to homemaking and childrearing. It was a woman's responsibility to foster a supportive, warm, and welcoming space to nurture their husbands and the future generations of American men and women (Sandage, 2005). Their work not only transformed the kitchen as a central place in the American household, but it also irrevocably redefined the meaning of womanhood (Goffigon, 2014).

The division between the domestic and public spheres also affected women's status in the industrial society. Beecher (1851) explored values and expectations of middle-class women, the skills necessary in keeping a healthy, clean, and moral home, and also advocated for the education, respect, and equality of women as important contributors to American society. Although the home was the female domain, women were not allowed to inherit property in Colonial America. In 1839, Mississippi was the first state that granted women the right to hold property in their own name, with their husbands' permission (Moncrief, 2008). Passed in 1848, New York's Married Women's Property Act was used by other states as a model for allowing women to own property (Chused, 1983). Immigrant and poor families did not fit into this nuclear family model ideal but instead lived in multiple-generation families due to the high costs of living in the mid-nineteenth century.

Love as Passion and Companionship

The change in the attitude to romantic love appeared in the eighteenth and nineteenth centuries (Kalyuga, 2012). Along with the ideal nuclear family structure, the notion of romantic love took hold in the nineteenth century among the urban, educated, and wealthy and middle classes, and later spread to rural areas. Due to economic growth, families had more leisure time and began to focus on the "pursuit of Happiness" set forth in the Declaration of Independence. Urban cities also made it easy for young adults to meet, interact, and fall in love. The notion of marriage soon shifted focus on love first and childrearing second.

Changing Family Life and Growing Family Diversity in the Twentieth Century

In 1913, Henry Ford dramatically increased the efficiency of his automobile factories by instituting large-scale use of the moving assembly line (Jaycox, 2005). The assembly line paved the way for mass production of almost all goods. By the 1880s, steam power shortened the journey to the United States. Successive waves of immigrants from southern and eastern Europe and East Asia arrived between the 1880s and the 1920s, seeking opportunities paved by industrialization. Families often immigrated together during this era, although men usually came first to find work.

Impact of the Progressive Movement on the Family

Around the time that immigrants began arriving from Europe, U.S. factory owners, hoping to earn greater profits, began hiring children as workers because they were cheaper and more manageable. In the early twentieth century, the number of child laborers in the United States peaked while rates of illiteracy rose, and many families faced problems such as **poverty**. The **Progressive Era** (from the 1890s to 1920s) was a response to the economic and social problems taking place at the time and was a period of widespread social activism and political reform across the United States (Buenker *et al.*, 1977). In 1904, the National Child Labor Committee aimed to end child labor with efforts to provide free **compulsory education** for all children. Most states passed compulsory school attendance laws, and children were therefore required to attend school up to a certain age. School attendance also increased as labor unions were successful in gaining bans on child labor, although many poor parents and businesses opposed these laws. After 1910, smaller cities began building high schools. By 1940, 50 percent of young adults had earned a high school diploma (Tyack, 1974). With the onset of compulsory education, public schools took over the job of teaching children, which was previously the domain of the family.

Progressives also sought to raise wages for male workers and promoted better housing and comfortable surroundings for families. The National American Woman Suffrage Association pushed for a constitutional amendment guaranteeing women's voting rights, and was instrumental in winning the ratification of the **Nineteenth Amendment** to the United States Constitution in 1920, which granted women the right to vote (McConnaughy, 2013). Retirement funds and pension plans were also provided so that old people were less dependent on their adult children. Worker compensation and unemployment insurance allowed families to survive with the loss of the breadwinner's income. All these changes in social policy supported more nuclear families, and working-class and middle-class families began to look similar in the early twentieth century. In the ideal nuclear family of the time, men went to work while women stayed at home and children attended school.

Companionate Marriage

During this time, the notion of marriage began to change. Traditionally, marriage had been characterized by male authority, sexual repression, and a focus on childbearing, kin, and property relations (Simmons, 2009). In the twentieth century, more young adults went to work and became economically independent. With the rise of individualism in society, more young adults called for freedom from parental control for coupling and marriage. The major processes that influenced the understanding of romantic love in the twentieth century were relaxation of sexual morals in Europe and the "sexual revolution" of the 1960s to early 1970s (Karandashev, 2015). Two transitions in the meaning of marriage that occurred in the United States during the twentieth century have created the social context for *deinstitutionalization*—a weakening of the social norms that define partners' behavior in marriage.

The first transition was from the institutional marriage to the companionate marriage (Cherlin, 2004). **Companionate marriage** refers to marriages based on affection, friendship, and sexual gratification (Hunt, 2016). This type of marriage viewed wedlock as the union of two individuals bonded through sexual love. Although controversial at the time, Lindsey and Evans (1927) proposed that young men and women should be able to live together in a trial marriage, where the couple could have a year to evaluate whether they were suitable for each other and divorce thereafter if no children were born. With companionate marriages, sexual aspects of love became glorified and wives were no longer expected to be guardians of sexual restraint. The breakdown of the old rules of a gendered institution such as marriage could create a more egalitarian relationship between wives and husbands (Cherlin, 2004) and also create new child-rearing patterns with more emphasis on nurturing and love (Hunt, 2016). Some scholars stress general spousal companionship and love as its defining characteristics and find companionate marriage even in seventeenth-century Europe (Simmons, 2015). Interestingly, in a study of the socially constructed ideal of companionate marriage in Elizabethan and Jacobean England through four dramas by Christopher Marlowe, William Shakespeare, John Fletcher, and Thomas Dekker & Thomas Middleton, Pierce (2018) suggested that these dramas served as proponents of the companionate marriage while dually challenging the persisting restrictive social norms that prevented prospective unions between religiously, socioeconomically, and/or racially divergent individuals.

During the twentieth century, *the second transition* was to the individualized marriage in which the emphasis on personal choice and self-development expanded (Cherlin, 2004). Companionate marriage later gave way to individualized marriages, which further challenged the notion of separate spheres, or the breadwinner–homemaker model, within families (Hunt, 2016).

Baby Boomer Family

World War II ended in 1945 and brought a baby boom to many countries. **Baby boomers** are people born between 1946 and 1964 during the demographic post-World War II baby boom. In the United States, baby boomers were divided into two cohorts. The first half of baby boomers, called the Leading-Edge Baby Boomers, was born between 1946 and 1955. The second cohort, called Late Boomers, was born between 1956 and 1964 (Green, 2006).

At the end of World War II, Congress passed the 1944 G.I. Bill of Rights, which helped millions of veterans return home and reintegrate into civilian life. This bill encouraged homeownership and investment in higher education through the distribution of loans at low or no-interest rates to veterans. The returning veterans pursued higher education, married, bought their houses in the suburbs of cities, and formed families. Thriving on the American Dream, a record number of babies were born. In 1954, annual births first topped 4 million and did not drop below that figure until 1965, when four out of ten Americans were under the age of twenty (Forman-Brunell, 2009).

The marriage rate rose from 12.2 per 1,000 people in 1945 to 16.4 in 1946. Age at first marriage dropped to 22.8 years for males and 20.3 for females in 1950, down from 24.3 for males and 21.5 for females in 1940 (U.S. Census Bureau, 2017b). The residential pattern was *neolocal*, meaning that newlywed couples moved out of their parents' homes to begin a new family.

In the 1950s, a young woman was likely to marry straight out of high school or take a job until she married. The wife would quit her job when she became pregnant and would stay home to care for her children while her husband had a steady job that paid enough to support the family. The 1950s was an era in which 60 percent of families consisted of a breadwinning father and a stay-at-home mother (Coontz, 2016).

The Dual-earner Family

The two-married-parents-with-kids model in the 1950s gradually eroded and dual-earner families became popular throughout the 1970s. A **dual-earner family** refers to both husband and wife providing income for the family, usually by working outside the home. Women sometimes sought work in the labor market to maintain the family's standard of living if the husband lost his job or if the husband's income failed to keep up with inflation. Also, dual-earner couples have less traditional gender ideologies and share household tasks more equally than single-earner couples. This increase in the husband's participation in housework promotes the wife's marital satisfaction (Dai, 2016). The dual-earner family became widely accepted in the twentieth century and rose from 35 percent in 1966 to 61 percent in 1994, replacing the traditional married-couple model of a "breadwinner" husband and "homemaker" wife (Winkler, 1998). The shift from single-earner to dual-earner family has affected the marital relationship, the well-being of family members, employers, and communities, with both positive and negative effects (Dai, 2016).

Contemporary Family in the Twenty-first Century Onward

The first Super Bowl was played on January 15, 1967 at Los Angeles Memorial Coliseum. Super Bowl 52 was played on February 4, 2018 at U.S. Bank Stadium in Minneapolis, MN. In the 1950s, most children spent their childhood in a family that was made up of a male breadwinner and a female homemaker. However, since that time, the dual-earner couple has become the norm. Between 1975 and 2016, the share of young women who were homemakers fell from 43 percent to 14 percent of all women age 25 to 34 (Vespa, 2017). From 1967 to 2018, the marriage rate in the United States declined; the age at first marriage rose; and family size decreased. Consider these statistics: the median age at first marriage was 23.1 for men and 20.6 for women in 1967, compared with 29.5 for men and 27.4 for women in 2017 (U.S. Census Bureau, 2017b), while family size was 3.28 people in 1967 and 2.54 people in 2016 (U.S. Census Bureau, 2016b). Young adults delaying marriage and putting off having children are reflected in demographic trends for the population as a whole (Vespa, 2017). Clearly, marriage and families were changing in the United States in the twenty-first century.

Diversity of Family Forms

Diverse family forms have become more popular, including cohabiting families, single-parent families, LGBT families, childfree families, and grandparent families. As one in two marriages ends in divorce, many people remarry and form stepfamilies. Many children also live in single-parent families. The number of live births to unmarried women was 1,569,796 (39.8 percent of all births to unmarried women) in 2016; the birth rate for unmarried women was 42.4 births per 1,000 unmarried women ages 15 to 44 in 2016 (CDC, 2017). Some people have decided to remain single throughout their entire life rather than marrying and forming a family at all. More than one in three young adults ages 18 to 36 lived in their parents' home in 2016, making it the most stable living arrangement for young adults (Vespa, 2017).

Factors Shaping Family Life

Factors such as education, income, gender, age, race, social class, and marital status contribute to the shaping of family life. Households with householders having lower levels of educational attainment, with widowed householders, or with older householders (age 65 and older) were more likely to remain in the bottom quintile or move into a lower quintile compared with their counterparts (Hisnanick *et al.*, 2017). One factor that has a large effect on families is household income, which has gone up from a median annual household income of $7,143 in 1967 (U.S. Census Bureau, 2016b) to $61,372 in 2017 (U.S. Census Bureau, 2018b). Household income varies by age cohort. Households maintained by householders ages 45 to 54 ($80,671) and ages 35 to 44 ($78,368) had the highest median income, while households maintained by householders ages 65 and older ($41,125) and ages 15 to 24 years ($40,093) had the lowest median income (U.S. Census Bureau, 2018b).

Annual homeownership rates also vary by age cohort. Figure 3.1 shows that 35 percent of people under age 35 owned homes in the United States annually compared to 80 percent of those ages 65 and over. Owning a home has long been a reflection of a family's economic prosperity. However, the financial and housing crisis of the late 2000s accelerated the decline in economic stability among the bottom 90 percent of families. Getting a home loan is difficult. So many families maintained by people under age 35 are now putting off buying a house. More Millennial households live in poverty than households headed by any other generation and dominate the ranks of the nation's renters. Millennials ages 18 to 35 (79.8 million in 2016) headed 18.4 million of the 45.9 million households that rent their homes, compared with 12.9 million Generation X households ages 36 to 51 and 10.4 million Boomer households ages 52 to 70 in 2016 (Fry, 2017).

Race also affects family life. In 2017, real median income in the United States was $81,331 for Asian

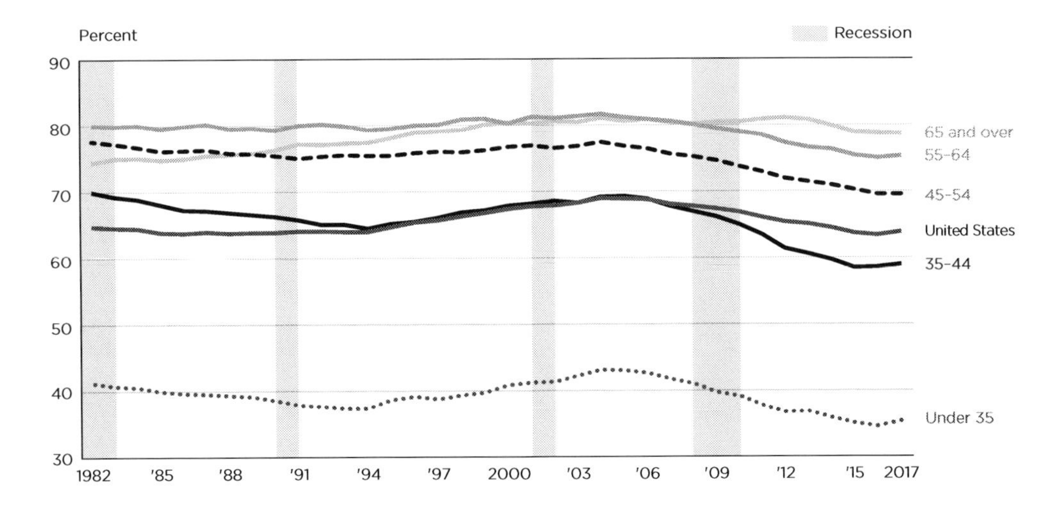

Figure 3.1 Annual Homeownership Rates for the United States by Age Group, 1982–2017
Source: U.S. Census Bureau, Current Population Survey/Housing Vacancy Survey, February 27, 2018; recession data from the National Bureau of Economic Research. https://www.census.gov/housing/hvs/data/charts/fig07.pdf

households, $68,145 for White households, $50,486 for Hispanic households, and $40,258 for Black households (U.S. Census Bureau, 2018b). Families with adequate resources likely enjoy good relationships and maintain family stability. Adolescents from Asian-American families (65 percent) are most likely to live with their married biological parents, compared with White teenagers (54 percent), Hispanic (43 percent), and African American adolescents (17 percent) (Zill & Wilcox, 2015). Black children in the United States enjoy less family stability than White children, experiencing close to twice as many family transitions—union dissolutions and partnership formations—as White children (DeRose, 2017).

Diversity Overview: Immigrant Families

Since the Colonial Era, the United States has been shaped by immigrants from many foreign countries. The U.S. Census Bureau (2018a) claims that someone in the world becomes a new migrant every 29 seconds. An **emigrant** is a person who has left his or her own country. **Push factors** such as famine and war make people leave their homelands. An **immigrant** is a person who enters a new country and takes up permanent residence. **Pull factors** such as freedom and opportunities bring immigrants to a new country. Immigration is what has made the United States such a diverse country.

Since 1965, a wave of Europeans, Asians, and Hispanics have arrived in the United States based on the Immigration and Naturalization Act of 1965, giving priority to people who already had family in the United States or had skills that were needed in the labor market. The lawful immigrants (33.8 million and 75.5 percent) and unauthorized immigrants (11.0 million and 24.5 percent) hit a record high of 44.8 million in 2015 (Cohn, 2017).

The Changes in Immigration Policies

The fluctuation of immigrant numbers over time to the United States has changed because of government policy. In 1740 the British Parliament passed the Naturalization Act of 1740, allowing any Protestant alien residing in any of its American colonies for seven years to become a natural-born subject of this Kingdom. The Naturalization Act of 1790 was the first naturalization act enacted by the newly created U.S. government, requiring two years of residency in the United States and one year in the state of residence when applying for citizenship. The Naturalization Act of 1795 increased the period of required residence from two to five years in the United States. The Naturalization Act of 1798 increased the period necessary for immigrants to become naturalized citizens in the United States from five to fourteen years. The Alien Act of 1800 reduced naturalization residency requirements from fourteen to five years.

Following the early settlers in the Colonial Era, by the 1880s immigrants poured in from Europe due to industrialization and steam power that shortened the journey to the United States. In 1924, Congress passed the National Origins Act that established a quota system for restricting immigrants by country of origin and determining how many immigrants could enter the United States. This Act gave the most quotas to northern and western Europe and restricted the number of immigrants from southern and eastern Europe. A downturn in the economy during the 1870s caused a backlash against Chinese immigrants in the workforce. The Chinese Exclusion Act of 1882 was implemented to prevent Chinese from immigrating to the United States. After the first Pacific railroad was built in 1869, Chinese immigrants returned home, and the quarter million Chinese immigrants previously in the country dwindled to 75,000 (Miller, 1901). Today, to restore the rule of law and secure our border, President Trump is committed to constructing a border wall and ensuring the swift removal of unlawful entrants (White House, 2018).

Challenges of Immigrant Families

Of 1,051,031 persons who became lawful permanent residents (LPR) in 2015, 65 percent were family members of a U.S. citizen or lawful permanent resident (Baugh & Witsman, 2017). Many immigrant families feel excited about a new life in the land of opportunity. However, many experience hardships in their homeland and in their new country.

Individuals who migrate experience multiple stresses that can impact their mental well-being, including the loss of cultural norms and changes in identity and concept of self (Bhugra & Becker, 2005). For example, children with language barriers are expected to learn at a rigorous pace and must take the same state tests given to native English-speaking students. They come to school at a disadvantage and are expected to close the achievement gap. These children also often serve as interpreters for their non-English-speaking parents. Another challenge for immigrant families is employment. Job growth has not come close to matching new immigration and natural population increases (Camarota, 2016). Figure 3.2 shows that between 2000 and 2015, the number of both native and immigrant working-age people increased by 25.9 million in the United States, but the number of those working increased by only 8.4 million. It means that there has been a dramatic decline in work, particularly among the young and less educated (Camarota & Zeigler, 2016).

At the heart of the immigration debate is the idea that there are not Americans available for work and therefore immigrants are needed to fuel the economy. Today, to protect American workers, President Trump supports ending chain migration, eliminating the Visa Lottery, and moving the country to a merit-based entry system (White House, 2018). In fact, immigrants made up 17.1 percent (27.6 million) workers out of

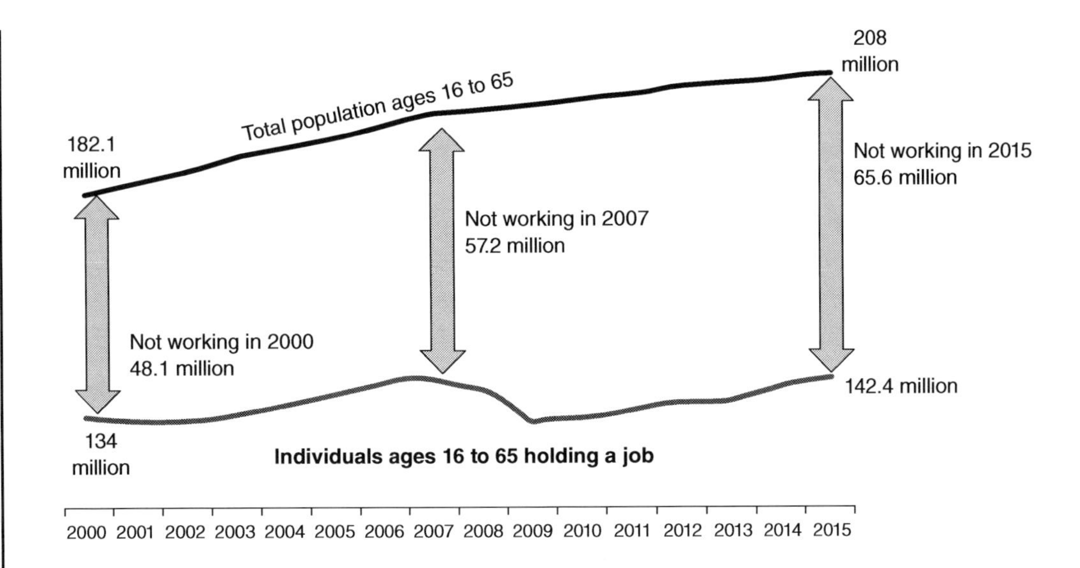

Figure 3.2 Natural Population Growth and New Immigration Growth Have Greatly Exceeded Employment Growth, 2000–2015

Source: Center for Immigration Studies 2016. https://cis.org/Impact-LargeScale-Immigration-American-Workers

the total U.S. workforce (161.4 million) in 2014. About 19.6 million were legal workers, or 12.1 percent of the total workforce and about 8 million, or 5 percent, were unauthorized without legal permission or had overstayed their visas (Desilver, 2017). As the U.S. baby boom generation heads into retirement, the increase in the potential labor force will slow, and immigrants will play the primary role in the future growth of the working-age population.

Indeed, U.S.-born children of immigrants will make up a growing share of working-age adults: 13 percent in 2035, compared with 6 percent in 2015 (Passel & Cohn, 2017).

Social Class Categories of American Families

Social class position of a child's parents matters. It matters for school success, and ultimately, occupational success (Lareau, 2011). **Social class** refers to a group of people with similar shares of resources in hierarchical social divisions such as upper, middle, and lower classes. Class divisions affect family life, and economic inequality has been one of the hottest topics in every U.S. presidential campaign. Income and opportunities are as unequally shared as they have ever been; and society is divided in terms of the different lives,

Upon the completion of this section, students should be able to:

LO7: Compare upper-upper-class families and lower-upper-class families.

LO8: Explain how middle-class household income affects family life.

LO9: Explain the effect of post-industrial society on working-class families.

LO10: Analyze the causes and consequences of working poor and underclass families.

hopes, and dreams that the rich and the poor have (Graham, 2017). Indeed, there has been rising income inequality across the class lines in the United States. Figure 3.3 shows that the incomes of the top 1 percent and 0.1 percent of U.S. families dwarf those of the bottom 99 percent. The top 1 percent of families captured 58 percent of total real income growth per family from 2009 to 2014, with the bottom 99 percent of families reaping only 42 percent. Meanwhile, in terms of annual family income in the United States, households in the bottom 90% had incomes of $33,068 or less while the top 0.1 percent of households had incomes of $6 million more. Let's look at some ways in which class affects American families.

Upper-class Families

The upper class comprises the capitalist class and entrepreneurs. Warner (1949) divides the upper class into the upper-upper class and the lower-upper class.

Upper-upper-class Families

The **upper-upper class** is the capitalist class, the wealthiest and most powerful class in the United States. Members of the upper-upper class own the means of production and have inherited this so-called old money over multiple generations. **Old money** is the accumulated and inherited wealth of established upper-class *families,* such as the Rockefeller family. Upper-upper-class families live off the income from their inherited wealth. With their multigenerational wealth, they pass high social status from generation to generation. They have substantial assets and have much leisure time for cultivating a variety of interests. Upper-upper-class families live in upscale homes and possess a high quality of life. Their children usually

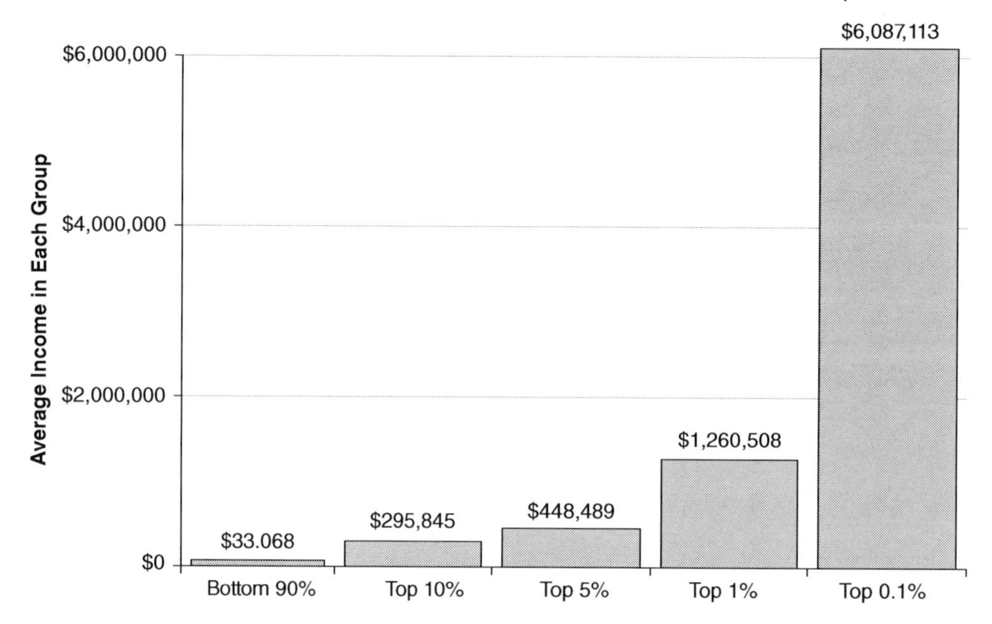

Figure 3.3 Outsized U.S. Income Inequality Persists in 2014

Source: Emmanuel Saez and Thomas Pikkety's analysis of IRS Data, Washington Center for Equitable Growth 2015. http://equitablegrowth.org/research-analysis/u-s-income-inequality-persists-amid-overall-growth-2014/

attend elite preparatory schools and prestigious universities.

Members of the upper-upper class are not only wealthy but also can shape policies and benefit from these policies. For example, President Donald Trump proposed new tax benefits for childcare aimed at bringing relief to working-class families. A recent analysis of his proposals shows that both in terms of dollars and as a share of household income, these benefits would be much larger for high-income families than for low- and middle-income families in America (Chikhale, 2017). The upper-upper class can influence political and economic institutions when they own mass media companies that allow them to influence the thinking of society. The upper-upper class is often referred to as the old upper class and is distinct from the lower-upper class.

Lower-upper-class Families

Industrial and post-industrial societies have provided opportunities for entrepreneurs to benefit from technology and build wealth. **Lower-upper-class families** are maintained by householders who have earned so-called new money from investments and business ventures within their own lifetimes. **New money** is recently acquired wealth rather than inherited wealth. Many entrepreneurs earned new money through investment and became self-made billionaires. *The World's Billionaires* has been published by the American business magazine *Forbes* since 1987. Many billionaires on the Forbes 400 list did not come from an affluent background and did not inherit family businesses but rather made their own fortune by applying their ideas and using new technologies in their industries. Microsoft's Bill Gates is one such example. Table 3.1 shows that his fortune was $90 billion in 2018. Indeed, entrepreneurs are increasingly prominent among the super-wealthy (Kaplan & Rauh, 2013).

Lower-upper-class families have much wealth. They live in exclusive neighborhoods, gather at expensive social clubs, and send their children to the finest schools. They serve on the boards of corporations and exert much influence on the decision-making process both nationally and globally. Although they command a good deal of wealth, some upper-class families engage in philanthropy. The **Giving Pledge** is a commitment made by some of the world's wealthiest individuals and families to dedicate the majority of their wealth to giving back. As of 2018, there are a total of 184 pledgers to the Giving Pledge (Giving Pledge, 2018).

Table 3.1 World's Billionaires in 2018

No.	*Name*	*Net Worth (USD)*	*Age*	*Citizenship*	*Source(s) of Wealth*
1	Jeff Bezos	$112 billion	54	United States	Amazon.com
2	Bill Gates	$90 billion	62	United States	Microsoft
3	Warren Buffett	$84 billion	87	United States	Berkshire Hathaway
4	Bernard Arnault	$72 billion	69	France	LVMH
5	Mark Zuckerberg	$71 billion	33	United States	Facebook
6	Amancio Ortega	$70 billion	81	Spain	Inditex, Zara
7	Carlos Slim	$67.1 billion	78	Mexico	America Movil, Grupo Carso
8	Charles Koch	$60 billion	82	United States	Koch Industries
8	David Koch	$60 billion	77	United States	Koch Industries
10	Larry Ellison	$58.5 billion	73	United States	Oracle Corporation

Source: Wikipedia, 2018: https://en.wikipedia.org/wiki/The_World%27s_Billionaires

Middle-class Families

In the great shift from manual skills to servicing people in post-industrial society in the 1950s, sociologist C. Wright Mills (1951) described the forming of a new class: the white-collar workers. "Being a member of the middle class can connote more than income, be it a college education, white-collar work, economic security, owning a home, or having certain social and political values" (Pew Research Center, 2015).

Members of the middle class typically have a comfortable standard of living, significant economic security, considerable work autonomy, and rely on their expertise to sustain themselves (Gilbert, 2002). The middle class is divided into the upper-middle class and the lower-middle class based on education, income, and occupation.

Upper-middle-class Families

Upper-middle-class families are maintained by householders who are mostly white-collar professionals who earn above-average income and have a high level of education (Thompson & Hickey, 2005) and a high degree of autonomy in their work (Eichar, 1989). Constituting roughly 15 percent of households, the upper or professional middle class is sometimes referred to as the "professional class" (Ehrenreich, 1989) and consists of highly educated, salaried professionals and managers (Gilbert, 2002).

The upper-middle class has been expanding since the United States became a post-industrial society in the 1960s. **Post-industrial societies** are based on computer technology that produces information, supports a service industry, and encourages consumption. Gilbert (1997) finds that the key to the success of the upper-middle class is the growing importance of educational certification. **Credentialism** refers to the practice of relying on earned credentials when hiring staff or assigning social status. Collins (1979) examines the connection between credentialism and stratification. Occupations that require high educational degrees such as a PhD, MBS, MD, or CPA are highly paid and are held in high status. Information technology (IT) occupations that require professional credentials are also likely to earn more in the post-industrial society. In 2014, the "highest earning IT occupations were computer and information research scientists, software developers, applications and system software, computer and information systems managers, and computer network architects, each with median earnings of $90,000 or more" (U.S. Census Bureau, 2016c). Figure 3.4 shows that the number of IT professionals has increased tenfold since 1970.

The main occupational tasks of upper-middle-class individuals tend to center on conceptualizing, consulting, and instruction (Ehrenreich, 1989). Florida (2014) noticed the rise of the "creative class," which composes up to 30 percent of the population of the United States. The **creative class** is a class of workers whose job is to create meaningful new forms. The people in this group make a living as artists and as participants in craft and related cultural events. The creative class is a key driving force for development of the service industry in post-industrial societies.

Student Accounts of Middle-class Family

My husband and I have worked very hard to even be considered middle class. With six kids, it has been a rough road. It can be tempting to let the kids go to my mom for things they want but I have stuck to my guns. My oldest son is eleven and recently bought his own iPad. The best part was him thanking me for not letting him waste every penny that he saved on crazy things like candy. He was very proud of himself and I was very proud of him. Now my eight-year-old son wants to buy his own iPad. He has $2 saved so far (personal communication, October 26, 2017).

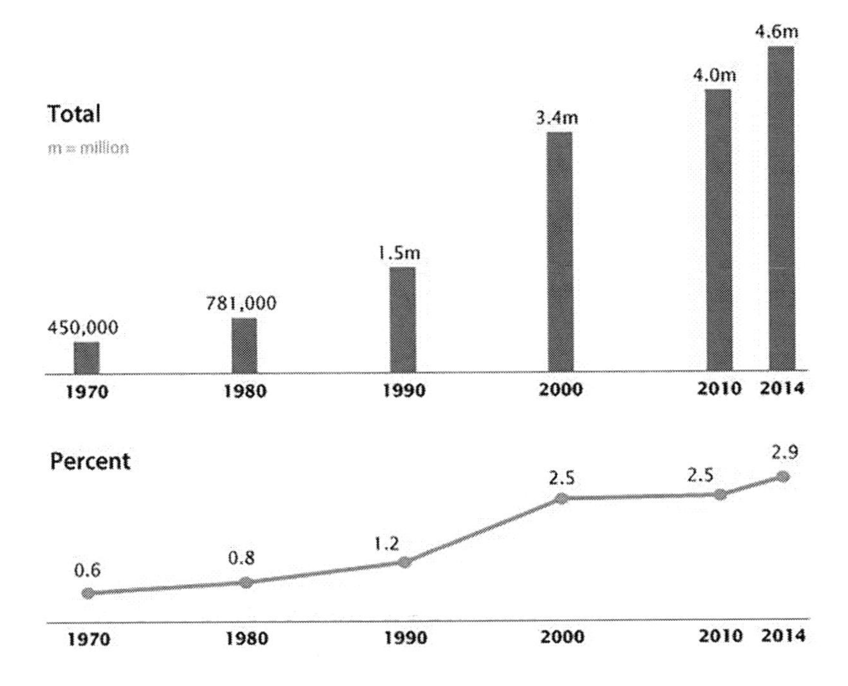

Figure 3.4 Number of IT Workers Has Increased Tenfold since 1970
Source: EEO Supplementary Reports from the 1970, 1980, 1990, 2000 decennial censuses and 2010 and 2014 American Community Survey, U.S. Census Bureau, August 16, 2016. https://www.census.gov/library/visualizations/2016/comm/cb16-139_itworkers.html

Lower-middle-class Families

The American lower-middle class is diverse and growing (Gilbert, 2002). **Lower-middle-class families** are maintained by householders who have white-collar jobs and college degrees. The lower-middle class has lower educational attainment, less autonomy in technical and lower-level management positions, and lower incomes than the upper-middle class. Given the growing service industry in our post-industrial society, about nine out of ten new jobs are projected to be added in the service-providing sector from 2016 to 2026, resulting in more than 10.5 million new jobs, or 0.8 percent annual growth (U.S. Bureau of Labor Statistics, 2017a). The lower-middle class is referred to as simply the middle class.

Middle-class family life is affected by household income and household size. Within the middle class, a three-person lower-income household earned $24,074 annually compared with $174,625 annually for an upper-income family of three in the middle class in 2014 (Pew Research Center, 2016).

Figure 3.5 shows that the income gap between upper-income and lower-income middle-class families widened substantially from 1970 to 2014.

Such middle-class income variations affect family life. For example, a family in which both partners are experienced high school teachers might earn $120,000 a year while an upper-middle-class family in which both partners are technical stock analysts might easily earn $400,000 (Murray, 2012). For example, Silicon Valley is a booming technical environment, but lower-middle-class families struggle to find affordable housing because of the affluent neighborhoods being built throughout the city.

Student Account of Working-class Family

My mom was a high school drop-out who had three kids and gave two up for adoption. I was the only one she kept. I learned how hard life would be if you did not finish school and lived a poor life. School was difficult as it was fun and fair game to harass the poor kid who had holes in his clothes and was in the free lunch program (personal communication, October 30, 2016).

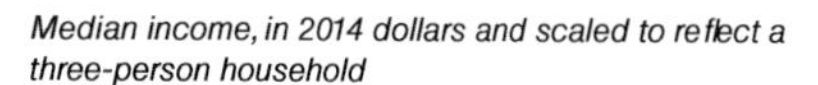

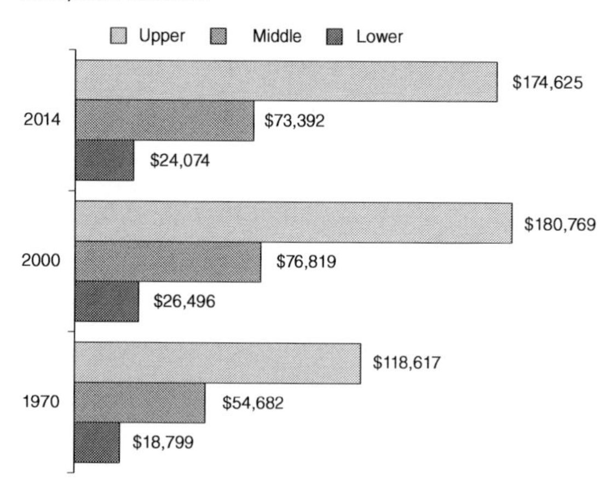

Figure 3.5 Growth in Income in Middle-income Households Is Less than Growth in Upper-income Households since 1970

Source: Pew Research Center December 8, 2015. http://www.pewsocialtrends.org/2015/12/09/the-american-middle-class-is-losing-ground/st_2015-12-09_middle-class-04/

Working-class Families

From a sociological perspective, the majority of Americans can be described as members of the working class (Vanneman & Cannon, 1988). The term *working class* is applicable if the position of householder in relation to the production of goods and services is the determinant of social class. A **working-class family** is maintained by householders who hold high school degrees and have blue-collar or service-work jobs in manual labor occupations. According to Gilbert (2002), those in the working class are commonly employed in manual labor and low-level white-collar occupations. Working-class occupations include industrial work and service-work jobs that are highly routinized. Skilled workers such as carpenters and plumbers are called blue-collar workers but may make more money than the lower-middle class. Gilbert and Kahl (1992) see the working class as the most populous in the United States. In 2013, the working class constituted nearly two-thirds (66.1 percent) of the civilian labor force in the United States between ages 18 and 64 (Wilson, 2016).

Effect of Post-industrial Society on Working-class Families

The manufacturing industry supported working-class families during the Industrial Revolution. National Manufacturing Day is observed on October 6. Developing technology has changed the way we approach manufacturing. American manufacturing lost roughly one-third of its jobs in the first decade of the new century (American Presidency Project, 2017).

Since the post-industrial society in the 1960s, millions of workers with manufacturing jobs in factories were laid off as their skill set was no longer a good fit for jobs in the information technology society. Due in large part to technology-driven growth, many blue-collar jobs in the traditional industries are disappearing. **Technological unemployment** is the loss of jobs caused by technological change in the process of invention, innovation, and diffusion of technology. Such change includes the introduction of "mechanical-muscle" machines or "mechanical-mind" processes (automation). With the introduction of mechanization and the assembly line in factories, the system of production was divided into smaller tasks. These tasks demanded less skilled workers. More workers in manufacturing industries were semi-skilled or unskilled. **De-skilling** is the process by which skilled labor within an industry is eliminated by the introduction of technologies operated by semi-skilled or unskilled workers. In de-skilling, technological development leads to the reduction of the scope of a worker's work to specialized tasks and thus work is fragmented. Workers lose the integrated skills and knowledge of the craftsperson. This results in cost savings due to lower investment in human capital, and reduces barriers to entry, weakening the bargaining power of the human capital (Braverman, 1974). Furthermore, more jobs were automated. In the twenty-first century, robots began to perform roles in manufacturing and other industries. The use of industrial robots may reduce employment in the economy. For example, a robot at Lowe's home improvement store in Sunnyvale, California now checks inventory instead of a real person. Acemoglu and Restrepo (2017) find that the employment impact for men is 1.5 to 2 times greater than for women, and the effects are concentrated in manufacturing industries. In addition, in many areas of work, corporations have access to labor around the world. Due to cheaper labor and material elsewhere, many manufacturing jobs are outsourced and sent overseas.

Since the 1960s, economic restructuring and occupational insecurity has become a major problem for American working-class families. The manufacturing sector has experienced the largest employment loss of any sector, and a projected loss of 814,100 jobs from 2014 to 2024 would reduce manufacturing employment to under 11.4 million (U.S. Bureau of Labor Statistics, 2015). The ten U.S. metropolitan areas with the greatest losses in economic status from 2000 to 2014, such as Springfield, Ohio, have one thing in common: a greater than average reliance on manufacturing industries (Pew Research Center, 2016). Figure 3.6 demonstrates that manufacturing establishments show gains in receipts but lower employment over time.

Wage stagnation is another pressing issue of working-class families. Income inequality has serious consequences for low-income families, particularly for children, many of whom underperform in school, thereby reducing their future employment prospects and perpetuating an intergenerational cycle of economic disadvantage (Cherlin, 2014). White blue-collar workers expected to live the American Dream and have a stable, middle-class existence but the typical two-parent household is disappearing. Blue-collar Whites are more likely to live in places where identities, friendships, and social support were traditionally tied to the mine or the factory (Graham, 2017). This makes it difficult to form other social support and take new jobs in service sectors such as health. Adopting policies to lift labor standards, broaden collective bargaining, and maintain genuine full employment would help ensure that there are good-quality jobs available for workers displaced by technology (Mishel & Bivens, 2017).

Working-poor Families

The **working poor** are people who spent at least 27 weeks in the labor force (that is, working or looking for work) but whose incomes still fall below the official poverty level (U.S. Bureau of Labor Statistics, 2017b). **Working-poor families** are maintained by householders with high school degrees who work in lower-paid jobs and live paycheck to paycheck. In 2015, there were 8.6 million working poor in the United States, down from 9.5 million in 2014. The working-poor rate was 5.6 percent, down 0.7 percentage point from 2014 (U.S. Bureau of Labor Statistics, 2017b). The **working-poor rate** is the ratio of the working poor to all individuals in the labor force for at least 27 weeks.

Demographic Characteristics of Working-poor Families

Among persons in the labor force for 27 weeks or more, 14.1 percent of part-time workers were

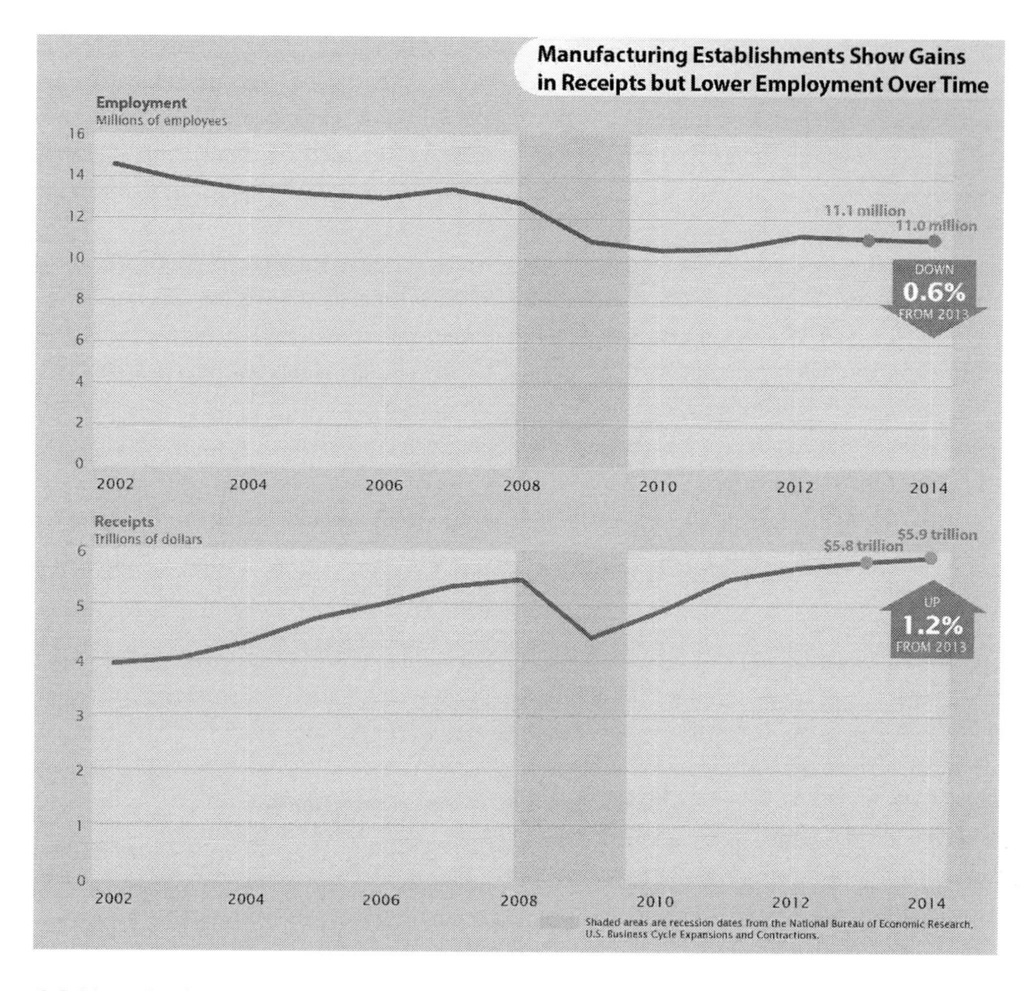

Figure 3.6 Measuring American Manufacturing in the United States, 2017
Source: 2014 Annual Survey of Manufactures (ASM), U.S. Census Bureau. https://www.census.gov/library/visualizations/2016/comm/manufacturing-in-the-us.html

classified as working poor, compared with 3.4 percent of those employed full time (U.S. Bureau of Labor Statistics, 2017b). Working-class and working-poor households may fall below the poverty line if an income earner becomes unemployed (Gilbert, 2002). For family members, the poverty threshold is determined by their family's total income. In 2016, 4.1 million families were living below the poverty level despite having at least one member in the labor force for half the year or more, down from 4.6 million (U.S. Bureau of Labor Statistics, 2018).

Working poor families vary by gender and race. Women were more likely than men to be among the working poor. In 2016, the number of women classified as working poor (4.1 million) is greater than the number of men (3.4 million) and women have a higher working-poor rate (5.8 percent) than men (4.2 percent). In 2016, the working-poor rates of Blacks and Hispanics were 8.7 percent and 8.5 percent, respectively, compared with 4.3 percent for Whites and 3.5 percent for Asians. Among Whites and Blacks, the working-poor rate was higher for women than for men. The working poor rates were 4.9 percent for White women and 3.9 percent for White men, 10.5 percent for Black women and 6.7 percent for Black men. Figure 3.7 illustrates the working-poor rates by gender and race in 2016.

Working-poor families vary by marital status and the makeup of the family. Among families, those with children in the household were much more likely to live below the poverty level (9.7 percent) than those without children (2.2 percent). Families maintained by women with children had a working-poor rate of 22.8 percent, more than double that of families maintained by men with children, at 11.2 percent. Married-couple families with only one member in the labor force in 2016 were less likely to be living below the poverty level, at 7.7 percent, than were families maintained by men, at 11.6 percent, and maintained by women, at 21.5 percent (U.S. Bureau of Labor Statistics, 2018).

Working-poor families struggle day-to-day to earn wages to meet basic needs such as food, shelter, health care, daily necessities. For example, the digital divide has been a central topic in tech circles for decades with researchers. A **digital divide** is an economic and social inequality with regard to access to, use of, or impact of information and communication technologies (ICT) (U.S. Department of Commerce, 1995). In 2016, one-fifth of adults living in households earning less than $30,000 a year were "smartphone-only" internet users—meaning they owned a smartphone but did not have broadband internet at home—compared with two-thirds of adults living in high-earning households having multiple devices (Anderson, 2017). Wealthy families are likely to have multiple devices such as home broadband services, a smartphone, a desktop or laptop computer, *and* a tablet that enable them to go online. **Broadband internet** refers to households who said

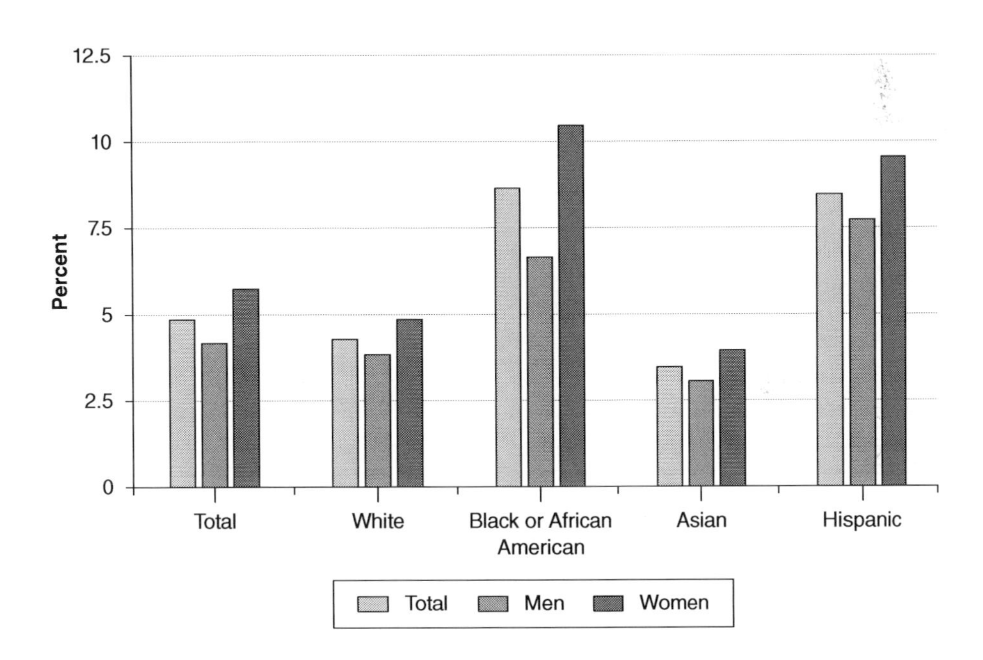

Figure 3.7 Working-poor Rates by Gender, Race, and Hispanic or Latino Ethnicity, 2016

Source: U.S. Bureau of Labor Statistics, Current Population Survey (CPS), Annual Social and Economic Supplement (ASEC). https://www.bls.gov/opub/reports/working-poor/2016/home.htm#chart2

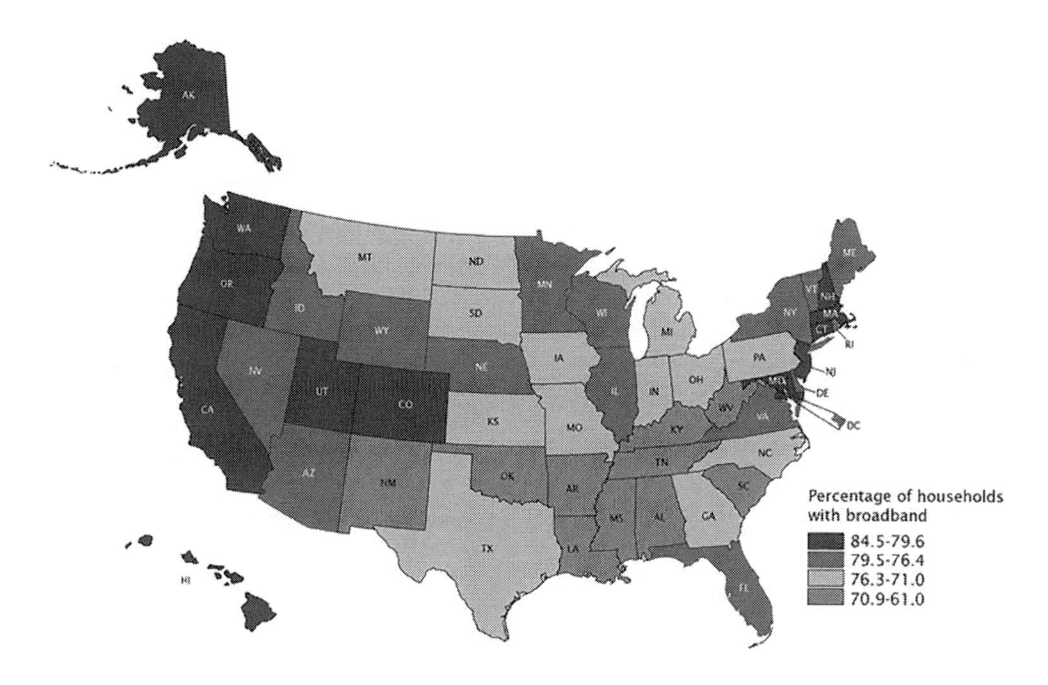

Figure 3.8 The Digital Divide: Percentage of Households with Broadband Internet Subscription by State
Source: 2015 American Community Survey. https://www.census.gov/library/visualizations/2017/comm/internet-map.html

"Yes" to subscriptions such as DSL, cable, fiber optic, mobile broadband, satellite, or fixed wireless. Figure 3.8 shows the digital divide by state.

Achieving higher levels of education and obtaining high-paying occupations reduce the incidence of living in working-poor families. Workers with an associate's degree and those with a bachelor's degree or higher had the lowest working-poor rates (5.1 percent and 1.4 percent, respectively). Workers in occupations such as management and professional-related occupations were least likely to be classified as working poor, at 1.6 percent in 2016. By contrast, service occupations, with 2.8 million working poor, accounted for 39 percent of all those classified as working poor as these jobs do not require high levels of education and are characterized by low earnings (U.S. Bureau of Labor Statistics, 2018). The Working Poor Families Project is a national initiative that seeks to strengthen state policies on behalf of low-income working families (WPFP, 2017).

Underclass Families

The underclass occupies the lowest possible position in the class hierarchy. Myrdal (1963) defines *underclass* as a class of the unemployed, unemployable, and underemployed who are more and more hopelessly set apart from the nation at large and do not share in its life, its ambitions, and its achievements. **Underclass families** are maintained by householders who are not well-educated and who are underemployed or unemployed. The term describes those in easily filled employment positions with little prestige or economic compensation, who lack a high school education and who, to some extent, are disenfranchised from mainstream society (Gilbert, 1997).

Families Living in Poverty

Many underclass families live in poverty. **Poverty** refers to a condition where a family's total income is

The Census Bureau releases two reports every year that describe who is poor in the United States. The first report calculates the nation's official poverty measure based on cash resources. The second report focuses on the Supplemental Poverty Measure (SPM) and takes into account cash resources and noncash benefits from government programs aimed at low-income families.

The Official Poverty Measure

The United States has an **official** measure of poverty. The current official poverty measure was developed in the early 1960s when President Lyndon Johnson declared war on poverty. This measure does not reflect the key government policies enacted since that time to help low-income individuals meet their needs.

Poverty Rate: 1959 to 2016
(In percent)

Recession

22.4 percent
Official
SPM
13.9 percent
12.7 percent

25
20
15
10
5
0

1959 '65 '70 '75 '80 '85 '90 '95 2000 '05 '10 2016

Note: The data points are placed at the midpoints of the respective years.
Source: U.S. Census Bureau, Current Population Survey, 1960 to 2017 Annual Social and Economic Supplements.

The Supplemental Poverty Measure

The **SPM** extends the official poverty measure by taking into account government benefits and necessary expenses, like taxes, that are not in the official measure. This second estimate of poverty has been released annually by the Census Bureau since 2011. In 2016, the SPM rate was slightly higher than the official measure identifying 44.6 million people as poor. This was 13.9 percent of the population.

Figure 3.9 Measuring America: How the U.S. Census Bureau Measures Poverty
Source: U.S. Census Bureau, American Community Survey 2011–2015, Current Population Survey, 1960–2017 Annual Social and Economic Supplements. https://www.census.gov/content/dam/Census/library/visualizations/2014/demo/poverty_measure-how.pdf

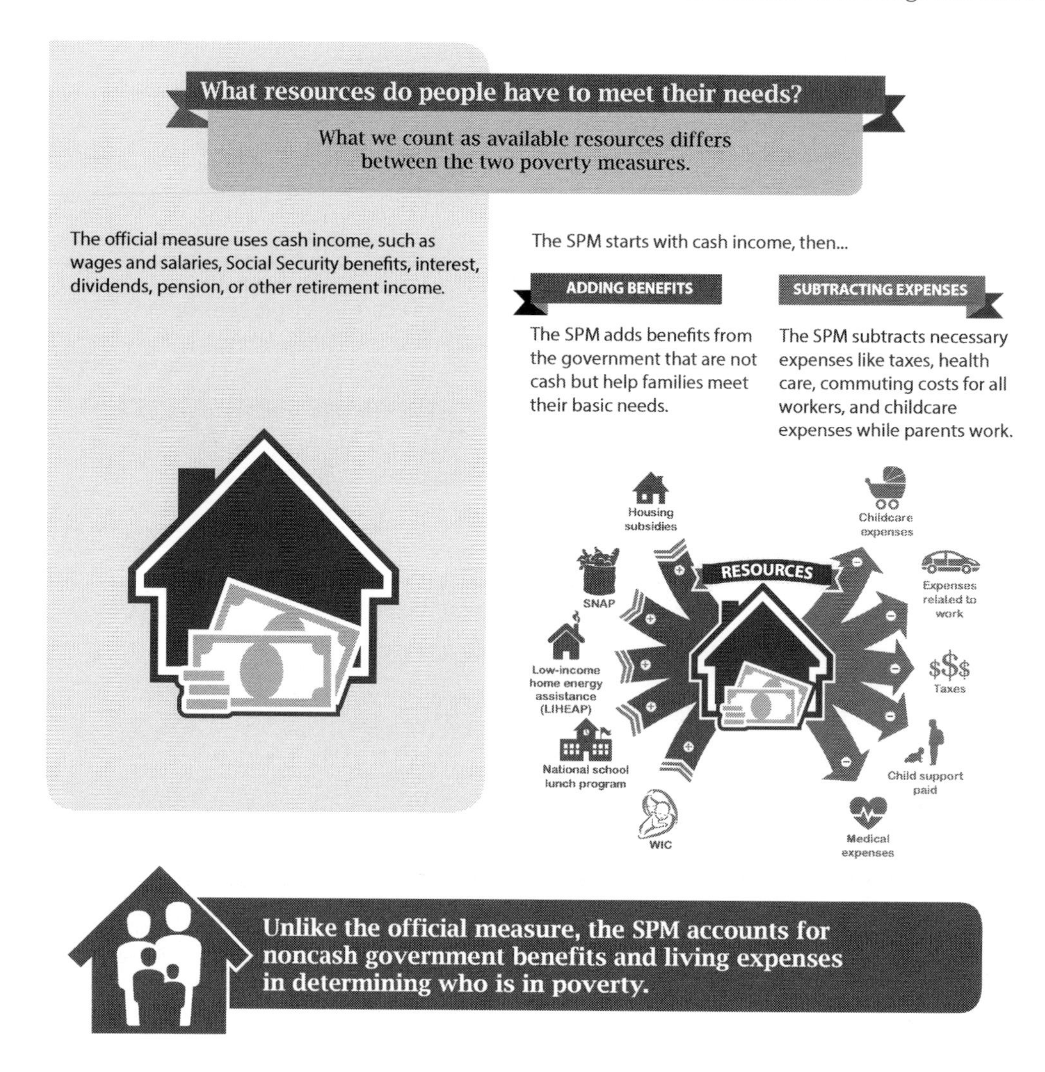

Figure 3.9 Continued

less than the official poverty threshold for a family of that size and composition and is not enough to meet basic needs. The current official poverty measure was developed after President Lyndon B. Johnson declared war on poverty on January 8, 1964. Molly Orshansky of the Social Security Administration developed and refined poverty thresholds and the Office of Economic Opportunity and other federal agencies started using them in 1965 (Fisher, 1997). **Poverty thresholds** are used to determine poverty status and are related to the size of the family and age of the members. In 2017, the weighted average poverty threshold for a family of four was $25,094; for a family of nine or more people, the threshold was $50,681; and for one person, it was $12,488 (Fontenot *et al.*, 2018). Figure 3.9 explains how the U.S. Census Bureau measures poverty based on cash resources and non-cash benefits from government programs aimed at low-income families. For both measures, individuals are considered in poverty if the resources they share with others in the household are not enough to meet basic needs.

The **poverty rate** is the ratio of the number of people who fall below the poverty line and the total population. The poverty rate for families in 2017 was 9.3 percent, representing 7.8 million families, a decline from 9.8 percent and 8.1 million families in 2016 (Fontenot *et al.*, 2018). The overall poverty rate in 2017 was 12.3 percent, down 0.4 percentage points from 12.6 percent in 2016 (Fontenot *et al.*, 2018). The poverty rate varies by age of the members: The poverty rate for people age 18 to 64 was 11.2 percent (22.2 million) in 2017, down from 11.6 percent (22.8 million) in 2016, compared with 17.5 percent (12.8 million) for children under age eighteen and 9.2 percent (4.7 million) for old adults ages 65 and older (see Figure 3.10).

Poverty rates vary by gender and age. Females age 18 to 64 (13.0 percent) had a higher poverty rate than males (9.4 percent). Female older adults age 65 and older had a higher poverty rate (10.5 percent) than male older adults (7.5 percent). Girls under age eighteen had a higher poverty rate (17.7 percent) than boys (17.3 percent) (Figure 3.11).

As noted earlier, the causes of the growing underclass relate to the impacts of age, income, gender, race, and behaviors on the perpetuation of poverty, along with education. Few people of this class finished high school, and they usually suffer from inadequate housing, food, and safety. For example, in 2017, 24.5 percent of people age 25 and older without high school diplomas lived in poverty, compared with people with a bachelor's degree or higher (4.8 percent) (Fontenot *et al.*, 2018).

Consequences of Social Class

Social class limits the access to resources that families need. Social class is associated with life expectancy. For example, life expectancies are six years higher in New York than in Detroit as New Yorkers live in affluent, educated cities with healthy behaviors (Chetty *et al.*, 2016). Class also affects life expectancies by gender. The richest American men live fifteen years longer than the poorest American men, while the richest American women live ten years longer than the poorest American women (HIP, 2016). Household income is also related to life expectancy. Hederos *et al.* (2017)

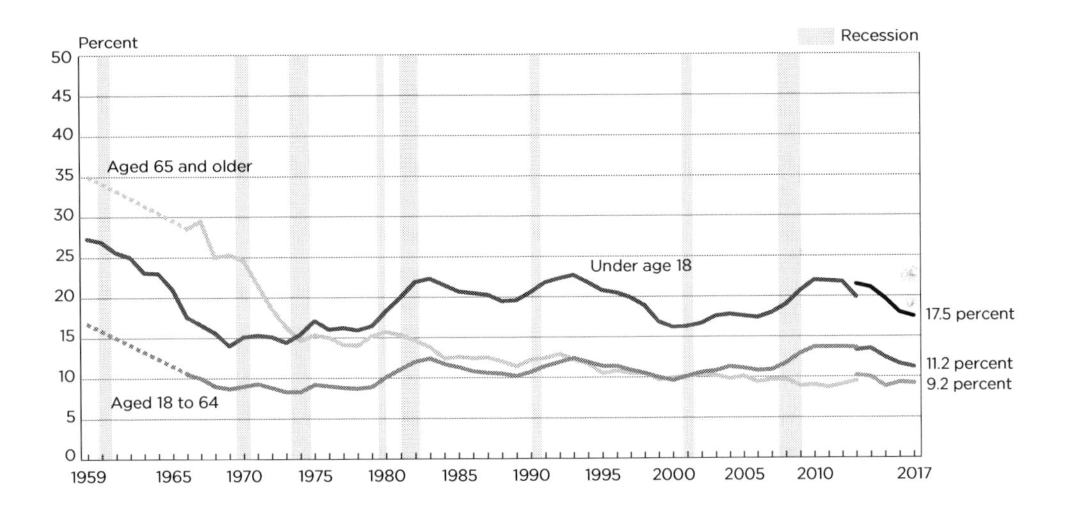

Figure 3.10 Poverty Rates by Age, 1959 to 2017

Source: U.S. Census Bureau, Current Population Survey, 1960–2018, Annual Social and Economic Supplements. https://www.census.gov/library/publications/2018/demo/p60-263.html

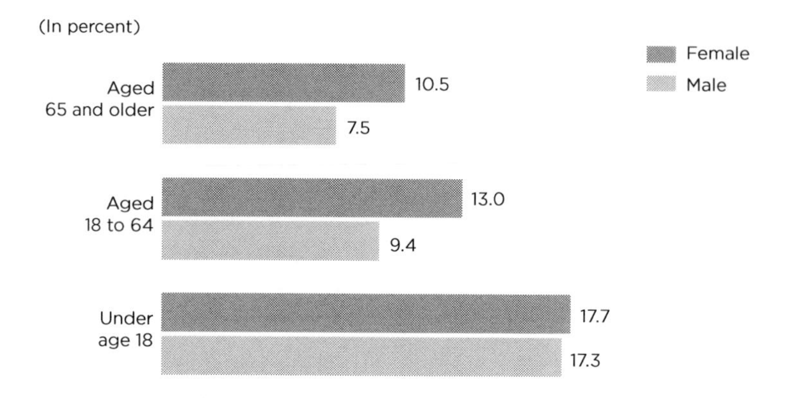

Figure 3.11 Poverty Rates by Age and Sex, 2017

Source: U.S. Census Bureau, Current Population Survey, 1960–2018, Annual Social and Economic Supplements. https://www.census.gov/library/publications/2018/demo/p60-263.html

find that the difference between the lowest and highest family income quintiles increased by about one year for women and by almost two years for men in Sweden between 1986 and 2007. This was the case both for individual and family income. Let's look at health inequality for families in the United States.

Class leads to likelihood of proximity to health inequality for underclass families. **Health-care access** is a supply side issue indicating the level of service which the health-care system offers individuals or families. Inequalities in health-care access persist in the United States. According to the inverse care law, the availability of good medical care tends to vary inversely with the need for it in the population served (Hart, 1971). Indeed, upper-class families have been the biggest buyers of high-quality health care. Government coverage continued to be most prevalent for the population in poverty (62.8 percent) and least prevalent for the population with income-to-poverty ratios at or above 400 percent of poverty (24.2 percent) in 2017 (Berchick *et al.*, 2018). Many people obtain health insurance coverage through a family member's plan. In 2017, people living in families had a higher health insurance coverage rate (91.7 percent) than unrelated individuals (88.8 percent) (Berchick *et al.*, 2018).

Meanwhile, health insurance coverage varies by metropolitan area and state. In 2017, the state with the lowest percentage of people without health insurance at the time of interview was Massachusetts (2.8 percent), while the state with the highest percentage was Texas (17.3 percent) (see Figure 3.12).

For adults age 19 to 64, the relationship between poverty status and change in the uninsured rate between 2015 and 2016 may be related to the state of residence and whether that state expanded Medicaid eligibility. Figure 3.13 shows that in states that expanded Medicaid eligibility on or before January 1, 2016 ("expansion states") and in states that did not expand Medicaid eligibility ("non-expansion states"), the uninsured rate decreased as the income to-poverty ratio increased for adults age 19 to 64.

Class leads to likelihood of retirement inequality for low-class families. Retirement inequality is greater than income inequality. Upper-income families hold a disproportionate share of retirement account balances within specific age groups, such as workers in their peak earning years (age 50–55) (Morrissey, 2016). Figure 3.14 shows the bottom 60 percent of families in this age group receive 23 percent of total income but hold only 14 percent of account balances.

Higher-paid workers have access to more retirement benefits than lower-paid workers. Figure 3.15 shows that among workers with an average wage in the highest 10 percent, 88 percent had access to employer-provided retirement plans in March 2016.

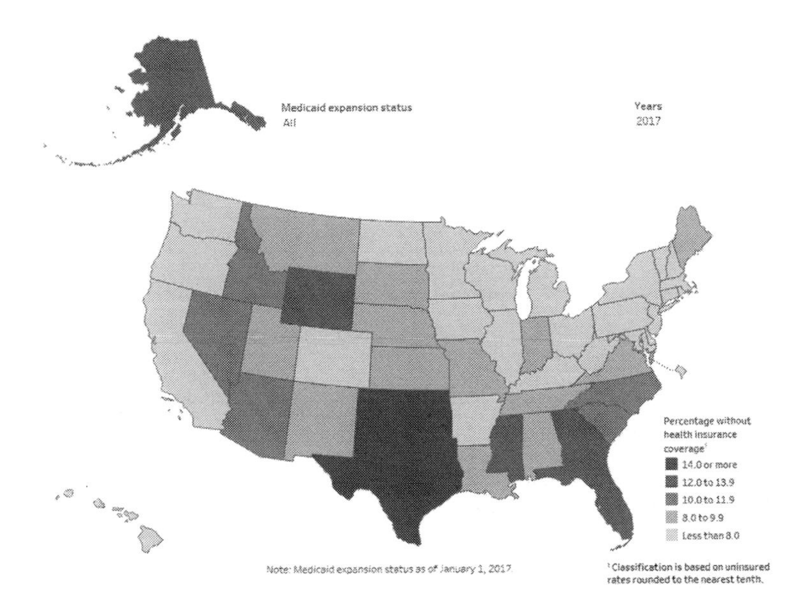

Figure 3.12 Uninsured Rate by State, 2017

Source: U.S. Census Bureau, 2008–2017, American Community Surveys, One-year Estimates. https://www.census.gov/library/publications/2018/demo/p60-264.html

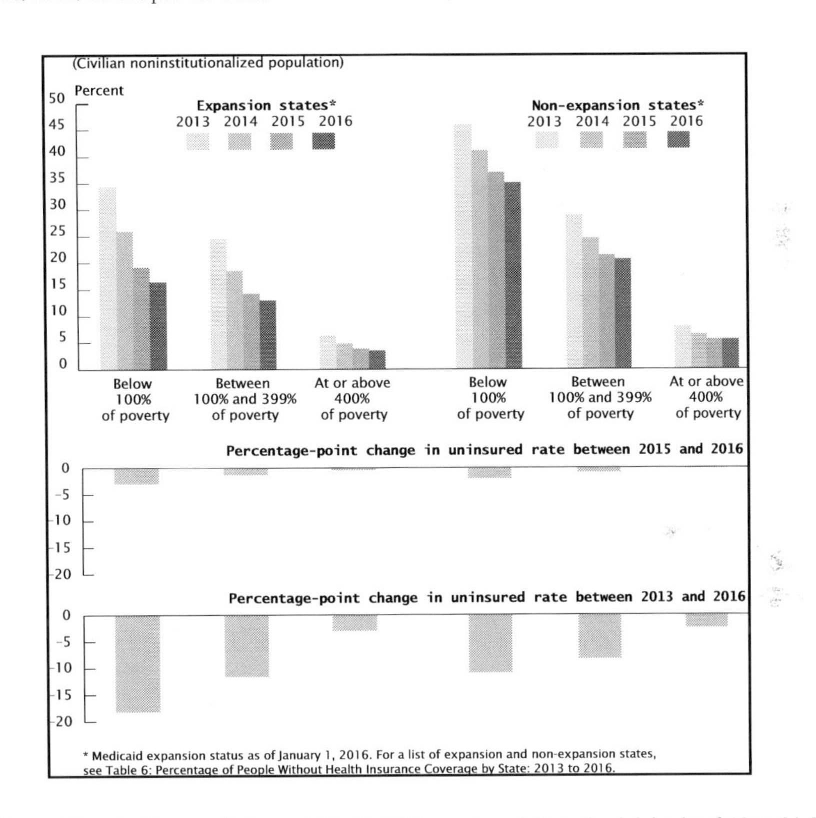

Figure 3.13 Uninsured Rate by Poverty Status and Medicaid Expansion of State for Adults Aged 19 to 64, 2013–2016

Source: U.S. Census Bureau, 2013–2016, One-Year American Community Surveys. https://census.gov/content/dam/Census/library/visualizations/2017/demo/p60-260/figure5.pdf

Note: For information on confidentiality protection, sampling error, nonsampling error, and definitions in the American Community Survey, see <www2.census.gov/programs-surveys/acs/tech_docs/accuracy/ACS_Accuracy_of_Data_2016.pdf>.

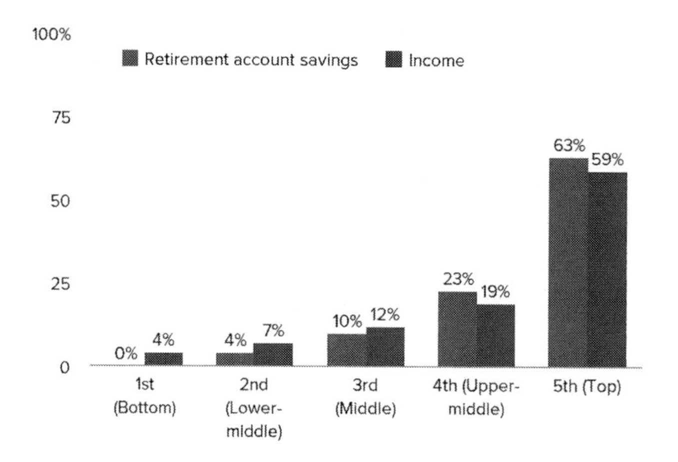

Figure 3.14 Retirement Inequality is Greater than Income Inequality Even in Peak-Earning Years
Source: EPI analysis of Survey of Consumer Finance data, 2013. http://www.epi.org/publication/retirement-in-america/

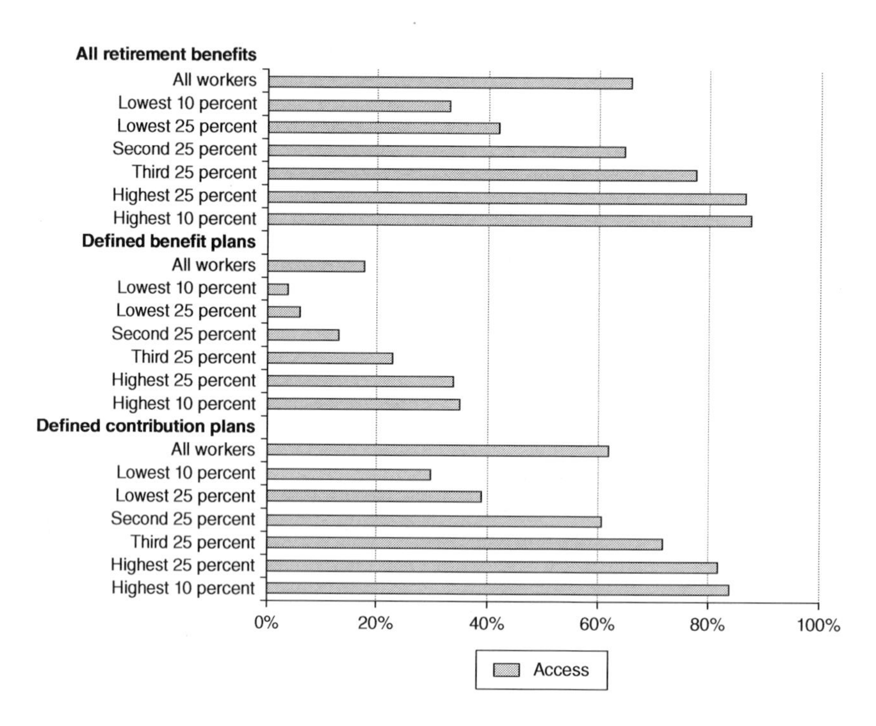

Figure 3.15 Percent of Private Industry Workers with Access to Retirement Benefits, by Type of Retirement Plan and Wage Category, March 2016
Source: U.S Bureau of Labor Statistics March 2017c. https://www.bls.gov/opub/ted/2017/higher-paid-workers-more-likely-to-have-access-to-retirement-benefits-than-lower-paid-workers.htm

Global Overview: World Families

International Day of Families, which is observed on May 15 every year by the United Nations, provides people with an opportunity to promote awareness of issues relating to families and to increase knowledge of the social, economic, and demographic processes affecting them (United Nations, 2017). One of the issues relating to families is **global stratification**, unequal distribution of resources on a global basis.

Global Wealthy and Poor Families

Sociologists have examined global stratification and unequal distribution of resources on a global basis. The GNI (Gross National Income) per capita reflects the average income of a country's citizens. The World Bank (2017a) defines low-income economies as those with a GNI per capita of $1,025 or less; lower-middle-income economies are those between $1,026 and $4,035; upper-middle-income economies are those between $4,036 and $12,475; and high-income economies are those $12,476 and more in 2015.

Large gaps between the wealthiest and poorest nations across the world shape family life. For example, Saudi Arabia has been run as an absolute monarchy by the al-Saud family from the eighteenth century. The entire House of Saud is composed of around 15,000 members and is one of the wealthiest families in the world (House of Saud, 2017). Based on 111 countries that account for 88 percent of the global population, Kochhar (2015) identified five income groups: poor (living on $2 or less daily), low income ($2.01–10), middle income ($10.01–20), upper-middle income ($20.01–50), and high income (more than $50). Figure 3.16 illustrates that on a global scale, more than half (56 percent) of Americans were high income by the global standard, living on more than $50 per day in 2011.

The world population is 7.5 billion (U.S. Census Bureau, 2018a). In 2016, just under 10 percent of the world's workers were living with their families on less than $1.90 per person per day, down from 28 percent in 2000, and in the least developed countries, nearly 38 percent of workers in 2016 were living below the poverty line (United Nations, 2017). The extreme poverty rate is the share of the population living on less than $1.90 a day in 2011 purchasing power parity terms. In 2013, over one in ten people were living under the international extreme poverty line of $1.90 a day, compared with almost four in ten in 1990 (World Bank, 2017b). In 2015, 702.1 million people were living in extreme poverty. Of these, about 347.1 million people

% of the U.S. and global populations by income in 2011

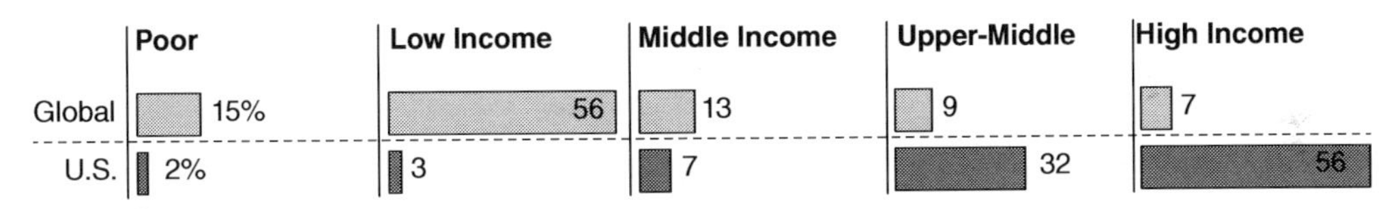

Figure 3.16 By a Global Standard, Majority of Americans are High Income

Source: Pew Research Center, 2015

http://www.pewresearch.org/fact-tank/2015/07/09/how-americans-compare-with-the-global-middle-class/ft_15-07-08_gmcglobal_vs_usa/

Note: The poor live on $2 or less daily, low income on $2.01–10, middle income on $10.01–20, upper-middle income on $20.01–50, and high income on more than $50; figures expressed in 2011 purchasing power parities in 2011 prices. People are grouped by the daily per capita income or consumption of their family, the choice of metric depending on how the source data for a country are collected.

lived in Sub-Saharan Africa (35.2 percent of the population) and 231.3 million lived in South Asia (13.5 percent of the population) (World Bank and International Monetary Fund, 2016). Poverty may increase the risk of disability through malnutrition, a polluted environment, health issues, and lack of access to safe water and sanitation. Poverty leads to the likelihood of proximity to environmental hazards, the risk of disability through malnutrition, and lack of access to safe water and sanitation. Research indicates that the quality of the neighborhood can influence children's development. Building the resilience of the poor and strengthening disaster risk reduction is a core development strategy for ending extreme poverty in the most afflicted countries (United Nations, 2017).

Global Middle-class Families

Middle-class values emphasize education, hard work, and thrift. The middle class emerged out of the bourgeoisie in the late fourteenth century and has been thought of as the source of entrepreneurship and innovation that make a modern economy thrive (Kharas, 2010). Figure 3.17 shows that we are witnessing the most rapid expansion of the global middle class, numbering about 3.2 billion in 2016.

Within a few years, based on current forecasts, a majority of the world's population could have middle-class or rich lifestyles for the first time ever (Kharas, 2017). A larger middle-class population and market has significant environmental and social implications. There is considerable evidence that a larger middle class will also imply a happier population, at least for new entrants into the middle class (Kahneman & Deaton, 2010). Middle-class households tend to invest more in their children's education and this, in turn, can reduce fertility rates and decrease the long-term population trajectory for the world (Kharas, 2017). Figure 3.18 shows that the middle-class expansion is expected to be broad based, but heavily concentrated in Asia. The vast majority (88 percent) of the next billion people in the middle class will be Asian.

The middle-class market in advanced economies has matured and is projected to grow at only 0.5 to 1 percent per year (Kharas, 2017). The middle class is a significant presence in western European countries. The share of adults living in middle income houses increased in France, the Netherlands, and the United

Figure 3.17 At a Global Level, We Are Witnessing the Most Rapid Expansion of the Middle Class the World Has Ever Seen

Source: Brookings. https://www.brookings.edu/wp-content/uploads/2017/02/global_20170228_global-middle-class-1.png

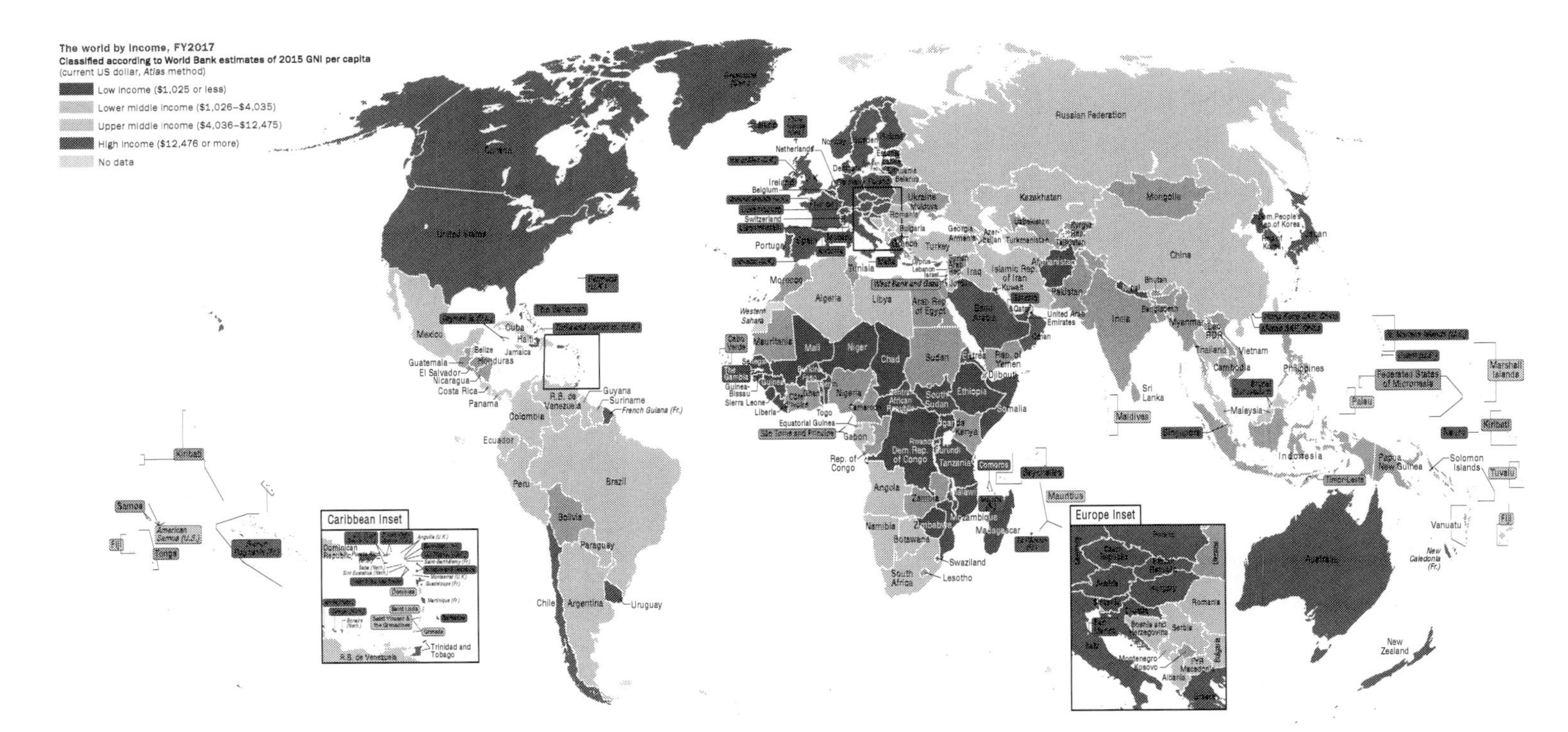

East Asia and Pacific

Country	Income group
American Samoa	Upper middle income
Australia	High income
Brunei Darussalam	High income
Cambodia	Lower middle income
China	Upper middle income
Fiji	Upper middle income
French Polynesia	High income
Guam	High income
Hong Kong SAR, China	High income
Indonesia	Lower middle income
Japan	High income
Kiribati	Lower middle income
Korea, Dem. People's Rep.	Low income
Korea, Rep.	High income
Lao PDR	Lower middle income
Macao SAR, China	High income
Malaysia	Upper middle income
Marshall Islands	Upper middle income
Micronesia, Fed. Sts.	Lower middle income
Mongolia	Lower middle income
Myanmar	Lower middle income
Nauru	High income
New Caledonia	High income
New Zealand	High income
Northern Mariana Islands	High income
Palau	Upper middle income
Papua New Guinea	Lower middle income
Philippines	Lower middle income
Samoa	Lower middle income
Singapore	High income
Solomon Islands	Lower middle income
Thailand	Upper middle income
Timor-Leste	Lower middle income
Tonga	Lower middle income
Tuvalu	Upper middle income
Vanuatu	Lower middle income
Vietnam	Lower middle income

Europe and Central Asia

Country	Income group
Albania	Upper middle income
Andorra	High income
Armenia	Lower middle income
Austria	High income
Azerbaijan	Upper middle income
Belarus	Upper middle income
Belgium	High income
Bosnia and Herzegovina	Upper middle income
Bulgaria	Upper middle income
Channel Islands	High income
Croatia	High income
Cyprus	High income
Czech Republic	High income
Denmark	High income
Estonia	High income
Faroe Islands	High income
Finland	High income
France	High income
Georgia	Upper middle income
Germany	High income
Gibraltar	High income
Greece	High income
Greenland	High income
Hungary	High income
Iceland	High income
Ireland	High income
Isle of Man	High income
Italy	High income
Kazakhstan	Upper middle income
Kosovo	Lower middle income
Kyrgyz Republic	Lower middle income
Latvia	High income
Liechtenstein	High income
Lithuania	High income
Luxembourg	High income
Macedonia, FYR	Upper middle income
Moldova	Lower middle income
Monaco	High income
Montenegro	Upper middle income
Netherlands	High income
Norway	High income
Poland	High income
Portugal	High income
Romania	Upper middle income
Russian Federation	Upper middle income
San Marino	High income
Serbia	Upper middle income
Slovak Republic	High income
Slovenia	High income
Spain	High income
Sweden	High income
Switzerland	High income
Tajikistan	Lower middle income
Turkey	Upper middle income
Turkmenistan	Upper middle income
Ukraine	Lower middle income
United Kingdom	High income
Uzbekistan	Lower middle income

Latin America and the Caribbean

Country	Income group
Antigua and Barbuda	High income
Argentina	Upper middle income
Aruba	High income
Bahamas, The	High income
Barbados	High income
Belize	Upper middle income
Bolivia	Lower middle income
Brazil	Upper middle income
British Virgin Islands	High income
Cayman Islands	High income
Chile	High income
Colombia	Upper middle income
Costa Rica	Upper middle income
Cuba	Upper middle income
Curaçao	High income
Dominica	Upper middle income
Dominican Republic	Upper middle income
Ecuador	Upper middle income
El Salvador	Lower middle income

World Bank estimates of 2015 GNI per capita

Figure 3.18 Eighty-eight Percent of the Next Billion Entrants into the Middle Class Will Be in Asia
Source: Brookings. https://www.brookings.edu/wp-content/uploads/2017/02/global_20170228_global-middle-class-3.png

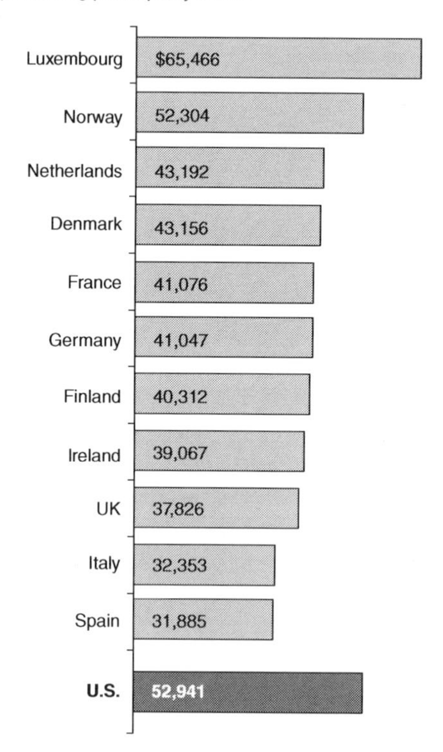

Figure 3.19 Luxembourg Households in All Income Tiers Earn More than Households in the U.S. and Other Western European Countries
Source: http://www.pewglobal.org/2017/04/24/middle-class-fortunes-in-western-europe/st_2017-04-24_western-europe-middle-class_2-01/

Kingdom, but they shrank in Germany, Italy, and Spain from 1991 to 2010 (Kochhar, 2017). Figure 3.19 illustrates that Luxembourg households had highest incomes and lived on $65,466 in 2010, compared to $52,941 in the United States. The middle class in Italy lived on a median income of $32,353. Financially, the American middle class is ahead of the middle classes in western European nations with the exception of Luxembourg (Kochhar, 2017).

Families over the world have transformed greatly in the past decades. The United Nations (2017) recognizes the family as the basic unit of society. International Day of Families focuses on the role of families and family-oriented policies in promoting education and the overall well-being of their members.

Summary

The United States was inhabited by the diverse Native tribes prior to the arrival of the Italian explorer Columbus. The colonial families settled down in the thirteen colonies and created a pioneer culture for the country. Slave families contributed to the building of the United States. Slave marriages were not legally recognized but slave breeding was seen as a way to acquire future slaves for slave owners. Slave families were ripped apart and family ties were hard to hold on to.

In the nineteenth century, families moved away from being a production and economic center to focusing on the reproduction and socialization of children. In the 1950s, the male breadwinner and stay-at-home mother was the ideal nuclear family model, and marriage was about mutual emotional and sexual satisfaction. In the twenty-first century onward, a young adult may marry late, remain single, or live in their parental home. Non-traditional forms of family such as single-parent family and a cohabiting family continued to rise due to the declining marriage rate, declining family size, and rising marriage age.

Social class has remained a visible factor that has shaped family life. Upper-upper-class families are members of the capitalist class and inheritors of wealth. Lower-upper-class families are self-made billionaires. Both upper-upper-class families with old money and lower-upper-class families with new money are exceptionally rich. Upper-middle-class families are headed by higher-status householders whose work is largely self-directed. Lower-middle-class families and working-class families tried hard to maintain an average standard of living. The working-poor families live paycheck to paycheck while underclass families live in poverty. From a diversity and global approach, the chapter examined immigrant families and world families along the lines of global stratification.

Key Terms

Baby boomers are people born between 1946 and 1964 during the demographic post-World War II baby boom.

Breadwinner was a term coined to describe the male role as provider.

Broadband internet refers to households who said "yes" to subscriptions such as DSL, cable, fiber optic, mobile broadband, satellite, or fixed wireless.

Colonial America refers to the history of European settlements from the colonization of America until their incorporation into the United States.

Companionate marriage refers to marriages based on affection, friendship, and sexual gratification.

Compulsory education is mandatory education imposed by law.

Creative class is a class of workers whose job is to create meaningful new forms.

Credentialism refers to the practice of relying on earned credentials when hiring staff or assigning social status.

De-skilling is the process by which skilled labor within an industry is eliminated by the introduction of technologies operated by semi-skilled or unskilled workers.

Digital divide is an economic and social inequality with regard to access to, use of, or impact of information and communication technologies (ICT).

Dual-earner family refers to both husband and wife providing income for the family, usually by working outside the home.

Emigrant refers to a person who leaves his or her own country.

Giving Pledge is a commitment by the world's wealthiest individuals and families to dedicate the majority of their wealth to giving back.

Global stratification refers to unequal distribution of resources on a global basis.

Health-care access is a supply side issue indicating the level of service which the health-care system offers individuals or families.

Homemaker is coined to describe the female role as homemaker.

Immigrant refers to a person who enters a new country and takes up permanent residence.

Indentured servant is a person under contract to work for another person for a period of time without pay but in exchange for free passage to the new country.

Industrialization is based on technology that mechanizes production to provide goods.

Industrial Revolution is the transition to manufacturing processes from hand production methods.

Lower-middle-class family is maintained by householders who have white-collar jobs and have associate's two-year college or bachelor's degrees.

Lower-upper-class family is maintained by householders who earn new money from investments and business ventures within their own lifetimes.

New money is recently acquired wealth rather than inherited wealth.

Nineteenth Amendment to the United States Constitution in 1920 granted women the right to vote.

Nuclear family consists of a father, a mother, and their children.

Old money is the accumulated and inherited wealth of established upper-class *families* such as the Rockefeller family.

Pilgrims are referred to as members of a Puritan separatist sect who set sail in the *Mayflower* bound for the Americas to establish a colony where they could enjoy religious freedom.

Post-industrial society is based on computer technology that produces information, supports a service industry, and encourages consumption.

Poverty refers to a condition where a family's total income is less than the official poverty threshold for a family of that size and composition and is not enough to meet basic needs.

Poverty rate is the ratio of the number of people who fall below the poverty line and the total population.

Poverty thresholds are used to determine poverty status and related to the size of the family and age of the members.

Progressive Era (from the 1890s to 1920s) was a response to the economic and social problems taking place at the time and was a period of widespread social activism and political reform across the United States.

Pull factors, such as freedom and opportunities, bring immigrants to a new country.

Push factors, such as famine and war, make people leave their country.

Slave breeding is the practice used by slave owners to reproduce slaves in order to increase the slave owner's wealth.

Slavery is a closed stratification system in which slaves are classified as property.

Social class refers to a group of people with similar shares of resources in hierarchical social divisions such as upper, middle, and lower classes.

Technological unemployment is the loss of jobs caused by technological change in the process of invention, innovation, and diffusion of technology.

Underclass family is maintained by householders who are lowly educated and are under-employed or unemployed.

Upper-middle-class family is maintained by householders who hold professional degrees and have high-paying professional occupations.

Upper-upper class is maintained by householders who have inherited old money over multiple generations.

Working-class family is maintained by householders who hold high school diplomas and have blue-collar jobs and service-work jobs in manual labor occupations.

Working poor are people who spent at least 27 weeks in the labor force (that is, working or looking for work) but whose incomes still fell below the official poverty level.

Working-poor family is maintained by householders with high school diplomas who work in lower paid jobs and live paycheck to paycheck.

Working-poor rate is the ratio of the working poor to all individuals in the labor force for at least 27 weeks.

Discussion Questions

- Discuss how every family is different and what makes a family a family.
- Think of a framework and use education, income, and occupation as a way to portray family life.
- Design a budget for a family based upon the combination of average costs and total income.

Suggested Film and Video

America's Forgotten Working Class, J.D. Vance: https://www.youtube.com/watch?v=iEy-xTbcr2A

America's Middle Class Is Shrinking but Still Rich Compared to Europe: http://fortune.com/2017/04/25/america-middle-class-shrinking/

Andrew Neil: Rise of the British Underclass – The Great British Class Survey, BBC Lab UK: https://www.youtube.com/watch?v=mIHeAOA13cw

Haley, A. (1977). *Roots.* Stan Margulies (Producers).

McQueen, S. (2013). *12 Years a Slave.* Brad Pitt *et al.* (Producers).

Internet Sources

Are You in the American Middle Class? Find Out with Our Income Calculator: http://www.pewresearch.org/fact-tank/2016/05/11/are-you-in-the-american-middle-class/

Anti-Poverty Family-Focused Policies in Developing Countries: http://www.un.org/esa/socdev/family/docs/BP_POVERTY.pdf

Family-oriented Anti-poverty Polices in Developed Countries: http://www.un.org/esa/socdev/family/docs/BACKGROUNDPAPERRICHARDSON.pdf

Fairness Initiative on Low-Wage Work: www.lowwagework.org

Organizations of Interest: http://www.workingpoorfamilies.org/resources/

Sustainable Development Goal 1: https://sustainabledevelopment.un.org/sdg1

The Health Inequality Project: https://healthinequality.org/

WDI 2017 Maps from the World Bank: http://data.worldbank.org/products/wdi-maps

Working Poor Families Project: http://www.workingpoorfamilies.org/

World Data Lab: http://www.worlddata.io/

World Poverty Clock: http://worldpoverty.io/

References

Acemoglu, D. & Restrepo, P. (2017). *Robots and Jobs: Evidence from US Labor Markets.* NBER Working Paper No. 23285. National Bureau of Economic Research. Retrieved on September 9, 2017 from http://www.nber.org/papers/w23285

American Presidency Project (2017). 670—Proclamation 9516—National Manufacturing Day, 2016. Retrieved on September 10, 2017 from http://www.presidency.ucsb.edu/ws/index.php?pid=119079

Anderson, M. (March 22, 2017). *Digital Divide Persists Even as Lower-Income Americans Make Gains in Tech Adoption.* Washington, DC: Pew Research Center. Retrieved on September 10, 2017 from http://www.pewresearch.org/fact-tank/2017/03/22/digital-divide-persists-even-as-lower-income-americans-make-gains-in-tech-adoption/

Baugh, R. & Witsman, K. (March 2017). *U.S. Lawful Permanent Residents: 2015. Annual Report.* Homeland Security. Retrieved on September 16, 2017 from https://www.dhs.gov/sites/default/files/publications/Lawful_Permanent_Residents_2015.pdf

Beauchamp, W. M. (1905). *A History of the New York Iroquois.* New York: New York State Education Department.

Beecher, C. (1851). *True Remedy for the Wrongs of Woman; with a History of an Enterprise Having That for Its Object.* Boston, MA: Phillips, Sampson, & Co.

Berchick, E. R., Hood, E. & Barnett, J. C. (September 2018). *Health Insurance Coverage in the United States: 2017.* U.S. Census Bureau. Retrieved on October 1, 2018 from https://www.census.gov/content/dam/Census/library/publications/2018/demo/p60-264.pdf

Berlin, I. & Rowland, L. (1998). *Families and Freedom: A Documentary History of African-American Kinship in the Civil War Era.* New York: The New Press.

Bhugra, D. & Becker, M. A. (2005). Migration, Cultural Bereavement and Cultural Identity. *World Psychiatry: Official Journal of the World Psychiatric Association (WPA),* 4(1), 18–24. Retrieved on September 8, 2017 from https://www.ncbi.nlm.nih.gov/pmc/articles/PMC1414713/

Boydston, J. (1994). *Home and Work: Housework, Wages, and the Ideology of Labor in the Early Republic*. New York: Oxford University Press.

Braverman, H. (1974). *Labor and Monopoly Capital*. New York: Monthly Review.

Broadberry, S. & Gupta, B. (2005). *Cotton Textiles and the Great Divergence: Lancashire, India and Shifting Competitive Advantage, 1600–1850*. International Institute of Social History, University of Warwick.

Brunner, B. (December 28, 2004). *TIME Almanac with Information Please 2005*. Boston, MA: Pearson Education Company.

Buenker, J. D., Burnham, J. C. & Crunden, R. M. (1977). *Progressivism*. Cambridge, MA: Schenkman Publication Company.

Camarota, S. A. (March 2016). *The Impact of Large-Scale Immigration on American Workers*. Center for Immigration Studies. Retrieved on November 15, 2016 from http://cis.org/Testimony/Camarota-The-Impact-of-Large-Scal-%20Immigration-on-American-Workers

Camarota, S. & Zeigler, K. (February 2016). The Employment Situation of Immigrants and Natives in the Fourth Quarter of 2015. Center for Immigration Studies *Backgrounder*, 2016.

CDC (Centers for Disease Control and Prevention) (March 31, 2017). *Unmarried Childbearing*. Retrieved on September 11, 2017 from https://www.cdc.gov/nchs/fastats/unmarried-childbearing.htm

Cherlin, A. (2004). The Deinstitutionalization of American Marriage. *Journal of Marriage and Family*, 66(4), 848–861. Retrieved on September 30, 2018 from http://www.jstor.org/stable/3600162

Cherlin, A. (2014). *Labor's Love Lost: The Rise and Fall of the Working-Class Family in America*. New York: Russell Sage Foundation.

Chetty, R., Hendren, N. & Katz, L. (2016). The Effects of Exposure to Better Neighborhoods on Children: New Evidence from the Moving to Opportunity Project. *American Economic Review*, 106(4).

Chikhale, N. (March 6, 2017). *A Childcare Plan for Wealthy Families in the United States*. Washington Center for Equitable Growth. Retrieved on September 9, 2017 from http://equitablegrowth.org/equitablog/value-added/a-childcare-plan-for-wealthy-families-in-the-united-states/

Chused, R. H. (1983). Married Women's Property Law: 1800–1850. *Georgetown Law Journal*, 71(1983), 1359, 1366.

Cohn, D.'V. (August, 2017). *5 Key Facts about U.S. Lawful Immigrants*. Washington, DC: Pew Research Center. Retrieved on September 16, 2017 from http://www.pewresearch.org/fact-tank/2017/08/03/5-key-facts-about-u-s-lawful-immigrants/

Cohn, D.'V. & Ruiz, N. G. (July 6, 2017). *More than Half of New Green Cards Go to People Already Living in the U.S.* Washington, DC: Pew Research Center. Retrieved on July 19, 2017 from http://www.pewresearch.org/fact-tank/2017/07/06/more-than-half-of-new-green-cards-go-to-people-already-living-in-the-u-s/

Collins, R. (April 1979). *The Credential Society: A Historical Sociology of Education and Stratification*. New York: Academic Press.

Coontz, S. (2016). *The Way We Never Were*. New York: Basic Books.

Dai, W. (2016). Dual-Earner Couples in the United States. In C. L. Shehan (Ed.), *Encyclopedia of Family Studies*. doi: 10.1002/9781119085621.wbefs406

DeRose, L. (September 5, 2017). *Race, Cohabitation, and Children's Family Stability*. Institute for Family Studies. Retrieved on September 11, 2017 from https://ifstudies.org/blog/race-cohabitation-and-childrens-family-stability

Desilver, D. (March, 2017). *Immigrants Don't Make Up a Majority of Workers in Any U.S. Industry*. Washington, DC: Pew Research Center. Retrieved on September 16, 2017 from http://www.pewresearch.org/fact-tank/2017/03/16/immigrants-dont-make-up-a-majority-of-workers-in-any-u-s-industry/

Ehrenreich, B. (1989). *Fear of Falling, The Inner Life of the Middle Class*. New York, NY: Harper Collins.

Eichar, D. (1989). *Occupation and Class Consciousness in America*. Westport, CT: Greenwood Press.

Fisher, G. M. (1997). *The Development of the Orshansky Poverty Thresholds and Their Subsequent History as the Official U.S. Poverty Measure*. U.S. Census Bureau. Retrieved on September 14, 2017 from https://www.census.gov/library/working-papers/1997/demo/fisher-02.html

Florida, R. (October 28, 2014). *The Rise of the Creative Class*. New York: Basic Books.

Fontenot, K., Semega, J. & Kollar, M. (2018). *Income and Poverty in the United States: 2017 Current Population Reports*. U.S. Census Bureau. Retrieved on September 29, 2018 from https://www.census.gov/content/dam/Census/library/publications/2018/demo/p60-263.pdf

Forman-Brunell, M. (2009). *Babysitter: An American History*. New York: New York University Press.

Fry, R. (September 6, 2017). *5 Facts about Millennial Households*. Washington, DC: Pew Research Center. Retrieved on September 15, 2017 from http://www.pewresearch.org/fact-tank/2017/09/06/5-facts-about-millennial-households/

Gilbert, D. (1997). *The American Class Structure: In An Age of Growing Inequality.* Belmont, CA: Wadsworth Publishing.
Gilbert, D. &. Kahl, J. A. (1992). *American Class Structure: A New Systhesis.* Belmont, CA: Wadsworth Publishing.
Giving Pledge (2018). *A Commitment to Philanthropy.* Retrieved on October 4, 2018 from https://givingpledge.org/
Goffigon, C. (2014). Make Me a Sandwich: A Cultural History of Domestic Kitchens in 19th Century America. *American Studies Honors Papers,* 6. http://digitalcommons.conncoll.edu/americanstudieshp/6
Graham, C. (July 10, 2017). *The Unhappiness of U.S. Working Class.* Washington, DC: Brookings Institution. Retrieved on September 18, 2017 from https://www.brookings.edu/opinions/the-unhappiness-of-the-us-working-class/
Green, B. (2006). *Marketing to Leading-Edge Baby Boomers: Perceptions, Principles, Practices, Predictions.* New York: Paramount Market Publishing.
Hart, J. T. (1971). The Inverse Care Law. *Lancet,* 1, 405–412.
HIP (Health Inequality Project) (2016). *How Can We Reduce Disparities in Health?* Retrieved on August 8, 2016 from https://healthinequality.org/
Hederos, K., Jäntti, M., Lindahl, L. & Torssander, J. (2017). Trends in Life Expectancy by Income and the Role of Specific Causes of Death. *Economica.* doi: 10.1111/ecca.12224
Higginbotham, A. L. (1978). *In the Matter of Color: Race and the American Legal Process: The Colonial Period.* New York: Oxford University Press.
Hisnanick, J. J., Giefer, K. G. & Williams, A. K. (July 2017). *Dynamics of Economic Well-Being: Fluctuations in the U.S. Income Distribution: 2009–2012.* U.S. Census Bureau. Retrieved on September 7, 2017 from https://www.census.gov/content/dam/Census/library/publications/2017/demo/p70-142.pdf
House of Saud (2017). *Royal Family Profiles.* Retrieved on July 17, 2017 from http://houseofsaud.com/
Hunt, A. N. (2016). Companionate Marriage. In *The International Encyclopedia of Human Sexuality.* https://doi.org/10.1002/9781119085621.wbefs255
Hunter, L. C. (1985). *A History of Industrial Power in the United States, 1730–1930, Vol. 2: Steam Power.* Charlottesville, VA: University Press of Virginia.
Jaycox, F. (2005). *The Progressive Era.* New York: Facts On File, Incorporated.
Kahneman, D. & Deaton, A. (2010). High Income Improves Evaluation of Life but Not Emotional Well-Being. *Proceedings of the National Academy of Sciences,* 107(38), 16489–16493.
Kalyuga, M. (2012). Vocabulary of Love. In M. A. Paludi (Ed.), *The Psychology of Love* (Vol. 3, pp. 75–87). Santa Barbara, CA: Praeger.
Kaplan, S. N. & Rauh, J. (May 2013). Family, Education, and Sources of Wealth among the Richest Americans, 1982–2012. *American Economic Review Papers & Proceedings.*
Karandashev, V. (2015). A Cultural Perspective on Romantic Love. *Online Readings in Psychology and Culture,* 5(4).
Kharas, H. (February 28, 2017). *The Unprecedented Expansion of the Global Middle Class.* Washington, DC: Brookings Institution. Retrieved on September 18, 2017 from https://www.brookings.edu/research/the-unprecedented-expansion-of-the-global-middle-class-2/
Kharas, H. (January 31, 2010). *The Emerging Middle Class in Developing Countries.* Washington, DC: Brookings Institution. Retrieved on September 18, 2017 from https://www.brookings.edu/research/the-emerging-middle-class-in-developing-countries/
Kochhar, R. (July 9, 2015). *How Americans Compare with the Global Middle Class.* Washington, DC: Pew Research Center. Retrieved on August 23, 2016 from http://www.pewresearch.org/fact-tank/2015/07/09/how-americans-compare-with-the-global-middle-class/
Kochhar, R. (April 24, 2017). *Middle Class Fortunes in Western Europe.* Washington, DC: Pew Research Center. Retrieved on July 17, 2017 from http://www.pewglobal.org/2017/04/24/middle-class-fortunes-in-western-europe/
Kochhar, R. & Fry, R. (February 14, 2015). *America's 'Middle' Holds Its Ground after the Great Recession.* Washington, DC: Pew Research Center. Retrieved on September 8, 2017 from http://www.pewresearch.org/fact-tank/2015/02/04/americas-middle-holds-its-ground-after-the-great-recession/
Kochhar, R. & Fry, R. (December 9, 2015). *The American Middle Class Is Losing Ground.* Washington, DC: Pew Research Center. Retrieved on August 7, 2016 http://www.pewsocialtrends.org/2015/12/09/the-american-middle-class-is-losing-ground/
Kochhar, R. & Morin, R. (January 27, 2014). *Despite Recovery, Fewer Americans Identify as Middle Class.* Washington, DC: Pew Research Center. Retrieved on September 9, 2017 from http://www.pewresearch.org/fact-tank/2014/01/27/despite-recovery-fewer-americans-identify-as-middle-class/
Krech III, S. (1999). *The Ecological Indian: Myth and History.* New York: W. W. Norton.
Lareau, A. (2011). *Unequal Childhoods, Class, Race, and Family Life, With an Update a Decade Later.* Berkeley, CA: University of California Press.
Library of Congress (April 25, 2017). *Primary Documents in American History.* Retrieved on September 7, 2017 from http://www.loc.gov/rr/program/bib/ourdocs/Indian.html

Library of Congress (May 15, 2017). *Primary Documents in American History.* Retrieved on September 11, 2017 from https://www.loc.gov/rr/program/bib/ourdocs/Homestead.html

Lindsey, B. B. & Evans, W. (1927). *The Companionate Marriage.* New York: Ayer Co Pub.

Loury, G. C. (March 1, 1998). *An American Tragedy: The Legacy of Slavery Lingers in Our Cities' Ghettos.* Washington, DC: Brookings Institution. Retrieved on September 1, 2017 from https://www.brookings.edu/articles/an-american-tragedy-the-legacy-of-slavery-lingers-in-our-cities-ghettos/

McConnaughy, C. M. (2013). *The Woman Suffrage Movement in America: A Reassessment.* New York: Cambridge University Press.

Miller, J. (Dec. 1901). The Chinese and the Exclusion Act. *The North American Review,* 173(541), 782–789.

Mills, C. W. (1951). *White Collar: The American Middle Classes.* New York: Oxford University Press.

Mishel, L. & Bivens, J. (May 24, 2017). *The Zombie Robot Argument Lurches On. There Is No Evidence that Automation Leads to Joblessness Or Inequality.* Economic Policy Institute. Retrieved on September 8, 2017 from http://www.epi.org/publication/the-zombie-robot-argument-lurches-on-there-is-no-evidence-that-automation-leads-to-joblessness-or-inequality/

Moncrief, S. (2008). *The Mississippi Married Women's Property Act of 1839.* Hancock Historical Society. Retrieved on August 19, 2017 from http://www.hancockcountyhistoricalsociety.com/vignettes/the-mississippi-married-womens-property-act-of-1839/

Morrissey, M. (March 3, 2016). *Retirement Inequality Chartbook.* Economic Policy Institute. Retrieved on November 12, 2016 from http://www.epi.org/publication/retirement-in-america/#chart1

Murray, C. (2012). *Coming Apart: The State of White America 1960–2010.* New York: Crown Forum.

Myrdal, G. (1963). *Challenge to Affluence.* New York: Random House.

National Women's History Project (NWHP) (2016). *Timeline of Legal History of Women in the United States.* Retrieved on April 26, 2016 from http://www.nwhp.org/resources/womens-rights-movement/detailed-timeline/

Padavic, I. & Reskin, B. (2002). *Women and Men at Work.* California: Sage Publications.

Passel, J. & Cohn, D'V. (March 8, 2017). *Immigration Projected to Drive Growth in U.S. Working-Age Population through at Least 2035.* Washington, DC: Pew Research Center. Retrieved on September 8, 2017 from http://www.pewresearch.org/fact-tank/2017/03/08/immigration-projected-to-drive-growth-in-u-s-working-age-population-through-at-least-2035/

Pew Research Center (2015). *The American Middle Class Is Losing Ground: No Longer the Majority and Falling Behind Financially.* Retrieved on November 24, 2018 from http://www.pewresearch.org/wp-content/uploads/sites/3/2015/12/2015-12-09_middle-class_FINAL-report.pdf

Pew Research Center (May 11, 2016). *America's Shrinking Middle Class: A Close Look at Changes within Metropolitan Areas.* Retrieved on November 24, 2018 from http://www.pewresearch.org/wp-content/uploads/sites/3/2016/05/Middle-Class-Metro-Areas-FINAL.pdf

Phillips, W. D. & Phillips, C. R. (1992). *The Worlds of Christopher Columbus.* Cambridge: Cambridge University Press, p. 9.

Pierce, M. L. (2018). *Tracking the Evolution of the Companionate Marriage Ideal in Early Modern Comedies.* Honors Theses and Capstones, 383. https://scholars.unh.edu/honors/383

Riedy, M. E. (Spring 1997). *Uncle Tom's Houses, the American Domestic Ideal 1840–1870.* Retrieved on March 26, 2016 from http://xroads.virginia.edu/~cap/utc/title.html

Roe, J. W. (1916). *English and American Tool Builders.* New Haven, CT: Yale University Press.

Sandage, S. A. (2005). *Born Losers: A History of Failure in America.* Cambridge, MA: Harvard University Press.

Simkin, J. (2015). *Spartacus Educational.* Retrieved on January 21, 2016 from http://spartacus-educational.com/USAES1920S.htm

Simmons, C. (2009). *Making Marriage Modern: Women's Sexuality from the Progressive Era to World War II.* New York: Oxford University Press.

Simmons, C. (2015). Companionate marriage. In A. Bolin and P. Whelehan (Eds.), *The International Encyclopedia of Human Sexuality.* doi: 10.1002/9781118896877.wbiehs096

Takaki, R. (2008). *A Different Mirror: A History of Multicultural America.* New York: Back Bay Books.

Thompson, W. & Hickey, J. (2005). *Society in Focus.* Boston, MA: Pearson.

Tolman, T. L. (March, 2011). *The Effects of Slavery and Emancipation on African-American Families and Family History Research.* Crossroads. Retrieved on May 7, 2016 from http://www.leaveafamilylegacy.com/African_American_Families.pdf

Tompkins, K. W. (2012). *Racial Indigestion: Eating Bodies in the 19th Century.* New York: NYU Press.

Tyack, D. B. (1974). *The One Best System: A History of American Urban Education*. Cambridge, MA: Harvard University Press.

Tyler, L. G. (Ed.) (April 1915). The F.F.V's of Virginia. *William and Mary College Quarterly Historical Magazine*, 23(4), 277.

United Nations (May 15, 2017). *2017 International Day of Families. Division for Social Policy and Development Family.* Retrieved on July 17, 2017 from https://www.un.org/development/desa/family/international-day-of-families/idf2017.html

U.S. Bureau of Labor Statistics (December 2015). *Monthly Labor Review*. Retrieved on October 26, 2016 from http://www.bls.gov/opub/mlr/2015/article/industry-employment-and-output-projections-to-2024.htm#BLS_table_footnotes

U.S. Bureau of Labor Statistics (October 24, 2017a). *Employment Projections—2016–26*. Retrieved on October 3, 2018 from https://www.bls.gov/news.release/pdf/ecopro.pdf

U.S. Bureau of Labor Statistics (April 2017b). *A Profile of the Working Poor, 2015*. Retrieved on July 15, 2017 from https://www.bls.gov/opub/reports/working-poor/2015/home.htm

U.S. Bureau of Labor Statistics (April 20, 2017c). *Employment Characteristics of Families Summary.* U.S. Department of Labor. Retrieved on July 12, 2017 from https://www.bls.gov/news.release/famee.nr0.htm

U.S. Bureau of Labor Statistics (July 2018). *A Profile of the Working Poor, 2016*. Retrieved on October 3, 2018 from https://www.bls.gov/opub/reports/working-poor/2016/home.htm#chart2

U.S. Census Bureau (1790). *1790 Overview.* Retrieved on November 01, 2016 from https://www.census.gov/history/www/through_the_decades/overview/1790.html

U.S. Census Bureau (October 04, 2016a). *FFF: Thanksgiving Day: Nov. 24, 2016*. Retrieved on October 31, 2016 from http://www.census.gov/newsroom/facts-for-features/2016/cb16-ff19.html

U.S. Census Bureau (January 29, 2016b). *Facts for Features: Super Bowl 50: Feb. 7, 2016*. Retrieved on June 25, 2017 from https://www.census.gov/newsroom/facts-for-features/2016/cb16-ff05.html

U.S. Census Bureau (August 16, 2016c). *Number of IT Workers Has Increased Tenfold Since 1970*. Retrieved on August 23, 2016 from https://www.census.gov/newsroom/press-releases/2016/cb16-139.html

U.S. Census Bureau (June 05, 2017a). *FFF: The Fourth of July: 2017*. Retrieved on July 1, 2017 from https://www.census.gov/newsroom/facts-for-features/2017/cb17-ff10-fourth-of-july.html

U.S. Census Bureau (November 2017b). *Historical Marital Status Tables*. Retrieved on October 2, 2018 from https://www.census.gov/data/tables/time-series/demo/families/marital.html

U.S. Census Bureau (October 2, 2018a). *U.S. and World Population Clock*. Retrieved on October 2, 2018 from https://www.census.gov/popclock/

U.S. Census Bureau (September 2018b). *Income and Poverty in the United States: 2017*. Retrieved on October 1, 2018 from https://www.census.gov/library/publications/2018/demo/p60-263.html

U.S. Department of Commerce, National Telecommunications and Information Administration (NTIA) (1995). *Falling Through the Net: A Survey of the Have Nots in Rural and Urban America*. Retrieved on September 10, 2017 from https://www.ntia.doc.gov/ntiahome/fallingthru.html

U.S. Government Publishing Office (GPO) (1953). *67 STAT. B132 – INDIANS*. Retrieved on September 5, 2017 from https://www.gpo.gov/fdsys/granule/STATUTE-67/STATUTE-67-PgB132/content-detail.html

Vanneman, R. & Cannon, L. W. (1988). *The American Perception of Class*. New York, NY: Temple University Press.

Vespa, J. (April 2017). *The Changing Economics and Demographics of Young Adulthood: 1975–2016*. U.S. Census Bureau. Retrieved on September 7, 2017 from https://www.census.gov/content/dam/Census/library/publications/2017/demo/p20-579.pdf

Waldman, C. (2006). *Encyclopedia of North American Indians*. New York: Checkmark Books.

Waldman, C. (2009). *Atlas of the North American Indian*. New York: Checkmark Books.

Warner, W. L. (1949). *Social Class in America: A Manual of Procedure for the Measurement of Social Status*. Chicago: Science Research Associates.

West, P. (2013). *Domesticating History: The Political Origins of America's House Museums*. Washington, DC: Smithsonian Books.

White House (2018). *Immigration*. Retrieved on October 3, 2018 from https://www.whitehouse.gov/search/?s=Immigration

Williams, G. L. (1947). The Legal Unity of Husband and Wife. *Modern Law Review*, 10(1), 16–31.

Williams, T. J., Collins, M. B., Rodrigues, K., Rink, W. J., Velchoff, N., Keen-Zebert, A., Gilmer, A., Frederick, C. D., Ayala, S. J. & Prewitt, E. R. (2018). Evidence of an Early Projectile Point Technology in North America at the Gault Site, Texas, USA. *Science Advances,* 4(7), eaar5954.

Wilson, V. (June 9, 2016). *People of Color Will Be a Majority of the American Working Class in 2032.* Economic Policy Institute. Retrieved on November 14, 2016 from http://www.epi.org/publication/the-changing-demographics-of-americas-working-class/

Winkler, A. E. (April 1998). Earnings of Husbands and Wives in Dual-Earner Families. *Monthly Labor Review.* U.S. Bureau of Labor Statistics. Retrieved on November 24, 2018 from https://www.bls.gov/opub/mlr/1998/04/art4full.pdf

WPFP (Working Poor Families Project) (2017). *Strengthening State Policies for Working Families.* Retrieved on November 7, 2016 from http://www.workingpoorfamilies.org/

World Bank (2017a). *WDI 2017 Maps.* Retrieved on July 17, 2017 from http://data.worldbank.org/products/wdi-maps

World Bank (2017b). *Understanding Poverty.* Retrieved on July 17, 2017 from http://www.worldbank.org/en/understanding-poverty

World Bank & International Monetary Fund (2016). *Development Goals in an Era of Demographic Change.* Retrieved on August 24, 2016 from http://pubdocs.worldbank.org/en/503001444058224597/Global-Monitoring-Report-2015.pdf

Zill, N. & Wilcox, B. W. (June 18, 2015). *Race, Ethnicity, and Family Stability in Red and Blue America.* Institute for Family Studies. Retrieved on September 6, 2017 from https://ifstudies.org/blog/race-ethnicity-and-family-stability-in-red-and-blue-america

4
RACE AND ETHNICITY IN FAMILIES

Learning Outcomes

Upon completion of the chapter, students should be able to:

1. describe some effects of racism on racial families
2. explain how the principle of third-generation interest can be applied to ethnic families
3. describe racial and ethnic families
4. describe factors that affect minority family life
5. describe the effects of prejudice and discrimination on minority families
6. describe the strategies that help reduce discrimination against minority families
7. explain the demographic status of White European Americans
8. explain the demographic status of Native American families
9. describe the main characteristics of African American families
10. describe the diversity of Hispanic American families
11. describe the diversity of Asian American families
12. explain the strengths of and challenges facing interracial families
13. describe the effect of transnational families on childrearing.

Brief Chapter Outline

Pre-test

Racial and Ethnic Family Diversity

Racial Family Diversity

Racial Classifications; Race as a Social Construct; The Sociology of Racism; Effects of Racism on Family Life

Ethnic Family Diversity

Family Ethnicity; Importance of Ethnicity

Racial and Ethnic Minority Families

Majority Group

Minority Group

Multidimensional Inequality for Minority Families

Prejudice and Discrimination against Minority Families

Prejudice against Minority Families

Types of Prejudice; Theories of Prejudice; Stereotypes

Discrimination against Minority Families

Effects of Discrimination on Minority Families; Combinations of Prejudice and Discrimination

Social Mobility of Minority Families

Types of Stratification Systems; Types of Social Mobility

Intragenerational mobility; Intergenerational mobility

Diversity Overview: Interracial Families

Racial and Ethnic Families in the United States

White American Families

Demographic Status of White American Families; White American Ethnicities

German American Families; Irish American Families; English American Families; Italian American Families; Polish American Families

Native American Families

Demographic Status of Native American Families; Family Life on Reservations

African American Families

The Great Migration of African American Families; Demographic Status of African American Families; Characteristics of African American Families

Hispanic American Families

Demographic Status of Hispanic American Families; Characteristics of Hispanic American Families

Asian American Families

Demographic Status of Asian American Families; Family Life and Marriage Patterns of Asian American Families

Global Overview: Transnational Families

Summary

Key Terms

Discussion Questions

Suggested Film and Video

Internet Sources

References

Pre-test

Engaged or active learning is a powerful strategy that leads to better learning outcomes. One way to become an active learner is to begin the chapter and try to answer the following true/false statements from the material as you read. You will find that you have ready answers to these questions upon the completion of the chapter.

Race is biologically determined but ethnicity is examined in terms of culture.	T	F
The minority group is the subordinate group.	T	F
The majority group is the dominant group.	T	F
Minority families are likely to be prejudiced and discriminated against.	T	F
Prejudice is a feeling or an attitude and discrimination is a treatment or an act.	T	F
An example of prejudiced nondiscriminatory action is when a hiring manager is prejudiced but offers a minority person the job due to affirmative action.	T	F
The largest European ancestries include German, Irish, English, Italian, and Polish.	T	F
Transnational families refer to families that have two separate living arrangements with close links in two or more countries.	T	F
According to Hansen, ethnicity is preserved among immigrants, weakens among their children, and returns with their grandchildren.	T	F
Institutional racism is expressed in the daily operations of social institutions.	T	F

Upon completion of this section, students should be able to:
LO1: Describe some effects of racism on family life.
LO2: Explain how the principle of third-generation interest can be applied to ethnic families.

White nationalism refers to the belief that White people seek to develop and maintain a White national identity (Conversi, 2004). White nationalist groups espouse **White separatism**—the pursuit of a "White-only state"—and **White supremacy**—the belief that White people are superior to non-Whites (Loftis, 2003). The violence in Charlottesville, Virginia, on August 12, 2017, attracted the attention of the entire world. White nationalists and counter-protesters clashed violently during a White nationalist march through the small college town. At the peak of the violence, a car plowed into a group of counter-protestors, killing one person and injuring nineteen others (Sotomayor *et al.*, 2017). Many people across the country were surprised that White nationalists had walked down an American street bearing torches and showing their faces in defense of something so abhorrent to so many. In response to the tragic events in Charlottesville, a multiracial coalition of student and community activists was organized to march against White supremacy. The group marched in the name of justice and equality from Charlottesville to Washington, DC, from August 28 to September 6.

Considering the persistence of racism in the United States, it is necessary to explore race and ethnicity in families. This section discusses the concepts of race and ethnicity and explores racial and ethnic family diversity. We discuss how race and ethnicity affect family life and describe the racism that many minority families face in the United States.

Racial and Ethnic Family Diversity

Have you ever been asked to check off your race or ethnicity in a box on an application form? Race and ethnicity are used in daily interactions as a way to categorize families as racially and ethnically distinct. Let's examine some aspects of racial and ethnic family diversity.

Racial Family Diversity

Race refers to a group of people who share similar biological characteristics such as skin color, hair texture, and eye shape. The meanings of race have changed over time. The term race was used to refer to a group of people based on ancestral ties and then to denote national affiliations. From the nineteenth century, the term race was used to denote genetically differentiated human populations defined by phenotype (Keita *et al.*, 2004). The scientific consensus is that race does not exist as a biological category among humans (Clair & Denis, 2015). Let's examine the origin of racial classifications.

Racial Classifications

Racial classifications were associated with the global distribution of humans. The traditional "out of Africa" model posits a dispersal of modern *Homo sapiens* across Eurasia as a single wave at ~60,000 years ago and the subsequent replacement of all indigenous populations. Bae *et al.* (2017) have discovered growing evidence for multiple dispersals predating 60,000 years ago in regions such as southern and eastern Asia. Western history is divided into classical antiquity, the Middle Ages or medieval period, and the modern period. During the *classical age* between the eighth century BCE and the fifth or sixth century CE centered on the Mediterranean Sea, ancient Greek and Roman society influenced Europe and other parts of the world. The term "race" referred to a group of people based on ancestral ties. Classical civilizations from Rome to China tended to invest the most importance in familial or tribal affiliation than in physical appearance (Goldenberg, 2003).

During the Middle Ages from the fifth to the fifteenth century, the English word "race" (derived from the Spanish *raza*, meaning "breed"), along with many of the ideas now associated with the term, were products of the European era of exploration (Smedley, 1999). During medieval travel (1241–1438), European expeditions crossed *Eurasia*—a combined continental landmass of Europe and Asia. Beginning with the *Silk Road* (i.e., an ancient network of the trade routes that connected the East and West), the Eurasian view of history sought to establish genetic, cultural, and linguistic links between European and Asian cultures of antiquity. In fact, much interest in this area lies with the presumed origin of the speakers of the *Proto-Indo-European language* (i.e., linguistic reconstruction of the hypothetical common ancestor of the Indo-European languages) and chariot warfare in Central Eurasia (Beckwith, 2009).

During the early modern period from the sixteenth to eighteenth century, the Age of Discovery marked the emergence of extensive overseas exploration, the beginning of globalization, and the adoption of colonialism. The term "race" was used to describe humans and societies in the way we understand national identity, or ethnic group. In this way, the idea of race came about during the historical process of exploration and conquest which brought Europeans into contact with groups from different continents (Marks, 2008). Biological and geological scientists are interested in natural history and arrange data from their explorations into categories according to certain criteria. For example, German and English scientists Bernhard Varen (1622–50) and John Ray (1627–1705) classified human populations into categories according to stature, shape, food habits, and skin color, along with any other distinguishing characteristics (Smedley, 1999). German anthropologist Johann Friedrich Blumenbach (1752–1840) divided the human species into the Caucasian, the Mongoloid, the Malayan, the Negroid, and the American Indian races in 1779. However, he did not propose hierarchy among the races (Graves, 2001) as he concluded that Africans were not inferior to the rest of mankind (Hitt, 2005). Many geographic, climatic, and historical factors have contributed to the patterns of human genetic variation in the world (Race, Ethnicity, and Genetics Working Group, 2005). No eighteenth-century dictionary defined "race" in the modern sense of a subdivision of the human species, identified by a shared appearance and other inherited traits (Hudson, 1996).

Starting from the nineteenth century, the term "race" was often used to denote genetically differentiated human populations defined by phenotype (Keita *et al.*, 2004). In Colonial America, the early Europeans encountered people from different continents and wondered whether social groups with physical and cultural differences could have corresponding mental differences. Swiss American biologist and geologist Louis Agassiz (1807–73) came up with **scientific racism** (race biology)—the pseudoscientific belief that empirical evidence exists to support or justify racism, racial inferiority, or racial superiority (Gould, 1981). Since the second half of the twentieth century, scientific racism has been criticized because it helped legitimate the domination of the globe by Whites (During, 2005).

Race as a Social Construct

Race only exists as a social construction within a network of force relations (i.e., social background of inequality against which all power interactions are played out) (Baldwin, 1963). Race is socially constructed and was redefined to justify **racial hierarchy**—a stratification system on the basis of the belief that some racial groups are superior to other racial groups. In Colonial America, for example, racial groups were placed into a hierarchy, with White or lighter-skinned people at the top, and Black and Natives at the bottom of the racial hierarchy. Therefore, the social construction of race allowed the early English to subjugate those who were not of Caucasian descent. In the United States, the root term **Caucasian** has often been used in a different, societal context as a synonym for "White" or of "European ancestry" (Bhopal & Donaldson, 1998; Baum, 2008). **Racial formation** refers to the process by which racial groups are created and manipulated. The decennial censuses conducted since 1790 in the United States created an incentive to establish racial categories and fit people into those categories (Nobles, 2000). In many parts of the world, the idea of race became a way of rigidly dividing groups by use of culture as well as physical appearances (Hannaford, 1996).

Comparing the United States with South Africa and Brazil, the National Research Council (2001) finds that historically, relations between African and European descendants fell along a continuum of legal racial domination. During the late nineteenth century and early twentieth century (1850–1930), the tension between some who supported racial superiority and others who pursued human equality was paramount. Three social and political developments (the coming of mass democracy, the age of imperialist expansion, and the impact of Nazism) that occurred during the period led to the decline in racial studies (Malik, 1996). For example, as a reaction to the rise of Nazi Germany and its prominent espousing of racist ideologies in the 1930s, Huxley and Haddon (1935) criticized the use of race to justify the politics of "superiority" and "inferiority." In 1978, the general assembly of the United Nations Educational, Scientific and Cultural Organization (UNESCO) published a collective "Declaration on Race and Racial Prejudice" and stated that "all human beings belong to a single species." Biologists, geneticists, and anthropologists long ago reached a common understanding that race is *not* a "scientific" concept rooted in discernible biological differences (National Research Council, 2001). However, skin tone is a status characteristic used in society to evaluate and rank the social position of minorities (Hargrove, 2018).

The Sociology of Racism

The sociology of racism is the study of the relationship between racism, racial discrimination, and racial inequality (Clair & Denis, 2015). **Racial discrimination** concerns the unequal treatment of racial groups. Under slavery in the United States, for example, a slave was counted as three-fifths of a person. The system of slavery was in place until the Emancipation Proclamation in 1863, but 100 years later, Martin

Luther King Jr., in his "I Have a Dream" speech, spoke to the enduring racism of the nation when he visualized a day when children were not judged by their skin color but by the content of their character. **Racial inequality** concerns unequal outcomes in terms of income, job, wealth, health, etc. Race determined where people could live and work, purchase a home, and attend school. Today we continue to reflect on the ongoing history of race relations. For example, race matters in hiring. The unemployment rate for Black college graduates sits at 4.0 percent, compared to 2.6 percent for White college graduates (Gould, 2017).

When you combine power with racial discrimination, the result is racism (Greene, 1998). **Racism** refers to the belief that one race is superior to other races. At root, racism is "an ideology of racial domination" (Wilson, 1999) in which the presumed biological or cultural superiority of one or more racial groups is used to justify or prescribe the inferior treatment or social position(s) of other racial groups (Clair & Denis, 2015). American sociologist W.E.B. Du Bois (1868–1963) identified the color line as the central social problem in the twentieth century (1903). He argued that skin color and hair texture both factored very heavily in terms of the privileges and opportunities American families are given. Contemporary approaches to racism explore persistence of racial inequality in the twenty-first century. Sociologists viewed racism as fundamentally rooted in political, economic, and/or status resource competition (Blalock, 1967). Indeed, racism has material effects on family life and occurs at a number of levels.

Effects of Racism on Family Life

With ideologies of racial domination as a direct cause, we can explain persistent racial inequalities in criminal sentencing, income, health, and wealth (Clair & Denis, 2015). **Institutional racism** is expressed in the daily operations of social institutions that maintain racial inequality. **Genocide**—intentional action to destroy an ethnic group—is a classic example of institutionalized racism which led to the death of millions of people based on race. The Holocaust was genocide during World War II in which Nazi Germany systematically murdered some 6 million European Jews between 1941 and 1945 (Gilbert, 1985).

In the United States, an educational institution is a place where people of different races gain an education. However, the college education entitlement of the 1944 G. I. Bill, although available to all veterans, imparted an unintended advantage along racial lines. Many minority veterans had not achieved a high school education, a requirement for college eligibility. Today the funds are not equally distributed in schools and racial minorities are more likely to attend inadequately financed schools in low-income communities. Over 526 teachers in Arizona have already quit this year due to low pay, a heavy workload, and not feeling supported (Roberts, 2017).

Using a census of all single-family tax-appraised homes in Harris County in Houston, Howell and Korver-Glenn (2018) have examined the influence of neighborhood racial composition on home values. The findings show that variation in appraisal methods coupled with appraisers' racialized perceptions of neighborhoods perpetuates neighborhood racial disparities in home value.

Contemporary approaches to racism also focus on the criminal justice system that accounts for unequal treatment and outcomes—the use of race as a criterion in the surveillance, questioning, searching, and arresting of suspects; jury selection; and differential punishment of minorities. Delgado (2018) has analyzed the state's guidelines for determining "reasonable suspicion" implemented by the Maricopa County Sheriff's Office (MCSO) in 2010 and 95 press releases from the desk of MCSO's head sheriff, Joe Arpaio, from 2011. The findings show that these discourses have enabled racial profiling, racial discrimination, and racial attacks on the Latino/a community in the Phoenix metropolitan area. Indeed, the use of this color-blind discourse masks state-sanctioned White supremacy perpetrated by the MCSO. In

2016, incarceration rates were higher for Black men (41.3 percent) than for White men (39 percent) (Bureau of Justice Statistics, 2018).

Environmental racism occurs when racial families are forced to live in proximity of a hazardous environment because they are not given access to healthy places in which to live. For example, in Chester, Pennsylvania, social, political, and economic forces shape the proportionate distribution of environmental hazards in poor communities of color (Cole & Foster, 2001). The slow government response to Hurricane Katrina in 2005, where a large number of those affected were African American, is another example of the vulnerability of racial families (Burns & Thomas, 2004). Racial minorities are more likely to have health problems because of their exposure to high-risk environments that are harmful to physical well-being and experience worse health than the majority population. The causal factors that contribute to the differences include socioeconomic position, health-risk behaviors, access to health care, culture and acculturation, genetic factors, and environmental and occupational exposures (National Research Council, 2001). Bravo *et al.* (2016) have explored *racial isolation* and air pollution—a measure of the extent to which minority racial/ethnic group members are exposed to only one another, and long-term particulate matter with an aerodynamic diameter of $< 2.5\ \mu$ (PM2.5)—in nonurban areas. The results show the strongest association between racial segregation and PM2.5 in Midwest rural communities. Indeed, long-term exposure to the pollutant is associated with racial segregation, with more highly segregated areas suffering higher levels of exposure. A new report from the Environmental Protection Agency finds that people of color are much more likely to live near polluters and breathe polluted air (Newkirk II, 2018).

Ethnic Family Diversity

The immigrants to the United States from different parts of Europe, Africa, and Asia began to mix among themselves and with Native Americans in the 1800s. Efforts to sort the increasingly mixed population of the United States into discrete categories generated many difficulties (Spickard, 1992). For example, many millions of children born in the United States have belonged to a different race than one of their biological parents. As the problems surrounding the word "race" became increasingly apparent during the twentieth century, the term "ethnicity" was promoted as a way of characterizing the differences between groups (Huxley & Haddon, 1935). **Ethnicity** refers to a group of people who identify with each other based on a shared cultural heritage or national origin.

Family Ethnicity

Family ethnicity is the sum total of our ancestry and cultural dimensions: how families collectively identify the core of their beings (McAdoo, 1999). Ethnicity emphasizes the cultural differences rather than genetic variation. For example, White is considered a "race" but Italians and Germans can be considered ethnicities. Cultural differences are what set ethnic families apart from one another. For example, Italian men are careful to remove their hats when they enter a church, whereas Jewish men are careful to cover their heads when entering a synagogue. Often, people of different ethnicities prefer certain types of foods that are a part of their cultural heritage. In neighborhoods with large populations of a certain group of people, such as Chinatown and Little Italy in New York City, families find shops and restaurants catering to their ethnicity. **Ethnic paradox** refers to the maintenance of one's ethnic ties that can assist with assimilation in society. Little Italy and Chinatown could serve as bridges between the mainstream and ethnic culture.

Importance of Ethnicity

Ethnicity is important because it is fundamental to the all-encompassing core of our identity and because it involves the unique family customs, proverbs, and stories that are passed on for generations

Upon completion of this section, students should be able to:
LO3: Describe racial and ethnic families.
LO4: Describe factors that affect minority family life.

(McAdoo, 1999). According to American historian Marcus Lee Hansen (1892–1938), ethnicity is preserved among immigrants, weakens among their children, and returns with their grandchildren (Hansen, 1938). Hansen's **principle of third-generation interest** maintains that the third generation (or the grandchildren of the original immigrants) will be more interested in their ethnicity than their parents. According to Hansen, the first generation of immigrants settle in ethnic enclaves; the second generation or American-born children try to distance themselves from their heritage; the third generation becomes more interested in their ancestry and ethnicity than their parents (Hansen, 1938). In this way, Hansen claims, "what the son wishes to forget, the grandson wishes to remember" (1938).

Racial and Ethnic Minority Families

Race and ethnicity are often used interchangeably (Oppenheimer, 2001). Ethnic groups can share a belief in a common ancestral origin (Cornell & Hartmann, 1998), which also can be a defining characteristic of a racial group (Race, Ethnicity, and Genetics Working Group, 2005). Race and ethnicity lead to the formation of racial and ethnic families who become grouped into families with minority status. Let's explore some concepts related to minority and majority status.

Majority Group

A **majority group** is a dominant group whose members have more control or power over their lives than members of a minority group. *White Anglo-Saxon Protestant (WASP)* or Old Stock Americans are people who are descended from the original settlers of the thirteen colonies, of mostly northwestern European ancestry, who immigrated in the seventeenth and eighteenth centuries (Hirschman, 2005). This high-status majority group once controlled disproportionate financial, political, and social power in the United States. Sociologists use the term very broadly to include all Protestant Americans of northern European or northwestern European ancestry regardless of their class or power (Glassman *et al.*, 2004).

In a sociological context, *majority* does not necessarily relate to having a numerical majority; rather, majority is examined in terms of power. Some groups may have greater numbers but are still considered a minority. For example, *apartheid* in South Africa was a system of racial segregation. The Black inhabitants of the country were a numerical majority but were oppressed by the small number of Whites. When it comes to gender, men have more power and rights than women and are considered to be the majority group even if women outnumber men. The number of females in the United States was 164 million and the number of men was 159.1 million in 2016 (U.S. Census Bureau, 2018a).

Minority Group

A **minority group** is a subordinate group whose members have less control or power over their lives than do members of the dominant group. A minority group has certain characteristics: (1) unequal treatment, (2) distinguishing physical or cultural traits, (3) involuntary membership, (4) awareness of subordination, and (5) in-group marriage (Wagley & Harris, 1958), although interracial marriage is accepted today.

The term minority is *multidefinitional* and can be examined in terms of race and ethnicity, gender, social class, age, sexual orientation, disability, religion, etc. When it comes to race, both historical and contemporary forces have shaped the four minority groups in the United States: African American, Hispanic American, Asian American, and Native American (Taylor, 2001).

Multidimensional Inequality for Minority Families

Intersectionality theory is a feminist sociological theory developed by Kimberlé Williams Crenshaw (1959–) that examines the experiences of minority people who are subjected to multiple forms of subordination in society. Intersectionality is a theoretical framework that posits that multiple social categories (e.g., race, ethnicity, gender, sexual orientation, religion, socioeconomic status) intersect at the micro level of individual experience to reflect multiple interlocking systems of privilege and oppression at the macro, social-structural level (e.g., racism, sexism, heterosexism) (Bowleg, 2012). The intersectionality theory explores how biological factors such as race and gender, social factors such as social class, and cultural factors such as ethnicity interact on multiple levels and provides a multidimensional approach to the study of *social inequality*—unequal outcomes (employment, income, wealth, health, etc.).

Let's examine how being a woman and therefore a member of the minority gender affects family life. Families headed by women are likely to be relegated to a minority position. In 2017, the median income of family households maintained by men with no wife present was $60,843, compared with $41,703 for those maintained by women with no husband present (Fontenot *et al.*, 2018). Meanwhile, the experience of minority women is more powerful than the sum of their race and sex (Crenshaw, 1989). Black women have to work seven months into 2017 to be paid the same as White men in 2016 (Wilson *et al.*, 2017). Latina women only make 54 cents for every dollar a White man makes (Bibler, 2015). In 2017, among women, Whites ($795) earned 88 percent as much as Asians ($903); Blacks ($657) earned 73 percent; and Hispanics ($603) earned 67 percent (U.S. Bureau of Labor Statistics, 2018a). Transgender women and disabled women are likely to experience other forms of minority status.

Social class is another marker of minority families. Lower-class families are less likely to accumulate assets, own a house, and protect their family from economic setbacks. Association with a minority religion is another basis for minority families. Protestantism is the dominant religion in the United States. Back in the 1880s, immigrant families from southern and eastern Europe had a different religion and cultural heritage from the English and received unequal treatment. For example, drawing on sources from classical antiquity and upon their own internal interactions, the hostility between the English and Irish powerfully influenced early European thinking about the differences between people (Takaki, 2008). Today, Muslims in America often are the subject of bullying by those who wrongly consider all Muslims as being affiliated with terrorists. According to Pew Research Center's 2017 survey, U.S. Muslims are concerned about their place in society, but continue to believe in the American Dream (Pew Research Center, 2017a). Based on combinations of race, gender, age, education, social class, disability, sexual orientation, etc., a poor, older, Arab American lesbian Muslim with a disability is more likely to be subjected to multiple forms of subordination.

Student Accounts of Prejudice

It has been a challenge living in the current political atmosphere, being a practicing Muslim woman. I feel the weight of eyes on me, questioning my patriotism to the United States that I call my home. The danger that lurks behind every stare, every slur that is yelled at me every day, makes me wonder if things can get any worse (personal communication, June 20, 2017).

Upon completion of this section, students should be able to:
LO5: Describe the effects of prejudice and discrimination on minority families.
LO6: Describe the strategies that help reduce discrimination against minority families.

Prejudice and Discrimination against Minority Families

Minority families are more likely to experience prejudice and discrimination. American sociologist Robert Merton (1910–2003) formulated a typology of prejudice and discrimination to show how both are related but not the same (1949).

Prejudice against Minority Families

In a diverse society, social interactions among people of different backgrounds are the norm. However, when we pre-judge someone or a group, we make up our minds about who they are before we actually get to know them. Prejudices are not based upon actual real-life interaction with a person or group. **Prejudice** refers to negative attitudes based on oversimplified generalizations about members of a certain group. For example, if a person believes that all people over 60 don't understand technology, this is a prejudice. **Subtle prejudice** refers to an implicit and indirect expression of negative beliefs. For example, a Black man sits next to a racist White woman and she moves over slightly to get comfortable in the seat. **Overt prejudice** refers to an explicit and direct expression of negative beliefs. As a universal attitude of tolerance toward all groups has yet to be adopted, it is becoming less acceptable in our society to overtly express prejudices (Dowden & Robinson, 1993).

Types of Prejudice

Types of prejudice include gender, racial and ethnic, age, sexual orientation, social class, and disability prejudice. For example, when a hiring manager believes that people of a certain race are in some way inferior to another, this is a racial prejudice or racism. When an elder-care facility or nursing home believes that older people do not have the ability to learn web navigation, this is an age prejudice or **ageism**—prejudice on the basis of a person's age. Prejudging people because of their sexual orientation is called **homophobia**. Sexual orientation prejudice is the idea of having negative opinions about the LGBTQ group. It is sometimes based on the stereotype that all gay men or lesbians are inferior. When the upper class of society believes that lower-class students are not entitled to quality education opportunities, this is a class prejudice or **classism**—the belief that those of a certain economic class are inferior to those of another class. When people believe that people with disability are not entitled to the rights of able-bodied people, this is a disability prejudice or **ableism**—the belief that those with a physical or mental disability are inferior to able-bodied people.

Theories of Prejudice

The **minimal group model** suggests that people automatically favor those similar to them, such as those of the same race, and reject those not similar to them. This theory of prejudice proposes that the determinant of a person or a group's attitude toward another is the similarity between the two persons or two groups' racial and ethnic background. Dissimilarity is presumed to lead to rejection. Several predictors of prejudice include social dominance orientation (Guimond *et al.*, 2003), religious fundamentalism (Altemeyer & Hunsberger, 1992), egalitarian beliefs (Dovidio & Gaertner, 1986), and a belief in fixed human character (Hong *et al.*, 2004). Individuals with higher levels of religious fundamentalism expressed more negative attitudes towards both *in-group* (i.e., group members with a shared interest or racial identity) and *out-group* people (i.e., those who do not belong to a specific in-group); individuals with higher levels of egalitarianism expressed more positive attitudes towards out-groups (Dudley & Mulvey, 2009). **Symbolic interactionist theory** emphasizes the role of social interaction in reducing racial and ethnic prejudice. The term **intergroup relationship (IGR)** refers to relationships among people from different social, economic, racial, and ethnic groups. American psychologist Gordon Allport (1897–1967) developed the contact hypothesis known as *intergroup contact theory* (Allport, 1954). The **contact hypothesis** suggests that interpersonal contact is one of the most effective ways to improve relations among groups who experience conflict and to reduce prejudice against minority group members. When it comes to formation of intergroup relationships, **structural functionalist theory** supported both assimilation and pluralism, two contrary pathways in response to family diversity. When people of different racial or ethnic origins come into contact with each other, such interracial or interethnic contact is likely to reduce prejudice. According to **conflict theory**, racial and ethnic prejudice is tied to social class conflict. Conflict theory argues that racial inequality among racial families must be reduced in order to lessen racial prejudice and class conflict in society.

Stereotypes

Prejudice stems from **stereotypes**—oversimplified generalizations about members of a certain group. Since the 9/11 attacks in 2001, for example, many Americans have viewed all Muslim Americans as terrorists. Based on the actions of others, the whole group of people gets stereotyped. The social consequences of expressing negative attitudes based on race are severe. **Racial profiling** refers to suspecting a person of a certain race based on a stereotype about their race. For example, in late 2014, a series of highly publicized police killings of unarmed Black male civilians in the United States prompted social turmoil (Hall *et al.*, 2016). Some states such as Arizona and Texas have issued immigration laws, allowing police officers to ask for papers just on suspicion that someone might be illegal in the United States. Many Hispanics face racial profiling when they are pulled over multiple times while driving.

Many people see fatal encounters with police as signs of a broader problem but racial groups differ in their views of police officers. The Pew Research Center (2017a) reports that about two-thirds of the public (64 percent) give officers a warm rating on the scale (between 51 and 100) but Whites and Blacks differ widely in their views. A majority of Whites (74 percent) give law enforcement warm ratings compared with African Americans (30 percent) and Hispanics (55 percent).

Discrimination against Minority Families

The process of favoring those similar to them and rejecting those not similar to them occurs on both an individual and group level (Hogg, 1987), and often results in discrimination (Mullin & Hogg, 1998). When a person acts on his or her prejudice, this action or unequal treatment of different group members on the basis of race, age, gender, etc. is **discrimination**. Therefore, prejudice is an attitude whereas discrimination is an action.

Different kinds of prejudice lead to different forms of discrimination. For example, if a nursing home refused to install internet technology because it believed that older people do not know how to learn web navigation, it would be acting on an age prejudice and would be discriminating against the older adults. If government limited the rights of people with disability to basic necessities that able-bodied people take for granted, such as employment and education, it would be acting on a disability prejudice and would be discriminating in favor of able-bodied people. Racial discrimination concerns the unequal treatment of races in terms of jobs, education, etc. If a hiring manager refused to hire those of a certain race because he believes they are in some way inferior to another, this person would be acting on a racial prejudice.

Effects of Discrimination on Minority Families

Discrimination affects family life in many ways. **Individual discrimination** takes place in one-to-one encounters and occurs when someone practices prejudicial treatment of another. Consider the following example: Jason is a young White father. He views a potential rental property and is quoted a price of $1,200 per month rent. However, when Maria, a Hispanic single mother, visits the same location, the property manager quotes a price of $2,000 per month rent for the unit. In this example, Maria is a victim of individual discrimination.

Institutional discrimination is the denial of opportunities and equal rights to minority families that results from day-to-day operations of social institutions. For example, White Anglo-Saxon Protestant (WASP) is an elite social class of powerful White Americans of British Protestant ancestry and is often used to describe their historical dominance over the financial and legal institutions of the United States. The term White Anglo-Saxon Protestant or Anglo-Americans is used to distinguish upscale WASPs from ordinary people of various White ethnic origins. In the nineteenth century, millions of immigrants from southern, central, and eastern Europe entered the United States. The English settlers came to define "White." **White ethnic** refers to White Americans who are not White Anglo-Saxon Protestant (Marger, 2008). These White ethnics had different cultures from the English settlers and experienced discrimination in employment more than did the White immigrant families from northwestern Europe.

There are reasons why Americans prefer homeownership to renting. Owning a home can serve as a vehicle for economic mobility or a marker of status attainment. Homeownership may deepen feelings of ontological security and enable families to move into more convenient neighborhoods (McCabe, 2018). When it comes to homeownership, the Home Owners' Loan Corporation denied mortgages for houses in minority neighborhoods in the 1930s (Simms *et al.*, 2009). Women did not get the right to open bank accounts and were denied the opportunity to build credit history in their own names. The Civil Rights Act of 1964 outlawed discrimination based on race, color, religion, sex, or national origin. However, racist ideas still linger today. For example, American Honda Finance Corporation's past practices between January 2011 and July 2015 resulted in thousands of African American, Hispanic, and Asian and Pacific Islander borrowers paying higher interest rates than White borrowers for their auto loans (CFPB, 2015).

Combinations of Prejudice and Discrimination

Merton (1949) identified four combinations of prejudice and discrimination. *Unprejudiced nondiscrimination* occurs when a hiring manager is not prejudiced and does not discriminate against job applicants. *Unprejudiced discrimination* occurs when a hiring manager has no personal prejudice but engages in discriminatory behavior due, for instance, to peer pressure. *Prejudiced nondiscrimination* occurs when a hiring manager is prejudiced but still offers the job, for example due to affirmative action policies. *Prejudiced discrimination* occurs when a hiring manager is sexist or racist and engages in discriminatory behavior.

In the past centuries, the United States has tried to move away from the situation in which prejudice and discrimination was manifested by genocide of Native Americans, slavery, Jim Crow laws (mandating racial segregation in all public facilities in the 1800s), and restrictive laws against immigrants. Yet still today, minority families are likely to encounter prejudice and discrimination in some way that reduces their chances of upward social mobility.

Social Mobility of Minority Families

Sociology examines the effects of stratification systems on social mobility of minority families. Every society has a stratification system that is either open or closed based on the possibility of social mobility.

Types of Stratification Systems

There are three social stratification systems on a global basis. A **slavery system** is a closed form of stratification in which slaves are owned by slave owners. For example, slavery was practiced throughout the American colonies in the seventeenth and eighteenth centuries. A **caste system** is a closed form of stratification in which people's status is determined at birth by their parent's ascribed status. The example of a caste system is the Hindu caste system. This caste system divided society into the Brahmins (priestly people), the Kshatriyas (e.g., rulers, administrators, and warriors), the Vaishyas (e.g., artisans, merchants, tradesmen, and farmers), and Shudras (e.g., laboring classes) (Fowler, 1997). Like a slavery system, a caste system does not allow a person to move up the ladder. For example, if the father was a farmer, the son would only be a farmer. A **class system** is an open form of stratification in which people's status is based on their control of resources and on the type of work they do. In a class system, as opposed to a caste system, it is possible for people to achieve a higher socioeconomic status than the ascribed status that was granted by their family. The United States has an open class system that allows people to move up the ladder.

Types of Social Mobility

Social mobility refers to the movement of individuals, families, or groups among stratified social positions from one stratum to another. In terms of social mobility, sociology distinguishes between intragenerational social mobility and intergenerational social mobility.

Intragenerational Mobility **Intragenerational mobility** is the social movement of individual family members within their own lifetime. Intragenerational upward social mobility occurs if a person born into poverty finds a way out of poverty and works his or her way up to a higher class. As education, income, and occupation are correlated, getting a college education is one way for people to escape poverty and achieve upward social mobility. An example of someone who experienced intragenerational upward social mobility is Bill Gates, who had humble beginnings but went on to become one of the world's wealthiest men through his founding of Microsoft.

Intergenerational Mobility **Intergenerational mobility** refers to changes in social class that occur across generations. An example of someone who experienced intergenerational upward social mobility is Former President Barack Obama. Obama's grandfather worked as a mission cook and as a local herbalist in Kenya. His father was a Kenyan senior governmental economist. In January 2008, President-elect Barack Obama and his family made history, becoming the first African American family to move into the White House. However, the Stanford Center on Poverty and Inequality (2017) finds that intergenerational mobility in the United States is racially asymmetrical as Blacks who grew up in more affluent households in the 1960s and 1970s had a greater chance of moving downward compared with Whites of similar origins.

The expectation that one's economic status will improve over one's parents' and grandparents' is salient in immigrant communities, in which the first generation often must work harder to overcome numerous cultural and economic challenges (Trevelyan *et al.*, 2016). The U.S. Census Bureau (2016a) find that 37.4 percent of second-generation immigrants (native-born children of a foreign-born parent) hold a bachelor's degree or higher and 14.9 percent hold a master's degree or higher; second-generation households have a higher median household income of $51,291 compared to $45,475 for first-generation households. Figure 4.1 shows that the second generation is likely college-educated and has higher incomes than their parents' generation.

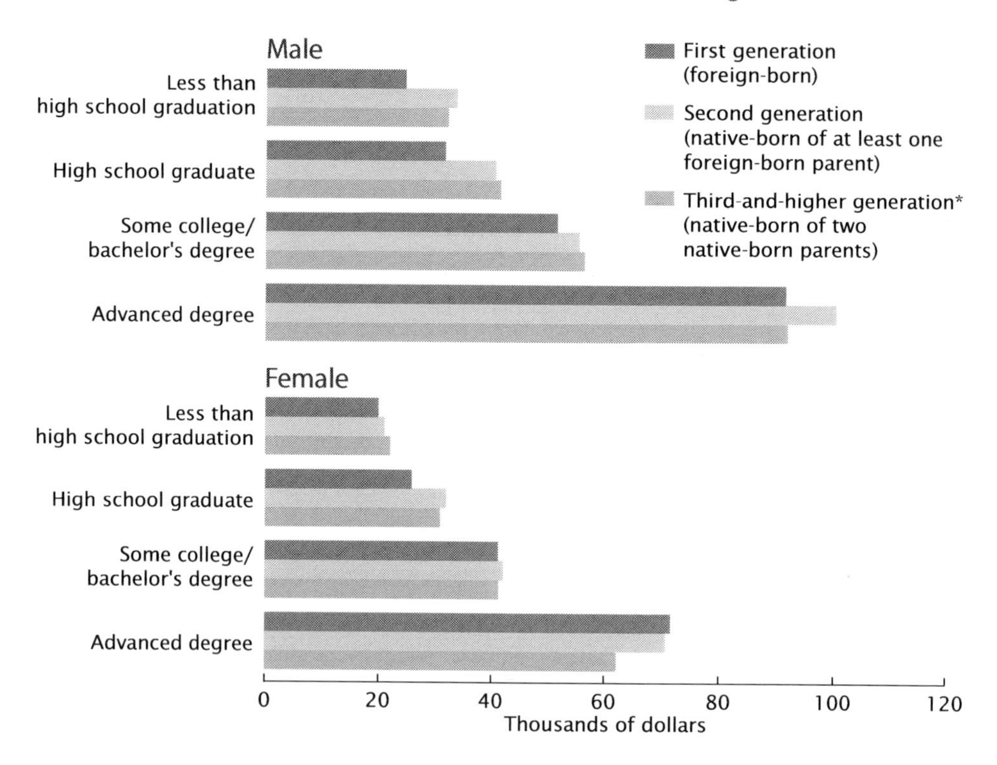

Figure 4.1 Generational Earnings and Education

Source: 2013 Current Population Survey, Annual Social and Economic Supplement, U.S. Census, November 30, 2016. https://www.census.gov/library/visualizations/2016/comm/cb16-203_earnings_education.html

The United States has tried to move toward American values of social equality and justice. **Affirmative action** is a set of procedures intended to eliminate unlawful discrimination among applicants on the basis of race, gender, or national origin. It is a policy of recruiting members of a subordinate or minority group for job and educational opportunities that they otherwise may not have had. Affirmative action was the outcome of the Civil Rights movement in the 1960s. It appeared in President Kennedy's *Executive Order 10925*, "ensuring equal opportunity for all qualified persons, without regard to race, creed, color, or national origin" (U.S. EEOC, 1961). Opponents claim that affirmative action leads to reverse

Upon completion of the next section, students should be able to:
LO7: Explain the demographic status of White European Americans.
LO8: Explain the demographic status of Native American families.
LO9: Describe the main characteristics of African American families.
LO10: Describe the diversity of Hispanic American families.
LO11: Describe the diversity of Asian American families.

Diversity Overview: Interracial Families

On June 12, 1967, the Supreme Court ruled unanimously that states cannot outlaw marriages between Whites and non-Whites. Since then, interracial marriages have increased. Today people from different races have relationships of all kinds. Using the National Longitudinal Study of Adolescent to Adult Health, Shiao (2018) has investigated the relationship between interracial friendship and interracial intimacy. The findings suggest that early interracial friendship remains a significant positive influence on the likelihood of subsequent interracial intimacy.

An **interracial family** is a form of family in which a couple identifies themselves as being of different races or ethnicities. In 2013, about one in eight new marriages (12 percent) were intermarried (Wang, 2015). In 2015, one in six of all U.S. newlyweds (17 percent) had a spouse of a different race or ethnicity, a fivefold increase from 3 percent in 1967 (Livingston & Brown, 2017). Intermarriage families vary by race. About three in ten Asian newlyweds (29 percent) married someone from another race in 2015, and the share was 27 percent among recently married Hispanics. Intermarriage families vary by state. Among states, more people who identified as being of two or more races lived in California (1.5 million) than in any other state, with an increase of 32,900 from 2015 (U.S. Census Bureau, 2017a).

There has also been a rise in multiracial and multiethnic children in the United States. One in seven U.S. infants (14 percent) was multiracial or multiethnic in 2015. In Hawaii, roughly four in ten babies (44 percent) are multiracial or multiethnic compared with 28 percent of babies in Oklahoma and Alaska and 4 percent of babies younger than one in Vermont (Livingston, 2017).

How do interracial couples choose to identify the race of their child on U.S. Census forms? Since 2000, the U.S. Census forms have allowed Americans to check more than one race box. Interracial couples living in the West, the region with the largest Asian and Pacific Islander population, were more likely to report that their child is Asian and Pacific Islander alone or in combination with another race (Cohn, 2016). A child of a White or Black male householder was more likely to be reported as the same race as the father.

discrimination. **Reverse discrimination** is discrimination against members of a dominant or majority group in favor of a historically disadvantaged group.

Racial and Ethnic Families in the United States

The United States was home to 2.5 million living people in July 1776 (U.S. Census Bureau, 2017b) and 328.7 million as of October 2018 (U.S. Census Bureau, 2018c). The United States has been built by diverse racial and ethnic families. This section explores the diversity of family life among racial and ethnic groups and their struggles to attain the equal rights that they deserve.

The racial categories included in the census questionnaire reflect a social definition of race and not an attempt to define race biologically, anthropologically, or genetically (U.S. Census Bureau, 2018d). The U.S. Census Bureau recognizes six racial and ethnic groups: American Indian or Alaska Native, White, Black or African American, Asian, Native Hawaiian and other Pacific Islander, and Hispanic American. Figure 4.2 shows a more diverse nation with diverse racial and ethnic families.

White American Families

White refers to an American who has his or her origins in Europe, the Middle East, or North Africa.

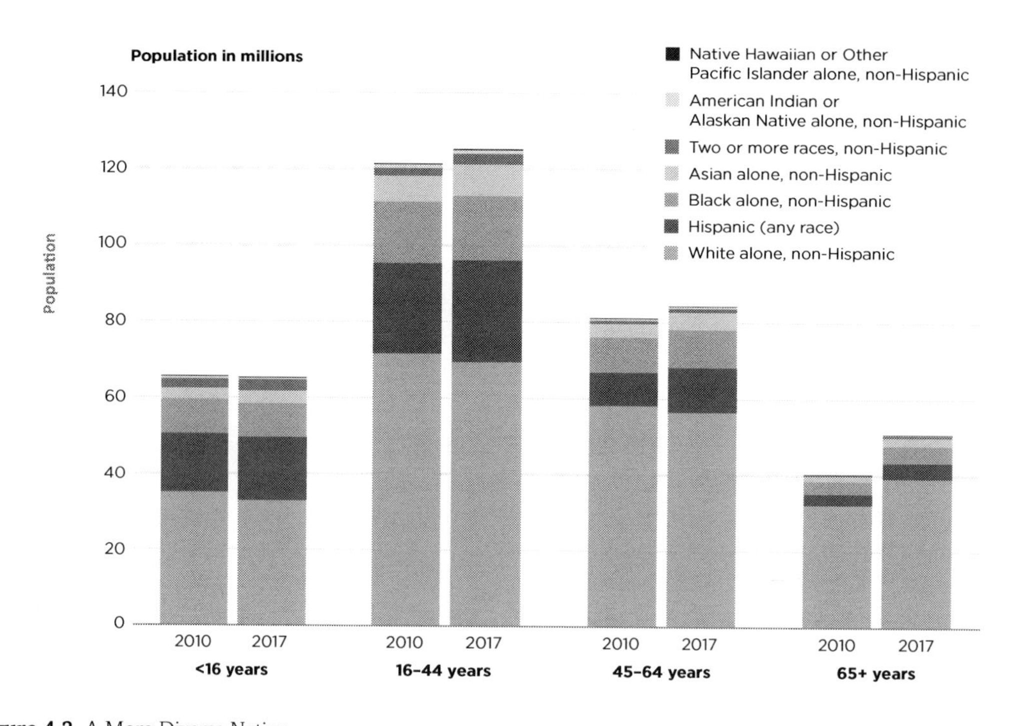

Figure 4.2 A More Diverse Nation

Source: Vintage 2017 Population Estimates. https://www.census.gov/library/visualizations/2018/comm/diverse-nation.html

White privilege refers to rights and social privileges that benefit people identified as White, compared to non-White people under the same circumstances. Social privileges are likely to benefit people identified as White. Race matters for the distribution of material resources. For example, the persistence of affluence has been stronger for Whites (Stanford Center on Poverty and Inequality, 2017).

Demographic Status of White American Families

White alone made up 76.6 percent of the total U.S. population (U.S. Census Bureau, 2017h), totaling 256 million as of July 1, 2017 (U.S. Census Bureau 2017a). In 2017, the share of White families with an employed member was 80.5 percent; White families remained less likely to have an unemployed member (5.2 percent) (U.S. Bureau of Labor Statistics, 2018b). In 2017, the poverty rate for Whites was 8.7 percent with 17 million people in poverty, compared with the official national poverty rate at 12.3 percent (U.S Census Bureau, 2018b). In 2017, 93.7 percent of Whites had health insurance coverage and were among the most likely to have private health insurance at 73.2 percent (Berchick *et al.*, 2018). Figure 4.3 shows that in 2017, the annual median household income of White Americans was $68,145, compared with $61,372 for all races.

The racial wealth gap is larger than the income gap. Social policy has created and maintained the racial wealth gap and has been likely to help White Americans with the materials to build wealth (Gould, 2017). Figure 4.4 shows that average wealth for

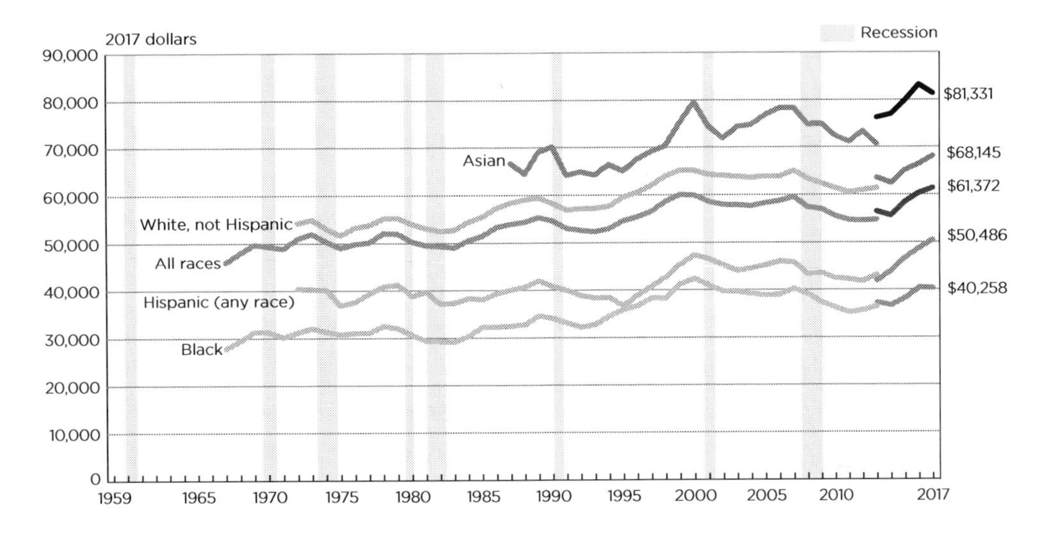

Figure 4.3 Real Median Household Income by Race and Hispanic Origin, 1967–2017
Source: U.S. Census Bureau, Current Population Survey, 1968–2018, Annual Social and Economic Supplements. https://www.census.gov/library/publications/2018/demo/p60-263.html

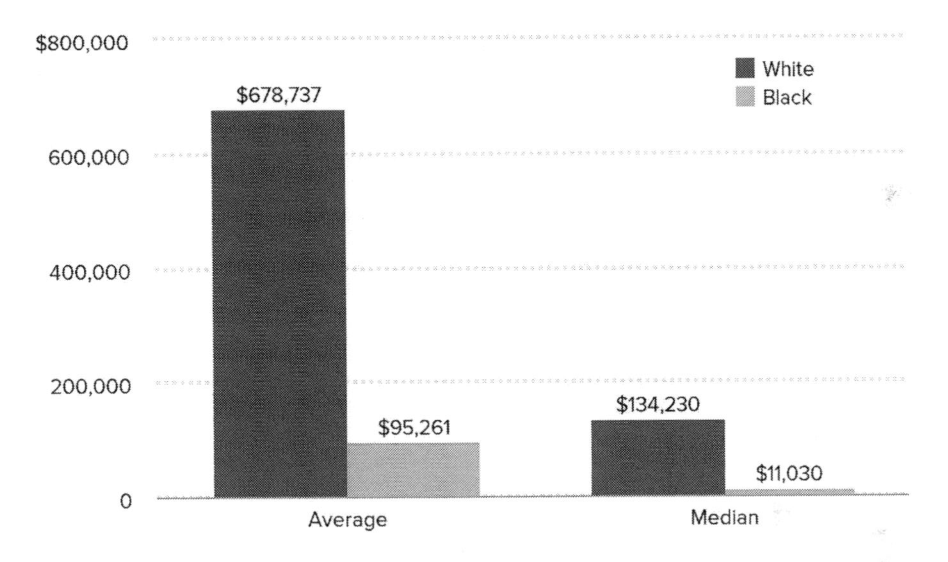

Figure 4.4 Median and Average Wealth, by Race
Source: Survey of Consumer Finance Combined Extract Data, 2013, Economic Policy Institute. http://www.epi.org/blog/the-racial-wealth-gap-how-african-americans-have-been-shortchanged-out-of-the-materials-to-build-wealth/

White families is seven times higher than average wealth for Black families. The median White wealth (wealth for the family in the exact middle of the overall distribution—wealthier than half of all families and less-wealthy than half) is *twelve times* higher than median Black wealth.

The racial wealth gaps are growing. Whites tend to earn more than non-Whites partly due to disparities in human capital and educational attainment (Stanford Center on Poverty and Inequality, 2017). However, even after taking education level and other factors into account, average wages for White college graduates are far higher ($31.83 per hour) than average wages for Black college graduates ($25.77) (Gould, 2017).

The White populations were more likely to hold a bachelor's degree or higher, 37.3 percent (U.S. Census Bureau, 2017c). Figure 4.5 shows a steady increase in high school and bachelor's degree attainment for the White population age 25 and over from 1974 to 2017.

White American Ethnicities

The terms "race," "ethnicity," and "ancestry" all describe just a small part of the complex web of biological and social connections that link individuals and groups to each other (Race, Ethnicity, and Genetics Working Group, 2005). White Americans have a variety of ethnicities. White includes Scandinavians, and eastern, western, northern, and southern Europeans. An alternative to the use of White Americans is to categorize individuals in terms of ancestry. **Ancestry** refers to a person's ethnic origin or descent, "roots," heritage, or the place of birth of the person or the person's parents or ancestors before their arrival in the United States (U.S. Census Bureau, 2017d). The five largest ancestries of White Americans are German, Irish, English, Italian, and Polish Americans.

German American Families People claiming German ancestry ranked first (43 million), making up 14.9 percent of the total American population (U.S. Census Bureau, 2017d/2017f). Germans were among the first settlers in the original thirteen colonies, bringing their talents and ideas across the ocean to a new and unfamiliar world (White House, 2014). German families initially landed in Philadelphia in the 1700s. Between 1820 and 1880, millions of German immigrants came to the United States following the

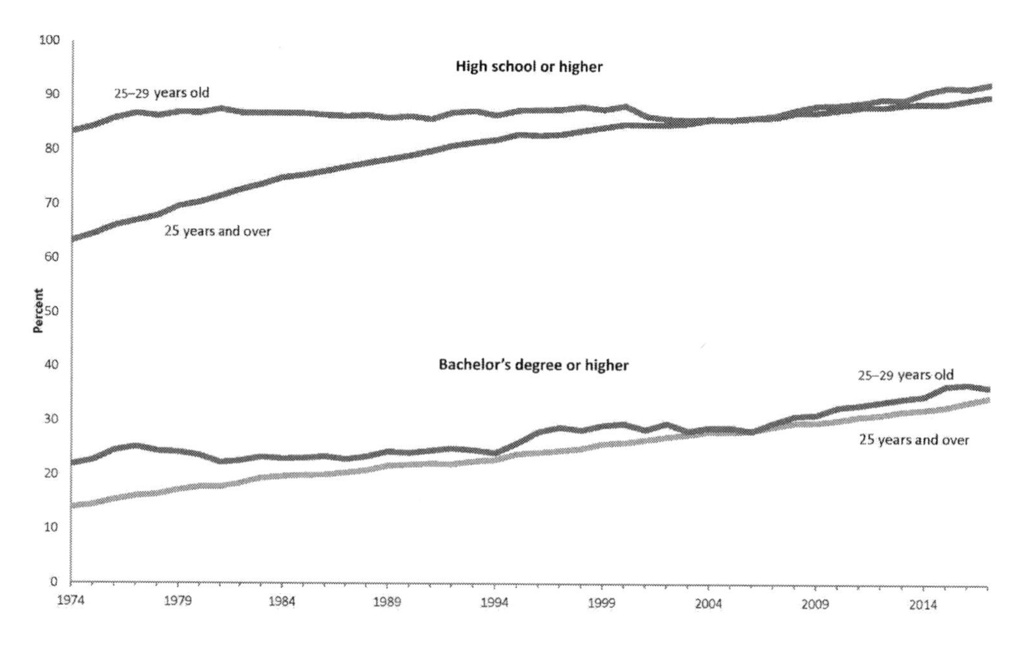

Figure 4.5 High School and Bachelor's Degree Attainment for White Population Age 25 and over, 1974–2017
Source: U.S. Census Bureau 1974–2002, March, Current Population Survey, 2003–2017 Annual Social and Economic Supplement to the Current Population Survey. https://www.census.gov/library/visualizations/time-series/demo/cps-historical-time-series.html

German revolutions of 1848–49. Contributions from German families to the United States are well-known. For example, German-born scientist Albert Einstein (1879–1955) developed the theory of relativity and received the Nobel Prize in physics in 1921. Einstein even predicted the existence of gravitational waves, which in 2016 scientists announced that they had detected, confirming the prediction Einstein made a century earlier. Former President Barack Obama proclaimed October 6, 2014, as German American Day.

Irish American Families Irish was the nation's second-most frequently reported European ancestry. Irish Americans (31.4 million) made up 10 percent of U.S. residents in 2017 (U.S. Census Bureau, 2018e/2017f). The Great Famine was a period of mass starvation, disease, and emigration in Ireland between 1845 and 1852 (Kinealy, 1994). Most Irish came to the United States to escape a devastating famine, as potato fields in their native land were riddled with disease that destroyed crops. Most Irish immigrants to the United States were Catholic (unlike the majority of English immigrants, who were Protestant) and they favored large cities such as Boston, Philadelphia, and New York (Diner, 1983). Their presence provoked a reaction among native-born Americans or nativists who denounced the Irish immigrants for their distinct culture and their Catholic religion. **Nativism** is the policy of favoring the native inhabitants over the immigrants. These Irish immigrant families took jobs left behind by German laborers. After a period of acculturation, Irish American families began to improve their economic and social positions. In 2017, the median income for households headed by an Irish American is $69, 944, and 68.4 percent of householders of Irish ancestry own the home in which they live, with the remainder renting in 2015 (U.S. Census Bureau, 2017e). Congress proclaimed March as Irish American Heritage Month in 1991. St. Patrick's Day, a religious holiday to honor St. Patrick, who introduced Christianity to Ireland in the fifth century, has evolved into a celebration of all things Irish (U.S. Census Bureau, 2018e).

English American Families People claiming English ancestry ranked third (23.0 million), making up 7.8 percent of the U.S. population in 2017 (U.S. Census Bureau, 2017b/2017f). Some are descendants of the original Plymouth colonists who participated in the autumn feast that is widely believed to be one of the first Thanksgivings (U.S. Census Bureau, 2016b). Some English Americans are White Anglo-Saxon Protestants (WASPs) who had ancestors in high-status and influential British Protestant families from northern Europe. Until at least the 1940s, WASPs formed the ruling class of the United States with a large role in Republican Party leadership, as well as finance, business, law, higher education, and especially high society (Kaufmann, 2004). Military, foreign affairs, and top political offices were once dominated by WASPs.

Italian American Families Italian Americans are the fourth largest ancestry group and are an ethnic group consisting of Americans who have ancestry from Italy. There were 16.6 million people claiming Italian ancestry in the United States in 2017, making up 5.5 percent of the U.S. population (U.S. Census Bureau, 2018f/2017f). Christopher Columbus set foot on American soil in 1492. Italian unification in 1861 caused economic conditions to worsen for many in the former Kingdom of the Two Sicilies, the largest of the states of Italy (De Sangro, 2003). Many Italian families began their journey to America. From 1880 to 1915, an estimated 13 million Italians migrated out of Italy, making Italy the scene of the largest voluntary emigration in recorded world history (Choate, 2008). These new arrivals thought of themselves as Neapolitans, Sicilians, Calabrians, or Syracuseans. They might not have understood each other's dialects, but on arrival in the United States they became Italian Americans (U.S. Census Bureau, 2018f). Most Italians began their new lives and worked as manual laborers in mining camps and on farms (Nelli, 1980). Many sought housing in *Little Italy*—a name for the ethnic enclave populated primarily by Italians or

people of Italian ancestry. The descendants of the Italian immigrants excelled in all fields of endeavor, and made substantial contributions in virtually all areas of American life.

Polish American Families Polish Americans are the fifth largest ancestry group. There were 9.0 million people claiming Polish ancestry in the United States in 2017, making up 3 percent of the U.S. population (U.S. Census Bureau, 2017f). At the turn of the twentieth century, imperial repression, land shortages, and chronic unemployment made life more and more untenable for the Poles of Europe, and they left for America by the thousands, then by the hundreds of thousands (U.S. Census Bureau, 2018g). The largest wave of Poles arrived from 1870 to 1914; the second wave came after World War II; and the third wave came after Poland's independence in 1989. Early Polish immigrants often accepted jobs that many other Americans did not want to take either because of the low pay or unsafe working conditions. Fifty-one percent of Polish Americans are now married and the median household income for Polish American families is $73,452 (U.S. Census Bureau, 2017g). Polish American Heritage Month is an annual event celebrated in October by Polish American communities.

Native American Families

Native Americans are also known as American Indians. **American Indian** or **Alaska Native** refers to a person having origins in any of the original peoples of North and South America (including Central America) and maintaining tribal affiliation or community attachment (U.S. Census Bureau, 2018d). The first American Indian Day was celebrated in May 1916 in New York.

Demographic Status of Native American Families

In 2016, American Indians and Alaska Natives totaled 6.7 million, making up about 2 percent of the total U.S. population (U.S. Census Bureau, 2017i). Native American families are the most disadvantaged minority in terms of education, occupation, and income. There were 841,943 American Indian and Alaska Native households in 2016; their median household income was $39,719, compared with $57,617 for the nation; 52.9 percent of Native American householders owned their own home compared with 63.1 percent for the nation as a whole; 7.2 percent of Native American grandparents were living with at least one of their grandchildren under the age of eighteen; 26.2 percent of Native Americans were in poverty compared with 14.7 percent for the nation as a whole (U.S. Census Bureau, 2017i). One major factor behind these high poverty rates and low wealth is the low rate of employment of Native Americans. Childhood poverty declines when working parents are able to find quality jobs with a decent wage and benefits including childcare and paid family leave (Jones, 2017b). Figure 4.6 shows that Native American children continue to face the highest poverty rates, hovering around 33.8 percent. Native American children are three times more likely to be in poverty than White children in 2016.

Racial and ethnic groups can exhibit substantial average differences in disease incidence, disease severity, disease progression, and response to treatment (LaVeist, 2002). Native Americans suffer from higher rates of diabetes, tuberculosis, pneumonia, influenza, and alcoholism than does the rest of the U.S. population (Mahoney & Michalek. 1998), plus 19.2 percent of American Indians and Alaska Natives lacked health insurance coverage in 2016 compared with 8.6 percent for the nation as a whole (U.S. Census Bureau, 2017i). In 2017, Native Americans had the lowest labor force participation rate and made up 1 percent of the labor force, compared with 78 percent for Whites (U.S. Bureau of Labor Statistics, 2018c). Figure 4.7 shows that the estimated number of American Indian and Alaska Native-owned employer firms was 26,757.

Family Life on Reservations

In 1876, the U.S. government forced Native American families to live on reservations—areas that have been set aside for the use of the Native American tribes and which can be legally described as communities. In 2017, there were 567 federally recognized Indian tribes in the United States and 326 federally recognized American Indian reservations (U.S. Census Bureau, 2017i); 22 percent of American Indians and Alaska Natives live on reservations or other trust lands (OMH, 2016). Of 841,943 households, 37.9 percent were married-couple families, including those with children. Of those living on reservations in 2016,

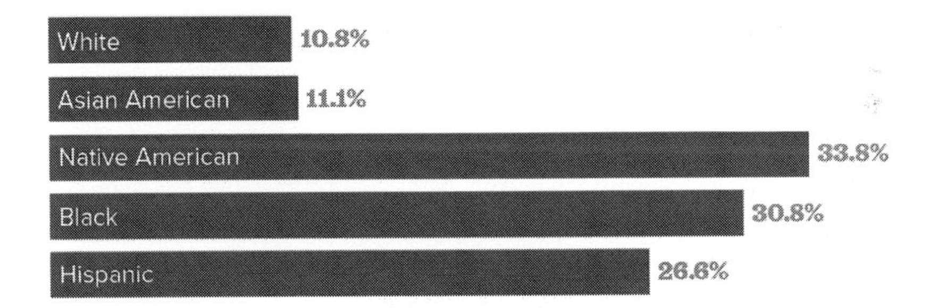

Figure 4.6 Share of Children in Poverty by Race and Ethnicity, 2016

Source: American Community Survey, 2015–2016, Economic Policy Institute. http://www.epi.org/publication/one-third-of-native-american-and-african-american-children-are-still-in-poverty/

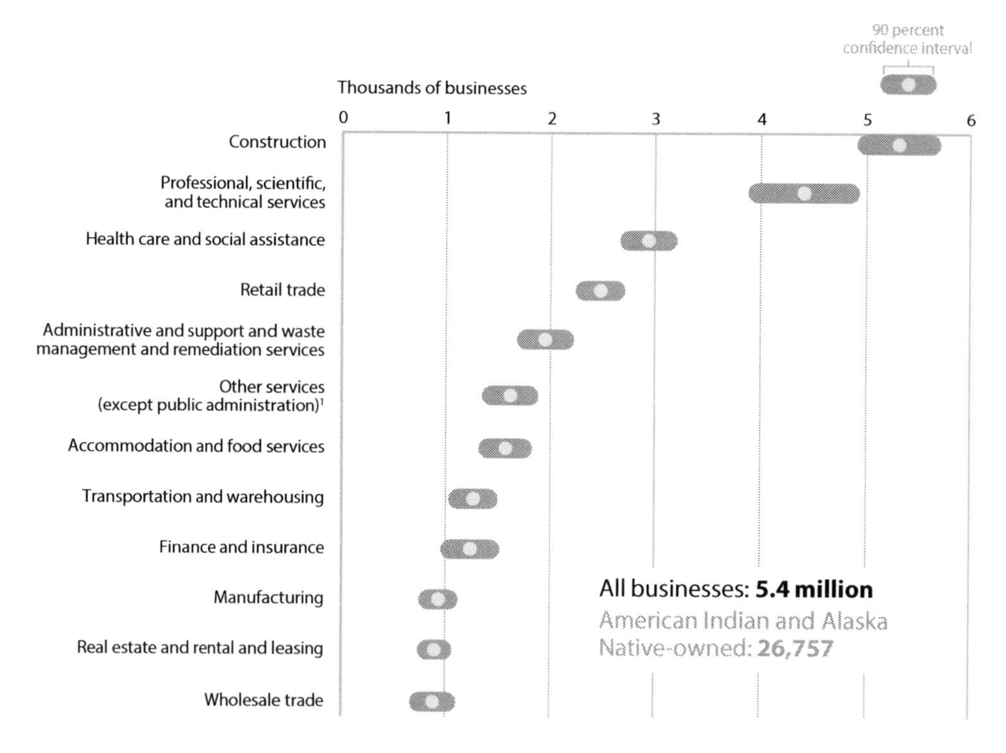

Figure 4.7 American Indian and Alaska Native-owned Businesses in the United States
Source: 2014 Annual Survey of Entrepreneurs. https://www.census.gov/content/dam/Census/library/visualizations/2016/comm/cb16-ff22_aian_graphic.pdf

27.0 percent speak their native language at home, compared with 21.6 percent for those who do not live on a reservation (U.S. Census Bureau, 2017i). Unfortunately, Native American reservations are often very poor. There are wait lists of up to years for families in need of housing. Meanwhile, many families live together in multigenerational families and must work together to survive.

African American Families

African American or **Black** refers to a person having origins in any of the Black racial groups of Africa. African American families contributed to the history of the United States from farming to sports and entertainment, from scientific inventions to the Civil Rights movement. Feagin (2016) emphasizes the centrality of African Americans in providing the massive amounts of labor over centuries that generated much of the past and present development and wealth of this country. Each year, the President of the United States proclaims February as National African American History Month.

The Great Migration of African American Families

Before 1910, most African American families lived in the South. The three states with the largest Black populations were Georgia, Mississippi, and Alabama. During the Great Migration between 1910 and 1970, an estimated 6 million Blacks left the South (U.S. Census Bureau, 2012). The **Great Migration** refers to the early-twentieth-century migration of Southern-born African Americans out of the rural South to largely urban locations in the North, Midwest, and West. The rise of manufacturing industries in the North and poor economic conditions in the South from 1910 to 1970 shaped the scale and timing of the Great Migration. The migration resulted in the creation of large African American residence areas across the Northeast, Midwest, and West. The city of Newark, New Jersey realized the largest increase in Black population, with the Black proportion of the city rising from 10 percent in 1940 to 54.2 percent in 1970 (U.S. Census Bureau, 2012).

The Great Migration changed African American family life as families moved away from agricultural pursuits and became wage earners in cities. This migration increased the mortality of African Americans born in the South in the early twentieth century (Black *et al.*, 2015). The National Research Council (2001) suggests Black migration from the South increased their risk of involvement with the criminal justice system and the increasing geographic marginalization of very poor Whites, which may insulate them from some forms of crime detection and social control found in large cities. Today, 100 years after the Great Migration, the nation is seeing more Blacks move into the South than move out (Toppo & Overberg, 2015). After nearly 100 years, the Great Migration seems to be reversing.

Demographic Status of African American Families

The African American population in the United States totaled 43.2 million in 2017 (U.S. Census Bureau, 2018h), making up 13.4 percent of the total U.S. population (U.S. Census Bureau, 2017h). Over the last several decades, Black workers have been offering more to the economy and the labor market to incredibly disappointing results in pay and unemployment (Jones & Wilson, 2017). While the unemployment rate for Black workers remained higher than for White workers (7.1 percent versus 3.8 percent) in 2017, compared with 16.8 percent versus 9.2 percent in 2010, the Black unemployment rate fell faster than overall unemployment over the year (Gould & Wilson, 2017). The share of African American families with an employed member was 78.7 percent; Black families remained more likely to have an unemployed member (9.5 percent) in 2017 (U.S. Bureau of Labor Statistics, 2018b).

Progress has been made in narrowing the wage gap for African American families since 1950 but wage differentials along racial lines are a permanent feature in the United States. In 2015, when including those out of the labor force, the median African American man age 25 to 54 earned $23,000, compared with the $44,000 earned by his White counterpart (Collins & Wanamaker, 2017). As of 2015, relative to the average hourly wages of White men with the same education, experience, metro status, and region of residence, Black men make 22.0 percent less, and Black women make 34.2 percent less (Gould, 2017). In 2017, the annual median income of African American households was $40,258, compared with $61,372 for the nation (see Figure 4.3). The persistence of poverty has been stronger for Blacks (Stanford Center on Poverty and Inequality, 2017). Approximately 21.2 percent of African Americans (9.0 million) lived in poverty in 2017, compared with 12.3 percent for the nation (Fontenot *et al.*, 2018). African American children are three times more likely to be in poverty than White children (see Figure 4.6). Poor Black children (81.1 percent) are much more likely to attend high-poverty schools than poor White children (53.5 percent) (García, 2017).

Occupation and income are correlated with education. African American ancestors struggled to get an education and during slavery most African slaves were denied formal education. Black schools originated in the Southern United States after the Civil War and Black colleges and universities (HBCUs) were set up when segregated colleges did not admit African Americans. Although African Americans lagged overall compared to White or Asian Americans, they advanced greatly in education attainment. Figure 4.8 shows that the Black population age 25 and over with a bachelor's degree or higher was on the rise at 22.5 percent in 2015 and at 23.3 percent in 2016. Research has explored the different roles of family

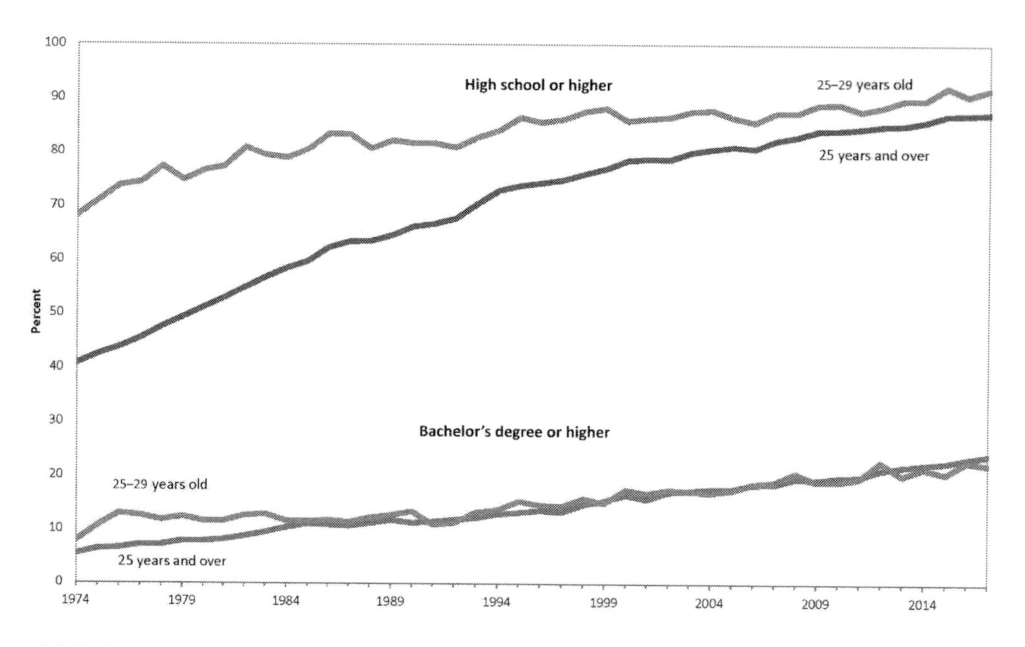

Figure 4.8 High School and Bachelor's Degree Attainment for Black Population Age 25 Years and Older, 1974–2017
Source: U.S. Census Bureau. 1974–2002, March, Current Population Survey, 2003–2017 Annual Social and Economic Supplement to the Current Population Survey. https://www.census.gov/library/visualizations/time-series/demo/cps-historical-time-series.html

support and college enrollment. The findings show that high parental income tends to decrease family support for Black adolescents. Black adolescents in high-income families tend to have lower chances of college enrollment than their White counterparts, leading to unequal mental health benefits for highly educated Blacks (Park, 2018).

African American families have been short-changed out of the materials to build wealth. This outcome holds for Black families regardless of the time and money spent on educational upgrading (Jones, 2017a). Figure 4.9 shows that median wealth for Black families whose head has a college degree is only one-eighth the wealth of the median White family whose head has a college degree. Even the typical Black family with a graduate or professional degree had more than $200,000 less wealth than a comparable White family.

Owning a home is a key to achieving the American Dream. The racial wealth gap is a housing wealth gap. African American families have been falling behind because of the systematic oppression of African Americans through government legislation and laws. Housing policies that prevented Blacks from acquiring land created redlining and restrictive covenants, and encouraged lending discrimination, reinforcing the racial wealth gap for decades (Jones, 2017b). For example, the Homestead Act, passed on May 20, 1862, gave an applicant ownership of land at little or no cost but barred families of color from acquiring free land. Following the National Housing Act of 1934, the Federal Housing Administration (FHA) established guidelines to steer private mortgage investors away from minority areas after 1935. Drawing on the National Housing Survey, McCabe (2018) recently finds that African Americans are more likely than Whites to identify the social status of ownership and the importance of building wealth as reasons to buy a home. While African Americans are more likely to pursue homeownership as a way

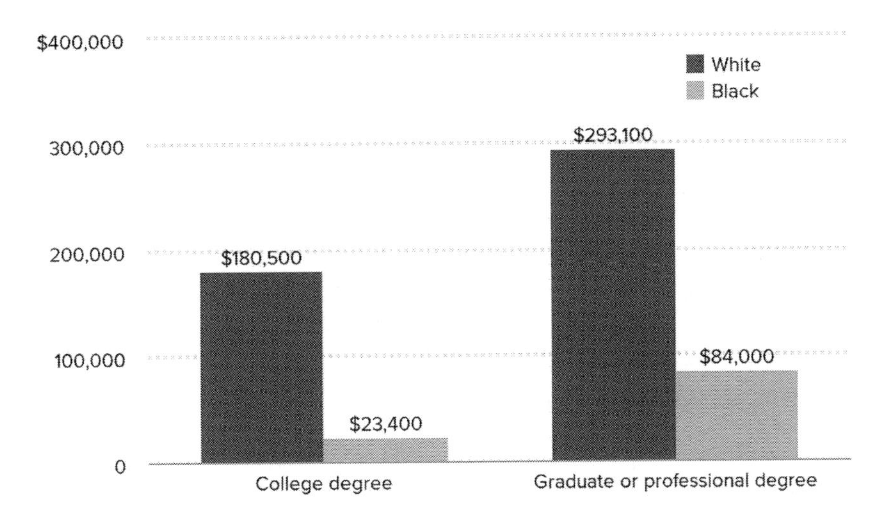

Figure 4.9 Median Wealth, by Degree and Race
Source: Survey of Consumer Finance Combined Extract Data, 2013, Economic Policy Institute. http://www.epi.org/blog/the-racial-wealth-gap-how-african-americans-have-been-shortchanged-out-of-the-materials-to-build-wealth/

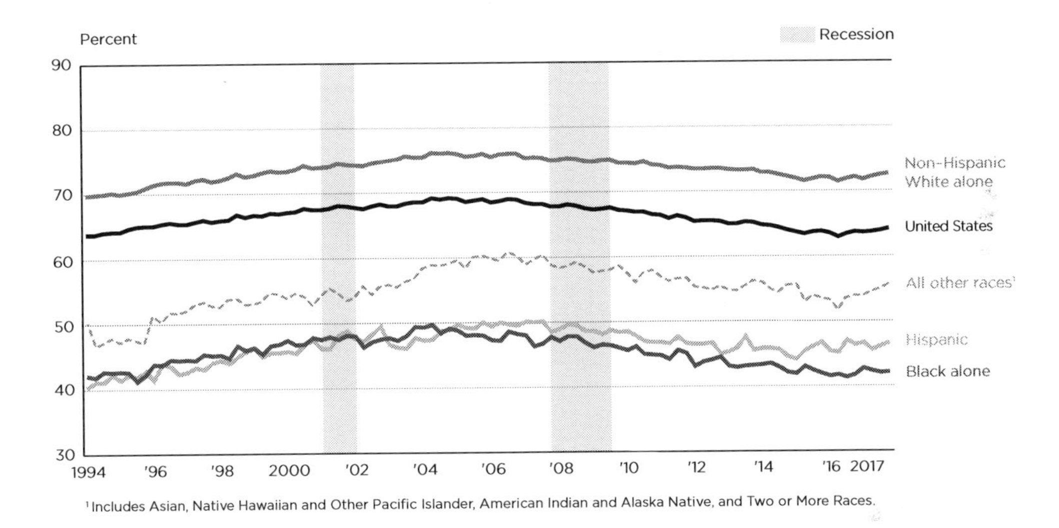

Figure 4.10 Quarterly Homeownership Rates by Race and Ethnicity of Householder for the United States, 1994–2017
Source: U.S. Census Bureau, Current Population Survey/Housing Vacancy Survey, February 21, 2017; recession data from the National Bureau of Economic Research. https://www.census.gov/housing/hvs/data/charts/fig08.pdf

to improve their housing quality, they are less likely to view ownership as a tool for accessing more convenient neighborhoods (McCabe, 2018). Figure 4.10 shows that African American families had the lowest quarterly homeownership rates (40 percent), compared with 70 percent of White families in 2017. African American families that have faced racial discrimination in employment and income for generations were more likely to be barred from accessing the housing market. In 2016, 58 percent of Black household heads were renting their homes (Ciliuffo & Geiger, 2017).

African Americans have higher rates of mortality than any other racial or ethnic group for eight of the top ten causes of death (Hummer *et al.*, 2004). In 2017, 89.4 percent of Blacks had health insurance coverage; the government coverage rate was the highest for Blacks, at 44.1 percent; and 56.5 percent of Blacks had private health insurance coverage (Berchick *et al.*, 2018). In a study of 1,680 African American adults about the relationship between skin color and health, Hargrove (2018) reports that dark-skinned women experience more physiological deterioration and self-report worse health than lighter-skinned women. These associations are not evident among men, and socioeconomic factors, stressors, and discrimination does not explain the dark–light disparity in physiological deterioration among women.

Characteristics of African American Families

In 2016, there were 9.8 million Black family households in the United States; 59.5 percent of households with a Black householder had at least one relative present; 45.1 percent of families with Black householders were married couples (U.S. Census Bureau, 2017j). Many African American adults remain unmarried. Figure 4.11 shows that Blacks have the highest proportion of living alone (35 percent) compared with their counterparts, and more African American families are headed by single mothers (20 percent) or are living as cohabiting couples (5 percent) in 2017.

A big challenge African Americans face is being a single parent. Even as violent crime rates have continued to fall, an African American child is six

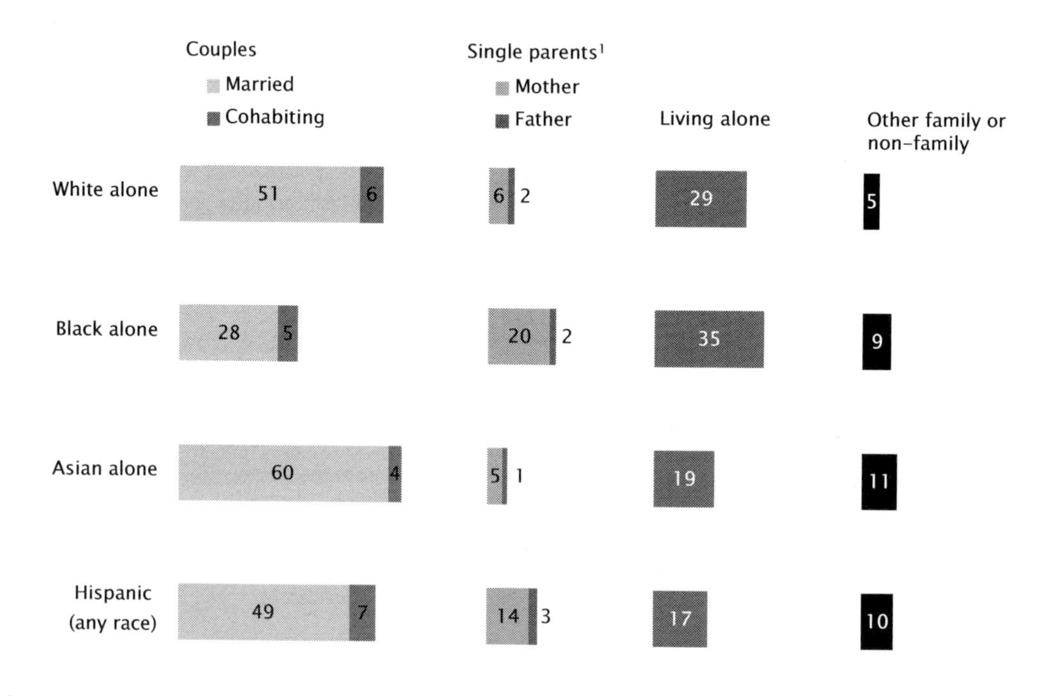

Figure 4.11 HH-7a Households and Family Forms by Race

Source: Current Population Survey, Annual Social and Economic Supplement, 2017. https://www.census.gov/content/dam/Census/library/visualizations/time-series/demo/families-and-households/hh-7a.pdf

times as likely as a White child to have or have had an incarcerated parent. These disparities in parental incarceration further perpetuate racial differences in achievement through an array of cognitive and non-cognitive outcomes linked to children's performance at school (Gould, 2017). Grandparents are often an important part of the kin network for African American families. The number of Black grandparents living with their grandchildren younger than eighteen in 2014 in the United States was 1.2 million (U.S. Census Bureau, 2017j). This kin network helps support the family and raise children.

African American working moms stand out as working moms play a larger economic role in families (Wilson, 2017). Figure 4.12 shows that more than two-thirds of all African American working mothers are single moms, making them the primary economic providers for their families—half of all African American female workers are moms.

Hispanic American Families

Hispanic or **Latino** refers to a person of Cuban, Mexican, Puerto Rican, South or Central American, or other Spanish culture or origin, regardless of race. Hispanic Heritage Month (September 15 to October 15) celebrates the culture and traditions of those who trace their roots to Spain, Mexico, and the Spanish-speaking nations of Central America, South America, and the Caribbean. Hispanic is a name that the American government gave in 1970 to make it easier to count people from those countries. The term *Latino* appeared for the first time in the U.S. Census in 2000.

Demographic Status of Hispanic American Families

The Hispanic or Latino population has increased ninefold, from 6.3 million in 1960 to 56.6 million in 2015, and grew to 58.9 million in 2017 (U.S. Census Bureau, 2018i), making up 18.1 percent of the total U.S. population (U.S. Census Bureau, 2017h) and making people of Hispanic origin the second largest racial or ethnic group behind Whites.

In the United States, Hispanic or Latino contains a variety of ethnic subgroups such as Mexican, Cuban, Puerto Rican, and South or Central Americans. In 2017, 63.2 percent of U.S. Hispanics were of Mexican origin (36.6 million), Puerto Rican (5.5 million), Cuban (2.3 million), Dominican (2.0 million), and Central American (5.5 million) (U.S. Census Bureau, 2018i). In 2016, 54.5 percent of the Hispanic residents with a population of 1 million or more lived in California, Arizona, Colorado, Florida, Illinois, New Jersey, New Mexico, and New York. California (15.3 million) had the largest Hispanic population of any state and Los Angeles County (4.9 million) had the largest Hispanic population of any county (U.S. Census Bureau, 2017k).

The share of Hispanic American families with an employed member was 86.9 percent; Hispanic families remained more likely to have an unemployed member (7.7 percent) in 2017 (U.S. Bureau of Labor Statistics, 2018b). In 2017, the median income of Hispanic households was $50,486, compared with $61,372 for the nation (see Figure 4.3). Approximately 18.3 percent of Hispanics (10.8 million) lived in poverty in 2017, compared with 12.3 percent for the nation (Fontenot *et al.*, 2018). Hispanics had the lowest rate of any health insurance coverage at 83.9 percent and also had the lowest rate of private coverage, at 53.5 percent (Berchick *et al.*, 2018). Drawing on the National Housing Survey, McCabe (2018) finds that Latinos are more deeply invested in the social status of homeownership, the importance of building wealth, and the promise of moving into a nicer home when they pursue ownership opportunities. In 2016, 54 percent of Hispanic household heads were renting their homes (Ciliuffo & Geiger, 2017). Figure 4.10 shows that 45 percent of Hispanics owned their home in 2016, compared with 70 percent of White families.

In terms of education, the Hispanic dropout rate remains higher, and college enrollment remains lower than for their counterparts. Some 66 percent of Hispanics cited the need to help support their family as

White mothers

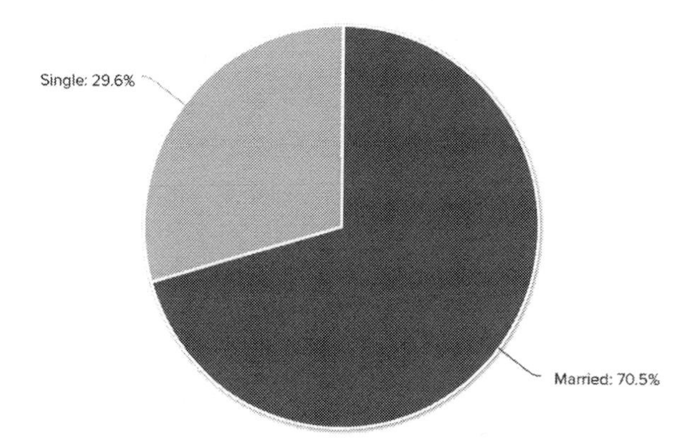

Black mothers

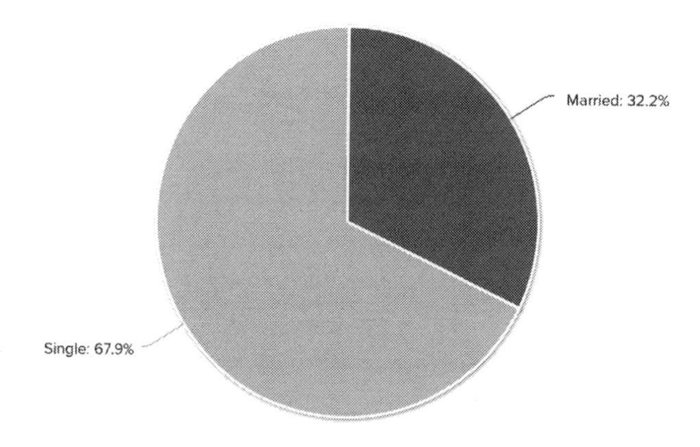

Hispanic mothers

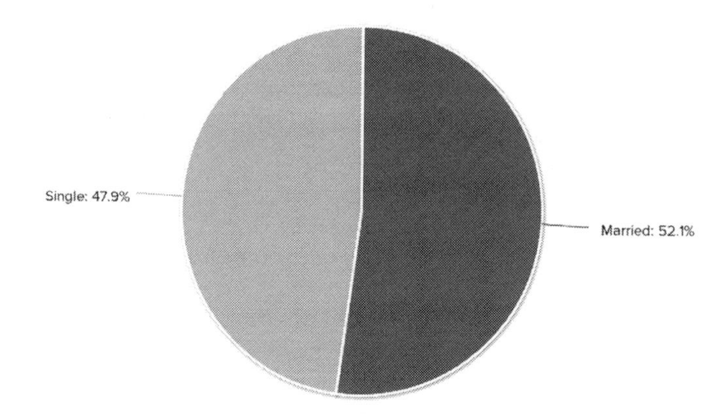

Figure 4.12 Working Mothers by Marital Status, by Race and Ethnicity, 2015

Source: EPI analysis of Current Population Survey Annual Social and Economic Supplement data, Economic Policy Institute. http://www.epi.org/blog/african-american-women-stand-out-as-working-moms-play-a-larger-economic-role-in-families/

a reason for not enrolling in college (Krogstad, 2015). Researchers attribute the educational gaps associated with ethnic minority status to factors such as fewer family resources, discrimination, teacher–student mismatch, English learner status, and social isolation at school. In fact, Salerno and Reynolds (2016) have found that school ethnic enclaves provide both academic and social support, help foster a positive ethnic self-image, and ultimately link ethnic minority status and heritage to success despite the educational challenges faced by Latina/o high school students at a public high school in the Southeast. A growing share of Hispanics have gone to college and almost 40 percent of Hispanics age 25 and older had any college experience in 2015, up from 30 percent in 2000 (Flores, 2017). Figure 4.13 shows Hispanic student enrollment on the rise in total enrollment by level of school from 1996 to 2016.

Characteristics of Hispanic American Families

Among the 16.7 million Hispanic U.S. households in 2016, 48.0 percent were married-couple households and 72.9 percent of Hispanics (40 million) age five and older spoke Spanish at home in 2016 (U.S. Census Bureau, 2017k). Mexican Americans are the largest group of Hispanics in the United States, accounting for 63.2 percent (36.6 million) of the nation's Hispanics in 2017 (Flores, 2017). These families trace their ancestry back to the merging of Spanish settlers with the Native Americans of Central America and Mexico. The 1848 Treaty of Guadalupe Hidalgo ended the Mexican–American war and resulted in

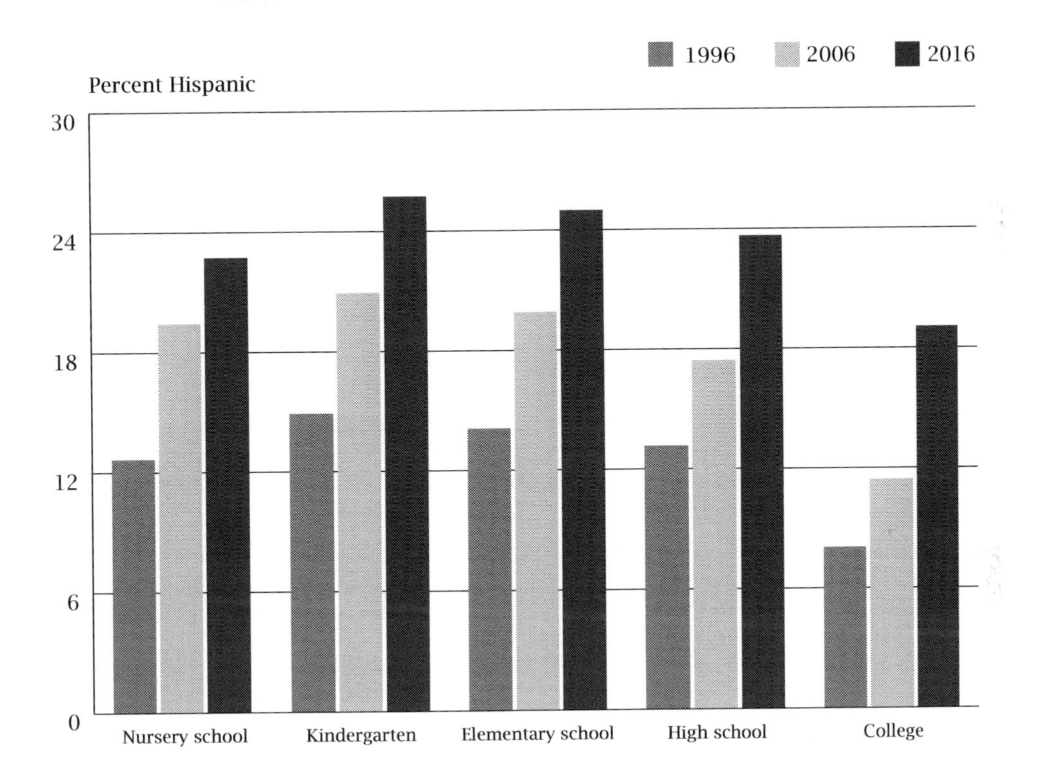

Figure 4.13 Hispanic Student Enrollment on the Rise: Percentage of Total Enrollment by Level of School

Source: Current Population Survey, October 2016. www.census.gov/programs-surveys/cps.html; https://census.gov/content/dam/Census/library/visualizations/2017/comm/hispanic-school.pdf

the United States gaining territory spanning across Arizona, California, Western Colorado, Nevada, New Mexico, Texas, and Utah.

Hispanic American families are family-oriented. For example, 49 percent of Hispanic American families were headed by two parents in 2017 (see Figure 4.11). *Familism* within the Hispanic community is associated with a sense of obligation to family members and the placement of family interests over individual desires. One way that Hispanic American families may demonstrate family closeness is by getting together every week for a family meal. Honoring ancestors by naming children after their grandparents and parents is common among Hispanic families. A record 40 million Hispanics age five and older speak Spanish at home, up from 25 million in 2000 (Flores, 2017). Extended families serve important roles, helping to provide childcare and eldercare. Grandparents help guide the family and raise their grandchildren. In turn, grandparents are taken care of by the younger generations.

Hispanic American families are often dedicated to their religion and do not encourage divorce. Being nominally Roman Catholic is another means of maintaining strong extended family ties. After interviewing 74 racially and religiously diverse highly religious families, Dollahite and Marks (2009) found that these families strive together to fulfill the sacred purposes suggested by their respective faiths. Identified processes include relying on God or God's word for support, guidance, and strength, sanctifying the family by living religion at home, and resolving conflict with prayer, repentance, and forgiveness. Cohabiting couples with or without children, marriage across ethnic lines, and other living arrangements were relatively uncommon among Hispanic groups but now are coming to resemble the pattern of non-Hispanics (see Figure 4.11).

Hispanic American families experienced a lot of issues with new U.S. immigration laws. These policies include pursuing a program of stepped-up arrests of unauthorized immigrants, cuts in legal immigration and refugee admissions, punitive sanctions for so-called "sanctuary" cities and states, and a request to fund construction of a wall along the United States–Mexico border (Reich, 2018). For example, a federal judge ruled in 2012 that Arizona enforce the "papers please" provision of its immigration law. In May 2017, Texas pushed to the forefront of debate over immigration when a "sanctuary city" ban was signed into law, allowing police officers to ask during routine stops whether someone is legally in the United States. The biggest fear that many Hispanic families face is deportation.

Asian American Families

Asian refers to a person having origins in any of the original peoples of the Far East, Southeast Asia, or the Indian subcontinent. Asian American and Pacific Islander Heritage Month celebrates the culture and traditions of those from these Asian countries and Native Hawaiians and other Pacific Islanders. The first ten days of May were chosen to coincide with two important milestones in Asian/Pacific American history: the arrival in the United States of the first Japanese immigrants (May 7, 1843) and the contributions of Chinese workers to the building of the transcontinental railroad, completed May 10, 1869 (U.S. Census Bureau, 2018j).

Demographic Status of Asian American Families

Asian American families are very diverse, as each Asian ethnic subgroup has its own history, culture, language, and pathway into the United States. The U.S. Asian population grew from 11.9 million in 2000 to 20.4 million in 2015, the fastest growth rate of any major racial or ethnic group (López *et al.*, 2017). As of July 1, 2017, the Asian population totaled 21.1 million (U.S. Census Bureau, 2018j), making up 5.8 percent of the U.S. population (U.S. Census Bureau, 2017h). California had the largest Asian population (6.5 million), followed by New York (1.8 million) (U.S. Census Bureau, 2017l).

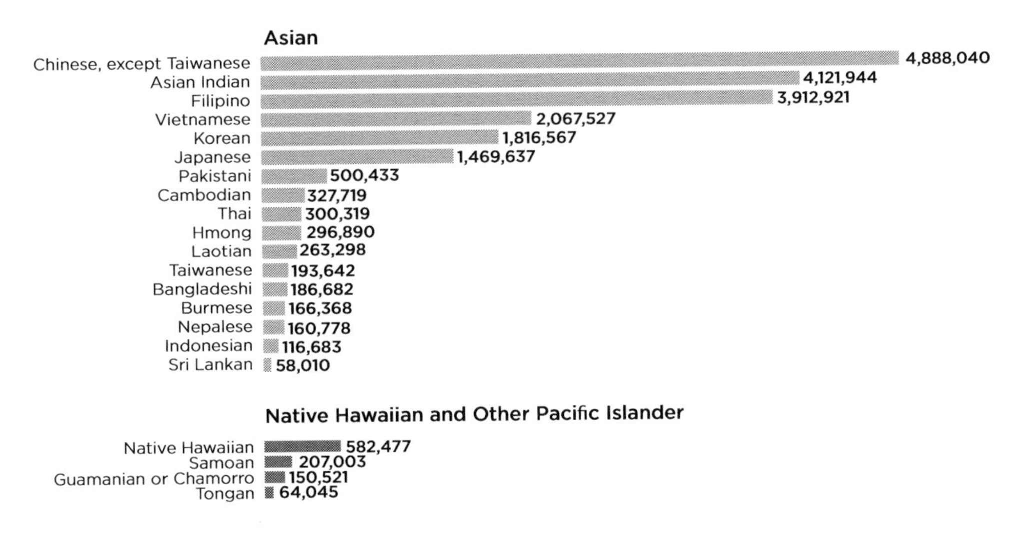

Figure 4.14 Asian and Pacific Islander Population in the United States

Source: 2016 American Community Survey One-Year Estimates, Table B02018: Asian Alone or in Any Combination, and Table B02019: Native Hawaiian and Other Pacific Islander Alone or in Any Combination. https://www.census.gov/content/dam/Census/library/visualizations/2018/comm/2018-api.pdf

Figure 4.14 shows the largest six ethnic subgroups are of Chinese, Asian Indian, Filipino, Vietnamese, Korean, and Japanese origin.

In 2016, Asian households had the highest median income ($81,331), compared with $61,372 for the nation as a whole (see Figure 4.3). Asians near the top of their income distribution (the 90th percentile) had incomes 10.7 times greater than the incomes of Asians near the bottom of their income distribution (the 10th percentile) (Kochhar & Cilluffo, 2018). This varies widely among Asian subgroups. Household incomes below the median household income for all Americans are Bangladeshi ($49,800), Hmong ($48,000), Nepalese ($43,500), and Burmese ($36,000); Asian Indian households have the highest median income ($100,000), followed by Filipinos ($80,000), and Japanese and Sri Lankans (each $74,000) (López *et al.*, 2017). In 2017, 92.7 percent of Asians had health insurance coverage and were among the most likely to have private health insurance at 72.2 percent (Berchick *et al.*, 2018).

About half of Asians age 25 and older (51 percent) have a bachelor's degree or more, compared with 30 percent of all Americans this age (López *et al.*, 2017). In 2015, 21.7 percent of the Asian population, alone or in combination, age 25 and older had a graduate or professional degree compared with 11.6 percent for all Americans (U.S. Census Bureau, 2017l). Those with advanced degrees are likely to earn more income. The average earnings for those with a bachelor's degree were $65,482, compared with $92,525 for those with an advanced degree and $35,615 for those with a high school diploma (U.S. Census Bureau, 2017c). The share of Asian families with an employed member was 88.6 percent; Asian families remained less likely to have an unemployed member (5.4 percent) in 2017 (U.S. Bureau of Labor Statistics, 2018b). Figure 4.15 shows that Asians are more likely to work in management, professional, and related occupations than their counterparts.

High-income households tend to have high levels of broadband internet use. Access to a computer and a broadband internet subscription have become increasingly important to American families in carrying out their day-to-day lives. This technology is used

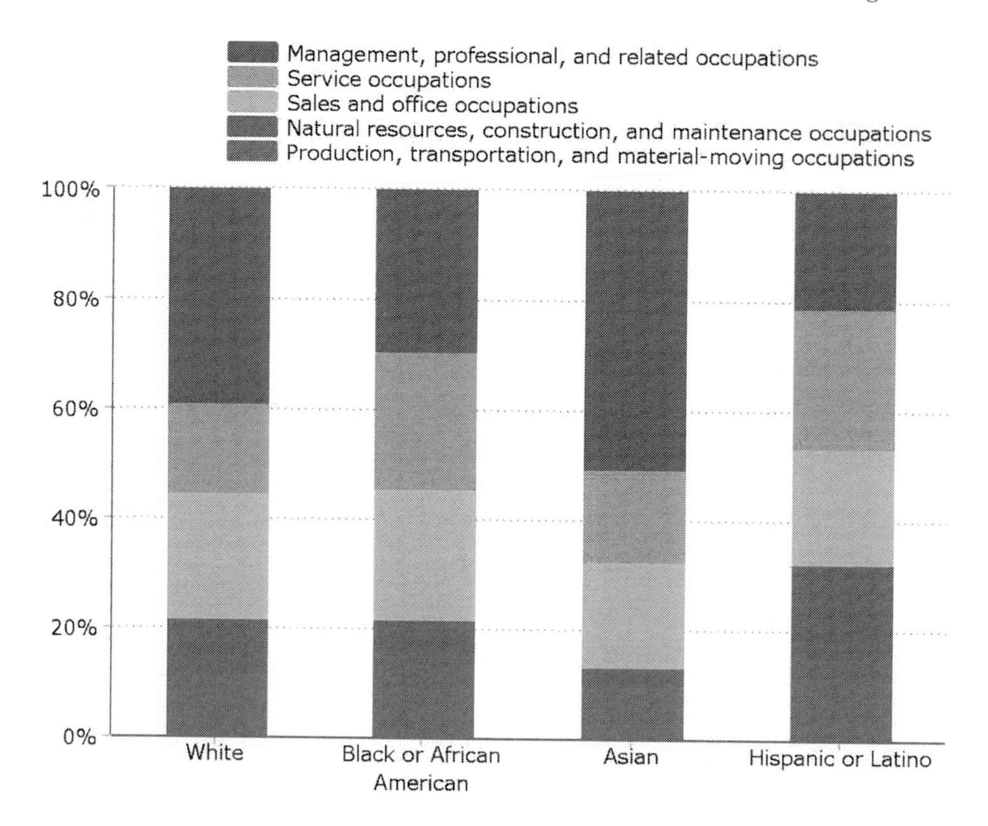

Figure 4.15 Employed People by Occupations, Race, and Hispanic or Latino Ethnicity, 2014, Annual Averages
Source: U.S. Bureau of Labor Statistics November 18, 2015. https://www.bls.gov/opub/ted/2015/educational-attainment-and-occupation-groups-by-race-and-ethnicity-in-2014.htm
Note: People whose ethnicity is identified as Hispanic or Latino may be of any race.
Data may not sum to 100 percent due to rounding.

for a variety of activities including accessing health information, online banking, choosing a place to live, applying for jobs, looking up government services, and taking classes (U.S. Census Bureau, 2017m). Figure 4.16 shows that Asian households were the most likely to have a desktop or laptop, handheld device, and broadband subscription. About 80 percent reported this combination of the three key items. Sixty-five percent of Whites reported all three items, compared with 55 percent of Hispanic households and 49 percent of African American households.

Asian American families were less likely to live in poverty but there are differences between Asian ethnic subgroups. The poverty rate for Asians was 10 percent (2 million), compared with 12.7 percent for the nation as a whole (Fontenot *et al.*, 2018). Hmong (28.3 percent), Bhutanese (33.3 percent), and Burmese (35.0 percent) had the highest poverty rates among Asian groups, compared with Filipinos (7.5 percent), Asian Indians (7.5 percent), and Japanese (8.4 percent) (López *et al.*, 2017).

Family Life and Marriage Patterns of Asian American Families

The traditional Asian American families were patriarchal, with males and elders having authority over women and children. Confucianism's notion of filial piety emphasizes the family above the individual.

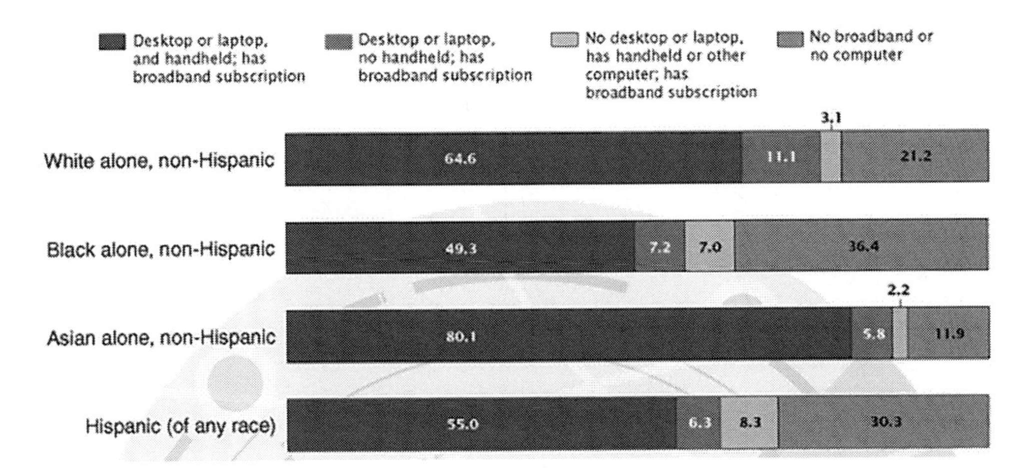

Figure 4.16 The Digital Divide: Percentage of Households by Broadband Internet Subscription, Computer Type, Race and Hispanic Origin

Source: 2015 American Community Survey. https://census.gov/library/visualizations/2017/comm/internet.html

Note: Estimates may not sum to 100 percent due to rounding.

Asian householders (60 percent) were married and less likely to be single parents than their counterparts (see Figure 4.11). About 26 percent of Asians live in multigenerational households, a higher share than for the United States overall (19 percent) (López *et al.*, 2017).

The first major wave of Asian immigration to the United States occurred between 1850 and the 1880s, primarily destined for Hawaiian sugar plantations, the California Gold Rush (1848–1855), and the transcontinental railroads in the 1860s. However, the distinct culture and religion that Asian immigrants embraced resulted in **xenophobia**, referred to as the fear of foreigners and strangers. Chinese families were rioted out of towns and driven towards establishing Chinatowns. Historically, Asian immigrants were concentrated in Hawaii and in states along the Pacific coast, settling in segregated ethnic enclaves such as Chinatown, Little Tokyo, and Little Manila (Zhou & Gatewood, 2000). Chinatown in San Francisco is the oldest in existence.

Research shows that three distinct immigrant family types emerged in different periods in response to particular political and economic conditions: split household, small producer, and dual-wage worker (Glenn, 1983). The passage of laws such as the Chinese Exclusion Act of 1882, the Gentlemen's Agreement of 1907 with Japan, and the Asian Exclusion Act of 1924 made the Chinese, Japanese, Koreans, and Filipinos face similar conditions of exclusion, forced many Asian families to remain separate, and forged a shared Asian experience of split-household family. A **split-household family** is a form of family in which one spouse breadwinner sojourns abroad and one spouse cares for children and elders in the home country.

Some Asian immigrant families saved money to open a small business. A **small business family** is a form of family engaged in a small business that relies on family members as free labor. Owners of particular businesses could obtain "merchant status," which enabled them to enter the United States and sponsor relatives (Lee, 2015). The small-producer family succeeded the split-household type around 1920 and became more common after the late 1940s when women were allowed to join their spouses (Glenn, 1983). Ethnic enclaves such as Chinatowns are the

Source: Christian Mehlführer, *Wikipedia*. https://en.wikipedia.org/wiki/Chinatown,_San_Francisco#/media/File: San_Francisco_China_Town_MC.jpg

central focus of family businesses. For example, Korean immigrant families were motivated to open small retail businesses through the *kye*—a private banking club to which Korean Americans pay contributions and from which they may get a loan to start a business. Due to occupational restrictions and limited capital, the small businesses were confined to laundries, restaurants, groceries, and sweatshops in the service sector. Between 1910 and 1920, the number of Chinese restaurants in New York City quadrupled, and then more than doubled again over the next ten years (Lee, 2015). Some Chinese immigrants started working as chefs or waitresses in Chinese restaurants and then saved enough money to open their own restaurants.

Historically, arranged marriages are not uncommon in many Asian countries. Usually, young people are socialized to expect such unions. For example, many parents from South Asia cling to the tradition of arranging the marriages of their adult children, but their children are growing up in a culture where most seek a marriage based on love. Even though Pakistani parents, especially fathers, are perceived to be resistant to change, Western values are playing a determining role in the process of mate-selection for second-generation Pakistani Muslim females (Zaidi & Shuraydi, 2002).

Global Overview: Transnational Families

The motivation behind the choice of migrating to another country is to pursue a better or different life. Because of globalization, it is increasingly common to see a family member moving to another country and leaving his or her family in the country of origin. **Transnational families** are "families where family members live some or most of the time separated from each other, yet hold together and create something that can be seen as a feeling of collective welfare and unity, namely 'familyhood', even across national borders" (COFACE, 2012). Transnational family arrangements are prevalent worldwide because of stringent migration policies as well as families' attempts to escape violent conflict or persecution (Mazzucato & Schans, 2011). A split-household family is a form of transnational family.

Transnational families are common in European countries. Citizens of European Union (EU) countries have the right to move between EU countries. As of 2015, nearly 20 million people, or about 4 percent of the EU's birth population, lived in a European country in which they were not born; the countries with the highest numbers of migrants from other EU countries are Germany (5.3 million), the United Kingdom (2.9 million), and France (2.3 million) (Pew Research Center, 2017b). Some families opt to live for long periods as a transnational family and apply for family reunification. Although EU citizenship is of greater advantage than non-EU citizenship for transnational families, many issues are common to both EU and non-EU families. For example, with the migration of one family member, family life inevitably changes in terms of parenthood, child development, and communication (COFACE, 2012).

Strengths and Challenges of Transnational Families

The studies on transnational families have shown that whether children benefit from their parents' migration depends not only on the outcomes that are studied (economic vs. psychological outcomes) but also on the characteristics of the parent and child (Mazzucato & Schans, 2011). An advantage of being a transnational family is that the family has ties in different countries. Migrant parents are perceived as having more economic resources than caregivers and therefore are expected to provide financial help to the caregiver and his or her family (Mazzucato & Schans, 2011).

In the past decades, scholars from different disciplines have engaged with the topic of families with members who live across national borders and the effects of such transnational living arrangements on children (Schmalzbauer, 2008). One disadvantage is that the family must go long periods of time without being together. Children might benefit from remittances while suffering emotionally from prolonged separation (Borraz, 2005). In this way, transnational families can be considered broken homes. Short-term and long-term separations may affect the relationship of married couples and the development of children. Separated husbands and wives may not be able to help the family with their children's education and family matters due to the distance between the two countries in which the family members live.

When it comes to child fostering, transnational families face challenges as some conditions that prevail in traditional fostering are not present. Long distances and travel costs make it difficult for migrant parents to visit and care for the families. Caregivers tend to live in urban contexts where they cannot rely on the help of extended family with the tasks of childrearing, as is done in a rural setting (Mazzucato & Schans, 2011). In migrant-sending countries, almost no policies target children who are left behind by their parents (Yeoh & Lam, 2006).

The increasing feminization of overseas labor migration in recent decades has prompted anxieties over a "crisis of care" when women and mothers leave (Parreñas, 2005). When mothers migrate, fathers assume "mothering" roles to a greater or lesser degree, and this may be an increasing trend in some parts of Asia where nuclear families are becoming more prevalent (Mazzucato & Schans, 2011). Migrant mothers themselves may assume dual roles, both as breadwinners and as nurturers from a distance (Hondagneu-Sotelo & Avila, 1997). Southeast Asia is a significant supplier of labor migrants. Graham *et al.* (2012) theorize the child's position in the transnational family nexus through the framework of the "care triangle," representing interactions between three subject groups—"left-behind" children, non-migrant parents/other carers, and migrant parent(s) in Indonesia and the Philippines.

Transnational families are able to stay in better touch through video technology such as Skype. Such communication technologies help people maintain virtual intimacy with faraway relatives, secure emotional support, and engage in transnational caregiving (Gonzalez & Katz, 2016).

Summary

Racial and ethnic diversity is what has shaped the United States into what it is today. Race refers to a group of people who share similar biological characteristics such as skin color, hair texture, and eye shape and color. Ethnicity refers to a group of people who identify with each other based on a shared cultural heritage or national origin. The terms *minority* and *subordinate group* are used interchangeably, as are the terms *majority* and *dominant group*. Families with majority and minority status are examined in terms of race and ethnicity, gender, religion, class, etc.

Racial and ethnic families often try to assimilate to the mainstream culture and at the same time maintain their ethnic cultures. Many cities and metropolitan areas have ethnic enclaves such as Chinatown and Little Italy that are inhabited by families of the same racial and ethnic background. This could affect the lifestyles of other families across the United States and could bring about prejudice and discrimination. Some people form prejudices against minority families based on stereotypes. Prejudice is an attitude toward people not based on actual experience. For example, lighter-skinned Americans may have prejudices against darker-skinned ones and vice versa. Prejudice may lead to discrimination. Prejudice is an attitude but discrimination is an action that affects family life. The chapter has explored the strengths and challenges of racial and ethnic minority families. From a diversity and global approach, the chapter has discussed interracial and transnational families.

Key Terms

Ableism refers to the belief that those with physical or mental disabilities or handicaps are inferior to able-bodied people.

Affirmative action is a set of procedures intended to eliminate unlawful discrimination among applicants on the basis of race, gender, or national origin.

African American is a person having origins in any of the Black racial groups of Africa.

Ageism refers to prejudice on the basis of a person's age.

Alaska Natives are indigenous peoples of Alaska, United States.

American Indian refers to a person having origins in any of the original peoples of North and South America (including Central America) and maintaining tribal affiliation or community attachment.

Ancestry refers to a person's ethnic origin or descent, "roots," heritage, or the place of birth of the person or the person's parents or ancestors.

Asian refers to a person having origins in any of the original peoples of the Far East, Southeast Asia, or the Indian subcontinent.

Black is a person having origins in any of the Black racial groups of Africa.

Caste system is a closed form of stratification in which people's status is determined at birth by their parent's ascribed status.

Caucasian has often been used in a different, societal context as a synonym for "White" or of "European ancestry."

Class system is an open form of stratification in which people's status is based on their control of resources and on the type of work they do.

Classism refers to the belief that those of a certain economic class are inferior to another class.

Conflict theory argues that racial inequality among racial families must be reduced in order to lessen racial prejudice and class conflict in society.

Contact hypothesis suggests that interpersonal contact is one of the effective ways to improve relations among groups who experience conflict and to reduce prejudice against minority group members.

Discrimination refers to unequal treatment of different groups of people on the basis of race, age, gender, etc.

Environmental racism occurs when racial families are forced to live in proximity of a hazardous environment because they are not given access to healthy places in which to live.

Ethnicity refers to a group of people who identify with each other based on a shared cultural heritage or national origin.

Ethnic paradox refers to the maintenance of one's ethnic ties that can assist with assimilation in society.

Genocide is intentional action to destroy an ethnic group.

Great Migration refers to the early-twentieth-century migration of Southern-born African Americans out of the rural South to largely urban locations in the North, Midwest, and West.

Hispanic refers to a person of Cuban, Mexican, Puerto Rican, South or Central American, or other Spanish culture or origin, regardless of race.

Homophobia is prejudging people because of their sexual orientation.

Individual discrimination takes place in one-to-one encounters and occurs when someone practices prejudicial treatment of another.

Institutional discrimination is the denial of opportunities and equal rights to minority families that results from day-to-day operations of social institutions.

Institutional racism is expressed in the daily operations of social institutions that maintain racial inequality.

Intergenerational mobility refers to changes in social class that occur across generations.

Intergroup relationship (IGR) refers to relationships among people from different social, economic, racial, and ethnic groups.

Interracial family is a form of family in which a couple identifies themselves as being of different races or ethnicities.

Intersectionality theory examines the experiences of minority people who are subjected to multiple forms of subordination in society.

Intragenerational mobility is the social movement of individual family members within their own lifetime.

Latino refers to a person of Cuban, Mexican, Puerto Rican, South or Central American, or other Spanish culture or origin, regardless of race.

Majority group is a dominant group whose members have more control or power over their lives than members of a minority group.

Minimal group model suggests that people automatically favor those similar to them, such as those of the same race, and reject those not similar to them.

Minority group is a subordinate group whose members have less control or power over their lives than do members of the dominant group.

Native Americans are also known as American Indians.

Nativism is the policy of favoring the native inhabitants over the immigrants.

Overt prejudice refers to an explicit and direct expression of negative beliefs.

Prejudice refers to negative attitudes based on oversimplified generalizations about members of a certain group.

Principle of third-generation interest maintains that the third generation (or the grandchildren of the original immigrants) will be more interested in their ethnicity than their parents.

Race refers to a group of people who share similar biological characteristics.

Racial discrimination concerns the unequal treatment of racial groups.

Racial formation refers to the process by which racial groups are created and manipulated.

Racial hierarchy is a stratification system on the basis of the belief that some racial groups are superior to other racial groups.

Racial inequality concerns unequal outcomes in terms of income, job, wealth, health, etc.

Racial profiling refers to suspecting a person of a certain race based on a stereotype about their race.

Racism refers to the belief that one race is superior to other races.

Reverse discrimination is discrimination against members of a dominant or majority group in favor of a historically disadvantaged group.

Scientific racism (race biology) is the pseudoscientific belief that empirical evidence exists to support or justify racism, racial inferiority, or racial superiority.

Slavery system is a closed form of stratification in which slaves are owned by slave owners.

Small business family is a form of family in which the family is engaged in a small business that relies on family members as free labor.

Social mobility refers to the movement of individuals, families, or groups among stratified social positions from one stratum to another.

Split-household family is a form of family in which one spouse breadwinner sojourns abroad and one spouse cares for children and elders in their home country.

Stereotypes refer to oversimplified generalizations about members of a certain group.

Structural functionalist theory supported both assimilation and pluralism, two contrary pathways in response to family diversity.

Subtle prejudice refers to an implicit and indirect expression of negative beliefs.

Symbolic interactionist theory emphasizes the role of social interaction in reducing racial and ethnic prejudice.

Transnational family is a form of family in which family members live some or most of the time separated from each other in two or more countries, yet hold together and create something that can be seen as a feeling of collective welfare and unity.

White refers to an American who has his or her origins in Europe, the Middle East, or North Africa.

White ethnic refers to White Americans who are not White Anglo-Saxon Protestant.

White nationalism refers to the belief that White people seek to develop and maintain a White national identity.

White privilege refers to rights and social privileges that benefit people identified as White, compared to non-White people under the same circumstances.

White separatism is the pursuit of a "White-only state."

White supremacy refers to the belief that White people are superior to non-Whites.

Xenophobia refers to the fear of foreigners and strangers.

Discussion Questions

- Propose strategies for fighting for justice and equality.
- Imagine how family life differs if one race is dominant over other races.

Suggested Film and Video

Asian Families (2014). Allen Blakey interviews Dr. Michael Kerr. [Video file]. Video posted to https://www.youtube.com/watch?v=spakYJ_0Mh4

Asian American Life: December 2015 [Video file]. Video posted to https://www.youtube.com/watch?v=T-aWAobejIE

From Our Lakota Hearts (2011). Pine Ridge Reservation SD. [Video file]. Video posted to https://www.youtube.com/watch?v=OTBd4tVkHaU

Harper, Hill (2012). *The African-American Family*. [Video file] Video posted to https://www.youtube.com/watch?v=KWnN8U1N1mQ

Hidden America: Children of the Plains [Video file] Video posted to https://www.youtube.com/watch?v=IJapHc7B8Xs

Hill Harper Discusses the African-American Family [Video file] Video posted to https://www.youtube.com/watch?v=KWnN8U1N1mQ

Identity: Being Asian American [Video file] Video posted to https://www.youtube.com/watch?v=TnbZAqLXIsQ

Mark, Diane and Onodera, Lisa. (Producer). (1995). *Picture Bride*. McJimsey, Mark. Producer. (1997). King of the Hill.

Mixed Families—Community Disapproval—Interracial Relationships [Video file] Video posted to https://www.youtube.com/watch?v=_EL8FsZbGKY

Pons, Lele. (2016). *My Big Fat Hispanic Family*. [Video file] Video posted to https://www.youtube.com/watch?v=8s_lMZ4LC-E

Internet Sources

American Family Immigration History Center: http://www.ellisisland.org/genealogy/ellis_island_visiting.asp

Bureau of Justice Statistics: https://www.bjs.gov/index.cfm?ty=tp&tid=11

Council on Contemporary Families (CCF): http://www.contemporaryfamilies.org/

Discover what Makes You Uniquely You: https://www.ancestry.com/

Minimum Wage Tracker: http://www.epi.org/minimum-wage-tracker/

National Indian Law Library. Tracing Native American Family Roots: http://www.narf.org/nill/resources/roots.html

Native American Resources: http://www-bcf.usc.edu/~cmmr/Native_American.html

Redefining Family at Colonial Williamsburg: http://www.history.org/Almanack/life/family/essay.cfm

The National Latino Fatherhood and Family Institute: http://www.nlffi.org/

References

Allport, G. W. (1954). *The Nature of Prejudice.* Reading, MA: Addison-Wesley.

Altemeyer, B. & Hunsberger, B. (1992). Authoritarianism, Religious Fundamentalism, Quest, and Prejudice. *International Journal for the Psychology of Religion,* 2, 113–133.

Bae, C. J., Douka, K. & Petraglia, M. D. (2017). On the origin of modern humans: Asian perspectives. *Science,* 358(6368), eaai9067.

Baldwin, J. (1963). *The Fire Next Time.* New York: Dial Press.

Baum, B. (2008). *The Rise and Fall of the Caucasian Race: A Political History of Racial Identity.* New York: New York University Press.

Beckwith, C. I. (2009). *Empires of the Silk Road: A History of Central Eurasia from the Bronze Age to the Present.* Princeton, New Jersey: Princeton University Press.

Berchick, E. R., Hood, E. & Barnett, J. C. (September 2018). *Health Insurance Coverage in the United States: 2017.* U.S. Census Bureau. Retrieved on October 1, 2018 from https://www.census.gov/content/dam/Census/library/publications/2018/demo/p60-264.pdf

Bhopal, R. & Donaldson, L. (1998). White, European, Western, Caucasian, or What? Inappropriate Labeling in Research on Race, Ethnicity, and Health. *American Journal of Public Health,* 88(9), 1303–1307.

Bibler, K. (July 21, 2015). The Pay Gap Is Even Worse for Black Women, and That's Everyone's Problem. *Economic Justice.* Retrieved on November 20, 2016 from http://www.aauw.org/2015/07/21/black-women-pay-gap/

Black, D. A., Sanders, S. G., Taylor, E. J. & Taylor, L. J. (2015). The Impact of the Great Migration on Mortality of African Americans: Evidence from the Deep South. *The American Economic Review,* 105(2), 477–503.

Blalock, H. M. (1967). *Toward a Theory of Minority-Group Relations.* Somerset, NJ: Wiley.

Blumenbach, J. F. (1779). *Handbuch der Naturgeschichte,* vol. 1. Verlag/Drucker: Johann Christian Dieterich.

Borraz, F. (2005). Assessing the Impact of Remittances on Schooling: The Mexican Experience. *Global Economy Journal,* 5(1), 1–32.

Bowleg, L. (2012). The Problem with the Phrase *Women and Minorities:* Intersectionality—An Important Theoretical Framework for Public Health. *American Journal of Public Health,* 102(7), 1267–1273.

Bravo, M. A., Anthopolos, R., Bell, M. L. & Miranda, M. L. (2016). Racial Isolation and Exposure to Airborne Particulate Matter and Ozone in Understudied US Populations: Environmental Justice Applications of Downscaled Numerical Model Output. *Environment International,* 92–93: 247–255.

Bureau of Justice Statistics (2018). *Prisoners in 2016.* Retrieved on September 26, 2018 from https://www.bjs.gov/content/pub/pdf/p16_sum.pdf

Burns, P. & Thomas, M. (2004). The Failure of the Nonregime: How Katrina exposed New Orleans as a Regimeless City. *Urban Affairs Review,* 41(4), 517–527.

CFPB (Consumer Financial Protection Bureau) (July 14, 2015). *CFPB and DOJ Reach Resolution with Honda to Address Discriminatory Auto Loan Pricing.* Retrieved on October 1, 2017 from https://www.consumerfinance.gov/about-us/newsroom/cfpb-and-doj-reach-resolution-with-honda-to-address-discriminatory-auto-loan-pricing/

Choate, M. (2008). *Emigrant Nation: The Making of Italy Abroad.* Cambridge, MA: Harvard University Press.

Ciliuffo, A. & Geiger, A. (July 19, 2017). *More U.S. Households Are Renting than at Any Point in 50 Years.* Washington, DC: Pew Research Center. Retrieved on August 14, 2017 from http://www.pewresearch.org/fact-tank/2017/07/19/more-u-s-households-are-renting-than-at-any-point-in-50-years/

Clair, M. & Denis, J. S. (2015). *Sociology of Racism.* James D. Wright (Ed.). *The International Encyclopedia of the Social and Behavioral Sciences,* 19, 857–863.

COFACE (Confederation of Family Organizations in the European Union) (2012). *Transnational Families and the Impact of Economic Migration on Families.* Retrieved on August 14, 2016 from http://coface-eu.org/en/upload/03_Policies_WG1/2012%20COFACE%20position%20on%20Transnational%20Families%20en.pdf

Cohn, D.'V. (April 6, 2016). *From Multiracial Children to Gender Identity, What Some Demographers Are Studying Now.* Washington, DC: Pew Research Center. Retrieved on July 28, 2016 from http://www.pewresearch.org/fact-tank/2016/04/08/from-multiracial-children-to-gender-identity-what-some-demographers-are-studying-now/

Cole, L. & Foster, S. (2001). *From the Ground Up: Environmental Racism and the Rise of the Environmental Justice Movement.* New York: New York University Press.

Collins, W. J. & Wanamaker, M. H. (June 13, 2017). *The Persistence of Earnings Differences among White and Black Men.* Washington Center for Equitable Growth. Retrieved on July 18, 2017 from http://equitablegrowth.org/research-analysis/the-persistence-of-earnings-differences-among-white-and-black-men/

Conversi, D. (July 2004). Can Nationalism Studies and Ethnic/Racial Studies Be Brought Together? *Journal of Ethnic and Migration Studies,* 30(4), 815–829.

Cornell, S. & Hartmann, D. (1998). *Ethnicity and Race: Making Identities in a Changing World.* Thousand Oaks, CA: Pine Forge Press.

Crenshaw, K. (1989). Demarginalizing the Intersection of Race and Sex: A Black Feminist Critique of Antidiscrimination Doctrine, Feminist Theory and Antiracist Politics. *The University of Chicago Legal Forum*, 140, 139–167.

Delgado, D. J. (2018). "My Deputies Arrest Anyone Who Breaks the Law": Understanding How Color-blind Discourse and Reasonable Suspicion Facilitate Racist Policing. *Sociology of Race and Ethnicity,* 4(4), 541–554.

De Sangro, M. (2003). *I Borboni nel Regno delle Due Sicilie* (in Italian). Lecce: Edizioni Caponi.

Diner, H. R. (1983). *Erin's Daughters in America: Irish Immigrant Women in the Nineteenth Century.* Baltimore, MD: Johns Hopkins University Press.

Dollahite, D. & Marks, L. (2009). A Conceptual Model of Family and Religious Processes in Highly Religious Families. *Review of Religious Research,* 50(4), 373–391. Retrieved on October 7, 2018 from http://www.jstor.org/stable/25593754

Dovidio, J. F. & Gaertner, S. L. (1986). Prejudice, Discrimination, and Racism: Historical Trends and Contemporary Approaches. In J. F. Dovidio & S. L. Gaertner (Eds.), *Prejudice, Discrimination, and Racism* (pp. 1–34). San Diego, CA: Academic Press.

Dowden, S. & Robinson, J. P. (1993). Age and Cohort Differences in American Racial Attitudes: The Generational Replacement Hypothesis Revisited. In P. M. Sniderman, P. E. Tetlock, & E. G. Carmines (Eds.), *Prejudice, Politics, and the American Dilemma* (pp. 86–103). Stanford, CA: Stanford University Press.

Du Bois, W. E. B. (1903). *The Souls of Black Folk.* Chicago: A.C. McClurg & Co.

Dudley, M. G. & Mulvey, D. (2009). Differentiating among Outgroups: Predictors of Congruent and Discordant Prejudice. *North American Journal of Psychology,* 11(1), 143–156.

During, S. (2005). *Cultural Studies: A Critical Introduction.* New York: Routledge.

Feagin, J. (2016). *How Blacks Build America: Labor, Culture, Freedom, and Democracy.* New York and London: Routledge.

Flores, A. (September 18, 2017). *How the U.S. Hispanic Population is Changing.* Washington, DC: Pew Research Center. Retrieved on September 30, 2017 from http://www.pewresearch.org/fact-tank/2017/09/18/how-the-u-s-hispanic-population-is-changing/

Fontenot, K., Semega, J. & Kollar, M. (2018). Income and Poverty in the United States: 2017 Current Population Reports. *U.S. Census Bureau.* Retrieved on September 29, 2018 from https://www.census.gov/content/dam/Census/library/publications/2018/demo/p60-263.pdf

Fowler, J. (1997). *Hinduism: Beliefs and Practices.* Sussex Academic Press.

García, E. (January 13, 2017). *Poor Black Children Are Much More Likely to Attend High-Poverty Schools than Poor White Children.* Economic Policy Institute. Retrieved on September 21, 2017 from http://www.epi.org/publication/poor-black-children-are-much-more-likely-to-attend-high-poverty-schools-than-poor-white-children/

Gilbert, M. (1985). *The Holocaust: A History of the Jews of Europe during the Second World War.* New York: Henry Holt.

Glassman, R. M., Swatos, W. H. Jr. & Demballs, B. J. (2004). *Social Problems in Global Perspective.* Lanham, MD: University Press of America.

Glenn, E. N. (1983). Split Household, Small Producer and Dual Wage Earner: An Analysis of Chinese-American Family Strategies. *Journal of Marriage and the Family,* 45(1), 35–46.

Goldenberg, D. M. (2003). *The Curse of Ham: Race and Slavery in Early Judaism, Christianity, and Islam.* Princeton: Princeton University Press.

Gonzalez, C. & Katz, V. S. (2016). Transnational Family Communication as a Driver of Technology Adoption. *International Journal of Communication,* 10(2016), 2683–2703.

Gould, E. (January 26, 2017). *Racial Gaps in Wages, Wealth, and More: A Quick Recap.* Economic Policy Institute. Retrieved on July 18, 2017 from http://www.epi.org/blog/racial-gaps-in-wages-wealth-and-more-a-quick-recap/

Gould, E. & Wilson, V. (July 7, 2017). *The Black Employment Rate Returns to Historic Low, but Not Really.* Economic Policy Institute. Retrieved on July 18, 2017 from http://www.epi.org/blog/the-black-unemployment-rate-returns-to-historic-low-but-not-really/

Gould, S. J. (1981). *The Mismeasure of Man* (pp. 28–29). New York: W. W. Norton.

Graham, E., Jordan, L. P., Yeoh, B.S.A., Lam, T., Asis, M. & Su-kamdi (2012). Transnational Families and the Family Nexus: Perspectives of Indonesian and Filipino Children Left Behind by Migrant Parent(s). *Environment & Planning A,* 44(4), 10.1068/a4445.

Graves, J. L. (2001). *The Emperor's New Clothes: Biological Theories of Race at the Millennium.* New Brunswick, NJ: Rutgers University Press.

Greene, K. R. (1998). Racism: Its Impact on the African American Family. *Leaven,* 6(2), Article 9. Retrieved on October 1, 2017 from http://digitalcommons.pepperdine.edu/leaven/vol6/iss2/9

Guimond, S., Dambrun, M., Michinov, N. & Duarte, S. (2003). Does Social Dominance Generate Prejudice? Integrating Individual and Contextual Determinants of Intergroup Cognitions. *Journal of Personality and Social Psychology,* 84, 697–721.

Hall, A. V., Hall, E. V. & Perry, J. L. (2016). Black and Blue: Exploring Racial Bias and Law Enforcement in the Killings of Unarmed Black Male Civilians. *American Psychologist,* 71(3), 175–186. Retrieved on September 22, 2017 from http://dx.doi.org/10.1037/a0040109

Hannaford, I. (1996). *Race: The History of an Idea in the West.* Baltimore, MD: Johns Hopkins University Press.

Hansen, M. L. (1938). *The Problem of the Third Generation Immigrant.* Rock Island, IL: Augustana Historical Society.

Hargrove, T. W. (Sept 20 2018). Light Privilege? Skin Tone Stratification in Health among African Americans. *Sociology of Race and Ethnicity.* https://doi.org/10.1177/2332649218793670

Hirschman, C. (2005). Immigration and the American Century. *Demography*, 42, 595–620. Retrieved on September 29, 2017 from https://www.ncbi.nlm.nih.gov/pubmed/16463913

Hitt, J. (2005). Mighty White of You: Racial Preferences Color America's Oldest Skulls and Bones. *Harper's Magazine.* Retrieved on October 4, 2018 from http://www.middlebury.edu/media/view/157971/original/jack_hitt_mighty_white_of_you.pdf

Hogg, M. A. (1987). Social Identity and Group Cohesiveness. In J. C. Turner, M. A. Hogg, P. J. Oakes, S. D. Reicher & M. S. Wetherell (Eds.), *Rediscovering the Social Group: A Self-categorization Theory* (pp. 89–116). Oxford, UK: Basil Blackwell.

Hondagneu-Sotelo, P. & Avila, E. (1997). "I'm Here but I'm There": The Meanings of Latina Transnational Motherhood. *Gender and Society*, 11, 548–571.

Hong, Y., Coleman, J., Chan, G., Wong, R.Y.M., Chiu, C., Hansen, I. G. *et al.* (2004). Predicting Intergroup Bias: The Interactive Effects of Implicit Theory and Social Identity. *Personality and Social Psychology Bulletin*, 30, 1035–1047.

Howell, J. & Korver-Glenn, E. (2018). Neighborhoods, Race, and the Twenty-first-century Housing Appraisal Industry. *Sociology of Race and Ethnicity*, 4(4), 473–490.

Hudson, N. (1996). From "Nation" to "Race": The Origin of Racial Classification in Eighteenth-Century Thought. *Eighteenth-Century Studies*, 29(3), 247–264. Retrieved from http://www.jstor.org/stable/30053821

Hummer, R. A., Benjamins, M.R. & Rogers, R. G. (2004). Racial and Ethnic Disparities in Health and Mortality among the U.S. Elderly Population. In N. B. Anderson, R. A. Bulatao & B. Cohen (Eds.), *Critical Perspectives on Racial and Ethnic Differences in Health in Later Life* (pp. 53–94). Washington, DC: National Academy Press.

Huxley, J. & Haddon, A. C. (1935). *We Europeans: A Survey of Racial Problems.* London: Jonathan Cape.

Jones, J. (February 13, 2017a). *The Racial Wealth Gap: How African-Americans Have Been Shortchanged Out of the Materials to Build Wealth.* Economic Policy Institute. Retrieved on September 24, 2017 from http://www.epi.org/publication/receiving-an-inheritance-helps-white-families-more-than-black-families/

Jones, J. (February 17, 2017b). *Receiving an Inheritance Helps White Families More than Black Families.* Economic Policy Institute. Retrieved on September 24, 2017 from http://www.epi.org/publication/receiving-an-inheritance-helps-white-families-more-than-black-families/

Jones, J. & Wilson, V. (March 27, 2017). *Low-wage African American Workers Have Increased Annual Work Hours Most since 1979.* Economic Policy Institute. Retrieved on September 21, 2017 from http://www.epi.org/blog/low-wage-african-american-workers-have-increased-annual-work-hours-most-since-1979/

Kaufmann, E. P. (2004). The Decline of the WASP in the United States and Canada. In E. P. Kaufmann (Ed.), *Rethinking Ethnicity: Majority Groups and Dominant Minorities* (pp. 54–73). New York: Routledge.

Keita, S.O.Y., Kittles, R. A., Royal, C.D.M., Bonney, G. E., Furbert-Harris, P., Dunston, G. M., Rotimi, C. N. *et al.* (2004). Conceptualizing Human Variation. *Nature Genetics*, 36(11s), S17–S20.

Kinealy, C. (1994). *This Great Calamity: The Irish Famine 1845–52.* Ireland: Gill & Macmillan.

Kochhar, R. & Cilluffo, A. (July 12, 2018). *Key Findings on the Rise in Income Inequality within America's Racial and Ethnic Groups.* Washington, DC: Pew Research Center. Retrieved on October 6, 2019 from http://www.pewresearch.org/fact-tank/2018/07/12/key-findings-on-the-rise-in-income-inequality-within-americas-racial-and-ethnic-groups/

Krogstad, J. (2015). *5 Facts about Latinos and Education.* Washington, DC: Pew Research Center. Retrieved on July 4, 2016 from http://www.pewresearch.org/fact-tank/2015/05/26/5-facts-about-latinos-and-education/

LaVeist, T. A. (Ed) (2002). *Race, Ethnicity, and Health.* San Francisco: Jossey-Bass.

Lee, H. R. (2015). *The Untold Story of Chinese Restaurants in America.* Scholars Strategy Network. Retrieved on August 27, 2016 from http://www.scholarsstrategynetwork.org/sites/default/files/ssn-basic-facts-lee-on-chinese-merchants-under-exclusion.pdf

Livingston, G. (June 6, 2017). *The Rise of Multiracial and Multiethnic Babies in the U.S.* Washington, DC: Pew Research Center. Retrieved on August 1, 2017 from http://www.pewresearch.org/fact-tank/2017/06/06/the-rise-of-multiracial-and-multiethnic-babies-in-the-u-s/

Livingston, G. & Brown, A. (May 18, 2017). *Intermarriage in the U.S. 50 Years after Loving v. Virginia.* Washington, DC: Pew Research Center. Retrieved on July 19, 2017 from http://www.pewsocialtrends.org/2017/05/18/intermarriage-in-the-u-s-50-years-after-loving-v-virginia/#fn-22844-2

Loftis, S. (April 11, 2003). Interviews Offer Unprecedented Look Into The World And Words Of The New White Nationalism. *Vanderbilt News*. Vanderbilt University.

López, G., Ruiz, N. G. & Patten, E. (September 8, 2017). *Key Facts about Asian Americans, a Diverse and Growing Population*. Washington, DC: Pew Research Center. Retrieved on September 30, 2017 from http://www.pewresearch.org/fact-tank/2017/09/08/key-facts-about-asian-americans/

Mahoney, M. C. & Michalek, A. M. (1998). Health Status of American Indians/Alaska Natives: General Patterns of Mortality. *Fam Med.*, 30, 190–195.

Malik, K. (1996). *The Meaning of Race*. New York: New York University Press.

Marger, M. N. (2008). *Race and Ethnic Relations: American and Global Perspectives* (8th edn). California: Cengage Learning.

Marks, J. (2008). Race: Past, Present and Future. In B. Koenig, S. Soo-Jin Lee & S. S. Richardson (Eds.), *Revisiting Race in a Genomic Age*. New Brunswick, NJ: Rutgers University Press.

Mazzucato, V. & Schans, D. (2011). Transnational Families and the Well-Being of Children: Conceptual and Methodological Challenges. *Journal of Marriage and the Family*, 73(4), 704–712.

McAdoo, H. P. (1999). *Family Ethnicity*. California: Sage Publishing.

McCabe, B J. (2018). Why Buy a Home? Race, Ethnicity, and Homeownership Preferences in the United States. *Sociology of Race and Ethnicity*, 4(4), 452–472.

Merton, R. K. (1949). Discrimination and the American Creed. In R. M. MacIver (Ed.), *Discrimination and National Welfare* (pp. 99–126). New York: Institute for Religious Studies.

Mullin, B. A. & Hogg, M. A. (1998). Dimensions of Subjective Uncertainty in Social Identification and Minimal Intergroup Discrimination. *British Journal of Social Psychology*, 37, 345–365.

National Research Council (2001). *America Becoming: Racial Trends and Their Consequences, Volume 1*. Washington, DC: The National Academies Press.

Nelli, H. S. (1980). Italians. In S. Thernstrom (Ed.), *Harvard Encyclopedia of American Ethnic Groups*. Cambridge, MA: Belknap Press.

Newkirk II, V. R. (Feb 28, 2018). Trump's EPA Concludes Environmental Racism Is Real. *The Atlantic*. Retrieved on October 4, 2018 from https://www.theatlantic.com/politics/archive/2018/02/the-trump-administration-finds-that-environmental-racism-is-real/554315/

Nobles, M. (2000). *Shades of Citizenship: Race and the Census in Modern Politics*. Stanford, CA: Stanford University Press.

OMH (Department of Health and Human Resources Office of Minority Health) (2016). *American Indian/Alaska Native Profile*. Retrieved on November 12, 2016 from http://minorityhealth.hhs.gov/omh/browse.aspx?lvl=3&lvlid=62

Oppenheimer, G. M. (2001). Paradigm Lost: Race, Ethnicity, and the Search for a New Population Taxonomy. *Am J Public Health*, 91, 1049–1055.

Park, K. (2018). Black–White Differences in the Relationship between Parental Income and Depression in Young Adulthood: The Different Roles of Family Support and College Enrollment among U.S. Adolescents. *Sociology of Race and Ethnicity*. https://doi.org/10.1177/2332649218776037

Parreñas, R. S. (2005). *Children of Global Migration: Transnational Families and Gendered Woes*. Stanford, CA: Stanford University Press.

Pew Research Center (July 26, 2017a). *Muslims are Concerned about Their Place in Society, but Continue to Believe in the American Dream*. Retrieved on October 5, 2018 from http://www.pewforum.org/2017/07/26/findings-from-pew-research-centers-2017-survey-of-us-muslims/

Pew Research Center (June 19, 2017b). *Origins and Destinations of European Union Migrants within the EU*. Retrieved on August 1, 2017 from http://www.pewglobal.org/interactives/origins-destinations-of-european-union-migrants-within-the-eu/

Race, Ethnicity, and Genetics Working Group (2005). The Use of Racial, Ethnic, and Ancestral Categories in Human Genetics Research. *American Journal of Human Genetics*, 77(4), 519–532. Retrieved on September 19, 2017 from https://www.ncbi.nlm.nih.gov/pmc/articles/PMC1275602/

Reich, G. (2018). Hitting a Wall? The Trump Administration Meets Immigration Federalism. *Publius: The Journal of Federalism*, 48(3), 372–395.

Roberts, L. (September 26, 2017). Roberts: 526 Arizona Teachers Have Already Quit This Year. *The Republic–AZ Central*. Retrieved on September 27, 2017 from http://www.azcentral.com/story/opinion/op-ed/laurieroberts/2017/09/26/arizona-teachers-quit-school-year/706419001/

Salerno, S. & Reynolds, J. R. (2016). Latina/o Students in Majority White Schools. How School Ethnic Enclaves Link Ethnicity with Success. *Sociology of Race and Ethnicity*, 3(1), 113–125.

Schmalzbauer, L. (2008). Family Divided: The Class Formation of Honduran Transnational Families. *Global Networks*, 8, 329–346.

Shiao, J. L. (2018). It Starts Early: Toward a Longitudinal Analysis of Interracial Intimacy. *Sociology of Race and Ethnicity*, 4(4), 508–526.

Simms, M. C., Fortuny, K. & Henderson, E. (August 2009). *Racial and Ethnic Disparities among Low-income Families*.

Retrieved on April 17, 2016 from http://www.urban.org/sites/default/files/alfresco/publication-pdfs/411936-Racial-and-Ethnic-Disparities-Among-Low-Income-Families.PDF

Smedley, A. (1999). *Race in North America: Origin and Evolution of a Worldview*, 2nd edn. Boulder, CO: Westview Press.

Sotomayor, M., Mccausland, P. & Brockington, A. (August 12, 2017). Charlottesville White Nationalist Rally Violence Prompts State of Emergency. NBC News. Retrieved on August 27, 2017 from https://www.nbcnews.com/news/us-news/torch-wielding-white-supremacists-march-university-virginia-n792021

Spickard, P. R. (1992). The Illogic of American Racial Categories. In M. P. P. Root (Ed.), *Racially Mixed People in America* (pp. 12–23). Newbury Park, California: Sage.

Stanford Center on Poverty and Inequality (2017). State of the Union: The Poverty and Inequality Report. Special Issue, *Pathways Magazine*. Retrieved on September 26, 2017 from http://inequality.stanford.edu/publications/pathway/state-union-2017

Takaki, R. (2008). *A Different Mirror: A History of Multicultural America*. New York: Back Bay Books.

Taylor, R. (2001). *Minority Families in the United States: A Multicultural Perspective*, 3rd edn. New Jersey: Pearson.

Toppo, G. & Overberg, P. (2015). After Nearly 100 years, Great Migration Begins Reversal. *USA Today*. Retrieved on August 26, 2016 from http://www.usatoday.com/story/news/nation/2015/02/02/census-great-migration-reversal/21818127/

Trevelyan, E., Gambino, C., Gryn, T., Larsen, L., Acosta, Y., Grieco, E., Harris, D. & Walters, N. (November 2016). *Characteristics of the U.S. Population by Generational Status: 2013*. Retrieved on July 19, 2017 from https://www.census.gov/content/dam/Census/library/publications/2016/demo/P23-214.pdf

U.S. Bureau of Labor Statistics (August 2018a). *Highlights of Women's Earnings in 2017*. Retrieved on October 5, 2018 from https://www.bls.gov/opub/reports/womens-earnings/2017/home.htm

U.S. Bureau of Labor Statistics (April 19, 2018b). *Employment Characteristics of Families Summary 2017*. Retrieved on October, 2018 from https://www.bls.gov/news.release/famee.nr0.htm

U.S. Bureau of Labor Statistics. (August 2018c). *Labor Force Characteristics by Race and Ethnicity, 2017*. Retrieved on October 6, 2018 from https://www.bls.gov/opub/reports/race-and-ethnicity/2017/home.htm

U.S. Census Bureau (September 13, 2012). *The Great Migration, 1910 to 1970*. Retrieved on August 26, 2016 from https://www.census.gov/dataviz/visualizations/020/

U.S. Census Bureau (November 30, 2016a). *Children of Foreign-Born Parents Generation More Likely to Be College-Educated Than Their Parents*, Census Bureau Reports. Retrieved on July 19, 2017 from https://www.census.gov/newsroom/press-releases/2016/cb16-203.html

U.S. Census Bureau (October 04, 2016b). *FFF: Thanksgiving Day: Nov. 24, 2016*. Retrieved on October 31, 2016 from http://www.census.gov/newsroom/facts-for-features/2016/cb16-ff19.html

U.S. Census Bureau (June 22, 2017a). *The Nation's Older Population Is Still Growing, Census Bureau Reports*. Retrieved on June 22, 2017 from https://census.gov/newsroom/press-releases/2017/cb17-100.html

U.S. Census Bureau (June 05, 2017b). *FFF: The Fourth of July: 2017*. Retrieved on June 26, 2017 from https://www.census.gov/newsroom/facts-for-features/2017/cb17-ff10-fourth-of-july.html

U.S. Census Bureau. (March 30, 2017c). *Highest Educational Levels Reached by Adults in the U.S. since 1940*. Retrieved on June 25, 2017 from https://www.census.gov/newsroom/press-releases/2017/cb17-51.html

U.S. Census Bureau (September 27, 2017d). *Stats for Stories: Ancestor Appreciation Day*. Retrieved on September 27, 2017 from https://www.census.gov/newsroom/stories/2017/september/ancestor.html

U.S. Census Bureau (February 21, 2017e). *FFF: Irish-American Heritage Month (March) and St. Patrick's Day (March 17) and 2017 American Community Survey*, Table S0201, Selected Population Profile for Irish in the United States. Retrieved on June 24, 2017 from https://www.census.gov/newsroom/facts-for-features/2017/cb17-ff05.html

U.S. Census Bureau (2017f). *C04006 Selected Population Profile in the United States 2017 American Community Survey 1-Year Estimates*. Retrieved on October 1, 2018 from https://factfinder.census.gov/faces/tableservices/jsf/pages/productview.xhtml?src=bkmk

U.S. Census Bureau (2017g). *S0201. Polish. Selected Population Profile in the United States 2017 American Community Survey 1-Year Estimates*. Retrieved on October 1, 2018 from https://factfinder.census.gov/faces/tableservices/jsf/pages/productview.xhtml?src=bkmk

U.S. Census Bureau (2017h). *QuickFacts: United States*. Retrieved on October 6, 2018 from https://www.census.gov/quickfacts/fact/table/US/PST045217#viewtop

U.S. Census Bureau (October 2017i). *American Indian and Alaska Native Heritage Month: November 2017*. Retrieved on October 6, 2018 from https://www.census.gov/newsroom/facts-for-features/2017/aian-month.html

U.S. Census Bureau (January 01, 2017j). *FFF: National African-American History Month: February 2017*. Retrieved on June 25, 2017 from https://www.census.gov/newsroom/facts-for-features/2017/cb17-ff01.html

U.S. Census Bureau (August 31, 2017k). *FFF: Hispanic Heritage Month 2017*. S0201 Selected Population Profile in the United States 2017 American Community Survey 1-Year Estimates. Retrieved on September 18, 2018 from https://www.census.gov/newsroom/facts-for-features/2017/hispanic-heritage.html

U.S. Census Bureau (March 14, 2017l). *FFF: Asian/Pacific American Heritage Month: May 17.* Retrieved on June 25, 2017 from https://www.census.gov/newsroom/facts-for-features/2017/cb17-ff07.html

U.S. Census Bureau (September 11, 2017m). *Pacific Coast & Most Northeast States Lead in Broadband Internet Use.* Retrieved on September 12, 2017 from https://census.gov/newsroom/press-releases/2017/internet-use.html

U.S. Census Bureau (February 15, 2018a). *Women's History Month: March 2018.* Retrieved on September 28, 2018 from https://www.census.gov/newsroom/facts-for-features/2018/womens-history.html

U.S. Census Bureau (September 2018b). *Income and Poverty in the United States: 2017.* Retrieved on October 1, 2018 from https://www.census.gov/library/publications/2018/demo/p60-263.html

U.S. Census Bureau (October 6, 2018c). *U.S. and World Population Clock.* Retrieved on October 2, 2018 from https://www.census.gov/popclock/

U.S. Census Bureau (January 23, 2018d). *About.* Retrieved on October 6, 2018 from https://www.census.gov/topics/population/race/about.html

U.S. Census Bureau (February 6, 2018e). *Irish-American Heritage Month (March) and St. Patrick's Day: March 2018 and 2017 American Community Survey.* Retrieved on October 6, 2018 from https://www.census.gov/newsroom/facts-for-features/2018/irish-american-month.html

U.S. Census Bureau (October 2018f). *Italian American Heritage and Culture Month: October 2018.* Retrieved on October 6, 2018 from https://www.census.gov/newsroom/stories/2018/italian-american.html

U.S. Census Bureau (October 2018g). *Polish-American Heritage Month: October 2018.* Retrieved on October 6, 2018 from https://www.census.gov/newsroom/stories/2018/polish-american.html

U.S. Census Bureau (Feb 2, 2018h). *Facts for Features: National African-American (Black) History Month: February 2018.* Retrieved on October 6, 2018 https://www.census.gov/newsroom/facts-for-features/2018/black-history-month.html

U.S. Census Bureau (Feb 2, 2018i). *Hispanic Heritage Month 2018.* Retrieved on October 6, 2018 from https://www.census.gov/newsroom/facts-for-features/2018/hispanic-heritage-month.html

U.S. Census Bureau (May 1, 2018j). *Asian and Pacific American Heritage Month: May 2018.* PEPSR5H- Annual Estimates of the Resident Population by Sex, Race Alone or in Combination, and Hispanic Origin for the United States, States, and Counties: April 1, 2010 to July 1, 2017. Population Estimates. Retrieved on October 7, 2018 from https://www.census.gov/newsroom/facts-for-features/2018/asian-american.html

U.S. Equal Employment Opportunity Commission (EEOC) (1961). *Executive Order 10925.* Retrieved on October 7, 2018 from https://www.eeoc.gov/eeoc/history/35th/thelaw/eo-10925.html

Wagley, C. & Harris, M. (1958). *Minorities in the New World: Six Case Studies.* New Jersey: Columbia University Press.

Wang, W. (June 12, 2015). *Interracial Marriage: Who is 'Marrying Out'?* Washington, DC: Pew Research Center. Retrieved on July 27, 2016 from http://www.pewresearch.org/fact-tank/2015/06/12/interracial-marriage-who-is-marrying-out/

White House (2014). *Presidential Proclamation—German-American Day, 2014.* Retrieved on October 1, 2017 from https://obamawhitehouse.archives.gov/the-press-office/2014/10/03/presidential-proclamation-german-american-day-2014

Wilson, V. (May 11, 2017). *African American Women Stand Out as Working Moms Play a Larger Economic Role in Families.* Economic Policy Institute. Retrieved on September 25, 2017 from http://www.epi.org/blog/african-american-women-stand-out-as-working-moms-play-a-larger-economic-role-in-families/

Wilson, V., Jones, J., Blado, K. & Gould, E. (July 28, 2017). *Black Women Have to Work 7 Months into 2017 to Be Paid the Same as White Men in 2016.* Economic Policy Institute. Retrieved on September 25, 2017 from http://www.epi.org/blog/black-women-have-to-work-7-months-into-2017-to-be-paid-the-same-as-white-men-in-2016/

Wilson, W. J. (1999). *The Bridge over the Racial Divide: Rising Inequality and Coalition Politics.* Berkeley, CA: University of California Press.

Yeoh, B. & Lam, T. (2006). The Cost of (Im)Mobility: Children Left Behind and Children Who Migrate with a Parent. Invited paper for ESCAP Regional Seminar on Strengthening the Capacity of National Machineries for Gender Equality to Shape Migration Policies and Protect Migrant Women, *UNESCAP,* 22–24 November 2006, Bangkok, Thailand.

Zaidi, A. U. & Shuraydi, M. (2002). Perceptions of Arranged Marriages by Young Pakistani Muslim Women Living in a Western Society. *Journal of Comparative Family Studies,* 33(4), 495–514.

Zhou, M. & Gatewood, J. V. (2000). *Contemporary Asian America: A Multidisciplinary Reader.* New York: New York University Press.

5
GENDER ROLES AND SOCIALIZATION

Learning Outcomes

Upon completion of the chapter, students should be able to:

1. compare the concepts of *gender, sex,* and *gender identities*
2. differentiate *instrumental roles* from *expressive roles*
3. compare men's and women's participation in household labor and childcare
4. define the concept of *gender socialization*
5. describe the social learning, cognitive development, and symbolic interaction perspectives of gender socialization
6. identify the agents of gender socialization
7. describe children's gender socialization in two-parent and single-parent headed families
8. identify the spousal stressors and systems of support in military families and families of children with disabilities
9. explain how sociocultural and geographic change may challenge long-standing norms and customs on gender and power in refugee and immigrant families
10. explain why some institutionalized support settings have been criticized as working against single homeless fathers
11. identify new life roles that may benefit midlife and older men and women
12. describe arguments for and against family care and medical leave policies.

Brief Chapter Outline

Pre-test

Engaged or active learning is a powerful strategy that leads to better learning outcomes. One way to become an active learner is to begin the chapter and try to answer the following true/false statements from the material as you read. You will find that you have ready answers to these questions upon the completion of the chapter.

Sex involves biological distinctions; *gender* is learned socially and culturally.	T	F
Talcott Parsons used the term *instrumental* to describe men's traditional roles and the term *emotional* to describe women's traditional roles.	T	F
Men with the highest and lowest incomes are most likely to have stay-at-home wives.	T	F
Men have become less involved with housework and childcare.	T	F
Gender socialization involves learning the separate roles, behaviors, attitudes, personalities, and representations that are socially attributed to men and women.	T	F
It is generally accepted that domestic labor performed by women constitutes "real" work.	T	F
Carla Pfeffer's study on housework and emotion work within transgendered families found their division of labor to be "strongly egalitarian."	T	F
In Sweden, the images found in children's textbooks are said to promote gender equity.	T	F
Cognitive development theory explains that both sexes enact behaviors based on symbols, roles, and meanings that are defined as either "masculine" or "feminine."	T	F
Men's concern for body image is growing to be equal to that of women.	T	F

Upon the completion of this section, students should be able to:
LO1: Compare the concepts of *gender, sex,* and *gender identities.*
LO2: Differentiate *instrumental roles* from *expressive roles.*
LO3: Compare men's and women's participation in household labor and childcare.

Barbara, a 22-year-old returning college student, recalls mornings at her home in rural Louisiana:

> By the time I made it to eleventh grade, I never needed anyone to get me out of bed for school. It was the noises and smells coming from around the house that did it for me. The "big one" was the smell of my mom cooking breakfast. I had to make sure the kitchen was cleaned ... Then, in the bathroom next to my bedroom, I heard the water faucet gurgling and my youngest brother acting cranky as usual. It was my dad's job to get him ready for daycare—comb his hair, make sure his teeth were brushed, stuff like that ... Dad would do this for the boys; mom did this in the mornings for me when I was little. My eldest brother had to feed the horses [before school], and it was my job to put the first load of clothes into the washer and dryer. I'd fold them after I got home from school ...This was before we all left the house [before] going to school and before my parents drove off to work (personal communication, October 1, 2016).

Gender Roles and the Family

Members of Barbara's family accurately might say that their roles in the family functioned to make the early morning more predictable and, thus, more organized. Yet, to some degree, their assigned family roles reflect expectations traditionally accorded to the social construct of gender—socially constructed roles, behaviors, activities, and attributes that a given society considers appropriate for men and women. It is Barbara's mother, for example, who cooks the morning meal while Barbara tidies up the kitchen and later does the laundry. Barbara's brother, however, is assigned no in-house chores but is required to care for the family's livestock outside the home. Put another way, it could be argued that some morning roles expected of some family members were gendered:

household assignments were structured around the obligations, characteristics, and behaviors that the larger society would expect of women and men.

Unlike **sex**—the *biological* distinction between females and males—**gender** is a *social* construct and not so fixed or innate. Rather than a biological fact, gender is more of a social performance that is contoured around men's and women's treatment and social evaluation because of their sex. Our **gender identity**—that is, our subjective sense of being a boy, girl, man, woman, or combination thereof—is developed early in our lives. We learn how to be "masculine" or "feminine," men or women, as we indirectly observe, directly experience, and come to associate different patterns of behavior with each sex. However, because gender and the roles traditionally attributed to men and women are learned, they also are less fixed and are subject to change. Recall that, in Barbara's story, it is her father who readies the youngest child, a preschooler, for daycare in the mornings. Moreover, Barbara's mother, a working woman with a career, leaves for work in the mornings alongside Barbara's father.

This section further discusses the concept of gender and how being female or male affects us all, especially with regard to our predicted and performed functions, responsibilities, and other behaviors related to family and social life. We begin with an examination of traditional gender roles in families.

Traditional Views on Gender Roles in Families

American sociologist Talcott Parsons (1902–79) characterizes the family as a basic social unit whose primary function is to meet critical human needs (Parsons & Bales, 1955). However, the family also is a social institution positioned within a larger, interconnected system of societal institutions. As a result, the family carries a reach so broad that the state of the

Source: Monkey Business Images/www.Shutterstock.com

family is consequential to the order of a society and vice versa. Parsons suggests that the different roles expected of men and women have a significant influence in maintaining societal order, especially when practiced within the traditional or nuclear family type.

How does Parsons relate traditional sex-specific roles and gender-based expectations in families to a more stable society? Parsons viewed men and women in nuclear families as having spheres of influence that are separate yet complementary. For example, men traditionally are expected to assume the position of family breadwinner. For Parsons, in this sexual division of labor, it would follow that males are socialized to demonstrate the confidence, rationality, and assertiveness required for leadership and success in the labor market. Parsons identified these masculine social roles as **instrumental roles**. Divergent yet in harmony with men's instrumental roles, Parsons argues, are women's feminine or **expressive roles**. Examples of expressive roles involve demonstrating affection, lovingness, supportiveness, selflessness, and the emotional support required in childrearing and managing family members' needs. For Parsons, women's nurturance in concert with men's economic protection was essential to the ordered operation of the home and a foundation for that same stability in the greater society.

Mothers Staying at Home

Men's and women's shares of labor, particularly the traditional husband-provider/wife-caregiver contrast as described by Parsons, has long been studied by sociologists. Although some experts would argue that men have had more choices in the kinds of paid labor they do, fewer would say that men have had a choice in *whether* to work. Women, particularly mothers, may experience more leniencies than men in whether to take on paid labor. Whether inside and/or outside the home, women's participation in labor has resulted in separate, if not competing identities for them as a group. For example, women who are not traditional "stay-at-home" wives and/or mothers may be viewed as "working mothers," who prioritize employment over caregiving and family life. Meanwhile, traditional "stay-at-home" wives and/or mothers may be viewed as "fully" committed to family obligations.

Certainly, the traditional role of being at home full-time is not a reality for most women and their families today. However, deciding not to engage in paid work and instead choosing to stay home has been a reality for a substantial portion of women with children. The trajectory of mothers' paid employment shows a notable increase followed by a plateau or leveling off in the rate of change. In 1970, only 30 percent of mothers of children under five had been employed in the previous year, which was followed by an increase to 46 percent in 1980 and 60 percent in 1990. The next decade, however, saw only a small increase in mothers' paid labor: a rate of 60 percent in 1990 to 65 percent in 2000 (Cotter *et al.*, 2008).

Why is it that women's employment had appeared to level off over time? Some believe that women's employment levels could not reach full parity with men's unless men took on a more equal share of childrearing. Others point out that employers could have adopted policies making it easier for parents to combine work and family responsibilities. A cultural backlash to the women's movement may be possible as well. For example, critics have long observed an antifeminist countermovement that is concurrent with feminist efforts to include women in professions historically dominated by men.

In 2014, researchers observed a reversal in the stall of stay-at-home mothers, something that had persisted for the last three decades of the twentieth century. The Pew Research Center found that the share of mothers who do not work outside the home rose to 29 percent in 2012, up from a modern-era low of 23 percent in 1999. The recent turnaround comes with a downturn in women's labor force participation and a continued ambivalence about the impact of working mothers on young children (Cohn *et al.*, 2014).

The conventional wisdom has been that married mothers will choose to stay home if their husbands' earnings were high enough to make staying at home affordable. Cotter *et al.* (2008) tested this assumption using longitudinal data spanning from 1968 to 2006. The sociologists were surprised to learn that between 1991 and 2006, the largest group of stay-at-home mothers were wives whose husbands were in the lowest 25 percent of the male earnings distribution. Women married to men in the highest 5 percent of the income distribution made up the next largest group, suggesting that the two groups of married mothers with the lowest employment rates are those with both the poorest and the richest spouses. The researchers speculate that low-income and high-income mothers have different reasons for their relatively low rate of employment. Mothers with low-earning husbands may have great economic need for jobs. However, these women often have low education and a lower earning potential, which hinders job pay above the cost of reliable childcare. On the other hand, mothers with the highest-earning husbands have little economic need to be employed.

Benefits and Drawbacks of Traditional Gender Roles in Families

Many women find value in their roles as stay-at-home mothers. For these women, foregoing paid labor and taking on traditional gender roles avoids the tension of navigating the workplace plus family demands. Staying at home also allows opportunities for mothers to maximize their time with their families.

There are stay-at-home mothers who are disabled, enrolled in school, or simply unable to find work; however, there are others who made the choice to leave paid work. In psychologist Adrienne Partridge's (2015) study of women who left paid careers to stay at home, some women became stay-at-home wives and mothers to end spousal conflict, including arguments with their husbands about their inaction in childcare and certain aspects of family life. Circumstances at work led others to feeling "pushed out" of their paid jobs, which made continuing to perform paid work difficult. All of Partridge's respondents had stay-at-home mothers when younger, something she suggests may have influenced their decisions to become adult homemakers. Also notable, the respondents were active community or school volunteers during their period of stay-at-home mothering; nonetheless, each missed some aspect about having a career.

Women who remain home to engage in unpaid household labor may find their contributions underappreciated. But why this drawback regarding women's traditional roles in family life? In a money-based economy, labor that yields a wage or salary is often viewed as "real" work or "real" economic activity, a characterization that effectively excludes household work. Some scholars have responded by arguing for an expanded definition of work, particularly one inclusive of unpaid sustenance labor as well as work performed in the home (Donahoe, 1999).

Changing Views on Traditional Gender Roles

Some things have changed for men and women over the decades. For example, did you know the following?

- In the 1960s, a bank could refuse to issue a credit card to an unmarried woman, and if she was married, her husband was required to co-sign. It was not until the Equal Credit Opportunity Act of 1974 that it became illegal to refuse a credit card to a woman based on her sex.
- Women were once kept out of jury pools because they were seen as too fragile and overly sympathetic to hear details and remain objective about certain criminal offenses. It was also believed that women's commitment was to home, not jury duty. It was not until 1973 that women could serve on juries in all 50 states.
- Men were forbidden to attend some state-supported nursing schools until 1982, when the Supreme Court found that denying men

admission also denied men's access to the career of their own choosing.

- It wasn't until the 1970s that hospitals and physicians began permitting men in the delivery room during the birth of their own child.

For some, it is difficult to identify with men and women's glaring, unequal positioning in the basic social processes shown in the above examples. Because of social movements and policy changes, some perhaps during your lifetime, expectations of conduct that once limited men and women have been challenged or overturned. Housework, to a degree, has also undergone a sort of gender "reboot."

Housework may involve everyday tasks like preparing meals, shopping for groceries, or caring for children; however, its implications are more than ordinary. For those who perform housework, the labor is ongoing, time intensive, sometimes physically demanding, but relevant to family well-being, maintenance, and order. For sociologists and other family scientists, housework is also a matter of gender. Household labor often is used as a measurement to conceptualize and test how responsibilities are divided between the sexes. How household labor is assigned among the sexes may help explain why some gender stereotypes tend to endure and reproduce. **Gender stereotypes** are culturally based beliefs about how people typically will look, act, feel, think, and behave based on their biological sex. Recent analyses by researchers have uncovered a growing participation by men in household labor and childcare.

Men's Growing Participation in Housework and Childcare

Increasingly, men's contribution to housework has become more involved and less traditional. In their 2008 report on men's growing participation in

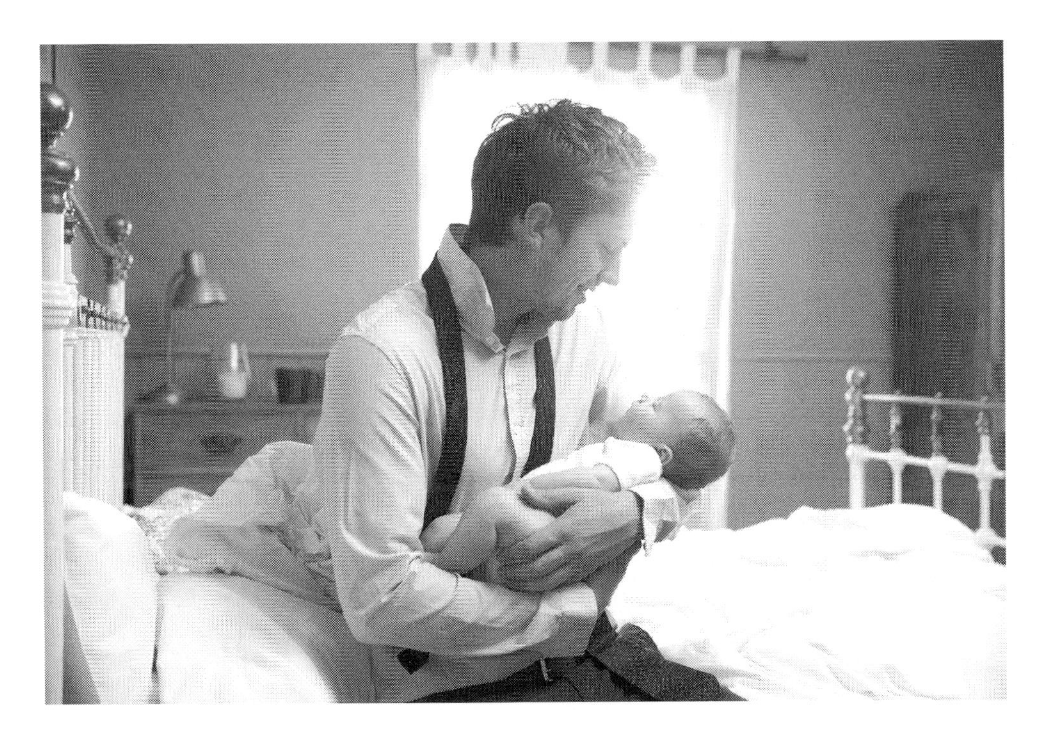

Source: Monkey Business Images/www.Shutterstock.com

domestic labor, sociologists Oriel Sullivan and Scott Coltrane (2008) observed that men's housework involvement doubled between the 1960s and the twenty-first century. In a more recent study by the Pew Research Center, it was found that fathers spent more than twice as much time doing housework as they did in the 1960s (ten hours vs. four hours per week), cutting mothers' housework time almost in half during the same period (18 hours vs. 32 hours per week) (Parker & Wang, 2013).

Fathers are also more involved in childcare than they were 50 years ago. In 2015, fathers reported spending seven hours a week on childcare on average—almost triple the time spent in 1965. Mothers reported an average of about fifteen hours a week on childcare and eighteen hours a week on housework that same year. Although fathers are more engaged with their children, many still perceive themselves as not doing enough: 48 percent of fathers feel that they spend too little time with their children, compared with 25 percent of mothers who say the same. Only 39 percent of fathers say that they are doing a "very good job" raising their children, compared with 51 percent of mothers (Parker & Livingston, 2016).

What is behind the movement of fathers and mothers sharing in childcare and housework? For one, creating and spending time with children seems to be distinctly taking hold as a quality of sound parenting. However, family scientists more frequently point to women entering the labor force in the 1970s and 1980s (Sullivan & Coltrane, 2015), making shared childcare and housework a practical response to employed women's absence from the home. There is also evidence that patterns of parent–child involvement may be predicted by having not one but two children of the same sex. A study of two-parent American families with at least two middle school children found that mothers spent more time with children than did fathers in families with two daughters. Conversely, fathers spent more time with children than did mothers in families with two sons (McHale & Crouter, 2003).

Women's Continued Participation in Housework and Childcare

Despite fathers' increase in childcare participation, mothers tend to spend about twice as much time with their children as fathers do. In 2011, the average childcare time was 7.3 hours per week for fathers but 13.5 hours per week for mothers (Parker & Wang, 2013). Because a proportion of women contribute to their families in both non-traditional and traditional ways, the standard "nine-to-five, Monday through Friday" work adage does not hold for many. In sociologist Arlie Hochschild's (1989) sample of two-income families, women continued their traditional roles as primary caregivers even when employed outside the home, creating what Hochschild calls a stressful "second shift" of unpaid domestic labor. Her study also found that women's unpaid family labor at home, including childcare and household maintenance, amounted to an extra month of work each year. Recent research adds that even in households with stay-at-home fathers and breadwinning mothers, the reversal of traditional gender roles may only go so far. When breadwinning wives and mothers returned home from their paid jobs, some shifted into traditional domestic responsibilities and allocated their husbands more time to pursue other activities. The study concludes that some parents continue to "do gender" in traditional ways during evenings and weekends, even when fathers specialize in caregiving and mothers specialize in breadwinning (Latshaw & Hale, 2016).

Sullivan and Coltrane (2015) affirm the growing convergence of gender roles, including a more equitable housework arrangement between mothers and fathers. Even so, they approach the subject with some caution, suggesting that movements toward gender equality tend to be met with counterprotest or forms of institutional backlash. They and other family scien-

Diversity Overview: Household Work and Women Partners of Transgender Men

Once almost exclusively relegated to sensationalistic portrayals on daytime talk shows, **transgender people** (those whose gender identity does not align with their biological sex) and **transsexual people** (a type of transgender identity in which individuals undergo anatomical and/or hormonal changes to their bodies, so that they are closer to their corresponding gender identity) are receiving more serious media depiction now than ever before. The lives of transgender or transsexual people can be viewed in films and documentaries such as *Boys Don't Cry* (1999), *Transamerica* (2005), and *Transparent* (2014). Transgender issues have been covered by every major U.S. television broadcast news network as well as the British Broadcasting Corporation (Pfeffer, 2010). One of the most recognized openly transgendered Americans in recent times is Caitlyn Jenner. Born Bruce, Jenner came out as Caitlyn, a transgender woman, during a 2015 television interview. In 1976, Jenner earned international status as a male Olympic gold-medal-winning decathlete. More than 30 years later, Jenner became a reality television personality on the popular show *Keeping Up with the Kardashians*.

Despite the group's growing media visibility, academic studies on transgender people and their families remain almost non-existent. In her qualitative study of 50 women whose intimate partners identify as transgender or transsexual men (trans), sociologist Carla A. Pfeffer (2010) exposes non-trans women's everyday experiences in transgender family life. Some of the study's major findings on women's perceptions of the division of household labor and **emotion work** (people's active engagement in the management of their own and others' emotions) within family relationships are provided below.

- Despite a strong feminist self-identification among respondents and their trans male partners, the women Pfeffer interviewed often spoke of an unequal division of household labor. Most framed non-egalitarian household labor practices as a personal preference and not concessions based on traditional gender role expectations. Pfeffer explains that given few cultural models of equality-sensitive male and trans male identities, findings showing an adherence to existing social models was unsurprising.
- The women in Pfeffer's (2010) study elaborated on attending to and being accountable for the detailed organization of their trans male partners' personal and emotional lives. Pfeffer suggests that such routines reflect traditional gender roles, specifically the expectation that women manage not only their emotions but also the emotions of others.

Pfeffer (2010) states that the division of household labor and emotion work within transgendered families "cannot simply be described as egalitarian" (p. 165). Her sample frequently acted "in critical ways to act to shape, support, reflect, and coproduce seemingly normative forms of masculinity and femininity" (p. 179). The respondent "Michele" offered one of the clearest examples of a woman partner's unequal investment in her relationship with her trans male partner. When asked how much of her life comprises taking care of her partner and issues related to his gender transition, Michelle replied, "I would say, percentage wise—and this is something I've been trying to change because I see it being a problem—I would say about 70% of my life. That's scaled back from what it was … like, 80%" (p. 174). Given her findings, Pfeffer cautions that we not make over-simplistic assumptions about the women in her sample. Without existing models for how to "do" transgender partnerships, she explains, these women and their partners are navigating uncharted territories.

Upon the completion of this section, students should be able to:
LO4: Define the concept of *gender socialization.*
LO5: Describe the social learning, cognitive development, and symbolic interaction perspectives of gender socialization.
LO6: Identify the agents of gender socialization.

tists also underscore that, despite fewer women being full-time homemakers compared to decades earlier, household labor remains a ceaseless commitment for women as a group. Either women are likely to do their own housework or they may hire a domestic worker, who is also likely to be a woman.

The Socialization of Gender Roles

Answer the following questions as truthfully as you can.

- One partner of a two-earner, two-career heterosexual family must stop working and stay home with their two-year old son, who has been diagnosed with a life-threatening illness. How should the couple decide which partner will quit work?
- Your child's school has just hired a man to teach preschool, and you learn that your child has been enrolled in his class. Would you feel uncomfortable with the decision to hire a man to teach preschool?

In whatever way you answer these questions, your responses are likely to reflect your gender socialization. **Gender socialization** is the process of learning the various roles, behaviors, attitudes, personalities, and representations that are socially associated with men and women. Gender socialization is an ongoing process that begins at birth and continues throughout our lives. But before we explore further

Source: blvdone/www.Shutterstock.com

the topic of gender and socialization, we start with its premise: **socialization**.

As members of a society, we cannot help but socially interact with others at some point. During these interactions, we learn and experience attitudes, ideas, and expectations about multiple areas of social life along the way. This ongoing process is socialization, in which individuals learn or adapt to patterns of thought and conduct in a manner approved by a group, society, or culture.

Social scientists argue that gender socialization facilitates the differences observed between males and females in their patterns of conduct, dress and appearance, and social acceptance and treatment, in addition to other factors contributing to their development as men and women. From this perspective, the countless distinctions we associate with being men and women, masculine or feminine, are a result of a lifetime of learning the social expectations and attitudes associated with one's sex. This is not to say that social scientists deny that hormones and other biological factors are involved in how males and females think, behave, and communicate. Rather, they are more likely to argue that, because gender socialization begins so early, it is difficult to pinpoint the distinction between biological and social influences. Let's begin with an examination of three theoretical perspectives on gender role socialization: the social learning, cognitive development, and symbolic interaction perspectives.

Theories of Gender Role Socialization

The **social learning perspective of gender role socialization** posits that males and females learn their gendered social roles through a system of socialization, which rewards their adherence to sex-typed norms and behaviors and which punishes their adherence to behaviors inconsistent with sex-typed norms and behaviors. A parent, for example, may heap praise (reward) upon a son who wants to play high school football. Yet that same parent may scold (punish) his or her daughter who aspires to do the same. In a different scenario, without a word about gender, the parent could try to sidetrack the daughter's enthusiasm by extolling the more "feminine" athleticism of cheerleading and promising to pay for next month's tryouts (reward). This system of reward and punishment—whether directly spoken or indirectly shown—makes up a socialization process in which individuals learn, acquire, and model expected gendered roles, norms, and behaviors through their interaction with and reactions from others.

Once behavioral standards are successfully socialized, individuals may model their own behavior against those standards, perhaps to where they no longer depend on external rewards or punishment for direction. The daughter, for example, may adopt the external message of "girls are not allowed to play football." Taking the parent's advice but still wanting to participate in sports, she may try out for cheerleading, making discouragement from her parent unnecessary. Indeed, not all individuals are swayed by the gendered messages of those around them. In fact, the theory's one-way effect of environment on behavior has been criticized as having a simplistic view of socialization and human development.

The **cognitive development perspective of gender role socialization** posits that individuals observe differences between women and men and then pattern their behaviors around gender categories aligned with one biological sex or another. An example may be for the aspiring football player son to observe the prevailing norm of boys playing football in his social environment and rationalize, "I am a teenage boy in high school with an athletic ability. It makes sense that I perform activities that adolescent athletic boys often do—like play football."

The cognitive development perspective removes social rewards or punishments as a primary source of our gender role socialization. Rather, the thought processes and social patterns that shape our beliefs about men and women are said to determine our perceptions of what is or what is not masculine or feminine. Consequently, not only does the perspective place importance on external or environmental contexts but it also emphasizes individuals' internal or mental frameworks.

The **symbolic interaction perspective of gender role socialization** posits that, through the process of socialization, men and women adapt to behaviors, character traits, self-identities, roles, symbols, and meanings that are socially identified as masculine or feminine/male or female. For example, let's revisit the example of the son and daughter who both expressed an interest in playing football. Say that not only their parents but also societal cues and symbols (male-centered football merchandizing, media advertisements, and local, regional, and national all-male football teams) present football as a "masculine" sport. The brother, who now plays high school football, may perceive himself as "masculine" partially on that basis. After all, he is "doing" a sport that is socially constructed as "masculine," thus confirming his own sense of maleness. Alternately, say that one of the parent's arguments against the daughter's football aspirations is that the sport is too "masculine" for a girl. The parent compensates for the daughter's disappointment by adding that cheerleading is also an athletic activity, although one of a more "feminine" quality and temperament. Say that the daughter adheres to the gendered expectations suggested by the parent. In doing so, the daughter may interpret: "As a cheerleader, I am able to develop my athleticism and compete, while I still retain my femininity." In this scenario, both siblings agree to codes of expectations that are socially endorsed as appropriate and allied with their sex: he enrolls in high school football; she signs up for high school cheerleading.

Agents of Gender Socialization

Who or what are the contexts—the people, groups, or other social institutions—through which gender socialization takes place? Let's now turn our attention to five influential mediators of our gender development: parents and family, schools, sports, peers, and the media. These mediators are called **agents of gender socialization**. We begin with what is often considered the most important socialization agent, the family unit.

Parents and Family

Parents exhibit and pass on their own beliefs about gender to their children, whether intentionally or not. In fact, the process of socialization is so persistent and gendered norms and expectations are ingrained so early that they often feel "natural" and not socially influenced.

Earlier studies on gender and children's toys found that parents often provided boys with vehicles, action figures, and sports equipment and encouraged sons to engage in more physical play. Daughters were provided dolls, miniature kitchen appliances, and dress-up toys and were guided toward games of cooking, cleaning, childcare, and family nurturing (Lytton & Romney, 1991). Parents have become more flexible regarding gender norms surrounding their children's play—but only to a point. Even parents with egalitarian attitudes about gender may act differently when it comes to their daughters and sons. For example, fathers may encourage the athletic participation of their daughters; however, few fathers or mothers encourage doll play for their sons (Leaper, 2014).

Current research has sought to uncover children's preferences for toys commercially targeted for one sex or the other. In a recent study of 101 boys and girls of three age groups (9 to 17 months, 18 to 23 months, and 24 to 32 months), children preferred to play with toys typed to their own gender, demonstrating that sex differences in toy preference appear early in development. The study also revealed a trend in which both boys and girls showed an increasing preference for toys stereotyped for boys as they increased in age (Todd *et al.*, 2017).

Social and behavioral scientists alike have questioned the long-term effects of socially endorsed gender-based norms, attitudes, behaviors, and identities on health and well-being. One focus has been the toll of **gender role conflict** among men, an intrapersonal and interpersonal stress resulting when a man does not or cannot conform to social expectations of male conduct. For example, not fulfilling masculinity standards may lead to feelings of low self-esteem for some. Even so, studies suggest that men who report greater gender role conflict are less willing to utilize counseling services, especially if uncomfortable with therapeutic processes or disclosing personal distress (Pederson & Vogel, 2007). In another study, college men who endorsed traditional male-role norms were less communicative about themselves to female friends and supported a one-way decision-making power with their intimate partners (Thompson *et al.*, 1985).

Parent-to-child interactions tend to socialize an emotive, more talkative but less assertive communication pattern with girls (see Rohlinger, 2007). Some caution that socializing non-assertiveness in girls may pose a liability in their future career paths as women, even if unconsciously. Psychologist Lois Frankel (2014) finds that women often miss opportunities for career-enhancing assignments or promotions because they are reluctant to highlight their abilities, hesitant to speak at meetings, and too overworked to build networks required for long-standing success. For Frankel, these and other inhibitors—wanting to be liked, speaking too softly, or asking for permission to produce an action—are residual impressions from women's early socialization as girls.

Schools

Many teachers say that they treat boys and girls equally; however, their attitudes may reflect subtle biases in classroom settings. Generally, boys experience more challenging student–teacher interactions and often dominate classroom activities and space. This tendency is related to boys receiving more attention than girls through their teachers' criticism, praise, and constructive feedback, regardless of whether the teacher is male or female (United Nations Educational, Scientific, and Cultural Organization, 2007).

Stereotypical gender imagery and bias have been documented in children's books. A study of top-selling and award-winning children's books in the United States found almost twice as many males as females among its sample of 200 book titles and main characters. Male characters appeared 53 percent more in illustrations; female main characters were presented as more nurturing and were seen in more indoor than outdoor scenes than males. Work roles and occupations also were stereotyped by gender, and more women than men appeared to have no paid occupation at all (Hamilton *et al.*, 2006). Experts emphasize textbooks in the United States now are less gender biased than those published more than twenty years ago. However, they also find that many children's textbooks continue to ignore the changes in women's societal position across recent decades (Blumberg, 2007).

A principal concern for experts is that both sexes' depictions may restrict girls' and boys' visions of who they are and what they can become. More to the point, will girls avoid mathematics and science and instead follow gender-stereotyped courses of study (literature, composition, foreign language) despite an interest or an ability to do otherwise? Will boys remain disadvantaged or unaided in reading and language skills, especially in countries where boys fail to show basic levels of reading literacy? High school course-taking patterns may foreshadow gender differences in colleges and universities, where degrees and specializations still show a high concentration of one sex or the other. In the United States, for example, women are likely to pursue degrees in education, English, nursing, and some social sciences and are less likely to pursue degrees in science, math, and engineering. Because these male-dominated fields are valued and demand a high salary, women's absence from them accounts for a great deal of the gender gap in pay.

Sports

Masculine-identified sports often demonstrate some aspect of aggressiveness and power. Specifically, there is an attempt to overpower the opponent by bodily contact, to project the body into or through space over distances, and to compete face-to-face in situations where bodily contact may happen (Metheny, 1965). A sport also may be perceived as "masculine" if it reinforces a sense of identity, camaraderie, and solidarity with men to the exclusion of women. Football has been pointed out as serving this function in the United States (Koivula, 2001).

Gender development comes early and with consequences for many young male athletes. Studies show that boys who are successful at sports are labeled as more "masculine" by their peers; boys who fail at sports may be labeled as "sissies" or "girls" (see Rohlinger, 2007), pejoratives taken as offensive or emasculating within heterosexual male culture. Sports environments that are fueled by hyper-masculine ideals may propagate sexist attitudes. In sociologist Steven Schacht's (1996) study of adult male rugby players, overt sexual harassment of women and homophobic comments were almost ritualistic throughout the athletes' gender identity development and social bonding. Denigrations like these internalize that "real" men detach themselves from "all aspects of femininity" which, as Schacht points out, can "foster a 'better than' hierarchical notion of men in relation to women" (p. 562).

Meanwhile, sports socially identified as "feminine" tend to allow female athletes to stay true to stereotyped expectations of femininity. "Feminine" athletic contests are those said either to demonstrate non-aggression ("flag" football) or some degree of gracefulness, aesthetics, or visual pleasure (figure skating and gymnastics) (Koivula, 2001). Even when embodying physical exertion, competitive sports that permit female participation may be viewed as insubordinate to men's competitions because they are not forceful enough. Sociologist Nancy Theberge (1997) offers women's ice hockey, a sport with little or no bodily contact, as a case in point. Given its low occurrence of bodily contact, some view women's ice hockey as less competitive, even inferior to the "real" game of men's full-body ice hockey. Yet parents have begun to encourage sports participation in their daughters, and sports commonly male-identified (soccer, boxing, football) have gained interest among female athletes.

Peers

Regardless of age or life course, we directly or indirectly learn about gendered norms, actions, and stereotypes from our peers, a group of individuals of similar age, status, and interests. Adults may gauge their movement from one age or status to another by observing the life courses of their friends. For example, adults may perceive and internalize when it is appropriate to get married or have children, or how their lifestyle or range of options therein should be at any time. Unconsciously or not, children watch and learn sex-typed attitudes, ideals, and behaviors as they interact with their peers. A boy, for example, may simply observe that males in his peer group can grow their hair to a certain length without negative reprisal from other peers. For that reason, he may deliberately choose to wear short hair. Although there is no expressed statement or negative feedback in this scenario that directly regulates the boy's style of hair, his peer influence is substantial enough to promote or reinforce his behavior regarding everyday personal appearance.

The Media

The search for the "perfect" body is documented as an ongoing issue among women; however, studies about men's anxieties over body image are scarcer. What studies there are suggest that adolescent, college-age, and older men are experiencing high levels of dissatisfaction with their physical appearance. Consequently, more researchers are examining the media's role in the socialization of men and boys, including their struggle with body image.

Men's concern over body image is growing to equal that of women. In a study by the University of West England's Centre for Appearance Research, four out of five men were unhappy about their bodies, almost 81 percent mentioned perceived imperfections in ways that promoted anxiety about their body image, and 38 percent stated that they would sacrifice at least a year of their lives in exchange for a perfect body. The study also found that 63 percent of the sample thought their arms or chests were not muscular enough, 29 percent thought about their appearance at least five times a day, and 18 percent were on a high-protein diet to increase muscle mass. Some respondents resorted to compulsive exercise, strict diets, or laxatives to lose weight or to achieve a more toned physique (Campbell, 2012).

Males tend to idealize physical images of high muscularity and low body fat. Like many women, men also compare their bodies against idealized media images, finding that they fail to resemble media codes of appearance. In fact, the proportion of undressed men in film and commercial advertisements has increased dramatically since the early 1980s (Pope *et al.*, 2001). Less recognized, however, is that use of anabolic steroids has made it possible for men to become more muscular by means other than exercise alone. As a result, the muscular media images that some men try to emulate may be developed unnaturally or even digitally modified.

Boys are just as vulnerable to judgment and self-criticism regarding their bodies. Findings show that boys' concerns and dissatisfaction may begin as early as second grade, and the effect has been noted among young children across racial and ethnic backgrounds (Heron *et al.*, 2013). Whether or to what extent male action figures or animated characters play a part in this trend is not known. That muscle-body measurements of male action figures exceed those of the largest bodybuilders is better-documented (Pope *et al.*, 1999). Male animated action characters also are more likely than their live-action counterparts to feature a large chest (15.4 percent vs. 4.9 percent), a small waist (18.4 percent vs. 4.3 percent), and an unrealistically large and muscularized physique (12.5 percent vs. 0.5 percent) (Smith & Cook, 2008).

Females tend to idealize thinness, especially a low waist-to-hip ratio. A number of scholars find that the "thin and sexy" cultural ideal has created a situation where the majority of women and girls are dissatisfied with their bodies (Hellmich, 2006). A desire for thinness has been found among girls as young as six years, and many children are well aware that dieting is a means toward achieving this ideal (Lowes & Tiggemann, 2003). The quest to meet culturally idealized codes of attractiveness is likely to continue into adulthood, sometimes with unhealthy results. In one study, women experienced body dissatisfaction, along with a short-term increase in depression, stress, guilt, shame, and insecurity, following their exposure to images of models having a thin-ideal body type (Stice & Shaw, 1994). Further, chasing or retaining the thin-body ideal can involve risks such as eating disorders, excessive exercising, and calorie counting, or a distorted view of personal value. Once restricted to the rich and famous, more middle-class adult women are re-contouring their physical appearance through corrective undergarments, cosmetic surgery (liposuction; face, breast, and buttock "lifts"; and "tummy tucks"), and other procedures (anti-wrinkle injections, chemical peels, and other skin treatments). Many incur substantial monetary debt in the process.

The complex relationship between standards of attractiveness and the self-perceptions of many women and girls is exacerbated by beauty ideals created by cultural media. An analysis of more than 4,000 characters depicted in 400 G, PG, PG-13, and R-rated movies revealed two dominant portrayals: the traditional woman and the hypersexual woman. Traditional depictions showed females in domestic or caregiving functions; hypersexual depictions showed an overemphasis of attractiveness and sexuality by way of clothing, thin body proportions, and an hourglass figure. Females in motion pictures were more than five times as likely as males to be depicted in clothing that enhanced, exaggerated, or called attention to parts of their bodies from neck to knees (Smith & Cook, 2008).

Upon the completion of this section, students should be able to:
LO7: Describe children's gender socialization in two-parent and single-parent headed families.
LO8: Identify the spousal stressors and systems of support in military families and families of children with disabilities.
LO9: Explain how sociocultural and geographic change may challenge long-standing norms and customs on gender and power in refugee and immigrant families.
LO10: Explain why some institutionalized support settings have been criticized as working against single homeless fathers.
LO11: Identify new life roles that may benefit midlife and older men and women.
LO12: Describe arguments for and against family care and medical leave policies.

Television content for children eleven and under also emphasizes media-stylized female body proportions. Females in animated stories are more likely to have small waists compared to females in live action stories (36.9 percent vs. 6.9 percent) and are more likely to have unrealistic body dimensions compared to live action females (22.7 percent vs. 1.2 percent). Researchers conclude that while all female portrayals need not be uplifting or inspirational and not all attractive females need to be eliminated from media, there needs to be a shift from how media most often represent women and girls "away from creating females as adornment, enticement, or with inclination to romance as the main or exclusive personality trait or motivator" (Smith & Cook, 2008, p. 12).

Gender Roles across Family Forms and Transitions

It is difficult if not impossible for a child to grow to adulthood without some exposure to gender role stereotyping. Beginning at birth, newborn daughters are more likely to be described as *little, beautiful,* and *pretty* compared to newborn sons. As boys develop, they may learn to get their parents' attention by crying, screaming, and showing aggression; girls are likely to use gestures, gentle touching, or non-demanding tones of voice (see Rohlinger, 2007). Gender-specific norms of appearance, play, and other behaviors are rewarded or reinforced. In fact, by the age of two, many girls and boys are aware of cultural differentiations regarding femininity and masculinity.

Parents who purchase toys and clothes for their children can attest to the rows, stacks, or web pages of gender-segregated items and their color-coded boundaries. In one study, girls' rooms contained more dolls, furniture, fictional characters, and yellow bedding; boys' rooms had more sports equipment, tools, vehicles, and blue bedding. Boys wore more blue, red, and white clothing and tended to have blue pacifiers; girls wore more pink and multicolored clothes, pink jewelry, and had pink pacifiers (Pomerleau *et al.*, 1990). Yet, the socialization of gender roles goes further than parents' socialization or the proverbial "pink or blue" consumerism. Other factors may be just as influential. Changing family types and consequential adjustments brought by poverty, illness, war or global conflict, military deployment, and other crises may compel family members to rethink and revise their gendered roles and identities.

In this section, we examine gender through the prism of various family complications and compositions, including single-parent, two-parent, and same-sex-couple families in addition to childfree families, military families, families of children with disabilities, immigrant and refugee families, homeless families, and families involving midlife and post-midlife members. Employee-based family care policies also are discussed.

Parenting and Children's Gender Socialization in Two-Parent Families

Some patterns of gender socialization in two-parent families may be easier to anticipate than others. For example, we may expect that there are parents with traditional attitudes about gender who expect their children—especially fathers with regard to their daughters—to perform sex-typed roles. There are two-parent families in which the mother, father, and children map out and perform an equal distribution of labor. But more complex and not so readily assumed are parents who endorse egalitarian views about certain domains for their children (areas of employment), but who are gender-specific about others (daughters' participation in certain family roles). There also are children (especially daughters) with fathers of traditional attitudes but egalitarian behaviors, who take note of their father's more egalitarian side and react against their mother's traditional role in domestic labor (Leaper, 2014; Marks *et al.*, 2009).

Of course, not all children internalize all gendered behaviors within their personal orbits. Nor do all children model the actions of their same-sex parent to every detail. Generally, children tend to be less traditional than their parents in the split of household labor. For example, fathers and mothers in Hilton and Haldeman's (1991) study were consistently traditional in dividing family household work. However, boys in two-parent families were the least sex-typed in household task behaviors and shared more in housework than all of the study's male participants, including adult fathers and boys from single-parent households. Boys in two-parent homes also were found to perform more household tasks compared to boys from single-parent households.

Still, there are certain domains of children's labor that continue to be modeled around parents' sometimes-traditional conventions. In the same study, for example, the principal variable distinguishing male and female children was time used for cleaning the home. Girls accumulated more time per day on house cleaning as well as food preparation and dishwashing than did boys. Boys logged more time compared to girls on a few chores often typified as female domestic labor, namely care of clothing and care of household linens. However, researchers also found that boys expended more time than girls on household management activities, a finding more consistent with conventional gender performance norms (Hilton & Haldeman, 1991).

Parenting and Children's Gender Socialization in Single-Parent-Headed Families

What does family composition have to do with single-parent households and the gender-role socialization of their children? Every day, sons and daughters are given work responsibilities within their single-parent homes; however, the manner and degree to which tasks are assigned can depend on a parent's marital and employment status, socioeconomic class, family structure, as well as social expectations about "men's work" and "women's work." For example, if an employed single parent works full-time—and especially if more than one child is involved—older children may be assigned caretaking duties to meet the family's needs (Weiss, 1979). Girls from single-mother headed homes in Hilton and Haldeman's (1991) study tended to model their mothers' household labor patterns. Girls washed the dishes, cared for clothing, played with children, helped with homework, discussed problems with family members, shared in social activities, and performed other non-physical care behaviors more intensively than boys. Like their single mothers, who balanced household responsibilities and additional demands, girls also performed responsibilities such as taking family members to the doctor and bathing and feeding family members. In fact, girls in single-parent families were less sex segregated in their household tasks than all other adults and children in the study.

That boys from single-parent families were less likely to participate in housework than girls from the same family type proved a surprising outcome for Hilton and Haldeman. The researchers posed several questions to explain the finding. For example, could it be that some single parents assign certain responsibilities to the most accommodating child in order to alleviate potential stress or conflict? Are daughters more compliant than sons? If so, could it be because girls are socialized to be so? Fathers' involvement in childcare was shown to have a negative correlation to gender stereotyping among their children. Are some single-women heads of households reinforcing traditional male roles for their sons to avoid "over-feminizing" them, given the absence of a resident male model?

What is more clearly known is that adult men's decision to act as caregivers may depend on the availability of a female surrogate. Adult sons of frail, elderly parents tend to become caregivers only in the absence of an available female sibling. They also rely on support from their own spouses and provide less overall "hands-on" assistance to their parents (Horowitz, 1985).

Children's Gender Socialization in Same-Sex Couple Families

The number of children with lesbian, gay, or bisexual parents has increased in recent decades. This includes homosexual men and women who are having children in the context of an already established lesbian or gay identity (Gates, 2013). Consider the following statistics.

- More than 125,000 same-sex couple households (19 percent) include nearly 220,000 children under age eighteen.
- Same-sex couples who are or consider themselves to be spouses are more than twice as likely to be raising biological, step, or adopted children (31 percent) when compared to same-sex couples who say that they are unmarried partners (14 percent).
- An estimated 3 million lesbian, gay, bisexual, or transgender Americans have had a child, and as many as 6 million American children and adults have a lesbian, gay, bisexual, or transgender parent.

Long before *Obergefell v. Hodges*, the 2015 United States Supreme Court decision that legalized same-sex marriage, many opponents of same-sex marriage raised the argument that children would experience social or psychological harm from having a gay or lesbian parent. Some resistance to this legal decision and growing family form remains in American culture. However, in their review of scholarship on same-sex parents and families, sociologists Timothy Biblarz and Everen Savci (2010) state that "evidence is strong that children raised in gay parent families enjoy high levels of psychological well-being and social adjustment" (p. 493).

Parenting and Children of Lesbian Mothers

A number of studies suggest that children of lesbian parents appear similar to children of opposite-sex parents with regard to psychological well-being, peer relationships, and social and behavioral adjustment. A study comparing 51 adolescents from two-mother families with 51 adolescents from intact mother–father families showed similar conclusions. For example, adolescents in both family types showed positive relationships with their parents, a measurement that was favorably associated with the children's emotional well-being. On the measurements of psychological adjustment and substance use, adolescents with lesbian mothers had higher levels of self-esteem and lower levels of conduct problems than those of heterosexual-headed families (Bos *et al.*, 2015).

Bos and colleagues (2005) conclude that lesbian and heterosexual families are "very much alike" (p. 263), adding that "although family functioning in lesbian families might be just as varied, challenging, comforting, amusing and frustrating as it is in heterosexual families, it is the stigma of lesbianism … that makes their family life different" (p. 273).

Gershon and colleagues (1999) found that children who perceived higher levels of stigma associated with having a lesbian mother tended to have lower self-esteem. However, adolescents who reported more decision-making coping skills showed higher scores on self-esteem than those who relied on coping skills involving high levels of social support. Altogether, the study suggests that the relationship between having a lesbian parent, perceived stigma, self-esteem, and well-being is moderated by effective coping strategies in response to homophobia. The same study also found that adolescents who were more open to close friends about their mother's lesbian identity had higher levels of self-esteem (Gershon *et al.*, 1999).

Despite findings of well-adjustment among many children of lesbian parents, the heterosexism or homophobia of others has been a noted cause for concern. Lesbian mothers know that teaching their sons to reject male dominance over women could mean ridicule or homophobic retaliations against their children. They know their son's social and economic achievement could mean accommodating cultural ideals of manhood that encourage success but involve sexism. In at least one study, a number of pre-adoptive lesbian parents who preferred to adopt girls did so because they believed that adoptive sons would experience more unfair treatment than adoptive daughters (see Biblarz & Savci, 2010).

Parenting and Children of Gay and Bisexual Male Fathers

Lesbians are more likely to enter parenthood through non-traditional means (donor insemination, for example). Gay men, however, may experience greater practical as well as emotional difficulties in becoming a parent. There are strong oppositional pressures for gay men not to parent. A gay man who chooses to do otherwise challenges gender and sexual norms within gay culture itself, in addition to traditional cultural beliefs defining how masculinity and paternity should be expressed (Stacey, 2006). Indeed, emergent family types have expanded the possibility for men to be both gay and a father. Yet, some men still equate "coming out" as gay with surrendering the decision to parent. For others, state-sanctioned legal barriers to parenthood are convincing enough either to postpone or to forgo child adoption.

For gay men who do become fathers, there is evidence to suggest that homosexual and heterosexual fathers share several important similarities. In a study of divorced, non-resident heterosexual and homosexual fathers, both groups reported similar reasons for wanting to become fathers and maintained similar degrees of involvement in their children's activities. However, gay fathers tended to be stricter in imparting standards for their children's behavior. Gay fathers also employed more reasoning strategies and were more responsive to their children's needs than heterosexual fathers (see Tasker, 2005).

Studies that analyze sons and daughters who live with their gay and bisexual fathers are scarce. However, Barrett and Tasker's (2001) study asked the views of more than 100 gay and bisexual fathers about their sons' and daughters' experiences. Daughters were more likely than sons to respond sympathetically to their fathers' disclosure about sexual orientation. Neither sons nor daughters were described as having different experiences insofar as problems or benefits developed from having a gay or bisexual father (Barrett & Tasker, 2001).

Psychologist Charlotte Patterson (2006) summarizes the social and developmental outcomes of children with same-sex parents this way: "Children of lesbian and gay parents may be exposed to prejudice against their parents in some settings, and this may be painful for them, but evidence for the idea that such encounters affect children's overall adjustment is lacking." Indeed, many adult children of lesbian, gay, and bisexual parents acknowledge that growing up with LGB parents fostered tolerance and open-mindedness about gender and sexuality. They are protective of their parents and the homosexual community at large. However, evidence also suggests that some struggle with issues of trust into adulthood, which, they say, is linked to experiences with bullying or parents' unexpected disclosure about their sexual orientation (Goldberg, 2007).

Gender and Childfree Families

Not all families involve children; some couples are childfree by choice. Among women, the reasons for voluntary childlessness generally involve a freedom to travel and an appreciation of solace. Others note the time, space, and greater ability to develop stronger relationships with other adults (Gillespie, 2003).

The decision to be childfree may be as much a matter of gender as it is a matter of preference for some, especially among partners who endorse low economic risk and a more equal division of household labor and work arrangements in the labor market. From this perspective, the experience of parenting, the establishment of future generations and namesakes, and other cultural ideals related to fertility are not the values underscoring some couples' decision to bear children. Values associated with gender equity as well as economic and relational stability also are given account when negotiating whether to parent or not (see Henz, 2008).

Gender and Military Families

Syndicated columnist Sarah Smiley (2013) has written about her experiences with imbalanced parental participation resulting from military life. A member of a military family from birth, her Navy father was deployed when he received news of her birth through telegram. By the time Smiley married a Navy pilot 22 years later, her father had accumulated eleven years of "sea time" and had been absent for half her life. However, Smiley writes, her mother did not recognize "the strangeness" of the thinly worded telegram announcing her birth until 23 years later. Nor, at the time, had it "seemed strange" that her father missed her birth and, later, most of her recitals and track meets. As a mother and wife of a serviceman, Smiley finds similarities between her past and her present family life. Although she and her husband were together for the birth of their son, he was deployed two weeks later and, subsequently, missed most of their son's first year. Two children later, Smiley's husband has been absent from seven family birthdays, two Thanksgivings, one Christmas, the couple's anniversary, Little League games, and their youngest son's first day of kindergarten.

Wives/mothers in military families traditionally have acted as caregivers after the deployment of a serving husband/father (Fitzsimons & Krause-Parello, 2009). But while more families have moved toward a shared household economy and more mothers work outside the home, military mothers/spouses may experience certain difficulties when trying to enter paid labor positions. Compared to their at-home civilian counterparts, more military spouses tend to earn less overall, are out of work, or work fewer hours than they would like. They may experience interruption in their careers, and employers may be hesitant to offer them jobs requiring a large investment in training or a long learning curve. Some service members and their families, particularly among the junior enlisted ranks, report substantial financial distress. Despite the fact that steady pay increases have allowed some service members to earn more than civilians with a comparable level of education, a handful of military families qualify for food stamps (Hosek & Wadsworth, 2013).

The issues affecting single-parent military families, who are likely to be headed by a female service member, are also difficult. Single military parents may have to arrange for childcare during extended training exercises and deployments. In addition, they can be stationed away from important extended family networks in which a single civilian parent otherwise would find support.

Not only stress but also resiliency is evidenced among military families. The spouses of military service men and women often utilize social supports to manage increased levels of stress and to develop stabilizing family routines. In a study of military children and protective factors against the stress of having a military parent, perceived social support from mothers was associated with fewer emotional symptoms among daughters and fewer conduct problems and overall problems among sons and daughters (Morris & Age, 2009).

It is not yet clear how traditional military culture might affect military civilian husbands and fathers. Nonetheless, irregular gender role expectations are well-recognized by the group. The traditional male/breadwinner paradigm is disrupted when a husband is a civilian and the wife works each day as military personnel. Further, civilian husbands are treated differently—more alienated—than civilian wives in social contexts involving other military spouses and personnel. To summarize one military sociologist: people generally possess an idea of how to regard or engage with civilian wives; people are less familiar with how to treat civilian husbands. Additionally, the low number of male civilian spouses in a military system dominated by male servicemen can compound issues, especially if the husband has no prior military experience (Ziezulewicz, 2009).

Gender and Families of Children with Disabilities

The care of children with disabilities can be intense, ranging from changing dressings to constant therapeutic attention to maintaining feeding and breathing technologies. Mothers who care for their disabled child may experience a reduction of work hours, less overall family income, or the end of paid labor altogether. The fathers of children with disabilities in sociologist Valerie Leiter's (2007) study performed some childcare work; however, almost one-fifth of mothers conducted carework for 20 hours or more per week (Leiter *et al.*, 2004). Studies also suggest that mothers are the parent most likely to daily experience the negative impact of their child's disabilities. In sociologist David Gray's (2003) sample of parents with a high-functioning autistic child, mothers primarily managed medical referrals and educational difficulties and more often endured emotional distress and career disruption. Because mothers acted as principal caregivers, their spouses and people outside the family more likely held mothers responsible for their child's behaviors.

To be sure, the fathers of disabled children undergo emotional distress and change. They also engage in tasks directly related to their child's condition, despite traditional gender role assignments. However, there are certain pressures that may limit some fathers' daily participation in their child's care. For example, fathers of disabled children report more stress and difficulty in areas such as child temperament, emotional attachment, and personal relationship with their child (Krauss, 1993). It is perhaps, then, not surprising that the husbands in Gray's sample were not the primary caregivers of their autistic children. Rather, the husbands saw themselves as a support system for their wives or someone on whom their wives could rely during periods of extreme stress. In fact, the most severe pressure the husbands experienced was through the stress experienced by their wives.

Gender, and Immigrant and Refugee Families

For centuries, families have been forced to flee their countries of origin due to socio-political opposition, massive economic strife, or threat of violence or death. Refugee families who leave their country to escape war, persecution, or natural disaster, in addition to families who permanently immigrate to an alternate country, often attest to substantial life and family adjustments in the process.

New sociocultural changes may be in opposition to prior norms and customs on gender and power. For example, refugee and immigrant women and adolescents may have to work in their country of destination, and their employment may be interpreted by immigrant men as contesting their roles as breadwinners. Further, Western women may exhibit an autonomy otherwise restricted for some people's countries of origin. The contrast may be particularly unsettling for husbands and fathers, especially those who held patriarchal privilege in their former regions. As one male respondent from Bosnia-Herzegovina stated, "Before I used to have a greater control over my wife and children. Now when I want to give advice to my children, they just ignore me. They tell me that my beliefs are old-fashioned, and that I should change and become more modern" (Weine *et al.*, 2004, p. 152).

Some immigrant or refugee families respond to their emergent cultural and economic challenges to patriarchal power by shifting family interactions in a more egalitarian direction. In a study of Mexican immigrants, wives managed the family's daily well-being after their husbands temporarily left their country of origin to work in the United States. Out of necessity and separation from their wives, immigrant husbands took on tasks typically assigned to their wives.

When the families reunified, Mexican wives expected their husbands to be more involved in household work; Mexican husbands, in some cases, reverted to traditional housework arrangements (Hondagneu-Sotelo, 1992). Still, some couples may retain distinct gendered boundaries, although in some ways nuanced by their new cultural surroundings. Interviews with Korean immigrant women show that wives wanted aspects of their husbands' hierarchal behaviors to change, but not insofar as challenging originating cultural norms on husband headship (Lim, 1997).

Gender and Homeless Families

Homeless families are mostly headed by highly disadvantaged women, making up 65 percent of homeless people who are members of households with children (National Coalition for the Homeless, 2009). However, the media have exposed an unprecedented incidence of single fathers living in some local homeless shelters. For example, the executive director of a California homeless shelter reported more homeless fathers in residence than any other demographic group. A survey of homeless shelters and food banks in nearby areas showed they also experienced an increase in services to single men with children (Scauzillo, 2012). Qualitative research on homeless fathers suggests that some fathers step into the single-parent role following intense and often chronic conflicts with their children's mother (see McArthur *et al.*, 2006).

Some institutionalized support settings have been criticized as working against single homeless men with children. Like single homeless mothers, single homeless fathers want to retain custody of their children and to keep their family intact. Yet homeless shelters tend to accommodate exclusively to single men and not to single fathers with children. Nearly all of Schindler and Coley's (2007) sample of homeless fathers emphasized a "loss of respect" not only by shelter staff but also by family members and the general public (p. 45). Discontent was exacerbated due to being homeless and male, something that began with their unemployment and was complicated by social expectations that men provide for their families. The researchers also point out that homeless single fathers might find themselves in parenting situations unlike ever before. Homeless shelters require that residents observe curfews, childcare guidelines, chore details, and restrictive visiting hours, which proved too structured for some respondents. However, the most unprecedented experience involved the men's assuming the roles of primary parent as well as provider. In fact, some fathers were negotiating for the first time the challenge of attending to their children throughout the day.

Like single homeless fathers, single homeless mothers find shelter life constraining to their authority over their children. One study described homeless mothers' loss of autonomy as an "out-of-order" relationship that "unravels" the mother's role and fosters a loss of confidence, parent–child bonding, and position as head of household (Boxill & Beaty, 1990). Meanwhile, both groups are aware of their stigmatization as inadequate parents, and they fear losing their children due to state-enforced child protection intervention. However, there are perceived dissimilarities between the two groups of homeless parents as well. Single fathers with experiences of living in homeless shelters with their children believe that shelters operate from the standpoint of homeless mothers. Single fathers also believe that the shelter experience of women is somehow different from that of men.

Gender Roles in Midlife and Later Life Cycles

For midlife and older individuals, increasing longevity allows longer periods in normal adult roles (longer parental involvement with their adult children)—plus the advantage of new, beneficial family opportunities. The fact that more family generations are alive at once allows the chance to maintain ties with not only children and grandchildren but also great-grandchildren and even great-great grandchildren.

Retirement from full-time employment also opens new possibilities for personal use of time. Some retirees take on new career paths or educational opportunities they once had postponed. Many integrate various work, leisure, and community participation into everyday family roles (Cox, 2015). For retired men, the need or desire to reorient time with family may be greater than for retired women. Why is this? Women will have spent 20 or 30 years of meeting spousal, parental, and other family needs by the time they reach the age of retirement. Retired men, whose traditional focus is family provider, will have largely interacted with their children and other kin through their wives.

One meaningful activity for both sexes has been to expand post-retirement roles as family helpers and to rekindle connections with friends. A study of high school students who graduated in the mid-twentieth century found that, as adults in their 50s and 60s, men and women reported very high levels of helping kin and non-kin alike. Women did more to assist elderly parents and to provide emotional support to others. Men offered more assistance than women with regard to housework, yard work, and repairs, especially as men transitioned from their 50s to their 60s. Retired midlife men also became significantly more involved in the care of their adult children and grandchildren, virtually eliminating any gender differences by the time they were in their 60s (Kahn *et al.*, 2011).

Older individuals may face chronic health problems or disabilities, making the need to be *cared for* a new reality for many men and women. Personal adjustments and family demands like these bring to light how gender role expectations may vary throughout the life course. A Pew Research Center study found that women are slightly more likely than men to be caring for a loved one, as are adults ages 50 to 64, compared with other age groups. Caregivers were more likely than others to report that they themselves were living with a disability (Fox & Brenner, 2012).

Employee and Family Care Policies

The 1993 Family and Medical Leave Act (FMLA) provides employees unpaid time off from work for up to twelve weeks per year due to certain family and medical conditions. Eligibility for FMLA provisions includes the following: the birth and care for an employee's newborn child or the placement of a child for adoption or foster care; the care of an employee's spouse, child, or parent with a serious health condition; or a medical leave if the employee is unable to work due to a serious health condition.

The FMLA applies to all public agencies, public and private elementary and secondary schools, and companies with 50 or more employees. A company may decide to absorb the costs of a longer absence or a paid leave from work. However, only 12 percent of U.S. private sector employees have access to paid family leave through their employer. In the low-wage job sector, only 5 percent of women have access to paid maternity leave (U.S. Department of Labor, 2015).

Globally, Papua New Guinea, Oman, and the United States are the only three countries with no paid maternity leave policy. The question of whether the United States should do so has fostered debate for and against paid maternity leave, in addition to other paid family leave policies. A chief position

Global Overview: Gender Roles in Children's Textbooks

As we just discussed, schools are a powerful agent of gender socialization. However, despite widespread revision efforts, patterns of gender inequality in children's textbooks continue worldwide.

The United Nations Educational, Scientific, and Cultural Organization (UNESCO) initiated a series of Education for All Global (EFA) Monitoring Reports in early 2000. Published between 2002 and 2015, the reports followed the progress of over 160 governments' commitment to reach targeted education goals by 2015. According to EFA reports, whether measured by lines of text, proportions of book characters, title mentions, or index citations, women and girls have remained underrepresented in school textbooks and curricula for more than 35 years. This outcome was consistent despite countries' level of income, development, and gender parity or the education level and subject matter assessed (UNESCO, 2007). Women, for example, were stereotypically portrayed as accommodating and girls were shown as passive conformists. Boys and men performed noble and exciting activities, and almost none of their actions or occupations were related to caring or acts socially defined as feminine (UNESCO, 2007). Generally, both sexes were depicted in highly stereotyped household and occupational roles, while performing stereotyped behaviors, attitudes, and traits. Some additional findings from the EFA series and accompanying support data included the following.

- In Chinese pre-primary and primary textbooks, males and females are represented disproportionately. The proportion of male characters rises from 48 percent in books for four-year-olds to 61 percent in those for six-year-olds (UNESCO, 2007). Females appear frequently only in reading materials for very young children (UNESCO, 2007). In social studies texts, 100 percent of scientists and soldiers are male; 100 percent of teachers and 75 percent of service personnel are female. Females represent only about one-fifth of the historical characters in the twelve-volume series of elementary Chinese textbooks and often appear "dull and lifeless" in comparison with males, who appear "more vibrant" (UNESCO, 2007).
- In a content analysis of 106 Romanian textbooks, illustrations depicting women and girls make up 24 percent in first- and second-grade textbooks. That proportion drops to 10 percent by twelfth grade (Blumberg, 2007).
- Men dominate textbook illustrations and activities in India, representing commercial, occupational, and marketing trades and situations in the six mathematics books used in primary schools. No women are depicted as merchants, executives, engineers, or entrepreneurs (UNESCO, 2007). More than half the illustrations in the average primary school English, Hindi, mathematics, science, and social studies textbooks depict only males; 6 percent depict only females (UNESCO, 2007).
- Anti-stereotyping initiatives have led to a far different vision of gender in Sweden, where the modern male may wear an apron as he stirs pots on the stove or conducts housework. Gender equality also is described or depicted in workplace divisions of labor (Blumberg, 2007).

among advocates is that strong paid family and medical leave policies help working families take time off for caregiving responsibilities, in addition to their own medical needs, without risking their economic security. Paid leave also encourages men to take paternity leave and serve as caregivers, which, advocates argue, has a number of positive effects on families. Opponents of family and medical leave policies argue that the cost for businesses to accommodate paid leave is significant (rehiring costs, costs for training temporary and replacement workers, and low productivity in the meantime). As a result, employers may reduce employee hiring or transition to a part-time only workforce, excluding family leave provisions for low-wage employees especially.

Summary

In this chapter, we focused on *gender socialization*: the ongoing process of learning the roles, behaviors, attitudes, personalities, and representations that are socially associated with either or both sexes. Certainly, sex-specific norms, behaviors, and other expectations have been challenged and perhaps are not as entrenched as they were centuries or decades prior. Still, for many, gender differences may "feel" natural or "feel" innate. The fact that gender roles and gender identities are *socially* appropriated early in life and subject to change may go unrecognized.

There are benefits and drawbacks to traditional gender roles and identities. For example, optioning out of paid work and staying at home can be a benefit that allows some women more time to interact with their families. Unfortunately, the often-narrow concept of what entails "real" work has undervalued the traditional stay-at-home status in our culture. At the same time, changing ideas, behaviors, and expectations about the sexes has facilitated a shift in how some women, men, and family members perceive and "do" gender. Although not on par with women, men's participation in household labor and childrearing—a traditionally female sphere—has increased over the decades.

To explain the differences in gender socialization between men and women, the social learning perspective posits that males and females learn gendered social roles through a system of socialization that rewards their adherence to sex-typed norms and behaviors and punishes them when they do not. The cognitive development perspective posits that individuals observe differences between women and men and then organize their behaviors around gender categories aligned with one biological sex or another. The symbolic interaction perspective posits that, through socialization, men and women adapt to behaviors, character traits, self-identities, roles, symbols, and meanings that are socially identified as masculine or feminine/male or female.

Another important question is who or what are the people, groups, institutions, and other contexts in which gender socialization takes place. Generally, there are five primary agents of gender socialization: family and parents, schools, sports, peers, and the media. Changing family composition, life adjustments, and challenges spurred by poverty, homelessness, rearing a child with a disability, displacement due to war or global conflict, military deployment, illness, and other crises may compel family members to rethink and revise their gendered roles and identities.

Key Terms

Agents of gender role socialization are individuals, groups, and institutions that comprise the social contexts in which gender socialization occurs.

Cognitive development perspective of gender role socialization posits that individuals cognitively seek out, organize, and enact attitudes and behaviors that are socially aligned with one's biological sex.

Emotion work refers to people's active engagement in the management of their own and others' emotions within family relationships.

Expressive roles refer to a sexual division of labor in which women demonstrate traditionally feminine social roles and behaviors.

Gender, a social construct, refers to the different behavioral and normative expectations that are associated with being female or male in a culture or society.

Gender identities comprise individuals' subjective sense of being a boy, girl, man, woman, or a combination thereof.

Gender socialization is the ongoing process of learning the roles, behaviors, attitudes, personalities, and representations that are socially associated with men and women.

Gender stereotypes are culturally based beliefs about how people typically will look, act, feel, think, and behave based on their biological sex.

Instrumental roles refer to a sexual division of labor in which men demonstrate traditionally masculine social roles and behaviors.

Sex is the biological distinction differentiating males and females.

Socialization is the ongoing process whereby individuals learn and adapt to norms and behaviors in a manner approved by a group, society, or culture.

Social learning perspective of gender role socialization posits that individuals learn gender roles through a system of socialization, which rewards their adherence to sex-typed norms and behaviors and which punishes their adherence to behaviors inconsistent with sex-typed norms and behaviors.

Symbolic interaction perspective of gender role socialization posits that, through the process of socialization, individuals adapt to behaviors, character traits, and a sense of self-reflective roles, symbols, and meanings that are socially defined as masculine or feminine.

Transgender people are those whose gender identity does not align with their biological sex.

Transsexual people are those who undergo anatomical and/or hormonal changes to their bodies.

Discussion Questions

- Discuss the many ways in which sex roles are internalized through the process of gender socialization.
- Examine the lyrics of five to ten of the most popular songs in a popular music genre. What cultural assumptions are conveyed with regard to gender? Are social expectations about men and women reproduced or challenged? Discuss your findings.
- Closely read the content of two magazines: one that targets a male audience; another that targets a female audience. Identify examples of text, images, advertisements, etc. that either reinforce or counter traditional gender roles and stereotypes. As you discuss your findings, be sure to focus on how the same companies may advertise in different ways, depending on the two different audiences. In addition, discuss how sex, independence, communication, and other social practices and values are depicted across the two sets of publications.

Suggested Film and Video

Earp, Jeremy. (Filmaker). *Tough Guise 2: Violence, Manhood, and American Culture.* (2013). (Film). Media Education Foundation.

Obaid-Chinoy, Sharmeen. (Filmaker). *A Girl in the River: The Price of Forgiveness.* (2015). (Film). HBO Documentary.

Newsom, Jennifer. (Filmaker). *Miss Representation.* (2011). (Film). Sundance Documentary.

Internet Sources

Gender Bias Learning Project. (2016) Center for Work Life Law, University of California, Hastings College of the Law. Retrieved on November 4, 2016 from http://www.genderbiasbingo.com/

The Gender Ads Project: Advertising, Education, and Activism. (2015). Scott A Lukas, Ph.D. Retrieved on November 4, 2016 from http://www.genderads.com/

The Transcending Gender Project: Celebrating Lives beyond Gender. The Transcending Gender Project. Retrieved on November 6, 2016 from http://www.transcendinggender.org/

References

Barrett, H. & Tasker, F. (2001). Growing Up with a Gay Parent: Views of 101 Gay Fathers on Their Sons' and

Daughters' Experiences. *Educational and Child Psychology*, 18, 62–77.

Biblarz, T. & Savci, E. (2010). Lesbian, Gay, Bisexual, and Transgender Families. *Journal of Marriage and Family*, 72, 480–497.

Blumberg, R. (2007). *Gender Bias in Textbooks: A Hidden Obstacle on the Road to Gender Equality in Education*. Paper commissioned for the Education for All Global Monitoring Report 2008: Will We Make It?

Bos, H., Van Balen, F. & van den Boom, D. (2005). Lesbian Families and Family Functioning: An Overview. *Patient Education and Counseling*, 59, 263–275.

Bos, H., van Gelderen, L. & Gartrell, N. (2015). Lesbian and Heterosexual Two-Parent Families: Adolescent–Parent Relationship Quality and Adolescent Well-Being. *Journal of Child and Family Studies*, 24, 1031–1046.

Boxill, N. & Beaty, A. (1990). Mother/Child Interaction among Homeless Women and Their Children in a Public Night Shelter in Atlanta, Georgia. *Child & Youth Services*, 14, 49–64.

Campbell, D. (January 5, 2012). Body Image Concerns More Men than Women, Research Finds. *The Guardian*. Retrieved from http://www.theguardian.com/ lifeandstyle/2012/jan/06/body-image-concerns-men-more-than-women

Cohn, D., Livingston, G. & Wang, W. (2014). *After Decades of Decline, a Rise in Stay-At-Home Mothers*. Washington, DC: Pew Research Center's Social & Demographic Trends Project. Retrieved from http://www.pewsocialtrends.org/2014/04/08/after-decades-of-decline-a-rise-in-stay-at-home-mothers/

Cotter, D., England, P. & Hermsen, J. (2008). *Moms and Jobs: Trends in Mothers' Employment and Which Mothers Stay Home*. Families as They Really Are: A Briefing Paper Prepared for the Council on Contemporary Families (pp. 416–424). Council on Contemporary Families.

Cox, H. (2015). *Later Life: The Realities of Aging*. New York: Routledge.

Donahoe, D. A. (1999). Measuring Women's Work in Developing Countries. *Population and Development Review*, 25, 543–576.

Fitzsimons, V. & Krause-Parello, C. (2009). Military Children: When Parents Are Deployed Overseas. *The Journal of School Nursing*, 25, 40–47.

Fox, S. and Brenner, J. (2012). *Family Caregivers Online*. Washington, DC: Pew Internet & American Life Project. Retrieved from http://www.pewinternet.org/files/old- media/Files/Reports/2012/PIP_Family_Caregivers_Online.pdf.

Frankel, L. (2014). *Nice Girls Don't Get the Corner Office: Unconscious Mistakes Women Make that Sabotage Their Careers*. UK: Hachette.

Gates, G. J. (2013). *LGBT Parenting in the United States*. The Williams Institute. University of California School of Law. Retrieved from http://www.ncbi.nlm.nih.gov/pmc/articles/PMC2807026/

Gershon, T., Tschann, J. & Jemerin, J. (1999). Stigmatization, Self-esteem, and Coping among the Adolescent Children of Lesbian Mothers. *Journal of Adolescent Health*, 24, 437–445.

Gillespie, R. (2003). Childfree and Feminine: Understanding the Gender Identity of Voluntarily Childless Women. *Gender & Society*, 17, 122–136.

Goldberg, A. (2007). (How) Does It Make a Difference? Perspectives of Adults with Lesbian, Gay, and Bisexual Parents. *American Journal of Orthopsychiatry*, 77, 550–562.

Gray, D. (2003). Gender and Coping: The Parents of Children with High-Functioning Autism. *Social Science & Medicine*. 56, 631–642.

Hamilton, M. C., Anderson, D., Broaddus, M. & Young, K. (2006). Gender Stereotyping and Under-representation of Female Characters in 200 Popular Children's Picture Books: A Twenty-First Century Update. *Sex Roles*, 55, 757–765.

Hellmich, N. (2006, September 26). Do Thin Models Warp Girls' Body Image? *USA Today*. Retrieved from http://www.usatoday.com/news/health/2006-09-25-thin-models_x.htm.

Henz, U. (2008). Gender Roles and Values of Children: Childless Couples in East and West Germany. *Demographic Research*, 19, 1451–1500.

Heron, K., Smyth, J., Akano, E. & Wonderlich, S. (2013). Assessing Body Image in Young Children. *SAGE Open*, 3, 1–7.

Hilton, J. & Haldeman, V. (1991). Gender Differences in the Performance of Household Tasks by Adults and Children in Single-Parent and Two-Parent, Two-Earner Families. *Journal of Family Issues*, 91, 114–130.

Hochschild, A. (1989). *The Second Shift: Working Parents and the Revolution at Home*. New York: Viking.

Hondagneu-Sotelo, P. (1992). Overcoming Patriarchal Constraints: The Reconstruction of Gender Relations among Mexican Immigrant Women and Men. *Gender & Society*, 6, 393–415.

Horowitz, A. (1985). Sons and Daughters as Caregivers to Older Parents: Differences in Role Performance and Consequences. *The Gerontologist*, 25, 612–617.

Hosek, J. & Wadsworth, S. (2013). Economic Conditions of Military Families. *The Future of Children*, 23, 41–59.

Kahn, J., McGill, B. & Bianchi, S. (2011). Help to Family and Friends: Are There Gender Differences at Older Ages? *Journal of Marriage and Family*, 73, 77–92.

Koivula, N. (2001). Perceived Characteristics of Sports Categorized as Gender-Neutral, Feminine and Masculine. *Journal of Sport Behavior*, 24, 377–393.

Krauss, M. (1993). Child-related and Parenting Stress: Similarities and Differences between Mothers and Fathers of Children with Disabilities. *American Journal on Mental Retardation*, 97, 393–405.

Latshaw, B. & Hale, S. (2016). 'The Domestic Handoff': Stay-At-Home Fathers' Time-Use in Female Breadwinner Families. *Journal of Family Studies*, 22, 97–120.

Leaper, C. (2014). Parents' Socialization of Gender in Children. *Encyclopedia of Early Childhood Socialization*, 6 [online]. Montreal, Quebec: Centre of Excellence for Early Childhood Development and Strategic Knowledge. Retrieved from http://www.child- encyclopedia.com/gender-early-socialization/according-experts/parents-socialization-gender-children.

Leiter, V. (2007). Nobody's Just Normal, You Know: The Social Creation of Developmental Disability. *Social Science and Medicine*, 65, 1630–1641.

Leiter, V., Krauss, M., Anderson, B. & Wells, N. (2004). The Consequences of Caring: Maternal Impacts of Having a Child with Special Needs. *Journal of Family Issues*, 25, 379–403.

Lim, I. (1997). Korean Immigrant Women's Challenge to Gender Inequality at Home: The Interplay of Economic Resources, Gender, and Family. *Gender & Society*, 11, 31–51.

Lowes, J. & Tiggemann, M. (2003). Body Dissatisfaction, Dieting Awareness and the Impact of Parental Influence in Young Children. *British Journal of Health Psychology*, 8, 135–147.

Lytton, H. & Romney, D. (1991). Parents' Differential Socialization of Boys and Girls: A Meta- Analysis. *Psychological Bulletin*, 109, 267–296.

Marks, J., Bun, L. & McHale, S. (2009). Family Patterns of Gender Role Attitudes. *Sex Roles*, 61, 221–234.

McArthur, M., Zubrzycki, J., Rochester, A. & Thomson, L. (2006). 'Dad, Where Are We Going To Live Now?': Exploring Fathers' Experiences Of Homelessness. *Australian Social Work*, 59, 288–300.

McHale, S. & Crouter A. (2003). How Do Children Exert an Impact on Family Life? In A. Crouter and A. Booth (Eds.), *Children's Influence on Family Dynamics: The Neglected Side of Family Relationships* (pp. 207–220). Mahwah, NJ: Lawrence Erlbaum Associates.

Metheny, E. (1965). *Connotations of Movement in Sport and Dance*. Dubuque, IA: Wm. C. Brown.

Morris, A. & Age, T. (2009). Adjustment among Youth in Military Families: The Protective Roles of Effortful Control and Maternal Social Support. *Journal of Applied Developmental Psychology*, 30, 695–707.

National Coalition for the Homeless (2009). Who is Homeless? [Fact Sheet]. Retrieved from http://www.national homeless.org/factsheets/Whois.pdf.

Parker, K. & Livingston, G. (2016). *Six Facts about American Fathers*. Washington, DC: Pew Research Center. Retrieved from www.pewresearch.org/fact-tank/2017/06/15/fathers-day-facts/

Parker, K. & Wang, W. (2013). *Modern Parenthood*. Washington, DC: Pew Research Center's Social & Demographic Trends Project. Retrieved from http://www.pewsocialtrends.org/2013/03/14/chapter-4-how-mothers-and-fathers-spend-their-time/#fn-ref-19095–15

Parsons, T. & Bales, R. (1955). *Family, Socialization, and Interaction Process*. Glencoe, IL: Free Press.

Partridge, A. (2015). *From Career Woman to Stay-At-Home Mother and Back Again: Understanding the Choices Mothers Make*. (Unpublished doctoral dissertation). San Francisco, CA: Alliant International University.

Patterson, C. (2006). Children of Lesbian and Gay Parents. *Current Directions in Psychological Science*, 15, 241–244.

Pederson, E. & Vogel, D. (2007). Male Gender Role Conflict and Willingness to Seek Counseling: Testing a Mediation Model on College-aged Men. *Journal of Counseling Psychology*, 54, 373–384.

Pfeffer, C. (2010). "Women's work"?: Women Partners of Transgender Men Doing Housework and Emotion Work. *Journal of Marriage and Family*, 1, 165–183.

Pomerleau, A., Bolduc, D., Malcuit, G. & Cossette, L. (1990). Pink or Blue: Environmental Gender Stereotypes in the First Two Years of Life. *Sex Roles*, 22(5–6), 359–367.

Pope, H. G., Olivardia, R., Gruber, A. & Borowiecki, J. (1999). Evolving Ideals of Male Body Image as Seen through Action Toys. *International Journal of Eating Disorders*, 26, 65–72.

Pope, H. G., Olivardia, R., Borowiecki III, J. & Cohane, G. H. (2001). The Growing Commercial Value of the Male Body: A Longitudinal Survey of Advertising in Women's Magazines. *Psychotherapy and Psychosomatics*, 70, 189–192.

Rohlinger, D. A. (2007). Socialization, Gender. In Ritzer, G. (Ed.), *Blackwell Encyclopedia of Sociology*. [Online]. Malden, MA: Blackwell Publications.

Scauzillo, S. (2012 June 6). Growing Number of Homeless Single Fathers Showing Up in Local Shelters. *San Gabriel Valley Tribune*. Retrieved from http://www.sgvtribune.com/article/ZZ/20120616/NEWS/120618571.

Schacht, S. (1996). Misogyny On and Off the "Pitch": The Gendered World of Male Rugby Players. *Gender & Society*, 10, 550–565.

Schindler, H. & Coley, R. (2007). A Qualitative Study of Homeless Fathers: Exploring Parenting and Gender Role Transitions. *Family Relations*, 56, 40–51.

Smiley, S. (2013, June 18). Changing Gender Roles Mean New Sacrifices for Military Dads. A Wife's Perspective on Life in the Military. *The Times Record*. Retrieved from http://www.timesrecord.com/news/2013-06-18/Family_(and)_Friends/Changing_gender_roles_mean_new_sacrifices_for_mili.html

Smith, S. & Cook, C. (2008). Gender Stereotypes: An Analysis of Popular Films and TV. *Conference*, 208, 12–23.

Stacey, J. (2006). Gay Parenthood and the Decline of Paternity as We Knew It. *Sexualities*, 9, 27–55.

Stice, E. & Shaw, H. (1994). Adverse Effects of the Media Portrayed Thin-Ideal on Women and Linkages to Bulimic Symptomatology. *Journal of Social and Clinical Psychology*, 13, 288–308.

Sullivan, O. & Coltrane, S. (2008). *Men's Changing Contribution to Housework and Childcare*. Chicago, IL: Council on Contemporary Families Briefing Paper. Retrieved from https://contemporaryfamilies.org/mens-changing-contribution-to-housework-and-childcare-brief-report/

Sullivan, O. & Coltrane, S. (2015). *The Continuing "Gender Revolution" in Housework and Care: Evidence from Long-term Time-use Trends*. Chicago, IL: Council on Contemporary Families Briefing Paper. Retrieved from https://contemporaryfamilies.org/continuing-gender-revolution-brief-report/

Tasker, F. (2005). Lesbian Mothers, Gay Fathers, and Their Children: A Review. *Journal of Developmental & Behavioral Pediatrics*, 26, 224–240.

Theberge, N. (1997). "It's Part of the Game": Physicality and the Production of Gender in Women's Hockey. *Gender & Society*, 11, 69–87.

Thompson Jr, E. H., Grisanti, C. & Pleck, J. H. (1985). Attitudes toward the Male Role and Their Correlates. *Sex Roles*, 13, 413–427.

Todd, B., Barry, J. & Thommessen, S. (2017). Preferences for "Gender-typed" Toys in Boys and Girls Aged 9 to 32 Months. *Infant and Child Development*, 26(3).

UNESCO (United Nations Educational, Scientific, and Cultural Organization) (2007). *Education for All by 2015: Will We Make It?* Education for All Global Monitoring Report. Paris, France: United Nations Educational, Scientific, and Cultural Organization.

U. S. Department of Labor. (2015). *DOL Factsheet: Paid Family and Medical Leave* [Fact Sheet]. Retrieved from https://www.dol.gov/wb/paidleave/PDF/PaidLeave.pdf.

Weine, S., Muzurovic, N., Kulauzovic, Y., Besic, S., Lezic, A., Mujagic, A. & Knafl, K. (2004). Family Consequences of Refugee Trauma. *Family Process*, 43, 147–160.

Weiss, R. (1979). Growing Up a Little Faster: The Experience of Growing Up in a Single Parent Household. *Journal of Social Issues*, 35, 97–111.

Ziezulewicz, G. (March 14, 2009). With their Wives—Not Them—Reporting for Duty, Military Husbands Work to Find Their Place. *Stars and Stripes*. Retrieved from http://www.stripes.com/news/with-their-wives-not-them-reporting-for-duty-military-husbands-work-to-find-their-place-1.89136

PART II

DEVELOPMENT AND MAINTENANCE OF RELATIONSHIPS

6
SEXUALITY AND SEXUAL RELATIONSHIPS THROUGHOUT LIFE

Learning Outcomes

Upon completion of the chapter, students should be able to:

1. apply sociological perspectives to the analysis of sexuality
2. explain sexual identity and its components: biological sex, gender roles, gender identity
3. describe types of sexual orientation
4. explain types of sexual behaviors and sexual behavior deviations
5. identify sexual response and barriers to sexual response
6. compare traditional sexual scripts with contemporary sexual scripts
7. describe sexuality in adolescence, young adulthood, middle adulthood, and late adulthood
8. describe sexually transmitted infections (STIs) and ways of preventing them
9. describe the contributing factors to healthy sexual relationships
10. explain the strengths and challenges of LGBTQ families
11. describe sexuality issues across the globe.

Brief Chapter Outline

Gender Identity
Gender Roles
Masculine and Feminine Gender Roles
Gender Expression; Cisgender; Transgender
Demographic Status of Transgender Persons; Cross-dressers; Genderqueer; Bigender; Third Gender; Two-spirit
Transsexuality

Sexual Orientation

Heterosexuality
Homosexuality
Same-sex But Non-sexual Relations; Situational Homosexuality; Preferential Homosexuality; Biological Roots of Homosexuality; The Cass Identity Model
Bisexuality
Pansexuality
Asexuality
Characteristics of Asexuality; Diversity of Asexuality
Romantic Orientation; Asexuality and Sexual Attraction; Asexuality and Transgender

Sexual Behaviors

Sexual Desire
Sexual Attraction
Most Common Lifetime Sexual Behaviors
Masturbation; Sexual Intercourse; Oral Sex; Anal Sex
Common Lifetime Sexual Behaviors
Flirting; Prostitution
Sociological Perspectives on Prostitution
Uncommon Sexual Behaviors
Sexual Behavior Deviations
Atypical Sexual Behaviors

Sexual Response

Gender Similarities and Differences
Barriers to Sexual Response
Categories of Sexual Dysfunction; Causes of Sexual Dysfunction

Diversity Overview: LGBT Families

Sexual Scripts

Traditional Sexual Scripts
Contemporary Sexual Scripts
Sexual Norms

Sexuality throughout the Life Cycle

Sexuality in Adolescence
Sexual Minority Youth
Sexuality in Young Adulthood
Sexuality in Middle Adulthood
Sexuality in Late Adulthood

Pre-test

Engaged or active learning is a powerful strategy that leads to better learning outcomes. One way to become an active learner is to begin the chapter and try to answer the following true/false statements from the material as you read. You will find that you have ready answers to these questions upon completion of the chapter.

Statement		
Another word for *heterosexual* is *straight*.	T	F
Gender identity is the extent to which one identifies as being either masculine or feminine.	T	F
Cisgender refers to a person whose gender identity is the same as his or her sex assigned at birth.	T	F
A homosexual is a person who is romantically or sexually attracted to people of the same sex.	T	F
Sexuality is viewed as a person's feelings and behaviors related to the pursuit of sexual pleasure.	T	F
Components of sexual identity are biological sex, gender role, gender identity, and sexual orientation.	T	F
Sexual behavior deviations are called *paraphilias*.	T	F
Traditional sexual scripts are characterized by a patriarchal pattern of male supremacy and female dependency.	T	F
Having sex regularly is one of the secrets of happy couples.	T	F
Condom use can reduce the risk of both sexually transmitted infections (STIs) and unintended pregnancies.	T	F

Upon completion of this section, students should be able to:
LO1: Apply the sociological perspectives to the analysis of sexuality.

Sexuality is a part of human life. **Sexuality** is the way people experience and express themselves as sexual beings. For example, people express their sexuality in the way they speak, sing, smile, sit, dress, dance, and cry. Sexuality has a history because it's a "cultural production" (Halperin, 1989). The first evidence of attitudes toward sexuality comes from Buddhist texts. German psychiatrist Richard von Krafft-Ebing (1840–1942) is considered to have established sexology as a scientific discipline (Hoenig, 1977). **Sexology** is the scientific study of human sexuality, sexual interests, sexual behaviors, and sexual functions. Just 27 years after Krafft-Ebing published *Psychopathia Sexualis* in 1886, the first academic association related to sex, Society for Sexology, was founded in 1913 (Kewenig, 1983).

Sociology of Sexuality

Studying sexuality is an interesting subfield of sociology because a sociological perspective transcends biological notions of sex and emphasizes the social and cultural bases of gender (ASA, 2017). Throughout time and place, the vast majority of human beings have participated in sexual relationships (Broude, 2003). Each society, however, interprets sexuality in different ways. For example, many societies around the world have different attitudes about premarital sex, the age of sexual consent, homosexuality, masturbation, and other sexual behaviors that are not consistent with universal cultural norms (Widmer *et al.*, 1998).

Sociological Perspectives on Sexuality

Social constructionism and postmodernism have been the most prominent approaches to the sociological study of human sexuality in the last two decades (Sanderson, 2003). The three major sociological perspectives fall within the paradigm of *social constructionism*—the belief that reality is socially constructed.

Structural Functionalism

When it comes to sexuality, structural functionalism examines the ways sexuality operates through social institutions such as the family. Structural functionalists stress the importance of regulating sexual behavior to ensure marital cohesion and family stability. **Marital cohesion** refers to the degree of connectedness, togetherness, and emotional bonding between marital partners (Doane, 2016). Talcott Parsons *et al.* (1955) argue that the regulation of sexual activity is an important function of the family, ensuring family stability. Most religions emphasize marital relationships as the appropriate context for sexual intimacy. From structural functionalist theory, the purpose of encouraging sexual activity in the confines of marriage is to intensify the bond between spouses and to ensure that procreation occurs within a legally recognized relationship.

Symbolic Interactionism

Symbolic interactionists suggest the importance of shared meanings and symbols in human sexuality. The two schools of thought are *situational symbolic interactionists*—behaviors based on the definition of situation—and *structural symbolic interactionists*—behaviors constructed through social institutions.

Situational symbolic interactionists focus on how individuals define situations and thereby construct the realities in which they live (Longmore, 1998). Sexuality is not biologically given but is produced by society through webs of social interaction and definition (Weeks, 1986). The interpersonal dimension opens the door for situational symbolic interactionism, where reality is defined by interacting people in a given situation (DeLamater & Hasday, 2007). For example, extramarital sex, defined as "abnormal" by one person, may be viewed otherwise by another. Without the proper elements of a sexual norm that defines the situation, names the actors, and plots the behavior, little is likely to happen (Longmore, 1998).

Structural symbolic interactionists focus on the ways in which social institutions influence the self's construction of sexuality. Each institution is associated with a sexual ideology or discourse (Foucault, 1998). For example, the family is associated with a discourse that emphasizes family functions of support and childrearing, norms of fidelity, and the incest taboo (DeLamater & Hasday, 2007). Medicine has become important in the conceptualization and control of sexuality, a trend referred to as the **medicalization of sexuality**—a process of the growing authority of medical experts over sexual experiences sponsored by the pharmaceutical industry (Tiefer, 2004).

In addition, **social exchange theory** explores how people determine the rewards and costs of the relationship for both partners. Rewards are defined as exchanged resources that are pleasurable and gratifying; costs are defined as exchanged resources that result in a loss. The theory is useful for understanding sexuality within a relational context, including why two people choose each other as sexual partners, which partner has more influence on what sexual activities they do together, sexual satisfaction, and the likelihood that one or both partners seek sexual activity outside the relationship (Sprecher, 1998). Therefore, social exchange theory is important in research on mate selection, relationship formation, and prediction of relationship dissolution.

Conflict Theory

Conflict theorists see power differentials as manifested throughout all aspects of a patriarchal society in which men hold power over women. Collins (1971) argues that men sexually overpower women and create the idea of women as sexual property. For them, sexual oppression based on sexual difference is omnipresent (Heasley & Crane, 2002). Belief that men have—or have the right to—more sexual urges than women creates a double standard. Reiss (1960) defines the **double standard** as prohibiting premarital sexual intercourse for women but allowing it for men. This standard has evolved into allowing women to engage in premarital sex only within committed love relationships, but allowing men to engage in sexual relationships with as many partners as they wish without condition (Milhausen & Herold, 1999).

Queer Theory

Queer theory is an analytical viewpoint within queer studies that challenges the putatively "socially constructed" categories of sexual identity (Branch, 2003). A central issue raised by queer theory is whether homosexuality, heterosexuality, and bisexuality are socially constructed or purely driven by biological forces (Pickett, 2015). **Queer theory** focuses on mismatches between sex, gender, and desire (Jagose, 1996) and constructedness of gendered and sexual identities and categorizations (Callis, 2009).

American academic scholar Eve Kosofsky Sedgwick (1950–2009) embraces the term *queer*, defining it as the open mesh of possibilities, gaps, overlaps, dissonances and resonances, and lapses and excesses of meaning when the constituent elements of anyone's gender, of anyone's sexuality, aren't made (or can't be made) to signify monolithically (1993). **Queer** is used as an umbrella term to describe individuals who don't identify as **cisgender**—a person whose gender

Upon completion of the next section, students should be able to:
LO2: Explain sexual identity and its components: biological sex, gender roles, gender identity.
LO3: Describe types of sexual orientation.

identity aligns with the assigned sex at birth. Sanderson (2003) argues that queer theory seeks to make homosexuality normal. **Sexual diversity** refers to all the diversities of sex characteristics, sexual orientations, and gender identity.

Components of Sexual Identity

Sexual identity is how people feel their gender identity aligns with or differs from their biological sex and how they choose to identify with those to whom they are romantically or sexually attracted. An individual's sexual identity can change throughout his or her life and may or may not align with his or her biological sex, sexual behavior, or actual sexual orientation (Rosario *et al.*, 2006).

Shively and DeCecco (1977) explore how biological sex, gender identity, gender role, and sexual orientation combine and conflict to form a person's sexual identity. This section examines the four components of sexual identity: biological sex, gender identity, gender role, and sexual orientation.

Biological Sex

Sex is biologically determined. As you learned in Chapter 5, **sex** refers to the biological traits associated with being male or female. Sex assignment is the determination of a newborn's sex at birth. Primary sexual characteristics are present at birth but secondary sexual characteristics emerge at puberty.

Primary Sex Characteristics

Primary sex characteristics are those a person is born with, including hormones, chromosomes, and the reproductive system used for reproduction. In the XY sex-determination system, a person's biological sex is determined by either of a pair of sex chromosomes. **Assigned sex** is a label that a person is given at birth based on a person's reproductive system. Being male means having testes where sperm are produced, testosterone, a penis, and an X chromosome and a Y chromosome (XY). Being female means having a vagina, uterus that holds the developing fetus, ovaries, and two X chromosomes (XX). In the majority of births, a relative, midwife, nurse, or physician inspects the genitalia when the baby is delivered, and sex is assigned, without the expectation of ambiguity (Reiner, 2002). For example, newborns with male genitalia or a penis are identified as males and newborns with female genitalia or a vagina are identified as females.

Secondary Sex Characteristics

Secondary sex characteristics are the non-reproductive physical features that develop later in life during puberty that distinguish males from females. For example, growth of breasts in females and growth of facial hair and beards in males are considered secondary sex characteristics. **Hormones** (the Greek word, meaning "set into motion") are chemical messengers that arouse cells and organs to specific activities and influence the way we look, feel, and behave (Hales, 2012). The development of primary and secondary sex characteristics is controlled by hormones produced by the body after the fetal stage.

Intersex

Someone may be born with atypical sex characteristics. Intersex people are born with biological sex characteristics (such as sexual anatomy, reproductive organs, hormonal patterns, and/or chromosomal patterns) that do not fit the typical definitions for male or female bodies (UN Office of Human Commissioner, 2016). For example, a person can be born with male genitalia or penis but have a functional female reproductive system inside. Intersex describes someone whose sexual organs are not strictly male or female (see image below).

In 1 in 100 births bodies differ from standard male or female; 1 or 2 in 1,000 births receive surgery to "normalize" genital appearance (Blackless *et al.*, 2000). Intersex persons routinely face forced medical

Biological sex refers to the objectively measurble organs, hormones, and chromosomes. Female = vagina, ovaries, XX chromosomes; male = penis, testes, XY chromosomes; intersex - a combination of the two.

Source: http://itspronouncedmetrosexual.com/2011/11/breaking-through-the-binary-gender-explained-using-continuums/#sthash.iX5eTSm7.3ZU6mWlH.dpbs

surgeries that are conducted at a young age without free or informed consent; these interventions jeopardize their physical integrity and ability to live freely (Kirby, 2016). Three former U.S. Surgeons-General issued a statement that calls for a moratorium on medically unnecessary surgeries on intersex children too young to participate in the decision (Elders *et al.*, 2017). Intersex children don't need to be "fixed" and they are perfect just as they are (UN Free & Equal, 2017).

Gender Identity

A person's sexual identity is also shaped by gender. As you learned in Chapter 5, gender refers to the social and cultural meanings and practices associated with masculinity and femininity. Gender examines a wide range of topics including gender identity, gender roles, gender inequality, and the interaction of gender with sexuality. **Gender identity** is one's personal perception and experience of being male, female, or combination of both, or neither. Gender identity is the extent to which one identifies as being either masculine or feminine (Diamond, 2002). A person can identify as male or female through gender roles.

Gender Roles

Gender roles are the expectations people have about attitudes and behaviors that go along with a person's assigned sex. **Gender roles** refer to the attitudes and behaviors considered socially and culturally appropriate for men or women and associated with masculinity and femininity. The meanings of masculinity and femininity play a role in how we understand gender roles. **Masculinity** is a social position, a set of practices, and the effects of the collective embodiment of those practices on individuals, relationships, institutional structures, and global relations of domination (Connell, 2000). Masculinity suggests that the attitudes and practices associated with masculinity are culturally valued above those associated with femininity. Connell (1995) distinguishes *hegemonic masculinity*—gender practices that guarantee the dominant position of men and the subordination of women—from *subordinate masculinity*—power relations between groups of men. Thus, hegemonic masculinity ensures male domination over women. An example of subordinate masculinity is the dominance of heterosexual men and the subordination of homosexual men.

Femininity refers to attributes and practices associated with women in distinction from men. For example, passive, submissive, and compassionate, nurturing behaviors toward others are traditionally considered feminine traits in comparison to masculine competitiveness and dominance. Pyke and Johnson (2003) examine the power relations among women and distinguish *hegemonic femininity*—dominance of White women—from *subordinated femininities*—subordination of non-White women (women of color). Whereas hegemonic masculinity is male domination over females, hegemonic femininity is confined to power relations among women (Pyke & Johnson, 2003). For example, a male/female wage gap (hegemonic masculinity) and a White women/women of color wage gap (hegemonic femininity) persist in the United States. Compared to a White male, a White woman working full-time loses $418,800 over a 40-year career. Compared to a White woman, a Black woman loses $840,040, a Native woman loses $934,240, and a Latina loses more than $1 million over a 40-year career (National Women Law Center, 2017).

Masculine and Feminine Gender Roles Masculine and feminine gender roles are learned through the process of socialization. The naturalness of masculinity/femininity, male/female, is an assumed part of American social institutions (LaMarre, 2007). The process begins before babies are born. For example, parents select gendered names on the basis of the biological sex of a fetus, and decorate the baby's room and select clothes in gendered ways that reflect cultural or societal expectations. Children are introduced to gender roles that are linked to their biological sex. For example, when it comes to toys, parents supply boys with trucks or toy guns while girls are usually given dolls that foster nurturing. At age four or five, most children are firmly entrenched in culturally appropriate gender roles (Kane, 1996) and they are able to become aware of the differences between boys and girls based on the behaviors and expectations of their parents. When it comes to occupational gender roles, women and men are likely to work according to the expectations of their gender roles.

Gender roles are therefore constructed and learned behaviors that are influenced by cultural factors because behaviors seen as appropriate in one culture may be viewed otherwise in another. Whether we are expressing our masculinity or femininity, we are *always* "doing gender" and thus gender is something we do or perform (West & Zimmerman, 1987). For example, gender identity has an effect on how people dress and present themselves.

Gender Expression

Our gender is demonstrated through the ways we dress and behave. **Gender expression** refers to the way a person expresses aspects of gender identity and shows his or her femininity, masculinity, or androgyny through physical appearance, clothing choice and accessories, hairstyles, voice or body language, and behaviors. Although social rules have restricted people's dress according to gender, gender expression has changed over time. For example, women wear dresses. Trousers were traditionally a male form of dress, frowned upon for women (Ewing & Mackrell, 2002). Starting around World War I, traditional gender roles blurred and fashion pioneers introduced trousers to women's fashion, which gave women an androgynous look (Köksal & Falierou, 2013). Gender expression may vary from country to country. For example, it is considered feminine to wear a dress or skirt but the kilt worn by Scottish males does not make them appear feminine in their culture. Gender expression may not reflect a person's gender. **Androgyny** refers to the combination of masculine and feminine characteristics. Androgynous describes an ambiguous form of expressing gender (see image below).

Cisgender

People perceive their gender identity based on gender and sex. **Cisgender** refers to a person whose gender identity aligns with the assigned sex at birth. If someone is born with male reproductive organs (biological sex), he is likely to identify as man (gender identity), express himself masculinely (gender expression), and grow up as a provider (gender role). We call this identity "cisgender" (when your biological sex aligns with how you identify) and it grants a lot of privilege (Killermann, 2013). **Cisgender identity privileges** include having your gender as an option on an application form and using public facilities without fear or anxiety. However, a person's gender identity can differ from his or her assigned sex at birth.

Gender Expression

Feminie **Androgynous** **Masculine**

Gender expression is how you demonstrate your gender (based on traditional gender roles) through the ways you act, dress, behave, and interact.

Source: http://itspronouncedmetrosexual.com/2011/11/breaking-through-the-binary-gender-explained-using-continuums/#sthash.iX5eTSm7.3ZU6mWIH.dpbs

Transgender

People tend to think of masculinity and femininity in dichotomous terms, viewing men and women as opposites. Most people feel that they are either male or female. However, the assignment of a biological sex may not align with how this person feels and how this person identifies. **Transgender persons** have a gender identity (masculine/feminine) that differs from their assigned sex at birth (male/female). For example, the assigned sex at birth is female, but gender identity may be male. A transgendered male may have an emotional connection to the feminine aspects of society and identify their gender as female.

Demographic Status of Transgender Persons
Approximately 0.6 percent of U.S. adults (1.4 million people) identify as transgender, ranging from 0.3 percent in North Dakota to 0.8 percent in Hawaii, California, Georgia, and New Mexico; an estimated 0.7 percent of adults ages 18 to 24, 0.6 percent of adults ages 25 to 64, and 0.5 percent of adults ages 65 and older identify as transgender (Flores *et al.*, 2016).

Transgender people are likely to suffer from poverty and need medical care. Compared with the general population, a national survey conducted in the United States in 2008 found that transgender individuals were four times more likely to live in extreme poverty (Grant *et al.*, 2011). In terms of health, transgender individuals had four times the rate of HIV infection and 28 percent postponed medical care because of discrimination (Bauer *et al.*, 2009). Factors that have put transgender people at risk for HIV infection include anal or vaginal sex without condoms or medicines to prevent HIV; injecting hormones or drugs with shared syringes; commercial sex work; mental health issues; incarceration; homelessness; unemployment, etc. (CDC, 2017a). Of 2,351 transgender people with newly diagnosed HIV infection during 2009–14, 84 percent were transgender women (male-to-female); 15.4 percent were transgender men (female-to-male), and 0.7 percent were additional gender identity (e.g., gender queer, bi-gender) (Clark *et al.*, 2016). Figure 6.1 shows that over half of both transgender women (50.8 percent; 1002/1974) and men (58.4 percent; 211/361) with newly diagnosed HIV infection were non-Hispanic Black/African American.

Student Account of Transgender
On the class roster, my name appears as "Ingrid Smith." However, I kindly ask for you to refer to me as England and direct me with male pronouns. I'm a transgender student and I am in the process of getting my name changed (personal communication, September 2017).

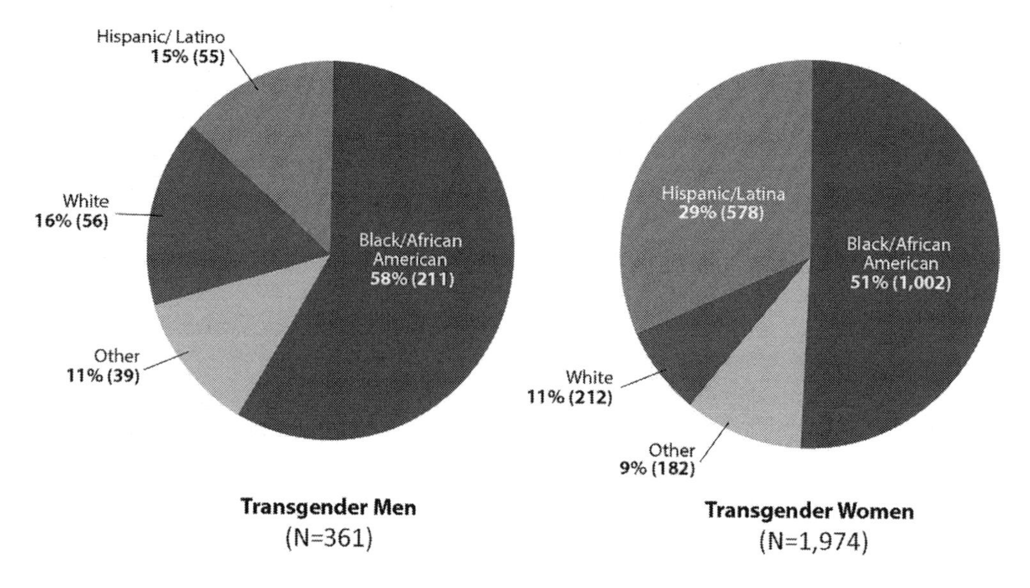

Figure 6.1 HIV Diagnoses among Transgender People in the United States
Source: Clark, H., Babu, A. S., Wiewel, E. W., Opoku, J., Crepaz, N. (2017). Diagnosed HIV Infection in Transgender Adults and Adolescents: Results from the National HIV Surveillance System, 2009–2014; 21(9), 2774–2783. Hispanics/Latinos can be of any race. https://www.cdc.gov/hiv/group/gender/transgender/index.html

Cross-dressers Transgender is used to include cross-dressers. A **cross-dresser** is a male who dresses as a woman or a female who dresses as a man but does not alter his or her reproductive organs or genitalia. For example, a cross-dresser with male reproductive organs (biological sex) identifies as female (gender identity) and expresses himself femininely (gender expression). Many transgendered children grow up hating their bodies, and this population can have high rates of drug abuse and suicide (Weiss, 2011). Others strive to express their gender identity through physical appearance, clothing choice, etc. Some begin dressing like, using names of, and wanting to be recognized as another gender. Cross-dressing men enjoy acting like women for a time, indulge their feminine persona, and fantasize about being transformed into a woman, but they still consider themselves to be men.

Genderqueer Some references use transgender broadly in such a way that it includes genderqueer (Cronn-Mills, 2014). The sixth- to twelfth-grade students in a national sample endorsed "genderqueer" more often than "transgender," and the number who endorsed "another gender" was almost half of those who endorsed "transgender" (Kosciw *et al.*, 2013). In a national survey of LGBT individuals in science, technology, engineering, and mathematics, more endorsed "genderqueer" than "transgender," whereas the number who endorsed "androgynous" was more than half of those who endorsed "transgender" (Yoder & Mattheis, 2016). **Genderqueer** (GQ) or non-binary (NB) is an inclusive term for gender identities that are not exclusively male or female. Genderqueer (see image below) is used for gender identity that is somewhere between woman and man.

Genderqueer people may identity themselves as being both men and women (bigender, binary) or having a fluctuating gender identity (genderfluid); having no gender, neither man nor woman (agender, genderless, genderfree, gender neutral); or being a third gender (Winter, 2010). Such gender identities outside of the binary of female and male are increasingly

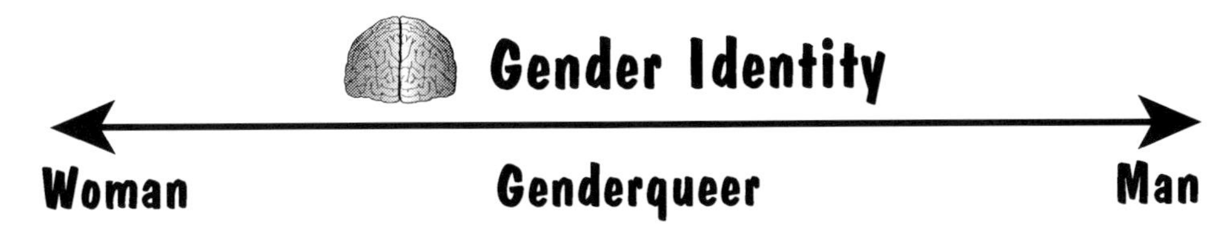

Gender identity is how you, in your head, think about yourself. It's the chemistry that composes you (e.g., hormonal levels) and how you interpret what that means.

Source: http://itspronouncedmetrosexual.com/2011/11/breaking-through-the-binary-gender-explained-using-continuums/#sthash.iX5eTSm7.3ZU6mWlH.dpbs

being recognized in legal, medical, and psychological systems and diagnostic classifications in line with the emerging presence and advocacy of these groups of people (Richards *et al.*, 2016).

Bigender **Bigender people** identify with both male and female and have two distinct gender identities, either at the same time, or at different times. A bigender person moves between feminine and masculine gender behavior and gender expression, depending on context. For example, a bigender person may play a dual role of female and male at the same time or may feel male one day and female the next. Among the transgender community, less than 3 percent of those who were assigned male at birth and less than 8 percent of those who were assigned female at birth identified as bigender (San Francisco Department of Public Health, 1999). Case and Ramachandran (2012) find that bigender people experience *alternating gender incongruity (AGI)*—involuntary switching of female and male states. When bigender people feel a change between their gender identities, it might have to do with a change in how they use parts of their brains between the left and right hemispheres (Case & Ramachandran, 2012).

Third Gender Some cultures have less distinct views of masculinity and femininity. Not all tribes/nations have rigid gender roles, but, among those that do, the most usual spectrum that has been documented is that of four genders: feminine woman, masculine woman, feminine man, and masculine man (Estrada, 2011). The term **third gender** is usually understood to mean "other" gender or a masculine woman, or a feminine man. The Gender Recognition Act of 2017 makes California the second state in the nation—following Oregon—to allow residents to be identified by a gender marker other than "F" or "M" on their driver's license and the first to allow a third gender marker on birth certificates (California State Senate Majority Caucus, 2017).

LaMarre (2007) explores how contemporary queer theorists have accounted for the role gender plays in becoming the primary marker of sexual identities. For example, Halberstam (1998) and Messerschmidt (2003) identify specific forms of female masculinity by looking at how women embody masculinity. In Samoa, located about halfway between Hawaii and New Zealand in the Polynesian region of the Pacific Ocean, *Fa'afafine*—the way of the woman, is used to describe individuals who are born biologically male but embody both masculine and feminine traits. In Native Hawaiian culture, **Māhū** refers to an individual who may be considered third-gendered with characteristics of both sexes, usually a male to female (University of Hawaii at Manoa Library, 2017).

Two-spirit **Two-spirit** is a gender identity that identifies people as combining the work and traits of both men and women in American Indian history. Two-spirit people fulfill a third-gender role in their cultures by performing work and wearing clothing associated with both men and women. Two-spirit within a traditional setting was a gender analysis and not a sexual orientation (Pruden & Edmo, 2016). The title *two-spirit* is a sacred, spiritual, and ceremonial role that is recognized and confirmed by the Elders of the two-spirit's ceremonial community (Estrada, 2011). The roles of the two-spirit include mediator, social worker, name giving, matchmaker, boy's and girl's puberty ceremony, and peace-maker for the tribe, etc. (Pruden & Edmo, 2016). Two-spirit is not interchangeable with LGBT Native American.

Both male- and female-bodied two-spirits have been documented in over 130 North American tribes, in every region of the continent (Roscoe, 1992). For example, the Navajos are among the tribes with links to the Grand Canyon. In the Navajo culture, there was a category of people called *berdaches*—anatomically normal men but defined as two-spirit to fall between male and female. American scholar Roscoe (1992) focuses on the life of We-Wha (1849–96), a *berdache*, and describes an alternative third-gender role (see image below). The term *two-spirit* was adopted in 1990 at an Indigenous lesbian and gay international gathering to replace the term *berdache* that was used from 1492 to 1990 (Pruden & Edmo, 2016).

Transsexuality

Some transgender people may cross-dress from their assigned sex but others may seek medical assistance

Source: Photographer John K. Hillers, 1843–1925. https://upload.wikimedia.org/wikipedia/commons/1/10/We-Wa%2C_a_Zuni_berdache%2C_weaving_-_NARA_-_523796.jpg

to transition from one sex to another. A medical diagnosis can be made if a person experiences discomfort as a result of a desire to be a member of the opposite sex. **Transsexuality** occurs when people identify with gender identities that do not match their assigned sex and want to undergo physical transitions from one sex to another. The medical interventions include gender reassignment surgeries and hormone therapy. A person born as male becomes recognizably female through the use of the hormones and/or surgical procedures. They can be known as male-to-female (MTF) or female-to-male (FTM) transsexuals. After transgender people alter the body through medical interventions, their biological sex can match their gender identity. However, transsexuals are unable to acquire the reproductive abilities of the birth sex to which they transition.

Sexual Orientation

Everyone has a gender identity and a sexual orientation, but a person's gender does not determine a person's sexual orientation (CDC, 2017a). What counts as sex, how it is socially regulated, and what people do in sex varies greatly between cultures and over historical time (Hawkes, 1996). In fact, sexual orientation is more multifaceted than any single number can reflect (Weinrich, 2014). **Sexual orientation** is an enduring pattern of emotional, romantic, or sexual attraction to persons of the opposite sex (heterosexuality), the same sex (homosexuality), or both sexes (bisexuality), or multiple sexes, or lack of sexual attraction to others. Let's explore the different forms of sexual orientation.

Heterosexuality

Heterosexuality is emotional, romantic, or sexual attraction between members of the opposite sex. A heterosexual person is therefore attracted to persons of the opposite sex. In 1886, German psychiatrist Richard von Krafft-Ebing used *heterosexual* in his book *Psychopathia Sexualis* (Ambrosino, 2017). This book was so popular that the term "heterosexual" became the most widely accepted term for sexual orientation. Another word for heterosexual is *straight*. For example, if a man is attracted to females, he is heterosexual or straight. Most men and women are heterosexual. Among adults ages 18 to 44, 92.3 percent of women and 95.1 percent of men said they were "heterosexual or straight" (Copen *et al.*, 2016).

The fact that the term "straight" is used to describe heterosexuals reflects **heteronormativity**—judging sexuality according to heterosexual norms such as sexual attraction between people of the opposite sex and alignment of the assigned sex and gender identity (masculine man/feminine woman). This naturalization of gender binaries is at the very foundation of society and shapes institutions such as marriage (Jackson, 2006). Gendered identities are reinforced by heteronormative society in which the media, political, and social institutions are constructed through the biased lens of assumed heterosexuality (Butler, 2004). The problem when looking at the heteronormative definitions of sexuality is that they assume that sexual desire, behavior, and identity are the same, that is, sexual desires and behaviors match up with overall sexual identity categories (LaMarre, 2007). **Homophobia** describes extreme prejudice directed at gays, lesbians, bisexuals, and others who are perceived as not being heterosexual.

Homosexuality

Homosexuality has existed throughout human history. **Homosexuality** is emotional, romantic, or sexual attraction between members of the same sex. A person who is **gay** is emotionally, romantically, or sexually attracted to members of the same gender; the term gay can be applied to both males and females, but it is more often applied to males. A **lesbian** is a woman who is emotionally, romantically, or sexually attracted to other women. Let us explore same-sex relations with a non-sexual nature (homosociality) and same-sex sexual relationships (homosexuality).

Same-sex but Non-sexual Relations

Sedgwick (1985) coined the term *homosocial* to demonstrate the immanence of men's same-sex bonds. **Homosociality** refers to same-sex relationships that are based on friendship or mentorship but are not of a romantic or sexual nature. The opposite of homosocial is **heterosocial**—non-sexual relations with the opposite sex. Homosocial relationships are not obliged to be sexual relationships and they are merely same-sex social interactions. For example, depending on the culture, family, and social structures, same-sex preferences have been found to develop between three and nine years of age (LaFreniere *et al.*, 1984). The sociological neologism "homosocial" was used to distinguish from "homosexual" and to connote a form of male bonding often accompanied by a fear or hatred of homosexuality (Yaeger, 1985). Therefore, homosociality does not imply homosexuality. Next, let us examine two forms of homosexuality—*situational homosexuality* and *preferential homosexuality*.

Situational Homosexuality

Sexuality in the ancient world was much more fluid and it wasn't uncommon for men to take both male and female sexual companions (Tyner, 2015). For example, pederasty seems to have developed in the late seventh century BCE as an aspect of ancient Greek homosocial culture (Hubbard, 2003). **Pederasty** (the Greek word meaning "love of boys") was a socially acknowledged erotic relationship between an adult male (the *erastes*) and a younger male (the *eromenos*) usually in his teens (Nissinen, 2004).

Situational homosexuality occurs when individuals engage in sexual relations with members of the same sex as a substitute for sexual relations with members of the opposite sex. The ancient Greeks practiced an elaborate form of *situational homosexuality*—the man–boy relationship (Cantarella, 1992; Percy, 1996). Adult men were expected to participate in mentorships with boys in order to introduce them to the general social order of men (Worthen, 2016). Greeks celebrated a social system in which mutual loving relationships between adult men and teenage boys were socially sanctioned and normalized (Kuefler, 2007). The Romans disapproved of sexual relationships between adult men and younger citizen boys (Worthen, 2016). Roman law dictated that those who took the passive (i.e., penetrated) role in a sex act between two men could lose their citizenship rights (Ormand, 2009). The Chinese also had a form of man–boy homosexuality that resembled the Greek pattern (Sanderson, 2003). For example, Ming Dynasty literature such as *Bian Er Chai* portrays homosexual relationships between men as more enjoyable and more "harmonious" than heterosexual relationships (Kang, 2009). Japan was another agrarian civilization with relative sexual openness and institutionalized man–boy homosexuality (Leupp, 1995). Situational homosexuality is widespread throughout Melanesia (Herdt, 1984). All of the patterns of situational homosexuality are ones engaged in only by men and are most commonly man–boy homosexuality (Sanderson, 2003). As in the case of the Greeks, the man–boy relationship is thought to contribute to the boy's masculinity (Harris, 1981).

Preferential Homosexuality

Preferential homosexuality implies that people have preferential erotic attraction to members of the same sex. The term "homosexuality" was coined in the late nineteenth century by a German psychologist, Karoly Maria Benkert (Pickett, 2015). The Victorian era is well known for its highly sexually repressive mores (Foucault, 1998). Laws dictated the criminality of same-sex behaviors in many Western countries (Worthen, 2016). Thus, "homosexuality" in Victorian society was locked up and hidden away in what contemporary theorists have described as "the closet," a place where same-sex desire is kept secret (Cocks, 2009). The degrees to which same-sex attraction was accepted varied by class, with the middle class taking the narrowest view, while the aristocracy and nobility often accepted public expressions of alternative sexualities (Pickett, 2015). Homosexuality became medicalized. Doctors were called in by courts to examine sex crime defendants (Greenberg, 1988). Homosexuality was outlawed in Colonial America.

Biological Roots of Homosexuality

It is estimated that about 2–4 percent of the populations of Western industrial societies are preferentially homosexual (Sanderson, 2003). After reviewing research on biological roots of homosexuality, sociologists Lee Ellis and Ashley Ames (1987) concluded that homosexuality develops when, during a critical period of fetal development, the brain receives an excess of the hormone(s) of the opposite sex. Male homosexuals thus have fetally "feminized" brains, whereas lesbians have fetally "masculinized" brains (Sanderson, 2003).

The Cass Identity Model

In the twentieth century, with the decline of prohibitions against sex for the sake of pleasure even outside of marriage, it became more difficult to argue against gay sex and these trends were strong in the 1960s (Pickett, 2015). Today more people come out of the closet and disclose their sexual orientation. **Coming out of the closet**, or coming out, is a metaphor for LGBT people's self-disclosure of their sexual orientation or gender identity. In late October 2017, Kevin Spacey (1959–), an American actor, film director, producer, screenwriter, and singer, came out as gay (Nordyke, 2017). Copen *et al.* (2016) found that 3.4 percent of single women and 4.5 percent of single men ages 18 to 44 said they were lesbian or gay; Hispanic women (11.2 percent) were less likely than White women (19.6 percent) and Black women (19.4 percent) to have had a same-sex sexual contact.

Sociologist Vivienne Cass (1979) developed the **Cass identity model**—a theory of gay and lesbian identity development. The coming-out process of homosexuals is described as a series of social stages that the individual is obliged to negotiate with others (Devor, 1997). The first stage is called identity confusion; the person attempts to deny their sexual identity as homosexual. In the identity comparison stage, the person examines the available identity options to see which one explains his or her self-identity. The third stage is called identity tolerance as the person recognizes "I probably am gay" and seeks out other gay people. In the identity acceptance stage, the person accepts or claims public acknowledgment of his or her sexual identity and continues to learn the gay and lesbian culture. In the identity pride stage, the person identifies gay culture as a source of support and acquires social connections. Finally, in the identity synthesis stage, the person views sexual orientation as his or her sexual identity and integrates sexual identity with his or her self-concept. Sociologists Kaufman and Johnson (2004) argue that this model does not take into account socio-cultural factors that can impact identity development.

American sexologist Alfred Kinsey (1894–1956) was among the first to conceptualize sexuality as a continuum rather than a dichotomy of straight or gay. The Kinsey scale (see image below) is used in research to describe a person's sexual orientation based on his or her experience or response at a given time, ranging from 0, meaning exclusively heterosexual, to 6, meaning exclusively homosexual.

Bisexuality

In Kinsey's scale, the rating of 3 indicates "equally heterosexual and homosexual" and means bisexual persons. In Kinsey reports, 11.6 percent of White males aged 20–35 were given a rating of 3 for this period of their lives (Kinsey *et al.*, 1948); 7 percent of single females aged 20–35 and 4 percent of previously

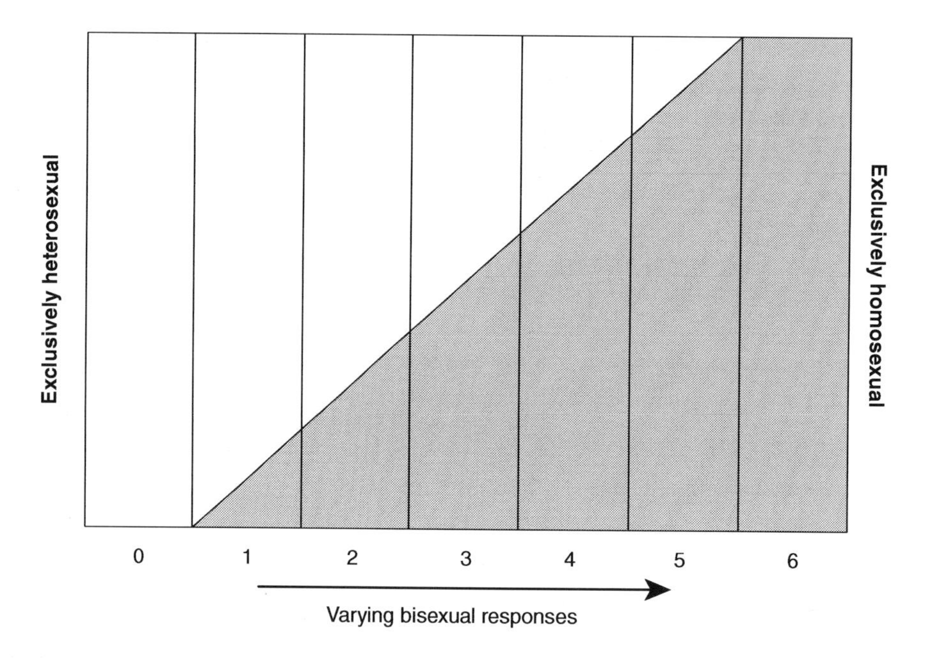

Source: Sexual Behavior in the Human Female (1953) by Alfred Kinsey. https://en.wikipedia.org/wiki/Kinsey_scale

married females aged 20–35 were given a rating of 3 for this period of their lives (Kinsey *et al.*, 1953).

Bisexuality is emotional, romantic, or sexual attraction to both the opposite and the same sex. Bisexuals are attracted sexually to both males and females and engage in sexual relationships with either males or females. For some, a bisexual identity serves as a transitional identity to a subsequent gay/lesbian identity (Rosario *et al.*, 2006). Bisexuals account for about 1.8 percent of the total U.S. adult population (Gates, 2011).

Bisexual identity, by its very existence, plays with categories of sexuality and gender (Callis, 2009). Bisexual women are more likely to share their sexual identity with their family and friends. According to the Pew Research Center survey in 2013, 40 percent of respondents said they were bisexual; bisexual women are much more likely than bisexual men to say most of their friends and family know about their sexuality (Brown, 2017). Copen *et al.* (2016) find some heterosexuals, married people, cohabiting couples, and singles are bisexual. Among heterosexual women and men, approximately 3.2 percent of currently married women and 1.7 percent of currently married men said they were bisexual; 5.9 percent of currently cohabiting women and 1.8 percent of currently cohabiting men said they were bisexual; and 7.2 percent of single women and 2.4 percent of single men said they were bisexual.

Pansexuality

Bisexuals believe gender to be a factor in their attraction to people. Pansexuals believe gender has nothing to do with sexual attraction. **Pansexuality** is emotional, romantic, or sexual attraction to people regardless of biological sex and gender identity. Pansexual people can be attracted toward any person such as cisgender people, transgender people, etc.

The term pansexualism denotes sexual instinct in human activity. Psychologist Sigmund Freud (1900) coined the term pansexualism to mean "the pervasion of all conduct and experience with sexual emotions." Pansexual people may refer to themselves as gender-blind, asserting that gender and sex are not determining factors in their romantic or sexual attraction to others (Diamond & Butterworth, 2008).

Asexuality

Asexuality refers to having no sexual attraction to a person of either sex (Bogaert, 2004). It means a lack of or low level of sexual attraction, sexual interest toward others, and sexual desire for sexual activity. Approximately 1 in 100 people is asexual (Decker, 2015). In the Kinsey scale (see Table 6.1), a separate category of "X" was created for those with "no socio-sexual contacts or reactions" (Kinsey *et al.*, 1948). The category of "X" explained people who "do not respond erotically to either heterosexual or homosexual stimuli, and do not have overt physical encounters with individuals of either sex" (Kinsey *et al.*, 1953). Storms (1980) proposed a two-dimensional map of erotic orientation showing four sexual orientation categories: homosexual, bisexual, asexual, and heterosexual.

Table 6.1 Table of the Kinsey Scale

Rating	*Description*
0	Exclusively heterosexual
1	Predominantly heterosexual, only incidentally homosexual
2	Predominantly heterosexual, but more than incidentally homosexual
3	Equally heterosexual and homosexual
4	Predominantly homosexual, but more than incidentally heterosexual
5	Predominantly homosexual, only incidentally heterosexual
6	Exclusively homosexual
x	No socio-sexual contacts or reactions

Source: https://en.wikipedia.org/wiki/Kinsey_scale; Kinsey *et al.* (1948)

Characteristics of Asexuality

Asexual people have the same emotional needs as anyone else (ALGBTICAL, 2017). Intimate relationship is as important to asexual people as it is to anyone. They enjoy indulging in non-sexual intimate practices such as kissing, hugging, and cuddling with close friends. In the absence of sex, these intimate practices are demonstrated by asexual people to close friends and vital to the endurance of bonds, like non-asexual people. Some asexuals engage in sexual activity to only satisfy their partners and regard it an act of altruistic love (Scott *et al.*, 2014). Many asexuals experience attraction, but do not act out their attraction sexually and unlike people who engage in abstinence, asexual people view sexual activity as an unnecessary part of their life (ALGBTICAL, 2017).

Empirical research on asexuality reveals significantly lower self-reported sexual desire and arousal and lower rates of sexual activity (Brotto & Yule, 2011). Research also shows that sexual desire and arousal are not absent in asexuals. For example, 38 women ages 19 to 55 (ten heterosexuals, ten bisexuals, eleven homosexuals, and seven asexuals) viewed neutral and erotic audiovisual stimuli while VPA (vaginal pulse amplitude) and self-reported sexual arousal and affect were measured (Brotto & Yule, 2011). When asexuals experience sexual arousal, they may masturbate but feel no desire for partnered sexuality (ALGBTICAL, 2017).

Diversity of Asexuality

Asexuality might be a fluid and changeable identity. **Fluid sexuality** refers to a person whose romantic and/or sexual attraction changes over time. Asexuality depends on individual experiences over the life course. Some may not realize their sexual orientation until adulthood. Among the asexual community, each asexual person experiences relationships, attraction, and arousal somewhat differently (ALGBTICAL, 2017).

Romantic Orientation One distinction between types of asexual people is whether they are romantically attracted to others or desire romantic relationships (Decker, 2015). **Romantic asexual people** are romantically attracted to others. Many described themselves as "romantic" asexuals, meaning that they wanted to have emotionally but not physically intimate relationships (Scott *et al.*, 2014). **Aromantic asexual people** are not romantically attracted to others but are often satisfied with non-romantic relationships such as closer friendships. Approximately 20 percent of asexual-identified survey respondents identified as aromantic (AVEN, 2014). **Grayromantic asexual people** are between romantic and aromantic and less likely to experience romantic attraction compared to most people. **Demiromantic asexual people** sometimes develop romantic attraction toward someone after becoming familiar with and emotionally fond of that person. Grayromantic and demiromantic people may or may not experience sexual attraction in normative ways, but find their romantic experiences to be somewhere between aromantic and romantic (Decker, 2015).

Asexuality and Sexual Attraction Another distinction between types of asexual people is whether they are sexually attracted to others or desire sexual relationships. "Aces" refers to the aggregate *Asexual, Demisexual, Grey-Asexual Spectrum* (AVEN, 2014). **Demiasexual people** only experience sexual attraction after forming a strong emotional connection, such as in the context of a "serious" committed relationship. Some people identify in the gray area between asexual and sexual. **Grayasexual people** do not normally experience sexual attraction but experience it sometimes.

Asexuality and Transgender Being transgender does not imply any specific sexual orientation. Like cisgender people, transgender people may identify as straight, gay, lesbian, bisexual, and so on (Human Rights Campaign, 2016). For example, in the 2014 survey, among ace women, 1.3 percent identified as transgender; among ace men, 17.5 percent identified as transgender; among non-binary aces, 30.5 percent identified as transgender (Siggy, 2015).

Just as people will rarely go from being straight to gay, asexual people will rarely become sexual and vice versa (AVEN, 2017). Asexuals are more likely to have low self-esteem and to be depressed than other sexual orientations (Nurius, 1983). The asexual woman is portrayed as invisible, "oppressed by a consensus that they are non-existent," and left behind by both the sexual revolution and feminist movement (Johnson, 1977). Some asexual people engage in political activism through groups (AVEN, 2017).

Sexual Behaviors

Sexual behavior refers to the manner in which people experience and express their sexuality. Sexual activity has sociological, emotional, and biological aspects. People tend to engage in sexual activities for personal pleasure or for reproduction, and arouse the sexual interest of another or attract partners during the courtship process. Levine (2003) views sexual desire as the sum of the forces that lean us toward and away from sexual behavior.

Sexual Desire

Sexual desire is often considered essential to romantic attraction and relationship development (Regan &

Upon completion of this section, students should be able to:
LO4: Explain types of sexual behaviors and sexual behavior deviations.

Atkins, 2006). It is clinically useful to think of sexual desire as consisting of drive (biological), motive (individual and relationship psychology), and wish (cultural) components (Levine, 2003). **Sexual desire** or libido is a motivational state and a drive to seek out sexual objects or engage in sexual activities (Regan & Atkins, 2006). Understood from a biological framework, sexual desire comes from an innate motivational force like "an instinct, drive, need, urge, wish, or want" (DeLamater & Sill, 2005).From a sociological perspective, sexual desire can be aroused by socio-cultural sources such as sexual fantasies and is affected by gender, age, education, and attitudes. For example, among women, the principal influences on strength of sexual desire are age, the importance of sex to the person, and the presence of a sexual partner; among men, age and education are factors in the importance of sex to the person; attitudes are more significant influences on sexual desire than biomedical factors (DeLamater & Sill, 2005). Men have higher sex drives and desire for sexual activity than women do; this also correlates with the finding that men report, on average, a greater total number of lifetime sexual partners (Ostovich & Sabini, 2004). Men think about sex an average of nineteen times per day, compared to ten times per day for women (Fisher *et al.*, 2011).

Sexual Attraction

Sexual attraction refers to a person's attraction to another person sexually on the basis of sexual desire. Sexual attraction is a factor in sexual selection and is diverse and culturally universal. It is a person's ability to attract the sexual interest of another person and it can be based on a physical trait or other qualities of a person. There is an extremely widespread, probably universal, desire on the part of men to be stimulated by the sight of female genitals (Sanderson, 2003). A person's sexual attractiveness also depends on another person's sexual orientation. For example, a gay or lesbian person would find a person of the same sex to be more attractive than one of the opposite sex. Sexual attraction varies by race and education. White women (79.6 percent) were less likely than Hispanic women (84.7 percent) and Black women (84.2 percent) to say they were attracted "only to the opposite sex"; women with a bachelor's degree or higher (84.7 percent) were more likely to say they were attracted "only to the opposite sex" compared with women with a high school diploma (79.1 percent) (Copen *et al.*, 2016).

Sexual attraction may be enhanced by a person's cultural traits such as customs or visual sexual stimuli (VSS). For example, technological innovations have resulted in greater access to sexually explicit material and greater ease of taking and sharing sexually explicit photographs and videos (Herbenick *et al.*, 2017). VSS are often used as pleasant, arousing stimuli that have an intrinsic positive value (Wierzba *et al.*, 2015). In most laboratory settings, VSS plays a role of reward, as evidenced by the experience of pleasure while watching VSS, possibly accompanied by genital reaction (Gola *et al.*, 2016). More hours viewing VSS such as erotic videos was related to stronger-experienced sexual responses to VSS in the laboratory and was related to stronger desire for sex with a partner (Prause & Pfaus, 2015). Lesbian women showed a significantly greater interest in VSS than heterosexual women (Bailey *et al.*, 1994). Men have been found to place more emphasis on VSS than women (Conley, 2011).

Most Common Lifetime Sexual Behaviors

People engage in many types of sexual behaviors, ranging from sexual behaviors performed alone, such as masturbation, to sexual behaviors performed with another person, such as sexual intercourse. Herbenick *et al.* (2017) find that more than 80 percent of people reported lifetime masturbation, vaginal sex, and oral sex.

Masturbation

Autoeroticism is the practice of becoming sexually stimulated through internal stimuli (Laplanche &

Pontalis, 1988). The most common autoerotic practice is **masturbation**—the sexual stimulation of one's own genitals for sexual arousal. Masturbation is more common than partnered sexual activities during adolescence and older age (70+) (Herbenick *et al.*, 2010). A survey of older adults in the United States found that 35 percent of men and 20 percent of women ages 60 to 69 reported masturbation and the principal correlate was frequency of sexual desire (DeLamater & Sill, 2005).

Sexual Intercourse

Most people engage in sexual intercourse, or vaginal intercourse, or vaginal sex. **Sexual intercourse** is the insertion and thrusting of the erect penis into the vagina for sexual pleasure, reproduction, or both. Vaginal sex is the mainstay of the heterosexual repertoire, experienced at least once in the past month by around 60 percent of all men and women aged 16–74 years (Mercer *et al.*, 2013). The proportion of adults who reported vaginal sex in the past year was highest among men ages 25 to 39 and for women ages 20 to 29 (Herbenick *et al.*, 2010). The 2011–2013 National Survey of Family Growth revealed that 94.2 percent of heterosexual women and 92.2 percent of heterosexual men ages 18 to 44 had ever had vaginal intercourse (Copen *et al.*, 2016). Forms of penetrative sexual intercourse also include oral sex, anal sex, sexual penetration by the fingers, and penetration with a sex toy.

Oral Sex

Most people engage in **oral sex**—a sexual behavior that involves contact between the mouth, lips, or tongue and the genitals. *Cunnilingus* is oral sex performed on a female. *Fellatio* is oral sex performed on a male. In 2009, 18.3 percent of males and 22.4 percent of females ages 16 to 17 performed oral sex; more than half of women and men ages 18 to 49 engaged in oral sex (Herbenick *et al.*, 2010). The 2011–2013 National Survey of Family Growth revealed that 86.2 percent of women and 87.4 percent of men had ever had oral sex (Copen *et al.*, 2016). Oral sex alone cannot result in pregnancy, and heterosexual couples may perform oral sex as their method of contraception (Crooks & Baur, 2010). Although oral sex has been recommended as a form of safe sex, oral sex is not necessarily an effective method of preventing sexually transmitted infections (STIs) (Fulbright, 2003).

Oral sex varies by race, education, and politics. Oral sex was reported more often among those with a bachelor's degree or higher (women 91.5 percent and men 91.7 percent); White women (91.9 percent), White men (91.0 percent), and Black men (90.4 percent) were more likely to have ever had oral sex than Hispanic men (78.6 percent) and other race groups (Copen *et al.*, 2016). Hatemi *et al.* (2016) examined the influence of political views on sexual behavior in the United States and found that conservatives were happy with fewer partners and one or two sex positions, while liberals tried a wide variety of sex positions and sex behavior (oral sex, anal sex) with more partners.

Anal Sex

Anal sex is the penetration of the anus by the erect penis for sexual pleasure. Other forms of anal sex include fingering and using sex toys for anal penetration. Anal sex without the protection of a condom is considered the riskiest form of sexual activity (Krasner, 2010). More than 20 percent of men ages 25 to 49 and women ages 20 to 39 reported anal sex in 2009 (Herbenick *et al.*, 2010). Lifetime anal sex was reported by 43 percent of men (insertive) and 37 percent of women (receptive) (Herbenick *et al.*, 2017). Although anal sex is associated with male homosexuality, research shows that not all gay males engage in anal sex and that it is not uncommon in heterosexual relationships (Wellings *et al.*, 2012).

Common Lifetime Sexual Behaviors

Common lifetime sexual behaviors included wearing sexy lingerie/underwear (75 percent women, 26 percent men), sending/receiving digital nude/semi-nude photos (54 percent women, 65 percent men), reading erotic stories (57 percent of participants), public sex (≥43 percent), role-playing (≥22 percent), tying/being tied up (≥20 percent), spanking (≥30 percent), and watching sexually explicit videos/DVDs (60 percent women, 82 percent men) (Herbenick *et al.*, 2017).

Flirting

Flirting can be considered a sexual behavior. **Flirting** (or **coquetry**) is sexual activity involving verbal and nonverbal language by one person to another. Examples of flirting include blowing a kiss, making eye contact, smiling, touching, and winking. Flirting behavior varies across cultures. For example, people hold eye contact to flirt in the United States but avoid eye contact to flirt in Japan. Japanese women are not expected to initiate eye contact as it is considered disrespectful. And a kiss is not viewed the same across cultures: A kiss can express sentiments of love, passion, affection, respect, greeting, friendship, peace, or good luck, depending on the culture.

Mobile phones and social media are a major vehicle for flirting and expressing interest in a potential partner. Lenhart *et al.* (2015) find that 55 percent of all teens ages 13 to 17 have flirted with someone in person to let them know they are interested; 50 percent of teens have let someone know they were interested in them romantically by friending them on a social media site. A more recent trend in flirting is **sexting**—the act of sending nude pictures or sexually explicit messages via text.

Prostitution

Prostitution is a universal feature of society. **Prostitution** is the practice of engaging in sexual behavior in exchange for payment. The sexual behavior may take place at the client's residence or hotel room or may take place at the prostitute or escort's residence or a hotel room rented for the occasion. In *street prostitution*, the prostitute solicits customers on the street.

Sociological Perspectives on Prostitution Structural functionalists suggest that prostitution is functional in society. For example, it provides a sexual outlet for men, such as the poor, who cannot compete in the marriage market and for men such as sailors who are away from home a lot (Davis, 1937). Prostitution was extremely prevalent in ancient Roman culture; men from all social standings could consort with prostitutes, to a reasonable degree, without much social censure (Tyner, 2015).

Conflict theorists view prostitution as a social problem of unequal distribution of resources. In traditional societies, women had few opportunities to support themselves and relied on economic support from men. Their young bodies were their most marketable resource. In ancient Rome, many prostitutes were slaves of wealthy businessmen forced to sell their body for the pleasure of other men (McGinn, 1998). Weitzer (2005) explores different kinds of power relations in contemporary prostitution between workers and managers and workers and customers. We need to repair the economic damage suffered globally by women as the result of low wages, limited occupational choices, restricted access to social resources, and inequitable distribution of wealth (Spanger & Skilbrei, 2017).

The *symbolic interactionist perspective* examines how prostitutes learn the trade and are trained to perform the prostitute's role. For example, prostitutes learn how to hustle—how to get the maximum amount of money for the minimum amount of work (Heyl, 1979). The *social exchange framework* focuses on the exchange of resources (material or symbolic) between or among people (Sprecher, 1998).

It is the domain of the states to permit or prohibit prostitution under the Tenth Amendment to the United States Constitution. Prostitution is illegal in the United States except for ten Nevada counties and on November 3, 2009, Rhode Island closed a legal loophole that had allowed indoor prostitution to exist since 1980 (ProCon.org, 2016).

Uncommon Sexual Behaviors

Lifetime group sex, sex parties, taking a sexuality class/workshop, and going to BDSM parties are uncommon (each <8 percent) (Herbenick *et al.*, 2017). **Group sex** is sexual behavior involving more than two participants between people of all sexual orientations and genders. Group sex commonly takes place at a private sex party. The film *The Ice Storm* (1997) centers around two dysfunctional families who attend a "key party" where married couples swap sexual partners by having wives select other husbands' keys from a bowl. A key party is a type of swinger sex event. **Swinging** (wife or partner swapping) is a non-monogamous sexual behavior in which singles and partners in a committed relationship engage in sexual activities with others as a recreational or social activity (Bergstrand & Williams, 2000). Key parties or swinger parties still exist in the twenty-first century. A diverse group of entrepreneurs, doctors, financiers and artists gather for the monthly *Bronze Party*—a lifestyle party, a modernized term for what many refer to as swingers' parties (Segall, 2015). Some couples see swinging as a healthy outlet and as a means to strengthen their relationship (Bergstrand & Williams, 2000).

BDSM refers to erotic or role-playing practices such as BD (bondage, discipline, dominance and submission) and SM (sadism and masochism). The two terms "sadism" and "masochism" incorporated into the compound BDSM were originally derived from the names of Marquis de Sade (1740–1814), who not only practiced sexual sadism, but also wrote novels about these practices, and Leopold von Sacher-Masoch (1836–95), who wrote novels expressing his masochistic fantasies (Hyde & DeLamater, 2017). The abbreviation SM is often used for *sadomasochism*—the giving or receiving of sexual pleasure from the infliction or reception of pain (see image below). The sexual sadist and masochist enjoy giving or receiving pain. Organizers often provide certain large pieces of BDSM equipment to which people can be bound or restrained while the partygoers usually bring their own whips, canes, and restraints (Moser, 2006).

Sexual Behavior Deviations

Sexual deviance refers to behaviors that involve individuals seeking erotic gratification through means that are considered different to most people. Sexual behaviors are redefined during periods of social change, For example, sexual behavior such as masturbation viewed as "deviant" before is seen as normal now in many cultures. So-called sexual deviance, therefore, is not a problem within the sexual field but an issue within the social field (de Lauretis, 2017).

Atypical Sexual Behaviors

Certain atypical sexual behaviors are considered sexual behavior deviations or **paraphilias** such as *exhibitionism, frotteurism, pedophilia, transvestic fetishism, voyeurism*, and *zoophilia*. **Exhibitionism** refers to the urge or act of exposing one's sex organs to nonconsenting others. **Frotteurism** refers to the urge or act of rubbing one's genitals against the body of an unfamiliar or nonconsenting person. **Pedophilia** involves sexual attraction or activity with those under the age of thirteen. Usually people with this disorder develop strategies to gain access to and the trust of children. **Transvestic fetishism** is a sexual deviation in which heterosexual men wear women's clothes in order to achieve sexual arousal. **Voyeurism** occurs when a person achieves sexual arousal by watching an unfamiliar person who is undressing or engaging in sexual activity. **Zoophilia** occurs when a person has sexual feelings for animals or engages in sexual

Source: Joe Shuster (1950s). https://en.wikipedia.org/wiki/Sadomasochism#/media/File:Shuster_Nights_of_Horror-18.jpg

Upon completion of this section, students should be able to:
LO5: Identify sexual response and barriers to sexual response.

behaviors involving animals. People with this disorder might focus on domestic animals like dogs, or farm animals such as sheep or goats (World Health Medicine, 2017).

Sexual Response

The **sexual response cycle** refers to the sequence of physical and emotional changes that occur as a person becomes sexually aroused and participates in sexual activities such as sexual intercourse. Austrian neurologist Sigmund Freud (1922) described the sexual response as a complicated sequence of related events ultimately resulting in the execution of the sex act. William Masters and Virginia Johnson (1966) developed the sexual response cycle (see image below), outlining the four phases to characterize the physical and emotional changes that most people experience when they engage in sexual intercourse: excitement, plateau, orgasm, and resolution. Although everyone has a unique sexual response, these phases are common among most people.

Gender Similarities and Differences

The duration and characteristics of each stage vary between men and women. For males, sexual intercourse begins with the *excitement phase*, in which physiologic signs of arousal are noticed, muscle tension increases, the heart rate quickens, and breathing is accelerated. Due to hormonal and psychological stimuli, men also begin secreting a lubricating liquid. This is followed by a *plateau phase* when more hormones are released, muscle tension increases, and heart rate increases. *Orgasm* is the climax of the sexual response cycle. Rhythmic contractions of the muscles at the base of the penis result in the ejaculation of semen. Orgasm leads to the *resolution phase*. In the resolution phase, the male is in an unaroused state. Men need recovery time after orgasm, called a *refractory period*, during which they cannot reach orgasm again. This rest period may increase with age (King, 2013).

A similar sexual response cycle occurs in females. In the excitement phase, as blood flow to the genitals increases, nipples become erect and vaginal lubrication begins. This is followed by the plateau phase in which the vagina continues to swell from increased blood flow and the vaginal walls turn a dark purple. During orgasm, the muscles of the vagina contract and the uterus undergoes rhythmic contractions. Muscle contractions in the vaginal area create a high level of pleasure, though all orgasms are centered in the clitoris (King, 2013).

Sexual responses in both men and women are similar; men's sexual response only differs in terms of duration while women can have responses that differ in both intensity and duration (Masters & Johnson, 1966). However, research shows that there are many differences between men and women in terms of their response. For example, Basson (2000) argues that not everyone fits this model and most women do not orgasm during penetrative sexual intercourse. Masters and Johnson (1966) equate a man's erection with a woman's vaginal lubrication during the excitement phase. Levin (2008) argues that a woman's clitoris is the anatomical parallel to a man's penis and, as a result, clitoral swelling would be the equivalent of a man's erection.

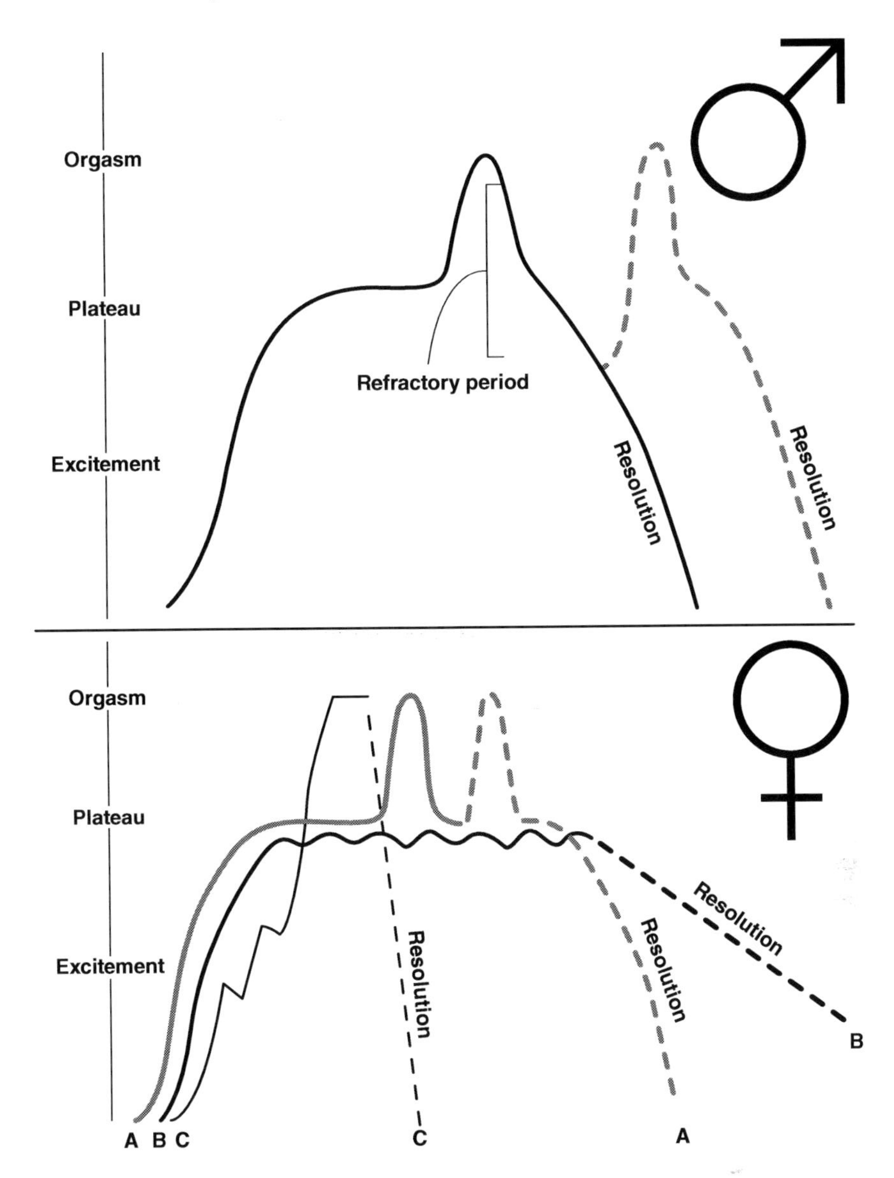

Sexual Response Cycle as the Sequence of Physical and Emotional Changes Occurring in Sexual Activities or as a Person becomes Sexually Aroused and Participates in Sexually Stimulating Activities.
Source: Avril 1975 [Public domain], via Wikimedia Commons. https://pbmo.files.wordpress.com/2012/04/sexual-response-cycle.png

Barriers to Sexual Response

The human sexual response cycle set the foundation for studying and categorizing sexual dysfunctions in men and women (Kaplan, 1974). Some people experience barriers to sexual response. **Sexual dysfunction** refers to a problem during any phase of the sexual response cycle that prevents a person from experiencing satisfaction from sexual activity. In a large American sample, the prevalence of sexual dysfunction is 43 percent in women and 31 percent in men (Sexual Diversity, 2017).

Categories of Sexual Dysfunction

There are three common sexual disorders for men including sexual desire, ejaculation disorder, and erectile dysfunction. The main symptoms of sexual dysfunction in men include low libido (lack of sex desire) and ejaculation problems. The four main categories of sexual dysfunctions for women include desire disorders, arousal disorders, orgasm disorders, and sexual pain disorders (King, 2013). Symptoms of sexual dysfunction in women include low libido and difficulty in achieving orgasm. **Hypoactive sexual desire disorder** (HSDD) refers to persistently or recurrently deficient or absent sexual fantasies and desire for sexual activity according to the *Diagnostic and Statistical Manual of Mental Disorders* (DSM III-R) in 1987. In patients with HSDD, 41 percent of women had at least one other sexual dysfunction and 18 percent had diagnoses in all three categories (that is, in desire, arousal, and orgasm disorders (Segraves & Segraves, 1991). Regardless of the adjustment in the 2013 DSM terminology from hypoactive sexual desire disorder (HSDD) to female sexual interest/excitement and arousal disorder (FSIAD), HSDD still exists as a center part of FSIAD (Jayne *et al.*, 2017).

Causes of Sexual Dysfunction

Sexual dysfunction can result from physical or psychological problems. For example, free testosterone is crucial as it affects key receptors both centrally and peripherally to promote a normal sexual response and desire (Jayne *et al.*, 2017). In women, testosterone acts centrally to promote desire as well as peripherally in the vulva and vagina to facilitate arousal by optimizing blood flow and facilitating lubrication (Traish *et al.*, 2007). Estrogen is derived from precursor hormones with a central and peripheral effect promoting the sexual response and desire in women (Cappelletti & Wallen, 2016).

Sexual dysfunction in men can also result from physical or psychological problems (WebMD, 2017). Many physical conditions such as diabetes, hormonal imbalances, and heart disease can cause problems with sexual function. Side-effects of medications such as antidepressant drugs can affect sexual desire and function. Psychological causes include stress, depression, anxiety, concern about sexual performance, relationship issues, or the effects of past sexual trauma (World Health Medicine, 2017). Sexual health and general health go hand in hand; the treatment of sexual dysfunction has global implications on health and well-being (Jayne *et al.*, 2017). Morrow (1994) argues that sexual dysfunctions are socially constructed as problems with reference to hegemonic masculine standards and patterns of normal sexual functioning.

Sexual Scripts

When studying sexuality, sociologists focus their attention on sexual scripts and practices. **Sexual script** is a set of expectations and guidelines for appropriate sexual behavior. Sociologists Gagnon and Simon (1973) introduced the concept of sexual scripts and bring social scientific tools to the study of sexualities. Socially learned sexual scripts tell people who to have sex with (e.g., race, gender), when and where it is appropriate to have sex, and what acts are appropriate (and in what order) once sexual behavior is initiated (DeLamater & Hasday, 2007).

Upon completion of this section, students should be able to:
LO6: Compare traditional sexual scripts with contemporary sexual scripts.

Diversity Overview: LGBT Families

LGBT family is a form of family in which lesbian, gay, bisexual, and/or transgender people form intimate relationships and raise children. LGBT History Month is observed in October every year to celebrate the achievements of lesbian, gay, bisexual, and transgender icons.

Changing Attitudes towards LGBT People

Out of 42 communities: homosexuality was accepted or ignored in nine; five communities had no concept of homosexuality; eleven considered it undesirable but did not set punishments; and seventeen strongly disapproved and punished (Broude & Greene, 1976). Today about six in ten Americans (62 percent) say they favor allowing gays and lesbians to marry legally (Pew Research Center, 2017a). The increase in the share of adults who favor same-sex marriage is due in part to generational change. For example, more than 56 percent of baby boomers favor allowing gays and lesbians to marry legally in 2017. Support for same-sex marriage also has grown among those in the Silent Generation in recent years (41 percent). Millennials (74 percent) and Generation X (65 percent) express higher levels of support for same-sex marriage in 2017.

Same-sex Couples on the Rise

Homosexuals account for about 1.7 percent of the total U.S. adult population (Gates, 2011). On June 26, 2015, the U.S. Supreme Court legalized gay marriage nationwide. Same-sex marriages are currently on the rise. As of 2017, more Americans support (62 percent) than oppose (32 percent) allowing gays and lesbians to marry legally (Masci *et al.*, 2017). However, many couples are still judged daily. For example, a Kentucky clerk denied same-sex couples licenses because of her religious beliefs (Blinder & Perez-Pena, 2015).

As of March 27, 2015, same-sex married workers were eligible for the Family and Medical Leave Act (FMLA) to care for a spouse, regardless of whether they reside in a state that recognizes their marriage (Movement Advancement Project, 2016). The application of stem cells to facilitate reproduction for gay and lesbian couples is widely reported in the news now. Several research laboratories are focused on the creation of cells that are genetically male but have been produced from eggs and alternately, they are also trying to create sperm from female eggs (Murnaghan, 2017). It means that they hold potential for the future of same-sex reproduction. How do LGBT families live their lives in the United States?

LGBT Family Life

There were 7.3 million unmarried-partner households in 2015 and of this number, 433,539 were same-sex households (U.S. Census Bureau, 2017). LGBT people are more likely to experience family crisis than their heterosexual peers. Gates (2011) finds that LGBT families are more likely to be headed by women and people of color: 41 percent of non-White women in same-sex couples have children under age eighteen, compared with 20 percent of non-White men; 39 percent of same-sex couples with children under age eighteen are people of color, compared to 36 percent White people. Same-sex families of color are likely to be poor. LGBT adults raising children are three times more likely than non-LGBT individuals to report household incomes near the poverty threshold (Gates, 2011). African American (42 percent), Hispanic (33 percent), and American Indian and Alaskan Native (32 percent) same-sex couples are likely to report not having enough money for the food that they or their families needed at some point in the last year (Brown *et al.*, 2016).

There are 2.4 million LGBT older adults over age 50 in the United States (Choi & Meyer, 2016). They have higher prevalence of mental and physical health problems than their heterosexual counterparts. Martos and colleagues (2017) identify 213 LGBT community health centers operating in 37 states and found that most provide wellness programs and services (72 percent), HIV/STI services (65 percent), and counseling services (52 percent).

Traditional Sexual Scripts

In sexual script's link to sexual activity, gender plays a role in what makes sexual scripts differ between males and females. **Traditional sexual scripts** are those that reflect a patriarchal pattern of male supremacy and female dependency. For example, ancient Rome was a society where a male citizen held power and established an identity for himself. Roman society developed a culturally consistent approach to systematically institutionalize sexuality in such a manner as to result in widespread sexual repression among all demographic subsets of Roman society (Tyner, 2015). A woman was dependent on the connection to her family for that identity; first her father, then her husband, then her children. As such, in terms of sexuality, a woman was viewed as serving as a childbearer, providing legitimate heirs to her husband (Goetting, 2017). In traditional societies, sexual behavior is considered acceptable only within the confines of marriage. Premarital sex and extramarital sex are considered taboo.

Traditional sexual scripts endorse different sexual behavior for women and men. Men are expected to initiate sexual encounters and are permitted to engage in and enjoy sexual behavior. For example, a man must always take the "active" role as opposed to the "passive" and the role of the Roman man in sex was to find sexual pleasure by penetrating the body of another whom he finds beautiful (Goetting, 2017). The overarching belief of how a proper man was to behave can be described as "impenetrable penetrators" (Hallet & Skinner, 1997). In traditional societies, women were not economically independent and wives were considered a commodity whose sexual purpose was to bear children. In such traditional sexual scripts, wives were not expected to initiate sex or express desire. The only women seen to be capable of doing so were prostitutes.

In ancient Rome, the view of man as impenetrable penetrator applied to sexual orientation (Goetting, 2017). It meant that it did not matter whether the other person was biologically male or female or heterosexual or homosexual. Our modern notions of classifying someone as gay or homosexual, straight or heterosexual, or bisexual would have little in common with the mindset of an ancient Roman (Clarke, 2003). Different cultures and subcultures perceived homosexuality differently. Many of these perceptions are influenced by religion or social norms.

Contemporary Sexual Scripts

Contemporary society is marked by a move away from the patriarchal pattern of male supremacy and female dependency towards gender equality in sexuality. **Contemporary sexual scripts** are sexual relationship-centered rather than male-centered. The **sexual revolution** was a social movement starting in the 1960s that challenged traditional codes of behavior related to sexuality throughout the West. The sexual revolution brought about social change in the attitudes and practices regarding premarital sex, birth control, pornography, kinds of sexual behavior, and the legalization of abortion. For example, in 1969, the first adult erotic film depicting explicit sex received wide theatrical release in the United States (Canby, 1969).

Contemporary sexual scripts are more likely to be egalitarian. For example, women and men both learn to take ownership of their sexual experiences, to communicate openly about their sexual feelings, and to meet each other's sexual desires. Sexual activities are a mutual exchange of erotic pleasure. Non-marital sex is acceptable within a relationship context. It allows either partner to initiate sex and accept orgasm through different types of stimulation. Gay, lesbian, and bisexual relationships are accepted.

Sexual Norms

When studying sexuality, sociologists focus their attention on sexual norms. **Sexual norms** refer to rules of accepted behaviors that are socially and culturally enforced. Certain sexual norms such as disapproval of incest are shared among most societies. Likewise, societies have certain sexual norms that reinforce their accepted social system of sexuality. In most societies, sexual norms vary from time to time and have been changing from a traditional society to modern society. For example, sexual norms have traditionally encouraged sexual activity within the family (marriage) and have discouraged premarital and extramarital sex. Today premarital sex once considered "abnormal" may be viewed as accepted in a certain culture. Historically, the family and religion has been the greatest influence on sexual behavior in most societies. Children are socialized to sexual attitudes by the family and religion.

Sexual norms regarding sexuality vary across cultures and societies. Deviance from normal sexual behavior can be classified in restrictive or permissive societies. In western Europe in the Victorian era, many sex acts were considered deviant including masturbation, adultery, and even sexual pleasure (Worthen, 2016). Such behaviors as extramarital sex may be viewed as "acceptable" in a permissive society. **Sociosexual orientation** is the individual difference in the willingness to engage in sexual activity outside of a committed relationship. Individual sexual norms are culturally enforced. Individuals with a more restricted sociosexual orientation prefer greater love, commitment, and emotional closeness before having sex with romantic partners while individuals with an unrestricted sociosexual orientation are more willing to engage in sex without love, commitment, or closeness (Simpson & Gangestad, 1991).

The United States moved toward a **permissive society**—a society in which social norms become increasingly liberal, especially with regard to sexual freedom (Ayto, 2006). Sexual freedom includes the freedom to take part in sexual activities, such as extramarital sex, which were previously considered unacceptable. Extramarital sex occurs outside marriage when a married person engages in sexual activity with someone other than his or her spouse. Three out of every four American adults believe that extramarital sex is always wrong, with only 3 percent of the population believing that extramarital sex isn't wrong at all (Wolfinger, 2017). Figure 6.2 tracks these attitudes between 2000 and 2016 for different age groups of General Social Survey (GSS) respondents. People today still disapprove of sex outside of wedlock. This shifting attitude toward extramarital sex has been greatest for older American people in their 60s, 70s, and 80s. People ages 50 to 59 had some of the highest rates of extramarital sex.

When an extramarital sexual relationship breaches a social norm, it can be referred to as

Upon completion of the next section, students should be able to:
LO7: Describe sexuality in adolescence, young adulthood, middle adulthood, and late adulthood.

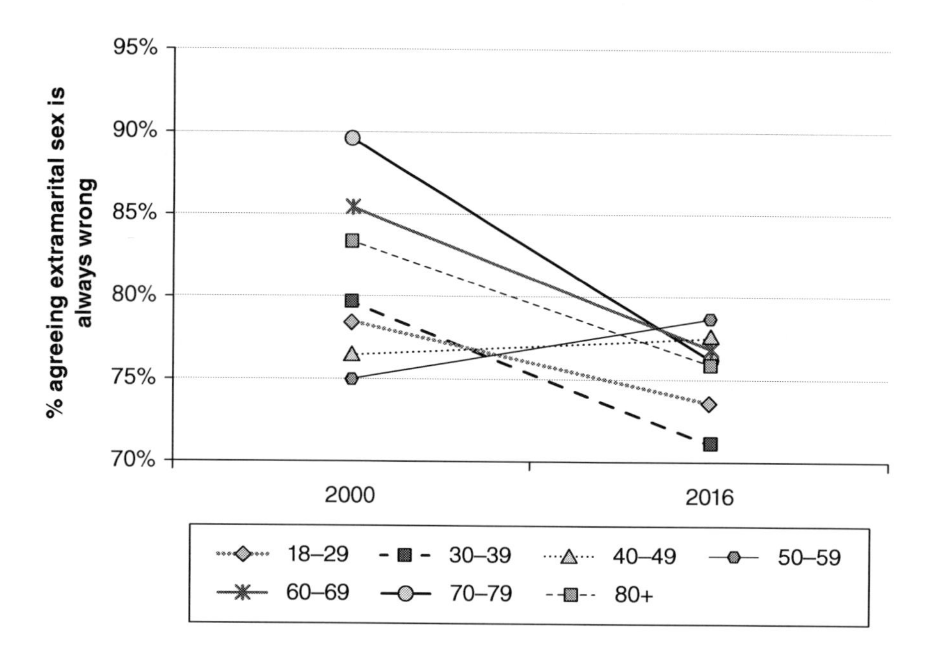

Figure 6.2 Shifting Attitudes toward Extramarital Sex

Source: General Social Survey, Institute for Family Studies 2017. https://ifstudies.org/blog/americas-generation-gap-in-extramarital-sex

infidelity. **Marital infidelity** is a violation of a married couple's emotional and sexual exclusivity. What constitutes an act of infidelity depends upon the exclusivity expectations within the relationship (Barta & Kiene, 2005).

Sexuality throughout the Life Cycle

Sexuality changes throughout the life cycle. Sexuality begins before birth and is a lifelong learning process. Preschool children are interested in everything about their world, including sexuality. Children are aware of differences in the male and female genitals. They start to imitate social and sexual behaviors such as kissing and hugging other people. Boys and girls can experience orgasm from masturbation although boys will not ejaculate until puberty. Researchers study sexuality throughout the life cycle, exploring the sexuality of adolescents through old adults. Let's look at sexuality in various life stages.

Sexuality in Adolescence

Adolescence is the transitional stage of physical and psychological human development that occurs during the period from puberty to adulthood. One large part of adolescence is that teenagers notice an increase in sexual feelings. The age at which adolescents become sexually active varies by culture and time.

Sexual experiences may help teenagers understand pleasure and establish their sexual orientation. Teenagers who had their first sexual experience at age sixteen revealed a higher well-being than those who were sexually inexperienced (Vrangalova & Savin-Williams, 2011). Oral sex is common among adolescent girls who fellate their boyfriends not only to preserve their virginity but also to create and maintain intimacy or to avoid pregnancy (Brady & Halpern-Felsher, 2007). From 2011 to 2013, among unmarried fifteen- to nineteen-year-olds in the United States, 44 percent of females and 49 percent of males

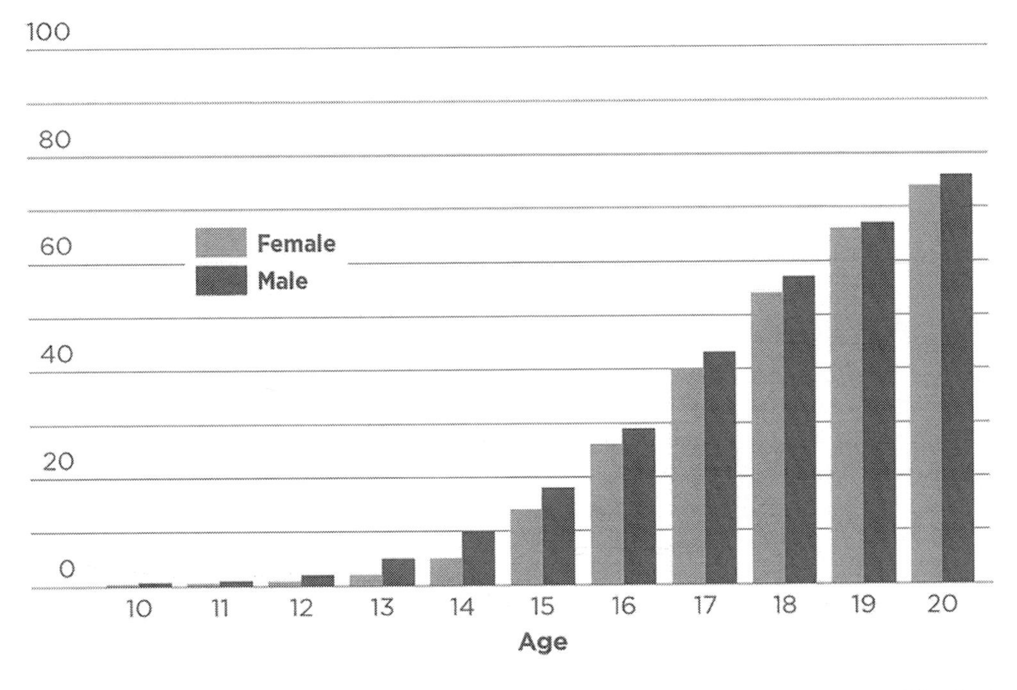

Figure 6.3 U.S. Teen Sexual Activity: The Proportion of Adolescents Who Have Had Sex Increases Rapidly with Age
Source: Guttmacher Institute, 2016. https://www.guttmacher.org/fact-sheet/american-teens-sexual-and-reproductive-health?gclid=COPb95z62NACFQ5YDQodU7lHjg#4a

had had sexual intercourse (Martinez & Abma, 2015). Figure 6.3 shows that the proportion of adolescents who have had sex increases rapidly with age. The proportion of U.S. females aged fifteen to nineteen who used contraceptives the first time they had sex has increased, from 48 percent in 1982 to 79 percent in 2011–2013 (Martinez & Abma, 2015). Still, more than three-quarters of births to teenagers are unintended (Karberg *et al.*, 2016).

Sexual Minority Youth

Sexual minority refers to a group of people whose sexual identity and practices differ from the majority of people in a society. Sexual minority youth are those who identify as gay or lesbian, bisexual, or unsure of their sexual identity, or youths who have only had sexual contact with persons of the same sex or with both sexes (Kann *et al.*, 2011). Health disparities exist

Student Account of Sexual Minority Youth

When I was in high school, I observed my friend Tim and many others endure a great deal of harassment because they were gay. The homosexual community fit the description of a subordinate group because their sexual orientation was not in agreement with the majority. Many of the young gay teens were called names, spat at, or attacked by other teens in the school (personal communication, November 20, 2017).

between sexual minority and non-sexual minority youth; 42.8 percent of gay, lesbian, or bisexual students and 31.9 percent of not-sure students had seriously considered attempting suicide, compared with 14.8 percent of heterosexual students during the previous twelve months (Kann *et al.*, 2016).

LGBTQ youth are four times as likely to report having been physically forced to have sex compared to their heterosexual peers (18 percent vs. 5 percent); twice as likely to experience sexual dating violence (23 percent vs. 9 percent); and twice as likely to report being bullied at school (34 percent vs. 19 percent); three out of ten report that they attempted suicide in 2015 (Kann *et al.*, 2016). For sexual minority youth, family and friend support promotes mental health and well-being and protects against psychological distress and depression (Child Trends Databank, 2014). Positive school climate and school connectedness play important roles in promoting LGBTQ teens' well-being. For example, gay–straight alliances, LGBTQ-inclusive curriculum, supportive staff, and anti-bullying policies can promote positive school climates for LGBTQ youth.

Sexuality in Young Adulthood

Many people are at their peak of sexual activity in young adulthood, which spans from age 18 to 39. Yet rates of sexual activity vary. For example, 15 percent of Millennials born in the 1980s and 1990s had no sexual partners, compared to 6 percent of GenXers born in the 1960s and 1970s (Twenge *et al.*, 2016). New parents tend to experience a decline in sexual activity. Pain following childbirth may reduce sexual interest. The decline in sexual activity for new parents is primarily due to the stresses associated with caring for a newborn, including lack of sleep and many other tasks that must be addressed with childcare (Van Anders *et al.*, 2013).

Many young adults engage in sex but do not wish to become pregnant. Pregnancy prevention programs have expanded their efforts to target men's contraceptive decision-making in addition to women's. Approximately 65 percent of women and 60 percent of men reported using contraception the last time they had sex; 50 percent of women and 45 percent of men reported using hormonal or long-acting reversible contraception (LARC) methods (Karberg *et al.*, 2016). Long-acting reversible contraception (LARC) methods including the intrauterine device (IUD) and birth control implant are effective in preventing pregnancy. Figure 6.4 shows more women's reports of contraceptive use compared with men. As is the case with many other aspects of social life, women are mistreated when it comes to sex; young women and teenage girls often face efforts by male partners to sabotage birth control (including damaging condoms) (Chan & Martin, 2009).

Sexuality in Middle Adulthood

Middle adulthood is the life stage between ages 40 and 64. As people age, some hormone levels decline. Women experience menopause that affects sex. For example, Avis and colleagues (2000) find that lower estrogen levels are related to pain with intercourse. However, after menopause, women may no longer fear an unwanted pregnancy and therefore may feel freer to enjoy sex more. In fact, singles over age 50 are very interested in being sexually active (OurTime, 2017). Figure 6.5 shows that starting after 2004, Americans over 55 began reporting rates of extramarital sex that were about five or six percentage points higher than were being offered by younger adults. In fact, by 2016, 20 percent of Americans in their 50s and 60s reported having extramarital sex while 14 percent of those under age 55 did so (Wolfinger, 2017).

Sexuality in Late Adulthood

One becomes an older adult at the age of 65. Normal physical changes sometimes affect the ability to have and enjoy sex in both men and women. For example, many women will have less vaginal lubrication.

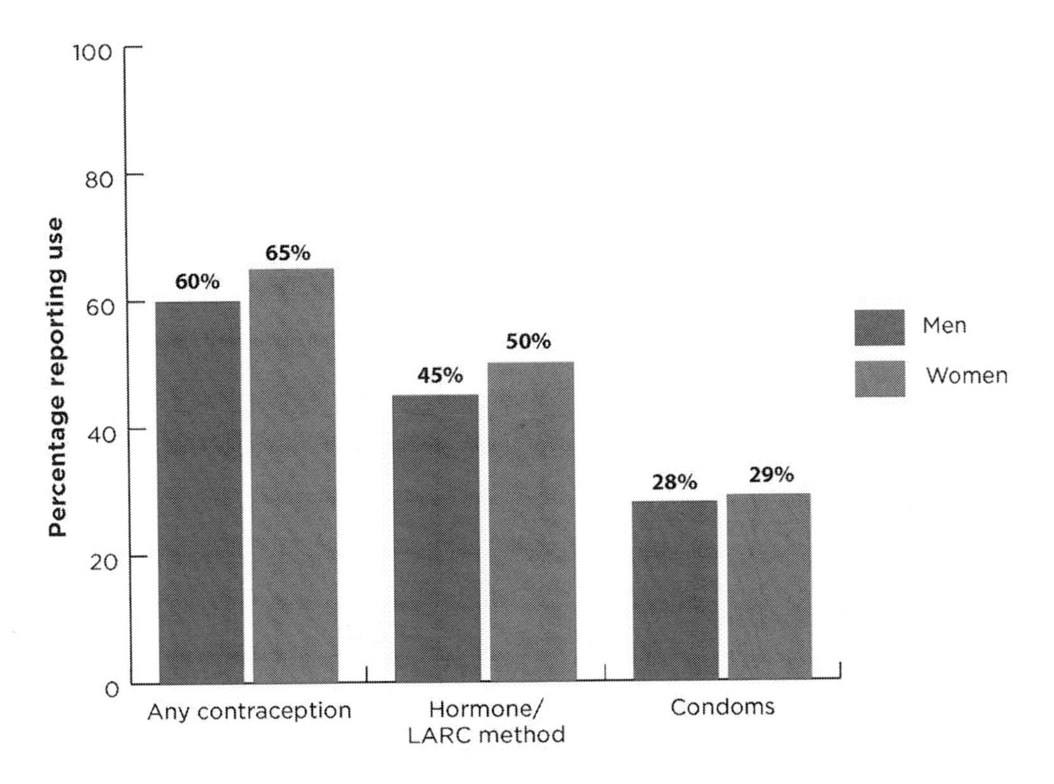

Figure 6.4 Men's and Women's Reports of Contraceptive Use at Last Sex, by Type

Source: Child Trends Databank November 21, 2016. http://www.childtrends.org/publications/intimate-inaccuracies-young-couples-dont-always-agree-contraceptive-use/

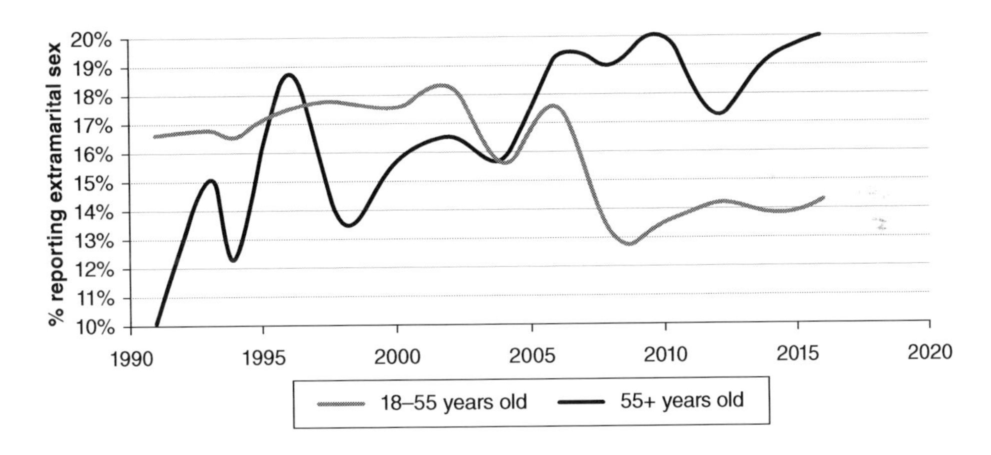

Figure 6.5 Divergent Trends in Extramarital Sex by Age

Source: Institute for Family Studies, 2017. https://ifstudies.org/blog/americas-generation-gap-in-extramarital-sex

Erectile dysfunction (ED)—the loss of ability to have and keep an erection for sexual intercourse—is common and may cause a man to take longer to have an erection. Health conditions associated with the aging process can impact sexuality. Some illnesses, disabilities, medicines, and surgeries can affect your ability to have and enjoy sex (National Institute of Health, 2017).

Sexual activity is important to adults age 65 and older. Given a state of reasonably good health and the availability of an interested partner, there was no absolute age at which sexual abilities disappeared (Masters & Johnson, 1966). Older people have fewer distractions, more time and privacy, no worries about getting pregnant, and greater intimacy with a lifelong partner (National Institute of Health, 2017). Many older couples find greater satisfaction in their sex life than they did when they were younger. Many older men and women are capable of excitement and orgasm into their 70s and beyond. In fact, 46 percent of men aged 70 to 80 years reported orgasm at least once a month; over 80 percent of all men who reported some level of erection stated that it was of importance to them to maintain the present level of erection stiffness (Helgason *et al.*, 1996).

Figure 6.6 shows older Americans even became sexually active outside marriage in recent years, although rates of extramarital sex decline once survey respondents enter their 70s and 80s. It also shows that sex outside of marriage is most likely to lead to divorce among the very same age groups that have the highest levels of extramarital sex.

Sexual Health and Healthy Sexual Relationships

Sexual health requires a positive and respectful approach to sexuality and sexual relationships, as well

Upon completion of this section, students should be able to:
LO8: Describe sexually transmitted infections and ways of preventing these diseases.
LO9: Describe the contributing factors to healthy sexual relationships.

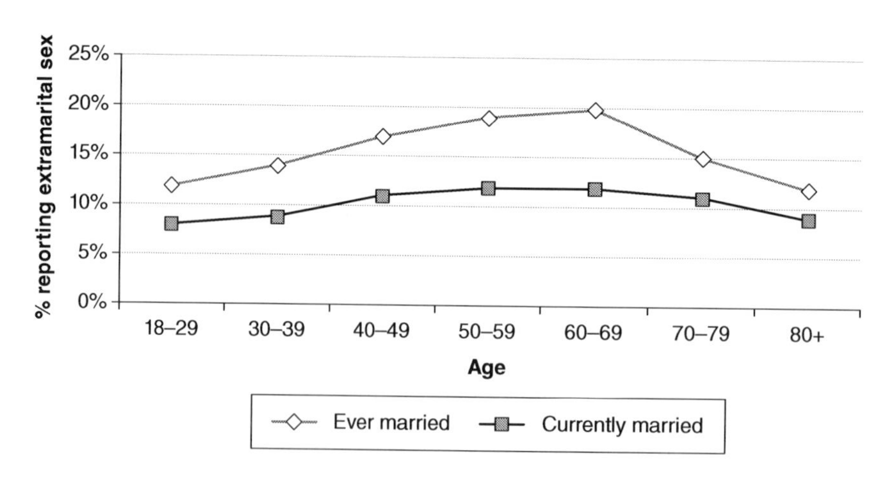

Figure 6.6 Extramarital Sex by Age and Marital Status
Source: Institute for Family Studies 2017. https://ifstudies.org/blog/americas-generation-gap-in-extramarital-sex

as the possibility of having pleasurable and safe sexual experiences (World Health Organization, 2017). Let us look at several topics related to sexual health, starting with sexuality education.

Sexuality Education

Comprehensive sexuality education is an instruction method in which students gain knowledge, attitudes, skills, and values to make appropriate and healthy choices in their sexual behavior (Loeber *et al.*, 2010). Some programs cover **sexual abstinence**—the practice of refraining from sexual activity for moral or religious reasons. The Bush administration introduced abstinence programs to encourage unmarried teens and young adults to remain abstinent until marriage. However, in 2009, the Obama administration removed most of the funding from sexual-abstinence education, claiming it had been proven to be ineffective (Dailard, 2006). Riggs and Bartholomaeus (2017) advocate for an approach to sexuality education that eschews the gendering of body parts and gametes so as to address the needs of transgender young people and provide cisgender young people with a more inclusive understanding of their own and other people's bodies and desires.

Sexual Health

Sexual health refers to the state of physical, emotional, mental, and social well-being in relation to sexuality (World Health Organization, 2017). Sexual health is increasingly recognized as not merely the absence of disease, but the ability to have pleasurable and safe sexual experiences, free from coercion (Mercer, 2014). Risks that arise from sexual activity include sexually transmitted infections (STIs), human immunodeficiency virus (HIV), and unintended pregnancy. For example, men who have sex with men (MSM) are at particularly high risk of acquiring HIV and STIs. Approximately 3 percent of men and women in Britain have had same-sex partner(s) in the past five years (Mercer *et al.*, 2013). Anal intercourse—especially when a condom is not used ("unprotected anal intercourse")—is the most significant risk activity. Approximately two-thirds of MSM report anal sex in 2003, and of these, 40 percent always and 25 percent sometimes used a condom for this practice (Mercer *et al.*, 2004).

Sexually Transmitted Infections (STIs)

There are more than 25 kinds of **sexually transmitted infections** (STIs) acquired primarily through sexual activity (CDC, 2000). In fact, half of all new STI cases each year are acquired by individuals between the ages of 15 and 24 (CDC, 2016). If untreated, chlamydia and gonorrhea, the two most common reportable STIs, can cause infertility, pregnancy complications, adverse pregnancy outcomes, pelvic inflammatory disease (PIV), and increased risk of HIV infection (CDC, 2017b). Figure 6.7 shows chlamydia, gonorrhea, and syphilis rates of fifteen- to nineteen-year-olds from 1996 to 2015.

Effects of STIs by Gender and Race

Adolescent females are much more likely than males to have reported a case of chlamydia (2,994 versus 768 cases per 100,000) or gonorrhea (442 versus 245 cases per 100,000) in 2015 (CDC, 2017b) (see Figure 6.8). However, adolescent boys were more likely than girls to have reported a case of syphilis in 2015 (8.0 versus 2.8 cases per 100,000).

Black youth are more likely than Hispanic and White fifteen- to nineteen-year-olds to have reported cases of chlamydia, gonorrhea, and syphilis (see Figure 6.9). For instance, in 2015, rates of chlamydia among Black adolescents were 4,201 per 100,000, compared with 1,067 for Hispanics and 775 per 100,000 for Whites (Child Trends Databank, 2017). Racial/ethnic differences were even greater for syphilis (although rates were relatively low for all adolescents).

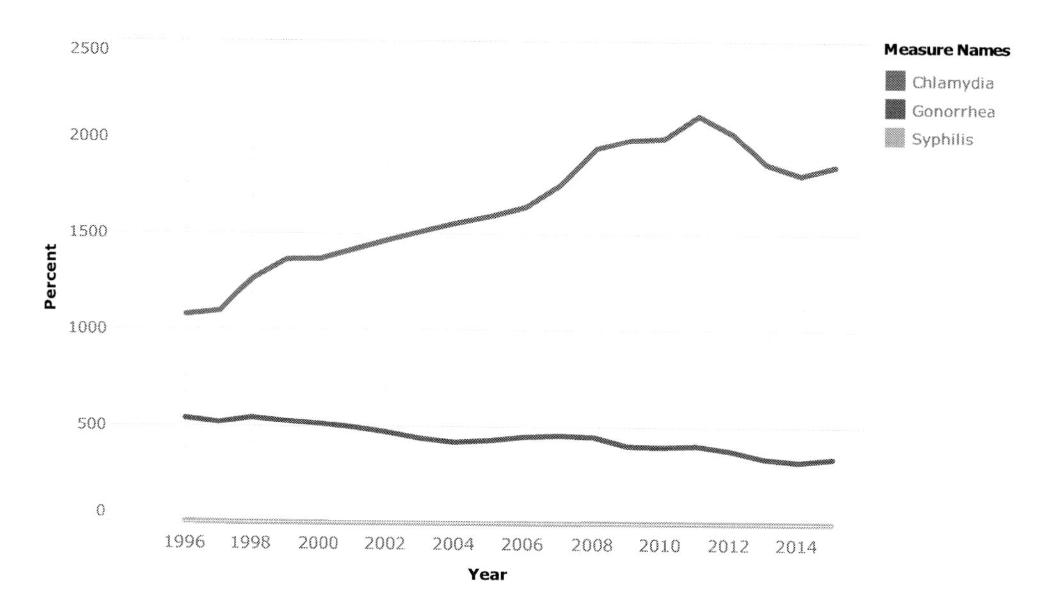

Figure 6.7 Chlamydia and Gonorrhea Rates per 100,000, Ages 15 to 19, 1996–2015
Source: CDC (2015) and Child Trends Databank March 2017. https://www.childtrends.org/indicators/sexually-transmitted-infections-stis/

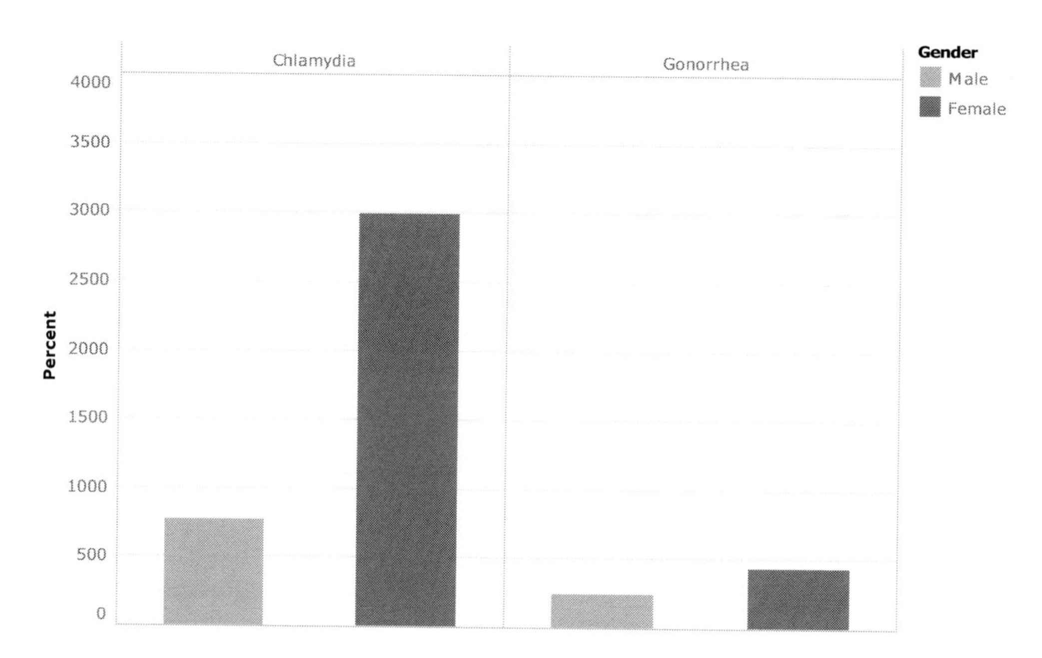

Figure 6.8 Chlamydia and Gonorrhea Rates per 100,000, Ages 15 to 19, by Gender, 2013
Source: CDC (2015) and Child Trends Databank March 2017. https://www.childtrends.org/indicators/sexually-transmitted-infections-stis/

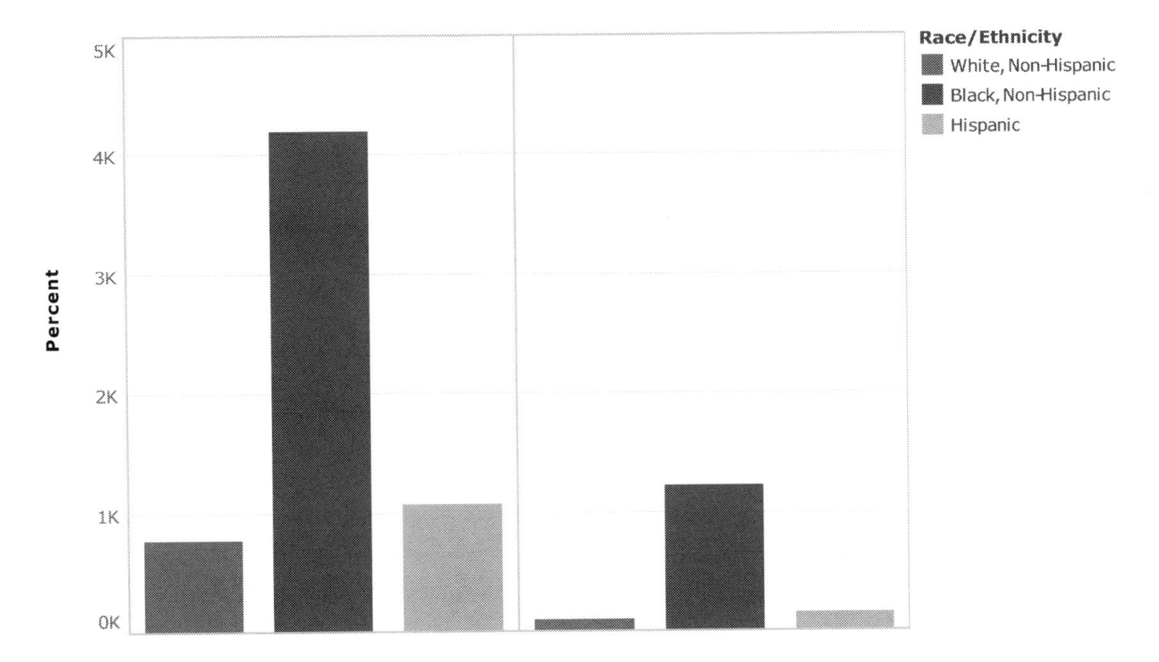

Figure 6.9 Chlamydia and Gonorrhea Rates per 100,000, Ages 15 to 19, by Race and Hispanic Origin, 2015
Source: CDC (2015) and Child Trends Databank March 2017. http://www.childtrends.org/?indicators=sexually-transmitted-infections-stis#_ednref3

Prevention of STIs and Pregnancy

STIs and unintended pregnancy are major health issues that can be consequences of unprotected sexual activity. Abstinence from vaginal, anal, and oral intercourse is the only 100 percent effective way to prevent STIs and pregnancy. Parental education, parental communication about contraception, and sex education courses can help increase condom use in sexual relationships. Condoms, if used correctly, can greatly reduce the risk of both STIs and unintended pregnancies (Child Trends Databank, 2016). In 2011–2013, 97 percent of sexually experienced female teens had used a condom at least once (Martinez & Abma, 2015). In 2015, 62 percent of male high school students used a condom during their most recent sexual intercourse, compared with 52 percent of females. Figure 6.10 shows that condom use, as reported by sexually active high school students, increased from 46 percent in 1991 to 63 percent in 2003 and decreased to 57 percent in 2015.

Black males were 27 percent more likely than Black females to report condom use at last sexual intercourse; Hispanic males were 14 percent and White males were 2 percent more likely than their female counterparts to report using a condom (Child Trends Databank, 2016). Figure 6.11 shows that Black male students (74 percent) were more likely than White male students (58 percent) to report condom use in 2015.

Effects of a Healthy Sex Life

In a healthy sexual relationship, both partners are able to express their feelings and show respect for each other regarding sex. There are physical, emotional, and mental benefits of having healthy sex. For example, couples who communicate and have a healthy sexual relationship are more likely to share financial responsibilities and stay faithful (Northrup *et al.*, 2013).

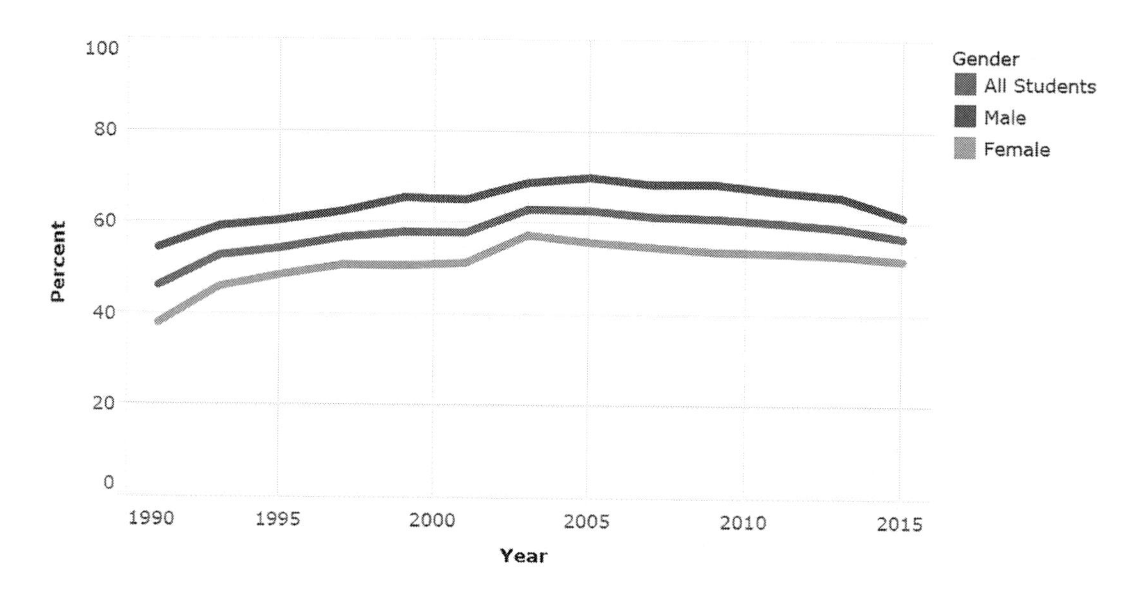

Figure 6.10 Condom Use as Reported by Sexually Active Students in Grades 9 through 12, by Gender: Selected Years, 1991–2015

Source: CDC (2016). 1991–2015 High School Youth Risk Behavior Survey Data and Child Trends Databank October 2016. https://www.childtrends.org/?indicators=condom-use

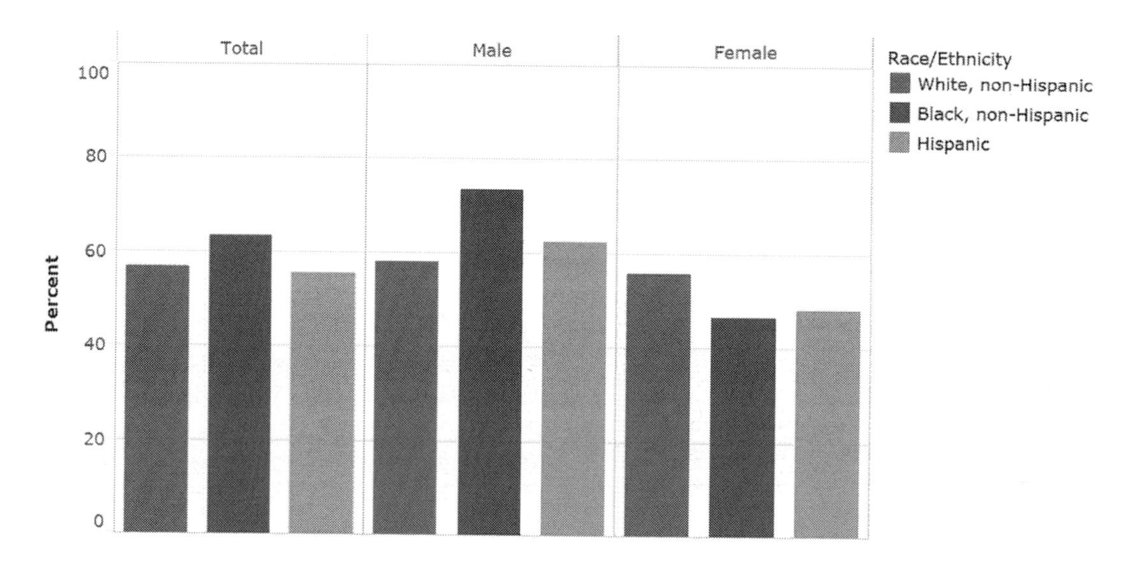

Figure 6.11 Condom Use as Reported by Sexually Active Students in Grades 9 through 12, by Race and Hispanic Origin, 2015

Source: CDC (2016). 1991–2015 High School Youth Risk Behavior Survey Data and Child Trends Databank October 2016. https://www.childtrends.org/?indicators=condom-use

Notes: At last sexual intercourse among sexually active students (students who had sex in the three months preceding this survey).

Estimates reported here only include respondents who selected one race category.

Intimacy

Men and women who enjoy sex are likely to have an intimate and long-term committed relationship. Endorphins and oxytocin are released during sex, and these feel-good hormones activate pleasure centers in the brain that create feelings of intimacy and relaxation and help stave off anxiety and depression (Berman, 2013). People who purchase prostitution services or have one-night sexual encounters derive pleasure with little intimacy. Because sexual behavior involves physical intimacy and can involve emotional intimacy, it can provide an important mechanism for regulating intimacy (Reis & Sprecher, 2009). Couples feel more intimate with each other when they have satisfying sex. Having sex regularly can do more than make you feel closer to your partner—it can make you physically healthier (Hutcherson, 2006).

Physical and Psychological Benefits

When couples are engaged in work and are struggling to raise children, sex might help ease stress. In fact, having sex can help people sleep more soundly. The same endorphins that help de-stress can also relax your mind and body (Meston & Buss, 2007). Cardiologists consider sexual activity comparable to a modest workout on a treadmill. Regular sex promotes the release of hormones, including testosterone and estrogen, which can keep the body looking young and vital (Meston & Buss, 2007).

Couples in dual-earner families play multiple roles, and sex is a small part of daily time allocation. Based on surveys about what makes couples happy, having sex regularly is one of the surprising secrets of happy couples. Three to four times a week was found to be the perfect amount for prime levels of happiness (Northrup *et al.*, 2013).

Global Overview: Sexuality

Many parents in many societies hardly talk about sexuality at home. Ayushi, a student at Banaras Hindu University, comes from a family where talking about sex is taboo but she was intent on attending a youth health workshop (UNESCO, 2017). Young people are facing many barriers when it comes to receiving unbiased and practical information on sexual health. Whether it is due to the family, community, or education system, they are facing similar problems in countries across the globe, making a solution to this issue a universal one (Urbanski, 2017).

Comprehensive Sexuality Education

Comprehensive sexuality education (CSE) is recognized as an "age-appropriate, culturally relevant approach to teaching about sexuality and relationships by providing scientifically accurate, realistic, non-judgmental information" (UNESCO, 2009). Evidence demonstrates that CSE contributes to HIV prevention and gender equality outcomes; almost 80 percent of countries have policies or strategies that support CSE (UNESCO, 2015). When it comes to sexuality education, research shows the importance of sexual diversity. Although social acceptance of gender and sexuality diversity is growing in schools in Australia, visibility and inclusion of knowledge pertaining to those who are gender- and/or sexuality-diverse, such as lesbians, gay men, and transgender people, remain marginalized (Ferfolja & Ullman, 2017). The corpus of research describes the challenges LGBT youth face in schools and points to the need for change (Francis, 2017). In a study of 27 publications on gender and sexuality diversity in South African schools, post-apartheid, Francis (2017) finds that schools proliferate compulsory heterosexuality and heteronormativity.

Permissive Approach and Abstinence Model

Sweden took the lead in the sexual revolution of the 1960s and the rest of Europe and the United States soon followed. Sweden is thought to have a permissive approach to sex and be the most liberal when it comes to attitudes about sex. Sex education is a compulsory part of Swedish school curricula. The introduction of the birth control pill by the mid-1960s gave people more freedom (Duiker & Spielvogel, 2017). In countries such as the UK, Sweden, or Germany, promoting a permissive approach has resulted in a smaller percentage of AIDS cases as well as significantly lower abortion rates (Urbanski, 2017).

In Poland, the sex education model chosen by the Polish Ministry of Education promotes abstinence. Due to religious, political, and other external factors, young people's access to sexual education is highly limited in Poland; the result is above EU average teenage birth rates (Urbanski, 2017). In most cultural, ethical, and religious contexts, societies are likely to be restrictive in their attitudes about chastity, a synonym for sexual abstinence. For example, a study of 37 countries reported that non-Western societies—like China, Iran, and India—valued chastity highly in a potential mate, while western European countries such as France, the Netherlands, and Sweden placed little value on prior sexual experiences (Buss, 1989).

Sex Education Also Belonging in the Home

Gordon (1999) suggests raising a child responsibly in a sexually permissive world. Many parents in the United States are likely to forbid discussion of the topic as they understand sex among teenagers as sensitive or to be feared. However, Dutch parents take a different approach, talking to their kids about sex and even encouraging them to have sex when they are ready (Schalet, 2011). Teenagers are encouraged to have sex in their parents' home and ensure that it is a safe and pleasant experience. Nine out of ten Dutch parents are okay with their teenage children having sex in their home, but in the United States, nine out of ten parents are opposed to it and teenage pregnancy rates are three times higher than in Holland (Schalet, 2011). There are lower rates of contraceptive use in the United States (72.7 percent) for women ages 15 to 44 in 2013–2015 than in Netherlands (73 percent) for women ages 18 to 45 in 2013 (CIA, 2018).

Genital Surgeries on Intersex Infants

Globally, intersex infants, children, and adolescents are subjected to medically unnecessary surgeries, hormonal treatments and other procedures in an attempt to forcibly change their appearance to be in line with societal expectations about female and male bodies. In recognition of Intersex Awareness Day on October 26, 2017, the United States stands in solidarity with intersex persons and their advocates around the world. At a young age, intersex persons routinely face forced medical surgeries without free or informed consent. These interventions jeopardize their physical integrity and ability to live freely (Nauert, 2017).

In 2017, intersex activists from Australia and New Zealand issued the "Darlington Statement" that calls for "the immediate prohibition as a criminal act of deferrable medical interventions, including surgical and hormonal interventions that alter the sex characteristics of children without personal consent." On March 30–31, 2017, in Vienna, European intersex activists issued the "Vienna Statement," which notes that, "until this day more than 50 times UN bodies, regional and national human rights bodies have called on governments, policy makers and stakeholders to put an end to human rights violations faced by intersex people" (OII Europe, 2017a). Activists convened the fourth annual International Intersex Forum in Amsterdam and discussed "infanticide, intersex genital mutilation and other harmful medical practices, lack of appropriate and consented health care as well as discrimination in access to education, other services and employment" (OII Europe, 2017b).

Same-sex Marriage

The first English statute against homosexuality was placed on the books by Parliament in 1533 under Henry VIII. The death penalty for homosexuality has historically been implemented worldwide. Colonial America had the laws of the United Kingdom. In 1776 male homosexuals in the original thirteen colonies were universally subject to the death penalty. After the Revolution, Pennsylvania took the lead, in 1786, in dropping the death penalty (Crompton, 1976). It is currently still in existence in a small number of countries or parts of countries. In 2017, countries that allow gay and lesbian marriage include Argentina (2010), Belgium (2003), Brazil (2013), Canada (2005), Colombia (2016), Denmark (2012), England/Wales (2013), France (2013), Germany (2017), Greenland (2015), Iceland (2010), Ireland (2015), Luxembourg (2014), Malta (2017), the Netherlands (2000), New Zealand (2013), Norway (2008), Portugal (2010), Scotland (2014), South Africa (2006), Spain (2005), Sweden (2009), the United States (2015), Uruguay (2013) (Pew Research Center, 2017b). Since the 1970s, a growing number of governments have been considering whether to grant legal recognition to same-sex marriages. However, today same-sex relationships are illegal in 72 countries. It should be noted that some of these states either have no law, or have such repressive regimes (like Egypt, Qatar, and Iraq) that same-sex sexual relations are functionally severely outlawed (ILGA & Carroll, 2016). ILGA (the International Lesbian, Gay, Bisexual, Trans and Intersex Association) is the world federation of national and local organizations dedicated to achieving equal rights for lesbian, gay, bisexual, trans, and intersex people (ILGA, 2017).

Summary

This chapter has explored the sociology of sexuality and sexual relationships, including sexual identity, sexual behavior, sexual response, sexual script, sexual health, and healthy sexual relationships. Sexual identity involves biological sex, gender role, gender identity, and sexual orientation. Sex refers to biological characteristics differentiating males and females, while gender denotes social and cultural characteristics of masculine and feminine behavior. Types of sexual orientation include homosexuality, heterosexuality, bisexuality, pansexuality, and asexuality. Most common sexual behaviors include masturbation, sexual intercourse, oral sex, and anal sex. Common sexual behaviors include flirting, wearing sexy underwear, sending/receiving digital nude/semi-nude photos, reading erotic stories, watching sexually explicit videos/DVDs, etc. Uncommon sexual behaviors include group sex, key parties, BDSM, etc. The sexual response cycle includes excitement, plateau, orgasm, and resolution.

Traditional sexual scripts are based on patriarchal sex. Contemporary sexual scripts are more egalitarian. Sexuality throughout the life cycle involves the stages of adolescence, young adulthood, middle adulthood, and older adulthood. Sexual health involves the state of physical, emotional, mental, and social well-being in relation to sexuality. Contraceptive methods help to keep sex healthy by avoiding unwanted pregnancy and STIs. Regular sex is one of the surprising secrets of happy couples. From a diversity and global perspective, this chapter has discussed LGBT families and topics on sexuality education, genital surgeries on intersex children, and same-sex marriage.

Key Terms

Adolescence is the transitional stage of physical and psychological human development that occurs during the period from puberty to adulthood.

Anal sex is the penetration of the anus by the erect penis for sexual pleasure.

Androgyny refers to the combination of masculine and feminine characteristics.

Aromantic asexual people are not romantically attracted to others but are often satisfied with non-romantic relationships such as closer friendships.

Asexuality refers to having no sexual attraction to a person of either sex.

Assigned sex is a label that a person is given at birth based on primary sex characteristics such as genitals.

BDSM refers to erotic or role-playing practices such as BD (bondage, discipline, dominance and submission) and SM (sadism and masochism).

Bigender people identify with both male and female and have two distinct gender identities, either at the same time, or at different times.

Bisexuality is emotional, romantic, or sexual attraction to both the opposite and the same sex (both males and females).

Cass identity model is a theory of gay and lesbian identity development.

Cisgender refers to a person whose gender identity aligns with the assigned sex at birth.

Cisgender identity privileges include having your gender as an option on an application form and using public facilities without fear or anxiety.

Coming out of the closet is a metaphor for LGBT people's self-disclosure of their sexual orientation or of their gender identity.

Comprehensive sexuality education is an instruction method in which students gain knowledge, attitudes, skills, and values to make appropriate and healthy choices in their sexual behavior.

Contemporary sexual scripts are sexual-relationship-centered rather than male-centered.

Cross-dresser is a male who dresses as a woman or a female who dresses as a man but does not alter his or her genitalia.

Demiasexual people only experience sexual attraction after forming a strong emotional connection such as in the context of a "serious" committed relationship.

Demiromantic asexual people sometimes develop romantic attraction toward someone after becoming familiar with and emotionally fond of that person.

Double standard is defined as prohibiting premarital sexual intercourse for women but allowing it for men.

Erectile dysfunction (ED) is the loss of ability to have and keep an erection for sexual intercourse. It is common and may cause a man to take longer to have an erection.

Exhibitionism refers to the urge or act of exposing one's sex organs to non-consenting others.

Extramarital sex occurs outside marriage when a married person engages in sexual activity with someone other than his or her spouse.

Femininity refers to attributes such as passive, submissive, and compassionate, nurturing practices associated with women in distinction from men.

Flirting (or **coquetry**) is sexual activity involving verbal and non-verbal language by one person to another.

Fluid sexuality refers to a person whose romantic and/or sexual attraction changes over time.

Frotteurism refers to the urge or act of rubbing one's genitals against the body of an unfamiliar or non-consenting person.

Gay is emotionally, romantically, or sexually attracted to members of the same gender and can be applied to both males and females, but it is more often applied to males.

Gender refers to the social and cultural meanings and practices associated with masculinity and femininity in a culture.

Gender expression refers to the way a person expresses aspects of gender identity and shows his or her femininity, masculinity, or androgyny through physical appearance, etc.

Gender identity is one's personal perception and experience of being male, female, or a combination of both, or neither.

Genderqueer (GQ) or non-binary (NB) is an inclusive term for gender identities that are not exclusively male or female.

Gender roles refer to the attitudes and behaviors considered socially and culturally appropriate for men or women and associated with masculinity and femininity.

Grayasexual people do not normally experience sexual attraction but experience it sometimes.

Grayromantic asexual people are between romantic and aromantic and less likely to experience romantic attraction than most people.

Group sex is sexual behavior involving more than two participants between people of all sexual orientations and genders.

Heteronormativity refers to judging sexuality according to heterosexual norms such as sexual attraction between people of the opposite sex and alignment of the biological sex, gender identity, and gender role.

Heterosexuality is emotional, romantic, or sexual attraction between members of the opposite sex.

Heterosociality refers to non-sexual relationships with the opposite sex.

Homophobia describes extreme prejudice directed at gays, lesbians, bisexuals, and others who are perceived as not being heterosexual.

Homosexuality is emotional, romantic, or sexual attraction between members of the same sex.

Homosociality refers to non-sexual relationships with the same sex.

Hormones are chemical messengers that arouse cells and organs to specific activities and influence the way we look, feel, and behave.

Hypoactive sexual desire disorder (HSDD) refers to persistently or recurrently deficient or absent sexual fantasies and desire for sexual activity.

Intersex people are born with biological sex characteristics (such as sexual anatomy, reproductive organs, hormonal patterns and/or chromosomal patterns) that do not fit the typical definitions for male or female bodies.

Lesbian is a woman who is emotionally, romantically, or sexually attracted to other women.

LGBT family is a form of family in which lesbian, gay, bisexual, and/or transgender people form intimate relationships and raise children.

Māhū in Native Hawaiian culture refers to an individual who may be considered third-gendered with characteristics of both sexes, usually a male to female.

Marital cohesion refers to the degree of connectedness, togetherness, and emotional bonding between marital partners.

Marital infidelity is a violation of a married couple's emotional and sexual exclusivity.

Masculinity is a social position, a set of practices, and the effects of the collective embodiment of those practices on individuals, relationships, institutional structures, and global relations of domination.

Masturbation refers to the sexual stimulation of one's own genitals for sexual arousal.

Medicalization of sexuality is a process of the growing authority of medical experts over sexual experiences sponsored by the pharmaceutical industry.

Oral sex is sexual behavior that involves contact between the mouth, lips, or tongue and the genitals.

Pansexuality is emotional, romantic, or sexual attraction to people regardless of biological sex and gender identity.

Paraphilias are considered certain atypical sexual behaviors or sexual behavior deviations.

Pederasty was a socially acknowledged erotic relationship between an adult male (the *erastes*) and a younger male (the *eromenos*) usually in his teens.

Pedophilia involves sexual attraction or activity with those under age thirteen.

Permissive society is a society in which social norms become increasingly liberal, especially with regard to sexual freedom.

Preferential homosexuality implies that people have preferential erotic attraction to members of the same sex.

Primary sex characteristics include hormones, chromosomes, and the reproductive organs used in reproduction.

Prostitution is the practice of engaging in sexual behavior in exchange for payment.

Queer is used as an umbrella term to describe individuals who don't identify as heterosexual or cisgender—a person whose gender identity differs from the sex that he or she was assigned at birth.

Queer theory focuses on mismatches between sex, gender, and desire.

Romantic asexual people are romantically attracted to others.

Secondary sex characteristics are the non-reproductive physical features that appear in humans during puberty that distinguish males from females.

Sex refers to the biological traits associated with being male or female.

Sexology is the scientific study of human sexuality, sexual interests, sexual behaviors, and sexual functions.

Sexting refers to the act of sending nude pictures or sexually explicit messages via text.

Sexual abstinence is the practice of refraining from aspects of sexual activity for moral or religious reasons.

Sexual attraction refers to a person's attraction to another person sexually on the basis of sexual desire.

Sexual behavior is the manner in which people experience and express their sexuality.

Sexual desire is a motivational state and a drive to seek out sexual objects or engage in sexual activities.

Sexual deviance refers to behaviors that involve individuals seeking erotic gratification through means that are considered different to most people.

Sexual diversity refers to all the diversities of sex characteristics, sexual orientations, and gender identity.

Sexual dysfunction refers to a problem during any phase of the sexual response cycle that prevents a person from experiencing satisfaction from sexual activity.

Sexual health refers to the state of physical, emotional, mental, and social well-being in relation to sexuality.

Sexual identity is how people feel their gender identity aligns with or differs from their biological sex and how they choose to identify with whom they are romantically or sexually attracted.

Sexual intercourse is the insertion and thrusting of the erect penis into the vagina for sexual pleasure, reproduction, or both.

Sexuality is the way people experience and express themselves as sexual beings.

Sexual minority refers to a group of people whose sexual identity and practices differ from the majority of people in a society.

Sexual norms refer to rules of accepted sexual behaviors that are socially enforced.

Sexual orientation is an enduring pattern of emotional, romantic, or sexual attraction to persons of the opposite sex (heterosexuality), the same sex (homosexuality), or both sexes (bisexuality), or multiple sexes, or lack of sexual attraction to others.

Sexual response cycle refers to the sequence of physical and emotional changes that occur as a person becomes sexually aroused and participates in sexually stimulating activities, including intercourse and masturbation.

Sexual revolution was a social movement starting in the 1960s that challenged traditional codes of behavior related to sexuality throughout the West.

Sexual script is a set of expectations and guidelines for appropriate sexual behavior.

Sexually transmitted infections (STIs) are infections acquired primarily through sexual activity.

Situational homosexuality occurs when individuals engage in man–boy sexual relations with members of the same sex as a substitute for sexual relations with members of the opposite sex.

Social exchange theory explores how people determine the rewards and costs of the relationship for both partners.

Sociosexual orientation is the individual difference in the willingness to engage in sexual activity outside of a committed relationship.

Swinging (wife or partner swapping) is a non-monogamous sexual behavior in which both singles and partners in a committed relationship engage in sexual activities with others as a recreational or social activity.

Third-gender is usually understood to mean "other" gender or a masculine female, or a feminine male.

Traditional sexual scripts are those that reflect a patriarchal pattern of male supremacy and female dependency.

Transgender persons have a gender identity (masculine/feminine) that differs from their assigned sex at birth (male/female).

Transsexuality occurs when people identify with gender identities that do not match their assigned sex and want to undergo physical transitions from one sex to another.

Transvestic fetishism is a sexual deviation in which heterosexual men wear women's clothes in order to achieve sexual arousal.

Two-spirit is a gender identity that identifies people as combining the work and traits of both men and women in American Indian history.

Voyeurism occurs when a person achieves sexual arousal by watching an unfamiliar person who is undressing, or engaging in sexual activity.

Zoophilia occurs when a person has sexual feelings for animals or engages in sexual behaviors involving animals.

Discussion Questions

- A traditional sexual script refers to patriarchal sex while a contemporary sexual script is more egalitarian. What future sexual scripts may guide our sexual behaviors?

Suggested Film and Video

Asexuality: An Overview: http://www.asexualawarenessweek.com/videos.html

Hope, Ted; Schamus, James; and Lee, Ang. (Producers). (1997). *The Ice Storm*. Fox Searchlight Pictures

Ossana, Diana and Schamus, James. (Producers). (2005). *Brokeback Mountain*. River Road Entertainment

Internet Sources

2016–2020 NIH Strategic Plan to Advance Research on the Health and Well-being of Sexual and Gender Minorities (PDF): https://www.edi.nih.gov/sites/default/files/EDI_Public_files/sgm-strategic-plan.pdf

Child Trends. 5 things to know about LGBT Youth: http://www.childtrends.org/child-trends-5/5-things-to-know-about-lgbtq-youth/

US: End Irreversible Genital Surgeries on Intersex Infants: https://www.hrw.org/report/2017/07/25/i-want-be-nature-made-me/medically-unnecessary-surgeries-intersex-children-us

References

ALGBTICAL (Association for Lesbian Gay Bisexual & Transgender Issues in Counseling of Alabama) (2017). *Love without Sex*. Retrieved on July 25, 2017 from http://www.algbtical.org/2A%20ASEXUAL.htm

Ambrosino, B. (March 16, 2017). *The Invention of Heterosexuality*. BBC. Retrieved on October 30, 2017 from http://www.bbc.com/future/story/20170315-the-invention-of-heterosexuality

ASA (American Sociological Association) (2017). *Gender and Sexuality*. Retrieved on October 3, 2017 from http://www.asanet.org/topics/gender-and-sexuality

AVEN (The Asexual Visibility and Education Network) (2014). *The 2014 AVEN Community Census: Preliminary Findings*. Retrieved on October 5, 2017 from https://asexualcensus.files.wordpress.com/2014/11/2014censuspreliminaryreport.pdf

AVEN (The Asexual Visibility and Education Network) (2017). *Overview*. Retrieved on October 5, 2017 from http://www.asexuality.org/?q=overview.html

Avis, N. E., Stellato, R., Crawford, S., Johannes, C. & Longcope, C. (2000). Is There an Association between Menopause Status and Sexual Functioning? *Menopause*, 7(5), 297–309. Retrieved on December 3, 2016 from https://www.nia.nih.gov/newsroom/2000/09/can-menopause-change-your-sex-life

Ayto, J. (2006). *Movers and Shakers: A Chronology of Words that Shaped Our Age*. New York: Oxford University Press.

Bailey, J. M., Gaulin, S., Agyei, Y. & Gladue, B. (1994). Effects of Gender and Sexual Orientation on Evolutionarily Relevant Aspects of Human Mating Psychology. *Journal of Personality and Social Psychology*, 66(6), 1081–1093. Retrieved on October 28, 2017 from https://www.ncbi.nlm.nih.gov/pubmed/8046578

Barta, W. D. & Kiene, S. M. (2005). Motivations for Infidelity in Heterosexual Dating Couples: The Roles of Gender, Personality Differences, and Sociosexual Orientation. *Journal of Social and Personal Relationships*, 22(3), 339–360.

Basson, R. (2000). The Female Sexual Response: A Different Model. *Journal of Sex and Marital Therapy*, 26, 51–65.

Bauer, G. R., Hammond, R., Travers, R., Kaay, M., Hohenadel, K. M. & Boyce, M. (2009). I Don't Think This Is Theoretical; This Is Our Lives: How Erasure Impacts Health Care for Transgender People. *J Assoc Nurses AIDS Care*, 20(5), 348–361. Retrieved on October 21, 2017 from http://www.nursesinaidscarejournal.org/article/S1055-3290(09)00107-1/fulltext

Bergstrand, C. & Williams, J. B. (October 10, 2000). Today's Alternative Marriage Styles: The Case of Swingers. *Electronic Journal of Human Sexuality*, 3. Retrieved on October 10, 2017 from http://www.ejhs.org/volume3/swing/body.htm

Berman, L. (2013). *It's Not Him, It's You!: How to Take Charge of Your Life and Create the Love and Intimacy You Deserve*. New York: DK.

Blackless, M., Charuvastra, A., Derryck, A., Fausto-Sterling, A., Lauzanne, K. & Lee, E. (2000). How Sexually Dimorphic Are We? Review and Synthesis. *American Journal of Human Biology*, 12, 151–166.

Blinder, A. & Perez-Pena, R. (September 1, 2015). Kentucky Clerk Denies Same-Sex Marriage Licenses, Defying Court. *The New York Times*. Retrieved on October 29, 2017 from https://www.nytimes.com/2015/09/02/us/same-sex-marriage-kentucky-kim-davis.html

Bogaert, A. F. (2004). Asexuality: Prevalence and Associated Factors in a National Probability Sample. *Journal of Sex Research*, 41(3), 279–287.

Brady, S. S. & Halpern-Felsher, B. L. (2007). Adolescents' Reported Consequences of Having Oral Sex Versus Vaginal Sex. *Pediatrics*, 119(2), 229–236.

Branch, M. A. (April 2003). Back in the Fold. *Yale Alumni Magazine*. Retrieved on October 14, 2017 fromYale-AlumniMagazine.com

Brotto, L. A. & Yule, M. A. (2011). Physiological and Subjective Sexual Arousal in Self-Identified Asexual Women. *Archives of Sexual Behavior*, 40(4), 699–712. Retrieved on October 6, 2017 from https://doi.org/10.1007/s10508-010-9671-7

Broude, G. J. (2003). Sexual Attitudes and Practices. In *Encyclopedia of Sex and Gender: Men and Women in the World's Cultures* Volume 1 (pp. 177–184). New York: Springer.

Broude, G. & Greene, S. J. (1976). Cross-cultural Codes on Twenty Sexual Attitudes and Practices. *Ethnology*, 15, 409–429.

Brown, A. (June 13, 2017). *5 key findings about LGBT Americans*. Washington, DC: Pew Research Center. Retrieved on August 11, 2017 from http://www.pewresearch.org/fact-tank/2017/06/13/5-key-findings-about-lgbt-americans/

Brown, T.N.T., Romero, A. P. & Gates, G. J. (July 2016). *Food Insecurity and SNAP Participation in the LGBT Community*. The Willman Institute. Retrieved on December 4, 2016 from http://williamsinstitute.law.ucla.edu/research/lgbt-food-insecurity-2016/

Buss, D. M. (1989). Sex Differences in Human Mate Preferences: Evolutionary Hypothesis Tested in 37 Cultures. *Behavioral and Brain Sciences*, 12(1), 1–49.

Butler, J. (2004). *Undoing Gender.* New York: Taylor & Francis.

California State Senate Majority Caucus (October 16, 2017). Celebration and Statement after Governor Brown's Signs SB 179 – The Gender Recognition Act. Retrieved on October 26, 2017 from http://sd11.senate.ca.gov/news/20171016-celebration-and-statement-after-governor-browns-signs-sb-179-%E2%80%93-gender-recognition-act

Callis, A. S. (2009). Playing with Butter and Foucault: Bisexuality and Queer Theory. *Journal of Bisexuality,* 9, 213–233. Retrieved on October 30, 2017 from http://www.tandfonline.com/doi/pdf/10.1080/15299710903316513

Canby, V. (August 10, 1969). Warhol's Red Hot and 'Blue' Movie. D1. Print. (behind paywall). *New York Times.*

Cantarella, E. (1992). *Bisexuality in the Ancient World.* Trans. Cormac O'Cuilleanain. New Haven, CT: Yale University Press.

Cappelletti, M. & Wallen K. (2016). Increasing Women's Sexual Desire: The Comparative Effectiveness of Estrogens and Androgens. *Hormones and Behavior,* 78, 178–193.

Case, L. K. & Ramachandran, V. S. (2012). Alternating Gender Incongruity: A New Neuropsychiatric Syndrome Providing Insight into the Dynamic Plasticity of Brain-sex. *Medical Hypotheses,* 78(5), 626–631. Retrieved on October 26, 2017 from https://www.ncbi.nlm.nih.gov/pubmed/22364652

Cass, V. (1979). Homosexual Identity Formation: A Theoretical Model. *Journal of Homosexuality,* 4(3), 219–235.

CDC (Centers for Disease Control and Prevention) (2000). *Tracking the Hidden Epidemics: Trends in STDs in the United States, 2000.* Atlanta: Division of STD Prevention, U.S. Department of Health and Human Services, Public Health Service.

CDC (Centers for Disease Control and Prevention) (2015). *Sexually Transmitted Disease Surveillance, (various years).* Atlanta, GA: Department of Health and Human Services. Retrieved on December 1, 2017 from https://www.cdc.gov/std/stats/

CDC (Centers for Disease Control and Prevention) (July 18, 2016). *Sexual Risk Behaviors: HIV, STD, & Teen Pregnancy Prevention.* Retrieved on December 1, 2017 from http://www.cdc.gov/healthyyouth/sexualbehaviors/

CDC (Centers for Disease Control and Prevention) (March 18, 2017a). *Transgender Persons. Lesbian, Gay, Bisexual, and Transgender Health.* Retrieved on October 21, 2017 from https://www.cdc.gov/lgbthealth/transgender.htm

CDC (Centers for Disease Control and Prevention) (2017b). *Sexually Transmitted Infections (STIs).* Retrieved on October 21, 2017 from https://www.childtrends.org/?indicators=sexually-transmitted-infections-stis

Chan, R. L., & Martin, S. L. (2009). Physical and Sexual Violence and Subsequent Contraception Use among Reproductive-aged Women. *Contraception,* 80, 276–281.

Child Trends Databank (September 11, 2014). *5 Things to Know about LGBT Youth.* Retrieved on November 29, 2016 from http://www.childtrends.org/child-trends-5/5-things-to-know-about-lgbtq-youth/

Child Trends Databank (October, 2016). *Condom Use.* Retrieved on November 29, 2016 from https://www.childtrends.org/?indicators=condom-use

Child Trends Databank (March 2017). *Sexually Transmitted Infections (STIs).* Retrieved on July 27, 2017 from https://www.childtrends.org/indicators/sexually-transmitted-infections-stis/

Choi, S. K. & Meyer, I. H. (August 2016). *LGBT Aging: A Review of Research Findings, Needs, and Policy Implications.* The William Institute. Retrieved on December 4, 2016 from http://williamsinstitute.law.ucla.edu/category/research/census-lgbt-demographics-studies/

CIA (Central Intelligence Agency) (2018). *The World Factbook.* Retrieved on November 24, 2018 from https://www.cia.gov/library/publications/resources/the-world-factbook/fields/357.html

Clark, H., Babu, A. S., Wiewel, E. W., Opoku, J. & Crepaz, N. (December 2016). Diagnosed HIV Infection in Transgender Adults and Adolescents: Results from the National HIV Surveillance System, 2009–2014. *AIDS and Behavior,* 21(9), 2774–2783. Retrieved on October 21, 2017 from https://www.ncbi.nlm.nih.gov/pubmed/28035497

Clarke, J. R. (2003). *Roman Sex.* New York: Harry N. Abrams, Inc.

Cocks, H. G. (2009). *Nameless Offences: Homosexual Desire in the 19th Century.* London and New York: I B Tauris.

Collins, R. (1971). A Conflict Theory of Sexual Stratification. *Social Problems,* 19(1), 3–21.

Conley, T. D. (2011). Perceived Proposer Personality Characteristics and Gender Differences in Acceptance of Casual Sex Offers. *Journal of Personality and Social Psychology,* 100(2), 309–329.

Connell, R. W. (1995). *Masculinities.* Berkeley: University of California Press.

Connell, R. W. (2000). *The Men and the Boys.* Berkeley, CA: University of California Press.

Copen, C. E., Chandra, A. & Febo-Vazquez, I. (January 2016). Sexual Behavior, Sexual Attraction, and Sexual Orientation among Adults aged 18–44 in the United States: Data from the 2011–2013 National Survey of Family Growth. *National Health Statistics Report,* 7(88), 1–14.

Crompton, L. (1976). Homosexuals and the Death Penalty in Colonial America. *Journal of Homosexuality,* 1(3).

Cronn-Mills, K. (2014). IV. Trans' spectrum. Identities. *Transgender Lives: Complex Stories, Complex Voices.* Minneapolis, MN: Twenty-First Century Books.

Crooks, R. & Baur, K. (2010). *Our Sexuality.* California: Cengage Learning.

Dailard, C. (2006). New Bush Administration Policy Promotes Abstinence Until Marriage Among People in their 20s. *Guttmacher Policy Review*, 9(4). Retrieved on November 30, 2016 from http://www.guttmacher.org/pubs/gpr/09/4/gpr090423.html

Davis, K. (1937). The Sociology of Prostitution. *American Sociological Review*, 2(5), 744–755.

Decker, J. S. (2015). *The Invisible Orientation: An Introduction to Asexuality.* New York: Skyhorse Publishing Books.

DeLamater, J. D. & Sill, M. (2005). Sexual Desire in Later Life. *The Journal of Sex Research,* 42(2). Retrieved on October 28, 2017 from http://www.tandfonline.com/doi/abs/10.1080/00224490509552267

DeLamater, J. & Hasday, M. (2007). The Sociology of Sexuality. In C. Bryant and D. L. Peck (Eds.), *21st Century Sociology: A Reference Handbook*, Vol. 1. Thousand Oaks, CA: Sage.

de Lauretis, T. (February 2017). The Queerness of the Drive. *Journal of Homosexuality,* 64(14).

Devor, A. (1997). *FTM: Female-to-Male Transsexuals in Society.* Bloomington, IN: Indiana University Press.

Diamond, L. M. & Butterworth, M. R. (2008). Questioning Gender and Sexual Identity: Dynamic Links Over Time. *Sex Roles*, 59, 365–376.

Diamond, M. (2002). Sex and Gender Are Different: Sexual Identity and Gender Identity Are Different. *Clinical Child Psychology & Psychiatry*, 7(3), 320–334.

Doane, M. J. (2016). Cohesion, Marital. *The Wiley Blackwell Encyclopedia of Family Studies*, 1–3. Retrieved on October 15, 2017 from http://onlinelibrary.wiley.com/doi/10.1002/9781119085621.wbefs289/abstract

Duiker, W. J. & Spielvogel, J. J. (2017). *The Essential World History, Volume II: Since 1500.* Boston, MA: Cengage Learning.

Elders, J. M., Satcher, D. & Carmona, R. (June 2017). Re-Thinking Genital Surgeries on Intersex Infants. *Palm Center.* Retrieved on October 22, 2017 from http://www.palmcenter.org/wp-content/uploads/2017/06/Re-Thinking-Genital-Surgeries-1.pdf

Ellis, L. & Ashley Ames, M. (1987). Neurohormonal Functioning and Sexual Orientation: A Theory of Homosexuality-Heterosexuality. *Psychological Bulletin*, 101, 233–258.

Estrada, G. (2011). Two Spirits, Nádleeh, and LGBTQ2 Navajo Gaze. *American Indian Culture and Research Journal,* 35(4), 167–190. Retrieved on October 14, 2017 from http://uclajournals.org/doi/10.17953/aicr.35.4.x500172017344j30?code=ucla-site

Ewing, E. & Mackrell, A. (2002). *History of Twentieth Century Fashion.* Los Angeles: Quite Specific Media Group Ltd.

Ferfolja, T. & Ullman, J. (2017). Gender and Sexuality Diversity and Schooling: Progressive Mothers Speak Out. *Journal of Sex Education,* 17(3), 348–362. Retrieved on October 14, 2017 from http://www.tandfonline.com/doi/abs/10.1080/14681811.2017.1285761

Fields, J. (2012). Sexuality Education in the United States: Shared Cultural Ideas across a Political Divide. *Sociology Compass*, 6(1), 1–14.

Fisher, T. D., Moore, Z. T. & Pittenger, M. (2011). Sex on the Brain?: An Examination of Frequency of Sexual Cognitions as a Function of Gender, Erotophilia, and Social Desirability. *The Journal of Sex Research*, 49(1), 69–77.

Flores, A. R., Herman. J. L., Gates, G. J., & Brown, T. N. T. (2016). *How Many Adults Identify as Transgender in the United States?* Los Angeles, CA: The Williams Institute.

Foucault, M. (1980). *The History of Sexuality (Volume 1: An Introduction).* Translated by Robert Hurley. New York: Vintage Books.

Foucault, M. (1998). *The History of Sexuality: The Will to Knowledge.* London: Penguin.

Francis, D. A. (June 2017). Homophobia and Sexuality Diversity in South African schools: A Review. *Journal of LGBT Youth,* 14(4), 359–379, DOI: 10.1080/19361653.2017.1326868

Freud, S. (1900). The Interpretation of Dreams. Volumes 4–5 of *The Standard Edition of the Complete Psychological Works.* London: Hogarth; New York: Macmillan.

Freud, S. (1922). *Beyond the Pleasure Principle.* London: International Psychoanalytical Press.

Fulbright, Y. K. (2003). *The Hot Guide to Safer Sex.* California: Hunter House.

Gagnon, J. H. and Simon, W. (1973). *Sexual Conduct: The Social Sources of Human Sexuality.* Chicago: Aldine Books.

Gates, G. J. (April 2011). *How Many People Are Lesbian, Gay, Bisexual, and Transgender?* The William Institute. Retrieved on August 11, 2017 from http://williamsinstitute.law.ucla.edu/wp-content/uploads/Gates-How-Many-People-LGBT-Apr-2011.pdf

Goetting, C. (2017). A Comparison of Ancient Roman and Greek Norms Regarding Sexuality and Gender. Honors Projects, 221. http://scholarworks.bgsu.edu/honorsprojects/221

Gola, M., Wordecha, M., Marchewka, A. & Sescousse, G. (2016). Visual Sexual Stimuli—Cue or Reward? A Perspective for Interpreting Brain Imaging Findings on Human Sexual Behaviors. *Frontiers in Human Neuroscience*, 10, 402. Retrieved on October 28, 2017 from https://www.ncbi.nlm.nih.gov/pmc/articles/PMC4983547/

Gordon, S. (1999). *Raising A Child Responsibly in A Sexually Permissive World.* Holbrook, MA: Adams Media.

Grant, J. M., Mottet, L. A., Tanis, J., Harrison, J., Herman, J. L. & Keisling, M. (2011). *Injustice at Every Turn: A Report of the National Transgender Discrimination Survey.* Washington, DC: National Center for Transgender Equality and National Gay and Lesbian Task Force.

Greenberg, D. F. (1988). *The Construction of Homosexuality.* Chicago: University of Chicago Press.

Guttmacher Institute (September 2016). *American Teens' Sexual and Reproductive Health.* Retrieved on December 3, 2016 from https://www.guttmacher.org/fact-sheet/american-teens-sexual-and-reproductive-health?gclid=COPb95z62NACFQ5YDQodU7IHjg

Halberstam, J. (1998). *Female Masculinity.* Durham, NC: Duke University Press.

Hales, D. (2012). *An Invitation to Health.* California: Cengage Learning.

Hallet, J. P. & Skinner, M. B, (1997). *Roman Sexualities.* Princeton, NJ: Princeton University Press.

Halperin, D. M. (1989). Is There a History of Sexuality? *History and Theory,* 28(3).

Harris, M. (1981). *America Now: The Anthropology of a Changing Culture.* New York: Simon & Schuster.

Hatemi, P. K., Crabtree, C. & McDermott, R. (2016). The Relationship between Sexual Preferences and Political Orientations: Do Positions in the Bedroom Affect Positions in the Ballot Box? *Personality and Individual Differences,* 105, 318–325.

Hawkes, G. (1996). *Sociology of Sex and Sexuality.* Massachusetts: Open University Press.

Heasley, R. & Crane, B. (Eds.) (2002). *Sexual Lives.* New York: McGraw-Hill.

Helgason, Á. R., Adolfsson, J. P., Dickman, P. W., Arver, S. T., Fredrikson, M., Göthberg, M. & Steineck, G. (1996). Sexual Desire, Erection, Orgasm and Ejaculatory Functions and Their Importance to Elderly Swedish Men: A Population-based Study. *Age and Ageing,* 25(4), 285–291.

Herbenick, D., Reece, M., Schick, V., Sanders, S. A., Dodge, B. & Fortenberry, J. D. (2010). Sexual Behavior in the United States: Results from a National Probability Sample of Men and Women Ages 14–94. *The Journal of Sexual Medicine,* 7(s5), 255–65. Retrieved on October 9, 2017 from https://www.ncbi.nlm.nih.gov/pubmed

Herbenick, D., Bowling, J., Fu, T.-C. (Jane), Dodge, B., Guerra-Reyes, L. & Sanders, S. (July 20, 2017). Sexual Diversity in the United States: Results from a Nationally Representative Probability Sample of Adult Women and Men. *PLOS ONE,* 12(7). Retrieved on October 9, 2017 from http://journals.plos.org/plosone/article?id=10.1371/journal.pone.0181198

Herdt, G. H. (1984). *Ritualized Homosexuality in Melanesia.* Berkeley, CA: University of California Press.

Heyl, B. (1979). Prostitution: An Extreme Case of Sex Stratification. In F. Adler & R. Simon (Eds.), *The Criminology of Deviant Women.* Boston, MA: Houghton Mifflin.

Hoenig, J. (1977). Dramatis Personae: Selected Biographical Sketches of 19th Century Pioneers in Sexology. In J. Money & H. Musaph (Eds.), *Handbook of Sexology* (pp. 21–43). Amsterdam: Elsevier.

Hope, D. A. (Ed.) (2009). Contemporary Perspectives on Lesbian, Gay, and Bisexual Identities. Nebraska Symposium on Motivation, 54.

Hubbard, T. K. (2003). *Homosexuality in Greece and Rome.* Berkeley, CA: University of California Press.

Human Rights Campaign (2016). *Sexual Orientations and Gender Identity Definitions.* Retrieved on December 1, 2016 from http://www.hrc.org/resources/sexual-orientation-and-gender-identity-terminology-and-definitions

Hutcherson, H. (2006). *Pleasure: A Woman's Guide to Getting the Sex You Want, Need and Deserve.* New York: G.P. Putnam's Sons.

Hyde, J. S. & DeLamater, J. D. (2017). *Understanding Human Sexuality.* New York: McGraw-Hill, Inc.

ILGA (2017). *About us.* Retrieved on October 16, 2017 from http://ilga.org/about-us/

ILGA & Carroll, A. (2016). International Lesbian, Gay, Bisexual, Trans and Intersex Association: *State Sponsored Homophobia 2016: A World Survey of Sexual Orientation Laws: Criminalisation, Protection and Recognition.* Retrieved on October 16, 2017 from http://ilga.org/downloads/02_ILGA_State_Sponsored_Homophobia_2016_ENG_WEB_150516.pdf

Jackson, S. (2006). Gender, Sexuality, and Heterosexuality: The Complexity (and Limits of) Heteronormativity (pp. 105–117). *Feminist Theory.* London: Sage.

Jagose, A. (1996). *Queer Theory: An Introduction.* New York: New York University Press

Jayne, C. J., Heard, M. J., Zubair, S. & Johnson, D. L. (2017). New Developments in the Treatment of Hypoactive Sexual Desire Disorder – A Focus on Flibanserin. *International Journal of Women's Health,* 9, 171–178. Retrieved on October 10, 2017 from https://www.ncbi.nlm.nih.gov/pmc/articles/PMC5396928/

Johnson, M. (1977). Asexual and Autoerotic Women: Two Invisible Groups. In H. Gorchros & J. Gochros (Eds.), *The Sexually Oppressed.* New York: Associated Press.

Kane, E. (1996). *Gender, Culture, and Learning.* Washington, DC: Academy for Educational Development.

Kang, W. (2009). *Obsession: Male Same-Sex Relations in China, 1900–1950.* Hong Kong: Hong Kong University Press.

Kann, L., Olsen, O. E., McManus, T., Kinchen, S., Chyen, D., Harris, W. A. & Wechsler, W. (2011). *Sexual Identity, Sex of Sexual Contacts, and Health-Risk Behaviors among Students in Grades 9–12—Youth Risk Behavior Surveillance, Selected Sites, United States, 2001–2009.* CDC. Retrieved on July 26, 2017 from https://www.cdc.gov/mmwr/preview/mmwrhtml/ss6007a1.htm

Kann, L., Olsen, O. E., McManus, T., Harris, W. A., Shanklin, S. L., Flint, K. H. *et al.* (August 12, 2016). Sexual Identity, Sex of Sexual Contacts, and Health-Related Behaviors Among Students in Grades 9–12—United States and Selected Sites, 2015. CDC. MMWR. *Surveillance Summaries*, 65(9). Retrieved on July 26, 2017 from https://www.cdc.gov/mmwr/volumes/65/ss/pdfs/ss6509.pdf

Kaplan, H. S. (1974). *The New Sex Therapy: Active Treatment of Sexual Dysfunctions.* New York: Brunner/Mazel, Publishers, Inc.

Karberg, E., Wildsmith, E. & Manlove, J. (November 21, 2016). Intimate Inaccuracies: Young Couples Don't Always Agree About Contraceptive Use. *Child Trends.* Retrieved on November 29, 2016 from http://www.childtrends.org/publications/intimate-inaccuracies-young-couples-dont-always-agree-contraceptive-use/

Kaufman, J. & Johnson, C. (2004). Stigmatized Individuals and the Process of Identity. The Sociological Quarterly, 45(4), 807–833.

Kewenig, W. A. (1983). Foreword. In E. J. Haeberle, *The Birth of Sexology: A Brief History in Documents.* Washington, DC: World Association for Sexology.

Killermann, S. (2013). *The Social Justice Advocate's Handbook: A Guide to Gender.* Austin, TX: Impetus Books.

King, B. M. (2013). *Human Sexuality Today.* New Jersey: Pearson.

Kinsey, A. C., Pomeroy, W. R. & Martin, C. E. (1948). *Sexual Behavior in the Human Male.* Philadelphia, PA: W. B. Saunders.

Kinsey, A., Pomeroy, W., Martin, C. & Gebhard, P. (1953). *Sexual Behavior in the Human Female.* Philadelphia, PA: W. B. Saunders.

Kirby, J. (2016). In Recognition of Intersex Awareness Day: Statement by Assistant Secretary and Department Spokesperson. U.S. Department of State. Retrieved on October 22, 2017 from http://www.palmcenter.org/wp-content/uploads/2017/06/Re-Thinking-Genital-Surgeries-1.pdf

Köksal, D. & Falierou, A. (2013). *A Social History of Late Ottoman Women: New Perspectives.* Leiden: Brill.

Kosciw, J. G., Greytak, E. A., Palmer, N. A. & Boesen, M. J. (2013). *The 2013 National School Climate Survey: The Experiences of Lesbian, Gay, Bisexual and Transgender Youth in Our Nation's Schools. 2014.* Retrieved on October 21, 2017 from http://www.glsen.org/article/2013-national-school-climate-survey.

Krasner, R. I. (2010). *The Microbial Challenge: Science, Disease and Public Health.* Burlington, MA: Jones & Bartlett Publishers.

Kuefler, M. (2007). *The History of Sexuality Sourcebook.* Toronto: University of Toronto Press.

LaFreriere, P., Strayer, F. F. & Gauthier, R. (1984). The Emergence of Same-Sex Preferences among Preschool Peers: A Developmental Ethological Perspective. *Child Development,* 55, 1958–1965.

LaMarre, N. (2007). Compulsory Heterosexuality and the Gendering of Sexual Identity: A Contemporary Analysis. *The New York Sociologist,* 2.

Laplanche, J. & Pontalis, J.-B. (1988). *The Language of Psycho-analysis. London: Karnac Books.*

Lenhart, A., Anderson, M. & Smith, A. (2015). *Teens, Technology and Romantic Relationships.* Washington, DC: Pew Research Center. Retrieved on October 22, 2017 from http://www.pewinternet.org/2015/10/01/teens-technology-and-romantic-relationships/

Leupp, G. P. (1995). *Male Colors: The Construction of Homosexuality in Tokugawa Japan.* Berkeley: University of California Press.

Levin, R. J. (2008). Critically Revising Aspects of the Human Sexual Response Cycle of Masters and Johnson: Correcting Errors and Suggesting Modifications. *Sexual and Relationship Therapy,* 23(4), 393–399.

Levine, S. B. (2003). The Nature of Sexual Desire: A Clinician's Perspective. *Archives of Sexual Behavior,* 32(3), 279–285. Retrieved on October 27, 2017 from https://www.ncbi.nlm.nih.gov/pubmed/12807300

Loeber, O., Reuter, S., Apter, D. & Lazdane, P. (June 2010). Aspects of Sexuality Education in Europe – Definitions, Differences and Developments. *European Journal of Contraception & Reproductive Health Care,* 15(3), 169–176.

Longmore, M. A. (1998). Symbolic Interactionism and the Study of Sexuality. *Journal of Sex Research,* 35, 44–57.

Martinez, G. M. & Abma, J. C. (2015). *Sexual Activity, Contraceptive Use, and Childbearing of Teenagers Aged 15–19 in the United States.* Data Brief No. 209. National Center for Health Statistics. Retrieved on November 29, 2016 from http://www.cdc.gov/nchs/data/databriefs/db209.pdf

Martos, A. J., Wilson, P. A. & Meyer, I. H. (2017). Lesbian, Gay, Bisexual, and Transgender (LGBT) Health Services in the United States: Origins, Evolution, and Contemporary Landscape. The Williams Institute. *PLOS ONE* 12(7), e0180544. https://doi.org/10.1371/journal.pone.0180544

Masci, D., Brown, A. & Kiley, J. (June 26, 2017). *5 Facts about Same-Sex Marriage.* Washington, DC: Pew Research Center. Retrieved on August 10, 2017 from http://www.pewresearch.org/fact-tank/2017/06/26/same-sex-marriage/

Masters, W. H. & Johnson, V. E. (1966). *Human Sexual Response.* Toronto; New York: Bantam Books.

McGinn, T. A. (1998). *Prostitution, Sexuality, and the Law in Ancient Rome.* New York: Oxford University Press.

Mercer, C. H. (2014). Sexual Behaviour. *Medicine,* 42(6), 291–293.

Mercer, C. H., Fenton, K. A. & Copas, A. J. (2004). Increasing Prevalence of Male Homosexual Partnerships and Practices in Britain 1990–2000: Evidence from National Probability Surveys. *AIDS,* 18, 1453–1458.

Mercer, C. H., Tanton, C. & Prah, P. (2013). Changes in Sexual Attitudes and Lifestyles in Britain through The Life Course and over Time: Findings from the Third National Survey of Sexual Attitudes and Lifestyles (Natsal-3). *Lancet,* 382, 1781–1794.

Messerschmidt, J. W. (2003). *Flesh and Blood: Adolescent Gender Diversity and Violence.* New York: Rowman & Littlefield.

Meston, C. M. & Buss, D. (2007). Why Humans Have Sex. *Archives of Sexual Behavior,* 36, 477–507.

Milhausen, R. & Herold, E. (1999). Does the Sexuality Double Standard Still Exist? Perceptions of University Women. *Journal of Sex Research,* 36(4), 361–368.

Morrow, R. (1994). The Sexological Construction of Sexual Dysfunction. *Journal of Sociology,* 30(1), 20–35.

Moser, C. (2006). Demystifying Alternative Sexual Behaviors. *Sexuality, Reproduction & Menopause,* 4(2).

Movement Advancement Project (MAP) (2016). *National LGBT Movement Report 2011.* Retrieved on December 4, 2016 from http://www.lgbtmap.org/lgbt-movement-overviews/2011-national-lgbt-movement-report

Murnaghan, I. (Sep. 13, 2017). Stem Cells and Same Sex Reproduction. *Explore Stem Cells.* Retrieved on October 7, 2017 from http://www.explorestemcells.co.uk/stem-cells-same-sex-reproduction.html

National Institute of Health (NIH) (2017). *Sexuality in Later Life.* National Institute on Aging. Retrieved on October 11, 2017 from https://www.nia.nih.gov/health/sexuality-later-life

National Women Law Center (March 2017). *Women and the Lifetime Wage Gap: How Many Women Years Does it Take to Equal 40 Man Years?* Retrieved on October 25, 2017 from https://nwlc.org/wp-content/uploads/2017/03/Women-and-the-Lifetime-Wage-Gap-2017-1.pdf

Nauert, H. (October 26, 2017). *In Recognition of Intersex Awareness Day.* U.S. Department of State. Retrieved on October 31, 2017 from https://www.state.gov/r/pa/prs/ps/2017/10/275098.htm

Nissinen, M. (2004). *Homoeroticism in the Biblical World: A Historical Perspective,* translated by Kirsi Stjerna. Minnesota: Augsburg Fortress.

Nordyke, K. (October 29, 2017). Star Trek Star Claims Kevin Spacey Made a Pass at Him at Age 14; Spacey Apologizes, Comes Out as Gay. *The Hollywood Reporter.* Los Angeles.

Northrup, C., Schwartz, P. & Witte, J. (2013). *The Normal Bar: The Surprising Secrets of Happy Couples and What They Reveal About Creating a New Normal in Your Relationship.* New York: Harmony.

Nurius, P. (1983). Mental Health Implications of Sexual Orientation. *The Journal of Sex Research,* 19(2), 119–136.

OII Europe (Organization Intersex International Europe) (March 2017a). *Statement of the 1st European Intersex Community Event (Vienna, March 30–31, 2017).* Retrieved on October 22, 2017 from https://oiieurope.org/statement-1st-european-intersex-community-event-vienna-30st-31st-march-2017/

OII Europe (Organization Intersex International Europe) (April 2017b). *4th International Intersex Forum – Media Statement.* Retrieved on October 22, 2017 from https://oiieurope.org/statement-1st-european-intersex-community-event-vienna-30st-31st-march-2017/

Ormand, K. (2009). *Controlling Desires: Sexuality in Ancient Greece and Rome.* Santa Barbara, CA: Praeger.

Ostovich, J. M. & Sabini, J. (2004). How are Sociosexuality, Sex Drive, and Lifetime Number of Sexual Partners Related? *Personality and Social Psychology Bulletin,* 30(10), 1255–1266.

OurTime. (2017). *Senior Dating Facts: Dating After 50 Survey Results.* Retrieved on July 29, 2017 from http://www.sexualdiversity.org/sexuality/love/700.php

Parsons, T., Bales, R. F., Olds, J., Zelditsch, M. & Slater, P. E. (1955). *Family, Socialization, and Interaction Process.* New York: Free Press.

Percy, W. A. III. (1996). *Pederasty and Pedagogy in Archaic Greece.* Urbana: University of Illinois Press.

Pew Research Center (October 5, 2017a). *Homosexuality, Gender and Religion.* Retrieved on October 5, 2017 from http://www.people-press.org/2017/10/05/5-homosexuality-gender-and-religion/

Pew Research Center (August 8, 2017b). *Gay Marriage around the World.* Retrieved on October 5, 2017 from http://www.pewforum.org/2017/08/08/gay-marriage-around-the-world-2013/

Pickett, B. (2015). Homosexuality. *The Stanford Encyclopedia of Philosophy* (Fall 2015 Edition), Edward N. Zalta (Ed.).

Retrieved on October 5, 2017 from https://plato.stanford.edu/archives/fall2015/entries/homosexuality/>.
Prause, N. & Pfaus, J. (2015). Viewing Sexual Stimuli Associated with Greater Sexual Responsiveness, Not Erectile Dysfunction. *Sexual Medicine*, 3(2), 90–98.
ProCon.org (2016, July 14). *US Federal and State Prostitution Laws and Related Punishments.* Retrieved on October 10, 2017 from http://prostitution.procon.org/view.resource.php?resourceID=000119
Pruden, H. & Edmo, S. (2016). *Two-Spirit People: Sex, Gender and Sexuality in Historic and Contemporary Native America.* National Congress of American Indians Policy Research Center. Retrieved on October 14, 2017 from http://www.ncai.org/policy-research-center/initiatives/Pruden-Edmo_TwoSpiritPeople.pdf
Pyke, K. D. & Johnson, D. L. (2003). Asian American Women and Racialized Femininities: 'Doing' Gender across Cultural Worlds. *Gender and Society,* 17(1), 33–53.
Rosario, M., Schrimshaw, E., Hunter, J. & Braun, L. (2006). Sexual Identity Development Among Lesbian, Gay, and Bisexual Youths: Consistency and Change Over Time. *Journal of Sex Research*, 43(1), 46–58.
Regan, P. C. & Atkins, L. (2006). Sex Differences and Similarities in Frequency and Intensity of Sexual Desire. *Social Behavior & Personality: An International Journal*, 34(1), 95–101.
Reiner, W. G. (2002). Gender Identity and Sex Assignment: A Reappraisal for the 21st Century. *Adv. Exp. Med. Biol. 511, 175–189.* Retrieved on October 26, 2017 from https://www.ncbi.nlm.nih.gov/pubmed/12575762
Reis, H. T. & Sprecher, S. (2009). *Encyclopedia of Human Relationships.* Thousand Oaks, CA: Sage.
Reiss, I. L. (1960). Premarital Sexual Standards in America. Glencoe, IL, The Free Press.
Richards, C., Bourman, W. P., Seal, L., Barker, M. J., Nieder, T. O. & T'Sjoen, G. (2016). Non-binary or Genderqueer Genders. *Journal International Review of Psychiatry,* 28(1), 95–102.
Riggs, D. W. & Bartholomaeus, C. (2017). Transgender Young People's Narratives of Intimacy and Sexual Health: Implications for Sexuality Education. *Journal of Sex Education,* 18(4), 376–390.
Roscoe, W. (1992). *The Zuni Man-Woman.* New Mexico: University of New Mexico Press.
Sanderson, S. K. (2003). *The Sociology of Human Sexuality A Darwinian Alternative to Social Constructionism and Postmodernism.* Retrieved on October 16, 2017 from http://stephenksanderson.com/documents/TheSociologyofHumanSexualityADarwinianAlternativetoSocialConstructionismandPostmodernism.pdf
San Francisco Department of Public Health (1999). *The Transgender Community Health Project.* Retrieved on October 22, 2017 from http://hivinsite.ucsf.edu/InSite?page=cftg-02-02#S4.8X
Schalet, A. T. (2011). Not under My Roof: Parents, Teens, and the Culture of Sex.
Scott, S., McDonnell, L. & Dawson, M. (2014). Asexual Lives: Social Relationships and Intimate Encounters. Discover Society. Retrieved on October 6, 2017 from http://discoversociety.org/2014/06/03/asexual-lives-social-relationships-and-intimate-encounters/
Sedgwick, E. K. (1985). *Between Men: English Literature and Male Homosocial Desire.* New York: Columbia University Press.
Sedgwick, E. K. (1993). *Tendencies.* Durham, NC and London: Duke University Press.
Segall, L. (January 25, 2015). Inside a High-tech San Francisco Swinger's Party. CNN. Retrieved on October 10, 2017 from http://money.cnn.com/2015/01/25/technology/swingers-silicon-valley/
Segraves, R. T. & Segraves, K. B. (1991). Hypoactive Sexual Desire Disorder: Prevalence and Comorbidity in 906 Subjects. *Journal of Sex and Marital Therapy*, 17, 55–58.
Sexual Diversity (2017). Sexual Dysfunction in Males and Females. Retrieved on July 26, 2017 from http://www.sexualdiversity.org/sexuality/health/dysfunction/
Shively, M. G. & De Cecco, J. P. (1977). Components of Sexual Identity. *Journal of Homosexuality*, 3(1). Retrieved on October 7, 2017 from http://www.tandfonline.com/doi/abs/10.1300/J082v03n01_04
Siggy (February 25, 2015). Transgender and Assigned Sex. *The Asexual Census.* Retrieved on October 8, 2017 from https://asexualcensus.wordpress.com/author/tlmiller/
Simpson, J. A. & Gangestad, S. W. (1991). Individual Differences in Sociosexuality: Evidence for Convergent and Discriminant Validity. *Journal of Personality and Social Psychology*, 60, 870–883. Retrieved on October 28, 2017 from http://psycnet.apa.org/doiLanding?doi=10.1037%2F0022-3514.60.6.870
Spanger, M. & Skilbrei, M.-L. (2017). *Prostitution Research in Context: Methodology, Representation and Power.* New York: Taylor & Francis.
Sprecher, S. (1998). Social Exchange Theories and Sexuality. *The Journal of Sex Research,* 35(1), 32–43. Retrieved from http://www.jstor.org/stable/3813163
Storms, M. D. (1980). Theories of Sexual Orientation. *Journal of Personality and Social Psychology,* 38, 783–792.
Tiefer, L. (2004). *Sex is Not a Natural Act and Other Essays* (2nd edn.). Boulder, CO: Westview Press.
Traish, A. M., Kim, S. W., Stankovic, M., Goldstein, I. & Kim, N. N. (2007). Testosterone Increases Blood Flow and Expression of Androgen and Estrogen Receptors in the Rat Vagina. *Journal of Sexual Medicine,* 4(3), 609–619.

Twenge, J. M., Sherman, R. A. & Wells, B. E. (2016). Sexual Inactivity during Young Adulthood Is More Common among U.S. Millennials and iGen: Age, Period, and Cohort Effects on Having No Sexual Partners after Age 18. *Arch Sex Behav.* doi: 10.1007/s10508–016–0798-z

Tyner, K. (2015). *Roman Social-Sexual Interactions: A Critical Examination of the Limitations of Roman Sexuality.* Undergraduate Honors Theses, Paper 849.

UNESCO (UN Educational, Scientific and Cultural Organization) (2009). *International Technical Guidance on Sexuality Education.* Retrieved on October 12 2017 from http://unesdoc.unesco.org/images/0018/001832/183281e.pdf

UNESCO (UN Educational, Scientific and Cultural Organization) (2015). *Emerging Evidence, Lessons and Practice in Comprehensive Sexuality Education: A Global Review.* Retrieved on October 12 2017 from http://unesdoc.unesco.org/images/0024/002431/243106e.pdf

UNESCO (UN Educational, Scientific and Cultural Organization) (March 2017). *Health Workshop Educates Youth in India.* Retrieved on October 12, 2017 from http://www.un.org/youthenvoy/2017/03/health-workshop-educates-youth-india-2/

UN Free & Equal (2017). *UN for Intersex Awareness.* Retrieved on October 25, 2017 from https://www.unfe.org/intersex-awareness/

University of Hawaii at Manoa Library (2017). *Gender Identity and Sexual Identity in the Pacific and Hawai'i: Introduction.* Retrieved on October 26, 2017 from https://guides.library.manoa.hawaii.edu/Pacificsexualidentity

UN Office of Human Commissioner (OHCHR) (2016). *Intersex Awareness Day – Wednesday 26 October.* Retrieved on October 22, 2017 from http://www.ohchr.org/EN/NewsEvents/Pages/DisplayNews.aspx?NewsID=20739&%3BLangID=E

Urbanski, P. (September 25, 2017). Summary of the Polish Sexual Education Model – Global Takeaways. *OpenIDEO.* Retrieved on October 15, 2017 from https://challenges.openideo.com/challenge/youth-srh/research/summary-of-the-polish-sexual-education-model-global-takeaways

U.S. Census Bureau (August 14, 2017). *FFF: Unmarried and Single Americans Week: Sept. 17–23, 2017.* Retrieved on August 15, 2017, 2016 from https://www.census.gov/newsroom/facts-for-features/2017/single-americans-week.html

Van Anders, S. M., Hipp, L. E. & Low, L. K. (2013). Exploring Co-Parent Experiences of Sexuality in the First 3 Months after Birth. *The Journal of Sexual Medicine,* 10(8), 1988–1999.

Vrangalova, Z. & Savin-Williams, R. C. (2011). Adolescent Sexuality and Positive Well-being: A Group-Norms Approach. *Journal of Youth Adolescence,* 40, 931–934.

WebMD (2017). *Sexual Problems in Men.* Retrieved on October 16, 2017 from https://www.webmd.com/men/guide/mental-health-male-sexual-problems#1

Weeks, J. (1986). *Sexuality.* London: Routledge.

Weinrich, J. D. (2014). Multidimensional Measurement of Sexual Orientation: Past. *Journal of Bisexuality,* 14, 3–4, 314–332.

Weiss, D. C. (2011). Report: 'Staggering' Rate of Attempted Suicides by Transgenders Highlight Injustices. *ABA Journal,* February 4.

Weitzer, R. (2005). New Directions in Research on Prostitution. *Crime, Law & Social Change,* 43, 211–235.

Wellings, K., Mitchell, K. & Collumbien, M. (2012). *Sexual Health: A Public Health Perspective.* New York: McGraw-Hill Education

Wesp, J. (1992). Polyamory Frequently asked Questions. [Online]. May 29. *Society for Human Sexuality.* Retrieved on October 10, 2017 from http://www.sexuality.org/l/polyamor/p olyfaq.html

West, C. & Zimmerman, D. H. (1987). Doing Gender. *Gender and Society,* 1(2), 125–151.

Widmer, E. D., Treas, J. & Newcomb, R. (1998). Attitudes toward Nonmarital Sex in 24 Countries. *Journal of Sex Research,* 35(4), 349–358.

Wierzba, M., Riegel, M., Pucz, A., Leśniewska, Z., Dragan, W., Gola, M., *et al.* (2015). Erotic Subset for the Nencki Affective Picture System (NAPS ERO): Cross-sexual Comparison Study. *Front. Psychol.,* 6, 1336. Retrieved on October 28, 2017 from https://www.ncbi.nlm.nih.gov/pubmed/26441715

Winter, C. R. (2010). *Understanding Transgender Diversity: A Sensible Explanation of Sexual and Gender Identities.* California: CreateSpace.

Wolfinger, N. H. (July 5, 2017). *America's Generation Gap in Extramarital Sex.* Institute for Family Studies. Retrieved on July 27, 2017 from https://ifstudies.org/blog/americas-generation-gap-in-extramarital-sex

World Health Medicine (2017). *Types of Sexual Behavior Deviation.* Retrieved on July 26, 2017 from http://kkkmedicine.blogspot.com/2010/05/10-types-of-sexual-behavior-deviation.html

World Health Organization (2017). *Sexual and Reproductive Health.* Retrieved on July 27, 2017 from http://www.who.int/reproductivehealth/topics/sexual_health/sh_definitions/en/

WorldPress (2012). *Asexuality Archive.* Retrieved on October 28, 2017 from http://www.asexualityarchive.com/things-that-are-not-asexuality/

Worthen, M. G. (2016). *Sexual Deviance and Society: A Sociological Examination.* New York: Routledge.

Yaeger, P. S. (December 1985). Eve Kosofsky Sedgwick, Between Men: English Literature and Male Homosocial Desire. *MLN*, 100(5), 1139–1144.

Yoder, J. B. & Mattheis, A. (2016). Queer in STEM: Workplace Experiences Reported in a National Survey of LGBTQA Individuals in Science, Technology, Engineering, and Mathematics Careers. *J. Homosex.*, 63(1), 1–27. Retrieved on October 21, 2017 from https://www.ncbi.nlm.nih.gov/pubmed/26241115?dopt=Abstract

7
LOVE

Learning Outcomes

Upon completion of the chapter, students should be able to:

1. define the concept of love
2. explain love as emotion and action
3. describe the four elements of love
4. identify the seven types of love
5. explain sociological theories of love
6. compare Sternberg's triangular theory of love with Lee's color wheel theory of love
7. explain the functions of love
8. describe problems of experiencing love
9. explore love problem causes and solutions
10. list tips for maintaining a long-term love relationship
11. explain expression of love from a diversity perspective
12. describe expressions of love from a global perspective.

Brief Chapter Outline

Pre-test

Love Involves Emotion and Action

Love as an Emotion

Emotions as Response to Events; Classification of Basic Emotions; Mind and Body Interaction in Shaping Emotions; Sociology of Emotions

Micro-level Approach to Emotions; Macro-level Approach to Emotions

Love Involves Actions

Measurement of Love; Behavioral Expectations

Elements of Love
- Intimacy
  - Sexual Intimacy; Emotional Intimacy; Psychological intimacy
- Passion
  - Passionate Love; Habituation
- Commitment
  - The Investment Model of Commitment Process
- Caring
  - Human Caregiving; Three Types of Affective Care Relations; Characteristics of Love Labor

Types of Love
- Self-love
  - Self-love before Loving Others; Self-growth; Self-care
- Parental Love
  - Maternal Love; Filial Love; Parental Love and Entrepreneurial Love
- Familial Love
  - Family Love Forever
- Friendship Love
  - Liking
- Romantic Love
  - Romantic Love as Attachment; Romantic Love as Caregiving; Romantic Love as Sexual Attraction; Adolescent Love; Expression of Romantic Love
- Companionate Love
- Compassionate Love
  - Social Support and Sacrifice; Altruistic Caregiving

Diversity Overview: Expressions of Love by Gender

Love from a Sociological Standpoint
- Conflict Theory on Love
- Structural Functionalism on Love
- Symbolic Interactionism on Love
- The Color Wheel Theory of Love
  - Primary Types of Love; Secondary Types of Love
- Sternberg's Triangular Theory of Love

Functions of Love
- Love as the Foundation for Interpersonal Relationships
- Love as a Theme in the Search for a Mate
- Love as Essential for Human Survival
- Love as a Provider of Health Benefits

Problems of Experiencing Love
- Jealousy
  - Causes and Effects of Jealousy; Stalking; Tips for Keeping the Stalker Away
- Unrequited Love

Pre-test

Engaged or active learning is a powerful strategy that leads to better learning outcomes. One way to become an active learner is to begin the chapter and try to answer the following true/false statements from the material as you read. You will find that you have ready answers to these questions upon completion of the chapter.

Love means romance only. Once romance is over, there is no love at all.	T	F
Parental love is unconditional and unbreakable.	T	F
Self-love is not centered on the self.	T	F
Love involves both emotion and action.	T	F
Love is expressed through intimacy, passion, commitment, and caregiving.	T	F
People with low self-esteem are more likely to be jealous.	T	F
Love brings happiness and preserves physical and psychological health.	T	F
Women are interested in sex more than men are.	T	F
Companionate love combines intimacy, commitment, and caregiving.	T	F
Compassionate love is altruistic love.	T	F

When Angela comes home from work, the dog wags its tail and licks her. It exhibits happiness. When she leaves for work the dog cries at the window. The dog grew up with a sibling and created a strong bond to that sibling. Angela remembers that when the sibling died, the dog went into depression and refused to eat. This could be interpreted as sadness. Happiness is the feeling love brings. Non-human animals can feel love as well, although it is less complex and less creative (Singer, 2009).

Human beings sing, dance, marry, cry, and may commit suicide because of love. Today people still sing The Beatles' "All You Need Is Love." Indeed, many people think there is nothing more important than love. After all, isn't true love what we all dream of and want to experience? Love makes us feel happy

Upon completion of this section, students should be able to:
LO1: Define the concept of love.
LO2: Explain love as emotion and action.

and makes our lives meaningful. Yet some people think that true love takes place only in the films—*The Notebook* projects that love can conquer all, the idealization of one's partner, soul mate/only one, and love at first sight (Hefner & Wilson, 2013). Strong love even involves sacrificing your own happiness and even sometimes your own life for the other person (Boeree, 2003). Does true love exist in the real world? No matter whether people like it or not, love in its many forms is always around us. Let's start this chapter by looking at the concept of love.

Love Involves Emotion and Action

How do you define love? Although love has existed since the beginning of humankind, the definition of love has changed and continues to change across time and place. For the purposes of our text, we'll define *love* as the physical, emotional, sexual, and/or social affection someone holds for another.

Love involves both emotion and action. Psychologist Frijda (1927–2015) never stopped wondering about human emotions, and human behaviors generally (Mesquita, 2016). Most theorists view love as an emotion and a set of behaviors. For example, in a study of word frequencies, love was listed 179 times as a noun and 145 times as a verb (Frances & Kucera, 1982). This is because we often use actions to express our love.

Love as an Emotion

Love is a human emotion with a complex intentional structure, having its own kind of reasons (Solomon, 2007) and is a universal emotion experienced by a majority of people (Karandashev, 2015). Indeed, love should be considered an emotion, like anger, sadness, happiness, and fear (Shaver *et al.*, 1996). Love is one of the most powerful emotions we can experience. The term *emotion* (from French *émouvoir*, meaning "to stir up") was introduced into academic discussion as a catch-all term for passions, sentiments, and affections (Dixon, 2003). When they were asked to name items belonging to the category of "emotion," most undergraduate students nominated love (Fehr & Russell, 1984). When studying the time children begin to name emotion-related words, such as hug and kiss, Bretherton and Beeghly (1982) found that children could use the word "love" when they were 25 months. Our empirically driven bottom-up approach suggests the usefulness of defining love as a linking emotion (Seebach & Núñez-Mosteo, 2016). **Emotions** refer to a positive or negative experience that is associated with physiological reactions to events or influenced by social relationships and social structure.

Emotions as Response to Events

Psychologists believe that emotions are responses to significant internal and external events (Schacter, 2011). American psychologist William James (1884) and Danish physiologist Carl Lange (1887) suggested that emotions are the results of physiological reactions to external events. Lange (1886) became a pioneer of psychophysiology with his contribution to the so-called **James–Lange theory of emotion** (Schioldann, 2011). The theory illustrates an event happening, followed by a person's physiological response, and then an emotion being experienced (see image below). When a person experiences physiological response during an event, the emotion is the interpretation of this physiological arousal. For example, when Angela is dating a man, her heart beats faster, her pupils dilate, her face blushes, and her love-related emotion is experienced. When we see the film *The Notebook* on TV, we might shed tears for the confounding sadness of brief true-love moments lost.

Classification of Basic Emotions

Psychologists believe that emotions are motivational states with the special function of producing adaptation to situational conditions. The theory assumes

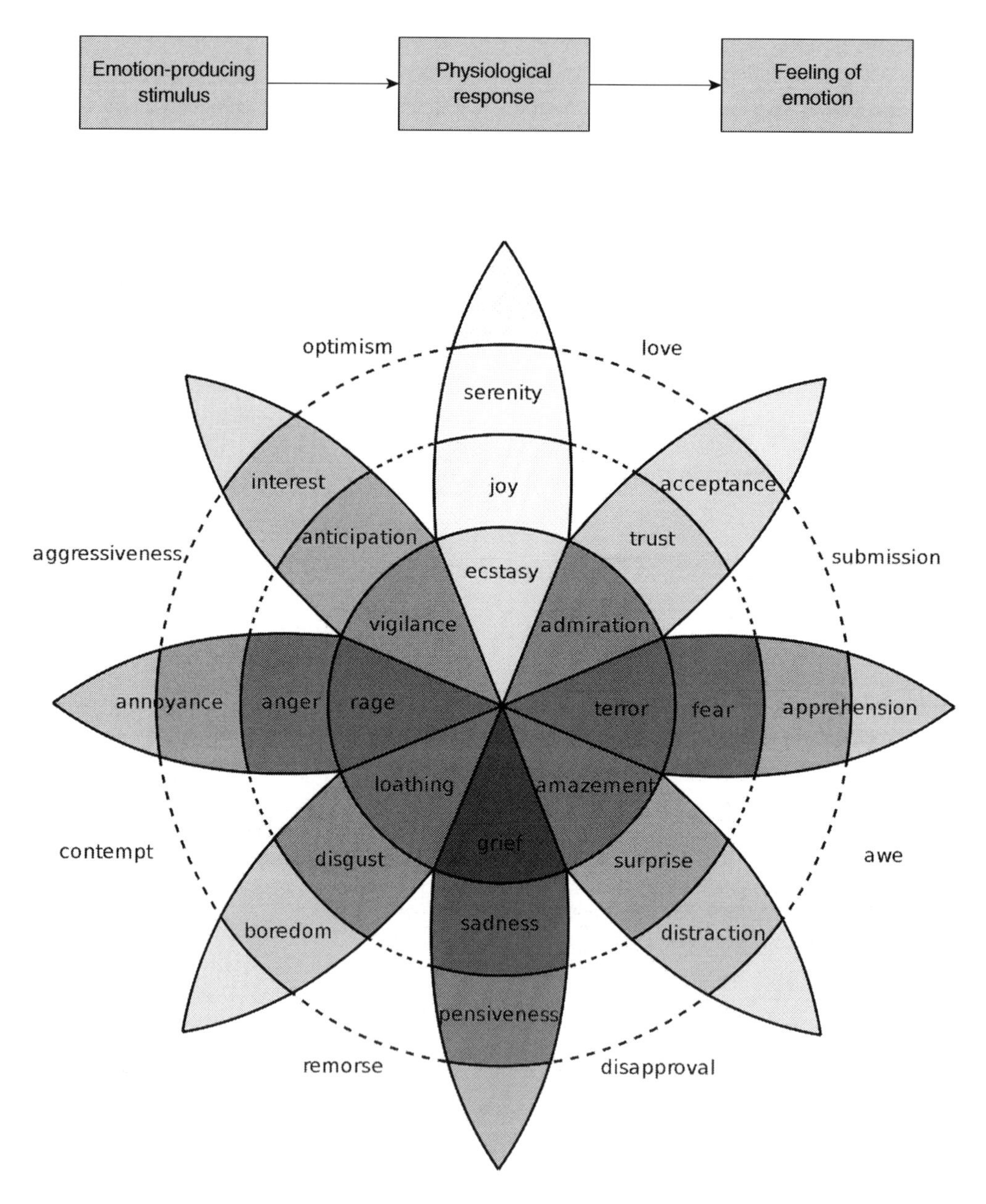

Psychologist Robert Plutchik's Wheel of Emotions Proposes Eight Basic Emotions to Describe how Emotions Related.
Source: Author: Machine Elf 1735. https://upload.wikimedia.org/wikipedia/commons/thumb/c/ce/Plutchik-wheel.svg/2000px-Plutchik-wheel.svg.png

that the emotional system in the central nervous system can change from one emotional state to another, producing only one emotion at a time (Brehm, 2016).

Psychologist Robert Plutchik (1927–2006) proposed eight basic emotions to describe how emotions are related. Eight primary bipolar emotions (joy versus sadness; anger versus fear; trust versus disgust; and surprise versus anticipation) can result in positive or negative influences. Eight primary emotions (see image on previous page), like colors, can mix with one another to form complex emotions. For example, interpersonal *joy* could mix with *trust* to produce *love*. Interpersonal *anger* and *disgust* could blend to form *contempt*.

Mind and Body Interaction in Shaping Emotions

Psychologists have long been aware that mind and body interact in shaping emotional experience. Both mind and body make indispensable contributions to emotion (Hatfield & Sprecher, 1986). Emotional stimuli in the external environment induce a specific mental state involving an early brain–body interaction as a function of emotional arousal (D'Hondt *et al.*, 2010). This link between emotions and bodily states is reflected in the way we speak of emotions (Kövecses, 2000): a young bride getting married next week may suddenly have "cold feet," severely disappointed lovers may be "heartbroken," and our favorite song may send "a shiver down our spine."

The connection between mind and body is powerful. Emotions are often felt in the body. In experiments, participants reported bodily sensations associated with six "basic" and seven nonbasic ("complex") emotions (see image below). The body maps show regions whose activation increased

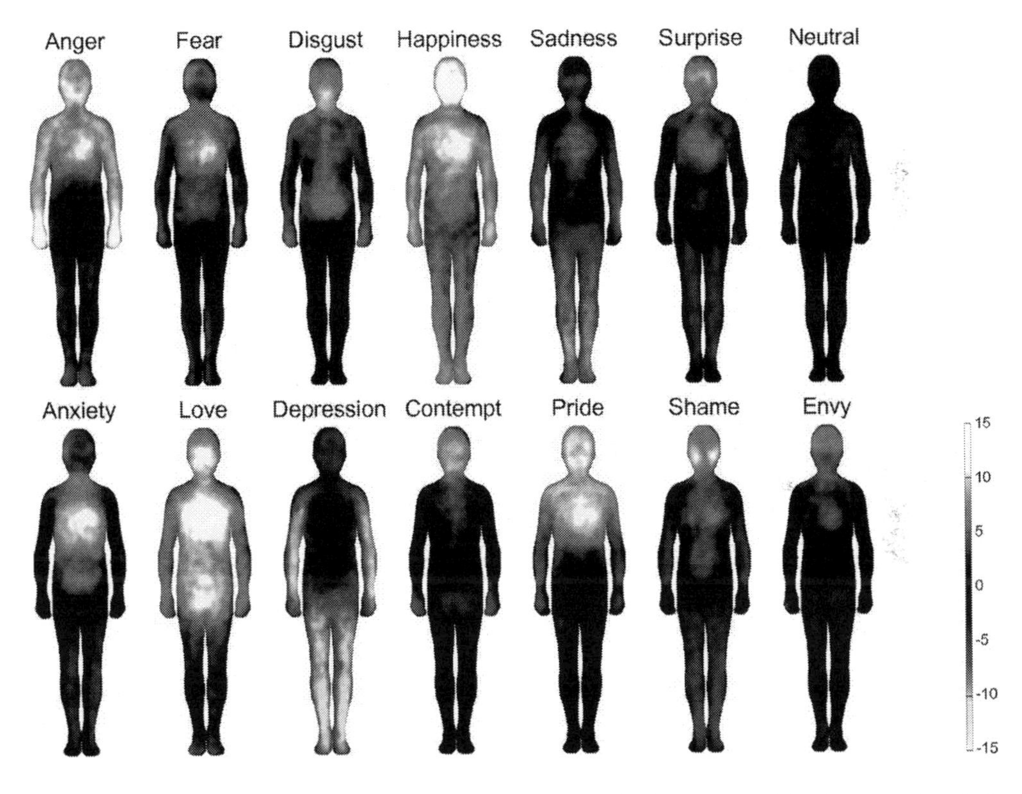

The Body Maps Showing Regions whose Activation Increased (warm colors) or Decreased (cool colors) when Feeling Each Emotion.
Source: PNAS (Proceedings of the National Academy of Sciences of the United States of America, 2013). http://www.pnas.org/content/111/2/646/F2.large.jpg

Student Accounts of Feelings of Love

Love is a strong and overwhelming feeling. It can be felt in your body, your stomach, heart, or head. It makes you feel special. It is almost like an "I am on top of the world" feeling. It is a combination of affection, attraction, enjoyment, and pleasure. These feelings can make your emotions heighten and make your excitement prosper (personal communication, October 22, 2017).

(warm colors) or decreased (cool colors) when feeling each emotion. If you copy URL https://www.pnas.org/content/111/2/646 into your browser, you are able to see the the changes in color shown in the image on the previous page. Perception of these emotion-triggered bodily changes may play a key role in generating consciously felt emotions (Nummenmaa *et al.*, 2013).

Sociology of Emotions

The **sociology of emotions** examines how human emotions influence, and are influenced by society. Bourdieu's emotion practice framework (1990) suggests that emotion emerges from the (mis)fit between the structured environment and the individual's internalized adaptive patterns (dispositions, tastes, values, etc.). This emotion practice framework draws attention to the ways that emotion operates at both the micro and macro levels of analysis (Cottingham *et al.*, 2017).

Micro-level Approach to Emotions Sociologists examine emotions on a micro level—treating emotions as outcomes of social relationships and social influences. *Social interaction theory* and *affect control theory* treat emotions as key mechanisms linking social structure to individual behavior and vice versa (Thoits, 1989). Kemper's social interaction theory (1978) suggests that a person's power and status dimensions of social relationships are universal emotion-elicitors. Increases or decreases in one's own and others' status or power generate specific emotions whose quality depends on the patterns of change (Kemper, 1978). For example, after dating a man, Angela develops love-related emotions towards him because this guy is handsome and CEO of a company. Possessing sufficient or adequate power produces feelings of security; receiving sufficient or adequate status produces happiness (Thoits, 1989).

Heise's affect control theory (1979) attempts to predict specific emotions from a person's powerful/weak and likable/dislikable dimensions of relationship. Affect control theory assumes that people are motivated to experience transient feelings that confirm fundamental sentiments (Thoits, 1989). Emotions can be viewed as culturally delineated types of feelings or affects. **Affects** mean positive and negative evaluations (liking/disliking) of an object, behavior, or idea (Heise, 1979). In response to the emotion, our bodies can evaluate and react. For example, in a scene from the film *Pretty Woman*, when rich businessman Edward Lewis gives Vivian Ward, a Hollywood hooker, a dazzling ruby diamond necklace worth $250,000 to wear out for the evening, Vivian's pupils dilate and a huge smile lights up her face. The audience has a positive impression of Edward, which generates the feeling of happiness towards Vivian.

Feelings are the representations of emotions. People can experience an emotion only if they have some feelings aroused physiologically. Indeed, when a person experiences physiological response during an event or interaction with someone, the *emotion* is the interpretation of this physiological arousal and the *feeling* is the representation of the emotion. You have no doubt experienced many of the positive feelings that love produces—the warm feeling you may have thinking about the love you have for your family or hot feeling you get when you fall in love with someone romantically. Classical Greek philosopher Aristotle (384–322 BCE) suggests that *eudaimonia* (Greek word, meaning happiness) is the highest quality a human life can acquire (Nordenfelt, 1994). **Happiness** is the emotion that we feel when we are overwhelmed with an experience that involves emotional stimuli ranging from contentment to intense joy. For example, when Angela is in love, she has feelings of happiness and never wants the feeling to end.

Macro-level Approach to Emotions Sociologists explore emotions on a macro level—treating emotions as outcome and mechanism of inequitable social arrangements. Like other resources such as income and power, emotions are unequally distributed to individuals and families. **Emotional stratification** means that unequal distribution of emotional labor and emotional energy underlies forms of social inequality in the areas of gender, social class, race, and age. **Emotional energy** (EE) is an aliveness of the mind, a happiness of the heart, and a spirit filled with hope (Kirshenbaum, 2003). The distribution of emotional energy corresponds with the ranking of social classes. Collins and McConnell (2015) use the concepts of emotional energy to explain social inequality. People who possess power and wealth are more likely to have high emotional energy. For example, a chief executive officer (CEO) holding power over employees is more likely to have high emotional energy. High-emotional-energy persons are confident and proactive; they are forward moving; they have a path to a goal (Collins & McConnell, 2015). Today working parents pour their emotional energy into home life, work life, and possibly school life. Their lives may be full but their energy gas tanks seem to be running on fumes (O' Connor, 2016). One out of eight adults is in the same low-energy state (Kirshenbaum, 2003).

Love Involves Actions

Love is not merely a feeling but is also action (Fromm, 1959). Have you ever done anything vulnerable, funny, or silly when you were in love? For sure, love can cause people to behave or act both rationally and irrationally.

Measurement of Love

One way to measure love is to look at behaviors that lovers engage in to express love. Social psychologist Chapman (2015) proposes five categories of behaviors that people would engage in to express love: (1) words of affirmation, (2) spending quality time, (3) giving gifts, (4) acts of service, and (5) physical touch. Goff *et al.* (2007) designed a series of questions to measure one of these behaviors and examine a lover's behavioral expectations such as "giving presents," "offers encouragement," "spends time with me," "holds my hand," and "does yard work."

British biologist Charles Darwin (1809–82) describes the main action tendency of love as "a strong desire to touch the beloved person." He describes facial expressions associated with maternal and romantic love (Darwin, 1872). For example, the maternal face is marked by "a gentle smile and tender eyes" whereas romantic lovers' "hearts beat quickly, their breathing is hurried, and their faces flush."

Emotions coordinate our behavior and physiological states during survival-salient events and pleasurable interactions (Nummenmaa *et al.*, 2013).

Student Accounts of Love Involving Action

We can express our love for someone with words, but actions speak louder than words. We can tell someone that we love them 100 times a day, but it might not mean anything unless the love is shown (personal communication, October 22, 2017).

A couple who loves each other is likely to act in a loving manner. For example, they are likely to spend time together and care deeply for each other. Romantic love can include acts such as kissing, sex, emotional contact, and companionship that contribute to happiness in relationships (Kim & Hatfield, 2004). Mutual gazing could momentarily increase feelings of romantic love (Kellerman *et al.*, 1989). College dating couples who love each other a great deal spend more time gazing into one another's eyes than couples who love each other to a lesser degree (Rubin, 1970). Friends show their love for each other by spending time together, laughing together, and doing things together.

Behavioral Expectations

People who are loved or love have behavioral expectations. And it's not just romantic love that involves action, of course. Lazarus (1991) views caregiving tendencies as part of love; "these manifestations of affection communicate to the loved one that he or she is valued and secure in the relationship."

People are likely to do anything for each other when they care deeply for one another. For example, parents often find themselves doing things they never thought for the children they love. Those who love their partners compassionately would be expected to respond to their partner's needs with behaviors that convey understanding and caring (Berscheid, 2006). Love involves the very real activities of "looking out for" and "looking after" the other, including the management of the tensions and conflict that are an integral part of love labor relations (McKie *et al.*, 2002). Truly loving people care deeply for their needs such as massaging the body, or giving financial help if needed. Those who love their partner compassionately perceive that their partner is responsive to them ($r = .56$) (Fehr *et al.*, 2010).

Elements of Love

Psychologist Robert Sternberg (1986) suggests that love has three components: intimacy, commitment, and passion. All couples in love will experience intimate, passionate, and committed love in the same patterns (Acker & Davis, 1992). In a cross-cultural study of love, the top features of love listed by subjects include commitment, honesty/sincerity, and caregiving (Wu & Shaver, 1993). In another cross-cultural study of love, devotion–care was the strongest predictor of relationship satisfaction within each culture, independent of demographics and perceived stress (Jackson *et al.*, 2006). This section discusses four elements of love: intimacy, passion, commitment, and caregiving. These four elements are essential when it comes down to love.

Intimacy

Intimacy is an important ingredient for any love relationship. Intimate love encompasses the close bonds of loving relationships and is called "warm" love because of the way it brings people closer together. A couple with intimate love deeply values each other (Levy, 2013). Erickson (1963) explains that either we satisfy our needs for intimacy or we remain socially and emotionally isolated. For example, if people do not meet their intimacy needs, they might feel lonely and depressed. Hook and colleagues (2003) posit that there are four key features that make up intimacy: the presence of love and affection, personal validation (feeling accepted or understood), trust, and

Upon completion of this section, students should be able to:
LO3: Describe the four elements of love.

self-disclosure. **Intimacy** refers to feelings of closeness and bondedness in loving relationships sexually, emotionally, and psychologically. Let us discuss three kinds of intimacy: sexual intimacy, emotional intimacy, and psychological intimacy.

Sexual Intimacy

When people hunger for sexual connection, the longing is physical. Romantic love includes sexual intimacy. **Sexual intimacy** or physical intimacy is a form of intimacy that involves physical affection. **Physical affection**—"any touch intended to arouse feelings of love in the giver and/or the recipient"—is an essential ingredient in sexual intimacy. Physical affection has been categorized into seven types of behaviors including holding hands, cuddling/holding, backrubs/massages, caressing/stroking, kissing on the face, hugging, and kissing on the lips (Gulledge *et al.*, 2003). Physical affection is associated with positive outcomes in romantic relationships (Mackey *et al.*, 2000). Physical affection has been related to the formation of attachment bonds. Sexual intimacy can take place with or without emotional intimacy.

Emotional Intimacy

When people seek an emotional bond, they want a connection with another person at an emotional level. **Emotional intimacy** is a form of intimacy that involves sharing the feelings of bondedness and demonstrating caring for emotional support. Emotional intimacy is demonstrated from a mother to a child in maternal love. Emotional intimacy does not require physical contact but it is associated with **attachment** (the emotional bond). Psychologist Bowlby (1969) uses the *attachment-behavioral system* to describe the experience of love. The most intense emotions arise when a person forms and maintains a bond, disrupts, and renews attachment relationships. For example, when forming a bond, a person falls in love. Similarly, threat of loss arouses anxiety and actual loss gives rise to sorrow. While each of these situations is likely to arouse anger, the unchallenged maintenance of a bond is experienced as a source of security and the renewal of a bond is experienced as a source of joy (Bowlby, 1969). Edelstein *et al.* (2012) have examined whether emotional intimacy increases *estradiol*—a steroid hormone associated with attachment and caregiving processes. The results reveal that among single participants, levels of estradiol increased in response to the emotionally intimate parent–child interaction video clip. Women with an avoidant attachment style showed smaller increases in estradiol after watching the emotionally intimate clip.

Psychological Intimacy

Psychological intimacy is a form of intimacy that involves being open and honest in talking with a partner about personal thoughts and feelings not usually expressed in other relationships. It has been used synonymously with personal disclosure which involves "putting aside the masks we wear in the rest of our lives" (Rubin, 1983). Psychologist Sidney Jourard (1971) investigated *disclosure*—telling others about the self. **Self-disclosure** means that people trust someone and feel comfortable exposing their true self without fear of rejection. Self-disclosure is a rewarding experience, comparable to those of food and sex (Tamir & Mitchell, 2012). People expect to receive rates of disclosure like those that they give to others (Jourard, 1971). In studying the characteristics of relationships that had lasted an average of 30 years, Mackey *et al.* (2000) reported that psychological intimacy emerged as a significant predictor of satisfaction between partners. Twice as many lesbian respondents compared to heterosexuals viewed their relationships as psychologically intimate.

Communication privacy theory explains the way people make decisions about revealing and concealing private information (Petronio, 2000). Indeed, in everyday life, we try to manage boundaries when we interact with different people with whom we choose to disclose. If there are different criteria for disclosure between partners, expectations can go unfulfilled, resulting in relational dissatisfaction (Jorgensen & Gaudy, 1980). For example, at Thanksgiving dinner party, Angela's boyfriend shares the pet name he has for Angela to everyone at the party. After the party, Angela informs her boyfriend that the pet name he calls her should be kept private between the two of them. The adjustment is essential for both the partners to have the same definition of information such as "pet name" and draw divisions between private information and public information.

Passion

Passion (from Latin *pati*, meaning to suffer) is an intense emotion. French philosopher Denis Diderot (1713–84) viewed passion as the pleasures and pains of the senses; pleasures of the mind or of the imagination; our perfection or our imperfection of virtues or vices; and pleasures and pains in the happiness or misfortunes of others (Diderot, 1765). These guiding principles of passion can cause people to have a strong sexual feeling for someone or act in an irrational way (Plutchik, 2002). **Passion** refers to an intense emotion that leads to romance, physical attraction, and sexual desire in loving relationships.

Passionate Love

Passionate love is considered "hot" love because of the strong presence of sexual arousal between two partners. As a state that promotes passion, sexual desire plays a role in initiating contact, motivating sexual interest, and seeking proximity (Gonzaga *et al.*, 2006). Sexual desire has been described as a "longing for sexual union" (Hatfield & Rapson, 1987). Passion is often expressed through touching, kissing, and sexual interactions.

In the beginning of romantic relationships, couples feel physically attracted to and respond to each other in an exciting state. Romantic love is associated, particularly in the early stages, with physiological, psychological, and behavioral indices (Fisher, 1998). The body is charged up because the sensation is thrilling. The neurons in the brain release *dopamine*—a feel-good hormone and neurotransmitter associated with euphoria. **Euphoria** is an affective state in which a person experiences pleasure and happiness (Bearn & O'Brien, 2015). Passionate love and components of the human sexual response cycle are associated with the induction of euphoria (Georgiadis & Kringelbach, 2012). Biologist Jeremy Griffith (2011) defines love as "unconditional selflessness." When asked whether they would seek a divorce if they got married and the passion disappeared from their marriage, 70 percent of college students stated that they would (Buri, 2009).

Habituation

Why does passion wane after couples get married? **Habituation** is a form of learning in which an organism decreases or ceases its responses to a stimulus after repeated presentations (Bouton, 2007). The body and mind may use up their energy resources. A progressive decline of a behavior in a habituation procedure may reflect nonspecific effects such as fatigue (Fennel, 2011). When couples are repeatedly exposed to each other, these intense emotions and electrifying feelings begin to lose their shine and couples become habituated.

Commitment

Commitment is a key factor for feeling safe in a relationship. **Commitment** refers to a decision that

Student Accounts of Commitment

People can be sexually intimate without being committed to each another. True intimacy requires personal commitment (personal communication, October 22, 2017).

a person loves another and/or maintains the love. Commitment is considered "cold" love because it does not require either intimacy or passion. However, commitment and its resulting feelings of safety are the most powerful and consistent predictors of marital satisfaction (Acker & Davis, 1992).

Short-term commitment involves a conscious decision to love someone. The strength of commitment to a romantic relationship is associated with feelings of satisfaction (Impett *et al.*, 2001) but the "feeling" of love is superficial in comparison to one's commitment to love via a series of loving actions over time (Fromm, 1956). Commitment is the expectation that the relationship is permanent and is "intent to persist in a relationship, long-term orientation toward the involvement and feelings of psychological attachment to it" (Arriaga & Agnew, 2001). Committed love is for lovers who are committed to being together for a long period of time and increases in intensity as the relationship grows (Sternberg, 1986; Acker & Davis, 1992).

The Investment Model of Commitment Process

The **investment model** suggests that commitment to a target is influenced by three factors: satisfaction level, quality of alternatives, and investment size (Rusbult *et al.*, 2011). *Satisfaction* refers to the level of "positive versus negative effect experienced in a relationship." Satisfaction level increases to the extent that a relationship gratifies the individual's most important needs, including needs for companionship, security, intimacy, sexuality, and belonging. *Investment size* refers to the magnitude and importance of the resources (i.e., investing their time and energy, giving things of value, self-disclosure) that are attached to a relationship. "The resources would decline in value if the relationship were to end" (Rusbult *et al.*, 1998). *Quality of alternatives* involves the perceptions of relational rivals or social threats (i.e., rejection and negative evaluation). Gere *et al.* (2013) found that stronger reward (i.e., connection and intimacy) perceptions were associated with higher commitment, investment, and satisfaction, as well as lower quality of alternatives in all studies. When a couple is satisfied, lack alternatives, and have made a considerable investment in their relationship, they are likely to have a committed love relationship. Many people end romantic relationships if they feel that mutual commitment is not increasing (Brown, 1995).

Caring

Another important element of love is caring. When a couple feels that their lives are intertwined, they are likely to show their care for each other by supporting, helping, and protecting each other in any kind of situation. **Caring** is the practice of providing care and support to an intimate relationship partner or family member. Caring is central to all types of love. For example, human parental caregiving is attuned to a child's needs, from proximity to safety to emotional security and beyond (LaRossa & LaRossa, 1981).

Human Caregiving

Caregiving is defined as the behaviors an individual evokes to meet the needs of another who is in distress or in need of assistance (Canterberry & Gillath, 2012). An attachment system has evolved across species to respond to this expansion of the caregiving system (Bell, 2012). In rats, caregiving includes retrieving offspring to the nest (Bowlby, 1969). Rabbits with a minimal cortex feed their offspring and huddle with them but will not retrieve them when they wander from the burrow (MacLean, 1990). Protective caregiving behavior directed toward young is performed not only by mothers in these species but also by other adult herd members (Bell, 2012). In felines and canines, the caregiving system has expanded beyond protection to teaching and emotional support (Lopez, 1978). After a hundred million years or so, caregiving began to show characteristics associated with the feelings of affection that we recognize in humans (Bell, 2012).

Human caregiving can be viewed as a motivation to meet the needs of a special other (Bell, 2012). All humans are born with the capacity and motivation to engage in caregiving behaviors. The two goals of the caregiving behavioral system are to protect close others from harm and decrease their suffering during times of threat, and to promote close others' personal growth and exploratory behavior (Bowlby, 1969).

Three Types of Affective Care Relations

Lynch (2007) divides affective relations into primary, secondary, and tertiary care relations. **Primary care relations** are produced through **love labor**—any love-producing activity voluntarily performed without consideration of reward between parents and children and between two partners, for instance. **Secondary care relations** involve *general care work* including affective bonds with neighbors, or work colleagues, or the caring given in nursing or paid caring. In the modern era, affective relations are productive, materialist human relations that constitute people mentally, emotionally, physically, and socially (Cantillon & Lynch, 2017). **Tertiary care relations** involve *solidarity care work,* where people work informally, politically, culturally, or economically through solidarity to challenge injustices such as gender inequality. Thus, mutuality, commitment, trust, and responsibility at the heart of *love labor* make it distinct from general care work and solidarity work (Lynch, 2007).

Characteristics of Love Labor

The first characteristic of love labor is that its principal goal is the enhancement of the love relationship itself (Cantillon & Lynch, 2017). The love labor that produces love through nurture variously involves physical, cognitive, and mental work as well as emotional work (Ruddick, 1989). For example, love labor is created when a caregiver is doing physical tasks such as cooking meals. Angela has learned how to care for her dog and keep the dog safe, healthy, and happy at home. For a caring parent, it is the full range of the child's needs that determines the parent's response (Bell, 2012). Also, love labor involves the emotional and other work oriented to the enrichment and enablement of others, and the bond between self and others (Lynch, 2009). Emotional support involves affirming, supporting, and challenging, as well as identifying with someone and supporting them emotionally at times of distress (Mattingly, 2014), and includes expressions of care, affection, sympathy, and encouragement and/or instrumental support (provision of information, advice, and tangible resources) (Collins *et al.*, 2006). The love laboring required to nurture primary care relations involves emotionally laden responsibilities, including holding the person in mind, planning, listening, attending, and making a commitment to the relationship itself, that do not apply to other care relations (Cantillon & Lynch, 2017).

Secondly, love labor is time consuming and endless. Since love relations are based on human needs and arise from certain vulnerabilities and dependencies, neither care nor love is ordered on clock time (Bryson, 2013). Unlike paid care time, love laboring time is person-specific, and it has a longer and more unpredictable trajectory as it is not tied to a time-bound contract (Cantillon & Lynch, 2017). The affective engagement in terms of time, responsibility, commitment, and emotional engagement is more intense and prolonged (Engster, 2005).

Finally, love laboring might be gendered. Bubeck (1995) argues that love laboring is unequally distributed between women and men, with the result that women's exploitation as love laborers is arguably the

Upon completion of the next section, students should be able to:
LO4: Identify the seven types of love.

Student Accounts of Types of Love

Love is defined as something different for each person. I love my family in a way that is different than how I love a friend. I love my husband in a way that is sexual and yet respectful. I love my children with a love that is unconditional. The love for my children can never be replaced in any way (personal communication, October 12, 2017).

principal form of exploitation that applies to them as women. The resolution of care-related inequality is fundamental to the attainment of equality for women (Lynch, 2009). Partners are expected to share love laboring in their love relationships. Love laboring is not entirely altruistic; the bonds that develop in love relationships have the potential to be mutually beneficial (Cantillon & Lynch, 2017).

Types of Love

On August 9, 2011, the 85-year-old Spanish Duchess of Alba made headlines when she married Alfonso Diez, a civil servant who was 24 years younger than she was. The billionaire duchess, who attracted headlines throughout her colorful life, died at age 88 on November 20, 2014, leaving no money to her husband. Did the duchess buy love with money? What type of love did they experience?

Source: AP Associated Press, October 5, 2011

People experience different types of love as they reach different levels of maturity in their lives. The ancient Greeks identified kinship or familiarity (*storge*), friendship and/or platonic desire (*philia*), sexual and/or romantic desire (*eros*), and self-emptying or divine love (*agape*) (Lewis, 1960). When participants were asked to list types of love, the following types of love appeared among the top examples in the United States and China: love between parent and child (also listed as motherly, fatherly, parental, and love for parents); family love (including sibling love); romantic love (including true love, everlasting love, and deep, unforgettable love); marital love (including committed love and love between spouses); friendship; and love of life (Wu & Shaver, 1993). This section examines seven types of love: self-love, parental love, familial love, friendship love, romantic love, companionate love, and compassionate love. All types of love are viewed as varying combinations of intimacy, passion, commitment, and caring.

Self-love

Self-love is a type of love through which we acknowledge our own good qualities and focus on our self-growth and well-being. French philosopher Jean-Jacques Rousseau (1712–78) suggested that *amour de soi* (from French, meaning "love of self") was lost during the transition from the pre-societal condition to society, but it can be restored through "good" institutions (Derathé, 1995). Self-love has often been a moral flaw, akin to vanity and *selfishness*—being concerned excessively or exclusively for one's own advantage, pleasure, or welfare, regardless of others (Kirkpatrick, 1998). American psychoanalyst Fromm (1956) re-evaluates self-love in a positive sense.

Student Account of Self-Love

Love brings happiness to the lives of those who are experiencing it. Happiness is possible with love, but love starts with loving yourself (personal communication, 2017).

Self-love before Loving Others

Sedikides and Gregg (2003) define self-love as "a person's subjective appraisal of himself or herself as intrinsically positive or negative." A person first needs to love oneself in the way of caring about oneself, taking responsibility for oneself, respecting oneself, and knowing oneself such as one's strengths and weaknesses in order to be able to truly love another person. Self-love is crucial to our lives and is a life-long journey. It allows us to understand and accept our own selves. Self-love is important because we must know ourselves well in order to bring partners into our life in a healthy, sustainable way. Recognizing our own positive qualities helps us identify them in others. In fact, the experience of self-love is a prerequisite for loving others. If we can take care of ourselves and nourish ourselves, we are more likely to be able to do the same for another person.

Self-growth

Self-love supports personal growth. Rath and Harter (2010) have identified five essential elements of well-being: career, social, financial, physical, and community well-being. We are likely to love ourselves and boost our well-being if we value education, pursue a career, have a rich social network, are secure financially, eat a balanced diet, do regular exercise, and take pride in community activities. However, note that self-love is not the same as being self-centered. If people are centered only on themselves, they are likely to dominate their partners and attempt to control love. This might lead to issues such as the controlling behaviors we discuss later in this chapter.

Self-care

We will take better care of our basic needs if we love ourselves. **Self-care** is any necessary human regulatory function which is under individual control, deliberate, and self-initiated (Segall & Goldstein, 1998). People high in self-love are more likely to nourish themselves daily through a healthy lifestyle such as fitness and intimate relationships with others. Sedikides & Gregg (2008) suggest that **self-enhancement** is a type of motivation that maintains and cultivates positive feelings of the self. For example, eating is used as an analogy to explain that self-enhancement is a fundamental part of human nature. Self-enhancement is thought to be the foremost motive in the perpetual search for self-knowledge (Sedikides, 1993). We are more likely to love ourselves when we participate in activities that help us grow physically and emotionally. For example, we can find new ways to practice self-care such as jogging. Tracking devices such as a Fitbit come together with the promotion of individual responsibility in health-care policy (Beck & Beck-Gernsheim, 2001) and "healthy living" (Burrows *et al.*, 1995). Fotopoulou and O'Riordan (2017) explore how people learn to self-care with wearable technologies through a series of micropractices that involve processes of mediation and the sharing of their own data via social networking.

Parental Love

Parental love is a type of love that parents have for their children, characterized by the presence of intimacy, commitment, and caregiving. Love is an important component in children's emotional life. When an infant is born, parents often have an immediate and unbreakable bond with the newborn. **Unconditional love** refers to affection without limits or conditions. It means that a person still loves another no matter what conditions occur. This unconditional love is healthy and essential for the child's development. Because parents are often willing to do anything to protect their children, children learn to trust through parental love. Bowlby (1969) chose to focus on attachment as the foreground in the parent–child relationship, relegating caregiving to the background. Caregiving from a parent or other caregiver precedes the development of the attachment system in the child (Bell, 2010). Parental love includes *maternal love* (love of a mother) and *paternal love* (love of a father).

Maternal Love

Maternal love is the love a mother feels for their child. Studying twenty types of love, Fehr and Russell (1991) find that maternal love is the type of love with the highest average rating. Studies of infants reveal that "the desire for relationship, pleasure in connection and the ability to make and maintain relationship are present at the onset of development" (Gilligan, 1995). Maternal love allows infants to form an attachment to their mothers. "Neurobiology of attachment" emerges and helps illuminate the importance of love (Damasio, 2006). *Oxytocin* is a nonapeptide hormone best known for its role in lactation (production of milk) and parturition (act of giving birth) (Lee *et al.*, 2009). In mothers, the prebirth production of oxytocin primes their caregiving system to proactively care about the child emotionally and to care for the child behaviorally (Uvnäs-Moberg *et al.*, 2005). Breastfeeding is a biological process and a culturally determined behavior (Macadam, 2017). Before pabulum, mothers chewed solid food and then transferred it to their children by putting their mouths directly on their child's mouth. Because mouth-mating was a universal experience, it was also universally recognized as a signal of positive emotions (Devon, 2015). Both tender physical contact and focused social interaction cause the production of oxytocin (Uvnäs-Moberg *et al.*, 2005). Maternal love ensures that mothers will care for children and contributes to the child's happiness and development. **Monogyny** (or male monogamy)—the practice of one male marrying one woman—is favored as means of increasing paternity (Fromhage *et al.*, 2007). **Monogynic love** encourages men to support the mother of their children. For fathers and others, physical contact and interaction contribute to the production of oxytocin to strengthen the caregiving system toward the child (Bell, 2012).

Filial Love

Filial love is the love a child has for a parent. The love a parent has for a child is usually unconditional but the love a child has for a parent is based on social norms. The Chinese philosopher Confucius (551–479 BCE) developed Confucian *role ethics*—morality based on a person's fulfillment of a role, such as that of a parent or a child. Filial piety is central to Confucian role ethics. **Filial piety** (Chinese: 孝, *xiào*) is a virtue of love and respect for one's parents, elders, and ancestors. In serving his parents, a filial son reveres them in daily life, nourishes them, takes anxious care of them in sickness, shows great sorrow over their death, and sacrifices to them with solemnity (Ikels, 2004). The term can also be applied to general obedience, and is used in religious titles in Christian churches, like "filial priest" or "filial vicar" for a cleric whose church is subordinate to a larger parish (Chang & Kalmanson, 2010).

Parental Love and Entrepreneurial Love

Sternberg (1995) views partners in close relationships as business partners. Entrepreneurs' emotional experience toward the venture has been suggested to resemble parental love (Mirabella, 1993; Cardon *et al.*, 2005). Acevedo and Aron (2009) conceptualize love as a "peak experience" that is an intense emotion. Peak experiences are particularly associated with high-growth ventures (Schindehutte *et al.*, 2006). Entrepreneurs' relationship with their firm is often characterized by unrealistic expectations that often cannot be met. Similarly, parents create idealistic images of their children and put their child on an unachievable pedestal (Cardon *et al.*, 2005). Both entrepreneurial and parental love seem to be supported by brain structures associated with reward

Student Account of Familial Love

My dad always said "Friends come and go, but family is forever." The older I get, the more I see how this may be true. I have a love for my brothers unlike any I have for my friends—it's based on our long-term shared experiences, maybe our shared genes, and for sure our knowledge that no matter what, until death do us part, we will have each other in our lives (Shannon, October 31, 2017).

and emotional processing as well as social understanding; both parentnal and entrepreneurial love activate the brain areas which fall within the reward network of the brain (Halko *et al.*, 2017).

Familial Love

Familial love is a type of love shared between parents and their offspring, siblings, spouses, and other family members based on kinship ties, characterized by the presence of intimacy, commitment, and caregiving. The Greek word *storge* (pronounced store-gae) is familial love for natural affection. Familial love includes the love of a parent towards children and vice versa. The attachment and caregiving systems can be activated in humans toward a wide variety of special others, including other adults, siblings, and pets (Bell, 2012). The primary caregiver can be any major attachment figure in the adult life and most attachment figures include spouses, grandparents, siblings, and/or close friends.

Family Love Forever

Familial love is based on intimacy, caregiving, and commitment to one another. The family is a primary group that provides physical, economic, and emotional support. Familial love is a type of love that we experience in our daily life. The bond that familial love creates can be stronger than other types of love. For example, children experience signs of anxiety when separated from their primary caregiver; and adults can be diagnosed with adult separation anxiety (Bell, 2012). Twins grow up together and share familial love. They may argue and fight but in the end, they still protect and care for each other because they have the sibling relationship and share a feeling of familial love. Both maternal grandmothers (the mother of the child's mother) and paternal grandmothers (the mother of the child's father or their in-law counterparts) may help provide an additional element of unconditional love. We feel and receive familial love no matter what type of family we live in—nuclear, extended, cohabiting, single parent, LGBTQ, or step-families. No matter the context, familial love is one of the most enduring types of love.

Friendship Love

Friendship is another important source of love that we experience in our daily lives. **Friendship love** is a type of love that close friends feel for each other, characterized by the presence of intimacy. The Greek word *philia* (pronounced fill-ee-ah) is the term for the love found between friends. Studying twenty types of love, Fehr and Russell (1991) find friendship love with the third highest average rating after maternal and paternal love.

Liking

Liking includes components of favorable evaluation and respect for the target person, as well as the

Student Account of Friendship Love

I am currently in a relationship with someone as my best friend. I believe that friendship is one of the building foundations to any relationship whether it becomes romantic (personal communication, November 10, 2017).

perception that the target is like oneself (Lindzey & Byrne, 1968). Friendship is a relationship between two people who view each other as equals and includes the elements of trust, respect, enjoyment, and acceptance (Davis, 1985). Allan (1998) examines the ways in which friendships are socially patterned in modern society. Class and status divisions are important for understanding the character of informal solidarities. Friendship love is based on shared interests and emotional support. Indeed, experiencing friendship love is associated with better health and well-being. Friends have a big effect on our physical, social, and psychological health—sometimes they have an even bigger impact than our family relationships (Giles *et al.*, 2005). Friendship is often the basis for a romantic love relationship. Friendship love relationships are built upon the consistent display of affection. For a healthy friendship to develop into a love relationship, passion and a level of intimacy must exist.

Romantic Love

Out of the four elements of love, intimacy and passion are the two highlighted components for romantic love. **Romantic love** is a type of love found between two lovers and characterized by the presence of intimacy, caregiving, and passion. *Eros* (pronounced air-os) has the root of the ancient Greek word for "erotic" with the notion of selfish love (Lewis, 1960).

Romantic love is seen by many as an important aspect of human experience. Romantic love takes center stage in music, literature, and films. Examples of romantic love can be seen in romance drama films such as William Shakespeare's *Romeo and Juliet*. Media influences the expectations we have of what love should be (Storey & McDonald, 2013). Young adults are the target population that is mainly influenced by an unrealistic idea of love that they see over and over in films. Intense romantic love is a cross-culturally universal phenomenon (Aron *et al.*, 2005). Jankowiak and Fischer (1992) "are able to document the occurrence of romantic love in 88.5 percent of the sampled cultures [147 out of 166], concluding that romantic love constitutes 'a human universal'." Psychologist John Bowlby (1907–90) analyzed three of the behavioral systems—*attachment*, *caregiving*, and *sexual attraction*—that play a role in romantic love.

Romantic Love as Attachment

Romantic love is associated with intimacy. When a person is romantically in love with another, it means "I am emotionally dependent on you for happiness, safety, and security; part of my identity is based on my attachment to you" (Bowlby, 1969). **Secure attachment** is attachment with neither fear of losing the relationship nor fear of entering relationships. Romantic love is associated with strong feelings of love and desire for a specific person. Aron *et al.* (2005) suggest that romantic love uses subcortical reward and motivation systems to focus on a specific individual. For example, romantic love can activate the *striatum*—the reward system of the human brain and a major projection site of midbrain dopamine cells (Bartels & Zeki, 2004), activated by a variety of rewards such as food and drink (Kelley, 2004) and money (Knutson *et al.*, 2001). Oxytocin ("love hormone") is associated with affiliative bonding in mammals (Insel *et al.*, 1997) that mediates social behavior, pair-bonding, and parental attachment across a variety of species (Carter, 1998). People in the first stages of romantic attachment had higher levels of oxytocin, compared with non-attached single people (Schneiderman *et al.*, 2012).

Romantic Love as Caregiving

Caregiving is central to romantic love. Providing sensitive and responsive care to one's partner has been shown to have important benefits for the quality of one's romantic relationships (Fitzgerald, 2017). Maslow (1955) makes the developmental sequence from attachment to caring in his analysis of love. When a person is romantically in love with another, it means "I get great pleasure from supporting, caring for, and taking care of you; from facilitating your progress, health, growth, and happiness" (Bowlby, 1969). Bell (2012) suggests that at base the strength of attachment should be conceptualized as the level of trust. *Trust* captures the focus of attachment on security. This trust is a trust in the other's caring and support, and not the self-interest-based predictability of others' actions (Molm *et al.*, 2009). Romantic love is conceptualized as a synthesis of prototypical "caring" love and noncrude attributes of sex (Manoharan & de Munck, 2015).

Romantic Love as Sexual Attraction

Romantic love is associated with passion. The major difference between romantic love and other kinds of love is the element of **eroticism**—sexual desire. Romantic love is often hormonally driven and can be motivated by fulfillment of one's own needs and desires (Underwood, 2009). When a person is romantically in love with another, it means "I am sexually attracted to you and can't get you out of my mind; You excite me, 'turn me on,' make me feel alive, complete my sense of wholeness; I want to see you, devour you, touch you, merge with you, lose myself in you" (Bowlby, 1969). The concept of "chemistry" as it applies to relationships is one that is widely recognized about romance (Fowler, 2007). Neurological studies have been able to describe the actual chemical processes that occur in the human body when a person is experiencing love. It is *phenylethylamine (PEA)*—an amphetamine-related compound that produces the mood—lifting and energizing the effects of romantic love (Liebowitz, 1983). Those experiencing romantic love are experiencing increased phenylethylamine (Levy, 2013). The crash that follows a break-up is much like amphetamine withdrawal.

Adolescent Love

Passionate love is fueled by adolescent hormonal changes and thus necessarily appears after puberty (Gadpaille, 1975). **Adolescent love** is characterized by the presence of physical attraction and passion. There are three stages of adolescence: early adolescence that ranges from twelve to fourteen years, middle adolescence from fifteen to seventeen years, and eighteen and over is classed as late adolescence (Woollaston, 2013). Romantic love and adolescence are described as intense, overwhelming, passionate, consuming, exciting, and confusing (Kephart, 1967). Adolescents usually find relationships with peers in high schools or colleges. They generally fall in love with a willing partner for the first time (Farber, 1980). Psychologist Maslow (1955) suggests that "*love hunger*" is a deficiency disease like salt hunger. Rubin (1970) argues that the attitudes and behaviors of romantic love may differ, depending on whether the most salient rewards exchanged are those of security or those of stimulation. As an individual grows from a child to a teen, **sexual love**—a person's strong passionate feelings—becomes more relevant in their life (Milligan, 2011). Sexual love is not love at first sight—it is basic human instinct and hormonal responses (Singer, 2009).

Expression of Romantic Love

Romantic love is often demonstrated through behavioral synchrony between partners—hugging, kissing, lifting each other, scuffling, sitting close to each other, confessing their love to each other and to others, talking about each other when apart, seeking each other out and excluding others, grief at being separated, giving of gifts, extending courtesies withheld from others, making sacrifices for each other, and feeling jealous (Carlson & Hatfield, 1992). Flushing or blushing occurs in the presence of the loved one, as well as dilation of the pupils and an increase in tear production that causes the eyes to glisten (Morris, 1971). With romantic love, the most passionate feelings are usually at the beginning of the relationship. Over time, these feelings may fade. Later in the chapter, we'll discuss how partners can keep romantic relationships alive.

Source: Rich Lam via Getty Images

Companionate Love

Companionate love is often the product of romantic relationship and over time develops into a relationship that involves less passion and more intimacy, commitment, and caregiving. **Companionate love** is a type of love found between two partners and characterized by the presence of intimacy, commitment, and caregiving. Research on companionate love focuses primarily on nonromantic contexts (e.g., family and friends, strangers, and even all of humanity) (Fehr *et al*., 2009). Companionate love is observed in long-term marriages where passion is no longer present (Ashford & Le Croy, 2009).

Companionate love is a type of long-term love and is known as *affectionate love*. Couples who have been married a long time often feel companionate love for one another. Love is ultimately not a feeling at all, but rather is a commitment to, and adherence to, loving actions towards another, oneself, or many others, over a sustained duration (Fromm, 2006). Emotional attachments to others and affective commitments (e.g. desires, attitudes, values, beliefs) influence a significant portion of human behavior (Etzioni, 1988; Hochschild, 1975). Through meta-analytic factor analysis, Graham (2010) finds that love is positively associated with relationship satisfaction and length. When a couple reaches this level of love, they feel committed to each other and care for each other.

Compassionate Love

True love is caring about the happiness of another person without requiring anything in return. Love can be a virtue representing *compassion*—the unselfish loyal and benevolent concern for the good of others (Aristotle, 2003). **Compassionate love** is altruistic love characterized by the presence of intimacy, commitment, and altruistic caregiving. **Altruism** or selflessness is the practice of concern for the welfare of others. French sociologist Auguste Comte coined the term *altruisme* for the opposite of *egoism*—an excessive sense of self-importance (Teske, 2009). Compassionate love refers to a collection of attitudes, cognitions, emotions, and actions related to selfless concern and giving of oneself for the well-being of others (Underwood, 2009). In Greek, *agape* (pronounced a-gah-pay) is altruistic and self-sacrificing love with the fulfillment of others' needs. Both compassionate love and agape love styles are focused on giving selflessly to benefit the well-being of another. Compassionate love is like unconditional love (Miller *et al.*, 2015) and plays an important role in the quality and stability of relationships (Fehr *et al.*, 2014). Hispanics (47 percent) and Blacks (32 percent) are more likely than Whites (24 percent) to say that every person has only one true love (Pew Research Center, 2011).

Social Support and Sacrifice

Underwood (2009) and Berscheid (2006) have developed models that delineate behaviors associated with compassionate love (e.g., social support and sacrifice). Scores on the Compassionate Love Scale (CLS) that assess the provision of social support (e.g., practical and emotional support) are strongly correlated (in the .50s) with scores completed with respect to close others in general (Sprecher & Fehr, 2005; Fehr *et al.*, 2014). Fehr and Harasymchuk (2013) reported a similar correlation between CLS scores and social support received from one's dating partner.

Given that women tend to be nurturers (Taylor, 2006), when the CLS is administered with respect to close others (family and friends) or strangers/humanity, women generally score higher than men (Sprecher & Fehr, 2005; Fehr & Sprecher, 2013). People who love their partner compassionately are expected to be willing to make more sacrifices for him or her than those who love their partner less compassionately. In empirical investigations, compassionate love is strongly associated with making sacrifices for a dating partner (Fehr & Harasymchuk, 2013). The religious or cultural socialization of an individual will affect the likelihood of being compassionate; motivations that focus on self-gain will impede the expression of compassionate love (Fehr *et al.*, 2014).

Altruistic Caregiving

Compassionate love is equivalent to altruistic or unlimited love with the notion of selfless love. Bioethicist Stephen Post (2003) placed greater emphasis on the extension of love to all human beings and views unlimited love as a subtype of compassionate love.

Compassionate love is associated with caregiving. It may describe compassionate and affectionate actions towards other humans, one's self or animals (Fromm, 2006). Fehr *et al.* (2010) found that correlations between compassionate love and the caregiving subscales were 0.37 for cooperation, 0.50 for sensitivity, 0.64 for proximity, and 0.18 for compulsive caregiving with scores on CLS. In their qualitative analysis of compassionate love and caregiving, Roberts *et al.* (2009) found that compassionate love took the form of providing physical and emotional care, healing and forgiving past transgressions, and letting go of the other. Mothers' compassionate love for their children supported positive parenting and buffered against adverse parenting under stress (Miller *et al.*, 2015). Fostering compassionate love may help mothers, and particularly those who experience strong physiological arousal during difficult parenting situations, to establish positive socialization contexts for their children (Miller *et al.*, 2015). AD (Alzheimer's Disease) individuals' and caregivers' compassionate love were associated with less burden and positive appraisals of caregiving (Monin *et al.*, 2015).

Compassionate love focuses on the other and brings benefits to people around the world. You often see examples of compassionate love following natural disasters, where people demonstrate love and compassion for strangers in often inspiring ways. Research with healthy young adults shows that compassionate goals and feeling compassionate love in relationships enhances positive emotions (Canevello & Crocker, 2010).

Diversity Overview: Expressions of Love by Gender

Have you ever heard of the book *Men Are from Mars, Women Are from Venus*? Written by John Gray (1994), the book contends that women and men are different biologically and emotionally, which is reflected in how they choose their partners and behave in their relationships. Ubando (2016) found that there was a slightly negative correlation between verbal emotional expression and relationship satisfaction in men. Since gender communication focuses on the way gendered beings communicate, let's look at some ways in which gender plays a role in love relationships.

The Feminization of Love

Many people believe that women are more skilled than men at expressing their feelings. Women tend to specialize in the "social-emotional" aspects of interaction (Strodtbeck & Mann, 1956). Intimate communication may be experienced differently by men and women. How do women express their feelings?

Emotional intimacy can be expressed in *verbal* communication. Women are more likely to express how much they love (Cancian, 1990) and to express love with words. The **feminization of love** means that people who express how much they love are women. Boys' and girls' love understanding was associated with maternal emotional expressiveness in different ways. Girls understand emotion better than boys and show better understanding of the complex emotion (Bosacki & Moore, 2004). The results of Brown and Dunn (1992) show that girls scored higher than boys did in decoding and explaining emotions. Women usually express their love through words—talking about their emotions and feelings (Know *et al.*, 2002). Women express emotions, share personal feelings, relate stories, and listen empathetically whereas men engage in competitive joking and assertive speech to win control of conversation; women tend to use talk to create connections whereas men tend to use it to emphasize status (Tannen, 2001). Women are also more likely to expect their partners to express their love with words. Ubando (2016) has explored how emotional expressivity affects perceived relationship satisfaction in undergraduate couples from ages 18 to 24. Women reported that they felt they shared more personal information with their partners yet they were less trusting of and comfortable with their partners than men.

Gender Differences in Self-disclosure

Research consistently indicates that women disclose more than men. Men have a greater need to control their privacy (Petronio *et al.*, 1984). Women self-disclose more than men in all types of friendship (Sheldon, 2013). Men and women use different criteria for deciding to open or close their boundaries. For example, women tend to talk about intimate or personal topics with each other and prefer disclosing to same-sex friends while men prefer to disclose while engaging in some activity (Caldwell & Peplau, 1982). When the target was friend, parent, or spouse, women disclosed more than men and when the target was a stranger, men reported that they disclosed similarly to women (Dindia & Allen, 1992).

Self-disclosure between Online and Offline Relationships

Today, social network users tend to rely more on self-disclosure to create a sense of closeness because they do not receive verbal and nonverbal cues that would be otherwise exchanged in face-to-face communications (Schafer, 2015). Women still disclose to their exclusive face-to-face and exclusive Facebook friends more than men (Sheldon, 2013). Since offline relationships are characterized by higher interdependence, and greater breadth and depth of self-disclosure (Cummings *et al.*, 2000), both men and women disclosed more to their exclusive face-to-face friend than exclusive Facebook friend (Sheldon, 2013).

Gender Differences Nonverbally

Emotional intimacy can be expressed in *nonverbal* communication. Women tend to make more eye contact than men in same-sex groups (Exline, 1963). In a study of four experimental groups, Rubin (1970) found that the women spent much more time looking at the men than the men did the women; *gazing* may serve as a vehicle of emotional expression for women and allow women to obtain cues from their male partners concerning the appropriateness of their behavior.

Women and men see gift-giving in different ways. Women are more likely than men to consider gifts as tangible proof of love. Christenson (2003) found that 62 percent of men say that "just spending time together" would be an ideal Valentine's Day celebration, whereas 53 percent of women would break up with someone who didn't give them a gift on Valentine's Day.

Men are more likely to express love through actions rather than words. Men are likely to express their feelings through sexual intimacy and emphasize behaviors such as sexual activities (Cancian, 1990). Meanwhile, men can be romantic but not see love as necessarily leading to marriage. Women sometimes belittle men for being "commitment dodgers," "commitment phobics," "paranoid about commitment," and "afraid of the M word" (Crittenden, 1999). Men self-reported higher for verbal and nonverbal affection while women self-reported higher for supportiveness (Ubando, 2016).

When it comes to expression of emotions, women report negative feelings, such as sadness and anxiety, more than men; men reported more positive feelings such as excitement than women (Simon & Nath, 2004). A new research study suggests that women are not universally more expressive across all facial actions. Women express more happiness and sadness while men express more anger. The pattern of findings was consistent across the five countries (US, Germany, UK, China, and France) in the study (McDuff *et al.*, 2017).

Gender Differences in the Process of Socialization

These gender differences can be attributed to the socialization process through which males and females have adopted different roles. For example, Macoby (1990) catalogued some of the interpersonal behaviors that men may learn through socialization: competitiveness, assertiveness, autonomy, self-confidence, instrumentality, and the tendency to not express intimate feelings. Noller (1993) describes some of the behaviors that women may learn through socialization: nurturance, emotional expressivity, verbal exploration of emotions, and warmth. These ideas have become part of our *stock of knowledge*—the "common-sense" facts given to us by family, friends, school, and other socializing agents (Farganis, 2008). Therefore, men may experience intimacy through shared activities and women experience intimacy through verbal self-disclosure and shared affect (Markman & Kraft, 1989).

Upon completion of the next section, students should be able to:
LO5: Explain sociological theories of love.
LO6: Compare Lee's color wheel theory of love with Sternberg's triangular theory of love.

Love from a Sociological Standpoint

Sociology is interested in the study of love because love has consequential importance in shaping individual lives and society; love can be shaped by society. Sociological theories have different explanations of how people fall in love. The theoretical conceptualizations of love are all embedded in wider theoretical constructions set up to account for the modernization process (Rusu, 2017).

Conflict Theory on Love

Conflict theory argues that love is shaped by social relationships; the capacity for love is not equally distributed. According to Karl Marx and Frederick Engels, the history of society is a history of class struggle. Both Marx and Engels appear to regard marriage as being an expression of such a class struggle (Fowler, 2007). For example, gender inequality has always been a global issue. Women must participate in public production and must fulfill their duties at home; but if they remain at home they are deprived of the possibility of earning their own living independent of men (Engels, 1988). Love is shaped and produced by concrete social relations and circulates in a marketplace of unequal competing actors (Illouz, 2013). For example, unequitable love and jealousy occur because partners with money and power are likely to dominate and control love.

Conflict theory explores how the economy of love privileges some types of love relationships. The capacity for love is not equally distributed and some people are fortunate enough to have a gift for love (Hacker, 2017). Society plays a big role in the way we see love based on social differences such as gender, race, economic status, religion, education, and ethnicity (Ridgeway, 2009). For example, richer people marry those with similar levels of wealth and income (Choi & Mare, 2012). Marriages that unite two people from different class backgrounds might seem to be more egalitarian but there are limitations to cross-class marriages. For example, couples often overlook class-based differences in beliefs, attitudes, and practices until they begin to cause conflict and tension (Streib, 2015).

Structural Functionalism on Love

Structural functionalists argue that love serves as a binding force of love relationships in society. The central tenet of structural functionalism is that society is made up of interrelated parts that work together and contribute to the stability of society. When love binds love relationships and performs functions, society is likely to be stable. Structural functionalists believe that love provides a foundation on which relationships are built. In a traditional society, the family performed multiple functions, such as creating an economic unit. In dual-earner families today, both husband and wife earn wages and raise the family. Parsons (1943) views love as a binding force whose social function is to "integrate the conjugal couple of the modern nuclear family in the absence of the external pressures exerted by the kinship network" (Rusu, 2017).

Love can function to meet people's needs and represent the path to extreme individualization (Beck & Beck-Gernsheim, 1995). For example, love played a part in women's emancipation as the rise of romantic love and its link with marriage gave women more choice in their future partners (Carter, 2015). With growing choice and freedom, love has become "confluent" and temporary subject to individuals' needs (Giddens, 1992). Romantic partners act as people's primary attachment figures (Doherty & Feeney, 2004) and play a vital role in meeting people's most fundamental socioeconomic needs (Fitzgerald, 2017).

Symbolic Interactionism on Love

The central tenet of symbolic interactionism is that symbols such as languages and meanings of the symbols are essential to comprehending how people communicate with one another. Newcomb (1960) assumes that love is an attitude held by a person toward another person, involving predispositions to think, feel, and behave in certain ways towards that person. This assumption means that love is an interpersonal attraction. People are likely attracted to those of the same race/ethnicity, religion, culture, or social class. **Homophily** is the tendency of individuals to socialize and bond with people of similar backgrounds. For a couple absolutely in love, love comes to symbolize romantic love. The couple expects love to supply long-term emotional highs. Carter (2013) explored the discourses and languages that women used to explain love and found that many women had difficulties talking about their feelings and love. Indeed, when a couple "drafted" into and out of a relationship, love was simultaneously loudly absent and quietly present.

Symbolic structuralists view love as primarily dependent on the definition of the situation, which varies across time and location. From a historical and psychological viewpoint, we can understand love only in terms of cultural conceptions of (a) the beloved, (b) the feelings that accompany love, (c) the thoughts that accompany love, and (d) the actions, or the relations one has with the beloved (Beall & Sternberg, 1995). Indeed, the definition and the emotional experience of love are contextually and culturally bound. A person's definition of a situation is determined not only by the physical space which they occupy but also by their status and role in society (Fowler, 2007). Although couples have similar backgrounds, the definitions and interpretations of love still vary from person to person. For instance, when it comes to her wedding, Angela wants to invite her whole family and close friends but her fiancé only wants to invite his parents. Angela does not understand the way that her fiancée defines the wedding and love because the way people express and define love is the product of socialization. Agents of socialization include parents, school, peers, and the media.

The Color Wheel Theory of Love

Canadian sociologist John Alan Lee (1973) explains how people fall in love and describes different love styles in terms of the traditional color wheel. The three primary types are *eros, ludus*, and *storge*, and the three secondary types are *mania, pragma*, and *agape*. If you copy URL https://en.m.wikipedia.org/wiki/Color_wheel_theory_of_love#/media/File%3AColour_Wheel_of_Love.jpeg into your browser, you can see the changes in color shown in the image on the next page.

Primary Types of Love

Eros is romantic love. Lee (1973) describes eros as a passionate and emotional love of wanting to satisfy, create sexual contentment, security, and aesthetic enjoyment for each other. Erotic lovers are more likely to say they fall in love at first sight than those of other love styles. **Ludus** is game-playing love or playful love. Ludus is uncommitted love characterized by the enjoyment of many sexual partners at one time without deep commitment in the early stages of a relationship. The acquisition of love and attention itself may be part of the game (Lee, 1973). Ludus is relevant to all stages of relationship development; ludic attitudes are associated with absence of concern for partner loyalty, short and uncommitted relationships, and positive feelings about relationship dissolution (Hammock & Richardson, 2011). **Storge** is companionate love and friendship love. Storge is the love between companions. This kind of love involves commitment. There is a love between siblings, spouses, cousins, parents, and children. Storge also grows slowly out of friendship and is based more on similar interests with no passion.

Secondary Types of Love

Mania (pronounced may-nee-uh) is obsessive love and possessive love with involvement of strong emotional intensity, extreme jealousy, and dependency. Mania comes from the Greek word for "madness." This kind of love is equivalent to roller-coaster manic love. It is represented by the color purple, as it is a mix between ludus and eros. Manic people typically are not attractive to individuals who have strong self-concept and high self-esteem (Lasswell & Lasswell, 1976).

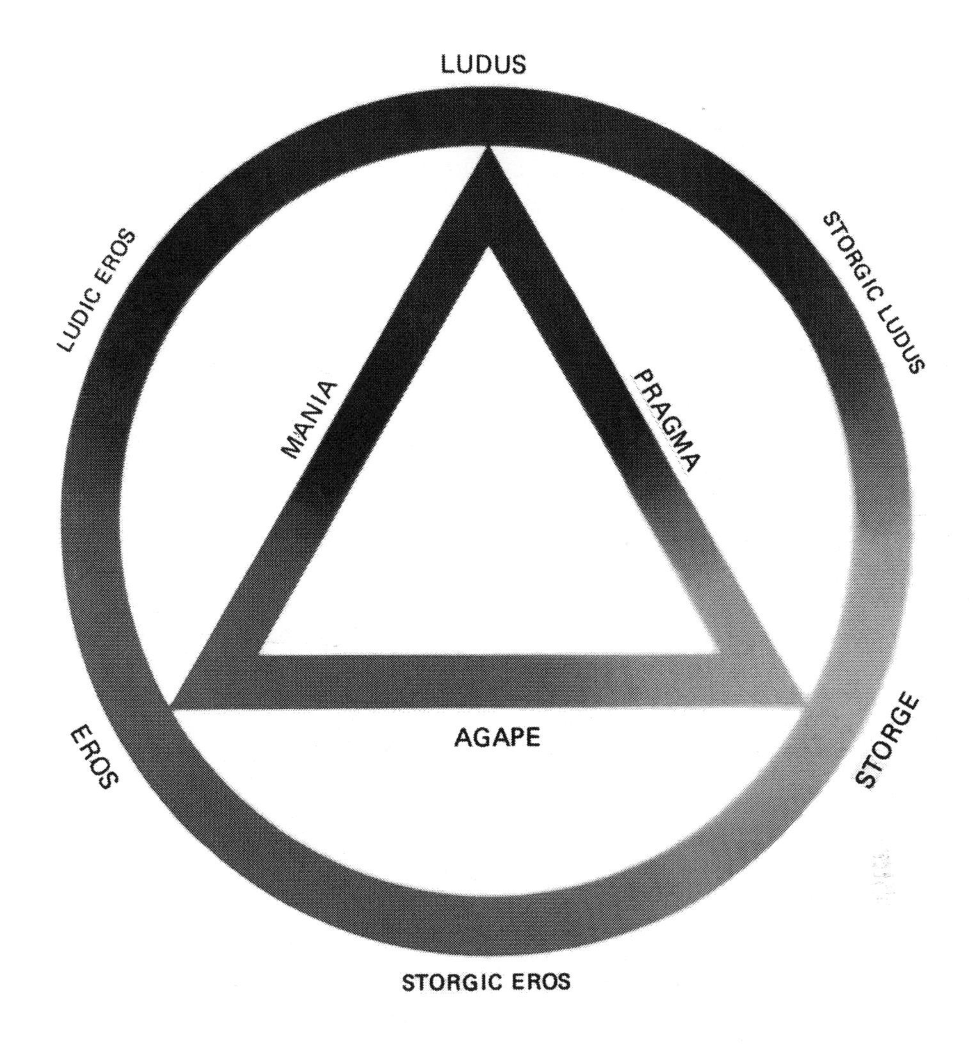

THE COLOUR WHEEL OF LOVE

Source: Kaitlindzurenko, Color wheel as according to John Alan Lee. https://upload.wikimedia.org/wikipedia/commons/f/f7/Colour_Wheel_of_Love.jpeg

Pragma (pronounced prag-ma) is practical love. Pragma is defined as a convenient and realistic type of love not necessarily derived out of true romantic love. Pragmatic lovers are pragmatic as they are lovers who are logical in their approach of looking for someone who meets their needs (Lee, 1973). Pragma is associated primarily with relationship initiation (i.e., selection of an appropriate partner) (Hammock & Richardson, 2011). **Agape** is altruistic love and unselfish love. Agape is the traditional Christian love that is chaste, patient, and undemanding (Lee, 1973). It is based on an unbreakable commitment and an unconditional, selfless love. Agape love is signified by the color orange.

Rohmann *et al.* (2016) examined the connection of Lee's six love styles with relationship satisfaction across cultures, including in Bosnia, Germany, Romania, Russia, and German immigrants from Russia and Turkey. Results show that the effects of love styles on relationship satisfaction were consistent across cultures. However, game-playing love negatively predicted relationship satisfaction for Germans, Romanians, and Turkish migrants but not for Bosnians, Russians, and Russian migrants.

Sternberg's Triangular Theory of Love

Psychologist Robert Sternberg's (1988) **triangular theory of love** suggests that partners express their love depending on combinations of intimacy, passion, and commitment (see image below). Love triangles emanate from stories, and almost all of us are exposed to large numbers of diverse stories that convey different conceptions of how love can be understood (Sternberg, 1995). In Sternberg's theory, **nonlove** is the absence of intimacy, passion, and commitment. This applies to our acquaintances or someone we are not attached with. Liking is a form of love that involves intimacy without passion or commitment. In the questionnaire study, Rubin (1970) found that respondents' estimates of the likelihood that they would marry their partners were more highly related to their love than to their liking for their partners.

Infatuated love involves passion but no intimacy and commitment. People with only sexual love manifest this category. This is considered "puppy love" or relationships that have not become serious yet (Rothwell, 2016). Without developing intimacy or commitment, infatuated love may disappear. **Empty love** is a form of love that involves only commitment without intimacy or passion. An example of empty love is an unhappy marriage characterized by the absence of passion and intimacy. In an arranged marriage, the spouses' relationship may begin as empty love and develop into another form (Sternberg & Hojjat, 1997). Romantic love involves passion and intimacy but no commitment. Romantic relationships often start out as infatuated love and become romantic love as intimacy develops over time. Companionate love involves intimacy and commitment but no passion. **Fatuous love** involves passion and commitment without intimacy. Finally, **consummate love** involves all components of intimacy, passion, and commitment (see image on p.288). Consummate love sits at the center of the triangle because it is an ideal type of love. However, consummate love may not be permanent. For example, if passion is lost over time, it may change into companionate love. Table 7.1 shows these eight combinations of love relationships.

People show their love differently. Results indicated mixed support for the triangular theory (Acker & Davis, 1992). For example, the predicted decline over time in passion emerged only for females; intimacy levels did not generally display the predicted decline for longer relationships; commitment was the most powerful and consistent predictor of relationship satisfaction, especially for the longest relationships.

Functions of Love

Love is such a powerful concept to understand whether it is an expression of emotion, or link with marriage, or societal glue. Structural functionalist theory suggests that love performs certain functions and

Upon completion of this section, students should be able to:
LO7: Explain the functions of love.

Table 7.1 Sternberg's Eight Combinations of Love

None of intimacy, passion and commitment	**Non-love:** A couple remains in a relationship without intimacy, passion, and commitment.
Intimacy only	**Liking/friendship:** A couple feels comfortable in a relationship of good friends.
Passion only	**Infatuated love:** A couple falls into love on first sight with intense emotional involvement.
Commitment only	**Empty Love:** A couple remains in a relationship with little or no passion.
Intimacy and passion	**Romantic love:** A couple develops into romantic love.
Intimacy and commitment	**Companionate love:** A couple remains in a long-term relationship with the same intensity in intimacy, and commitment, but with little or no passion.
Passion and commitment	**Fatuous Love:** A couple moves from meeting to marriage with emotional response.
Intimacy, passion, commitment	**Consummate love:** A couple remains in a relationship with the same intensity in intimacy, passion, commitment.

Combinations of intimacy, passion, commitment

	Intimacy	*Passion*	*Commitment*
Non-love			
Liking/friendship	x		
Infatuated love		x	
Empty love			x
Romantic love	x	x	
Companionate love	x		x
Fatuous love		x	x
Consummate love	x	x	x

Source: Sternberg, R. J. (1986) A triangular theory of love. *Psychological Review*, 93(2), 119–135. https://en.wikipedia.org/wiki/Triangular_theory_of_love

contributes to the smooth functioning of society. Love ensures human survival and enhances our physical and emotional health. Love allows people to attribute a sense of purpose for living (Määttä, 2013).

Love as the Foundation for Interpersonal Relationships

Love is experienced in the individual social interactions of everyday life. Lazarus (1991) acknowledges that "love commonly means a social relationship rather than an emotional process." He argues that when "love" means an emotion, it is a "momentary state, a reaction that comes and goes like anger, guilt, shame, and jealousy." Love is one of the processes through which human beings become attracted to one another (Fowler, 2007).

Love is first produced in the primary world of intimate relations where there is strong attachment,

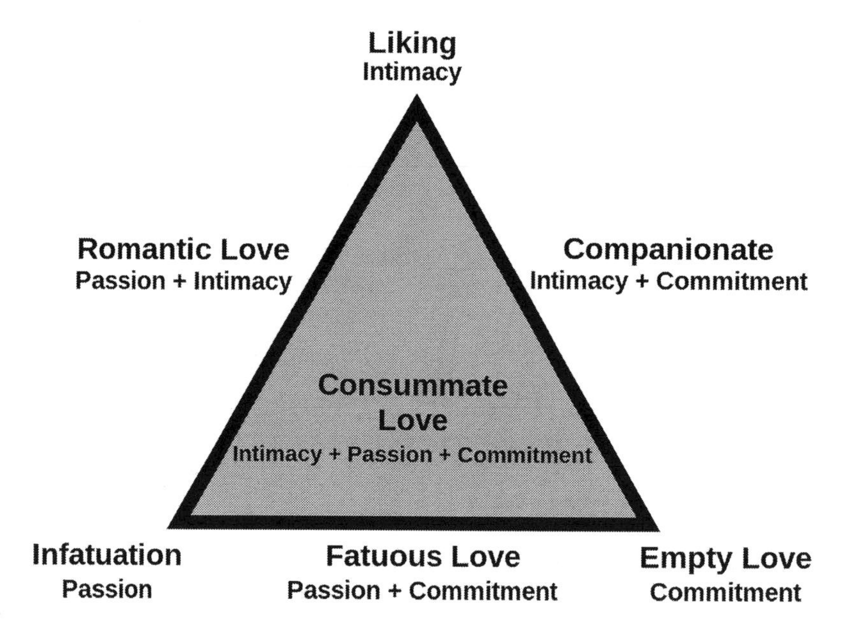

Partners Express their Love Depending on Combinations of Intimacy, Passion, and Commitment.
Source: https://commons.wikimedia.org/wiki/File:Triangular_Theory_of_Love.svg

interdependence, deep engagement, and intensity (Gürtler, 2005). From the moment of birth, relationships are made between mother and child, father and child, grandparent and child, and sibling and child. As the child grows older, enters schools, goes to work, and gets married, the number of relationships such as friendships has grown. These relationships are maintained by love. For a person in love, life is never without meaning (Singer, 2009).

Interpersonal love is closely associated with interpersonal relationships (Fromm, 1959). Love acts as a facilitator of interpersonal relationships in different forms. Such love exists between family members, friends, and couples. Without love, a couple lacks a reason to continue a relationship, and social stability would therefore be at risk. The association between love and marriage gives it a unique status as a link between the individual and the structure of society (Rubin, 1970). Love was positively and obsession was negatively associated with relationship satisfaction and length (Graham, 2010). Also love may act as a function to keep human beings together against menaces. Love is essential to life and serves as a binding force that drives the family and partners to stick together and overcome difficulties during periods of family crisis.

Love as a Theme in the Search for a Mate

In traditional societies, individual happiness was subordinate to the well-being of the family, and arranged marriage was the norm. Today, love has become a

Student Accounts of the Functions of Love

Love is the ultimate high that makes us at peace with everything around us. It brings out the best of us; it purifies our souls; it puts a smile on our face; it makes us laugh; it makes us vulnerable yet full of trust; it makes us forgive and forget; it is an essential food to all (personal communication, October 28, 2017).

central theme in the search for a mate. Love has been identified by relationship scientists as a major force in the development of romantic relationships (Fehr *et al.*, 2014). Romantic love is regarded as a universal human phenomenon and serves as an honest signal of one's deep emotional involvement, thus influencing mate-choice (Schiefenhövel, 2009).

Marriage or cohabitation can happen when two people fall in love and decide to share their life together. Indeed, love remains Americans' top reason to marry, with 88 percent of Americans citing love as a very important reason to get married, ahead of making a lifelong commitment (81 percent) and companionship (76 percent) (Livingston & Caumont, 2017).

Love as Essential for Human Survival

Sociologists contend that love is essential to the survival of human beings. Sorokin (2002) affirms that love conquers all. Loving someone and being loved ensures the survival of our species. Love brings together people such as couples, parents, children, and friends, and shapes our existence. Parental love assures the survival of the human species.

Compassionate love brings benefits to people around the globe. Love offers a unique opportunity; it is a path to salvation (Jackson, 1993). As a humanitarian organization, the Tzu Chi Foundation (2017) strives to expand the love we all have for our families to society at large. Since 1966, the foundation has provided international relief work in 70 countries to people suffering from disasters, believing firmly that love is the solution to human problems. In addition to hurricanes and wildfires, in 2017 the Red Cross mobilized more than 56,000 disaster workers to provide help to people affected by 242 significant disasters in 45 states and three territories (American Red Cross, 2017).

Love as a Provider of Health Benefits

We are informed that love will insure long life, health (both mental and physical), security, appreciation, and power, and bring about the regeneration of civilization (Sorokin, 2002).

Love can serve as a buffer against stress and preserve our physical and psychological well-being. Research indicates that love plays an important role in mental health treatment. For example, the continuum of love involves autistic love (all about me), empathetic love (all about you), and integral love (all about us); the cycle of love manifests as the three pillars of recovery: awareness, acceptance, and integration (Sleeth, 2011). Our social relationships—both in terms of quantity and quality—affect our mental health, health behavior, physical health, and mortality risk (Umberson & Karas Montez, 2010). **Social isolation** refers to the absence of social relationships.

Love is an important predictor of happiness, satisfaction, and positive emotions (Diener & Lucas, 2000). Love can produce the feeling of happiness. Kissing, sex, emotional closeness, and companionship found in love relationships contribute to happiness (Ross *et al.*, 1990). The feelings love brings—happiness, empathy, mutual respect, a sense of purpose—can lead to stronger motivation, less stress, a positive outlook on life, and hope (Määttä, 2013). Research on compassionate love investigates health effects from engaging in compassionate love and suggests primarily positive evidence (Post, 2007). Harvard's 75-year-long study of adult life suggests that our relationships

Upon completion of the next section, students should be able to:
LO8: Describe problems of experiencing love.
LO9: Identify love problem causes and solutions.

and how happy we are in our relationships has a powerful influence on our health (Mineo, 2017).

Problems of Experiencing Love

Sociologists focus not just on the functions of love in society but also on the potential problems it may cause, including jealousy, controlling behaviors, and infidelity. This section discusses the negative issues associated with love.

Jealousy

When two groups in a cross-cultural study were asked about negative features of love, both American and Chinese subjects listed similar problems, including pain, jealousy, unrequited love, separation, loss, betrayal/desertion (Wu & Shaver, 1993). People in love exhibit signs of *emotional dependency* on their relationship, including possessiveness, jealousy, fear of rejection, and separation anxiety. **Jealousy** is an emotional response to a real or perceived rival who is believed to be a threat to a partner. Jealousy is most likely to occur when a partner believes that a rival is competing for a desired partner. Jealousy varies across cultures, and the socialization process can influence the situations that trigger jealousy and the way in which jealousy is expressed.

Causes and Effects of Jealousy

Insecurity about a relationship can be related to jealousy. Jealousy is usually an unhealthy manifestation of insecurity, low self-confidence, and possessiveness (Buunk & Dijkstra, 2001). People who feel insecure about relationships are afraid of losing the relationship and are likely to become jealous lovers. The greater your insecurity about the relationship, the greater the chances you will feel jealousy (Marelich *et al.*, 2003). Jealousy can lead to controlling or otherwise negative behaviors. For example, people might engage in various types of abuse (verbal, physical, and sexual) when they are jealous.

People with low self-esteem are likely to be jealous. Some people are even jealous when their partner spends time with family members or relatives or in pursuing hobbies (Hanna, 2003). Love is expected to be reciprocal and mutual. If a partner tries to control love, a love relationship is more likely to diminish. Sometimes, to test the relationship, some people attempt to make their partners jealous.

Jealousy is emotionally and physically damaging. It can destroy a relationship if it leads to suspicion and other negative behaviors. When jealousy leads to behaviors such as verbal, physical, and sexual abuse, love relationships are likely to diminish or end. If people feel jealous in a relationship, it is important that they learn how to deal with it. For example, the persons experiencing jealousy can discuss their feelings with their partners and attempt to behave more positively in the love relationship.

Stalking

Taken to the extreme, jealous people may engage in stalking. **Stalking** refers to controlling or obsessive behaviors that intrude on a person's privacy or personal space and cause the victim to fear for personal safety (including the safety of another person) or suffer mental distress. Stalking behaviors include threatening the victim, intruding on the victim's property, or using any method or device to monitor or follow the victim. Estimates suggest that 19.3 million women and 5.1 million men have been stalked in their lifetime in the United States (Black *et al.*, 2011). California was the first state in the United States that passed an antistalking law in 1990. Since the mid-1990s, all 50 states have enacted similar legislation.

Tips for Keeping the Stalker Away

Stalking is not just a violation of privacy; it is also a crime. If someone is being stalked on campus or is in immediate danger, the victim of the stalking should act immediately. Some tips include contacting the police or campus authorities. The victim should let friends and family know about the situation. The victim could get a court order to keep the stalker away. The victims of stalking should save all communications and record when and where each contact occurred. The victim can try to devise a safety plan and consider ways to change their driving route home or to class, and/or walk with a group of friends.

Unrequited Love

Most people who give love would like it to be reciprocated. However, unrequited love is not. **Unrequited love** (UL) refers to a relationship in which one person loves another who does not reciprocate the romantic affection. Is unrequited love a type of romantic love? Romantic love involves equal love. Both women and men tend to link love and sex in their romantic relationships (Hendrick & Hendrick, 1992). Unrequited love is not a good simulation of true romantic love, but an inferior approximation of that ideal (Bringle *et al.*, 2013).

There are many reasons for unrequited love. Baumeister *et al.* (1993) described several ways in which UL can develop (e.g., growing out of friendship, loving from afar). UL has been found to be more prevalent among individuals who reported an anxious/ambivalent attachment style (Aron *et al.*, 1998) and who were low on defensiveness (Dion & Dion, 1975). There are often large disparities between the two persons in their desirability and the desirability of their alternatives, decreasing the prospect of a reciprocal, romantic relationship (Berscheid & Reis, 1998).

Bringle *et al.* (2013) delineate five types of UL on a continuum from lower to greater levels of interdependence: crush on someone unavailable, crush on someone nearby, pursuing a love object, longing for a past lover, and an unequal love relationship. For example, someone can have a crush on a love object who exists, but who is seldom proximal (e.g., a rock star, a movie star). All types of UL relationships were less emotionally intense than equal love and four times more frequent than equal love during a two-year period; all types of UL were less intense than equal love on passion, sacrifice, dependency, commitment, and practical love, but more intense than equal love on turmoil (Bringle *et al.*, 2013).

Unrequited love can cause pain and heartache to those who love and are not loved in return. Anderson (2014) found that the grief of being spurned in a romantic relationship can create an *abandonment* trauma powerful enough to implant an emotional drain deep within the self. If left unresolved, it leeches self-esteem and creates self-sabotage (McAdams *et al.*, 2009). The grief from unrequited love is different from other types of bereavement in its ability to leave residual damage to a person's self-esteem (Anderson, 2014). If love goes unrequited for those who want to receive love and if the unrequited love is upsetting, the best solution is to let go of the unrequited love and develop relationships with people who return love.

Lost Love

Loss of love can cause the feeling of pain or an acute sensation in the stomach. Losing a love due to a break-up can be a devastating experience. For a new study nationally, Knopp *et al.* (2017) found that the average duration of the first relationships was 38.8 months while the average duration of the second was 29.6 months. In fact, 40 percent of people who had been broken up with by their partner in the previous eight weeks experienced clinical depression and 12 percent experienced severe depression (McManamy, 2011). Even more startling, 50 to 70 percent of female homicides are committed by lovers and spouses, often following a break-up.

Broken Heart

How does it feel to have a broken heart? **Broken heart** is a metaphor for the intense emotional pain one feels when experiencing great longing. People with broken heart syndrome may have sudden chest pain and exhibit different levels of emotions to love or broken relationships. Research indicates that physical pain, such as a broken heart, and social pain overlap in their underlying neural circuitry and computational processes. Indeed, the anterior cingulate cortex plays a key role in the physical–social pain overlap (Eisenberger & Lieberman, 2004). Functional magnetic resonance imaging (fMRI) was used to study women and men who had recently been rejected by a partner and reported they were still intensely "in love" (Fisher *et al.*, 2010). The severely stressful life event of losing a partner was followed by a transiently increased risk of atrial fibrillation lasting for one year, especially for the least predicted losses (Graff *et al.*, 2016).

A broken heart is a major stressor and has been found to precipitate episodes of major depression. In one study (death of a spouse), 24 percent of mourners were depressed at two months, 23 percent at seven months, 16 percent at thirteen months and 14 percent at 25 months (Zisook & Shuchter, 1991). Romantic rejection causes a profound sense of loss and negative affect; and it can induce clinical depression and in extreme cases lead to suicide and/or homicide (Fisher *et al.*, 2010). For example, Millennials spend their 20s (and sometimes 30s) in pointless dating, uncertain relationships, and painful break-ups; this "cycle of meeting someone, falling in love, and breaking up is a formula for anxiety and depression" (Twenge, 2014).

It is important to know that the pain of loss is a natural part of the human existence and broken heart syndrome is usually treatable, and most people make a full recovery. To stay healthy, it's important to find ways to reduce stress and cope with particularly upsetting situations. Learning how to relax and cope with problems can improve your emotional and physical health (NHLBI, 2014).

Infidelity and ESI

An overwhelming majority of people have the expectation of fidelity of sexual and, often, emotional connection in monogamous relationships (Stanley, 2017). With the monogamous sexual couple at the center of personal life, infidelity is regarded as a threat, revealing wider limitations to claims about the extent to which relationships have been detraditionalized (van Hooff, 2017). **Infidelity** refers to unfaithful behavior on the part of a spouse in a married relationship. If a couple is in a committed or exclusive relationship, infidelity can be considered a betrayal and may cause serious pain. Love is an important part of life, but finding a lasting love relationship can be a challenge. Extra-dyadic sexual involvement (ESI) refers to having sexual relations with someone else in unmarried, serious, and romantic relationships. Forty-four percent of the national sample reported having had sex with someone other than their present partner in one or both relationships studied (Knopp *et al.*, 2017). If unmarried couples don't declare a relationship exclusive and if there was no implicit promise of sexual exclusivity, there might be no violation.

Serial infidelity refers to the practice of engaging in a succession of sexual relationships with someone other than your partner. Knopp *et al.* (2017) found that those who reported engaging in ESI in the first relationship were three times more likely to report engaging in ESI in their next relationship compared to those who did not engage in ESI in the first relationship. Thus, prior infidelity emerged as an important risk factor for serial infidelity in next relationships.

Why are people unfaithful? There are many reasons ranging from pure selfishness to boredom. Variables that are biological (e.g., differences in proneness

Upon completion of the next section, students should be able to:
LO11: List tips for maintaining a long-term love relationship.

to sexual excitement) or cultural (and thus impacting individual values) are in the mix (Stanley, 2017). Some people are unfaithful as a way of leaving a relationship. Commitment ends when repetitive hurt such as **cheating** comes in. An affair in a dating relationship is likely to be the beginning of the end.

Keeping Love Alive

As a couple enters a romantic relationship, their focus is on each other. However, as their relationship becomes a long-term one, their focus might turn to their careers, children, and friends and less on just each other. This may cause the relationship to wither.

The relationships need to be continually nurtured (Sternberg, 1995). Nurturing love is what produces people in their relational humanity as mentally healthy, warm, and considerate human beings (Cantillon & Lynch, 2017). A combination of passion, intimacy, caregiving, and commitment allows love to flourish and develop into a long-term love. There are several ways for established couples to keep their love alive, which we will discuss here.

Evaluating a Potential Love Relationship

Love is created by two partners. Therefore, it is important that before engaging in a long-term love relationship, a couple gets to know each other to determine whether they have common interests, values, and goals. Lee (1973) believes that to have a mutually satisfying love affair, a person must find a partner who shares the same style and definition of love. Otherwise, the more different two people are in their styles of loving, the less chance they will share in understanding each other's love.

Social exchange theory posits that partners in a love relationship give and receive items of value from each other based on their costs such as pain and benefits such as happiness. The cohesiveness or strength of a social bond is determined by the degree of reinforcement that members of the group receive (Farganis, 2008). A person is attracted to another if they perceive a reward from the relationship that they establish and when there is mutual attraction based on intrinsic benefits, such as the pleasure of the other's company. Then, a relationship between lovers is likely to develop (Fowler, 2007). Therefore, each partner in the relationship must determine honestly whether engaging in the relationship provides enough positive benefits to outweigh the long-term costs.

Fostering Intimacy

Love is based on intimacy and feelings of closeness. Love relationship satisfaction is rooted in intimacy—"individuals' subjective experiences of closeness and connectedness with their romantic partners, which emerge from couple relationship processes that involve self-disclosure, mutual trust, empathy, and acceptance" (Yoo *et al.*, 2013).

To sustain a love relationship, it is important for a couple to remain intimate with each other and to spend time together. When a couple spends time together, they learn more about each other and grow with each other through experience. Love manifests the function of intimacy as a factor in facilitating information exchange and learning (Restivo, 1977). A couple can maintain intimacy verbally. **Verbal intimacy** involves having someone to talk to and share your thoughts with. Intimacy allows a couple to communicate their feelings and learn from each other to show how much they care for each other. Emotional expressivity can be used as a tool in measuring intimacy. Ubando (2016) notes that if each partner has different expectations of emotional expression within their relationship, conflict will ensue. For example, males who reported high verbal emotional expressivity had lower levels of relationship satisfaction than women. It means that communication style is an essential ingredient in intimacy. A positive communication style led to greater feelings of intimacy as well as relationship satisfaction (Yoo *et al.*, 2013). Emotional intimacy involves disclosing thoughts to offer mutual and emotional support. Mutual self-disclosures create trust. Love blooms when it is based on mutual respect and trust. Self- disclosure is a two-step process. First, a person must make a self-disclosure that is neither too general nor too intimate. Second, the self-disclosure must be received with empathy, caring, and respect (Schafer, 2015).

Note that fostering intimacy is not the same thing as avoiding conflict. For example, when men and women enter a love or marital relationship, the couple often must change their personal rules to coordinate with their partners and agree on ways to mutually manage their shared boundary, otherwise, conflict might erupt. Healthy conflict in a relationship (rather than ignoring problems) can be a sign of intimacy. Couples who don't fight are usually the ones who have affairs (Weil, 2017). **Relationship talk** refers to talking with a partner about the relationship. "We need to talk" are the four words people may love or hate. If a woman says it to a man, he might dread what is to come. When a guy says that to a woman, she may be anxious, but she is likely to welcome the opportunity for discussion (Forsythe, 2015). Satisfaction with the frequency of relationship talk was associated with lower baseline distress for patients and partners (Badr *et al.*, 2008).

We know about the golden rule, which is to treat others the way we want to be treated. People who are truly in love tend to focus on the positive qualities of their beloved, while overlooking his or her negative traits. Chapman (2015) suggests the five love languages: words of affirmation, quality time, receiving gifts, acts of service, and physical touch. If your primary language is physical touch, you are likely to associate physical closeness with love. For example, you may demonstrate it with hand-holding in public, or back rubs when watching TV together.

Giving space is another way to maintain intimacy. If a couple is with each other constantly, it might drive both too close. With this type of intimacy, there is so much togetherness that the space between the two partners disappears (Lyman, 2011). If a couple spends every minute with each other, they do not have time to enjoy others. As the saying goes, "absence makes the heart grow fonder." "I would learn to respect my partner's boundaries; not everyone is comfortable with sharing every aspect of their lives, which is why I would refrain from pushing them to open if we hit a topic that's off limits" (Arbeit *et al.*, 2016).

Keeping Passion Alive

A passionate love may deteriorate into empty love. Sternberg (1986) cautions that maintaining a consummate love may be even harder than achieving it. One way to keep passion alive is to *be* passionate. Strgar (2017) recommends broadening your definition of passion and broadening your horizons to a more realistic kind of love. A couple can keep a relationship exciting by trying new things together. For example, they can take a road trip and a romantic weekend getaway. In a survey about what helps people *stay* married, married adults say having shared interests (64 percent) and a satisfying sexual relationship (61 percent) are very important to a successful marriage (Livingston & Caumont, 2017). Without excitement, a relationship may become routine. When couples become bored with one another, it can lead to infidelity.

Another way to keep passion alive is nonverbal interaction. Eye contact is essential to maintaining intimacy in a relationship (Schnarch, 1997). Hugging and kissing are also important to maintaining intimacy. Long hugs (lasting 10 to 30 seconds) increase the body's production of oxytocin ("love hormone"), which is the chemical that increases our sense of connectedness (Lyman, 2011).

Another powerful way to keep passion alive is the "relationship checkup." **Relationship checkup** occurs when a couple regularly sets aside a specific date to look systematically at how the two partners are doing individually and as a couple (Staheli & Schwartz, 2015). American dramatist and screenwriter Lillian Hellman (1905–84) said, "People change and forget to tell each other." When women are breastfeeding or as they age, couples should talk about the physical changes that affect the sexual aspect of their relationship (Rock, 2017). Making sex work—embracing your erotic soul and deepening the intimacy in your life—is a consequence of deep presence and its gift (Strgar, 2017).

Taking Care of the Loved One

In a healthy love relationship, each partner serves as caregiver for the other partner, and both partners care for each other consistently. The caregiving system is activated when an individual detects that another is distressed or needs help and the system is deactivated when the care recipient's need is met or his/her sense of security is reestablished (Fitzgerald, 2017).

One way to express caring is to be a good listener and provide emotional support. In addition, being equal partners is another way to demonstrate care for each other in a long-term love relationship. For example, sharing household chores and childrearing can foster the sense of caring. When it comes to the caregiving system, people can engage in one of the three broad caregiving strategies: (1) sensitive and responsive caregiving (delivered in an appropriate manner that meets the care recipient's specific needs), (2) hyperactivating caregiving (persistent in nature and delivered in a way that usually intensifies the care recipients' distress), or (3) deactivating caregiving (distant, minimal, and lacking in emotional content) (Canterberry & Gillath, 2012).

At the mental level, love labor involves holding the person and their interests in mind, keeping them "present" in mental planning, and anticipating and prioritizing their needs and interests (Cantillon & Lynch, 2017). Mindfulness contributes to greater intimate relationship satisfaction by fostering more relationally skillful emotion repertoires (Wachs & Córdova, 2007). In social exchange theory, caregiving behavior is viewed as a rational choice where two partners seek to maximize benefit and minimize cost. If the loving laboring toward each other is reciprocated, the social bond will be strong.

Maintaining a Sense of Commitment

Loving relationships need commitment to thrive and grow. Maintaining a sense of commitment is another important component of sustaining a healthy relationship.

The ingredient for finding lasting love is understanding commitment. Men and women find themselves in half-committed relationships that lead to frustration, sadness, and, in many cases, divorce (Stanley, 2005). Commitment to each other is important even if it is difficult to articulate. It is easy to form relationships with friends. What this points to is the de-formalization of relationship practices—there is more freedom now to begin and end relationships and drift into and out of them (Carter, 2015).

Part of being committed to the relationship is being loyal and honest with each other. One way to be honest is to communicate freely with each other. The most valuable approach to a healthy marriage reflects the fundamental premise that love and commitment thrive most within the context of social and economic opportunity (Randles, 2016). When a couple loves and commits to each other, they are likely to make sacrifices and become one. Greater commitment was associated with greater satisfaction, greater investments, and poorer quality alternatives (Rusbult, 1983). Women, both lesbians and heterosexuals, reported that they had invested more in their relationships and were more committed to maintaining their relationships than did men (Duffy & Rusbult, 1986).

Stanley (2005) offers a five-step plan for understanding commitment: learning to handle the pressures of everyday life; moving through the pain of unfulfilled dreams and hopes; overcoming attraction to others that might endanger a marriage; transforming your thinking from "me versus you" to "we" and "us"; and capturing the beauty and mystery of lifelong devotion, loyalty, teamwork, and building a lasting vision for the future.

Global Overview: Expressions of Love

Valentine's Day is an annual holiday celebrated in the United States in which people express their love for each other. In fact, in 2016, Americans spent an estimated $131 million on flowers alone (U.S. Census Bureau, 2017).

Transitions in the Meaning of Romantic Love

Valentine's Day originated as a Western Christian liturgical feast day honoring an early saint named Valentinus. The day became associated with courtly love, which had emerged in thirteenth-century Europe (Shaver *et al.*, 1988) and was based on ideals of male chivalry and the idealization of women. Courtly love was practiced across Europe during the Middle Ages and allowed knights and ladies to show their admiration for each other regardless of their marital state (see image on next page). Usually, a married lady gave a token to a knight of her choice to be worn during a medieval tournament. Some illicit court romances were fueled by the practice of courtly love.

Romantic love today is equivalent to courtly love. The art of courtly love can be found in the paintings and literary works of artists in the Middle Ages. This courtly love was a combination of aesthetic love and selfless love (Hendrick & Hendrick, 1992). In eighteenth-century England, Valentine's Day became associated with the courtly love when lovers expressed their love for each other by presenting flowers and sending greeting cards.

Note that at that time, romantic love was not associated with marriage as marriages were arranged. The notion of romantic love had historically been reserved for sexual relationships outside of marriage. Romantic love that occurred outside of marriage has replaced other social benefits such as the formation of a nuclear family as the basis of marriage; this change is just one example of the transitions in the meaning of love and the individualization apparent in society today (Fowler, 2007). As the interconnection between love and marriage grew, so did the condemnation of extra-marital relationships (Carter, 2015).

Expression of Love across the World

People express love with cultural objects. Geertz (1959) views culture as the symbols and meanings people use for communicating and perpetuating their knowledge about life. People use symbols to express love. Symbolic internationalist theory suggests that people make sense of the world with the symbols they create and use. Although symbols of love vary from culture to culture, it is common to associate red hearts with love and use a bouquet of red roses to express love. In some cultures, the color red symbolizes love whereas yellow may symbolize friendship. An apple in some European countries such as Greece symbolizes love. A maple leaf can be seen in Asian cultures as a symbol of love. Celtic love knots may symbolize love in Irish cultures.

Igbo Romantic Pet Names

A person's name is mostly symbolic in meaning. Igbo is an ethnic group from southeast Nigeria. Some names have strong undertones as Igbo people believe in appraising the strength and power of God; surnames are used for sustaining the family heritage; women do not change their last names after marriage (Momjunction, 2017). In Igbo, people use romantic pet names in local languages to keep their relationships fresh and fun. Romantic names, in Igbo worldview, go beyond marriage and companionship and connote intimacy, beauty, fondness, appreciation, love, admiration, and lust. Girls may call their boyfriends *Ademi* (my

Source: John William Waterhouse (1849–1917). http://www.tandfonline.com/doi/full/10.1080/09505431.2013.809412

crown) when in love. Romantic pet names in Igbo feature linguistic elements that portray meaning beyond sex and privacy (Anyanwu & Oha, 2017).

Public Displays of Affection

Displays of love are different based on societal norms (Rosenblatt, 1967). **Public displays of affection** (PDA) refer to displays of affection between lovers and acts of physical intimacy in the view of others. Displays of affection are present in everyday life. For example, a mother kisses her child goodbye as she leaves home to go to school. Close friends greet with a kiss on the cheek. Romantic love is popularized in almost every culture, and among lovers physical affection has been defined as "any touch intended to arouse feelings of love in the giver and/or the recipient" (Gulledge *et al.*, 2003). However, people express love to their partners with cultural practices. For example, in France, people show their love by holding hands, kissing, and initiating sexual relationships. The perfect expression of Argentine romanticism can be found in the tango—the world's most sensuous dance—but is detectable in everything from heady red wines to the sublime football skills of stars such as Diego Maradona and Lionel Messi (Forgan, 2012). In Latin American countries, there are two phrases, "Te quiero" and "Te amo"; the first is used casually by friends, family members, and couples (similar to "I love you") whereas the latter carries a more dramatic connotation that only applies to an amour or lover (Shapley, 2017). Muslims see smiling as a sign of sexual attraction, so women have learned not to smile at men (Robbins *et al.*, 2009). In the United States, the phrase "I love you" can be used among friends but it would *not* be acceptable in the Czech Republic (Shapley, 2017).

People may express their love in a subtle way. For example, self-deprecation and not self-aggrandizement is generally the way they do things in Ireland (Forgan, 2012). In traditional China, if a woman fell in love with a man and wanted to express her love, she might give him a present such as a handkerchief with an embroidered mandarin duck to symbolize an affectionate couple. Today, people often express their love verbally or with gifts. In Japan, public displays of affection are discouraged and draw stares, especially among older adults. Instead, men carry their partners' bags, while women might make an *aisai bento* ("beloved-wife lunch box") for work. Public displays of affection can easily be expressed online on social media. Emery *et al.* (2014) found that avoidant individuals showed low desire for relationship visibility, whereas anxious individuals reported high desired visibility.

Summary

Chapter 7 has examined love as emotion and action. Love exists in many forms such as self-love, parental love, friendship love, familial love, romantic love, companionate love, and compassionate love. The elements of love are intimacy, passion, caregiving, and commitment. The chapter has explored the sociology of love, Lee's color wheel of love, and Sternberg's triangular theory of love. The chapter has also discussed functions of love and problems of experiencing love. Finally, the chapter has examined ways to keep love alive in the long-term, through intimacy, passion, caregiving, and commitment. Using diversity and global perspectives, the chapter has delved into expressions of love by gender and in different cultures.

Key Terms

Adolescent love is characterized by the presence of physical attraction and passion.

Affects mean positive and negative evaluations (liking/disliking) of an object, behavior, or idea.

Agape is altruistic love and unselfish love.

Altruism or selflessness is the practice of concern for the welfare of others.

Attachment is conceptualized as characteristic of the dependent, weaker member of an affectionally bonded dyad.

Broken heart is a metaphor for the intense emotional pain one feels when experiencing great longing.

Caregiving is defined as the behaviors an individual evokes in order to meet the needs of another who is in distress or in need of assistance.

Caring is the practice of providing care and support to an intimate relationship partner or family member.

Cheating refers to unfaithful behavior to a spouse or lover.

Commitment refers to a decision that a person loves another and/or maintains the love.

Communication privacy theory explains the way people make decisions about revealing and concealing private information.

Companionate love is a type of love found between two partners and characterized by the presence of intimacy, commitment, and caring.

Compassionate love is altruistic love characterized by the presence of intimacy, commitment, and altruistic caregiving.

Consummate love involves all components of intimacy, passion, and commitment.

Emotional energy (EE) is an aliveness of the mind, a happiness of the heart, and a spirit filled with hope.

Emotional intimacy is a form of intimacy that involves sharing the feelings of bondedness and demonstrating caring for emotional support.

Emotional stratification refers to the unequal distribution of emotional labor and emotional energy underlying forms of social inequality in the areas of gender, social class, race, and age.

Emotions refer to a positive or negative experience that is associated with physiological reactions to events or influenced by social relationships and social structure.

Empty love is a form of love that involves only commitment without intimacy or passion.

Eros is romantic love.

Eroticism refers to sexual desire.

Euphoria is an affective state in which a person experiences pleasure and happiness.

Extra-dyadic sexual involvement (ESI) refers to having sexual relations with someone else in unmarried, serious, and romantic relationships.

Familial love is an emotional bond between family members and is based on kinship ties.

Fatuous love involves passion and commitment without intimacy.

Feelings are the representations of emotions. People can experience an emotion only if they have some feelings aroused physiologically.

Feminization of love means that people who express how much they love are women.

Filial love is the love a child has for a parent.

Filial piety is a virtue of love and respect for one's parents, elders, and ancestors.

Friendship love is a type of love that close friends feel for each other, characterized by the presence of intimacy.

Habituation refers to a decrease in response to an emotional stimulus after repeated presentations.

Happiness is the emotion that we feel when we are overwhelmed with an experience that involves emotional stimuli ranging from contentment to intense joy.

Homophily is the tendency of individuals to socialize and bond with people of similar backgrounds.

Infatuated love involves passion but no intimacy and commitment.

Infidelity refers to unfaithful behavior on the part of a spouse in married relationships.

Intimacy is the ability to be close to others as a lover, a friend, and as a participant in society.

Investment model suggests that commitment to a target is influenced by three factors: satisfaction level, quality of alternatives, and investment size.

James–Lange theory of emotion suggests that emotions be the results of physiological reactions to external events.

Jealousy is an emotional response to a real or perceived rival who is believed to be a threat to a partner.

Liking includes components of favorable evaluation and respect for the target person.

Love is the physical, emotional, sexual, social affection someone holds for another.

Love labor refers to any love-producing activity voluntarily performed without consideration of reward.

Ludus is game-playing love or playful love.

Mania is obsessive love and possessive love with the involvement of strong emotional intensity, extreme jealousy, and dependency.

Maternal love is the love a mother feels for their child.

Monogynic love encourages men to support the mother of their children.

Monogyny is the practice of one male marrying one woman.

Nonlove is the absence of intimacy, passion, and commitment.

Parental love is unconditional love of parents towards their children.

Passion refers to an intense emotion that leads to romance, physical attraction, and sexual desire in loving relationships.

Physical affection is an essential ingredient in sexual intimacy.

Pragma is practical love.

Primary care relations are produced through love labor—any love-producing activity voluntarily performed without consideration of reward between parents and children or between two partners.

Psychological intimacy is a form of intimacy that involves being open and honest in talking with a partner about personal thoughts.

Public displays of affection (PDA) refer to displays of affection between lovers and acts of physical intimacy in the view of others.

Relationship talk refers to talking with a partner about the relationship.

Relationship checkup occurs when a couple regularly sets aside a specific date to look systematically at how the two partners are doing individually and as a couple.

Romantic love is a type of love found between two lovers and characterized by the presence of intimacy, caregiving, and passion.

Secondary care relations involve general care work including affective bonds with neighbors, or work colleagues, or the caring given in nursing or paid caring.

Secure attachment is attachment with neither fear of losing the relationship nor fear of entering relationships.

Self-care is any necessary human regulatory function which is under individual control, deliberate and self-initiated.

Self-disclosure means that people trust someone and feel comfortable exposing their true self without fear of rejection.

Self-enhancement is a type of motivation that maintains and cultivates positive feelings of the self.

Self-love is a type of love in which we acknowledge our own good qualities and focus on our self-growth and well-being.

Serial infidelity refers to the practice of engaging in a succession of sexual relationships with someone other than your partner.

Sexual intimacy or physical intimacy is a form of intimacy that involves physical affection.

Sexual love is basic human instinct and hormonal responses.

Social isolation refers to the absence of social relationships.

Sociology of emotions examines how human emotions influence, and are influenced by, society.

Stalking refers to controlling or obsessive behaviors that intrude on a person's privacy or personal space and causes the victim to fear for personal safety (including the safety of another person) or suffer mental distress.

Storge is companionate love and friendship love.

Tertiary care relations involve solidarity care work, where people work informally, politically, culturally, or economically through solidarity to challenge injustices such as gender inequality.

Triangular theory of love suggests that partners express their love depending on combinations of intimacy, passion, and commitment.

True love or real love is an unconditional love of someone towards a "special person."

Unconditional love refers to affection without limits or conditions.

Unrequited love refers to a relationship in which one person loves another who does not reciprocate the romantic affection.

Verbal intimacy involves having someone to talk to and share your thoughts with.

Discussion Questions

- Do you believe in love at first sight? Why or why not?
- How does romantic love occur when friendship escalates?
- How do you think it is best to deal with unrequited love?
- What tips do you have for maintaining a long-term love relationship?
- Why does friendship help us develop a love relationship and become more intimate? What factors contribute to maintaining satisfaction in a love relationship over time?

Suggested Film and Video

Dove Self-Esteem Program: https://vimeo.com/151480167
Minsky, Howard G. (Producer). (1970). *Love Story*. Paramount Pictures.
Somatosensory Tracts: https://www.khanacademy.org/science/health-and-medicine/human-anatomy-and-physiology/nervous-system-introduction/v/somatosensory-tracts

Internet Sources

Family Caregiving: https://www.aarp.org/caregiving/
National Center for Family and Marriage Research: http://www.bgsu.edu/ncfmr.html

References

Acevedo, B. P. & Aron, A. (2009). Does a Long-term Relationship Kill Romantic Love? *Review of General Psychology*, 13, 59–65.

Acker, M. & Davis, M. H. (1992). Intimacy, Passion, and Commitment in Adult Romantic Relationships: A Test of the Triangular Theory of Love. *Journal of Social and Personal Relationships*, 9(1), 21–50. Retrieved on November 19, 2017 from http://journals.sagepub.com/doi/abs/10.1177/0265407592091002

Allan, G. (1998). Friendship, Sociology and Social Structure. *Journal of Social and Personal Relationships*, 15(5).

American Red Cross (December 7, 2017). *2017 Year in Review: Red Cross Delivers More Food, Relief Items and Shelter Stays than Last 4 Years Combined*. Retrieved on December 10, 2017 from http://www.redcross.org/news/press-release/2017-Year-in-Review-Red-Cross-Delivers-More-Food-Relief-Items-and-Shelter-Stays-than-Last-4-Years-Combined

Anderson, S. (2014). *The Journey from Abandonment to Healing: Surviving Through and Recovering from the Five Stages That Accompany the Loss of Love*. Berkley Trade.

Anyanwu, B. & Oha, A. C. (2017). Gender and Meaning in Igbo Romantic Pet Names. *South African Journal of African Languages*, 37(1).

Arbeit, M. R., Hershberg, R. M., Rubin, R. O., DeSouza, L. M. & Lerner, J. V. (2016). "I'm Hoping that I Can Have Better Relationships": Exploring Interpersonal Connection for Young Men. *Qualitative Psychology*, 3(1), 79–97.

Aristotle (2003). *Nicomachean Ethics*. Hugh Treddenick (Ed.). London: Penguin.

Aron, A., Aron, E. N. & Allen, J. (1998). Motivation for Unreciprocated Love. *Personality and Social Psychology Bulletin*, 24, 787–796.

Aron, A., Fisher, H., Mashek, D. J., Strong, G., Li, H. & Brown, L. L. (2005). Reward, Motivation, and Emotion Systems Associated With Early-Stage Intense Romantic Love. *Journal of Neurophysiology*, 94(1), 327–337. Retrieved on November 4, 2017 from http://jn.physiology.org/content/94/1/327.long

Arriaga, X. B. & Agnew, C. R. (2001). Being Committed: Affective, Cognitive, and Conative Components of Relationship Commitment. *Personality and Social Psychology Bulletin*, 27, 1190–1203.

Ashford, J. B. & LeCroy, C. W. (2009). *Human Behavior in the Social Environment*. San Francisco, CA: Cengage Learning.

Badr, H., Acitelli, L. K. & Taylor, C. L. (2008). Does Talking about Their Relationship Affect Couples' Marital and Psychological Adjustment to Lung Cancer? *J Cancer Surviv.*, 2(1), 53–64. Retrieved on November 5, 2017 from https://www.ncbi.nlm.nih.gov/pubmed/18648987

Bartels, A. & Zeki, S. (2004). The Neural Correlates of Maternal and Romantic Love. *Neuroimage*, 21(3), 1155–1166.

Baumeister, R. F., Wotman, S. R. & Stillwell, A. M. (1993). Unrequited Love: On Heart-break, Anger, Guilt, Scriptlessness, and Humiliation. *Journal of Personality and Social Psychology*, 64, 377–394.

Beall, A. E. & Sternberg, R. J. (1995). The Social Construction of Love. *Journal of Social and Personal Relations*, 12(3).

Bearn, J. & O'Brien, M. (2015). Addicted to Euphoria: The History, Clinical Presentation, and Management of Party Drug Misuse. *Int. Rev. Neurobiol.*, 120, 205–233.

Beck, U. & Beck-Gernsheim, E. (1995). *The Normal Chaos of Love*. Cambridge: Polity Press.

Beck, U. & Beck-Gernsheim, E. (2001). *Individualization: Institutionalized Individualism and its Social and Political Consequences*. London: Sage.

Bell, D. C. (2010). *The Dynamics of Connection: How Evolution and Biology Create Caregiving and Attachment*. New York: Lexington Books.

Bell, D. C. (2012). Next Steps in Attachment Theory. *Journal of Family Theory & Review*, 4(4), 275–281.

Berscheid, E. (2006). Searching for the Meaning of "Love". In R. J. Sternberg & K. Weis (Eds.), *The New Psychology of Love* (pp. 171–183). New Haven, CT: Yale University Press.

Berscheid, E., & Reis, H. T. (1998). Attraction and Close Relationships. In D. T. Gilbert, S. T. Fiske, & G. Lindzey (Eds.), *The Handbook of Social Psychology* (4th edn, Vol. 2, pp. 193–281). New York: McGraw-Hill.

Black, M. C., Basile, K. C., Breiding, M. J., Smith, S. G., Walters, M. L., Merrick, M. T., Chen, J. & Stevens, M. R. (2011). *The National Intimate Partner and Sexual Violence Survey (NISVS): 2010 Summary Report*. Atlanta, GA: National Center for Injury Prevention and Control, Centers for Disease Control and Prevention.

Boeree, C. G. (2003). *General Psychology.* Retrieved on October 27, 2011 from http://webspace.ship.edu/cgboer/genpsylove.html

Bosacki, S. L. & Moore, C. (2004). Preschoolers' Understanding of Simple and Complex Emotions: Links with Gender and Language. *Sex Roles*, 50, 659–675.

Bourdieu P. (1990). *The Logic of Practice.* Nice R. (Trans.). Stanford, CA: Stanford University Press.

Bouton, M. E. (2007). *Learning and Behavior: A Contemporary Synthesis.* Sunderland: MA Sinauer.

Bowlby, J. (1969). *Attachment and Loss: Vol. 1. Attachment* (2nd edn.). New York: Basic Books.

Brehm, J. W. (2016). The Intensity of Emotion. *Personality and Social Psychology Review,* 3(1), 2–22.

Bretherton, I. & Beeghly, M. (1982). Talking about Internal States: The Acquisition of an Explicit Theory of Mind. *Developmental Psychology,* 18, 906–921.

Bringle, R. G., Winnick, T. & Rydell, R. J. (2013). The Prevalence and Nature of Unrequited Love. *Sage Open*, 3(2).

Brown, P. M. (1995). *The Death of Intimacy: Barriers to Meaningful Interpersonal Relationships.* New York: Haworth.

Brown, J. R. & Dunn, J. (1992). Talk with Your Mother or Your Sibling? Developmental Changes in Early Family Conversations about Feelings. *Child Development*, 63, 336–349.

Bryson, V. (2013). *Time to Love. In Love: A Question for Feminism in the Twenty-first Century.* A. G. Jónasdóttir & A. Ferguson (Eds.). London: Routledge.

Bubeck, D. E. (1995). *Care, Justice and Gender.* Oxford: Oxford University Press.

Buri, J. R. (November 03, 2009). Mistaking Passion for Love. *Psychology Today.* Retrieved on September 27, 2011 from http://www.psychologytoday.com/blog/love-bytes/200911/mistaking-passion-love

Burrows, R., Nettleton, S. & Bunton, R. (1995). Sociology and Health Promotion. Health, Risk and Consumption under Late Modernism. *The Sociology of Health Promotion*, 1–12.

Buunk, B. P. & Dijkstra, P. (2001). Evidence for a Sex-Based Rival-Oriented Mechanism: Jealousy as a Function of a Rival's Physical Attractiveness and Dominance in a Homosexual Sample. *Personal Relationships*, 8, 391–406.

Caldwell, M. A. & Peplau, L. A. (1982). Sex Differences in Same-sex Friendship. *Sex Roles*, 8, 721–732.

Cancian, F. M. (1990). *Love in America: Gender and Self Development.* Cambridge, UK: Cambridge University Press.

Canevello A. & Crocker J. (2010). Creating Good Relationships: Responsiveness, Relationship Quality, and Interpersonal Goals. *Journal of Personality and Social Psychology,* 99, 78–106. Retrieved on November 5, 2017 from https://www.ncbi.nlm.nih.gov/pubmed/20565187

Canterberry, M. & Gillath, O. (2012). Attachment and Caregiving. In P. Noller & G. C. Karantzas (Eds.), *The Wiley-Blackwell Handbook of Couples and Family Relationships* (pp. 207–219). New York: Wiley- Blackwell.

Cantillon, S. & Lynch, K. (2017). Affective Equality: Love Matters. *Hypatia*, 32, 169–186.

Cardon, M. S., Zietsma, C., Saparito, P., Matherne, B. & Davis, C. (2005). A Tale of Passion: New Insights into Entrepreneurship from a Parenthood Metaphor. *Journal of Business Venturing*, 20, 23–45.

Carlson, J. G. & Hatfield, E. (1992). *Psychology of Emotion.* Fort Worth, TX: Harcourt Brace Jovanovich.

Carter, C. S. (1998). Neuroendocrine Perspectives on Social Attachment and Love. *Psychoneuroendocrinology,* 23, 779–818. Retrieved on December 10, 2017 from https://www.ncbi.nlm.nih.gov/pubmed/9924738

Carter, J. (2013). The Curious Absence of Love Stories in Women's Talk. *The Sociological Review,* 61(4), 728–744.

Carter, J. (February 2015). The Sociology of Love. *The Sociological Review.* Retrieved on December 5, 2017 from https://www.thesociologicalreview.com/blog/the-sociology-of-love.html

Chang, W. & Kalmanson, L. (November 8, 2010). *Confucianism in Context: Classic Philosophy and Contemporary Issues, East Asia and Beyond.* Albany, NY: State University of New York Press.

Chapman, G. D. (2015). *The Five Love Languages for Men: Tools for Making a Good Relationship Great.* Chicago, IL: Northfield Publishing.

Choi, K. H. & Mare, R. D. (2012). International Migration and Educational Assortative Mating in Mexico and the United States. *Demography*, 49(2), 449–476.

Christenson, E. (February 17, 2003). What Women Want. *Newsweek.*

Collins, N. L., Guichard, A. C., Ford, M. B. & Feeney, B. C. (2006). An Attachment-theoretical Approach to Caregiving in Romantic Relationships. In M. Mikulincer & G. S. Goodman (Eds.), *Dynamics of Romantic Love: Attachment, Caregiving, and Sex* (pp. 149–189). New York: Guilford Press.

Collins, R. & McConnell, M. (2015). *Napoleon Never Slept: How Great Leaders Leverage Emotional Energy.* Los Angeles, CA: Maren Ink.

Cottingham, M. D., Johnson, A. H. & Erickson, R. J. (November 2, 2017). "I Can Never Be Too Comfortable": Race, Gender, and Emotion at the Hospital Bedside. *Qualitative Health Research.*

Crittenden, D. (1999). *What Our Mothers Didn't Tell Us: Why Happiness Eludes the Mother Women.* New York, NY: Simon & Schuster.

Cummings, J., Butler, B. & Kraut, B. (2000). The Quality of Online Social Relationships. *Communications of the ACM,* 45, 103–108.

Damasio, A. (2006). *Descartes' Error: Emotion, Reason and the Human Brain.* London: Vintage Books.

Darwin, C. (1872/1965). *The Expression of the Emotions in Man and Other Animals.* Chicago: University of Chicago Press.

Davis, K. E. (1985). Near and Dear: Friendship and Love Compared. *Psychology Today,* 19(2), 22–30.

Derathé, R. (1995). *Jean-Jacques Rousseau et la science politique de son temps* (pp. 135–141). Paris, France: Librairie Philosophique J. Vrin.

D'Hondt, F., Lassonde, M., Collignon, O., Dubarry, A.-S., Robert, M., Rigoulot, S. & Sequeira, H. (2010). Early Brain-Body Impact of Emotional Arousal. *Frontiers in Human Neuroscience,* 4, 33.

Devon, M. (2015). *The Origin of Emotions.* Lexington, KY: CreateSpace Independent Publishing Platform.

Diderot, D. (1765). *Passions. The Encyclopedia of Diderot & d'Alembert Collaborative Translation Project.* Translated by Timothy L. Wilkerson. Ann Arbor: Michigan Publishing, University of Michigan Library, 2004. Retrieved on November 6, 2017 from http://hdl.handle.net/2027/spo.did2222.0000.248>. Trans. of *Passions, Encyclopédie ou Dictionnaire raisonné des sciences, des arts et des métiers,* vol. 12. Paris, 1765.

Diener, E. & Lucas, R. (2000). Subjective Emotional Wellbeing. In M. Lewis & J. M. Haviland-Jones (Eds.), *Handbook of Emotions,* 2nd edn (pp. 325–337). New York: Guilford.

Dindia, K. & Allen, M. (1992). Sex Differences in Self-disclosure: A Meta-analysis. *Psychological Bulletin,* 112, 106–124.

Dion, K. K. & Dion, K. L. (1975). Self-esteem and Romantic Love. *Journal of Personality,* 43, 39–57.

Dixon, T. (2003). *From Passions to Emotions: The Creation of a Secular Psychological Category.* Cambridge, UK: Cambridge University Press.

Doherty, N. A. & Feeney, J. A. (2004). The Composition of Attachment Networks throughout Adult Years. *Personal Relationships,* 11(4), 469–488.

Duffy, S. M. & Rusbult, C. E. (1986). Satisfaction and Commitment in Homosexual and Heterosexual Relationships. *J. Homosex.* 1985–1986 Winter; 12(2), 1–23. Retrieved on November 4, 2017 from https://www.ncbi.nlm.nih.gov/pubmed/3835198

Edelstein, R. S., Kean, E. & Chopik, W. J. (2012). Women with an Avoidant Attachment Style Show Attenuated Estradiol Responses to Emotionally Intimate Stimuli. *Horm Behav.,* 61(2), 167–175. Retrieved on November 4, 2017 from https://www.ncbi.nlm.nih.gov/pubmed/22154613

Eisenberger, N. I. & Lieberman, M. D. (2004). Why Rejection Hurts: A Common Neural Alarm System for Physical and Social Pain. *Trends Cogn Sci.,* 8(7), 294–300.

Emery, L. F., Muise, A., Dix, E. L. & Le, B. (2014). Can You Tell That I'm in a Relationship? Attachment and Relationship Visibility on Facebook. *Personality and Social Psychology Bulletin,* 40(11), 1466–1479.

Engels, F. (1988). Engels on the Origin and Evolution of the Family. *Population and Development Review,* 14(4) (Dec. 1988), 705–729.

Engster, D. (2005). Rethinking Care Theory: The Practice of Caring and the Obligation to Care. *Hypatia,* 20(3), 50–74.

Erickson, E. (1963). *Childhood and Society,* 2nd edn. New York: Norton.

Etzioni, A. (1988). Normative-affective Factors: Toward a New Decision-making Model. *J. Econ. Psych.,* 9, 125–150.

Exline, R. V. (1963). Explorations in the Process of Person Perception: Visual Interaction in Relation to Competition, Sex, and Need for Affiliation. *Journal of Personality,* 31, 1–20.

Farber, B. A. (1980). Adolescence. In *On Love and Loving* (pp. 44–60). Pope, K. S. (Ed.). San Francisco: Jossey-Bass.

Farganis, J. (2008). *Readings in Social Theory,* 5th edn. Boston, MA: McGraw Hill.

Fehr, B. & Harasymchuk, C. (2013). *Compassionate Love in Intimate Relationships.* Unpublished manuscript, University of Winnipeg.

Fehr, B. & Russell, J. A. (1984). Concept of Emotion Viewed from a Prototype Perspective. *Journal of Experimental Psychology: General,* 113, 464–486.

Fehr, B. & Russell, J. A. (1991). Concept of Love Viewed from a Prototype Perspective. *Journal of Personality and Social Psychology,* 60, 425–438.

Fehr, B. & Sprecher, S. (2013). Compassionate Love: What We Know So Far. In M. Hojjat & D. Cramer (Eds.), *Positive Psychology of Love* (pp. 106–120). New York: Oxford University Press.

Fehr, B., Sprecher, S. & Underwood, L. (Eds.) (2009). *The Science of Compassionate Love: Theory, Research, and Applications.* Malden, MA: Wiley-Blackwell.

Fehr, B., Harasymchuk, C. & Gouriluk J. (2010, June). Validation of the Quadrumvirate Model of Love. Invited paper presented at the Canadian Psychological Association Conference (Social Psychology Faculty Symposium on Romantic Relationships), Winnipeg, MB.

Fehr, B., Harasymchuk, C. & Sprecher, S. (2014). Compassionate Love in Romantic Relationships: A Review and Some New Findings. *Journal of Social and Personal Relationships,* 31(5), 575–600.

Fennel, C. T. (2011). Habituation Procedures. In E. Hoff, *Research Methods in Child Language: A Practical Guide* (pdf). Hoboken, NJ: John Wiley & Sons.

Fisher, H. E. (1998). Lust, Attraction, and Attachment in Mammalian Reproduction. *Human Nature,* 9, 23–52.

Fisher, H. E., Brown, L. L., Aron, A., Strong, G. & Mashek, D. (2010). Reward, Addiction, and Emotion Regulation

Systems Associated With Rejection in Love. *Journal of Neurophysiology*, 104(1), 51–60.

Fitzgerald, J. (Ed.) (2017). *Foundations for Couples' Therapy: Research for the Real World*. New York: Taylor & Francis.

Forgan, D. (February 8, 2012). *World's Most Romantic Nationalities*. CNN. Retrieved on November 2, 2017 from http://travel.cnn.com/explorations/life/valentines/most-romantic-nationalities-807423/

Forsythe, K. (January 9, 2015). Tips To Ace The "Relationship Talk." *Huffington Post*. Retrieved on November 30, 2017 from https://www.huffingtonpost.com/katherine-forsythe-msw/the-relationship-talk_b_6084602.html

Fotopoulou, A. & O'Riordan, K. (2017). Training to Self-care: Fitness Tracking, Biopedagogy and the Healthy Consumer. *Journal of Health Sociology Review*, 26(1).

Fowler, A. C. (2007). Love and Marriage: Through the Lens of Sociological Theories, *Human Architecture. Journal of the Sociology of Self-Knowledge*, 5(2), Article 6. Retrieved on September 30, 2016 from http://scholarworks.umb.edu/humanarchitecture/vol5/iss2/6

Frances, W. N. & Kucera, H. (1982). *Frequency Analysis of English Usage: Lexicon and Grammar*. Boston, MA: Houghton Mifflin.

Fromhage, L., Jacobs, K. & Schneider, J. M. (2007). Monogynous Mating Behaviour and its Ecological Basis in the Golden Orb Spider *Nephila fenestrata*. *Ethology*, 113, 813–820.

Fromm, E. (1959). *Sigmund Freud's Mission: An Analysis of his Personality and Influence*. Oxford, UK: Harper.

Fromm, E. (2006). *The Art of Loving*. New York: Harper & Row.

Gadpaille, W. (1975). *The Cycles of Sex*. New York: Bantam.

Geertz, H. (1959). The Vocabulary of Emotion: A study of Javanese Socialization Processes. *Psychiatry: Journal for the Study of Interpersonal Processes*, 22, 225–237.

Georgiadis, J. R. & Kringelbach, M. L. (2012). The Human Sexual Response Cycle: Brain Imaging Evidence Linking Sex to Other Pleasures. *Progress in Neurobiology*, 98(1), 49–81.

Gere, J., MacDonald, G., Joel, S., Spielmann, S. S. & Impett, E. A. (2013). The Independent Contributions of Social Reward and Threat Perceptions to Romantic Commitment. *Journal of Personality and Social Psychology*, 105(6), 961–977. Retrieved on November 4, 2017 from https://www.ncbi.nlm.nih.gov/pubmed/23915039

Giddens, A. (1992). *The Transformation of Intimacy: Sexuality, Love, and Eroticism in Modern Societies*. Redwood City, CA: Stanford University Press.

Giles, L. C., Glonek, G. V. F., Luszcz, M. A. & Andrews, G. R. (May 2005). Effect of Social Networks on 10-year Survival in Very Old Australians: The Australian Longitudinal Study of Aging. *Journal of Epidemiology and Community Health*, 59, 574–579.

Gilligan, C. (1995). Hearing the Difference: Theorizing Connection. *Hypatia*, 10(2), 120–127.

Goff, B. G., Goddard, H. W., Pointer, L. & Jackson, G. B. (2007). Measures of Expressions of Love. *Psychological Reports*, 101, 357–360.

Gonzaga, G. C., Turner, R. A., Keltner, D., Campos, B. & Altemus, M. (2006). Romantic Love and Sexual Desire in Close Relationships. *Emotion*, 6(2), 163–179.

Graff, S., FengerGrøn, M., Christensen, B. *et al.* (2016). Long-term Risk of Atrial Fibrillation after the Death of a Partner. *Open Heart*, 3, e000367.

Graham, J. M. (2010). Measuring Love in Romantic Relationships: A Meta-analysis. *Journal of Social and Personal Relationships*, 28(6), 748–771.

Gray, J. (1994). *Men Are from Mars, Women Are from Venus: The Classic Guide to Understanding the Opposite Sex*. New York: HarperCollins Publishers.

Griffith, J. (2011). *The Book of Real Answers to Everything!* Australia: WTM Publishing & Communications Pty Ltd.

Gulledge, A. K., Gulledge, M. H. & Stahmann, R. F. (2003). Romantic Physical Affection Types and Relationship Satisfaction. *The American Journal of Family Therapy*, 31(4), 233–242.

Gürtler, S. (2005). The Ethical Dimension of Work: A Feminist Perspective. Trans. A. Smith. *Hypatia*, 20(2), 119–134.

Hacker, P. (2017). *The Passions: A Study of Human Nature*. New York: Wiley-Blackwell.

Halko, M., Lahti, T., Hytönen, K. & Jääskeläinen, I. P. (2017). Entrepreneurial and Parental Love—Are They the Same? *Human Brain Mapping*, 38(6), 2923–2938.

Hammock, G. & Richardson, D. S. (2011). Love Attitudes and Relationship Experience. *The Journal of Social Psychology*, 151(5). Retrieved on November 5, 2017 from http://www.tandfonline.com/doi/abs/10.1080/00224545.2010.522618?src=recsys&journalCode=vsoc20

Hanna, S. L. (2003). *Person to Person: Positive Relationships Don't Just Happen*, 4th edn. Upper Saddle River, NJ: Prentice Hall.

Hatfield, E. & Rapson, R. L. (1987). Passionate Love/Sexual Desire: Can the Same Paradigm Explain Both? *Archives of Sexual Behavior*, 16(3), 259–278.

Hatfield, E. & Sprecher, S. (1986). Measuring Passionate Love in Intimate Relations. *Journal of Adolescence*, 9, 383–410.

Hefner, V. & Wilson, B. J. (2013). From Love at First Sight to Soul Mate: The Influence of Romantic Ideals in Popular Films on Young People's Beliefs about Relationships. *Communication Monographs*, 80(2).

Heise, D. (1979). *Understanding Events: Affect and the Construction of Social Action*. New York: Cambridge University Press.

Hendrick, S. & Hendrick, C. (1992). *Romantic Love*. Newbury Park, CA: Sage.

Hochschild, A. R. (1975). The Sociology of Feeling and Emotion: Selected Possibilities. In M. Millman & R. M. Kanter (Eds.), *Another Voice: Feminist Perspectives on Social Life and Social Science* (pp. 280–307). New York: Anchor.

Hook, M., Gerstein, L., Detterich, L. & Gridley, B. (2003). How Close Are We? Measuring Intimacy and Examining Gender Differences. *Journal of Counseling and Development*, 81, 462–472.

Ikels, C. (2004). *Filial Piety: Practice and Discourse in Contemporary East Asia*. Redwood City, CA: Stanford University Press.

Illouz, E. (2013). *Why Love Hurts: A Sociological Explanation*. Cambridge, UK: Polity Press.

Impett, E. A., Beals, K. P. & Peplau, L. A. (2001). Testing the Investment Model of Relationship Commitment and Stability in a Longitudinal Study of Married Couples. *Current Psychology: Developmental, Learning, Personality, Social.*, 20(4), 312–326.

Insel, T. R., Young, L. & Wang, Z. (1997). Molecular Aspects of Monogamy. *Ann NY Acad Sci.*, 807, 302–316. Retrieved on December 10, 2017 from https://www.ncbi.nlm.nih.gov/pubmed/9071359

Jackson, S. (1993). Even Sociologists Fall in Love: An Exploration in the Sociology of Emotions. *Sociology*, 27(2), 201–222.

Jackson, T., Chen, H., Guo, C. & Gao, X. (2006). Stories We Love by: Conceptions of Love among Couples from the People's Republic of China and the United States. *Journal of Cross-Cultural Psychology*, 37(4), 446–464.

James, W. (1884). What is an Emotion? *Mind*, 9, 188–205.

Jankowiak, W. & Fischer, E. (1992). A Cross-Cultural Perspective on Romantic Love. *Ethnology*, 31(2), 149–155.

Jorgensen, S. R. & Gaudy, J. C. (1980). Self-disclosure and Satisfaction in Marriage: The Relation Examined. *Family Relations*, 29, 281–287.

Jourard, S. (1971). *The Transparent Self.* New York: Van Nostrand.

Karandashev, V. (2015). A Cultural Perspective on Romantic Love. *Online Readings in Psychology and Culture*, 5(4).

Kellerman, J., Lewis, J. & Laird, J. D. (1989). Looking and Loving: The Effect of Mutual Gaze on Feelings of Romantic Love. *Journal of Research in Personality*, 23, 145–161.

Kelley, A. E. (2004). Ventral Striatal Control of Appetitive Motivation: Role in Ingestive Behavior and Reward–Related Learning. *Neurosci Biobehav Rev.*, 27, 765–776.

Kemper, T. D. (1978). *A Social Interactional Theory of Emotions*. New York: Wiley.

Kephart, W. (1967). Some Correlates of Romantic Love. *Journal of Marriage and the Family*, 29, 470–474.

Kim, J. & Hatfield, E. (2004). Love Types and Subjective Well-Being: A Cross-Cultural Study. *Social Behavior and Personality: An International Journal*, 32(2), 173.

Kirkpatrick, B. (Ed.). (1998). *Roget's Thesaurus of English Words and Phrases*. Harmondsworth, UK: Penguin.

Kirshenbaum, M. (2003). The Emotional Energy Factor: The Secrets High-Energy People Use to Beat Emotional Fatigue. New York: Bantam Dell.

Knopp, K., Scott, S., Ritchie, L. *et al.* (2017). Once a Cheater, Always a Cheater? Serial Infidelity across Subsequent Relationships. *Arch Sex Behav.*, 46, 2301.

Know, D., Zusman, M. & Daniels, V. (March 2002). College Students Attitudes toward Interreligious Marriage. *College Student Journal*, 36(1), 84–87.

Knutson, B., Adams, C. M., Fong, G. W. & Hommer, D. (2001). Anticipation of Increasing Monetary Reward Selectively Recruits Nucleus Accumbens. *J Neurosci.*, 21, U1–U5.

Kövecses, Z. (2000). *Metaphor and Emotion: Language, Culture, and Body in Human Feeling*. Cambridge, UK: Cambridge University Press.

Lange, C. (1886). *On Periodical Depressions and their Pathogenesis* (Om Periodiske Depressionstistande og deres Patogenese). Copenhagen: Lunds Forlag.

LaRossa, R. & LaRossa, M. M. (1981). *Transition to Parenthood: How Infants Change Families*. Beverly Hills, CA: Sage.

Lasswell, T. E. & Lasswell, M. E. (1976). I Love You but I'm not in Love with You. *Journal of Marriage and Family Counseling*, 38, 211–224.

Lazarus, R. S. (1991). *Emotion and Adaptation*. New York: Oxford University Press.

Lee, H.-J., Macbeth, A. H., Pagani, J. & Young, W. S. (2009). Oxytocin: The Great Facilitator of Life. *Progress in Neurobiology*, 88(2), 127–151.

Lee, J. A. (1973). *Colors of Love: An Exploration of the Ways of Loving*. Toronto, CA: New Press.

Levy, P. E. (2013). *Industrial Organizational Psychology*, 4th edn (pp. 316–317). New York: Worth.

Lewis, C. S. (1960). *The Four Loves*. New York: Harcourt Brace Jovanovich.

Liebowitz, M. R. (1983). *The Chemistry of Love*. Boston: Little, Brown & Co.

Lindzey, G. & Byrne, D. (1968). Measurement of Social Choice and Interpersonal Attractiveness. In G. Lindzey & E. Aronson (Eds.), *Handbook of Social Psychology, Vol. 2.*, 2nd edn. Reading, MA: Addison-Wesley.

Livingston, G. & Caumont, A. (February 13, 2017). *5 Facts on Love and Marriage in America*. Washington, DC: Pew Research Center. Retrieved on July 20, 2017 from http://www.pewresearch.org/fact-tank/2017/02/13/5-facts-about-love-and-marriage/

Lopez, B. H. (1978). *Of Wolves and Men*. New York: Charles Scribner's Sons.

Lyman, J. (2011). Five Tips for Creating Intimacy. *Divine Caroline*. Retrieved on November 17, 2011 from http://www.divinecaroline.com/22072/109666-five-tips-creating-intimacy/2

Lynch, K. (2007). Love Labor as a Distinct and Non-Commodifiable Form of Care Labor. *Sociological Review*, 54(3), 550–570.

Lynch, K. (2009). Affective Equality: Who Cares? *Development*, 52(3), 410–415. Retrieved on November 24, 2017 from https://doi.org/10.1057/dev.2009.38

Määttä, K (2013). *Many Faces of Love*. The Netherlands: Sense Publishers.

Macadam, P. S. (2017). *Breastfeeding: Biocultural Perspectives*. New York: Routledge.

Mackey, R. A., Diemer, M. A. & O'Brien, B. A. (2000). Psychological Intimacy in the Lasting Relationships of Heterosexual and Same-Gender Couples. *Sex Roles*, 43(3/4), 201–227.

MacLean, P. D. (1990). *The Triune Brain in Evolution: Role in Paleocerebral Functions*. New York: Plenum.

Macoby, E. E. (1990). Gender and Relationships. *American Psychologist*, 45, 513–520.

Manoharan, C. & de Munck, V. (2015). The Conceptual Relationship Between Love, Romantic Love, and Sex: A Free List and Prototype Study of Semantic Association. *Journal of Mixed Methods Research*, 11(2), 248–265.

Marelich, W. D., Gaines, S. O. & Banzet, M. R. (2003). Commitment, Insecurity, and Arousability: Testing a Transactional Model of Jealousy. *Representative Research in Social Psychology*, 27, 23–31.

Markman, H. J. & Kraft, S. A. (1989). Men and Women in Marriage: Dealing with Gender Differences in Marital Therapy. *Behavior Therapist*, 12, 51–56.

Maslow, A. H. (1955). *Deficiency Motivation and Growth Motivation*. Nebraska Symposium on Motivation, 1955, 2.

Mattingly, C. (2014). Love's Imperfection—Moral Becoming, Friendship and Family Life. *Suomen Anthropologies: Journal of the Finnish Anthropological Society*, 39(1), 53–67.

McAdams, III, C. R., Foster, V. A., Dotson-Blake, K. & Brendel, J. M. (2009). Dysfunctional Family Structures and Aggression in Children: A Case for School-Based, Systemic Approaches with Violent Students. *Journal of School Counseling*, 7(9).

McDuff, D., Kodra, E., Kaliouby, R. & LaFrance, M. (2017). A Large-Scale Analysis of Sex Differences in Facial Expressions. *PLoS ONE*, 12(4), e0173942.

McKie, L., Gregory, S. & Bowlby, S. (2002). Shadow Times: The Temporal and Spatial Frameworks and Experiences of Caring and Working. *Sociology*, 36(4), 897–924.

McManamy, J. (January 2011). *The Brain in Love and Lust*. Retrieved on January 30, 2012 from http://www.mcmanweb.com/love_lust.html

Mesquita, B. (2016). The Legacy of Nico H. Frijda (1927–2015). *Journal of Cognition and Emotion*, 30(4).

Miller, J. G., Kahle, S., Lopez, M. & Hastings, P. D. (2015). Compassionate Love Buffers Stress – Reactive Mothers. From Fight-or-Flight Parenting. *Developmental Psychology*, 51(1), 36–43. Retrieved on November 5, 2017 from https://www.ncbi.nlm.nih.gov/pubmed/25329554

Milligan, T. (2011). *The Art of Living: Love*. New York: Routledge.

Mineo, L. (2017). Good Genes Are Nice, But Joy Is Better. *Harvard Gazette*. Retrieved on December 3, 2017 from https://news.harvard.edu/gazette/story/2017/04/over-nearly-80-years-harvard-study-has-been-showing-how-to-live-a-healthy-and-happy-life/

Mirabella, A. (1993). Publishing pro takes on top job at Smart Money. *Crain's N.Y. Bus*, 9, 13.

Molm, L. D., Schaefer, D. R. & Collett, J. L. (2009). Fragile and Resilient Trust: Risk and Uncertainty in Negotiated and Reciprocal Exchange. *Sociological Theory*, 27(1), 1–32.

Momjunction. (2017). 176 Igbo Baby Names with Meanings. Retrieved on November 25, 2018 from https://www.momjunction.com/baby-names/igbo/#gref

Monin, J. K., Schulz, R. & Feeney, B. C. (2015). Compassionate Love in Individuals with Alzheimer's Disease and Their Spousal Caregivers: Associations With Caregivers' Psychological Health. *The Gerontologist*, 55(6), 981–989. Retrieved on November 5, 2017 from https://www.ncbi.nlm.nih.gov/pmc/articles/PMC4668762/

Morris, D. (1971). *Intimate Behavior*. New York: Random House.

Newcomb, T. M. (1960). The Varieties of Interpersonal Attraction. In D. Cartwright & A. Zander (Eds.), *Group Dynamics*, 2nd edn. Evanston, IL: Row, Peterson.

NHLBI (National Heart, Lung, Blood Institute) (2014). How is Broken Heart Syndrome Treated? Retrieved on December 10, 2017 from https://www.nhlbi.nih.gov/health/health-topics/topics/broken-heart-syndrome/treatment

Noller, P. (1993). Gender and Emotional Communication in Marriage. *Journal of Language and Social Psychology*, 12, 132–154.

Nordenfelt, L. Y. (1994). *Concepts and Measurement of Quality of Life in Health Care*. The Netherlands: Springer.

Nummenmaa, L., Glerean, E., Hari, R. & Hietanen, J. K. (2013). Bodily Maps of Emotions. *PNAS*, 111(2), 646–651. Retrieved on November 14, 2017 from http://www.pnas.org/content/111/2/646.full

O'Connor, M. (October 27, 2016). The Importance of Emotional Energy. *Huffpost*. Retrieved on November 15, 2017.

Parsons. T. (1943). The Kinship System of the Contemporary United States. *American Anthropologist*, 45(1), 22–38.
Petronio, S. (Ed.) (2000). *The Boundaries of Privacy: Praxis of Everyday Life in Balancing the Secrets of Private Disclosures.* Hillsdale, NJ: Lawrence Erlbaum Associates.
Petronio, S., Martin, J. & Littlefield, R. (1984). Prerequisite Conditions for Self-disclosing: A Gender Issue. *Communication Monographs*, 51, 268–273.
Pew Research Center (January 6, 2011). *Only One True Love.* Retrieved on December 10, 2017 from http://www.pewresearch.org/fact-tank/2011/01/06/only-one-true-love/
Plutchik, R. (2002). Nature of Emotions. *American Scientist*, 89(4), 349.
Post, S. G. (2003). *Unlimited Love: Altruism, Compassion and Service.* Philadelphia: Templeton Foundation Press.
Post, S. G. (Ed.) (2007). *Altruism and Health: Perspectives from Empirical Research.* New York: Oxford University Press.
Randles, J. M. (December 2016). *Proposing Prosperity? Marriage Education Policy and Inequality in America.* New York: Columbia University Press.
Rath, T. & Harter, J. (2010). *Well Being: The Five Essential Elements.* New York: Gallup Press.
Restivo, S. (1977). An Evolutionary Sociology of Love. *International Journal of Sociology of the Family,* 7(2), 233–245. Retrieved from http://www.jstor.org/stable/23027993
Ridgeway, C. (2009). How Easily Does a Social Difference Become a Status Distinction? Gender Matters. *American Sociological Review,* 74(1).
Robbins, S. P., Judge, T. A. & Odendaal, A. (2009). *Organizational Behavior: Global and Southern African Perspectives.* Cape Town, South Africa: Pearson Holdings Southern Africa.
Roberts, L. J., Wise, M. & Du Benske, L. L. (2009). Compassionate Family Caregiving in the Light and Shadow of Death. In B. Fehr, S. Sprecher & L. Underwood (Eds.), *The Science of Compassionate Love: Research, Theory and Application* (pp. 311–344). Malden, MA: Blackwell Publishers.
Rock, A. (2017). An Annual Relationship Checkup Can Keep the Spark Alive. *Health Central.* Retrieved on December 4, 2017 from https://www.healthcentral.com/article/an-annual-relationship-checkup-can-keep-the-spark-alive
Rohmann, E., Führer, A. & Bierhoff, H.-W. (2016). Relationship Satisfaction across European Cultures. The Role of Love Styles. *Cross-Cultural Research,* 50(2).
Rosenblatt, P. (1967). Marital Residence and the Function of Romantic Love. *Ethnology,* 6, 471–480.
Ross, C. E., Mirowsky, J. & Goldsteen, K. (1990). The Impact of the Family on Health: The Description in Review. *Journal of Marriage and the Family*, 52, 1059–1078.
Rothwell, J. D. (2016). *In the Company of Others.* Oxford, UK: Oxford University Press.
Rubin, L. B. (1983). *Intimate Strangers.* New York: Harper & Row.
Rubin, Z. (1970). Measurement of Romantic Love. *Journal of Personality and Social Psychology,* 16(2), 265–273. Retrieved on November 4, 2017 from http://citeseerx.ist.psu.edu/viewdoc/download?doi=10.1.1.452.3207&rep=rep1&type=pdf
Ruddick, S. (1989). *Maternal Thinking: Toward a Politics of Peace.* New York: Ballantine.
Rusbult, C. E. (1983). A Longitudinal Test of the Investment Model: The Development (and Deterioration) of Satisfaction and Commitment in Heterosexual Involvements. *Journal of Personality and Social Psychology*, 45, 101–117.
Rusbult, C. E., Martz, J. M. & Agnew, C. R. (1998). The Investment Model Scale: Measuring Commitment Level, Satisfaction Level, Quality of Alternative, and Investment Size. *Personal Relationships*, 5, 357–391.
Rusbult, C. E., Agnew, C. & Arriaga, X. (2011). The Investment Model of Commitment Processes. *Department of Psychological Sciences Faculty Publications.* Paper 26. http://docs.lib.purdue.edu/psychpubs/26
Rusu, M. S. (April 5, 2017). Theorising Love in Sociological Thought: Classical Contributions to a Sociology of Love. *Journal of Classic Sociology,* 18(1), 3–20.
Schacter, D. L., Gilbert, D. T., Wegner, D. M. & Hood, B. M. (2011). *Psychology.* Basingstoke: Palgrave Macmillan.
Schafer, J. (2015). Self-disclosures Increase Attraction. *Psychology Today.* Retrieved on December 6, 2017 from https://www.psychologytoday.com/blog/let-their-words-do-the-talking/201503/self-disclosures-increase-attraction
Schiefenhövel, W. (2009), Romantic Love. A Human Universal and Possible Honest Signal. *Hum Ontogenet,* 3, 39–50.
Schindehutte, M., Morris, M. & Allen, J. (2006). Beyond Achievement: Entrepreneurship as Extreme Experience. *Small Business Economics,* 27, 349–368.
Schioldann, J. (2011). On Periodical Depressions and their Pathogenesis by Carl Lange (1886). *Hist Psychiatry,* 22(85, Pt 1), 108–130.
Schnarch, D. (1997). *Passionate Marriage: Sex, Love, and Intimacy in Emotionally Committed Relationships.* New York: W.W. Norton & Company.
Schneiderman, I., Zagoory-Sharon, O., Leckman, J. F. & Feldman, R. (2012). Oxytocin during the Initial Stages of Romantic Attachment: Relations to Couples' Interactive Reciprocity. *Psychoneuroendocrinology,* 37(8), 1277–1285.

Sedikides, C. (1993). Assessment, Enhancement, and Verification Determinants of the Self Evaluation Process. *Journal of Personality and Social Psychology*, 65, 317–338.

Sedikides, C. & Gregg, A. P. (2003). Portraits of the self. In M. A. Hogg & J. Cooper (Eds.), *The Sage Handbook of Social Psychology* (pp. 92–122). Thousand Oaks, CA: Sage.

Sedikides, C. & Gregg, A. P. (2008). Self-Enhancement Food for Thought. *Perspectives on Psychological Science*, 3(2).

Seebach, S. & Núñez-Mosteo, F. (2016). Is Romantic Love a Linking Emotion? *Sociological Research Online*, 21(1), 1–12.

Segall, A. & Goldstein, J. (1998). Exploring the Correlates of Self Provided Health Care Behaviour. In D. Coburn, A. D'Arcy, & G. M. Torrance (Eds.), *Health and Canadian Society: Sociological Perspectives* (pp. 279–280). Toronto: University of Toronto Press.

Shapley, H. (2017). Saying I Love You Around the World. Retrieved on November 2, 2017 from https://www.match.com/cp.aspx?cpp=/cppp/magazine/article0.html&articleid=12952

Shaver, P., Hazan, C. & Bradshaw, D. (1988). Love as Attachment: The Integration of Three Behavioral Systems. In R.J.B. Sternberg & M. L. Barnes (Eds.), *The Psychology of Love* (pp. 68–99). New Haven, CT: Yale University Press.

Shaver, P. R., Morgan, H. & Wu, S. (1996). Is Love a "Basic" Emotion? *Personal Relationships*, 3.

Sheldon, P. (2013). Examining Gender Differences in Self-disclosure on Facebook versus Face-to-Face. *The Journal of Social Media in Society*, 2(1).

Simon, R. W. & Nath, L. E. (2004). Gender and Emotion in the United States: Do Men and Women Differ in Self-Reports of Feelings and Expressive Behavior? *American Journal of Sociology*, 109(5), 1137–1176.

Singer, I. (2009). *Meaning of Life: The Pursuit of Love*. Cambridge, MA: MIT Press.

Sleeth, D. B. (2011). Three Pillars of Recovery: The Role of Integral Love in Clinical Practice. *Journal of Humanistic Psychology*, 53(1), 5–25.

Solomon, R. C. (2007). Lessons of Love (and Plato's Symposium). In R. C. Solomon (Ed.), *True to Our Feelings* (pp. 51–62). Oxford: Oxford University Press.

Sorokin, P. A. (2002). *The Ways and Power of Love: Types, Factors, and Techniques of Moral Transformation*. Philadelphia, PA: Templeton Foundation Press.

Sprecher, S. & Fehr, B. (2005). Compassionate Love for Close Others and Humanity. *Journal of Social and Personal Relationships*, 22, 629–652.

Staheli, L. & Schwartz, P. (2015). *Snap Strategies for Couples: 40 Fast Fixes for Everyday Relationship Pitfalls*. Berkeley, CA: Seal Press.

Stanley, S. (2005). *The Power of Commitment: A Guide to Active, Lifelong Love*. San Francisco, CA: Jossey-Bass.

Stanley, S. (September 26, 2017). *Cheating Then and Again*. Institute for Family Studies. Retrieved on December 3, 2017 from https://ifstudies.org/blog/cheating-then-and-again

Sternberg, R. J. (1986). A Triangular Theory of Love. *Psychological Review*, 93(2), 119–135.

Sternberg, R. J. (1988). *The Triangle of Love*. New York: Basic Books.

Sternberg, R. J. (1995). Love as a Story. *Journal of Social and Personal Relationships*, 12(4).

Sternberg, R. J. & Hojjat, M. (1997). *Satisfaction in Close Relationships*. New York: Guilford Press.

Storey, J. & McDonald, K. (2013). Love's Best Habit: The Uses of Media in Romantic Relationships. *International Journal of Culture Studies*, 17(2), 113–125.

Strgar, W. (2017). *Sex That Works: An Intimate Guide to Awakening Your Erotic Life*. Louisville, CO: Sounds True.

Streib, J. (2015). *The Power of the Past: Understanding Cross-Class Marriages*. Oxford, UK: Oxford University Press.

Strodtbeck, F. L. & Mann, F. (1956). Sex Role Differentiation in Jury Deliberations. *Sociometry*, 19, 3–11.

Tamir, D. I. & Mitchell, J. P. (2012). Disclosing Information about the Self is Intrinsically Rewarding. *PNAS*, 109(21), 8038–8043.

Tannen, D. (2001). *Talking from 9 to 5: Women and Men at Work*. New York: William Morrow.

Taylor, S. (2006). Tend and Befriend: Biobehavioral Bases of Affiliation under Stress. *Current Directions in Psychological Science*, 15, 273–277.

Teske, N. (2009). *Political Activists in America: The Identity Construction Model of Political Participation*. University Park, PA: Pennsylvania State University Press.

Thoits, P. A. (1989). The Sociology of Emotions. *Annual Review of Sociology*, 15, 317–342.

Twenge, J. M. (2014). *Generation Me – Revised and Updated: Why Today's Young Americans Are More Confident, Assertive, Entitled – and More Miserable Than Ever Before*. New York: Atria Books.

Tzu Chi USA (2017). *After Disasters in 2016 Bringing Relief with Love*. Retrieved https://www.tzuchi.us/story/after-disasters-in-2016/

Ubando, M. (2016). Gender Differences in Intimacy, Emotional Expressivity, and Relationship Satisfaction. *Pepperdine Journal of Communication Research*, 4, Article 13. Available at: http://digitalcommons.pepperdine.edu/pjcr/vol4/iss1/13

Umberson, D. & Karas Montez, J. (2010). Social Relationships and Health: A Flashpoint for Health Policy. *J Health Soc Behav.*, 51(Suppl), S54–S66.

Underwood, L. G. (2009). Compassionate Love: A Framework for Research. In B. Fehr, S. Sprecher & L. G. Underwood (Eds.), *The Science of Compassionate Love: Theory, Research, and Applications* (pp. 3–25). Malden, MA: Blackwell Press.

U.S. Census Bureau (January 23, 2017). *FFF: Valentine's Day 2017: Feb. 14.* Retrieved on July 21, 2017 from https://www.census.gov/newsroom/facts-for-features/2017/cb17-ff02.html

Uvänas-Moberg, K., Arn, I. & Magnusson, D. (2005). The Psychobiology of Emotion: The Role of the Oxytocinergic System. *International Journal of Behavioral Medicine,* 12(2), 59–65.

van Hooff, J. (2017). An Everyday Affair: Deciphering the Sociological Significance of Women's Attitudes towards Infidelity. *The Sociological Review,* 65(4), 850–864.

Wachs, K. & Córdova, J. V. (2007). Mindful Relating: Exploring Mindfulness and Emotion Repertoires in Intimate Relationships. *Journal of Marital and Family Therapy,* 33(4), 464–481.

Weil, B. E. (2017). *Make Up, Don't Break Up: Finding and Keeping Love for Singles and Couples.* New York: Waterfront Digital Press.

Woollaston, V. (2013). An Adult at 18? Not Any More: Adolescence Now Ends at 25 to Prevent Young People Getting an Inferiority Complex. *Daily Mail.* Retrieved on December 5, 2017 from http://www.dailymail.co.uk/health/article-2430573/An-adult-18-Not-Adolescence-ends-25-prevent-young-people-getting-inferiority-complex.html

Wu, S. & Shaver, P. R. (1993, August). American and Chinese Love Conceptions: Variations on a Universal Theme. Poster presented at the 101st convention of the American Psychological Association, Toronto, Ontario.

Yoo, H., Bartle-Haring, S., Day, R. D. & Gangamma, R. (2013). Couple Communication, Emotional and Sexual Intimacy, and Relationship Satisfaction. *Journal of Sex & Marital Therapy,* 40(4), 275–293.

Zisook, S. & Shuchter, S. R. (1991). Depression throughout the First Year after the Death of a Spouse. *American Journal of Psychiatry* 148, 1346–1352.

8
COURTSHIP, DATING, AND MATE SELECTION

Learning Outcomes

Upon completion of the chapter, students should be able to:

1. identify two types of courtship systems
2. describe different types of arranged marriage
3. compare arranged marriage with love marriage
4. explain dating goals, strategies, and scripts
5. describe peer-influenced modern dating
6. evaluate types of contemporary dating services
7. identify types of dating violence
8. describe reasons for and effects of dating violence
9. describe strategies for preventing dating violence
10. identify factors that affect mate selection in the marriage market
11. analyze two rules of mate selection
12. explain how theories of mate selection can be applied to reality
13. compare male selection criteria between men and women
14. explain traditional and modern mate selection around the world.

Brief Chapter Outline

Pre-test

Courtship

Closed Courtship System: Arranged Marriage

Reasons for Arranged Marriage

Types of Arranged Marriages

Traditional Arranged Marriages; Modern Arranged Marriages; Westernized Version of Arranged Marriage; Advantages and Disadvantages of Arranged Marriage

Pre-test

Engaged or active learning is a powerful strategy that leads to better learning outcomes. One way to become an active learner is to begin the chapter and try to answer the following true/false statements from the material as you read. You will find that you have ready answers to these questions upon the completion of the chapter.

Courtship is dating with the goal of marriage.	T	F
When people date, they are not necessarily committed to marrying the person who they are dating.	T	F
In forced arranged marriages, the bride and groom are forced to marry despite their opposition to the marriage.	T	F
Unlike courtship, the end goal of dating is not necessarily marriage.	T	F
A dinner date is considered a traditional dating behavior.	T	F
Online dating is the practice of searching for a romantic or sexual partner on the internet via a dedicated website.	T	F
The marriage market is a marketplace where eligible partners try to find a mate with equal assets and liabilities.	T	F
Acquaintance rape refers to sexual assault or rape of a victim who knows or is familiar with his or her rapist.	T	F
Homogamy is the practice of choosing a mate with similar characteristics.	T	F
According to filter theory, social structure sifts the pool of eligible candidates for a mate according to specific criteria.	T	F

Upon completion of this section, students should be able to:
LO1: Identify two types of courtship systems.
LO2: Describe different types of arranged marriage.
LO3: Compare arranged marriage with love marriage.

Although mate selection is universal, the process by which partners are selected is determined by cultural and social factors (Bejanyan *et al.*, 2015). *Collectivist cultures* such as those of Asia and Africa emphasize family goals above individual needs. With globalization, there is a growing trend in collectivist cultures for young adults to exercise greater personal choice in their mate selection and engage in dating (Buunk *et al.*, 2010) in spite of parents' disapproval (Dugsin, 2001). *Individualist cultures* such as those of Western countries, with their emphasis on personal desires and independence (Markus & Kitayama,

1991), expect individuals to exercise personal control over their own partner selection and relationship maintenance (Bejanyan *et al.*, 2015).

Two types of courtship systems co-exist globally: "love" marriages in the West and "arranged" marriages in many parts of Asia and Africa (Penn, 2011).

Courtship

Let's start this chapter with a few definitions. **Courtship** is dating with the intent of marriage. During courtship, a couple learns about one another and determines whether they will become engaged and marry each other. Sometimes, in their search for love, people attempt to find a prospective mate through assistance. A **matchmaker** is a person who arranges relationships and marriages between others, either informally in cultural communities or formally in a matchmaking business. Dating services can serve as matchmakers and help people find eligible mates. Members of a person's **social network** online and offline may serve the role of matchmakers for people to meet socially and form relationships that may develop into marriage. Let's now take a closer look at courtship systems, starting with the closed courtship system.

Closed Courtship System: Arranged Marriage

In a **closed courtship system**, the parents of adult children choose eligible mates for their sons and daughters to form an arranged marriage. **Arranged marriage** is a type of marital union where the bride and groom are selected by their families. A marriage is proposed and arranged by parents, matchmakers, matrimonial sites. A matchmaker often serves the role of a go-between for families during the mate selection process.

Reasons for Arranged Marriage

An arranged marriage is a product of culture. The first major transition in heterosexual mating was around 10,000 to 15,000 years ago in the agricultural revolution when we became less migratory and more settled, leading to the establishment of marriage as a cultural contract (Garcia & Heywood, 2016). Parents had a great deal of influence over their offspring's choice of a mate because children were economically dependent on and had strong ties to the family. *Family allocentrism* (collectivist personality trait)—the strength of closeness and devotion between family members (Lay *et al.*, 1998)—can potentially influence the willingness for children to take their parents' opinions into consideration when selecting a mate (Bejanyan *et al.*, 2015). It was the norm for adult children to marry for pragmatic reasons. Instead of placing emphasis on the romantic connection between two partners, parents encourage adult children to assign more weight to pragmatic qualities in a prospective partner such as economic resources, social and religious status, and positive interactions between the two families (Myers *et al.*, 2005).

Arranged marriages were practiced among royal families and ordinary families. In royal families, one purpose of arranged marriage was to keep the royal bloodlines pure. Royal families also used arranged marriages to form political and economic alliances and to maintain territorial stability. Marriage for political, economic, or diplomatic reasons was a pattern seen for centuries among European rulers (Fleming, 1973). A **marriage of state** is diplomatic marriage between two members of different nations or between two power coalitions to secure peace between two parties. In the film *Helen of Troy*, the story of the Trojan War in 1100 BCE, Paris of Troy sails to Sparta to secure a peace treaty between the two city-states. In nonroyal families, the choice of marriage was made by the parents based on economic considerations (Murstein, 1971). Parents are considered to have the authority and experience to select an appropriate mate for their adult children. An arranged marriage was the only acceptable beginning of a new family, since sex was something that couples engaged in after marriage for procreation.

Types of Arranged Marriages

Broude and Greene (1983) have reported that 130 cultures have elements of arranged marriage after studying 142 cultures worldwide. Arranged marriages were very common throughout the world until the eighteenth century (O'Brien, 2008) and they are still a frequently used method of matchmaking in many areas.

Traditional Arranged Marriages

An extreme form of loveless marriage is a **forced arranged marriage** in which the two partners are married without their consent or against their wishes. In the nineteenth and twentieth centuries, people in India defined self and nation through a system of arranged marriage, in which parents claimed the authority to give final consent to a match (Black, 2017). Parents relied on the intuition of village elders, family members, or friends to select which pairs of unacquainted singles would be compatible (Finkel *et al.*, 2012). When the parents found the match to be a good one, they would formalize and present the match to the bride and groom. Couples would marry despite their opposition and had no right to reject the arranged marriage. Courtship can be omitted in forced arranged marriages where the couple does not meet before the wedding. Globally, 11 million girls were forced to marry before the age of eighteen, including girls (46.4 percent) in South Asia and girls (42 percent) in Africa (Statistical Brain, 2016).

A less extreme form of arranged marriage is **arranged marriages without force,** in which parents choose a mate but do not force their adult child to marry that person. An arranged marriage in Japan would begin with courtship that would allow love to develop and would lead into an arranged marriage (Lebra, 1978). While this may be the case, passionate love is seldom encouraged outside of movies and stories (Derne, 1995). If parents saw the value of an arranged marriage, they would often persuade their adult child to marry the selected candidate. Because of parental influence, individuals sometimes date in secrecy, exercising temporary liberties over their own partner choice until they are expected to abide by parental expectations and choose a marital partner congruent with their parents' standards (Netting, 2006).

Modern Arranged Marriages

Passionate love may be a cross-cultural phenomenon and not just confined to the West (Hatfield & Rapson, 2002). Western notions of romance and passion are increasingly influencing collectivists' perceptions of relationships (Netting, 2010). Many cultures practice **modern arranged marriages** in which parents choose a mate and respect their adult children's opinions. In *modern arranged marriages with courtship***,** the parents select a prospective mate and allow the couple to court and develop a love relationship. The couple has the freedom to get to know one another and continue courtship until they decide to marry each other. Dating and engaging in sexual activity prior to marriage are often considered inappropriate in collectivist cultures (Lau *et al.*, 2009) because it contradicts many traditional cultural practices regarding mate selection (Myers *et al.*, 2005). The parents may arrange meetings or chaperone dates between the couple when they feel that their children's actions can pose a greater liability to the family (Nesteruk & Gramescu, 2012).

Westernized Version of Arranged Marriage

The institution of the old arranged marriage is subject to numerous pressures, such as declining social networks, high geographical mobility, and growing complexity in the choice of a marital partner (Agrawal, 2015). For example, many young adults leave their home countries for Western countries to pursue higher education and to stay there to work. They seek love but have difficulties finding the right partners. "There were certain expectations that I wanted my life partner to have but I just never found anyone who was close enough to those expectations" (Toledo, 2009). Many turned to tradition and asked their parents to arrange a match for them. They have entered **Westernized versions of arranged marriages**—mate selection endorsed by parents through media usage and free choice of partners. Indian matrimonial websites have become popular hunting grounds for prospective partners in India's marriage market as of the late 1990s (Titzmann, 2016). Instead of selecting potential mates in their inner circle, many parents place matrimonial ads on the internet on their adult children's behalf. The use of internet-mediated services becomes a means of undertaking such "kin work" (Agrawal, 2015).

Advantages and Disadvantages of Arranged Marriage

Most modern arranged marriages have high levels of martial stability. Statistic Brain (2016) reports that 53.25 percent of marriages in the world are arranged and the global divorce rate for arranged marriages is 6 percent, compared with 50 percent of love marriages ending in divorce in the United States.

Why is this so? A key difference between arranged marriage and love marriage is that an arranged marriage is more of a practical partnership than a romantic fantasy. People generally stay in the relationship because arranged marriages are more about the needs of the families than the individuals (Regan *et al.*, 2012). As marriage is a serious life-long commitment, parents carefully check a candidate's background and suitability as a partner. Adult children often respect their parents' decision about a partner and seek family support in the case of a relationship issue. Parents support the relationship and take responsibility for keeping it alive.

In arranged marriages, people marry without loving each other at the beginning but they know that the feeling of love can develop. As people say in India, "First comes marriage, then comes love." Love is expected to grow as the spouses learn more about each other as the years go by (Myers *et al.*, 2005). A study of couples from twelve countries and six religions revealed that self-reported love grew from a mean of 3.9 to 8.5 on a 10-point scale in the first study (n=30) and that love grew from a mean of 5.1 to 9.2 in the second study (n=22) (Epstein *et al.*, 2013). Indeed, commitment was found to be the most important factor that contributed to the growth of love. If the feeling of love does not develop, they will form a partnership and committed relationship.

The disadvantage of forced arranged marriages includes denying a couple's freedom and individual choice. Adult children may experience excessive pressure from parents to act pragmatically, suppress any feelings of passion for a romantic partner (Hamid *et al.*, 2011), and sacrifice their own happiness for the sake of keeping their parents happy. Even if couples are not forced into marriage, familial pressure still makes them feel like it. Another disadvantage is that the couple might experience lack of privacy in their relationship if parents chaperone dates between them. The dowry system could be a disadvantage as the bride's family is expected to pay enormous amounts of money or give expensive things to the groom's family (see Chapter 9).

There is a growing trend in almost every culture for young adults to exercise both free mate choice and arrangement in their mate selection, for instance, in the *Yali*—a native Papuan (Indonesian) tribe (Sorokowski *et al.*, 2017). India's marriages (90 percent) are still termed as "arranged" but increasing social, physical, and media mobility serve as key factors in social change around women and marriage (Titzmann, 2016). *Elopement*, known as love marriage, has become dominant in some parts of India (Keera, 2013).

Open Courtship System: Love Marriage

Free mate choice is the version of boy-meets-girl that we see in films and TV shows. In an **open courtship system**, people choose their partners, get to know each other, move on to being in a serious relationship, and make their own decisions regarding marriage. Traditionally, it is the role of a male to court and encourage a female to accept a *marriage proposal*—an event where one person in a relationship asks for the other's hand in marriage (Schlesinger, 2008). If it is accepted, it marks the initiation of *engagement*—a mutual promise of later marriage. Open courtship systems produce **love marriages** in which the marital union is based on love and is the sole decision of the couple.

Love marriage suggests (1) individual freedom to choose a life partner, (2) the importance of romantic companionship, and (3) the changing status of women (Black, 2017). Love marriages are a product of modern society, in which young adults enter into the labor force and become economically independent from their parents. As they become educated and financially independent, women no longer need to depend on their husbands for economic and social security. As such autonomy in the mate selection process increased, the old-fashioned way of close courtship under parental supervision began to fade in individualist cultures. By the end of the nineteenth century, love was regarded as the most important prerequisite to marriage and is still Americans' top reason to marry today (Livingston & Caumont, 2017).

Advantages and Disadvantages of Love Marriage

Love marriage is associated with individualism and freedom from parental intervention and control. Individualist cultures such as that of the United States value self-sufficiency and the development of personal identity (Madathil & Benshoff, 2008). Individual happiness is considered the chief purpose of marriage (Beigel, 1951). The couple takes control over the mate selection process. In some cases, this happens without the knowledge of their parents. One reason for the beneficial effects of free choice mating might be that these pairs are more compatible than arranged pairs (Ihle *et al.*, 2015). Love marriage is a union between two individuals who are attracted to each other. After a period of courtship, they may decide to become life partners.

Why is the divorce rate so high if love marriages work? In a love marriage, partners tend to use screening criteria that emphasize attractiveness. "Attractiveness matching" gives heavy weight to the physical attractiveness of potential partners. Sternberg's *triangle theory of love* (1986) affirms *intimacy* (i.e., feelings of closeness) and *passion* (i.e., feelings of sexual desire) as the foundation for love marriages. Passion is touted as the essence of love—the basis upon which love is cultivated (Hatfield & Rapson, 2002). Americans (71 percent) believe in love at first sight (Statistic Brain, 2017b). What does love at first sight feel like? When you meet someone, your heart starts racing and your palms get sweaty. People experience intense euphoria and elevated levels of testosterone and estrogen at the beginning of a relationship. This kind of *atypical* form of love relationships is called *limerence*—an involuntary interpersonal state that involves an acute longing for emotional reciprocation (Tennov, 1979) similar to substance dependence and obsessive-compulsive disorder (Wakin & Vo, 2011).

Student Accounts of Courtship

Finding a date all depends on your age group. Most of the younger crowd still goes out to clubs, malls, and find dates at school. The internet is also a big dating source for all ages when they are finding love. Outside of that, I just think it is being at the right place at the right time, even if that means bumping into someone at a supermarket. You never know when you are about to find love; sometimes love finds you. And if it is meant to be, then it will be. Everything happens for a reason (personal communication, September 23, 2017).

Passion is the strongest element in the early stages of relationship formation (Myers *et al.*, 2005) and such feelings begin to fade by as much as a half in eighteen months (Epstein *et al.*, 2013).

Passion is likely rooted in the idealization of one's partner (Sternberg, 1986). *Idealization* occurs when lovers generate positive illusions by maximizing virtues and minimizing flaws. When idealized images give way to realistic ones and when the intense romance of early marriage weakens, marriage partners are disappointed by the changes (Niehuis *et al.*, 2006). When a couple feels that the passion that they fell in love for is gone, they may consider their marriage a failure and seek the next idealized mate.

Love changes in time if the couple is not committed to the relationship. Westerners leave their love lives to chance or fate whereas those in other cultures look for more than just passion (Epstein *et al.*, 2013). It is commitment that allows a love relationship to be sustained during fluctuations of passion and intimacy (Sternberg, 1988). Commitment and sacrifice that strengthen love in arranged marriages could be introduced into autonomous marriages in Western cultures, where love normally weakens over time (Epstein *et al.*, 2013).

Courtship Issues in Love Marriage

"The basic cause of divorce ... is faulty mate selection and inadequate preparation for a companionship type of marriage" (Waller, 1938). Scholars have detected two courtship issues (Huston *et al.*, 2001; Niehuis & Huston, 2000).

Event-driven Courtship Surra and Huston (1987) distinguish between event-driven and relationship-driven courtships. **Event-driven courtship** refers to a relationship that escalates based on external factors, such as moving in together to save on rent, that have little to do with a couple's true level of intimacy. The courtship issue here is that premarital partners have a very long courtship characterized by very little passion (Niehuis *et al.*, 2006). *The early exiters* are those couples who had a long but passionate-less courtship and divorced after two to seven years of marriage. Couples in such unions report more conflict and greater uncertainty about the relationship (Surra & Huston, 1987) but don't want to disrupt the courtship (Huston & Burgess, 1979). They attempt to rekindle an unpredictable romance through the wedding itself. Loss of love and affection early in marriage has its roots in the couple's premarital courtship (Niehuis *et al.*, 2006).

Length of Courtship and Relationship-driven Courtship **Length of courtship** refers to the amount of time between the couple's first date and the decision to get married. Grover *et al.* (1985) explored three factors that contributed to marital satisfaction: "length of courtship, age at marriage, and whether they broke up at least once while dating." The only factor that consistently correlated with marital satisfaction was the length of courtship. Hansen (2006) found a positive correlation between length of courtship and marital satisfaction. He reported the highest divorce rates for couples who spent less than six months dating. With longer periods of acquaintance, individuals can screen out incompatible partners. Happily married couples dated for an average of 25 months (Huston *et al.*, 2001).

In courtship experience, some partners may blindly rush into marriage. *The delayed-action divorcers* are those couples who had passionate but short courtships of less than twelve months (Niehuis *et al.*, 2006) and divorced after at least seven years of marriage. Some couples focus on the wedding rather than the marriage (Hawkins *et al.*, 2004) and follow the romantic ideal but have the least consensus about commitment (Niehuis *et al.*, 2006). Such early highly romantic courtships make people stick it out longer (Huston *et al.*, 2000) when they became less affectionate over the first two years of marriage. Moderate feelings of passion and an average length of courtship may be characteristic of stably married couples (Niehuis & Huston, 2000).

Upon completion of this section, students should be able to:
LO4: Explain dating goals, strategies, and scripts.
LO5: Describe peer-influenced modern dating.
LO6: Evaluate types of contemporary dating services.

Dating

Courtship occurs when two people are prepared to make a commitment to marriage. Where courtship had been customary, dating became a new institution during the transition to an industrial society. The practice of dating has a long history in the United States, going back about a century (Bailey, 1988). Many agricultural workers flocked to cities to find job opportunities. When parents worked away from the home, teenagers spent an increasing amount of time free from parental supervision. Later, when young adults found jobs in cities and became financially independent, they were likely to search for romantic partners. Courtship encounters were soon mobilized by the average citizens' increased access to automobiles and women's increasing role in the public sphere (Eaton & Rose, 2011). By the mid-1920s, going on "dates" had become a "universal custom" for young men and women in the United States and the dominant script for romantic interactions between singles (Bailey, 1988). **Dating** is a form of courtship where an individual chooses to meet socially with another person with the aim of each assessing the other's suitability either as a prospective partner in a relationship or marriage.

Dating is considered a means by which people become coupled. A Harvard study reveals that Millennials are looking for guidance on how to form loving relationships (Talalas, 2017). The respondents (70 percent) wish that they had received more information from their parents about "how to have a more mature relationship" (38 percent), "how to deal with breakups" (36 percent), "how to avoid getting hurt in a relationship" (34 percent), or "how to begin a relationship" (27 percent) (Weissbourd *et al.*, 2017). Let's take a closer look at dating as a route to coupling.

Where Do People Couple Up?

We used to live in a tight-knit inner circle. In the 1930s in the United States, most couples lived within ten blocks of each other when they met (Vandenberg, 1972). At that time, the sheltered courtship practice of "calling" was closely monitored by the family and community, and took place in the bachelorette's home (Rothman, 1984). Improved transportation and high geographical mobility widened our circles for meeting people and "moved courtship from the home to public locations, such as movie theaters, dance halls, and restaurants" (Eaton & Rose, 2011). Americans met their partners in the last decades, listed by importance: through mutual friends, in bars, at work, in educational institutions, at church, through their families, or neighbors (Rosenfeld & Thomas, 2012).

Friends are the primary introducers. It is a great way to meet people through friends. We would probably not marry our best friends, but we are likely to end up marrying a friend of a friend or someone we coincided with in the past (Ortega & Hergovich, 2017). Married couples (63 percent) reported finding their mates through a friend; only 9 percent of women found relationships at a bar (Statistic Brain, 2017a). Figure 8.1 shows that respondents have met their partners through friends (39 percent), followed by at work (15 percent) or in bars or other public areas (12 percent), sport/religion/hobby events (9 percent), internet/dating apps (8 percent), family (7 percent), school (6 percent), and speeding date (1 percent).

Dating Goals and Strategies

Unlike courtship, the end goal of dating is not necessarily marriage. Dating is no longer the direct path to

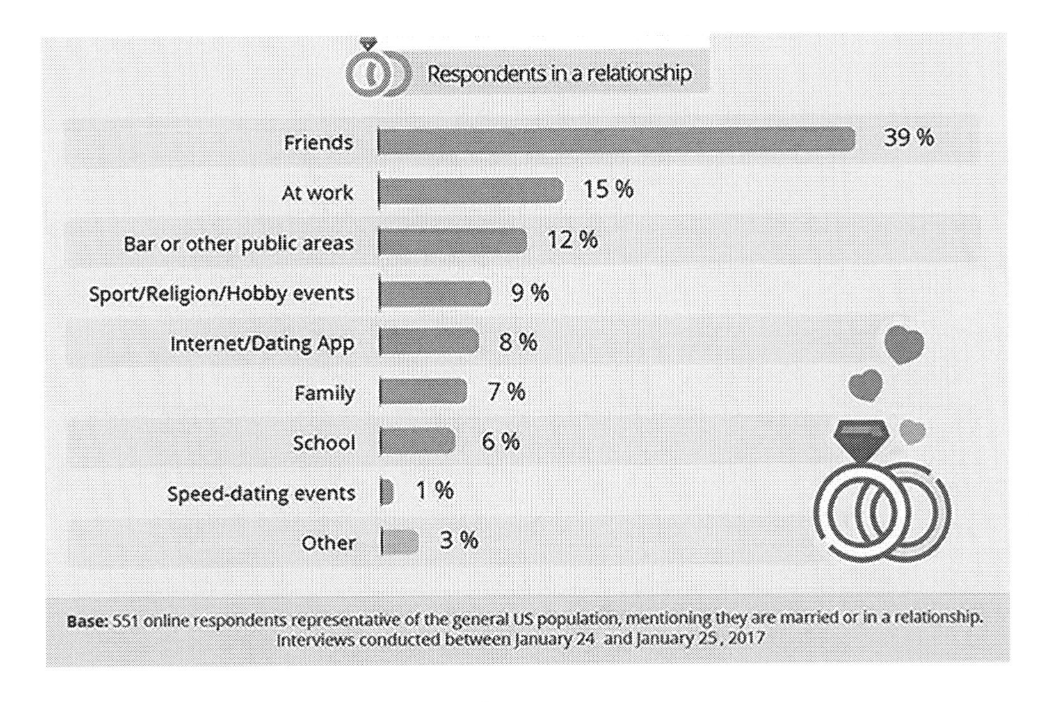

Figure 8.1 How Did You Meet Your Spouse or Partner?
Source: ReportLinker. https://www.reportlinker.com/insight/finding-love-online.html

marriage that it once was (Libby, 1976). If the question posed by courtship is, "Would this be a man or woman I would want to be my husband or wife?" then the question posed by dating is, "Where could I find a man or woman who I want to spend time with in a romantic way?"

Dating Goals

People are dating for many purposes. Mongeau (2004) identified six purposes for dating: recreation (to have fun), socialization (to get to know the partner), status grading (increasing social status by dating an attractive partner), companionship (finding a friend to do things with), mate selection/courtship (finding a spouse), and intimacy (establishing a meaningful relationship). Dating goals may change over the course of a date. For example, a person may go into a date with the purpose of having fun but during the date decide to pursue a love relationship.

First Date Expectations

A **first date** is an initial meeting between two persons with the goals of screening potential dating candidates or producing a relationship outcome such as a sexual partnership, friendship, romantic relationship, or life partnership. Figure 8.2 shows that expectations for a date can be formed by the communicator, relationship, and context factors (Morr Serewicz & Gale, 2008). *The communicator factor* involves the communication style that allows the individual to predict how the partner will communicate in a dating situation. Greater communication and greater disclosure predicted first date success (Sharabi & Caughlin, 2017). Most dating couples were found to use a female-demand/male-withdraw style as their predominant communication pattern and its use increased in response to difficult discussions (Vogel *et al.*, 1999). *The relationship factor* involves characteristics such as the degree of attraction or similarity that describes

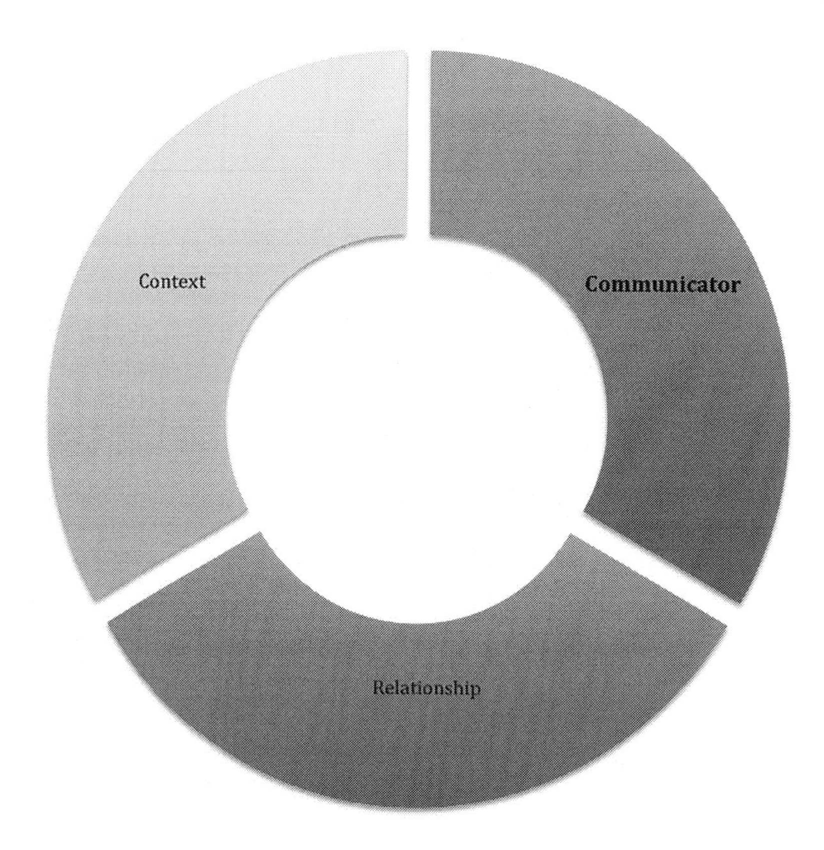

Figure 8.2 The Expectations for a Date Can Be Formed Based on the Communicator, the Relationship, and the Context
Source: Natalie Stevenson. https://commons.wikimedia.org/wiki/File:Context,Communication%26Relationship.jpg

the relationship between the two partners. During the initial relationship stage, a couple's main task is to establish trust and construct a stable satisfying relationship structure (Zhou *et al.*, 2017). *The context factor* involves situations such as the environment that can enhance or diminish the interaction on the date and help structure one's goals (Mongeau *et al.*, 2004). Women develop short-term dating strategies such as using temporal context to assess the long-term potential of a current partner (Buss & Shmitt, 1993).

Dating Goals by Gender

The three most common goals that people have on first dates are to reduce uncertainty, achieve relational escalation, and have fun (Mongeau *et al.*, 2004). Initial romantic encounters are known as being vehicles for uncertainty reduction (Afifi & Lucas, 2008). More women than men go on first dates to reduce uncertainty (see Figure 8.3).

Men and women may have different goals to get where they want to go with love. Women are more likely to express companionship, friendship, and romantic relationship goals and consider the first date in terms of its relational implications than men (Mongeau *et al.*, 2004). Women look for certain cues on a first date. For example, they will observe if the man comes to pick her up, opens doors for her, pays for the date, or compliments her on how she looks. If her expectations are not met, she may decide to end the date. If they are met, she may wait for a second date in

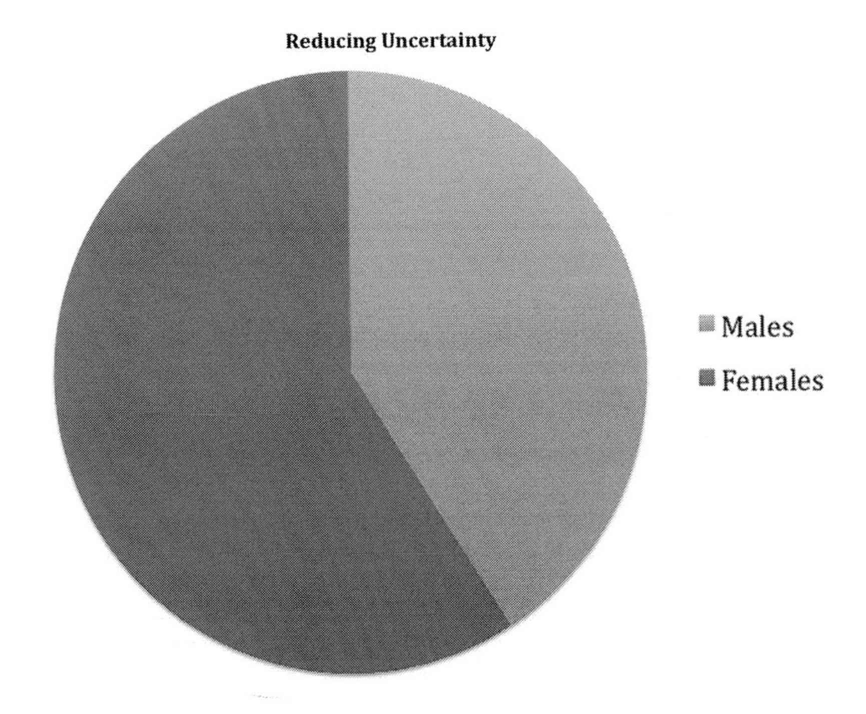

Figure 8.3 Men vs. Women Who Go on First Dates to Reduce Uncertainty
Source: Natalie Stevenson. https://upload.wikimedia.org/wikipedia/commons/1/14/Reducing_Uncertainty.jpg

the future. Women are more likely to insist that sexual intercourse occur in relationships that involve affection and marital potential (Townsend & Levy, 1990).

Men have a higher expectation for sexual goals. Men perceive relationships in more of a sexualized manner than women do (Abbey, 1987), report willingness to engage in sexual relations after any length of acquaintance from one hour up to five years, have a greater interest in uncommitted sex (Buss & Shmitt, 1993), are more sexually autonomous (Peplau *et al.*, 1977), and express more interest in having sex with hypothetical partners (Epstein *et al.*, 2007). College male students have higher sexual expectations than do female students; their sexual expectations are heightened when alcohol is available (Morr & Mongeau, 2004).

Dating Strategies

Sexual strategies theory suggests that both men and women have evolved different mechanisms that underlie short-term and long-term mating strategies. Dating is a means of determining whether potential romantic partners will meet their short- or long-term relationship needs. Short-term mating strategies seem to work for both men and women (Buss, 2018). Men do not want to commit, and pursue a short-term mating strategy. Many young women are focusing on their education and are career-oriented. Dating apps like Tinder offer them information about potential mates. Thus, men are making the whole mating system shift towards short-term dating, and women are forced to go along with it in order to mate at all (Buss, 2018). Men have positive control in a sexual encounter (using available strategies to initiate sex) while women have negative control (using strategies to avoid having sex) (Ehrmann, 1959).

Gray (1997) explored **five stages of dating**—attraction, uncertainty, exclusivity, intimacy, and engagement that moves romantic partners toward more committed relationships. Baxter and Bullis (1986) investigated turning points that are related to positive and negative changes in romantic relationships. A **turning point** refers to an event that is associated with a change in a relationship. First dates represent an important early event in the development of dating relationships (Mongeau *et al.*, 2004). Both men and women are looking for physical attractiveness that fits their taste and style. For example, women were likely to use cosmetics when preparing for a date and believed that sexual attractiveness could be altered using cosmetics (Franzoi, 2001). Participants identified the first date (i.e., "get to know you time" events) with positive relationship consequences. First dates can represent "turning points" where a relationship might move from platonic to romantic (Morr & Mongeau, 2004).

Knapp (1984) suggested a **relationship phase model** in terms of three phases—coming together, maintenance, and coming apart. In the *coming together phase*, partners meet, exchange information about each other, spend time together, and become a couple (Alder & Rodman, 2003). A person's personality is related to the formation of the person's relationship satisfaction; this is called the *actor effect* (Zhou *et al.*, 2017). The advantage of *extraverts*—being social and expressive in the initial stages of relationships—is supported by studies (Vazire, 2010). Also, the personality is related to the perceived relationship quality of the person's partner; this is called the *partner effect* (Kenny *et al.*, 2006). For example, a partner's personality will affect the person's evaluation of relationship quality through daily interaction, emotional sharing, and conflict resolution (Zhou *et al.*, 2017). Women consistently expressed a greater desire for relationship support than men (Perrin *et al.*, 2010) and would like their partner to do things such as "be a good listener to me," "be sympathetic to my feelings," and "remember my birthday."

Dating Scripts

The significance of a successful first date to relationship escalation is highlighted when first *date scripts* are considered (McDaniel, 2005). **Dating scripts** are what allow people to predict the actions of others and serve as guidelines for their own decisions on how to react to the other person (Laner & Ventrone, 2000). Because society is invested in the outcomes of courtship (i.e., marriage), and because dating is a public act, it has always come with a host of prescribed rules and expectations (Gilligan, 1982). These scripts are used to organize, interpret, and predict the behavior of individuals in dating encounters (Simon & Gagnon, 1986). From the 1980s to the 2000s, heterosexual dating in the United States remained highly gender-typed in terms of cultural scripts (e.g., beliefs, ideals, and expectations) and interpersonal scripts (e.g., the use of a specific cultural scenario by an individual) (Eaton & Rose, 2011).

Cultural Scripts

Cultural scripts are "collective guides" for situational norms, values, and practices that are characteristic of and accessible to cultural insiders (Goddard & Wierzbicka, 2004). If a person displays *dating scripts* and behaviors deemed to be negative, the other person may decide to end the date or the activity they are engaged in. Thus, a first date might be highly scripted because it helps to create a good impression (Simon & Gagnon, 1986).

First dates tend to be more "traditional" (male-dominated) but in certain cases, such as gay dating, the circumstances are different (Laner & Ventrone, 2000). In dating, a couple sets a specific date, time, and place to meet socially. According to old-fashioned dating etiquette, the male had to get the permission of the female's parents to take their daughter on a date. After the male spoke with the female's parents, he came to pick her up at her house and promised to bring her home before a set curfew, opened the car door and took her out, paid the bill, took the girl home, and said goodnight. What did dating look like?

> In those days, trends emerged like "rating and dating," meaning men with more money got

> the prettiest ladies and more of them. And "petting and paying" referred to a man paying for a woman—and then expecting her to return to his car later to "pet," anything but going all the way was fair game.
>
> (Persch, 2008)

The hypothetical first-date scripts heavily emphasized gender roles. In a study of 21 dating scripts that college students expected to occur on the first date, Morr Serewicz and Gale (2008) found a trend toward traditional gender roles for date partners. The gender roles indicate that maintaining the traditional gender-power ratio is a significant aspect of creating a positive impression (Lipman-Blumen, 1984). Four dating advice manuals (Titus & Fadal, 2008; Casey, 2009; Miller, 2004; McGraw, 2005) were gender-typed. For example, women are reminded to indirectly manage their relationship from behind the scenes, given that men "need to feel like the leader in relationships" (McGraw, 2005). Men are expected to initiate, plan, and pay for the date, and are the sexual aggressors while women are supposed to assume a subordinate role by being alluring, facilitating the conversation, and limiting sexual activity (Rose & Frieze, 1993). For example, gender-typed media versions of dating may have a strong influence on novice daters. Undergraduate students who watched *reality dating programs* were more likely to endorse a *double standard of sexual behavior*—"a man should be more sexually experienced than his wife" and "a woman who initiates sex is too aggressive" (Eaton & Rose, 2011).

The norm of female thinness persisted as a theme in the cultural dating script (Zurbriggen & Morgan, 2006). Even sixth-grade girls expressed concern about their appearance and about their weight (Gershon *et al.*, 2004). Both women and men rated an ideal woman as thinner than an ideal man (Stake & Lauer, 1987). A study of 5,810 heterosexuals who placed profiles on Yahoo revealed that men were five times more likely than women to prefer to date only those with fit or toned bodies; White men preferred a thin and toned woman; African American and Latino men accepted a thin and toned woman and one with a thicker body (Glasser *et al.*, 2009); Mexican American women expected that Mexican American men would prefer less achieving women as potential partners (Gonzalez, 1988).

Interpersonal Scripts

Interpersonal scripts are the behavioral enactment of a specific cultural script by an individual (Simon & Gagnon, 1986). For example, women were seen as sexual objects and emotional facilitators but men as planners, economic providers, and sexual initiators (Rose & Frieze, 1989). On a first date, a man is supposed to control the public domain (i.e., making plans, transporting the date, and paying the bill) whereas women are supposed to control the private sphere (i.e., level of sexual intimacy and charm) (LaPlante et al., 1980). Women reported more often expressing both positive and negative emotions in dating relationships than men (Sprecher & Sedikides, 1993).

Interpersonal script appears useful in the analysis of *unrequited love*—dating situations in which a person pursues someone who does not return their interest. Sinclair and Frieze (2005) found that both female and male undergraduates believed that when they were the pursuer, the object of their affection was more accepting than when they themselves were the object of unwanted pursuit. Research indicates racial hierarchies in decisions about whom to date. The college-educated Latinas stated openness to dating men of all racial/ethnic backgrounds but found that White men seemed reluctant to date members of minoritized groups (Muro & Martinez, 2018).

Alksnis *et al.* (1996) used dating scripts to determine what constitutes a bad or good date and found five date events unique to women's bad date script such as "date made sexual advances too early, repeatedly tells you how sexy you look, stares at you, leans in close to you whenever you are sitting together, and repeatedly touches you." In general, those who are more knowledgeable about an event have more well-developed scripts (Chase & Simon, 1973) and may feel freer to modify standard ways of behaving (Rose & Frieze, 1989).

The availability of alcohol sold in restaurants and in all licensed bars could influence dating scripts. Alcohol can reduce anxiety and increase self-disclosure for men but women are more concerned that their actions might be misinterpreted if alcohol is involved (George & Norris, 1991). Individuals were likely to approve the friendship goal when alcohol was not involved (Mongeau *et al.*, 2004).

Who Pays on a Date?

If the initial interaction is successful, potential partners might move on to a dinner date. A **dinner date** is a traditional date to eat together in a restaurant. Who pays for dates? Heterosexual dating scripts remain quite traditional, with the man expected to ask a woman out and to pay for the date (Emmers-Sommer *et al.*, 2010). A survey of 12,899 heterosexuals on 21 campuses revealed that men (63 percent) on a "recent date" paid; both women and men (19 percent) paid; and woman (2 percent) treated (England & Bearak, 2013). Most men (82 percent) and women (58 percent) reported that men paid for most expenses, even after dating for a while (Lever *et al.*, 2015).

Why Do Men Likely Pay for the Date?

When the bill came, the server was likely to place it in front of the man. The man-initiated and man-paid date script was the dominant cultural script. The practices are still resistant to change for several reasons. Men pay on a date because traditional dating scripts stem from *patriarchal society*—male dominance. Tradition goes all the way back to a time when men were breadwinners. Gender provides "a clear framework of cultural beliefs that differentiate men and women" (Ridgeway, 2011). Men's paying reflects their historical domination of financial resources and reinforces the gender stereotype of "male as provider" (Lever *et al.*, 2015). West and Zimmerman (1991) recognized the social construction of differences and conceptualized the achievement as "doing gender."

Men who hold on to conventional notions of chivalry are likely to pay for dating expenses. **Chivalry** is the idea that men, to show they cherish and protect women, engage in acts specifically for women that they may not do for other men (Lever *et al.*, 2015). Examples of chivalry include picking the woman up, opening the door, and paying for the dates. Many women valued chivalry as sign of a respectful and caring man. Some men pay because they feel socially obligated to do so, and may feel guilty if they fail to live up to these gendered expectations (Lamont, 2014). Chivalry benefits men because the early stages of dating are fraught with uncertainties and ambiguities; men who pay on a date can be positively evaluated and this provides one incentive for men to continue to pay for dates (Lever *et al.*, 2015). Men who do not pay may be afraid of being viewed as lacking economic resources.

Research indicates the link between money and sexual intimacy. The party who pays for a first date shapes the expectations of what is going to happen next (Patrick, 2017). Over dinner, a couple enjoys engaging conversation and chemistry. Men who pay for a date are more likely to be attracted to their date (Cohen, 2016) and have higher first date sexual expectations than women (Emmers-Sommer *et al.*, 2010).

Egalitarian Dating

Today, the bill may be placed in the middle of the table. The arrival of the check can spark a showdown if one picks it up or they split the bill or alternate. **Egalitarian dating** is a form of courtship that is not structured by gender role norms and instead favors and promotes equality between partners and the sharing of power and responsibilities (Fay & Eaton, 2014).

Feminist Perspective on Dating

Dating is a gendered part of heterosexual romantic relationships that supports men's power and gender stereotypes (Mahoney & Knudson-Martin, 2009). If gender roles and norms are used to stabilize and structure early relationship interactions, they may establish a trajectory for future interactions that contributes to the perpetuation of gender-differentiated behavior (Eaton & Rose, 2011). **Feminist theory** focuses on analyzing gender inequality. Dating is a way of examining progress towards gender equality. Egalitarian practices suggest that gender should not determine who pays for dating expenses.

Sharing dating expenses or woman-initiated dates implies that alternative scripts were developed in the 1990s. Most men (72 percent) had been on at least one date where the woman paid all the date expenses (Lottes, 1993). Men and women spend similar amounts of actual effort (e.g., money) on one-night stands (Pedersen *et al.*, 2010). Along with young men and women in their twenties, younger women with college degrees were most likely to endorse egalitarian practices (Lever *et al.*, 2015). Therefore, traditional gender stereotypes are loosening their hold, where the performers by their social actions may be "undoing gender" (Risman, 2009).

Why Do Women Share Dating Expenses?

For many women, sharing expenses made them feel equal. One important motive to share dating expenses is that both men and women want their personal actions to be consistent with their professed beliefs (Lever *et al.*, 2015). More women are economically independent and would like to share date expenses (Korman, 1983). More women (10.3 percent) than men (8.7 percent) have received master's degrees (U.S. Census Bureau, 2017b). Among couples where both partners worked, wives (9.9 percent) earned at least $30,000 more than their husbands in 2016 (U.S. Census Bureau, 2017c). These social and economic changes in the public sphere have been accompanied by documented social and economic changes in couples' private sphere (Ridgeway, 2011).

The link between money and sexuality explains why some women choose to pay for dates. Women (64 percent) experienced the incidence of *unwanted sex on a date* when the man initiated and paid for the date (Hannon *et al.*, 1995). Women aged 18–25 (22 percent) and women aged 56–65 (46 percent) reported paying for dates to avoid the pressure to be sexual on a date (Lever *et al.*, 2015).

Paying Attitudes and Behaviors by Gender

A study of 17,607 unmarried heterosexuals reveals that men (64 percent) expect some degree of financial contribution from women; young men (44 percent) would stop dating a woman who never offers to pay any expenses on a date (Lever *et al.*, 2015). As both men and women share breadwinning responsibilities in many dual-earner families, some men use women's paying behavior to see whether a partner for transitional dating demonstrates potential to be a long-term mate (Wills, 2000). If she has not offered to pay over a month of incurring shared dating expenses, it is not a good sign for the future. In other cases, most men still believe that they should pay on dates. The internal experience of shame, guilt, or regret is elicited when people have violated social expectations (Tangney *et al.*, 2007).

Some women are resisting a change that is associated with loss of female privilege (Goode, 1980). Those who choose not to pay view sex as men's reward for paying. Women (39 percent) hope that men will reject their offers to pay; women (44 percent) were bothered when men expected them to pay; and this was truer of older women than younger women (Lever *et al.*, 2015).

Friendship Script

Cultural and interpersonal scripts from the 1980s to the 2000s remain highly influenced by gender roles that emphasize men's power. Eaton and Rose (2011) proposed a **friendship script**—an equal-power dating script characterized by mutual responsiveness and shared responsibility for date events, from asking for the date and paying for it, to monitoring the date conversation. In the absence of cultural and institutional obligations, friendships are created and maintained in the context of mutual support, equality, and fairness (Rawlins, 1992).

"Friendship-based love" or "compassionate love" already exists in intimate heterosexual relationships (Fehr *et al.*, 2009). Also, gay and lesbian partners are using friendship scripts as the basis for early and long-lasting egalitarian romantic relationships (Rose & Zand, 2000). Committed gay and lesbian couples tend to be more egalitarian than heterosexual couples (Connolly & Sicola, 2005) in the division of responsibilities and in terms of relationship maintenance efforts (Peplau & Fingerhut, 2007).

World of Modern Dating

The days of going to an expensive restaurant on a first date might give way to more affordable locations. People often meet for first dates in parks, coffee shops, bars, bookstores, or on a picnic. People (16 percent) said that no money was spent on a date (England & Bearak, 2013). Along with dating, there are other means through which people move from being single to being coupled. People may move away from one-on-one dating toward peer-influenced dating.

Hanging Out

Hanging out is a form of get together with friends or acquaintances that can lead to a relationship. For example, going to see a movie with friends can be considered hanging out. Instead of going on a formal one-on-one date, people want to socialize with friends, get together at work, school, the gym, or elsewhere, and explore whether they are romantically interested in another member of the group. With the rise of modern entertainment venues and the availability of automobile use, high school and college students like to hang out on the weekends or after school. They connect via text messaging or on social media. From there, they hang out at places on school campus, at a friend's house, or at the mall. The activities allow them to get to know each other and possibly form a romantic relationship. The term is widely used on college campuses; "hanging out" together long enough leads to defining themselves as boyfriend and girlfriend (Kuperberg & Padgett, 2015).

Group Dating

Many people go on "group dates," in which a handful of young men and women meet at common gathering places for having fun with the potential for dyadic relationship initiation (Bredow *et al.*, 2008). **Group dating** is a modern form of dating where a group of men and women organize a night out, with the hope of forming romantic partnerships. Group dating is known as *gōkon* in Japan. A single man and woman who know each other may bring some eligible friends to a restaurant. Everyone gives a self-introduction before chatting and drinking.

For those who are shy on a one-on-one date, group dating might be an alternative. It helps ease tension because it does not put the pressure on two partners to keep the conversation going. Group dating may be a solution to the problem some people have in finding a partner. It is a learning experience since everyone can observe how people interact with each other. Group dating is an alternative to single dating. If people are not committed to spending the whole day or night with one partner, they can interact with other people there. In a group setting, it is necessary to lay down some ground rules and avoid conflict. For example, people may not get to talk if someone dominates conversations. Competition might occur if two people like the same woman or man. Group dating becomes more popular with group dating apps such as Grouper.

Hooking Up

Since the 1920s, there has been a transition from an age of dating to an era of hookup culture (Bogle,

2008). **Hooking up** is a form of casual sexual relationship and experience (CSRE) in which uncommitted sexual encounters between casual acquaintances or strangers typically last just one night (England *et al.*, 2007). Hook up has become the "primary currency of social interaction" between the sexes in high schools and colleges (Stepp, 2008). People often begin with hookups (Armstrong *et al.*, 2009) and expect their encounters to develop into a relationship and replace hooking up with formal "dating" (Bogle, 2008). By their senior year, heterosexual students (69 percent) had been in a relationship of at least six months (Armstrong *et al.*, 2010).

Causes and Consequences of Hookup Culture

Hookups (uncommitted sexual encounters) are becoming progressively more engrained in popular culture, reflecting evolved changing social and sexual scripts (Garcia *et al.*, 2012). The rise of hookups is "related to the increased prevalence and social acceptance of premarital sex" (Currier, 2013). Also, human beings gravitate toward polyamorous relationships by the trends developing around dating apps (Ryan & Jetha, 2012). Adolescent females (70 percent) and adolescent males (65 percent) have had sex by age nineteen (Abma *et al.*, 2004). Adult hookups have become more apparent. For example, most people (95 percent) had premarital sex by age 44 (Finer, 2007). In a study of 33,380 adults, Twenge *et al.* (2015) found that they had more sexual partners, were more likely to have had sex with a casual date, and accepted most premarital sex such as teen sex. Furthermore, the rise of hookups is related to "increased dating and sexual initiating behavior among women" (Currier, 2013). Women (33 percent) had sex on the first online dating encounter (Statistic Brain, 2017b). A survey of 14,000 students from nineteen universities and colleges revealed that both men and women (72 percent) reported at least one hookup by their senior year (England *et al.*, 2007).

The problem in navigating the hookup culture is gender inequality. Men "gain status" by having sexual encounters without commitment and treating women like "hookup material" (Armstrong *et al.*, 2010). Women prefer relationships, whereas men prefer casual encounters. Women are less satisfied with hookup culture than their male peers (Bogle, 2008). Hookup sexual behaviors such as penetrative intercourse often transpire without any desire for a romantic relationship (Garcia *et al.*, 2012). The *sexual scripts* (guidelines for appropriate sexual behavior) assume *cunnilingus* (oral sex performed on a female) in relationships, but not in hookups (Backstrom *et al.*, 2012). Women had orgasms more often in relationships than in hookups (Armstrong *et al.*, 2012).

Hook ups place women at risk of "low self-esteem, depression, alcoholism, and eating disorders" (Stepp, 2008). Random hookups have shown to cause feelings of pressure and anxiety (Garcia *et al.*, 2013) and women experience more regret than men (Eshbaugh & Gute, 2008). Finer (2007) suggested that education and interventions provide the skills and information people need to protect themselves from unintended pregnancy and sexually transmitted diseases.

Contemporary Dating Services

Hanging out, group dating, and hookups allow people to have face-to-face interactions but require people to meet in person. Most people live busy lives and need assistance finding an eligible date or mate. Many turn to non-face-to-face forms of meeting others for potential romantic relationships. Technology can serve in the role of matchmaker through various dating services. In fact, the dating service industry has been a growing one.

Online Dating

Technology has changed the game of love, making it possible for people to find and meet prospective mates on the internet. The major transition in mating is with the rise of the internet and now internet is changing so much about the way we act both romantically and sexually (Garcia & Heywood, 2016). The most popular dating innovation is online dating. **Online dating** (also called *e-dating, cyberdating*, or *internet dating*) is the practice of searching for a romantic partner or date on the internet via online dating websites such as Match.com and eHarmony. The internet allows people to form relationships with perfect strangers (Ortega & Hergovich, 2017).

Online Dating by Gender and Age There are more than 54 million single people in the United States and more than 49 million have tried online dating (Statistic Brain, 2017b), whether it is for a hookup, date, relationship, or marriage (Brooks, 2017).

Gender and age affect dating behaviors. Online dating attracts more men (52.4 percent) than women (47.6 percent) (Statistic Brain, 2017b). Most same-sex people meet each other online (Ortega & Hergovich, 2017). Online dating is gaining wider acceptance across most age ranges, tripling among people aged 18–24 from 10 percent to 27 percent between 2013 and 2015 (Livingston & Caumont, 2017) and doubling among older singles aged 55–64 from 6 percent in 2013 to 12 percent in 2015 (see Figure 8.4).

Advantages of Online Dating Most Americans say that online dating is a good way to meet people (Smith & Anderson, 2016). Online dating has advantages over offline dating in terms of *access, communication,* and *matching* services provided by online dating sites (Finkel *et al.*, 2012). **Access** refers to users' exposure to and opportunity to evaluate potential romantic partners they are otherwise unlikely to encounter. The online dating customer spends about $243 annually (Statistic Brain, 2017b) and can choose potential mates from millions of profiles on online dating sites. Internet dating services offer unprecedented levels of access to potential partners (Finkel *et al.*, 2012) and could increase connections between people outside their social groups (Ortega & Hergovich, 2017).

Online dating services facilitate communication and foster affection between online daters. **Communication** refers to users' opportunity to use various forms of computer-mediated communication (CMC) to interact with potential partners through the dating site before meeting face to face. The ability to find out more information ahead of time versus even meeting

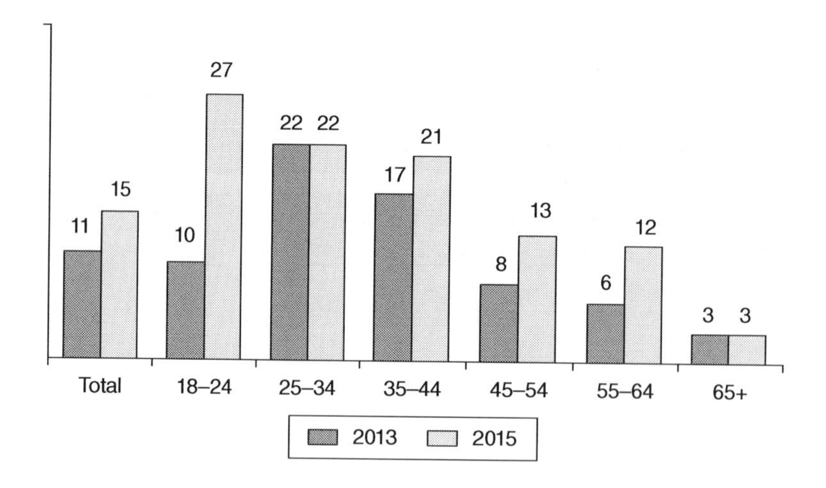

Figure 8.4 Use of Online Dating Sites or Mobile Apps by Age
Source: Pew Research Center. http://www.pewresearch.org/fact-tank/2016/02/29/5-facts-about-online-dating/

a stranger at a party is an advantage that online dating has over conventional dating (Sharabi & Caughlin, 2017). **Matching** refers to a site's use of a mathematical algorithm to select potential partners for users. Certain dating sites may collect data and banish from the dating pool people who are likely to be poor relationship partners (Finkel *et al.*, 2012). Married couples (30 percent) who dated online reported greater marital satisfaction (Cacioppo *et al.*, 2013).

Online dating has become an efficient means of seeking partners. The average length of courtship for marriage partners who met online was 18.5 months, compared with 42 months offline (Statistic Brain, 2017b). Online dating works for people who do not prefer face-to-face interaction due to a lack of social skills. Those without the confidence to approach someone might not feel shy online.

Disadvantages of Online Dating More than half of Americans (61 percent) who are married or in a relationship say that they have a negative view of online dating (ReportLinker, 2017). The ways that online dating sites implement the services of access, communication, and matching do not always improve romantic outcomes (Finkel *et al.*, 2012). For example, it may take months of browsing hundreds of invitations before a single response materializes (Hall *et al.*, 2010). The act of browsing and comparing large numbers of profiles can make online daters reduce their willingness to commit to any one person. Online daters (53 percent) say that they have dated more than one person simultaneously (Statistical Brain, 2017b). As CMC lacks the experiential richness of a face-to-face encounter, some important information about potential partners is impossible to glean from CMC alone (Finkel *et al.*, 2012).

Inaccuracy of the personal information posted and lack of privacy are potential disadvantages of online dating. Online dating services require users to provide information such as a user's age, gender, location, interests, and a picture. This allows other users to search by some criteria, but it can be risky to post personal information in a public domain. Some people can maintain a degree of anonymity. Men are likely to lie most about age, height, and income whereas women are more likely to lie most about weight, physical build, and age (Statistical Brain, 2017b). The first date is usually disappointing because expectations are idealized in the absence of accurate information about the potential partner, "as indicated by having less attraction after meeting than during online engagement" (Sharabi & Caughlin, 2017).

Face-ism in Online Dating **Face-ism**, or facial prominence, is the relative prominence of the face or the proportion of the head relative to the body in the portrayal of men and women. In a study of 6,286 profile photos from online dating sites in seven countries (Austria, Denmark, Hungary, Japan, the Netherlands, Sweden, and the United States), Prieler and Kohlbacher (2017) found that male pictures focused more on their facial features while female photographs focused more on their body than on their facial features.

Why did the media depict men with more facial prominence and women with greater body prominence? Photographs high in face-ism received higher ratings on dominance (Zuckerman, 1986). Photographs of men tend to be higher in facial prominence across cultures and elicited more favorable impressions than photographs low in face-ism (Archer *et al.*, 1983). Men were shown more facially prominent than women in U.S. periodicals (Levesque & Lowe, 1999). There were gender and age differences in facial prominence. Young women aged 18–24 had a higher facial prominence than men; men aged 41 and older had higher facial prominence than women (Prieler & Kohlbacher, 2017).

Converting Online Dating into a Successful First Date What factors help convert online contact into a first face-to-face date? Online communication was most effective in leading to an in-person meeting if there were a genuine interest, a rapid turnaround, and reciprocity in self-disclosure (Khan & Chaudhry, 2015).

Source: Wen Ren from *Café Glass*. https://www.tribecafilm.com/stories/watch-this-short-cafe-glass-directed-by-wen-ren

Creating an attractive profile has an influence on likeability. Men are more attracted to screen names that indicate physical attractiveness (e.g., Blondie, Cutie), whereas women are more attracted to screen names that indicate intelligence (e.g., cultured) (Whitty & Buchanan, 2010). Screen names starting with a letter near the top of the alphabet are presented first. When it comes to photo, attire and physical appearance in still photos have a powerful influence on likeability (Berry & McArthur, 1986). A genuine smile will make a good first impression (Frank *et al.*, 1993). Overall attractiveness of the text is positively correlated with photo attractiveness (Brand *et al.*, 2012). Simple language in the headline message makes information-processing easier and increases likeability (Khan & Chaudhry, 2015). Group photos showing other people having a good time in your company are desirable. Women find a man more attractive when they see other women smiling at him (Jones *et al.*, 2007). Capitalizing on the center-stage effect by selecting photos where you are in the middle creates a sense of importance (Raghubir & Valenzuela, 2006). This can be further enhanced in group photos where you are shown touching another person (confining this to the upper arm to be socially acceptable) (Major & Heslin, 1982) because a toucher is perceived to be of higher status than the one touched (Summerhayes & Suchner, 1978). Dynamic video clips can be more realistic than still photos and may promote familiarity at the first face-to-face encounter (Mierke *et al.*, 2011).

A good description of personal traits increased likeability when it showed who the dater was and what they were looking for in a 70:30 ratio (Khan & Chaudhry, 2015). Men prefer physical fitness in women gained via yoga, aerobics, and gym, not via rugby and bodybuilding, while women prefer bravery, courage, and a willingness to take risks rather than kindness and altruism in their partners (Bassett & Moss, 2004).

What factors set the stage for a successful first date? There is a difference between how it feels faceless and what it feels like face-to-face on a date. First date success was predicted by perceived similarity, expressed similarity, lower uncertainty, and greater information seeking (Sharabi & Caughlin, 2017). Disclosure of personal information to each other will make you feel close (Collins & Miller, 1994). The prospect of ending with a face-to-face meeting is best met through a profile closer to reality (Wiseman, 2007). When online daters are similar to each other and communicate more with each other prior to meeting face to face, the first date is more likely to be successful because doing so reduces uncertainty.

Maturity—the ability to respond to the environment in an appropriate manner—is another factor when it comes to successful online dating. Maturity encompasses being aware of the correct time and place to behave and knowing when to act, according to the circumstances and the culture of the society one lives in (Wechsler, 1950). Mature online daters tend to be more successful than their younger counterparts (Davis, 2013) and are more honest than the younger ones, which is the basis for their success (Statistic Brain, 2016).

Social Network Sites

Online dating sites require a fee but social network sites are free. **Social network sites** (SNSs) are internet-based applications that allow users to create and share their profiles and to view and interact with their online connections made by others in the user's network (Amichai-Hamburger & Hayat, 2017). Like online dating services, SNSs provide an online platform that people can use to build social relations. Around seven in ten Americans use social networking sites; Facebook (68 percent) and YouTube (73 percent) dominate the landscape, compared with Instagram (35 percent), Pinterest (29 percent), Snapchat (27 percent), LinkedIn (25 percent), Twitter (24 percent), and WhatsApp (22 percent) (Smith & Anderson, 2018).

Social networking services provide another online venue for navigating dating partners for a one-time date, or short-term or long-term relationships. Young adults (86 percent) used SNSs in 2016, up from 41 percent in 2006. They tend to friend their date on Facebook, follow up after a first date by texting or emailing each other, and talk over the phone if they are interested in a second date (Pew Research Center, 2017).

Companies such as Skype provide internet phone services that help people get connected across the country and around the globe at no cost. It fulfills many of the same functions that social media sites perform. Through webcams, people can chat via voice or video call. This audio- and video-based technology can help people initiate a date or decide whether they want to meet physically and further the relationship.

Mobile Dating

Advances in telecommunication have given rise to mobile devices such as smartphones that allow people to communicate with potential mates. **Mobile dating services** (or cell phone dating) allow individuals to chat and become romantically involved by means of text messaging, mobile chatting, and *the mobile web*—browser-based internet services accessed from mobile devices through a mobile or other wireless network. As mobile devices become integrated into daily life, people use smartphones to find a date or a potential mate.

Mobile dating services allow their users to provide a short profile that is stored in their phones. These services are free to use but standard text messaging fees may still apply. With the advent of mobile phones with *GPS* (Global Positioning System)—a device that can receive information from GPS satellites and calculate the device's geographical position—proximity dating becomes popular. Dating apps are intermediaries through which individuals engage in strategic performances in pursuit of love, sex, and intimacy (Hobbs *et al.*, 2016) and can take people into an era of anytime, anywhere dating.

> Everyone is Tindering. The tables are filled with young women and men who've been chasing money and deals on Wall Street all day, and now

they're out looking for hookups. Everyone is drinking, peering into their screens, and swiping on the faces of strangers they may have sex with later that evening. [See image below.]

(Sales, 2015)

Speed Dating

Speed dating refers to organized matchmaking events that allow multiple men to meet multiple women of similar age for brief encounters one after the other and assess mutual interest before subsequent dates. This design allows researchers to separate *actor effects* (how do I behave towards others in general?) from *partner effects* (which behavior do I evoke in others?) in the dyadic interaction where one participant is interacting with only one dating partner (Kenny *et al.*, 2006). Actor and partner effects can be estimated because speed daters get access to a dating partner's address only in the case of matching (reciprocal choices). After following 382 participants aged 18–54 over a period of one year, Asendorpf *et al.* (2011) reported the chance (6 percent) for mating with a partner (an increase by men's short-term mating interest) and the chance (4 percent) for relating (an increase by women's long-term mating interest). According to Back *et al.* (2011), actual mate choices are not reciprocal but flirting behavior is strongly reciprocal.

Advantages of speed dating include efficiency and low cost. However, because the interactions are so brief, people may not have sufficient time to assess whether those they interact with could be a potential partner.

Personal Classified Advertisements

People searching for a mate may use classified ads that are published in "Personals" or "Eligibles" sections in magazines and newspapers, either in print or

Source: Justin Bishop. https://www.vanityfair.com/culture/2015/08/tinder-hook-up-culture-end-of-dating

online. Craigslist is a classified advertisement website with a section devoted to personals that enable people to meet others either platonically or romantically.

Research has indicated that male and female personal ads were aligned with a gender-typed cultural script. Men sought physical beauty and thinness in a partner, whereas women sought an understanding partner (Smith *et al.*, 1990). Men offered status and sought attractiveness in partners, and women offered attractiveness and sought status in partners (Willis & Carlson, 1993). Almost three times as many men responded to a personal ad placed by a fictional "beautiful waitress" than one placed by an "average-looking woman lawyer" and more women replied to an ad by an "average-looking man lawyer" than a "handsome cabdriver" (Goode, 1996). A study of 600 personal ads from newspapers and websites across the United States revealed that advertisers (47 percent) indicated their race, including White (60 percent) and Black (29 percent) (Smith *et al.*, 2011).

Advantages of classified ads include low cost, allowing for anonymous posting, and helping people meet others quickly. However, as with online dating sites, people may provide inaccurate information about their socioeconomic status, age, appearance, or other factors.

Professional Matchmaking Services

Professional matchmaking services exist because people often need assistance in finding a mate. There were 399 dating service establishments nationwide as of 2012 (U.S. Census Bureau, 2017a). Professional matchmakers charge a fee and offer social interactions among a diverse population. For the sake of privacy, matchmakers use case numbers to label applicants. People who seek a mate are required to describe the criteria that they seek in a partner in an application form. A comparative study of heterosexuals and lesbians revealed that heterosexual men offered financial security in their ads and sought attractiveness in a partner; heterosexual women offered their physical attractiveness and sought financial security in a partner; lesbians offered and requested partner attributes such as honesty above all other personality traits (Smith *et al.*, 2011).

Mail-order Brides

A **mail-order bride** is a woman from a developing country who advertises herself in a catalog or an online mail-order bride agency to find and marry a man in a developed nation. Initially, the mail-order bride system consisted of printed catalogs filled with pictures of women and information on their age, height, weight, and interests. Today, most mail-order bride systems exist online. The advertisements and catalogs are part of an expanding multi-million-dollar industry that markets women from developing countries as potential brides to men in Western industrialized nations (Chun, 1996).

Dating Violence

Dating violence can happen to any teen and person in a romantic, dating, or sexual relationship, anytime, anywhere. **Dating violence** refers to an act of violence by one or more members on the other member in the context of dating or courtship. Gender and gender role stereotypes are likely to affect dating violence.

Upon completion of this section, students should be able to:
LO7: Identify types of dating violence.
LO8: Describe reasons for and effects of dating violence.
LO9: Describe strategies for preventing dating violence.

Types of Dating Violence

There are two forms of violence against women by *an intimate partner* and *non-partner.* Globally, women (7.2 percent) reported ever having experienced non-partner sexual violence (WHO, 2013a). In the United States, a national survey of people aged 18–25 revealed that women (87 percent) reported having experienced being catcalled (55 percent), touched without permission by a stranger (41 percent), and insulted with sexualized words (e.g., bitch) by a man (47 percent) (Weissbourd *et al.*, 2017).

Dating violence includes physical, verbal/emotional, sexual, and digital abuse, or a combination thereof. **Physical dating violence** is the use of physical force with the intent to cause fear or injury. This occurs when a partner is pinched, hit, shoved, slapped, punched, or kicked. A national survey revealed that one in ten teens reported being hit or physically hurt on purpose by a boyfriend or girlfriend at least once in the past twelve months (CDC, 2018). High school students (10 percent) report experiencing physical dating violence in the previous twelve months (NCSL, 2017). **Psychological/emotional dating violence** is the use of words with the intent to cause fear or harm his or her sense of self-worth. It is exemplified by bullying, calling names, embarrassing on purpose, or threatening to hurt someone, or keeping him/her away from friends. One in three women worldwide has experienced emotional dating violence by a non-partner (WHO, 2017).

Dating violence can take place electronically. **Digital abuse** refers to the use of technologies and/or social media networking as a tool to intimidate, harass, or threaten a current or ex-dating partner. Digital abuse includes demanding passwords, cyberbullying, stalking on social media, repeated texting, or posting sexual pictures of a partner online. Sex offenders (10 percent) use online dating to meet people (Statistic Brain, 2017b). Americans (41 percent) have been personally subjected to harassing behavior online and an even larger share (66 percent) has witnessed these behaviors directed at others. Men (30 percent) and women (23 percent) have been called offensive names online whereas women receive sexualized forms of online abuse at much higher rates than men (Duggan, 2017b). **Acquaintance rape** or **date rape** refers to a sexual assault or a rape of a victim who knows or is familiar with the rapist. Most victims of acquaintance rape are females.

Sexual dating violence is forcing a partner to engage in a sex act when he or she does not or cannot consent. It includes such behaviors as unwanted touching and kissing, forcing someone to have sex or engage in other sexual activities, and restricting access to birth control. During the previous twelve months, one in ten teens reported having been kissed, touched, or physically forced to have sexual intercourse when they did not want to at least once by someone they were dating (CDC, 2018). High school students (10 percent) reported experiencing sexual dating violence in the previous twelve months (NCSL, 2017).

Dating Violence by Gender

Women are more likely than men to be victims of dating violence. **Gender-based violence** (GBV) refers to any act that results in physical, sexual, or psychological harm or suffering to women. GBV is a major public health problem in all societies, and a violation of human rights.

More women were represented as victims of dating violence. Hannon *et al.* (1996) found that lifetime incidence of unwanted sex dates was 64 percent for women and 35 percent for men. Youths involved in same-sex dating are just as likely to experience dating violence as youths involved in opposite sex dating (Young *et al.*, 2004). In the United States, one in five women and one in seventy-one men have been raped in their lifetime (Black *et al.*, 2011). In response to rape vignettes, rejection of coercive strategies increased as the level of force increased (Struckman-Johnson & Struckman-Johnson, 1991). Participants rated a date rape initiated by a woman as more justifiable, more understandable, and less aggressive and inappropriate than a date rape initiated by a man (Hannon *et al.*, 2000). Both women and men evaluated a man who was sexually coerced by a woman as being more responsible and as more in control of the situation than a woman who was sexually coerced by a man (Katz *et al.*, 2007).

Reasons for Dating Violence

Heise (1998) conceptualizes gender-based violence (GBV) as a multifaceted phenomenon grounded in interplay among personal, situational, and sociocultural factors. A history of family violence can be a factor that contributes to dating violence. Lower maternal acceptance and higher exposure to marital conflict in early adolescence were both independently associated with involvement in teen dating violence (National Institute of Justice, 2017).

Gender role expectations can be another factor that contributes to dating violence. The distinct gender-related social norms and behaviors influence inequalities between men and women and build the foundation for gender-based violence (Darj *et al.*, 2017). Violence against women is a manifestation of historically unequal power relations between men and women and is one of the crucial social mechanisms by which women are forced into a subordinate position (United Nations, 1993). A cross-sectional study of 1,568 male participants in Sri Lanka revealed that most men agreed with statements such as: "There are times when a woman deserves to be beaten" and "The men connected manhood to dominance and violence" (de Mel & Gomez, 2013).

Peer pressure can be another factor that leads to dating violence. Heavy drinking (59 percent), the casual hookup culture (43 percent), and fraternities and sororities (42 percent) are among the top five causes of sexual assault during the start of the school year (Palumbo, 2017). Women who used alcohol and drugs were more likely to be victimized (Yeater *et al.*, 2008). Both men and women were more likely to blame a woman victim of a hypothetical date rape when she was wearing a short skirt rather than a long skirt (Workman & Freeburg, 1999).

Effects of Dating Violence on Victims

Dating violence is a widespread issue that has serious long-term and short-term effects on victims (see Figure 8.5). The health effects of non-partner sexual violence include mental health effects such as depression and anxiety disorders. Women (52 percent) who have been physically and/or sexually abused have experienced injuries because of that violence (WHO, 2013a). Youth who experience dating violence are more likely to experience symptoms of depression and anxiety, engagement in unhealthy behaviors, such as tobacco, alcohol, and drug use, involvement in antisocial behaviors, and thoughts about suicide (CDC, 2017).

Women who experienced psychological aggression from men in dating relationships were more likely to express bulimic symptoms and attempt to diet (Skomorovsky *et al.*, 2006). Women who have experienced non-partner sexual violence are 2.3 times more likely to have alcohol use disorders and 2.6 times more likely to have depression or anxiety than women who have not experienced non-partner sexual violence (WHO, 2013a). Youth (50 percent) who experienced both dating violence and rape reported attempting suicide, compared to non-abused girls (12.5 percent) and non-abused boys (5.4 percent) (NCADV, 2015). Women who were raped had higher rates of use of medical care than women who were not raped, even years after the event (Golding *et al.*, 1988).

Gender role stereotypes continue to affect individual judgments of female victims of rape and dating violence. Men were more likely to perceive the woman in a date rape scenario as having more interest in sex, and were less likely to label the scenario as "rape" (Maurer & Robinson, 2008). Men were more likely to blame the victim/woman and less likely to blame the perpetrator/man (Brown & Testa, 2008) and were more likely to say that a hypothetical rape

Women exposed to intimate partner violence are →

Mental Health

TWICE as likely to experience depression

ALMOST TWICE as likely to have alcohol use disorders

Sexual and Reproductive Health

16% more likely to have a low birth-weight baby

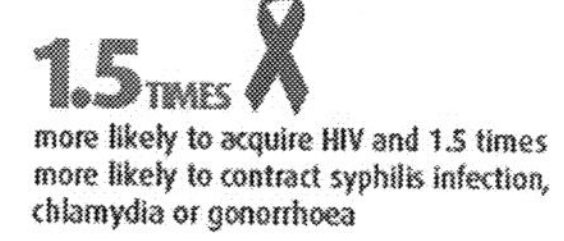

more likely to acquire HIV and 1.5 times more likely to contract syphilis infection, chlamydia or gonorrhoea

Death and Injury

42% of women who have experienced physical or sexual violence at the hands of a partner have experienced injuries as a result

38% of all murders of women globally were reported as being committed by their intimate partners

Figure 8.5 Violence against Women: Health Impact

Source: World Health Organization, 2013. http://www.who.int/reproductivehealth/publications/violence/VAW_health_impact.jpeg?ua=1

was provoked by the woman victim (Cowan, 2000). Both men and women undergraduates rated dating violence in which the perpetrator was a man and the victim was a woman as more violent than any other gender dyad combination, including "female-on-male," "male-on-male," and "female-on-female" (Hamby & Jackson, 2010).

Strategies for Preventing Dating Violence

The goal of prevention and intervention is to stop dating violence before it begins. The most effective tools for preventing dating violence include *primary prevention*—involving early interventions within families (Devries *et al.*, 2015) and *secondary prevention*—involving awareness of gender-based violence and its consequences in schools and workplaces (Darj *et al.*, 2017).

Many prevention strategies are proven to prevent dating violence. At the individual level, programs that focus on improving parents' mental health and parenting skills may prove to be beneficial (National Institute of Justice, 2014). At a community level, some effective initiatives include having stricter policies that strengthen penalties and more options to make it easier to report (Palumbo, 2017). It is important to reduce levels of childhood exposures to dating violence, reform discriminatory family law, and strengthen women's economic and legal rights, etc. (WHO, 2013b).

Services need to be provided for those who have experienced dating violence. Access to comprehensive post-rape care is essential, and must ideally happen within 72 hours (WHO, 2013b). Current state actions and state laws that address dating violence can be viewed through the National Conference of State Legislatures (NCSL, 2017). Victims of dating violence can call the National Dating Abuse Helpline at 1-866-331-9474 for help.

Upon completion of the next section, students should be able to:
LO10: Identify factors that affect mate selection in the marriage market.
LO11: Analyze two rules of mate selection.
LO12: Explain how theories of mate selection can be applied to reality.

Student Account of Mate Selection

I am getting married on September 3 to a wonderful man. He is twenty years older than I am. Some of his family members do not approve. Often, I feel incredibly left out and isolated at family parties because they stare and whisper when I walk in the room (personal communication, September 3, 2017).

Mate Selection

What do you look for in a mate? **Mate selection** is a process in which we consider criteria that we prefer in a mate, from physical appearance to cultural background to education and socioeconomic standing. Let's look at some factors related to mate selection.

The Marriage Marketplace

When people reach a marriageable age, they often begin to pay attention to the spousal supply and demand system in which coupling and marriage occur. Sociologists use the term **marriage marketplace** to describe the mate selection process where people compare the assets and liabilities of eligible candidates and choose the best available mate. Young age, high-paying occupation, and physical attractiveness are examples of potential assets or resources.

Mating Market Approach

In the marriage marketplace, prospective partners compare the personal, social, and financial resources of eligible mates and then bargain for the best they can get (Coltrance, 1998). **Mating market approach** suggests that people with desirable traits have a stronger "bargaining hand" and can be more selective when choosing partners. A survey of 27,605 heterosexuals revealed that people with higher incomes or with more education had stronger preferences for good-looking and slender partners (Fales *et al.*, 2016). Li and Kenrick (2006) distinguish between *necessities* (what people find essential in a partner) and *luxuries* (what people prefer in a partner, but could live without). For example, with long-term mates, physical attractiveness was considered a necessity by men, and resources or social status was a necessity for women. If women use short-term mating to assess or attain potential long-term relationships, then we would expect women to prioritize status/resources and kindness (Li *et al.*, 2002)—and to treat physical attractiveness as more of a luxury (Li & Kenrick, 2006).

Sex Ratio

The sex ratio is an important social indicator that affects the availability of eligible candidates in the marriage market. **Sex ratio** is the number of males per 1,000 females in a population. A sex ratio of 1 means there are equal numbers of females and males. A sex ratio above 1 means there are more males than females, whereas a sex ratio below 1 means there are more females than males. Significant sex discrepancies exist in societies due to biological differences, infant mortality rates, and birth control policies. In the United States, the sex ratio was 1.00 for people aged 25–54 in 2016 (CIA, 2018a). However, other countries do not have a sex ratio of 1. For example, the sex ratio in China in 2016 was 1.04 for people aged 25–54 (CIA, 2018b). It means that there are more males than females in China. A country's sex ratio affects its **marriage rate**, which is the number of marriages per 1,000 people in a year. If there are more single men than eligible women, a significant portion of men must remain single or use other avenues for finding mates elsewhere.

A country's sex ratio affects its **fertility rate**. The total fertility rate in the United States in 2016 was 1.87 children born per woman, compared with 1.6 children born per woman in China (CIA, 2018a). A country's sex ratio is also influenced by **sex-selective abortion**—the practice of terminating a pregnancy based upon the predicted sex of the infant. The one child per family policy from the 1970s to 2010s in China controlled the population growth. However, selective abortion of female fetuses occurred because cultural norms prefer male children to female children. The fact that there are more men than women in China represents a **gender imbalance**, a disparity between males and females in a population. During the 1960s in the United States, single men married at an increasing rate, and single women at a decreasing rate. These trends can be explained by **marriage squeeze**—an imbalance between the number of men and women available to marry in a certain society (Akers, 1967). In 2000, a new marriage squeeze was observed when African American women found it difficult to meet and marry desirable and eligible men (Crowder & Tolnay, 2000).

Rules of Mate Selection

There are two rules an individual may use to select and marry someone: homogamy and heterogamy, choosing eligible candidates who are similar to or different from oneself. *Assortative mating* occurs when individuals exhibit a preference for those who are either similar (homogamy) or dissimilar (heterogamy) to themselves (Domingue *et al.*, 2014).

Homogamy

Homogamy is the tendency to select a mate with similar characteristics. In choosing marriage partners, we are likely to look for someone with similar characteristics to our own and mate with someone of a similar socioeconomic status, ethnicity, race, age, education level, and/or religion. Couples (70–90 percent) are similar regarding racial, education, age, and religion (Laumann *et al.*, 1994); examples include Barack and Michelle Obama and Bill and Melinda Gates.

Racial Homogamy Race and ethnicity play a role in who selects and marries whom. If communities are not racially integrated, people may choose to marry a mate with similar racial features. **Spousal racial homogamy** refers to marriages between spouses with the same racial background. **Intraracial marriage**—the practice of marrying someone of a similar race—remains the dominant marriage pattern among newlyweds (83 percent) and married couples (90 percent) in 2016 (Wu, 2018). Figure 8.6 shows that all Black and White marriages were racially homogamous in 1964.

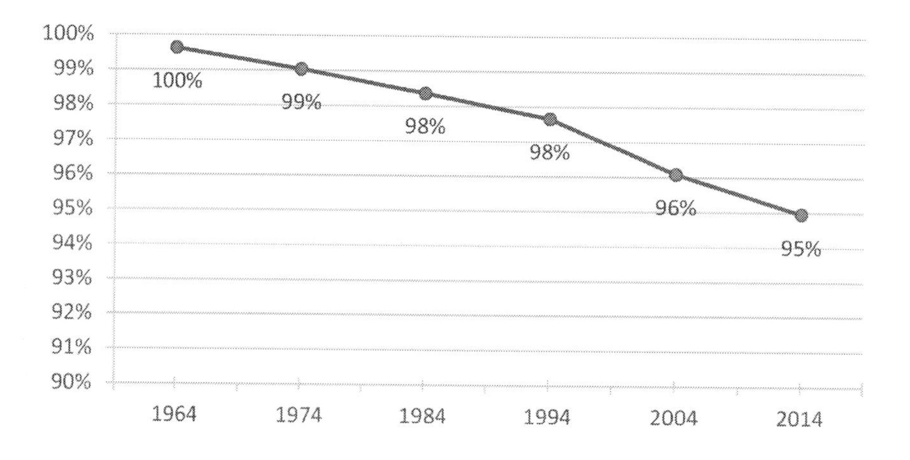

Figure 8.6 Trends in Racial Homogamy among Black and White Couples in the United States, 1964–2014
Source: U.S. Census Bureau, Current Population Survey, 1964–2014 and NCFMR. https://magic.piktochart.com/output/7230569-lamidi-brown-manning-racial-homogamy-fp-15-16web

Reasons for Homogamy To preserve the cultural heritage and maintain solidarity within the familial unit, parents may strive to cultivate family allocentrism (i.e., the strength of closeness and devotion between family members) (Uskul *et al.*, 2010). Those who were higher in family allocentrism showed greater adherence to their heritage, cultural customs and group membership (Lay *et al.*, 1998). Anthropologists use the term **endogamy** to refer to the practice of selecting a marriage mate within the same ethnic or cultural group. (Note that homogamy is a broader term than endogamy.) Endogamy involves cultural pressures to marry in one's social group and carries cultural sanctions against marrying outside of one's own group (Winch, 1958).

People often want to marry others of similar characteristics because they tend to feel comfortable with others who are like themselves. Education, social class, age, and religion facilitate meeting. People are likely to meet those who go to the same school and attend the same church. The parents of Hindu families may want their sons or daughters to marry other Hindus, while the parents of Jewish families may want their sons or daughters to marry other Jews. People are drawn to others of the same social class because they are likely to have similar lifestyles. These people have a higher chance of common personal tastes, opinions, and values with one another, making it easier to establish affinitive relations (Degenne & Forse, 1999).

Age Homogamy **Age homogamy** refers to age-similar marriage or *coeval*—a person of roughly the same age as oneself. Economic development generally leads to a rise in age homogamy. The family in pre-industrial societies is characterized by a relatively large age gap between an older breadwinner husband and a younger wife with limited non-domestic labor participation (Van Poppel *et al.*, 2001). In the second half of the nineteenth century, there was a cultural shift towards the rise of love marriage and the increase of the preference for age-similar marriage. Marriages with age gaps of less than two years (two excluded) rose by 7–20 percent for the Dutch regions between 1812 and 1913 (Van de Putte *et al.*, 2009). During the past three decades, 17 out of 24 low-income countries experienced a similar increase in low-age-gap marriages with a gap of 0–5 years (0 and 5 included) (Casterline *et al.*, 2010). U.S. marriages with 0 to 4-year spousal age gaps (4 included) increased from 37.1 percent in 1900 to 63.3 percent in 1960 and 69.9 percent in 1980 (Atkinson & Glass, 1985). Husbands were just over two years older in 2000 (Rolf & Ferrie, 2008). Table 8.1 shows that couples (33.9 percent) have an age difference of 1 year. Data in Australia and the United Kingdom (Wilson & Smallwood, 2014) show an almost identical pattern.

Table 8.1 Age Difference in Heterosexual Married Couples, 2017 Survey

Age difference	*Percentage of All Married Couples*
Husband 20+ years older than wife	1.0
Husband 15–19 years older than wife	1.6
Husband 10–14 years older than wife	5.0
Husband 6–9 years older than wife	11.2
Husband 4–5 years older than wife	12.8
Husband 2–3 years older than wife	19.6
Husband and wife within 1 year	33.9
Wife 2–3 years older than husband	6.9
Wife 4–5 years older than husband	3.4
Wife 6–9 years older than husband	2.8
Wife 10–14 years older than husband	1.1
Wife 15–19 years older than husband	0.3
Wife 20+ years older than husband	0.4

Source: U.S. Census Bureau, Current Population Survey, 2017 Annual Social and Economic Supplement, Internet Release Date: November 2017. https://census.gov/data/tables/2017/demo/families/cps-2017.html

Heterogamy

People may mate with someone of a different socio-economic status, ethnicity, race, age, education level, and/or religion. **Heterogamy** is the tendency to select a marriage mate with different characteristics.

Racial Heterogamy **Racial heterogamy** refers to marriages involving two partners of different racial or ethnic groups. Anthropologists use the term **exogamy** to refer to the practice of marrying someone outside one's own ethnic or cultural group. For example, Indonesia has more than 300 different ethnic groups, with substantial evidence of both exogamous and endogamous marriage norms (Heaton *et al.*, 2001). In Bali, Indonesia, marriage is exogamous and a bride price is paid by the groom (Rammohan & Robertson, 2012).

The local marriage marketplace can influence mate selection. For example, racially mixed communities might produce more mixed-race couples and mixed nativity couples are more likely to intermarry than same nativity couples (Choi & Tienda, 2014). The importance of racial homogamy in marital relationships is decreasing (Feldman, 2011) due to **interracial marriage**—the practice of marrying someone of a different race. For example, Facebook founder Mark Zuckerberg and Priscilla Chan have an interracial marriage.

The connections people make online are expanding their own social circles and help "nearly complete racial integration" (Ortega & Hergovich, 2017). After the introduction of the first dating websites like Match.com in 1995, interracial marriages increased in the 2000s from 10.68 percent to 15.54 percent, jumped to 17.24 percent after the creation of Tinder in 2014, and remained the same in 2015 (Ortega & Hergovich, 2017). The United States is experiencing a decline in spousal racial homogamy across generations. Figure 8.7 shows that those in the Greatest Generation (Americans born before 1928) (100 percent) were racially homogamous, compared with those in the Millennial generation (94 percent). College-educated couples experienced a greater decline in racial homogamy, from 99 percent in 1964 to 94 percent in 2014 (Lamidi *et al.*, 2015). In 2016, one in six newlyweds (17 percent) was racially homogamous (Wu, 2018).

Age Heterogamy People say that love is ageless. **Age heterogamy** or age-dissimilar marriage refers to the extent of the age difference between the partners. Age differences between spouses are more than peculiarities of individual preference but in most marriages, husbands are older than wives at the time of marriage (Abel & Kruger, 2008). "May–September" and "May–December" unions are marriages in which there is a big age gap between the partners (Vera *et al.*, 1985).

Sociologists assumed that age-heterogamous marriages were most prevalent among the upper class and believed that such marriages exhibited poorer marital quality than age-similar unions. However, Vera *et al.* (1985) found that the age-heterogamous unions were more prevalent among lower classes; age-dissimilar couples did not exhibit poorer marital quality than age-similar unions. Age heterogamy is also associated with remarriages (Berardo *et al.*, 1993) and with family income, arranged marriage norms, patrilocal residence, and autocratic family authority in rural Bangladesh (Uddin *et al.*, 2017). When it comes to the age heterogamy effect on longevity, Abel and Kruger (2008) confirm that men and women married to younger spouses live longer than those married to spouses that are the same age at time of marriage. In the United States, in 2016, one in four marrieds had a husband who was at least two years older than his wife (Wu, 2018).

Hypogamy and Hypergamy People can move up or down the social ladder through mate selection. **Hypogamy** involves marrying down the social ladder. For example, the Duke of Windsor gave up the British throne in 1936 to marry an American divor-

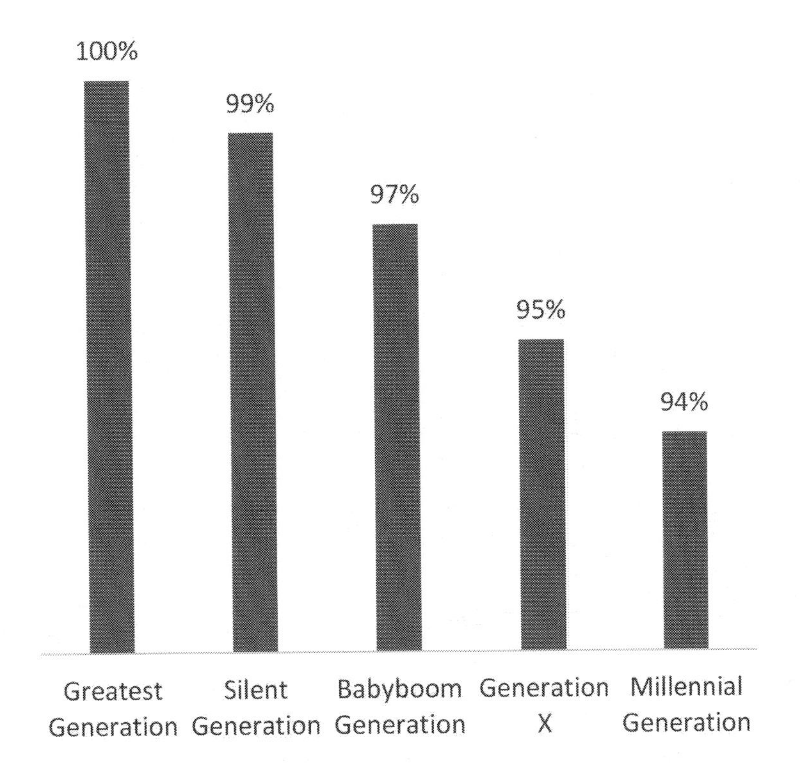

Figure 8.7 Generational Differences in the Share of Racially Homogamous Black and White Couples in the United States, 1964–2014

Source: U.S. Census Bureau, Current Population Survey, 1964–2014 and NCFMR. https://magic.piktochart.com/output/7230569-lamidi-brown-manning-racial-homogamy-fp-15-16web

cee and commoner named Wallis Simpson. **Hypergamy** involves marrying up the social ladder and improving one's social status. A commoner marrying into royalty is another example of hypergamy. For example, when Kate Middleton married Britain's Prince William in 2011, she elevated her social status and became known as the Duchess of Cambridge (see image on next page). **Status** is a social position held by a person within a group. **Status inconsistency** occurs when people have a mismatch between their statuses. People from different family backgrounds are socialized to fulfill built-in expectations that guide their behaviors. Status inconsistency that deviates from expected behaviors can lead to social discomfort.

Educational Heterogamy **Educational heterogamy** refers to marriages involving two partners with different levels of education. Sociologists are interested in educational heterogamy's effect on marital satisfaction, health, and childrearing. Interestingly, when husbands had more education than their wives, both partners reported less than happy marriages but when the wife had more education, both partners reported more satisfaction with the marriage (Tynes, 1990).

Social capital theory suggests that average health is better in countries with more educational heterogamy. The European Social Survey of 59,314 respondents in 2002, 2004, and 2006, revealed that the degree of educational heterogamy did not influence the average level of self-assessed health

Source: Wikipedia 2016

(Huijts *et al.*, 2010). Eeckhaut *et al.* (2013) found a positive link between educational heterogamy and differences in childrearing values and behaviors between partners. The opposite of educational heterogamy is **educational homogamy**—marriages involving two partners with the same level of education. In the United States, in 2016, newlyweds (53 percent) and marrieds (54 percent) were in educationally homogamous marriages; it was more common for the wife to have more education than the husband among newlyweds (30 percent) compared with those (24 percent) who had been married at least a year (Wu, 2018).

Theories of Mate Selection

Desired aspects in a partner vary by culture. In individualist cultures, love-marriage selection criteria reflect individual freedom and personal concerns. In collectivist cultures, arranged-marriage selection criteria reflect concerns of the total family unit. These personal and family concerns include socioeconomic status, personality, and similar characteristics of the prospective spouse. Yet, the similarities in characteristics between the two partners are consistent in both marriages (Yalom & Carstensen, 2002). This section introduces filter theory in male selection.

Filter Theory

Filter theory proposes that our social structure sifts and narrows the pool of eligible candidates for a mate according to specific criteria. Theoretical explanations have been provided for why similarity leads to attraction and satisfaction. Similarity (in attitudes and beliefs) is consensually validating and reinforcing (Clore & Byrne, 1974), leading to uncertainty reduction and

Diversity Overview: Mate Selection Criteria by Gender

The criteria that people use to choose a mate are social in origin. Miller and Perlman (2009) identify warmth and loyalty, attractiveness and vitality, and status and resources as criteria with which people evaluate potential mates in the world. Let's look at gender selection criteria for short-term and long-term partners.

Male and Female Criteria for Short-term Partners

For short-term mates, a preference for physical attractiveness has been identified consistently (Regan, 1998). **Physical attractiveness** is the degree to which a person's physical features are considered aesthetically pleasing or beautiful. Both sexes prioritize physical attractiveness in short-term mates (Li & Kenrick, 2006), desire more physical attractiveness as relationship duration shortens (Buunk *et al.*, 2002), favor a short-term/fling partner who is high on attractiveness/vitality over one who has high warmth/trustworthiness (Fletcher *et al.*, 2004). Adolescents (Regan & Joshi, 2003) as well as homosexual men and homosexual women (Regan *et al.*, 2001) value attractiveness more in short- than in long-term partners.

Men value beauty more in choosing a partner in a hypothetical dating game (Hetsroni, 2000) and are more concerned about the appearance of their dates than women (Hatfield & Sprecher, 1986). Male college students considered it important to have higher levels of physical attractiveness for romantic partners than for friends (Sprecher & Regan, 2002). A study of participants who were attracted to same-sex individuals shows that the attractiveness–desire association was stronger for women than men (Eastwick & Smith, 2018). For long-term mates, both sexes had relatively high standards for various characteristics (Kenrick *et al.*, 1993). Let's look at gender selection criteria for long-term mates.

Male Mate-selection Criteria

The physical attractiveness of the sexual partner is much more important to men than to women (Sanderson, 2003) and the association of physical attractiveness with romantic evaluations is stronger for heterosexual men than for women in photograph-rating contexts (Eastwick & Smith, 2018). Heterosexual males (92 percent) and females (84 percent) indicated that "their potential partner was good-looking"; male (84 percent) and female (58 percent) indicated that "their potential partner had a slender body" (Fales *et al.*, 2016).

Why is physical attractiveness important? Physical attractiveness may serve as a gatekeeper directing us toward partners who are healthy, age appropriate, and able to reproduce (Weeden & Sabini, 2005). For males, physical attractiveness plays a major role in sexual attraction, with males using such indicators of high female fecundity as health, complexion, cleanliness, and condition of the skin (Symons, 1995). Men are attracted to a low *waist-to-hip ratio* (WHR) that is associated with women's health, fecundity, and cognitive ability (Singh, 1993). Attractiveness and health are related to WHR and male preference for female body shape is adaptive (Kościński, 2014). Attractive individuals are expected to be happier and to have more rewarding life experience than unattractive individuals (Griffin & Langlois, 2006).

Men also prefer to marry younger women. Males everywhere show an extremely strong desire for young females (Sanderson, 2003). Younger females are more likely to become pregnant and to produce strong, healthy offspring. Men (80 percent) date women who are at least five years younger than them (Statistic Brain, 2017a). The preference of men for younger women is indicated by studies of what biologists call *neoteny*, or retention of juvenile characteristics in adults (Sanderson, 2003). Men are inclined to value physical features such as full lips, soft hair, smooth skin, colorful cheeks, good muscle tone, and secondary sexual characteristics including breasts and buttocks, which tend to be cues to youth, sexual maturity, and fecundity (Symons, 1995).

Female Mate-selection Criteria

Two important aspects of American society's definition of manhood are *financial success* and *masculine personality attributes* such as confidence and determination (Brannon, 1976) and both affect women's romantic attraction to men (Kenrick *et al.*, 1990). Women prefer masculine men as potential mates (Kimlicka *et al.*, 1982) and prefer mates who have a high earning capacity, who are educated, self-confident, and intelligent, and who have a higher social status than themselves (Buunk *et al.*, 2002).

Women place more emphasis on a partner's socioeconomic status (SES) in relationships. Women (88 percent) find money to be very important in a relationship (Statistic Brain, 2017a). Females (95 percent) and males (75 percent) indicated that "their potential partner had a steady income"; females (69 percent) and males (47 percent) indicated that "their potential partner made/will make a lot of money" (Fales *et al.*, 2016). Their ability to invest affection and resources in relationships may often outweigh the effects of their physical attractiveness in women's actual selection of partners (Townsend & Levy, 1990). Because increasing gender equality reduces gender differences in mate selection, the strategies men and women use to choose mates may not be as hardwired as scientists originally thought (Zentner & Mitura, 2012).

predictability (Berger & Calabrese, 1975), and leading to enjoyable interactions (Fehr, 2001).

Filtering Mechanism The major filtering mechanism is *homogamy*—similar individuals marrying each other. Filter theory starts with the base of all people in the marriage market and removes married couples and keeps eligible partners. The compatibility filter removes all people who are not attracted to each other. "Compatible matches" are associated with the similarity principle (Sprecher, 2011). People are physically attracted to those who look like themselves.

People are similar to each other on a number of dimensions, including attitudes and beliefs, personality, leisure interests, communication styles, education, and sociocultural background factors (Baxter & West, 2003). Assortative mating involves *genetic assortative mating* (genetically similar spouse)—mate choice based on genetic expression—and *social assortative mating*—mate choice based on cultural or socioeconomic or other factors. For example, males prefer female faces that resemble their own (Guo *et al.*, 2014) but this genetic assortative mating occurs because people are more likely to marry those who live in the same ethnic subgroups (Abdellaoui *et al.*, 2014). Online daters (64 percent) say that common interests are the most important reason to date (Statistical Brain, 2017b). A couple with the same social class background may have similar interests, hobbies, and goals in life. Streib (2015) has found that marriages between someone with a middle-class background and someone with a working-class background can involve differing views on all sorts of important things—childrearing, money management, career advancement, how to spend leisure time.

The Attraction–Similarity Model

Social scientists distinguish between *perceived similarity* (the degree to which similarity is perceived with the other) and *actual similarity* (the degree to which two people are similar). The **attraction–similarity model** suggests that people in existing relationships are motivated to perceive similarity to achieve balance and satisfaction over time (Morry *et al.*, 2011).

According to the attraction–similarity model, perceived similarity increases over time in a relationship. Perceived similarity changes in a corresponding way if actual similarity changes over the course

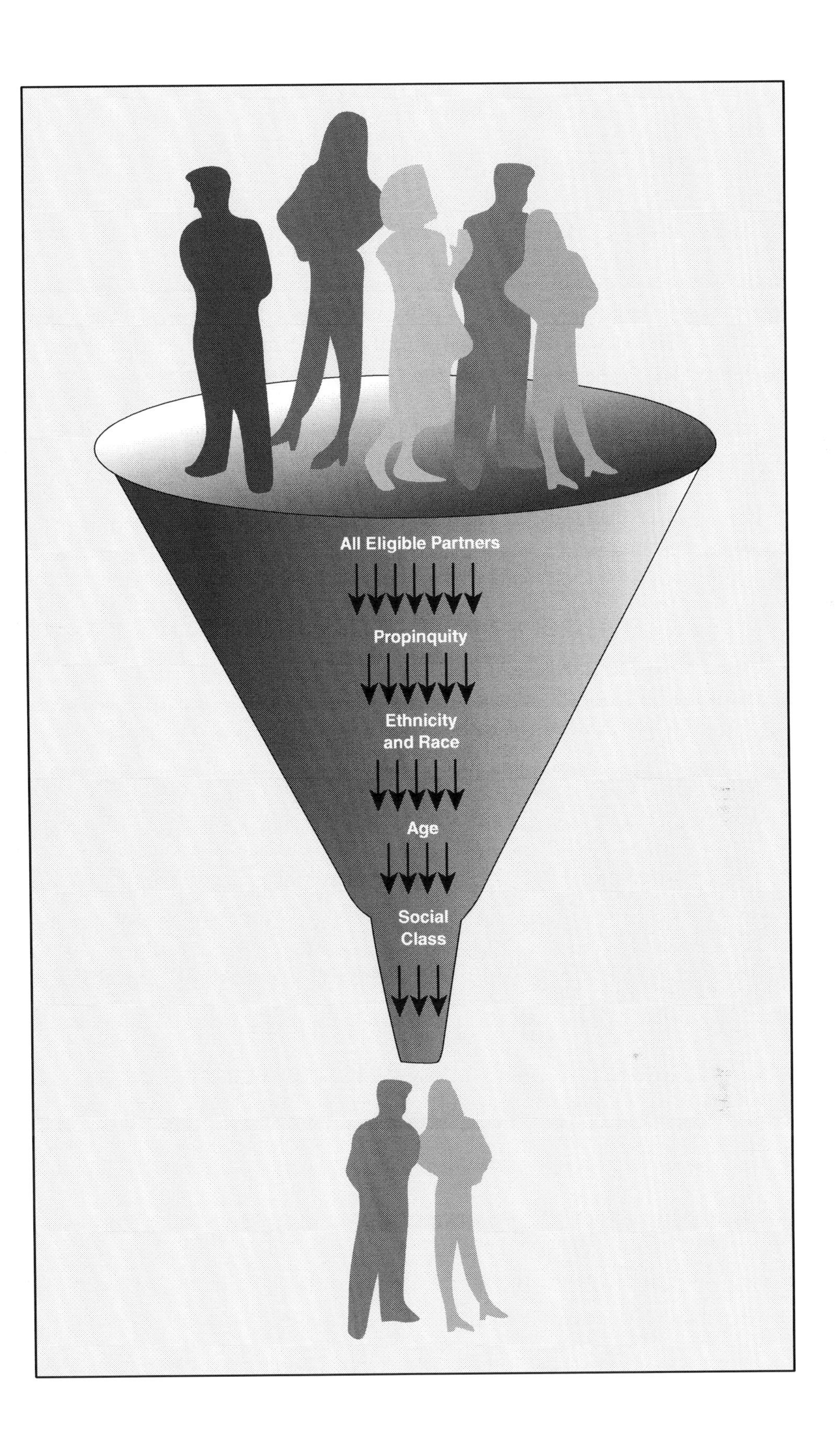
All Eligible Partners
Propinquity
Ethnicity
and Race
Age
Social
Class

of a relationship (Sprecher *et al.*, 2013). For example, as pairs become interdependent, engage in the same activities, and live in the same environment, they are likely to develop similar interests (Gonzaga, 2009) and even begin to look alike because of being influenced by each other's diets and exercise patterns (Christakis & Fowler, 2007). When the couple spends more time together over time, they are likely to develop more similar attitudes. **Attitude alignment** refers to "the tendency of interacting partners to modify their attitudes in such a manner as to achieve attitudinal congruence" (Davis & Rusbult, 2001). A couple that is more satisfied in their relationship perceives greater similarity. Thus, perceived similarity is more important than actual similarity in generating attraction and relationship satisfaction (Klohnen & Luo, 2003) and as a predictor of satisfaction and attraction (Montoya *et al.*, 2008).

There were inconsistent research findings on the similarity effect on marital relationships. For example, research considers the implications of online dating for homogamy (similar partners marrying each other). It is assumed that if a couple shares things in common, they are less likely to have problems. However, this is not the case over the course of a lifetime (Finkel *et al.*, 2012). For example, a study of 23,000 married couples reveals that the similarity of spouses accounted for less than 0.5 percent of spousal satisfaction (Lehrer, 2016). In speed-dating, partner similarity had weak effects on dating success (Asendorpf *et al.*, 2011).

Global Overview: Mate Selection

People today choose mates based on love, but in traditional societies people developed courtship systems for matching couples. Let's look at some traditional and contemporary mate selection processes around the globe.

Source: Wikipedia, 2016

Bride Purchase

Bride purchase—in which a female was purchased as property—was once practiced as a courtship system. Because traditionally women were subordinate to men and were economically dependent on them, marriage was a transfer of property from the father's side to the husband's side. Purchasing a bride is therefore associated with the concept of women as property. *The Babylonian Marriage Market* (see image on previous page) is an 1875 painting by the British painter Edwin Long that vividly portrays women being auctioned into marriage. Although the practice still takes place in some areas of the world, it is illegal in most countries.

Picture Bride

The advance of photography brought about a new form of mate selection: the **picture bride** industry. Like the mail-order bride industry today, the picture bride industry helped Japanese and Korean immigrants living in Hawaii and the U.S. West Coast in the early twentieth century to find brides from their native countries via a matchmaker. Brides would be chosen based on a photograph or recommendation only (Chun, 1996).

The Blind Date: Making a Comeback in Contemporary China

In China, arranged marriages, sometimes called blind marriages, were the norm before the mid-twentieth century. In arranged marriages in traditional China, a marriage was a negotiation and decision between parents and other older members of two families. A matchmaker would set up a meeting between the groom's family and the bride's family to confirm their attitude toward the marriage. However, since the Chinese Marriage Law was passed in 1950, people have moved away from arranged marriages to love marriages in mate selection.

Yet the blind date is making a comeback in contemporary China. A **blind date** is a social engagement between two persons who have not previously met; the date is usually set up by a mutual acquaintance. Because most young Chinese adults today are education and career oriented, they often postpone marriage and parenthood. Women who are unmarried by the age of 30 face the stigma of being labeled "*left over*" or *Shengnu*—the Chinese equivalent of a spinster (Suliman, 2016). These single women often face huge family pressure to get married.

When these adults reach marriage age but are not yet married, their parents often begin to try to find a spouse for them. They meet at spouse-hunting fairs or try their luck at finding eligible candidates for their adult sons and daughters elsewhere. Parents have even spearheaded blind date corners in public parks. For example, many parents frequent the blind date corner at People's Park in Shanghai on Saturdays and Sundays from noon to 5 pm (see image below). They use an umbrella to advertise that they are seeking eligible candidates to meet with their adult children. In the Chinese mate selection market, physical attributes and socioeconomic status are strongly prioritized (Lange *et al.*, 2014).

Source: Wen Ren

Summary

Courtship is dating with the goal of marriage. In closed courtship systems, arranged marriages take place in which parents select mates and arrange the marriage. An open courtship system is characterized by love marriages, where individuals are granted freedom of choice in mate selection. Traditional dating behaviors include dinner dates. Modern peer-influenced dating includes hanging out, group dating, and hooking up. In contemporary society, people often use dating services. One risk of dating is dating violence. Factors that influence mate selection vary from sex ratio to socioeconomic standing. Homogamy is the tendency to select a marriage mate with similar characteristics. Heterogamy is the tendency to select a marriage mate with different characteristics. Sociological theories explain the mate selection process, including filter theory which narrows the pool of eligible candidates for a mate according to specific criteria. The chapter has also discussed gender differences in mate selection and explored mate selection practices across countries.

Key Terms

Access refers to users' exposure to and opportunity to evaluate potential romantic partners they are otherwise unlikely to encounter.

Acquaintance rape or **date rape** refers to a sexual assault or a rape of a victim who knows or is familiar with the rapist.

Age heterogamy or age-dissimilar marriage (ADM) refers to the extent of the age difference between the partners.

Age homogamy refers to age-similar marriage or *coeval*—a person of roughly the same age as oneself.

Arranged marriage is a type of marital union where the bride and groom are selected by their families.

Arranged marriages without force occur when parents choose a mate but do not force their adult child to marry that person.

Attitude alignment refers to the tendency of interacting partners to modify their attitudes in such a manner as to achieve attitudinal congruence.

Attraction–similarity model suggests that people in existing relationships are motivated to perceive similarity to achieve balance and satisfaction over time.

Blind date is a social engagement between two persons who have not previously met; the date is usually set up by a mutual acquaintance.

Bride purchase occurs when a female is purchased as property and was once practiced as a courtship system.

Chivalry is the idea that men engage in acts specifically for women that they may not do for other men.

Closed courtship system occurs when the parents of adult children choose eligible mates for their sons and daughters to form an arranged marriage.

Communication refers to users' opportunity to use various forms of computer-mediated communication (CMC) to interact with potential partners through a dating site before meeting face to face.

Courtship is dating with the intent of marriage.

Cultural scripts are "collective guides" for situational norms, values, and practices that are characteristic of and accessible to cultural insiders.

Dating is a form of courtship where an individual chooses to meet socially with another person with the aim of each assessing the other's suitability either as a prospective partner in a relationship or marriage.

Dating scripts are what allow people to predict the actions of others and serve as guidelines for their own decisions on how to react to the other person.

Dating violence refers to an act of violence by one or more members on the other member in the context of dating or courtship.

Digital abuse refers to the use of technologies and/or social media networking as a tool to intimidate, harass, or threaten a current or ex-dating partner.

Dinner date is a traditional date to eat together in a restaurant.

Educational heterogamy refers to marriages involving two partners with different levels of education.

Educational homogamy refers to marriages involving two partners with the same level of education.

Egalitarian dating is a form of courtship that is not structured by gender role norms but instead favors and promotes equality between partners and the sharing of power and responsibilities.

Endogamy is the practice of selecting a marriage mate within the same ethnic or cultural group.

Event-driven courtship refers to a relationship that escalates based on external factors that have little to do with a couple's true level of intimacy.

Exogamy is the practice of marrying someone outside one's ethnic or cultural group.

Face-ism, or facial prominence, is the relative prominence of the face or the proportion of the head relative to the body in the portrayal of men and women.

Feminist theory focuses on analyzing gender inequality.

Fertility rate means the number of live births in women over a specific length of time.

Filter theory proposes that our social structure sifts and narrows the pool of eligible candidates for a mate according to specific criteria.

First date is an initial meeting between two persons with the goals of screening potential dating candidates or producing a relationship outcome.

Five stages of dating includes attraction, uncertainty, exclusivity, intimacy, and engagement that moves romantic partners toward more committed relationships.

Forced arranged marriage occurs when the two partners are married without their consent or against their wishes.

Friendship script is an equal-power dating script characterized by mutual responsiveness and shared responsibility for date events between partners.

Gender-based violence (GBV) refers to any act that results in physical, sexual, or psychological harm or suffering to women.

Gender imbalance is a disparity between males and females in a population.

Group dating is a modern form of dating where a group of men and women organize a night out, with the hope of forming romantic partnerships.

Hanging out is a form of get-together with friends or acquaintances that can lead to a relationship.

Heterogamy is the tendency to select a marriage mate with different characteristics.

Homogamy is the tendency to select a mate with similar characteristics.

Hooking up is a form of casual sexual relationship and experience (CSRE) in which uncommitted sexual encounters between casual acquaintances or strangers last just one night.

Hypergamy involves marrying up the social ladder and improving one's social status.

Hypogamy involves marrying down the social ladder.

Interpersonal scripts are the behavioral enactment of a specific cultural script by an individual.

Interracial marriage is the practice of marrying someone of a different race.

Intracial marriage is the practice of marrying someone of a similar race.

Length of courtship refers to the amount of time between the couple's first date and the decision to get married.

Love marriage is a marital union based on love and is the sole decision of the couple.

Mail-order bride is a woman from a developing country who advertises herself in a catalog or an online mail-order bride agency to find and marry a man in a developed nation.

Marriage of state is diplomatic marriage between two members of different nations or between two power coalitions to secure peace between two parties.

Marriage marketplace describes the mate selection process where people compare the assets and liabilities of eligible candidates and choose the best available mate.

Marriage rate is the number of marriages per 1,000 people in a year.

Marriage squeeze refers to an imbalance between the number of men and women available to marry in a certain society.

Matching refers to a site's use of a mathematical algorithm to select potential partners for users.

Matchmaker is a person who arranges relationships and marriages between others, either informally in cultural communities or formally in a matchmaking business.

Mate selection is a process in which we consider criteria that we prefer in a mate, from physical appearance to cultural background to education and socioeconomic standing.

Mating market approach suggests that people with desirable traits have a stronger "bargaining hand" and can be more selective when choosing partners.

Maturity is the ability to respond to the environment in an appropriate manner.

Mobile dating services allow individuals to chat and become romantically involved by means of text messaging, mobile chatting, and the mobile web.

Modern arranged marriages with courtship occur when the parents select a prospective mate and allow the couple to court and develop a love relationship.

Online dating is the practice of searching for a romantic partner or date on the internet via online dating websites.

Open courtship system occurs when people choose their partners, get to know to each other, move on to being in a serious relationship, and make their own decisions regarding marriage.

Physical attractiveness is the degree to which a person's physical features are considered aesthetically pleasing or beautiful.

Physical dating violence is the use of physical force with the intent to cause fear or injury.

Psychological/emotional dating violence is the use of words with the intent to cause fear or harm his or her sense of self-worth.

Racial heterogamy refers to marriages involving two partners of different racial groups.

Relationship phase model suggests three phases—coming together, maintenance, and coming apart.

Sex ratio is the number of males per 1,000 females in a population.

Sex-selective abortion is the practice of terminating a pregnancy based upon the predicted sex of the infant.

Sexual dating violence is forcing a partner to engage in a sex act when he or she does not or cannot consent.

Sexual strategies theory suggests that both men and women have evolved different mechanisms that underlie short-term and long-term mating strategies.

Social capital theory suggests that average health is better in countries with more educational heterogamy.

Social network may serve the role of matchmakers for people to meet socially and form relationships that may develop into marriage.

Social network sites (SNSs) are internet-based applications that allow users to create and share their profiles and to view and interact with online connections made by others in the user's network.

Speed dating refers to organized matchmaking events that allow multiple men to meet multiple women of similar age for brief encounters one after the other and assess mutual interest before subsequent dates.

Spousal racial homogamy refers to marriages between spouses with the same racial background.

Status is a social position held by a person within a group.

Status inconsistency occurs when people have a mismatch between their statuses.

Turning point refers to an event that is associated with a change in a relationship.

Westernized versions of arranged marriages are forms of mate selection endorsed by parents through media usage and free choice of partners.

Discussion Questions

- Why are marriages sometimes arranged by parents or grandparents?
- How does dating differ from courtship?
- Are there any alternative dating options in contemporary society?
- Why do Americans select partners who are often similar to themselves?

Suggested Film and Video

Apatow, J. (Producer). (2017). *The Big Sick*. Amazon Studios Lionsgate.

Ren, W. (Lana Del Rey). (2013). *Every 15 Minutes 2013 (City of La Mirada)* (Video). Retrieve from https://www.youtube.com/watch?v=8Hmk3-NBzCM

Ren, W. (Wen Ren). (2014). Café Glass. (Short Film). Retrieved from https://www.youtube.com/watch?v=CbNjzBsIf78

Subash, K. (Vikram Krishna). (2001). *Love Marriage*. (Film). India.

Internet Sources

Centers for Disease Control and Prevention: https://www.cdc.gov/violenceprevention/datingmatters/

The National Center for Victims of Crime. (2012). *National Hotlines and Help Links*: http://www.victimsofcrime.org/help-for-crime-victims/national-hotlines-and-helpful-links

The National Sexual Violence Resources Center (NSVRC): https://www.nsvrc.org/

References

Abbey, A. (1987). Misperceptions of Friendly Behavior as Sexual Interest: A Survey of Naturally Occurring Incidents. *Psychology of Women Quarterly*, 11(2), 173.

Abdellaoui, A., Verweij, K. J. H. & Zietsch, B. P. (2014). No Evidence for Genetic Assortative Mating Beyond That Due to Population Stratification. *Proceedings of the National Academy of Sciences*, 111(40), E4137–E4137.

Abel, E. L. & Kruger, M. L. (2008). Age Heterogamy and Longevity: Evidence from Jewish and Christian Cemeteries. *Journal of Biodemography and Social Biology*, 54(1), https://doi.org/10.1080/19485565.2008.9989128

Abma, J. C., Martinez, G. M., Mosher, W. D. & Dawson, B. S. (2004). Teenagers in the United States: Sexual Activity, Contraceptive Use, and Childbearing, 2002. *Vital Health Stat*, 23(24), 1–48.

Afifi, W. A. & Lucas, A. A. (2008). Information Seeking in Initial Stages of Relational Development. In S. Sprecher, A. Wenzel & J. Harvey (Eds.), *Handbook of Relationship Initiation* (pp. 197–215). New York: Psychology Press.

Agrawal, A. (2015). Cyber-matchmaking among Indians: Re-arranging Marriage and Doing 'Kin Work.' *South Asian Popular Culture*, 13(1).

Akers, D. S. (1967). On Measuring the Marriage Squeeze. *Demography*, 4(2), 907–924.

Alder, R. B. & Rodman, G. (2003). *Understanding Human Communication*, 8th edn. New York: Oxford University Press.

Alksnis, C., Desmarais, S. & Wood, E. (1996). Gender Differences in Scripts for Different Types of Dates. *Sex Roles*, 34, 499–509.

Allendorf, K. (2013). Schemas of Marital Change: From Arranged Marriages to Eloping for Love. *Journal of Marriage and the Family*, 75(2), 453–469.

Amichai-Hamburger, Y. & Hayat, T. (2017). Social Networking. *The International Encyclopedia of Media Effects*, 1–12.

Archer, D., Iritani, B., Kimes, D. & Barrios, M. (1983). Face-ism: Five Studies of Sex Differences in Facial Prominence. *Journal of Personality and Social Psychology*, 45, 725–735.

Armstrong, E. A., England, P. & Fogarty, A. C. K. (2009). Orgasm in College Hookups and Relationships. In B. J. Risman (Ed.), *Families as They Really Are* (pp. 362–377). New York: Norton.

Armstrong, E. A., Hamilton, L. & England, P. (2010). Is Hooking up Bad for Young Women? *Contexts*, 9(3), 22–27.

Armstrong, E. A., England, P. & Fogarty, A. C. K. (2012). Accounting for Women's Orgasm and Sexual Enjoyment in College Hookups and Relationships. *American Sociological Review*, 77(3), 435–462.

Asendorpf, J. B., Penke, L. & Back, M. D. (2011). From Dating to Mating and Relating: Predictors of Initial and Long-Term Outcomes of Speed-Dating in a Community Sample. *Eur. J. Pers.*, 25, 16–30.

Atkinson, M. & Glass, B. (1985). Marital Age Heterogamy and Homogamy, 1900 to 1980. *Journal of Marriage and Family*, 47, 685–691.

Back, M. D., Penke., L, Schmukle, S. C., Sachse, K., Borkenau, P. & Asendorpf, J. B. (2011). Why Mate Choices Are Not as Reciprocal as We Assume: the Role of Personality, Flirting and Physical Attractiveness. *European Journal of Personality*, 25(2), 120–132.

Backstrom, L., Armstrong, E. A. & Puentes, J. (2012). Women's Negotiation of Cunnilingus in College Hookups and Relationships. *Journal of Sex Research*, 49(1), 1–12.

Bailey, B. L. (1988). *From Front Porch to Back Seat: Courtship in Twentieth-Century America*. Baltimore, MD: Johns Hopkins University Press.

Bassett, J. F. & Moss, B. (2004). Men and Women Prefer Risk Takers as Romantic and Nonromantic Partners. *Curr Res Soc Psychol*, 9, 135–144.

Baxter, L. A. & Bullis, C. (1986). Turning Points in Developing Romantic Relationships. *Human Communication Research*, 12, 469–493.

Baxter, L. A. & West, L. (2003). Couple Perceptions of their Similarities and Differences: A Dialectical Perspective. *Journal of Social and Personal Relationships*, 20, 491–514.

Becker, G. S., Hubbard, W. H. J. & Murphy, K. M. (2010). Explaining the Worldwide Boom in Higher Education of Women. *Journal of Human Capital*, 3, 203–241.

Beigel, H. G. (1951). Romantic Love. *American Sociological Review*, 16(3).

Bejanyan, K., Marshall, T. C. & Ferenczi, N. (2015). Associations of Collectivism with Relationship Commitment, Passion, and Mate Preferences: Opposing Roles of Parental Influence and Family Allocentrism. *PLoS ONE*, 10(2), e0117374.

Beland, N. (March 2003). The Ultimate Guide to The Girl Next Door. *Men's Health*, 18, 132–140. Retrieved November 7, 2003, from EBSCOHost databases.

Belsey, C. (1994). *Desire: Love Stories in Western Culture*. Oxford: Blackwell.

Berardo, F. M., Appel, J. & Berardo, D. H. (1993). Age Dissimilar Marriages: Review and Assessment. *Journal of Aging Studies*, 7(1), 93–106.

Berger, C. R. & Calabrese, R. J. (1975). Some Explorations in Initial Interaction and Beyond: Toward a Developmental Theory of Interpersonal Communication. *Human Communication Research*, 1, 99–112.

Berry, D. S. & McArthur, L. Z. (1986). Perceiving Character in Faces: The Impact of Age-Related Craniofacial Changes on Social Perception. *Psychol Bull*, 100, 3–18.

Bertrand, M., Kamenica, E. & Pan, J. (2015). Gender Identity and Relative Income with Households. *The Quarterly Journal of Economics*, 571–614. doi:10.1093/qje/qjv001. Advance Access publication on January 29, 2015.

Beth, M. (June 19, 2014). Colleges Get New Rules on Dating Violence. *USA Today*.

Black, M. C., Basile, K. C., Breiding, M. J., Smith, S. G., Walters, M. L., Merrick, M. T. & Stevens, M. R. (2011). *The National Intimate Partner and Sexual Violence Survey: 2010 summary report*. Centers for Disease Control and Prevention. Retrieved on November 26, 2018 from http://www.cdc.gov/ViolencePrevention/pdf/NISVS_Report2010-a.pdf

Black, S. (2017). Love Marriage. *South Asia: Journal of South Asian Studies*, 40(2).

Blackwell, D. L. & Lichter, D. T. (2004). Homogamy among Dating, Cohabiting, and Married Couples. *Sociol. Q.*, 45, 719–737.

Blossfeld, H. (2009). Educational Assortative Marriage in Comparative Perspective. *Annual Review of Sociology*, 35, 513–530.

Bogle, K. A. (2007). The Shift from Dating to Hooking Up in College: What Scholars Have Missed. *Sociology Compass*, 1(2), 775–788.

Bogle, K. A. (2008). *Hooking up: Sex, Dating, and Relationships on Campus*. New York: New York University Press.

Bradshaw, C., Kahn, A. S. & Saville, B. K. (2010). To Hook Up or Date: Which Gender Benefits? *Sex Roles*, 62, 661–669.

Brand, R. J., Bonatsos, A., D'Orazio, R. *et al.* (2012). What Is Beautiful Is Good, Even Online: Correlations between Photo Attractiveness and Text Attractiveness in Men's Online Dating Profiles. *Comput Human Behav.*, 28, 166–170.

Brannon, R. (1976). The Male Sex Role: Our Culture's Blueprint of Manhood, and What It's Done for Us Lately. In D. S. David & R. Brannon (Eds.), *The Forty-nine Percent Majority: The Male Sex Role*. Reading, MA: Addison-Wesley.

Break the Cycle (2012). Dating Violence 101. Retrieved on July 29, 2014 from http://www.breakthecycle.org/what-is-dating-violence

Bredow, C. A., Cate, R. M. & Huston, T. L. (2008). Have We Met Before? A Conceptual Model of First Romantic Encounters. In S. Sprecher, A. Wenzel & J. Harvey (Eds.), *Handbook of Relationship Initiation* (pp. 3–28). New York: Psychology Press.

Brooks, A. (December 4, 2017). Amazing Online Dating Statistics—(The Good, Bad & Weird Facts). Dating Advice. Retrieved on December 16, 2017 from http://www.datingadvice.com/online-dating/online-dating-statistics

Broude, G. J. & Greene, S. J. (1983). Cross-cultural Codes on Husband–Wife Relationships. *Ethnology*, 22(3), 263–280.

Brown, A. L. & Testa, M. (2008). Social Influence on Judgments of Rape Victims: The Role of Negative and Positive Social Reactions of Others. *Sex Roles*, 58, 490–500.

Brown, P. H. (2009). Dowry and Intrahousehold Bargaining Evidence from China. *Journal of Human Resources*, 44(1), 25–46.

Buettner, R. (2017). Getting a Job via Career-Oriented Social Networking Markets. *Electron Markets*, 27, 371.

Burnett, P. (1987). Assessing Marital Adjustment and Satisfaction: A Review. Measurements and Evaluation. *Counseling and Development*, 20(3), 113–121.

Buss, D. M. (2018). *The Evolution of Desire: Strategies of Human Mating*. Audiobook, CD. Hachette Audio and Blackstone Audio.

Buss, D. M. & Shmitt, D. P. (1993). Sexual Strategies Theory: A Contextual Evolutionary Analysis of Human Mating. *Psychological Review*, 100, 204–232.

Buss, D. M., Abbott, M., Angleitner, A., Asherian, A., Biaggio, A., Blanco-Villasenor, A., *et al.* (1990). International Preferences in Selecting Mates: A Study of 37 Cultures. *Journal of Cross-Cultural Psychology*, 21, 5–47.

Buunk, B. P., Dijkstra, P., Fetchenhauer, D. & Kenrick, D. T. (2002). Age and Gender Differences in Mate Selection Criteria for Various Involvement Levels. *Personal Relationships*, 9, 271–278.

Buunk, A. P., Park, J. H. & Duncan, L. A. (2010). Cultural Variation in Parental Influence on Mate Choice. *Cross-Cultural Research*, 44, 23–40.

Cacioppo, J. T., Cacioppo, S., Gonzaga, G. C., Ogburn, E. L. & VanderWeele, T. J. (2013). Marital Satisfaction and Break-Ups Differ Across On-Line and Off-line Meeting Venues. *PNAS*, June 18, 110(25).

Campolo, T. (2009). *Choose Love, Not Power: How to Right the World's Wrongs from a Place of Weakness*. Ventura, CA: Regal Books.

Casey, W. (2009). *The Man Plan: Drive Men Wild ... Not Away*. New York: Penguin.

Casterline, J., Qian, Z. & Liu, J. (2010). Assortative Mating on Age: Trends in the Spousal Age Difference. Paper presented at annual meeting of the Population Association of America; Dallas, TX.

Castro, F. N. & de Araujo Lopes, F. (2011). Romantic Preferences in Brazilian Undergraduate Students: From the Short Term to the Long Term. *J. Sex Res.*, 48, 479–485.

Cate, R. & Lloyd, S. (1992). *Courtship*. Newbury Park, CA: Sage.

CDC (Centers for Disease Control and Prevention) (2016) *Understanding Teen Dating Violence*. National Center for Injury Prevention and Control.

CDC (Centers for Disease Control and Prevention) (December 05, 2017). *Keeping you Safe 7/24*. Washington: The Department of Health and Human Services. Retrieved on November 27, 2018 from https://www.cdc.gov/about/24-7/

CDC (Centers for Disease Control and Prevention) (June 11, 2018). *Teen Dating Violence*. GA: National Center for Injury Prevention and Control. Retrieved on November 27, 2018 from https://www.cdc.gov/violenceprevention/intimatepartnerviolence/teen_dating_violence.html

Chase, W. G. & Simon, H. A. (1973). The Mind's Eye in Chess. In W. G. Chase (Ed.), *Visual Information Processing* (pp. 215–282). New York: Academic Press.

Chen, D. (Apr 12, 2017). 4 Lessons from the Longest-Running Study on Happiness. TED Ideas.

Choi, K. H. & Mare, R. D. (2012). International Migration and Educational Assortative Mating in Mexico and the United States. *Demography*, 49, 449.

Choi, K. H. & Tienda, M. (2014). *Intermarriage: Bringing Marriage Markets Back*. Retrieved on January 17, 2018 from http://paa2015.princeton.edu/papers/152771

Christakis, N. A. & Fowler, J. H. (2007). The Spread of Obesity in a Large Social Network over 32 Years. *The New England Journal of Medicine*, 357, 370–379.

Chun, C. S. Y. (1996). The Mail-Order Bride Industry: The Perpetuation of Transnational Economic Inequalities and Stereotypes. *University of Pennsylvania Journal of International Economic Law*, 17(4).

CIA (Central Intelligence Agency) (January 4, 2018a). United States. *The World Factbook*. Retrieved on January 9, 2018 from https://www.cia.gov/library/publications/resources/the-world-factbook/geos/us.html

CIA (Central Intelligence Agency) (January 3, 2018b). China. *The World Factbook*. Retrieved on January 9, 2018 from https://www.cia.gov/library/publications/resources/the-world-factbook/geos/ch.html

Clore, G. L. & Byrne, D. (1974). A Reinforcement-Affect Model of Attraction. In T. L. Huston (Ed.), *Foundations of Interpersonal Attraction* (pp. 143–170). New York: Academic Press.

Clutton-Brock, T. H. (1991). *The Evolution of Parental Care*. Princeton, NJ: Princeton University Press.

Cohen, M. T. (2016). It's Not You, It's Me ... No, Actually It's You: Perceptions of What Makes a First Date Successful or Not. *Sexuality & Culture: An Interdisciplinary Quarterly*, 20(1), 173–191.

Collins, N. L. & Miller, L. C. (1994). Self-Disclosure and Liking: A Meta-analytic Review. *Psychol Bull.*, 116, 457–475.

Coltrance, S. (1998). *Gender and Families*. Thousand Oaks, CA: Pine Forge Press.

Conner, B. H., Peters, K. & Nagasawa, R. H. (1975). Person and Costume: Effects on the Formation of First Impressions. *Home Econ Res J.*, 4, 32–41.

Connolly, C. M. & Sicola, M. K. (2005). Listening to Lesbian Couples: Communication Competence in Long-term Relationships. *Journal of GLBT Family Studies*, 1, 143–167.

Cowan, G. (2000). Beliefs about the Causes of Four Types of Rape. *Sex Roles*, 42, 807–823.

Crowder, K. D. & Tolnay, S. E. (2000). A New Marriage Squeeze for Black Women: The Role of Racial Intermarriage by Black Men. *Journal of Marriage and Family*, 62(3), 792–807.

Currier, D. M. (2013). Strategic Ambiguity: Protecting Emphasized Femininity and Hegemonic Masculinity in the Hookup Culture. *Gender & Society*, 27(5), 704–727.

Darj, E., Wijewardena, K., Lindmark, G. & Axemo, P. (2017). Even Though a Man Takes the Major Role, He Has No Right to Abuse: Future Male Leaders' Views on Gender-Based Violence in Sri Lanka. *Global Health Action*, 10(1), 1348692.

Davis, J. L. & Rusbult, C. E. (2001). Attitude Alignment in Close Relationships. *Journal of Personality and Social Psychology*, 81, 65–84.

Davis, L. (2013). *Love at First Click: The Ultimate Guide to Online Dating.* Upper Saddle River, NJ: Simon and Schuster.

Degenne, A. & Forse, M. (1999). *Introducing Social Networks.* London: Sage.

de Mel, N. P. P. & Gomez, S. (2013). Why Masculinities Matter: Attitudes, Practice and Gender Based Violence in Four Districts in Sri Lanka. Columbo: CARE International Sri Lanka.

Demos, V. (2007). The Intersection of Gender, Class and Nationality and the Agency of Kytherian Greek Women. Paper presented at the annual meeting of the American Sociological Association, August 11.

Derne, S. (1995). *Culture in Action: Family Life, Emotion, and Male Dominance in Banaras, India.* Albany, NY: State University of New York Press.

Devries, K. M., Knight, L., Child, J. C. *et al.* (2015). The Good School Toolkit for Reducing Physical Violence from School Staff to Primary School Students: A Cluster-Randomised Controlled Trial in Uganda. *Lancet Globe Health*, e378–379.

Dholakia, U. (November 24, 2015). Why Are So Many Indian Arranged Marriages Successful? *Psychology Today*, New York: Sussex Publishers, LLC. Retrieved on November 27, 2018 from https://www.psychologytoday.com/us/blog/the-science-behind-behavior/201511/why-are-so-many-indian-arranged-marriages-successful

Domingue, B. W., Fletcher, J., Conley, D. & Boardman, J. D. (2014). Genetic and Educational Assortative Mating among US Adults. *PNAS*, May 19, 2014.

Doosje, B., Rojahn, K. & Fischer, A. (1999). Partner Preferences as a Function of Gender, Age, Political Orientation, and Level of Education. *Sex Roles*, 40, 45–60.

Duggan, M. (June 26, 2017a). *Same-Sex Marriage Detailed Table, 2017 Total.* Washington, DC: Pew Research Center.

Duggan, M. (July 14, 2017b). *Men, Women Experience and View Online Harassment Differently.* Washington, DC: Pew Research Center.

Dugsin, R. (2001). Conflict and Healing in Family Experience of Second-Generation Emigrants from India Living in North America. *Fam Process*, Summer; 40(2), 233–241.

Dutton, D. G. & Aron, A. P. (1974). Some Evidence for Heightened Sexual Attraction under Conditions of High Anxiety. *Journal of Personality and Social Psychology*, 30(4), 510–517.

Dyrenforth, P. S., Kashy, D. A., Donnellan, M. B. & Lucas, R. E. (2010). Predicting Relationship and Life Satisfaction from Personality in Nationally Representative Samples from Three Countries: The Relative Importance of Actor, Partner, and Similarity Effects. *J. Pers. Soc. Psychol.*, 99, 690–702.

Eastwick, P. W. & Smith, L. K. (2018) Sex-differentiated Effects of Physical Attractiveness on Romantic Desire: A Highly Powered, Preregistered Study in a Photograph Evaluation Context, Comprehensive Results. *Social Psychology*, 3(1), 1–27.

Eaton, A. A. & Rose, S. (2011). Has Dating Become More Egalitarian? A 35 Year Review Using Sexual Roles. *Sex Roles*, 64, 843.

Eeckhaut, M. C. W., Van de Putte, B., Gerris, J. R. M. & Vermulst, A. (2013). Educational Heterogamy: Does it Lead to Cultural Differences in Child-rearing? *Journal of Social and Personal Relationships*, 31(6), 729–750.

Ehrmann, W. (1959). *Premarital Dating Behavior.* New York: Henry Holt.

Emmers-Sommer, T. M., Farrell, J., Gentry, A., Stevens, S., Eckstein, J., Battocletti, J. & Gardener, C. (2010). First Date Sexual Expectations: The Effects of Who Asked, Who Paid, Date Location, and Gender. *Communication Studies*, 61(3), 339–355.

Emond, A. & Eduljee, N. B. (2014). Gender Differences: What We Seek in Romantic and Sexual Partners. *Universal Journal of Psychology*, 2(2), 90–94.

England, P. & Bearak, J. (2013, April 11–13). Is There a War of the Sexes in College? Gender, Meanings, and Casual Sex. Paper prepared for presentation at the Population Association of America, New Orleans, LA.

England, P., Shafer, E. F. & Fogarty, A. C. K. (2007). Hooking Up and Forming Romantic Relationships on Today's College Campuses. In M. Kimmel (Ed.), *The Gendered Society Reader* (pp. 531–547). New York: Oxford University Press.

Epstein, J., Klinkenberg, W. D., Scandell, D. J., Faulkner, K. & Claus, R. E. (2007). Perceived Physical Attractiveness, Sexual History, and Sexual Intentions: An Internet Study. *Sex Roles*, 56, 23–31.

Epstein, R., Pandit, M. & Thakar, M. (2013). How Love Emerges in Arranged Marriages: Two Cross-cultural Studies. *Journal of Comparative Family Studies*, 44(3), 341–360. Retrieved from http://www.jstor.org/stable/23644606

Eshbaugh, E. M. & Gute, G. (2008). Hookups and Sexual Regret among College Women. *The Journal of Social Psychology*, 148, 77–89.

Faier, L. (2007). Filipina Migrants in Rural Japan and Their Professions of Love. *American Ethnologist*, 34, 148–162.

Fales, M. R., Frederick, D. A., Garcia, J. R., Gildersleeve, K. A., Haselton, M. G. & Fisher, H. E. (2016). Mating Markets and Bargaining Hands: Predictors of Mate Preferences for Attractiveness and Resources in Two National U.S. Studies. *Personality and Individual Differences*, 88, 78–87.

Fay, B. & Eaton, A. A. (2014) Egalitarian Dating over 35 Years. In A. C. Michalos (Ed.), *Encyclopedia of Quality of Life and Well-Being Research*. Dordrecht: Springer.

Fehr, B. (2001). The Life Cycle of Friendship. In C. Hendrick & S. S. Hendrick (Eds.), *Close Relationships: A Sourcebook* (pp. 71–82). Thousand Oaks, CA: Sage.

Fehr, B. (2008). Friendship Formation. In S. Sprecher, A. Wenzel, & J. Harvey (Eds.), *Handbook of Relationship Initiation* (pp. 29–54). New York: Psychology Press.

Fehr, B., Sprecher, S. & Underwood, L. (Eds.) (2009). *Compassionate Love: Theory, Research, and Applications*. Malden: WileyBlackwell.

Feldman, R. S. (2011). *Gender and Sexuality. Life Span Development: A Topical Approach*. Upper Saddle River, NJ: Prentice Hall/Pearson.

Finer, L. B. (2007). Trends in Premarital Sex in the United States, 1954–2003. *Public Health Reports*, 122(1), 73–78.

Finkel, E. J., Eastwick, P. W., Karney, B. R., Reis, H. T. & Sprecher, S. (2012). Online Dating: A Critical Analysis from the Perspective of Psychological Science. *Psychological Science in the Public Interest*, 13(1), 3–66.

Fisher, T. & Johnson, P. (2008). *The Power of Prestige: Why Young Men Report Having More Sex Partners than Young Women*. Published online: 19 July 2008, Springer Science and Business Media, LLC.

Fleming, P. H. (June 1973). The Politics of Marriage among Non-Catholic European Royalty. *Current Anthropology*, 14(3), 231–249.

Fletcher, G. J. O., Tither, J. M., O'Loughlin, C., Friesen, M. & Overall, N. (2004). Warm and Homely or Cold and Beautiful? Sex Differences in Trading Off Traits in Mate Selection. *Personality & Social Psychology Bulletin*, 30, 659–672.

Foshee, V. A., McNaughton Reyes, H. L., Ennett, S. T., Cance, J. D., Bauman, K. E. & Bowling, J. M. (2012). Assessing the Effects of Families for Safe Dates, a Family-based Teen Dating Abuse Prevention Program. *Journal of Adolescent Health*, 51(4), 349–356.

Frank, M. G., Ekman, P. & Friesen, W. V. (1993). Behavioral Markers and Recognizability of the Smile of Enjoyment. *J Pers Soc Psychol*, 64, 83–93.

Franzoi, S. L. (2001). Is Female Body Esteem Shaped by Benevolent Sexism? *Sex Roles*, 44, 177–188.

Friedman, S. L. (2005). The Intimacy of State Power: Marriage, Liberation, and Socialist Subjects in Southeastern China. *The Journal of the American Ethnological Society*, 32(2).

Garcia, J. R. & Heywood, L. L. (2016). Moving toward Integrative Feminist Evolutionary Behavioral Sciences. *Feminism & Psychology*, 26(3), 327–334.

Garcia, J. R., Reiber, C., Massey, S. G. & Merriwether, A. M. (2012). Sexual Hookup Culture: A Review. *Review of General Psychology*, 16(2), 161–176.

Garcia, J. R., Reiber, C., Massey, S. G. & Merriwether, A. M. (February 2013). Sexual Hook-up Culture. *APA (American Psychological Association) Monitor on Psychology*, 44(2), 60–67.

Garcia, J. R., Seibold-Simpson, S. M., Massey, S. G. & Merriwether, A. M. (2015). Casual Sex: Integrating Social, Behavioral, and Sexual Health Research. In J. DeLamater & R. F. Plante (Eds.), *Handbook of the Sociology of Sexualities* (pp. 203–222). Dordrecht, Netherlands: Springer.

George, W. H. & Norris, J. (1991). Alcohol, Disinhibition, Sexual Arousal, and Deviant Sexual Behavior. *Alcohol Health and Research World*, 15, 133–138.

Gershon, A., Gowen, L. K., Compian, L. & Hayward, R. C. (2004). Gender-stereotyped Imagined Dates and Weight Concerns in Sixth Grade Girls. *Sex Roles*, 50, 515–523.

Gilligan, C. (1982). *In a Different Voice: Psychological Theory and Women's Development*. Cambridge, MA: Harvard University Press.

Ginsburg, G. P. (1988). Rules, Scripts and Prototypes in Personal Relationships. In S. Duck, D. F. Hay, S. E. Hobfoll, W. Ickes, & B. M. Montgomery (Eds.), *Handbook of Personal Relationships: Theory, Research and Interventions* (pp. 23–39). Oxford, UK: John Wiley & Sons.

Gladwell, M. (2000). *The Tipping Point: How Little Things Can Make A Big Difference*. New York: Little Brown.

Glasser, C. L., Robnett, B. & Feliciano, C. (2009). Internet Daters' Body Type Preferences: Race–Ethnic and Gender Differences. *Sex Roles*, 61, 14–33.

Glenn, N. & Marquardt, E. (2001). *Hooking Up, Hanging Out, and Hoping for Mr. Right: College Women on Dating and Mating Today*. New York: Institute for American Values.

Goddard, C. & Wierzbicka, A. (2004). Cultural Scripts: What Are They and What Are They Good For? *Intercultural Pragmatics*, 1(2), 153–166.

Golding, J. M., Stein, J. A., Siegel, J. M., Burnam, M. A. & Sorenson, S. B. (1988). Sexual Assault History and Use of Health and Mental Health Services. *American Journal of Community Psychology,* 16(5), 625–644.

Gonzaga, G. C. (2009). Similarity Principle of Attraction. In H. Reis & S. Sprecher (Eds.), *Encyclopedia of Human Relationships* (Vol. 3, pp. 1496–1500). Thousand Oaks, CA: Sage.

Gonzalez, J. T. (1988). Dilemmas of the High-achieving Chicana: The Double-bind Factor in Male/Female Relationships. *Sex Roles,* 18, 367–380.

Goode, E. (1996). Gender and Courtship Entitlement: Responses to Personal Ads. *Sex Roles,* 34, 141–169.

Goode, W. J. (1980). Why Men Resist. *Dissent,* 27, 181–193.

Gottman, J. M., Levenson, R. W., Swanson, C., Swanson, K., Tyson, R. & Yoshimoto, D. (2003). Observing Gay, Lesbian and Heterosexual Couples' Relationships: Mathematical Modeling of Conflict Interaction. *Journal of Homosexuality,* 45, 65–91.

Gray, J. (1997). *Mars and Venus on a Date: A Guide for Navigating the 5 Stages of Dating to Create a Loving and Lasting Relationship.* New York: Harper Paperbacks.

Griffin, A. M. & Langlois, J. H. (2006). Stereotype Directionality and Attractiveness Stereotyping: Is Beauty Good or Is Ugly Bad? *Social Cognition,* 24(2), 187–206.

Grover, K., Russell, C., Schumm, W. & Paff-Bergen, L. (1985). Mate Selection Processes and Marital Satisfaction. *Family Relations,* 34(3), 383–386.

Guo, G., Wang, L., Liu, H. & Randall, T. (2014). Genomic Assortative Mating in Marriages in the United States. *PLoS ONE,* 9(11), e112322.

Hall, J. A., Park, N., Song, H., & Cody, M. J. (2010). Strategic Misrepresentation in Online Dating: The Effects of Gender, Self-Monitoring, and Personality Traits. *Journal of Social and Personal Relationships,* 27(1), 117–135.

Hamby, S. & Jackson, A. (2010). Size Does Matter: The Effects of Gender on Perceptions of Dating Violence. *Sex Roles,* 63, 324–331.

Hamid, S., Johansson, E. & Rubenson, B. (2011). Good Parents Strive to Raise 'Innocent Daughter'. *Cult Health Sex,* Aug; 13(7), 841–851.

Hannon, R., Hall, D. S., Kuntz, T., Van Laar, S. & Williams, J. (1995). Dating Characteristics Leading to Unwanted vs. Wanted Sexual Behavior. *Sex Roles,* 33, 767–783.

Hannon, R., Kuntz, T., Van Laar, S., Williams, J. & Hall, D. S. (1996). College Students' Judgments Regarding Sexual Aggression during a Date. *Sex Roles,* 35, 765–773.

Hannon, R. K., Hall, D. S., Nash, H., Formati, J. & Hopson, T. (2000). Judgments Regarding Sexual Aggression as a Function of Sex of Aggressor and Victim. *Sex Roles,* 43, 311–322.

Hansen, S. R. (2006). *Courtship Duration as a Correlate of Marital Satisfaction and Stability.* Alliant International University, California School of Professional Psychology, San Diego.

Hatfield, E. & Rapson, R. L. (2002). Passionate Love and Sexual Desire: Cross-Cultural and Historical Perspectives In A. Vangelisti, H. T. Reis & M. A. Fitzpatrick (Eds.), *Stability and Change in Relationships* (pp. 306–324). Cambridge, UK: Cambridge University Press.

Hatfield, E. & Sprecher, S. (1986). *Mirror, Mirror… The Importance of Looks in Everyday Life.* Albany: State University of New York Press.

Hawkins, A. J., Carroll, W. D. & Willoughby, B. (2004). A Comprehensive Framework for Marriage Education. *Family Relations,* 53, 547–558.

Heaton, T., Cammack, M. & Young, L. (2001) Why Is the Divorce Rate Declining in Indonesia? *Journal of Marriage and the Family,* 63(2), 480–490.

Heise, L. L. (1998). Violence against Women: An Integrated, Ecological Framework. *Violence Against Women,* 4, 262–290.

Hernan, V., Berardo, D. H. and Berardo, F. M. (1985). Age Heterogamy in Marriage. *Journal of Marriage and Family,* 47(3), 553–566.

Hetsroni, A. (2000). Choosing a Mate in Television Dating Games: The Influence of Setting, Culture and Gender. *Sex Roles,* 42, 83–106.

Hobbs, M, Owen, A. & Gerber, L. (2016). Liquid Love? Dating Apps, Sex, Relationships and the Digital Transformation of Intimacy. *Journal of Sociology,* 53(2), 271–284.

Holden, C. J., Zeigler-Hill, V., Pham, M. N. & Shackelford, T. K. (2014). Personality Features and Mate Retention Strategies: Honesty–Humility and the Willingness to Manipulate, Deceive, and Exploit Romantic Partners. *Pers. Individ. Dif.,* 57, 31–36.

Howard, J. A., Blumstein, P. & Schwartz, P. (1987). Social or Evolutionary Theories? Some Observations on Preferences in Human Mate Selection. *Journal of Personality and Social Psychology,* 53, 194–200.

Huda, S. (2006). Dowry in Bangladesh: Compromising Women's Rights. *South Asian Research,* 26(3), 249–268.

Huijts, T., Monden, C. W. S. & Kraaykamp, G. (2010). Education, Educational Heterogamy, and Self-Assessed Health in Europe: A Multilevel Study of Spousal Effects in 29 European Countries. *European Sociological Review,* 26(3), 261–276.

Huston, T. L. & Burgess, R. L. (1979). Social Exchange in Developing Relationships: An Overview. In R. L. Burgess & T. L. Huston (Eds.), *Social Exchange in Developing Relationships.* New York: Academic Press.

Huston, T. L., Niehuis, S. & Smith, S. E. (2000). Courtship and the Newlywed Years: What They Tell Us about the Future of a Marriage. *Revista de Psicologia Social y Personalidad*, 16(2), 155–178.

Huston, T. L., Caughlin, J. P., Houts, R. M., Smith, S. E. & George, L. J. (2001). The Connubial Crucible: Newlywed Years as Predictors of Marital Delight, Distress, and Divorce. *Journal of Personality and Social Psychology*, 80, 237–252.

Ianzito, C. (2018). 8 May–December Celebrity Marriages. *AARP.* Retrieved on January 21, 2018 from https://www.aarp.org/home-family/sex-intimacy/info-2015/celebrity-marriage-may-december-photo.html

Ihle, M., Kempenaers, B. & Forstmeier, W. (2015) Fitness Benefits of Mate Choice for Compatibility in a Socially Monogamous Species. *PLoS Biology*, 13(9), e1002248.

Japan for the Uninvited (June 23, 2006). Compa Parties (Group Dating). Retrieved on December 30, 2017 from http://www.japanfortheuninvited.com/articles/compa-parties.html

Jauk, E., Neubauer, A. C., Mairunteregger, T., Pemp, S., Sieber, K. P. & Rauthmann, J. F. (2016). How Alluring Are Dark Personalities? The Dark Triad and Attractiveness in Speed Dating. *Eur. J. Pers.*, 30, 125–138.

Jones, B. C., DeBruine, L. M., Little, A. C., Burriss, R. P. & Feinberg, D. R. (2007). Social Transmission of Face Preferences among Humans. *Proc Biol Sci*, 274, 899–903.

Katz, J., Moore, J. A. & Tkachuk, S. (2007). Verbal Sexual Coercion and Perceived Victim Responsibility: Mediating Effects of Perceived Control. *Sex Roles*, 57, 235–247.

Kelleher International (2015). In *Wikipedia online*. Retrieved on April 5, 2015 from https://en.wikipedia.org/wiki/Kelleher_International

Kennedy, P. (2013). Who Made Speed Dating? *The New York Times*. Retrieved on March 18, 2015 from http://www.nytimes.com/2013/09/29/magazine/who-made-speed-dating.html?_r=0

Kenny, D. A., Kashy, D. A. & Cook, W. L. (2006). *Dyadic Data Analysis*. New York: Guilford.

Kenrick, D. T., Sadalla, E. K., Groth, G. & Trost, M. R. (1990). Evolution, Traits, and the Stages of Human Courtship: Qualifying the Parental Investment Model. *Journal of Personality*, 58, 97–116.

Kenrick, D. T., Groth, G. E., Trost, M. R. & Sadalla, E. K. (1993). Integrating Evolutionary and Social Exchange Perspectives on Relationship: Effects of Gender, Self-Appraisal, and Involvement Level on Mate Selection Criteria. *Journal of Personality and Social Psychology*, 64, 951–969.

Khan, K. S. & Chaudhry, S. (2015). An Evidence-based Approach to an Ancient Pursuit: Systematic Review on Converting Online Contact into a First Date. *BMJ Evidence-based Medicine*. Published online first: 12 February 2015. doi: 10.1136/ebmed-2014–110101

Kimlicka, T. A., Wakefield, J. A. & Goad, N. A. (1982). Sex-roles of Ideal Opposite Sexed Persons for College Males and Females. *Journal of Personality Assessment*, 46, 519–521.

Klohnen, E. C. & Luo, S. (2003). Interpersonal Attraction and Personality: What Is Attractive—Self Similarity, Ideal Similarity, Complementarity or Attachment Security? *Journal of Personality and Social Psychology*, 85, 709–722.

Knapp, M. (1984). *Interpersonal Communication and Human Relationships*. Boston: Allyn and Bacon.

Konrath, S., Au, J. & Ramsey, L. R. (2012). Cultural Differences in Face-ism. Male Politicians Have Bigger Heads in More Gender-Equal Cultures. *Psychology of Women Quarterly*, 36(4).

Korman, S. (1983). Nontraditional Dating Behavior: Date-initiation and Date Expense-sharing among Feminists and Nonfeminsts. *Family Relations*, 32(4), 575–581.

Kościński, K. (2014). Assessment of Waist-to-Hip Ratio Attractiveness in Women: An Anthropometric Analysis of Digital Silhouettes. *Archives of Sexual Behavior*, 43(5), 989–997.

Kreider, R. M. & Fields, J. M. (2002). *Number, Timing, and Duration of Marriages and Divorces: 1996. In Current Population Reports (No. P70–80)*. Washington, DC: U. S. Census Bureau.

Kuperberg, A. & Padgett, J. E. (2015). Dating and Hooking Up in College: Meeting Contexts, Sex, and Variation by Gender, Partner's Gender, and Class Standing. *Journal of Sex Research*, 52, 517–531.

Lamidi, E., Brown, S. L., & Manning, W. D. (2015). Assortative mating: Racial Homogamy in U.S. Marriages, 1964–2014. *Family Profiles*, FP-15–16. Bowling Green, OH: National Center for Family & Marriage Research. https://www.bgsu.edu/ncfmr/resources/data/family-profiles/lamidibrown-manning-assortative-mating-racial-homogamy-fp-15-16.html

Lamont, E. (2014). Negotiating Courtship: Reconciling Egalitarian Ideals with Traditional Gender Norms. *Gender & Society*, 28, 189–211.

Laner, M. R. & Ventrone, N. A. (2000). Dating Scripts Revisited. *Journal of Family Issues*, 21, 488–500.

Lange, R., Houran, J. & Li, S. (2014). Dyadic Relationship Values in Chinese Online Daters: Love American Style? *Sex. Cult.*, 19, 190–215.

LaPlante, M., McCormick, N., & Brannigan, G. (1980). Living the Sexual Script: College Students' Views of Influence in Sexual Encounters, *Journal of Sex Research*, 16, 338–355.

Lau, M., Markham, C., Lin, H., Flores, G. & Chacko, M. R. (2009). Dating and Sexual Attitudes in Asian-American Adolescents. *Journal of Adolescent Research*, 24, 91–113.

Laumann, E. O., Gagnon, J. H., Michael, R. T. & Michaels, S. (1994). *The Social Organization Of Sexuality: Sexual Practices in the United States*. Chicago: University of Chicago Press.

Lay, C., Fairlie, P., Jackson, S., Ricci, T., Eisenberg, J. *et al.* (1998) Domain-specific Allocentrism-idiocentrism: A measure of Family Connectedness. *Journal of Cross-Cultural Psychology*, 29, 434–460.

Lebra, T. S. (1978). Japanese Women and Marital Strain. *Ethos*, 6(1), 22–41.

Lehrer, J. (2016). *A Book About Love*. New York: Simon & Schuster.

Lever, J., Frederick, D. A. & Hertz, R. (2015). Who Pays for Dates? Following Versus Challenging Gender Norms. *Sage Open*, 5(4).

Levesque, M. J. & Lowe, C. A. (1999). Face-ism as a Determinant of Interpersonal Perceptions: The Influence of Context on Facial Prominence Effects. *Sex Roles*, 41, 241.

Lewis, R. & Spanier, G. (1979). Theorizing about the Quality and Stability of Marriage. In W. R. Burr, R. Hill, F. I. Nye & I. Reiss (Eds.), *Contemporary Theories about the Family*, Vol. 2. New York: Free Press.

Li, N. P., Bailey, J. M., Kenrick, D. T. & Linsenmeier, J. A. W. (2002). The Necessities and Luxuries of Mate Preferences: Testing the Tradeoffs. *Journal of Personality and Social Psychology*, 82, 947–955.

Li, N. P., & Kenrick, D. T. (2006). Sex Similarities and Differences in Preferences for Short-Term Mates: What, Whether, and Why. *Journal of Personality and Social Psychology*, 90(3), 468–489. Available at: http://ink.library.smu.edu.sg/soss_research/722

Libby, R. (1976). Social Scripts for Sexual Relationships. In S. Gordon & R. Libby (Eds.), *Sexuality Today and Tomorrow: Contemporary Issues in Human Sexuality* (pp. 172–173). N. Scituate: Duxbury.

Lipman-Blumen, J. (1984). *Gender Roles and Power*. Englewood Cliffs, NJ: Prentice-Hall.

Livingston, G. & Caumont, A. (February 13, 2017). *5 Facts on Love and Marriage in America*. Washington, DC: Pew Research Center.

Lottes, I. L. (1993). Nontraditional Gender Roles and the Sexual Experiences of Heterosexual College Students. *Sex Roles*, 29, 645–669.

Louis, R. & Copeland, D. (1998). *How to Succeed With Women*. New Jersey: Reward Books.

Luo, B. (2008). Striving for Comfort: "Positive" Construction of Dating Cultures among Second-Generation Chinese American Youths. *Journal of Social and Personal Relationships*, 25, 867–888.

Luo, S. & Zhang, G. (2009). What Leads to Romantic Attraction: Similarity, Reciprocity, Security, or Beauty? Evidence from a Speed-Dating Study. *Journal of Personality*, 77, 933–964.

MacColl, G. & Wallace, C. M. (2012). *To Marry An English Lord: Takes of Wealth and Marriage, Sex and Snobbery*. New York: Workman Publishing.

Madathil, J. & Benshoff, J. M. (2008). Importance of Marital Characteristics and Marital Satisfaction: A Comparison of Asian Indians in Arranged Marriages and Americans in Marriages of Choice. *The Family Journal*, 16, 222–230.

Mahoney, A. R. & Knudson-Martin, C. (2009). Gender Equality in Intimate Relationships. In C. Knudson-Martin & A. R. Mahoney (Eds.), *Couples, Gender, and Power: Creating Change in Intimate Relationships* (pp. 3–16). New York: Springer.

Major, B. & Heslin, R. (1982). Perceptions of Cross-Sex and Same-Sex Nonreciprocal Touch: It Is Better to Give than to Receive. *J Nonverbal Behav.*, 6, 148–162.

Malamuth, N. M. (Ed.), *Sex, Power, Conflict: Evolutionary and Feminist Perspectives*. New York: Oxford University Press.

Mannan, M. A. & Chaudhuri, Z. S. (2009). *An Inventory and Statistics on Violence against Women in Bangladesh: Who is Doing What and Where*. Dhaka: Bangladesh Institute of Development Studies (BIDS).

Markus, H. R. & Kitayama, S. (1991). Culture and the Self: Implications for Cognition, Emotion, and Motivation. *Psychology Review*, 2, 224–253.

Masci, D. (February 10, 2017). *On Darwin Day, 6 Facts about the Evolution Debate*. Washington, DC: Pew Research Center.

Maurer, T. W. & Robinson, D. W. (2008). Effects of Attire, Alcohol, and Gender on Perceptions Of Date Rape. *Sex Roles*, 58, 423–434.

McDaniel, A. K. (2005). Young Women's Dating Behavior: Why/Why Not Date a Nice Guy? *Sex Roles*, 53, 347–359.

McGraw, P. (2005). *Love Smart: Find the One You Want–Fix the One You Got*. New York: Free Press.

Medora, N. P., Larson, J. H., Hortacsu, N. & Dave, P. (2002). Perceived Attitudes towards Romanticism: A Cross-Cultural Study of American, Asian-Indian, and Turkish Young Adults. *Journal of Comparative Family Studies*, 33, 155–178.

Mensinger, J. L., Bonifazi, D. Z. & LaRosa, J. (2007). Perceived Gender Role Prescriptions in Schools, the Superwoman Ideal, and Disordered Eating among Adolescent Girls. *Sex Roles*, 57, 557–568.

Menski, W. (Ed.) (1999). *South Asians and the Dowry Problem*. Stoke-on-Trent, UK: Trentham Books.

Mierke, K., Aretz, W., Nowack, A., Wilmsen, R. & Heinemann, T. (2011). Impression Formation in Online-Dating-Situations: Effects of Media Richness and Physical Attractiveness Information. *J Bus Media Psychol.*, 2, 49–56.

Miller, E, Tancredi, D. J., McCauley, H. L., Decker, M. R., Virata, C. D. M., Anderson, H. A., O'Connor, B. & Silverman, J. G. (2013). One-Year Follow-up of a Coach-delivered Dating Violence Prevention Program: A Cluster Randomized Controlled Trial. *American Journal of Preventive Medicine*, 45, 108–112.

Miller, M. (1971). A Comparison of the Duration of Interracial with Intraracial Marriage in Hawaii. *International Journal of Sociology of the Family*, 1(2), 197–201. Retrieved from http://www.jstor.org/stable/23027023

Miller, R. (2004). *Understanding Women: The Definitive Guide to Meeting, Dating and Dumping, if Necessary.* New York: The Book Factory.

Miller, R. S. & Perlman, D. (2009). *Intimate Relationships*, 5th edn. New York: McGraw-Hill.

Mongeau, P., Serewicz, M. C. & Therrien, L. F. (2004). Goals for Cross-Sex First Dates: Identification, Measurement, and the Influence of Contextual Factors. *Communication Monographs*, 71(2), 121–147.

Mongeau, P. A., Jacobsen, J. & Donnerstein, C. (2007). Defining Dates and First Date Goals: Generalizing from Undergraduates to Single Adults. *Communication Research*, 34(5), 526–547.

Montoya, R. M., Horton, R. S. & Kirchner, J. (2008). Is Actual Similarity Necessary for Attraction? A Meta-Analysis of Actual and Perceived Similarity. *Journal of Social and Personal Relationships*, 25, 889–922.

Moore, M. & Gould, J. (2001). *Date Like a Man: What Men Know about Dating and Are Afraid You'll Find Out.* New York: Quill.

Morr, M. C. & Mongeau, P. A. (2004). First-Date Expectations: The Impact of Sex of Initiator, Alcohol Consumption, and Relationship Type. *Communication Research*, 31(1), 3–35.

Morr Serewicz, M. C. & Gale, E. (2008). First-Date Scripts: Gender Roles, Context, and Relationship. *Sex Roles*, 58, 149.

Morry, M. M., Kito, M. & Ortiz, L. (2011). The Attraction-Similarity Model and Dating Couples: Projection, Perceived Similarity, and Psychological Benefits. *Personal Relationships*, 18, 125–143.

Muro, J. A. & Martinez, L. M. (2018). Is Love Color-blind? Racial Blind Spots and Latinas' Romantic Relationships. *Sociology of Race and Ethnicity*, 4(4), 527–540.

Murstein, B. I. (1971). *Theories of Attraction and Love.* New York: Springer.

Myers, J. E., Madathil, J. & Tingle, L. R. (2005). Marriage Satisfaction and Wellness in India and the United States: A Preliminary Comparison of Arranged Marriages and Marriages of Choice. *Journal of Counseling & Development*, 83, 183–190.

Nada, S. (2000). Arranging a Marriage in India. In P. Devita (Ed.), *Stumbling Toward Truth: Anthropologists at Work* (pp. 196–204). Long Grove, IL: Waveland Press.

National Conference of State Legislatures (NCSL) (May 9, 2017). *Teen Dating Violence.* Retrieved on July 23, 2017 from http://www.ncsl.org/research/health/teen-dating-violence.aspx

National Domestic Violence (2017). Hotline. Retrieved on January 28, 2018 from http://www.thehotline.org/2013/02/25/dating-abuse-resources-for-teens/

National Institute of Justice (NIJ) (February 13, 2014). Prevention and Intervention of Teen Dating Violence. Washington: NIJ.gov: https://nij.gov/topics/crime/intimate-partner-violence/teen-dating-violence/Pages/prevention-intervention.aspx

National Institute of Justice (NIJ) (April 4, 2017). Family Context Is an Important Element in the Development of Teen Dating Violence and Should Be Considered in Prevention and Intervention. Washington: NIJ.gov: https://nij.gov/topics/crime/intimate-partner-violence/teen-dating-violence/Pages/family-context-in-development-of-teen-dating-violence.aspx

NCADV (National Coalition against Domestic Violence) (2015). *Facts about Dating Abuse and Teen Violence.* Retrieved on July 22, 2017 from https://ncadv.org/files/Dating%20Abuse%20and%20Teen%20Violence%20NCADV.pdf

Nesteruk, O. & Gramescu, A. (2012). Dating and Mate Selection among Young Adults from Immigrant Families. *Marriage & Family Review*, 48, 40–58.

Netting, N. (2006). Two-Lives, One Partner: Indo-Canadian Youth between Love and Arranged Marriages. *Journal of Comparative Family Studies*, 37(1), 129–147.

Netting, N. (2010). Marital Ideoscapes in 21st-Century India: Creative Combination of Love and Responsibility. *Journal of Family Issues*, 31, 707–726.

Neyer, F. J. & Voigt, D. (2004). Personality and Social Network Effects on Romantic Relationships: A Dyadic Approach. *Eur. J. Pers.*, 18, 279–299.

Niehuis, S. & Huston, T. L. (2000). A Longitudinal Study of Courtship Predictors of Marital Distress and Divorce. Paper presented at the 10th International Conference on Personal Relationships, Brisbane, Australia.

Niehuis, S. & Huston, T. L. (2002). The Premarital Roots of Disillusionment Early in Marriage. Paper presented at the International Conference on Personal Relationships, Halifax, Nova Scotia, Canada.

Niehuis, S., Skogrand, L., Huston, T. L. (2006). When Marriages Die: Premarital and Early Marriage Precursors to Divorce. *The Forum for Family and Consumer Issues.* Retrieved on November 27, 2018 from https://www.researchgate.net/publication/280625117_When_marriages_die_Premarital_and_early_marriage_precursors_to_divorce

NSVRC (The National Sexual Violence Resources Center) (2017). *April is Sexual Assault Awareness Month.*

Retrieved on July 23, 2017 from https://www.nsvrc.org/news/3756

O'Brien, J. (2008). *Encyclopedia of Gender and Society,* Volume 1 (pp. 40–42). California: Sage.

Ortega, J. & Hergovich, P. (2017). *The Strength of Absent Ties: Social Integration via Online Dating.* Retrieved on January 15, 2018 from https://arxiv.org/pdf/1709.10478.pdf

Orth, U. (2013). How Large Are Actor and Partner Effects of Personality on Relationship Satisfaction? The Importance of Controlling for Shared Method Variance. *Pers. Soc. Psychol. Bull.*, 39, 1359–1372.

Palumbo, L. (September 27, 2017). New Survey Finds that Adults are Most Likely to Believe Stricter Campus Policies are Most Effective Approach to Decrease Campus Sexual Assault. PA: National Sexual Violence Resource Center. Retrieved on November 27, 2018 from https://www.nsvrc.org/sites/default/files/2018-03/news_nsvrc_press-release-new-survey-finds-adults-more-likely-to-believe-stricter-campus-policies-most-effective_0.pdf

Parru, S. (October 2, 2017). Love and Marriage, South Asian Style. *The New York Times.*

Patrick, W. L. (September 2, 2017). Who Pays For a First Date? Why It Matters- How Paying for a First Date Could Affect Expectations of What Comes Next. New York: Sussex Publishers, LLC: *Psychology Today.* Retrieved on November 27, 2018 from https://www.psychologytoday.com/us/blog/why-bad-looks-good/201709/who-pays-first-date-why-it-matters

Paul, E. L. (2006). Beer Goggles, Catching Feelings, and the Walk of Shame: The Myths and Realities of the Hookup Experience. In D. C. Kirkpatrick, S. Duck & M. K. Foley (Eds.), *Relating Difficulty* (pp. 141–160). Mahwah, NJ: Lawrence Erlbaum Associates.

Paul, E. L., McManus, B. & Hayes, A. (2000). "Hookups": Characteristics and Correlates of College Students' Spontaneous and Anonymous Sexual Experiences. *Journal of Sex Research*, 37, 76–88.

Paulhus, D. L. & Williams, K. M. (2002). The Dark Triad of Personality. *Journal of Research in Personality*, 36, 556–563.

Pedersen, W. C., Putcha-Bhagavatula, A. & Miller, L. C. (2010). Are Men and Women Really that Different? Examining Some of Sexual Strategies Theory (SST)'s Key Assumptions about Sex-Distinct Mating Mechanisms. *Sex Roles.* Advance online publication. doi:10.1007/s11199-010-9811-5.

Penn, R. (2011). Arranged Marriages in Western Europe: Media Representations and Social Reality. *Journal of Comparative Family Studies*, 42(5), 637–650.

Peplau, L. A. & Fingerhut, A. W. (2007). The Close Relationships of Lesbians and Gay Men. *Annual Review of Psychology,* 58, 405–424.

Peplau, L. A., Rubin, Z. & Hill, C. T. (1977). Sexual Intimacy in Dating Relationships. *Journal of Social Issues,* 33, 86–109.

Perrin, P. B., Heesacker, M., Tiegs, T. J., Swan, L. K., Lawrence, A. W., Smith, M. B., *et al.* (2010). Aligning Mars and Venus: The Social Construction and Instability of Gender Differences in Romantic Relationships. *Sex Roles.* Advance online publication. doi:10.1007/s11199-010-9804-4.

Persch, J. A. (2008). It's Complicated: Who Pays on Dates. NBC News. Retrieved on March 28, 2015 from http://www.nbcnews.com/id/23244363/ns/business-personal_finance/t/its-complicated-who-pays-dates/#.VRbKvPnF_-4

Pew Research Center (January 12, 2017). *Social Media Fact Sheet.* Retrieved on July 21, 2017 from http://www.pewinternet.org/fact-sheet/social-media/

Prieler, M. & Kohlbacher, F. (2017). Face-ism from an International Perspective: Gendered Self-Presentation in Online Dating Sites Across Seven Countries. *Sex Roles,* 77, 604.

Raghubir, P. & Valenzuela, A. (2006). Center-of-inattention: Position Biases in Decision-making. *Organ Behav Human Decis Processes,* 99, 66–80.

Rammohan, A. & Robertson, P. (2012). Do Kinship Norms Influence Female Education? Evidence from Indonesia. *Oxford Development Studies,* 40(3), 283–304.

Rawlins, W. K. (1992). *Friendship Matters: Communication, Dialectics, and the Life Course.* Hawthorne: Aldine.

Regan, P. C. (1998). What if You Can't Get What You Want? Willingness to Compromise Ideal Mate Selection Standards as a Function of Sex, Mate Value, and Relationship Context. *Personality and Social Psychology Bulletin,* 24(12), 1294–1303.

Regan, P. C. & Joshi, A. (2003). Ideal Partner Preferences among Adolescents. *Social Behavior and Personality,* 31, 13–20.

Regan, P. C., Medina, R. & Joshi, A. (2001). Partner Preferences among Homosexual Men and Women: What Is Desirable in a Sex Partner Is Not Necessarily Desirable in a Romantic Partner. *Social Behavior and Personality,* 29, 625–634.

Regan, P. C., Lakhanpal, S. & Anguiano, C. (2012). Relationship Outcomes in Indian-American Love-based and Arranged Marriages. *Psychol. Rep.*, 110, 915–924.

Reilly, M. E. & Lynch, J. M. (1990). Power-sharing in Lesbian Partnerships. *Journal of Homosexuality,* 19, 1–30.

ReportLinker (2017). What We Talk About When We Talk About Finding Love Online. Retrieved on January 8, 2018 from https://www.reportlinker.com/insight/finding-love-online.html

Ridgeway, C. L. (2011). *Framed by Gender: How Gender Inequality Persists in the Modern World.* New York: Oxford University Press.

Risman, B. J. (2009). From Doing to Undoing: Gender As We Know It. *Gender & Society,* 23, 81–84.

Rolf, K. & Ferrie, J. P. (2008). The May–December Relationship since 1850: Age Homogamy in the U.S. Retrieved on January 17, 2018 from http://paa2008.princeton.edu/papers/80695

Rose, S. & Frieze, I. H. (1989). Young Singles' Scripts for a First Date. *Gender & Society,* 3, 258–268.

Rose, S. & Frieze, I. H. (1993). Young Singles' Contemporary Dating Scripts. *Sex Roles,* 28, 499–509.

Rose, S. & Zand, D. (2000). Lesbian Dating and Courtship from Young Adulthood to Midlife. *Journal of Gay and Lesbian Social Services,* 11, 77–104.

Rosenfeld, M. J. & Thomas, R. J. (2012). Searching for a Mate: The Rise of the Internet as a Social Intermediary. *American Sociological Review,* 77(4), 523–547.

Rothman, E. K. (1984). *Hands and Hearts: A History of Courtship in America.* New York: Basic Books.

Rubin, Z. (1970). Measurement of Romantic Love. *Journal of Personality and Social Psychology,* 16, 265–273.

Ryan, C. & Jetha, C. (2012). *Sex at Dawn. How We Mate, Why We Stray, and What It Means for Modern Relationships.* New York: Harper Perennial.

Ryan, K. M. & Mohr, S. (2005). Gender Differences in Playful Aggression during Courtship in College Students. *Sex Roles,* 53, 591–601.

Sales, N. J. (September 2015). Tinder and the Dawn of the "Dating Apocalypse." As Romance Gets Swiped from the Screen, Some Twenty Somethings Aren't Liking What They See. *Vanity Fair.* Retrieved on November 27, 2018 from https://www.vanityfair.com/culture/2015/08/tinder-hook-up-culture-end-of-dating

Sanderson, S. K. (2003). The Sociology of Human Sexuality: A Darwinian Alternative to Social Constructionism and Postmodernism. In Annual Meeting of the American Sociological Association. Atlanta, Georgia.

Schlesinger, H. J. (2008). *Promises, Oaths, and Vows: On the Psychology of Promising.* New York: Routledge.

Schmitt, D. P., Couden, A. & Baker, M. (2001). The Effects of Sex and Temporal Context on Feelings of Romantic Desire: An Experimental Evaluation of Sexual Strategies Theory. *Personality and Social Psychology Bulletin,* 27, 833–847.

Schramm, D., Marshall, J., Harris, V. & Lee, T. (2012). Religiosity, Homogamy, and Marital Adjustment: An Examination of Newlyweds in First Marriages and Remarriages. *Journal of Family Issues,* 33(2), 246–268.

Schroer, J. W. (2016) Generations X, Y, Z and the Others. *The Social Librarian.* Retrieved on November 26, 2015 from http://www.socialmarketing.org/newsletter/features/generation3.htm

Seth, R. (2008). *First Comes Marriage: Modern Relationship Advice from the Wisdom of Arranged Marriages.* New York: Touchstone.

Sharabi, L. L. & Caughlin, J. P. (2017). What Predicts First Date Success: A Study of Modality Switching in Online Dating. *Personal Relationships: Journal of the International Association for Relationship Research.* April 11.

Simon, W. & Gagnon, J. H. (1986). Sexual Scripts: Permanence and Change. *Archives of Sexual Behavior,* 15(2), 97–120.

Simpson, J. A., Farrell, A. K., Ori-a, M. M. & Rothman, A. J. (2015). Power and Social Influence in Relationships. In M. Mikulincer and P. R. Shaver (Eds.), *APA Handbook of Personality and Social Psychology,* Vol. 3 (pp. 393–420). American Psychological Association.

Sinclair, H. C. (2012). Stalking Myth-attributions: Examining the Role of Individual, Cultural, and Contextual Variables on Judgments of Unwanted Pursuit Scenarios. *Sex Roles: A Journal of Research,* 66(5–6), 378–391.

Sinclair, H. C. & Frieze, I. H. (2005). When Courtship Persistence Becomes Intrusive Pursuit: A Comparison of Rejecter and Pursuer Perspectives of Unrequited Attraction. *Sex Roles,* 52, 839–852.

Singh, D. (1993). Adaptive Significance of Female Physical Attractiveness: Role of Waist-to-hip Ratio. *Journal of Personality and Social Psychology,* 65, 293–307.

Skomorovsky, A., Matheson, K. & Anisman, H. (2006). The Buffering Role of Social Support Perceptions in Relation to Eating Disturbances among Women in Abusive Dating Relationships. *Sex Roles,* 54, 627–638.

Smith, A. & Anderson, M. (March 2018). *Social Media Use in 2018. A Majority of Americans Use Facebook and YouTube, but Young Adults Are Especially Heavy Users of Snapchat and Instagram.* Washington, DC: Pew Research Center.

Smith, C. A., Konik, J. A. & Tuve, M. V. (2011). In Search of Looks, Status, or Something Else? Partner Preferences among Butch and Femme Lesbians and Heterosexual Men and Women. *Sex Roles,* 64, 658. https://doi.org/10.1007/s11199-010-9861-8

Smith, J. E., Waldorf, V. A. & Trembath, D. L. (1990). Single White Male Looking for Thin, Very Attractive ... *Sex Roles,* 23, 675–685.

Sorokowski, P., Groyecka, A., Karwowski, M., Manral, U., Kumar, A., Niemczyk, A. & Pawłowski, B. (2017). Free Mate Choice Does Not Influence Reproductive Success in Humans. *Scientific Reports,* 7, 10127.

Spoon, T. R., Millam, J. R. & Owings, D. H. (2006). The Importance of Mate Behavioural Compatibility in Parenting and Reproductive Success by Cockatiels, Nymphicus Hollandicus. *Anim. Behav.,* 71, 315–326.

Sprecher, S. (2011). Relationship Compatibility, Compatible Matches, and Compatibility Matching. *Psychological Records Journal,* 1, 187–215.

Sprecher, S. (2013). Correlates of Couples' Perceived Similarity at the Initiation Stage and Currently. *Interpersonal:*

An International Journal on Personal Relationships, [S.1.], 7(2), 180–195.

Sprecher, S. & Regan, P. C. (2002). Liking Some Things (in Some People) More than Others: Partner Preferences in Romantic Relationships and Friendships. *Journal of Social and Personal Relationships*, 19, 463–481.

Sprecher, S. & Sedikides, C. (1993). Gender Differences in Perception of Emotionality: The Case of Close Heterosexual Relationships. *Sex Roles*, 28, 511–530.

Sprecher, S., Treger, S., Hilaire, N., Fisher, A., & Hatfield, E. (2013). You Validate Me, You Like Me, You're Fun, You Expand Me: "I'm Yours!" *Current Research in Social Psychology*, 22, 22–34.

Stake, J. & Lauer, M. L. (1987). The Consequences of Being Overweight: A Controlled Study of Gender Differences. *Sex Roles*, 17, 31–47.

Statistic Brain (August 16, 2016). Arranged/Forced Marriage Statistics. Statistic Brain Research Institute.

Statistical Brain (2017a). Online/Relationship Statistics. Statistic Brain Reserch Institute.

Statistic Brain (2017b). Online Dating Statistics. Statistic Brain Reserch Institute.

Stepler, R. (2017). *Led by Baby Boomers, Divorce Rates Climb for America's 50+ Population.* Washington, DC: Pew Research Center.

Stepp, L. S. (2008). *Unhooked: How Young Women Pursue Sex, Delay Love, and Lose at Both.* New York: Riverhead Books.

Sternberg, R. J. (1986). A Triangular Theory of Love. *Psychol. Rev*, 93, 119–135.

Sternberg, R. J. (1988). *The Triangle of Love: Intimacy, Passion, Commitment.* New York: Basic Books.

Strauss, A. & Corbin, J. (1990). *Basics of Qualitative Research: Grounded Theory Procedures and Techniques.* Newbury Park, CA: Sage.

Streib, J. (2015). *The Power of the Past: Understanding Cross-Class Marriages.* Oxford, UK: Oxford University Press.

Struckman-Johnson, D. & Struckman-Johnson, C. (1991). Men and Women's Acceptance of Coercive Sexual Strategies Varied by Initiator Gender and Couple Intimacy. *Sex Roles*, 25, 661–676.

Suliman, A. (August 03, 2016). China's "Leftover Women" Are a Thing of the Past. New York: Vice Media LLC. Retrieved on November 27, 2018 from https://www.vice.com/en_us/article/nnkv8q/the-rise-of-unmarried-women-in-china

Summerhayes, D. L. & Suchner, R. W. (1978). Power Implications of Touch in Male–Female Relationships. *Sex Roles*, 4, 103–110.

Surra, C. A. & Huston, T. L. (1987). Mate Selection as a Social Transition. In D. Perlman & S. Duck (Eds.), *Intimate Relationships: Development, Dynamics, and Deterioration* (pp. 88–120). Newbury Park, CA: Sage.

Sutin, A. R., Costa, P. T., Wethington, E. & Eaton, W. (2010). Perceptions of Stressful Life Events as Turning Points Are Associated with Self-rated Health and Psychological Distress. *Anxiety, Stress, and Coping*, 23(5), 479–492.

Symons, D. (1995). Beauty is in the Adaptations of the Beholder: The Evolutionary Psychology of Human Female Sexual Attractiveness. In P. R. Abramson and S. D. Pinkerton (Eds.), *Sexual Nature/Sexual Culture* (pp. 80–120). Chicago: University of Chicago Press.

Talalas, K. (2017). *Millennials Are Looking for Parental Guidance on Love.* VA: Institute for Family Studies. Retrieved on November 27, 2018 from https://ifstudies.org/blog/millennials-are-looking-for-parental-guidance-on-love

Tangney J. P., Stuewig J. & Mashek D. J. (2007). Moral Emotions and Moral Behavior. *Annual Review of Psychology*, 58, 345–372.

Taylor, B. G., Stein, N. D., Mumford, E. A. & Woods, D. (2013). Shifting Boundaries: An Experimental Evaluation of a Dating Violence Prevention Program in Middle Schools. *Prevention Science*, 14, 64–76.

Tennov, D. (1979). *Love and Limerence: The Experience of Being in Love.* New York: Stein and Day.

Thompson, E. R. (2008). Development and Validation of an International English Big-Five Mini-Markers. *Personality and Individual Differences*, 45(6), 542–548. doi:10.1016/j.paid.2008.06.013.

Titus, M. & Fadal, T. (2008). *Why Hasn't He Called? How Guys Really Think and How to Get the Right One Interested in You.* New York: McGraw-Hill.

Titzmann, F.-M. (2016). Changing Patterns of Matchmaking: The Indian Online Matrimonial Market. *Asian Journal of Women's Studies*, 19(4).

Toledo, M. (2009). First Comes Marriage, Then Comes Love. ABC News. Retrieved on December 14, 2017 from http://abcnews.go.com/2020/story?id=6762309&page=1

Townsend, J. M. (1987). Sex Differences in Sexuality among Medical Students: Effects of Increasing Socioeconomic Status. *Archives of Sexual Behavior*, 16, 427–446.

Townsend, J. M. & Levy, G. D. (1990). Effects of Potential Partners' Physical Attractiveness and Socioeconomic Status on Sexuality and Partner Selection. *Archives Sexual Behavior*, 19(2), 149–164.

Tracy, J. (March 22, 2012). How Many Online Dating Sites Are There? *Online Dating Magazine.* Retrieved on November 27, 2018 from https://www.onlinedatingmagazine.com/faq/howmanyonlinedatingsitesarethere.html

Trivers, R. (1972). Parental Investment and Sexual Selection. In B. Campbell (Ed.), *Sexual Selection and the Descent of Man, 1871–1971* (pp. 136–179). Chicago, IL: Aldine.

Truman, J. & Langton, L. (2014). *Criminal Victimization, 2013.* Bureau of Justice Statistics. Retrieved on March 29, 2015 from http://www.bjs.gov/content/pub/pdf/cv13.pdf

Twenge, J. M., Sherman, R. A. & Wells, B. E. (2015). Changes in American Adults' Sexual Behavior and Attitudes, 1972–2012. *Archives of Sexual Behavior,* 44(8), 2273–2285.

Tynes, S. R. (1990). Educational Heterogamy and Marital Satisfaction between Spouses. *Social Science Research,* 19(2), 153–174.

Uddin, E., Hoque, N. & Islam, R. (2017). Familial Factors Influencing Age-Heterogamy vs. Age-Homogamy in Marriage in Bangladesh: Implication for Social Policy Practice. *Global Social Welfare,* 4, 127.

United Nations (1993). *Declaration on the Elimination of Violence against Women.* General Assembly. Retrieved on November 27, 2018 from http://www.un.org/documents/ga/res/48/a48r104.htm

Urbaniak, G. & Kilmann, P. R. (2003). Physical Attractiveness and the "Nice Guy Paradox": Do Nice Guys Really Finish Last? *Sex Roles,* 49, 413–426.

U.S. Census Bureau (January 23, 2017a). *FFF: Valentine's Day 2017: Feb. 14.*

U.S. Census Bureau (December 14, 2017b). *Educational Attainment in the United States: 2017. Table 3.*

U.S. Census Bureau (2017c). Married Couple Family Groups, By Presence of Own Children Under 18, And Age, Earnings, Education, And Race And Hispanic Origin Of Both Spouses. Current Population Survey, 2017. Annual Social and Economic Supplement.

Uskul, A. K., Lalonde, R. N. & Konanur, S. (2010). The Role of Culture in Intergenerational Value Discrepancies Regarding Intergroup Dating. *Journal of Cross-Cultural Psychology,* 42, 1165–1178.

Van Poppel, F., Liefbroer, A., Vermunt, J. & Smeenk, W. (2001). Love, Necessity and Opportunity: Changing Patterns of Marital Age Homogamy in the Netherlands, 1850–1993. *Population Studies,* 55, 1–13.

Vandenberg, S .G. (1972). Assortative Mating or Who Marries Whom? *Behavior Genetics,* 2, 127–158.

Van de Putte, B., Van Poppel, F., Vanassche, S., Sanchez, M., Jidkova, S., Eeckhaut, M., Oris, M. & Matthijs, K. (2009). The Rise of Age Homogamy in 19th Century Western Europe. *Journal of Marriage and Family,* 71, 1234–1253.

Vazire, S. (2010). Who Know What About a Person? The Self-Other Knowledge Asymmetry (SOKA) Model. *J. Pers. Soc. Psychol.,* 98, 281–300.

Vera, H., Berardo, D. H. & Berardo, F. M. (1985). Age Heterogamy in Marriage. *Journal of Marriage and Family,* 47(3), 553–566.

Vogel, D., Wester, S. & Heesacker, M. (1999). Dating Relationships and the Demand/Withdraw Pattern of Communication. *Sex Roles,* 41, 297–306.

Volokh, E. (January 23, 2015). Adult Incest and the Law. *The Washington Post.* Retrieved on https://www.washingtonpost.com/news/volokh-conspiracy/wp/2015/01/23/adult-incest-and-the-law/?utm_term=.1701a13b073b

Wade, L. (April 8, 2008). The Marriage Market. *Sociological Images.* Retrieved on July 23, 2017 from https://thesocietypages.org/socimages/2008/04/08/the-marriage-market/

Wade, N. (2015). *A Troublesome Inheritance: Genes, Race and Human History.* New York: Penguin Random House.

Wakin, A. & Vo, D. B. (2011). *Love-Variant: The Wakin-Vo I.D.R. Model of Limerence.* Retrieved on November 26, 2018 from http://citeseerx.ist.psu.edu/viewdoc/download?doi=10.1.1.729.1932&rep=rep1&type=pdf

Waller, W. W. (1938). *The Family, A Dynamic Interpretation.* New York: Warner Books.

Watson, D., Hubbard, B. & Wiese, D. (2000). General Traits of Personality and Affectivity as Predictors of Satisfaction in Intimate Relationships: Evidence from Self- and Partner-Ratings. *J. Pers.,* 68, 413–449.

Watson, D., Klohnen, E. C., Casillas, A., Simms, E. N., Haig, J. & Berry, D. S. (2004). Match Makers and Deal Breakers: Analyses of Assortative Mating in Newlywed Couples. *J. Pers.,* 72, 1029–1068.

Wechsler, D. (1 March 1950). Intellectual Development and Psychological Maturity. *Child Development,* 21(1), 45.

Weeden, J. & Sabini, J. (2005). Physical Attractiveness and Health in Western Societies: A Review. *Psychological Bulletin,* 131(5), 635–653.

Weissbourd, R., Anderson, T. R., Cashin, A. & McIntyre, J. (2017). The Talk: How Adults Can Promote Young People's Healthy Relationships and Prevent Misogyny and Sexual Harassment. Harvard Graduate School of Education. Retrieved on November 27, 2018 from https://static1.squarespace.com/static/5b-7c56e255b02c683659fe43/t/5bd51a0324a69425bd079b59/1540692500558/mcc_the_talk_final.pdf

West, C. & Zimmerman, D. H. (1991). Doing Gender. In J. Lorber & S. A. Farrell (Eds.), *The Social Construction of Gender* (pp. 13–37). Newbury Park: Sage.

Whitty, M. T. & Buchanan, T. (2010). What's in a Screen Name? Attractiveness of Different Types of Screen Names Used by Online Daters. *Int J Internet Sci.,* 5, 5–19.

WHO (World Health Organization) (2013a). *Global and Regional Estimates of Violence against Women: Prevalence and Health Effects of Intimate Partner Violence and Non-Partner Sexual Violence.* Geneva: WHO.

WHO (World Health Organization). (2013b). *Responding to Intimate Partner Violence and Sexual Violence against Women: WHO Clinical And Policy Guidelines.* Geneva: WHO.

WHO (World Health Organization) (2017). WHO Launches New Manual to Strengthen Health Systems To Better Respond To Women Survivors Of Violence. Geneva: WHO.

Wiederman, M. W. & Allgeier, E. R. (1992). Gender Differences in Mate Selection Criteria: Sociobiological or Socioeconomic Explanation? *Ethology and Sociobiology*, 13(2), 115–124.

Williams, A. (1999). *The Wet Spot: Why Don't Women Date Nice Guys?* Retrieved November 7, 2003, from http://www.pottymouth.org/wetspot/wetspot014.html

Willis, F. N. & Carlson, R. A. (1993). Singles Ads: Gender, Social Class, and Time. *Sex Roles*, 29, 387–404.

Wills, R. (2000). *Nice Guys and Players: Becoming the Man Women Want.* Arlington, VA: E.R.L. Publishing.

Wilson, B. & Smallwood, S. (2014). *Age Differences at Marriage and Divorce.* Office for National Statistics.

Winch, R. F. (1946). Interrelations between Certain Social Background and Parent–Son Factors in a Study of Courtship among College Men. *American Sociological Review*, 11, 333–343.

Winch, R. (1958). *Mate-Selection: A Study of Complementary Needs.* New York: Harper & Brothers.

Wiseman, R. (2007). *Quirkology: How to Discover Big Truths in Small Things.* New York: Basic Books.

Workman, J. E. & Freeburg, E. W. (1999). An Examination of Date Rape, Victim Dress and Perceiver Variables within the Context of Attribution Theory. *Sex Roles*, 41, 261–277.

Wu, H. (2018). Homogamy in U.S. Marriages, 2016. *Family Profiles*, FP-18–18. Bowling Green, OH: National Center for Family & Marriage Research. https://doi.org/10.25035/ncfmr/fp-18-18

Yalom, M. & Carstensen, L. L. (Eds.) (2002). *Inside the American Couple: New Thinking, New Challenges.* Berkeley, CA: University of California Press.

Yeater, E. A., Lenberg, K. L., Avina, C., Rinehart, J. K. & O'Donohue, W. T. (2008). When Dating Situations Take a Turn for the Worse: Situational and Interpersonal Risk Factors for Sexual Aggression. *Sex Roles*, 59, 151–163.

Young, M. L., Waller, M. W., Martin, S. L. & Kupper, L. L. (2004). Prevalence of Partner Violence in Same-sex Romantic and Sexual Relationships in a National Sample of Adolescents. *Journal of Adolescent Health*, 35(2), 124–131.

Zentner, M. & Mitura, K. (2012). Stepping out of the Caveman's Shadow: Nations' Gender Gap Predicts Degree of Sex Differentiation in Mate Preferences. *Psychological Science*, 23, 1176–1185.

Zhou, Y., Wang, K., Chen, S., Zhang, J. & Zhou, M. (2017). The Actor, Partner, Similarity Effects of Personality, and Interactions with Gender and Relationship Duration among Chinese Emerging Adults. *Front. Psychol.*, 8, 1698.

Zuckerman, M. (1986). On the Meaning and Implications of Facial Prominence. *Journal of Nonverbal Behavior*, 10, 215–229.

Zurbriggen, E. L. & Morgan, E. M. (2006). Who Wants to Marry a Millionaire? Reality Dating Television Programs, Attitudes toward Sex, and Sexual Behaviors. *Sex Roles*, 54, 1–17.

9
MARRIAGE AND MARITAL RELATIONSHIPS

Learning Outcomes

Upon completion of the chapter, students should be able to:

1. describe reasons for and effects of the declining marriage rate and rising first marriage age
2. analyze some predictors of a successful marriage
3. describe engagement, marriage contracts, bride price, and dowry
4. explain the benefits and requirements of marriage
5. describe different wedding rituals
6. describe marital adjustment in young adulthood
7. explain how having children affects a marriage
8. describe marital relationships in middle adulthood
9. explain marital relationships in late adulthood
10. describe how social exchange theory explains marital relationships
11. apply resource theory to the analysis of marital power
12. explain expression of love from a diversity perspective
13. describe diverse ethnic wedding customs
14. explain collective marriages in the world.

Brief Chapter Outline

Pre-test

The State of Marriage

Marriage Rate

Marriage Rate by Education, Gender, Geography, and Race

Marriage Age

Marriage Age by Education and Race

Pre-test

Engaged or active learning is a powerful strategy that leads to better learning outcomes. One way to become an active learner is to begin the chapter and try to answer the following true/false statements from the material as you read. You will find that you have ready answers to these questions upon completion of the chapter.

Marriage is a legally recognized or socially approved union between two partners that establishes their rights and obligations to each other.	T	F
The marriage rate is declining and the age at first marriage is rising in the United States	T	F
Marital satisfaction refers to an individual's global evaluation of the marital relationship.	T	F
Research indicates that communication is the most important predictor of a happy marriage.	T	F
Marital equity refers to the degree of balance of authority, power, or influence between spouses.	T	F
The sandwich generation refers to those adults who both (a) are caring for an aging parent age 65 or older and who (b) are raising a child under age eighteen or are supporting a grown child.	T	F
Marital adjustment is the accommodation of husband and wife to each other at a given time.	T	F
Activity theory proposes that successful aging occurs when older adults stay active and maintain social interactions.	T	F
Social exchange theory is based on reward and cost analysis.	T	F
Collective marriage is often referred to as a mass marriage.	T	F

People usually follow the sequence: career, marriage, and parenthood. This idea was introduced by American feminist Betty Friedan (1921–2006). But today, people are free to do what works for them and forge their own path. Many have resources to make their own life with or without a life partner. Some have

Student Account of Common Milestones

Apparently my sequence went in the following order: motherhood, marriage, career. I was nineteen when I had my first child. Therefore I had to drop out of college and go to work. Two and a half years later I had another child. Now I'm starting over going back to school. I don't regret it. Actually they're my biggest motivation but I do believe that it would have been a whole lot easier if I did it in that order. I like the quote that my husband says, "He is the muscle and I am the brain." I think that is best for my family (personal communication, February 10, 2018).

Upon completion of this section, students should be able to:
LO1: Describe reasons for and effects of the declining marriage rate and rising first marriage age.

career, marriage, and parenthood at the same time and thus get married while working on their career and growing their family.

The State of Marriage

Marriage is a legally recognized and/or socially approved union between two partners that establishes rights and obligations. Marriage is a part of adulthood. Throughout human history, young people have aspired to achieve adulthood and but now the concept of adulthood is in jeopardy (Mohler, 2016). A set of four common milestones signify the transition to adulthood: leaving the parental home, securing a job, getting married, and becoming a parent (Shanahan *et al.*, 2005). Today, only 24 percent of all 25- to 34-year-olds live away from their parents, are in the labor force, are married, and have had a child. This number compares to 45 percent of all 25- to 34-year-olds who had completed all four of these milestones in 1975 (Vespa, 2017) (see Figure 9.1). Another 22 percent had accomplished three of the four milestones in 1975 except they did not work (many were married mothers who, in 1975, were not working outside the home). In other words, in 1975, there was a good deal of uniformity in what people experienced by their early thirties: the two most common sets of milestones described the experiences of two-thirds of all 25- to 34-year-olds at that time. Today, the experiences of people ages 25 to 34 are more diverse, with an increasing number of people waiting to get married and have children.

Marriage Rate

Half of Americans ages eighteen and older were married in 2016, a share that has remained stable in recent years but is down nine percentage points over the past quarter-century (Geiger & Livingston, 2018). The **marriage rate** is the number of marriages per 1,000 unmarried women over the age of fifteen in a given period. In 2016, there were 31.9 marriages per 1,000 unmarried women, compared with 76.5 marriages in 1970 (Hemez, 2016). In 1960, it was unusual for a woman to reach age 25 without marrying: Only 10 percent of women ages 25 to 29 had never married.

Today, the marital rate is on the decline, with many Americans putting off starting a family until they are older (Vespa, 2017). Millennials (born 1981–1997) are less likely than previous generations of young adults to be married. In 2015, Millennials (40 percent) ages 25 to 35 were married, compared with Baby Boomers (68 percent) (born 1946–1964) ages 25 to 34 in 1980 (Anderson, 2017). In 2016, Millennials (42 percent) were married, compared with their Silent Generation counterparts (82 percent) (born mid-1920s–mid-1940s) in 1963 (Fry, 2017).

Figure 9.2 shows that half of American adults are married, down from over two-thirds in 1967.

Marriage Rate by Education, Gender, Geography, and Race

Marriage rates are linked to socioeconomic status. The education gap in marital status has continued to grow. In 2015, among adults ages 25 and older, those with a four-year college degree have the highest marriage rate (65 percent), compared with those with some college education (55 percent) and those with no education beyond high school (50 percent) (Geiger & Livingston, 2018).

Looking at men and women separately, the marital status for men continued to decline from 70 percent

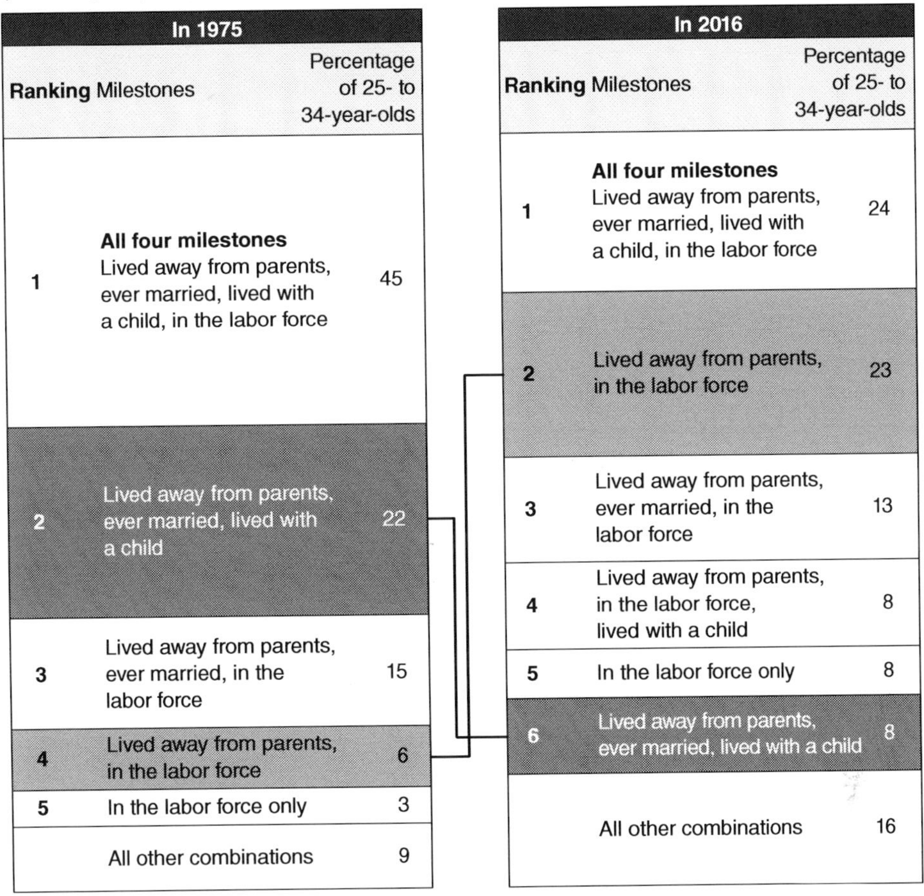

Figure 9.1 Four Common Milestones of Adulthood—Living Independently, Working, Getting Married, and Having Children—from 1975 to 2016

Source: U.S. Census Bureau, 1975 and 2016 Current Population Survey Annual Social and Economic Supplement. https://www.census.gov/newsroom/press-kits/2017/young_adulthood.html

in 1960 to 60 percent in 1990. In 2016, 46.8 percent of men age eighteen and older were unmarried (U.S. Census Bureau, 2017a). About 55 percent of men were married in 2017. Figure 9.3 shows the decline of marital status for men in the United States from 1960 to 2017.

Meanwhile, 53.2 percent of women age eighteen and older were unmarried in 2016 (U.S. Census Bureau, 2017a). Within marriage markets, when a randomly chosen woman becomes more likely to earn more than a randomly chosen man, marriage rates decline (Bertrand *et al.*, 2015). Figure 9.4 shows that women's marriage rates have been on the decline from almost 70 percent in 1960 to 50 percent in 2017.

In the United States, all states in the Northeast exhibited low marriage rates with Rhode Island having the lowest marriage rate, while most states in the Southern and Western region had high marriage rates

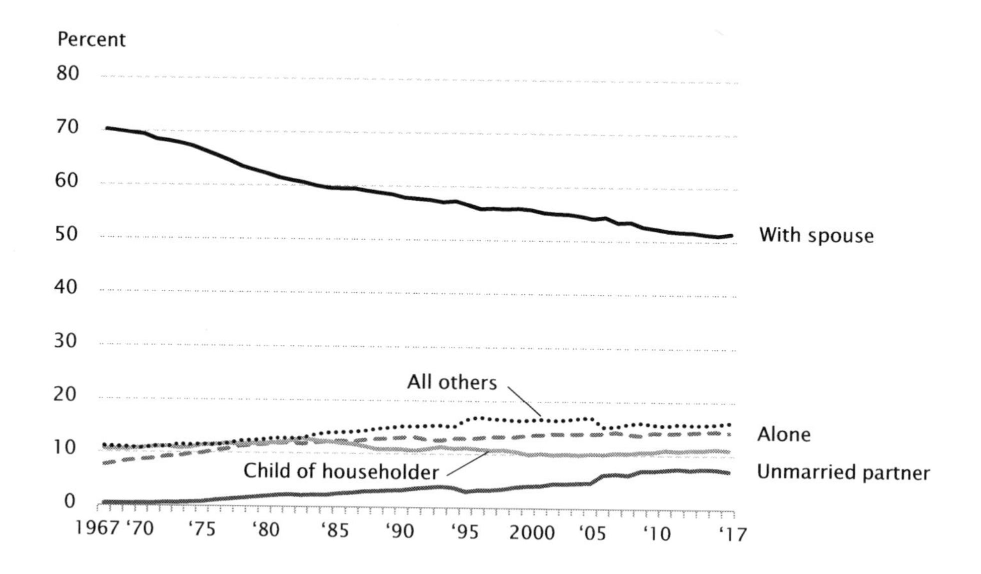

Figure 9.2 Living Arrangements of U.S. Adults Age 18 and Over

Source: Current Population Survey, Annual Social and Economic Supplement, 1967–2017. https://www.census.gov/data/tables/time-series/demo/families/adults.html

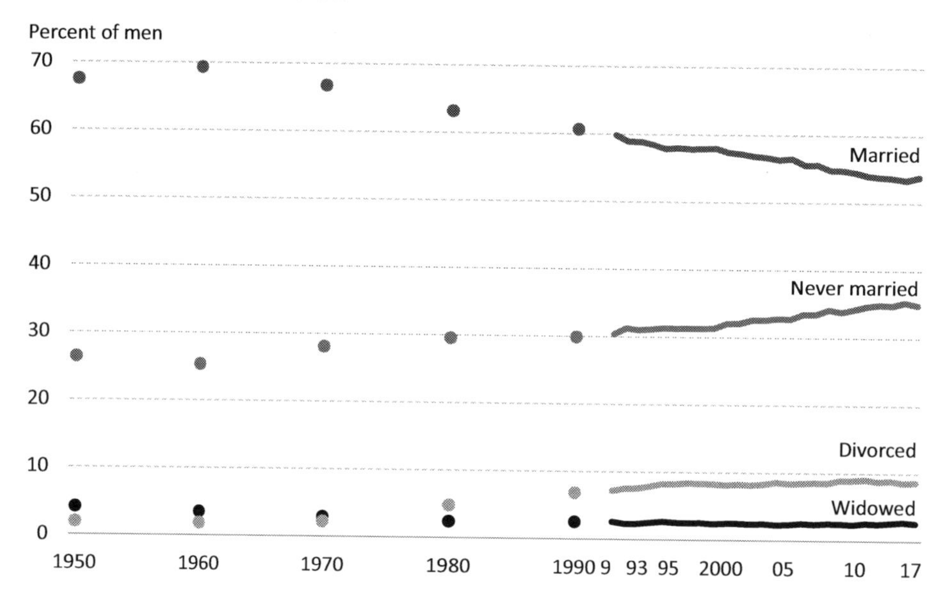

Figure 9.3 Men's Marital Status in the United States, 1950–2017

Source: U.S. Census Bureau, Decennial Censuses, 1950–1990, and Current Population Survey, Annual Social and Economic Supplements, 1993–2017. https://www.census.gov/data/tables/time-series/demo/families/marital.html

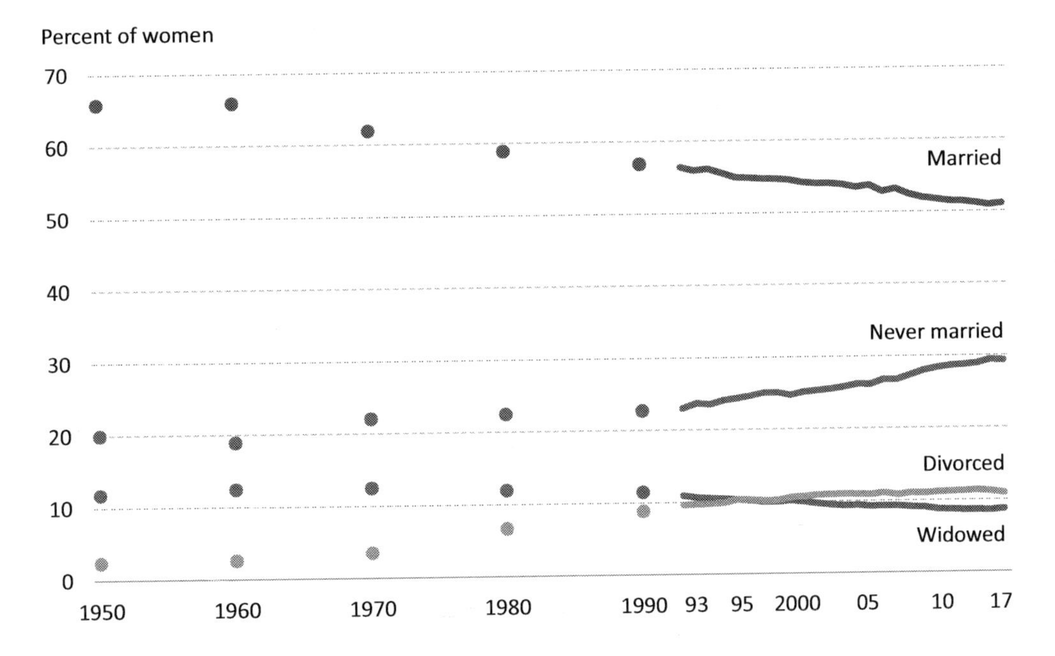

Figure 9.4 Women's Marriage Rate in the United States, 1950–2017
Source: U.S. Census Bureau, Decennial Censuses, 1950–1990, and Current Population Survey, Annual Social and Economic Supplements, 1993–2017. https://www.census.gov/data/tables/time-series/demo/families/marital.html

(Hemez, 2016). In Rhode Island, about 24 per 1,000 unmarried women over the age of fifteen got married, compared with 58 marriages in Utah in 2016. In terms of race, the marriage rate of all racial groups in the United States declined after 1980. In 2015, Asians ages eighteen and older had the highest marriage rate (61 percent), compared with Whites (54 percent), Hispanics (46 percent) and Blacks (30 percent) (Parker & Stepler, 2017a). Marriage rates continue to vary by race and ethnicity. Figure 9.5 shows that in 2017, 70 percent of White people were married, compared with 60 percent of Hispanics and 35 percent of African Americans.

Marriage Age

The decline in the share of married adults can be explained by the fact that Americans are marrying later in life (Parker & Stepler, 2017a). The age of first marriage in the United States is on the rise. In 1890, the median age at first marriage was 22 for women and 26 for men; it was 20 for women and nearly 23 for men in 1956 and 25 for women and 27 for men in 1999 (U.S. Census Bureau, 2004). In 2017, the median age of first marriage had reached its highest point on record: 27.4 for women and 29.5 for men (U.S. Census Bureau, 2017b). In the 1970s, eight in ten people had gotten married by the time they turned 30. Today, eight in ten people have gotten married by the time they turned 45 (Vespa, 2017). Figure 9.6 shows the median age at first marriage from 1890 to 2017.

Marriage Age by Education and Race

Many young adults in the United States today are putting off marriage in order to pursue higher education and career. People with a college degree or higher have the highest median age at first marriage, at age 29.3 for women and age 30.8 for men (Anderson & Payne, 2016). In terms of race, Black men (31.1) and

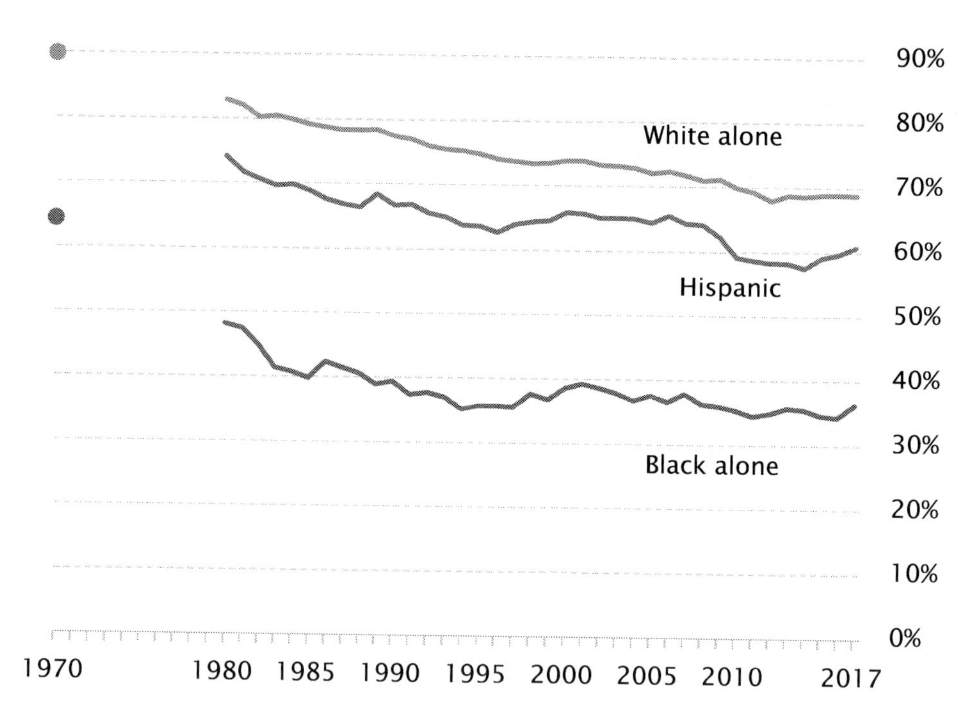

Figure 9.5 Percent of Family Groups with Children under 18 That Are Married

Source: U.S. Census Bureau, Current Population Survey, Annual Social and Economic Supplements 1970 and 1980–2017. https://www.census.gov/data/tables/time-series/demo/families/families.html

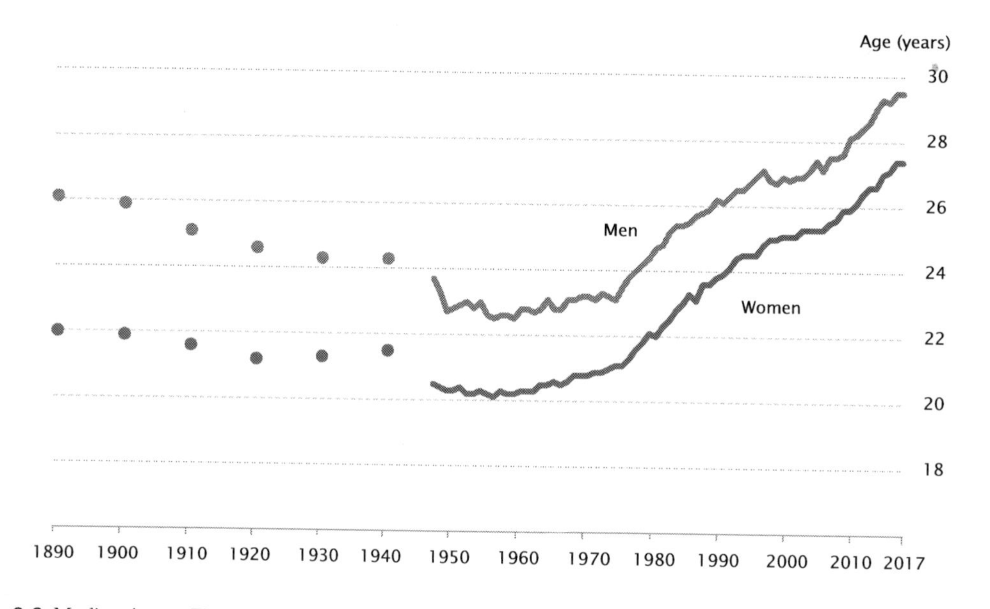

Figure 9.6 Median Age at First Marriage, 1890–2017

Source: U.S. Census Bureau, Decennial Censuses, 1890–1940, and Current Population Survey, Annual Social and Economic Supplements, 1947–2017. https://www.census.gov/data/tables/time-series/demo/families/marital.html

Black women (30.3) have the highest median ages at first marriage, compared with Asian men (30.1), Hispanic men (29.4), White men (29.3), Asian women (28.2), and White and Hispanic women (27.6) (Anderson & Payne, 2016). Socioeconomic indicators associated with labor force participation, wages, poverty, and housing (e.g., housing costs and living arrangements) all relate to marriage rates for young adults ages 18 to 34 (Currentz, 2018).

Deinstitutionalization of Marriage

"Love and marriage go together like a horse and carriage" are the lyrics from a song called "Love and Marriage" by American songwriter Sammy Cahn (1913–93). Marriage has undergone a process of *deinstitutionalization*—a weakening of the social norms that define partners' behavior—over the past few decades (Cherlin, 2004). Indeed, the role of marriage has changed, as 50 percent of survey respondents say that society is well off if people have priorities other than marriage and children (Fry, 2016).

The Changing Role of Marriage

Marriage has existed for around 5,000 years and served economic, political, and social functions; the individual needs and wishes of its members (especially women and children) took second place (Coontz, 2004). Prior to the 1960s, when the older, institutional model of marriage dominated popular consciousness, marriage was the only legitimate venue for having sex, bearing and raising children, and enjoying an intimate relationship (Wilcox, 2009). Although the idea of marrying for love started to emerge in the late eighteenth century, several factors acted to stabilize marriage. Traditional marriage enabled both the husband and wife to gain the benefit of each other's abilities. The husband took on the role as provider. The wife fulfilled the role of homemaker and was economically dependent on the husband. Sex was something that couples engaged in after marriage for procreation. There were penalties for illegitimacy such as marital infidelity. Reliable contraception and divorce were not readily available.

These restraining factors fell away in the last 200 years. The transition from agricultural subsistence to wage labor and the eventual joining of married women to the workforce were instrumental in changing the meaning of marriage (Cherlin, 2004). As economic and political institutions took over the roles played by marriage, people began to see marriage as a personal relationship that fulfills their emotional and sexual desires. In the 1950s, married couples were supposed to be partners, friends, and lovers; marriage transitioned "from an institution to a companionship" (Burgess & Locke, 1945). The contraceptive pill gave unmarried women a degree of sexual freedom that the sex radicals of the 1920s could only have dreamed of (Coontz, 2004). In the 1960s, the rise in the number of single young adults as they went through college and started their careers, as well as the rise in childbearing outside marriage, divorce rates, same-sex unions, and a greater acceptance of cohabitation before marriage, led to the second great change in the meaning of marriage for society (Fowler, 2007). By the end of the 1970s, women had access to legal rights, education, birth control, and decent jobs (Coontz, 2004). As more married couples chose not to have children, childless marriages weakened the connection between marriage and parenthood, eroding the traditional justifications for elevating marriage over all other relationships (Coontz, 2004). As love became the main reason to marry, loveless marriages were seen as a problem. By the end of the eighteenth century, Sweden, Prussia, France, and Denmark had legalized divorce on the grounds of incompatibility (Phillips, 1991). In 1969, Governor Ronald Reagan of California signed the nation's first no-fault divorce bill; New York was the last state to pass a no-fault divorce law in 2010 (Wilcox, 2009). *No-fault divorce* is a divorce in which one spouse can dissolve a marriage for any reason or no reason and gutted marriage of its legal power to bind husband and wife (Wilcox, 2009). During the 1980s and 1990s, all these changes came together to irrevocably transform the role of marriage (Coontz, 2004). Marriage transitioned from the companionate model of marriage to "individualized marriage" (Cherlin, 2004)—a shift from concern with playing a role to concern with self-development and emotional fulfillment (Cancian, 1987).

"Beta" Marriage

Almost half of Millennials (43 percent) would support a marriage model that involved a two-year trial—at which point the union could be either formalized or dissolved, no divorce or paperwork required (Bennett, 2014). **Beta marriage** refers to a union that you can test, renew, or abandon without consequence.

Where did that idea come from? American paleontologist E. D. Cope (1888) suggested that marriages should start with a five-year contract that either spouse could end or renew with a further ten- or fifteen-year contract and permanent contract if all still went well after that. British sexologist Havelock Ellis (1920) explored a temporary union of varying levels of commitment that allowed people to have sex, access birth control, and have an easy divorce if desired, as long as no children were involved. Gadoua and Larson (2014) took a look at the modern shape of marriages to help people open their minds to marrying creatively, from a parenting marriage for the sake of raising and nurturing children to a safety marriage for financial security or companionship. Some people want a marriage for life. Others want a temporary marriage so that they are free to determine how often they would consider renewing, renegotiating, or ending their marital contract based on their goals. Sutherland (2014) questioned how temporary, 100-percent negotiable, low-commitment relationships deserve the term "marriage."

Predictors of a Successful Marriage

People get married in the hope of having a happy marriage, which is conceptualized as a successful marriage (Alder, 2010). **Marital quality** is the subjective evaluation of a marital relationship on a number of dimensions (Spanier & Lewis, 1980) such as marital happiness, marital conflict, marital commitment, social support, marital interaction, marital discord, forgiveness, and domestic violence (Stanley, 2007). The terms "satisfaction, adjustment, success, happiness" have been used in describing the quality of marriage (Fincham *et al.*, 1997). Both marital happiness and satisfaction refer to positive feelings that a spouse derives from a marriage (Campbell *et al.*, 1976).

Throughout the modern world, most married couples devote effort to striving for a *happy* and *satisfying* marriage. **Marital satisfaction** refers to an individual's global evaluation of the marital relationship (Hinde, 1997). The highest amount of marital satisfaction is among the spouses who are compatible with each other concerning philosophy of life, their perceptions of sexual satisfaction, the amount of time they spend with each other, and how they spend leisure time with each other (Kaslow & Robison, 1996). Marital satisfaction has been found to correlate with communication, couple closeness, couple flexibility, personality issues, and conflict resolution in order of *importance* (Olson *et al.*, 2008), social and personal resources, satisfaction with lifestyle, and rewards from spousal interaction (Kendrick & Drentea, 2016), marital interaction, the presence of children, household income, egalitarian attitudes, and traditional marital attitudes (Amato *et al.*, 2007), sexual and interpersonal factors, mental health, sociodemographic factors such

Upon completion of this section, students should be able to:
LO2: Analyze some predictors of a successful marriage.

as occupation, length of marriage, and number of children (Zaheri *et al.*, 2016), education and age (Alder, 2010), background and context, individual traits and behaviors, and couple interactional processes (Larson & Holman, 1994). Let's take a look at some research findings that social scientists believe to be the factors that affect marital success and satisfaction.

Financial Stability

Financial stability is one of the key factors that determine whether a marriage will be successful. Lehrer (2008) investigated stabilizing forces from higher levels of education and older ages. According to a 2017 survey by the Pew Research Center, seven in ten respondents (71 percent) said it is very important for a man to be able to support a family financially to be a good husband or partner (Geiger & Livingston, 2018). A woman's job along with a man's occupation and income can help improve life quality and marital satisfaction (Rajabi *et al.*, 2013), because when both partners are employed, families are far more likely to be middle or upper-middle class (Rosen, 2016). Usually, marital stress is generated by financial instabilities or economic hardships (Martin, 2006). Economic aspects of life are crucial predictors of conflict and dissolution in both married and cohabiting couples (Hardie & Lucas, 2010). Thus, if people have attained some level of independence before marriage, they are more likely to be prepared to deal with hardships.

You might follow the sequence: career, marriage, and parenthood and wait until you are at least age 25 or over to get married. Doing so will pay dividends over the years ahead. Education and experience that come with age will contribute to the success of a marriage. A study of 6,850 couples has revealed that increased age at marriage has a strong effect on the success of the marriage (Lehrer & Chen, 2013). Marriage at a young age is becoming increasingly difficult due to increasing rates of unemployment and financial instability among younger populations (Martin, 2006). Also, adults who divorce at a young age tend to have lower assets than adults after 50 (Spangler *et al.*, 2016). Most studies suggest that suitable marriages for men and women are made in the age ranges of 18–25 and 24–30 years, respectively (Rezaeanlangroodi *et al.*, 2011) and are influenced by culture (Shahhosseini *et al.*, 2014).

Relationship Factors

Over the course of life, we learn to establish different relationships. The **Actor–Partner Interdependence Model** (APIM) is the most popular model used to measure the influence that members of a dyad have on each other (Fitzpatrick *et al.*, 2016). The APIM focuses on *actor–partner* effects of the marital relationship on marital satisfaction. For example, a person's personality affects marital happiness; this is called the *actor effect*. Individual persons who are satisfied with their own ability to communicate are generally more satisfied with themselves than with their partner (Brown, 2006). The APIM also examines the perceived relationship quality of the partner; this is called the *partner effect* (Kenny *et al.*, 2006). In a study of sexual agreement for HIV risk, Mitchell *et al.* (2013) reported that participants' likelihood of having unprotected anal intercourse (UAI) decreased with *actor effects* of commitment to a sexual agreement and *partner effects* of age, trust, and investment of relationship commitment. When examining actor–partner effects of sexual satisfaction, Jones (2016) found that male sexual satisfaction was related to female marital satisfaction. This partner effect predicted a female outcome. In other words, as men reported greater sexual satisfaction, women reported more satisfaction in their relationship. Yoo *et al.* (2014) found that communication and sexual satisfaction were the most important predictors of marital satisfaction.

Relationship Outcomes in Marriage Forms

Dinna (2005) reported that couples of *love marriages* are more satisfied with their marriages than couples of *arranged marriages*. Although evidence shows that couples are more satisfied, engaged, and less prone to domestic violence within love-based than arranged marriages (Şahin *et al.*, 2010), research results are not consistent (Regan *et al.*, 2012). For example, the satisfaction level of couples in arranged marriages in India was higher than that of Indian Americans in love marriages (Yelsma & Athappilly, 1988); the satisfaction level of Indian Americans in arranged marriages was higher than that of couples in arranged marriages in India and those of Indian Americans in love marriages (Madathil & Benshoff, 2008). There was no difference in marital satisfaction between love and arranged marriages in Japan (Walsh & Taylor, 1982), in the United States (Myers *et al.*, 2005), and among Indian Americans living in contemporary American society (Regan *et al.*, 2012). The couples were more satisfied in both arranged marriages and love marriages with parental acceptance than those in love marriages without parental acceptance (Arif & Fatima, 2015). Interestingly, some studies have shown fitness benefits of mating with partners of free choice (Pogány *et al.*, 2014) in terms of number of offspring (Edward & Chapman, 2012) and their growth rate (Sandvik *et al.*, 2000). However, a study of Tsimane, Yali, and Bhotiya tribes revealed that free choice marriages did not have more children and did not lower child mortality rates any more than arranged marriages (Sorokowski *et al.*, 2017).

Research has suggested that both institutional and companionate marriages would have the greatest likelihood of success in the long term (Kamp Dush & Taylor, 2012). *Institutional marriage*—characterized by traditional gender roles and supported by religion, law, and local community—may promote marital happiness because first, such marriages are involved in religious institutions that provide social support for marriage, and second, they have a more altruistic mindset toward the marriage that makes them less likely to seek their own interests (Wilcox & Nock, 2006). For example, women do the majority of housework and childcare (Bianchi *et al.*, 2006), and gender attitudes reinforcing this behavior may help wives cope with the inequity (Kamp Dush & Taylor, 2012). A study of 20-year data from the Marital Instability over the Life Course Study has supported the suppositions of the institutional model of marriage (Kamp Dush & Taylor, 2012).

Companionate marriage emphasizes affection and marital equality in family roles and has higher marital quality and lower marital conflict (Wilcox & Nock, 2006) when both partners share equally in the division of household labor and decision-making in the household (Kamp Dush & Taylor, 2012). The growth of "companionate" marriage is viewed as a sure indicator that marriage would increasingly become a relationship of equality (Clark *et al.*, 1991). Individuals who began the 20-year study period with companionate marriage were more likely to stay happily married with average or little conflict (Kamp Dush & Taylor, 2012).

Communication Factors

Communication is a vital factor in determining the tenor and perceived closeness of intimate relationships (Holland *et al.*, 2017). For example, conversation is a key point in long-distance relationships. A long-distance relationship can be stronger in comparison to a normal one, because it forces the two partners to enhance the conversation process (Jurkane-Hobein, 2015).

Much of relationship research has focused on general relationships. Jones (2016) argues that marital satisfaction, general communication, sexual communication, and sexual satisfaction are interrelated.

General Communication

Happy couples with marital satisfaction engage in more positive and more effective communication (Lavner *et al.*, 2016), use active listening skills, agree, approve, assent, and often use laughter and humor (Noller & Fitzpatrick, 1991). In fact, 95 percent of happily married couples agreed that communication was the most important predictor of a happy marriage

(Olson *et al.*, 2008). Satisfied couples maintain a five-to-one ratio of positive to negative exchanges in interactions (Gottman, 1995), strengthen their relationship, and increase well-being on a daily basis (Gable *et al.*, 2004).

Sexual Communication

Sexual communication refers to the communication, self-disclosure, and communication processes around sexual topics and problems. Partners may have different opinions about discussing sexuality, due to differing expectations, desires, experiences, or beliefs, which in turn may affect the relationship (Khoury & Findlay, 2014). Sex may also be considered taboo due to social and cultural influences (Moyer-Gusé *et al.*, 2011).

Sexual communication is found to be a key factor identified in successful marriages. For example, to decrease unintended pregnancies, couples must mutually agree on family planning. A study of 40 Latino couples revealed a positive association between both husbands' and wives' general communication and sexual communication, and a negative association between husbands' views on family planning and their wives' sexual communication (Matsuda, 2017). Higher disclosure to one's partner about sexual preferences and desires is positively correlated with sexual satisfaction and marital quality (Montesi *et al.*, 2011). Couples who discussed sex more were more likely to be relationally and sexually satisfied.

Sexual Satisfaction

Sexual satisfaction refers to satisfaction with the frequency, variety, and quality of various aspects of one's sexual life, including functioning and connection (Jones, 2016). Happily married couples experience greater sexual satisfaction, increased sexual activity, and enjoy higher rates of fidelity (Robles & Kiecolt-Glaser, 2003). In 2017, married adults (61 percent) said that a satisfying sexual relationship is very important to a successful marriage (Geiger & Livingston, 2018). A study of 1,310 couples living in Louisiana also revealed positive effects of sexual satisfaction on marital satisfaction (Dzara, 2010).

Different relational factors that influence sexual satisfaction include length of the relationship, frequency of sex, number of children at home, attitudes, desire discrepancy, spouses' ages and education levels (Christopher & Sprecher, 2000). A study of 387 married couples revealed that general communication and sexual satisfaction independently predict marital satisfaction (Litzinger & Gordon, 2005). If couples have difficulty communicating but are sexually satisfied, they will still experience greater marital satisfaction than if they have a less satisfying sexual relationship. Thus, sexual satisfaction may compensate for the negative effects of poor communication on marital satisfaction (Litzinger & Gordon, 2005).

Gender role may influence sexual satisfaction. When the wife is satisfied with the division of labor, the couple engages in sex more often (Carter, 2011). Using data from Wave II of the National Survey of Families and Households, Kornrich *et al.* (2013) reported that couples with more traditional housework arrangements reported higher sexual frequency, suggesting the importance of gender display rather than marital exchange for sex between heterosexual married partners. Men tended to overestimate their partner's sexual satisfaction, whereas women were more accurate in estimating their partner's sexual satisfaction (Fallis *et al.*, 2014).

Interestingly, research shows that having more premarital partners and younger age at first sexual experience tended to lower marital satisfaction (Legkauskas & Stankevičienė, 2009) and negatively affected marital quality (Jose *et al.*, 2010). Figure 9.7 shows that the lowest divorce rates are associated with married women with zero premarital partners in the 2000s.

Religious women without premarital sexual partners tended to increase marital satisfaction and lower divorce rates. Figure 9.8 illustrates virgin marriages and the low divorce rate for religious brides between the 1980s and 2000s.

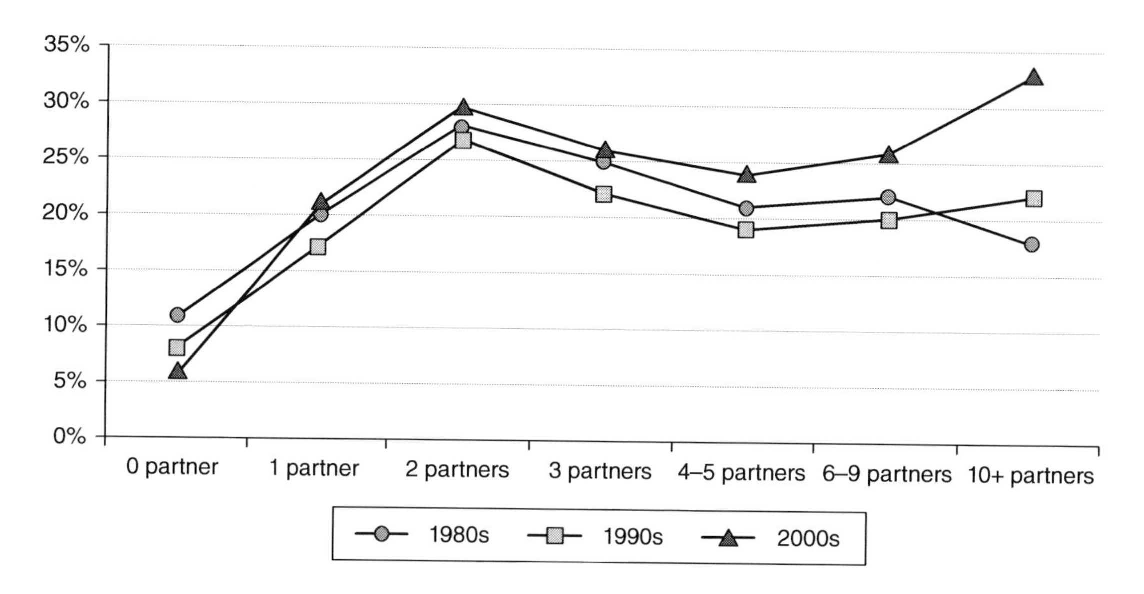

Figure 9.7 Chances of Divorce after Five Years of Marriage by Marriage Cohort and Number of Premarital Sexual Partners

Source: NSFG, 2002–2013, Wolfinger (2016). https://ifstudies.org/blog/counterintuitive-trends-in-the-link-between-premarital-sex-and-marital-stability/

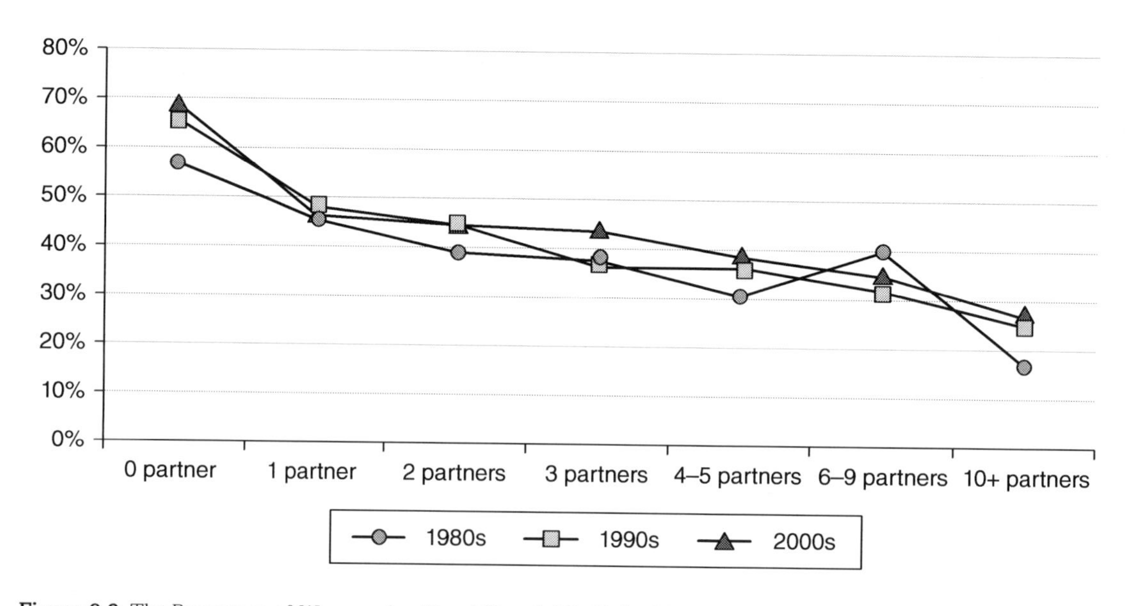

Figure 9.8 The Percentage of Women who Attend Church Weekly by Marriage Cohort and Number of Premarital Sexual Partners

Source: NSFG, 2002–2013, Wolfinger (2016). https://ifstudies.org/blog/counterintuitive-trends-in-the-link-between-premarital-sex-and-marital-stability/

Personality Traits

Research has indicated associations between personality traits and marital satisfaction. *Personality* is a stable and fundamental psychological construct (Donnellan *et al.*, 2005). After three decades of studying romantic relationships, Fisher (2010) discovered that your dominant personality type guides not only who you are but also who you love. Women more often than men selected personality traits as a reason to choose a partner (McDaniel, 2005). Igbo *et al.* (2015) affirmed a strong relationship between the Big Five personality factors and conflict resolution strategies of spouses. The British Household Survey (n=4,169), the Divorce in Flanders study (n=4,377), and the German Socio-Economic Panel (n=8,155) all have revealed the associations between the Big Five personality traits and divorce (Boertien & Mortelmans, 2017). Let's take a look at **Five Factor Model** (FFM) that is composed of five personality traits (see image below): *openness to experience, conscientiousness, extraversion, agreeableness, neuroticism.*

Openness involves active imagination, aesthetic sensitivity, attentiveness to inner feelings, preference

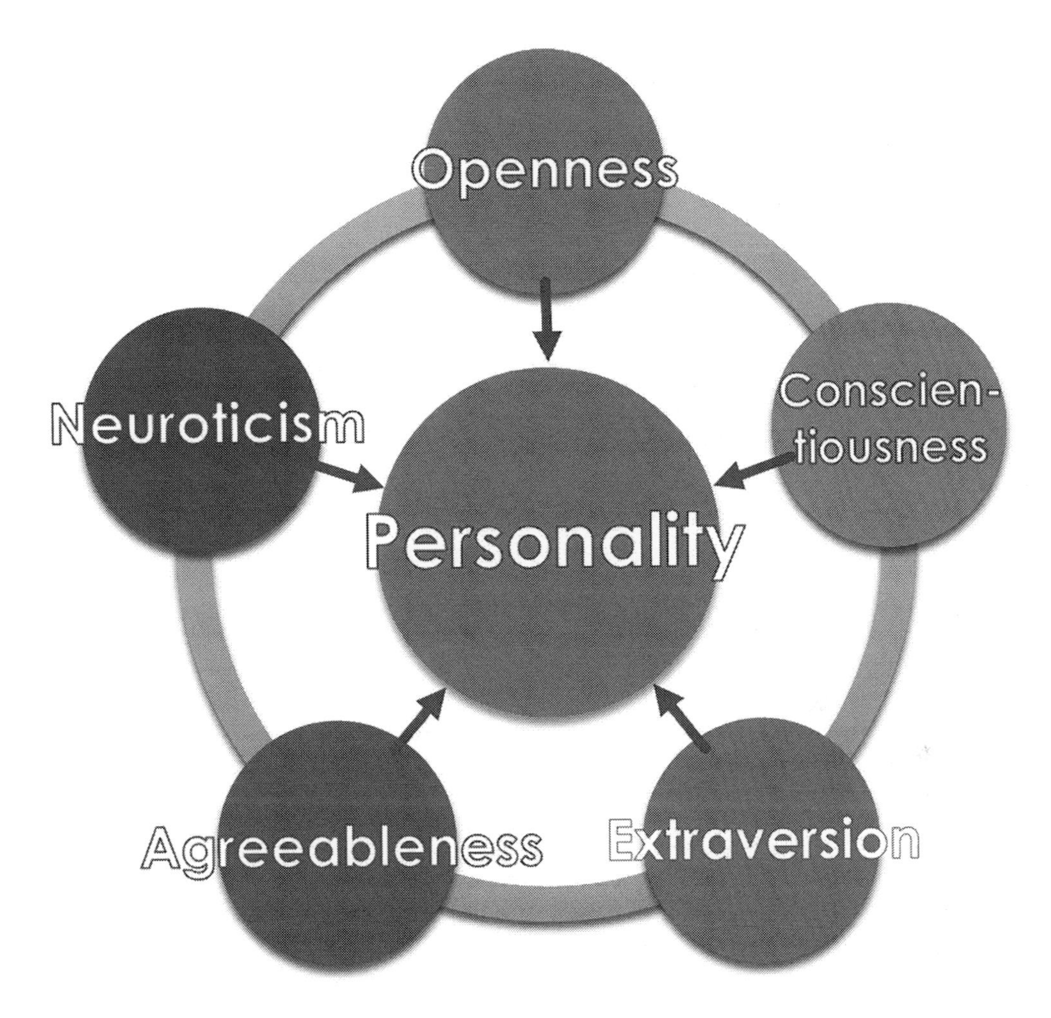

Five Factor Model (FFM)
Source: Eganos. https://upload.wikimedia.org/wikipedia/commons/archive/8/86/20170929033341%21Bigfive_en.png

for variety, and intellectual curiosity (Costa & McCrae, 1992). When compared to closed people, those who are open to experience are intellectually curious, open to emotion, and willing to try new things. Openness to experience correlates with intelligence (Moutafi *et al.*, 2006). In speed dating, women based their choices additionally on men's openness to experience (Asendorpf *et al.*, 2011). Across Britain, Flanders, and Germany, people who behave with high openness to experience are less likely to divorce (Boertien & Mortelmans, 2017). In terms of sexual communication, openness was affected by traditional gender roles. For example, men had more influence in sexual negotiations than women; more traditional couples were less sexually self-disclosing, less communicative, and less effective at sexual negotiations than less traditional couples (Greene & Faulkner, 2005).

Conscientiousness is manifested in doing a task carefully and taking obligations to others seriously. Conscientious people tend to be efficient and organized as opposed to easy-going and disorderly people. Conscientiousness similarity is higher in friends-first relations (Barelds & Barelds-Dijkstra, 2007). Comparisons between dating and married couples showed that the most consistent effect was *conscientiousness* (Cupach & Metts, 1986). Highly conscientious people are less likely to get divorced and are positively associated with partners' relationship quality (Roberts *et al.*, 2009). Across Britain, Flanders, and Germany, low conscientious people who do not keep up social relations are more likely to get divorced than highly conscientious people (Boertien & Mortelmans, 2017).

Extraversion is manifested in outgoing, talkative, energetic behavior, whereas **introversion** is manifested in more reserved and solitary behavior (Thompson, 2008). *Extraverts* correlated with arousal and pleasantness (Kuppens, 2008); *introverts*—being thoughtful and contemplative—took the slow penetration strategy to develop relations (Nelson & Thorne, 2012). Although men's high extraversion led to their popularity in speed-dating contexts (Back *et al.*, 2011), husbands' extraversion predicted low marital satisfaction (Belsky & Hsieh, 1998). Wives' extraversion was positively related to husbands' marital satisfaction (Chan et al., 2007).

Agreeableness is manifested in kind, sympathetic, cooperative, warm, and considerate behavior (Thompson, 2008). Agreeable people value getting along with others and tend to compromise their interests with others (Rothmann & Coetzer, 2003). In terms of casual sexual encounters, cute macho guys (e.g., less agreeable/more attractive) were more successful than nice guys (e.g., more agreeable/less attractive) (Urbaniak & Kilmann, 2006).

Neuroticism is the tendency to experience negative emotions such as anger, anxiety, or depression (Jeronimus *et al.*, 2014). As neuroticism among couples increases, marital satisfaction decreases (Rogge *et al.*, 2006). A study of 400 couples revealed a positive correlation between neuroticism and negative interactions for marriage (Khalatbari *et al.*, 2013). Neurotic extraverts experience high levels of both positive and negative emotional states, a kind of emotional rollercoaster (Passer & Smith, 2009). Thus, neuroticism seems problematic to marital satisfaction and people who scored higher on agreeableness, conscientiousness, and openness enjoyed better marital quality (Zhou *et al.*, 2017b).

Personality traits vary by gender and geography. Females are more likely to experience negative emotions than men. Cross-cultural studies reported higher levels of female neuroticism across all nations (Lippa, 2010). Another study of gender differences in personality traits in 55 nations revealed that women were significantly higher than men in average neuroticism, extraversion, agreeableness, and conscientiousness (Schmitt *et al.*, 2008). In the United States, neuroticism is highest in the Middle Atlantic states and declines westward while openness to experience is highest in ethnically diverse regions of the mid-Atlantic, New England, the West Coast, and cities (Rentfrow & Jokela, 2016).

The Dark Triad

The Five Factor Model has associations with the **dark triad of personality**—three interrelated higher-order personality constructs (i.e., *psychopathy, narcissism, and Machiavellianism*). Use of the term "dark" implies that people possessing these traits have malevolent qualities (Paulhus & Williams, 2002). **Psychopathy** is characterized by continuing antisocial behavior, impulsivity, selfishness, callousness, and remorselessness (Skeem *et al.*, 2011) and has the strongest correlations with low dutifulness and deliberation aspects of conscientiousness (Furnham *et al.*, 2013). **Narcissism** is characterized by grandiosity, pride, egotism, and a lack of empathy (Kohut, 1977) and derives from the psychodynamic formulations such as a pathological form of self-love (Freud, 1914). Ryan *et al.* (2008) reported that men higher in covert narcissism (i.e., "associated with conscious shame and unconscious grandiosity") often physically assaulted their partners; women high in sexual narcissism (i.e., "sexual behavior that involves low self-esteem and an inflated sense of sexual entitlement") often sexually coerced their partners. **Machiavellianism** (MACH) refers to interpersonal strategies that advocate self-interest, deception, and manipulation (Jakobwitz & Egan, 2006). Persons high in MACH are likely to exploit others and are less likely to be concerned about other people beyond their own self-interest (Barnett & Thompson, 1985). People with dark triad personalities have more sex partners and more favorable attitudes towards casual sex (Jonason *et al.*, 2009), lower standards in their short-term mates (Jonason *et al.*, 2011), tend to steal mates from others (Jonason *et al.*, 2010), and have a pragmatic and game-playing love style (Jonason & Kavanagh, 2010). As people with dark triad traits put more effort into their appearance (Holtzman, 2012), people with dark triad personalities are judged as slightly better-looking than average on first sight (Carter *et al.*, 2014).

How a couple's similar or dissimilar personalities influence marital relationship can be explained by *similarity effect* or *complementary effect*. Let us take a look at personality similarity first.

Personality Similarity

The similarity hypothesis poses that similarity in some important areas will make people evaluate other people as more attractive and perceive higher levels of relationship satisfaction (Lucas *et al.*, 2004). Research has shown a positive similarity impact on marital relationship. For example, personality similarity among newlyweds predicted marital satisfaction (Asoodeh *et al.*, 2010) and could help newlywed couples maintain good relationship quality (Gonzaga *et al.*, 2007). A study of 248 married couples (Gaunt, 2006) and a study of 1,608 romantic couples (Furler *et al.*, 2013) both revealed a strong association between spousal personality similarity and life satisfaction. Compared to similarities in a single trait, profile personality similarity (the couple-centered approach) is a stronger and more consistent influence on marital satisfaction (Zhou *et al.*, 2017a).

Meta-emotions Research has indicated that the type of personality similarity that matters for marital satisfaction is **meta-emotions**—encompassing both partners' feelings and thoughts about emotion. A couple who shares their meta-emotional style is likely to give them a common emotional template (Lehrer, 2016). You might want to marry a partner who handles emotions the same way you do because meta-emotion can help resolve marital conflict (Gottman *et al.*, 1997). When you have compatible meta-emotional styles—when people agree on how feelings should be expressed—you are able to diffuse these tensions before they get too big and dangerous (Lehrer, 2016). In other words, a couple should share happiness and suppress anger the same way. For example, trivial annoyances may lead to huge fights because one partner expresses anger and rage but the other does not think that it is an issue at all. There is also a significant relationship between marital satisfaction and **emotional stability**—a person's ability to remain stable and balanced. People who scored higher on emotional stability enjoyed better relationship quality (Zhou *et al.*, 2017b).

The Theory of Complementary Needs

The expression of "opposites attract" reflects the theory of complementary needs. The **complementary hypothesis** suggests that complementary partners are more beneficial to the relationship and have a higher possibility of meeting individual needs (De Raad & Doddema-Winsemius, 1992). Some people choose and marry those with dissimilar personalities. In fact, partners who are complementary in extraversion, emotional stability, and openness are more likely to fall in love at first sight (Barelds & Barelds-Dijkstra, 2007). Closed people are more inclined to seek open people to balance themselves, enhance the stability of the relationship, and obtain comfort (Gurtman, 1995). Couples who were dissimilar in their profile personality had better marital quality in a long-term relationship (Zhou *et al.*, 2017a).

The social role of segmentation highlights the necessity of complementation in a smoothly functioning relationship (Eagly *et al.*, 2000). For example, couples with greater discrepancy of conscientiousness perceive better relationship quality in the long term (Zhou *et al.*, 2017b). This phenomenon is especially observed in organizations in which conscientious complementation enhances interaction quality (Chuang & Hsu, 2013). For Chinese dating young adults, openness in complementary matching dyads corresponds to better relationship quality than openness in moderate matching dyads (Zhou *et al.*, 2017b).

Genetic and Environmental Influence on Personality Traits

Both genetic and non-shared environmental factors influence all personality traits (Jang *et al.*, 1996) and account for personality changes (Hopwood *et al.*, 2011). Openness to experience (57 percent), extraversion (54 percent), conscientiousness (49 percent), neuroticism (48 percent), and agreeableness (42 percent) was estimated to have a genetic influence (Bouchard *et al.*, 2003). Sociologists suggest that the most important environmental factor during childhood development is *socialization*—the ongoing process whereby individuals learn to become a functioning member of society through the agents of socialization such as parents, school, or peers. Personality traits are inherited through observed behavior during the process of socialization. For example, where a person lives shapes and forms who he becomes because his environment is the main determinant of his personalities and individual characteristics (Holmes, 2017).

The transition to adulthood between the ages of 18 and 30 involves significant psychological development with regard to intimacy, identity, work, and parenthood (Arnett, 2007). These changes are accompanied by both stability and change in personality traits (Blonigen *et al.*, 2008). Genetic factors were responsible for differential stability while differential change was influenced by the non-shared environmental factors among twins assessed twice around the ages of 20 and 30 (Blonigen *et al.*, 2008).

Goal Theory of Marital Satisfaction

The **goal theory of marital satisfaction** argues that to reach marital satisfaction, people have to achieve three marital goals—*personal growth, instrumental goals, and companionship goals* (Li & Fung, 2011). **Personal growth goals** are based on the improvement and development of the newlywed couple with the help of each other within the marriage. When the newlyweds meet these goals and help each other adjust to married life, they are likely to reach marital satisfaction and feel capable of future challenges. **Instrumental goals** focus on the tasks that occur throughout life that include using the spouse's physical and mental resources. Usually, middle-aged

Upon completion of this section, students should be able to:
LO3: Describe engagement, marriage contracts, bride price, and dowry.
LO4: Explain the benefits and requirements of marriage.
LO5: Describe different wedding rituals.

couples prioritize the instrumental goals. **Companionship goals** emphasize the bonding and emotional goals that a spouse needs with the other spouse. Older couples focus on the companionship goals and realize that it takes time and effort to build commitment and achieve these goals. Indeed, many relationships do not last long because people fail to reach the marital goals and make sacrifices for their partners (Li & Fung, 2011). The **ideal–real gap theory** proposes that discrepancies between what one wants or considers to be ideal or desirable and what one actually has will affect marital satisfaction (Michalos, 1986). A study of 300 couples revealed that happy couples do not differentiate between the realities of life and their ideals (Asoodeh *et al.*, 2010).

Getting Married

Marriage is a social or cultural, spiritual, or legal union of individuals. This union may be called *matrimony*, while the ceremony that marks its beginning is called a *wedding* and the married status created is called *wedlock* (Ghaffar *et al.*, 2010). When you are planning on getting married, you might check marriage requirements and wedding ceremonies in your state. You may consider the costs of the wedding. Let's discuss engagement first.

Engagement

An **engagement** is a promise to wed and is the period of time between a marriage proposal and marriage. A **fiancée** is a future bride or a wife-to-be, while a **fiancé** is a future groom or husband-to-be. An engagement is often announced at an engagement party. In a study of engaged couples, Stewart and Olson (1990) reported that if both sets of parents were positive about the marriage, the majority of engaged couples experienced a positive premarital relationship.

Customs regarding engagement rings vary by culture. The Western practice of giving a ring to symbolize an engagement began in 1477 when the Holy Roman Emperor Maximilian I gave Mary of Burgundy a diamond ring as an engagement present (Collings, 2009). Wearing a ring indicates that she has promised to marry and committed herself to her future spouse. During an engagement period, the couple makes preparations for their marriage. Weeks before the wedding, *the maid of honor*—the bride's chief support before and after the wedding—plans a *wedding shower*, where the bride-to-be receives gifts from family and friends.

Marriage Contract

In some cultures, an engagement ring accompanies a marriage contract. A **marriage contract** is a special form of contract whereby a duty of good faith is placed on both parties during negotiations. Marriage contracts are also called *premarital* or *prenuptial* agreements that have long been recognized as valid in several European countries such as France and Belgium (see image below).

Marriage contracts are just one of several legal tools that can be used to address your specific needs (Ausman, 2003). Both parties should have lawyers to represent them to ensure that the agreement is enforceable. Instead of making blanket rules for every couple, marriage contracts are best tailored to the needs and wishes of the parties and specify the expectations of both spouses in regard to the

Source: The Web Gallery of Art (WGA). https://upload.wikimedia.org/wikipedia/commons/b/bc/Jan_Josef_Horemans_%28II%29_-_The_Marriage_Contract_-_1768.jpg

rights, responsibilities, and obligations of the marriage (Ausman, 2003). If a marriage contract is not legally binding in nature, it can serve as a communication tool that sets up the wedding, discussions on infidelity, etc. If couples don't have conversations about exclusivity and expectations about fidelity, the door to greater fallout remains open because they will undoubtedly default to dishonesty (Finkel, 2017). When you get married, your property and finances might merge with those of your spouse. The success of a marriage may hinge on how well the couple deals with issues such as financial assets, communication, conflict resolution, parenting, expectations, etc. All this could be discussed in the marriage contract. Establishing parameters in a marriage contract ensures that the couple knows what they can and can't do to contribute to a happy marriage and a healthy family. A cohabitation agreement is ideal for couples choosing to live together (Ausman, 2003).

Bride Price and Dowry

In many societies, marriage often involves some kind of social and economic exchange. In many cultures and in parts of China, both dowry and bride price were practiced from ancient eras to the twentieth century (Adrian, 2004) and continue to be practiced.

Bride Price

Bride price is similar to the engagement ring. **Bride price**, also called bridewealth, is a payment of money, valuable goods, or property by the groom or his family to the bride's family for permission to marry her. Bridewealth paid at marriage has functions in different societies: to indemnify the girl's family for the loss of her services, to solidify the new affinal bonds created by marriage, and to legitimize children born to the union (Dalton, 1966). In *patrilocal societies*, the bride will live with her husband's family, which deprives her family of her labor and she also benefits her new household with offspring. Different from the purchase of a woman, bride price is a purely symbolic gesture acknowledging (but never paying off) the husband's permanent debt to the wife's parents (Graeber, 2011). Thus, her family is compensated for their loss when the groom pays a bride price (William *et al.*, 2013).

The tradition of paying a bride price is still practiced in contemporary Asia and Africa. The bride price is a centuries-old tradition in China. Factors such as the shortage of women and economic boom over the last two decades have bumped up the "bride price" to hundreds of thousands of Chinese Yuan. News of a bride who got married after her family received a 3.8 million Chinese Yuan ($550,000) bride price caused controversy (Spooky, 2017).

Dowry

A **dowry** is payment of a woman's inheritance, by her parents, at the time of her marriage either to her or to her husband (William *et al.*, 2013). The custom of dowry was part of traditional arranged marriages in Europe, Asia, Africa, and other parts of the world. In patrilocal societies, a bride is required to provide a dowry—a transfer of parental property, gifts, or money at the marriage of a daughter (Goody, 1976).

The types of dowry vary depending on the economic circumstances of the families involved and the customary expectations of the society. In the Indian subcontinent, dowry is said to originate from Hindu customs (Jhabvala, 1981). The Romans practiced two types of dowry—*dos profectitia* given by the father or father's father of the bride and *dos adventitia* given by others (Mackeldey, 1883). The English dowry system allowed most noble families to marry off their daughters and thus gain extended kin and patronage ties. The custom of dowry was brought to the United States by colonists from Europe and existed in certain Native American tribes.

Functions of Dowry The custom of giving dowries performs some functions. First, a dowry may form an important part of the economic arrangements for a marriage and affirms an alliance between two families united by marriage. For example, Pocahontas (1596–1617), daughter of the Chief of the Powhatan Indian confederacy, brought a dowry to her marriage that included a large amount of land when marrying English tobacco planter John Rolfe in Jamestown, Virginia in 1613 (Mirza, 2007). The marriage ensured peace between the Jamestown settlers and the Powhatan Indians for several years. A dowry is also an indicator of the bride's family status and it is vital for the security of the bride after her marriage. For example, dowry was the more prestigious form and associated with the Brahmanic (priestly) caste (Tambiah, 1989). The bride's father offered his daughter adorned property; such voluntary gifts were to provide economic security in her new home (Rabby *et al.*, 2013). A woman who brought a large dowry was considered more virtuous in Chinese culture than one who did not (Mann, 2008). A young couple may also use the dowry to set up their nuclear family. For example, the daughters of wealthy nineteenth-century industrialists were given "dowries" such as large amounts of money and property by their fathers to marry European aristocrats who held a title but had little wealth. The mutual exchange of title and wealth raised the status of both bride and groom (Ferraro & Andreatta, 2009). Finally, a bride may rely upon her dowry for economic support if the groom or husband leaves her. A bride with a large dowry is likely to be treated well by the groom's family and the dowry may also have served as a form of protection for the wife against the possibility of ill treatment by her husband and his family (Wilson, 2005). Thus, when families give dowry, they not only ensure their daughter's economic security, they also "buy" the best possible husband for her, and son-in-law for themselves (Schlegel & Eloul, 1988).

Effects of Dowry on Women Dowry is the property that a woman brings during her marriage (Nazzari, 1991). In recent years, dowry levels have bumped up to previously unforeseen levels; among Hindus in north India they can amount to three or four times a family's total assets (Luciana *et al.*, 2004). Dowry has both positive and negative impact on a bride's marital life. If a bride arrives with a large amount of dowry, then she receives a warm welcome and fair treatment from the groom and in-laws' family. However, when the amount of dowry is inadequate or not as much as demanded by the groom's party, then it can have an adverse and damaging effect on the marital life of the newly married women (Rabby *et al.*, 2013). The practice of dowry inevitably leads to discrimination in different areas against daughters and makes them vulnerable to various forms of violence (Singh, 2013). In India, incidents of bride burning, physical abuse, and mental torture by husbands and their families make the headlines of newspapers daily (Ahmad *et al.*, 2014). **Femicide** or feminicide refers to "the intentional killing of females (women or girls) because they are females" (Russell & Harmes, 2011) and is a sex-based hate crime. Rabby *et al.* (2013) recommend specific intervention programs to explore possible ways to reduce and eliminate dowry.

Benefits of Marriage

People desire love, companionship, commitment, and children. Having a family is a high priority among many Americans (Nadelson & Notman, 1981). Increased marital satisfaction can improve physical health, psychological health, economic development, job satisfaction, and overall life satisfaction (Omran *et al.*, 2016). Indeed, marriage provides many benefits.

The ability of the relationship to provide "each partner a dependable companion" appears to be the most beneficial (Myers, 1992). Marriage also allows two people to merge resources, divide tasks, and accumulate more capital than they could as singles (Coontz, 2004). Larger shares of married-parent families at the state level are linked to greater economic mobility, higher family incomes, and less child poverty (Elhage, 2016). College-educated couples are more likely to reap the benefits of marriage, including better education, higher incomes, and family stability for their children (Wilcox *et al.*, 2015). The **marriage premium** is the difference in median personal income between married and unmarried men (Anderson & Guzzo, 2016). Figure 9.9 shows that men with a bachelor's degree have the largest marriage premium ($19,400), and those with less than a high school education have the smallest ($3,000).

Marriage is also beneficial to health and well-being. In fact, it is the marital satisfaction that provides the benefits (Dush *et al.*, 2008). Marriage can increase life expectancy for both genders, but this benefit is

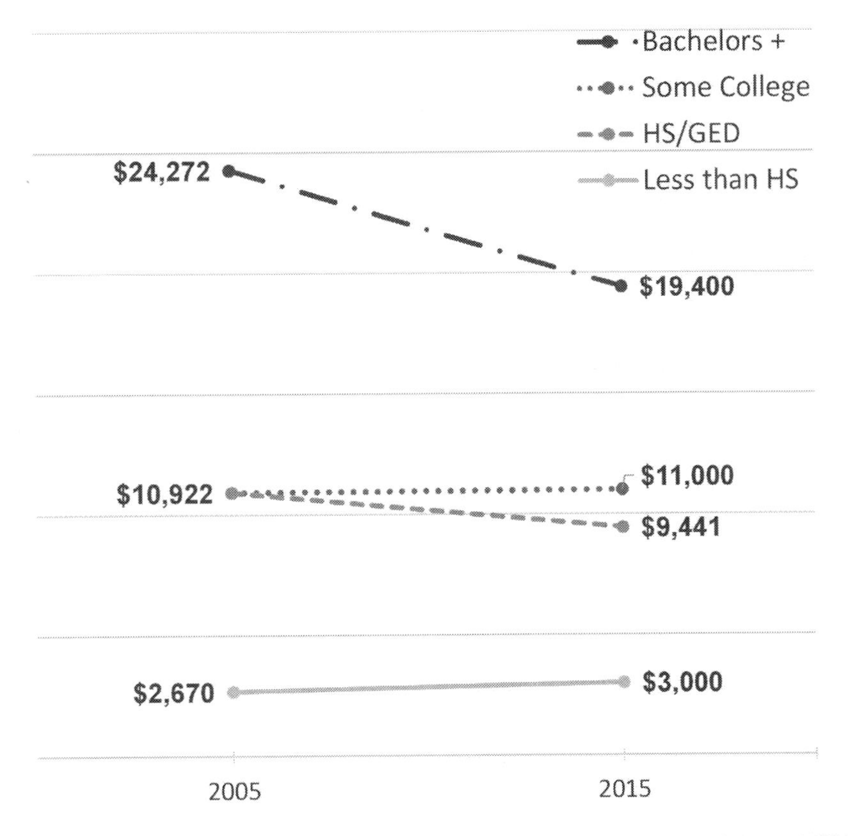

Figure 9.9 Difference in Median Personal Income between Full-Time Employed Married and Unmarried Men Ages 25–44 in the United States, 2005 and 2015

Source: Current Population Survey, Annual Social and Economic Supplement 2005 & 2015 and NCFMR 2016. https://www.bgsu.edu/ncfmr/resources/data/family-profiles/anderson-guzzo-men-employment-marriage-fp-16-06.html

five times stronger for men than for women (Robles & Kiecolt-Glaser, 2003). Good marriages specifically were a main ingredient to longevity, with those in stable and loving relationships enjoying greater well-being than those who are unattached (Myers, 1992). Marital happiness is positively related to measures of global health (Hetherington, 1993), indices of better immunity (antibody titers to various viral agents) (Kiecolt-Glaser *et al.*, 1988), and cardiovascular system functioning (Ewart *et al.*, 1991). Satisfying marital functioning protects against the development of psychological distress and depression (Trudel & Goldfarb, 2010). Compared to married people, unmarried people suffer higher rates of depression, anxiety, hostility, general distress, low self-esteem, less positive relationships with others, and less sense of life purpose and personal mastery (Cave, 2004).

Reasons for marriage also include social legitimacy, social pressure, the desire for a high social status, economic security, or validation of an unplanned pregnancy (Nadelson & Notman, 1981). Marriage is an institution in which interpersonal relationships such as spouse are acknowledged. Marriage determines rights and obligations connected to sexuality, gender roles, relationships with in-laws, and the legitimacy of children, and allows the property and status of the couple to be passed down to the next generation in an orderly manner (Coontz, 2004). Legal marriage provides rights and protections to spouses, including access to health care, tax breaks, and other benefits. Federal law does not regulate state marriage law but provides rights and responsibilities of married couples that differ from those of unmarried couples. There are 1,138 statutory provisions in which marital status determines benefits, rights, and privileges (GAO, 2004).

Marital status is a powerful predictor of employment behavior. Marriage is motivated by the values of love and family in addition to influencers, such as economic situation, family background, and children (Zhou *et al.*, 2017b). An American man ages 25 to 54 was likely to be employed in 2015 if he was married and had children (Eberstadt, 2016). A single or cohabiting mother is more likely to receive welfare benefits if she remains unmarried rather than marrying a partner who is earning a steady income (Wilcox *et al.*, 2015). A longitudinal data set spanning seventeen years reported that happier singles opt for marriage and actual division of labor contributes to spouses' well-being, especially for women raising young children (Stutzer & Frey, 2006).

Marriage System

Marriage systems are culturally inherited. For example, marriage norms govern who can marry whom, pays for the marriage ritual, gets the children in the event of the groom's or bride's death, and can inherit property and titles (Henrich *et al.*, 2012). Failure to conform to marriage norms results in reputational damage, loss of status, and various forms of sanctioning (Chudek & Henrich, 2010). Marriage norms also specify rules about the form of marriage. Today, *monogamy*—the practice of marrying one partner during their lifetime—is the only accepted form of marriage sanctioned by law in the United States.

In history, approximately 85 percent of societies in the anthropological record have permitted men to marry multiple wives (i.e., *polygynous marriage*) (White *et al.*, 1988) because taking wives is always positively associated with status, wealth, or nobility (Cashdan, 1996). After the origins of agriculture, as human societies grew in size, complexity, and inequality, levels of polygynous marriage intensified, reaching extremes in the earliest empires whose rulers assembled immense harems (Scheidel, 2009b). Today, however, with absolute wealth gaps greater than any seen in human history, monogamous marriage is both normative and legally enforced in most of the world's highly developed countries (Henrich *et al.*, 2012). Although the roots of the modern monogamous marriage can be traced back to classical Greece and Rome (Scheidel, 2009a), the global spread of this marriage system has occurred only in recent centuries, as other societies sought to emulate the West, with laws prohibiting polygyny arriving in 1880 in Japan, 1953 in China, 1955 in India, and 1963 in Nepal (Henrich *et al.*, 2012). By shifting male efforts from seeking wives to paternal investment, monogamy increases savings, child investment, and economic productivity and also reduces the intensity of intrasexual competition and intra-household conflict (Henrich *et al.*, 2012).

Who Can Marry?

Marriage laws vary by state. Each state is different when it comes to marriage-related issues such as the legal requirements for marriage and the specifics of pre-marital agreements. As long as a couple meets a certain set of requirements, they can receive state recognition of their marriage and apply for a **marriage license**—a document that grants a couple permission to marry—at any county clerk office in the state in which they want to be married. After the wedding ceremony, the person who performs the ceremony can file a **marriage certificate**—a document that proves that a couple is married—in the county office within a few days. It is signed by the married couple, the person who officiates the wedding, and two witnesses. Since June 26, 2015, same-sex marriage has been established in all 50 states. Same-sex couples have the right to marry on the same terms and conditions as opposite-sex couples.

In the United States, people have the freedom to choose whoever they want to marry. However, a couple who is eligible for marriage needs to be eighteen. Every state allows exceptions to their age of marriage. A judge might consent to an underage marriage for various reasons, such as if a female is pregnant, but this often requires proof that the couple can financially support their family in accordance with law. People who are already married cannot get married again until they divorce their former spouse. Proof of divorce or death of a previous spouse is required to show termination of any prior marriage.

A couple must have the mental capacity to enter into a marriage contract. If a partner does not understand what it means to be married because of mental illness, drug or alcohol use, or other issues that affect judgment, then the person lacks the mental capacity to consent to the marriage.

Marriage requirements might also include blood tests, residency requirements, and more. The couple makes sure that both have fulfilled all marriage requirements.

Costs of Weddings

Weddings in the United States are a $70 billion a year industry and are therefore a significant component of the American economy (Howard, 2008). The wedding industry began to take off in the United States between the 1920s and the 1950s (Howard, 2008), promoting goods such as diamond rings and wedding fashions as well as new services such as gift registries and wedding packages that standardized wedding ceremonies and receptions. Wedding costs may include those for the couple's attire, the venue, the photographer/videographer, flowers, and catering. In 2017, the average cost of a U.S. wedding was $25,764 (The Wedding Report, 2018).

Marriage Ceremony

A marriage ceremony is known as a **wedding** in which two partners are united in marriage. A wedding day is one of the most memorable events for the **bride** (a woman) and **groom** (a man), or just before and after the event. The wedding ceremony symbolizes that the couple is making a public promise to each other and formalizing their relationship in marriage.

Most states have legal requirements pertaining to the marriage ceremony, including who may perform and officiate the marriage ceremony and whether any witness to the ceremony is required. Civil unions are performed by a judge or a court clerk. Religious wedding ceremonies are conducted by a priest or minister. Wedding ceremonies may take place in a religious building such as a church or outdoors or in a county office or other location.

Traditional Wedding Rituals

A number of wedding rituals are prevalent in the United States. A **white wedding** is a traditional formal Western wedding in which the bride wears a white wedding dress. The color white became popular for brides after Queen Victoria wore a white gown when she married Prince Albert in 1840. By the 1950s, the wedding industry had made the formal white wedding a part of the modern American Dream (Howard, 2008).

Traditionally, the couple exchanges wedding rings of valuable metals like platinum, gold, or silver to symbolize their love and commitment to one another. Women may receive two rings, one upon engagement and another at the wedding ceremony. Weddings often incorporate flowers such as red roses to symbolize everlasting love. It is customary for the father of the bride to walk the bride down the aisle to the groom. During the ceremony, the bride and groom vow their love and commitment for one another, exchange rings, and become husband and wife. The newlyweds often share a kiss to seal their union.

Those with close relationships to the couple are often selected to be part of the wedding party as *bridesmaids* and *groomsmen*. A *maid of honor* and a *best man* are the principal bridesmaid and groomsman, and deliver a toast at the reception that follows

Diversity Overview: Ethnic Wedding Customs

Many Americans may consider incorporating *ethnic wedding traditions* or customs into their wedding ceremony. Let's examine some customs that stem from the diverse cultures that make up the United States.

African Wedding

Because a wedding in African culture involves a union between two families, many traditional African American wedding rituals center on family. For example, *knocking* was a traditional ritual in which a potential groom visits the house of a potential bride alongside his elder male relatives and requests the bride's family's permission to marry her. *Jumping the broom* is another marriage ritual that originated in Africa and symbolized the beginning of making a home together and "sweeping away" the past. Jumping the broom was used as a marriage ritual for slaves in the Southern United States in the 1840s and 1850s. Its revival in twentieth-century African American culture is due to the publication of the novel *Roots* (Parry, 2011).

Jewish Wedding

A **Jewish wedding** is a wedding ceremony that follows Jewish law and traditions. When a Jewish couple gets engaged, families traditionally announce the engagement and the wedding date at an engagement party. In Jewish tradition, the two ceremonies of engagement and marriage usually take place up to a year apart. The *ketubah* is a Jewish wedding contract that testifies that the husband guarantees to his wife that he will meet certain minimum human and financial conditions of marriage. The wife agrees only to accept the husband's proposal of marriage and it is not a mutual agreement (Lamm, 2017). The origins of European engagement are found in Jewish law, where marrying without such an agreement is considered immoral (Kaplan, 1983). Traditionally, both the bride and the groom walk down the aisle with both of their parents. A *huppah* is used as a wedding canopy that the bride and groom stand beneath during the ceremony. Breaking a glass at Jewish weddings can symbolize that the smashing of the glass is irrevocable, as is marriage.

Chinese Wedding

In Chinese culture, an auspicious date is selected to propose marriage (meaning 提亲 in Chinese or tí qīn in Pinyin), where both families meet to discuss the wedding and the groom's family presents gifts such as jewelry and bride price (China Bridal, 2003). The couple is considered officially engaged. It is popular to combine traditional Chinese wedding rituals with a white wedding. The wedding date is selected carefully as odd dates are considered unlucky. The bride often changes outfits throughout the wedding ceremony and reception, while the groom often wears a suit and a traditional Chinese costume. When it comes to the bride's dress during the reception, red is the color of choice as it represents good luck and is believed to ward off evil spirits. During a traditional Chinese wedding, a *tea ceremony* takes place. The newlyweds kneel down and present a cup of tea, bowing to Heaven and Earth, to the groom's parents and ancestors and finally to each other. The bride serves tea to the guests in order of seniority. The parents and guests pass on monetary gifts wrapped in red envelopes to the newlyweds to welcome them to the family or to show respect for their parents.

Native American Wedding

Native American families traditionally believe in the Great Spirit that is manifested in natural elements such as the sun, water, and fire. The sun is considered the most powerful element. Native Americans therefore traditionally hold wedding ceremonies at dusk, feasting, dancing, and celebrating until the sun rises the next day. Because Native Americans believe that water is Mother Earth's blood and is used as a symbol of purification and cleansing, the bride and groom engage in a ceremonial washing of hands to wash away past wrongdoings and old memories. Fire is considered to be a sacred element for some tribes. Therefore, a fire circle is created using stones and types of wood. Two small fires represent the bride and groom's individual lives. The bride and groom each offer a prayer and then push the two small fires into the large stack of wood in the center which catches fire. All sing praises to the Creator as the two lives are merged into one holy union (Manataka American Indian Council, 2017).

the wedding. During the wedding reception, other rituals often take place, including the practice of the bride throwing a bouquet behind her head to the single women attending the wedding and the groom throwing a garter to the single men in attendance. The idea is that whoever catches the bouquet or garter will marry next. Other rituals include the mother–son dance, the father–daughter dance, and the cutting of the cake. When the newlywed leaves the reception, the family and friends throw rice or blow bubbles at the married couple, symbolizing fertility. After the wedding, the newlyweds may go on a honeymoon to celebrate their marriage before settling down to daily life.

Marital Relationships throughout the Life Cycle

National Spouses Day is observed annually on January 26. The unofficial holiday encourages people to let their spouses know how much they are appreciated. **Spouse** refers to a person married to and living with a householder who is of the opposite sex of the householder (U.S. Census Bureau, 2018). Marriage marks a new stage because both spouses feel secure in the relationship after taking a binding and permanent vow and adjust to one another as their love demands (Sutherland, 2014). The **family life cycle perspective** addresses the events related to their ongoing structural entrances and exits using a framework of family developmental transitional periods (Carter & McGoldrick, 1989). A **transition** is defined as the passage from one ending state to another beginning state (Bridges, 1980). The family life cycle transitions such as marriage, parenting, and retirement are anticipated but couples and family members still experience a great amount of stress and difficulty in managing these transitions (Carter & McGoldrick, 1989).

Studies have shown that marital relationship quality is associated with marital adjustment (Madahia *et al.*, 2013) since it is an aspect of the relationship between spouses. **Marital adjustment** is the accommodation of husband and wife to each other at a given time (Locke & Wallace, 1959). Well-adjusted couples are expected to have long-lasting,

Upon completion of this section, students should be able to:
LO6: Describe marital adjustment in young adulthood.
LO7: Explain how having children affects a marriage.
LO8: Describe marital relationships in middle adulthood.
LO9: Explain marital relationships in late adulthood.

stable marriages, whereas poorly adjusted and maladjusted marriages are expected to experience instability and/or to end in divorce (Kendrick & Drentea, 2016) and unhappily married couples were distinguished by their failure to manage conflict and initiate repair activities (Mace, 1989). Factors contributing to marital adjustment include marital satisfaction, cohesion, affection, and marital conflict (Kendrick & Drentea, 2016). Satisfaction involves three levels: (a) the satisfaction with one's spouse, (b) satisfaction with family life, (c) general satisfaction with life (Garcia, 1999).

Marital Relationships in Young Adulthood

Young adulthood is the period of adulthood encompassing the ages of 18 to 39. After the wedding ceremony is over, a period of marital adjustment occurs in which each partner makes accommodations for the other, and spousal lives become integrated. If marital adjustment takes place successfully, the two personalities are not merely merged but interact to complement each other to achieve mutual satisfaction and common objectives (Burgess & Cottrell, 1939). In fact, newlywed couples are likely to experience the greatest level of marital satisfaction.

Marital Intimacy

Newlyweds must place a higher priority on the marital relationship and differentiate it from some of the close attachments they may have formed with parents, children, siblings, and relatives (Mace, 1989). For example, establishing new couple boundaries between the couple and their families is a critical task during the transition to marriage (Morris, 1999). **Marital intimacy** is a form of intimacy that involves feelings of closeness in relationships with spouses. Spanier (1979) found a positive correlation between intimacy and marital satisfaction. Some characteristics of marital intimacy are commitment, purpose, communication, loyalty, faithfulness, safety, caring, and mindfulness/empathy (Wachs & Cordova, 2007). **Mindfulness** is a process of openly attending, with awareness, to one's present moment experience (Creswell, 2017). The capacity to be mindful is associated with higher well-being in daily life (Brown & Ryan, 2003), while mind-wandering predicts subsequent unhappiness (Killingsworth & Gilbert, 2010). Mindfulness techniques such as positive emotional exchanges have a positive impact on overall marital satisfaction (Burpee & Langer, 2005) and have been shown to be particularly beneficial for regulating emotions (Dubert *et al.*, 2016).

Another important expectation of marriage is that married partners will meet each other's need for love, intimacy, and affection (Morris, 1999). Failure to fulfill one's partner's expectations about intimacy in the relationship predicts marital dissatisfaction (Kelley & Burgoon, 1991). Romance is the most important ingredient in the newlywed relationship and is kept alive by frequent interactions, spending time together, and openly disclosing one's thoughts and feelings (Gottman, 1995). A study of 610 newly married couples revealed that couples who spend more time together reported higher levels of marital satisfaction than those who don't spend as much time together (Johnson & Anderson, 2013). A study of 193 couples revealed that a partner's behaviors such as attachment security could enhance the marital relationship (Feeney, 2002). Attachment avoidance was a characteristic of distressed couples (Mondor *et al.*, 2011).

Marital Role and Adjustment

After tying the knot, couples should continue to negotiate practical matters like work and household chores as necessary (Sutherland, 2014). **Role theory** suggests that our everyday activities consist of acting out socially defined roles associated with a given status and it allows us to understand the forms of adjustment that newlywed couples in their marital roles need to make relative to each other.

The forming of a new family constitutes a major change in the behavior patterns of the young adults since they play an entirely new role of husband and wife (Dyer, 1962). The category of "husband" or "wife" includes people in formal marriages as well as in common-law marriages (marriages without marriage license) (U.S. Census Bureau, 2018). Traditionally, wives were expected to tend to family responsibilities while husbands were expected to provide for the family. Today, although traditional gender roles have changed, Americans still place a higher value on a man's role as financial provider (Parker & Stepler, 2017b). While forming a new family, couples may experience differences in needs and values over issues such as family leadership, loyalty, money, power, sex, privacy, and children (Holman & Li, 1997). For example, women's quality of life is affected by family functions and marital adjustment in family (Basharpoor & Sheykholeslami, 2015) if women take the sole household responsibilities. If the expected roles are defined before a marriage, marital relationships and adjustment tend to function well.

Are there any gender role differences in marital relationships between family of origin and marital adjustment? Newlywed couples are learning how to perform marital roles, using the skills they learned throughout childhood and adolescence. What they have observed constitutes their basic attitudes and behaviors about family life. If the newlywed couple has been raised in the same culture or race, they would share the same norms and reported higher levels of marriage satisfaction in the areas of personality issues, communication, conflict resolution, leisure activities, spousal role consensus, and personal habit tolerance (Schwartz, 1994). Similarly, if couples were exposed to the ideological representations about family roles in post-modern society, family of origin had no significant effect upon marital adjustment for either women or men (Muraru & Turliuc, 2013). The only difference between traditional and nontraditional couples was in family management (Asoodeh *et al.*, 2010). For example, both husband and wife in the traditional marriage believed that the man should make the financial decisions. However, when the husband and wife are directed by non-shared normative patterns (Dyer, 1962), both partners must learn to adapt to challenges in their new family of two. Intercultural couples face similar difficulties in their marital relationships. Cultural differences in acculturation, language and communication, gender roles, and childrearing may potentially amplify the difficulties (Skowroński *et al.*, 2014). Since conflicts and disagreements may arise out of behavior directed by unmet expectations or non-shared norms, adjustment of these disparate normative orientations is important in marital relationships (Dyer, 1962).

Marital Communication and Adjustment

Marital communication is a process in which couples can interact in their thoughts and feelings in verbal and nonverbal forms (Abbasi & Afsharinia, 2015). Three communication patterns (*mutual constructive communication, demand–withdraw, mutual avoidance*) affect marital satisfaction. **Mutual constructive communication** refers to mutual expression and discussion of feelings about the problem and problem resolution. The couples "see with other eyes, hear with other ears and feel with other heart" (Ansbacher & Ansbacher, 1956). Mutual constructive communication proved to be the first factor predicting marital satisfaction (Amiri *et al.*, 2011), led to less distress and more marital satisfaction (Manne *et al.*, 2006), correlated with sexual health promotion and HIV prevention among male couples (Mitchell & Gamarel, 2018), and had the highest correlation with marital satisfaction in both women and men (Abbasi & Afsharinia, 2015).

Demand–withdraw communication occurs when one partner attempts to discuss a problem while the other avoids the issue or ends the discussion (Christensen, 1988). For many possible reasons, wives demand due to a depressive belief that change will only happen through nagging or shouting, or wives withdraw due to a depressive belief that change is not possible (Byrne *et al.*, 2004). The demand–

withdraw pattern ranked among the most destructive and least effective interaction patterns in couples' problem-solving communication repertoires (Heavey *et al.*, 1993), correlated with marital relationship dysfunction (Eldridge *et al.*, 2007), led to marital dissatisfaction (Madahia *et al.*, 2013), and lowered marital satisfaction (Manne *et al.*, 2006). A study of 116 U.S. couples revealed that both husband demand–wife withdraw and wife demand–husband withdraw predicted negative emotions, lower levels of conflict resolution, and spousal depression (Papp *et al.*, 2009).

Mutual avoidance occurs when a couple avoids discussing problems and stressors. In a study of 147 patients and 127 partners during cancer treatment, Manne *et al.* (2006) found that mutual avoidance was associated with more distress and marital dissatisfaction for both patient and partner. Abbasi and Afsharinia (2015) also reported a negative correlation between marital satisfaction and mutual avoidance, suggesting that couples change unhealthy communication styles through education, counseling, and other therapeutic interventions.

Marital Conflict and Adjustment

Becoming a couple is one of the most complex and difficult transitions of the family life cycle as marital conflict is present in every relationship. **Marital conflict** is the pursuit of incompatible interest and goals by two partners (Best, 2006). Conflict is characterized in most cases by quarrels, fighting, severe anger, aggression, violence, bitterness, and hatred and can weaken the spousal relationship (Igbo *et al.*, 2015).

Marital conflict is a primary source of marital distress between husband and wife due to factors such as unmet expectations. For example, disagreement over the division of household responsibilities has emerged as a major cause of trouble in U.S. marriages (Berk, 1985). Marital conflict is also associated with increased risk for a major depressive episode (Whisman & Bruce, 1999), abuse of partners (O'Leary & Cano, 2001), and alcohol problems (Murphy & O'Farrell, 1994). Four behaviors that are most corrosive to a happy and successful marriage are *criticism, contempt, defensiveness,* and *stonewalling*. **Criticism** "is any statement that implies that there is something globally wrong with one's partner"; **defensiveness** is "any attempt to defend oneself from a perceived attack"; **contempt** is "any statement or nonverbal behavior that puts oneself on a higher plane than one's partner"; **stonewalling** "occurs when the listener withdraws from the conversation" (Gottman & Silver, 1999). When these behaviors and reactions irritate and conflict with or otherwise disturb the marital partner, some adjustment is necessary (Dyer, 1962).

Marital conflict resolution is an act of finding a solution to a conflict (Burton, 1990) and to a disagreement whereby the two parties to a conflict are mutually satisfied with the outcome of a settlement (Igbo *et al.*, 2015). **Marital conflict resolution strategies** include avoiding the conflict, giving in, standing your ground, compromise, collaboration, mediation, and voting. These conflict resolution strategies can be affected by gender and personality traits (Igbo *et al.*, 2015). For example, females are more likely to use solution-oriented strategies while males indicate use of control variables. Interestingly, Driver and Gottman (2004) reported that the ability to use humor and affection in conflict discussions led to higher marital satisfaction than if there was no humor or affection during conflict. Also, married couples who are flexible are more likely to make adjustments in their marital relationship and maintain marital happiness. **Couple flexibility** refers to the amount of change in marital relationships that occurs. Fincham *et al.* (2007) suggest forgiveness and sacrifice as transformative processes in marital research. Couples who manage to resolve marital conflict in a mutually satisfying way become happier with their marriages.

Marital Relationships and Parenthood

Many newlywed couples choose to have children. When their first child comes into the world, the couple must learn how to adapt to the challenges that a newborn poses. The emotional maturity of the couple, the stability of their relationship, and the couple's financial stability are some factors that affect how well the couple weathers the storms of parenthood.

Parental Roles and Marital Adjustment

In addition to the marital roles of the new husband and wife, the situation of adjustment is further complicated if the couple starts to play parental roles. Marital satisfaction often declines following the birth of children. There were significant statistical differences between existence of children and marital satisfaction (Omran *et al.*, 2016). Sometimes disagreement over childcare or family matters can lead to frequent or less interaction. Chores, division of labor, and finances are areas of potential spousal disagreement during the transition to parenthood that may affect the marital relationship (Belsky & Kelly, 1994). The lack of complete agreement over important issues like raising children and relationship with relatives in the middle of married life will lead to marital dissatisfaction (Abbasi *et al.*, 2010). The amount of communication between the couple decreases during the transition to parenthood, with reduced communication associated with decreased marital satisfaction (Cowan & Cowan, 1988). **Marital distress** occurs when a married or unmarried couple experiences emotional or physical conflicts that threaten to end the relationship. Marital distress can lead to higher levels of depression and anxiety, can negatively affect children of the union, and can lead to negative outcomes later in life (Bradbury *et al.*, 2000).

How could the couple perform the parental role and maintain a good marital relationship? **Dramaturgical analysis** (Goffman, 1959) is the study of social interaction that compares everyday life to a theatrical presentation. As on the stage in the theater, the couples live their lives by talking about love verbally and nonverbally and playing a parental role to give an impression to each other. Goffman described their performance as *the presentation of self*—an individual's effort to create favorable impressions in the minds of others. This analysis allows a couple to look at the parental role they play. Our selves are made up of all of the different roles we perform in our daily lives (Fowler, 2007). Sociologists differentiate between *role expectations*—the expectations required of a role—and *role performance*—the actual delivery of those expectations. Role performance is the heart of marital or parental interactions. For example, the role expectation attached to parental status is that both parents care and provide for their child. The role performance would be their actual day-to-day parenting activities. Usually, the feeling of emotional stability occurs when a couple can significantly meet the needs and expectations of their spouse in the marital relationship (Khalatbari *et al.*, 2013). Otherwise, disagreement over the division of household responsibilities between the husband and wife could emerge as a major cause of marital conflict. For example, female labor force participation not only improves family economic well-being (White & Rogers, 2000) but also increases work–family conflict. Husbands' lack of participation in housework when wives are employed full time (Bianchi *et al.*, 2006) is associated with tension, conflict, and poor marital quality (Frisco & Williams, 2003). When dissatisfactions come around, the couple tends to blame one another for failing to meet the expectations. Thus, inaccurate or unmet expectations and role performance lead to adjustment problems.

Predictors of marital adjustment include length of time married, and number of children (Kendrick & Drentea, 2016). Many couples are dissatisfied with the balance of attention given to their marriage versus the attention given to the newborn child (Olson *et al.*, 2008). This may be related to expectations the couple have about having a child and the roles they play in the infant's care (Kach & McGhee, 1982). Lawrence *et al.* (2008b) suggested that family planning and pre-pregnancy marital satisfaction generally protect marriages from these declines. In families with many children, husbands' participation in housework and joint decision-making are required (Twenge *et al.*, 2003). If couples share equally in parenting duties, they can keep it from interfering with marital satisfaction (Belsky & Kelly, 1994). Thus, there is a parallel between marital and parent–child relationships.

Marital Equality

Equality in a marriage emerges out of the interactions of both partners. **Equity theory** is based on the idea that individuals are motivated by fairness. Fairness is important in intimate relationships and spouses in equitable relationships—which may be characterized by joint decision-making—have been found to be more satisfied (Walster *et al.*, 1978). Harris (2006) supported marital equality after examining the stories that people told about their equal and unequal marriages. **Marital equality** refers to an even distribution of domestic, paid work, and/or childcare, equal decision-making and entitlement to express and meet needs, mutual reciprocation of attention and accommodation to each other, and burden-sharing (MFRI, 2018). Couples who believed in lifelong marriage, and shared decision-making and housework were likely to have a happy and low-conflict marriage (Kamp Dush & Taylor, 2012).

The husband's provision of support was found to predict marital satisfaction. Providing invisible and visible support regarding division of labor can lead to healthy marital adjustment. **Invisible support** originates outside the recipient's awareness. For example, one partner may make the other breakfast without being asked to do so. A study of 275 married couples revealed that husbands should try to provide more support without being solicited (Lawrence *et al.*, 2008a). **Visible support** occurs when both partners are aware of the supportive actions. For example, it may occur if one partner performs a task, such as changing a diaper, when the other partner requests it. Lawrence *et al.* (2008a) suggested that wives should try to solicit more support in order to try to increase marital satisfaction. Couples who work to support each other and grow together, cope with stress, deal with family hardships, and work on their relationship problems are more likely to have a successful marriage (Vanover, 2016). Couples who had a long-term supportive relationship perceived better relationship quality (Zhou *et al.*, 2017a).

Marital Relationships in Middle Adulthood

Middle adulthood is the period of adulthood that spans ages 40 to 64. In terms of marriage, research results indicated a U-shaped pattern of marital happiness over the life course, with marital happiness being high in the beginning of marriage, declining in middle adulthood, and then rising again in later years. In a study of 1,988 couples, Kamp Dush and Taylor (2012) reported that the overall mean trajectory was a slightly U-shaped curve with stable high, middle, and low marital happiness trajectories across 20 years (see Figure 9.10).

Stressful Life Events and Marital Adjustment

From a life-cycle perspective, couples strive for stability but must adapt to stressful life events (Wethington, 2005). A study of 1,038 participants revealed that 885 (93 percent) experienced a stressful event in the last ten years: 390 (44 percent) about the death of a loved one, 122 (14 percent) about their own physical health, 94 (11 percent) about a close other's physical health, 56 (6 percent) about problems with a child, 54 (6 percent) about separation/divorce, 54 (6 percent) about stress over employment/finances, 37 (4 percent) about accidents/natural disasters, and 78 (9 percent) about other stressful events (Sutin *et al.*, 2010). These stressful events may be turning points associated with a major life change. For example, a heart attack or widowhood may be a turning point because it is associated with declines in physical health or increased distress. Turning points are often triggered by a severe event or long-term difficulty (Wethington *et al.*, 2004), and when resolved can lead to lower distress and

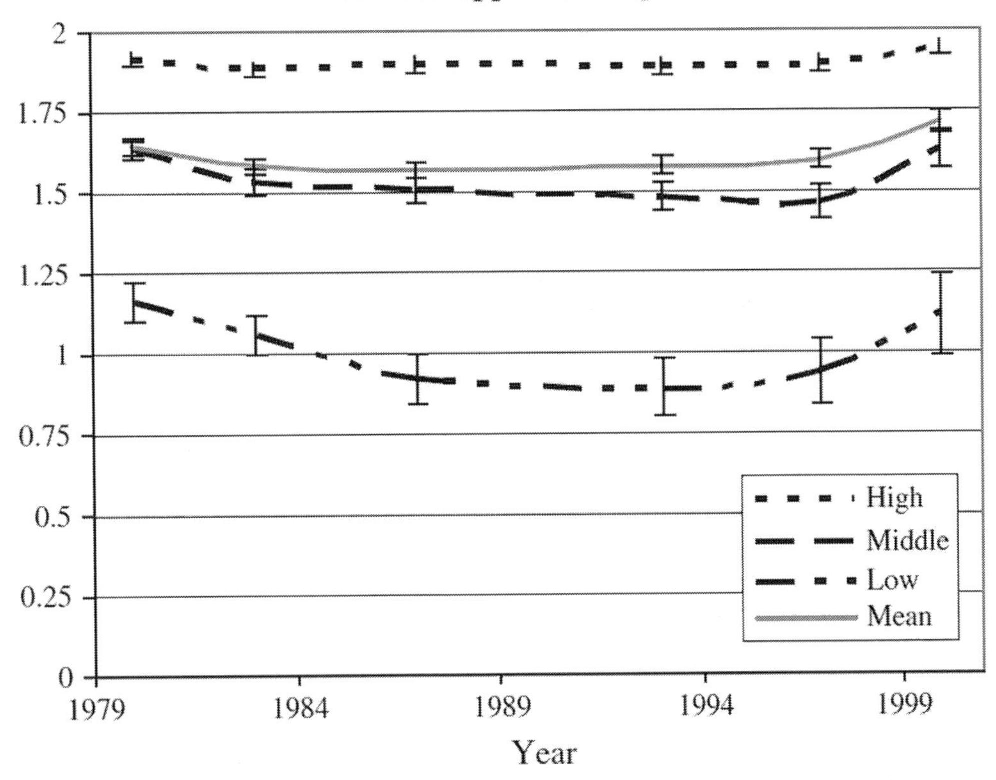

Figure 9.10 Marital Happiness Trajectories, 1980–2000
Source: Kamp Dush & Taylor (2012). https://www.ncbi.nlm.nih.gov/pmc/articles/PMC3274734/figure/F1/

greater psychological growth (Wethington, 2003). Those who can step back from the stressful event, evaluate their actions, and learn from the experience tend to be better adjusted than those who have difficulty evaluating the event in such a way (Blagov & Singer, 2004).

Why did marital happiness decline in middle age? In part, the decline may be related to the conflicting demands placed on the couple to support both a family and their career (Van Landeghem, 2012). **Sandwich generation** refers to adults who care for an aging parent ages 65 and older as well as raising a child under age eighteen or supporting a grown child. One in seven middle-aged adults (15 percent) are providing financial support to both an aging parent and a child; 36 percent of married adults are sandwiched between their parents and their children, compared with 13 percent of those who are unmarried (Parker & Patten, 2013).

Stressful events are related to physical or psychological health (Sutin *et al.*, 2010). As individuals age, changes in health status may be considered a life transition (Liang *et al.*, 2008).

For women, **menopause** is a process of a natural declines in reproductive hormones experienced in their 40s or 50s. Psychological symptoms include anxiety and less interest in sexual activity, while physical symptoms include lack of energy and hot flashes. These mental and physical changes may affect marital satisfaction. Men also experience a decline in the production of the male hormone testosterone with aging. Along with the decline in testosterone, some men experience symptoms that include fatigue, weakness, depression, and sexual problems (WebDC, 2017). As such, these health declines may have an effect on construals of stressful events, such as the designation of a turning point (Sutin *et al.*, 2010). Middle-aged adults experience signs of aging, which may in turn affect their level of marital satisfaction. Figure 9.11 shows a high marital conflict group (23 percent of respondents) where marital conflict had a slight upside-down U-shape, gradually increasing across the first eight years and decreasing across the final twelve years (Kamp Dush & Taylor, 2012).

When individuals experience more stress on a given day, they are more likely to describe their interactions with their partners as negative (Bolger *et al.*, 1989). The **Vulnerability-Stress-Adaptation Model** (VSA) posits that couples who have few enduring vulnerabilities, encounter few stressors, and employ effective adaptive processes are likely to experience high marital quality and stability. Employing the VAS model in a study of 103 couples over the first three years of marriage, Langer *et al.* (2008) found that personality traits of husbands predicted both the husband and wife's physical aggression and stress level. Edwards *et al.* (2018) found that individual attachment and perceived relational equity influenced

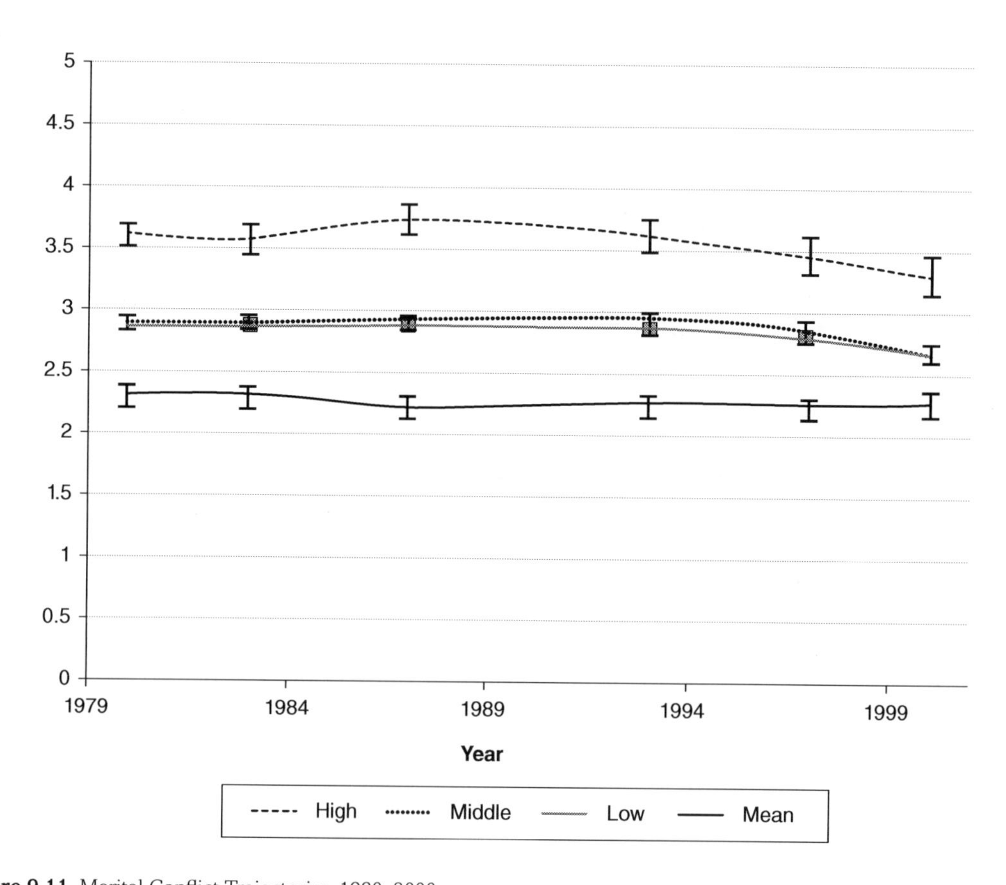

Figure 9.11 Marital Conflict Trajectories, 1980–2000

Source: Kamp Dush & Taylor (2012) and used by permission of the publisher, SAGE Publications. https://www.ncbi.nlm.nih.gov/pmc/articles/PMC3274734/figure/F2/

how the couples used forgiveness strategies to adapt to a relational transgression in marital relationships.

Another change in the family that may affect marital happiness occurs when the couple's children leave home to attend college or take a job elsewhere. An **empty nest** is a parental home that grown children have left to begin their adult lives. **Empty nest syndrome** is a phenomenon in which parents experience feelings of sadness and loss when the last child leaves home. However, not all parents experience a sense of loss when their children leave the home, and some may even see their marital happiness increase once they are no longer caring for their children. When children become separated and independent of family, marital satisfaction is more than when they live with their family (Abbasi *et al.*, 2010). The empty nest can be a time to renew the marital relationship. Despite the potential for possible marital satisfaction decline in middle adulthood, couples in middle age may see their marital satisfaction increase after their finances have stabilized and parenting responsibilities have ended (Sasser, 2005).

Spiritual Intimacy

Predictors of marital adjustment include the way that couples adapt to stressful events in determining marital satisfaction. A shared religion is seen as an important way to a successful marriage. **Spiritual intimacy** is a form of intimacy that reveals your spirituality to your partner (spiritual disclosure) and listening to your partner's disclosures in a supportive (spiritual support) manner (BGSU, 2017). The perception of spiritual intimacy has been positively linked to marital satisfaction through a path of emotional intimacy (Hatch *et al.*, 1986). A study of 63 older married couples revealed that wives' greater appreciation of the sacred qualities of marriage was positively linked to marital satisfaction (Sabey *et al.*, 2014). Some 47 percent of U.S. adults say that shared religious beliefs are "very important" for a successful marriage (Masci, 2016). Why is it so?

Spiritual intimacy is about learning how to connect with your partner through your faith. Indeed, your relationship with God has a positive effect on your marital intimacy and well-being (Holland *et al.*, 2017). Spouses with a strong religious orientation tend to espouse the idea that marriage is forever and report more marital happiness than those with a weak religious orientation (Heaton & Pratt, 1990). Successfully married couples trust and consult each other, believe in God, make decisions together, are committed to each other, and have a friendly relationship (Asoodeh *et al.*, 2010). The best predictor of perceived intimacy with God for young people is the use of prayers of praise, whereas for older people it is prayers of thanksgiving that become more selfless and frequent as people age (Hayward & Kraus, 2013).

Marital Relationships in Late Adulthood

Late adulthood is the period of adulthood beginning at age 65. Older adults are a growing population segment in the United States as the country is experiencing **population aging**—an increase in the median age of the population due to rising life expectancy. In terms of marriage, approximately 26 percent of older adults ages 70 and older have been married for at least 50 years; 24 percent have marriages of shorter durations; and 49 percent are currently unmarried (Wu & Brown, 2016). Approximately 54 percent of older Whites have a marital duration of at least 50 years, compared with 45 percent of older Hispanics, 40 percent of older African Americans, and 38 percent of older Asian Americans. Pursuing higher education sometimes delays the age at marriage, leading to a shorter duration of marriage. This is reflected in marital statistics, with more older married couples with a high school diploma/GED (57 percent) having a marital duration of at least 50 years, compared to those with a bachelor's degree (45 percent). The state with the highest proportion of older adults married at least 50 years is Nebraska with 65.3 percent, while the state with the lowest proportion is Alaska with 35.8 percent (Wu & Brown, 2016). Length of marriage was found to significantly influence satisfaction in first marriages (Mirecki *et al.*, 2013). Although some older adults face physical and social issues as they advance in age, including potential declines in physical and cognitive abilities, loneliness following retirement, and the effects of ageism, meaningful activities can lead to adjustment in late adulthood and can help sustain satisfaction in a marriage.

Adjustment to Retirement

One theory of successful aging is activity theory. Based on the "use it or lose it" assumption, **activity theory** suggests that no matter what age, people who remain active physically, mentally, and socially adjust better to the aging process. Ekerdt (1986) suggested that activity enables older adults to adjust to retirement. For example, older adults who face role loss (i.e., employment) can substitute former roles with alternatives. This may take the form of an older adult replacing a former professional role with volunteer work after retirement. Activity theory suggests that older adults who socialize with others increase their feelings of self-worth and better adjust to late adulthood. This in turn may increase marital satisfaction. It can also help older adults who experience the death of their spouse. Satisfaction in old age depended on active maintenance of personal relationships and endeavors (Bengtson & Norella, 2009). Activity theory rose in opposing response to *disengagement theory*—claiming that it is natural and acceptable for older adults to withdraw from society

Self-disclosure

For marital couples in close relationships, there exists a critical need to manage shared private information because it plays a functional role in the relationship (Derlega, 1984). **Self-disclosure** was defined as telling others about oneself. Research has explored the relationship between *self-disclosure output* (i.e., disclosure to spouse) and marital satisfaction (Webb, 1972) and between *self-disclosure input* (i.e., disclosure received from spouse) and marital satisfaction. Couples with discrepancies in amount of disclosure output reported less marital satisfaction than couples who reported bilaterally low or bilaterally high amounts of disclosure output (Davidson, 1981). Husbands' disclosure to wives was positively related to husbands' marital satisfaction and, similarly, wives' disclosure to husbands was positively related to wives' marital satisfaction (Hansen & Schuldt, 1984).

The decision-making behind the act of disclosing private information can be a complicated process. For example, soliciting disclosive information about a partner's health—such as asking about his or her level of pain—can actually increase the severity of pain a partner feels (Cutrona, 1996). That is, when disclosing about their discomfort, the sick spouse pays more attention to the chronic pain. West (2017) suggested that self-disclosure may be used to disconfirm the patient's negative sense of self or to induce a positive sense of self.

Marital Relationships from a Sociological Standpoint

Marital satisfaction is gaining increasing concern in modern society (Li & Fung, 2011). Marriage relationships require ongoing repair and maintenance. Let's look at explanations of marital relationships from social exchange theory and resource theory.

Upon completion of this section, students should be able to:
LO10: Describe how social exchange theory explains marital relationships.
LO11: Apply resource theory to the analysis of marital power.

Social Exchange Theory and Marital Relationships

Social exchange theory suggests that a marital relationship is the result of a benefit and cost exchange process between two partners. *Benefits* are reward elements of a relationship that include love, socioeconomic status, and sexual compatibility. *Costs* are the expenditure elements of a relationship that include effort, labor, and sacrifice. Social exchange theory argues that couples calculate the worth of a relationship by subtracting its costs from the rewards it provides. A good marriage depends on how couples are satisfied with their share of the rewards and costs. All relationships are based on give and take but the balance of this exchange is not always equal.

The intervening mechanisms in social exchange are the norms of fairness. Indeed, marital couples who receive favorable reward/cost outcomes from each other—the proportion of rewards and costs are distributed in a fair ratio to each other—are more likely to be satisfied with their marriage (Homans, 1974). In successful marriages, the ratio of positive to negative behaviors is approximately 5:1, whereas in unsuccessful marriages the ratio is 1:1 or less (Gottman & Levenson, 1992). For example, if a spouse plays multiple roles (i.e., wage earner, homemaker, caregiver, etc.) without support from the partner, the spouse is less likely to be happy and more likely to terminate the relationship.

Marital Cohesion

Social exchange theory serves to establish bonds between partners (Nakonezny & Denton, 2008). **Marital cohesion** refers to the degree of connectedness, togetherness, and emotional bonding between marital partners (Doane, 2016) and is related to positive marital relationship between partners. Marital cohesion of a social bond involves the dyadic adjustment process. For example, if the loving behavior that someone is expressing toward her fiancé is reciprocated through his loving behavior towards her, their social bond is likely to be stronger and they are likely to have marital satisfaction. However, extremely high levels of cohesion may be unhealthy for marriages, reflecting too much dependence between partners (Nakonezny & Denton, 2008). Decreased relational solidarity results in a "fragmented marriage in which individuals are cut loose from their marital moorings" (Durkheim, 1893/1933). Stoker (2004) found that marital cohesion and adaptability enhanced marital satisfaction in couples with a child with cerebral palsy (i.e., permanent movement disorders in early childhood).

Social exchange theory is a powerful tool used to examine the marital relationship on a reward and cost basis. However, as the calculation of rewards and costs often occurs on a post hoc basis, the partners may not know their prevailing profit status until they reflect back on previous behaviors (Bem, 1972). Moreover, social exchange theory ignores the means in which partners use communication to define rewards and punishments within the marital relationship (Nakonezny & Denton, 2008).

Resource Theory and Marital Relationships

Marital inequality and the distribution of power within the relationship have long been a concern within sociology of the family. For example, ideas about historic shifts in the dominance of husbands/fathers within families have vied with feminist-inspired views of the continuing significance of patriarchal control in both public and private spheres (Allan, 2007).

Marital Power

Resource theory is a theoretical framework for the study of family power. **Resources** refer to material goods (e.g., money and physical possessions) and nonmaterial goods (i.e., love, status, knowledge, and power) that can be accessed and used in social actions. Resources are viewed as "ability, possession, or other attribute of a partner giving the capacity to reward or punish another partner" (Emerson, 1976). The resources that spouses administer to each other are a principal source of marital power. **Marital power** is defined as a partner's ability to have his/her will prevail in the marriage relationship (Treas & Kim, 2016) and is based on a spouse's ability to reward his or her partner (Nakonezny & Denton, 2008). For example, in dual-earner marriages, women are likely to reward their husbands if they take sole household responsibilities. *Structural functionalist theory* emphasizes cooperation. *Conflict theory* regards power as creating and perpetuating marital inequality. The resource differential, thus, produces relationship asymmetry. The asymmetrical nature of the relationship allows for the emergence of exploitation in the marital relationship (Blau, 1964). Such exploitation is often used in dysfunctional couples to maintain the power imbalance between partners and to increase the bargaining power of one partner relative to the other partner (Nakonezny & Denton, 2008). With the increase in women's participation in employment and men's growing contribution to housework, there is support for *bargaining theory*—partners' relative resources determining their power relations (Treas & Kim, 2016)—and *marital equity*—the degree of balance of authority, power, or influence between spouses.

Marital power is apportioned and based on the relative resources (i.e., income, occupational and educational attainment) that each partner contributes to the family. In a study of White middle-class wives in Detroit, Michigan, Blood and Wolfe (1960) reported that the greater the men's resources, the greater the men's perceived power within the family. Marital power is also measured as decision-making outcomes. Husbands who contribute more resources such as income and educational attainment to the family are more likely to take the leadership role at home and have the final say. A further source of marital power is derived from the level of commitment to the relationship or to the extent that the self is identified with the marital relationship (Nakonezny & Denton, 2008). For example, if a wife's level of commitment increases relative to her husband, it means that the husband gains power and status relative to his wife. Thus, resource theory provides a valuable lens for understanding and addressing social class, inequality, and injustice (Törnblom & Kazemi, 2012).

Global Overview: Collective Marriages

Palm Bean County Clerk Sharon Bock officiates a free group wedding every year at the National Croquet Center in West Palm Beach, Florida (Jeffries, 2018). A **collective wedding** is a single marriage ceremony in which numerous couples are married at the same time. Collective marriages are a universal phenomenon and known as mass marriages (Ghaffar *et al.*, 2010). In 324 BCE, Alexander the Great married and in the same ceremony, he wed many of his leading officers and outstanding soldiers to other Persian women (Stadter, 1967). In Christianity Sun Myung Moon introduced mass marriages to make couples true parents without any sin and considered mass marriages a ritual of salvation and restoration (Wolfe, 1997). Mass marriages held in Buddhism remind the newlyweds of the teachings of the Buddha on the role of both husband and wife (Chye, 2007).

Societies have adopted collective marriages for many reasons. For example, Assyrians celebrated their new year with mass marriages (Abraham, 2002). The Galicnik wedding festival was organized for those men who migrated to other cities to earn money and came back in the summer (Seeney, 2007). In Belgium, mass marriages were organized to eliminate racial discrimination and create tolerance in the society (Wiki News, 2007). Many Muslim countries have the tradition of mass marriages. In Palestine, many hundreds of women had been widowed as a result of war and mass marriages have been organized to support the widows of war (El-Khodary, 2008). Collective marriages held in many cities of India provide financial assistance to the poor in that they save money, and families can spend the saved amount on other useful purposes (Bareth, 2002).

Mass weddings are preferred for economic and social reasons, such as the reduction of costs for the venue, officiants, and decorations, as well as the celebrations afterwards (McBee, 2000).

Let's take a look at collective marriages in the Punjab—a region in the northern part of the Indian subcontinent, comprising areas of eastern Pakistan and northern India. As dowry increased the expenditures of parents on the marriages of their daughters (Heyer, 1992), it became really difficult for poor parents to arrange a dowry for their daughters in Pakistan. The government of Pakistan realized this problem and adopted the concept of collective marriages from 2004 to 2007 (Ghaffar *et al.*, 2010). The collective marriages were organized by the Governor of the Punjab in Lahore, the capital city of the Pakistani province of Punjab. In a study of a random sample of fifteen couples and brides' parents, Ghaffar *et al.* (2010) found that the majority of beneficiaries of collective marriages were illiterate people; only healthy men got married through collective marriages after medical screening; the government provided to all couples a dowry—furniture (bed, chairs, and tables), a sewing machine, pedestal fans, cash gifts, necessary utensils, and other household goods; 70 percent of parents and couples who were part of a collective marriage were fully satisfied with the government efforts of collective marriages on a scale from 0 to 1. As collective marriages were successfully organized and gave economic benefits to many people, it was suggested that they continue (Ghaffar *et al.*, 2010).

Summary

Out of marriage, family is formed. Marriage and the family tend to be two sides of the same coin. Marriage is a legally recognized union between two partners that establishes their obligations and rights. The marriage rate has declined and the marriage age has risen. The chapter has explored the benefits of marriage, marital adjustment, and satisfaction in early adulthood, parenthood, middle adulthood, and late adulthood. Activity theory argues that older adults are likely to maintain happy marital relationships if they are involved in social activities. Social exchange theory examines marital satisfaction when the two partners receive benefits that outweigh their costs. Resource theory explores marital power based on the relative resources that each partner contributes to the family. From diversity and global perspectives, the chapter discussed wedding rituals and collective marriage.

Key Terms

Activity theory suggests that no matter what age, people who remain active physically, mentally, and socially adjust better to the aging process.

Actor–Partner Interdependence Model (APIM) is the most popular model used to measure the influence that members of a dyad have on each other.

Agreeableness is manifested in kind, sympathetic, cooperative, warm, and considerate behavior.

Beta marriage refers to a union that you can test, renew, or abandon without consequence.

Bride is a woman just before and after her wedding.

Bride price, also called bridewealth, is a payment such as money, valuable goods, or property by the groom or his family to the bride's family for permission to marry her.

Collective wedding is a single marriage ceremony in which numerous couples are married at the same time.

Companionship goals emphasize the bonding and emotional goals that a spouse needs with the other spouse.

Complementary hypothesis suggests that complementary partners are more beneficial to the relationship and have a higher possibility of meeting individual needs.

Conscientiousness is manifested in doing a task carefully and taking obligations to others seriously.

Contempt is any statement or nonverbal behavior that puts oneself on a higher plane than one's partner.

Couple flexibility refers to the amount of change that occurs in marital relationships.

Criticism is any statement that implies that there is something globally wrong with one's partner.

Dark triad of personality refers to three interrelated higher-order personality constructs (i.e., psychopathy, narcissism, and Machiavellianism).

Defensiveness is any attempt to defend oneself from a perceived attack.

Demand–withdraw communication occurs when one partner attempts to discuss a problem, while the other avoids the issue or ends the discussion.

Dowry is payment of a woman's inheritance, by her parents, at the time of her marriage either to her or to her husband.

Dramaturgical analysis is the study of social interaction that compares everyday life to a theatrical presentation.

Emotional stability is a person's ability to remain stable and balanced.

Empty nest is a parental home that grown children have left to begin their adult lives.

Empty nest syndrome is a phenomenon in which parents experience feelings of sadness and loss when the last child leaves home.

Engagement is a promise to wed and is the period of time between a marriage proposal and marriage.

Equity theory is based in the idea that individuals are motivated by fairness.

Extraversion is manifested in outgoing, talkative, energetic behavior.

Family life cycle perspective addresses the events related to their ongoing structural entrances and exits using a framework of family developmental transitional periods.

Femicide or feminicide refers to the intentional killing of females (women or girls) because they are females.

Fiancé is a future groom or husband-to-be.

Fiancée is a future bride or a wife-to-be.

Five Factor Model (FFM) is comprised of five personality traits: openness to experience, conscientiousness, extraversion, agreeableness, neuroticism.

Goal theory of marital satisfaction argues that to reach marital satisfaction, people have to achieve three marital goals—personal growth, instrumental goals, and companionship goals.

Groom is a man just before and after his wedding.

Ideal–real gap theory proposes that discrepancies between what one wants or considers to be ideal or desirable and what one actually has will affect marital satisfaction.

Instrumental goals focus on the tasks that occur throughout life that include using the spouse's physical and mental resources. Usually, middle-aged couples prioritize the instrumental goals.

Introversion is manifested in more reserved and solitary behavior.

Invisible support originates outside the recipient's awareness.

Jewish wedding is a wedding ceremony that follows Jewish law and traditions.

Machiavellianism (MACH) refers to interpersonal strategies that advocate self-interest, deception, and manipulation.

Marital adjustment is the accommodation of husband and wife to each other at a given time.

Marital cohesion refers to the degree of connectedness, togetherness, and emotional bonding between marital partners.

Marital communication is a process by which couples can interact in their thoughts and feelings in verbal and nonverbal forms.

Marital conflict is the pursuit of incompatible interests and goals by two partners.

Marital conflict resolution is an act of finding a solution to a conflict or a disagreement whereby the two parties are mutually satisfied with the outcome of a settlement.

Marital conflict resolution strategies include avoiding conflict, giving in, standing your ground, compromise, collaboration, mediation, and voting.

Marital distress occurs when a married or unmarried couple experiences emotional or physical conflicts that threaten to end the relationship.

Marital equality refers to an even distribution of domestic, paid work, and/or childcare, equal decision-making, equal entitlement to express and meet needs, mutual reciprocation of attention and accommodation to each other, and burden-sharing.

Marital intimacy is a form of intimacy that involves feelings of closeness in relationships with spouses.

Marital power is defined as a partner' ability to have his/her will prevail in the marriage relationship.

Marital quality is the subjective evaluation of the marital relationship on a number of dimensions such as marital satisfaction.

Marital satisfaction refers to an individual's global evaluation of the marital relationship.

Marriage is a legally recognized and/or socially approved union between two partners that establishes rights and obligations.

Marriage certificate is a document that proves that a couple is married.

Marriage contracts are a special form of contract whereby a duty of good faith is placed on both parties during negotiations.

Marriage license is a document that grants permission to marry.

Marriage premium is the difference in median personal income between married and unmarried men.

Marriage rate is the number of marriages per 1,000 unmarried women aged fifteen years and older over a given period.

Menopause is a process of natural declines in reproductive hormones that woman experience in their 40s or 50s.

Meta-emotions encompass both partners' feelings and thoughts about emotion.

Mindfulness is a process of openly attending to one's present-moment experience.

Mutual avoidance occurs when a couple avoids discussing problems and stressors.

Mutual constructive communication refers to mutual expression and discussion of feelings about the problem and problem resolution.

Narcissism is characterized by grandiosity, pride, egotism, and a lack of empathy.

Neuroticism is the tendency to experience negative emotions such as anger, anxiety, and depression.

Openness involves active imagination, aesthetic sensitivity, attentiveness to inner feelings, preference for variety, and intellectual curiosity.

Personal growth goals are based on the improvement and development of the newlywed couple with the help of each other within the marriage.

Population aging refers to an increase in the median age of the population due to rising life expectancy.

Psychopathy is characterized by continuing antisocial behavior, impulsivity, selfishness, callousness, and remorselessness.

Resources refer to material goods (e.g., money and physical possessions) and nonmaterial goods such as love and status that can be accessed and used in social actions.

Resource theory is a theoretical framework for the study of family power.

Role theory suggests that our everyday activities consist of acting out socially defined roles associated with a given status.

Sandwich generation refers to adults who care for an aging parent age 65 or older as well as raising a child under age eighteen or supporting a grown child.

Self-disclosure is defined as telling others about oneself.

Sexual communication refers to the communication, self-disclosure, and communication processes around sexual topics and problems.

Sexual satisfaction refers to one's satisfaction with the frequency, variety, and quality of various aspects of one's sexual life, including functioning and connection.

Similarity hypothesis poses that similarity in some important areas will make people evaluate other people as more attractive.

Social exchange theory suggests that a marital relationship is the result of a benefit and cost exchange process between two partners.

Spiritual intimacy is a form of intimacy that reveals your spirituality to your partner.

Spouse refers to a person married to and living with a householder who is of the opposite sex to the householder.

Stonewalling occurs when the listener withdraws from the conversation.

Transition is defined as the passage from one ending state to another beginning state.

Visible support occurs when both partners are aware of the supportive actions.

Vulnerability-Stress-Adaptation Model (VSA) posits that couples who have few enduring vulnerabilities, encounter few stressors, and employ effective adaptive processes are likely to experience high marital quality and stability.

Wedding is a marriage ceremony in which two partners are united in marriage.

White wedding is a traditional formal Western wedding in which the bride wears a white wedding dress.

Discussion Questions

- Conduct a survey in class that asks students whether they plan to get married and why.
- Compose an informal marriage contract specifying the rights, obligations, duties, and other related terms you would expect in your marriage.

Suggested Film and Video

Goetzman, G. *et al.* (Producer). *My Big Fat Greek Wedding.* (Movie). United States.

Lee, A. (Producer). (1993). *The Wedding Banquet.* (Movie). United States, Taiwan.

Internet Sources

Marriage Laws around the World, Pew Research Center: http://assets.pewresearch.org/wp-content/uploads/sites/12/2016/09/FT_Marriage_Age_Appendix_2016_09_08.pdf

Marriage Rate in the U.S.: Geographic Variation: https://magic.piktochart.com/output/8810838-wu-marrt-us-2014-fp-15-20-web

References

Abbasi, F. & Afsharinia, K. (2015). Relationship between Couples Communication Patterns and Marital Satisfaction. *Int. J. Econ. Manag. Soc. Sci.*, 4(3), 369–372.

Abbasi, M., Dehghani, M., Mazaheri, M. A., Ansarinejad, F., Fadaie, Z., Nikparvar, F. *et al.* (2010). Trend Analysis of Changes in Marital Satisfaction and Related Dimensions across Family Life Cycle. *Journal of Family Research,* 6(21), 5–22.

Abraham, S. (April 2, 2002). Assyrians Celebrated their New Year with Mass Wedding. Associated Press. Retrieved on November 29, 2018 from http://www.atour.com/news/assyria/20020402a.html

Adrian, B. (2004). The Camera's Positioning: Brides, Grooms, and Their Photographers in Taipei's Bridal Industry. *Ethos,* 32(2), 140–163.

Ahmad, N., Hussain, A., Tariq, M. S. & Raza, M. A. (2014). Role of Dowry in Successful Marital Life: A Case Study of District Dera Ghazi Khan, Pakistan. *Arabian Journal of Business and Management Review* (Nigerian Chapter), 2(11).

Alder, E. S. (2010). *Age, Education Level, and Length of Courtship in Relation to Marital Satisfaction* (Master's thesis, Pacific University). Retrieved on February 14, 2018 from http://commons.pacificu.edu/spp/145

Allan, G. (2007). Marital Power/Resource Theory. In G. Ritzer (Ed.), *The Blackwell Encyclopedia of Sociology.* Malden, MA: Blackwell Publishing.

Amato, P. R., Booth, A., Johnson, D. R. & Rogers, S. J. (2007). *Alone Together: How Marriage in America is Changing.* Cambridge, MA: Harvard University Press.

Amiri, M., Farhoodi, F., Abdolvandc, N. & Bidakhavidi, A. R. (2011). A Study of the Relationship between Big-Five Personality Traits and Communication Styles with Marital Satisfaction of Married Students Majoring in Public Universities of Tehran. *Procedia: Social and Behavioral Sciences,* 30, 685–689.

Anderson, L. R. (2017). Families and Households of Baby Boomers and Millennials. *Family Profiles.* FP-17–07. Bowling Green, OH: NCFMR.

Anderson, L. & Guzzo, K. B. (2016). Trends in Men's Economic Characteristics and Marriage. *Family Profiles.* FP-16–06. Bowling Green, OH: NCFMR.

Anderson, L. & Payne, K. K. (2016). Median Age at First Marriage, 2014. *Family Profiles.* FP-16–07. Bowling Green, OH: NCFMR.

Ansbacher, H. L. & Ansbacher, R. R. (Eds) (1956). *The Individual Psychology of Alfred Adler: A Systematic Presentation in Selection from His Writings.* New York: Harper & Row.

Arif, N. & Fatima, I. (2015). Marital Satisfaction in Different Types of Marriage, Pakistan. *Journal of Social and Clinical Psychology,* 13(1), 36–40.

Arnett, J. J. (2007). Emerging Adulthood: What Is It, and What Is It Good For? *Child Development Perspectives,* 1, 68–73.

Asendorpf, J. B., Penke, L. & Back, M. D. (2011). From Dating to Mating and Relating: Predictors of Initial and Long-term Outcomes of Speed-Dating in a Community Sample. *Eur. J. Pers.*, 25, 16–30.

Asoodeh, M. H., Khalili, S., Daneshpour, M. & Lavasani, M. G. (2010). Factors of Successful Marriage: Accounts from Self-described Happy Couples. *Procedia: Social and Behavioral Sciences,* 5, 2042–2046.

Ausman, B. (2003). Affairs of the Heart—Marriage Contracts for Professionals. *The Canadian Veterinary Journal,* 44(1), 83–86.

Back, M. D., Penke, L., Schmukle, S. C., Sachse, K., Borkenau, P. & Asendorpf, J. B. (2011). Why Mate Choices Are Not as Reciprocal as We Assume: The Role of Personality, Flirting and Physical Attractiveness. *Eur. J. Pers.*, 25, 120–132.

Barelds, D. P. H. & Barelds-Dijkstra, P. (2007). Love at First Sight or Friends First? Ties among Partner Personality Trait Similarity, Relationship Onset, Relationship Quality, and Love. *J. Soc. Pers. Relat.*, 24, 479–496.

Bareth, N. (January 9, 2002). Mass Marriages of Rajasthan Muslims. BBC News. Retrieved on November 29, 2018 from http://news.bbc.co.uk/2/hi/south_asia/1750769.stm

Barnett, M. A. & Thompson, S. (1985). The Role of Perspective Taking and Empathy in Children's Machiavellianism,

Prosocial Behaviour and Motive for Helping. *Journal of Genetic Psychology,* 146, 295–305.

Basharpoor, S. & Sheykholeslami, A. (2015). The Relation of Marital Adjustment and Family Functions with Quality of Life in Women. *Europe's Journal of Psychology,* 11(3), 432–441.

Belsky, J. & Kelly, J. (1994). *The Transition to Parenthood.* New York: Delacorte Press.

Bem, D. (1972). Self-perception Theory. In L. Berkowitz (Ed.), *Advances in Experimental Social Psychology* (vol. 6, pp. 2–62). New York: Academic Press.

Bengtson, V. L., Gans, D., Norella, P. & Silverstein, M. (2009). *Handbook of Theories of Aging.* New York: Springer.

Bennett, J. (July 25, 2014). The Beta Marriage: How Millennials Approach "I Do." *Time.* Retrieved on February 18, 2018 from https://ifstudies.org/blog/a-beta-marriage-is-no-marriage-at-all

Berk, S. F. (1985). *The Gender Factory.* New York: Plenum.

Bertrand, M., Kamenica, E. & Pan, J. (2015). Gender Identity and Relative Income with Households. *The Quarterly Journal of Economics,* 571–614.

Best, A. (2006). *Introduction to Peace and Conflict Studies in West Africa.* Ibadan: Spectrum Books Limited.

BGSU (Bowling Green State University) (2017). *Spiritual Intimacy: Talking As "Soul Mates."* Retrieved on November 29, 2018 from https://www.bgsu.edu/arts-and-sciences/psychology/graduate-program/clinical/the-psychology-of-spirituality-and-family/research-findings/marriage-couples/spiritual-intimacy.html

Bianchi, S. M., Robinson, J. P. & Milkie, M. A. (2006). *Changing Rhythms of American Family Life.* New York: Russell Sage.

Blagov, P. S & Singer, J. A. (2004). Four Dimensions of Self-Defining Memories (Specificity, Meaning, Content, and Affect) and Their Relationships to Self-Restraint, Distress, and Repressive Defensiveness. *J Pers.,* 72(3), 481–511.

Blau, P. M. (1964). *Exchange and Power in Social Life.* New York: John Wiley.

Blonigen, D. M., Carlson, M. D., Hicks, B. M., Krueger, R. F. & Iacono, W. G. (2008). Stability and Change In Personality Traits From Late Adolescence To Early Adulthood: A Longitudinal Twin Study. *J Pers Soc Psychol.,* 76(2), 229–266.

Blood, R. O. &Wolfe, D. M. (1960). *Husbands and Wives: The Dynamics of Married Living.* Glencoe, IL: Free Press.

Boertien, D. & Mortelmans, D. (2017). Does the Relationship between Personality and Divorce Change over Time? A Cross-Country Comparison of Marriage Cohorts. *Acta Sociologica.* Retrieved on February 11, 2018 from http://journals.sagepub.com/doi/pdf/10.1177/0001699317709048#articleCitationDownloadContainer

Bolger, N., DeLongis, A., Kessler, R. C. & Wethington, E. (1989). The Contagion of Stress across Multiple Roles. *Journal of Marriage and the Family,* 51, 175–183.

Bouchard, T. J. & McGue, M. (2003). Genetic and Environmental Influences on Human Psychological Differences. *Journal of Neurobiology,* 54(1), 4–45.

Bradbury, T. N., Fincham, F. D. & Beach, S. R. H. (2000). Research on the Nature and Determinants of Marital Satisfaction: A Decade in Review. *Journal of Marriage and the Family,* 62, 954–980.

Bridges, W. (1980). *Transitions.* Reading, MA: Addison-Wesley.

Brown, J. D. (2006). *Ethnic Identity in Health Care: Intercultural Health Communication and Physician–Patient Satisfaction* (doctoral dissertation). Texas Tech University, Lubbock, Texas.

Brown, K. W. & Ryan, R. M. (2003). The Benefits of Being Present: Mindfulness and Its Role in Psychological Well-being. *J. Pers. Soc. Psychol.,* 84(4), 822–848.

Burgess, E. W. & Cottrell, L. S. (August, 1939). Predicting Success or Failure in Marriage. *Journal of Marriage and Family,* 62(3), 849–852.

Burgess, E. W. & Locke, H. J. (1945). *The Family: From Institution to Companionship.* New York: The American Book Company.

Burpee, L. C. & Langer, E. (2005). Mindfulness and Marital Satisfaction. *Journal of Adult Development,* 12(10).

Burton, S. (1990). *Conflict Resolution and Prevention.* London: Macmillan.

Byrne, M., Carr, A. & Clark, M. (2004). Power in Relationships of Women with Depression. *Journal of Family Therapy,* 26, 407–429.

Campbell, A., Converse, P. E. & Rodgers, W. L. (1976). *The Quality of American Life: Perceptions, Evaluations, and Satisfactions.* New York: Russell Sage Foundation.

Cancian, F. M. (1987). *Love in America: Gender and Self Development.* Cambridge, UK: Cambridge University Press.

Carter, B. & McGoldrick, M. (1989). Overview: The Changing Family Life Cycle: A Framework for Family Therapy. In B. Carter & M. McGoldrick (Eds.), *The Changing Family Life Cycle: A Framework for Family Therapy,* 2nd edn. (pp. 3–28). Boston, MA: Allyn and Bacon.

Carter, C. (2011). *Raising Happiness: 10 Simple Steps for More Joyful Kids and Happier Parents.* New York: Ballantine Books.

Carter, G. L., Campbell, A. C. & Muncer, S. (2014). The Dark Triad Personality: Attractiveness to Women. *Personality and Individual Differences,* 56, 57–61.

Cashdan, E. (1996). Women's Mating Strategies. *Evol. Anthropol.*: Issues, News, Rev. 5, 134–143.
Cave, E. (2004). Harm Prevention and the Benefits of Marriage. *Journal of Social Philosophy*, 35(2), 233–243.
Cherlin, A. J. (2004). The Deinstitutionalization of American Marriage. *Journal of Marriage and Family*, 66(4), 848–861. http://www.jstor.org/stable/3600162
China Bridal (2003). *Complete Guide to Chinese Wedding*. ChinaBridal.com. Retrieved on February 13, 2018 from http://www.chinabridal.com/etiquette/guide.htm
Christensen, A. (1988). Dysfunctional Interaction Patterns in Couples. In P. Noller & M. A. Fitzpatrick (Eds.), *Perspectives on Marital Interaction* (pp. 31–52). Philadelphia: Multilingual Matters.
Christopher, F. S. & Sprecher, S. (2000). Sexuality in Marriage, Dating, and Other Relationships: A Decade Review. *Journal of Marriage and Family*, 62(4), 999–1017.
Chuang, A. & Hsu, R. S. (2013). Conscientiousness Match in Vertical Dyads: The Moderation of Gender Match. *Academy of Management Proceedings*, 2013(1).
Chudek, M. & Henrich, J. (2010). Culture–Gene Coevolution, Norm-psychology, and the Emergence of Human Prosociality. *Trends Cogn. Sci.* 15, 218–226.
Chye, H. Y. (October 01, 2007). *Buddhist Mass Wedding*. Than Hsiang Foundation. Retrieved on November 29, 2018 from http://www.thanhsiang.org/en/buddhist-mass-wedding
Clark, L. A., Simms, L. J., Wu, K. D. & Casillas, A. (1991). *Schedule for Nonadaptive and Adaptive Personality*. Minneapolis: University of Minnesota.
Collings, M. R. (2009). *Gemlore: An Introduction to Precious and Semi-Precious Stones. Cabin John*, MD: Wildside Press LLC.
Coontz, S. (2004). *Marriage, a History: How Love Conquered Marriage*. New York: Penguin Books.
Cope, E. D. (1888). *The Marriage Problem*. Chicago, IL: Open Court Publishing Company.
Costa, P. T. & McCrae, R. R. (1992). *NEO Personality Inventory Professional Manual*. Odessa, FL: Psychological Assessment Resources.
Cowan, P. A. & Cowan, C. P. (1988). Changes in Marriage during the Transition to Parenthood: Must We Blame the Baby? In G. Y. Michaels & W. A. Goldberg (Eds.), *The Transition to Parenthood: Current Theory and Research*. New York: Cambridge University Press.
Creswell, J. D. (2017). Mindfulness Interventions. *Annual Review of Psychology*, 68(1), 491–516.
Cupach, W. R. & Metts, S. (1986). Accounts of Relational Dissolution: A Comparison of Marital and Non-marital Relationships. *Commun. Monogr.* 53, 311–334.
Currentz, B. (2018). For Young Adults, Economic Security Matters for Marriage. *U.S. Census Bureau*.
Cutrona, C. E. (1996). *Social Support in Couples*. Thousand Oaks, CA: Sage.
Dalton, G. (1966). Brief Communications: "Bridewealth" vs. "Brideprice". *American Anthropologist*, 68(3), 732–737.
Davidson, B. (1981). The Relations between Partners' Levels of Affective Self-Disclosure and Marital Adjustment. *Dissertation Abstracts International*, 41(10-B), 3931–3932.
De Raad, B. & Doddema-Winsemius, M. (1992). Factors in the Assortment of Human Mates: Differential Preferences in Germany and the Netherlands. *Personality and Individual Differences*, 13, 103–114.
Derlega, V. J. (1984). Self-disclosure and Intimate Relationships. In V. J. Derlega (Ed.), *Communication, Intimacy, and Close Relationship*. Orlando, FL: Academic Press.
Dinna, M. (2005). Marital Satisfaction in Autonomous and Arranged Marriages: South African Indian sample, MA dissertation, University of Pretoria, Pretoria, viewed December 17, 2017, http://hdl.handle.net/2263/30227
Doane, M. J. (2016). Cohesion, Marital. *The Wiley Blackwell Encyclopedia of Family Studies*, 1–3.
Donnellan, M. B., Larsen-Rife, D. & Conger, R. D. (2005). Personality, Family History, and Competence in Early Adult Romantic Relationships. *J. Pers. Soc. Psychol.*, 88, 562–576.
Driver, J. L. & Gottman, J. M. (2004). Daily Marital Interactions and Positive Affect During Marital Conflict Among Newlywed Couples. *Family Process*, 43(3).
Dubert, C. J., Schumacher, A. M., Locker, L., Gutierrez, A. P. & Barnes, V. A. (2016). Mindfulness and Emotion Regulation among Nursing Students: Investigating the Mediation Effect of Working Memory Capacity. *Mindfulness*, 7(5), 1061–1070.
Durkheim, E. (1893/1933). *The Division of Labor in Society* (G. Simpson, Trans.). New York: Free Press.
Dush, C. M. K., Taylor, M. G. & Kroeger, R. A. (2008). Marital Happiness and Psychological Well-being across the Life Course. *Family Relations*, 57(2), 211–226.
Dyer, W. (1962). Analyzing Marital Adjustment Using Role Theory. *Marriage and Family Living*, 24(4), 371–375.
Dzara, K. (2010). Assessing the Effect of Marital Sexuality on Marital Disruption. *Social Science Research*, 39(5), 715–724.
Eagly, A. H., Wood, W. & Diekman, A. B. (2000). Social Role Theory of Sex Differences and Similarities: A Current Appraisal. In T. Eckes, & H. M. Trautner (Eds.), *The Developmental Social Psychology of Gender* (pp. 123–174). Mahwah, NJ: Erlbaum.
Eberstadt, N. (October 5, 2016). *A Portrait of the Un-Working American Man*. Charlottesville, VA: Institute for Family Studies.
Edward, D. A. & Chapman, T. (2012). Measuring the Fitness Benefits of Male Mate Choice in Drosophilia Melanogaster. *Evolution*, 66, 2646–2653.

Edwards, T., Pask, E. B., Whitbred, R. & Neuendorf, K. A. (2018). The Influence of Personal, Relational, and Contextual Factors on Forgiveness Communication following Transgressions. *Personal Relationships,* 25, 4–21.

Ekerdt, D. J. (1986). The Busy Ethic: Moral Continuity between Work and Retirement. *The Gerontologist,* 26(3), 239–244.

Eldridge, K. A., Sevier, M., Jones, J., Atkins, D. C. & Christensen, A. (2007). Demand–Withdraw Communication in Severely Distressed, Moderately Distressed, and Nondistressed Couples: Rigidity and Polarity during Relationship and Personal Problem Discussions. *J Fam Psychol.,* 21(2), 218–226.

ElHage, A. (November 9, 2016). *Mr. President: A Healthy Marriage Culture is Vital to America's Success.* Charlottesville, VA: Institute for Family Studies.

El-Khodary, A. (October 30, 2008). Mass Wedding Celebration in Gaza. *New York Times.* Retrieved on November 29, 2018 from https://www.nytimes.com/2008/10/31/world/middleeast/31gaza.html

Ellis, H. (1920). *Studies in the Psychology of Sex – Sex in Relation to Society.* Philadelphia: F.A. Davis Company.

Emerson, R. M. (1976). Social Exchange Theory. In A. Inkeles, J. Coleman & N. Smelser (Eds.), *Annual Review of Sociology,* Vol. 2 (pp. 335–362). Palo Alto, CA: Annual Reviews.

Ewart, C. K., Taylor, C. B., Kraemer, H. C. & Agras, W. S. (1991). High Blood Pressure and Marital Discord: Not Being Nasty Matters More than Being Nice. *Health Psychol.,* 10(3), 155–163.

Fallis, E. E., Rehman, U. S. & Purdon, C. (2014). Perceptions of Partner Sexual Satisfaction in Heterosexual Committed Relationships. *Archives of Sexual Behavior,* 43(3), 541–550.

Feeney, J. A. (2002). Attachment, Marital Interaction, and Relationship Satisfaction: A Diary Study. *Personal Relationships,* 9(1), 39.

Ferraro, G. P. & Andreatta, S. (2009). *Cultural Anthropology: An Applied Perspective.* Stamford, CT: Cengage Learning.

Fincham, F. D., Beach, S. R. & Kemp-Fincham, S. I. (1997). Marital Quality: A New Theoretical Perspective. In R. J. Sternberg & M. Hojjat (Eds.), *Satisfaction in Close Relationships* (pp. 275–304). New York: Guilford Press.

Fincham, F. D., Stanley, S. M. & Beach, S. R. H. (2007). Transformative Processes in Marriage: An Analysis of Emerging Trends. *Journal of Marriage and the Family,* 69(2), 275–292.

Finkel, E. J. (2017). *The All-or-Nothing Marriage: How the Best Marriages Work.* Penguin Audio.

Fisher, H. (2010). *Why Him? Why Her?: How to Find and Keep Lasting Love.* New York: Holt Paperbacks.

Fitzpatrick, J., Gareau, A., Lafontaine, M.-F. & Gaudreau, P. (2016). How to Use the Actor-Partner Interdependence Model (APIM) To Estimate Different Dyadic Patterns in MPLUS: A Step-by-Step Tutorial. *The Quantitative Methods for Psychology,* 12(1), 74–86.

Fowler, A. C. (2007). Love and Marriage: Through the Lens of Sociological Theories. *Human Architecture: Journal of the Sociology of Self-Knowledge,* 5(2), Article 6. Available at: http://scholarworks.umb.edu/humanarchitecture/vol5/iss2/6

Freud, S. (1914). On Narcissism: An Introduction. In *Complete Psychological Works* (pp. 30–59). London: Hogarth Press.

Frisco, M. L. & Williams, K. L. (2003). Perceived Housework Equity, Marital Happiness, and Divorce in Dual-Earner Households. *Journal of Family Issues,* 24, 51–73.

Fry, R. (May 24, 2016). *For First Time in Modern Era, Living With Parents Edges Out Other Living Arrangements for 18–34-Year-Olds.* Washington, DC: Pew Research Center.

Fry, R. (February 13, 2017). *Americans are Moving at Historically Low Rates, in Part because Millennials Are Staying Put.* Washington, DC: Pew Research Center.

Furler, K., Gomez, V. & Grob, A. (2013). Personality Similarity and Life Satisfaction in Couples. *Journal of Research in Personality,* 47(4).

Furnham, A., Richards, S. & Paulhus, D. (2013). The Dark Triad of Personality: A 10-Year Review. *Social and Personality Psychology Compass,* 7, 199–216.

Gable, S. L., Reis, H. T., Impett, E. & Asher, E. R. (2004). What Do You Do When Things Go Right? The Intrapersonal and Interpersonal Benefits of Sharing Positive Events. *Journal of Personality and Social Psychology,* 87, 228–245.

Gadoua, S. P. & Larson, V. (2014). *The New I Do: Reshaping Marriage for Skeptics, Realists and Rebels.* Berkeley, CA: Seal Press.

GAO (General Accounting Office) (2004). *Defense of Marriage Act: Update to Prior Report.* Retrieved on February 27, 2018 from https://www.gao.gov/new.items/d04353r.pdf

Garcia, S. D. (1999). Perceptions of Hispanic and African-American Couples at the Friendship or Engagement Stage of a Relationship. *Journal of Social and Personal Relationship,* 16, 65–86.

Gaunt, R. (2006). Couple Similarity and Marital Satisfaction: Are Similar Spouses Happier? *Journal of Personality,* 74, 1401–1420.

Geiger, A. & Livingston, G. (2018). *8 Facts about Love and Marriage in America.* Washington, DC: Pew Research Center. Retrieved on March 5, 2018 from http://www.pewresearch.org/fact-tank/2018/02/13/8-facts-about-love-and-marriage/

Ghaffar, F., Shenaz, Z. & Town, J. (2010). Social and Economic Factors of Collective Marriage in Lahore. *The Journal of Animal & Plant Sciences,* 20(2), 132–135.

Goffman, E. (1959). *The Presentation of Self in Everyday Life.* New York: Doubleday.

Gonzaga, G. C., Campos, B. & Bradbury, T. (2007). Similarity, Convergence, and Relationship Satisfaction in Dating and Married Couples. *J. Pers. Soc. Psychol.*, 93, 34–48.

Goody, J. (1976). *Production and Reproduction: A Comparative Study of the Domestic Domain.* Cambridge, UK: Cambridge University Press.

Gottman, J. (1995). *Why Marriages Succeed or Fail.* New York: Simon & Schuster.

Gottman, J. M. & Levenson, R. W. (1992). Marital Processes Predictive of Later Dissolution: Behavior, Physiology, and Health. *Journal of Personality and Social Psychology*, 63(2), 221–233.

Gottman, J. M. & Silver, N. (1999). *The Seven Principles for Making Marriages Work.* New York: Three Rivers Press.

Gottman, J. M., Katz, L. F. & Hooven, C. (1997). *Meta-emotion: How Families Communicate Emotionally.* Hillsdale, NJ: Lawrence Erlbaum Associates.

Graeber, D. (2011). *Debt: The First 5,000 Years.* New York: Melville House.

Greene, K. & Faulkner, S. L. (2005). Gender, Belief in the Sexual Double Standard, and Sexual Talk in Heterosexual Dating Relationships. *Sex Roles*, 53, 239–251.

Gurtman, M. B. (1995). Personality Structure and Interpersonal Problems: A Theoretically-Guided Tem Analysis of the Inventory of Interpersonal Problems. *Assessment*, 2, 343–361.

Hansen, J. & Schuldt, W. (1984). Marital Self-Disclosure and Marital Satisfaction. *Journal of Marriage and Family*, 46(4), 923–926.

Hardie, J. & Lucas, A. (2010). Economic Factors and Relationship Quality among Young Couples: Comparing Cohabitation and Marriage. *Journal of Marriage and Family*, 72(5), 1141–1154.

Harris, S. (2006). *The Meanings of Marital Equality.* Albany, NY: State University of New York Press.

Hatch, R., James, D. & Schumm, W. (1986). Spiritual Intimacy and Marital Satisfaction. *Family Relations*, 35, 539–545.

Hayward, R. D. & Kraus, N. (2013). Patterns of Change in Prayer Activity, Expectancies, and Contents during Older Adulthood. *Journal for the Scientific Study of Religion*, 52(1), 17–34.

Heaton, T. B. & Pratt, E. L. (1990). The Effects of Religious Homogamy on Marital Satisfaction and Stability. *Journal of Family Issues*, 11, 191–207.

Heavey, C. L., Layne, C. & Christensen, A. (1993). Gender and Conflict Structure in Marital Interaction: A Replication and Extension. *Journal of Consulting and Clinical Psychology*, 61, 16–27.

Hemez, P. (2016). Marriage Rate in the U.S.: Geographic Variation, 2015. *Family Profiles.* (FP-16–22). Bowling Green, OH: National Center for Family & Marriage Research (NCFMR).

Henrich, J., Boyd, R. & Richerson, P. J. (2012). The Puzzle of Monogamous Marriage. *Phil. Trans. R. Soc. B*, 367(1589), 657–669.

Hetherington, E. M. (1993). An Overview of the Virginia Longitudinal Study of Divorce and Remarriage with a Focus on Early Adolescence. *Journal of Family Psychology*, 7, 39–56.

Heyer, J. (1992). Roles of Dowries and Daughters: Marriages in Accumulation and Disturbance of Capital in a South Indian Community. *Journal of International Development*, 4(4), 419–436.

Hinde, R. A. (1997). *Relationships: A Dialectical Perspective.* East Sussex, UK: Psychology Press.

Holman, T. & Li, B. (1997). Premarital Factors Influencing Perceived Readiness for Marriage. *Journal of Family Issues*, 18, 124–144.

Holmes, T. (2017). Environmental Traits in Humans. *Synonym.* Retrieved on February 19, 2018 from http://classroom.synonym.com/environmental-traits-in-humans-12083411.html

Holtzman, N. (2012). People With Dark Personalities Tend to Create a Physically Attractive Veneer. *Social Psychological and Personality Science*, 4, 461–467.

Homans, G. C. (1974). *Social Behavior: Its Elementary Forms* (rev. edn). New York: Harcourt Brace Jovanovich.

Hopwood, C. J., Donnellan, M. B., Blonigen, D. M., Krueger, R. F., McGue, M., Iacono, W. G. & Burt, S. A. (2011). Genetic and Environmental Influences on Personality Trait Stability and Growth during the Transition to Adulthood: A Three Wave Longitudinal Study. *Journal of Personality and Social Psychology*, 100(3), 545–556.

Howard, V. (2008). *Brides, Inc. American Weddings and the Business of Tradition.* Philadelphia, PA: University of Pennsylvania Press.

Igbo, H. I., Awopetu, R. G. & Ekoja, O. C. (2015). Relationship between Duration of Marriage, Personality Trait, Gender and Conflict Resolution Strategies of Spouses. *Procedia: Social and Behavioral Sciences*, 190, 490–496.

Jakobwitz, S. & Egan, V. (2006). The "Dark Triad" and Normal Personality Traits. *Personality and Individual Differences*, 40(2), 331–339.

Jang, K., Livesley, W. J. & Vemon, P. A. (1996). Heritability of the Big Five Personality Dimensions and Their Facets: A Twin Study. *Journal of Personality*, 64(3), 577–591.

Jeffries, C. (Jan 12, 2018). This Might Be Your Chance to Get Married ... For Free. *PalmBeach Post.* Retrieved on October 13, 2018 from https://www.palmbeachpost.

com/entertainment/this-might-your-chance-get-married-for-free/vQQaUbbhiF4oQkvqsG9d5O/

Jeronimus, B. F., Riese, H., Sanderman, R. & Ormel, J. (2014). Mutual Reinforcement between Neuroticism and Life Experiences: A Five-Wave, 16-Year Study to Test Reciprocal Causation. *Journal of Personality and Social Psychology*, 107(4), 751–764.

Jhabvala, N. H. (1981). *Principles of Hindu Law.* Bombay: Jamnadas and Co.

Johnson, M. D. & Anderson, J. R. (2013). The Longitudinal Association of Marital Confidence, Time Spent Together, and Marital Satisfaction. *Family Process,* 52(2), 244–256.

Jonason, P. K. & Kavanagh, P. (2010). The Dark Side of Love: The Dark Triad and Love Styles. *Personality and Individual Differences,* 49, 606–610.

Jonason, P. K., Li, N. P., Webster, G. W. & Schmitt, D. P. (2009). The Dark Triad: Facilitating Short-term Mating in Men. *European Journal of Personality,* 23, 5–18.

Jonason, P. K., Li, N. P. & Buss, D. M. (2010). The Costs and Benefits of the Dark Triad: Implications for Mate Poaching and Mate Retention Tactics. *Personality and Individual Differences,* 48(4), 373–378.

Jonason, P. K., Valentine, K. A., Li, N. P. & Harbeson, C. L. (2011). Mate-selection and the Dark Triad: Facilitating a Short-Term Mating Strategy and Creating a Volatile Environment. *Personality and Individual Differences,* 51(6), 759–763.

Jones, A. C. (2016). The Role of Sexual Communication in Committed Relationships. All Graduate Theses and Dissertations, 4994. Retrieved on March 6, 2018 from https://digitalcommons.usu.edu/etd/4994

Jose, A., O'Leary, K. D. & Moyer, A. (2010). Does Premarital Cohabitation Predict Subsequent Marital Stability and Marital Quality? A Meta-Analysis. *Journal of Marriage and Family,* 72(1), 105–116.

Jurkane-Hobein, I. (2015). Imagining the Absent Partner: Intimacy and Imagination in Long-distance Relationships. *Innovative Issues and Approaches in Social Sciences,* 8(1), 223–241.

Kach, J. & Mcghee, P. (1982). Adjustment to Early Parenthood: The Role of Accuracy of Pre Parenthood Expectations. *Journal of Family Issues,* 3, 361–374.

Kamp Dush, C. M. & Taylor, M. G. (2012). Trajectories of Marital Conflict across the Life Course: Predictors and Interactions with Marital Happiness Trajectories. *Journal of Family Issues,* 33(3), 341–368.

Kamp Dush, C. M., Taylor, M. G. & Kroeger, R. A. (2008). Marital Happiness and Psychological Well-Being across the Life Course. *Family Relations,* 57(2).

Kaplan, R. A. (1983). *Made in Heaven, A Jewish Wedding Guide.* New York; Jerusalem: Moznaim Publishing.

Kaslow, F. & Robison, J. (1996). Long-term Satisfying Marriage: Perceptions of Contributing Factors. *The American Journal of Family Therapy,* 124, 153–170.

Kelley, D. L. & Burgoon, J. K. (1991). Understanding Marital Satisfaction and Couple Type as Functions of Relational Expectations. *Human Communication Research,* 18, 40–69.

Kendrick, H. M. & Drentea, P. (2016). Marital Adjustment. *The Wiley Blackwell Encyclopedia of Family Studies,* 1–2.

Kenny, D. A., Kashy, D. A. and Cook, W. L. (2006). *Dyadic Data Analysis.* New York: Guilford.

Khalatbari, J., Ghorbanshiroudi, S., Azari, K. N., Bazleh, N. & Safaryazdi (2013). The Relationship between Marital Satisfaction (Based on Religious Criteria) and Emotional Stability. *Procedia: Social and Behavioral Sciences,* 84, 869–873.

Khoury, C. B. & Findlay, B. M. (2014). What Makes for Good Sex? The Associations among Attachment Style, Inhibited Communication and Sexual Satisfaction. *Journal of Relationships Research,* 5. Retrieved on March 6, 2018 from http://doi.org/10.1017/jrr.2014.7

Kiecolt-Glaser, J. K., Kennedy, S., Malkoff, S., Fisher, L., Speicher, C. E. & Glaser, R. (1988). Marital Discord and Immunity in Males. *Psychosom Med.,* 50(3), 213–229.

Killingsworth, M. A. & Gilbert, D. T. (2010). A Wandering Mind Is an Unhappy Mind. *Science,* 330(6006), 932.

Kohut, H. (1977). *The Restoration of the Self.* New York: International Universities Press.

Kornrich, S., Brines, J. & Leupp, K. (2013). Egalitarianism, Housework, and Sexual Frequency in Marriage. *American Sociological Review,* 78(1), 26–50.

Kuppens, P. (2008). Individual Differences in the Relationship between Pleasure and Arousal. *Journal of Research in Personality,* 42(4), 1053–1059.

Lamm, M. (2017). The Marriage Contract (Ketubah). *Chabad.org.* Retrieved on February 13, 2018 from https://www.chabad.org/library/article_cdo/aid/465168/jewish/The-Marriage-Contract-Ketubah.htm

Langer, A., Lawrence, E. & Barry, R. A. (2008). Using a Vulnerability-Stress-Adaptation Framework to Predict Physical Aggression Trajectories in Newlywed Marriage. *Journal of Consulting and Clinical Psychology,* 76(5), 756.

Larson, J. & Holman, T. (1994). Premarital Predictors of Marital Quality and Stability. *Family Relations,* 43(2), 228–237.

Lavner, J. A., Karney, B. R. & Bradbury, T. N. (2016). Does Couples' Communication Predict Marital Satisfaction, or Does Marital Satisfaction Predict Communication? *Journal of Marriage and Family,* 78(3), 680–694.

Lawrence, E., Bunde, M., Barry, R. A., Brock, R. L., Sullivan, K. T., Pasch, L. A., White, G. A., Dowd, C. E. & Adams,

E. E. (2008a). Partner Support and Marital Satisfaction: Support Amount, Adequacy, Provision, and Solicitation. *Personal Relationships*, 15(4), 445–463.

Lawrence, E., Cobb, R. J., Rothman, A. D., Rothman, M. T. & Bradbury, T. N. (2008b). Marital Satisfaction across the Transition to Parenthood. *Journal of Family Psychology: Journal of the Division of Family Psychology of the American Psychological Association (Division 43*), 22(1), 41–50. https://www.ncbi.nlm.nih.gov/pmc/articles/PMC2367106/

Legkauskas, V. & Stankevičienė, D. Ž. (2009). Premarital Sex and Marital Satisfaction of Middle Aged Men and Women: A Study of Married Lithuanian Couples. *Sex Roles*, 60(1), 21–32.

Lehrer, E. L. (2008). Age at Marriage and Marital Instability: Revisiting the Becker–Landes–Michael Hypothesis. *Journal of Population Economics*, 21(2), 463–484.

Lehrer, E. L. & Chen, Y. (2013). Delayed Entry into First Marriage and Marital Stability: Further Evidence on the Becker–Landes–Michael Hypothesis. *Demographic Research*, 29, 521–541.

Lehrer, J. (2016). *A Book About Love*. New York: Simon & Schuster.

Li, T. & Fung, H. H. (2011). The Dynamic Goal Theory of Marital Satisfaction. *Review of General Psychology*, 15(3), 246–254.

Liang, J., Bennett, J. M., Shaw, B. A., Quiñones, A. R., Ye, W., Xu, X. & Ofstedal, M. B. (2008). Gender Differences in Functional Status in Middle and Older Age: Are There Any Age Variations? *J Gerontol B Psychol Sci Soc Sci.*, Sep; 63(5), S282–S292.

Lippa, R. A. (2010). Gender Differences in Personality and Interests: When, Where, and Why? *Social and Personality Psychology Compass*, 4(11), 1098–1110.

Litzinger, S. & Gordon, K. C. (2005). Exploring Relationships among Communication, Sexual Satisfaction, and Marital Satisfaction, *Journal of Sex & Marital Therapy*, 31(5).

Locke, H. J. & Wallace, K. M. (1959). Short Marital Adjustment and Prediction Tests: Their Reliability and Validity. *Marriage and Family Living*, 21(3), 251–255.

Lucas, T. W., Wendorf, C. A., Imamoglu, E. O., Shen, J., Parkhill, M. R., Weisfeld, C. C., *et al.* (2004). Marital Satisfaction in Four Cultures as a Function of Homogamy, Male Dominance and Female Attractiveness. *Sexual. Evol. Gender*, 6, 97–130.

Luciana, S., Sajada. A., Lopita, H. & Kobita, C. (2004). *Does Dowry Improve Life for Brides? A Test of the Bequest Theory of Dowry in Rural Bangladesh*. New York: Population Council.

Mace, D. (1989). Three Ways of Helping Married Couples. *Journal of Marriage and Family Therapy*, 13, 179–185.

Mackeldey, F. (1883). *Handbook of the Roman Law*, Two Volumes in One. Philadelphia, PA: T. & J. W. Johnson & Co.

Madahia, M. E., Samadzadehb, M. & Javidi, N. (2013). The Communication Patterns and Satisfaction in Married Students. *Procedia: Social and Behavioral Sciences*, 84, 1190–1193.

Madathil, J. & Benshoff, J. M. (2008). Importance of Marital Characteristics and Marital Satisfaction: A Comparison of Asian Indians in Arranged Marriages and Americans in Marriages of Choice. *The Family Journal*, 16, 222–230.

Manataka American Indian Council (2017). *Fire Ceremony*. Retrieved on February 14, 2018 from https://www.manataka.org/page25.html#FIRE CEREMONY

Mann, S. (2008). Dowry Wealth and Wifely Virtue in Mid-Qing Gentry Households. *Late Imperial China*, 29(1S), 64–76.

Manne, S. L., Ostroff, J. S., Norton, T. R., Fox, K., Goldstein, L. & Grana, G. (2006). Cancer-related Relationship Communication in Couples Coping with Early Stage Breast Cancer. *Psychooncology*, 15(3), 234–247.

Martin, S. (2006). Trends in Marital Dissolution by Women's Education in the United States. *Demographic Research*, 15(20), 537–559.

Masci, D. (October 27 2016). *Shared Religious Beliefs in Marriage Important to Some, But Not All, Married Americans*. Washington, DC: Pew Research Center.

Matsuda, Y. (2017). Actor–Partner Interdependence Model Analysis of Sexual Communication and Relationship/Family Planning Factors among Immigrant Latino Couples in the United States. *Health Communication*, 32(5).

McBee, R. D. (2000). *Dance Hall Days: Intimacy and Leisure among Working-Class Immigrants in the United States* (pp. 222–228). New York: New York University Press.

McDaniel, A. (2005). Young Women's Dating Behavior: Why/Why Not Date a Nice Guy? *Sex Roles*, 53, 347–359.

MFRI (Marriage and Family Research Institute) (2018). *Marital Equality: Gender and Power in Couples Therapy*. Retrieved on December 1, 2018 from http://www.marriageandfamilyresearchinstitute.com/Marital-Equality.html

Michalos, A. C. (1986). Job Satisfaction, Marital Satisfaction, and the Quality of Life: A Review and a Preview. In Andrews, F. M. (Ed.), *Research on the Quality of Life* (pp. 57–83). Ann Arbor, MI: University of Michigan Press.

Mirecki, R. M., Chou, J. L., Elliott, M. & Schneider, C. M. (2013). *Journal of Divorce & Remarriage*, 54(1).

Mirza, R. M. (2007). *The Rise and Fall of the American Empire: A Re-Interpretation of History, Economics and Philosophy: 1492–2006*. Victoria, BC: Trafford Publishing.

Mitchell, J. W. & Gamarel, K. E. (2018). Constructive Communication Patterns and Associated Factors among

Male Couples. *Journal of Couple & Relationship Therapy,* 17(2), 79–96.

Mitchell, J. W., Champeau, D. & Harvey, S. M. (2013). Actor–Partner Effects of Demographic and Relationship Factors Associated with HIV Risk within Gay Male Couples. *Archives of Sexual Behavior*, 42(7), 1337–1345.

Mohler, A. (2016). *Marriage as a Part of Adulthood.* Colorado Springs, CO: Focus on the Family.

Mondor, J., McDuff, P., Lussier, Y. & Wright, J. (2011). Couples in Therapy: Actor–Partner Analyses of the Relationships between Adult Romantic Attachment and Marital Satisfaction. *American Journal of Family Therapy,* 39(2), 112–123.

Montesi, J. L., Fauber, R. L., Gordon, E. A. & Heimberg, R. G. (2011). The Specific Importance of Communicating about Sex to Couples' Sexual and Overall Relationship Satisfaction. *Journal of Social & Personal Relationships,* 28(5), 591–609.

Morris, M. L. (1999). Transition to Marriage: A Literature Review. *Journal of Family and Consumer Sciences Education,* 17(1), 1–21.

Moutafi, J., Furnham, A. & Crump, J. (2006). What Facets of Openness and Conscientiousness Predict Fluid Intelligence Score? *Learning and Individual Differences,* 16, 31–42.

Moyer-Gusé, E., Chung, A. H. & Jain, P. (2011). Identification with Characters and Discussion of Taboo Topics after Exposure to an Entertainment Narrative about Sexual Health. *Journal of Communication*, 61(3), 387–406.

Muraru, A. & Turliuc, M. (2013). Predictors of Marital Adjustment: Are There Any Differences Between Women and Men? *Europe's Journal of Psychology*, North America, 9, Aug. 2013.

Murphy, C. M. & O'Farrell, T. J. (1994). Factors Associated with Marital Aggression in Male Alcoholics. *Journal of Family Psychology*, 8, 321–335.

Myers, D. (1992). *The Pursuit of Happiness.* New York: Avon Books.

Myers, J. E., Madathil, J. & Tingle, L. R. (2005). Marriage Satisfaction and Wellness in India and the United States: A Preliminary Comparison of Arranged Marriages and Marriages of Choice. *Journal of Counseling & Development*, 83, 183–190.

Nadelson, C. & Notman, M. (1981). To Marry or Not to Marry: A Choice. *The American Journal of Psychiatry,* 138(10), 1352–1356.

Nakonezny, P. A. & Denton, W. H. (2008). Marital Relationships: A Social Exchange Theory Perspective. *The American Journal of Family Therapy,* 36, 402–412.

Nazzari, M. (1991). Disappearance of the Dowry, Women, Families and Social Change in São Paulo, Brazil, 1600–1900. Retrieved on December 18, 2017 from http://www.saopaulo.org/pdfs/SU.html

Nelson, P. A. & Thorne, A. (2012). Personality and Metaphor Use: How Extraverted and Introverted Young Adults Experience Becoming Friends. *Eur. J. Pers.,* 26, 600–612.

Noller, P. & Fitzpatrick, M. A. (1991). Marital Communication in the Eighties. In A. Booth (Ed.), *Contemporary Families: Looking Forward, Looking Back.* Minneapolis: National Council on Family Relations.

O'Leary, K. D. & Cano, A. (2001). Marital Discord and Partner Abuse: Correlates and Causes of Depression. In S. R. H. Beach (Ed.), *Marital and Family Processes in Depression* (pp. 163–182). Washington, DC: American Psychological Association.

Olson, D. H., Olson-Sigg, A. & Larson, P. J. (2008). *National Survey of Married Couples.* Retrieved on July 11, 2016 from https://www.prepare-enrich.com/pe/pdf/research/2011/national_survey_of_married_couples_2008.pdf

Omran, A. S, Shaikhaleslami, F., Tabari, R., Kazemnejad, L. E. & Pariad, E. (2016). Role of Career Factors on Marital Satisfaction of Nurses. *Nursing and Midwifery Community,* 25(78), 102–109.

Papp, L. M., Kouros, C. D. & Cummings, E. M. (2009). Demand–Withdraw Patterns in Marital Conflict in the Home. *Personal Relationships,* 16(2), 285–300.

Parker, K. & Patten, P. (2013). *The Sandwich Generation: Rising Financial Burdens for Middle-Aged Americans.* Washington, DC: Pew Research Center.

Parker, K. & Stepler, R. (2017a). *As U.S. Marriage Rate Hovers at 50%, Education Gap in Marital Status Widens.* Washington, DC: Pew Research Center.

Parker, K. & Stepler, R. (2017b). *Americans See Men as the Financial Providers, Even as Women's Contributions Grow.* Washington, DC: Pew Research Center.

Parry, T. (2011). An Irregular Union: Exploring the Welsh Connection to a Popular African American Wedding Ritual. In A. L. Becker & K. Noone (Eds.), *Welsh Mythology and Folklore in Popular Culture: Essays on Adaptations in Literature, Film, Television and Digital Media* (pp. 109–110, 123–124). Jefferson, NC: McFarland and Company.

Passer, M. W. & Smith, R. E. (2009). *Psychology: The Science of Mind and Behavior.* New York: McGraw-Hill Higher Education.

Paulhus, D. L. & Williams, K. M. (2002). The Dark Triad of Personality: Narcissism, Machiavellianism, and Psychopathy. *Journal of Research in Personality,* 36(6), 556–563.

Phillips, R. (1991). *Untying the Knot: Divorce in Western Society.* Cambridge: Cambridge University Press.

Pogány, Á., Barta, Z., Szurovecz, Z., Székely, T. & Vincze, E. (2014). Mate Preference Does Not Influence Reproductive Motivation and Parental Cooperation in Female Zebra Finches. *Behaviour,* 151, 1885–1901.

Rabby, A. A., Raihman, U., Hossen, A. & Sultana, M. (2013). Dowry and its Impact on Women's Marital Life: A Study in Sylhet City, Bangladesh. *South Asian Anthropologist*, 13(1), 17–25.

Rajabi, G., Sarvestani, Y., Aslani, K. & Khojastemehr, R. (2013). Predicators of Marital Satisfaction in Married Female Nurses. *IJN*, 26(82), 23–33.

Regan, P. C., Lakhanpal, S. & Anguiano, C. (2012). Relationship Outcomes in Indian-American Love-based and Arranged Marriages. *Psychol. Rep.*, 110, 915–924.

Rentfrow, P. J. & Jokela, M. (December 2016). Geographical Psychology. *Current Directions in Psychological Science*, 25(6), 393–398.

Rezaeanlangroodi, R., Azizinazahad, M. & Hashemi, M. (2011). The Investigation of Biological and Psychological Marriage Criteria from the Viewpoints of Iranian University Students with that of Their Parents. *Procedia: Social and Behavioral Sciences*, 28, 406–410.

Roberts, B. W., Jackson, J. J., Fayard, J. V., Edmonds, G. & Meints, J (2009). Conscientiousness. In M. R. Leary & R. H. Hoyle, *Handbook of Individual Differences in Social Behavior* (pp. 257–273). New York, London: Guildford Press.

Robles, T. F. and Kiecolt-Glaser, J. K. (2003). The Physiology of Marriage: Pathways to Health. *Physiology and Behavior*, 79, 409–416.

Rogge, R. D., Bradbury, T. N., Hahlweg, K., Engl, J. & Thurmaier, F. (2006). Predicting Marital Distress and Dissolution: Refining the Two Factor Hypothesis. *Journal of Family Psychology*, 20(1), 156–159.

Rosen, R. (2016). Marriage Will Not Fix Poverty. *The Atlantic*. Retrieved on February 18, 2018 from https://www.theatlantic.com/business/archive/2016/03/marriage-poverty/473019/

Rothmann, S. & Coetzer, E. P. (2003). The Big Five Personality Dimensions and Job Performance. *SA Journal of Industrial Psychology*, 29.

Russell, D. E. H. & Harmes, R. A. (Eds.) (2011). *Femicide in Global Perspective*. New York: Teachers College Press.

Ryan, K. M., Weikel, K. & Sprechini, G. (2008). Gender Differences in Narcissism and Courtship Violence in Dating Couples. *Sex Roles*, 58, 802–813.

Sabey, A. K., Rauer, A. J. & Jensen, J. F. (2014). Compassionate Love as a Mechanism Linking Sacred Qualities of Marriage to Older Couples' Marital Satisfaction. *J Fam Psychol.*, Oct; 28(5), 594–603.

Şahin, N. H., Sermin, T., Ergin, A. B., Taşpinar, A., Balkaya, N. A. & Çubukçu, S. (2010). Childhood Trauma, Type of Marriage and Self-Esteem as Correlates of Domestic Violence in Married Women in Turkey. *J. Fam. Violence*, 25, 661–668.

Sandvik, M., Rosenqvist, G. & Berglund, A. (2000). Male and Female Mate Choice Affects Offspring Quality in a Sex-Role-Reversed Pipefish. *Proc. R. Soc. B Biol. Sci.*, 267, 2151–2155.

Sasser, L. (2005). *Empty Nest Syndrome*. Colorado Springs, CO: Focus on the Family.

Scheidel, W. (2009a). A Peculiar Institution? Greco-Roman Monogamy in Global Context. *History Family*, 14, 280–291.

Scheidel, W. (2009b). Sex and Empire: A Darwinian Perspective. In I. Morris & W. Scheidel (Eds.), *The Dynamics of Ancient Empires: State Power from Assyria to Byzantium* (pp. 255–324). Oxford, England: Oxford University Press.

Schlegel, A. & Eloul, R. (1988). Marriage Transactions: Labor, Property, Status. *American Anthropologist*, 90, 291–309.

Schmitt, D. P., Realo, A., Voracek, M. & Allik, J. (2008). Why Can't a Man Be More Like a Woman? Sex Differences in Big Five Personality Traits across 55 Cultures. *Journal of Personality and Social Psychology*, 94(1), 168–182.

Schwartz, P. (1994). *Peer Marriage: How Love between Equals Really Works*. New York: Free Press.

Seeney, H. (August 2007). Macedonia: Attending the Galicnik Wedding Festival. Retrieved on November 29, 2018 from https://www.dw.com/en/macedonia-attending-the-gali%C4%8Dnik-wedding-festival/a-2773483

Shahhosseini, Z., Hamzehgardeshi, Z. & Kardan Souraki, M. (2014). The Effects of Premarital Relationship Enrichment Programs on Marriage Strength: A Narrative Review Article. *Journal of Nursing and Midwifery Sciences*, 1(3), 62–72.

Shanahan, M. J., Porfeli, E. J., Mortimer, J. T., Erickson, L. D. (2005). Subjective Age Identity and the Transition to Adulthood: When Do Adolescents Become Adults? In R. Settersten Jr., F. Furstenberg, Jr. & R. Rumbaut (Eds.), *On The Frontier of Adulthood: Theory, Research, and Public Policy* (pp. 225–255). Chicago: University of Chicago Press.

Sharon Bock, Clerk & Comptroller Palm Beach County. (August 2017). *Valentine's Day Group Wedding Ceremony 2017*.

Singh, K. (2013). *Laws and Son Preference in India: A Reality Check*. United Nations Population Fund (UNFPA) – India.

Skeem, J. L., Polaschek, D. L. L., Patrick, C. J. & Lilienfeld, S. O. (2011). Psychopathic Personality: Bridging the Gap between Scientific Evidence and Public Policy. *Psychological Science in the Public Interest*, 12(3), 95–162.

Skowroński, D. P., Othman, A. B., Siang, D. T. W., Han, G. L. W., Yang, J. W. J. & Waszyńska, K. (2014). The Outline of Selected Marital Satisfaction Factors in the Intercultural Couples based on the Westerner and non-Westerner relationships. *Polish Psychological Bulletin* 45(3), 346–356.

Snyder, D. K., Heyman, R. E. & Haynes, S. N. (2005). Evidence-based Approaches to Assessing Couple Distress. *Psychological Assessment,* 17(3), 288–307.

Sorokowski, P., Groyecka, A., Karwowski, M., Manral, U., Kumar, A., Niemczyk, A. & Pawłowski, B. (2017). Free Mate Choice Does Not Influence Reproductive Success in Humans. *Scientific Reports,* 7, 10127.

Spangler, A., Brown, S. L., Lin, I.-F., Hammersmith, A. & Wright, M. (2016). Divorce Timing and Economic Well-being. *Family Profiles.* FP-16–01. Bowling Green, OH: NCFMR.

Spanier, G. B. (1979). The Measurement of Marital Quality. *J. Sex Marital Ther.,* Fall; 5(3), 288–300.

Spanier, G. B. & Lewis, R. A. (1980). Marital Quality: A Review of the Seventies. *Journal of Marriage and Family,* 42, 825–839.

Spooky (2017). China's Increasing "Bride Price" Makes Marriage Virtually Impossible for Poor Bachelors. *Odditycentral.*

Stadter, P. A. (1967). Flavius Arrianus: The New Xenophon. *Roman, and Byzantine Studies,* 8, 155–161.

Stanley, S. M. (2007). Assessing Couple and Marital Relationships: Beyond Form and Toward a Deeper Knowledge of Function. In L. M. Casper & S. L. Hoffereth (Eds.), *Handbook of Measurement Issues in Family Research* (pp. 85–100). Mahwah, NJ: Lawrence Erlbaum & Associates.

Stewart, K. L. & Olson, D. H. (1990). *Predicting Premarital Satisfaction on PREPARE using Background Factors.* Unpublished manuscript. Minneapolis, MN: PREPARE/ENRICH, Inc.

Stoker, S. L. (2004). *Factors that Influence Marital Satisfaction in Couples Raising a Child with Cerebral Palsy.* All Graduate Theses and Dissertations, 2612. https://digitalcommons.usu.edu/etd/2612

Stutzer, A. & Frey, B. S. (2006). Does Marriage Make People Happy, or Do Happy People Get Married? *The Journal of Socio-Economics,* 35(2), 326–347.

Sutherland, A. (July 30, 2014). A "Beta Marriage" Is No Marriage at All. *IFS.* Retrieved on February 18, 2018 from https://ifstudies.org/blog/a-beta-marriage-is-no-marriage-at-all

Sutin, A. R., Costa, P. T., Wethington, E. & Eaton, W. (2010). Perceptions of Stressful Life Events as Turning Points Are Associated with Self-rated Health and Psychological Distress. *Anxiety, Stress, and Coping,* 23(5), 479–492.

Tambiah, S. J. (1989). Bridewealth and Dowry Revisited: The Position of Women in Sub-Saharan Africa and North India. *Current Anthropology,* 30(4), 426.

The Wedding Report (2018). Wedding Statistics for the United States. Retrieved on July 30, 2017 from https://www.theweddingreport.com/

Thompson, E. R. (2008). Development and Validation of an International English Big-Five Mini-Markers. *Personality and Individual Differences,* 45(6), 542–548.

Thorne, A., McLean, K. C. & Lawrence, A. M. (2004). When Remembering Is Not Enough: Reflecting on Self-Defining Memories in Late Adolescence. *J. Pers.,* Jun; 72(3), 513–541.

Törnblom, K. & Kazemi, A. (Eds.) (2012). *Handbook of Social Resource Theory: Theoretical Extensions, Empirical Insights, and Social Applications.* Critical Issues in Social Justice. New York: Springer.

Townsend, J. M. & Levy, G. D. (1990). Effects of Potential Partners' Physical Attractiveness and Socioeconomic Status on Sexuality and Partner Selection. *Arch Sex Behav.,* Apr; 19(2), 149–164.

Treas, J. & Kim, J. (2016). *Marital Power.* The Wiley Blackwell Encyclopedia of Family Studies, 1–5.

Trudel, G. & Goldfarb, M. R. (2010). Marital and Sexual Functioning and Dysfunctioning, Depression and Anxiety. *Sexologies,* 19(3), 137–142.

Twenge, J. M., Campbell, W. K. & Foster, C. A. (2003). Parenthood and Marital Satisfaction: A Meta-analytic Review. *Journal of Marriage and Family,* 65, 574–583.

Tyler, L. G. (1907). *The Cradle of the Republic.* Richmond: Hermitage Press, Inc.

Urbaniak, G. C. & Kilmann, P. R. (2006). Niceness and Dating Success: A Further Test of the Nice Guy Stereotype. *Sex Roles,* 55, 209–224.

U.S. Census Bureau (2000). *Estimated Median Age at First Marriage, by Sex: 1890 to the Present.* Accessed online at: www.census.gov/population/socdemo/ms-la/tabms-2.txt, on Oct. 11, 2000.

U.S. Census Bureau (September 15, 2004). *Estimated Median Age at First Marriage, by Sex: 1890 to the Present.*

U.S. Census Bureau (August 14, 2017a). *FFF: Unmarried and Single Americans Week: Sept. 17–23, 2017.*

U.S. Census Bureau (November 15, 2017b). Table MS-2. Estimated Median Age at First Marriage: 1890 to Present.

U.S. Census Bureau (2018). *Stats for Stories: National Spouses Day.* Retrieved on March 5, 2018 from https://www.census.gov/newsroom/stories/2018/spouses.html

Van Landeghem, B. (2012). A Test for the Convexity of Human Well-Being over the Life Cycle: Evidence from a 20-Year Panel. *Journal of Economic Behavior and Organization,* 81, 571–582.

Vanover, B. (2016). Important Factors in Marital Success and Satisfaction: Marriage Counselors' Perspectives. *Master of Social Work Clinical Research Papers.* Paper 685. Retrieved on November 2016 from http://sophia.stkate.edu/msw_papers/685

Vespa, J. (April 2017). *Changing Economics and Demographics of Young Adulthood: 1975–2016.* Current Population Reports P20–579. U.S. Census Bureau.

Wachs, K. & Cordova, J. (2007). Mindful Relating: Exploring Mindfulness and Emotion Repertoires in Intimate Relationships. *Journal of Marital and Family Therapy*, 33(4), 464–481.

Wallerstein, J. S. & Blakeslee, S. (1996). *The Good Marriage: How and Why Love Lasts.* New York: Grand Central Publishing.

Walsh, M. & Taylor, J. (1982). Understanding in Japanese Marriages. *Journal of Social Psychology*, 118, 67–76.

Walster, E., Walster, G. W. & Berscheid, E. (1978). *Equity: Theory and Research.* Boston, MA: Allyn & Bacon.

Webb, D. G. (1972). Relationship of Self-Acceptance and Self-Disclosure to Empathy and Marital Need Satisfaction (Doctoral Dissertation, United States International University). *Dissertation Abstracts International*, 33, 432–438.

WebDC (2017). *Male Menopause.* Retrieved on February 15, 2018 from https://www.webmd.com/men/guide/male-menopause

Weiser, D. A. & Weigel, D. J. (2017). Exploring Intergenerational Patterns of Infidelity. *Pers Relationship*, 24, 933–952.

West, M. (2017). Self-disclosure, Trauma and the Pressures on the Analyst. *J. Anal Psychol.* 62(4), 585–601.

Wethington, E. (2003). Turning Points as Opportunities for Psychological Growth. In C. L. Keyes & J. Haidt (Eds.), *Flourishing: Positive Psychology and the Life Well-lived* (pp. 37–53). Washington, DC: American Psychological Association.

Wethington, E. (2005). An Overview of the Life Course Perspective: Implications for Health and Nutrition. *J. Nutr Educ Behav.*, May–Jun; 37(3), 115–120.

Wethington, E., Kessler, R. C. & Pixley, J. E. (2004). Turning Points in Adulthood. In O. G. Brim, C. D. Ryff, R. C. Kessler (Eds.), *How Healthy Are We? A National Study of Well-being at Midlife* (pp. 586–613). Chicago: University of Chicago Press.

Whisman, M. A. & Bruce, M. L. (1999). Marital Dissatisfaction and Incidence of Major Depressive Episode in a Community Sample. *J. Abnorm Psychol.*, Nov; 108(4), 674–678.

White, D. R., Betzig, L., Borgerhoff Mulder, M., Chick, G., Hartung, J., Irons, W., Low, B. S. & Otterbein, K. F. (1988). Rethinking Polygyny: Co-wives, Codes, and Cultural Systems (includes comments and author's reply). *Curr. Anthropol.*, 29(4), 529.

White, L. & Rogers, S. J. (2000). Economic Circumstances and Family Outcomes: A Review of the 90s (Working Paper No. 00–03). *Journal of Marriage and Family*, 62(4), 1035–1051.

Wiki News (2007). Mass Wedding Held against Racism in Belgium. Retrieved on October 2018 from http://en.wikinews.org/wiki/Mass_wedding_held_against_racism_in_Belgium

Wilcox, B. (2009). The Evolution of Divorce. *National Affairs.* Retrieved on February 19, 2018 from https://www.nationalaffairs.com/publications/detail/the-evolution-of-divorce

Wilcox, B. & Wang, W. (June 14, 2017). *The Millennial Success Sequence: Marriage, Kids, and the "Success Sequence" Among Young Adults.* Washington, DC: American Enterprise Institute.

Wilcox, B., Lerman, R. I. & Price, J. (2015). *Strong Families, Prosperous States: Do Healthy Families Affect the Wealth of States.* Washington, DC: American Enterprise Institute.

Wilcox, W. B. (2004). *Soft Patriarchs, New Men: How Christianity Shapes Fathers and Husbands.* Chicago: University of Chicago Press.

Wilcox, W. B. & Nock, S. L. (2006). What's Love Got to Do with it? Equality, Equity, Commitment and Women's Marital Quality. *Social Forces*, 84, 1321–1345.

William, A., Haviland, H. E. L., Prins, D. & McBride, W. B. (2013). *Telecourse Study Guide for Haviland/Prins/Walrath/McBride's Anthropology: The Human Challenge*, 14th edn. Belmont, CA: Cengage Learning.

Wilson, N. (2005). *Encyclopedia of Ancient Greece* (Encyclopedias of the Middle Ages). New York: Routledge.

Wolfe, D. (1997). Emboldened Rev Moon to Bring His Blessing Ceremony to our Nation's Capital. Retrieved on November 29, 2018 from http://www.newcovpub.com/unification/masswedding.html

Wolfinger, N. H. (2016). *Counterintuitive Trends in the Link between Premarital Sex and Marital Stability.* Charlottesville, VA: Institute for Family Studies.

Wu, H. (2018). Homogamy in U.S. Marriage, 2016. *Family Profiles*, FP-18–18. Bowling Green, OH: National Center for Family & Family Research.

Wu, H. & Brown, S. L. (2016). Long-term Marriage among Older Adults. *Family Profiles*, FP-16–08. Bowling Green, OH: National Center for Family & Family Research.

Yazdanpanah, F., Khalili, M. & Keshtkaran, Z. (2015). Level of Marital Satisfaction in Couples Living in Iran. *Indian Journal of Research*, 4(4), 4–7.

Yelsma, P. & Athappilly, K. (1988). Marriage Satisfaction and Communication Practices: Comparisons among Indian and American Couples. *Journal of Comparative Family Studies*, 19, 37–54.

Yoo, H., Bartle-Haring, S., Day, R. D. & Gangamma, R. (2014). Couple Communication, Emotional and Sexual Intimacy, and Relationship Satisfaction. *Journal of Sex & Marital Therapy*, 40(4), 275–293.

Young, M., Riggs, S. & Kaminski, P. (July 21, 2017). Role of Marital Adjustment in Associations between Romantic Attachment and Coparenting. *Family Relations Interdisciplinary Journal of Applied Family Science*, 66(2), 331–345.

Zaheri, F., Dolatian, M., Shariati, M., Simbar, M., Ebadi, A. & Azghadi, S. B. H. (2016). Effective Factors in Marital Satisfaction in Perspective of Iranian Women and Men: A Systematic Review. *Electronic Physician*, 8(12), 3369–3377.

Zhou, Y., Wang, K., Chen, S., Zhang, J. & Zhou, M. (2017a). The Actor, Partner, Similarity Effects of Personality, and Interactions with Gender and Relationship Duration among Chinese Emerging Adults. *Front. Psychol.*, 8, 1698.

Zhou, Y., Wang, K., Chen, S., Zhang, J. & Zhou, M. (2017b). An Exploratory Investigation of the Role of Openness in Relationship Quality among Emerging Adult Chinese Couples. *Front. Psychol.*, 8, 382.

10
PREPARING FOR CHILDREN AND CHILD ADOPTION

Learning Outcomes

Upon completion of the chapter, students should be able to:

1. describe fertility patterns in the United States
2. explain factors linked to the likelihood of rising maternal age
3. apply demographic transition theories to the analysis of fertility patterns
4. describe reasons for choosing to have or not to have a child
5. describe how a pre-baby checklist helps people calculate the potential cost of raising a child
6. explain five birth control methods
7. describe three ways of handling an unintended pregnancy
8. describe how prenatal care prevents risks during every stage of pregnancy and in childbirth
9. explain how postnatal care helps prevent risks after childbirth
10. describe causes and treatment of infertility
11. explain types of adoption and experiences of adoptive families
12. describe the characteristics of childfree families
13. explain sex selection and consequences on a global basis.

Brief Chapter Outline

Pre-test

Fertility Patterns in the United States

- Declining Fertility Rates
    - The Birth Rate; The General Fertility Rate; The Total Fertility Rate
- Teen Births
- Unmarried Childbearing
- Multiple Births
    - Twins; Triplet and Higher-order Multiple Births

Rising Mean Age of Mother

Mean Age by Birth Order; Mean Age by Birth Rate

Increasing Multiracial and Minority Babies

Shrinking Family Size

Factors Influencing Family Size and Fertility

Demographic Transition Theory

Second Demographic Transition

Diversity Overview: Childfree/Childless Families

Deciding Whether to Have a Child

Reasons People Choose to Have a Child

Reasons People Choose to Not Have a Child

Costs of Having a Child

Calculating Cost: The Pre-baby Checklist

Preventing Pregnancy

Birth Control Methods

Unintended Pregnancy

Abortion; Placing a Child for Adoption; Parenting

Pregnancy

Prenatal Care

Postnatal Care

Stages of Pregnancy

Stages of Labor

Method of Delivery

Self-control over Method of Delivery

Period of Gestation

Early-, Full- and Late-term Birth; Preterm Birth; Causes, Consequences, and Prevention of Preterm Birth

Infertility

Causes and Treatment of Infertility

Adoption

Types of Adoption

Domestic Adoption

Types of Domestic Adoption

International Adoption

Adoptive Families

Global Overview: Sex Selection

Summary

Key Terms

Discussion Questions

Suggested Film and Video

Internet Sources

References

Pre-test

Engaged or active learning is a powerful strategy that leads to better learning outcomes. One way to become an active learner is to begin the chapter and try to answer the following true/false statements from the material as you read. You will find that you have ready answers to these questions upon the completion of the chapter.

The general fertility rate is total births per 1,000 women ages 15 to 44 per year.	T	F
The average age of mothers at first birth has risen.	T	F
Involuntary childlessness refers to couples who want children but cannot conceive or are past childbearing age.	T	F
The United States has completed the demographic transition.	T	F
A triplet birth refers to an individual live birth in a triplet pregnancy.	T	F
A childfree family is a form of family in which couples choose not to have kids.	T	F
Options for handling an unplanned pregnancy include abortion, placing a child for adoption, and parenting.	T	F
Open adoption allows identifying information to be communicated between adoptive and biological parents.	T	F
Prenatal care improves pregnancy outcomes.	T	F
The postpartum period is the period beginning after the birth of a child.	T	F

After completion of this section, students should be able to:
LO1: Describe fertility patterns in the United States.
LO2: Describe factors linked to rising maternal age.
LO3: Apply demographic transition theories to the analysis of fertility patterns.

According to structural functionalist theory, one function of the family is to bear and rear children and therefore provide society with productive new members. In fact, the main reason the institution of marriage developed was to promote successful childrearing (Popenoe, 2009). Of course, not all couples decide to have children and, as we will see, patterns of childbirth are changing in the United States. Let's start our chapter by looking at fertility in the United States.

Fertility Patterns in the United States

Three factors that affect changes in a nation's population are *fertility*, *mortality*, and *migration*. In the United States, for example, there is one birth every 8 seconds, one death every 12 seconds, and one international migrant (net) every 29 seconds (U.S. Census Bureau, 2018). Let's take a look at fertility patterns in the United States.

Declining Fertility Rates

Today young adults are experiencing traditional milestones such as getting a job, marrying, and having children at a later age than their parents (Vespa, 2017). **Fertility** is the natural ability to conceive offspring. Tracking trends in fertility is essential in planning for current and future populations. High fertility has profound effects on their families and on socioeconomic development from national to local levels, on supplies of resources, and on environmental sustainability

(UNFPA, 2009). For example, when the baby boomer generation (born between 1946 and 1964) hit school age, existing schools were not able to accommodate the large numbers of incoming students. Conversely, low fertility rates may lead to issues concerning lack of future labor due to *depopulation*—substantial reduction in the population of an area. Let's look at three common ways to measure fertility.

The Birth Rate

The **birth rate** (or natality)—the number of live births per 1,000 population people in a year—is a basic measure of fertility. It can be calculated by dividing the total population and multiplying by 1,000 (birth rate = number of births ÷ total population × 1,000). In 2017, the global birth rate was 18.5 births per 1,000 population (CIA, 2017a), compared with 44.2 births in Angola and 6.6 births in Monaco. In the United States, the birth rate was 12.5 births in 2017 (CIA, 2017b), down from 2010 (13.0) (see Table 10.1).

Specific birth rate is used relative to a specific criterion such as age, gender, and race. Birth rates have fallen for seven specific age groups since 1957 (see Figure 10.1). For example, the birth rate declined to record lows for women aged 20–24 from 1957 (260.6) to 2015 (76.8).

Birth rates have fallen for all racial and ethnic groups, but rates among Hispanic (16.0) and Native Hawaiian or other Pacific Islander (16.8) remain higher than those for White (10.5) in 2016 (see Table 10.1). In 1990–2015, declines in birth rates were seen for American Indian or Alaska Native from

Table 10.1 Births and Birth Rates: United States, 2010–2016, and by Race, 2016

All Races and Origins	*Number*	*Birth Rate (births per 1,000 population)*	*General Fertility Rate (the number of births per 1,000 females aged 15–44)*
2016	3,945,875	12.2	62.0
2015	3,978,497	12.4	62.5
2014	3,988,076	12.5	62.9
2013	3,932,181	12.4	62.5
2012	3,952,841	12.6	63.0
2011	3,953,590	12.7	63.2
2010	3,999,386	13.0	64.1
Single Race, 2016	*Number*	*Birth Rate*	*General Fertility Rate*
White	2,056,332	10.5	58.8
Black	558,622	14.0	63.3
American Indian or Alaska Native (AIAN)	31,452	13.3	62.7
Asian	254,471	14.6	61.1
Native Hawaiian or Other Pacific Islander (NHOPI)	9,342	16.8	72.9
Hispanic (all persons of Hispanic origin of any race)	918,447	16.0	70.6

Source: NCHS, National Vital Statistics System, Natality, Martin *et al.* (2018a). https://www.cdc.gov/nchs/data/nvsr/nvsr67/nvsr67_01.pdf

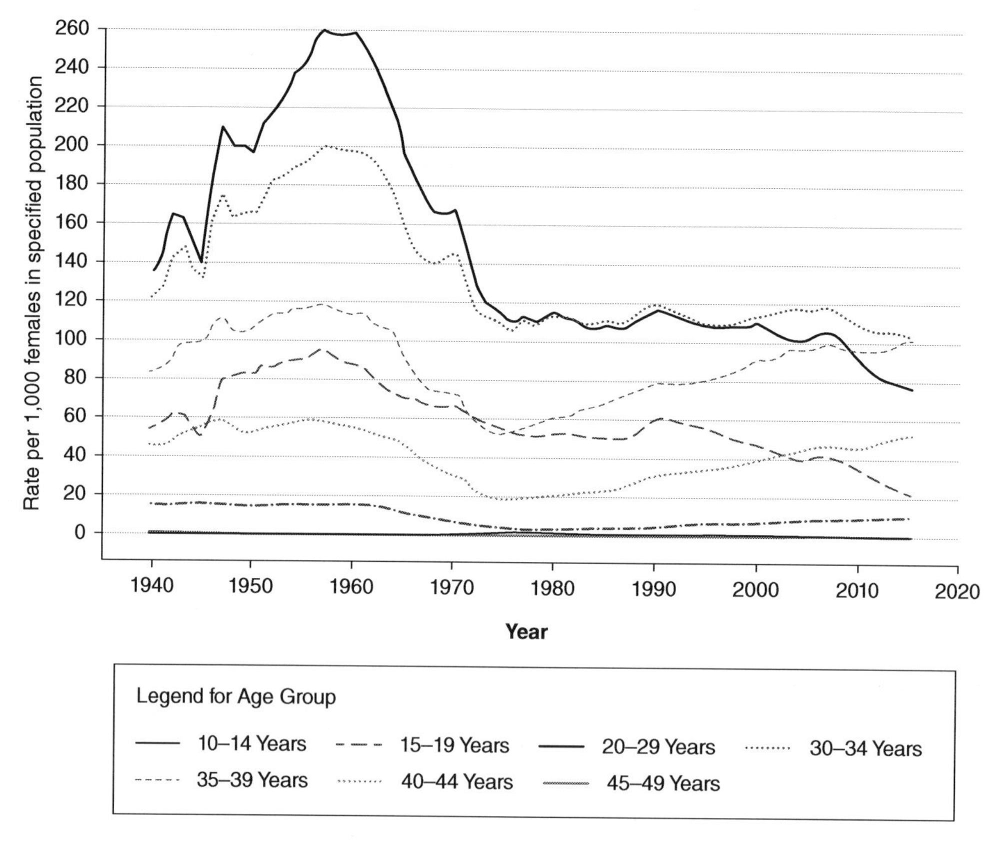

Figure 10.1 Birth Rates by Age Groups, 1940–2015
Sources: CDC/NCHS, National Vital Statistics System, birth data; Hamilton *et al.*, (2017). https://www.cdc.gov/nchs/data-visualization/natality-trends/index.htm

18.9 to 9.7, for White from 14.4 to 10.7, for Asian/Pacific Islander from 19.0 to 14.0, for Black from 23.0 to 14.2, and for Hispanic women from 26.7 to 16.3 (see Figure 10.2).

The changes in personal income and initial unemployment claims were related to changes in fertility rates (Livingston, 2011). In the United States, record low points occurred during periods of serious economic crisis (PRB, 2009). The number of births declined from 1921 (3,055,000) to 1933 (2,307,000) during the Great Depression in the 1920s and from 4,300,000 in 1957 to 3,136,965 in 1973 during the traumatic "oil shock" inflationary period of the 1970s. During the Great Recession from December 2007 to June 2009, the unemployment rate rose from 5 percent to 10.1 percent and housing prices plummeted by 20 percent (Duque *et al.*, 2018). The number of births declined from 4,326,233 in 2007 to 3,932,181 in 2013 (see Figure 10.3).

The General Fertility Rate

A more precise measure of fertility patterns is the **general fertility rate**—the number of births per 1,000 women aged 15–44 in a population in a year. The general fertility rate can be obtained by dividing the number of births by the number of women aged 15 through 44 and multiplying by 1,000 (general fertility rate = total number of live births ÷ females aged

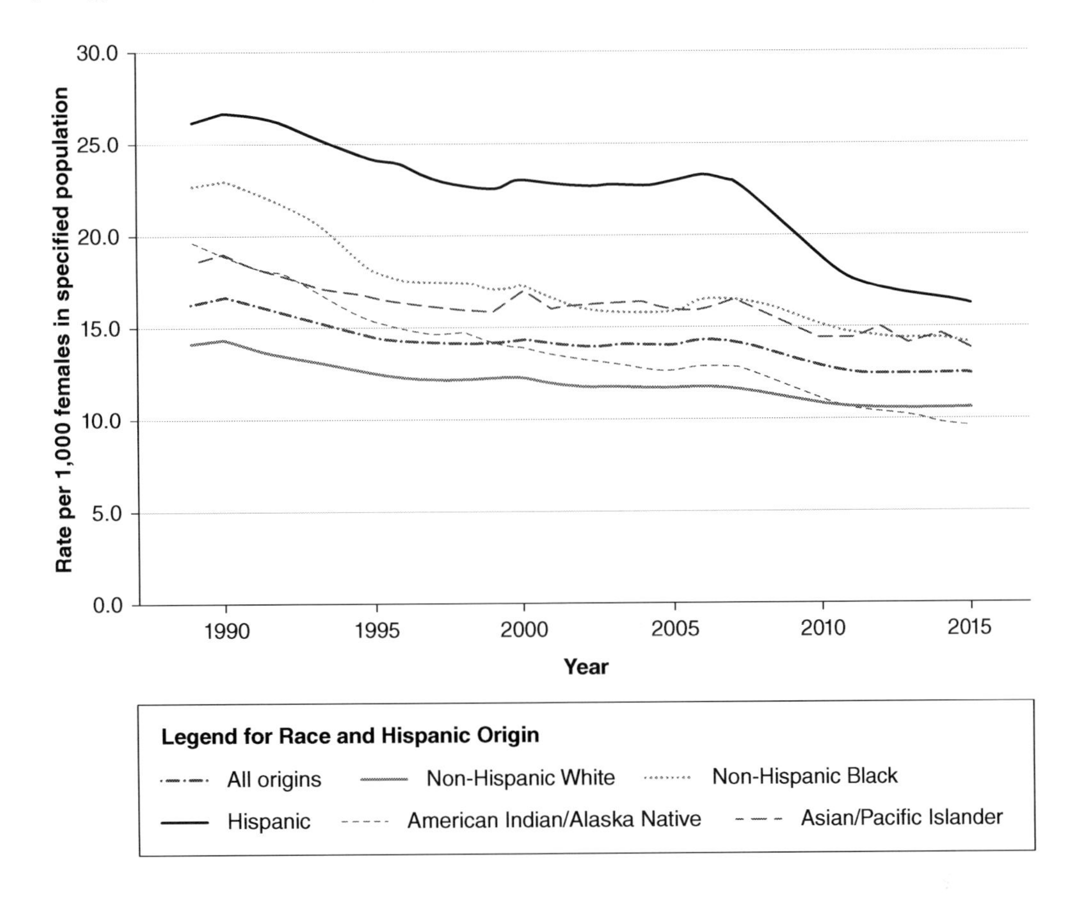

Figure 10.2 Birth Rates by Race and Hispanic Origin, the United States, 1960–2015
Sources: CDC/NCHS, National Vital Statistics System, birth data, Hamilton *et al.*, (2017). https://www.cdc.gov/nchs/data-visualization/natality-trends/index.htm

15–49 × 1000). In the United States, the general fertility rates declined from 1957 (122.9) to 2015 (62.5) per 1,000 females (see Figure 10.3) and declined from 2016 (62) to 2017 (60.3) (Martin *et al.*, 2018b).

The general fertility rate varies among racial groups. Native Hawaiian or other Pacific Islanders (NHOPI) had the highest fertility rate (72.9), compared with Hispanic (70.6), Black (63.3), American Indian or Alaska Native (AIAN) (62.7), and Asian (61.1) in 2016 (see Table 10.1). Between 2016 and 2017, the largest declines were for Asian and AIAN females, down 5 percent each, to 58.0 and 59.5, respectively (see Figure 10.4).

The Total Fertility Rate

The **total fertility rate** refers to the average number of children born per woman if all women survive their childbearing years and is a more direct measure of the level of fertility than the birth rate since it refers to births per woman (CIA, 2018a). The total fertility rate can be calculated by adding up the age-specific birth rates within a five-year increment during the reproductive years 10 through 49 (see Figure 10.1) and multiplying by 5 (total fertility rate = $\sum$ [five-year age-specific birth rates for females aged 10–49] × 5).

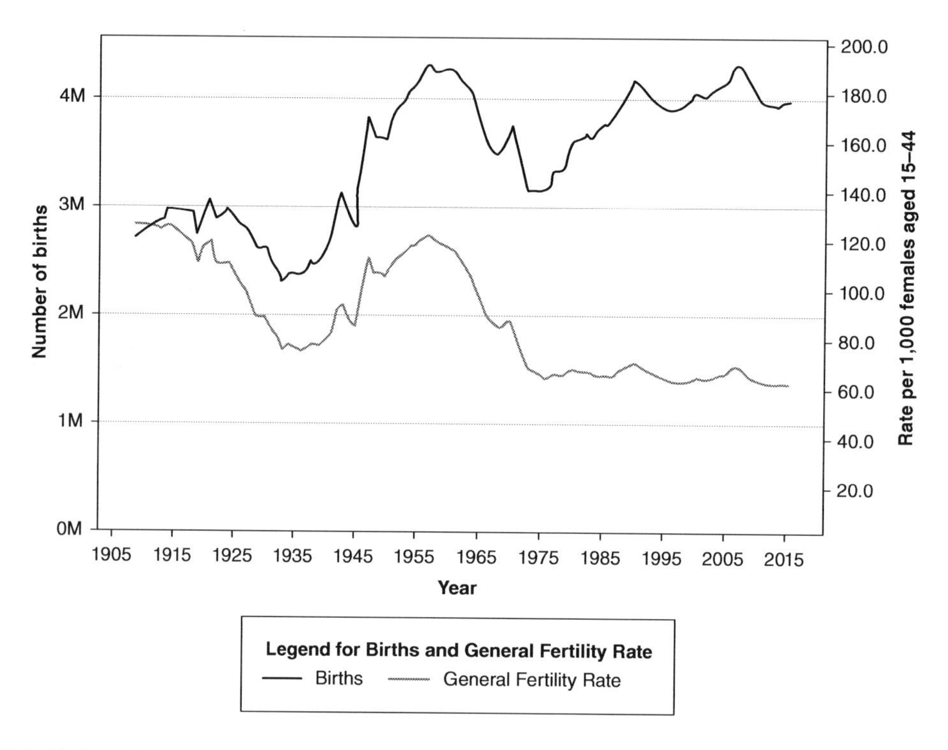

Figure 10.3 Births and General Fertility Rates, 1909–2015

Sources: CDC/NCHS, National Vital Statistics System, birth data; Hamilton *et al.*, (2017). https://www.cdc.gov/nchs/data-visualization/natality-trends/index.htm

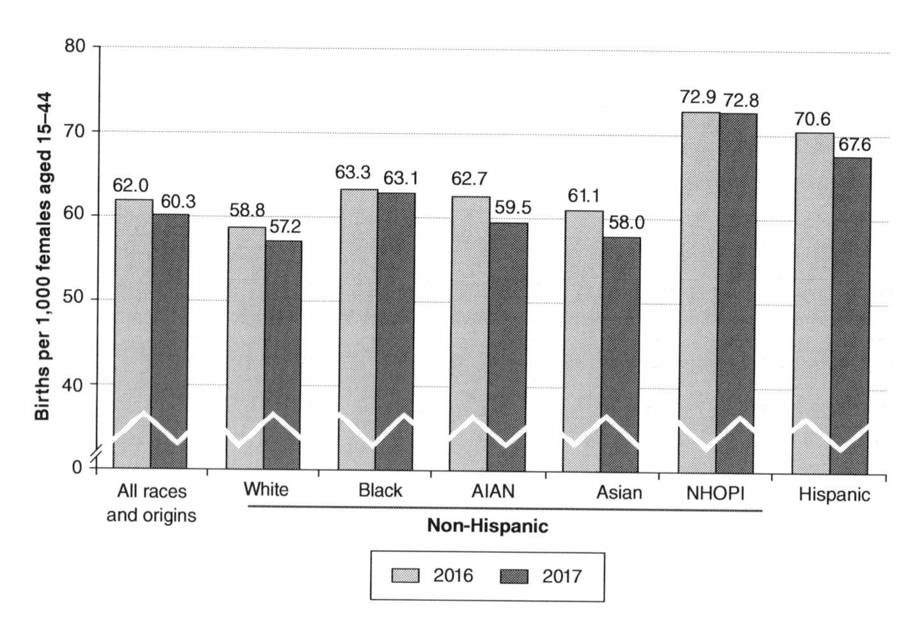

Figure 10.4 General Fertility Rates, by Race and Hispanic Origin of Mother: United States, 2016 and 2017

Source: NCHS, National Vital Statistics System, Natality. https://www.cdc.gov/nchs/products/databriefs/db318.htm

In the United States, total fertility rates have seen ups and downs. In 1936, the total fertility rate fell to 2.1 children per woman in the wake of the stock market crash of 1929, rose to three children born per woman during the baby boom between 1946 and 1964 (PRB, 2009), fell to a record low of 1.738 in 1976, remained at about 1.8 for the first half of the 1980s, rose between 1989 (2.014) and 1994 (2.002), and fell to 1.87 in 2017 (CIA, 2018a).

Total fertility rates vary among age and racial groups. Women aged 40–44 had 2.07 children in 2014, up from 2006 (1.86) (Livingston, 2018). Hispanic (2,092.5) and Native Hawaiian or Pacific Islander (2,076.5) had the highest total fertility rate, compared with Black (1,832.5), White (1,719.0), and Asian (1,690.5) per 1,000 women (see Table 10.2).

Replacement level fertility (total fertility rate of 2.1 children born per woman) is the level of fertility at which a population exactly replaces itself from one generation to the next (Craig, 1994). A total fertility rate of 2.1 is equivalent to the replacement rate of 1, which means that on average one woman in a community would produce an average of one *girl* (to replace her mother) throughout her reproductive life (Yip *et al.*, 2006). The global total fertility rate is 2.42 children born per woman in 2016 (CIA, 2017a), compared with 6.49 in Niger and 0.83 in Singapore, Macau (0.95), Japan (1.41), Greece (1.43), Italy (1.44), Germany (1.45), China (1.60), Canada (1.60), Brazil (1.75), Australia (1.77), Belgium (1.78), Norway (1.85), United Kingdom (1.88), and Sweden (1.88) below replacement fertility level (CIA, 2017b). Replacement level fertility will lead to zero population growth only if mortality rates remain constant and migration has no effect (Craig, 1994).

Table 10.2 Total Fertility Rates in United States, 2010–2016 and by Race

All Races and Origins	*Total Fertility Rate (births per 1,000 women)*
2016	1,820.5
2015	1,843.5
2014	1,862.5
2013	1,857.5
2012	1,880.5
2011	1,894.5
2010	1,931.0
Single Race, 2016	*Total Fertility Rate (births per 1,000 women)*
White	1,719.0
Black	1,832.5
American Indian or Alaska Native (AIAN)	1,794.5
Asian	1,690.5
Native Hawaiian or Other Pacific Islander (NHOPI)	2,076.5
Hispanic (all persons of Hispanic origin of any race)	2,092.5

Source: NCHS, National Vital Statistics System, Natality, Martin *et al.* (2018a). https://www.cdc.gov/nchs/data/nvsr/nvsr67/nvsr67_01.pdf

Teen Births

The change in fertility rates has been driven in part by declines in births to teens. In the United States, the **teen birth rate** (i.e., the number of births per 1,000 women aged 15–19) declined due to the combination of an increased percentage of adolescents who are waiting to have sexual intercourse and the increased use of contraceptives by teens (Kost & Henshaw, 2014). Figure 10.1 shows that teen birth rates fell to 22.3 births for teenagers aged 15–19 in 2015, down 8 percent from 2014, down 46 percent since 2007, and down 64 percent since 1991 (Hamilton & Mathews, 2016). Teen birth rates declined in 40 states, with declines ranging from 4 percent for Oklahoma to 15 percent for New Hampshire in 2016 (Martin *et al.*, 2018a). The largest decline in the teen birth rate from 2016 to 2017 was for Asian females, down 15 percent to 3.3 births per 1,000 (see Figure 10.5).

Still, the U.S. teen birth rate is higher than that of other developed countries such as Canada and the

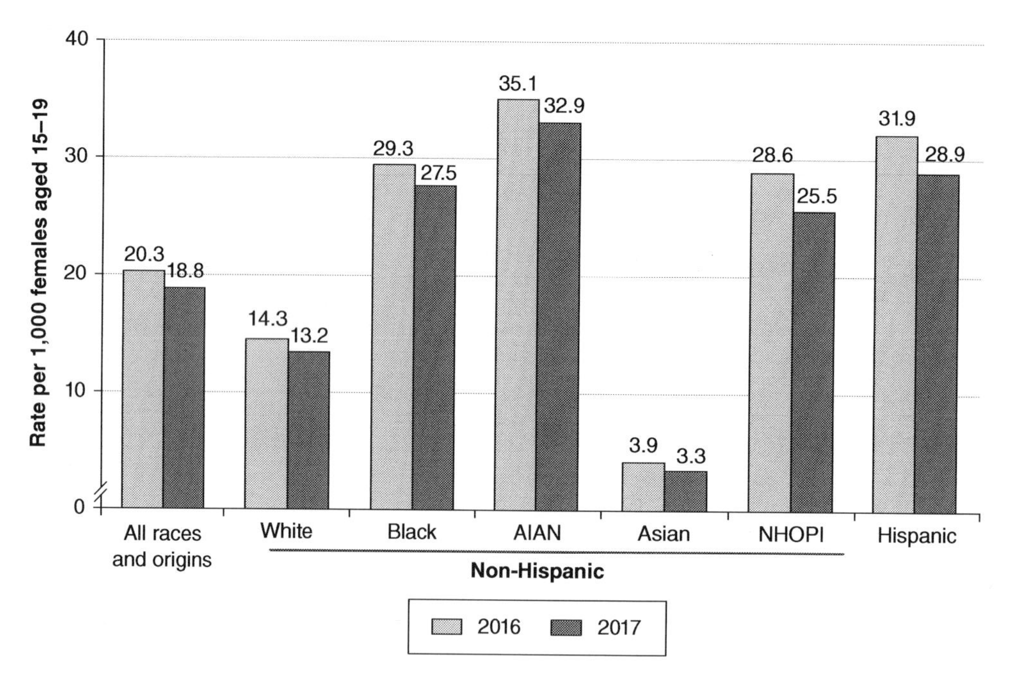

Figure 10.5 Birth Rates for Females Aged 15–19, by Race and Hispanic Origin of Mother, United States, 2016 and 2017
Source: NCHS, National Vital Statistics System, Natality. https://www.cdc.gov/nchs/products/databriefs/db318.htm

United Kingdom (UNSD, 2015). Individual, family, and community characteristics have been linked to teen births. Adolescents who are enrolled in school and performing well are less likely to have a baby than other adolescents (Kirby & Lepore, 2007). At the family level, adolescents whose mothers gave birth as teens and had a high school degree are more likely to have a baby than teens whose mothers were older at their birth or who attended college; adolescents living with both biological parents at age fourteen are associated with a lower risk of a teen birth (Martinez *et al.*, 2011). At the community level, adolescents who live in wealthier neighborhoods with strong levels of employment are less likely to have a baby than those in neighborhoods with limited employment opportunities (Kirby & Lepore, 2007).

Unmarried Childbearing

The **non-marital birth rate** refers to the number of births per 1,000 unmarried women (aged 15–44). In 1940, non-marital births were rare but today two-fifths of all births are to unmarried women; this has resulted from a complicated combination of moral and behavioral changes such as less taboo of premarital sex and more acceptable unmarried childbearing (VerBruggen, 2017). The number and rate of non-marital childbearing declined from the 2008 peak (see Figure 10.6) and fell to 42.4 births (1,569,796) in 2016, compared with 43.4 births (1,601,527) in 2015, 44.3 births (1,605,643) in 2013, and 51.8 births (1,726,566) in 2007–2008 (Martin *et al.*, 2018a).

The non-marital birth rates declined for all racial groups, with Hispanic women having the greatest decline between 2007 (102 per 1,000) and 2012 (73). Declines in non-marital birth rates were seen for Black women between 2007 (71) and 2012 (63), White women between 2007 (34) and 2012 (32), and Asian women between 2007 (24) and 2012 (23) (see Figure 10.7). In 2016, the decline in non-marital birth rates continued for Hispanic (52.0), White (28.5), and

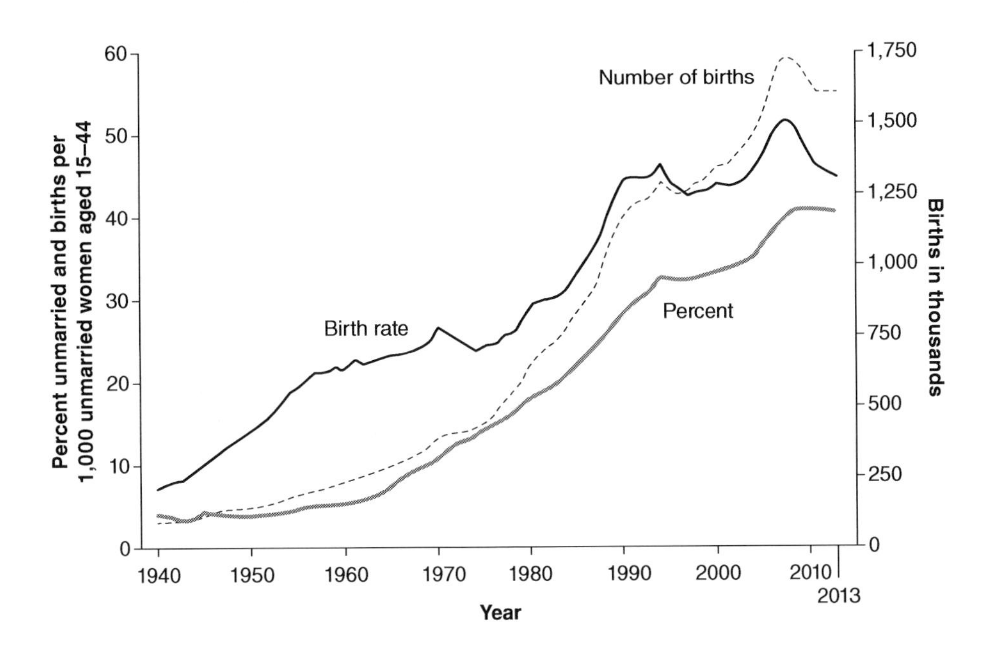

Figure 10.6 Number of Births, Birth Rate, and Percentage of Births to Unmarried Women, United States, 1940–2013
Source: CDC/NCHS, National Vital Statistics System; Curtin *et al.* (2014); https://www.cdc.gov/nchs/data/databriefs/db162.htm

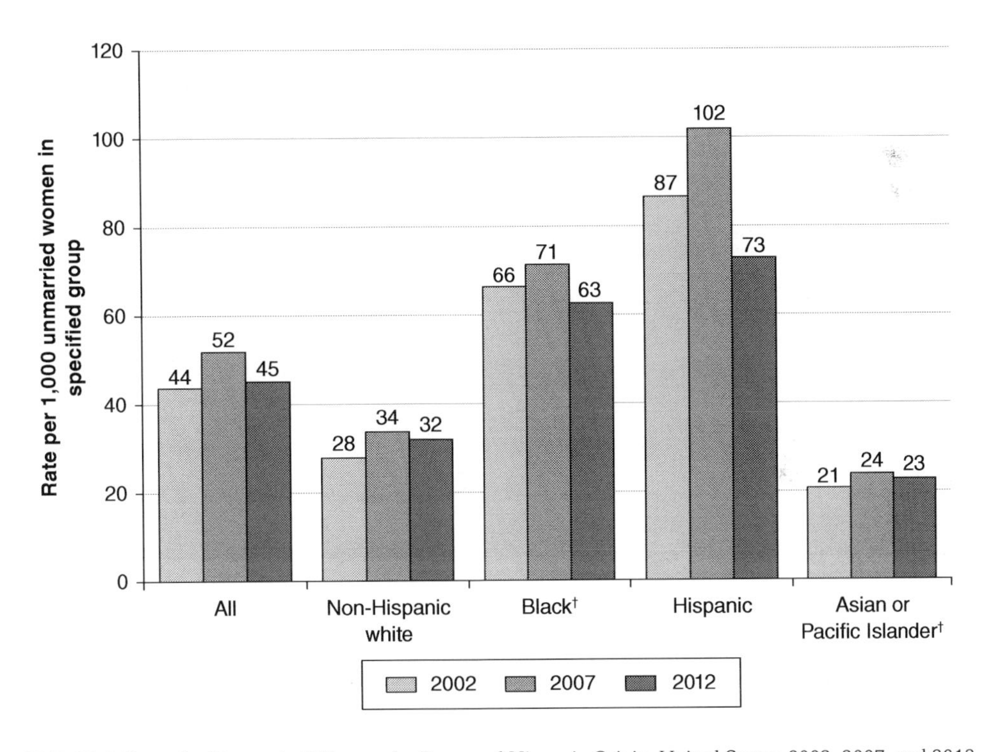

Figure 10.7 Birth Rates for Unmarried Women, by Race and Hispanic Origin, United States, 2002, 2007, and 2012
Source: CDC/NCHS, National Vital Statistics System, Curtin *et al.* (2014); https://www.cdc.gov/nchs/data/databriefs/db162.htm
Note: †Trend data for black and Asian or Pacific Islander mothers by Hispanic origin are not available.

Asian (12.0) but the rate for Black mothers increased (69.8) (Martin *et al.*, 2018a).

Young couples today do not feel the need to marry when a baby enters the picture (VerBruggen, 2017). **Shotgun marriage** is a form of marriage in which vows are taken out of a sense of obligation after an accidental pregnancy has occurred. The custom of shotgun marriage has plummeted as well: 43 percent of unwed pregnancies resulted in a shotgun marriage in the early 1960s, down to 9 percent today (VerBruggen, 2017). Shotgun marriage was once the norm for White women but now it is rare for all racial groups (VerBruggen, 2018).

Multiple Births

A **multiple birth** is the culmination of one multiple pregnancy, wherein the mother delivers two or more offspring. Cases of twins (two), triplets (three), quadruplets (four), quintuplets (five), sextuplets (six), septuplets (seven), and octuplets (eight) have been recorded. **Assisted reproductive technology** (ART) includes all fertility treatments in which either eggs or embryos are handled in the laboratory (i.e., in vitro fertilization [IVF] and related procedures). **In vitro fertilization** (IVF) involves extracting a woman's eggs, fertilizing the eggs in the laboratory, and then transferring the resulting embryos into the woman's uterus through the cervix (CDC, 2017d). Although the majority of infants conceived through ART are **singletons** (i.e., a child that is the only one born at one birth), women who undergo ART procedures are more likely than women who conceive naturally to deliver multiple-birth infants (Sunderam *et al.*, 2018). Maternal age and hereditary components are also the best-defined determinants for spontaneous multiple births (Bortolus *et al.*, 1999). In 2011, approximately 19 percent of twin births and 45 percent of triplet or higher-order births in the United States were attributable to non-IVF fertility treatments whereas 17 percent of twin births and 32 percent of triplet or higher-order births were attributable to IVF fertility treatments (Kulkarni *et al.*, 2013). In 2016, triplet and higher-order multiple births included 3,755 triplets, 217 quadruplets, and 31 quintuplets and higher-order multiple births, totaling 4,003, the lowest number since 1992 (Martin *et al.*, 2018a).

Twins

The most common form of multiple births is twins. **Twins** are two offspring produced by the same pregnancy. Twins form in one of two ways. **Fraternal twins** occur when two, separate eggs are fertilized by two, separate sperm (see image below). Fraternal twins occur when two fertilized eggs are implanted in the uterus wall at the same time. Fraternal twins are male–female and are more common for older mothers, with twinning rates doubling in mothers over the age of 35 (Bortolus *et al.*, 1999). **Mixed twins** are fraternal twins from biracial couples and exhibit differing racial features. One such pairing was born in Germany in 2008 to a white father from Germany and a black mother from Ghana (Moorhead, 2011).

Identical twins occur when a single fertilized egg splits into two (see image below). Identical twins are formed after a blastocyst collapses, splitting the progenitor cells (those that contain the body's fundamental genetic material) in half, leaving the same genetic material divided in two on opposite sides of the embryo (Bortolus *et al.*, 1999). Identical twins look almost exactly alike and share the exact same genes (OWH, 2018a).

Identical twins, although genetically very similar, are not exactly the same (Li *et al.*, 2014). For example, they do not have the same fingerprints because the fetuses touch different parts of their environment even within the confines of the womb, giving rise to small variations in their corresponding prints and thus making them unique (Patwari & Lee, 2008). Another cause of difference between identical twins is *epigenetic modification*—heritable alterations not caused by changes in DNA sequence but by differing environmental influences throughout their lives (Handy *et al.*,

Fraternal Twins Occur when Two, Separate Eggs are Fertilized by Two, Separate Sperm while Identical Twins Occur when a Single Fertilized Egg Splits into Two.

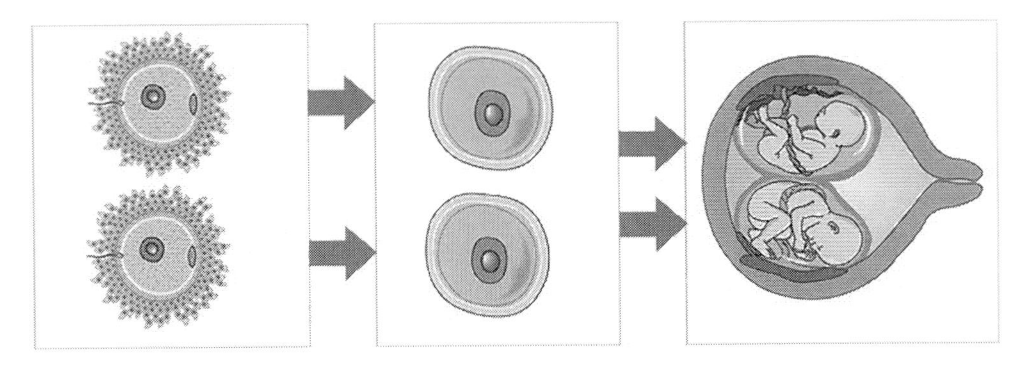

Source: Office on Women's Health, 2018. https://www.womenshealth.gov/pregnancy/youre-pregnant-now-what/twins-triplets-and-other-multiples

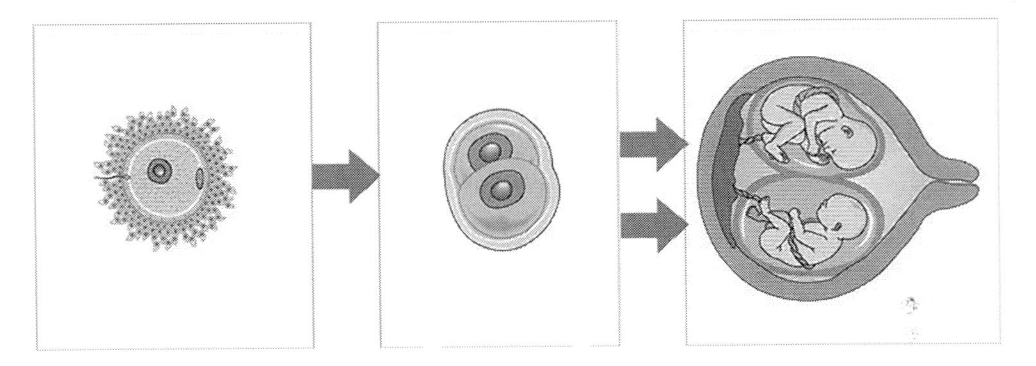

Source: Office on Women's Health in 2018. https://www.womenshealth.gov/pregnancy/youre-pregnant-now-what/twins-triplets-and-other-multiples

2011). For example, twins who spent their lives apart after they were adopted by two different sets of parents at birth had the greatest difference (Fraga *et al.*, 2005).

ART-conceived twins accounted for approximately 96.1 percent (22,491 of 23,413) of all ART-conceived infants born in multiple-birth deliveries in 2015 (Sunderam *et al.*, 2018). The **twin birth rate** (i.e., number of live births in twin deliveries per 1,000 live births) was 33.4 in 2016, down from 2015 (33.5) and from 2014 (33.9). In 2016, twin birth rates were highest among Black (39.9), followed by White (35.7), Hispanic (24.6), and Native Hawaiian or Pacific Islander women (24.4) (Martin *et al.*, 2018a).

Triplet and Higher-order Multiple Births

Triplet and higher-order multiple births refer to the number of triplets, quadruplets, and quintuplets and other higher-order multiples per 100,000 births. Triplet birth refers to individual live birth in a triplet pregnancy. Multiple births can be fraternal, identical, or a combination; multiples associated with fertility treatments are mainly fraternal (OWH, 2018a). After the first U.S. infant conceived with assisted reproductive technology (ART) was born in 1981, triplet and higher-order multiple births rose fourfold with increasing maternal age and use of fertility therapies in the 1980s to 1990s, reaching the highest rate (193.5) in 1998 (Martin *et al.*, 2016), and then declining to 113.5 in 2014, down to 103.6 in 2015, and down to 101.4 per 100,000 live births in 2016 (Martin *et al.*, 2018a).

Triplet and higher-order multiple births declined and varied widely by race, state, and age. White (121.7) and Black (112.4) women were about twice as likely to have multiple births as Hispanic (58.6) women in 2016 (Martin *et al.*, 2018a). In Connecticut, Illinois, Minnesota, Massachusetts, New Hampshire, New Jersey, and Rhode Island, triplet and higher-order birth rates in 2012–2014 were 50 percent lower than for those of the earlier period (1998–2000) (Martin *et al.*, 2016). Declines in triplet and higher-order multiple birth rates were seen for women aged 20 and over from 1998 to 2016, with the largest declines among women aged 30–39 (down 64 percent) from 1998 (376.3) to 2016 (135.3) and women aged 40 and over (down 55 percent) from 1998 (517.6) to 2016 (232.4) (see Figure 10.8).

The declines in triplet and higher-order multiple birth rates were linked to changes in assisted reproductive technologies (ART) practices, such as reducing the number of embryos transferred in ART procedures (Stern *et al.*, 2007). Also, multiple births pose substantial risks for both mothers and infants, including obstetric complications, preterm delivery (<37 weeks), and low birthweight (<2,500 g) infants (Sunderam *et al.*, 2018). For example, 7 percent of triplets and higher-order births in 2013 did not survive their first year of life, compared with 0.5 percent of singletons (Martin *et al.*, 2016). Those who survived were more likely to suffer long-term morbidities (Luke & Brown, 2007).

Rising Mean Age of Mother

One factor driving down annual fertility rates is that women are becoming mothers later in life. Maternal age is on the rise in the United States, with both

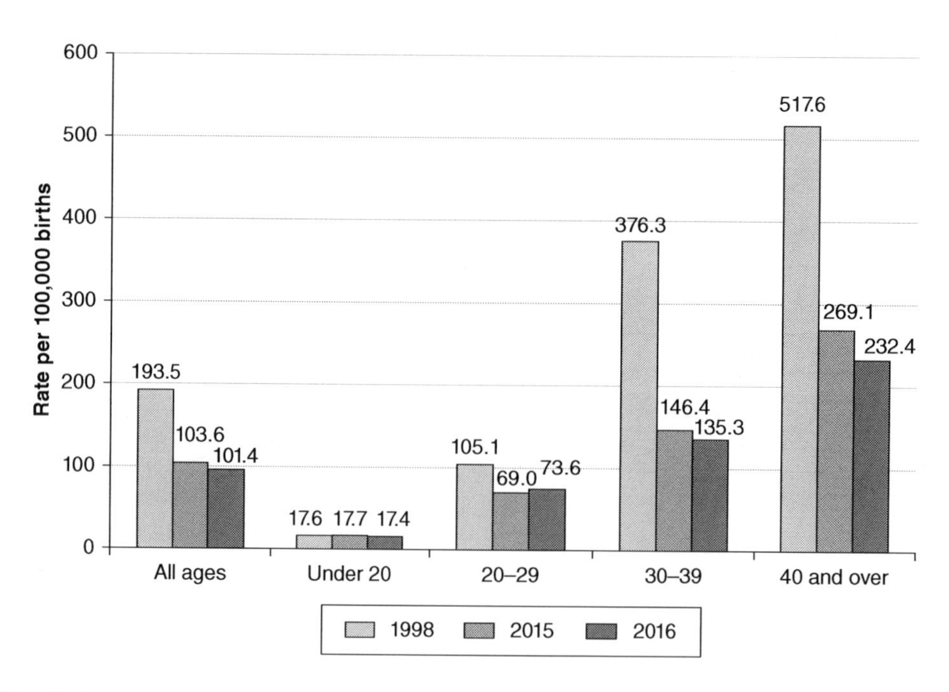

Figure 10.8 Triplet and Higher-order Multiple Birth Rates, by Age of Mother: United States, 1998, 2015, and 2016
Source: NCHS, National Vital Statistics System, Natality. https://www.cdc.gov/nchs/products/databriefs/db287.htm

men and women continuing to delay entry into parenthood (Eickmeyer, 2016). The **mean age at first birth** (i.e., the average age of mothers at birth of first child) was 26.6 years in 2016, compared with 26.4 in 2015 (Martin *et al.*, 2018a), 26.3 in 2014, 26 in 2013, 24.9 in 2000 (Mathews & Hamilton, 2016), 23 in 1994 (Livingston, 2018), and 21 in 1970 (Blalik, 2017).

Mean Age by Birth Order

The mean age of mothers increased for all birth orders, with age at first birth having the largest increase. The largest factor in the rising mean age at first birth is the decline in the proportion of first births to mothers under age 20, down 42 percent from 2000 to 2014 (Mathews & Hamilton, 2016) and the increase for women aged 30 and over. The **first-birth rate** (i.e., the number of first births per 1,000 women) was 23.7 births per 1,000 women aged 15–44 in 2016, down 2 percent from 2015, down 8 percent for women aged 15–19, down 3 percent for women aged 20–24, down 2 percent for women aged 25–29, and up 2 percent for women aged 30–39, and up 0.3 percent for women in their 40s in 2016 (Martin *et al.*, 2018a). Trends in mean age of mother for higher-order births were similar to those for first births. Figure 10.9 illustrates the trend in rising mean age of the mother by birth order.

The mean age at first birth increased for all racial groups from 2000 to 2016. Asian or Pacific Islander mothers had the oldest average age at first birth (29.5), compared with White (27.0), Black (24.2), Mexican (23.7), and American Indian and Alaska Native (23.1) women. In 2016, Asian women continued to have the oldest average age at first birth (30.1), compared with White (27.4), Black (24.8), Mexican (24.1), and American Indian and Alaska Native (23.2) women in 2014 (see Figure 10.10) (Martin *et al.*, 2018).

Mean Age by Birth Rate

Birth rates declined for women aged 15–29 and increased for those aged 30–49 (Martin *et al.*, 2018a). In 2016, the birth rate was 73.8 births per 1,000 women aged 20–24, down 4 percent from 2015 (76.8) and was 102.1 births per 1,000 women aged 25–29, down 2 percent from 2015 (104.3).

The birth rates increased for women in their 30s and 40s (see Figures 10.1 and 10.11). In 2016, the

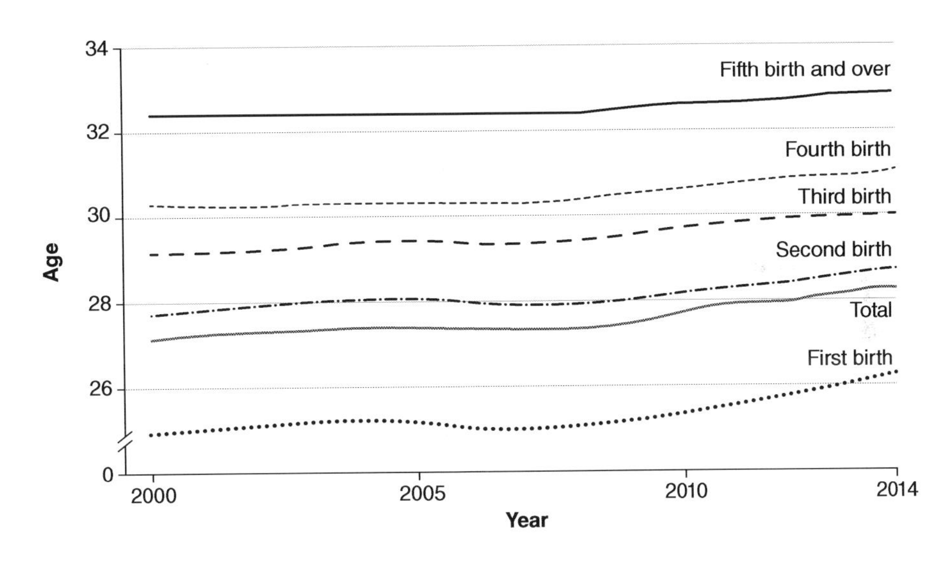

Figure 10.9 Mean Age, by Birth Order: United States, 2000–2014
Source: CDC/NCHS, National Vital Statistics System, 2016. https://www.cdc.gov/nchs/products/databriefs/db232.htm

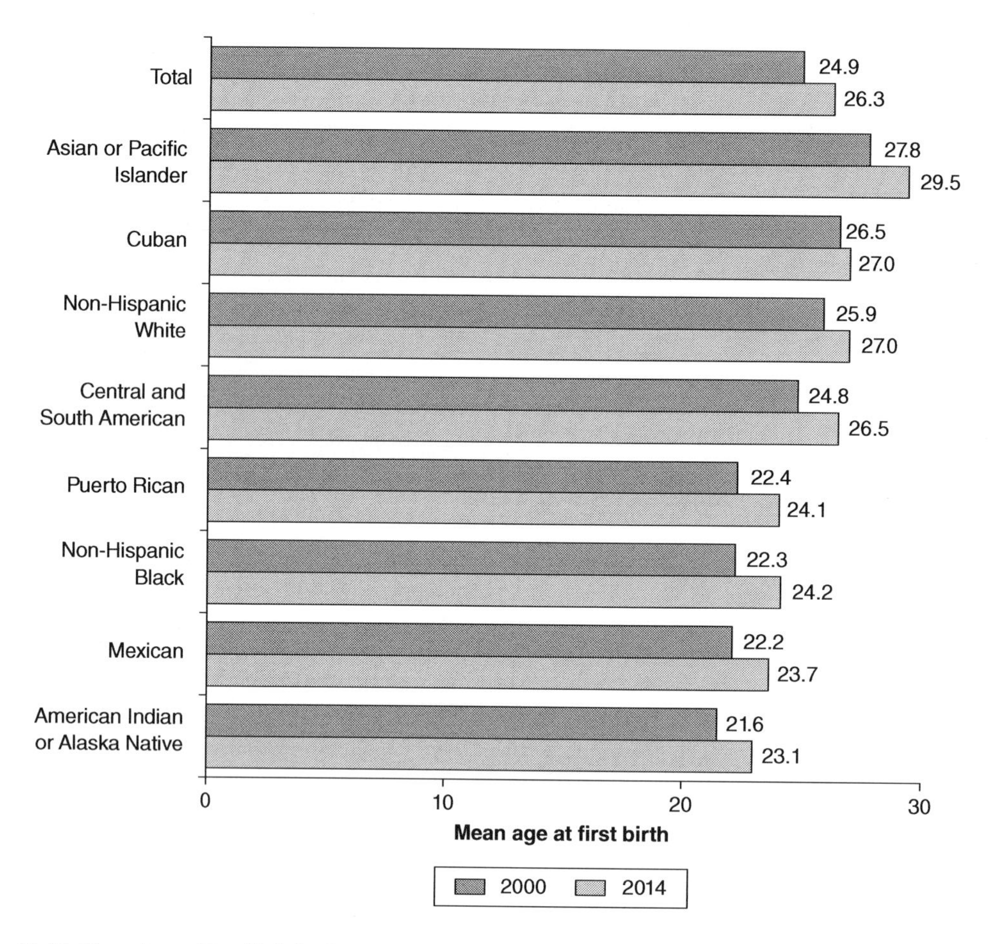

Figure 10.10 Mean Age at First Birth, by Race and Hispanic Origin of Mother: United States, 2000 and 2014
Source: CDC/NCHS, National Vital Statistics System, 2016. https://www.cdc.gov/nchs/products/databriefs/db232.htm

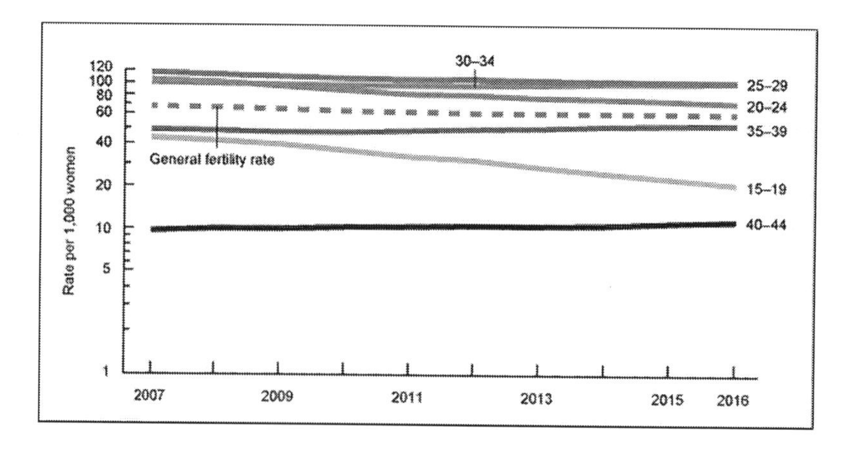

Figure 10.11 General Fertility and Age-specific Birth Rates, by Age of Mother: United States, 2007–2016
Source: NCHS, National Vital Statistics System, Natality; Martin *et al.* (2017a). https://www.cdc.gov/nchs/products/databriefs/db287.htm

birth rates include 102.7 births to women aged 30–34, up from 2015 (101.5) and up from 1975 (52.3), 52.7 births to women aged 35–39, up from 2015 (51.8) and up from 1975 (19.5), 11.4 births to women aged 40–44, up from 2015 (11.0) and up from 1975 (4.6), 0.9 births to women aged 45–49, up from 2015 (0.8) and up from 1975 (0.3) (Martin *et al.*, 2018a).

Increasing Multiracial and Minority Babies

There is a rapid rise of multiracial or multiethnic babies. One in seven U.S. infants (14 percent) were multiracial or multiethnic in 2015 (Livingston, 2017). The U.S. Census has shown that the fastest-growing racial group who claim two or more races is projected to grow 200 percent by 2060 (Vespa *et al.*, 2018). The minority population is projected to rise to 56 percent of the total in 2060, compared with 38 percent in 2014 (see Figure 10.12). In 2045, Whites will comprise 49.7 percent of the population in contrast to 24.6 percent for Hispanics, 13.1 percent for Blacks, 7.9 percent for Asians, and 3.8 percent for the multiracial population (Frey, 2018).

Non-Hispanic Whites May No Longer Comprise Over 50 Percent of the U.S. Population by 2044

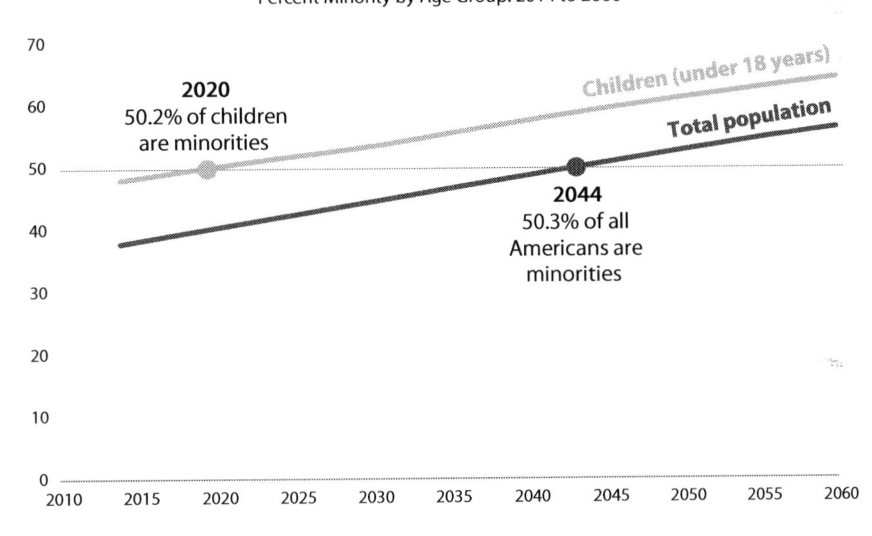

Figure 10.12 Projecting Majority-Minority: Non-Hispanic Whites May No Longer Comprise Over 50 Percent of the U.S. Population by 2044

Source: 2014 National Projections. https://www.census.gov/content/dam/Census/newsroom/releases/2015/cb15-tps16_graphic.pdf

Note: Minority is defined in this figure as any group other than non-Hispanic white.

The shift is the result of two trends. Between 2018 and 2060, minority populations are growing; the aging White population will see a gain through 2023 and then experience a long-term decline through 2060, a consequence of more deaths than births (Frey, 2018).

Shrinking Family Size

Average household size (i.e., the average number of persons per household) is calculated by dividing the total household population by the number of households in a country or area (UNPD, 2017a). In addition to fertility, trends in household size are influenced by trends in health, longevity, and migration; cultural patterns surrounding intergenerational coresidence, cohabitation, marriage, and divorce; and socioeconomic factors that shape trends in education, employment, and housing markets (UNPD, 2017b).

In the United States, household sizes are shrinking, with the average household size decreasing from 1940 (3.7) to 2017 (2.5) per household (see Figure 10.13). In 2014, 22.3 percent of women aged 15–50 had given birth to two children, 42.4 percent had no children, 17 percent had one child, 11.7 percent had three children, and about 6.8 percent had four or more children (U.S. Census Bureau, 2017).

The declines in fertility led Americans to have smaller families. In 1976, 40 percent of mothers aged 40–44 had four or more children and 24 percent had two children, while by 2014, 41 percent of mothers had two children and only 14 percent had four or more (Blalik, 2017). Figure 10.14 shows the shrinking U.S. family size and declining average number of children born to married couples from 1955 to 2017.

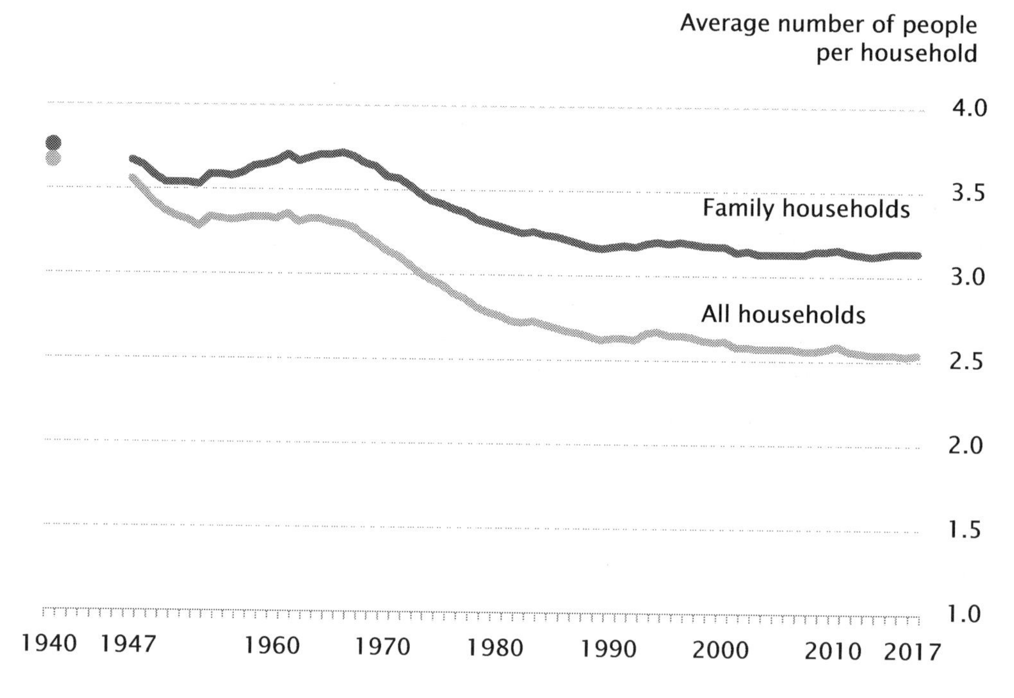

Figure 10.13 Changes in Household Size

Source: U.S. Census Bureau, Current Population Survey, Annual Social and Economic Supplements, 1940 and 1947–2017. https://www.census.gov/data/tables/time-series/demo/families/households.html

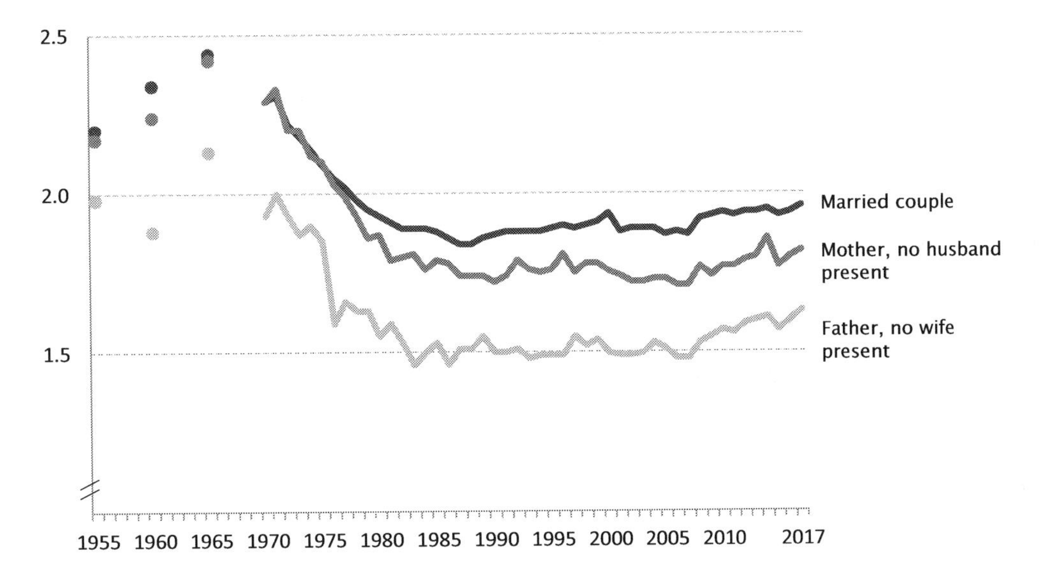

Figure 10.14 Average Number of Own Children per Family (for Families with Children under 18)
Source: U.S. Census Bureau, Current Population Survey, Annual Social and Economic Supplements, 1955, 1960, 1965 and 1970–2017. https://www.census.gov/data/tables/time-series/demo/families/families.html

Average household size across the globe ranges from two in most of Europe and northern America to nine persons per household in the countries of Africa and Asia. For example, households in Monaco and Serbia averaged 1.9 and 2.9 persons, respectively, compared with the largest household sizes found in Senegal and Oman, averaging 9.0 and 8.0 persons, respectively (UNPD, 2017b). For countries of Europe and the United States, multigenerational households with both a child (under age fifteen years) and an older person (aged 60 or over) accounted for 2 percent of all households, compared with 14 and 13 percent, respectively, in the countries of Africa and Asia (UNPD, 2017b).

Factors Influencing Family Size and Fertility

Most Americans believe that educational and economic accomplishments are extremely important milestones of adulthood. In contrast, marriage and parenthood rank low: over half believe that marrying and having children are not important parts of becoming an adult (Vespa, 2017). Women's increasing educational attainment and labor force participation as well as improvements in contraception are all factors in shrinking family size (Parker *et al.*, 2015). Seven in ten moms with kids younger than eighteen were in the labor force in 2014, up 47 percent from 1975, and mothers are the primary breadwinners in four in ten U.S. families (Blalik, 2017).

Sociologists have examined the impact of *uncertainty* on the entry into parenthood in a cross-national context. Mills and Blossfeld (2005) described three types of uncertainty: economic, temporal, and employment relationship. Youth postponed parenthood due to economic uncertainty (e.g., lower earnings, unemployment) (Oppenheimer, 1988), temporal uncertainty (i.e., often in the form of temporary or fixed-term contracts) (Breen, 1997), and lower employment relationship uncertainty (e.g., dependent workers versus self-employed or contract workers) (Mills & Blossfeld, 2005).

Research also has explored factors such as the individual decision-making process that influences the formation and change in fertility intentions. The **theory of planned behavior** (TPB) suggests that fertility intentions are the culmination of a combination of attitudes (i.e., perceived costs and benefits), subjective norms (e.g., influence of close friends and relatives), and perceived control over behavior (i.e., extent to which behavior is perceived as subject to control by the individual) (Ajzen, 1991). For example, the partner's fertility intentions play an important role in the realization of an individual's intentions, since generally childbearing in advanced societies is a joint couple decision (Balbo *et al.*, 2013). If there is a disagreement about childbearing expectations within the couple, the positive fertility intentions of one of the partners are less likely to be realized (Thomson, 2002). The main factors causing variations in fertility intentions are partner's expectations (Iacovou & Tavares, 2011) and changes in partnership status, activity status, and actual fertility events (Liefbroer, 2009). For example, partnership status is a strong predictor, with those who are not in a stable relationship being less likely to have a child (Hobcraft & Kiernan, 1995). A study of opportunity cost of having children for women revealed that since raising children requires parental and especially maternal time, fertility is more costly for higher-income mothers, who are therefore expected to have fewer children (Kravdal, 1992).

Another important factor influencing fertility is the gendered division of domestic labor of couples within the household (Balbo *et al.*, 2013). **Gendered fertility theory** suggests that very low fertility is the result of a hiatus of sustained gender inequity in family-oriented social institutions. In a comparative study of the Netherlands and Italy, Mills *et al.* (2008) report that an unequal division of household labor significantly impacts women's fertility intentions when they already have a heavy load (more work hours, children), which is particularly salient for working women in Italy. The unequal distribution of household labor lowered men's fertility intentions in Austria (Tazi-Preve *et al.*, 2004).

Demographic Transition Theory

With dramatic declines in fertility taking place throughout the world, it is increasingly important to understand the demographic transition as a global process (Reher, 2004). In 1945, American demographer Frank W. Notestein (1902–83) developed a formal demographic transition theory proposed by Warren Thompson (1887–1973) in 1929 (Weeks, 2014). Based on historical population trends of two demographic characteristics (i.e., birth and death rates), **demographic transition theory** suggests that societies are moving from high birth and death rates to low birth and death rates because of technological developments that have taken place. The existence of demographic transition is accepted in the social sciences because of the well-established historical correlation linking dropping fertility to social and economic development (Myrskylä *et al.*, 2009). Demographic transition relates to five stages (see image below).

At Stage 1, or pre-transition in the pre-industrial society in the eighteenth and nineteenth centuries, there was high death rate due to a lack of advanced medicine and natural events such as famine or disease, and a high birth rate due to lack of contraception and a demand for family labor. Historically, high birth rates were attributed to pre-industrial societies that relied heavily on agricultural activities. Larger families meant a larger workforce. In the United States, 94 percent of the population lived in rural areas and women in the rural economy had a higher fertility rate than those in urban areas in 1800; the average White woman gave birth to seven children (Greenwood & Seshadri, 2002). A high birth rate is often the response to a high death rate as societies with low levels of economic development seek to maintain replacement levels (total fertility rate at 2.1 births per woman) (Grover, 2014). **Infant mortality** (i.e., infants who die during the first year of life) is high when medicine and maternal care are insufficient in these countries. In countries with high infant and child

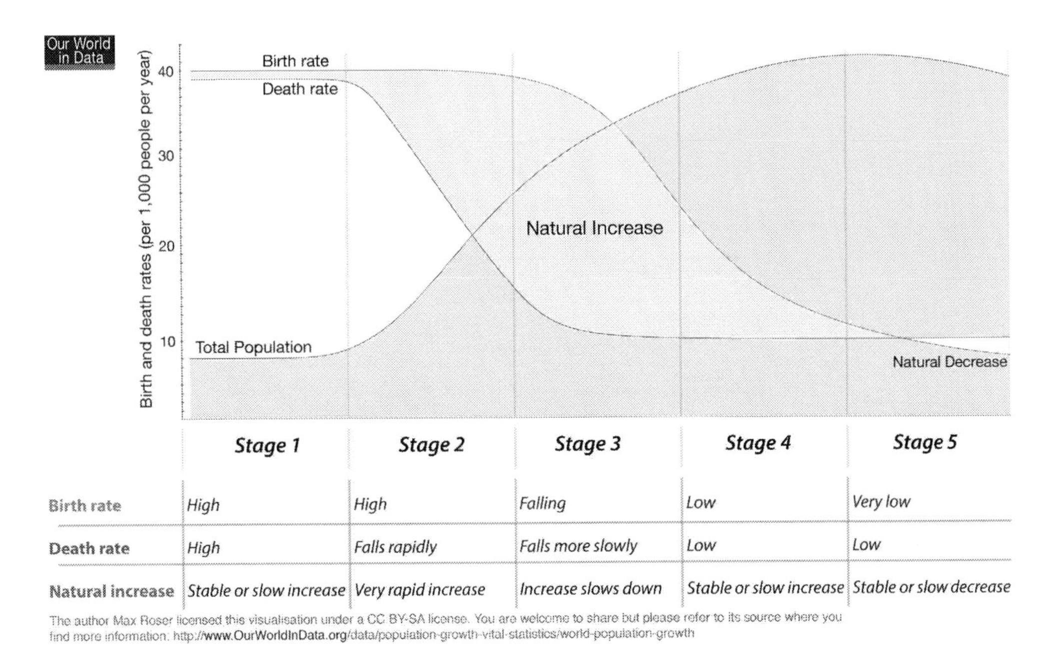

	Stage 1	Stage 2	Stage 3	Stage 4	Stage 5
Birth rate	*High*	*High*	*Falling*	*Low*	*Very low*
Death rate	*High*	*Falls rapidly*	*Falls more slowly*	*Low*	*Low*
Natural increase	*Stable or slow increase*	*Very rapid increase*	*Increase slows down*	*Stable or slow increase*	*Stable or slow decrease*

Demographic Transition Related to Five Stages when Societies Move from High Birth and Death Rates to Low Birth and Death Rates because of Technological Developments.
Source: Max Roser. https://upload.wikimedia.org/wikipedia/commons/8/8c/Demographic-TransitionOWID.png

mortality rates, the average number of births may need to be much higher (Craig, 1994).

At Stage 2, that of a developing country and early transition, there is a rapid decrease in a country's death rate while the birth rate remains high. The global demographic transition began in the nineteeenth century in the now economically developed parts of the world (the North) with declines in death rates (Bongaarts, 2009). The Industrial Revolution helped countries make the transition from an agricultural society to an industrial one. Along with public health care, mass education, women's education attainment, and technological advances in food supply worked to decrease the death rate. For example, for infectious diseases, the death rate has fallen from about 11 per 1,000 to less than 1 per 1,000 in England (Osborn, 1992).

The decline in death rate and high birth rate at Stage 2 may transform the age structure of the population. The second stage of the demographic transition implies a rise in child dependency and creates a youth bulge in the population structure (Weeks, 2014). **Population pyramid** is a graphic with a broad base that comes to a point at the top, illustrating the age and sex structure of a developing country's population (UNFPA, 2013). The decline in death rates entails the increasing survival of children and a growing population. The age structure of the population becomes increasingly youthful and has big families. The age pyramid narrows toward the top due to declining births for many years and due to higher death rates at older ages.

The demographic transitions in Africa, Asia, and Latin America started later and are still underway (Bongaarts, 2009). The now-industrial society experienced a decrease in the death rate due to improvements in public health and an increase in the birth rate due to demand for a labor force. Population size is projected to increase from 5.3 to 7.9 billion between 2005 and 2050 in the South (i.e. Africa, developing Asia, and Latin America), whereas population size is growing slightly from 1.22 to 1.25 billion in the North

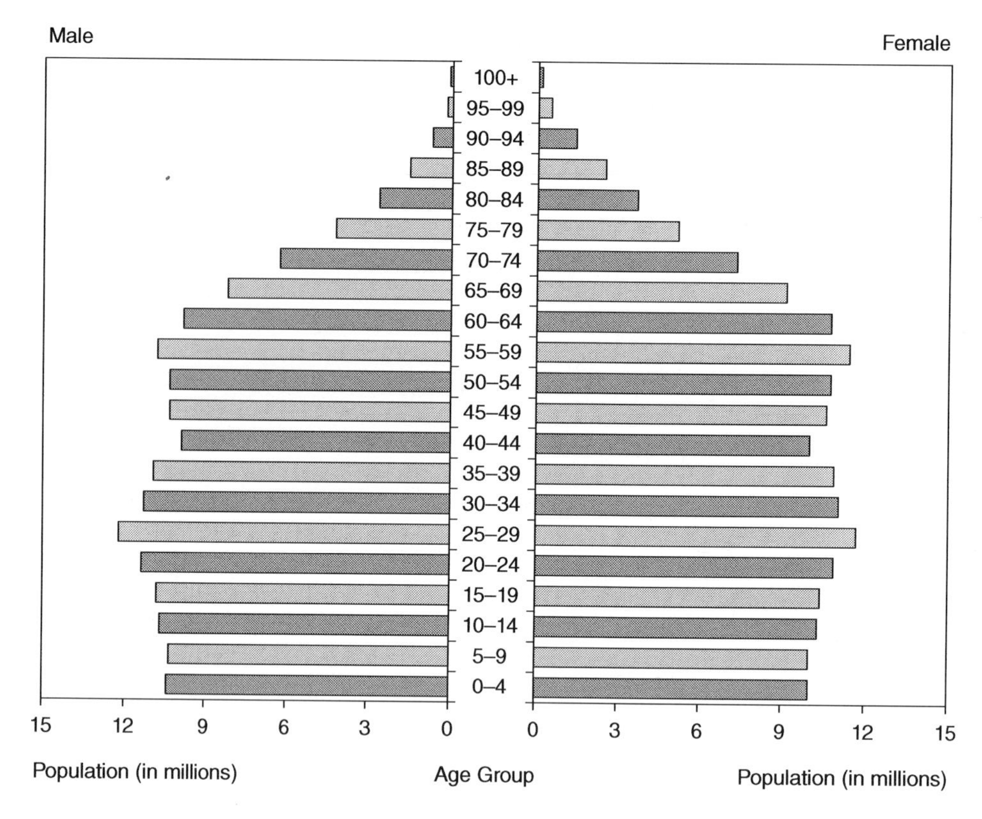

Figure 10.15 Population Pyramid Graph, Custom Region, United States, 2018
Source: U.S. Census Bureau. https://www.census.gov/data-tools/demo/idb/region.php?N=%20Results%20&T=12&A=separate&RT=0&Y=2018&R=-1&C=US

(i.e., western Europe, northern America, developed parts of Asia and Australia/New Zealand) (Bongaarts, 2009).

Since the mid-twentieth century, most of the world's countries have continued to progress to Stage 3 but a number of countries such as Afghanistan remain in Stage 2 or Stage 1 for a variety of social and economic reasons (Grover, 2014). Fertility in these developing countries did not seem to follow the expected pattern (Reher, 2004) as population growth continues at a rapid pace in sub-Saharan Africa, where the AIDS epidemic is most severe and the birth rate is expected to remain higher than the elevated death rate in the future (Bongaarts, 2009). Afghanistan still has high birth and death rates. The total fertility rate was 5.12 children per woman in 2017 (CIA, 2018b), compared with the world (2.42) in 2016 (CIA, 2017a); the death rate was 13.4 deaths per 1,000 population in 2017 (CIA, 2018b), compared with the world (7.8) (CIA, 2017a). In Afghanistan, the economy before 2014 had sustained a decade of strong growth, because of international assistance, but since 2014 has slowed because of the withdrawal of 100,000 foreign troops that had inflated the country's economic growth (CIA, 2018b).

At Stage 3, the industrial society saw a decrease in the death rate and birth rate due to improved economic conditions, access to contraception, and women's educational attainment. Education and career have been linked to delaying women's childbearing years, providing opportunities to women outside the home. Indeed, decline in birth rate is caused by changes in values about childbearing and improvements in contraceptive technology. Two more reasons such as cost of childrearing and industrialization may explain declining fertility in the United States. Between 1800 and 1940, real wages grew about five-fold and this increased the time cost of children in terms of consumption goods; the role of agriculture in the economy declined and families might not need a larger workforce (Greenwood & Seshadri, 2002). The driver of these changes was industrialization that, in turn, increased motivation for reduced family size but did not undermine the universal expectation of marriage and parenthood (Zaidil & Morgan, 2017).

At Stage 4, which has occurred in post-industrial societies, birth rates and death rates have continued to decline and become stable. These countries tend to have stronger economies, higher levels of education, better health care, a higher proportion of working women, and a total fertility rate around two children per woman (Grover, 2014). As a result of this close connection between development and fertility decline, more than half of the global population now lives in regions with below-replacement fertility (less than 2.1 children per woman) (Wilson, 2004). For example, in the United States, only 43 percent of the population lived in rural areas, and the average White woman birthed two children in 1940 (Greenwood & Seshadri, 2002). Most developed countries have completed the demographic transition and have low birth rates; most developing countries are in the process of this transition (Caldwell *et al.*, 2006). Between 1800 and 1940, the Unites States experienced a demographic shift from a mostly rural population with high fertility to one with low fertility (Greenwood & Seshadri, 2002). Between about 1750 and 1975, England and Wales experienced the demographic transition from

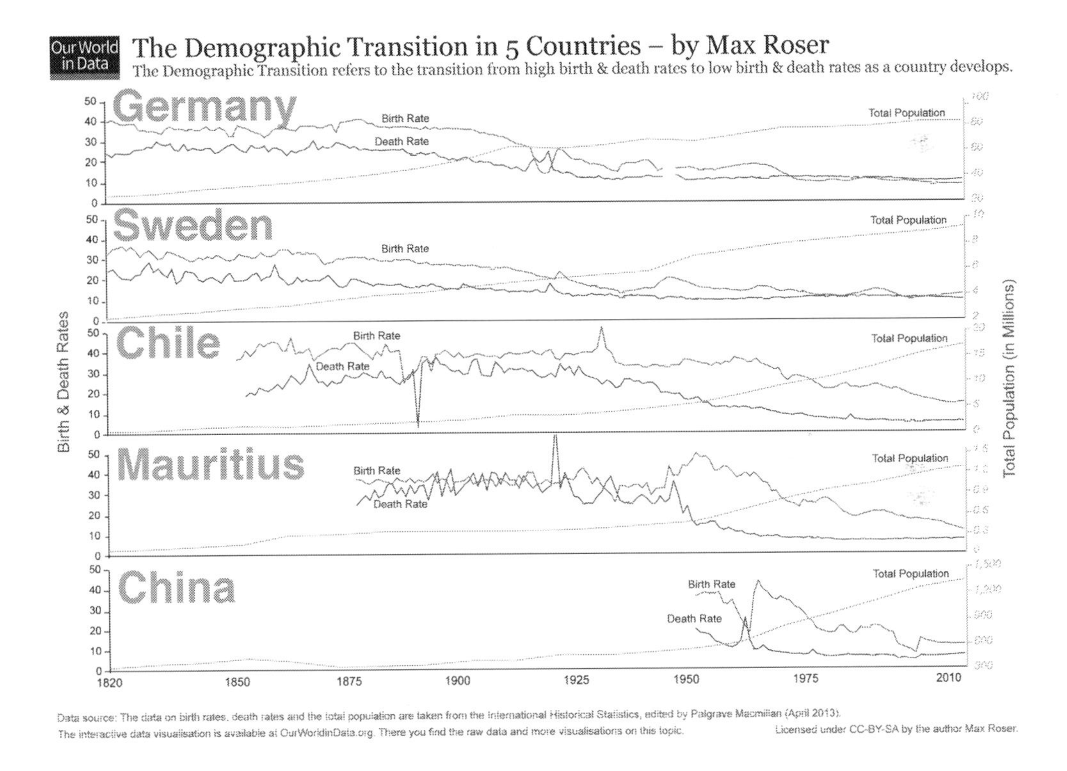

Source: Max Roser. https://upload.wikimedia.org/wikipedia/commons/2/20/Demographic-Transition-5-countries.png

high levels of both mortality and fertility to low levels (Osborn, 1992). Demographic change in East and Southeast Asia from the 1960s to 1990 can be seen as a byproduct of social and economic development together with, in some cases, strong governmental pressures (McNicoll, 2006).

Stage 5 has been proposed in addition to the original four stages. Both more-fertile and less-fertile futures have been claimed at Stage 5. Since the early 1950s, average fertility has declined to 2.0 births per woman in northern America and to 1.4 births per woman in Europe (see Figure 10.16). The decline in death and birth rates that occurs during the demographic transition may transform the age structure of the population. Further declines in both mortality and fertility will result in an aging population. As the population ages, the **old-age dependency ratio** (i.e., the ratio of older adults to working-age adults) is projected to rise. By 2030, all baby boomers will be older than age 65 and this will expand the size of the older population (Livingston, 2018). An increase of the old-age dependency ratio often indicates that a population has reached below-replacement levels of fertility, and as result does not have enough people in the working ages to support the economy, and the growing dependent population (Weeks, 2014). Some countries in eastern and southern Europe have reached a negative rate of *natural increase* (i.e., birth rate minus death rate) as their death rates are higher than their birth rates (Grover, 2014). For example, in Germany, the death rate (11.7 per 1,000) was higher than the birth rate (8.6 per 1,000) in 2017 (CIA, 2018c).

In the 1950s, the total fertility rate in the South was high at around six births per woman; this high level of fertility reflected a near absence of birth control (Bongaarts, 2009). Fertility levels started to decline in Asia and Latin America in the late 1960s and varied widely among regions from as high as 5 births per woman in Africa, to 2.5 births per woman in Asia and Latin America in 2000–2005.

Second Demographic Transition

Lesthaeghe and van de Kaa (1986) proposed a second demographic transition (SDT) that began in Europe after World War II. Lesthaeghe (1995) described three phases of second demographic transition: increasing

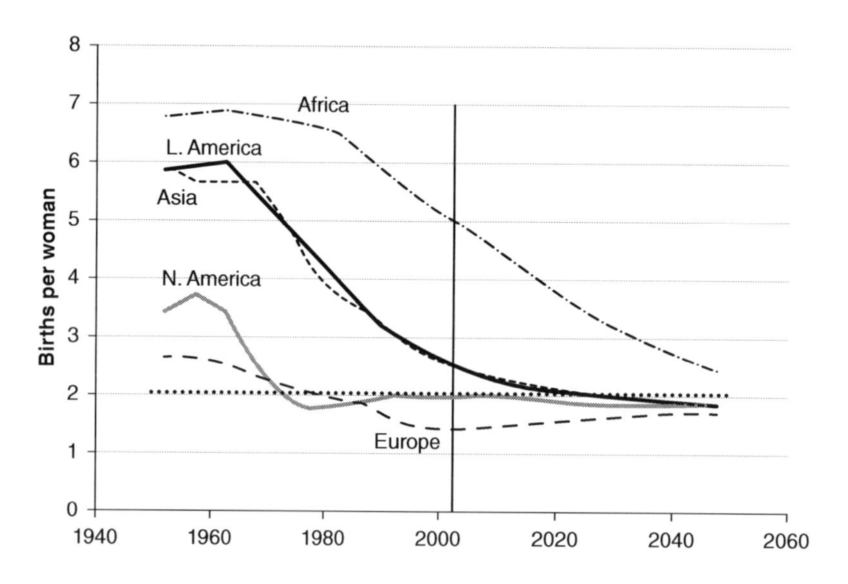

Figure 10.16 Trends in the Total Fertility Rate by Region

Source: John Bongaarts (2009). https://www.ncbi.nlm.nih.gov/pmc/articles/PMC2781829/figure/RSTB20090137F2/

divorce, fertility decline, contraceptive revolution, stop in declining age at marriage in Phase I (1955–1970); rise in premarital cohabitation, rise in non-marital fertility in Phase II (1970–1985); divorce rates plateau, decline in remarriage, recuperation of 30+ fertility in Phase III (1985 onward). The second demographic transition entails "sustained sub-replacement fertility, a multitude of living arrangements, a disconnection

Diversity Overview: Childfree/Childless Families

In most parts of Europe, childlessness and non-marriage were common phenomena during the course of the demographic transition (Rowland, 2007). For example, nearly one in five women in England and Wales born in 1971 has no children at all, compared to one in ten of their mother's generation (Farncombe, 2018). A **childfree family** is a form of family in which couples choose not to have kids.

How do we differentiate between childless and childfree? Sociologists focused on heterosexual women and utilized a childless framework in the 1970s and saw a shift to voluntarily childless couples after the 1980s (Baber & Dreyer, 1986) or "childless-by-choice" or "childfree" framework (Blackstone & Stewart, 2012). "Childlessness" conflates those who want kids and those who do not (Blackstone, 2014a). **Involuntary childlessness** refers to couples who want children but cannot conceive or are past childbearing age. **Voluntary childlessness,** also described by some as being *childfree*, is the lifelong voluntary choice to not have biological, step, or adopted children. Traditional perspectives see "femininity" and "motherhood" as inseparable (Balmer, 1994). Childfree, leading from the childless, involves the separation of "femininity" from "motherhood" (Basten, 2009). Today, the terms "childless-by-choice" and "childfree" are more commonly seen than "childless" when referring to individuals who have made the intentional choice not to have or rear children (Blackstone & Stewart, 2012).

Childlessness has been a central topic of research as families without kids have grown in developed nations since the 1970s (Blackstone, 2014b). Voluntary childlessness is characterized as less traditional and conventional in their gender roles (Callan, 1986), urban residency (DeOllos & Kapinus, 2002), greater financial stability and professional employment (Cwikel *et al.*, 2006), higher levels of education (Kneale & Joshi, 2008), the highest income, work experience, and lowest religiosity (Abma & Martinez, 2006).

Voluntary childless couples expressed similar levels of marital happiness and relationship satisfaction as couples with children (Kreyenfeld & Konietzka, 2017). However, voluntary childfree couples have unique characteristics and problems from couples with children, including developing ways to escape the pressure from society and others to reproduce (Pelton & Hertlein, 2011). For example, some opponents of the childfree choice perceive it as selfish, cold, and materialistic (Kelly, 2009). The "emerging childfree movement" demonstrates that childfree adults recognize their status as stigmatized and challenge the characterization of their choice as a deviant one (Park, 2005). Houseknecht (1977) noted early on the importance of forming a supportive reference group as one strategy for managing the stigma associated with the childfree status. The internet and social networking sites have facilitated the efforts of the childfree to connect with like-minded others (Basten, 2009). The techniques that childfree women and men use to manage stigma range from defensive or reactive (i.e., "identity substitution") to proactive (i.e., redefining "childlessness" in a positive way) (Park, 2005). Proponents of child-freedom posit that choosing to have children may be the more selfish choice, especially when poor parenting risks creating many long-term problems for both the children themselves and society at large (Leone, 1989).

After completion of this section, students should be able to:
LO4: Describe reasons for choosing to have or not to have a child.
LO5: Describe how a pre-baby checklist helps people calculate the potential cost of raising a child.

between marriage and procreation" (Lesthaeghe, 1995). Examining demographic change in 30 European countries, van de Kaa (1987) suggested that "the principal demographic feature of this second transition is the decline in fertility from above the 'replacement' level of 2.1 births per woman to a level well below replacement." For example, the total fertility rate in Hong Kong dropped from 1976 (2.48 births per woman) to 2005 (0.966). The increase in the number of spinsters, the delay in marriage, and the reduction of family size among married couples has increased in the past three decades, which has had a significant effect on the total fertility rate of Hong Kong (Yip *et al.*, 2006).

The United States is well on its way to a second demographic transition with rising ages at marriage, growing rates of cohabitation, increases in single-person households, declining remarriage rates, a trend towards fertility postponement, and higher rates of childlessness (Lesthaeghe & Neidert, 2006).

Deciding Whether to Have a Child

New motivations underlying family formation and parenthood distinguished the second demographic transition from the first transition. The watershed between the first demographic transition and the second demographic transition is the shift in norms from *altruistic* to *individualistic* (Lesthaeghe, 1995, 2014; van de Kaa, 2002). Greater female emancipation and individual autonomy were more central to the second demographic transition than they were to the first transition (Lesthaeghe, 1995). Let's look at factors involved in a couple deciding to have a child.

Reasons People Choose to Have a Child

In traditional societies, parents often wanted to have children so that they could serve as a labor force to help on family farms. In addition, people at the time tended to have large families because child mortality rates were high. Today, with far fewer children working on farms and a far lower infant mortality rate, the reasons people have for wanting children are diverse. Many people choose to have a child because they place a high value on the experience of parenthood. Others choose to have children to carry on their family name. For example, in Asia, traditional families often want a boy child to carry on the family name. Some couples choose to have a child because their parents want a grandchild. Still others choose to have children because they want to be taken care of when they get old. Unfortunately, couples sometime choose to have a child to "save their marriage." However, doing so is misguided, as creating a new life involves changing roles, responsibilities, and challenges. At the heart of the decision to bring a child into the world often lies the parents' own desires (to enjoy childrearing or perpetuate one's legacy/genes), rather than the potential person's interests (Benatar, 2006). It is therefore recommended that before a couple decides to bring a new life into the world, they should think carefully about their decision.

In the United States, women have waited longer to have children than a decade ago. Millennial women (those born from 1981 to 1997) appear to be waiting longer to become parents, compared with prior generations. Around four in ten Millennial women (42 percent) aged 18–33 were moms in 2014, compared with moms (49 percent) of Generation X (born between 1965 and 1980) of the same age range (Blalik, 2017).

Women aged 40–44 are likely to choose to have children before the end of their childbearing age. In 2016, some 86 percent of women aged 40–44 were mothers, compared with 80 percent in 2006 (Livingston, 2018). Women with high levels of education in their early 40s were likely to choose to have children. For example, in 2014, 82 percent of women in their 40s with a bachelor's degree were mothers, compared with 76 percent of their counterparts in 1994; 80 percent of women with a PhD or professional degree were mothers, compared with 65 percent in 1994 (Livingston, 2018). White and Black never-married women in their 40s were likely to choose to have children. In 2014, 37 percent were White mothers, compared with 13 percent in 1994; 75 percent were Black mothers in 2014, compared with 69 percent in 1994 (Livingston, 2018).

Interestingly, research indicates significant fitness benefits of free mate choice. In a comparative study of arranged and non-arranged marriages in terms of number of offspring, Sorokowski *et al.* (2017a) reported that free mate choice did not influence reproductive success in humans. Love is formed as a triangle with intimacy, passion, and commitment (Sternberg, 1986). A study of how love is related to biological fitness revealed that passion and commitment are the key factors that increase biological fitness or the ability to produce offspring (Sorokowski *et al.*, 2017b).

Reasons People Choose to Not Have a Child

There is an agreed-upon assumption that "family" includes children, but these expectations are at odds with changes in the way people are living (Baer, 2016). Today, some choose not to have children and stick with this decision; others may postpone childbearing decisions for such a period of time that eventually biological reproduction is no longer an option; others may intend to bear children but never do so (Blackstone & Stewart, 2012).

An increasing number of people have opted out of parenthood. In the United States, the total percentage of childless women aged 15–44 increased from 2002 (40.1) to 2015 (45.1). Voluntary childlessness grew from 1982 (5 percent) to 1995 (9 percent) (Abma & Martinez, 2006) and increased from 2002 (6.2 percent) to 2015 (7.4 percent). In 2002, 18 percent of women aged 40–44 had never had a child (Osborne, 2003), which rose to 20 percent in 2006 (Dye, 2008). There are racial and ethnic differences in childlessness in the United States. In the late 1980s, 15 percent of White women aged 40–44 were childless, compared with 14 percent of Black women and 11 percent of Hispanic women (Livingston, 2015), and for much of the twentieth century, childlessness was higher among Blacks than among Whites (Frejka, 2017).

At macro-level analysis, explanations for increasing rates of voluntary childlessness focus on social changes such as the feminist movement in the 1970s, increased reproductive choice, and women's increasing labor force participation (Gillespie, 2003). Baudin *et al.* (2017) proposed two types of childlessness: opportunity driven and poverty driven. *Opportunity-driven childlessness* is the main reason for the high levels of childlessness in developed countries (Baudin *et al.*, 2017). Highly educated women who earn high wages and face a high opportunity cost should they leave work to have a child may factor this into their decision (Aaronson *et al.*, 2014). For example, a year of delayed motherhood increased women's earnings by 9 percent, their work experience by 6 percent, and average wage rates by 3 percent (Miller, 2010). *Lack of financial resources* is a reason couples decide to remain childfree because they can't afford to have a baby. Many women saw parenting as conflicting with career and leisured identities and men rejected reproduction because of its perceived financial sacrifices (Park, 2005). Indeed, childfree couples have more free time and financial resources than couples with children. In the United States, factors that shape childlessness include high costs of childrearing, inadequate childcare infrastructure, insecurity of employment and income, and concern for the well-being of children (Frejka, 2017). For example, couples with children spend 60 percent more on entertainment, 79 percent more on food, and 101 percent more on dining out (Burkett, 2000). In terms of financial expenditure, married couples without kids have more discretionary income than households with children (Paul, 2001).

At micro-level analysis, research has explored individuals' motives for and consequences of remaining childfree. In a study of 25 voluntarily childless women, Gillespie (2003) identified two distinct and interrelated factors contributing toward choosing to be childfree. *The "pull" towards being childless* is characterized by increased freedom and better relationships with partners. Women attributed the benefits to enhanced autonomy and the ability to make choices about education, career, or personal development (Settle & Brumley, 2014). The most common reason for not having children was "freedom from childcare responsibility and greater opportunity for self-fulfillment and spontaneous mobility" (Houseknecht, 1987). Voluntarily childfree couples want to put their relationship first (Scott, 2009) and displayed higher levels of dyadic cohesion and dyadic satisfaction (Somers, 1993). *The "push" away from motherhood* involves a loss of identity and a rejection of the activities associated with motherhood (Basten, 2009). Women believed that the costs and constraints of motherhood outweighed the benefits (Settle & Brumley, 2014). Houseknecht (1979) observed less stressful "marital adjustment" among those without children. Another explanation for a growing rate of voluntary childlessness is that Millennials choose pets over babies. Young Americans may be less likely to be parents of children or homeowners, but they are leading in their rate of pet ownership (Ledbetter, 2017). Singles often see pets as "family members"; for some owners, pets replace kids; for many, the companionship provided by a pet replaces a spouse (Stone, 2017).

Not having children is a deliberate choice (Blackstone, 2014a). Women expressed concern they would not be "good" mothers because they did not have that "mother instinct" (Park, 2005). Family background factors, self-other attitudes, and reference groups play a central role in the decision to remain voluntarily childless (Houseknecht, 1978). In an interview of 21 women and 10 men who have chosen not to have children, Blackstone and Stewart (2016) reported that most respondents said that (1) the decision was a conscious decision and that (2) the decision occurred as a process as opposed to a singular event. For many, it is an ongoing decision as individuals have to check in with their partners or their family over the years. In fact, women aged 15–44 who choose to be temporarily childless increased from 2002 (31.5 percent) to 2015 (35.6 percent) (see Table 10.3).

Table 10.3 Percent Distribution of Childlessness Status among Women 15 to 44 Years of Age

	2002[1]	*2006–2010*[1]	*2011–2015*[2]
Total percent of women who are childless (0 births)	40.1%	42.6%	45.1% (0.90)
Percent who are temporarily childless (expect 1 or more)	31.5%	34.3%	35.6% (0.80)
Percent who are voluntarily childless (expect none, physically able to have children)	6.2%	6.0%	7.4% (0.44)
Percent who are involuntarily childless (expect none, physically unable to have children)	2.5%	2.3%	2.1% (0.18)

Source: CDC, March 1, 2018. https://www.cdc.gov/nchs/nsfg/key_statistics/c.htm#childlessness

Note: The percentages may not add to the total percent childless due to rounding.

In 2016, 30.8 percent of women aged 30–34 in the United States were childless, up 4 percent from 2006 (26.2 percent) (see Figure 10.17). This increase does not necessarily mean that women are choosing not to be mothers (Monte, 2017b) and they may go on to have children later.

What are the characteristics of older childless persons? A cross-country study of childlessness in old age revealed that childless, never-married women had consistently higher education levels than other women; and among men, marriage rather than parenthood was consistently linked with higher socioeconomic status in Australia, Finland, Germany, the Netherlands, the United Kingdom, and the United States, with the exception of Japan (Koropeckyj-Cox & Call, 2007).

The aging population of adults without children became a point of focus for researchers in the 2000s (Blackstone & Stewart, 2012). By 2030, 16 percent of women aged 80–84 will be childless, compared with 11.6 percent in 2010, but the **caregiver support ratio** (i.e., the number of potential caregivers aged 45–64 for each person aged 80 and older) is projected to decline from seven potential caregivers for one person to three for one by 2050 (Redfoot *et al.*, 2013). Elderly childless couples appear to have less social support and emotional ties and are subject to less benevolent surveillance and crisis help (Koropeckyj-Cox, 1988). Childless women in midlife and old age reported lower life satisfaction and self-esteem than both mothers with residential children and empty nest mothers (Hansen *et al.*, 2009).

Studies have shown that the elderly childless create bonds to fulfill many of the same functions that families with children fulfill (Blackstone, 2014b) and often have more extensive social support systems than people with children owing to greater social involvement (Stephens *et al.*, 1978). For example, single childless people fare better than married couples, as they are more likely to be involved with family and

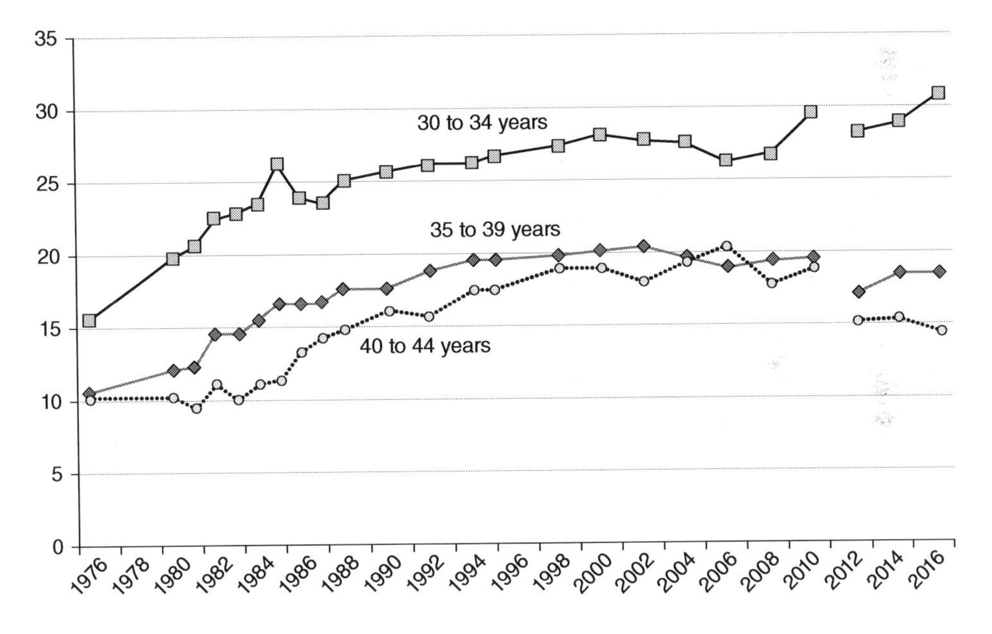

Figure 10.17 Percent Childless, Women aged 30–44, 1976–2016

Source: Current Population Survey, Fertility Supplement, 1976–2016. https://www.census.gov/library/stories/2017/11/women-early-thirties.html

friends (Rubinstein, 1987). Their support networks included stronger links with a broader range of relatives as well as friends and other non-relatives, and they tended to be more actively engaged in charity work than their counterparts with children, thus contributing to their communities in unique and important ways (Albertini & Kohli, 2009). In Wales, childfree older adults adapted their lives to accommodate their needs and fill "voids" that might be presumed to exist by relying on other family members or on engagement in community endeavors (Wenger, 2009).

Costs of Having a Child

In 2015, a middle-income ($59,200 to $107,400), two-child, married-couple family in the United States spent approximately $12,980 annually per child (Lino *et al.*, 2017). Indeed, housing, food, and childcare cost money. For middle-income families, housing accounts for the largest share, at 29 percent of total childrearing costs; food is second at 18 percent; and childcare/education is third at 16 percent (Lino, 2017). Expenses vary depending on the age of the child. Overall annual expenses averaged about $300 less for children from birth to two years old and averaged $900 more for teenagers aged fifteen to seventeen. Middle-income, married-couple parents of a child born in 2015 could expect to spend $233,610 for food, shelter, and other necessities to raise a child through age seventeen (see Figure 10.18).

Calculating Cost: The Pre-baby Checklist

To calculate the cost a child may have on a family, people sometimes create a pre-baby checklist. One item on the checklist is quality health care, which is often expensive. Maternity care is also expensive. On average, paying for maternity care out of pocket would cost more than a woman's annual premiums or rent, and roughly half her income. A nationwide study in 2010 found that for women with employer-sponsored insurance, the average payment for a vaginal birth was around $12,500 or $15,300 in today's cost after adjusting for inflation; for the one-third of U.S. births that involve a caesarean section, those payments were $16,700 or $20,400 in today's cost (Guttmacher

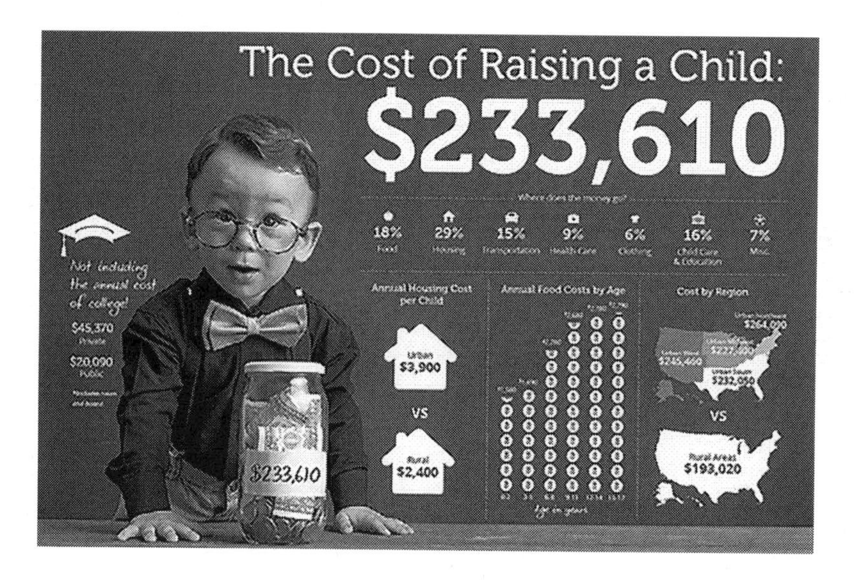

Figure 10.18 The Cost of Raising a Child
Source: U.S. Department of Agriculture (USDA) 2017. https://www.usda.gov/media/blog/2017/01/13/cost-raising-child

Institute, 2017a). Compared to a singleton (i.e., a child that is the only one born at one birth), multiple pregnancies are a major public health concern due to the higher health risks for mothers and infants and the health-care costs. Pregnancies with the delivery of twins cost approximately five times as much when compared with singleton pregnancies and pregnancies with delivery of triplets or more cost nearly twenty times as much (Lemo *et al.*, 2013). In 2013, the mean health-care cost to patients and insurers was estimated to be $26,922 for ART-conceived singleton deliveries, $115,238 for ART-conceived twins, and $434,668 for ART-conceived triplets and higher-order infants (Sunderam *et al.*, 2018).

The principal source of payment for the delivery of most births in 2016 was either private insurance (49.4 percent) or Medicaid (42.6 percent); the remainder of births includes uninsured deliveries or self-pay (4.1 percent) and other insurance (3.9 percent) such as federal, state, or local government, and charity (Martin *et al.*, 2018a). Source of payment for caesarean delivery varies by race. Medicaid coverage ranged from 24.9 percent for Asian women to 66.9 percent for American Indian or Alaska Native women. Private insurance ranged from 20 percent for American Indian or Alaska Native women (19.7 percent) to 60 percent for White (63.3 percent) and Asian (65.4 percent) women (Martin *et al.*, 2018a).

In addition, it is important that people understand their employer's maternity policy. In the United States, the Family and Medical Leave Act (FMLA) guarantees twelve weeks of unpaid leave after the natural birth or adoption of a child. This sometimes means that people wishing to have a child must save up several months of pay ahead of its birth.

A plan for childcare is another item on a pre-baby checklist. Low-income and rural communities may have limited or no access to quality childcare. Meanwhile, the average cost of center-based infant care exceeds more than 40 percent of the median income for single mothers and 27 percent of the median income for Millennials (CCAoA, 2017). The pre-baby checklist also often includes items such as a crib, bed, dresser, strollers, car seats, bottles, swings, bouncers, clothing, a baby bath, and diapers.

Preventing Pregnancy

Pregnancies that occur too early or too late in a woman's life, or that are spaced too closely, negatively affect maternal health and increase the risk of prematurity and low birth weight (Guttmacher Institute, 2011). People who wish to not have children can prevent unwanted pregnancies by engaging in abstinence and by using contraceptives. **Contraception** refers to birth control or family planning methods of preventing pregnancy.

Birth Control Methods

Non-permanent methods of contraception are available for both men and women. Many hormonal methods (the pill, vaginal ring, patch, implant, and IUD) offer a number of health benefits in addition to contraceptive effectiveness, such as treatment for excessive menstrual bleeding, menstrual pain, and acne (Jones, 2011). Birth control pills keep a woman's eggs from leaving her ovaries and must be taken every day at the same time to work properly. Four of every five sexually experienced women have used the pill (Daniels *et*

After completion of this section, students should be able to:
LO6: Explain five birth control methods.
LO7: Describe three ways of handling an unintended pregnancy.

al., 2013). Hormonal vaginal contraceptive rings are inserted inside the vagina and release hormones that prevent ovulation. Diaphragms are inserted inside the vagina and over the cervix to prevent semen from contacting an egg. Another option is a skin patch that blocks pregnancy and is worn on the lower abdomen, buttocks, or upper body (but not on the breasts) (CDC, 2017a). An intrauterine device (IUD) is a T-shaped piece of plastic inserted into the uterus to prevent pregnancy. Condoms are options for both males and females. Male condoms are worn on the penis during intercourse to keep semen from coming into contact with eggs. Female condoms are shaped like a ring with a pouch and are used to collect semen during intercourse. An added benefit of condoms is that they protect both parties from some sexually transmitted diseases (STDs).

For people who are certain that they no longer wish to have children, *sterilization* is the most effective family planning method. Female sterilization surgery is conducted to close off the fallopian tubes in a process called tubal ligation. Men can choose to undergo a vasectomy, which blocks their vas deferens, through which sperm flows, preventing the sperm from mixing with semen and thus avoiding pregnancy. Sterilization is most common among Blacks and Hispanics, women aged 35 or older, ever-married women, women with two or more children, women living below 150 percent of the federal poverty level, women with less than a college education, and women living outside of metropolitan areas (Jones *et al.*, 2012).

Emergency contraception is a way to prevent pregnancy after unprotected sex or contraceptive failure and it has no effect on an established pregnancy (Trussell & Raymond, 2018). Emergency contraceptives available in the United States include emergency contraceptive pills (ECPs) and the Copper T intrauterine contraceptive (IUC) (Stewart *et al.*, 2007; Glasier, 1997; Hatcher *et al.*, 1995). Nonhormonal copper IUDs, inserted up to five days after unprotected intercourse, can act as emergency contraception (Trussell & Raymond, 2018). A study of the IUD and oral levonorgestrel (LNG) users for emergency contraception revealed that women who chose the IUD had fewer pregnancies in the following year than those who chose oral LNG (Turok *et al.*, 2016).

More than 99 percent of women ages 15–44 who have ever had sexual intercourse have used at least one contraceptive method (Daniels *et al.*, 2013). In 2011–2013, the most common contraceptive methods currently being used were the oral contraceptive pill (16.0 percent), female sterilization (15.5 percent), condoms (9.4 percent), and long-acting reversible contraceptives (7.2 percent) (Daniels *et al.*, 2014). Some 8 percent of women of reproductive age use multiple contraceptive methods, often the condom combined with another method (Eisenberg *et al.*, 2012). When used correctly, modern contraceptives can effectively prevent pregnancy (Guttmacher Institute, 2016).

Unintended Pregnancy

An **unintended pregnancy** is one that was either mistimed or unwanted. In 2010–2014, an estimated 44 percent of pregnancies worldwide were unintended (Bearak *et al.*, 2018). From 1990–1994 to 2010–2014, the unintended pregnancy rate dropped less sharply in developing regions (16 percent) (90 percent UI 5–24) than in developed regions (30 percent) (90 percent UI 21–39) (see Figure 10.19). The high unintended pregnancy rate in developing regions corresponds with a substantial unmet need for contraception in these parts of the world (Bearak *et al.*, 2018). It is critical for governments to prioritize making the full range of contraceptive services available so that women and couples are able to choose the method that best fits their needs to prevent unintended pregnancy (Guttmacher Institute, 2018b).

Contraceptive services and supplies can be costly. In the United States, the federal and state governments provide funding for family planning services and supplies to help women meet these challenges

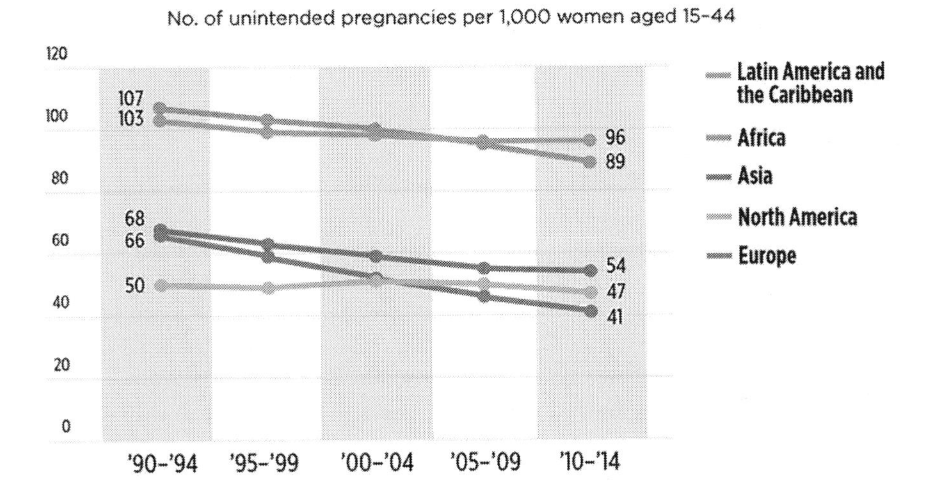

Figure 10.19 Unintended Pregnancy Rates Have Declined Worldwide but Vary Widely across Regions
Source: Guttmacher Institute (2018a); Bearak *et al.* (2018). https://www.guttmacher.org/infographic/2018/changes-unintended-pregnancy-rates-world-region

(Frost *et al.*, 2016). Federal law requires health insurance coverage for the full range of contraceptive methods used by women, including counseling and related services, without out-of-pocket costs (Guttmacher Institute, 2018b). These publicly funded contraceptive services helped women prevent 1.9 million unintended pregnancies. For example, the teen pregnancy rate declined by 51 percent from 116.9 to 57.4 pregnancies per 1,000 girls between 1990 and 2010 (Kost & Henshaw, 2014), down to 24.2 births in 2014 (Hamilton *et al.*, 2015), down to 22.3 in 2015, and down to 20.3 births in 2016 (Hamilton *et al.*, 2017). Couples who do not use any method of contraception have an approximately 85 percent chance of experiencing a pregnancy over the course of a year (Trussell, 2011). Three options that most people choose when facing an unplanned pregnancy are abortion, adoption, and parenting.

Abortion

Abortion is the termination of an embryo or fetus, naturally or via medical methods, often performed during the first 27 weeks of pregnancy. Women may choose to have an abortion because of birth control failure, an inability to support or care for a child, a pregnancy resulting from rape or incest, or physical or mental conditions that endanger the woman's health if the pregnancy is continued. In the world, 56 percent of unintended pregnancies ended in abortion in 2010–2014 and more women terminated their unintended pregnancies in developing countries than in developed countries (Guttmacher Institute, 2018b).

Since the Supreme Court handed down its 1973 decisions in *Roe v. Wade* and *Doe v. Bolton*, states have constructed abortion law, codifying, regulating, and limiting whether, when, and under what circumstances a woman may obtain an abortion. For example, 37 states require parental involvement in a minor's decision to have an abortion (Guttmacher Institute, 2017b). Abortion did rise in the 1970s, but it is lower today than it was when *Roe v. Wade* was decided (VerBruggen, 2017). A total of 664,435 abortions were reported in 2013 (Jatlaoui *et al.*, 2016). The majority of abortion patients (60 percent) were in their 20s, and three-fourths of abortion patients were poor and low income (see Figure 10.20).

Abortion varied widely among racial groups. Figure 10.21 shows that 39 percent of those who obtained an abortion were White, 28 percent were Black, 25 percent were Hispanic, 6 percent were Asian or Pacific Islander, and 3 percent were of some other race or ethnicity in 2016.

Placing a Child for Adoption

Some people consider placing a child for adoption. In general, any person who has the right to make decisions about a child's care and custody may place that child for adoption (CWIG, 2016a). Such persons include the birth parents or the child's legal guardian or legal entities such as state departments of social services or licensed child-placing agencies. There are tens of thousands of birth parents who placed their children for adoption in New Jersey as far back as 1920 (State of New Jersey, 2018). Adoption often benefits everyone in the adoption triad. For example, if birth mothers wish to finish school or attend college, adoption can bring a positive ending to a problematic situation. In turn, adoptive families experience the joy of adding a child to their family.

Parenting

Some choose to marry and proceed with having the baby after the discovery of an unplanned pregnancy. The highly educated are more likely than the less educated to marry after an unwed pregnancy (Ver-Bruggen, 2018). If the two partners recognize that the timing is not right for marriage but they both are interested in having the baby, joint custody is an option. Single parenting occurs when one partner chooses to participate in caring for the child. Child support is expected and required of the other parent.

Pregnancy

Conception refers to the moment when the sperm fertilizes the egg. **Pregnancy** is a term used to

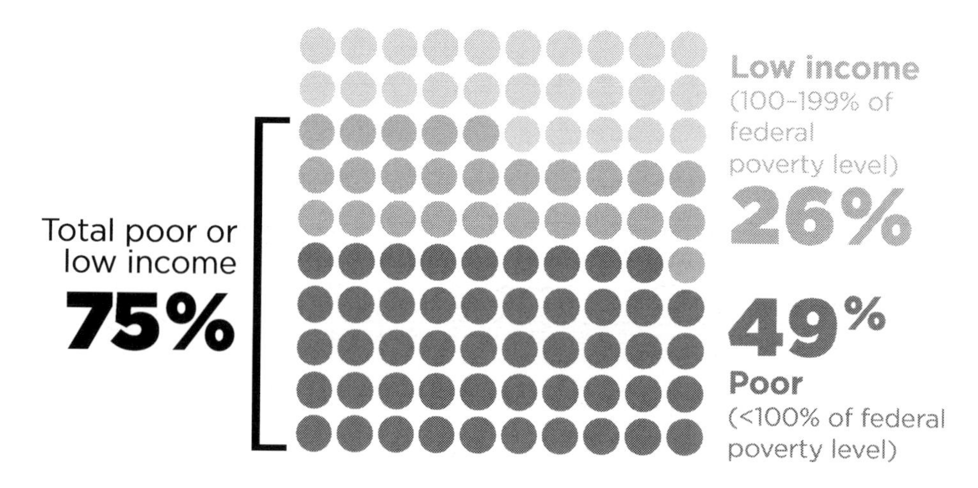

Figure 10.20 Abortion Patients Are Disproportionately Poor and Low Income
Source: New York: Guttmacher Institute 2016. https://www.guttmacher.org/infographic/2016/abortion-patients-are-disproportionately-poor-and-low-income

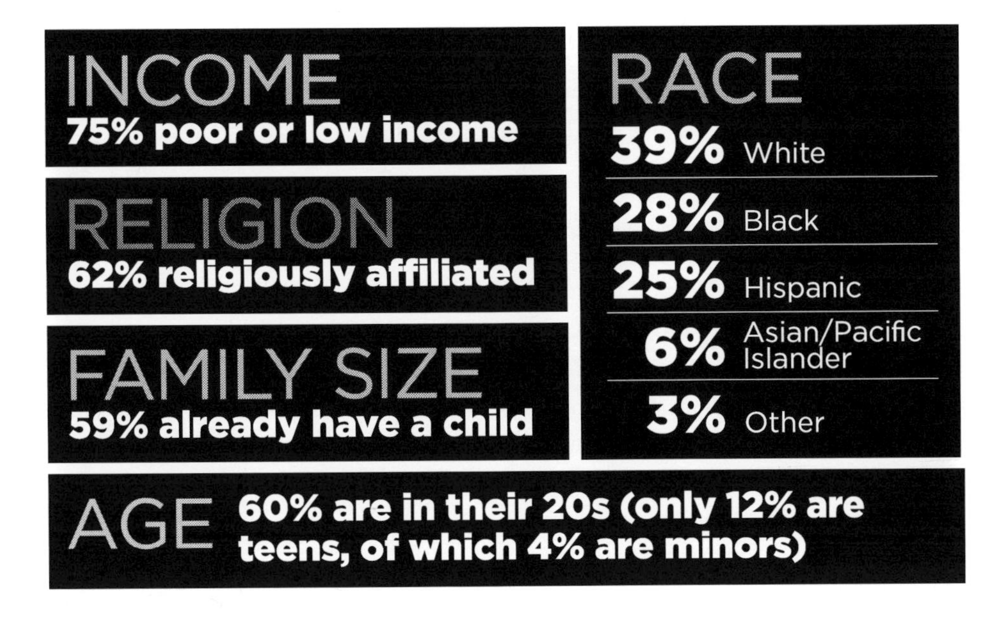

Figure 10.21 U.S. Abortion Patients, 2016
Source: New York: Guttmacher Institute, 2016. https://www.guttmacher.org/infographic/2016/us-abortion-patients

describe the period in which a fetus develops inside a woman's womb or uterus (NIH, 2017). It is important to learn what pregnant women can do before, during, and after pregnancy to give a baby a healthy start to life.

Prenatal Care

Gestational age is a measure of the age of a pregnancy that is taken from the woman's last menstrual period (LMP) to the current date. It is also used to describe pregnancy such as prenatal development, estimation of due date (expected date of delivery), birth classification into preterm, term or postterm, and estimation of various risk factors (see image below).

Quality care before, between, and during pregnancies will ensure that women have a positive pregnancy experience (WHO, 2018). Each year, approximately 500 women die as a result of pregnancy-related complications and an estimated 500 additional such deaths that are not identified as being caused by pregnancy occur (CDC, 2017b). Many of these adverse outcomes can be prevented by prenatal care during pregnancy and childbirth. **Prenatal care**, or antenatal care, is a type of preventive health care with the goal of providing regular

After completion of this section, students should be able to:
LO8: Describe how prenatal care prevents risks during every stage of pregnancy.
LO9: Explain how postnatal care helps prevent risks after childbirth.

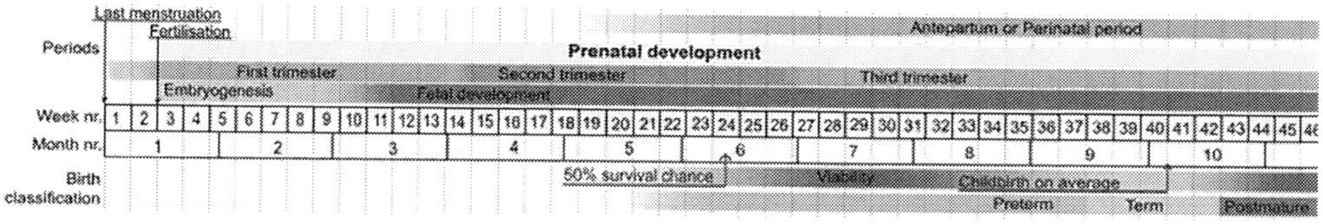

Last menstruation
Fertilisation
Periods
Antepartum or Perinatal period
Prenatal development
First trimester
Second trimester
Third trimester
Embryogenesis
Fetal development
Week nr. 1 2 3 4 5 6 7 8 9 10 11 12 13 14 15 16 17 18 19 20 21 22 23 24 25 26 27 28 29 30 31 32 33 34 35 36 37 38 39 40 41 42 43 44 45 46
Month nr. 1 2 3 4 5 6 7 8 9 10
Birth classification
50% survival chance
Viability
Childbirth on average
Preterm
Term
Postmature

Prenatal Development Table used to Describe Prenatal Development, Estimation of Due Date, Birth Classification into Preterm, Term or Postterm, Estimation of Various Risk Factors.
Source: Mikael Häggström. https://commons.wikimedia.org/wiki/File:Prenatal_development_table.svg

check-ups that allow doctors to treat and prevent potential health problems throughout the course of the pregnancy. Prenatal care improves pregnancy outcomes and benefits for both mother and child. Women carrying more than one baby are at higher risks for both mothers and infants. More frequent prenatal visits help the doctor to monitor both mother's and baby's health. With close monitoring, babies will have the best chance of being born near term and at a healthy weight (OWH, 2018a). Nutrition during pregnancy is important to ensure the healthy growth of the fetus. The doctor will tell how much weight to gain, if a pregnant woman needs to take extra vitamins, and how much activity is safe. The CDC (2017b) suggests taking folic acid before conception, eating a well-balanced diet, limiting caffeine, and avoiding alcohol, cigarettes, and drugs.

In the United States, prenatal care varies by stage of pregnancy, race, and age of mother (Martin *et al.*, 2018a). In 2016, 77.1 percent of pregnant women began prenatal care in the first trimester of pregnancy; 6.2 percent began in the third trimester or had no prenatal care. Women under age twenty (61.2 percent) and women aged 30–34 (82.1 percent) received first trimester prenatal care while women under age twenty (11.2 percent) and women in their 30s (4.8 percent) received late or no prenatal care. The majority of White (82.3 percent) and Asian (80.6 percent) women began prenatal care in the first trimester, compared with Hispanic (72.0 percent), Black (66.5 percent), American Indian or Alaska Native (63.0 percent), and Native Hawaiian or Pacific Islander women (51.9 percent) (Martin *et al.*, 2018a).

Postnatal Care

The **postpartum period** or postnatal period is the period beginning immediately after childbirth and extending for about six weeks. The postnatal period is considered the most critical but the most neglected period in the lives of mothers and babies, with most deaths of mothers and babies occurring during the postnatal period on a worldwide basis (WHO, 2013b). Risks during the postpartum period include hemorrhaging, high blood pressure, blood clots, and postpartum depression. **Postpartum hemorrhage** (PPH) occurs when a woman has heavy bleeding after delivery. The rate of PPH with procedures to control hemorrhage increased from 4.3 in 1993 to 21.2 in 2014 and the rate of PPH with blood transfusions also increased from 7.9 in 1993 to 39.7 in 2014 (CDC, 2017g). Another complication of birth is **deep vein thrombosis** that happens when a blood clot forms in a deep vein. The most serious complication of deep vein thrombosis is when part of the clot breaks off and travels to the lungs, causing a blockage called *pulmonary embolism*. The rate of pulmonary embolism doubled over time from 1.1 in 1993 to 2.2 in 2014 (CDC, 2017g). Figure 10.22 shows the rate of deep vein thrombosis and pulmonary embolism per 1,000 delivery hospitalizations from 1994 through 2014 in the United States.

Many women have the "baby blues" in the days after childbirth when they have mood swings, feel

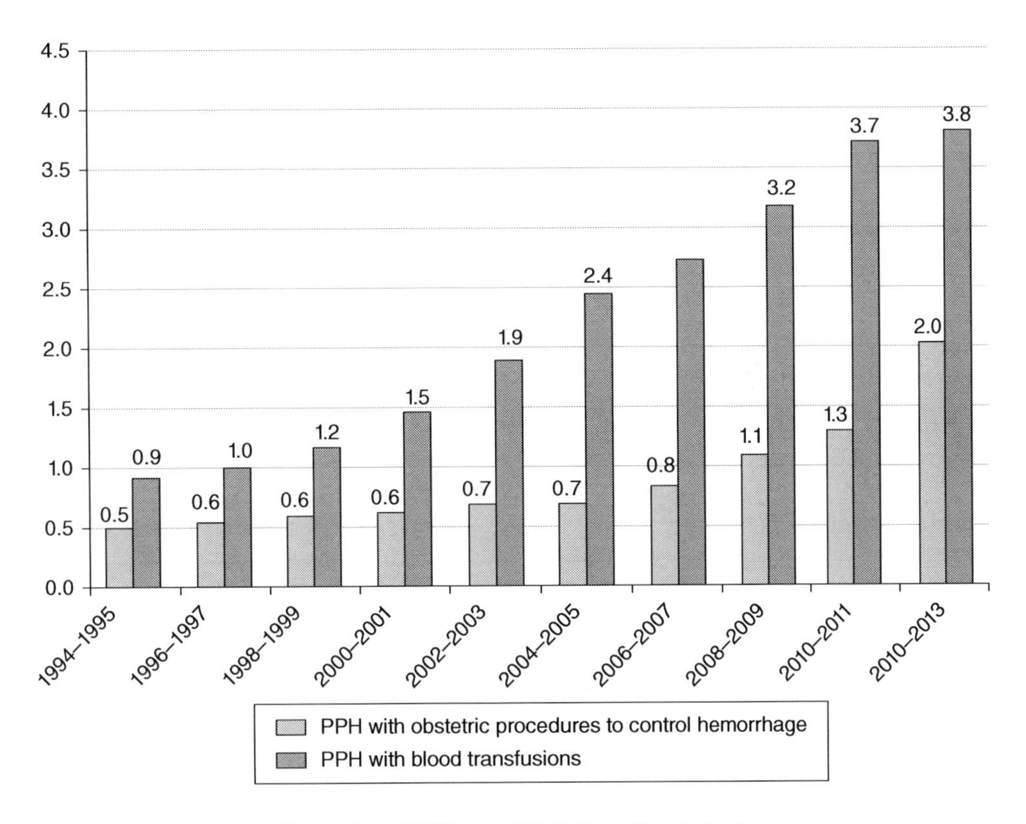

Figure 10.22 Rate of Postpartum Hemorrhage (PPH) per 1,000 Delivery Hospitalizations
Source: CDC October 19, 2016. https://www.cdc.gov/reproductivehealth/maternalinfanthealth/pregnancy-complications-data.htm

sad, anxious, or overwhelmed, or have trouble sleeping. The baby blues most often go away within a few days or a week. However, **postpartum depression** (PPD) is a type of mood disorder associated with childbirth, including symptoms such as sadness, thoughts of hurting the baby, thoughts of hurting yourself, and not having any interest in the baby, and needs to be treated by a doctor (OWH, 2017).

Stages of Pregnancy

The German obstetrician Franz Karl Naegele (1778–1851) devised a standard way of calculating the due date for a pregnancy when assuming a gestational age of 280 days (40 weeks) at childbirth. Naegele's rule estimates the expected date of delivery (EDD) by adding a year, subtracting three months, and adding seven days to the origin of gestational age (woman's last menstrual period).

Pregnancy is divided into three trimesters. The **first trimester** is from week 1 through 12 and includes conception in which a sperm penetrates an egg. The fertilized egg travels through the woman's fallopian tube to the uterus, where it implants itself in the uterine wall (NIH, 2017). During the first trimester, hormonal changes affect almost every organ system in the body and can trigger symptoms even in the very first weeks of pregnancy (OWH, 2018b). Pregnant women with multiples have more severe body changes, including rapid weight gain in the first trimester, intense nausea and vomiting, or extreme breast tenderness (OWH, 2018b). The **second trimester** is from week 13 through 28. Women might experience more noticeable changes to their body. For example, the abdomen will expand as the baby continues to grow (OWH, 2018b). Around the middle of the second trimester, a pregnant woman can feel the movement of the fetus. At 24 weeks, footprints and fingerprints have formed and the fetus sleeps and wakes regularly (NIH, 2017). The **third trimester** is from 29 weeks through 40 weeks. At 32 weeks, the bones are soft and yet almost fully formed, and the eyes can open and close (NIH, 2017). Many women find breathing difficult and notice that they have to go to the bathroom even more often. This is because the baby is getting bigger and it is putting more pressure on their organs (OWH, 2018b). **Childbirth** (or labor and delivery) is the ending of a pregnancy by one or more babies leaving a woman's uterus by vaginal passage or caesarean section.

Stages of Labor

In the last month of pregnancy, contractions of the uterus become noticeable, and the baby settles into the pelvis (Lothian, 2000). Labor occurs in three stages. The first stage begins with the onset of labor and ends when the cervix is fully opened (OWH, 2018c). The pain of labor is what most women worry about. Whether a woman gives birth in a hospital, birthing center, or at home, she is able to use a wide variety of comfort measures, moving freely, listening to music, taking a shower or bath, and having her feet and hands massaged (Lothian, 2000). If women are at the hospital, doctors will monitor the progress of their labor by checking their cervix, the baby's position, and location in the birth canal.

The second stage involves pushing and delivery of the baby (OWH, 2018c). Women will push hard during contractions, and rest between contractions. A woman can give birth in positions such as squatting, sitting, kneeling, or lying back (OWH, 2018c). The third stage involves delivery of the placenta (i.e., a large organ that develops during pregnancy) and contractions will begin 5 to 30 minutes after birth, signaling that it is time to deliver the placenta (OWH, 2018c). Labor is over once the placenta is delivered. If necessary, the doctor will repair any episiotomy (i.e., a surgical cut made at the opening of the vagina during childbirth).

Method of Delivery

Methods of delivery include virginal birth and caesarean delivery. Women are inherently capable of giving birth, have a deep, intuitive instinct about birth, and are able to give birth without interventions (Lothian, 2000). Most healthy pregnant women with no risk factors for problems during labor or delivery have their babies vaginally. **Caesarean delivery**, also called c-section, is surgery to deliver and take out the baby through the mother's abdomen. Most caesarean births result in healthy babies and mothers but a c-section may carry risks. The benefits of having a c-section may outweigh the risks when the mother carries more than one baby (twins, triplets, etc.) or has problems with health issues or with the position of the baby (OWH, 2018c). Caesarean delivery increased with advancing maternal age. Women aged 40 and over were more than twice as likely to deliver by caesarean as women under age 20 (Martin *et al.*, 2018a).

The caesarean delivery rate (i.e., the number of births delivered by caesarean per 100 births) declined for the fourth year in a row to 31.9 percent in 2016, down from 32.0 percent in 2015 and from 32.9 percent in 2009 in the United States. By race, the caesarean delivery rate was lowest among American Indian or Alaska Native women (28.0 percent) and highest among Black women (35.9 percent); for Hispanic subgroups, caesarean delivery ranged from 30.3 percent for Mexican women to 47.0 percent for Cuban women (Martin *et al.*, 2018a). From 2015 to 2016, caesarean delivery rates declined for all maternal age groups, down 13.0 percent for mothers aged 20 and under (20.2 percent), down 6 percent for mothers aged 20–29 (28.5 percent), down 5 percent for mothers aged 30–39 (36.3 percent), and down 3 percent for mothers aged 40 and over (47.9 percent) from the 2009 peaks (see Figure 10.23).

The **primary caesarean delivery rate** measures caesarean deliveries among women who have not had a previous caesarean delivery. In 2016, the primary caesarean delivery rate was 21.8 percent (about one of every five women who had not had a previous baby delivered by caesarean) (Martin *et al.*, 2018a). **Vaginal delivery after c-section** (VBAC) is a vaginal childbirth by a woman who has delivered previous babies by c-section and would like to have their next baby vaginally. Women give many reasons for wanting a VBAC. Some want to avoid the risks and long recovery of surgery and others want to experience vaginal delivery (OWH, 2018c). In 2016, 12.4 percent of women with a previous caesarean delivered vaginally (Martin *et al.*, 2018a). Favorable initial pelvic examination, spontaneous labor, and a lack of oxytocin use are associated with successful VBAC in women (Durnwald & Mercer, 2010).

The **low-risk caesarean delivery rate**—caesarean delivery among nulliparous (first birth), term (37 or more completed weeks based on the obstetric estimate), singleton (one fetus), cephalic (head first) births—declined from 2015 (25.8 percent) to 2016 (25.7 percent) (Martin *et al.*, 2018a).

Self-control over Method of Delivery

Choosing a caesarean section is seen as having a sense of control (Davis-Floyd, 1994). Caesarean sections have increased in developed countries undergoing modernization (Barros *et al.*, 1991) and occurred in developing countries. Social class plays a significant role in how women perceive childbearing and the extent to which they wish to have control over their pregnancies and births (Nelson, 1983). For example, middle-class women are able to choose where to give birth, have access to private care, and actively seek

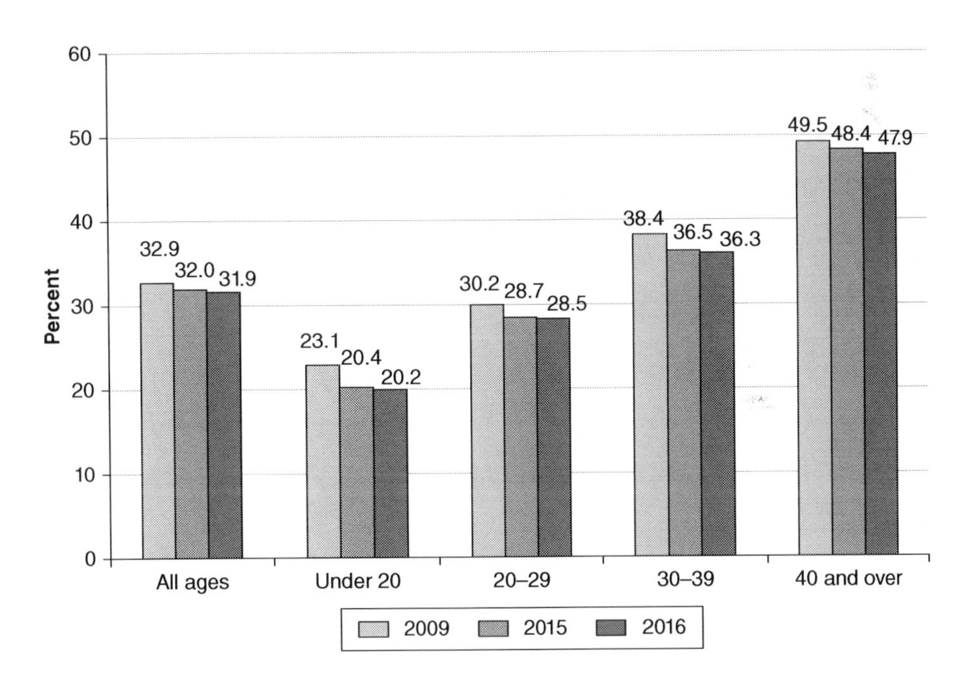

Figure 10.23 Caesarean Delivery Rates, by Age of Mother: United States, 2009, 2015, and 2016

Source: NCHS, National Vital Statistics System, Natality; Martin *et al.* (2017a). https://www.cdc.gov/nchs/products/databriefs/db287.htm

medical technology as a way to have control over their birth, but these choices were denied to the rural poor women in northern Thailand (Liamputtong, 2005). In Australia, middle-class women were able to exercise choice and control more than working-class women because the middle-class women's material resources "enable choice" (Zadoroznyj, 1999).

The notion of risk has penetrated many areas of the postmodern world (Beck, 1992), including the world of childbearing (Lane, 1995). Women's social positions influence the extent to which they passively accept or actively reject medical technologies (Kabakian-Khasholian *et al.*, 2000). For example, middle-class women sought control of themselves as they labored and gave birth and resisted medical control, whereas working-class women rejected the idea of self-control but focused more on their lived experience of childbirth such as the intensity or length of labor pain (Martin, 1992).

Period of Gestation

Gestation in singleton pregnancies lasts an average of 40 weeks (280 days) from the first day of the last menstrual period to the estimated date of delivery. Using the last menstrual period (LMP) method, a full-term human pregnancy is considered to be 40 weeks (280 days), though pregnancy lengths between 38 and 42 weeks are considered normal. Full-term infants have better health outcomes than do infants born earlier (NIH, 2017). A fetus born prior to the 37th week of gestation is considered to be preterm. Using gestational age, births can be classified into broad categories (see Table 10.4).

Early-, Full- and Late-term Birth

Early-term birth rate (i.e., number of births delivered between 37 weeks and 38 weeks 6 days of gestation per 100 births) declined from 2007 to 2014, went up from 2015 (24.99 percent) to 2016 (25.47 percent) (Martin *et al.*, 2018a), and went up again from 2017 Q1 (25.60 percent) to 2018 Q1 (26.13 percent) (CDC, 2018a). **Late-term birth rate** (i.e., number of births delivered between 41 weeks and 41 weeks 6 days of gestation per 100 births) declined from 2017 Q1 (6.34 percent) to 2018 Q1 (6.21 percent) (CDC, 2018a). **Full-term birth rate** (i.e., number of births delivered between 39 weeks and 40 weeks 6 days of gestation per 100 births) increased from 2007 to 2014, and declined from 58.47 percent in 2015 to 57.94 percent in 2016 (Martin *et al.*, 2017b), down to 57.84 percent in 2017 Q1 and 57.41 percent in 2018 Q1 (CDC, 2018a). **Post-term birth rate** (i.e., number of births delivered between 42 weeks and more days of gestation per 100 births) declined from 2015 to 2016 (Martin *et al.*, 2018a).

Preterm Birth

A developing baby goes through important growth throughout pregnancy. The final weeks of pregnancy are critical for the brain, lungs, and liver to fully develop. **Preterm birth** (or premature birth) occurs when a baby is born before 37 weeks of pregnancy are completed. In the United States, **preterm birth rates** (i.e., number of births delivered before 37 completed weeks of gestation per 100 births) rose 1 percent in 2017 to 9.93 percent of all births, from 9.85 percent in 2016 (see Figure 10.24). Preterm birth

Table 10.4 Birth Classification

Gestational Age in Weeks	*Classification*
< 37 0/7	Preterm Early preterm (before 34) Late preterm (34 to 36 weeks)
37 0/7–38 6/7	Early term
39 0/7–40 6/7	Full term
41 0/7–41 6/7	Late term
> 42 0/7	Postterm

Source: The American College of Obstetricians and Gynecologists (2013). https://www.acog.org/Clinical-Guidance-and-Publications/Committee-Opinions/Committee-on-Obstetric-Practice/Definition-of-Term-Pregnancy; CDC (2017). https://www.cdc.gov/nchs/products/databriefs/db287.htm

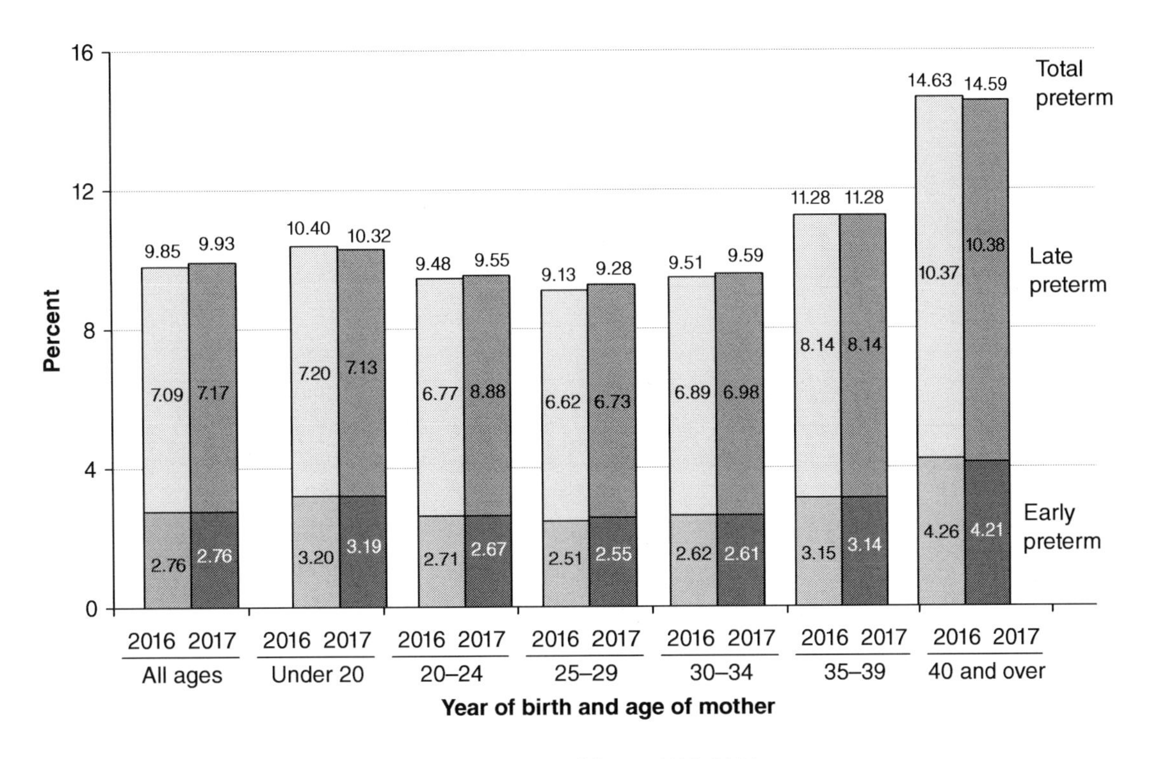

Figure 10.24 Preterm Birth Rates, by Age of Mother, United States, 2016–2017
Source: NCHS, National Vital Statistics System, Natality, CDC 2018. https://www.cdc.gov/nchs/products/databriefs/db318.htm

rate among African American women (14 percent) was 50 percent higher than that among White women (9 percent) (CDC, 2018c) and increased in California, Georgia, Idaho, Indiana, Kentucky, Michigan, Minnesota, Mississippi, New York, Oregon, Tennessee, Texas, and Virginia from 2015 to 2016 (Martin *et al.*, 2018a).

Preterm birth includes early preterm (less than 34 weeks) and late preterm (34 to 36 weeks). **Early preterm birth rate** (i.e., number of births delivered prior to 34 completed weeks of gestation per 100 births) was unchanged from 2017 Q1 to 2018 Q1 (2.75 percent) (CDC, 2018b). **Late preterm birth rate** (i.e., number of births delivered between 34 and 36 completed weeks of gestation per 100 births) increased from 2017 Q1 (7.12 percent) to 2018 Q1 (7.18 percent) (CDC, 2018b).

Causes, Consequences, and Prevention of Preterm Birth

Every year, an estimated 15 million babies (more than one in ten babies) are born preterm (before 37 weeks of gestation); the majority (60 percent) occur in Africa and South Asia (WHO, 2018). Common causes of preterm birth include multiple pregnancies, infections, chronic conditions such as diabetes and high blood pressure, or genetic influence (WHO, 2018). Women carrying more than one baby are at higher risk of preterm birth (OWH, 2018c). In low-income countries, half of the babies born at or below 32 weeks (two months early) die due to a lack of cost-effective care for infections and breathing difficulties, while in high-income countries almost all of these babies survive (WHO, 2018).

After completion of this section, students should be able to:
LO10: Describe causes and treatment of infertility.

Preterm births are a leading cause of infant mortality (Sunderam *et al.*, 2018) and globally it is the leading cause of death in children under the age of five years (WHO, 2018). Many survivors face a lifetime of disability, including learning disabilities and visual and hearing problems (WHO, 2018). It takes an emotional toll and is a financial burden for families (CDC, 2018c).

More than three-quarters of premature babies can be saved with essential care during childbirth and in the postnatal period for every mother and baby. WHO (2018) has developed key interventions for improving the chances of survival and health outcomes for preterm infants, including steroid injections (given to pregnant women at risk of preterm labor and under set criteria to strengthen the babies' lungs) and antibiotics to treat newborn infections. Prenatal care helps prevent preterm birth, including counseling on healthy diet and optimal nutrition, advice on tobacco and substance use, and a minimum of eight contacts with health professionals throughout pregnancy to identify and manage other risk factors, such as infections (WHO, 2018).

Infertility

Involuntary childlessness occurs when couples want children but cannot conceive or are past childbearing age. **Infertility** refers to the inability of couples to achieve, maintain, or carry a pregnancy to live birth after one year of trying. Infertility is divided into **primary infertility** (i.e., infertility in a woman who has never had a pregnancy) and **secondary infertility** (i.e., infertility in a woman who has had one or more pregnancies, but cannot become pregnant again) (WHO, 2013a). From 2011 to 2015, 6.7 percent of married women aged 15–44 in the United States were infertile; 7.3 million women aged 15–44 used infertility services; 12.1 percent of women aged 15–44 had *impaired fecundity*—infertility in which a woman has difficulty getting pregnant or carrying a pregnancy to term (CDC, 2016).

Causes and Treatment of Infertility

Both men and women can have underlying conditions that cause infertility. Infertility may result from a problem with any or several of these steps: a woman's body must release an egg from one of her ovaries (ovulation); a man's sperm must join with the egg along the way (fertilize); the fertilized egg must go through a fallopian tube toward the uterus (womb); the fertilized egg must attach to the inside of the uterus (implantation) (CDC, 2017c). Most cases of female infertility are caused by problems with ovulation. Without ovulation, there are no eggs to be fertilized. Age can contribute to infertility for women. In addition, smoking, poor diet, excess alcohol use, sexually transmitted diseases, and being overweight or underweight can also affect fertility for women (HHS, 2018).

Infertility in men is most often caused by a problem called *varicocele* (vair-ih-koh-seel). This happens when the veins on a man's testicle(s) are too large, heating the testicles and affecting the number or shape of the sperm (HHS, 2018). Making too few sperm or none can contribute to infertility for men. In addition, smoking, heavy alcohol use, drugs, age, environmental toxins, and medicine can also affect fertility for men.

For females, many cases are treated with medication or surgery (HHS, 2018). **Intrauterine**

After completion of this section, students should be able to:
LO11: Explain types of adoption and experiences of adoptive families.

insemination (IUI) is an infertility treatment or artificial insemination that involves placing sperm inside a woman's uterus to facilitate fertilization. In addition, assisted reproductive technology (ART) is used to treat infertility. Male infertility is often treated with medications that increase sperm production, antibiotics to heal an infection, and/or hormones to improve a hormone imbalance. Health-care providers may conduct a semen analysis to check the number, shape, and movement of sperm (HHS, 2018).

Adoption

Adoption is a mechanism through which an adult legalizes the parental relationship to a non-biological child and assumes all rights and responsibilities for rearing the child. Adoption is practiced globally. In the United States, 6.4 million (11.5 percent) females aged 18–44 have ever cared for a non-biological child from 2011 to 2015 (CDC, 2017e). The number of adoptions that finalized each year has remained flat, ranging between 50,700 and 53,600 over the last decade. The fiscal year (FY) 2016 represents a second year of increases in adoptions to 57,200, matching the number of adoptions in FY 2009 (Children's Bureau, 2017).

Types of Adoption

Adoption is an institution that fulfills purposes in society (Jones, 2008). For example, infertility is one reason parents seek to adopt children. In fact, infertility accounted for 80 percent of unrelated infant adoptions and half of adoptions through foster care (Berry, 1996). Others choose adoption (a legal process) in order to grow their family or from a desire to help provide a loving home to a child in need. Let's take a look at types of adoption that most people pursue.

Domestic Adoption

Domestic adoption occurs when the adoptive parents, birth parents, and the child live within the United States. All states, the District of Columbia, American Samoa, Guam, Puerto Rico, the Northern Mariana Islands, and the Virgin Islands have laws that specify the persons who are eligible to adopt and the persons who can be adopted (CWIG, 2013).

Types of Domestic Adoption

The adoption process includes initial placement, background checks, consent, and finalization. **Agency adoption** is a form of adoption in which an adoption agency completes the entire adoption process from beginning to end. Agency adoption is broken down into private and public agency. *Public agency*—any agency of the United States, a state, or a political subdivision of a state or any interstate governmental agency (U.S. Department of Labor, 2011)—handles only special-needs adoptions. These children have special needs (child abuse, born to drug-addicted mothers, etc.) and have been placed under the care of the state. For example, the Office of Adoption Operations is a licensed adoption agency and is approved to provide adoption services to children in the public child welfare system, which includes placing children into adoptive homes as well as providing other adoption-related services in New Jersey (State of New Jersey, 2018). People looking to adopt a newborn baby may want to look into private agencies. *Private agency*—supported by private funds and licensed or approved by the state in which it operates—handles both domestic and international adoptions. Private agencies use screening factors to choose parents they like to work with and find infants or pregnant women who want to place their babies for adoption. Birth parents often relinquish their parental rights to the private agency at the birth of their child.

Independent adoption occurs when the birth parents relinquish their parental rights directly to the adoptive parents through an attorney. For independent adoption, the adoptive family finds a pregnant mother through advertising or networking and uses an attorney to legally complete the adoption. Understanding the costs associated with different types of adoption may help parents decide which type of adoption to pursue. In many cases, tax credits, subsidies, employer benefits, and loans or grants can help with adoption costs. The average total cost of adoption via agency was $39,966, compared with $34,093 for independent adoption (American Adoptions, 2018). Forty-seven U.S. states have laws that provide regulation of the fees and expenses that adoptive parents are expected to pay in private domestic adoption (CWIG, 2017a).

Closed adoption is the practice of sealing all identifying information, preventing disclosure of the adoptive family and birth family, and having no contact between the families. **Open adoption** allows identifying information to be communicated between adoptive and biological parents. The adoptive and biological parents may enter into a legal agreement concerning visitation or interaction regarding the child. In the United States, most states have instituted procedures by which parties to an adoption may obtain information from an adoption record and protect the interests of all parties. *Non-identifying information* is limited to descriptive details about an adopted person and the adopted person's birth relatives and *identifying information* is from the disclosure of adoption records or elsewhere that may lead to the positive identification of birth parents, the adopted person, or other birth relatives. When an adoption is finalized, a new birth certificate for the child is customarily issued to the adoptive parents. The original birth certificate is then sealed and kept confidential by the state registrar of vital records (CWIG, 2016b). Under a new law in New Jersey, adult adoptees can request access to their birth records beginning January 1, 2017. A birth parent can notify the New Jersey Department of Health's Office of Vital Records that he or she wants his or her identifying information redacted from the birth certificate (State of New Jersey, 2018).

Stepparent adoption, or the legal adoption of a child by the spouse of the child's birth or previous adoptive parent, is the most common form of adoption in the United States and is governed by state law. Steps include checking out a state's laws on stepparent adoptions, contacting the court in the county that handles adoptions, submitting required legal forms, going to the preliminary hearing, finalizing the adoption, and applying for a new birth certificate. After the stepparent adoption occurs, the biological parent no longer has any rights or responsibilities for the child (CWIG, 2013). More men than women aged 18–44 adopted a stepchild in 2002 (Jones, 2008) because children are more likely to live in households with their biological mothers than with their biological fathers when parents divorce (Kreider & Fields, 2005).

Among adopters, 17 percent of women and 6 percent of men were never married in 2002 (Jones, 2008). **Second-parent adoption** is the adoption of a child by a second parent in the home who is not married to the legal parent of the child. A second-parent adoption allows a second parent to give the child two legal parents and gives both parents legal rights.

International Adoption

International adoption refers to the adoption of a child from another country. The U.S. Department of State (DOS) is committed to maintaining intercountry adoption as an option around the world for children who cannot find permanent families in their countries of birth (U.S. DOS, 2018a) and uses bilateral and multilateral cooperation and stakeholder outreach to develop a more stable global environment for intercountry adoption (U.S. DOS, 2018c). Intercountry adoptions are governed by three sets of laws: U.S. federal law, the laws of the child's country of residence, and the laws of your U.S. state of residence. Prospective adoptive parents must meet the legal adoption requirements of their country of residence and those of the country whose nationality the child holds. Procedures include a pre-travel visit for prospective adoptive parents, overseas medical examinations of the adopted child, and follow-up medical examinations after arrival in the United States. Generally, it takes one to four years and depends on the country of origin's procedures, and the specific circumstances of the adoption (U.S. DOS, 2018d). The Child Citizen Act of 2000 allows certain foreign-born, biological, and adopted children of American citizens to acquire American citizenship automatically. These children did not acquire American citizenship at birth, but they are granted citizenship when they enter the United States as lawful permanent residents (LPRs) (U.S. DOS, 2018e). Post-adoption services include support groups, education, social, and cultural activities, camps, therapists, medical resources, assistance in the search for family, and access to files (U.S. DOS, 2018b).

The number of international adoptions has increased from the 1990s, partially offsetting the decline in the number of infants relinquished at birth and available domestically to be adopted (Jones, 2008). In the past fifteen years, more than 250,000 children have come to the United States to join their families through international adoption (CDC, 2017f). In 2017, the number of adoptions to the United States increased in 42 countries and the overall number of adoptions to the United States was 4,714, a decline of 658 from 2016 (U.S. DOS, 2018c). Adoption service providers (ASPs) disclose all expected costs associated with adoption and reported charging between $0 and $64,357 for all adoption services, with half charging less than $30,400 and half charging more in 2017 (U.S. DOS, 2018c).

Many international adoptions are transracial adoptions. Some parents adopt a child of the same racial background as themselves. Others choose to adopt a child of a different race or ethnicity. **Transracial adoption** (or interracial adoption) refers to the adoption of a child that is of a different race than that of the adoptive parents.

Adoptive Families

Adoptive family is a form of family in which people become parents by adoption. Adoptive families are diverse. **Transracial families** consist of children of one race or culture being raised by parents of a different race or culture.

Research has explored the challenges and strengths of adoptive families. For example, Korean-born transracial international adoptees experienced both benefits (i.e., greater learning and advantages of belonging to multiple cultures) and challenges (i.e., racism, alienation from their White families and peers, unfamiliarity with Korean culture, and identity confusion) (Godon-Decoteau & Ramsey, 2018). In a study of 357 adoptive families, Hamilton *et al.* (2015) found some differences in general identity development and adjustment. Chen (2016) suggests that the identity formation of internationally adopted children and the adjustments of adoptive families reside in the processes of cross-cultural and intercultural communication. For example, it is important to teach them about their ethnicity and develop a sense of identity and appreciation for cultural diversity (Patton-Imani, 2012). Nelson and Colaner (2018) found that adoptive parents utilized narrating, naming, ritualizing, discussing, normalizing, and praising to communicatively negotiate racial and cultural identity differences. In a study of a longitudinal sample of 519 adoptive families, Taraban *et al.* (2018) suggested the importance of accounting for broader family context in predicting early childhood parenting and child outcomes. Following four infant girls from China during the first year post-adoption, Hwa-Froelich and Matsuoh (2017) reported that each child demonstrated a significant drop in communication and symbolic behavior scores during the sixth, seventh, and eight month post-adoption. Rolock *et al.* (2018) suggest that as adoptive children reach adolescence, families need support and services that help them reduce external forces affecting the adoptive family. For example, there are many public and private non-profit organizations that provide such post-adoption resources and services (U.S. DOS, 2018b). Indeed, the life-changing bond is formed when parents and

Global Overview: Sex Selection

Sex selection is the attempt to control the sex of the offspring to achieve a desired sex. Though illegal in most parts of the world, sex-selective infanticide and abandonment are still practiced. Today ultrasounds have enabled parents to detect the sex of a fetus during prenatal screenings. Prenatal diagnosis and methods such as pre-implantation and sperm sorting can be used for social sex selection. Those who prefer sons may therefore arrange to abort female fetuses, or kill children of the unwanted sex (i.e., *sex-selective infanticide*), or abandon children of the unwanted sex (i.e., *sex-selective abandonment*), or place children of the unwanted sex up for adoption (i.e., *sex-selective adoption*). Adoption may afford families that have a gender preference a legal means of choosing offspring of a particular sex. Tragically, 117 million women are believed to be "missing" in those countries where son preference is prevalent (UNPFA, 2017).

There are a couple of reasons for sex selection. In patriarchal (male-dominant) cultures, parents often prefer sons because they believe that sons will ensure that the family name is carried on and because they believe that sons are more capable of providing for the families. The dowry system in India and other countries is also responsible for a strong son preference. Having a son ensures that families are more economically secure by not having to provide dowry payments, but rather being on the receiving end of this practice (Seager, 2009). Girls and women have a marginal social, economic, and symbolic position, and consequently enjoy fewer rights. In countries where there are discriminatory practices regarding women inheriting, owning, or controlling land by law, having a son ensures that the family will not have to worry about the legal aftermath if something were to happen to them (Seager, 2009). Also, local fertility restrictions and spontaneous rapid fertility decline below replacement levels tend to compel parents who want both a son and a small family size to resort to sex selection (UNPFA, 2012). For example, a *one child one family policy* was introduced in China between 1978 and 1980 in order to control population growth, but contributed to the sex imbalance. Most families were subject to a strict one-child restriction, with some in rural areas being allowed to have a second child if the first child was a girl and disabled. This family policy lowered the total fertility rate (1.6 children born/woman) but led to an unbalanced sex ratio of 1.06 male(s) per female in 2017 (CIA, 2018d). Son preference is most prevalent in an arc of countries from East Asia through South Asia to the Middle East and North Africa (Arnold, 1987). Sex selection leads to imbalanced sex ratios throughout Asia and eastern Europe (UNFPA, 2017), reaching between 110 and 120 male births per 100 female births in several countries, pointing to the intensity of gender discrimination and son preference (UNFPA, 2012).

Sex selection in favor of boys is a symptom of pervasive social, cultural, political, and economic injustices against women, and a manifest violation of women's human rights (UNFPA, 2012). In March 2017, the United Nations Population Fund (2017) launched the Global Program to prevent son preference and gender-biased sex selection, hoping to end sex selection.

children transcend all boundaries to form a family to call their own (Younes & Klein, 2014).

Summary

Fertility rates have declined, the maternal age has increased, and family size has declined in the United States. People wish to remain childfree for different reasons. Couples who choose not to have children can use birth control methods to prevent pregnancy. Three options for handling an unintended pregnancy are abortion, adoption, and parenting. Male and female infertility can be treated through a thorough physical exam. Those who decide to have children often face high costs. Prenatal care has improved pregnancy outcomes and postnatal care also helps ensure children and mothers are healthy. People pursue adoption for different reasons. Adoption involves domestic and international adoption. Types of domestic adoption include agency adoption, independent adoption, open adoption and closed adoption, stepparent adoption, and second-parent adoption. From a diversity and global approach, the chapter explored childfree families and sex selection.

Key Terms

Abortion is the termination of an embryo or fetus, naturally or via medical methods, often performed during the first 28 weeks of pregnancy.

Adoption is a mechanism through which an adult legalizes the parental relationship to a non-biological child and assumes all rights and responsibilities for rearing the child.

Adoptive family is a form of family in which people become parents by adoption.

Agency adoption is a form of adoption in which an adoption agency completes the entire adoption process from beginning to end.

Assisted reproductive technology (ART) includes all fertility treatments in which either eggs or embryos are handled in the laboratory (i.e., in vitro fertilization [IVF] and related procedures).

Average household size is the average number of persons per household.

Birth rate is the number of live births per 1,000 population in a year.

Caesarean delivery, also called c-section, is surgery to deliver and take out the baby through the mother's abdomen.

Caesarean delivery rate is the number of births delivered by caesarean per 100 births.

Caregiver support ratio is the number of potential caregivers aged 45–64 for each person aged 80 and older.

Childbirth (or labor and delivery) is the ending of a pregnancy by one or more babies leaving a woman's uterus by vaginal passage or caesarean section.

Childfree family is a form of family in which couples choose not to have kids.

Closed adoption is the practice of sealing all identifying information, preventing disclosure of the adoptive family and birth family, and having no contact between the families.

Conception refers to the moment when the sperm fertilizes the egg.

Contraception refers to birth control or family planning methods of preventing pregnancy.

Deep vein thrombosis happens when a blood clot forms in a deep vein.

Demographic transition theory suggests that societies are moving from high birth and death rates to low birth and death rates because of technological developments that have taken place.

Domestic adoption occurs when the adoptive parents, birth parents, and the child live within the United States.

Early preterm birth rate refers to the number of births delivered prior to 34 completed weeks of gestation per 100 births.

Early-term birth rate is the number of births delivered between 37 weeks and 38 weeks 6 days of gestation per 100 births.

Emergency contraception is a way to prevent pregnancy after unprotected sex or contraceptive failure and it has no effect on an established pregnancy.

Fertility is the natural ability to conceive offspring.

First-birth rate is the number of first births per 1,000 women.

First trimester is from week 1 through 12 of pregnancy and includes conception.

Fraternal twins occur when two, separate eggs are fertilized by two, separate sperm.

Full-term birth rate is the number of births delivered between 39 weeks and 40 weeks 6 days of gestation per 100 births.

Gendered fertility theory suggests that very low fertility is the result of a hiatus of sustained gender inequity in family-oriented social institutions.

General fertility rate is the number of births per 1,000 women aged 15–44 in a population in a year.

Gestational age is a measure of the age of a pregnancy that is taken from the woman's last menstrual period (LMP) to the current date.

Identical twins occur when a single fertilized egg splits into two.

Independent adoption occurs when the birth parents relinquish their parental rights directly to the adoptive parents through an attorney.

Infant mortality refers to infants who die during the first year of life.

Infertility refers to the inability of couples to achieve, maintain, or carry a pregnancy to live birth after one year of trying.

International adoption refers to the adoption of a child from another country.

Intrauterine insemination (IUI) is an infertility treatment or artificial insemination that involves placing sperm inside a woman's uterus to facilitate fertilization.

In vitro fertilization (IVF) involves extracting a woman's eggs, fertilizing the eggs in the laboratory, and then transferring the resulting embryos into the woman's uterus through the cervix.

Involuntary childlessness refers to couples who want children but cannot conceive or are past childbearing age.

Late preterm birth rate refers to the number of births delivered between 34 and 36 completed weeks of gestation per 100 births.

Late-term birth rate is the number of births delivered between 41 weeks and 41 weeks 6 days of gestation per 100 births.

Low-risk caesarean delivery rate is caesarean delivery among nulliparous (first birth), term (37 or more completed weeks based on the obstetric estimate), singleton (one fetus), and cephalic (head first) births.

Mean age at first birth is the average age of mothers at birth of first child and is computed from the frequency of first births by age of mothers of all marital statuses.

Mixed twins are fraternal twins from biracial couples and exhibit differing racial features.

Multiple birth is the culmination of one multiple pregnancy, wherein the mother delivers two or more offspring.

Non-marital birth rate is the number of births per 1,000 unmarried women ages 15–44.

Old-age dependency ratio is the ratio of older adults to working-age adults.

Open adoption allows identifying information to be communicated between adoptive and biological parents.

Population pyramid is a graphic with a broad base that comes to a point at the top, illustrating the age and sex structure of a developing country's population.

Postpartum depression (PPD) is a type of mood disorder associated with childbirth.

Postpartum hemorrhage (PPH) occurs when a woman has heavy bleeding after delivery.

Postpartum period or postnatal period is the period beginning immediately after childbirth and extending for about six weeks.

Post-term birth rate is the number of births delivered between 42 weeks and more days of gestation per 100 births.

Pregnancy is the time during which one or more offspring develop inside a woman.

Prenatal care, or antenatal care, is a type of preventive health care with the goal of providing regular check-ups that allow doctors to treat and prevent potential health problems throughout the course of the pregnancy.

Preterm birth (or premature birth) occurs when a baby is born before 37 weeks of pregnancy are completed.

Preterm birth rate is the number of births delivered before 37 completed weeks of gestation per 100 births.

Primary caesarean delivery rate measures caesarean deliveries among women who have not had a previous caesarean delivery.

Primary infertility is infertility in a woman who has never had a pregnancy.

Replacement level fertility (total fertility rate of 2.1 children per woman) is the level of fertility at which a population exactly replaces itself from one generation to the next.

Secondary infertility is infertility in a woman who has had one or more pregnancies, but cannot become pregnant again.

Second-parent adoption is the adoption of a child by a second parent in the home who is not married to the legal parent of the child.

Second trimester is from week 13 through 28 of pregnancy.

Sex selection is the attempt to control the sex of the offspring to achieve a desired sex.

Shotgun marriage is a form of marriage in which vows are taken out of a sense of obligation after an accidental pregnancy has occurred.

Singleton is a child that is the only one born at one birth.

Stepparent adoption is the legal adoption of a child by the spouse of the child's birth or previous adoptive parent.

Teen birth rate is the number of births per 1,000 women aged 15–19.

Theory of planned behavior (TPB) suggests that fertility intentions are the culmination of a combination of attitudes, subjective norms, and perceived control over behavior.

Third trimester is from 29 weeks through 40 weeks of pregnancy.

Total fertility rate refers to the average number of children that would be born per woman if all women survive their childbearing years.

Transracial adoption (or interracial adoption) refers to the adoption of a child that is of a different race to that of the adoptive parents.

Transracial families consist of children of one race or culture being raised by parents of a different race or culture.

Triplet and higher-order multiple births refer to the number of triplets, quadruplets, and quintuplets and other higher-order multiples per 100,000 births.

Twin birth rate is the number of live births in twin deliveries per 1,000 births.

Twins are two offspring produced by the same pregnancy.

Unintended pregnancy is one that was either mistimed or unwanted.

Vaginal delivery after c-section (VBAC) is a vaginal childbirth by a woman who has delivered previous babies by c-section but would like to have their next baby vaginally.

Voluntary childlessness, also described by some as being *childfree*, is the lifelong voluntary choice to not have biological, step, or adopted children.

Discussion Questions

- What factors might affect your decision about having children?
- What are some advantages and disadvantages of choosing to be childfree family?
- According to the USDA Cost of Raising a Child Calculator, what is the cost of raising a child? (See www.cnpp.usda.gov/tools/crc_calculator)

Suggested Film and Video

National Adoption Month: https://www.childwelfare.gov/topics/adoption/nam/spread-the-word/watch/

Watch CDC Grand Rounds Video: Time for Public Health Action on Infertility: https://www.cdc.gov/reproductivehealth/infertility/publichealth.htm

Internet Sources

Adoption Options: Where Do I Start?: https://www.childwelfare.gov/pubs/f-adoptoption/

How Effective Are Birth Control Methods?: https://www.cdc.gov/reproductivehealth/contraception/index.htm

USDA Cost of Raising a Child Calculator: https://www.cnpp.usda.gov/calculatorintro

References

Aaronson, D., Lange, F. & Mazumder, B. (2014). Fertility Transitions along the Extensive and Intensive Margins. *American Economic Review*, 104(11), 3701–3724.

Abma, J. C. & Martinez, G. M. (2006). Childlessness among Older Women in the United States: Trends and Profiles. *Journal of Marriage and Family.* Retrieved on April 14, 2018 from https://doi.org/10.1111/j.1741-3737.2006.00312.x

Ajzen, I. (1991). The Theory of Planned Behavior. *Organizational Behavior and Human Decision Processes,* 50(2), 179–211.

Albertini, M. & Kohli, M. (2009). What Childless Older People Give: Is the Generational Link Broken? *Ageing & Society*, 29, 1261–1274.

American Adoptions (2018). *Comparing the Costs of Domestic, International and Foster Care Adoption.* Retrieved on April 11, 2018 from https://www.americanadoptions.com/adopt/the_costs_of_adopting

Arnold, F. (1987). The Effect of Sex Preference on Fertility and Family Planning: Empirical Evidence. *Popul Bull UN*, 23–24, 44–55.

Baber, K. M. & Dreyer, A. S. (1986). Gender-role Orientations in Older Child-free and Expectant Couples. *Sex Roles*, 14, 501–512.

Baer, D. (August 15, 2016). How Child-free People Decide Not to Have Kids. *The Cut.* Retrieved on March 21, 2018 from https://www.thecut.com/2016/08/how-childfree-people-decide-not-to-have-kids.html

Balbo, N., Billari, F. C. & Mills, M. (2013). Fertility in Advanced Societies: A Review of Research: La fécondité dans les sociétés avancées: un examen des recherches. *European Journal of Population*, 29(1), 1–38.

Balmer, R. (1994). American Fundamentalism: The Ideal of Femininity. In J. S. Hawley (Ed.), *Fundamentalism and Gender* (pp. 47–62). New York: Oxford University Press.

Barros, F. C., Vaughan, J. P., Victora, C. G. & Huttly, S. R. A. (1991). Epidemic of Caesarean Sections in Brazil. *Lancet*, 338, 167–169.

Basten, S. (2009). *Voluntary Childlessness and Being Childfree*. Oxford: University of Oxford.

Baudin, T., de la Croix, D. & Gobbi, P. (2017). *Endogenous Childlessness and Stages of Development*. London, Centre for Economic Policy Research. https://cepr.org/active/publications/discussion_papers/dp.php?dpno=12071

Bearak, J., Popinchalk, A., Alkema, L. & Sedgh, G. (2018). Global, Regional, and Subregional Trends in Unintended Pregnancy and Its Outcomes from 1990 to 2014: Estimates from a Bayesian Hierarchical Model. *Lancet Global Health*, 6(4), e380–e389.

Beck, U. (1992). *Risk Society: Towards a New Modernity*. Thousand Oaks, CA: Sage.

Benatar, D. (2006). *Better Never to Have Been*. Oxford, UK: Oxford University Press.

Berry, M. (1996). Preparation, Support and Satisfaction of Adoptive Families in Agency and Independent Adoptions, Table 2. *Child and Adolescent Social Work Journal*, 13(2) (April 1996).

Blackstone, A. (2014a). Childless … or Childfree. *SAGE Journals*, 13(4), 68–70.

Blackstone, A. (2014b). Doing Family without Having Kids. *Sociology Compass*, 8(1), 52–62.

Blackstone, A. & Stewart, M. D. (2012). Choosing to be Childfree: Research on the Decision Not to Parent. *Sociology School Faculty Scholarship*. 5. https://digitalcommons.library.umaine.edu/soc_facpub/5

Blackstone, A. & Stewart, M. D. (2016). "There's More Thinking to Decide": How the Childfree Decide Not to Parent. *The Family Journal*, 24(3), 296–303.

Blalik, K. (May 11, 2017). *6 Facts about U.S. Mothers*. Washington, DC: Pew Research Center.

Bongaarts, J. (2009). Human Population Growth and the Demographic Transition. *Philosophical Transactions of the Royal Society B: Biological Sciences*, 364(1532), 2985–2990.

Bortolus, R., Parazzini, F., Chatenoud, L., Benzi, B., Bianchi, M. M. & Marini, A. (1999). The Epidemiology of Multiple Births. *Human Reproduction Update*, 5(2), 179–187.

Breen, R. R. (1997). Recommodification and Stratification. *Sociology*, 31(3), 473–489.

Burkett, E. (2000). *The Baby Boon: How Family-Friendly America Cheats the Childless*. New York: The Free Press.

Caldwell, J. C., Caldwell, B. K., Caldwell, P., McDonald, P. F. & Schindlmayr, T. (2006). *Demographic Transition Theory*. Dordrecht, Netherlands: Springer.

Callan, V. J. (1986). Single Women, Voluntary Childlessness and Perceptions about Life and Marriage. *Journal of Biosocial Science*, 18(4), 479–487.

Centers for Disease Control and Prevention (CDC) (July 15, 2016). *Infertility*. Retrieved on August 4, 2017 from https://www.cdc.gov/nchs/fastats/infertility.htm

Centers for Disease Control and Prevention (CDC) (February 9, 2017a). *How Effective Are Birth Control Methods?* Retrieved on December 1, 2018 from https://www.cdc.gov/reproductivehealth/contraception/index.htm

Centers for Disease Control and Prevention (CDC) (September 15, 2017b). *Pregnancy and Prenatal Care*. Retrieved on December 1, 2018 from https://www.cdc.gov/healthcommunication/toolstemplates/entertainmented/tips/PregnancyPrenatalCare.html

Centers for Disease Control and Prevention (CDC) (March 30, 2017c). *Infertility FAQs*. Retrieved on December 1, 2018 from https://www.cdc.gov/reproductivehealth/infertility/index.htm

Centers for Disease Control and Prevention (CDC) (November 1, 2017d). *ART Success Rate*. Retrieved on April 15, 2018 from https://www.cdc.gov/art/artdata/index.html

Centers for Disease Control and Prevention (CDC) (June 6, 2017e). *Adoption and Nonbiological Parenting*. Retrieved on December 1, 2018 from https://www.cdc.gov/nchs/nsfg/key_statistics/a.htm#adoption

Centers for Disease Control and Prevention (CDC) (June 12, 2017f). *International Adoption*. Retrieved on December 1, 2018 from https://wwwnc.cdc.gov/travel/yellowbook/2018/international-travel-with-infants-children/international-adoption

Centers for Disease Control and Prevention (CDC) (August 2017g). *Data on Selected Pregnancy Complications in the United States*. Retrieved on December 1, 2018 from https://www.cdc.gov/reproductivehealth/maternalinfanthealth/pregnancy-complications-data.htm

Centers for Disease Control and Prevention (CDC) (July 31, 2018a). *Term Birth Rates*.

Centers for Disease Control and Prevention (CDC) (July 31, 2018b). *Preterm Birth Rates*.

Centers for Disease Control and Prevention (CDC) (April 24, 2018c). *Preterm Birth*. Retrieved on December 1, 2018 from https://www.cdc.gov/reproductivehealth/maternal infanthealth/pretermbirth.htm

Central Intelligence Agency (CIA) (2017a). *World.* Retrieved on March 17, 2018 from https://www.cia.gov/library/publications/resources/the-world-factbook/geos/xx.html

Central Intelligence Agency (CIA) (2017b). *Birth Rate.* Retrieved on April 6, 2018 from https://www.cia.gov/library/publications/resources/the-world-factbook/rankorder/2054rank.html

Central Intelligence Agency (CIA) (2018a). *United States.* Retrieved on March 14, 2018 from https://www.cia.gov/library/publications/the-world-factbook/rankorder/2127rank.html

Central Intelligence Agency (CIA) (2018b). *Afghanistan.* Retrieved on March 19, 2018 from https://www.cia.gov/library/publications/the-world-factbook/geos/af.html

Central Intelligence Agency (CIA). (2018c). *Germany.* Retrieved on April 17, 2018 from https://www.cia.gov/library/publications/resources/the-world-factbook/geos/gm.html

Central Intelligence Agency (CIA) (2018d). *China.* Retrieved on April 17, 2018 from https://www.cia.gov/library/publications/resources/the-world-factbook/geos/ch.html

Chen, Z. J. (2016). Exploring the Relationship between Adoptive Parents and International Adoptees: From the Perspectives of Cross-Cultural Communication and Adaptation. *Adoption Quarterly,* 19(2), 119–144.

Child Care Aware of America (CCAoA) (2017). *2017 Report.* Retrieved on August 4, 2017 from http://usa.childcareaware.org/wp-content/uploads/2017/07/FINAL_SFS_REPORT.pdf

Children's Bureau (2017). *Trends in Foster Care and Adoption.* Retrieved on October 13, 2018 from https://www.acf.hhs.gov/cb/resource/trends-in-foster-care-and-adoption

Child Welfare Information Gateway (CWIG) (2013). *Stepparent Adoption.* Washington, DC: U.S. Department of Health and Human Services, Children's Bureau.

Child Welfare Information Gateway (CWIG) (2016a). *Who May Adopt, Be Adopted, or Place a Child for Adoption.* Washington, DC: U.S. Department of Health and Human Services, Children's Bureau.

Child Welfare Information Gateway (2016b). *Access to Adoption Records.* Washington, DC: U.S. Department of Health and Human Services, Children's Bureau.

Child Welfare Information Gateway (CWIG) (March 2017a). *Regulation of Private Domestic Adoption Expenses.* Washington, DC: U.S. Department of Health and Human Services, Children's Bureau.

Child Welfare Information Gateway (CWIG) (2017b). *Transracial Adoption.* Retrieved on August 6, 2017 from https://www.childwelfare.gov/topics/adoption/adopt-parenting/foster/transracial/

Craig, J. (1994). Replacement Level Fertility and Future Population Growth. *Population Trends,* Winter (78), 20–22.

Curtin, S. C., Ventura, S. J. & Martinez, G. M. (2014). *Recent Declines in Nonmarital Childbearing in the United States. NCHS* Data Brief, no 162. Hyattsville, MD: National Center for Health Statistics.

Cwikel, J., Gramotnev, H. & Lee, C. (2006). Never-married Childless Women in Australia: Health and Social Circumstances in Older Age. *Social Science and Medicine,* 62(8), 1991–2001.

Daniels, K., Mosher, W. D. & Jones, J. (2013). Contraceptive Methods Women Have Ever Used: United States, 1982–2010. *National Health Statistics Reports,* 2013, No. 62.

Daniels, K., Daugherty, J. & Jones, J. (2014). Current Contraceptive Status among Women Aged 15–44: United States, 2011–2013. *National Health Statistics Reports,* No. 173.

Davis-Floyd, R.E. (1994). The Technocratic Body: American Childbirth as Cultural Expression, *Social Science and Medicine,* 38, 8, 1125–40.

DeOllos, I. Y. & Kapinus, C. A. (2002). Aging Childless Individuals and Couples: Suggestions for New Directions in Research. *Sociological Inquiry,* 72(1), 72–80.

Duque, V., Pilkauskas, N. V. & Garfinkel, I. (2018). Assets among Low-income Families in the Great Recession. *PLoS ONE,* 13(2), e0192370.

Durnwald, Celeste P. & Mercer, Brian M. (2010). Vaginal Birth after Cesarean Delivery: Predicting Success, Risks of Failure. *The Journal of Maternal-Fetal & Neonatal Medicine,* 15(6), 388–393, DOI: 10.1080/14767050410001724290

Dye, J. L. (2008). *Fertility of American Women: 2006.* Washington, DC: U.S. Census Bureau.

Eickmeyer, K. J. (2016). Over 25 Years of Change in Men's Entry into Fatherhood, 1987–2013. *Family Profiles.* FP-16–10. Bowling Green, OH: National Center for Family and Marriage Research.

Eisenberg, D. L., Allsworth, J. A., Zhao, Q. & Peipert, J. F. (2012). Correlates of Dual-Method Contraceptive Use: An Analysis of the National Survey of Family Growth (2006–2008). *Infectious Diseases in Obstetrics and Gynecology,* Article ID 717163. doi:10.1155/2012/717163.

Farncombe, V. (2018). Fifty and Childless: Finding a Way Forward. *BBC.* Retrieved on April 14, 2018 from http://www.bbc.com/news/education-42647471

Fraga, M. F., Ballestar, E., Paz, M. F., Ropero, S. & Setien, F. *et al.* (2005). Epigenetic Differences Arise during the

Lifetime of Monozygotic Twins. *Proc. Natl. Acad. Sci. U.S.A.*, 102 (30), 10604–10609.

Frejka, T. (2017). Childlessness in the United States. In M. Kreyenfeld & D. Konietzka (Eds.), *Childlessness in Europe: Contexts, Causes, and Consequences.* Demographic Research Monographs (a series of the Max Planck Institute for Demographic Research). Cham, Switzerland: Springer.

Frey, W. H. (March 14, 2018). The U.S. Will Become "Minority White" in 2045, Census Projects. Youthful Minorities Are the Engine of Future Growth. Washington, DC: Brookings Institution.

Frost, J. J., Frohwirth, L. & Zolna, M. R. (2016). *Contraceptive Needs and Services, 2014 Update.* New York, New York: Guttmacher Institute.

Gillespie, R. (2003). Childfree and Feminine. Understanding the Gender Identity of Voluntarily Childless Women. *Gender and Society,* 17 (1), 122–136.

Glasier, A. (1997). Emergency Postcoital Contraception. *N Engl J Med.*, 337, 1058–1064.

Godon-Decoteau, D. & Ramsey, P. G. (2018). Positive and Negative Aspects of Transracial Adoption: An Exploratory Study from Korean Transracial Adoptees' Perspectives. *Adoption Quarterly*, 21(1), 17–40.

Greenwood, J. & Seshadri, A. (2002). The U.S. Demographic Transition. *Economic Development across Time and Space,* 92(2).

Grover, D. (2014). What is the Demographic Transition Model? *Population Education.* Retrieved on March 29, 2018 from https://populationeducation.org/what-demographic-transition-model/

Guttmacher Institute. (2011). *Testimony of Guttmacher Institute: Submitted to the Committee on Preventive Services for Women, Institute of Medicine.* Retrieved on April 4, 2018 from http://www.guttmacher.org/pubs/CPSW-testimony.pdf.

Guttmacher Institute (2016). *Contraceptive Use in the United States. Who Needs Contraceptives?* Retrieved on April 3, 2018 from https://www.guttmacher.org/fact-sheet/contraceptive-use-united-states

Guttmacher Institute (August, 2017a). *Insurance Coverage of Contraceptives.* Retrieved on August 4, 2017 from https://www.guttmacher.org/state-policy/explore/insurance-coverage-contraceptives

Guttmacher Institute (August, 2017b). *An Overview of Abortion Laws.* Retrieved on August 4, 2017 from https://www.guttmacher.org/state-policy/explore/overview-abortion-laws

Guttmacher Institute (August 1, 2017c). *Parental Involvement.* Retrieved on August 4, 2017 from https://www.guttmacher.org/state-policy/explore/parental-involvement-minors-abortions

Guttmacher Institute (March 5, 2018a). *Unintended Pregnancy Rates Declined Globally from 1990 to 2014.* Retrieved on December 1, 2018 from https://www.guttmacher.org/news-release/2018/unintended-pregnancy-rates-declined-globally-1990-2014

Guttmacher Institute (April 1, 2018b). *Insurance Coverage of Contraceptives, State Laws and Policies.* Retrieved on April 4, 2018 from https://www.guttmacher.org/state-policy/explore/insurance-coverage-contr.

Hamilton B. E. & Mathews, T. J. (2016). *Continued Declines in Teen Births in the United States, 2015.* NCHS Data Brief, no 259. Hyattsville, MD: National Center for Health Statistics. Available from: https://www.cdc.gov/nchs/data/databriefs/db259.pdf.

Hamilton, B. E., Martin, J. A., Osterman, M. J. K. & Curtin, S. C. (2015). *Births: Final Data for 2014.* Hyattsville, MD: National Center for Health Statistics. Retrieved March 21, 2018 from http://www.cdc.gov/nchs/data/nvsr/nvsr64/nvsr64_12.pdf- PDF

Hamilton, B. E., Lu, L., Chong, Y. *et al.* (August 18, 2017). Natality Trends in the United States, 1909–2015. Atlanta, Georgia: National Center for Health Statistics. Retrieved on March 18, 2018 from https://www.cdc.gov/nchs/data-visualization/natality-trends/index.htm

Hamilton, E., Samek, D. R., Keyes, M., McGue, M. K. & Iacono, W. G. (2015). Identity Development in a Transracial Environment: Racial/Ethnic Minority Adoptees in Minnesota. *Adoption Quarterly,* 18(3), 217–233.

Handy, D. E., Castro, R. & Loscalzo, J. (2011). Epigenetic Modifications: Basic Mechanisms and Role in Cardiovascular Disease. *Circulation,* 123(19), 2145–2156.

Hatcher, R. A., Trussell, J., Stewart, F., Howells, S., Russell, C. R. & Kowal, D. (1995). *Emergency Contraception: The Nation's Best Kept Secret.* Decatur, GA: Bridging the Gap Communications.

Hobcraft, J. & Kiernan, K. (1995). Becoming a Parent in Europe, Vol. 1. In EAPS-IUSSP, (Ed.), *European Population Conference* (pp. 27–65). Milan: Franco Angeli.

Houseknecht, S. K. (1977). Reference Group Support for Voluntary Childlessness: Evidence for Conformity. *Journal of Marriage and the Family,* 39, 285–292.

Houseknecht, S. K. (1978). Voluntary Childlessness. *Alternative Lifestyles,* 1, 379.

Houseknecht, S. K. (1979). Childlessness and Marital Adjustment. *Journal of Marriage and the Family,* 41(2), 259–265.

Houseknecht, S. K. (1987). Voluntary Childessness. In M. B. Sussman & S. K. Steinmetz (Eds.), *Handbook of Marriage and the Family.* New York: Plenum Press.

Hwa-Froelich, D. A. & Matsuoh, H. (2017). Cross-Cultural Adaptation of Internationally Adopted Chinese Children, Communication and Symbolic Behavior Development. *Communication Disorders Quarterly*, 29(3), 149–165.

Iacovou, M. & Tavares, L. P. (2011). Yearning, Learning, and Conceding: Reasons Men and Women Change Their Childbearing Intentions. *Popul Dev Rev.*, 37(1), 89–123.

Jatlaoui, T. C., Ewing, A., Mandel, M. G. *et al.* (2016). Abortion Surveillance—United States, 2013. *MMWR Surveill Summ*, 65(No. SS-12), 1–44.

Jones, J. (2008). Adoption Experiences of Women and Men and Demand for Children to Adopt by Women 18–44 Years of Age in the United States, 2002. National Center for Health Statistics. *Vital Health Stat*, 23(27).

Jones, J., Mosher, W. D. & Daniels, K. (2012). Current Contraceptive Use in the United States, 2006–2010, and Changes in Patterns of Use since 1995. *National Health Statistics Reports*, 2012, No. 60. http://www.cdc.gov/nchs/data/nhsr/nhsr060.pdf.

Jones, R. K. (2011). *Beyond Birth Control: The Overlooked Benefits of Oral Contraceptive Pills.* New York: Guttmacher Institute.

Kabakian-Khasholian, T., Campbell, O., Shediac-Rizkallah, M. & Ghorayeb, F. (2000). Women's Experiences of Maternity Care: Satisfaction or Passivity? *Social Science and Medicine*, 51, 103–113.

Kelly, M. (2009). Women's Voluntary Childlessness: A Radical Rejection of Motherhood? *Women's Studies Quarterly*, 37, 157–172.

Kirby, D. & Lepore, G. (2007). *Sexual Risk and Protective Factors: Factors Affecting Teen Sexual Behavior, Pregnancy, Childbearing and Sexually Transmitted Disease.* Washington, DC: ETR Associates and the National Campaign to Prevent Teen and Unplanned Pregnancy.

Kneale, D. & Joshi, H. (2008). Postponement and Childlessness: Evidence from Two British Cohorts. *Demographic Research*, 19, 1935–1964.

Koropeckyj-Cox, T. (1988). Loneliness and Depression in Middle and Old Age: Are the Childless More Vulnerable? *Journals of Gerontology Series, B*, 53B, S303–S312.

Koropeckyj-Cox, T. & Call, V. R. A. (2007). Characteristics of Older Childless Persons and Parents Cross-National Comparisons. *SAGE Journals*, 28(10), 1362–1414.

Kost, K. & Henshaw, S. (2014). *U.S. Teenage Pregnancies, Births and Abortions, 2010: National Trends by Age, Race and Ethnicity.* New York: Guttmacher Institute. Retrieved on December 1, 2018 from https://www.guttmacher.org/report/us-teen-pregnancy-state-trends-2011

Kravdal, Ø. (1992). The Emergence of a Positive Relation between Education and Third Birth Rates in Norway with Supportive Evidence from the United States. *Population Studies*, 46(3), 459–475.

Kreider, R. M. & Fields, J. (2005). Living Arrangements of Children: 2001. *Current Population Reports P70–104.* Washington, DC: U.S. Census Bureau.

Kreyenfeld, M. & Konietzka, D. (Eds.) (2017). Childlessness in Europe: Contexts, Causes, and Consequences. *Demographic Research Monographs,* doi: 10.1007/978-3-319-44667-7_2

Kulkarni, A. D., Jamieson, D. J., Jones, H. W. Jr. *et al.* (2013). Fertility Treatments and Multiple Births in the United States. *N Engl J Med,* 369, 2218–2225.

Lane, K. (1995). The Medical Model of the Body as a Site of Risk: A Case Study of Childbirth. In J. Gabe (Ed.), *Medicine, Health and Risk: Sociological Approach.* Oxford: Blackwell.

Ledbetter, O. (2017). Pets vs. Parenthood: Why Millennials Are Owning Pets Instead of Having Kids. *Millennial Marketing.* Retrieved on March 18, 2018 from http://www.millennialmarketing.com/2017/07/pets-vs-parenthood-why-millennials-are-owning-pets-instead-of-having-kids/

Leone, C. (December 1989). *Fairness, Freedom and Responsibility: The Dilemma of Fertility Choice in America.* Dissertation. Pullman, Washington: Washington State University.

Lesthaeghe, R. (1995). The Second Demographic Transition in Western Countries: An Interpretation. In K. O. Mason & A. M. Jensen (Eds.), *Gender and Family Change in Industrialized Countries* (pp. 17–62). Oxford, UK: Clarendon Press.

Lesthaeghe, R. & Neidert, L. (2006). The Second Demographic Transition in the United States: Exception or Textbook Example? *Population and Development Review,* 32(4), 669–698.

Lesthaeghe, R. & van de Kaa, D. J. (1986). Twee demografische transities? In R. Lesthaeghe & D. van de Kaa (Eds.), *Bevolking: Groei of Krimp?* (pp. 9–24). Deventer, Netherlands: Van Loghum-Slaterus.

Li, R., Montpetit, A., Rousseau, M., Wu, S., Greenwood, C. M., Spector, T. D., Pollak, M., Polychronakos, C. & Richards, J. B. (2014). Somatic Point Mutations Occurring Early in Development: a Monozygotic Twin Study. *Journal of Medical Genetics,* 51(1), 28–34.

Liamputtong, P. (2005). Birth and Social Class: Northern Thai Women's Lived Experiences of Caesarean and Vaginal Birth. *Sociology of Health & Illness,* 27(2), 243–70.

Liefbroer, A. C. (2009). Changes in Family Size Intentions across Young Adulthood: A Life-Course Perspective. *Eur J Popul.,* 25(4), 363–386.

Lino, M. (January 13, 2017). *The Cost of Raising a Child.* U.S. Department of Agriculture (USDA). Retrieved on August 3, 2017 from https://www.usda.gov/media/blog/2017/01/13/cost-raising-child

Lino, M., Kuczynski, K., Rodriguez, N. & Schap, T. R. (March 2017). *Expenditures on Children by Families, 2015.*. Miscellaneous Publication No. 1528–2015. U.S. Department of Agriculture, Center for Nutrition Policy and Promotion. Retrieved on December 1, 2018 from https://www.cnpp.usda.gov/sites/default/files/crc2015.pdf

Livingston, G. (October 12, 2011). *In a Down Economy, Fewer Births.* Washington, DC: Pew Research Center.

Livingston, G. (May 7, 2015). *Childlessness Falls, Family Size Grows Among Highly Educated Women.* Washington, DC: Pew Research Center.

Livingston, G. (June 7, 2017). *The Rise of Multiracial and Multiethnic Babies in the U.S.* Washington, DC: Pew Research Center.

Livingston, G. (January 18, 2018). *They're Waiting Longer, but U.S. Women Today More Likely to Have Children than a Decade Ago.* Washington, DC: Pew Research Center.

Lothian, J. A. (2000). Why Natural Childbirth? *The Journal of Perinatal Education,* 9(4), 44–46.

Luke, B. & Brown, M. B. (2007). Contemporary Risks of Maternal Morbidity and Adverse Outcomes with Increasing Maternal Age and Plurality. *Fertil Steril.*, 88(2), 283–293.

Martin, E. (1992). *The Woman in the Body: A Cultural Analysis of Reproduction.* Boston: Beacon Press.

Martin, J. A., Osterman, M. J. K. & Thoma, M. E. (April 2016). *Declines in Triplet and Higher-order Multiple Births in the United States, 1998–2014.* Centers for Disease Control and Prevention.

Martin, J. A., Hamilton, B. E. & Osterman, M. J. K. (2017a). *Births in the United States, 2016.* NCHS data brief, no 287. Hyattsville, MD: National Center for Health Statistics.

Martin, J. A., Hamilton, B. E., Osterman, M. J. K., Driscoll, A. K. & Mathews, T. J. (2017b). Births: Final Data for 2015. *National Vital Statistics Reports,* 66(1). Hyattsville, MD: National Center for Health Statistics. Retrieved on December 1, 2018 from https://www.cdc.gov/nchs/data/nvsr/nvsr66/nvsr66_01.pdf

Martin, J. A., Hamilton, B. E., Osterman, M. J. K., Driscoll, A. K. & Drake, P. (2018a). Births: Final data for 2016. *National Vital Statistics Reports,* 67(1). Hyattsville, MD: National Center for Health Statistics.

Martin, J. A., Hamilton, B. E. & Osterman, M. J. K., (2018b). *Births in the United States, 2017.* NCHS Data Brief, no 318. Hyattsville, MD: National Center for Health Statistics.

Martinez, G., Copen, C. E., & Abma, J. C. (2011). Teenagers in the United States: Sexual Activity, Contraceptive Use, and Childbearing, 2006–2010 National Survey of Family Growth. *Vital Health Statistics,* 23(31).

Mathews, T. J. & Hamilton, B. E. (January 2016). *Mean Age of Mothers is on the Rise: United States, 2000–2014.* NCHS Data Brief No. 232. Hyattsville, MD: National Center for Health Statistics. Retrieved on August 2, 2016 from http://www.cdc.gov/nchs/data/databriefs/db232.htm

McNicoll, G. (2006). Policy Lessons of the East Asian Demographic Transition. *Population and Development Review,* March; 32(1).

Miller, A. R. (2010). The Effect of Motherhood Timing on Career Path. *Journal of Population Economics,* 24(3), 1071–1100.

Mills, M. & Blossfeld, H-P. (2005). Globalization, Uncertainty and the Early Life Course: A Theoretical Framework. In H.-P. Blossfeld, E. Klijzing, M. Mills & K, Kurz (Eds.), *Globalization, Uncertainty and Youth in Society* (pp. 1–24). London; New York: Routledge.

Mills, M., L., Mencarini, M., Tanturri, L. & Begall, K. (2008). Gender Equity and Fertility Intentions in Italy and the Netherlands. *Demographic Research,* 18, 1–26.

Monte, L. M. (November 2017b). *Some Delay Childbearing, Others Opt Out.* Washington, DC: U.S. Census Bureau.

Moorhead, J. (2011). Black and White Twins. *Guardian.* Retrieved on March 13, 2018 from https://www.theguardian.com/lifeandstyle/2011/sep/24/twins-black-white?INTCMP=SRCH

Myrskylä, M., Kohler, H.-P. & Billari, F. C. (2009). Advances in Development Reverse Fertility Declines. *Nature,* 460(7256), 741–743. Retrieved on March 25, 2018 from https://www.nature.com/articles/nature08230

National Institutes of Health (NIH) (2017). *About Pregnancy.* US Department of Health and Human Services. Retrieved on December 1, 2018 from https://www.nichd.nih.gov/health/topics/pregnancy/conditioninfo

Nelson, L. R. & Colaner, C. W. (2018). Becoming a Transracial Family: Communicatively Negotiating Divergent Identities in Families Formed Through Transracial Adoption. *Journal of Family Communication,* 18(1), 51–67.

Nelson, M. K. (1983). Working-class Women, Middle-class Women, and Models of Childbirth. *Social Problems,* 30(3), 284–297.

Office on Women's Health (OWH) (2017). *Depression During and After Pregnancy.* Retrieved on April 15, 2018 from https://www.womenshealth.gov/a-z-topics/depression-during-and-after-pregnancy

Office on Women's Health (OWH) (2018a). *Twins, Triplets, and Other Multiples.* Retrieved on April 14, 2018 from https://www.womenshealth.gov/pregnancy/youre-pregnant-now-what/twins-triplets-and-other-multiples

Office on Women's Health (OWH) (2018b). *Stages of Pregnancy.* Retrieved on April 15, 2018 from https://www.womenshealth.gov/pregnancy/youre-pregnant-now-what/stages-pregnancy

Office on Women's Health (OWH) (2018c). *Labor and Birth*. Retrieved on April 15, 2018 from https://www.womenshealth.gov/pregnancy/childbirth-and-beyond/labor-and-birth

Oppenheimer, V. K. (1988). A Theory of Marriage Timing. *American Journal of Sociology*, 94, 563–591.

Osborn, J. (1992). The Change in the Causes of Mortality in England and Wales during the Last 150 Years. *Med Secoli.*, 4(3), 43–61.

Osborne, R. S. (2003). *Percentage of Childless Women 40 to 44 Years Old Increases since 1976, Census Bureau Reports*. U.S. Census Bureau Press Release.

Park, K. (2005). Choosing Childlessness: Weber's Typology of Action and Motives of the Voluntarily Childless. *Sociological Inquiry*, 75(3), 372–402.

Parker, K., Horowitz, J. M. & Rohal, M. (December 17, 2015). *Parenting in America Outlook, Worries, Aspirations are Strongly Linked to Financial Situation*. Washington, DC: Pew Research Center.

Patton-Imani, S. (2012). Orphan Sunday: Narratives of Salvation in Transnational Adoption. *Dialog: A Journey of Theology*, 51(4), 301.

Patwari, P. & Lee, R. T. (2008). Mechanical Control of Tissue Morphogenesis. *Circulation Research*, 103(3), 234–243.

Paul, P. (2001). Childless by Choice. *American Demographics*, 23, 45–50.

Pelton, S. L. & Hertlein, K. M. (2011). A Proposed Life Cycle for Voluntary Childfree Couples. *Journal of Feminist Family Therapy*, 23(1), 39–53.

Popenoe, D. (2009). *Families without Fathers: Fathers, Marriage and Children in American Society.* Piscataway, NJ: Transaction Publishers.

PRB (Population Reference Bureau) (2009). *The U.S. Recession and the Birth Rate*. Retrieved on March 25, 2018 from https://www.prb.org/usrecessionandbirthrate/

Redfoot, D., Feinberg, L. & Houser, A. (2013). *The Aging of the Baby Boom and the Growing Care Gap: A Look at Future Declines in the Availability of Family Caregivers. AARP Public Policy Institute.*

Reher, D. S. (2004). The Demographic Transition Revisited as a Global Process. *Population, Space, and Place*, 10(1).

Rolock, N., Blakey, J. M., Wahl, M. & Devine, A. (2018). The Evolution of Challenges for Adoptive Families: The Impact of Age as a Framework for Differentiation. *Journal of Adolescent and Family Health*, 9(1), Article 3.

Rowland, D. T. (2007). Historical Trends in Childlessness. *Journal of Family Issues*, 28, 1311–1337.

Rubinstein, R. L. (1987). Childless Elderly: Theoretical Perspectives and Practical Concerns. *Journal of Cross-Cultural Gerontology*, 2(1), 1–14.

Scott, L. S. (2009). *Two is Enough: A Couple's Guide to Living Childless by Choice*. Berkeley, CA: Seal Press.

Seager, J. (2009). *The Penguin Atlas of Women in the World*. New York: Penguin Group

Settle, B. & Brumley, K. (2014). "It's the Choices You Make That Get You There": Decision-making Pathways of Childfree Women. *Michigan Family Review*, 18(1), 1–22.

Somers, M. D. (1993). A Comparison of Voluntarily Child-free Adults and Parents. *Journal of Marriage and the Family*, 55(3), 643–650.

Sorokowski, P., Groyecka, A., Karwowski, M., Manral, U., Kumar, A., Niemczyk, A. & Pawłowski, B. (2017a). Free Mate Choice Does Not Influence Reproductive Success in Humans. *Scientific Reports*, 7, 10127.

Sorokowski, P., Sorokowska, A., Butovskaya, M., Karwowski, M., Groyecka, A., Wojciszke, B. & Pawłowski, B. (2017b). Love Influences Reproductive Success in Humans. *Frontiers in Psychology*, 8, 1922.

State of New Jersey (2018). *Adoption*. Retrieved on April 26, 2018 from http://www.state.nj.us/njfosteradopt/adoption/birthrecords/

Stephens, R. C., Smith Blau, Z., Oser, G. T., & Millar, M. D. (1978). Aging, Social Support Systems, and Social Policy. *Journal of Gerontological Social Work*, 1(1), 33–45.

Stern, J. E., Cedars, M. I., Jain, T., Klein, N. A., Beaird, C. M., Grainger, D. A. *et al.* (2007). Assisted Reproductive Technology Practice Patterns and the Impact of Embryo Transfer Guidelines in the United States. *Fertil Steril.*, 88(2), 275–82.

Sternberg, R. J. (1986). A Triangular Theory of Love. *Psychol. Rev.*, 93, 119–135.

Stewart, F., Trussell, J. & Van Look, P. F. A. (2007). Emergency Contraception. In R. A. Hatcher, J. Trussell, A. Nelson, W. Cates, F. Guest, F. Stewart & D. Kowal (Eds.), *Contraceptive Technology*, 19th revised edn. New York: Ardent Media.

Stone, L. (2017). *Fewer Babies, More Pets? Parenthood, Marriage, and Pet Ownership in America*. Charlottesville, VA: Institute for Family Studies.

Sunderam, S., Kissin, D. M., Crawford, S. B. *et al.* (2018). Assisted Reproductive Technology Surveillance—United States, 2015. *MMWR Surveill Summ.*, 67(SS-3), 1–28.

Taraban, L., Shaw, D. S., Leve, L. D., Natsuaki, M. N., Ganiban, J. M., Reiss, D. & Neiderhiser, J. M. (February 20, 2018). Parental Depression, Overreactive Parenting, and Early Childhood Externalizing Problems: Moderation by Social Support. *Child Development*. Retrieved on April 21, 2018 from https://doi.org/10.1111/cdev.13027

Tazi-Preve, I., Bichlbauer, D. & Goujon, A. (2004). Gender Trouble and Its Impact on Fertility Intentions. *Yearbook of Population Research in Finland*, 40, 5–24.

Thomson, E. (2002). Motherhood, Fatherhood and Family Values. In R. Lesthaeghe (Ed.), *Meaning and Choice: Value Orientations and Life Course Decisions* (pp. 251–272). The Hague: NIDI/CBGS Publications No. 37.
Trussell, J. (2011). Contraceptive Failure in the United States, *Contraception*, 83(5), 397–404.
Trussell, J. & Raymond, E. G. (2018). *Emergency Contraception: A Last Chance to Prevent Unintended Pregnancy.* Princeton, NJ: Princeton University Press.
Turok, D. T., Sanders, J. N., Thompson, I. S., Royer, P. A., Eggebroten, J. & Gawron, L. (2016). Preference for and Efficacy of Oral Levonorgestrel for Emergency Contraception with Concomitant Placement of a Levonorgestrel IUD: A Prospective Cohort Study. *Contraception,* 93, 526–532.
United Nations Population Fund (UNFPA) (2009). *State of World Population 2009, Facing a Changing World: Women, Population and Climate.* Retrieved on December 1, 2018 from https://www.unfpa.org/sites/default/files/pub-pdf/state_of_world_population_2009.pdf
United Nations Population Fund (UNFPA) (2012). *Sex Imbalances at Birth.* Retrieved on December 1, 2018 from https://www.unfpa.org/publications/sex-imbalances-birth
United Nations Population Fund (UNFPA) (2013). *From Population Pyramids to Pillars.* Retrieved on March 31, 2018 from https://www.prb.org/population-pyramids/#!/
United Nations Population Fund (UNFPA) (March 15, 2017). *Sex Selection Leads to Dangerous Gender Imbalance; New Program to Tackle Root Causes.* Retrieved on December 1, 2018 from https://www.unfpa.org/news/sex-selection-leads-dangerous-gender-imbalance-new-programme-tackle-root-causes
United Nations, Department of Economic and Social Affairs, Population Division (UNPD) (2017a). *Household Size and Composition around the World 2017.* New York: Data Booklet (ST/ESA/ SER.A/405). Retrieved on December 1, 2018 from http://www.un.org/en/development/desa/population/publications/pdf/ageing/household_size_and_composition_around_the_world_2017_data_booklet.pdf
United Nations, Department of Economic and Social Affairs, Population Division (UNPD) (October 2017b). *Household Size and Composition around the World.* Population Facts, No. 2017/2. Retrieved on December 1, 2018 from http://www.un.org/en/development/desa/population/publications/pdf/popfacts/PopFacts_2017-2.pdf
United Nations Population Fund (UNFPA) (n.a.). *The Demographic Transition.* Retrieved on March 31, 2018 from http://papp.iussp.org/sessions/papp101_s01/PAPP101_s01_090_010.html
United Nations Statistics Division (UNSD). (2015). *Demographic Yearbook 2013.* New York: United Nations.
U.S. Census Bureau (2005). *International Data Base: Online Demographic Aggregation.* Retrieved on July 17, 2016 from www.census.gov/ipc/www/idbnew.html
U.S. Census Bureau (May 12, 2017). *Facts for Features: Mother's Day: May 14, 2017.* Washington, DC: U.S. Census Bureau.
U.S. Census Bureau (October 13, 2018). *U.S. and World Population Clock.* Retrieved on October 13, 2018 from https://www.census.gov/popclock/
U.S. Department of Health and Human Services (HHS) (2018). *Infertility.* Retrieved on April 10, 2018 from https://www.womenshealth.gov/a-z-topics/infertility#b
U.S. Department of Labor (2011). *Fact Sheet #7: State and Local Governments under the Fair Labor Standards Act (FLSA).* Retrieved on April 26, 2018 from https://www.dol.gov/whd/regs/compliance/whdfs7.pdf
U.S. Department of State (DOS) (March 22, 2018a). *Message from the Assistant Secretary – Intercountry Adoption.*
U.S. Department of State (DOS) (2018b). *What to Expect After Adoption.* Retrieved on April 8, 2018 from https://travel.state.gov/content/travel/en/Intercountry-Adoption/post-adoption/what-to-expect-after-adoption.html
U.S. Department of State (DOS) (March 23, 2018c). *FY 2017 Annual Report on Intercountry Adoption.*
U.S. Department of State (DOS) (2018d). *How to Adopt?* Retrieved on April 24, 2018 from https://travel.state.gov/content/travel/en/Intercountry-Adoption/Adoption-Process/how-to-adopt.html
U.S. Department of State (DOS) (2018e). *FAQs: The Child Citizen Act of 2000.* Retrieved on April 24, 2018 from https://travel.state.gov/content/travel/en/Intercountry-Adoption/adopt_ref/adoption-FAQs/child-citizenship-act-of-2000.html
van De Kaa, D. J. (1987). Europe's Second Demographic Transition. *Popul Bull.,* Mar; 42(1), 1–59. Retrieved on March 31, 2018 from https://www.ncbi.nlm.nih.gov/pubmed/12268395
van de Kaa, D. J. (2002). The Idea of a Second Demographic Transition in Industrialized Countries. Paper presented at the Sixth Welfare Policy Seminar of the National Institute of Population and Social Security, Tokyo, Japan.
VerBruggen, R. (2017). *How We Ended Up With 40 Percent of Children Born Out of Wedlock.* Charlottesville, VA: Institute for Family Studies.
VerBruggen, R. (2018). *Trends in Unmarried Childbearing Point to a Coming Apart.* Charlottesville, VA: Institute for Family Studies.
Vespa, J. (2017). *Most Popular Living Arrangement is Living with Parents.* Washington, DC: U.S. Census Bureau.

Vespa, J., Armstrong, D. M. & Medina, L. (2018). *Demographic Turning Points for the United States: Population Projections for 2020 to 2060*, Current Population Reports, P25–1144. Washington, DC: U.S. Census Bureau.

Weeks, J. (2014). *Population, an Introduction to Concepts and Issues*. Belmont, CA: Cengage Learning.

Wenger, G. C. (2009). Childlessness at the End of Life: Evidence from Rural Wales. *Ageing & Society*, 29, 1243–1259.

WHO (World Health Organization) (2013a). *Sexual and Reproductive Health: Infertility Definitions and Terminology*. Retrieved on December 1, 2018 from http://www.who.int/reproductivehealth/topics/infertility/definitions/en/

WHO (World Health Organization) (2013b). *WHO Recommendations on Postnatal Care of the Mother And Newborn*. Retrieved on June 16, 2017 from http://www.who.int/maternal_child_adolescent/documents/postnatal-care-recommendations/en/

WHO (World Health Organization) (2018). *Preterm Birth*. Retrieved on March 15, 2018 from http://www.who.int/mediacentre/factsheets/fs363/en/

Wilson, C. (2004). Fertility below Replacement Level. *Science*, 304, 207–209.

Yip, P. S. F., Li, B. Y. G., Xie, K. S. Y. & Lam, E. (2006). *An Analysis of the Lowest Total Fertility Rate in Hong Kong SAR*, Discussion Paper 289, Center for Intergenerational Studies, Institute of Economic Research, Hitotsubashi University.

Younes, M. N. & Klein, S. A. (2014). The International Adoption Experience: Do They Live Happily Ever After? *Adoption Quarterly*, 17(1), 65–83.

Zadoroznyj, M. (1999). Social Class, Social Selves and Social Control in Childbirth. *Sociology of Health and Illness*, 21(3), 267–289.

Zaidil, B. & Morgan, S. P. (2017). The Second Demographic Transition Theory: A Review and Appraisal. *Annu. Rev. Sociol.*, 43, 473–492.

PART III

CHALLENGES TO MARRIAGE AND THE FAMILY

11
PARENTING

Learning Outcomes

Upon completion of the chapter, students should be able to:

1. describe how family developmental theory explains the transition to successful parenthood
2. describe how the alternative model explains the transition to successful coparenthood
3. describe four distinct types of intelligence that a successful person can possess
4. analyze parental roles in forms of family
5. describe challenges and forms of parenthood
6. explain how the sociology of childhood explores a variety of childhoods
7. describe ways of parenting a child with high IQ, EI, and SI
8. describe ways of parenting a teen with high IQ, EI, SI, and FI
9. describe three parenting styles and their effects
10. describe parenting experiences of children in family diversity
11. explain the benefits and downsides of being third-culture children.

Brief Chapter Outline

Pre-test

Successful Parenthood

- The Transition to Successful Parenting
- Parenting a Child with a High IQ, EI, SI, and FI

Contemporary Parental Roles

- Forms of Parenthood
    - Two Married Parents
        - Coparenthood
    - Cohabiting Parents; LGBT Parents; Single Parents
        - Single Mothers; Single Fathers

Pre-test

Engaged or active learning is a powerful strategy that leads to better learning outcomes. One way to become an active learner is to begin the chapter and try to answer the following true/false statements from the material as you read. You will find that you have ready answers to these questions upon the completion of the chapter.

Statement		
Parenting is the process of supporting the development of a child from infancy to childhood.	T	F
RPM3 refers to parenting roles that involve responding, preventing, monitoring, mentoring, and modeling.	T	F
A coparenting relationship is the extent to which parents work with or against each other to care for their child.	T	F
Parentification is the process of role reversal whereby a child is obliged to act as parent to their own parent.	T	F
Attachment parenting is a parenting method that promotes the attachment of mother and infant through parental responsiveness and closeness.	T	F
African American children often experience communal parenting.	T	F
The permissive parenting style is responsive but not demanding.	T	F
The authoritarian parenting style is not responsive and demanding.	T	F
The authoritative parenting is both demanding and responsive.	T	F
Third-culture kids move among three cultures and countries internationally.	T	F

Upon completion of this section, students should be able to:
LO1: Describe how family developmental theory explains the transition to successful parenthood.
LO2: Describe how the alternative model explains the transition to successful coparenthood.
LO3: Describe four distinct types of intelligence that a successful person can possess.

Parenting actually starts before the child is born. The human life cycle involves several stages, including pregnancy, early childhood, preteen, adolescence, adulthood, and old adulthood. Following conception, a baby begins as a single cell and then multiplies into many cells that form the body parts and organs that create a new human life. Birth takes place approximately 40 weeks after conception, at which time the baby exits the mother's uterus and enters society.

Successful Parenthood

The transition to parenthood is a major family developmental milestone in many people's lives. Let's look at how people may experience the transition to parenthood and prepare to play their role as a parent.

The Transition to Successful Parenting

Family developmental theory suggests that the transition to parenthood involves developmental life changes from one stage such as spouse to another such as parent. Indeed, many people view the addition of a child into their marriage as a period of transition that is somewhat stressful (Hobbs, 1965). Stressful events such as the birth of a child bring about more changes than those occurring at any other developmental stage of the family life cycle (Priel & Besser, 2002). *The sociology of parenthood* focuses on coping with the transition to parenthood, combining paid work, parenting and household duties, the concept of "commitment" to children and paid work, experiences of mothers and fathers in relation to equal opportunities at work, and the implications for policy (Gatrell, 2005). Indeed, parents today play multiple roles. *Roles* are "socially constructed patterns of behavior and sets of expectations that provide us a position in our families" (Turner & West, 2000). Parenting is an example of a family role. Family development theory argues that as long as they prepare for the transition to parenthood and define parental roles, a couple is likely to grow and be happy when their first child is born.

Parenting is the process of supporting the emotional, intellectual, and financial development of a child from infancy to adulthood. Good parents take on their parenting roles, develop positive parenting skills, accept responsibility for the healthy development of their child, act as a positive role model, and guide their child through childhood to a successful adulthood (American SPCC, 2017). One set of guidelines for successful parenting is **RPM3** (NICHD, 2018). With RPM3, a parent is ready to respond to children, prevent risky behaviors, monitor children's contact with their surrounding world, mentor them to encourage desired behaviors, and model their own behavior to provide a consistent and positive example for them. **Parental responsiveness** (sensitivity or sensitive responsiveness) is characterized by warm acceptance of children's needs and interests (Merz *et al.*, 2015). Warm, responsive, and sensitive parenting promotes children's emotional, social, and even cognitive development during the preschool period (Landry *et al.*, 2001) whereas unresponsive and rejecting parenting is strongly tied to poor socioemotional development and externalizing problems (Rothbaum & Weisz, 1994). Parents function as caregivers and protect their children from birth through adulthood. Parents mentor their children through example and behaviors into successful roles of their own. Children learn an intense amount from people in their family as a model and children do what their family members do (Goleman, 2005). **Social learning theory** suggests that children acquire new behaviors by observing and imitating others. Social learning theorists have identified the pervasive roles that *conditioning and modeling* play as children acquire associations that subsequently form the basis for their culturally constructed selves (Bornstein, 2012). For example, parental infidelity was positively associated with offspring infidelity (Weiser & Weigel, 2017).

Parenting a Child with a High IQ, EI, SI, and FI

The goals of parenting are to ensure children's health and to prepare them to be functioning members of society. Nearly all parents regardless of culture seek to lead a happy, healthy, fulfilled parenthood and rear happy, healthy, fulfilled children (Bornstein, 2012). **Intelligence** is the aggregate or global capacity of the individual to act purposefully, to think rationally, and to deal effectively with their environment (Wechsler, 1958). *Four types of intelligence* can be used to measure a successful person.

Intellectual intelligence is considered a precursor to academic success and is measured using *intellectual quotient* (IQ)—a score derived from standardized tests designed to assess a person's academic intelligence (Lau, 2016). Research indicates that maternal intellectual ability is linked to VPT (born very preterm) and FT (born full-term) children's intellectual and language outcomes. Parental involvement in activities can help children master new skills and may promote cognitive development in VPT children born to mothers of lower intellectual ability (Lean *et al.*, 2018). Education institutions focus on improving children's intellectual ability.

Succeeding at school is not all about intellectual quotient. Rather, emotional intelligence is the key to consistent academic achievement (Olivo, 2016) and it is the foundation for a host of social skills and impacts most everything you do and say each day (Bradberry & Greaves, 2009).

A more important measure of a successful person than IQ (Goleman, 2005) is **emotional intelligence** (EI)—the ability to perceive and understand emotions, and to regulate emotions so as to promote emotional and intellectual growth (Salovey *et al.*, 2004). *Emotional quotient* (EQ) is used to assess the level of a person's emotional intelligence. Emotionally intelligent children enjoy increased self-confidence, greater physical health, better performance in school, and healthier social relationships (Gottman *et al.*, 1998). Children with poor emotional competence and who lack social skills have more difficulty forming peer relationships and benefit less from the educational environment of school than do children with stronger emotional and social skills (Kaiser *et al.*, 2000).

Teaching emotional intelligence to children has proven to be effective (Brackett *et al.*, 2015). Emotional intelligence is a protective factor for suicidal behavior (Cha & Nowak, 2009). **Suicide** is death caused by injuring oneself with the intent to die and is a growing public health problem (CDC, 2018a). In the United States, suicide rates have been rising in nearly every state and in 2016, 45,000 Americans age ten or older died by suicide, with one death every twelve minutes (CDC, 2018b). Suicidal behaviors are prevalent and dangerous behavior problems around the world, particularly among adolescents (Nock *et al.*, 2008). Research on suicidal behaviors has primarily aimed at identifying risk factors, such as the experience of childhood maltreatment (Gould *et al.*, 2003). People engage in suicidal behaviors due to an inability to tolerate the experience of negative affect (Lynch *et al.*, 2004). Those who are adept at perceiving, integrating, understanding, and managing their emotions would be at reduced risk for suicidal behaviors in response to stressful life events (Cha & Nowak, 2009).

It is through parents' socialization of emotion that children can learn how to handle their emotions. **Emotion** is characterized by intense mental activity and a high degree of pleasure or displeasure (Cabanac, 2002). The **sociology of emotions** also examines how emotions are differentiated, socialized, and managed socially and culturally (McCarthy, 1994). Women and men are socialized to adopt different cultural gender roles for emotion (Chaplin *et al.*, 2005). For example, women are expected to be more relationship-oriented than men, whereas men are expected to be assertive and even overtly aggressive if needed (Brody & Hall, 2000). Consistent with these roles, females may be more likely than males to express emotions that support relationships, mute emotions that assert one's own interests over the interests of others, and engage in more covert, relational forms of aggression (Zahn-Waxler, 2010). *Emotion-related parenting behaviors* are conceptualized as part of parents' socialization of emotion (Mirabile, 2010). Parents socialize children's emotions through their responses to children's emotions, their discussion of emotion, and by providing models of how to express and regulate emotions (Morris et al., 2002).

The sociology of emotion also explores the expression of emotions across cultures. Culture helps construct parents and parenting, and is maintained and transmitted by influencing parental cognitions that in turn are thought to shape parenting practices (Bornstein & Lansford, 2010). *Emotion cultures* consist of more than feeling and expression rules (Gordon, 1989) and are learned through the process of emotional socialization (Thoits, 1989). People from all over the world were found to interpret negative and positive emotions the same way (Robbins *et al.*, 2009). Indeed, the bodily maps of the sensation patterns for different emotions were consistent across different West European and East Asian cultures (Nummenmaa *et al.*, 2014).

The **sociology of depression** looks at the cultural context in which people live and the social stressors that people encounter. Some cultures may view symptoms of depression as normal emotional responses to particular life events and others may expect the grief process to last longer than the one-year period in the West (Nemade, 2018). Maternal depressive symptoms have been related to negative parenting behaviors (Nelson *et al.*, 2003). For example, depressed mothers have been described as more likely to report difficulties managing their infants' crying and demands (Seeley *et al.*, 1996); more withdrawn, with less rapid, consistent, and positive responsiveness to their children's actions (Burt *et al.*, 2005); and more intrusive, coercive, or hostile in their play or interaction with their child (Field, 1992).

The third important skill that parents can help their children learn is **social intelligence** (i.e., people skills)—the ability to get along well with others, and to get them to cooperate with you (Albrecht, 2003), including the abilities to understand and navigate social situations and maintain our relationships, to understand and manage people (Thorndike & Stein, 1937), and to manipulate the responses of others (Weinstein, 1969). In social intelligence, Goleman (2007) explores an emerging new science with startling implications for our interpersonal world. Social intelligence is equated with *interpersonal intelligence*. In contemporary society, we rely on a variety of methods to communicate with others, from traditional face-to-face interactions with known individuals to a much wider social network of known and unknown individuals in our social media distribution lists (Lau, 2016). In a diverse society today, there is an increasing need to successfully interact with people from different backgrounds and countries. Thus the development of social intelligence is more important now than ever before (Lau, 2016).

If a parent raises a socially intelligent child, the child is likely to know how to navigate a complex social world. Parents have an opportunity to shape their child's social intelligence by modeling their own responses to their life's challenges and by encouraging children to discuss theirs (Goleman, 2007). Those with high interpersonal intelligence communicate effectively and empathize easily with others, and may be either leaders or followers (Gardner, 1995). *Social competence* refers to children's ability to engage in social interaction, attain social goals, make and maintain friendships, and achieve peer acceptance in early childhood (Rubin *et al.*, 2006).

In an increasingly automated and globalized workplace, we are seeing a trend in businesses placing more value on social intelligence skills. People with higher levels of social intelligence have the potential to earn more than their colleagues (Lau, 2016). Equally, people with a high degree of emotional intelligence make more money—an average of $29,000 more per year than people with a low degree of emotional intelligence (Bradberry & Greaves, 2009). **Financial intelligence** (FI) is the ability to build wealth and manage a person's life by understanding how money works. Those with high financial intelligence are able to accumulate and manage wealth. *The financial quotient* (FQ) or financial intelligence quotient (FiQ) score is measured using six categories (i.e., spending, credit and debt, career and income, investing, financial planning, and risk and protection) (Yu & Zhang, 2015) and can be used to analyze an individual's behavior such as using credit cards (Nik *et al.*, 2014) and to improve financial literacy education (Russia G20 & OECD, 2013), or financial independence of women as a prerequisite to women's empowerment (Sujan, 2016). In personal and family life, money is a huge worry and a common cause of conflict, separation, or divorce in relationships. With the right attitude and financial intelligence, couples can become financially compatible and live happily. What parents say and do about money has a profound influence on their children.

Upon completion of this section, students should be able to:
LO4: Analyze parental roles in forms of family.
LO5: Describe challenges and forms of parenthood.

The way parents raise their children will have a lifelong impact on their emotional health and social behavior. Having a high IQ is not enough because other forms of intelligence—social intelligence and emotional intelligence, financial intelligence—are also important for success. These four types of intelligence (intellectual, emotional, social, and financial intelligence) can be taught through the process of parenting and are equally important to a successful person through a lifelong learning process.

Contemporary Parental Roles

This section examines children living with two married parents, cohabiting parents, LGBT parents, single parents, non-residential parents, grandparents, foster parents, and digital parents.

Most women and men become parents at some point in their adult lives. Mother and father are two primary roles in the family. Women who carry a pregnancy and deliver and/or raise children experience motherhood (the state of being a mother). A **mother** is the female parent of a child. Women who raise children (adoptive mother or other caregivers), or agree to carry a pregnancy for another person (surrogacy mother), can be considered mothers. **Biological mother** is the female genetic contributor to the creation of a child. In the United States, about 69 percent of women were the biological mother of at least one child in 2014 (Monte, 2017).

Men who raise children experience fatherhood (the state of being a father). A **father** is the male parent of a child. In the United States, there were 72.2 million fathers across the nation in 2014 (U.S. Census Bureau, 2018c). **Biological father** is the male genetic contributor to the creation of the baby, through sexual intercourse or sperm donation. About 59 percent of biological fathers had at least one child in 2014 (Monte, 2017) and roughly one in four men were grandfathers (U.S. Census Bureau, 2018c). Today a non-biological father such as adoptive father or stepfather (male married to biological mother) fills in the role of childrearing. Figure 11.1 shows that of the 121 million men age fifteen and over, about 75 million (six in ten men) were fathers to biological, step, or adopted children.

Forms of Parenthood

Today, children live with different forms of family. Children living in a two-parent family stood at 46 percent in 2013, down from 61 percent in 1980 and 73 percent in 1960. Let's examine different forms that parenthood may take beyond the traditional two-married-parent family. The types of parenthood categories are not mutually exclusive (Arendell, 2000).

Two Married Parents

Having two parents is a great benefit to a child's life. From structural functionalist theory, a father plays an instrumental role and a mother plays an expressive role, complementing each other in raising a child. Parsons (1954) illustrated this through his focus on the expressive traits of mothers concerned with the personal and intimate aspects of social life, compared to the instrumental traits of fathers concerned with aspects outside the family (Leonard, 2016). The maternal bond and paternal bond that develop with children impact children throughout their lifetime. Research suggests that having two resident biological caregivers predicts greater time investments in children and that shared parenting may be an important

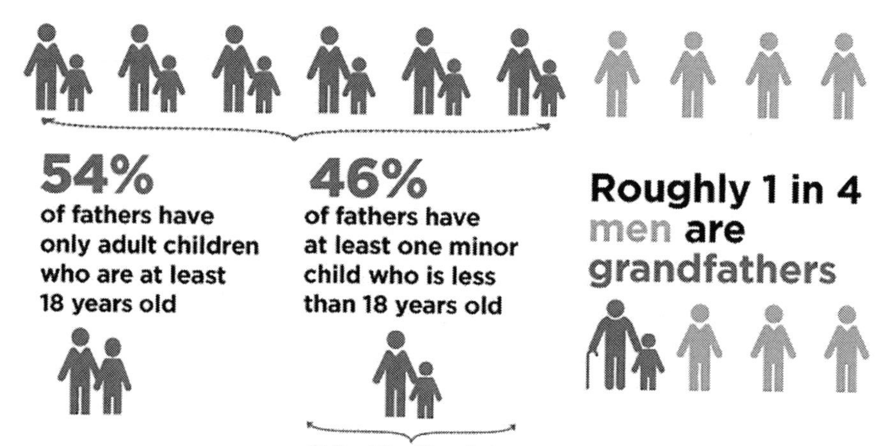

Figure 11.1 Fatherly Figures: A Snapshot of Dads Today
Source: Survey of Income and Program Participation, 2014 Panel, Wave 1, U.S. Census Bureau (Stats for Stories: Father's Day: June 17, 2018). https://www.census.gov/newsroom/stories/2018/fathers-day.html

dimension of family structure. For example, children who live with married biological parents receive the greatest share of caregiving time in the form of shared caregiving, compared with children in all other family structures (Kalil *et al.*, 2014). The long-term goal of equal opportunity for women in American society will never be achieved without serious and meaningful recognition of the significance, interest, and responsibility of fathers in children's lives (Levine, 1976). In 2017, almost 70 percent of children lived with two parents in the United States (see Figure 11.2).

Both the mother and father play important parental roles in the growth and development of children. Within this family type, the child would see how adults have learned distinctive patterns of behavior. "Fathers don't mother and mothers don't father" (Erickson, 2015). For instance, mothers and fathers speak to their toddlers in very different ways (Bayley, 2006). Men and women's brains operate differently (Moore *et al.*, 2002) and they may ask different questions and get children to talk differently in interaction. Children living with two married adults have better health, greater access to health care, and fewer emotional or behavioral problems than children living in other types of families (Blackwell, 2010). For example, high school graduation rates were higher in Florida counties with more children growing up in married-couple families (Wilcox & Zill, 2016). Children have more stable family lives when born within marriage regardless of their mother's educational background (DeRose, 2017). Those living with no biological parents were found to be less likely to exhibit behavioral self-control and more likely to be exposed to high levels of problematic parenting than children with two biological parents (Manning & Lamb, 2003). In a study of 17,000 children born in England, Scotland, and Wales, Eirini and Buchanan (2014) found that both father's and mother's involvement in families of lower socioeconomic status were highly protective against an adult experience of homelessness in sons.

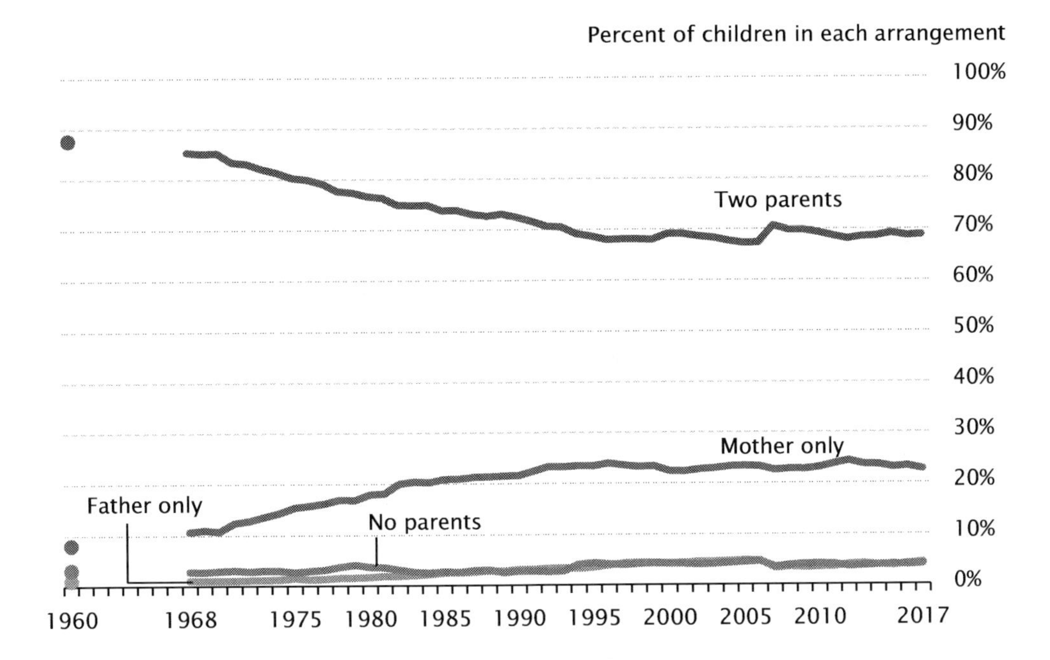

Figure 11.2 Living Arrangements of Children, 1960–2017
Source: U.S. Census Bureau, Decennial Census, 1960, and Current Population Survey, Annual Social and Economic Supplements, 1968–2017. https://www.census.gov/data/tables/time-series/demo/families/children.html

Coparenthood Studies of coparenting have gained in importance as researchers increasingly focus on both mothers' and fathers' roles in the family system (Feinberg, 2002). **Coparenting** has been defined as a unique component of the marital relationship in which parents work together, or, alternatively, struggle against each other when it comes to childrearing (McHale, 2007), and has been described as "the center around which the family process evolves" (Weissman & Cohen, 1985). As they become parents across the transition to parenthood, couples established their *coparenting relationship*—the extent to which they work with or against each other to care for their child (Murphy *et al.*, 2017). Each partner's involvement in making parenting decisions and each partner's support of the other's parenting decisions are likely to differ by *parent gender* (Coltrane & Shih, 2010). In a study of 125 first-time parents with a 24-month-old child, Murphy *et al.* (2017) found that fathers showed greater instances of support for their partner than did mothers, and fathers' higher support of mothers and higher involvement in parenting decisions was related to higher cooperative coparenting, whereas mothers demonstrated higher levels of involvement in parenting decisions than did fathers, and mothers' higher support of fathers' parenting was related negatively to competitive coparenting and positively to fathers' involvement. **Self-efficacy** refers to a personal judgment of how well one can execute courses of action to deal with prospective situations (Bandura, 1982). A study of 182 dual-earner couples to examine the role of supportive coparenting showed that mothers' perceptions of greater supportive coparenting were associated with lower parenting stress only when their parenting self-efficacy was low, but fathers' perceptions of greater supportive coparenting were associated with greater parenting satisfaction only when their parenting self-efficacy was high (Schoppe-Sullivan *et al.*, 2016).

Coparenting has been viewed as the intersection between the marital and parent–child relationship (McHale & Sullivan, 2008). The **alternative model** suggests that the coparenting alliance predicts both marital relationships and parenting practices. A **coparenting alliance** is characterized by (1) both parents' investment in the child, (2) valuing each other's involvement with the child, (3) respect for each other's judgment about childrearing, and (4) having a desire to communicate child-related information (Weissman & Cohen, 1985). For example, parental investment is any parental expenditure (time, energy, resources) that benefits one offspring at a cost to parents' ability to invest in other components of fitness such as parents' future sexual reproduction (Clutton-Brock, 1991). Indeed, a coparenting alliance has been found to equip parents with resources to parent effectively and to coordinate their parenting roles in ways that benefit their children (Behnke *et al.*, 2008), link to parent–child relationships (Bonds & Gondoli, 2007), and predict both marital quality and parenting practices for wives and husbands (Morrill, 2010).

Cohabiting Parents

Today, more primary caregivers are raising children together with their cohabiting partner or biological parents. A study of 68 countries shows that most countries have experienced a rise in cohabiting births and a decline in the share of children living with both natural parents (DeRose, 2017). In the United States, unmarried partner households increased from 6 million in 2006 to 7 million in 2016 (U.S. Census Bureau, 2018a) and one in four American parents are unmarried in 2017 (Livingston, 2018a). The share of American children living with an unmarried parent has doubled, jumping from 13 percent in 1968 to 32 percent in 2017 (Livingston, 2018a). Figure 11.3 shows that more children live with cohabiting parents in 2017.

Cohabiting parents are younger, less educated, and less likely to have ever been married than single parents. The median age of cohabiting parents is 34 years, compared with 38 among single parents and 40 among married parents; cohabiting parents (54 percent) have a high school diploma or less education, compared with single parents (45 percent) and married parents (31 percent) (Livingston, 2018a). Cohabitation tends to be associated with instability and suited to those struggling with unemployment, temporary jobs, or general life uncertainty (Perelli-Harris & Styrc, 2017). Unfortunately, the growth of cohabitation leads to fewer children having the benefits of living with two married parents. Individual children born to cohabiting couples suffered more family instability than children born to married couples across the globe (DeRose, 2017). In the United States, two-thirds of children born to cohabiting couples experienced family instability before age twelve, compared with about one-fourth of children born to married couples (Manning, 2015). When examining whether cohabitation and marriage influence mental well-being in midlife, Perelli-Harris and Styrc (2017) found that matching on childhood characteristics eliminates differences in well-being between marriage and cohabitation for men and women.

LGBT Parents

An estimated 3 million LGBT Americans have a child (Gates, 2013). **LGBT parenting** refers to lesbian, gay, bisexual, and transgender (LGBT) people raising children as same-sex couples, single LGBT parents, etc. In 2016, 15.7 percent of 887,456 same-sex couples had their own children in the household, compared with 39.1 percent of 56,480,804 married opposite-sex couples (U.S. Census Bureau, 2018b). Figure 11.4 shows an increase in same-sex households from 2008 to 2014.

LGBT couples become parents through former relationships, adoption, coparenting, donor insemination, surrogacy, or foster parenting. As of June 26, 2015, adoption by LGBT individuals is legal in every

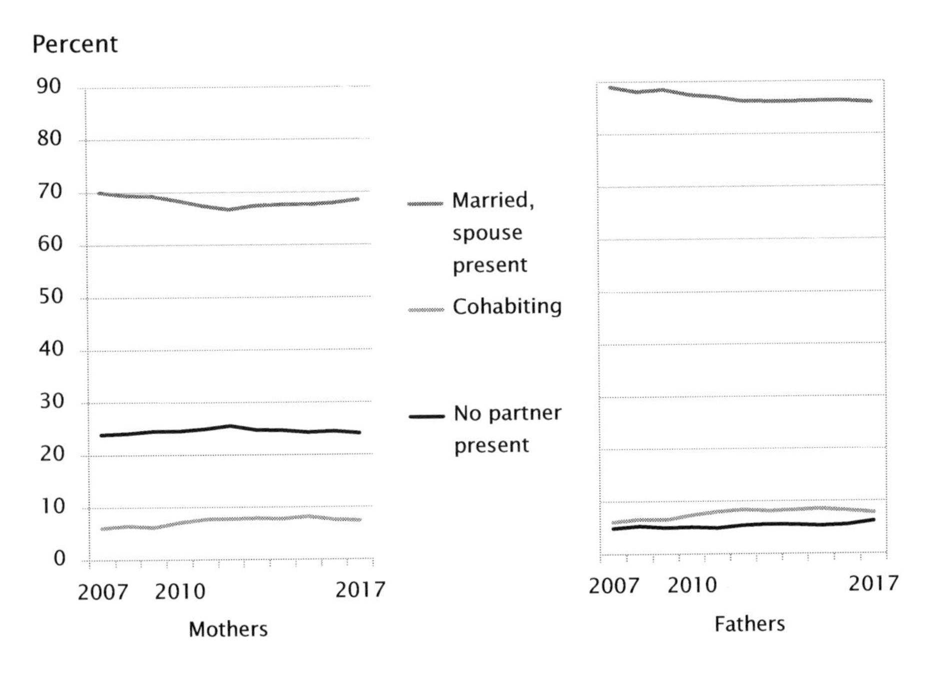

Figure 11.3 Parents of Coresident Children under 18, by Living Arrangement
Source: Current Population Survey, Annual Social and Economic Supplement, 2007–2017. https://www.census.gov/data/tables/time-series/demo/families/adults.html

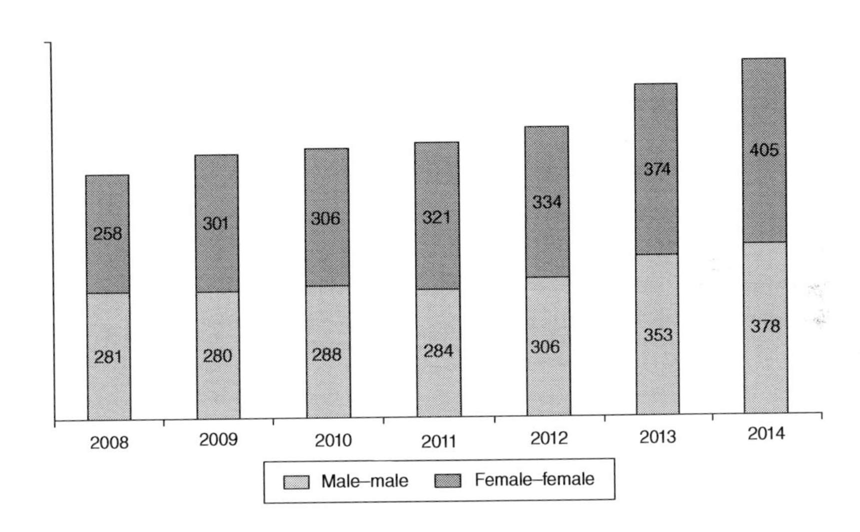

Figure 11.4 Estimates of Same-Sex Couple Households by Sex of Couple, 2008–2014
Source: U.S. Census Bureau, American Community Survey, 2008–2014 and Stats for Stories: LGBT Pride Month: June 2018. https://www.census.gov/newsroom/stories/2018/lgbt.html

jurisdiction in the United States. Joint adoption laws permit same-sex couples to adopt a child together and both prospective parents become that child's legal parents. Following the introduction of in vitro fertilization, lesbian couples are able to have the biological offspring of one of the partners. With advances in reproductive technology, an egg donor and a sperm donor can create a baby who has both biological parents and custodial parents. Male same-sex couples can become parents through surrogacy if they want a biological child of one of the partners. **Surrogacy**—the process of giving birth as a surrogate mother—enables a woman to carry a baby through a pregnancy and enables a couple elsewhere to raise the child. In a study of 40 gay father families created through surrogacy and 55 lesbian mother families created through donor insemination with a child ages three to nine, Golombok *et al.* (2017) found that children in both family types showed high levels of adjustment with lower levels of children's internalizing problems. In Italy, gay father families formed through surrogacy did not differ from lesbian mother families formed through donor insemination (Baiocco *et al.*, 2015).

Research has consistently shown that children in lesbian mother families do not differ from children in comparable groups of heterosexual parent families in terms of psychological adjustment (Golombok *et al.*, 2017). A study of 3,500 children with same-sex parents found no significant differences between households headed by same-sex and opposite-sex parents (Rosenfeld, 2010). Children with gay and lesbian parents and those with heterosexual parents showed a similar level of emotion regulation and psychological well-being (Baiocco *et al.*, 2015).

Gay fathers may be exposed to greater stigmatization because they possess the additional non-traditional feature of being in families headed solely by men (Golombok & Tasker, 2010) than are lesbian mothers (Goldberg & Smith, 2011). Children whose parents perceived greater stigmatization and children who experienced negative parenting showed higher levels of parent-reported externalizing (Golombok *et al.*, 2017). Since families with same-sex parents continue to experience stigma in society, there is reason to believe that these parents engage in *cultural socialization*—the processes by which parents communicate cultural values, beliefs, customs, and behaviors to their children. A study of 52 same-sex fathers and 43 mothers revealed that the majority of parents endorsed behaviors designed to promote children's awareness of diverse family structures and prepare them for potential stigma-related barriers along three dimensions: cultural socialization, preparation for bias, and proactive parenting (Oakley *et al.*, 2017). A study of 160 lesbian mothers and 106 gay fathers showed that lesbian and gay parents endorse the importance of unique socialization practices and feel confident engaging in these practices (Battalen *et al.*, 2018).

Single Parents

Single parent is the only male or female parent to a child, responsible for all the child's financial and emotional needs. Single parenthood historically came about in different ways, such as when one parent would die from a disease or a war. In the early years of the twentieth century, it was not uncommon for children to live with only one parent because of the death of one of their parents (Uhlenberg, 1980). At the time, almost one-third of single mothers living with children under age eighteen were widows. By the end of the 1970s, only 11 percent of single mothers were widowed, whereas two-thirds were divorced or separated (Bianchi, 1995). From 1960 to 2016, the percentage of children under age eighteen living with only their mother nearly tripled from 8 percent to 23 percent, and the percentage of children living with only their father increased from 1 percent to 4 percent (see Figure 11.5). In 2017, there were 11.6 million single parents living with children younger than age eighteen (U.S. Census Bureau, 2018c).

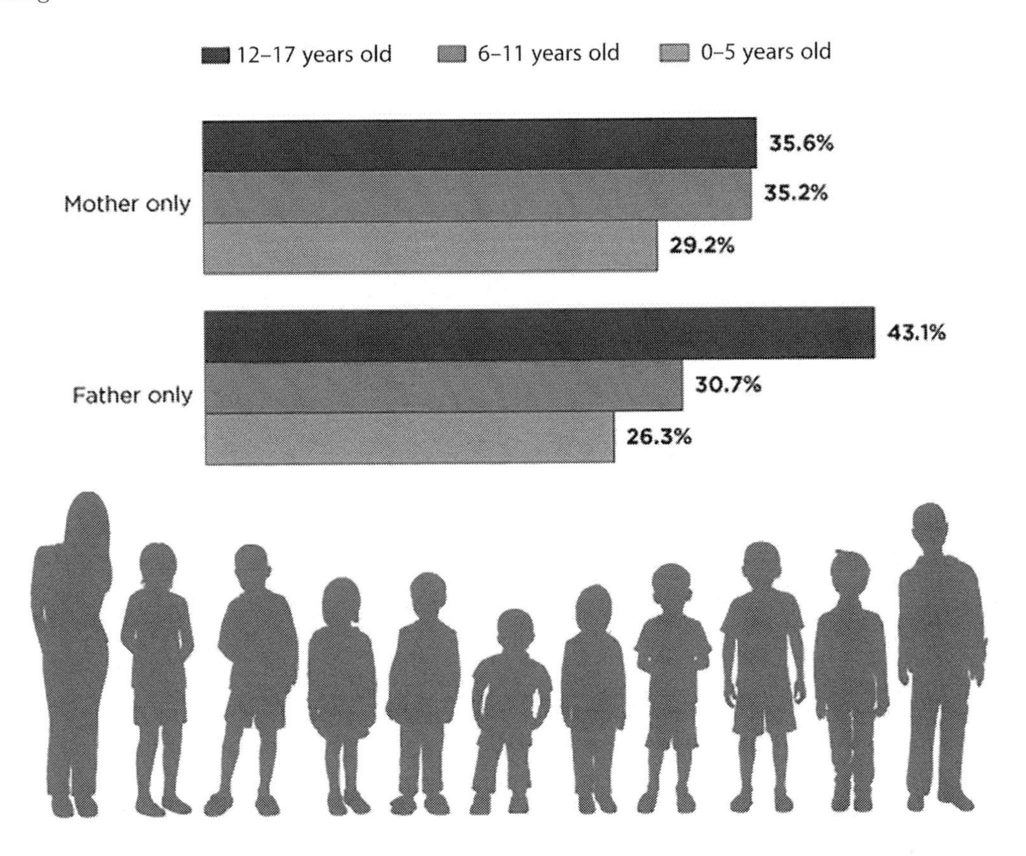

Figure 11.5 Children under 18 Living with One Parent
Source: 2017 Current Population Survey Annual Social and Economic Supplement. https://www.census.gov/library/visualizations/2017/comm/children-living-with-parent.html

Single Mothers In the United States, one in five children are living with a solo mom (Livingston, 2018b) and 9.5 million single mothers lived with their children under eighteen in 2017 (U.S. Census Bureau, 2018c), compared with 9.8 million in 2016, up from 7.7 million in 1985 (U.S. Census Bureau, 2017a). Single mothers varied by race, marital status, and education. Over twice as many children living with single mothers were identified as Black than children living with single fathers (Eickmeyer, 2017). Figure 11.6 shows that more than 50 percent of children under eighteen lived with their African American mothers in 2017, compared with their counterparts.

In 2017, 50.3 percent of single mothers living with their own children under 18 were never married; 29.3 percent were divorced; 16.8 percent were separated; and 3.6 percent were widowed (U.S. Census Bureau, 2018c). Single mothers who were never married have shown an increase from 1960 to 2014 (see Figure 11.7).

Single mothers were less likely to have a college degree, compared with single fathers. Fewer single mothers (18 percent) than fathers (23 percent) reported having at least a bachelor's degree. The share of single mothers without a high school degree was higher than single fathers (Eickmeyer, 2017). Single fathers varied by race and marital status. A majority of children living with single fathers identified as White (56 percent), whereas children living with single mothers identified as

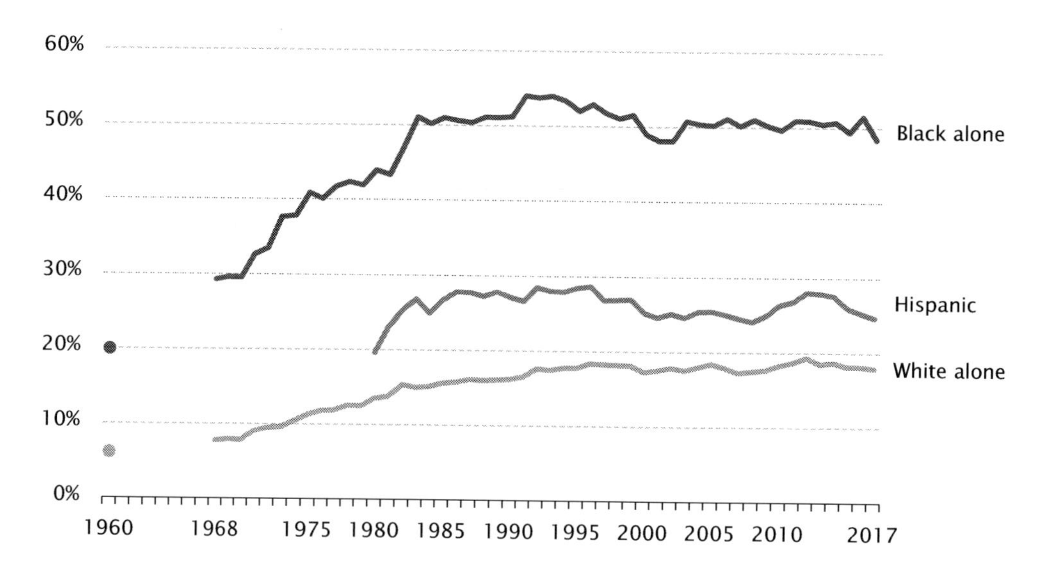

Figure 11.6 Percent of Children under 18 who Live with Their Mother Only, 2017
Source: U.S. Census Bureau, Decennial Census 1960, and Current Population Survey, Annual Social and Economic Supplements, 1968–2017. https://www.census.gov/content/dam/Census/library/visualizations/time-series/demo/families-and-households/ch-2-3-4.pdf

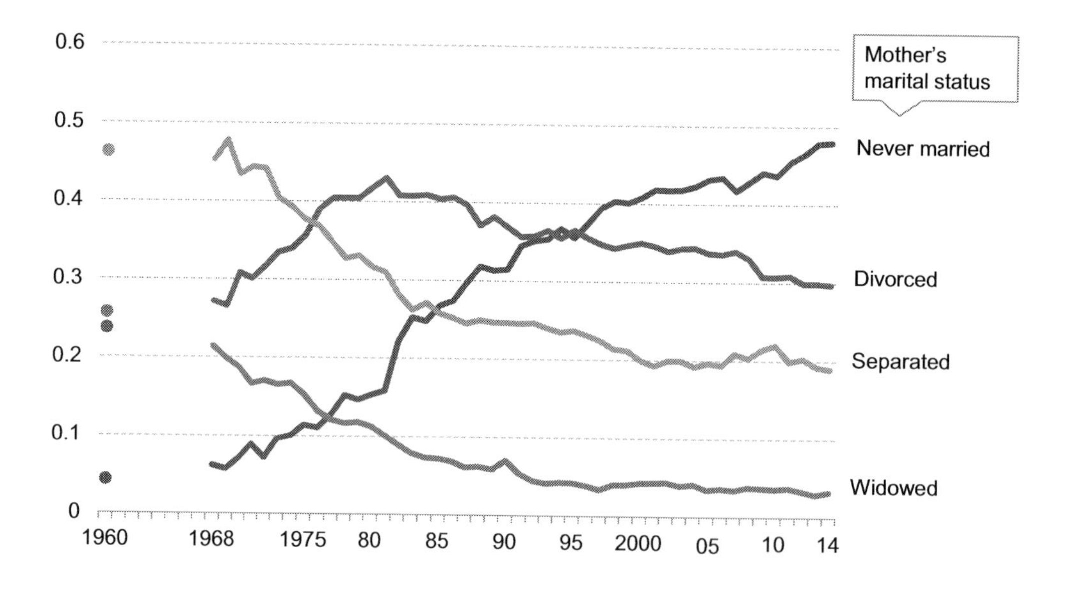

Figure 11.7 Children under 18 Living with their Mother Only
Sources: U.S. Census Bureau, Decennial Census, 1960, and Current Population Survey, Annual Social and Economic Supplements, 1968–2014, and Historical Living Arrangements of Children, November 2017. https://www.census.gov/data/tables/time-series/demo/families-and-households/ch-5.pdf

White (33 percent) or Black (32 percent) in 2016 (see Figure 11.8).

In the United States, three in ten solo mothers are living in poverty; 30 percent of single-mother families were living in poverty, compared with 17 percent of single-father families and 16 percent of cohabiting families and 8 percent of married-couple families (Livingston, 2018b). A higher percentage of girls than boys lived with a single mother (Dufur *et al.*, 2018). Single mothers who want to work often struggle with childcare costs, and children, especially boys, suffer from the distance or absence of their fathers (Reeves, 2015). Despite the challenges of single parenthood, many single mothers are able to provide positive experiences for their children, for example, through communal parenting (see Diversity Overview).

Single Fathers Single fathers without the mother present have shown an increase. In 2017, there were 2.1 million single fathers living with their children under age eighteen (U.S. Census Bureau, 2018c), up from 2 million in 2016 (U.S. Census Bureau, 2017b). Figure 11.9 shows the increase of single fathers of children under age eighteen from 1950 to 2017 in the United States.

More single fathers today are divorced or never married than are separated or widowed (see Figure 11.10). In 2017, 40.9 percent of single fathers living with their own children under eighteen were divorced; 37.5 percent were never married; 17.5 percent were separated; and 4 percent were widowed (U.S. Census Bureau, 2018c).

Single-father families are better off financially than single-mother families. In 2017, the poverty rate for never-married single fathers was 45.8 percent (145,000), compared with 56.3 percent for never-married single mothers (1.7 million) (U.S. Census Bureau, 2018c). Because single-father families are more likely to be in poverty than married families, they benefit from access to government services and aid (Brown, 2013). With fewer opportunities for personal

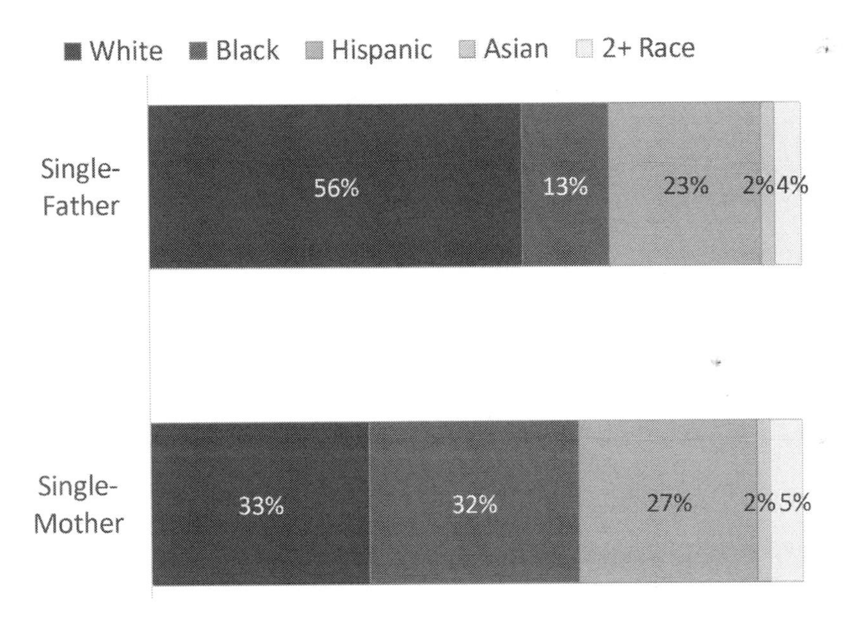

Figure 11.8 Variation in Racial and Ethnic Composition among Children in Single-parent Families, 2016

Source: Current Population Survey, 2016; Family Profiles FP-17–17 NCFMR. https://create.piktochart.com/output/23456881-eickmeyer-single-parent-families-fp-17-17

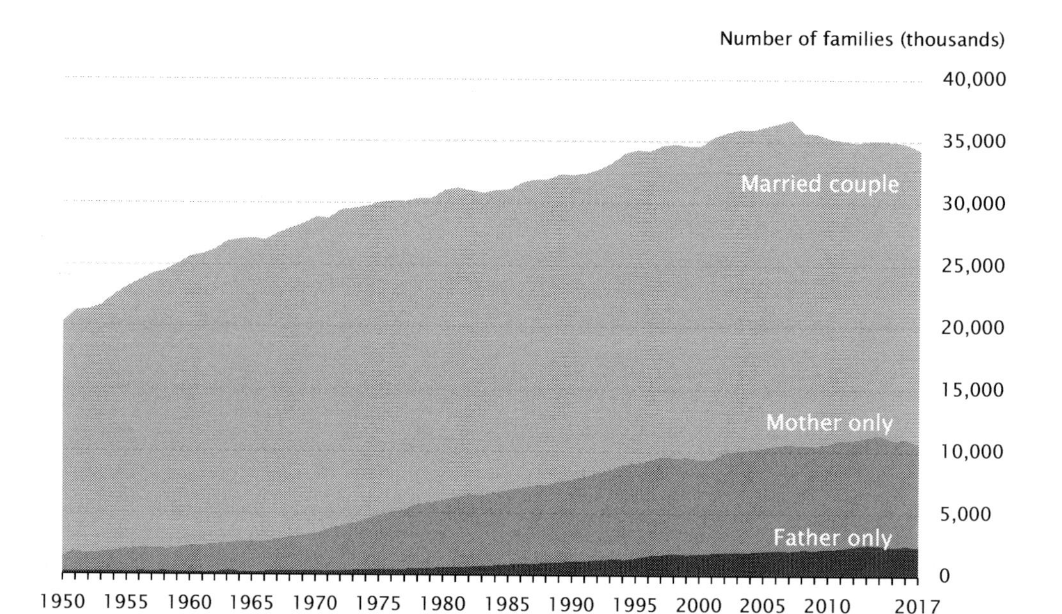

Figure 11.9 Families with Children under 18

Source: U.S. Census Bureau, Current Population Survey, Annual Social and Economic Supplements 1950 to 2017. https://www.census.gov/data/tables/time-series/demo/families/families.html

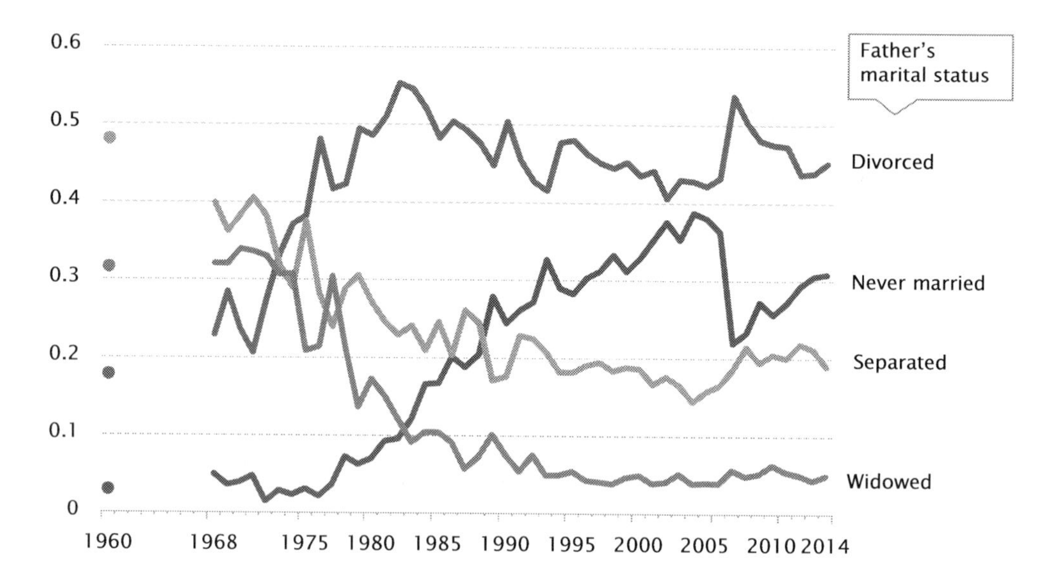

Figure 11.10 Children under 18 Living with their Father Only

Sources: U.S. Census Bureau, Decennial Census, 1960, and Current Population Survey, Annual Social and Economic Supplements, 1968–2014, and Historical Living Arrangements of Children, November 2017. https://www.census.gov/data/tables/time-series/demo/families/children.html

achievement, low-income fathers may be especially invested in the parental role (Edin *et al.*, 2011). Some fathers' poor mental health, however, is a risk factor for low father involvement (Paulson *et al.*, 2011). For example, *depressive symptoms*—poor mood, lack of energy, irritability, anger, or difficulty concentrating (Costello, 1993)—reduce father involvement regardless of father residence (Paulson *et al.*, 2011). In a sample of resident and non-resident fathers with three-year-old children, Lyons-Ruth *et al.* (2000) found that depressive symptoms were associated with reductions in play, reading, and displays of affection, and increases in negative parenting behaviors such as aggravation, yelling, and negative discipline techniques.

Single fathers represent a unique and growing share of single-parent families for children (ElHage, 2017). While they differ in important ways from single mothers, single dads need some of the same supports from the other parent and their network (e.g. faith community and direct-service non-profits) to raise their children as effective coparents (Brown, 2013). With a few possible exceptions, the children of single fathers do about as well in terms of academic performance but appear to be more likely to participate in substance use (Coles, 2015). Using data from the National Longitudinal Study of Adolescent to Adult Health (n=2570) on youth ages fifteen to nineteen, Dufur *et al.* (2018) found that youth in single-father families were more motivated to use birth control as they were more likely to report being diagnosed with an STD.

Non-residential Parents

A **custodial parent** is a parent with whom the child(ren) lived when their other parent(s) lived outside the household. In 2016, about four of every five (80.4 percent) of the 13.6 million custodial parents were mothers, while one of every five custodial parents were fathers (19.6 percent); custodial mothers were more likely to receive full child support payments in 2015 (44.9 percent) than were custodial fathers (35.5 percent) (Grall, 2018). **Non-residential parents** are defined as parents who do not live with one or more of their biological children all or most of the time and they may or may not have custodial rights (Braithwaite & Baxter, 2006). Parents become non-residential after a divorce or after the end of their cohabitating relationship, or because their children were born outside of a marital relationship. About 22.4 million children had a parent who lived outside their household, representing more than one-fourth (27.0 percent) of all children under age 21 (Grall, 2018).

Research on non-residential parenting has focused on patterns of contact and visitation (Braver *et al.*, 1993), child support (Natalier & Hewitt, 2010), child adjustment (Falci, 2006), father involvement (Carlson & McLanahan, 2010), levels of coparenting (Carlson & Högnäs, 2011), incarceration and absenteeism (Geller *et al.*, 2012), paternal engagement (Zhang & Fuller, 2012), and non-residential parent adjustment (Anderson & Letiecq, 2005). Indeed, non-residential parents experience a decrease in the amount of face-to-face and daily interaction they share with their children (Wilbur & Wilbur, 1988). As a result of these changes, non-residential parents often report feeling less parental control over the lives of their children (Braver *et al.*, 1993). Parental control involves *primary control*—the ability to control the environment in ways that are consistent with one's own wishes—and *secondary control*—the ability to adjust to an environment over which they have little to no control (Braver *et al.*, 1993). Residential parents have a high degree of primary control over their own lives and the lives of their children, whereas non-residential parents often have limited, secondary control over the lives of their children and must learn to adapt to this new situation (Kartch, 2013), and commonly report feelings of loss, self-doubt, and depression as they feel their parental role diminishing (Arditti, 1995).

The Role of Non-residential Fathers and Mothers The benefits of father involvement for children are widely acknowledged. Braver *et al.* (1993) defined *parental involvement* as a mixture of the "payment of child support and the frequency and emotional quality of the relationship with the child." Using a "good provider" framework, *the role of non-residential father* has been operationalized through child support payments (Arditti, 1995). Since most non-resident parents are fathers, there is a tendency for policymakers to see them primarily in terms of their financial obligations (Reeves, 2015). However, many of these men are in no position to make serious financial contributions because 41 percent of poor non-resident fathers have been out of paid work for at least a year (Sorensen, 2002) and because non-residential fathers (9 million) were less educated (15 percent high school dropout and 38 percent high school graduate) and unemployed (2.35 million) (Reeves, 2015). One-half (50.2 percent) of all 13.6 million custodial parents had a court order, child support award, or some other type of agreement to receive financial support from the non-custodial parent(s) in 2016 (Grall, 2018).

Due to the evolution of good fathers who are expected to visit their children (Arditti, 1995) in the 1970s and 1980s, the non-residential father role has been operationalized through frequency of visitation (Kartch, 2013). *Generative care*—caring for and developing the next generation—is primarily accomplished through the nurturing of one's own children (Paulson *et al.*, 2011). Fathers who embrace the generative aspect of fatherhood maintain close relationships with their children in an effort to foster their children's personal and social development (Kotila & Kamp Dush, 2013). Indeed, low-income and non-resident fathers place a high premium on the *generative aspect* of fatherhood—valuing stability, mentoring, engagement, and emotional support in their relationships with their children; these characteristics encompass their understanding of what it takes to be a "good father" (Summers *et al.*, 2006). Children in stepfather families and multigenerational households receive substantial time investments from *non-resident biological fathers* and grandparents, respectively (Kalil *et al.*, 2014). Using symbolic interactionalist theory to study father involvement in their children's lives, Ihinger-Tallman *et al.* (1993) suggest that non-residential fathers who identify their fatherly role as salient were more involved with their children than fathers who do not identify as strongly with their parental role. However, low-income and non-resident fathers may face challenges in remaining involved with their children as their lives are often characterized by multiple sources of stress, such as under- or unemployment, unreliable transportation, and no permanent housing (Anderson *et al.*, 2005). As a result, non-resident fathers were at risk for increased depressive symptoms and reduced father involvement (Cheadle *et al.*, 2010). Mothers limited father contact when fathers were depressed (Kotila & Kamp Dush, 2013).

Non-residential mothers claimed that their role in intensive mothering had to change because they no longer interacted with their children face-to-face on a daily basis (Bemiller, 2005). They enacted one of two strategies to cope with role changes. Some non-residential mothers were referred to as "accommodators" as they attempted to accommodate the traditional maternal role and continued to enact intensive mothering whenever they were with their children. These mothers reported lower levels of adjustment to being non-residential, higher levels of guilt, and increased prevalence of depression, and others were referred to as "resistors" as they attempted to reconceptualize their parental role, resist dominant, cultural role definitions of mother, and enact mothering in ways that fit within their new parental role (Bemiller. 2005). In a study of 40 non-residential parents (twenty mothers and twenty fathers), Kartch (2013) identified eight non-residential parent roles (limited role, active participant, nurturer, provider–tangible, teacher, sole parent, coparent, and disciplinarian).

Grandparents

In the 1960s and early 1970s, grandparent–grandchild relationships consisted of extended stays of grandchildren over holidays and summers at their grandparents' house (Grey *et al.*, 2013) but in the 1990s and 2000s were defined by grandparents' proximity to grandchildren in order to provide support and childcare because of parental relationship breakdowns or financial hardship (Ellis & Simmons, 2014). In 2015, 7.3 million grandparents served as primary caregivers for grandchildren under age eighteen and 2.6 million grandparents were responsible for the basic needs of one or more grandchild under age eighteen living with them. Of these caregivers, 1.6 million were grandmothers and 1.0 million were grandfathers (U.S. Census Bureau, 2017c).

One in ten grandparents lived with their grandchildren in 2014 and increases in grandparents living with grandchildren indicated that the grandparent role has changed.

Foster Parents

Foster care is a system in which a minor is placed into a ward or private home of a state-certified caregiver. The child's care is arranged through the government, a social service agency, or a school. The institution or foster parent is compensated for its expenses. A caregiver of this kind is referred to as a **foster parent**.

Foster care in the United States can trace its history back to the efforts of American philanthropist Charles Loring Brace (1826–90). In mid-nineteenth-century New York, both orphans and runaways filled the streets, and Brace proposed a radical solution to the problem by creating the Children's Aid Society (O'Connor, 2001). In 1854, the Society opened homes for homeless boys that provided them with room and board at low cost.

Foster care today consists of children staying in the care of the foster system while the child's birth family completes a case plan of services targeted at rehabilitation and prevention of future child maltreatment (Vandivere & Malm, 2015). In 2015, 269,509 children were removed from their families for neglect (61 percent), drug abuse by a parent (32 percent), and physical abuse (13 percent) and placed in the U.S. foster care system (Child Trends Databank, 2017). Foster care is intended to be a short-term solution until a permanent placement can be made.

Foster parents become certified caregivers after they are trained to learn about their rights and responsibilities as a foster parent. Foster parents play a role in parenting foster children and providing guidance and financial support that directly relates to their well-being. For example, social media helps youth in foster care fit in with their peers and fosters normalcy but foster parents should provide guidance and boundaries to help youth in their care use social media safely (Child Welfare Information Gateway, 2017).

Digital Parents

In 2018, the vast majority of Americans (95 percent) own a cellphone (Pew Research Center, 2018) and most U.S. households contain at least one of these devices: smartphone, desktop/laptop computer, tablet or streaming media device (Olmstead, 2017). **Digital parent** refers to a parent who uses one or more digital applications and devices on a daily basis, particularly in parenting. Digital parents are likely to shape how their children use the devices and give rise to online children. As digital media is one of the socialization agents, children and teens can access online material and explore the world. Social media, games, and free resources are available at the touch of a tablet. This leads to digital parenting by tablet, television, or electronics. Parents put their little ones in front of a device and walk away, letting the electronics teach and parent their child for them (Reynolds, 2017).

Digital connectivity offers many potential benefits from connecting with peers to accessing educational content but parents have voiced concerns about the behaviors children engage in online (Anderson, 2016). For example, babies learn an enormous amount from their parents or caregivers when interacting with them but when the same material is delivered through audio-visual media much less is learned (Bavelier *et al.*, 2010). Also parents were aware of digital dangers that children face. Examples of digital dangers include adult content, cyber bullying, texting, online addiction, and online grooming. **Online grooming** occurs when someone with a sexual interest in children uses social media, online chat, messaging apps, or text messages to build a relationship with a young person in order to meet and abuse the child physically. The vast majority of parents try to take a proactive approach to preventing problems by speaking with their teen about what constitutes acceptable and unacceptable online behavior (Anderson, 2016). Parents may set up rules with regard to their children's digital usage. Some serve as **helicopter parents**—paying close attention to a child's experiences and problems—and watch digital media together with children.

Research has linked the time spent using technology with "problems such as paying attention in class and deficits in attention, visual memory, imagination and sleep" (Bavelier *et al.*, 2010), and greater chances of obesity and depression (Reeves *et al.*, 2008). In a study of the effects of video games on children's school performance over a four-month-long study, Weis and Cerankosky (2009) found that children who play video games spent 30 percent less time reading and 34 percent less time completing homework than children who do not play video games. At preschool age, the interest in board games and computer games constantly increases and the interest in outdoor games gradually reduces (Sobkina & Skobeltsinaa, 2014). Increased use of technology showed greater chances of obesity, which in turn showed a greater chance of depression. Reeves *et al.* (2008) found that 30 percent of youth with major depressive symptoms used the internet for at least three hours a day and had decreased motivation. Sedentary behavior is a key factor in obesity in children and changes in sedentary activity go hand in hand with the reduction of weight loss in children (Epstein *et al.*, 1995).

Technology is part of a child's life. Today family members stay connected more through technology. **Networked family** is a form of family in which family members multitask and stay more connected through information and communication technology (ICT). A variety of devices enable multigenerational families to stay connected. Smartphones are one of the most quickly adopted consumer technologies today. In 2018, the Silent Generation (30 percent, ages 73 to 90) and boomers (67 percent, ages 54 to 72) own a smartphone, compared with Millennials (92 percent, ages 22 to 37) and Gen X (85 percent, ages 38 to 53) (Jiang, 2018). Many grandparents are actively learning how to use the devices in order to communicate better with their online grandchildren. In networked families, family time may become networked time. Even when family members spent time together face-to-face, they may focus more on device or internet use.

Parenting over the Life Course

The demands of parenting change over the life cycle. This section explores how to parent an infant, child, and teenager. Childhood is the building block upon which adolescence and adulthood is built. First, let's look at the sociology of childhood.

Upon completion of this section, students should be able to:
LO6: Explain how the sociology of childhood explores a variety of childhoods.
LO7: Describe ways of parenting a child with high IQ, EI, and SI.
LO8: Describe ways of parenting a teen with high IQ, EI, SI, and FI.

Sociology of Childhood

Historically, the majority of children worked in farming and in factories. During the 1600s, the concept of childhood began to emerge in Europe (Ariès, 1960). Children were seen as innocent and in need of protection and training by the adults around them. The modern concept of childhood didn't begin to emerge until the late 1800s and early 1900s when the Victorian middle and upper classes emphasized the role of the family and the sanctity of the child (Jordan, 1998) and when child labor laws began being passed in the United States and compulsory education laws took hold.

Child development was seen as "an inevitable and invariable process driven by a biologically rooted structure which the child inherits" (Archard, 1993). In psychology, child development is a body of knowledge constructed by adults for other adults to use in order to regulate and promote children's lives and learning (Woodhead & Faulkner, 2000). These concepts were incorporated into sociological theorizing on childhood during the 1950s in the form of socialization, through which children acquire and internalize the norms and values of the society into which they are born (Leonard, 2016) and through which children are shaped and trained in order to eventually become competent and contributing members of society (Corsaro, 2005). Developmental psychology and socialization theory positioned children in particular ways with reference to how they were seen as passive dupes of biology and/or socialization (Leonard, 2016) and were viewed as "less than fully human, unfinished or incomplete" (Jenks, 1996) and as helpless victims (Tisdall & Punch, 2012).

> Today, a four year-old who can tie his or her shoes is impressive. In colonial times, four-year-old girls knitted stockings and mittens and could produce intricate embroidery: at age six they spun wool. A good, industrious little girl was called 'Mrs.' instead of 'Miss' in appreciation of her contribution to the family economy: she was not, strictly speaking, a child.
>
> (Ehrenreich & English, 2005)

The **sociology of childhood** is a discipline within the field of childhood studies and explores the wide variety of childhoods across the world. In social research into children, the influence of the sociology of childhood is extensive (Kampmann, 2003) with a focus on children as social actors, the generational order, and on the interdisciplinary study of childhood. The **children as a social actor approach** views children as active members of society from the beginning (James & Prout, 2015) and as active actors who are shaped by the world around them and actively shape and change that world (Tisdall & Punch, 2012). The (new) sociology of childhood emerged in the 1980s and 1990s as a reaction against prevailing views of the child in developmental psychology and traditional socialization theory (James *et al.*, 1998). It sought to give "conceptual autonomy" (Thorne, 1987) to children and see them in their own right without reference to adulthood (Leonard, 2016), and perceived them as social actors and holders of rights rather than as passive and dependent in the private family (Qvortrup, 1994). James and Prout (1990) defined *children* as social actors who are "active in the construction of their own lives, the lives of those around them and of the societies in which they live." Therefore, the sociology of childhood has begun to shift from children being objects of adult work, to being competent and contributing social actors (Mayall, 2000) such as child soldiers (Rosen, 2007), child prostitutes (Montgomery, 2009), and street children (Hecht, 1998).

The **generational order approach** focuses on age cohorts and allows us to recognize that child–adult relations take place between groups of people subject to differing constellations of social, historical, and political ideas (Mayall, 2000). *A generation*—groups of people born over a fifteen- to twenty-year span—has a sense of belonging and shares similar assumptions

(Corsten, 1999). For example, the trend in opinion on legalizing marijuana highlights how societal forces can shift attitudes and how people may be influenced by those forces at different age cohorts (Pew Research Center, 2015a). In 2015, 68 percent of Millennials or Generation Y (born in 1981 and 1996) favored the legal use of marijuana, compared with Generation X (born in 1965 and 1980) (52 percent), boomers (born in 1946 and 1964) (50 percent), and the Silent Generation (born in 1926 and 1945) (29 percent).

Within the generational order approach *childhood* is defined as *inferior* to adulthood. Children learn to view adults as superior and to accept this superiority (Leonard, 2016). A case in point is schooling, where children's days are largely controlled by adult agendas (Mayall, 2000). Within school, children learn to accept the authority of adults and this ensures the smooth functioning of present and future society (Parsons, 1954).

Children's Rights

Given that children emerge both as competent actors and as heavily controlled and subordinated, the rights of children becomes not only of crucial importance to the quality of childhood, but also problematic (Mayall, 2000). The appearance and expansion of the sociology of childhood was simultaneous with the growing worldwide interest in children's human rights (Freeman, 1998). Traditional socialization theory recognizes society as a powerful determinant of individual behavior (Qvortrup, 1994) and legitimizes the exertion of adult power over children (Mayall, 2002). Like groups of adults such as women, colored adults and disabled adults, children have been denied rights on the grounds that they do not possess the competence and rationality that is necessary to have and exercise rights (Naughton *et al.*, 2007). Particularly in educational settings, research into children's rights can be attributed to children both as full-status humans in a sociopolitically contextual present, and as continually growing and changing, immature and dependent humans (Quennerstedt & Quennerstedt, 2013). Globally, millions of children are subjected to exploitation, including deprivation of education, child labor, forced military service, or imprisonment in institutions or detention centers where they endure poor conditions and violence (Human Rights Watch, 2018).

The United Nations Convention on the Rights of the Child (UNCRC) (1989) defines a *child* as "any human being below the age of eighteen years" and outlines the civil, political, economic, social, health, and cultural rights of children. Education is established as a human right in UNCRC and in other human rights instruments (UN, 1948). Children's rights are the human rights of children with attention to *child protection*—preventing and responding to violence, child abuse, commercial sexual exploitation, trafficking, child labor, and harmful traditional practices, such as female genital mutilation/cutting and child marriage (UNICEF, 2006).

The **multidisciplinary approach** makes the interdisciplinary study of childhood a range of ways of thinking spatially about children's lives (Seymour *et al.*, 2015) and makes connections with related disciplines such as developmental psychology, biology, social policy, anthropology, history, education, and health (Gabriel, 2017). For example, Prout (2005) examined how childhoods are constructed socially and materially through toys, food, and medicines. Lee and Motzkau (2011) have analyzed "life, resource, and voice" as three "multiplicities" in which the "entanglements" between children and technologies have become ever-more important to the regulation of children's lives. Research also has examined how childhoods are produced and experienced through complex intersections of emotion, affect, embodiment, and materiality (Kraftl, 2015).

Parenting a Child from Infanthood through Early Childhood

Childhood is the stage of life that ranges from birth to age twelve, including the development stages of

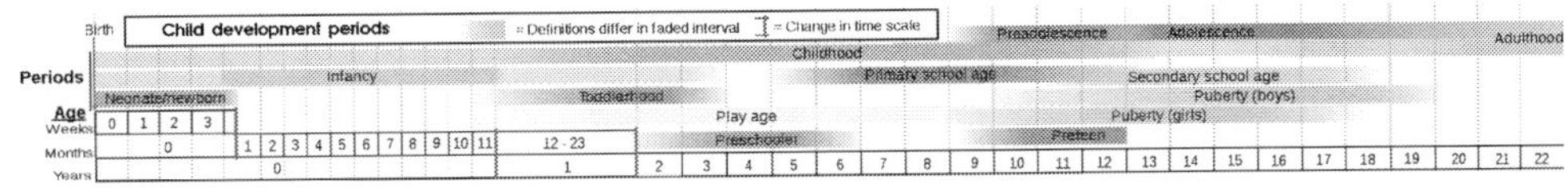

Child Development Stages Approximately Outlining Development Periods in Postnatal Human Development until Adulthood. *Source*: Mikael Häggström. https://commons.wikimedia.org/wiki/File:Child_development_stages.svg

early childhood (birth to eight years) (Farrell *et al.*, 2015), and preadolescence (ages nine to twelve). Some age-related development periods of early childhood include newborn (ages zero to five weeks), infant (ages five weeks to one year), toddler (ages one to three), preschooler (ages three to five), and middle childhood (ages six to eight). Let's look at some ways in which parents can have an influence on their children and adolescents (ages twelve to eighteen) (Kail, 2011).

Parenting an Infant

Early childhood begins with the infancy stage. **Newborn** is an infant who is only hours, days, or up to five weeks old. **Infant** is a child between the ages of one month and twelve months. At birth, babies are born in total dependency on their mother or carers (Gabriel, 2017). A healthy infant can breathe without the assistance of the mother but is completely dependent upon parents or caretakers for survival. Food given to children during infancy has a major impact on their health, development, and growth. For example, parents' choice criteria for infant food brands were significantly correlated to brand familiarity, satisfaction, and loyalty (Romána *et al.*, 2018). An infant is absorbing everything as the brain grows. Parents can stimulate cognitive development with all kinds of brain-boosting activities geared to helping babies develop intellectually. Sun (2011) found no Asian American advantage in cognitive skills at nine months but noted that East Asian American (Chinese, Japanese, and Korean) children made greater gains between nine months and two years of age than White children. Indeed, toys, games, interactive activities, and reading a picture book to an infant are activities for intellectual development. Parenting an infant is exciting but challenging.

Parental Influence on an Infant's Social Intelligence

The building blocks of *social intelligence* are present very early in life (Lau, 2016). Babies are born with a natural tendency to look at human faces, particularly their mother's (Sorce *et al.*, 1985). Infants cry as a form of basic instinctive communication (Chicot, 2015). At this stage, children learn through observing and communicating with parents or caregivers. At infancy and early childhood, if parents are caring and attentive towards their children, those children will be more prone to secure attachment (Aronoff, 2012). A crying infant may try to express feelings such as hunger, discomfort, or loneliness. Infants with mothers who have difficulty responding appropriately to their mental states may feel less recognized by their mothers, predicting high infant disorganized attachment behavior (Bigelow *et al.*, 2018).

Attachment Parenting Sociologists of childhood integrate the biological and social aspects of early children's development and discuss **attachment theory**—attempting to describe the dynamics of long-term and short-term interpersonal relationships between humans (Gabriel, 2017). The infant or child seeks proximity to a specified attachment figure in situations of alarm or distress for the purpose of survival (Tronick *et al.*, 1992). Attachment behavior anticipates a response by the attachment figure which will remove threat or discomfort (Bowlby, 1960).

American pediatrician William Sears (1993) developed **attachment parenting**—a parenting philosophy that promotes the attachment of mother and child by parental responsiveness and closeness. Parental responsiveness is theorized to foster a secure parent–child attachment relationship and keep emotions and stress reactivity at manageable levels, which in turn allows the child to explore the environment and engage in learning activities (Bowlby, 1982). For example, attachment parenting advocated extended breastfeeding and responsiveness to infants. Indeed, breastfeeding supports infants' immunologic, nutritional, physical, and cognitive development (Gartner & Eidelman, 2005). Mothers are encouraged to breastfeed and teach their infants because eating and teaching are both sources of happiness. In the United States, the proportion of infants who continued to be breastfed at twelve months increased from 16 percent in 2000 to 31 percent in 2014 (Child Trends Databank, 2016). Breastfeeding varied by race. Figure 11.11 shows that infants born in 2013 to Black mothers (66 percent) were less likely than infants born to White (84 percent), Hispanic (83 percent), Asian (84 percent), or American Indian mothers (68 percent) to ever be breastfed.

Forming strong attachments with caregivers is necessary to healthy child development. **Primary attachment** refers to an attachment between a principal or primary caregiver and a child. Any caregiver is likely to become the principal attachment figure if they provide most of the childcare and related social interaction (Holmes, 1993). The quality of the infant–parent attachment is a powerful predictor of a child's later social and emotional outcome (Benoit, 2004). **Secondary attachment** refers to an attachment between a child and a non-primary caregiver, such as a childcare provider, teacher, grandparent, sibling, or other adult (Mayseless, 2005). Babies gradually develop secondary attachments to other people around them—siblings, grandparents, other relatives and friends, teachers, or classmates (Gabriel, 2017). Children with multiple caregivers grow up not only feeling secure, but also develop "more enhanced capacities to view the world from multiple perspectives" (Hrdy, 2009).

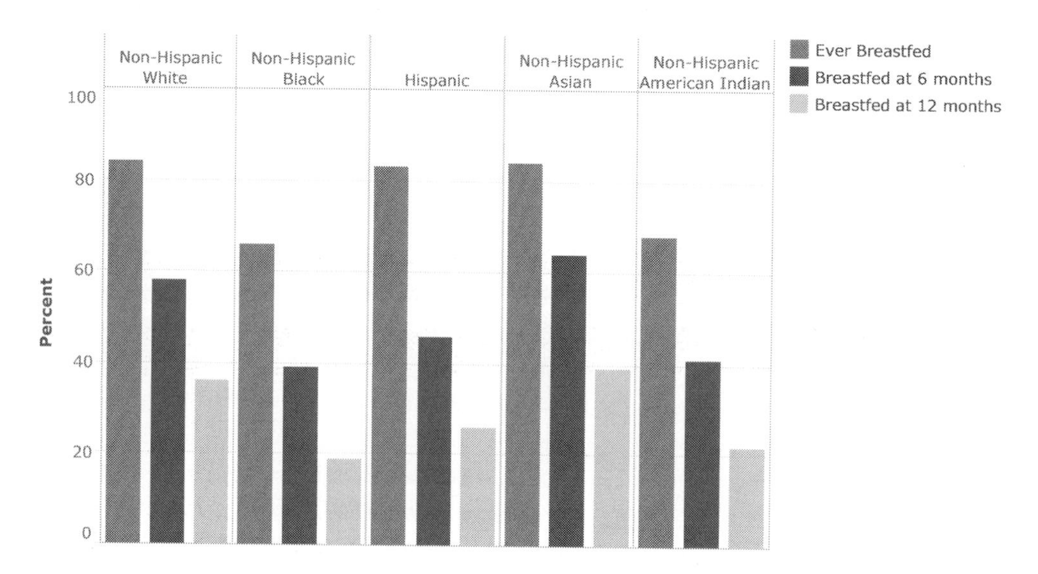

Figure 11.11 Percentage of Mothers Breastfeeding Their Infants by Race and Hispanic Origin, 2013>

Source: U.S. Department of Health and Human Services. Breastfeeding among U.S. Children Born 2001–2013, CDC National Immunization Survey. https://www.childtrends.org/indicators/breastfeeding/

Anxiety, fear, illness, and fatigue will cause a child to increase attachment behaviors (Karen, 1998). Eight-week-old infants smile, babble, and cry to attract the attention of potential caregivers (Prior & Glaser, 2006). During the first six months of life, the caregiver who responded promptly and warmly most of the time to the baby's cries will have created secure, organized attachment with all of the associated benefits (Benoit, 2004). A two- to six-month-old infant discriminates between familiar and unfamiliar adults, becoming more responsive toward the caregiver. The infant's behavior toward the caregiver becomes organized on a goal-directed basis to achieve the conditions that make it feel secure (Prior & Glaser, 2006). By the end of the first year, the infant is able to display a range of attachment behaviors designed to maintain proximity. These manifest as protesting the caregiver's departure, greeting the caregiver's return, and clinging when frightened (Karen, 1998). By the end of the first year, they can reliably discriminate between different facial expressions and use the different emotional expressions of their parents to guide behavior (Sorce *et al.*, 1985).

Touch Research indicates the importance of touch during infancy, including for social, cognitive, and physical development (Greenough, 1990). Babies are helpless at birth and a nurturing touch helps meet a baby's intense need for physical contact and secure attachment. **Touch** is the action of an object on the skin and the registration of information by the sensory systems of the skin (i.e., feeling) and is capable of communicating emotions and specific information (Hertenstein, 2002).

Touch is an important modality through which infants and mothers communicate (Moszkowski & Stack, 2007) and generates positive emotional displays, as well as modulates negative ones (Stack & Muir, 1992) through the specific movements such as rubbing, holding, or squeezing. These movements provide three types of stimuli to the brain: cutaneous (stimulation of the surface receptors of the skin), proprioceptive (stimulation of end-organs in muscles, tendons, and joints), and vestibular (stimulation of semicircular canals due to spatial displacements) (Weiss *et al.*, 2001). The four-month-old infant who receives touch along with other stimuli shows more smiling and vocalizing and less crying (Pelaez-Nogueras *et al.*, 1997). Infants who have been the recipients of positive touch experience benefit more as they develop emotionally and socially (Field, 2002). Negative touch such as harsh touch has been associated with later emotional behavioral problems (Weiss *et al.*, 2001).

In addition to emotional content, touch may transmit *specific information* to the infant (Tronick, 1995). For example, touch may communicate the presence or absence of a caregiver or the identity of the person who touches the infant (Hertenstein, 2002). Attachment theorists have long regarded the quality of parent–infant physical touch as a central feature of the *responsive* and *available* caregiving environment that is necessary to foster an infant's sense of security (Ainsworth *et al.*, 1978). Different types of touch such as maternal stroking, tickling, kissing, and pinching can convey "very specific messages to the infant, as specific messages as facial expressions" (Tronick, 1995). Infants gain meaning from touch through direct perception, learning, and cognitive processes. The **theory of perception** holds that organisms directly perceive meaning from stimulus arrays (Gibson, 1979) and **learning explanation** suggests that the infant makes associations between touch and environmental events to impute meaning on tactile patterns. Touch is a vital means through which infants can self-regulate and explore their surroundings (Moszkowski & Stack, 2007). For example, infants may learn the meaning of specific types of touch by observing how others are touched and how they behave in relation to the touch they receive (Bandura, 1965). **Memory processes** can influence how infants gain meaning from the touch that their caregivers administer. Infants may encode and store in their memory a particular meaning of a touch in the context of receiving the tactile stimulus from their caregiver (Hertenstein, 2002).

Parental Influence on an Infant's Emotional Intelligence

Emotional development includes expressions, attachment, and personality (Doherty & Hughes, 2009). During the infancy stage, a strong emotional bond is created between the child and their parents or care providers. Parent–child relationships are the first context in which children learn about emotions and serve as a rehearsal stage for children's developing emotional skills (Mirabile, 2010). These parent–baby interactions during the first twelve months of infant life are crucial for their emotional development. For example, when parents are singing and making faces, they are communicating with their infants. Through this ongoing interaction, a parent gets to know when her baby is angry or happy. At the same time, the baby feels that someone is caring for him or her. The most important thing during the first year is building **trust** (i.e., believing that the person who is trusted will do what is expected) that is the critical foundation for emotional well-being throughout a person's life (Schiller, 2005). A young child needs a sensitive and responsive mother (or caretaker) for the development of emotional security (Bowlby, 2005). **Emotional investment**—having emotionally bonded with an infant—is an integral part of early years practice because the work involves strong feelings towards young children and their families (Dahlberg & Moss, 2005).

Parenting a Toddler

Toddlerhood starts at around twelve months when toddlers begin learning how to walk and talk and ends at 36 months (Barker, 2001). As they grow into toddlers, they begin to grasp that other people have minds and feelings which are separate from what they themselves think, feel, and know (Buttelmann *et al.*, 2009).

Parental Influence on a Toddler's Emotional Intelligence

Any parent who has followed an active toddler around for a day knows that a child of this age is a whirlwind of explosive, contradictory, and ever-changing emotions (Lieberman, 1995). Toddlers tend to have *temper tantrums* (i.e., emotional outburst) because they have such strong emotions but do not know how to express themselves the way that older children and adults do (Kuersten-Hogan & McHale, 1998). A toddler might use different ways to release their toddler energy such as playing with a toy and whacking their parent with it. The first step in helping toddlers develop emotional intelligence is to help them learn to manage their emotions, to develop empathy for others, and to express their needs and feelings without attacking others (Markham, 2018).

Self-regulation **Self-regulation** refers to the ability to control or redirect one's disruptive emotions and behaviors in response to a particular situation (Goleman, 2005). For example, one of the reasons children fight is that they want the same toy. Instead of forcing children to share, parents can use self-regulation to allow the child who has the toy to play and decide how long he or she needs it. When children wait their turns, or share, they are practicing self-regulation. Children become more generous by having the experience of giving to others and learning how good that feels (Eisenberg, 2009).

An important dimension of self-regulation is **effortful control**—"the ability to inhibit a dominant response to perform a subdominant response" (Rothbart & Bates, 2006). Measures of effortful control involve *attentional control* (e.g., the ability to voluntarily focus attention as needed) and/or *behavioral control* (e.g., the ability to inhibit behavior upon command, called inhibitory control) (Apps *et al.*, 2016) as needed to adapt when the child does not want to do so (Posner & Rothbart, 2007). Attentional control shows modest advances in the early months of life and at eighteen months, becomes considerably better in the third year of life (Ruff & Rothbart, 1996), and continues to develop in childhood (Murphy *et al.*, 1999) and into adulthood (Williams *et al.*, 1999). Although infants are very limited in voluntary behavioral control, improvement in inhibitory control is seen in the first year (Putnam & Stifter, 2002) and between 22 and 33 months and at about 44 months (Reed *et al.*, 1984), and considerably in the third year of life (Kochanska, 2000). The toddler and preschool years are a time in which temperamentally based effortful control emerges rapidly and provides the basis for the emergence of self-regulation (Eisenberg, 2012). The abilities to focus attention when there are distractions, sit still in church or class, and force oneself to do an unpleasant task are aspects of effortful control that underlie the emergence of self-regulation, a major milestone in children's development (Kopp & Neufeld, 2003). **Self-regulated learning** is aligned closely with educational aims (Panadero, 2017) and is critical because it affects the quality of children's social interactions and their capacity for learning (Eisenberg, 2012). Children with strong self-regulatory skills (e.g., attentional skills involved in effortful control) are better prepared for school and have stronger social and academic outcomes than their peers who struggle to master these abilities (National Center for Education Statistics, 2001).

Individual differences in effortful control are associated with *heredity* and the quality of mother–child interactions (Eisenberg, 2012). Twin studies confirm that there is a genetic basis to effortful control (Goldsmith *et al.*, 2008). Research indicates that *the anterior cingulate cortex* (ACC) of the brain is associated with response selection and avoidance learning (van Veen & Carter, 2002), decision-making (Mulert *et al.*, 2008), emotion (Jackson *et al.*, 2006), and persistent antisocial behavior (Aharoni *et al.*, 2013). The quality of parenting has also been associated with individual differences in effortful control. A study of toddlers who are required to switch attention and inhibit behavior accordingly revealed that children showed significant improvement in performance by 30 months of age and performed with high accuracy by 36 to 38 months of age (Posner & Rothbart, 1998). In general, young children's self-regulation (behaviors that reflect effortful control) has been positively associated with maternal support and sensitivity, and negatively related to a directive and controlling caregiving style (Eisenberg *et al.*, 2011).

Parent self-regulatory competencies influence the quality of parent–child interactions and parenting behavior. For example, a depressed mother's inability to model effective self-regulatory strategies and to provide regulatory assistance may increase the stressful nature of mother–infant interaction and fail to provide the infant with effective coping strategies in distressing situations (Ashman & Dawsons, 2002).

Parental Influence on a Toddler's Social Intelligence

Infants and toddlers experience life more holistically than any other age group (Grotewell & Burton, 2008). The toddler years are a time of great cognitive, emotional, and social development. Social development includes the ability to interact with the world through playing with others. Indeed, eighteen-month-old infants are expected to share the toy they worked together to attain, and to possess at least a basic understanding of cooperation well before their second birthday (Wang & Henderson, 2018).

Parenting a Preschool Child

Preschool is the stage of childhood that ranges from three to five years old. The human brain undergoes its most constant and dramatic transformation during the preschool years. A range of early cognitive skills are important to children's school readiness and later academic success, including language, emergent literacy, and emergent math skills (Duncan *et al.*, 2007). Family income and socioeconomic status (SES), childcare quality, and aspects of parenting during early childhood have each been shown to be positively related to early development and school readiness (Denham *et al.*, 2012).

Preschooler's Intellectual Development

From birth to age five, it is not what children learn but how they learn that counts (Perlmutter & Colman, 2008). Curiosity plays a big part in preschoolers' lives and when young children ask "why" questions, they are motivated by a desire for explanation (Society for Research in Child Development, 2009). Studying of preschoolers aged two to four and three to five showed that young children were motivated to seek causal information actively and used specific conversational strategies to obtain it (Frazier *et al.*, 2009). When preschoolers got an explanation, they seemed satisfied but when they got answers that were not explanations, they would repeat their original question or provide an alternative explanation (Frazier *et al.*, 2009). Attachments are usually formed when children ask questions and parents listen to their concerns and respond to them thoughtfully. The home environment and center-based childcare can provide children with intellectual ability to do well in school.

The home environment has a greater impact on future outcomes than preschool (Scottish Government, 2008). Children's early language skills include a variety of expressive and receptive abilities such as understanding and responding appropriately to different word and sentence structures and breadth of vocabulary (Chapman, 2000), and have been found to predict later reading achievement (Muter *et al.*, 2004). Parental responsiveness during early childhood is found to play an important role in early cognitive development (Landry *et al.*, 2001), in language development (Pungello *et al.*, 2009), in emergent literacy and math (Hirsh-Pasek & Burchinal, 2006), and in early literacy and early math. By age four, East Asian American (Chinese, Japanese, and Korean) children outperformed peers from all other racial/ethnic groups in math and literacy by a large margin (Sun, 2011). The advantages of East Asian children were explained by *parenting resources* such as socioeconomic status, parent–child communication (e.g., reading books together, telling stories), and maternal care (e.g., spoke to the child spontaneously twice or more and kept the child in view). *Parental language input* (i.e., the quantity of language spoken to the child) is theorized to facilitate children's development of cognitive reasoning skills needed to learn about reading, math, and emotions (Snow, 1991). Investigators of young children's linguistic and cognitive development emphasized the importance of *inferential language input* (e.g., talk about the past or future or hypothetical or abstract situations) as children need to develop in order to support inferential reasoning skills needed to learn school readiness concepts (van Kleeck, 2008). Longitudinal analyses revealed that for children with stronger initial language skills, higher levels of parental inferential language input facilitated greater vocabulary development (Merz *et al.*, 2015). Parents with higher education and wages allocate more time to direct childcare (Welsch & Zimmer, 2008). Working women with a college degree spend about 70 percent more time caring for their children than working women with less than a high school degree; fathers with a college degree spend more than double the time on childcare than high school dropout fathers (Guryan *et al.*, 2008).

Quality childcare for young children provides a foundation for future academic success. Working parents are in need of *center-based childcare*—the caring for and supervision of a child (preschool) in a group setting. Childcare providers include licensed childcare centers, preschools and nursery schools, summer camps, etc. In the United States, 13 percent of children from birth to one year of age regularly attended a center-based program and this rose to 66 percent for four-year-olds; graduate and professional families pay more than twice that of parents with less than a high school diploma; more affluent families spend more on center-based programs (Whitehurst, 2018). Economically disadvantaged families send their children to a federally funded Head Start program and their children have shown especially strong improvements after attending preschool (Child Trends Databank, 2018). During early childhood, children exhibit variation in their development of early cognitive and social-emotional skills, which set the stage for individual differences in school readiness as well as later academic performance (Duncan *et al.*, 2007).

Emotional Competence during Preschool Years

When students struggle in academics or have a difficult time managing study habits, children and their families can often take a hit with their emotional regulation (Olivo, 2016). The preschool period is marked by a number of cognitive, motor, and language achievements that coincide with greater expectations from parents for autonomous regulation of emotion and behavior (Kopp & Neufeld, 2003). The acquisition of emotion knowledge during the preschool years contributes to adaptive social-emotional functioning in a school setting (Denham *et al.*, 2003). During the preschool period, emotional competence involves the ability to recognize and understand one's emotions and the emotions of others as well as the ability to regulate, express, and use one's emotions in socially appropriate, adaptive ways (Saarni *et al.*, 2006). Let's discuss three components of emotional competence: emotional expression, emotion regulation, and emotional understanding.

Emotional Expression **Emotional expression** refers to the frequency, intensity, latency, and duration of expression of both negative (e.g., anger, fear, sadness) and positive (e.g., happiness, contentment) emotions (Dougherty, 2006). The preschool period is marked by children's tendency to intensively express emotions such as happiness, sadness, anger, fear, surprise, and interest (Denham, 2007). Higher parent–child attachment security has been associated with more advanced emotion knowledge in the preschool years (Raikes & Thompson, 2006). The ways to build emotional intelligence with the child include helping the child recognize and express their own emotions.

One way to classify emotions is whether they are *positive* (i.e., expression of favorable feelings) or *negative* (i.e., expression of unfavorable feelings) *emotions* (Watson *et al.*, 1999). Parental love is the positive emotion that a child can feel when a parent hugs his or her child. Consider another example of children gathered for a birthday party, with one child opening gifts as the guests observe. The child opening gifts understands that certain emotional expressions (e.g., smiling or laughing) convey their pleasure or excitement (Mirabile, 2010). It is important to have a positive emotional experience as mental efficiency increases, memory becomes more acute, our understanding increases, and we make better decisions (Robbins *et al.*, 2009). Children who are able to read the emotional expressions of others accurately and predict a likely emotional reaction in a given social situation can use this information to negotiate interpersonal relationships with teachers and peers (Izard *et al.*, 2001). Negative emotions affect health. For example, losing a loved one can trigger intense feelings of grief that can lead to depression or make underlying depression worse (Healthline, 2017). Children who are able to adaptively regulate negative emotions and balance their expression of positive and negative emotions can maintain social relationships better (Fabes & Eisenberg, 1992) and are viewed as friendlier, less aggressive, and less sad by their teachers (Denham & Burger, 1991). Children's ability to display positive emotions and modulate the expression of negative emotions is critical for building and maintaining relationships with peers (Fabes *et al.*, 1999).

Gender differences in children's emotion expression have been observed in preschool years, with girls being less likely to show anger, and more likely to show sadness than boys (Brody, 1999). *Submissive emotions* (e.g., sadness and anxiety) do not threaten interpersonal interaction in most cases, whereas *disharmonious emotions* (e.g., anger, laughter), at the expense of another, have the potential to threaten relationships or interactions (Barrett & Campos, 1987). A study of 60 preschool children and their mothers and fathers revealed that girls expressed more submissive emotion than boys and fathers attended more to girls' submissive emotions than to boys' and attended more to boys' disharmonious emotions than to girls' (Chaplin *et al.*, 2005).

Emotion Regulation **Emotion regulation** is the ability to effectively manage and respond to an emotional experience and it involves changes in valence, intensity, and timing of emotions that occur as a result of intra-individual or inter-individual processes (Thompson, 1994). Emotionally well-regulated children are able to modulate their emotional experience and expression to fit contextual demands and their own goals (Grolnick *et al.*, 1996). Consider a situation in which the birthday-child is disappointed with a certain gift, yet keeps that disappointment from showing. This case illustrates the overlap between children's expression and regulation of their emotions (Mirabile, 2010). During stressful interactions, such as conflict with peers, some preschool-aged children often find it difficult to regulate their emotions and behaviors (Denham *et al.*, 2003). Higher levels of parental responsiveness may lead to more shared experiences around emotion regulation which give the child opportunities to learn about emotion in a supportive context (Raikes & Thompson, 2006). For example, parents who respond to children's emotion and encourage them to express emotions will promote children's effective emotion regulation (Gottman *et al.*, 1997).

Parents can teach effective regulatory strategies in helping their children reduce emotional distress. *Ineffective strategies* include *avoidance* because children do not directly cope with their emotions (Krohne *et al.*, 2002), *strict attentional focusing* in the absence of information gathering because children increase their attention to a distressing stimulus while failing to address or change the distressing situation (Gilliom *et al.*, 2002), *aggression and venting* because such strategies fail to reduce and may amplify children's distress (Eisenberg *et al.*, 1994), and *emotional suppression* because it leads to increases in physiological arousal, strains interpersonal relations, and impairs memory (Gross, 2002). Emotionally competent children know *effective strategies* (e.g., self-soothing and comfort and support-seeking behaviors) to regulate emotional expression (Eisenberg *et al.*, 1998), redirect their focus away from the source of their distress, and reduce their negative emotionality (Calkins *et al.*, 1999). Children's information-gathering behavior, such as finding out how long they must wait before they can have a desired object, effectively reduces distress because children understand the parameters associated with an event (Gilliom *et al.*, 2002). As children become proficient in their use of these strategies, they are expected to transition to advanced strategies such as *cognitive restructuring* (i.e., technique to replace irrational thoughts that trigger anxiety) (Murphy *et al.*, 1999).

Emotional Understanding **Emotional understanding** is the ability to recognize and name emotions and understand the causes and consequences of emotions (Saarni *et al.*, 2006) that facilitate social interaction, especially with peers and caregivers (Halberstadt *et al.*, 2001). **Self-awareness** is the ability to recognize and understand personal moods, emotions, and drives. Little kids experience a mysterious flood of feelings but they do not understand why they feel that way (Goleman, 2005). Parents can explain what a child is feeling and what that feeling is called. Emotion understanding abilities increase from ages three to six (Fabes *et al.*, 1991) when children can use these skills to recognize their peers' emotions and discrepancies between what their peers say and how they truly feel (Miller *et al.*, 2005).

Parents can socialize emotional understanding *directly*. Eisenberg *et al.* (1998) identified three primary ways in which parents socialize their children's emotional development: emotion discussion, parent reactions to children's emotion, and parent expressiveness. *Emotion discussions* (e.g., labeling the emotion and describing the causes and consequences of emotions) occurred during or following emotion-laden situations and provided parents with opportunities to teach children about emotions (Brown & Dunn, 1992), and contributed to children's awareness and understanding of emotion (Denham *et al.*, 1992). *Parent reactions* to children's negative emotions allow children to understand and explore the emotions surrounding them, to learn how to regulate emotions and cope with troublesome situations, and in turn underlie emergent capacities for emotion regulation and social competence (Denham, 2007). Promoting Alternative Thinking Strategies (PATHS) for Preschool is a preschool-based program that aims to increase social competence, emotional knowledge, and problem-solving skills in young children through direct instruction (Domitrovich, 2017). Teachers and parents reported children exposed to PATHS as more socially competent than their peers (Morris *et al.*, 2014).

Parents can socialize their children's emotional competence *indirectly*. Emotionally competent parents can *model* adaptive emotional expression during their interactions with their children. Children are likely to learn about emotions by watching how parents handle their own emotions (Denham, 2007). Some activities that can nurture emotional intelligence include role-playing games, playing with puzzles and blocks, coloring with crayons, and learning songs (Perlmutter & Colman, 2008). When children play out their perceptions of a particular role, they learn to appreciate what others must be feeling based on their behavior.

Parental Influence on a Preschooler's Social Intelligence

Children begin kindergarten at this age to start their social lives but some may find it difficult to connect with others. **Social skill** is social competence facilitating interaction and communication with others and managing relationships to move people in the desired direction. To raise a child's social intelligence, Baron-Cohen (2000) suggested that parents teach the social skills about how to join a group, how to choose friends, and how to handle angry feelings. The preschool period involves the use of negotiation and bargaining (Waters *et al.*, 1991). For example, four-year-olds are not distressed by separation if they and their caregiver have already negotiated a shared plan for the separation and reunion (Marvin & Britner, 2008). Throughout childhood, children are increasingly able to use their understanding of others to shape their own behavior in simple acts such as helping others when they are in distress (Staub, 1979) or more complex decisions such as choosing friends who share their values, rules, and characteristics (Bigelow & La Gaipa, 1980). Preschoolers are generally interested in other kids, able to verbalize their feelings, and are quick to adopt social norms (Markham, 2018).

Play Play is essential to the cognitive, physical, social, and emotional well-being of children (Ginsburg, 2009) and it offers children opportunities for physical (running, jumping, climbing, etc.), intellectual (general knowledge), social (friendships and conflict resolution), and emotional (happiness, anger, etc.) development. Usually, play activities of parents with their preschool child ages five to seven take a leading place. More parents (65.1 percent) play with children ages four to five, compared with those (59.1 percent) with children ages five to seven; boys are likely to prefer such games as computer games, building and construction, and military games, whereas girls prefer board games, playing in "family" and in various professions (Sobkina & Skobeltsinaa, 2014).

In the family, the involvement of parents in joint play may depend on the child's culture. Parents in some societies think that it is senseless for parents to play with infants, whereas parents in other societies play with babies and see them as interactive partners (Bornstein, 2007). Children are conceptualized as active explorers and playful agents while creating knowledge, skills, and understandings of themselves and their life-worlds (Gurholt & Sanderud, 2016). When play is controlled by adults, children acquiesce to adult rules and concerns and lose some of the benefits that play offers them; undirected play allows children to learn how to work in groups, to share, to negotiate, to resolve conflicts, and to learn self-advocacy skills (Ginsburg, 2007). *Curious play* is a theoretical framework to understand and communicate children's experiences of free play in nature (Gurholt & Sanderud, 20016). Gray (2013) argued that play is how children learn to take control of their lives and learn all they need to know if free to pursue their own interests. Play is recognized by the United Nations Commission on Human Rights as a right of every child.

Parenting a Primary School Child

Middle childhood begins at age five or primary school age and ends at around age eight, as opposed to the time children are generally considered to reach pre-adolescence (Freeman, 2016). Middle childhood is characterized by a child's readiness for school and being able to wait, to follow directions, and get along with other children (Goleman, 2005). The child tries to move toward a degree of independence as attachment behaviors such as clinging decline and autonomy such as self-reliance increases.

Parental Influence on a Primary School Child's School Performance

Schiller (2005) suggested that a sense of *trust* affects a child's intellectual development. For example, if a child does not feel safe in an environment (e.g., home, school), their brain goes into protection mode and then the curiosity that drives learning cannot work to its fullest potential. Two major agents of socialization in children's lives are the family environment and formal educational institutions (Lemish, 2007). The family functions as the primary socializer of the child that predominates in the first five years of life (Cooper, 1971) and continues throughout childhood.

The school may take over from the family as a key site for intellectual skill. Testing is a fundamental aspect of school organization and ability can be developed and strengthened through hard work (Leonard, 2016) and parental involvement. Parental involvement in a child's education includes monitoring the child's progress, establishing an open communication with the school, and helping the child with homework (Juvonen *et al.*, 2004). *Modeling, reinforcement, and direct instructions* are the three procedures in a parent's involvement in their child's education (Hoover-Dempsey & Sandler, 1997). Modeling predicts that children will emulate their parents' behavior. Parents can influence their children's educational outcomes by giving their children praises and rewards related to positive attitudes toward learning and by providing direct instruction. Family involvement practices that provide direct instruction and support are more prevalent in the elementary school years (Gouyon, 2004). Children with involved fathers are more likely to get good grades in school (Flouri & Buchanan, 2004). Many parents turn to tutors or after-school math and reading programs to help their children develop academically.

Gender, class, and ethnicity, among other variables, impact on how childhood is experienced (Leonard, 2016). Some parents were found to overestimate sons' abilities in math and science compared to daughters'. For example, parents tend to enroll their daughters in more cultural activities than their sons (e.g. art classes, dance classes, and musical instrument lessons), and tend to be more invested in school-related parent involvement programs for their sons than their daughters (Raley & Bianchi, 2006). In fact, parents have higher academic expectations for girls than they do for boys, and this gender difference becomes apparent as early as sixth grade (Child Trends Databank, 2015). The academic achievement of children varies across socioeconomic backgrounds. Children from stable middle-class backgrounds have the highest levels of academic achievement, compared with working-class families (Roksa & Potter, 2011). Financial strain in single-mother families has been shown to influence adolescent school performance through its effect on parental emotional health/depressive symptoms and parenting behavior (Jackson *et al.*, 2000). McCallum and Demie (2010) provided evidence of substantial differences between the backgrounds of pupils attending different schools and of a strong relationship between these differences and differences in the GCSE (General Certificate of Secondary Education) performance of schools in an inner London borough. In the United States, the lives of many children today are better than the lives of children 50 years ago but serious racial and ethnic inequities persist (Emig, 2018). Indeed, within school, children learn to accept the unequal distribution of reward and status in wider society (Parsons, 1954). Lareau (1987) explained this unequal distribution of reward and levels of academic achievement by socio-economic status from the three sociological theories. *The culture of poverty theory*—poverty resulting from the transmission of cultural values from generation to generation—explains that working-class families do not value education as highly as middle- and upper-class families. *The institutional theory* argues that schools are more welcoming to middle-class families and subtly discriminate against working-class families. *Cultural capital theory* suggests that schools are largely middle-class institutions with middle-class values.

Parental Influence on a Primary School Child's Emotional Intelligence

Emotional issues such as anxiety make academic struggles all the harder and the effort of coping with learning issues leads to increased stress or sadness. In either case, parents will find it helpful to supplement academic coaching with emotional support (Olivo, 2016). With the parent as the initial external regulator, emotional and behavioral regulation gradually is appropriated by the child (Cassidy, 1994).

Some middle children come to school seeking a home from home, a stable emotional situation in which they can exercise their own emotional lability, a group of which they can gradually become a part (Winnicott, 2000). *Emotional lability* (i.e., sudden strong shifts in emotion) commonly occurs in youth with attention-deficit/hyperactivity disorder (ADHD) (Posner *et al.*, 2014) and affected 3–10 percent of school-age children (Barkley, 2005). Children with high levels of emotional lability have been shown to tolerate frustration poorly, have high levels of irritability, and demonstrate frequent crying spells or tantrums (Sobanski *et al.*, 2010); this is linked to an over-expression of positive emotions such as exuberance and excitability that can be off-putting to peers (Wehmeier *et al.*, 2010). A study of 165 mothers and their third-grade children revealed that middle children's emotional lability was affected by multiple forms of parental socialization such as lack of contempt for children's emotions and lack of emotional suppression of mothers (Rogers *et al.*, 2016).

Emotion Socialization Behaviors Parents with an *emotion-coaching* philosophy are comfortable with emotion in the parent–child relationship and see child emotion as an opportunity for intimacy, teaching, encouragement of expression, and exploration of emotion (Baker *et al.*, 2011). Gottman *et al.* (1998) suggested a five-step "emotion coaching" process that teaches parents how to (1) be aware of a child's emotions, (2) recognize emotional expression as an opportunity for intimacy and teaching, (3) listen empathetically and validate a child's feelings, (4) label emotions in words a child can understand, and (5) help a child come up with an appropriate way to solve a problem. For example, children may be afraid to discuss challenges such as a physical difference, an absent father or mother, a learning disability, etc. Markham (2018) suggested that parents first need to overcome their own discomfort with the issue and learn how to turn tough conversations into learning opportunities and help their children develop the ability to cope with emotional stress. To raise an emotionally intelligent person, parents must recognize the importance of a strong parent–child attachment. Children are most likely to experience healthy social-emotional development when they are secure in their attachment to their parents (Erickson, 2015). For example, when a father plays catch with his child, the father and child are not only playing but also interacting and forming emotional bonds. Having consistent schedules around bedtimes and family mealtimes helps young children learn to manage their energy levels and their emotions (Bernier *et al.*, 2010). Engaging children in positive ways helps build brain connections that lead to emotional intelligence.

Emotion socialization behaviors varied by mothers and fathers. Fathers reported less awareness of emotion and held lower emotion coaching attitudes than did mothers (Gottman *et al.*, 1996); they engaged in more punitive responses and less supportive reactions to child emotion than mothers (McElwain *et al.*, 2007), and reported less optimal emotion-related attitudes and behaviors than did mothers (Baker *et al.*, 2011). Adolescents also reported receiving fewer coaching behaviors from their fathers (Stocker *et al.*, 2007). With regard to emotion coaching, mothers tend to endorse attitudes consonant with coaching behaviors more than fathers (Gottman *et al.*, 1997).

Gender, social class, and culture may impact on how emotional socialization is experienced. Gender differences in positive emotions varied across studies, with some studies reporting greater positive emotions for women and with others for men (Lucas & Gohm, 2000). Parents who respond to a boy's angry outbursts with supportive responses may contribute to more intense angry displays and further contribute to the development of conduct problems (Cole *et al.*, 1994). A study of 48 countries revealed that people desired high levels of happiness and fearlessness for their children and happiness was more desirable for girls than boys (Diener & Lucas, 2004). In this way, women expressed facial actions more frequently than men (McDuff *et al.*, 2017). Also, women experience emotions more intensely, tend to "hold onto" emotions longer than men, display more frequent expressions of both positive and negative emotions, except anger (Brody & Hall, 2000), and are more likely to express these emotions based on the experiences at hand, whereas men are more likely to misuse alcohol to help regulate their emotions (Nolen-Hoeksema, 2012). In terms of social class, low-income children are at elevated risk for emotion-related problems because emotion socialization may differ for low-income versus middle-income families (Chaplin *et al.*, 2010).

Parental Influence on a Primary School Child's Social Intelligence

Of all the skills we encourage our children to develop, social intelligence may be the most essential for predicting a successful life and parents' influence is most profound because their every word, move, and gesture is studied (Lansbury, 2014). In middle childhood, children meet a wider circle of people within school (Leonard, 2016). When children enter school, institutional settings set the tone for the relationships between individual children, *dyads* (child–caregiver, peer–peer friendships and playmates), and *group interactions* between young children and their caregivers (Gabriel, 2017). They gain social skills, make new friends, and learn how to become independent. When young children spend time together over a long period, as they do in preschool, they develop **social capital**—the networks of relationships among people who live and work in a particular society—in their own peer cultures (Gabriel, 2017).

Parental Influence on a Primary School Child's Financial Intelligence

Parents can teach children financial intelligence and build their confidence along the way. "It is not so much a matter of what you do as a parent; it's who you are" (Levitt & Dubner, 2005). Secure parental employment is a major factor in the financial well-being of children as it is associated with higher family income and access to wealth. It has also been linked to a number of positive outcomes for children, including better health, education, and social/emotional development (Cauthen, 2002). To build financial intelligence in their children, parents may try to model positive spending behaviors and give practical activities to teach kids basic money skills and management (Burkett & Osborne, 1996). In a growingly complex economic environment, children today need to develop financial intelligence more than ever and the best way for them to do this is through the family (Yu & Zhang, 2015). For example, many children witness financial transactions when they go shopping or banking with their parents. As first graders to fourth graders, children are exposed to everyday finance as they have seen money such as cash or credit cards. One of the ways to teach children financial literacy is by giving them an allowance. Parents can teach children how to save money and pay for things out of their allowance. Gallo and Gallo (2005) suggest that parents can help raise financially responsible children through parenting behaviors such as teaching financial literacy and a work ethic.

Parenting a Preteen and Adolescent

Preadolescence is the period from nine to twelve years (Corsaro, 2005)—the stage of child development following early childhood and preceding adolescence. Biologically, puberty begins in the preadolescence years (Appelbaum, 2016). *Adolescence* or teenage is a transitional stage of physical and psychological development between the ages of thirteen and seventeen. During adolescence, boys' voices change and girls begin menstruating. Both become more sexually aware beings and begin to become more independent. Adolescence is usually associated with a growth spurt, but physical growth (particularly in males) and cognitive development can extend into the early twenties (Dorn & Biro, 2011).

Parental Influence on an Adolescent's Academic Success

Parents have a strong impact on their teens' educational attainment. However, parents tend to become less involved in their children's educational careers over time, especially after elementary school, and they might feel they have grown enough to start handling schoolwork on their own (Crosnoe, 2001). For example, for children at age nine, about 75 percent of American parents report high or moderate involvement in school-related activities, but when children reach age fourteen the rate of parent involvement has dropped to 55 percent. The rate continues to drop throughout high school (U.S. Department of Education, 2003). In France, homework practice declines steadily with age (Gouyon, 2004). At the same time, *less* instrumental forms of involvement (monitoring school performance and progress, discussing plans for higher education, and maintaining high expectations) become more *common* in adolescence (Catsambis & Garland, 1997).

Parental Involvement in Education Epstein and Dauber (1991) distinguished six types of involvement: basic obligations at home (e.g., the provision of school supplies, general support and supervision at home); school-to-home and home-to-school communications; assistance at the school (e.g., volunteering); assistance in learning activities at home; involvement in school decision-making, governance, and advocacy; and collaboration and exchange with community organizations. When adolescents feel that parents care about them, they are more likely to do well academically and less likely to get involved in risky behaviors, including substance use, early sexual activity, and suicide attempts (McNeely & Falci, 2004).

Sociologists differentiate between *home-based activities* (e.g., helping children with homework, discussing their children's experiences at school) and *school-based activities* (e.g., participation in school activities) (Sui-Chu & Willms, 1996). Research indicates that parental involvement contributed to social class variation in students' experiences and maintained inequality beyond the K-12 education. A study of 41 families reveals that affluent parents serve as a "college concierge," using class resources to provide youth with academic, social, and career support and access to exclusive university infrastructure, whereas less affluent parents describe themselves as "outsiders" who are unable to help their offspring and unresponsive to their needs (Hamilton *et al.*, 2018). Socioeconomic explanations of the Asian American achievement advantage (e.g., GPA, high school completion rates, college entrance exams, and enrollment rates at prestigious universities) attribute the pattern to better-educated parents with more resources to invest in their children's education (Sakamoto *et al.*, 2009; Lee & Zhou, 2014). Research indicates that women earn better grades than men across levels of education (Quadlin, 2018).

Parental Influence on an Adolescent's Social Intelligence

The teenage years are a crucial time for the development of social intelligence (Lau, 2016). The transition from childhood to adolescence engenders changes in the individual, social context, and social norms that serve to elevate the importance of peers (Brown & Larson, 2009). Teens go to secondary school and spend more time with friends from different backgrounds and across social media networks.

Teen development is about separating from parents and gaining peer acceptance. **Peer relations**—associations among equals—become more salient in adolescence and encompass a wide variety of affiliations, including romantic relationships, sexually based interests and activities, etc. (Brown & Larson, 2009). In fact, teens can be ranked in terms of status as well as power or prestige, differed substantially on various emotional and behavioral outcomes (Cillessen & Mayeux, 2004), and must negotiate peer relationships and issues on a broader set of levels than they did in childhood (Brown & Larson, 2009). Openness to and success in romantic relationships are contingent on experiences within the friendship group (Connolly *et al.*, 2000). Teens with social intelligence are better adjusted than those with poor social skills (Brown & Larson, 2009). In fact, most adolescents named a peer from their own racial group as a best friend (Kao & Joyner, 2004). Socially withdrawn and victimized youth may pursue friendships with a peer who is socially skilled enough to avoid victimization (Guroglu *et al.*, 2007).

Preadolescents may use socially interactive technologies (SITs), cell phones/text messaging, and instant messaging (IM) as a learning resource (Lemish, 2007). Online usage does not impact on social intelligence levels during adolescence but can be linked to *loneliness* in later life (Lau, 2016). If online rejection occurs, such as "de-friending," symptoms of depression may be the result (Geddes, 2018). Research suggests that positive relationships between teens and their parents are important socially and emotionally, especially when young teens go through developmental changes and settle into high school (Juvonen *et al.*, 2004). The adolescents with stable *secure peer attachment* reported lower overall levels of depression and loneliness (Holt *et al.*, 2018). Those who transitioned from lower to higher *parental attachment* showed significant declines in loneliness (Holt *et al.*, 2018).

Today, employers value social intelligence more than intellectual quotience. A combination of school and out-of-school activities can improve a young person's social intelligence levels, with parents, schools. and extracurricular activities playing an important role (Lau, 2016). These positive social exchanges play an important role in maintaining good adult mental well-being (Coyne & Downey, 1991). Sleigh and Ritzer (2004) suggest that applicants' resumes reflect their social intelligence such as demonstrating with examples their ability to work in groups and maintain interpersonal relationships, and resolve conflicts.

Parental Influence on an Adolescent's Emotional Intelligence

The teen years are a period of intense growth physically, emotionally, and intellectually. Teenage behavior problems such as moodiness, quickness to anger, fatigue, and miscommunication are linked to physical changes and growth in the adolescent brain from childhood into adolescence and adulthood (Walsh & Bennett, 2005). A study of adults who had been diagnosed with a mental illness at some point during their lives revealed that half reported experiencing symptoms by their mid-teens (Kessler *et al.*, 2005). Social concerns and worries in teens can have negative impacts on mental health and functioning such as school performance and recreational activities (Erath *et al.*, 2007).

The teenage years are often portrayed as stressful for both parents and teens (Kopko, 2007). At this life stage, adolescents tend to assert their independence while parents often believe they still need a good deal of guidance and protection. Parents still remain a child's primary caregivers and continue to provide physical and emotional security even as the child's peer relationships become important. To promote emotional intelligence, parents need to stay connected to their teens and discuss with them how to overcome stressful events. To avoid parent–teen conflicts, parents must establish clear rules and allow their teens the room to become independent. Nurturing parent–teen relationships are essential to maintaining mental health (Afifi, 2009). Adolescents with consistently *secure parental and peer attachment* evidenced the best academic, social, and emotional functioning overall (Holt *et al.*, 2018).

Social Anxiety Teens gain more independence from their family and can have **social anxiety** (i.e., the fear of interaction with other people) about building good peer-to-peer relationships in this period (Haller *et al.*, 2013). A survey of more than 123,000 students at 153 colleges reveals that more than half experience overwhelming anxiety, and about a third feel intense depression during the school year (American College Health Association, 2013). **Shyness** is the feeling of awkwardness or lack of comfort when a person is around other people and **social phobia** (i.e., social anxiety disorder) is defined by the core feature of excessive fear of embarrassment, which is often accompanied by avoidance of social or public situations (Schneier *et al.*, 2002). A study of the frequency of shyness and its overlap with social phobia in a nationally representative adolescent sample revealed that half think of themselves as "shy" and more say that they were "shy" as children (Burstein *et al.*, 2011). In fact, among all youth who endorsed shyness, 12.4 percent met criteria for lifetime social phobia (see image below).

SNS Use and Self-esteem Social networking sites (SNSs) have revolutionized the way people interact. Individuals with social anxiety (not comfortable talking with others in person) may benefit from SNS use. For example, social anxiety was correlated positively

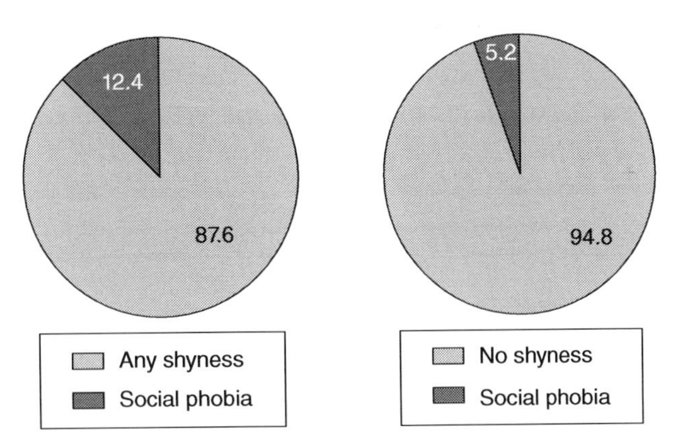

Proportion of Adolescents with Lifetime Social Phobia Among those with (n = 4749) and without (n = 5374) Shyness According to Adolescent Reports. Among all Youth who Endorsed Shyness, 12.4 Percent Meeting Criteria for Lifetime Social Phobia.
Source: Burstein *et al.* (2011). https://www.ncbi.nlm.nih.gov/pmc/articles/PMC3208958/figure/F1/

with feelings of comfort online (Prizant-Passal *et al.*, 2016). People are motivated to use Facebook for two primary reasons: (1) a need to belong and (2) a need for self-presentation (Nadkarni & Hofmann, 2012). Facebook profiles help satisfy individuals' need for self-worth and self-integrity (Toma & Hancock, 2013). Women were more likely to use social networking sites for comparing themselves with others and acquiring information, whereas men primarily used social networking sites to look at other people's profiles to find friends (Haferkamp *et al.*, 2012).

Studies have investigated the relationships between SNS use and **self-esteem**—the positive or negative emotional evaluations of the self (Smith & Mackie, 2007)—and have had mixed results. Facebook may be beneficial for individuals with low self-esteem by helping these individuals bridge social capital, gain acceptance, and adapt to a new culture especially within a university setting (Ellison *et al.*, 2007). A study of 852 adolescents aged ten to fifteen found that self-esteem longitudinally predicted higher SNS use (Valkenburg *et al.*, 2017). A study of 200 women in India indicated a negative correlation between Facebook use and self-esteem. The users usually show the good things and hide the negative things. As the person goes away from the real world, their self-respect lowers and they are not satisfied with self (Ashtaputre-Sisode, 2018).

Social comparison theory believes that there is a drive within individuals to gain accurate self-evaluations (Festinger, 1954). **Downward social comparison** occurs when people compare themselves to those who are less proficient than they are and can elevate self-esteem (Gibbons, 1986). **Upward social comparison** occurs when people compare themselves to people who are better than they are and can lower self-regard (Tesser *et al.*, 1988). *Facebook depression* is defined as depression that develops when preteens and teens spend a great deal of time on social media sites and then begin to exhibit classic symptoms of depression (O'Keeffe *et al.*, 2011). If a teen becomes moody and withdrawn after spending time on Facebook and comparing their lives to the posts of other people, it could be a sign of Facebook depression. A study of 190 adolescents at the mean age of 18.9 found no evidence supporting a relationship between internet use and clinical depression (Jelenchick *et al.*, 2013). The positive association between time spent on Facebook and depressive symptoms was found only among those high in *neuroticism*—the tendency to worry extensively and experience a pervasive sense of insecurity (Chow & Wan, 2017).

Parents might moderate the chances of their teens developing Facebook depression and help them develop *high self-esteem*—predicting success and well-being in life domains such as relationships, work, and health (Orth & Robins, 2014). One way to help teens develop self-esteem is altruistic behavior (e.g., voluntary behavior) intended to benefit another and society (Eisenberg, 2007). A study of 500 adolescents aged twelve to twenty revealed that altruistic behavior might raise teens' feelings of self-worth (Padilla-Walker *et al.*, 2017).

Impact of Parental Role on an Adolescent's Financial Intelligence

Financially intelligent parents can help their teens learn how to build their financial future and how to accumulate savings and build wealth. The strategies for raising financially intelligent children proposed by Gallo and Gallo (2005) include encouraging a work ethic, facilitating financial reflection, and teaching financial literacy. A work ethic is a learned behavior. For example, parents can teach children how to do chores around the house, teach them the value of work, and allow them to earn additional money or allowance. Many teens may have a part-time job, earn income, and possibly file a tax return. Parents can instill this behavior into their children through the socialization process so that children can carry it on later in life. Parents can encourage their teens to engage in financial reflection regarding issues such as spending or saving money. Many teens may have learned how to deposit and use money through

Diversity Overview: Parenting Experiences of Children in Family Diversity

Studies compared time invested in children across family structures as a means to understand differences in children's development. In a study of six family structures—married biological parents, cohabiting biological parents, mother and stepfather, mother and cohabiting boyfriend, single mother only, and multigenerational households—Kalil *et al.* (2014) found that children received little time investment from resident non-biological fathers; and children in single-mother, cohabiting-boyfriend, and multigenerational households received little time investment from their non-resident biological fathers. Factors such as family type, social class, culture, and race embody different characteristics that are deemed essential or advantageous and disadvantageous to their children.

Communal Parenting

Multigenerational families are defined as those consisting of more than two generations living under the same roof. In 2016, 64 million people (20 percent of the U.S. population) lived with multiple generations under one roof (Cohn & Passel, 2018). It means that one in five Americans lived in a multigenerational household in 2016.

Growing racial and ethnic diversity in the U.S. population helps explain some of the rise in multigenerational living (Cohn & Passel, 2018). Among Asians living in the U.S., 29 percent lived in multigenerational family households in 2016, compared with Hispanics (27 percent) and Blacks (26 percent), and Whites (16 percent) respectively. Households with three or more generations—for example, a grandparent, an adult child, and a grandchild of any age—housed 28.4 million people in 2016, up from 26.5 million in 2012 (Cohn & Passel, 2018).

A multigenerational household involves the kinship network of social relationships related by descent and by marriage, as well as members of the community. **Communal parenting** means that kinship networks or extended families serve as a backup to parents and share their responsibilities. For example, if single parents are unable to raise their biological children, parenting becomes communal. Older siblings, grandparents, relatives, and friends help monitor and share the responsibilities. Since the beginning of slavery in the United States, African Americans have viewed their families in terms of kinship networks (Tolman, 2011). Certain cultural resources—spirituality, ethnic identity, and mutual support—contribute greatly to the resilience of African American families (Lincoln & Mamiya, 1990). Supportive kinship networks play key roles in helping raise children. People with or without blood ties contribute to the task of bringing up a child, sharing the parental role, and thus can be sources of multiple attachments. This communal parenting throughout history would have significant implications for the evolution of multiple attachments (Quinn & Mageo, 2013).

Racial Differences in Childhood Adversity by Immigration History and Income

Childhood adversity comprises acts of commission of sexual, physical, and emotional abuse, as well as acts of omission such as emotional and physical neglect and witnessing intimate partner violence or child maltreatment (Schilling *et al.*, 2016). This has been associated with multiple adverse somatic and mental diseases in adulthood, maladjustment and an unhealthy life style (Grabe *et al.*, 2012), and reduced adjustment, social support, and resilience (Beutel *et al.*, 2017).

Childhood maltreatment has been increasingly recognized as a major public health problem in high-income countries (Bethell *et al.*, 2016). In the United States, for example, 18 percent of Michigan third graders have been formally investigated by Child Protective Services (CPS) for possible exposure to maltreatment; early childhood maltreatment is associated with significantly poorer school districts and lower academic outcomes (Jacob & Ryan, 2018). Parents with histories of maltreatment or violence exposure may have particular difficulty in responding sensitively and in a developmentally appropriate manner to their toddlers' tantrums and thus may benefit from parent–child mental health consultations (Gudmundson & Leerkes, 2011). In a study of adverse children from birth to age seventeen (n=84,837), Slopen *et al.* (2016) found that Black and Hispanic children were exposed to more adversities, compared with White children.

The Effects of Immigrant Family and Race on Parenting Aggravation

Among many hardships faced by immigrant families, those related to economic conditions, health insurance, acculturation, access of public benefit programs, and English proficiency are likely the most challenging (Beiser, 2005). These may all contribute to stress and aggravation in the family lives of immigrants. **Parenting aggravation** is a measurement of stress experienced by parents associated with caring for children (Yu & Singh, 2012). High parenting aggravation has previously been associated with maternal and paternal depression, autism, learning disabilities, child obesity, maternal chronic illness, paternal alcoholism, single parenthood, and family transitions (Annunziato *et al.*, 2007).

Yu and Singh (2012) explored elevated parenting aggravation among immigrant and minority families. All foreign-born parents, regardless of race/ethnicity, reported elevated parenting aggravation. Twenty-six percent of foreign-born parents with foreign-born children were highly aggravated, followed by 22 percent of foreign-born parents with U.S.-born children and 11 percent of U.S.-born parents.

Parenting Issues in Dysfunctional Families

Dysfunctional family is a form of family that has destructive and harmful parenting and lack of concern for the child. If the distress caused to the child is severe and long-lasting s/he may develop a psychiatric condition such as post-traumatic stress disorder (PTSD), which if not properly treated may adversely affect the rest of his/her life (Hosier, 2016). Let's take a look at some instances or factors that contribute to a dysfunctional family.

Parentification is the process of role reversal whereby a child is obliged to act as parent to their own parent. In extreme cases, the child is used to fill the void of the alienating parent's emotional life (Gardner *et al.*, 2006). **Instrumental parentification** involves the child completing physical tasks for the family, such as looking after a sick relative, or providing assistance to younger siblings that would normally be provided by a parent. **Emotional parentification** occurs when a child must take on the role of a confidant or mediator for (or between) parents or family members (Jurkovic, 1998). For example, if a mother discusses her worries with her child, the child would be counseling his or her mother and likely become emotionally parentified. Such parentification of the child represents unequivocal emotional abuse (Hosier, 2016). Parental disclosures involving sensitive, inappropriate information can also be detrimental to children's mental health (Afifi, 2009).

Parental alienation is the process, and the result, of emotional manipulation of a child into showing unwarranted fear, disrespect, or hostility towards a parent (Lorandos *et al.*, 2013). Bow *et al.* (2009) examined the role of an alienating parent. The primary cause is a parent wishing to exclude another parent from the life of their child that occurs in association with family separation or divorce (Warshak, 2015). The behaviors of all family members such as the alienated parent may lead to family dysfunction and the rejection of a parent (Ackerman, 2001). Bala *et al.* (2007) focused on the alienated child, and the relationship dynamics that contribute to the alienation. It often leads to adverse childhood experience and increased lifetime risks of both mental and physical illness (Bernet *et al.*, 2016). When a child is estranged from a parent, it is necessary to evaluate all contributing factors and all possible remedies to the estrangement (Bala *et al.*, 2007). Within the context of relationship dynamics, alienation is seen as a breakdown of attachment between parent and child. Behaviors that result in parental alienation may reflect other mental health disorders, both on the part of the alienating parent and the alienated parent, which may be relevant to a custody determination (Baker & Eichler, 2016).

Upon completion of this section, students should be able to:
LO9: Describe three parenting styles and their effects.

savings and credit. It is also a great way to teach teens how to manage and plan their spending out of an allowance or income. Some teens may be interested in possibly starting a business. Parents can help them understand how they make decisions regarding money and investment.

Parenting Styles

There are recurrent and occasionally heated political debates about youth crime and truancy, to which inadequate parenting is often seen as a contributing factor (Chan & Koo, 2008). Part of that learning process is **parenting styles**—practices that parents use in the raising of their children. Psychologist Diana Baumrind (1967) identified parenting styles based on two aspects of parenting behavior: control and warmth. **Parental control** refers to the degree to which parents manage their children's behavior from being very controlling to setting few rules and demands (Kopko, 2007) and is known as **demandingness**—"the claims parents make on children to become integrated into the family whole, by their maturity demands, supervision, disciplinary efforts and willingness to confront the child who disobeys" (Baumrind, 1991). **Parental warmth** refers to the degree to which parents are accepting and responsive to their children's behavior and is known as responsiveness—"the extent to which parents intentionally foster individuality, self-regulation, and self-assertion by being attuned, supportive, and acquiescent to children's special needs and demands" (Baumrind, 1991).

Parenting styles often mirror society. In a traditional patriarchal family, for example, a father held authority over the family and sought to raise obedient children using physical discipline such as spankings. The Civil Rights movement in the 1960s promoted human rights, freedom, and individualism. This created a change in views in parenting, and behaviors such as physical discipline once seen as commonplace in the raising of a child began being viewed as less acceptable. Since then, parents have become more sensitive to a child's needs and more concerned with their parenting styles. Let's explore *authoritarian*, *authoritative*, and *permissive* parenting styles in a bit more detail. Note that these parenting styles are not sequential and can be practiced at the same time.

Authoritarian Parenting Style

An **authoritarian parenting style** (high control and low warmth) is demanding and not responsive. Parents are highly controlling and demanding but display little warmth. Children are required to obey their parents' orders without question and may be punished if they fail to follow the rules. For authoritarian parents, "because I said so" is a viable explanation of why children should follow their rules. Verbal give-and-take between a parent and a child is discouraged. Authoritarian parents demand obedience and evaluate their children's behaviors according to rules and enforce discipline through threats, timeouts, yelling, or punishment such as spanking.

Sociologists examined parenting styles by race and social class within a cultural context. For example, European American and Puerto Rican mothers of toddlers believe in the differential value of individual autonomy versus connected interdependence (Harwood *et al.*, 1999). Puerto Rican mothers use more direct means of structuring, such as commands, physical positioning and restraints, and direct attempts to recruit their children's attention. Asian parents are considered to be authoritarian parents. Gibbs *et al.* (2017) separates *parenting* into abstract ideals and tangible behaviors. *Abstract ideals* are beliefs parents have about the importance of schooling and the kinds of skills necessary for school success, while *tangible behaviors* encompass specific and observable parenting behaviors that reflect how abstract ideals are put into action. These two dimensions of parenting represent what parents think (attitudes) and do (behaviors). Asian American parents were found to have high educational expectations compared with Whites but are less engaged in traditional measures of parenting (e.g., reading to the child, maternal warmth, parent–child relationship) (Gibbs *et al.*, 2017). Poor and working-class parents were found to practice *accomplishment of natural growth parenting* (Lareau, 2011). For example, these authoritarian parents gave orders to their children, rather than soliciting their opinions as they believed that they should care for their children. The differences in parenting styles might be linked to ethnic cultures. Children's experiences with their parents within a cultural context consequently scaffold them to become culturally competent members of their society (Bornstein, 2012). Parenting styles that are congruent with cultural norms appear to be effective in transmitting values from parents to children, perhaps because parenting practices that approach the cultural norm result in a childrearing environment that is more positive, consistent, and predictable (Bornstein, 2012). An uncooperative, immature, and irresponsible teen may be more likely to elicit an authoritarian parenting style (Kopko, 2007). In contemporary UK, the parenting style in single-parent families and stepfamilies is more likely to be authoritarian or permissive rather than authoritative (Chan & Koo, 2008).

Effects of Authoritarian Parenting on Children

Research has looked at the impact of authoritarian parenting styles on child and adolescent development. An authoritarian parenting style leads to positive outcomes in African American and Hong Kong Chinese school children (Leung *et al.*, 1998). Unlike Western authoritarian parents, Chinese authoritarian parents have closer relationships to their kids, and closeness is a predictor of higher school achievement (Chao, 2001). People learn better from positive feedback than from negative feedback, and this may be especially true for kids (Schmittmann *et al.*, 2006). Studying and comparing American high schools revealed that authoritarian schools had the worst rate of dropouts (Pellerin, 2005). For every region of the globe, associations of authoritarian parenting with academic achievement were less negative in Hispanic families than White families (Pinquart & Kauser, 2018).

The authoritarian parenting style has a number of negative effects on children. Kids with authoritarian parents are more likely to use and abuse alcohol (Calafat *et al.*, 2014) and are less advanced when it comes to self-regulation and moral reasoning (Piotrowski *et al.*, 2013). High school students with authoritarian parents were more likely to be involved in bullying, particularly if their parents attempted to control them through the use of punitive discipline in Spain (Gómez-Ortiz *et al.*, 2016). College students raised by authoritarian parents were more likely to engage in acts of bullying (Luk *et al.*, 2016). A study of 231 young adolescents about their cultural values and experiences with peers in Cyprus revealed that kids from authoritarian homes were more likely to have experienced bullying—both as victims and perpetrators (Georgiou *et al.*, 2013). A study of American middle and high school studies over eighteen months revealed that kids with authoritarian parents were more likely to reject their parents as legitimate authority figures (Trinkner *et al.*, 2012). A meta-analysis of more than 1,400 published studies found that harsh control and psychological control were actually the biggest predictors of worsening behavior problems over time (Pinquart, 2017). Risky behaviors are more common among teenagers with authoritarian parents in the UK (Chan & Koo, 2008).

Authoritarian parenting may put children at greater risk of anxiety, low self-esteem, and depression. Authoritarian parenting contributes to the risk of major depression later in life (Long *et al.*, 2015). Adolescents and adults are more likely to suffer depressive symptoms if they characterize their parents as having used authoritarian practices in the United States (King *et al.*, 2016) and in Caribbean countries (Lipps *et al.*, 2012). In China, kids with authoritarian fathers were more likely to suffer from a psychiatric disorder—even after accounting for the influence of genes (Yin *et al.*, 2016).

Authoritative Parenting Style

Authoritative parenting style (high control and high warmth) is both demanding and responsive. The authoritative parent takes a moderate approach that emphasizes standards, offers emotional support, and shows respect for children as independent social actors. Like authoritarian parents, authoritative parents enforce rules and standards of conduct. Unlike authoritarian parents, authoritative parents emphasize the reasons for rules and allow children to explore freely. Authoritative parents may offer options for their child to choose from and discuss the consequences of the child's actions but are less likely to control their children through harsh punishment. The authoritative parenting style reflects a balance between freedom and responsibility.

Authoritative parenting styles vary by race and culture. In Western countries like the United States, the authoritative parenting style includes democratic practices such as encouraging children to express their own opinions. European American mothers use suggestions (rather than commands) and other indirect means of structuring their children's behavior (Harwood *et al.*, 1999). The United States and Japan are both child-centered modern societies with high standards of living and so forth, but American and Japanese parents value different childrearing goals, which they express in different ways (Morelli & Rothbaum, 2007). For example, American mothers try to promote autonomy, assertiveness, verbal competence, and self-actualization in their children, whereas Japanese mothers try to promote emotional maturity, self-control, social courtesy, and interdependence in theirs (Bornstein, 2012).

The parenting styles cut across class lines. In a study of 88 families from Black, White, middle-class, working-class, and poor backgrounds, Lareau (2011) found that middle-class families raised their children in a different way than working-class and poor families. Middle-class parents practice *parenting concerted cultivation* that teaches children how to challenge authority, navigate bureaucracy, and manage their time (Lareau, 2011). Middle-class kids are driven to soccer practice and band recitals, are involved in family debates at dinner time, and are told to ask their teacher why they received a B on a French exam (Mckenna, 2012). Therefore, middle-class kids are raised in a way that provides them with the skills necessary to remain in the middle class (Lareau, 2011). Authoritative parenting was found to be more common among two-parent families (53 percent) than among single-parent families (32 percent) or stepfamilies (40 percent) in contemporary UK (Chan & Koo, 2008). A cooperative, motivated, and responsible teen may be more likely to have parents who exercise an authoritative parenting style (Kopko, 2007). Although authoritative parenting is less common in ethnic minority families, this parenting style has been linked with adolescent competence across a wide range of families (Steinberg & Silk, 2002).

Effects of Authoritative Parenting on Children

Authoritative parenting (supervision and autonomy) leads to better adolescent school performance and stronger school engagement (Steinberg *et al.*, 1992) and leads to positive outcomes in European American school children (Bornstein, 2012). In a comparative study of American high schools, Pellerin (2005) found that children from authoritative families may find it easier to fit in when the school climate is authoritative. Parental involvement is much more likely to promote adolescent school success when it occurs in the context of an authoritative home environment (Steinberg *et al.*, 1992). In Western countries, associations of authoritative parenting with academic achievement were stronger in White families than in Asian minorities (Pinquart & Kauser, 2018).

The authoritative parenting style (parental supervision) promotes the development of morality (Patrick & Gibbs, 2016) and moral reasoning skills (Kerr *et al.*, 2004), helps kids become more empathic, helpful, conscientious, and kind to others (Knafo & Plomin, 2008), promotes teen's exposure to positive activities, reduces teens' opportunities for engaging in delinquent, risky behaviors (Wargo, 2007), helps prevent children from developing aggressive behavior problems and reduces peer problems in preschoolers (Arsenio & Ramos-Marcuse, 2014). The children of authoritative parents are less likely to be engaged in drug and alcohol use, juvenile delinquency, or other antisocial behavior (Luyckx *et al.*, 2011). Authoritative parents seek to encourage independence in their children, which has been linked with more self-reliance, better problem solving, and improved emotional health (Luyckx *et al.*, 2011). Pinquart and Kauser (2018) suggest that parents across the globe could behave authoritatively.

Permissive Parenting Style

Permissive parenting style (low control and high warmth) is not demanding and responsive. Unlike authoritative parents, permissive parents do not enforce rules (Calafat *et al.*, 2014) and may explain rules but not enforce them. Permissive parents don't present themselves as authority figures or role models and avoid exercising overt power (Baumrind. 1966). Children are allowed to make their own choices or decisions with minimal restrictions. An example of this style would be a parent allowing their children to eat or play whatever and whenever they want without any limitations. Permissive parents may let their children get away with leaving chores or homework unfinished. A quarter of parents say that they give their children too much freedom (Pew Research Center, 2015b). Like authoritative parents, permissive parents are emotionally supportive and responsive to their children's needs and wishes. Thus, permissive parenting is non-directive and lenient with no behavioral expectations for a child.

Effects of Permissive Parenting on Children

Permissive parents are called *indulgent parents*—rejecting the notion of setting rules for their children or keeping them under control. When children are granted autonomy, they may make bad choices and end up staying up too late at night and eating too many nutrient-poor, energy-dense treats, for instance (Hennessy *et al.*, 2012). In a study of British ten- to eleven-year-olds, Jago *et al.* (2011) reported that kids with permissive parents had five times the risk of watching more than four hours of television per day. Research indicates links between permissive parenting and higher rates of school misconduct and lower levels of academic achievement (Lamborn *et al.*, 1991). Adolescents with permissive parents are less positively oriented to school than adolescents who are raised by authoritative and authoritarian parents (Hoskins, 2014).

Children of permissive parents tend to have low emotional regulation, are rebellious and defiant, have some antisocial behavior, and low motivation to complete tasks (Baumrind, 1967). A study of more than 1,000 young American kids ages two to eight revealed that preschoolers with permissive parents had more problems with self-control, management of their moods, and executing plans (Piotrowski *et al.*, 2013). Children with permissive parents were more likely to increase levels of aggression (Underwood *et al.*, 2009). Tracking American kids for over ten years revealed that preschoolers were more likely to develop anxiety and depression if they were raised by permissive parents (Williams *et al.*, 2009). Adolescents with permissive parents reported high frequencies of substance abuse (Hoskins, 2014).

Global Overview: Third-culture Kids

Through the process of globalization where countries become interconnected, families are choosing to expose their children to other cultures and are welcoming the opportunity to spend several years outside of their passport countries for financial gain and business opportunities (Moore & Barker, 2012). Useem *et al.* (1963) identified the blending of a host culture and home culture (the passport culture) to form a third culture in American children raised overseas and coined **third-culture kids** (TCKs) to describe people who have spent a significant part of their developmental years outside their parents' culture. A child's *first culture* refers to the culture of the home country (the passport country) from which his or her parents originated; a child's *second culture* refers to the host culture in which the family currently resides; and a child's *third culture* refers to the amalgamation of both the passport and host cultures (Useem, 1963). Examples of third-culture kids include children of parents who work in the military, missionary kids or children of government diplomats or international business people, etc. Two well-known third-culture kids are news reporter Abby Huntsman (who is the daughter of a former U.S. Ambassador to China) and former president George Bush (who is the son of a former U.S. Ambassador to China). Because third-culture kids move among three countries internationally, they are referred to as "global nomads." As they have multicultural experience, they are also called "cultural hybrids."

There are benefits to being a third-culture kid. Third-culture kids are bilingual or multilingual. In addition, they are often multicultural as they are exposed to diverse languages, cultures, and cultural customs. Because they are brought up in multiple cultures, third-culture kids have the capacity to function effectively across national, ethnic, and organizational cultures.

Third-culture kids can experience confusion growing up in different cultures. A longer period of time within a host culture provides opportunity to challenge beliefs of the passport culture and develop autonomy informed by cultural norms of the host country (Arthur, 2003). They might experience value conflict. For example, Americans tend to value individualism whereas the Japanese tend to value collectivism. Significant periods within the host culture also alter previous attachments (Terrazas-Carrillo *et al.*, 2014) and affect the individual's feelings about the return to family and friends within the passport culture (Arthur, 2003). Third-culture kids who move back and forth among multiple cultures may struggle to develop a personal and cultural identity. Third-culture kids are between cultures, stating that the third culture is developed by the child to explain an identity that is different from that of the host country or the parents' home country (Pollock & Van Reken, 2001).

Espada-Campos (2018) explored the undergraduate college experiences of adult third-culture kids (ATCKs) to understand the risk and protective factors associated with repatriation and collegiate engagement. Repatriation to pursue higher education was one of three primary reasons identified by the UN DESA (UN, 2013) for international migrant re-entry to the passport country. International schools and government agencies that facilitate the transition back to the passport country have identified several areas of concern specific to TCKs. For example, TCKs report grief and loss, feelings of cultural homelessness, interpersonal relationship struggles, self-identity versus cultural identity crises, and mental health issues as recurring themes (Morales, 2015). Risk factors were defined as "individuals, school, peer, family, and community influences that increase the likelihood that a young person will experience behavioral, social, or emotional health problems" (Jenson & Fraser, 2011). For students who participate in campus organizations that focus on cultural identification with a group, ethnic identity may serve as a protective factor that exacerbates stress and discourages social engagement (Forrest-Bank & Jenson, 2015).

Espada-Campos (2018) suggested the need for culturally informed administrative practices to mitigate psychosocial challenges associated with academic engagement. Interventions related to student identification procedures, supportive resources, and campus life programs should be incorporated to support multicultural students, starting at the time of application and continuing through to graduation.

Summary

The chapter has examined the transition to successful parenting and explored the forms parenthood may take today, including two married parents, cohabiting parents, LGBT parents, single parents, non-residential parents, foster parents, and grandparents. We have also examined how the roles of parents change as they parent an infant, a child, and a teen and the three parenting styles identified by Baumrind (authoritarian—requires obedience and enforces rules; authoritative—provides guidance and involves democratic elements; and permissive—promotes indulgent parenting). The chapter also explored parenting experiences of children in family diversity and third-culture kids from a diversity and global perspective.

Key Terms

Alternative model suggests that the coparenting alliance predicts both marital relationships and parenting practices.

Attachment parenting is a parenting philosophy that promotes the attachment of mother and child by parental responsiveness and closeness.

Attachment theory—attempting to describe the dynamics of long-term and short-term interpersonal relationships between humans.

Authoritarian parenting style (low warmth and high control) is demanding and not responsive.

Authoritative parenting style (high warmth and high control) is both responsive and demanding.

Biological father is the male genetic contributor to the creation of the baby, through sexual intercourse or sperm donation.

Biological mother is the female genetic contributor to the creation of the baby.

Childhood is the age span ranging from birth to preadolescence between the ages of one and twelve.

Childhood adversity is maltreatment, exposure to domestic violence, or living with another person with serious mental illness.

Children as a social actor approach views children as active members of society from the beginning.

Communal parenting means that kinship networks or extended families serve as a backup to parents and share their responsibilities.

Coparenting has been defined as a unique component of the marital relationship in which parents work together, or, alternatively, struggle against each other, when it comes to childrearing.

Coparenting alliance is characterized by both parents' investment in the child, valuing each other's involvement with the child, respect for each other's judgment about childrearing, and having a desire to communicate child-related information.

Custodial parent is the parent with whom the child(ren) lived when their other parent(s) lived outside the household.

Demandingness refers to the claims parents make on children to become integrated into the family whole.

Digital parent refers to a parent who uses one or more digital applications and devices on a daily basis, particularly in parenting.

Downward social comparison occurs when people compare themselves to those who are less proficient than they are.

Dysfunctional family is a form of family that has destructive and harmful parenting and lack of concern for the child.

Effortful control is the ability to inhibit a dominant response to perform a subdominant response.

Emotion is characterized by intense mental activity and a high degree of pleasure or displeasure.

Emotional expression refers to the frequency, intensity, latency, and duration of expression of both negative (e.g., anger, fear, sadness) and positive emotions.

Emotional intelligence (EI) is the ability to perceive, understand, and regulate emotions so as to promote emotional and intellectual growth.

Emotional investment involves having emotionally bonded with an infant.

Emotional parentification occurs when a child must take on the role of a confidant or mediator for (or between) parents or family members.

Emotional understanding is the ability to recognize and name emotions and understand the causes and consequences of emotions.

Emotion regulation is the ability to effectively manage and respond to an emotional experience.

Family developmental theory suggests that the transition to parenthood involves developmental life changes from one stage, such as spouse, to another, such as parent.

Father is the male parent of a child.

Financial intelligence (FI) is the ability to build wealth and manage a person's life by understanding how money works.

Foster care is a system in which a minor is placed into a ward or private home of a state-certified caregiver.

Foster parent is a caregiver of a child arranged through the government or a social service agency.

Generational order approach focuses on age cohorts and allows us to recognize that child–adult relations take place between groups of people subject to differing constellations of social, historical, and political ideas.

Helicopter parent pays close attention to a child's experiences and problems.

Infant is a young child between the ages of one month and twelve months.

Instrumental parentification involves the child completing physical tasks for the family, such as looking after a sick relative.

Intellectual intelligence is considered a precursor to academic success.

Intelligence is the aggregate or global capacity of the individual to act purposefully, to think rationally, and to deal effectively with their environment.

Learning explanation suggests that the infant makes associations between touch and environmental events to impute meaning on tactile patterns.

LGBT parenting refers to lesbian, gay, bisexual, and transgender (LGBT) people raising children.

Memory processes can influence how infants gain meaning from the touch that their caregivers administer.

Mother is the female parent of a child.

Multidisciplinary approach makes the interdisciplinary study of childhood a range of ways of thinking spatially about children's lives.

Networked family is a form of family in which family members multitask and stay more connected through information and communication technology.

Newborn is an infant who is only hours, days, or up to a few weeks old.

Non-residential parents are defined as parents who do not live with one or more of their biological children all or most of the time.

Online grooming occurs when someone with a sexual interest in children uses social media, online chat, messaging apps, or text messages to build a relationship with a young person in order to meet and abuse the child physically.

Parent is the caretaker and caregiver of a child.

Parental alienation is the process, and the result, of emotional manipulation of a child into showing unwarranted fear, disrespect, or hostility towards a parent.

Parental control refers to the degree to which parents manage their children's behavior from being very controlling to setting few rules and demands.

Parental responsiveness (sensitivity or sensitive responsiveness) is characterized by warm acceptance of children's needs and interests.

Parental warmth refers to the degree to which parents are accepting and responsive to their children's behavior.

Parentification is the process of role reversal whereby a child is obliged to act as parent to their own parent.

Parenting is the process of supporting the physical, psychological, social, financial, and intellectual development of a child from infancy to adulthood.

Parenting aggravation is a measurement of stress experienced by parents associated with caring for children.

Parenting style refers to practices that parents use in the raising of their children.

Peer relations refer to associations among equals, including romantic relationships.

Permissive parenting style (high warmth and low control) is responsive and not demanding.

Primary attachment refers to an attachment between a significant or primary caregiver and a child.

RPM3 involves responding to children, preventing risky behaviors, monitoring children's contact with their surrounding world, mentoring them to encourage desired behaviors, and modeling behavior to provide a consistent and positive example for them.

Secondary attachment refers to an attachment between a child and a non-primary caregiver, such as a childcare provider.

Self-awareness is the ability to recognize and understand personal moods, emotions, and drives.

Self-efficacy refers to a personal judgment of how well one can execute courses of action to deal with prospective situations.

Self-esteem is the positive or negative emotional evaluations of the self.

Self-regulated learning is a child's capacity for learning and is aligned closely with educational aims.

Self-regulation is the ability to control or redirect one's disruptive emotions and behaviors in response to a particular situation.

Shyness is the feeling of awkwardness or lack of comfort when a person is around other people.

Single parent is the only male or female parent to the child, responsible for all financial and emotional needs.

Social anxiety is the fear of interaction with other people.

Social capital refers to the networks of relationships among people who live and work in a particular society.

Social comparison theory believes that there is a drive within individuals to gain accurate self-evaluations.

Social intelligence is the ability to get along well with others, and to get them to cooperate with you.

Social learning theory suggests that children acquire new behaviors by observing and imitating others.

Social phobia (i.e., social anxiety disorder) is defined by the core feature of excessive fear of embarrassment.

Social skill is social competence facilitating interaction and communication with others and managing relationships to move people in the desired direction.

Sociology of childhood is a discipline within the field of childhood studies and explores the wide variety of childhoods across the world.

Sociology of depression looks at the cultural context in which people live and the social stressors that people encounter.

Sociology of emotions examines how emotions are differentiated, socialized, and managed socially and culturally.

Suicide is death caused by injuring oneself with the intent to die.

Surrogacy is the process of giving birth as a surrogate mother.

Third-culture kid refers to a child who is raised in a culture outside of his or her parents' culture for part of his or her developmental years.

Toddlerhood starts at around twelve months when toddlers begin learning how to walk and talk and ends at 36 months.

Touch is the action of an object on the skin and the registration of information by the sensory systems of the skin.

Trust believes that the person who is trusted will do what is expected.

Upward social comparison occurs when people compare themselves to people who are better than they are.

Discussion Questions

- What parental behaviors make up a successful child?
- Explain four distinct types of intelligence that a successful person can possess.
- How can parents use role-modeling to shape teenage behavior?

Suggested Film and Video

Grazer, B. (Producer). (1989). *Parenthood*. Production Company.
Parenting videos: https://www.cdc.gov/parents/essentials/videos/

Internet Sources

Child Care Aware of America: http://childcareaware.org/
Important Milestones: Your Baby by Six Months: https://www.cdc.gov/ncbddd/actearly/milestones/milestones-6mo.html
Your Baby at Two Months: https://www.cdc.gov/ncbddd/actearly/pdf/checklists/all_checklists.pdf

References

Ackerman, M. J. (2001). *Clinician's Guide to Child Custody Evaluations*. New York: Wiley.

Afifi, T. D. (2009). Adolescents' Physiological Reactions to Their Parents' Negative Disclosure About the Other Parent in Divorced and Nondivorced Families. *Journal of Divorce & Remarriage*, 517–540.

Aharoni, E., Vincent, G. M., Harenski, C. L., Calhoun, V. D., Sinnott-Armstrong, W., Gazzaniga, M. S. & Kiehl, K. A. (2013). Neuroprediction of Future Rearrest. *PNAS*, 110(15), 6223–6228.

Ainsworth, M. D., Blehar, M. C., Waters, E. & Wall, S. (1978). *Patterns of Attachment: A Psychological Study of the Strange Situation*. Hillsdale, NJ: Lawrence Erlbaum Associates.

Albrecht, K. (2003). *Social Intelligence: The New Science of Success*. Hoboken, NJ: John Wiley & Sons.

American College Health Association (2013). *American College Health Association, National College Health Assessment II: Reference Group Executive Summary Spring 2013*. Hanover, MD: American College Health Association.

American Society for the Positive Care of Children (American SPCC) (2017). *Positive Parenting*. Retrieved on December 2, 2018 from https://americanspcc.org/positive-parenting/

Anderson, E. A., Kohler, J. K. & Letiecq, B. L. (2005). Predictors of Depression among Low-income, Nonresidential Fathers. *Journal of Family Issues*, 26(5), 547–567.

Anderson, M. (2016). *Parents, Teens and Digital Monitoring*. Washington, DC: Pew Research Center.

Annunziato, R. A., Rakotomihamina, V. & Rubacka, J. (2007). Examining the Effects of Maternal Chronic Illness on Child Well-Being in Single Parent Families. *J Dev Behav Pediatr.*, 28(5), 386–391.

Appelbaum, H. L. (2016). *Abnormal Female Puberty: A Clinical Casebook*. New York: Springer.

Apps, M. A. J., Rushworth, M. F. S. & Chang, S. W. C. (2016). The Anterior Cingulate Gyrus and Social Cognition: Tracking the Motivation of Others. *Neuron*, 90(4), 692–707.

Archard, D. (1993). *Children: Rights and Childhood*. London: Routledge.

Arditti, J. A. (1995). Noncustodial Parents: Emergent Issues of Diversity and Process. *Marriage & Family Review*, 20, 283–304.

Arendell, T. (2000). Conceiving and Investigating Motherhood: The Decade's Scholarship. *Journal of Marriage and Family*, 62(4), 1192–1207.

Ariès, P. (1960). *Centuries of Childhood: A Social History of Family Life*. Robert Baldick (Trans.). New York: Vintage Books.

Aronoff, J. (2012). Parental Nurturance in the Standard Cross-Cultural Sample: Theory, Coding, and Scores. *Cross-Cultural Research*, 46(4), 315–347.

Arsenio, W. & Ramos-Marcuse, F. (2014). Children's Moral Emotions, Narratives, and Aggression: Relations with Maternal Discipline and Support. *J Genet Psychol.*, 175(5–6), 528–546.

Arthur, N. (2003). Preparing International Students for the Re-entry Transition. *Canadian Journal of Counseling*, 37(3), 173–185.

Ashman, S. B. & Dawsons, G. (2002). Maternal Depression, Infant Psychobiological Development, and Risk for Depression. In S. H. Goodman & I. H. Gotlib (Eds.), Children of Depressed Parents: Mechanisms of Risk and Implications for Treatment (pp. 37–58). Washington, DC: American Psychological Association.

Ashtaputre-Sisode, A. (2018). Facebook and Self-esteem. *The International Journal of Indian Psychology*, 6(1).

Baiocco, R., Santamaria, F., Ioverno, S., Fontanesi, L., Baumgartner, E., Laghi, F. & Lingiardi, V. (2015). Lesbian Mother Families and Gay Father Families in Italy: Family Functioning, Dyadic Satisfaction, and Child Well-being. *Sexuality Research and Social Policy*, 12, 202–212.

Baker, A. J. L. & Eichler, A. (2016). The Linkage between Parental Alienation Behaviors and Child Alienation. *Journal of Divorce and Remarriage,* 57, 475–484.

Baker, J. K., Fenning, R. M. & Crnic, K. A. (2011). Emotion Socialization by Mothers and Fathers: Coherence among Behaviors and Associations with Parent Attitudes and Children's Social Competence. *Social Development,* 20(2), 412–430.

Bala, N., Fidler, B., Goldberg, D. & Houston, C. (2007). Alienated Children and Parental Separation: Legal Responses from Canada's Family Courts. *Queens Law Journal,* 33, 79–138.

Bandura, A. (1965). Influence of Models' Reinforcement Contingencies on the Acquisition of Imitative Responses. *Journal of Personality and Social Psychology,* 1, 589–595.

Bandura, A. (1982). Self-efficacy Mechanism in Human Agency. *American Psychologist,* 37(2), 122–147.

Barker, R. (2001). *The Mighty Toddler: The Essential Guide to the Toddler Years.* Sydney, Australia: Pan Macmillan.

Barkley, R. (2005). *Attention-Deficit Hyperactivity Disorder: A Handbook for Diagnosis and Treatment, 3.* New York: Guilford Press.

Baron-Cohen, S. (2000). Theory of Mind and Autism: A Fifteen Year Review. In S. Baron-Cohen, H. Tager-Flusberg & D. J. Cohen (Eds.), *Understanding Other Minds: Perspectives from Developmental Cognitive Neuroscience* (pp. 3–21). Oxford, UK: Oxford University Press.

Barrett, K. C. & Campos, J. J. (1987). Perspectives on Emotional Development II: A Functionalist Approach to Emotions. In J. Osofsky (Ed.), *Handbook of Infant Development* (pp. 555–578). New York: Wiley.

Battalen, A. W., Farr, R. H., Brodzinsky, D. M. & McRoy, R. G. (2018). Socializing Children about Family Structure: Perspectives of Lesbian and Gay Adoptive Parents. *Journal of GLBT Family Studies,* doi: 10.1080/1550428X.2018.1465875

Baumrind, D. (1966). Effects of Authoritative Parental Control on Child Behavior. *Child Development,* 37(4), 887–907.

Baumrind, D. (1967). Child Care Practices Anteceding Three Patterns of Preschool Behavior. *Genetic Psychology Monographs,* 75(1), 43–88.

Baumrind, D. (1991). The Influence of Parenting Style on Adolescent Competence and Substance Use. *Journal of Early Adolescence,* 11(1), 56–95.

Bavelier, D., Green, C. S. & Dye, M. W. (2010). Children, Wired: For Better and for Worse. *Neuron,* 67(5), 692–701.

Bayley, N. (2006). *Bayley Scales of Infant and Toddler Development.* San Antonio, TX: Pearson.

Behnke, A. O., MacDermid, S. M., Coltrane, S. L., Parke, R. D., Duffy, S. & Widaman, K. F. (2008). Family Cohesion in the Lives of Mexican American and European American parents. *Journal of Marriage and Family,* 70, 1045–1059.

Beiser, M. (2005). The Health of Immigrants and Refugees in Canada. *Can J Public Health,* 96(Suppl 2), S30–S44.

Bemiller, M. L. (2005). *Mothering on the Margins: The Experience of Noncustodial Mothers.* (Unpublished doctoral dissertation). University of Akron, Akron, Ohio.

Benoit, D. (2004). Infant–Parent Attachment: Definition, Types, Antecedents, Measurement and Outcome. *Paediatrics & Child Health,* 9(8), 541–545.

Bernet, W., Wamboldt, M. Z. & Narrow, W. E. (2016). Child Affected by Parental Relationship Distress. *Journal of the American Academy of Child & Adolescent Psychiatry,* 55(7), 571–579.

Bernier, A., Carlson, S. M., Bordeleau, S. & Carrier, J. (2010). Relations between Physiological and Cognitive Regulatory Systems: Infant Sleep Regulation and Subsequent Executive Functioning. *Child Development,* 81(6), 1739–1752.

Bethell, C., Gombojav, N., Solloway, M. & Wissow, L. (2016). Adverse Childhood Experiences, Resilience and Mindfulness-Based Approaches: Common Denominator Issues for Children with Emotional, Mental, or Behavioral Problems. *Child and Adolescent Psychiatric Clinics of North America,* 25(2), 139–156.

Beutel, M. E., Tibubos, A. N., Klein, E. M., Schmutzer, G., Reiner, I., Kocalevent, R.-D. & Brähler, E. (2017). Childhood Adversities and Distress – The Role of Resilience in a Representative Sample. *PLoS ONE,* 12(3), e0173826.

Bianchi, S. M. (1995). The Changing Demographic and Socioeconomic Characteristics of Single-Parent Families. *Marriage and Family Review,* 20(1–2), 71–97.

Bigelow, A. E., Beebe, B., Power, M., Stafford, A.-L., Ewing, J., Egleson, A. & Kaminer, T. (2018). Longitudinal Relations among Maternal Depressive Symptoms, Maternal Mind-Mindedness, and Infant Attachment Behavior. *Infant Behavior and Development,* 51, 33–44.

Bigelow, B. J., & La Gaipa, J. J. (1980). The Development of Friendship Values and Choice. In H. C. Foot, A. J. Chapman & J. R. Smith (Eds.), *Friendship and Social Relations in Children* (pp. 15–44). New York: John Wiley & Sons.

Blackwell, D. L. (2010). Family Structure and Children's Health in the United States: Findings from the National Health Interview Survey, 2001–2007. National Center for Health Statistics. *Vital Health Statistics,* 10(246).

Bonds, D. D. & Gondoli, D. M. (2007). Examining the Process by Which Marital Adjustment Affects Maternal Warmth: The Role of Coparenting Support as a Mediator. *J Fam Psychol.,* 21(2), 288–296.

Bornstein, M. H. (2007). On the Significance of Social Relationships in the Development of Children's Earliest Symbolic Play: An Ecological Perspective. In A. Göncü & S. Gaskins (Eds.), *Play and Development* (pp. 101–129). Mahwah, NJ: Erlbaum.

Bornstein, M. H. (2012). Cultural Approaches to Parenting. *Parenting, Science and Practice*, 12(2–3), 212–221.

Bornstein, M. H. & Lansford, J. E. (2010). Parenting. In M. H. Bornstein (Ed.), *The Handbook of Cross-cultural Developmental Science* (pp. 259–277). New York: Taylor & Francis.

Bow, J. N., Gould, J. W. & Flens, J. R. (2009). Examining Parental Alienation in Child Custody Cases: A Survey of Mental Health and Legal Professionals. *The American Journal of Family Therapy*, 37(2), 127–145.

Bowlby, J. (1960). Separation Anxiety. *International Journal of Psychoanalysis*, 41, 89–113.

Bowlby, J. (1982). *Attachment and Loss, Vol. 1. Attachment.* London: Hogarth Press.

Bowlby, J. (2005). *A Secure Base.* London: Routledge.

Brackett, M., Divecha, D. & Stern, R. (2015). *Teaching Teenagers to Develop Their Emotional Intelligence.* Boston, MA: Harvard Business Review. Retrieved on December 2, 2018 from https://hbr.org/2015/05/teaching-teenagers-to-develop-their-emotional-intelligence

Bradberry, T. & Greaves, J. (2009). *Emotional Intelligence 2.0.* TalentSmart; Enhanced.

Braithwaite, D. O. & Baxter, L. A. (2006). You're My Parent but You're Not: Dialectical Tensions in Stepchildren's Perceptions about Communicating with the Nonresidential Parent. *Journal of Applied Communication*, 34, 30–48.

Braver, S. L., Wolchik, S. A., Sandler, I. R., Sheets, V. L., Fogas, B. & Bay, C. (1993). A Longitudinal Study of Noncustodial Parents: Parents without Children. *Journal of Family Psychology*, 7, 9–23.

Brody, L. R. (1999). *Gender, Emotion, and the Family.* Cambridge, MA: Harvard University Press.

Brody, L. R. & Hall, J. A. (2000). Gender, Emotion, and Expression. In M. Lewis & J. M. Haviland-Jones (Eds.), *Handbook of Emotions*, 2nd edn. (pp. 338–349). New York: Guilford

Brown, B. B. & Larson, J. (2009). 3 Peer Relationships in Adolescence. In In R. M. Lerner & L. Steinberg (Eds.). *Contextual Influences on Adolescent Development, Volume 2* (pp. 74–103). Hoboken, NJ: John Wiley & Sons Inc.

Brown, C. A. (2013). The Father Doctor. Germantown, MD: National Fatherhood Initiative. Retrieved on December 2, 2018 from https://www.fatherhood.org/bid/183377/4-great-resources-for-single-dads

Brown, J. R. & Dunn, J. (1992). Talk with Your Mother or Your Sibling? Developmental Changes in Early Family Conversations About Feelings. *Child Development*, 63, 336–349.

Burkett, L. & Osborne, R. (1996). *Financial Parenting.* Colorado Springs, CO: Chariot Victor Pub.

Burstein, M., Ameli-Grillon, L. & Merikangas, K. R. (2011). Shyness versus Social Phobia in US Youth. *Pediatrics*, 128(5), 917–925.

Burt, K. B., Van Dulmen, M. H. M., Carlivati, J., Egeland, B., Sroufe, L. A., Forman, D. R., Appleyard, K. & Carlson, E. A. (2005). Mediating Links between Maternal Depression and Offspring Psychopathology: The Importance of Independent Data. *Journal of Child Psychology and Psychiatry*, 46, 490–499.

Buttelmann, D. Carpenter, M. & Tomasello, M. (2009). Eighteen-month-old Infants Show False Belief Understanding in an Active Helping Paradigm. *Cognition*, 112, 337–342.

Cabanac, M. (2002). What is Emotion? *Behavioural Processes*, 60, 69–83.

Calafat, A., García, F., Juan, M., Becoña, E. & Fernández-Hermida, J. R. (2014). Which Parenting Style is More Protective against Adolescent Substance Use? Evidence within the European Context. *Drug Alcohol Depend*, 138, 185–192.

Calkins, S. D., Gill, K. L., Johnson, M. C. & Smith, C. L. (1999). Emotional Reactivity and Emotional Regulation Strategies as Predictors of Social Behavior with Peers during Toddlerhood. *Social Development*, 8, 310–334.

Carlson, M. J. & Högnäs, R. S. (2011). Coparenting in Fragile Families: Understanding How Parents Work Together after a Nonmarital Birth. In J. P. McHale & K. M. Lindahl (Eds.), *Coparenting: A Conceptual and Clinical Examination of Family Systems.* Washington, DC: American Psychological Association.

Carlson, M. J. & McLanahan, S. S. (2010). Fathers in Fragile Families. In M. E. Lamb (Ed.), *The Role of the Father in Child Development* (pp. 241–269). Hoboken, NJ: John Wiley & Sons.

Cassidy, J. (1994). Emotion Regulation: Influence of Attachment Relationships. *Monographs of the Society for Research in Child Development*, 59(2/3, The Development of Emotion Regulation: Biological and Behavioral Considerations), 228–249.

Catsambis, S., & Garland, J. E. (1997). *Parental Involvement in Students' Education During Middle School and High School.* Center for Research on the Education of Students Placed At Risk (CRESPAR), Report no. 18.

Cauthen, N. (2002). *Improving Children's Economic Security: Research Findings about Increasing Family Income through Employment: Policies that Improve Family Income Matter for Children.* New York: National Center for Children in Poverty.

Centers for Disease Control and Prevention (CDC) (2018a). *Preventing Suicide.* Retrieved on June 9, 2018 from https://www.cdc.gov/violenceprevention/pdf/suicide-factsheet.pdf

Centers for Disease Control and Prevention (CDC) (2018b). *Suicide Rates Rising across the U.S.* Retrieved on June 9, 2018 from https://www.cdc.gov/media/releases/2018/p0607-suicide-prevention.html

Centers for Disease Control and Prevention (CDC) (2018c). *Vital Signs.* Retrieved on June 9, 2018 from https://www.cdc.gov/vitalsigns/

Cha, C. B. & Nowak, M. K. (2009). Emotional Intelligence is a Protective Factor for Suicidal Behavior. *Journal of the American Academy of Child and Adolescent Psychiatry,* 48, 422–430.

Chan, T. W. & Koo, A. (2008). *Parenting Style and Youth Outcome in the UK.* Retrieved on June 10, 2018 from https://www.sociology.ox.ac.uk/materials/papers/2008-06.pdf

Chao, R. (2001). Extending Research on the Consequences of Parenting Style for Chinese Americans and European Americans. *Child Development,* 72, 1832–1843.

Chaplin, T. M., Cole, P. M. & Zahn-Waxler, C. (2005). Parental Socialization of Emotion Expression: Gender Differences and Relations to Child Adjustment. *Emotion,* 5(1), 80–88.

Chaplin, T. M., Casey, J., Sinha, R. & Mayes, L. C. (2010). Gender Differences in Caregiver Emotion Socialization of Low-Income Toddlers. *New Directions for Child and Adolescent Development,* 128, 11–27.

Chapman, R. S. (2000). Children's Language Learning: An Interactionist Perspective. *Journal of Child Psychology and Psychiatry,* 41(1), 33–54.

Cheadle, J., Amato, P. & King, V. (2010). Patterns of Nonresident Father Contact. *Demography,* 47(1), 205–225.

Chicot, R. (2015). *The Calm and Happy Toddler: Gentle Solutions to Tantrums, Night Waking, Potty Training and More.* New York: Random House.

Child Trends Databank (2015). *Parental Expectations for Their Children's Academic Attainment.* Retrieved on December 2, 2018 from https://www.childtrends.org/indicators/parental-expectations-for-their-childrens-academic-attainment

Child Trends Databank (November, 2016). *Breastfeeding.* Retrieved on August 24, 2017 from https://www.childtrends.org/?indicators=breastfeeding

Child Trends Databank (February 15, 2017). *Supporting Youth in Foster Care: Research-Based Policy Recommendations for Executive and Legislative Officials in 2017.* Retrieved on December 2, 2018 from https://www.childtrends.org/publications/supporting-youth-foster-care-research-based-policy-recommendations-executive-legislative-officials-2017

Child Trends Databank (April 2018). *High-Quality Preschool can Support Healthy Development and Learning.* Retrieved on December 2, 2018 from https://www.childtrends.org/wp-content/uploads/2018/05/PreschoolFadeOutFactSheet_ChildTrends_April2018.pdf

Child Welfare Information Gateway (2017). *Social Media: Tips for Foster Parents and Caregivers.* Washington, DC: U.S. Department of Health and Human Services, Children's Bureau.

Chow, T. S. & Wan, H. Y. (2017). Is There Any 'Facebook Depression'? Exploring the Moderating Roles of Neuroticism, Facebook Social Comparison and Envy. *Personality and Individual Differences,* 119277–119282.

Cillessen, A. H. N. & Mayeux, L. (2004). Sociometric Status and Peer Group Behavior: Previous Findings and Current Directions. In J. B. Kupersmidt & K. A. Dodge (Eds.), *Children's Peer Relations: From Development to Intervention* (pp. 3–20). Washington, DC: American Psychological Association.

Clutton-Brock, T. H. (1991). *The Evolution of Parental Care.* Princeton, NJ: Princeton University Press.

Cohn, D.'V. and Passel, J. S. (2018). *A Record 64 Million Americans Live in Multigenerational Households.* Washington, DC: Pew Research Center.

Cole, P. M., Michel, M. K. & Teti, L. O. (1994). The Development of Emotion Regulation and Dysregulation: A Clinical Perspective. *Monogr Soc Res Child Dev.,* 59(2–3), 73–100.

Coles, R. L. (2015). Single-father Families: A Review of the Literature. *Journal of Family Theory & Review,* 7(2), 144–166.

Coltrane S. & Shih, K. Y. (2010). Gender and the Division of Labor. In J. Chrisler & D. McCreary (Eds.), *Handbook of Gender Research in Psychology.* New York: Springer.

Connolly, J., Furman, W. & Konarski, R. (2000). The Role of Peers in the Emergence of Heterosexual Romantic Relationships in Adolescence. *Child Development,* 71, 1395–1408.

Cooper, D. (1971). *The Death of the Family.* Harmondsworth, London: Penguin.

Corsaro, W. (2005). *The Sociology of Childhood.* London; Thousand Oaks; New Delhi: Sage.

Corsten, M. (1999). The Time of Generations. *Time and Society,* 8(2), 249–272.

Costello, C. G. (1993). *Symptoms of Depression.* Oxford, UK: John Wiley & Sons.

Coyne, J. C. & Downey, G. (1991). Social Factors and Psychopathology: Stress, Social Support and Coping Processes. *Annual Review of Psychology,* 42, 401–425.

Crosnoe, R. (2001). Academic Orientation and Parental Involvement in Education during High School. *Sociology of Education,* 74(3).

Dahlberg, G. & Moss, P. (2005). *Ethics and Politics in Early Childhood Education.* London: Routledge Falmer.

Denham, S. A. (2007). Dealing with Feelings: How Children Negotiate the Worlds of Emotions and Social Relationships. *Cognition, Brain, Behavior,* 11, 1–48.

Denham, S. A. & Burger, C. (1991). Observational Validation of Ratings of Preschoolers' Social Competence and Behavior Problems. *Child Study Journal,* 21, 185–201.

Denham, S. A., Cook, M. & Zoller, D. (1992). Baby Looks Very Sad: Implications of Conversations about Feelings between Mother and Preschooler. *British Journal of Developmental Psychology,* 10, 301–315.

Denham, S. A., Blair, K. A., DeMulder, E., Levitas, J., Sawyer, K. S., Auerbach-Major, S. T. & Queenan, T. (2003). Preschoolers' Emotional Competence: Pathway to Social Competence? *Child Development,* 74, 238–256.

Denham, S. A., Bassett, H. H., Way, E., Mincic, M., Zinsser, K. & Graling, K. (2012). Preschoolers' Emotion Knowledge: Self-Regulatory Foundations, and Predictions of Early School Success. *Cognition & Emotion,* 26(4), 667–679.

DeRose, L. (2017). *Cohabitation Contributes to Family Instability across the Globe.* Charlottesville, VA: Institute for Family Studies.

Diener, M. L. & Lucas, R. E. (2004). Adults Desires for Children's Emotions across 48 Countries Associations with Individual and National Characteristics. *Journal of Cross-Cultural Psychology,* 35(5), 525–547.

Doherty, J. & Hughes, M. (2009). *Child Development; Theory into Practice 0–11.* Harlow, Essex: Pearson.

Domitrovich, C. E. (2017). Promoting Alternative Thinking Strategies (PATHS) for Preschool. Bethesda, MD: Child Trends. Retrieved on December 2, 2018 from: https://www.childtrends.org/programs/promoting-alternative-thinking-strategies-paths-for-preschool

Dorn, L. D. & Biro, F. M. (2011). Puberty and Its Measurement: A Decade in Review. [Review]. *Journal of Research on Adolescence,* 21(1), 180–195.

Dougherty, L. R. (2006). Children's Emotionality and Social Status: A Meta-Analytic Review. *Social Development,* 15, 394–417.

Dufur, M. J., Hoffmann, J. P. & Erickson, L. D. (2018). Single Parenthood and Adolescent Sexual Outcomes. *Journal of Child and Family Studies,* 27(3), 802–815.

Duncan, G. J., Dowsett, C. J., Claessens, A., Magnuson, K., Huston, A. C., Klebanov, P. & Japel, C. (2007). School Readiness and Later Achievement. *Developmental Psychology,* 43(6), 1428–1446.

Edin, K., Nelson, T. & Reed, J. M. (2011). Daddy Baby; Momma, Maybe: Low Income Urban Fathers and the "Package Deal" of Family Life (pp. 85–107). In M. Carlson & P. England (Eds.), *Social Class and Changing Families in an Unequal America.* Stanford, CA: Stanford University Press.

Ehrenreich, B. & English, D. (2005). *For Her Own Good: Two Centuries of the Experts Advice to Women.* New York: Random House.

Eickmeyer, K. J. (2017). American Children's Family Structure: Single Parent Families. *Family Profiles.* FP- 17–17. Bowling Green, OH: NCFMR.

Eirini, F. & Buchanan, A. (2014). Childhood Families of Homeless and Poor Adults in Britain: A Prospective Study. *Journal of Economic Psychology,* 25(1), 1–14.

Eisenberg, N. (2007). Empathy-Related Responding: Its Role in Positive Development and Socialization Correlates. In R. K. Silbereisen & R. M. Lerner (Eds.), *Approaches to Positive Youth Development* (pp. 75–91). London, UK: Sage.

Eisenberg, N. (2009). Eight Tips to Developing Caring Kids. In D. Streight (Ed.), *Good Things To Do: Expert Suggestions for Fostering Goodness in Kids.* Portland, OR: The Center for Spiritual and Ethical Education.

Eisenberg, N. (2012). Temperamental Effortful Control (Self-Regulation). In R. E. Tremblay, M. Boivin, R De V. Peters & M. K. Rothbart (Eds.), *Encyclopedia on Early Childhood Development.* Retrieved on December 2, 2018 from http://www.child-encyclopedia.com/temperament/according-experts/temperamental-effortful-control-self-regulation

Eisenberg, N., Fabes, R. A., Nyman, M., Bernzweig, J. & Pinuelas, A. (1994). The Relations of Emotionality and Regulation to Children's Anger-related Reactions. *Child Development,* 65, 109–128.

Eisenberg, N., Cumberland, A. & Spinrad, T. L. (1998). Parental Socialization of Emotion. *Psychological Inquiry,* 9(4), 241–273.

Eisenberg, N., Smith, C. & Spinrad, T. L. (2011). Effortful Control: Relations with Emotion Regulation, Adjustment, and Socialization in Childhood. In R. F. Baumeister & K. D. Vohs (Eds.), *Handbook of Self-Regulation: Research, Theory, and Applications* (pp. 263–283). New York: Guilford.

ElHage, A. (2017). *Five Facts about Today's Single Fathers.* Charlottesville, VA: Institute for Family Studies.

Ellis, R. R. & Simmons, T. (2014). *Coresident Grandparents and Their Grandchildren: 2012.* U.S. Census Bureau.

Ellison, N. B., Steinfield, C. & Lampe, C. (2007). The Benefits of Facebook "Friends": Social Capital and College Students' Use of Online Social Network Sites. *Journal of Computer Mediated Communication,* 12, 1143–1168.

Emig, C. (2018). Kerner and Kids: Work Remains. In F. Harris. & A. Curtis (Eds.), *Healing Our Divided Society: Investing in America Fifty Years after the Kerner Report.* Philadelphia, PA: Temple University Press.

Epstein, J. L. (1986). Parents' Reactions to Teacher Practices of Parent Involvement. *The Elementary School Journal*, 86, 278–294.

Epstein, J. L. & Dauber, S. L. (1991). School Programs and Teacher Practices of Parent Involvement in Inner-City Elementary and Middle Schools. *The Elementary School Journal*, 91(3), 289–305.

Epstein, L. H., Valoski, A. M., Vara, L. S., Mccurely, J., Winiewski, L., Kalarchain, M. & Shrager, L. R. (1995). Effects of Decreasing Sedentary Behavior and Increasing Activity on Weight Change in Obese Children. *American Psychological Association*, 14(2), 109–115.

Erath, S. A., Flanagan, K. S. & Bierman, K. L. (2007). Social Anxiety and Peer Relations in Early Adolescence: Behavioral and Cognitive Factors. *Journal of Abnormal Child Psychology*, 35(3), 405–416.

Erickson, J. (2015). Fathers Don't Mother and Mothers Don't Father: What Social Science Research Indicates about the Distinctive Contributions of Mothers and Fathers to Children's Development (April 18, 2015). *SSRN*.

Espada-Campos, S. (2018). *Third Culture Kids (TCKs) Go to College: A Retrospective Narrative Inquiry of International Upbringing and College Engagement*. Doctorate in Social Work (DSW) Dissertations, 113. https://repository.upenn.edu/edissertations_sp2/113

Eunice Kennedy Shriver National Institute of Child Health and Human Development, NIH, DHHS (2001). *Adventures in Parenting* (00–4842). Washington, DC: U.S. Government Printing Office.

Fabes, R. A. & Eisenberg, N. (1992). Young Children's Coping with Interpersonal Anger. *Child Development*, 63, 116–128.

Fabes, R. A., Eisenberg, N., Nyman, M. & Michealieu, Q. (1991). Young Children's Appraisals of Others' Spontaneous Emotional Reactions. *Developmental Psychology*, 27, 858–866.

Fabes, R. A., Eisenberg, N., Jones, S., Smith, M., Guthrie, I., Poulin, R., Shepard, S. & Friedman, J. (1999). Regulation, Emotionality, and Preschoolers' Socially Competent Peer Interactions. *Child Development*, 70, 432–442.

Falci, C. (2006). Family Structure, Closeness to Residential and Nonresidential Parents, and Psychological Distress in Early and Middle Adolescence. *The Sociological Quarterly*, 47, 123–146.

Farrell, A., Kagan, S. L. & Tisdall, E. K. M. (2015). *The SAGE Handbook of Early Childhood Research*. London: Sage.

Feinberg, M. E. (2002). Coparenting and the Transition to Parenthood: A Framework for Prevention. *Clinical Child and Family Psychology Review*, 5, 173–195.

Festinger, L (1954). A Theory of Social Comparison Processes. *Human Relations*, 7(2), 117–140.

Field, T. (1992). Infants of Depressed Mothers. *Development and Psychopathology*, 4, 49–66.

Field, T. (2002). Infants' Need for Touch. *Human Development*, 45(2), 100–103.

Flouri, E. & Buchanan, A. (June, 2004). Early Father's And Mother's Involvement and Child's Later Educational Outcomes. *British Journal of Educational Psychology*, 74(2), 141–153.

Forrest-Bank, S. S. & Jenson, J. M. (2015). The Relationship among Childhood Risk and Protective Factors, Racial Microaggression and Ethnic Identity, and Academic Self Efficacy and Anti-Social Behavior in Young Adulthood. *Children and Youth Services Review*, 50, 64–74.

Frazier, B. N., Gelman, S. A. & Wellman, H. M. (2009). Preschoolers' Search for Explanatory Information within Adult–Child Conversation. *Child Development*, 80(6), 1592.

Freeman, D. C. (2016). *Essays in Modern Stylistics*. New York: Routledge.

Freeman, M. (1998). The Sociology of Childhood and Children's Rights. *The International Journal of Children's Rights*, 6(4), 33–44.

Gabriel, N. (2017). *The Sociology of Early Childhood: Critical Perspectives*. Los Angeles, CA: Sage.

Gallo, E. & Gallo, J. (2005). *The Financially Intelligent Parent: 8 Steps to Raising Successful, Generous, Responsible Children*. New York: NAL Trade.

Gardner, H. (1995). Reflections on Multiple Intelligence: Myths and Messages. *Phi Delta Kappa*, 77(3), 200–203 .

Gardner, R. A., Sauber, R. S. & Lorandos, D. (2006). *The International Handbook of Parental Alienation Syndrome: Conceptual, Clinical and Legal Considerations* (American Series in Behavioral Science and Law). Springfield, IL: Charles C Thomas Pub Ltd.

Gartner, L. M. & Eidelman, A. (2005). Breastfeeding and the Use of Human Milk. *Pediatrics*, 115(2), 496–506.

Gates, G. J. (February 2013). *GBT Parenting in the United States*. Los Angeles, CA: The William Institute.

Gatrell, C. (2005). Hard Labour: The Sociology of Parenthood, *Gender and Society*, 21(2), 298–300.

Geddes, J. K. (2018). The Secret Life of Kids Online: What You Need to Know. *Parenting*. Retrieved on May 13, 2018 from https://www.parenting.com/article/kids-social-networking

Geller, A., Cooper, C. E., Garfinkel, I., Schwartz-Soicher, O. & Mincy, R. B. (2012). Beyond Absenteeism: Father Incarceration and Child Development. *Demography*, 49, 49–76.

Georgiou, S. N., Fousiani, K., Michaelides, M. & Stavrinides, P. (2013). Cultural Value Orientation and Authoritarian

Parenting as Parameters of Bullying and Victimization at School. *Int J Psychol.*, 48(1), 69–78.

Gibbons, F. X. (1986). Social Comparison and Depression: Company's Effect on Misery. *Journal of Personality and Social Psychology,* 51(1), 140–148.

Gibbs, B. G., Shah, P., Downey, D. B. & Jarvis, J. A. (2017). The Asian American Advantage in Math among Young Children: The Complex Role of Parenting. *Sociological Perspectives*, 60(2), 315–337.

Gibson, J. J. (1979). *The Ecological Approach to Visual Perception.* Boston, MA: Houghton Mifflin.

Gilliom, M., Shaw, D. S., Beck, J. E., Schonberg, M. A. & Lukon, J. L. (2002). Anger Regulation in Disadvantaged Preschool Boys: Strategies, Antecedents, and the Development of Self-Control. *Developmental Psychology,* 38, 222–235.

Ginsburg, K. R. (2007). The Importance of Play in Promoting Healthy Child Development and Maintaining Strong Parent–Child Bonds. *American Academy of Pediatrics,* 119(1), 182–191.

Goldberg, A. E. & Smith, J. Z. (2011). Stigma, Social Context, and Mental Health: Lesbian and Gay Couples across the Transition to Adoptive Parenthood. *Journal of Counseling Psychology*, 58, 139–150.

Goldsmith, H. H., Pollak, S. D. & Davidson, R. J. (2008). Developmental Neuroscience Perspectives on Emotion Regulation. *Child Development Perspectives*, 2, 132–140.

Goleman, D. (2005). *Emotional Intelligence: Why It Can Matter More Than IQ.* New York: Bantam Books.

Goleman, D. (2007). *Social Intelligence: The New Science of Human Relationships.* New York: Bantam Books.

Golombok, S. & Tasker, F. (2010). Gay Fathers. In M. Lamb (Ed.), *The Role of the Father in Child Development* (pp. 319–340). Hoboken, NJ: Wiley.

Golombok, S., Blake, L., Slutsky, J., Raffanello, E., Roman, G. D. & Ehrhardt, A. (2017). Parenting and the Adjustment of Children Born to Gay Fathers through Surrogacy. *Child Development,* 89(4), 1223–1233.

Gómez-Ortiz, O., Romera, E. M. & Ortega-Ruiz, R. (2016). Parenting Styles and Bullying. The Mediating Role of Parental Psychological Aggression and Physical Punishment. *Child Abuse Negl.*, 51, 132–143.

Gordon, S. L. (1989). The Socialization of Children's Emotions: Emotional Culture, Competence, and Exposure. In C. I. Saarni & P. Harris (Eds.), *Children's Understanding of Emotion.* New York: Cambridge University Press.

Gottman, J. M., Katz, L. F. & Hooven, C. (1996). Parental Meta-emotion Philosophy and the Emotional Life of Families: Theoretical Models and Preliminary Data. *Journal of Family Psychology*, 10, 243–268.

Gottman, J. M., Katz, L. F. & Hooven, C. (1997). *Meta-emotion: How Families Communicate Emotionally.* Hillsdale, NJ: Lawrence Erlbaum Associates.

Gottman, J., Declaire, J. & Goleman, D. (1998). *Raising An Emotionally Intelligent Child: The Heart of Parenting.* New York: Simon & Schuster.

Gould, M. S., Greenberg, T., Velting, D. M. & Shaffer, D. (2003). Youth Suicide Risk and Preventive Interventions: A Review of the Past 10 Years. *J Am Acad Child Adolesc Psychiatry*, 42(4), 386–405.

Gouyon, M. (2004). L'aide aux devoirs apportée par les parents. INSEE première n. 996.

Grabe, H. J., Schulz, A., Schmidt, C. O., Appel, K., Driessen, M., Wingenfeld. K. *et al.* (2012). A Brief Instrument for the Assessment of Childhood Abuse and Neglect: The Childhood Trauma Screener (CTS). *Psychiatrische Praxis,* 39(3), 109–115.

Grall, T. (2018). *Custodial Mothers and Fathers and Their Child Support: 2015.* U.S. Census Bureau.

Gray, P. (2013). *Free to Learn: Why Unleashing the Instinct to Play Will Make Our Children Happier, More Self-Reliant, and Better Students for Life.* New York: Basic Books.

Greenoug, W. T. (1990). Brain Storage of Information from Cutaneous and Other Modalities in Development and Adulthood. In K. E. Barnard & T. B. Brazelton (Eds.), *Touch: The Foundation of Experience* (pp. 97–126). Madison, CT: International Universities Press.

Grey, J., Geraghty, R. & Ralph, D. (2013). Young Grandchildren and Their Grandparents: Continuity and Change across Four Birth Cohorts. *Families, Relationships and Societies*, 2, 2.

Grolnick, W. S., Bridges, L. J. & Connell, J. P. (1996). Emotion Regulation in Two-year Olds: Strategies and Emotional Expression in Four Contexts. *Child Development,* 67, 928–941.

Gross, J. J. (2002). Emotion Regulation: Affective, Cognitive, and Social Consequences. *Psychophysiology,* 39, 281–291.

Grotewell, P. & Burton, Y. (2008). *Early Childhood Education: Issues and Developments.* New York: Nova Sciences Publishers, Inc.

Gudmundson, J. A. & Leerkes, E. M. (2011). Links between Mothers' Coping Styles, Toddler Reactivity, and Sensitivity to Toddler's Negative Emotions. *Infant Behavior and Development,* 35(1), 158–166.

Gurholt, K. P. & Sanderud, J. R. (2016). Curious Play: Children's Exploration of Nature. *Journal of Adventure Education and Outdoor Learning,* 16(4), 318–329.

Guroglu, B., van Lieshout, C. F. M., Haselager, G. J. T. & Scholte, R. H. J. (2007). Similarity and Complementarity of Behavioral Profiles of Friendship Types and Types of

Friends: Friendships and Psychosocial Adjustment. *Journal of Research on Adolescence*, 17, 357–386.

Guryan, J., Hurst, E. & Kearney, M. (2008), Parental Education and Parental Time with Children. *Journal of Economic Perspectives*, 22(3), 23–46.

Haferkamp, N., Eimler, S. C., Papadakis, A. & Kruck, J. V. (2012). Men Are from Mars, Women Are from Venus? Examining Gender Differences in Self-Presentation on Social Networking Sites. *Cyberpsychology, Behavior, and Social Networking*, 15(2), 91–98.

Halberstadt, A., Denham, S. & Dunsmore, J. (2001). Affective Social Competence. *Social Development*, 10, 79–119.

Haller, S. P. W., Cohen Kadosh, K., Scerif, G. & Lau, J. Y. F. (2013). Social Anxiety Disorder in Adolescence: How Developmental Cognitive Neuroscience Findings May Shape Understanding and Interventions for Psychopathology, *Developmental Cognitive Neuroscience*, 13, 11–20.

Hamilton, L., Roksa, J. & Nielsen, K. (2018). Providing a "Leg Up": Parental Involvement and Opportunity Hoarding in College. *SAGE Journals*, 91(2), 111–131.

Harwood, R. L., Schoelmerich, A., Schulze, P. A. & Gonzalez, Z. (1999). Cultural Differences in Maternal Beliefs and Behaviors. *Child Development*, 70, 1005–1016.

Healthline (June 27, 2017). *Coping with Depression after a Loved One's Death*. Retrieved on December 2, 2018 from https://www.healthline.com/health/depression/death-loved-one

Hecht, T. (1998). *At Home in the Street: Street Children of Northeast Brazil*. Cambridge, UK: Cambridge University Press.

Hennessy, E., Hughes, S. O., Goldberg, J. P., Hyatt, R. R. & Economos, C. D. (2012). Permissive Parental Feeding Behavior is Associated with an Increase in Intake of Low-Nutrient-Dense Foods among American Children Living in Rural Communities. *J Acad Nutr Diet*, 112(1), 142–148.

Hertenstein, M. J. (2002). Touch: Its Communicative Functions in Infancy. *Human Development*, 45, 70–94.

Hirsh-Pasek, K. & Burchinal, M. (2006). Mother and Caregiver Sensitivity over Time: Predicting Language and Academic Outcomes with Variable- and Person-centered Approaches. *Merrill-Palmer Quarterly*, 449–485.

Hobbs, D. F. (1965). Parenthood as Crisis: A Third Study. *Journal of Marriage and the Family*, 27, 367–372.

Holmes, J. (1993). *John Bowlby and Attachment Theory. Makers of Modern Psychotherapy.* London: Routledge.

Holt, L. J., Mattanah, J. F. & Long, M. W. (2018). Change in Parental and Peer Relationship Quality during Emerging Adulthood: Implications for Academic, Social, and Emotional Functioning. *Sage Journals*, 35(5), 743–769.

Hoover-Dempsey, K. V. & Sandler, H. M. (1997). Why Do Parents Become Involved in Their Children's Education? *Review of Educational Research*, 67(1).

Hosier, D. (2016). *Childhood Trauma and Its Link to Borderline Personality Disorder*. childtraumarecovery.com publications.

Hoskins, D. H. (2014). Consequences of Parenting on Adolescence Outcomes. *Societies*, 4, 506–531.

Hrdy, S. B. (2009). *Mothers and Others: The Evolutionary Origins of Mutual Understanding*. The Belknap Press of Harvard University Press.

Human Rights Watch (2018). *Children's Rights*. Retrieved on May 10, 2018 from https://www.hrw.org/topic/childrens-rights

Ihinger-Tallman, M., Pasley, K. & Buehler, C. (1993). Developing a Middle-Range Theory of Father Involvement Postdivorce. *Journal of Family Issues*, 14, 550–571.

Izard, C., Fine, S., Schultz, D., Mostow, A., Ackerman, B. & Youngstrom, E. (2001). Emotion Knowledge as a Predictor of Social Behavior and Academic Competence in Children at Risk. *Psychological Science*, 12(1), 18–23.

Jackson, A. P., Brooks-Gunn, J., Huang, C. & Glassman, M. (2000). Single Mothers in Low-wage Jobs: Financial Strain, Parenting, and Preschoolers' Outcomes. *Child Development*, 71, 14–23.

Jackson, P. L., Brunet, E., Meltzoff, A. N. & Decety, J. (2006). Empathy Examined through the Neural Mechanisms Involved in Imagining How I Feel Versus How You Feel Pain: An Event-related fMRI Study. *Neuropsychologia*, 44, 752–61.

Jacob, B. A. & Ryan, J. (2018). *How Life Outside of a School Affects Student Performance in School*. Retrieved on June 13, 2018 from https://www.brookings.edu/research/how-life-outside-of-a-school-affects-student-performance-in-school/

Jago, R., Davison, K. K., Thompson, J. L., Page, A. S., Brockman, R. & Fox, K. R. (2011). Parental Sedentary Restriction, Maternal Parenting Style, and Television Viewing Among 10- to 11-Year-Olds. *Pediatrics*. Aug 22. [Epub ahead of print]

James, A. & Prout, A. (1990). *Constructing and Reconstructing Childhood*. London: Falmer.

James, A. & Prout, A. (2015). *Constructing and Reconstructing Childhood, Contemporary Issues in the Sociological Study of Childhood*. New York: Routledge.

James, A., Jenks, C. & Prout, A. (1998). *Theorizing Childhood*. Cambridge: Polity Press.

Jelenchick, L. A., Eickhoff, J. C. & Moreno, M. A. (2013). Facebook Depression? Social Networking Site Use and Depression in Older Adolescents. *The Journal of Adolescent Health*, 52(1), 128–130.

Jenks, C. (1996). *Childhood.* Abingdon: Routledge.

Jenson, J. M. & Fraser, M. W. (2011). *Social Policy for Children and Families: A Risk and Resilience Perspective*, 2nd edn. Thousand Oaks, CA: Sage.

Jiang, J. (2018). *Millennials Stand Out for Their Technology Use, but Older Generations also Embrace Digital Life.* Washington, DC: Pew Research Center.

Jordan, T. E. (1998). *Victorian Child Savers and Their Culture: A Thematic Evaluation.* Lewiston, NY: Edwin Mellen Press Ltd.

Jurkovic, G. J. (1998). Destructive Parentification in Families. In L. L'Abate (Ed.), *Family Psychopathology: The Relational Roots of Dysfunctional Behavior* (pp. 237–255). New York: The Guilford Press.

Juvonen, J., Le, V. N., Kaganoff, T., Augustine, C. & Constant, L. (2004). *Parental Involvement. Focus on the Wonder Years: Challenges Facing the American Middle School.* Santa Monica, CA: RAND Corporation.

Kail, R. V (2011). *Children and Their Development*, 6th edn. (My developmentlab Series). Englewood Cliffs, NJ: Prentice Hall.

Kaiser, A. P., Hancock, T. B., Cai, X., Foster, E. M. & Hester, P. P. (2000). Parent Reported Behavioral and Language Delays in Boys and Girls Enrolled in Head Start Classrooms. *Behavioral Disorders*, 26, 26–41.

Kalil, A., Ryan, R. & Chor, E. (2014). Time Investments in Children across Family Structures. *SAGE Journals*, 654(1), 150–168.

Kampmann, J. (2003). Barndomssociologi – fra marginaliseret provokatør til mainstream leverandør [Sociology of childhood – from marginalised provocateur to mainstream supplier]. *Dansk Sociologi*, 14(2), 79–93.

Kao, G. & Joyner, K. (2004). Do Race and Ethnicity Matter among Friends? Activities among Interracial, Interethnic, and Intraethnic Adolescent Friends. *Sociological Quarterly*, 45, 557–573.

Karen, R. (1998). *Becoming Attached: First Relationships and How They Shape Our Capacity to Love.* Oxford; New York: Oxford University Press.

Kartch, F. F. (2013). *Nonresidential Parenting: Parental Roles and Parent/Child Relationships.* Theses and Dissertations, Paper 122.

Kerr, D. C., Lopez, N. L., Olson, S. L. & Sameroff, A. J. (2004). Parental Discipline and Externalizing Behavior Problems in Early Childhood: The Roles of Moral Regulation and Child Gender. *J Abnorm Child Psychol.*, 32(4), 369–383.

Kessler, R. C., Berglund, P., Demler, O., Jin, R., Merikangas, K. R. & Walters, E. E. (2005). Lifetime Prevalence and Age-of Onset Distributions of DSM-IV Disorders in the National Comorbidity Survey Replication. *Archives of General Psychiatry*, 62(6), 593.

King, K. A., Vidourek, R. A. & Merianos, A. L. (2016). Authoritarian Parenting and Youth Depression: Results from a National Study. *J Prev Interv Community*, 44(2), 130–139.

Knafo, A. & Plomin, R. (2008). Prosocial Behavior from Early to Middle Childhood: Genetic and Environmental Influences on Stability and Change. *Developmental Psychology*, 42(5), 771–786.

Kochanska, G., Murray, K. & Harlan, E. T. (2000). Effortful Control in Early Childhood: Continuity and Change, Antecedents, and Implications for Social Development. *Developmental Psychology*, 36(2), 220–232.

Kopko, K. (2007). *Parenting Styles and Adolescents.* Retrieved on June 9, 2018 from https://www.human.cornell.edu/sites/default/files/PAM/Parenting/Parenting-20Styles-20and-20Adolescents.pdf

Kopp, C. B. & Neufeld, S. J. (2003). Emotional Development during Infancy. In R. J. Davidson, K. R. Scherer & H. H. Goldsmith (Eds.), *Handbook of Affective Sciences* (pp. 347–374). Series in Affective Science. Oxford, UK: Oxford University Press.

Kotila, L. E. & Kamp Dush, C. M. (2013). Involvement with Children and Low-Income Fathers' Psychological Well-Being. *Fathering*, 11(3), 306–326.

Kraftl, P. (2015). Alter-Childhoods: Biopolitics and Childhoods in Alternative Education Spaces. *Annals of the Association of American Geographers*, 105(1), 219–237.

Krohne, H. W., Pieper, M., Knoll, N. & Breimer, N. (2002). The Cognitive Regulation of Emotions: The Role of Success versus Failure Experience and Coping Dispositions. *Cognition and Emotion*, 16, 217–243.

Kuersten-Hogan, R. & McHale, J. P. (1998). Talking About Emotions during the Toddler Years and Beyond: Mothers' and Fathers' Coaching of Children's Emotion Understanding. *Infant Behavior and Development*, 21, 514.

Lamborn, S. D., Mounts, N. S., Steinberg, L. & Dornbusch, S. M. (1991). Patterns of Competence and Adjustment among Adolescents from Authoritative, Authoritarian, Indulgent, and Neglectful Families. *Child Development*, 62, 1049–1065.

Landry, S. H., Smith, K. E., Swank, P. R., Assel, M. A. & Vellet, S. (2001). Does Early Responsive Parenting Have a Special Importance for Children's Development or Is Consistency across Early Childhood Necessary? *Developmental Psychology*, 37, 387–403.

Lansbury, J. (2014). *No Bad Kids: Toddler Discipline without Shame.* Scotts Valley, CA: CreateSpace Independent Publishing Platform.

Lareau, A. (1987). Social Class Differences in Family–School Relationships: The Importance of Cultural Capital. *Sociology of Education*, 60(2), 73–85.

Lareau, A. (2011). *Unequal Childhoods: Class, Race, and Family Life.* Oakland, CA: University of California Press.

Lau, J. (2016). *Social Intelligence and the Next Generation.* King's College London and NCS. Retrieved on May 13, 2018 from http://www.ncsyes.co.uk/sites/default/files/Social%20Intelligence%20Report%20FINAL.PDF

Lean, R. E., Paul, R. A., Smyser, C. D. & Rogers, C. E. (2018). Maternal Intelligence Quotient (IQ) Predicts IQ and Language in Very Preterm Children at Age 5 Years. *J Child Psychol Psychiatry,* 59(2), 150–159.

Lee, J. & Zhou, M. (2014). The Success Frame and Achievement Paradox: The Costs and Consequences for Asian Americans. *Race and Social Problems,* 6, 38–55.

Lee, N. & Motzkau, J. (2011). Navigating the Bio-politics of Childhood. *Childhood,* 18(1), 7–19.

Lemish, D. (2007). *Children and Television.* Malden, MA: Wiley-Blackwell.

Leonard, M. (2016). The Sociology of Children, Childhood, and Generation. *An International Journal,* 12(3), 313–316.

Leung, K., Lau, S. & Lam, W. (1998). Parenting Styles and Academic Achievement: A Cross-cultural Study. *Merrill-Palmer Quarterly,* 44, 157–172.

Levine, J. A. (1976). *Who Will Raise the Children? New Options for Fathers (and Mothers).* New York: Lippincott.

Levitt, S. D & Dubner, S. J. (2005). *Freakonomics: A Rogue Economist Explores the Hidden Side of Everything.* New York: HarperCollins.

Lieberman, A. F. (1995). *The Emotional Life of the Toddler.* New York: Free Press.

Lincoln, C. E., & Mamiya, L. H. (1990). *The Black Church in the African American Experience.* Durham, NC: Duke University Press.

Lipps, G., Lowe, G. A., Gibson, R. C., Halliday, S., Morris, A., Clarke, N. & Wilson, R. N. (2012). Parenting and Depressive Symptoms among Adolescents in Four Caribbean Societies. *Child Adolesc Psychiatry Ment Health,* 6(1), 31.

Livingston, G. (2018a). *The Changing Profile of Unmarried Parents.* Washington, DC: Pew Research Center.

Livingston, G. (2018b). *About One-Third of U.S. Children Are Living with an Unmarried Parent.* Washington, DC: Pew Research Center.

Long, E. C., Aggen, S. H., Gardner, C. & Kendler, K. S. (2015). Differential Parenting and Risk for Psychopathology: A Monozygotic Twin Difference Approach. *Soc Psychiatry Psychiatr Epidemiol.,* 50(10), 1569–1576.

Lorandos, D., Bernet, W. & Sauber, S. R. (2013). Overview of Parental Alienation. In D. Lorandos, W. Bernet & S. R. Sauber (Eds.), *Parental Alienation: The Handbook for Mental Health and Legal Professionals.* Springfield, IL: Charles C. Thomas Publisher.

Lucas, R. E. & Gohm, C. L. (2000). Age and Sex Differences in Subjective Well-Being across Cultures. In E. Diener & E. M. Suh (Eds.), *Culture and Subjective Well-being* (pp. 291–318). Cambridge, MA: MIT Press.

Luk, J. W., Patock-Peckham, J. A., Medina, M., Terrell, N., Belton, D. & King, K. M. (2016). Bullying Perpetration and Victimization as Externalizing and Internalizing Pathways: A Retrospective Study Linking Parenting Styles and Self-Esteem to Depression, Alcohol Use, and Alcohol-Related Problems. *Subst Use Misuse,* 51(1), 113–125.

Luyckx, K., Tildesley, E. A., Soenens, B., Andrews, J. A., Hampson, S. E., Peterson, M. & Duriez, B. (2011). Parenting and Trajectories of Children's Maladaptive Behaviors: A 12-Year Prospective Community Study. *J Clin Child Adolesc Psychol.,* 40(3), 468–478.

Lynch, T. R., Cheavens, J. S., Morse, J. Q. & Rosenthal, M. Z. (2004). A Model Predicting Suicidal Ideation and Hopelessness in Depressed Older Adults: The Impact of Emotion Inhibition and Affect Intensity. *Aging & Mental Health,* 8(6), 486–497.

Lyons-Ruth, K., Wolfe, R. & Lyubchik, A. (2000). Depression and the Parenting of Young Children: Making the Case for Early Preventive Mental Health Services. *Harvard Review of Psychiatry,* 8, 148–153.

Manning, W. D. (2015). Cohabitation and Child Wellbeing. *The Future of Children,* 25(2), 51–66.

Manning, W. D. & Lamb, K. A. (2003). Adolescent Well-Being in Cohabiting, Married, and Single-parent Families. *Journal of Marriage and the Family,* 65(4), 876–893.

Markham, L. (2018). Social Intelligence for Toddlers. *AHA, Parenting,* Retrieved on May 2, 2018 from http://www.ahaparenting.com/parenting-tools/social-intelligence/toddlers

Marvin, R. S. & Britner, P. A. (2008). Normative Development: The Ontogeny of Attachment. In J. Cassidy & P. R. Shaver (Eds.), *Handbook of Attachment: Theory, Research and Clinical Applications* (pp. 269–94). New York; London: Guilford Press.

Mayall, B. (2000). The Sociology of Childhood in Relation to Children's Rights. *The International Journal of Children's Rights,* 8, 243–259.

Mayseless, O. (2005). Ontogeny of Attachment in Middle Childhood: Conceptualization of Normative Changes. In K. A. Kerns & R. A. Richardson (Eds.), *Attachment in Middle Childhood* (pp. 1–23). New York: Guilford Press.

McCallum, I. & Demie, F. (2010). Social Class, Ethnicity and Educational Performance. *Educational Research,* 43, 2, 147–159.

McCarthy, E. D. (1994). The Social Construction of Emotion: New Directions from Culture Theory. *Sodal Perspectives on Emotion,* 2, 267–279.

McDuff, D., Kodra, E., Kaliouby, R. & LaFrance, M. (2017). A Large-scale Analysis of Sex Differences in Facial Expressions. *PLoS ONE*, 12(4), e0173942.

McElwain, N. L., Halberstadt, A. G. & Volling, B. L. (2007). Mother- and Father-reported Reactions to Children's Negative Emotions: Relations to Young Children's Emotional Understanding and Friendship Quality. *Child Dev.*, 78(5), 1407–1425.

McHale, J. P. (2007). *Charting the Bumpy Road of Coparenthood: Understanding the Challenges of Family Life*. Washington, DC: Zero to Three.

McHale, J. P., & Sullivan, M. J. (2008). Family Systems. In M. Hersen & A. M. Gross (Eds.), *Handbook of Clinical Psychology* (pp. 192–226). Hoboken, NJ: Wiley.

Mckenna, L. (2012). Explaining Annette Lareau, or, Why Parenting Style Ensures Inequality. *The Atlantic*. Retrieved on June 9, 2018 from https://www.theatlantic.com/health/archive/2012/02/explaining-annette-lareau-or-why-parenting-style-ensures-inequality/253156/

McNeely, C. & Falci, C. (2004). School Connectedness and the Transition Into and Out of Health-Risk Behavior among Adolescents: A Comparison of Social Belonging and Teacher Support. *Journal of School Health*, 74(7), 284–292.

Merz, E. C., Zucker, T. A., Landry, S. H., Williams, J. M., Assel, M. & Taylor, H. B., (2015). The School Readiness Research Consortium. Parenting Predictors of Cognitive Skills and Emotion Knowledge in Socioeconomically Disadvantaged Preschoolers. *Journal of Experimental Child Psychology*, 132, 14–31.

Miller, A. L., Gouley, K. K., Seifer, R., Zakariski, A., Eugia, M. & Vergnani, M. (2005). Emotion Knowledge Skills in Low-Income Elementary School Children: Associations with Social Status and Peer Experiences. *Social Development*, 14, 637–651.

Mirabile, S. P. (2010). *Emotion Socialization, Emotional Competence, and Social Competence and Maladjustment in Early Childhood*. University of New Orleans Theses and Dissertations, 1159. https://scholarworks.uno.edu/td/1159

Monte, L. M. (March 2017). *Fertility Research Brief, Current Population Reports, P70BR-147*. Washington, DC: U.S. Census Bureau.

Montgomery, H. (2009). *An Introduction to Childhood: Anthropological Perspectives on Children's Lives*. Chichester: Wiley-Blackwell.

Moore, A. M. & Barker, G. G. (2012). Confused or Multicultural: Third Culture Individuals' Cultural Identity. *International Journal of Intercultural Relations*, 36(4), 553–562.

Moore, K. A., Jekielek, S. M. & Emig, C. (2002). *Marriage from a Child's Perspective: How Does Family Structure Affect Children, and What Can We Do About It? Child Trends*. Retrieved on June 19, 2016 from http://www.childtrends.org/wp-content/uploads/2013/03/MarriageRB602.pdf

Morales, A. (2015). Factors Affecting Third Culture Kids' (TCKs) Transition. *Journal of International Education Research*, 11(1), 51–56.

Morelli, G. A. & Rothbaum, F. (2007). Situating the Child in Context: Attachment Relationships and Self-Regulation in Different Cultures. In S. Kitayama & D. Cohen (Eds.), *Handbook of Cultural Psychology* (pp. 500–527). New York: Guilford Press.

Morrill, M. (2010). Pathways between Marriage and Parenting for Wives and Husbands: The Role of Coparenting. *Family Process*, 49(1), 59–73.

Morris, A., Silk, J. S., Steinberg, L., Sessa, F., Avenevoli, S. & Essex, M. J. (2002). Temperamental Vulnerability and Negative Parenting as Interacting Predictors of Child Adjustment. *Journal of Marriage and Family*, 64(2), 461–471.

Morris, P., Mattera, S. K., Castells, N., Bangser, M., Bierman, K. & Raver, C. (2014). *Impact Findings from the Head Start CARES Demonstration: National Evaluation of Three Approaches to Improving Preschoolers' Social and Emotional Competence*. OPRE Report 2014–44. Washington, DC: Office of Planning, Research, and Evaluation, Administration for Children and Families, U.S. Department of Health and Human Services.

Moszkowski, R. J., & Stack, D. M. (2007). Infant Touching Behavior during Mother–Infant Face-to-Face Interactions. *Wiley Online Library*, 16(3), 307–319.

Mulert, C., Seifert, C., Leicht, G., Kirsch, V., Ertl, M., Karch, S. *et al.* (2008). Single-trial Coupling of EEG and FMRI Reveals the Involvement of Early Anterior Cingulate Cortex Activation in Effortful Decision Making. *Neuroimage*, 42, 158–168.

Murphy, B. C., Eisenberg, N., Fabes, R. A., Shepard, S. & Guthrie, I. K. (1999). Consistency and Change in Children's Emotionality and Regulation: A Longitudinal Study. *Merrill-Palmer Quarterly*, 45, 413–444.

Murphy, S. E., Gallegos, M. I., Jacobvitz, D. B. & Hazen, N. L. (2017). Coparenting Dynamics: Mothers' and Fathers' Differential Support and Involvement. *Personal Relationships*, 24, 917–932.

Muter, V., Hulme, C., Snowling, M. J. & Stevenson, J. (2004). Phonemes, Rimes, Vocabulary, and Grammatical Skills as Foundations of Early Reading Development: Evidence from a Longitudinal Study. *Developmental Psychology*, 40(5), 665–681.

Nadkarni, A. & Hofmann, S. G. (2012). Why Do People Use Facebook? *Personality and Individual Differences*, 52(3), 243–249.

Natalier, K. & Hewitt, B. (2010). "It's Not Just About Money": Non-Resident Fathers' Perspectives on Paying Child Support. *Sociology,* 44, 489–505.

National Center for Education Statistics (NCES) (2001). *Entering Kindergarten.* Retrieved on June 13, 2018 from https://nces.ed.gov/pubs2001/2001035.pdf

Naughton, G., Hughes, P. & Smith, K. (2007). Early Childhood Professionals and Children's Rights: Tensions and Possibilities around the United Nations General Comment No. 7 on Children's Rights. *International Journal of Early Years Education,* 15(2), 161–170.

Nelson, D. R., Hammen, C., Brennan, P. A. & Ullman, J. B. (2003). The Impact of Maternal Depression on Adolescent Adjustment: The Role of Expressed Emotion. *Journal of Consulting and Clinical Psychology,* 71(5), 935–944.

Nemade, R. (2018). Sociology of Depression – Effects of Culture. *Gracepoint.* Retrieved on June 12, 2018 from https://www.gracepointwellness.org/5-depression-depression-related-conditions/article/13009-sociology-of-depression-effects-of-culture

Nik, S., Kamila, S., Musab, R. & Zaleha Sahak, S. (2014). Examining the Role of Financial Intelligence Quotient (FIQ) in Explaining Credit Card Usage Behavior: A Conceptual Framework. *Procedia: Social and Behavioral Sciences,* 130, 568–576.

Nock, M. K., Borges, G., Bromet, E. J. *et al.* (2008). Cross-national Prevalence and Risk Factors for Suicidal Ideation, Plans, and Attempts. *Br J Psychiatry,* 192, 98–105.

Nolen-Hoeksema, S. (2012). Emotion Regulation and Psychopathology: The Role of Gender. *Annual Review of Clinical Psychology,* 8(1), 161–187.

Nummenmaa, L., Glerean, E., Hari, R. & Hietanen, J. K. (2014). *Proceedings of the National Academy of Sciences,* 111(2), 646–651.

Oakley, M., Farr, R. H. & Scherer, D. G. (2017). Same-sex Parent Socialization: Understanding Gay and Lesbian Parenting Practices as Cultural Socialization. *Journal of GLBT Family Studies,* 13(1), 56–75.

O'Connor, S. (2001). *Orphan Trains: The Story of Charles Loring Brace and the Children He Saved.* Chicago: University of Chicago Press.

O'Keeffe, G. S., Clarke-Pearson, K. and Council on Families Clinical Report (2011). Clinical Report–The Impact of Social Media on Children, Adolescents, and Families. *American Academy of Pediatrics,* 127(4), 601–603.

Olivo, S. T. (2016). Effective Treatments for Emotional Regulation in Children. Needham, MA: Beyond Booksmart. Retrieved on June 13, 2018 from https://www.beyondbooksmart.com/executive-functioning-strategies-blog/effective-treatments-for-emotional-regulation-in-children

Olmstead, K. (May 25, 2017). *A Third of Americans Live in a Household with Three or More Smartphones.* Washington, DC: Pew Research Center.

Orth, U. & Robins, R. W. (2014). The Development of Self-Esteem. *SAGE Journals,* 23(5), 381–387.

Padilla-Walker, L., Carlo, G. & Memmott-Elison, K. M. (2017). Longitudinal Change in Adolescents' Prosocial Behavior toward Strangers, Friends, and Family. *Journal of Research on Adolescence.* doi: 10.1111/jora.12362.

Panadero, E. (2017). A Review of Self-Regulated Learning: Six Models and Four Directions for Research. *Frontiers in Psychology,* 8(442).

Parsons, T. (1954). *Essays in Sociological Theory.* Glencoe, IL: Free Press.

Patrick, R. B. & Gibbs, J. C. (2016). Maternal Acceptance: Its Contribution to Children's Favorable Perceptions of Discipline and Moral Identity. *J Genet Psychol.,* 177(3), 73–84.

Paulson, J. F., Dauber, S. E. & Leiferman, J. A. (2011). Parental Depression, Relationship Quality, and Nonresident Father Involvement with Their Infants. *Journal of Family Issues,* 32(4), 528–549.

Pelaez-Nogueras, M., Field, T., Gewirtz, J. L., Cigales, M., Gonzalez, A., Sanchez, A. & Richardson, S. C. (1997). The Effects of Systematic Stroking versus Tickling and Poking on Infant Behavior. *Journal of Applied Developmental Psychology,* 18, 169–178.

Pellerin, L. A. (2005). Applying Baumrind's Parenting Typology to High Schools: Toward a Middle-range Theory of Authoritative Socialization. *Social Science Research,* 34, 283–303.

Perelli-Harris, B. & Styrc, M. (2017). Mental Well-Being Differences in Cohabitation and Marriage: The Role of Childhood Selection. *Journal of Marriage and Family,* 80, 239–255.

Perlmutter, D. & Colman, C. (2008). *Raise a Smarter Child by Kindergarten: Raise IQ by up to 30 Points and Turn On Your Child's Smart Genes.* New York: Harmony.

Pew Research Center (2015a). *The Whys and Hows of Generations Research.* Retrieved on June 23, 2018 from http://www.people-press.org/2015/09/03/the-whys-and-hows-of-generations-research/#key-differences-between-the-generations

Pew Research Center (December 17, 2015b). *Parenting in America: Outlook, Worries, Aspirations Are Strongly Linked to Financial Situation.* Retrieved on August 11, 2017 from http://www.pewresearch.org/fact-tank/2015/12/17/key-takeaways-about-parenting/

Pew Research Center (2018). *Mobile Fact Sheet.* Retrieved on June 25, 2018 from http://www.pewinternet.org/fact-sheet/mobile/

Pinquart, M. (2017). Associations of Parenting Dimensions and Styles with Externalizing Problems of Children and Adolescents: An Updated Meta-analysis. *Dev Psychol.*, 53(5), 873–932.

Pinquart, M. & Kauser, R. (2018). Do the Associations of Parenting Styles with Behavior Problems and Academic Achievement Vary by Culture? Results from a Meta-analysis. *Cultural Diversity and Ethnic Minority Psychology,* 24(1), 75–100.

Piotrowski, J. T., Lapierre, M. A. & Linebarger, D. L. (2013). Investigating Correlates of Self-Regulation in Early Childhood with a Representative Sample of English-speaking American Families. *J Child Fam Stud.*, 22(3), 423–436.

Pollock, D. C. & Van Reken, R. E. (2001). *Third Culture Kids: The Experience of Growing up among Worlds.* London: Nicholas Brealey Publishing.

Posner, J., Kass, E. & Hulvershorn, L. (2014). Using Stimulants to Treat ADHD-related Emotional Lability. *Current Psychiatry Reports,* 16(10), 478.

Posner, M. I. & Rothbart, M. K. (1998). Attention, Self-regulation and Consciousness. *Philosophical Transactions of the Royal Society of London,* 353(1377), 1915–1927.

Posner, M. I. & Rothbart, M. K. (2007). Research on Attention Networks as a Model for the Integration of Psychological Science. *Annual Review of Psychology,* 58, 1–23.

Priel, B. & Besser, A. (2002). Perceptions of Early Relationships during the Transition to Motherhood: The Mediating Role of Social Support. *Infant Mental Health Journal,* 23, 343–360.

Prior, V. & Glaser, D. (2006). *Understanding Attachment and Attachment Disorders: Theory, Evidence and Practice.* Child and Adolescent Mental Health, RCPRTU. London; Philadelphia: Jessica Kingsley Publishers.

Prizant-Passal, S., Shechner, T. & Aderka, I. M. (2016). Social Anxiety and Internet Use – A Meta-analysis: What Do We Know? What Are We Missing? *Computers in Human Behavior,* 62, 221–229.

Prout, A. (2000). Childhood Bodies: Construction, Agency and Hybridity. In A. Prout & J. Campling (Eds.), *The Body, Childhood and Society.* London: Palgrave Macmillan.

Prout, A. (2005). *The Future of Childhood.* London: Routledge.

Pungello, E. P., Iruka, I. U., Dotterer, A. M., Mills-Koonce, R. & Reznick, J. S. (2009). The Effects of Socioeconomic Status, Race, and Parenting on Language Development in Early Childhood. Developmental Psychology, 45(2), 544–557.

Putnam, S. P. & Stifter, C. A. (2002). Development of Approach and Inhibition in the First Year: Parallel Findings from Motor Behavior, Temperament Ratings, and Directional Cardiac Response. Developmental Science, 5, 441–451.

Quadlin, N. (2018). The Mark of a Woman's Record: Gender and Academic Performance in Hiring. *SAGE Journals,* 83(2), 331–360.

Quennerstedt, A. & Quennerstedt, M. (2013). Researching Children's Rights in Education: Sociology of Childhood Encountering Educational Theory. *British Journal of Sociology of Education,* 35(1), 115–132.

Quinn, N. & Mageo, J. M. (2013). *Attachment Reconsidered: Cultural Perspectives on a Western Theory.* New York: Palgrave Macmillan.

Qvortrup, J. (1994). Childhood Matters: An Introduction. In J. Qvortrup, M. Bardy, G. Sgritta, & H. Wintersberger (Eds.), *Childhood Matters. Social Theory, Practice and Politics* (pp. 1–24). Aldershot: Avebury.

Raikes, H. A. & Thompson, R. A. (2006). Family Emotional Climate, Attachment Security and Young Children's Emotion Knowledge in a High-risk Sample. *British Journal of Developmental Psychology,* 24(1), 89–104.

Raley, S. & Bianchi S. (2006). Sons, Daughters, and Family Processes: Does Gender of Children Matter? *Annual Review of Sociology,* 32(1), 401–421.

Reed, M. A., Pien, D. L. & Rothbart, M. K. (1984). Inhibitory Self-control in Preschool children. *Merrill-Palmer Quarterly,* 30, 131–147.

Reeves, G. M., Postolache, T. T., & Snitker, S. (2008). Childhood Obesity and Depression: Connection between these Growing Problems in Growing Children. *NCBI,* 103–114.

Reeves, R. V. (2015). *Non-resident Fathers: An Untapped Childcare Army?* Washington, DC: Brookings Institution. Retrieved on December 3, 2018 from https://www.brookings.edu/opinions/non-resident-fathers-an-untapped-childcare-army/

Reynolds, R. (2017). *Using Sociological Perspectives to Analyze the Changes in Parenting.* New York: Odyssey.

Robbins, S. P., Judge, T. A. & Odendaal, A. (2009). *Organizational Behavior: Global and Southern African Perspectives.* Cape Town, South Africa: Pearson.

Rogers, M. L., Halberstadt, A. G., Castro, V. L., MacCormack, J. K. & Garrett-Peters, P. (2016). Maternal Emotion Socialization Differentially Predicts Third-Grade Children's Emotion Regulation and Lability. *Emotion,* 16(2), 280–291.

Roksa, J. & Potter, D. (2011). Parenting and Academic Achievement: Intergenerational Transmission of Educational Advantage. *Sociology of Education,* 84(4), 299–321.

Romána, S. & Sánchez-Siles, L. M. (2018). Parents' Choice Criteria for Infant Food Brands: A Scale Development and Validation. *Food Quality and Preference,* 64, 1–10.

Rosen, D. (2007). Child Soldiers, International Humanitarian Law, and the Globalization of Childhood. *American Anthropologist,* 109(2), 296–306.

Rosenfeld, M. J. (2010). Nontraditional Families and Childhood Progress through School. *Demography,* Aug.; 47(3), 755–775.

Rothbart, M. K. & Bates, J. E. (2006). Temperament. In W. Damon, R. Lerner, & N. Eisenberg (Eds.), *Handbook of Child Psychology: Social, Emotional, and Personality Development,* 6th edn, Vol. 3 (pp. 99–106). New York: Wiley.

Rothbaum, F. & Weisz, J. R. (1994). Parental Caregiving and Child Externalizing Behavior in Nonclinical Samples: A Meta-analysis. *Psychological Bulletin,* 116, 55–74.

Rubin, K., Bukowski, W. & Parker, J. (2006). Peer Interactions, Relationships, and Groups. In W. Damon, R. Lerner & N. Eisenberg (Eds.), *Handbook of Child Psychology, Vol. 3. Social, Emotional, and Personality Development,* 6th edn (pp. 571–645). New York: Wiley.

Ruff, H. A. & Rothbart, M. K. (1996). *Attention in Early Development: Themes and Variations.* London: Oxford University Press.

Russia's G20 Presidency and the OECD (2013). Advancing National Strategies for Financial Education: A Joint Publication. Paris, France: Organization for Economic Co-operation and Development (OECD)

Ryan, K. W. (2011). The New Wave of Childhood Studies: Breaking the Grip of Bio-social Dualism? *SAGE Journals,* 19(4), 439–452.

Saarni, C., Campos, J. J., Camras, L. & Witherington, D. (2006). Emotional Development: Action, Communication, and Understanding. In W. Damon & N. Eisenberg (Eds.), *Handbook of Child Psychology,* 6th edn (Vol. 3). New York: John Wiley & Sons.

Sakamoto, A., Goyette, K. A. & Kim, C. (2009). Socioeconomic Attainments of Asian Americans. *Annual Review of Sociology,* 35, 255–276.

Salovey, P., Mayer, J. & Caruso, D. (2004). Emotional Intelligence: Theory, Findings, and Implications, *Psychological Inquiry,* 15(3), 197–215.

Schiller, P. (2005). *The Complete Resource Book for Infants: Over 700 Experiences for Children from Birth to 18 Months* (Complete Resource Series). Beltsville, MD: Gryphon House.

Schilling, C., Weidner, K., Brähler, E., Glaesmer, H., Häuser, W. & Pöhlmann, K. (2016). Patterns of Childhood Abuse and Neglect in a Representative German Population Sample. *PloS One.,* 11(7), e0159510.

Schmittmann, V. D., Visser, I. & Raijmakers, M. E. J. (2006). Multiple Learning Modes in the Development of Performance on a Rule-Based Category Learning Task. *Neuropsychologia,* 44, 2079–2091.

Schneier, F. R., Blanco, C., Antia, S. X. & Liebowitz, M. R. (2002). The Social Anxiety Spectrum. *Psychiatric Clinics of North America,* 25(4), 757–774.

Schoppe-Sullivan, S. J., Settle, T., Lee, J.-K. & Kamp Dush, C. M. (2016). Supportive Coparenting Relationships as a Haven of Psychological Safety at the Transition to Parenthood. *Research in Human Development,* 13(1), 32–48.

Scottish Government (2008). *The Early Years Framework.* Retrieved on May 11, 2018 from http://www.gov.scot/Resource/Doc/257007/0076309.pdf

Sears, W. & Sears, M. (1993). *The Baby Book: Everything You Need to Know about Your Baby: From Birth to Age Two.* Boston: Little, Brown.

Seeley, S., Murray, L. & Cooper, P. J. (1996). The Detection and Treatment of Postnatal Depression by Health Visitors. *Health Visit,* 64, 135–138.

Seymour, J., Hackett, A. & Procter, L. (2015). *Children's Spatialities, Embodiment, Emotion and Agency.* New York: Palgrave Macmillan.

Sleigh, M. J. & Ritzer, D. R. (2004). Beyond the Classroom: Developing Students' Professional Social Skills. *Psychological Science Observer,* 17(9).

Slopen, N., Shonkoff, J. P., Albert, M. A., Yoshikawa, H., Jacobs, A., Stoltz, R. & Williams, D. R. (2016).Racial Disparities in Child Adversity in the U.S.: Interactions With Family Immigration History and Income. *Am J Prev Med.,* 50(1), 47–56.

Smith, E. R. & Mackie, D. M. (2007). *Social Psychology.* Hove, UK: Psychology Press.

Snow, C. E. (1991). The Theoretical Basis for Relationships between Language and Literacy in Development. *Journal of Research in Childhood Education,* 6, 5–10.

Sobanski, E., Banaschewski, T., Asherson, P., Buitelaar, J., Chen, W., Franke, B. *et al.* (2010). Emotional Lability in Children and Adolescents with Attention Deficit/Hyperactivity Disorder (ADHD): Clinical Correlates and Familial Prevalence. *J. Child Psychol Psychiatry,* 51(8), 915–923.

Sobkina, V. S. & Skobeltsinaa, K. N. (2014). Sociology Of Preschool Childhood: Age Dynamics of the Child's Play. *Procedia: Social and Behavioral Sciences,* 146, 450–455.

Society for Research in Child Development (2009, November 13). When Preschoolers Ask Questions, They Want Explanations. *ScienceDaily.*

Sorce, J., Emde, R., Campos, J. & Klinnert, M. (1985). Maternal Emotional Signaling: Its Effect on Visual Cliff Behavior of 1-Year Olds. *Developmental Psychology,* 21, 195–200.

Sorensen, E. (2002). *Helping Poor Nonresident Dads Do More.* Washington, DC: The Urban Institute.

Stack, D. M. & Muir, D. W. (1992). Adult Tactile Stimulation during Face-to-Face Interactions Modulates Five-Month-Olds' Affect and Attention. *Child Development,* 63, 1509–1525.

Staub, E. (1979). *Positive Social Behavior and Morality: Socialization and Development (Vol. 2).* New York: Academic Press.

Steinberg, L. D. & Silk, J. S. (2002). Parenting Adolescents. In M. Bornstein (Ed.), *Handbook of Parenting,* 2nd edn, Vol. 1. Mahwah, NJ: Erlbaum.

Steinberg, L., Lamborn, S. D., Dornbusch, S. M. & Darling, N. (1992). Impact of Parenting Practices on Adolescent Achievement: Authoritative Parenting, School Involvement, and Encouragement to Succeed. *Child Development,* 63, 1266–1281.

Stocker, C. M., Richmond, M. K., Rhoades, G. K. & Kiang, L. (2007). Family Emotional Processes and Adolescents' Adjustment. *Social Development,* 16, 310–325.

Sui-chu, E. H. & Willms, J. D. (1996). Effects of Parental Involvement on Eighth-Grade Achievement. *Sociology of Education,* 69, 126–141.

Sujan, N. (2016). Financial Independence – A Must for Women Empowerment. *International Education and Research Journal,* 2(3).

Summers, J. A., Boller, K., Schiffman, R. & Raikes, H. H. (2006). The Meaning of "Good Fatherhood": Low Income Fathers' Social Constructions of Their Roles. *Parenting: Science & Practice,* 6, 145–165.

Sun, Y. (2011). Cognitive Advantages of East Asian American Children: When Do Such Advantages Emerge and What Explains Them? *Sociological Perspectives,* 54, 377–402.

Terrazas-Carrillo, E. C., Hong, J. Y. & Pace, T. M. (2014). Adjusting to New Places: International Student Adjustment and Place Attachment. *Journal of College Student Development,* 55(7), 693–706.

Tesser, A., Millar, M. & Moore, J. (1988). Some Affective Consequences of Social Comparison and Reflection Processes: The Pain and Pleasure of Being Close. *Journal of Personality and Social Psychology,* 54(1), 49–61.

Thoits, P. A. (1989). The Sociology of Emotions. *Annual Review of Sociology,* 15, 317–342.

Thompson, R. A. (1994). Emotion Regulation: A Theme in Search of Definition. *Monographs of the Society for Research in Child Development,* 59, 25–52.

Thorndike, R. L. & Stein, S. (1937). An Evaluation of the Attempts to Measure Social Intelligence. *Psychological Bulletin,* 34, 275–284.

Thorne, B. (1987). Re-Visioning Women and Social Change: Where are the Children? *Gender and Society,* 1(1), 85–109. http://www.jstor.org/stable/190088

Tisdall, E. K. M. & Punch, S. (2012). Not So 'New'?: Looking Critically at Childhood Studies. *Children's Geographies,* 10(3), 249–264.

Tolman, T. L. (March 2011). The Effects of Slavery and Emancipation on African-American Families and Family History Research. *Crossroads Journal.* Retrieved on August 4, 2016 from http://www.leaveafamilylegacy.com/African_American_Families.pdf

Toma, C. L. & Hancock, J. T. (2013). Self-Affirmation Underlies Facebook Use. *Personality and Social Psychology Bulletin,* 39, 321–331.

Trinkner, R., Cohn, E. S., Rebellon, C. J. & Van Gundy, K. (2012). Don't Trust Anyone over 30: Parental Legitimacy as a Mediator between Parenting Style and Changes in Delinquent Behavior over Time. *J Adolesc.,* 35(1), 119–132.

Tronick, E. Z. (1995). Touch in Mother–Infant Interaction. In T. M. Field (Ed.), *Touch in Early Development* (pp. 53–65). Mahwah, NJ: Erlbaum.

Tronick, E. Z., Morelli, G. A. & Ivey, P. K. (1992). The Efe Forager Infant and Toddler's Pattern of Social Relationships: Multiple and Simultaneous. *Developmental Psychology,* 28(4), 568–577.

Turner, L. H. & West, R. L. (2000). *Perspectives on Family Communication.* Boston, MA: McGraw Hill.

Uhlenberg, P. (1980). Death and the Family, *Journal of Family History,* Fall; 313–320.

Underwood, M. K., Beron, K. J. & Rosen, L. H. (2009). Continuity and Change in Social and Physical Aggression from Middle Childhood through Early Adolescence. *Aggress Behav.,* 35(5), 357–375.

UN Human Rights (UNHR) (2018). *Convention on the Rights of the Child.* General Assembly resolution 44/25, 20 Nov. 1989. U.N. Doc. A/RES/44/25.

UNICEF (UN Children's Fund) (2006). *What is Child Protection?*

United Nations (UN) (1948). *Universal Declaration of Human Rights.* General Assembly Resolution 217 A (III), 10 Dec. 1948. U.N. Doc. A/RES/3/217/A.

United Nations (UN) (2013). *World Youth Report 2013.* New York: Youth Department of Economic and Social Affairs (DESA). Retrieved on December 3, 2018 2, 2018 from https://www.un.org/development/desa/youth/world-youth-report/2013-world-youth-report-3.html.

U.S. Census Bureau (May 12, 2017a). *Facts for Features: Mother's Day: May 14, 2017.* Table B13004. Retrieved on August 8, 2017 from https://www.census.gov/newsroom/facts-for-features/2017/cb17-ff09-mothers-day.html

U.S. Census Bureau (June 8, 2017b). *Facts for Features: Father's Day: June 18, 2017.*

U.S. Census Bureau (July 12, 2017c). *FFF: National Grandparents Day 2017: Sept. 10.*

U.S. Census Bureau (2018a). *Unmarried-Partner Households by Sex of Partner.* Universe: Households. 2016 American Community Survey 1-Year Estimates. Table B11009.

U.S. Census Bureau (2018b). *Characteristics of Same-Sex Couple Households: 2005 to Present.*

U.S. Census Bureau (2018c). *Stats for Stories: Father's Day: June 17, 2018.* Table FG6. Retrieved on June 15, 2018 from https://www.census.gov/newsroom/stories/2018/fathers-day.html

U.S. Department of Education (2003). *Parent Involvement – Helping Your Child through Early Adolescence.*

Useem, J., Donoghue, J. D. & Useem, R. H. (1963). Men in the Middle of the Third Culture: The Roles of American and Non-Western People in Cross-Cultural Administration. *Human Organization,* 22(3), 169–179.

Valkenburg, P. M., Koutamanis, M. & Vossen, H. G. M. (2017). The Concurrent and Longitudinal Relationships between Adolescents' Use of Social Network Sites and Their Social Self-Esteem. *Computers in Human Behavior,* 76, 35–41.

Vandivere, S. & Malm, K. (2015). Family Finding Evaluations: A Summary of Recent Findings. *Child Trends.* Retrieved on September 21, 2016 from http://www.childtrends.org/wp-content/uploads/2015/01/2015-01Family_Finding_Eval_Summary.pdf

van Kleeck, A. (2008). Providing Preschool Foundations for Later Reading Comprehension: The Importance of and Ideas for Targeting Inferencing in Storybook-Sharing Interventions. *Psychology in the Schools,* 45(7), 627–643.

van Veen, V. & Carter, C. S. (2002). The Anterior Cingulate as a Conflict Monitor: fMRI and ERP Studies. *Physiol Behav.,* 77(4–5), 477–482.

Walsh, D. & Bennett, N. (2005). *Why Do They Act That Way? A Survival Guide to the Adolescent Brain for You and Your Teen.* New York: Atria Books.

Wang, Y. & Henderson, A. M. E. (2018). We Cooperated So... Now What? Infants Expect Cooperative Partners to Share Resources. *Infant Behavior and Development,* 52, 9–13.

Wargo, E. (2007, September). *Adolescents and Risk: Helping Young People Make Better Choices.* Ithaca, NY: ACT for Youth Center of Excellence: Research Facts and Findings. Retrieved on June 13, 2018 from http://www.actforyouth.net/resources/rf/rf_risk_0907.pdf

Warshak, R. A. (2015). Parental Alienation: Overview, Management, Intervention, and Practice. *Journal of the American Academy of Matrimonial Lawyers.* Retrieved on May 5, 2018 from http://aaml.org/sites/default/files/MAT107_7.pdf

Waters, E., Kondo-Ikemura, K., Posada, G. & Richters, J. (1991). *Learning to Love: Mechanisms and Milestones.* In M. Gunnar & L. Sroufe (Eds.), *Minnesota Symposia on Child Psychology: Vol. 23. Self Processes and Development* (pp. 217–255). Hillsdale, NJ: Erlbaum.

Watson, D., Wiese, D., Vaidya, J. & Tellegen, A. (1999). The Two General Activation Systems of Affect: Structural Findings, Evolutionary Considerations, and Psychobiological Evidence. *Journal of Personality and Social Psychology,* 76, 820–838.

Wechsler, D. (1958). *The Measurement and Appraisal of Adult Intelligence.* Baltimore, MD: Williams & Wilkins.

Wehmeier, P. M., Schacht, A. & Barkley, R. A. (2010). Social and Emotional Impairment in Children and Adolescents with ADHD and the Impact on Quality Of Life. *J Adolesc Health,* 46(3), 209–217.

Weinstein, E. A. (1969). The Development of Interpersonal Competence. In D. A. Goslin (Ed.), *Handbook of Socialization Theory and Research.* Chicago: Rand McNally.

Weis, R. & Cerankosky, B. C. (2009). Effects of Video-Game Ownership on Young Boys' Academic and Behavioral Functioning: A Randomized, Controlled Study. *APS Association for Psychological Science,* 463–470.

Weiser, D. A. & Weigel, D. J. (2017). Exploring Intergenerational Patterns of Infidelity. *Personal Relationships,* 24(4), 933–952.

Weiss, S. J., Wilson, P., Seed, J. & Paul, S. M. (2001). Early Tactile Experience of Low Birth Weight Children: Links to Later Mental Health and Social Adaptation. *Infant and Child Development,* 10, 93–115.

Weissman, S. & Cohen, R. S. (1985). The Parenting Alliance and Adolescence. *Adolesc Psychiatry,* 12, 24–45.

Welsch, D. M. & Zimmer, D. M. (2008), After-school Supervision and Children's Cognitive Achievement. *The B. E. Journal of Economic Analysis & Policy,* 8, art. 49.

Whitehurst, G. (Russ). (2018). *What is the Market Price of Daycare and Preschool?* Washington, DC: Brookings Institution. Retrieved on June 13, 2018 from https://www.brookings.edu/wp-content/uploads/2018/04/report.pdf

Wilbur, J. R. & Wilbur, M. (1988). The Noncustodial Parent: Dilemmas and Interventions. *Journal of Counseling and Development,* 66, 434–437.

Wilcox, W. B. & Zill, N. (2016). *Strong Families, Successful Schools.* Charlottesville, VA: Institute for Family Studies.

Williams, B. R., Ponesse, J. S., Schachar, R. J., Logan, G. D. & Tannock, R. (1999). Development of Inhibitory Control across the Life Span. *Developmental Psychology,* 35, 205–213.

Williams, L. R., Degnan, K. A., Perez-Edgar, K. E., Henderson, H. A., Rubin, K. H., Pine, D. S., Steinberg, L. & Fox, N. A. (2009). Impact of Behavioral Inhibition and Parenting Style on Internalizing and Externalizing Problems from Early Childhood through Adolescence. *J Abnorm Child Psychol.*, 37(8), 1063–1075.

Winnicott, D. W. (2000). *The Child, the Family, and the Outside World.* New York: Penguin.

Woodhead, M. & Faulkner, D. (2000). Subjects, Objects or Participants: Dilemmas of Psychological Research with Children. In A. James & P. Christensen (Eds.), *Research with Children: Perspectives and Practices* (pp. 9–35). London: Falmer Press.

Yin, P., Hou, X., Qin, Q., Deng, W., Hu, H. *et al.* (2016). Genetic and Environmental Influences on the Mental Health of Children: A Twin Study. *J Psychosoc Nurs Ment Health Serv.*, 54(8), 29–34.

Yu, C. & Hong, Z. (2015). *Financial Intelligence for Parents and Children (FIFPAC): Financial Intelligence Quotient (FQ) Test.* Piscataway, NJ: Institute of Financial Intelligence.

Yu, S. M. & Singh, G. K. (2012). High Parenting Aggravation among US Immigrant Families. *American Journal of Public Health*, 102(11), 2102–2108.

Zahn-Waxler, C. (2010). Socialization of Emotion: Who Influences Whom and How? *New Dir Child Adolesc Dev.*, 128, 101–109.

Zhang, S. & Fuller, T. (2012). Neighborhood Disorder and Paternal Involvement of Nonresident and Resident Fathers. *Family Relations*, 61, 501–513.

12
BALANCING PAID WORK, FAMILY WORK, AND SCHOOL WORK

Learning Outcomes

Upon completion of the chapter, students should be able to:

1. describe components of work
2. explain the impact of Agricultural and Industrial Revolutions on work
3. apply labor market segmentation theory to the analysis of the labor force
4. explain how occupations and jobs are categorized in the United States
5. describe how the division of paid work and unpaid work led to gender inequality
6. compare work experience and amount of earnings between women and men
7. compare the amount of housework and caring activity between women and men
8. explain factors affecting unequal division of household work between women and men
9. describe the benefits and challenges of dual-earner families and multigenerational families
10. apply structural functionalist theory to the analysis of stay-at-home-parent families
11. apply role theory to the analysis of the multiple roles of working parents
12. explain strategies for balancing paid work, family life, and school work
13. explore military families from a diversity approach
14. examine the future of work and work–family balance across countries.

Brief Chapter Outline

Pre-test

Work

Components of Work

Impact of Agricultural and Industrial Revolutions on Work

Labor Market

Labor Market Segmentation Theory

Pre-test

Engaged or active learning is a powerful strategy that leads to better learning outcomes. One way to become an active learner is to begin the chapter and try to answer the following true/false statements from the material as you read. You will find that you have ready answers to these questions upon the completion of the chapter.

Work encompasses paid work, unpaid work, and coerced work.	T	F
Paid work generates income and is considered "real work," whereas unpaid work is devalued and often invisible.	T	F
Work refers to any activity involving mental or physical effort done in order to achieve a purpose.	T	F
People are only considered contingent workers if their jobs are temporary and not expected to last.	T	F
Gig work is an emerging mode of work.	T	F
Role conflict refers to a conflict among the roles corresponding to more status.	T	F
Within-group inequality refers to wage inequality occurring among workers who are similar in education, age, and other characteristics.	T	F
Fathers have nearly tripled the time they spend with children since 1965.	T	F
Dual-military marriages refer to military members who are married to other military members.	T	F
Work provides a family with a certain socioeconomic status.	T	F

Upon completion of this section, students should be able to:
LO1: Describe components of work.
LO2: Explain the impact of Agricultural and Industrial Revolutions on work.
LO3: Apply labor market segmentation theory to the analysis of the labor force.
LO4: Explain how occupations and jobs are categorized in the United States.

Work

The American Dream has long provided many in this country with the opportunity to make a better life for themselves and their families. The price for that better life is work. **Work** refers to any activity involving mental or physical effort done in order to achieve a purpose. As adults, we spend most of our lives at work. Even if we are sitting back and enjoying *Hamilton*, a musical about the life of American founding father Alexander Hamilton, we are consuming the work of others: playwrights, composers, producers, directors, administrators, managers, advertising agents, newscasters, actors, musicians, electrical technicians, computer operators, and maintenance workers, to name a few.

Indeed, life has involved work since human history began. Work has many functions: it provides a means by which people can support a family; it provides people with a sense of self-worth, meaning, and connection with others; and it provides a family with a certain socioeconomic status. However, the world of work has undergone a great deal of change recently due to globalization, automation, and technology. We'll discuss some of these issues in this chapter.

Components of Work

When it comes to work, most people think of paid work. However, every family is occupied with other kinds of work, such as unpaid housework and childcare, on a daily basis. This section discusses types of work we do every day, including paid work, unpaid work, and coercive work.

In pre-industrial societies, families lived on farms; they were rarely paid for their labor but rather consumed all that they produced. **Paid work** (or market work) generates an income. Paid work emerged in the industrial society as people began to be hired by factory owners and became wage earners. Because paid work generates income, it is considered visible "real work."

Unpaid work refers to non-market work that people voluntarily perform for their families and others. Housework is one form of unpaid work that people do in their homes; it involves grocery shopping, cooking, cleaning, and caring tasks such as childcare and eldercare. Unpaid work also includes **volunteer work**—activities and tasks performed that mainly benefit others without direct reward in money or in kind. People choose to volunteer for a variety of reasons. Research has explored economic returns to volunteer work. Qvist and Munk (2018) find that for labor market entrants and for people in the early stages of their working life in Denmark, an additional year of volunteer work experience yields a significant positive return, but the economic returns to volunteer work experience decrease as a function of professional labor market experience. In the United States, about 62.6 million people, or 24.9 percent of the population age sixteen and older, volunteered through or for an organization at least once between 2014 and 2015 (U.S. Bureau of Labor Statistics, 2016a). The volunteer rate for men was 21.8 percent for the year ending in September 2015, and the rate for women was 27.8 percent that same year (see Figure 12.1).

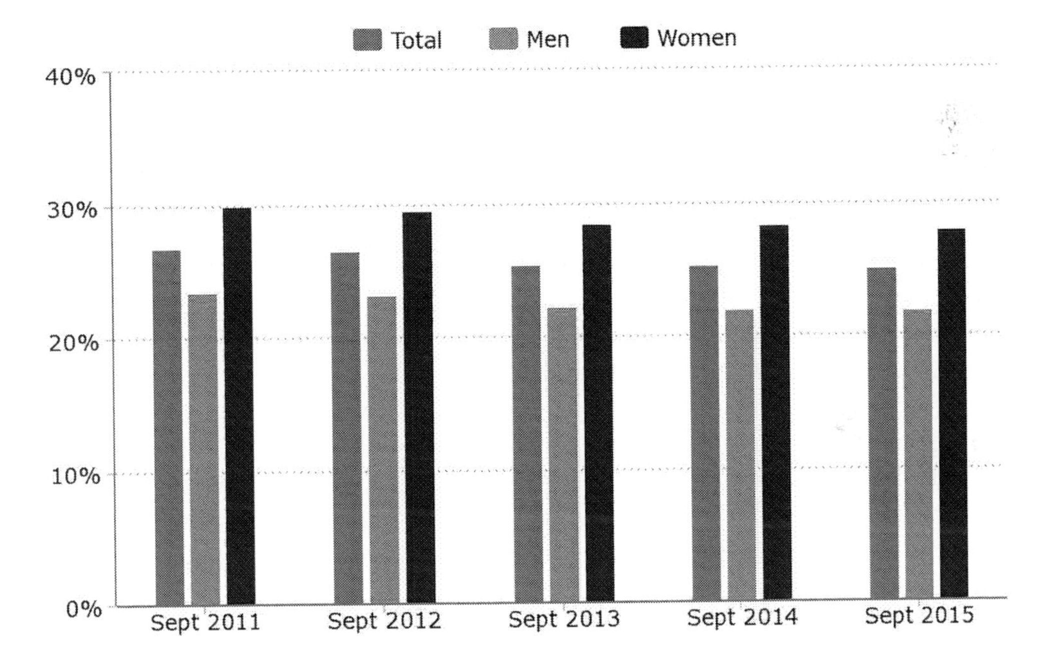

Figure 12.1 Volunteer Rate for the U.S. Population, 16 Years and Older, Men and Women, September 2011–September 2015

Source: U.S. Bureau of Labor Statistics, 2016a. https://www.bls.gov/opub/ted/2016/volunteer-rate-down-slightly-for-the-year-ending-in-september-2015.htm

Coercive work refers to activities that people are forced to do against their will and with little or no pay. It is exemplified by work done by slaves or prisoners. Forced labor is one form of coercive work. It refers to all work or service that is exacted from any person under the menace of any penalty and for which one has not offered oneself voluntarily (International Labor Organization, 2016). Forced labor is prevalent in five sectors of the U.S. economy: prostitution and sex services (46 percent), domestic service (27 percent), agriculture (10 percent), sweatshop/factory work (5 percent), and restaurant and hotel work (4 percent) (Human Rights Center, 2004). Workplace coercion may involve using power to force employees to follow orders. Coercion uses threats to influence the behavior of another and makes the other choose to comply (Bratton, 2012).

Impact of Agricultural and Industrial Revolutions on Work

We stand on the brink of a technological revolution that will fundamentally alter the way we live, work, and relate to one another (Schwab, 2016). Let's take a look at the Agricultural Revolution and several Industrial Revolutions that have transformed the world of work.

For many centuries, families were engaged in agricultural pursuits and lived on their farms in rural areas. Prior to industrialization in the eighteenth century in the West, most people worked in agriculture, either as serfs who farmed land held by members of the nobility or, later, as peasants who owned small parcels of land (Hanawalt, 1986). The Agricultural Revolution between 1650 and 1880 was the result of the complex interaction of social, economic, and farming technology changes (Mingay, 1977). The Norfolk System, for instance, rotates crops so that different crops are planted with the result that different kinds and quantities of nutrients are taken from the soil as the plants grow (Overton, 1996). The British Agricultural Revolution increased agricultural production in Britain due to increases in labor and land productivity. This increase in the food supply contributed to the rapid growth of population in England and Wales, from 5.5 million in 1700 to over 9 million by 1801 (Richards & Hunt, 1983). In China, intensive cultivation (including multiple annual cropping in many areas) had been practiced for many centuries and resulted in an increase in overall productivity per hectare of agricultural area (Merson, 1990). In the United States, new machinery—self-propelled combines and mechanical cotton pickers—reduced labor requirements in harvesting and significant advances occurred in the use of fertilizers, herbicides, insecticides, fungicides, antibiotics, growth hormones, and genetically modified organisms (GMOs) (Conkin, 2008). The rise in productivity accelerated the decline of the agricultural share of the labor force and the increase of the industrial workforce. In the early eighteenth century, most textile production was centered on *cottage industry* (small workshops). The spinning of cotton into threads for weaving into cloth had traditionally taken place in the homes of textile workers (White, 2009). A *cotton mill* is a factory housing powered spinning or weaving machinery for the production of yarn or cloth from cotton and contributed to the development of the factory system during the Industrial Revolution (Wadsworth *et al.*, 1931).

The **First Industrial Revolution**—using *water* and *steam* power to mechanize production (Schwab, 2016)—took place from the eighteenth to the nineteenth centuries in Europe and America. The key carrier technology of the First Industrial Revolution (1770–1850) was based on water-powered mechanization (Freeman & Louçã, 2002) and is widely taken to be the shift from our reliance on animals, human effort, and biomass as primary sources of energy to the use of fossil fuels such as coal and the mechanical power this enabled (Davis, 2016). The development of steam engines in the 1780s led to the growth of steam-powered mills and allowed them to be concentrated in urban mill towns (Thompson, 1993). The textile industries were developed. Family production was replaced by market production in which capitalists paid workers wages to produce goods in factories (Padavic & Reskin, 2002).

The **Second Industrial Revolution** or Technological Revolution—using *electric power* to create mass production (Schwab, 2016)—occurred between the end of the nineteenth century and the first two decades of the twentieth century and brought major breakthroughs in the form of electricity distribution, both wireless and wired communication, the synthesis of ammonia and new forms of power generation (Davis, 2016). British inventor Henry Bessemer and American inventor William Kelly both discovered an efficient way to convert iron ore into steel (Wagner, 2008).

The iron industries developed and grew alongside the demands for steel needed for railroad rails tracks, train parts, machines, and engines of all types. Chemical synthesis is a purposeful execution of chemical reactions to obtain a product, or several products, such as synthetic fabric, dyes, and fertilizer (Vogel *et al.*, 1996). In communication and transportation, major technological advances included the telegraph and telephone, and the automobile and the plane, at the beginning of the twentieth century.

The **Third Industrial Revolution** or Digital Revolution—using *electronics and information technology* to automate production (Schwab, 2016)—began in the 1950s with the development of digital systems, communication, and rapid advances in computing power, which have enabled new ways of generating, processing, and sharing information (Davis, 2016). Indeed, in information and communications technology society, people have used digital technology such as computers, the internet, and cellular phones to provide services and process information or store, retrieve, transmit, and manipulate information.

The **Fourth Industrial Revolution** refers to new technologies that are fusing the physical, digital, and biological worlds that alter how we live and work in societies. Schwab (2017) sees as part of this revolution "emerging technology breakthroughs" in fields such as artificial intelligence, Internet of Things, Big Data/data science. **Artificial intelligence** (AI) or machine intelligence refers to intelligence demonstrated by machines, in contrast to natural intelligence (NI) displayed by humans and other animals. Capabilities, creativity, and skills generally classified as AI include understanding human speech (Russell & Norvig, 2009), competing at the highest level in strategic game systems such as chess, robotics, intelligent routing in content delivery networks and military simulations or war games, and medical diagnosis. For example, human activities such as driving a vehicle or performing surgery or managing a project will be automated. A pair of recently developed AI systems can diagnose lung cancer (Patel *et al.*, 2017) and heart disease more accurately than human doctors (Leary, 2018). Researchers at Google have developed a model that can predict a patient's heath and death risk with 95 percent accuracy using AI (Rajkomar *et al.*, 2018). *The Second Machine Age* refers to a society in which technologies start to perform the cognitive tasks once the exclusive domain of humans and thus alters how we think about issues of technological, societal, and economic progress (Brynjolfsson & McAfee, 2016).

As robots and other computer-assisted technologies take over tasks previously performed by labor, there is increasing concern about the future of jobs and wages (Acemoglu & Restrepo, 2017). In fact, during the initial Industrial Revolution, German sociologist Karl Marx described the automation of the working class as a necessary feature of capitalism. During the Second Industrial Revolution, Keynes (1931) predicted that such innovations would lead to an increase in material prosperity but also to "technological unemployment." Leontief (1952) foretold similar problems as more and more workers will be replaced by machines. Indeed, in the United States, there are an estimated 1.5–1.75 million robots in operation with the number expected to increase to 4–6 million by 2025 and large decreases are found within blue-collar jobs that have routine manual operations, such as assembly workers, transportation workers, and machinists (Works, 2017). In analyzing the effect of the increase in industrial robot usage between 1990 and 2007 on U.S. local labor markets, Acemoglu and Restrepo (2017) estimate that one more robot per thousand workers reduces the employment-to-population ratio (the proportion of the country's working-age population) by about 0.18–0.34 percentage points and reduces wages between 0.25 and 0.5 percent. Approximately, one new robot replaces around three workers.

As you can see, throughout the course of history, new technologies have revolutionized work, allowing it to be performed anytime, anywhere. Technological innovation will lead to a supply-side miracle, with long-term gains in efficiency and productivity (Schwab, 2016) and technology will empower people rather than replace them (Schwab, 2017). To empower people, Brynjolfsson and McAfee (2016) suggest revamping education and preparing for the next economy, designing new collaborations that pair brute processing power with human ingenuity, and embracing policies that make sense in a radically transformed landscape.

Labor Market

Labor is a measure of the work done by people. In moving the production of goods from home to factory, industrialization created a new institution: a labor force (Padavic & Reskin, 2002). People who engage in paid work make up the **labor force**—the sum of the employed, or the labor pool, in employment and unemployed persons in a given country. **Employed persons** consist of persons who did any work for pay or profit during the survey reference week; persons who did at least fifteen hours of unpaid work in a family-operated enterprise; and persons who were temporarily absent from their regular jobs because of illness, vacation, bad weather, industrial dispute, or various personal reasons (U.S. Bureau of Labor Statistics, 2018b). In June 2018, 162 million people ages sixteen and over were in the labor force, compared with 160 million in June 2017 (U.S. Bureau of Labor Statistics, 2018d). **Unemployed persons** consist of persons who do not have a job, have actively looked for work in the prior four weeks, and are currently available for work, and persons who were not working and were waiting to be recalled to a job from which they had been temporarily laid off (U.S. Bureau of Labor Statistics, 2018b). In the United States, the unemployment rates for adult men and women ages sixteen and over was 3.8 percent, with rates for teenagers (13.3 percent), Whites (3.4 percent), Blacks (6.1 percent), Asians (2.7 percent), and Hispanics (4.5 percent) varying greatly in the second quarter of 2018 (U.S. Bureau of Labor Statistics, 2018c).

Labor market or job market refers to the *supply* and *demand* for labor, in which employees provide the supply and employers provide the demand. The number of employed and unemployed can be measured by the **labor force participation rate**—the percentage of the population that is either employed or unemployed (that is, either working or actively seeking work). Labor force levels and participation rates provide information on the *supply of labor* in an economy (U.S. Bureau of Labor Statistics, 2012). In June 2018, the U.S. labor force participation rate (seasonally adjusted) was 62.9 percent, compared with 62.8 percent in June 2017; labor participation rate for men ages sixteen and over was 69.1 percent, compared with 57.2 percent for women ages sixteen and over (U.S. Bureau of Labor Statistics. 2018d). Globally, labor force participation rates were higher for men than women in all selected countries (see Figure 12.2). China had the highest participation rate for women and India had the lowest in the labor market.

Labor Market Segmentation Theory

The labor market functions through the interaction of employers and workers. *Employment* is a relationship between two parties, usually based on a contract where work is paid for, where one party, which may be a corporation, for-profit organization, not-for-profit organization, co-operative or other entity, is the

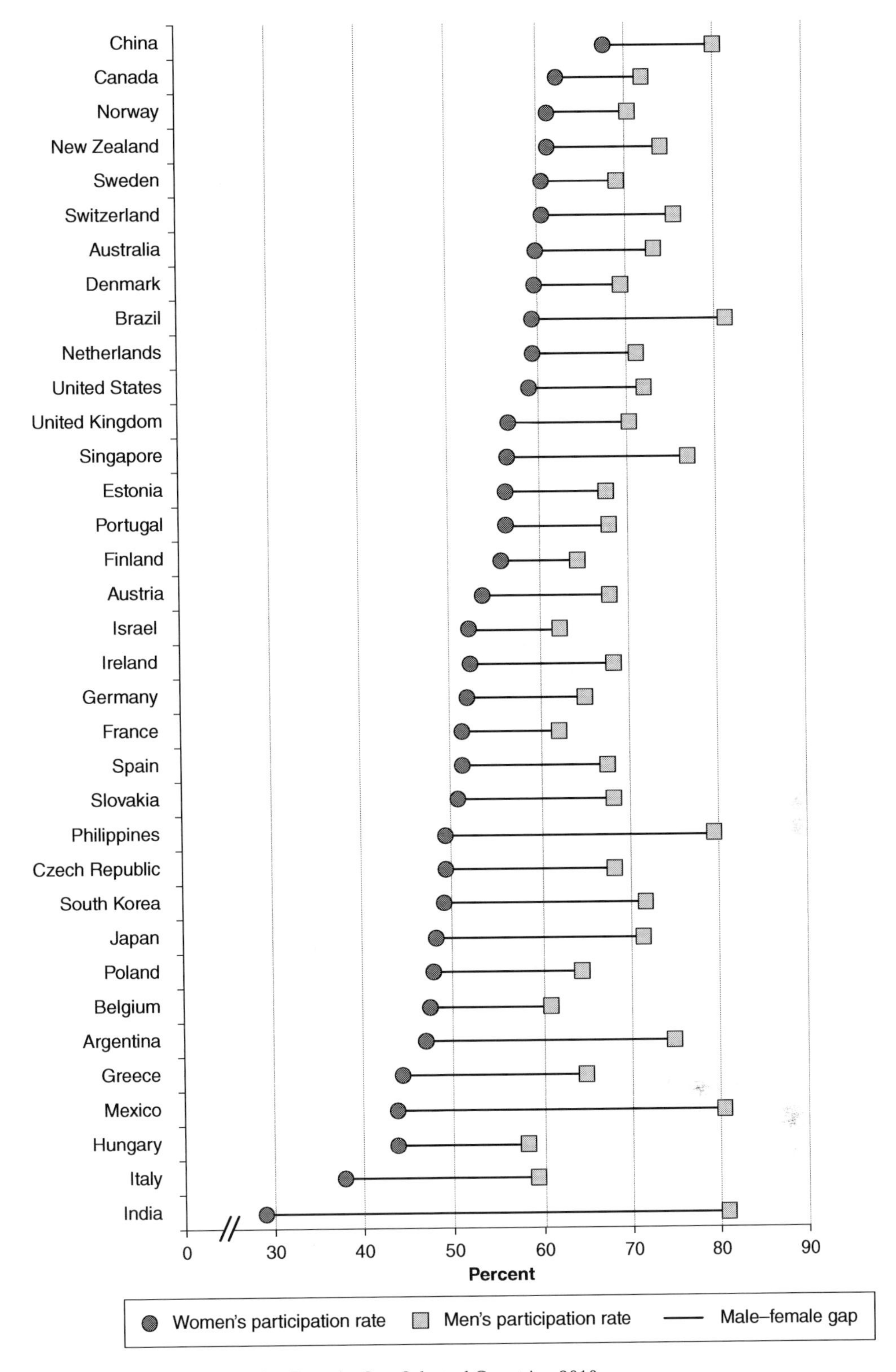

Figure 12.2 Labor Force Participation Rates by Sex, Selected Countries, 2010

Source: U.S. Bureau of Labor Statistics and International Labor Organization. https://www.bls.gov/fls/chartbook/2012/section2a.htm#chart2.2

employer and the other is the employee (Dakin & Armstrong, 1989). Both employer and workers are operating on a highly rational basis being solely interested in maximizing their income or selling their workforce for the best prize (Gerke & Evers, 1993). The various factors influencing wages and employment are neatly separated into two distinct groups: those affecting supply and those affecting demand. *Labor demand* relates to the characteristics of jobs, such as skill and educational demands, employment stability, wages, etc. (Gordon *et al.*, 1982). On the demand side, employers actively and consciously foster labor market segmentation in order to "divide and conquer" the labor force (Reich *et al.*, 1973). The *supply side* refers to attributes of labor, such as education, job skills, occupational preferences, etc. (Bauder, 2001). Persistent inequalities in the labor market competition led to the development of labor market segmentation theory in the 1970s (Wilkinson, 1981). **Labor market segmentation theory** argues for a dual split between primary (or independent) and secondary (or subordinate) segments (Ryan, 1981). Segmentation theorists have argued that labor market segments are able to function independently because both jobs and workers are divided by demand-side and supply-side processes (Bauder, 2001). "The rules governing the behavior of labor market actors differ from one segment of the labor market to the other" (Peck, 1996). Let's take a look at the primary labor market and secondary labor market.

The Primary Labor Market

A job within the labor market is expected to meet certain employment norms. Hodson and Sullivan (2002) suggest that a job should (1) be legal; (2) be covered by government work regulations, such as minimum standards of pay, working conditions, and safety standards; (3) be relatively permanent; and (4) provide adequate pay with sufficient hours of work each week to make a living. The **primary labor market** is the labor market consisting of high-wage-paying jobs, benefits, and longer lasting careers. The employees try to prove themselves to their employers by portraying their skills and educational credentials. The better paid and higher valued jobs in the labor market can be found in the primary sector (Cain, 1976). Those who work in the primary labor market are likely to be full-time employees. **Full-time work** refers to employment in which an employee works a minimum number of defined hours (35 hours or more per week) and receives benefits provided by an employer. In the second quarter of 2018 (not seasonally adjusted), there were 115.8 million full-time wage and salary workers and their median weekly earnings were $876 (U.S. Bureau of Labor Statistics, 2018a). The average workweek for all employees on private non-farm payrolls was 34.5 hours in June 2018 (U.S. Bureau of Labor Statistics, 2018d). The average hours full-time workers spent working by day of the week were 8.56 (see Figure 12.3).

The Secondary Labor Market

The **secondary labor market** is the labor market consisting of high-turnover, low-pay, part-time, or temporary work with few benefits and little job security or possibility for future advancement. The majority of the service sector, light manufacturing, and retail jobs are considered secondary labor (Bewley, 1995). People in the secondary labor market include part-timers, contingent workers, alternative arrangement workers, marginally attached and discouraged workers, workers marginally attached to the labor force.

Part-time work is a form of employment in which an employee works fewer hours per week than a full-time job and receives no or few benefits provided by an employer. Average hours part-time workers spent working by day of week were 5.35 hours, compared with 8.45 hours for multiple jobholders (see Figure 12.3). Workers work part-time for economic (involuntary) or non-economic (voluntary) reasons (Dunn, 2018). *Involuntary part-time workers* want to work 35 or more hours a week or had preferred full-time employment but were working 1 to 34 hours a week because their hours

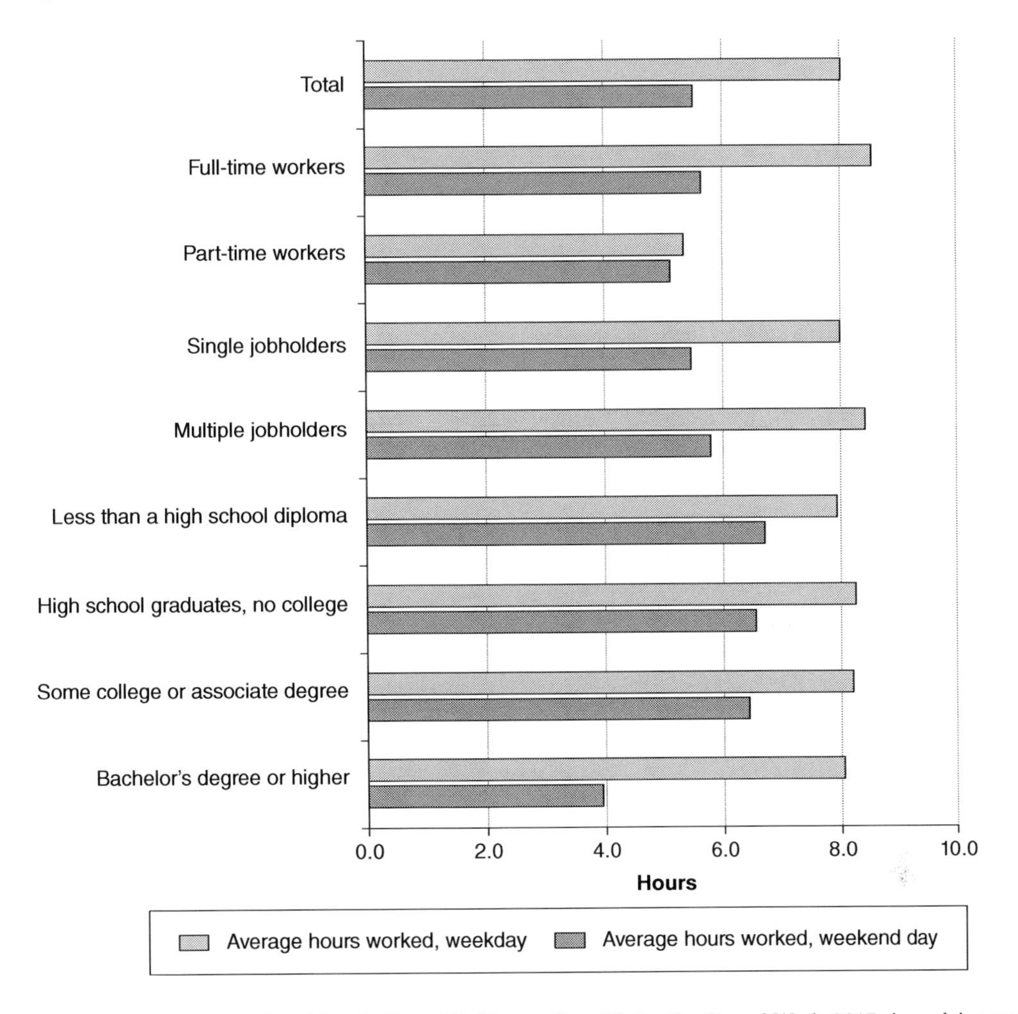

Figure 12.3 Average Hours Employed People Spent Working on Days Worked by Day of Week, 2017, Annual Averages
Source: U.S. Bureau of Labor Statistics, 2017d. https://www.bls.gov/charts/american-time-use/emp-by-ftpt-job-edu-h.htm
Note: Data for educational attainment refer to persons 25 years and over.

had been cut back or because they were unable to find full-time jobs. In June 2018, the number of involuntary part-time workers was 4.7 million, compared with 5.3 million in July 2017 (U.S. Bureau of Labor Statistics, 2018d). *Voluntary part-time workers* work 1 to 34 hours a week but do not want to work 35 or more hours a week. People work part-time voluntarily for a variety of non-economic (voluntary) reasons, including childcare problems, health problems, or a full-time workweek that is less than 35 hours. In 2016, 27.7 million people worked part-time voluntarily, including about one in five women, four in five teenagers ages sixteen to seventeen, one-third of older workers ages 65 and over, and White workers ages sixteen and older (14.6 percent versus 11.7 percent for Black, 12.1 percent for Asian, 12.4 percent for Hispanic) (Dunn, 2018). In 2016, median weekly earnings of all voluntary part-time workers were $245; women earned about 10 percent more than their male counterparts ($252 versus $230) but this earnings difference between male and female voluntary part-time workers is opposite to that of full-time workers, which is discussed earlier and more widely reported—female full-time workers had weekly earnings about 18 percent less than their male counterparts in 2016 (Dunn, 2018).

Contingent workers are people who do not expect their jobs to last or who report that their jobs are temporary and they do not have an implicit or explicit contract for ongoing employment (U.S. Bureau of Labor Statistics, 2018e). Three increasingly broad estimates of contingent work are (1) wage and salary workers who worked at their current job one year or less; (2) all workers including the self-employed and independent contractors who worked at their current job one year or less; and (3) wage and salary workers who do not expect their job to last and self-employed workers and independent contractors who have been self-employed for one year or less. Contingent workers were more than twice as likely as non-contingent workers to be under 25 years old (28 percent versus 12 percent). Of these young workers, three in five contingent workers were enrolled in school. Contingent workers remained slightly less likely than non-contingent workers to be White (76 percent versus 79 percent) and much more likely to be Hispanic (22 percent versus 16 percent) (U.S. Bureau of Labor Statistics, 2018f).

Contingent workers were more likely to earn less and receive fewer benefits than their non-contingent counterparts. In May 2017, median weekly earnings for contingent workers ($685) were 77 percent of those of non-contingent workers ($886); one-fourth of contingent workers had employer-provided health insurance, compared with half of non-contingent workers; and 23 percent of contingent workers were eligible for employer-provided pension or retirement plans, compared with 48 percent of non-contingent workers (U.S. Bureau of Labor Statistics, 2018f). In May 2017, 3.8 percent of workers (5.9 million) held contingent jobs, compared with 4.1 percent in 2005 and 4.9 percent of employment in 1995 (see Figure 12.4).

U.S. Bureau of Labor Statistics identifies four types of **alternative employment arrangements**: independent contractors, on-call workers, temporary help agency workers, and workers provided by contract firms (see Table 12.1). In May 2017, there were 10.6 million independent contractors (6.9 percent of total employment), 2.6 million on-call

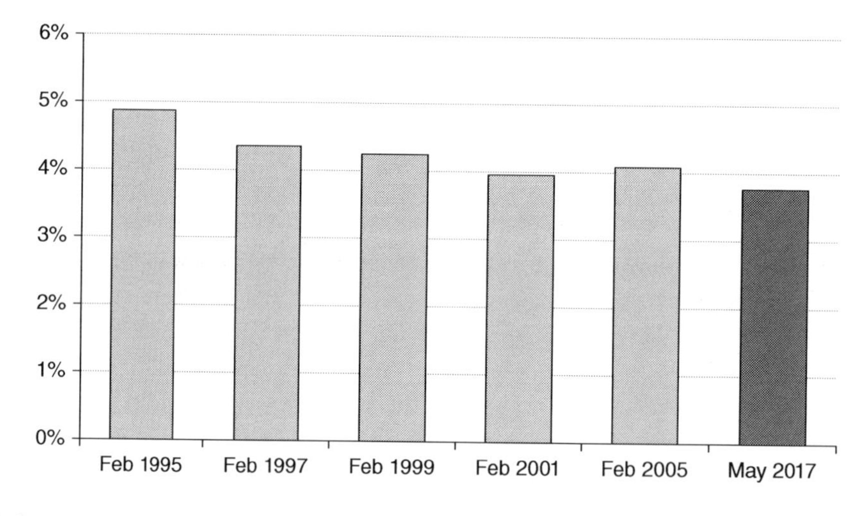

Figure 12.4 Contingent Workers as Percent of Total Employed, 1995–2017

Source: U.S. Bureau of Labor Statistics. https://www.bls.gov/opub/ted/2018/3-point-8-percent-of-workers-were-contingent-in-may-2017.htm

Table 12.1 Definitions of Alternative Employment Arrangements

Alternative employment arrangements	
Type of alternative work arrangement	*Definition*
Independent contractors	Independent contractors, consultants, and freelance workers, regardless of whether they are self-employed or wage and salary workers
On-call workers	People who are called into work only when they are needed, although they can be scheduled to work for several days or weeks in a row
Temporary help agency workers	Workers who are paid by a temporary help agency, whether or not their job was temporary
Workers provided by contract firms	Workers who are employed by a company that provides them or their services to others under contract; they are usually assigned to only one customer and usually work at that customer's worksite

Source: U.S. Bureau of Labor Statistics. https://www.bls.gov/cps/contingent-and-alternative-arrangements-faqs.htm#gig

workers (1.7 percent of total employment), 1.4 million temporary help agency workers (0.9 percent of total employment), and 933,000 workers provided by contract firms (0.6 percent of total employment) (U.S. Bureau of Labor Statistics, 2018f).

These specific types of alternative arrangements have evolved differently since 1995 (Shambaugh *et al.*, 2018). Independent contractors are more likely to be older and men than those in other alternative and traditional arrangements. More than one in three independent contractors were age 55 or older, compared with one in four workers in traditional arrangements and about two-thirds of independent contractors were men in May 2017. In terms of industry, independent contractors were more likely than traditional workers to be employed in construction and in professional and business services (U.S. Bureau of Labor Statistics, 2018f). In May 2017, on-call workers were more likely to work in education and health services and in construction. Temporary help agency workers were more likely to be Black or Hispanic and were heavily concentrated in the production, transportation, and material-moving occupations and in manufacturing industries (Shambaugh *et al.*, 2018). Contract company workers were much more likely to be employed in public administration than those in other alternative or traditional arrangements (U.S. Bureau of Labor Statistics, 2018f).

People in alternative employment arrangements earned less income than workers in a traditional arrangement. In May 2017, median weekly earnings were $1,077 for contract company workers, $851 for independent contractors, $797 for on-call workers, and $521 for temporary help agency workers. Workers in alternative arrangements remained less likely than workers in traditional arrangements to have employer-provided health insurance and retirement benefits. In May 2017, on-call workers (28 percent), temporary help agency workers (13 percent), and contract company workers (41 percent) had employer-provided health insurance; temporary help agency workers (13 percent) and on-call workers (35 percent) were less likely to be eligible for employer-provided plans than were contract company workers (48 percent) or those in traditional arrangements (51 percent) (U.S. Bureau of Labor Statistics, 2018f). Workers in alternative arrangements with independent contractors made up 6.9 percent of employment in May 2017 (see Figure 12.5).

The U.S. Bureau of Labor Statistics measures contingent work and alternative employment arrangements separately. A person in an alternative

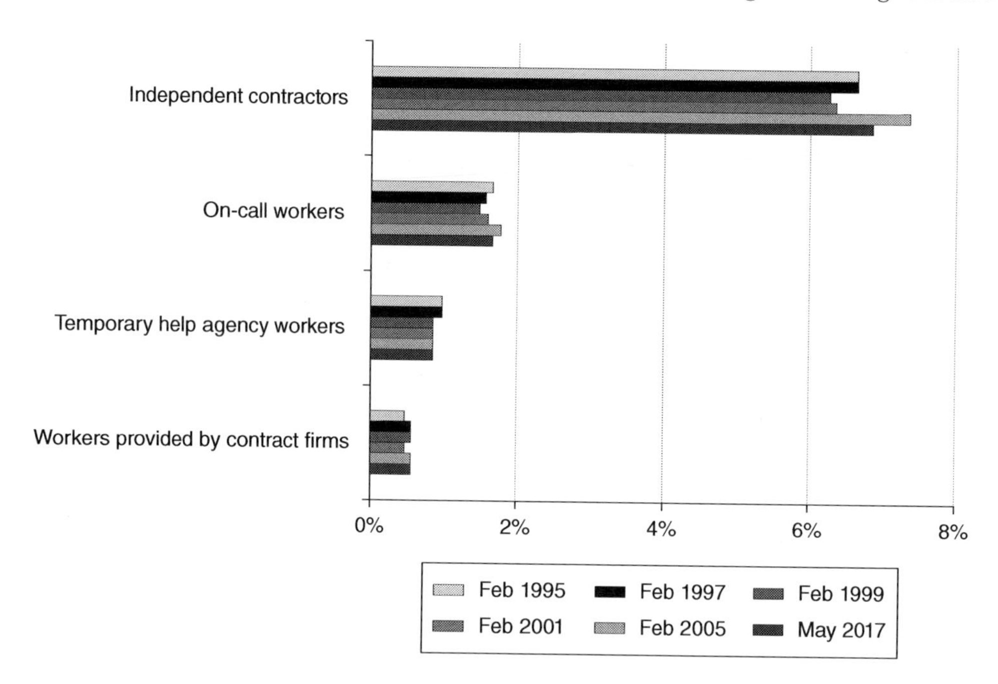

Figure 12.5 Workers in Alternative Arrangements as Percent of Total Employed, 1995–2017
Source: U.S. Bureau of Labor Statistics. https://www.bls.gov/opub/ted/2018/independent-contractors-made-up-6-point-9-percent-of-employment-in-may-2017.htm

employment arrangement may or may not be a contingent worker. For example, "Julio started working for a temporary help agency last week. He has been placed at a law firm preparing legal documents. The firm has told him that if his work is satisfactory, he can be assigned to work for them by the temporary help agency for as long as he wants. In this case, Julio is a temporary help agency worker. However, he is not a contingent worker because his job is not temporary as he can work at his job as long as he wishes" (U.S. Bureau of Labor Statistics, 2018e).

Gig Work

Have you heard "gig work" and "gig workers"? These definitions may overlap with contingent workers and some of those in alternative employment arrangements (U.S. Bureau of Labor Statistics, 2018e) and also overlap with self-employment. **Gig work** refers to technology-enabled contingent work provided by agencies or contract firms to independent contractors, freelancers, or on-call workers (Katz & Krueger, 2016). Gig workers often get individual gigs using a website or mobile app that helps to match them with customers (Torpey & Hogan, 2016). For example, people use their home cars to provide transportation to passengers who communicate with them using ridesharing apps on their cell phones. Airbnb matches owners of spare space with those seeking temporary quarters.

Gig workers are spreading among diverse occupation groups (Torpey & Hogan, 2016). Any occupation in which workers may be hired for on-demand jobs has the potential for gig employment. The advantages of gig work include flexibility, autonomy, and availability of various jobs. The disadvantages include inconsistency and lack of benefits. Gig workers do not have an explicit contract for long-term employment. It can therefore be stressful for gig workers to ensure that they have consistent income (Torpey & Hogan, 2016).

Lower-income Americans with annual household incomes of $30,000 or less are more than twice as likely to engage in technology-enabled gig work, compared with those living in households earning $75,000 per year or more (Smith, 2016). Figure 12.6 shows that in 2015, 18- to 29-year-olds (16 percent) earned money from online gig work platforms, compared with those ages 50 and older (6 percent).

In June 2018, there were 1.4 million **marginally attached workers** (persons who were not in the labor force, wanted and were available for work, and had looked for a job sometime in the prior twelve months), 359,000 **discouraged workers** (persons not currently looking for work because they believe no jobs are available for them), and 1.1 million **persons marginally attached to the labor force** (persons who had not searched for work for reasons such as school attendance or family responsibilities) (U.S. Bureau of Labor Statistics, 2018d).

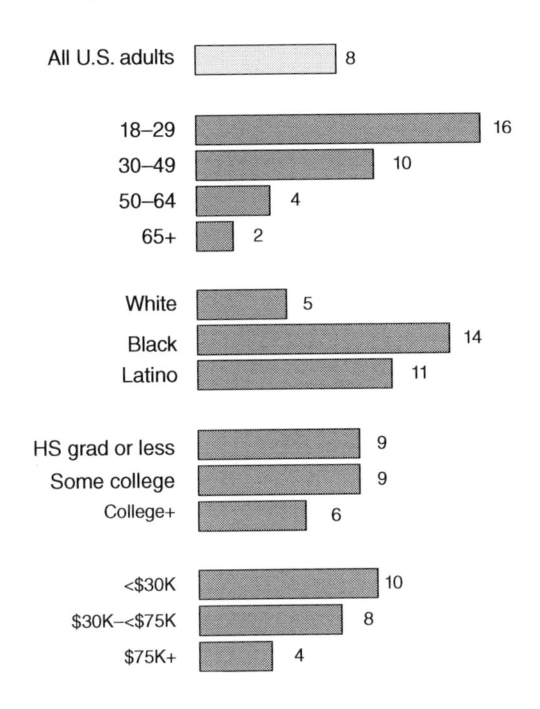

Figure 12.6 Young Adults and Non-Whites Are Especially Likely to get Work via Online Gig Platforms, 2015

Source: Pew Research Center, 2016. http://www.pewinternet.org/2016/11/17/labor-platforms-technology-enabled-gig-work/

As discussed earlier in the chapter, the secondary labor market is characterized by low wages, poor working conditions, lack of job security, and low status (Cain, 1976). Bohle *et al.* (2001) and Sverke *et al.* (2002) have confirmed the effect of work insecurity on health. Job insecurity and poor job quality might affect mental health almost as much as job loss (Winefield *et al.*, 1988). The highest excess risk is experienced by those whose contracts have no fixed term, or who have no formal employment contract (Artazcoz *et al.*, 2005). Bartley (2005) suggests that reductions in job insecurity should be a point of intervention for government policies aimed at improving population health and reducing health inequalities.

Occupational Segregation

Labor market segmentation theory explains the economic marginalization of ethnic minorities, lower classes, and women (Clairmont *et al.*, 1983). **Occupational segregation** refers to the concentration of one group to an occupation. Social scientists have long observed the tendency of societies to form ranks in hierarchies which are based on a person's gender, age, family connections, race/ethnicity, social class, religion, etc. In Colonial America, for example, the early settlers used race as the basis for job assignments and physical work. Enslaved men, women, and children as well as immigrant women worked on farms, mills, and mines. Those who constructed transcontinental railways included minority families. The use of immigrant labor on dairy farms is another example of occupational segregation with respect to Latinos, largely male, to the manual labor on farms.

Looking at the labor market and distribution of people among jobs, we can observe the extension of the nature of existing social relations (Gerke & Evers,

1993). At the core of labor market segmentation are *social groups* and *labor institutions* (Piore, 1983). The processes governing allocation and pricing within internal labor markets are *social*. Indeed, the social and economic status of an individual or group can be measured as a blend of wealth, income, occupation, and education (Noël, 2018). Over the 2014–16 period, 70 percent of Asian households had a member with a bachelor's degree or higher, compared with 23 percent of Hispanic households (Noël, 2018). Thus, educational attainment can have a strong impact on income distribution (Noël, 2018). *Labor institutions* lead to networks of access to job opportunities and also largely determine working wage setting and other types of enumeration (Gerke & Evers, 1993), and widen the gap between those with high and low incomes in the United States. In segmented labor markets, "wages do not equilibrate for similarly endowed workers in different parts of the market due to different patterns of wage setting" (Jones & Roemer, 1989). For example, in the second quarter of 2018, persons employed full-time in management or professional occupations had the highest median weekly earnings—$1,463 for men and $1,080 for women, compared with $615 for men and $512 for women employed in service jobs (U.S. Bureau of Labor Statistics, 2018a).

Classifications of Occupations

The U.S. Department of Labor identifies categories of occupations that reflect the jobs associated with the work that people do. **Jobs** are specific positions (e.g., help desk associate) in specific establishments in which workers perform specific activities. **Occupations** are categories of jobs that involve similar activities at different work sites (Padavic & Reskin, 2002). Based on the 2018 Standard Occupational Classification (SOC) system, all workers are classified into one of 867 detailed occupations, and detailed occupations are combined to form 459 broad occupations, 98 minor groups, and 23 major groups (U.S. Bureau of Labor Statistics, 2018g).

Major groups of industries include the *goods-producing industries* and *services-producing industries*, and each industry sector and subsector is placed into these groups. Knowing which industries are projected to grow or decline helps jobseekers make more informed career decisions (U.S. Bureau of Labor Statistics, 2015). Figure 12.7 illustrates projected general employment trends and percent change in employment of wage and salary workers from 2014 to 2024.

Occupations and Work in Goods-producing Industries

Goods refer to tangible creations that people make and use. Goods-producing industries consist of sectors such as natural resources and mining, construction, and manufacturing (U.S. Bureau of Labor Statistics, 2018h). In June 2018, employment in goods-producing industries was 20.6 million; average hourly earnings of all employees were $28.15 and average weekly hours of all employees were 40.5 (U.S. Bureau of Labor Statistics, 2018h).

Agriculture Work

Agricultural workers usually perform their duties outdoors in all kinds of weather and maintain the quality of farms, crops, and livestock by operating machinery and doing physical labor under the supervision of agricultural managers. Activities associated with the sector of natural resources include the subsector of *agriculture, forestry, fishing, and hunting*—establishments primarily engaged in growing crops, raising animals, harvesting timber, and harvesting fish and other animals from a farm, ranch, or their natural habitats. In 2017, the unemployment rate was 7.2 percent and the annual mean pay was $24,700 for farm workers and laborers, crop, nursery, and greenhouse workers—the largest occupation (U.S. Bureau of Labor Statistics, 2018j) (see Figure 12.8). Despite increased demand for crops and other agricultural products, employment growth is expected to be tempered as agricultural establishments continue to use technologies that increase output per farmworker (U.S. Bureau of Labor Statistics, 2018k).

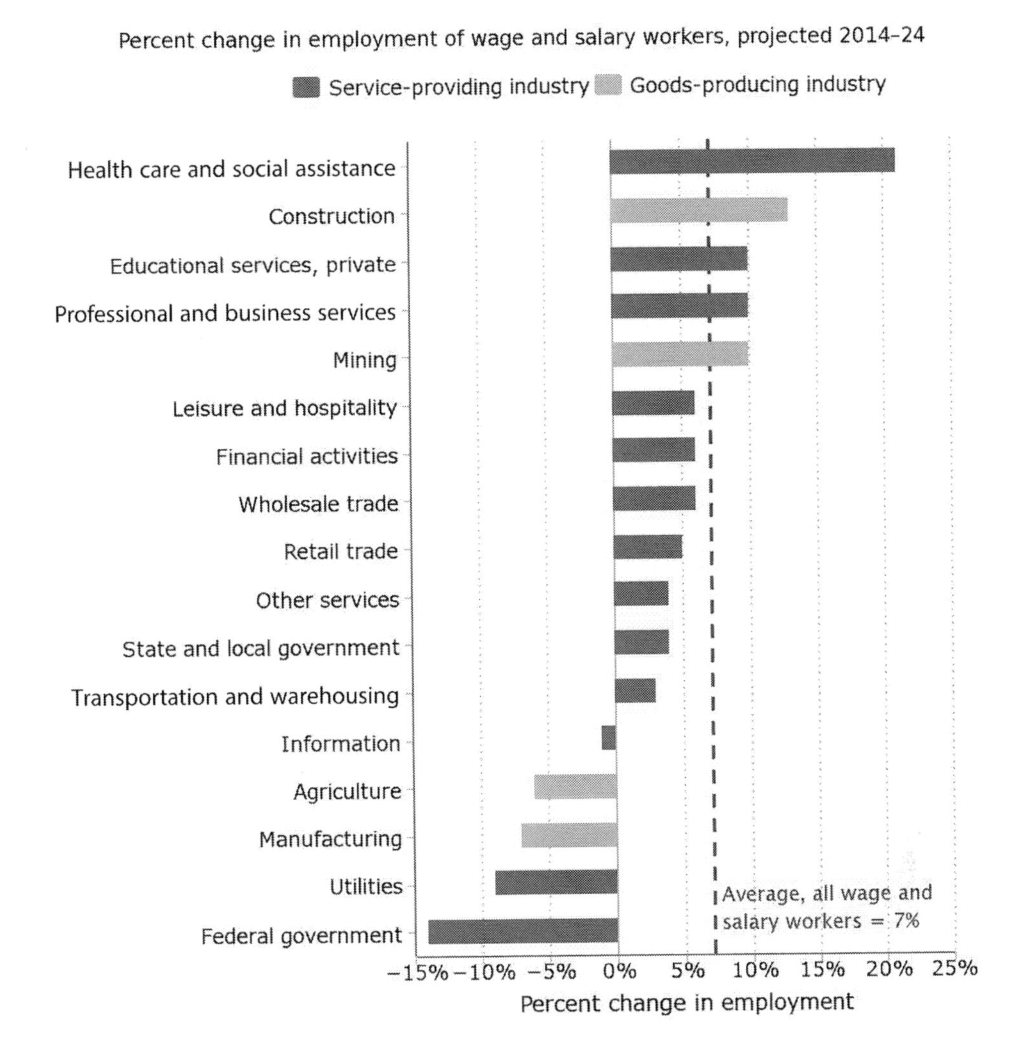

Figure 12.7 Growth by Major Industry Sector
Source: U.S. Bureau of Labor Statistics. https://www.bls.gov/careeroutlook/2015/article/projections-industry.htm

Manufacture Work

The initial Industrial Revolution impacted the transformation of society from rural communities based upon agriculture to industrial urban communities based on the *factory system*—a method of manufacturing using machinery, technology, and division of labor. People who work in factories and perform manual labor are often referred to as *blue-collar workers* owing to the fact that manual workers often wear navy and blue-colored clothes.

The manufacturing sector comprises establishments engaged in the mechanical, physical, or chemical transformation of materials into new products, and consists of subsectors such as food, apparel, leather, wood products, paper, petroleum and coal products, chemical, plastics and rubber products, primary metal, machinery, computer and electronic products, electrical equipment, and furniture. In 1990, the manufacturing industry was the leading employer in most U.S. states, followed by retail trade, but

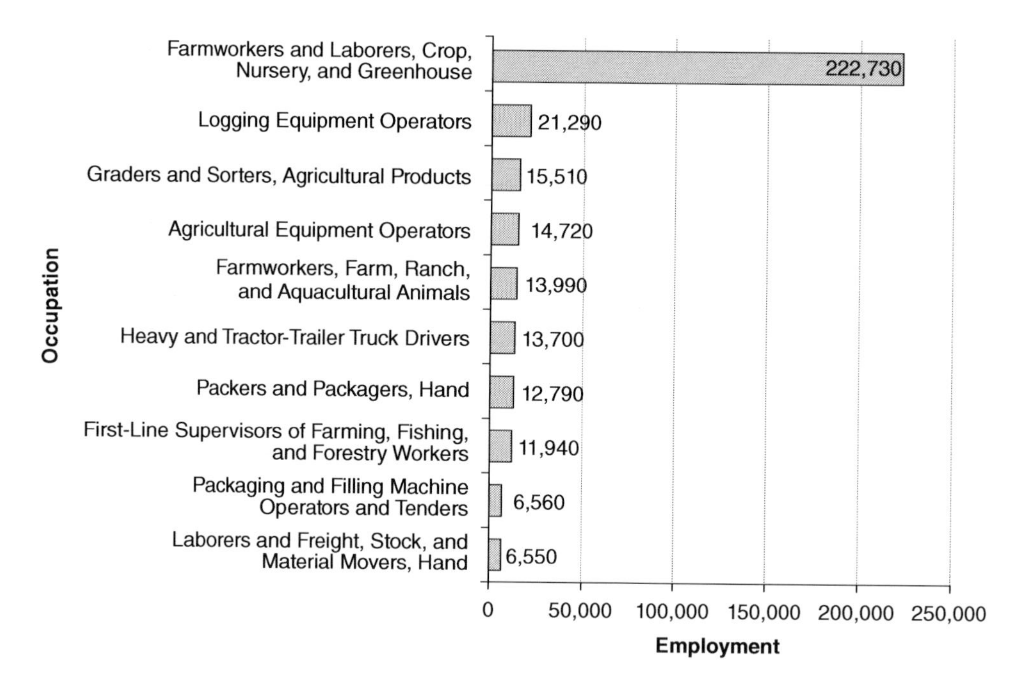

Figure 12.8 Largest Occupations in Sector 11—Agriculture, Forestry, Fishing and Hunting, 2017
Source: Bureau of Labor Statistics. https://www.bls.gov/oes/current/ind_emp_chart/ind_emp_chart.htm

employment in manufacturing declined steadily over the 1990–2013 period (U.S. Bureau of Labor Statistics, 2016b). During the process of globalization, many manufacturing jobs have been offshored to developing nations that pay their workers lower wages. Due to automation and robotics, employment in manufacturing experienced a steady decline of the workforce. Still, manufacturing is the fifth largest employer, with 11.6 million employees producing goods that we consume domestically or export abroad (U.S. Census Bureau, 2018b). Figure 12.9 shows the largest occupations in manufacturing in May 2017.

Occupations and Work in Service-producing Industries

The service industries involve the provision of services to businesses and consumers and grow fast in industrial societies. **Services** refer to intangible creations provided by others. The term *white-collar* refers to the white-collared shirts worn by office workers, as opposed to the blue-collar shirts worn by many manual workers. For example, white-collar workers perform health-care, professional, managerial, or administrative work in an office environment. Many administrators work for governmental bureaucracies or organizations dealing with health, education, or welfare. In June 2018, employment in service-producing industries was 128.2 million (U.S. Bureau of Labor Statistics, 2018i) in sectors such as trade, transportation, utilities, information, financial activities, professional and business services, education and health services, leisure and hospitality, and other services.

Care Work

Care work refers to occupations that provide services that help people develop their abilities to pursue the valuable aspects of their lives. Examples of these occupations include childcare, all levels of teaching (from preschool through university professors), and health care of all types (nurses, doctors, physical

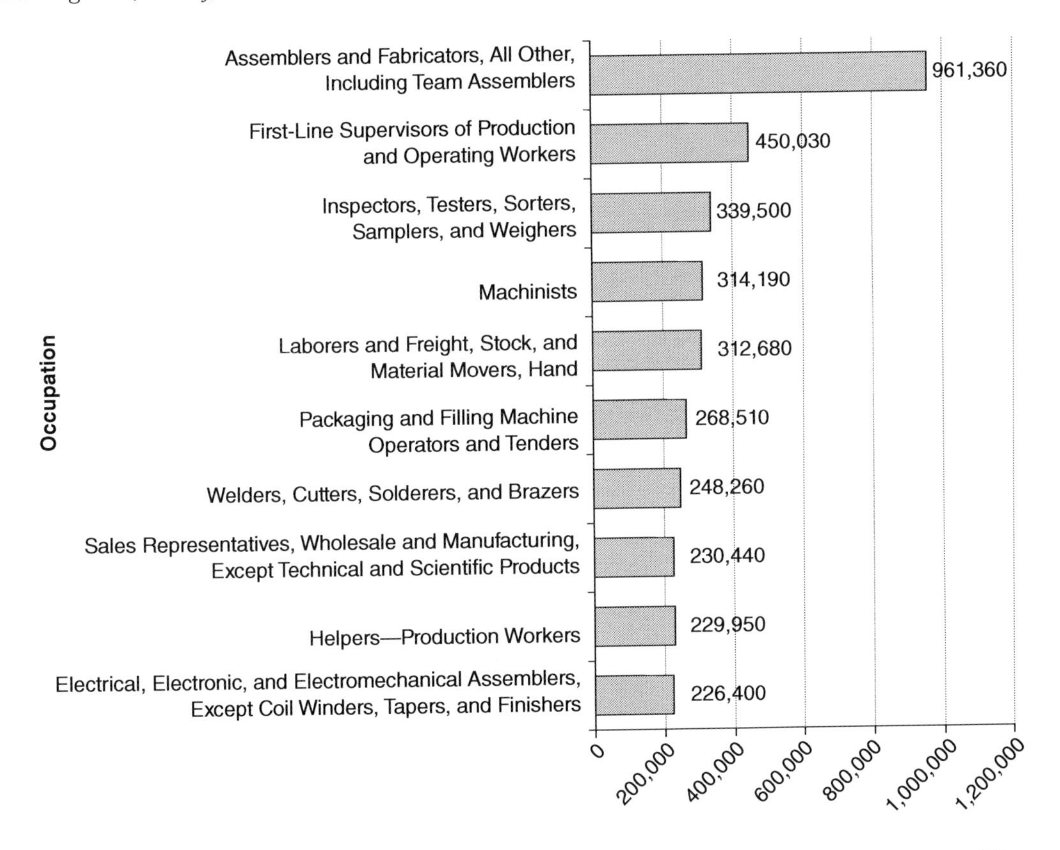

Figure 12.9 Largest Occupations in Sectors 31, 32, and 33—Manufacturing, 2017
Source: Bureau of Labor Statistics. https://www.bls.gov/oes/current/ind_emp_chart/ind_emp_chart.htm

therapists, and psychologists) (England, 2005). Care work also refers to any work done in the immediate service of others, regardless of the recipient's dependent or non-dependent status, and can even extend to animals and things (O'Hara, 2014). Home care work includes unpaid domestic work that is often performed by family members and others. The health-care and social assistance sector is projected to have the fastest job growth. For example, registered nurses are the fastest-growing occupation followed by personal care aides (see Figure 12.10). Personal care aides and home health aides help people with disabilities, chronic illnesses, or cognitive impairment by assisting in their daily living activities; demand for these services is expected to grow as the elderly population increases; the annual median pay was $23,130 in 2017 (U.S. Bureau of Labor Statistics, 2018l).

Professional, Scientific, and Technical Services

The professional, scientific, and technical services sector performs activities including legal advice and representation; accounting, bookkeeping, and payroll services; architectural, engineering, and specialized design services; computer services; consulting services; research services; advertising services; photographic services; translation and interpretation services; veterinary services; and other professional, scientific, and technical services (U.S. Bureau of Labor Statistics, 2018m). The contemporary professional middle class includes most doctors, natural

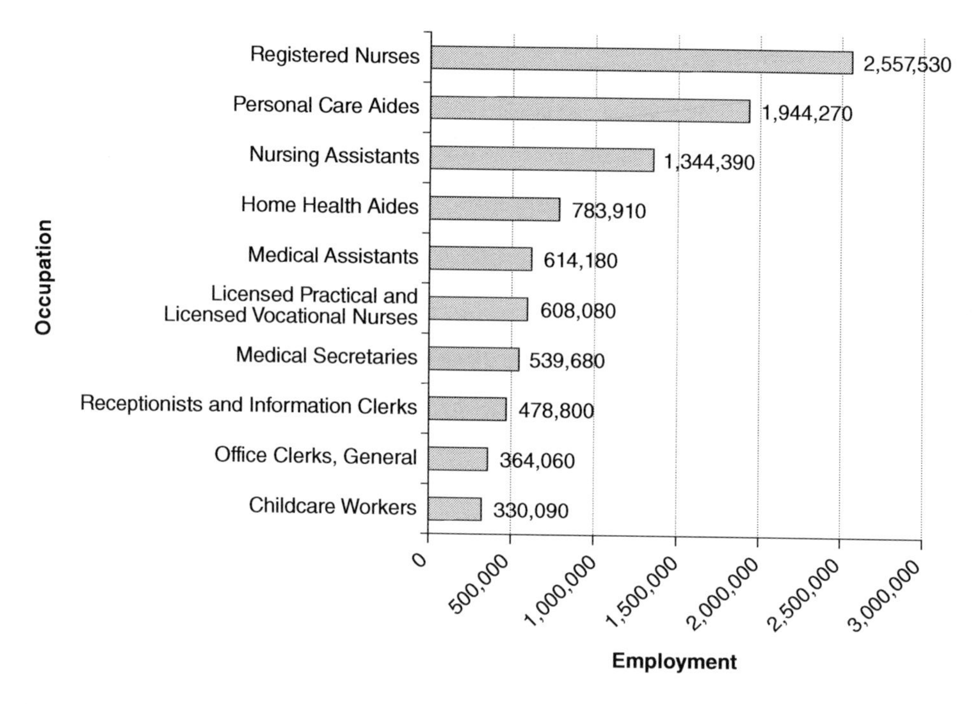

Figure 12.10 Largest Occupations in Sector 62—Health Care and Social Assistance, May 2017
Source: U.S. Bureau of Labor Statistics. https://www.bls.gov/oes/current/ind_emp_chart/ind_emp_chart.htm

scientists, engineers, computer scientists, certified public accountants, economists, social scientists, psychotherapists, lawyers, policy experts of various sorts, professors, at least some journalists and editors, some clergy, and some artists and writers (Brint, 1994). These activities require a high degree of expertise and training. The average hourly earnings were $41.31 and the average weekly hours were 36.7 in May 2018 (U.S. Bureau of Labor Statistics, 2018m). Figure 12.11 shows the largest occupations in professional, scientific, and technical services in May 2017.

Management Work

A variety of occupations are classified as "management" positions. A *manager* has control or direction of an institution or business. The management of companies and enterprises sector comprises (1) establishments that hold the securities of companies and enterprises for the purpose of owning a controlling interest or (2) establishments (except government establishments) that administer, oversee, and manage establishments of the company or enterprise and that normally undertake the strategic or organizational planning and decision-making role of the company or enterprise (U.S. Bureau of Labor Statistics, 2018n). The Chief Executive Officer (CEO) determines policies and provides overall direction of a company within guidelines set up by a board of directors; they plan, direct, or coordinate operational activities at the highest level of management with the help of subordinate executives and staff managers (U.S. Bureau of Labor Statistics, 2018o). Managers are often paid more as they have control over workers and employees in organizations. The mean hourly wage of national estimate for this occupation was $94.25 and the mean annual wage was $196,050, which was the highest wage of all the major occupational groups in May 2017 (U.S. Bureau of Labor Statistics, 2018o). Figure 12.12 shows the largest occupations in management of companies and enterprises in May 2017.

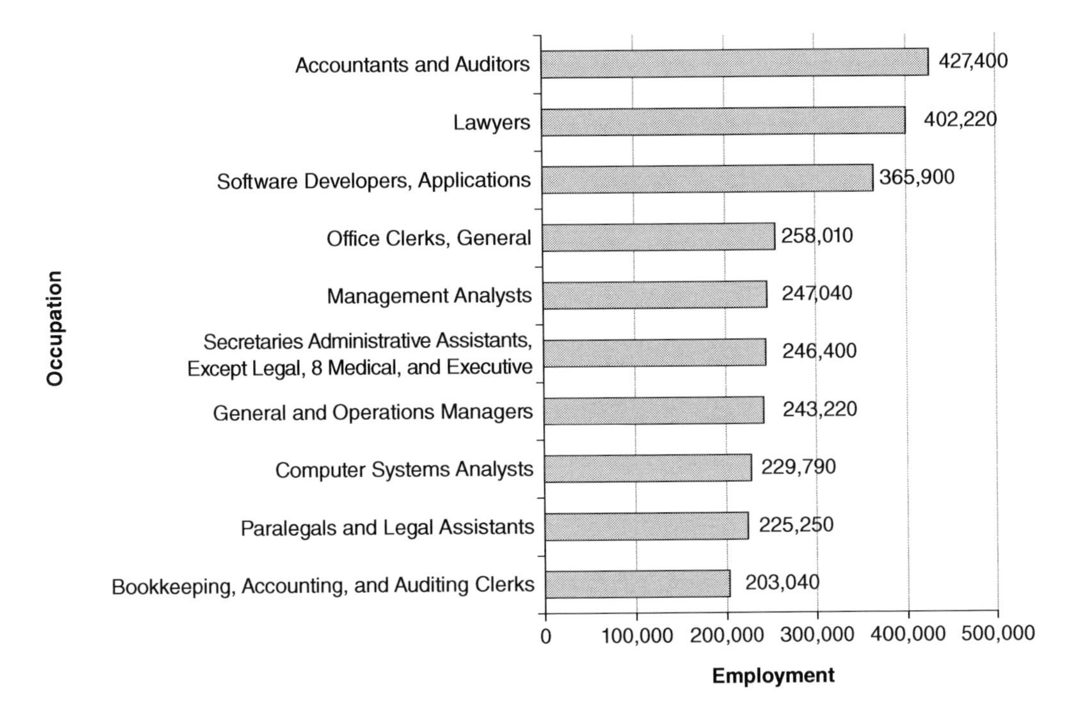

Figure 12.11 Largest Occupations in Sector 54: Professional, Scientific, and Technical Services, May 2017
Source: U.S. Bureau of Labor Statistics. https://www.bls.gov/oes/current/ind_emp_chart/ind_emp_chart.htm

Military Work

Members of the U.S. military service maintain the U.S. national defense. *Active Duty Service* branches include the DoD's (Department of Defense) Army, Navy, Marine Corps, and Air Force; the DHS's (Department of Homeland Security) Coast Guard; *Reserve* components include the DoD's Army National Guard, Army Reserve, Navy Reserve, Marine Corps Reserve, Air National Guard, and Air Force Reserve, and the DHS's Coast Guard Reserve (Military OneSource, 2018).

The military employs people in numerous occupational specialties, many of which are similar to civilian occupations such as nurses, doctors, and lawyers. The military distinguishes between enlisted and officer careers. Enlisted personnel make up about 82 percent of the Armed Forces and carry out military operations. The remaining 18 percent are officers—military leaders who manage operations and enlisted personnel. Members serve in the Army, Navy, Air Force, Marine Corps, or Coast Guard, or in the Reserve components of these branches, and in the Air National Guard and Army National Guard (U.S. Bureau of Labor Statistics, 2018p).

Self-employed Workers

Self-employment is the state of working for oneself as a freelancer or the ownership of a business rather than for an employer. The U.S. Bureau of Labor Statistics classifies self-employed workers as **incorporated workers** such as small-business owners who have established a legal corporation and employed others and as **unincorporated workers** such as freelancers who have not established a corporation and often operate alone. Of 15.0 million (10.1 percent of total U.S. employment) self-employed workers, 9.5 million (6 in 10) were unincorporated and 5.5 million were incorporated in 2015 (Hipple & Hammond, 2016). The unincorporated self-employment rate

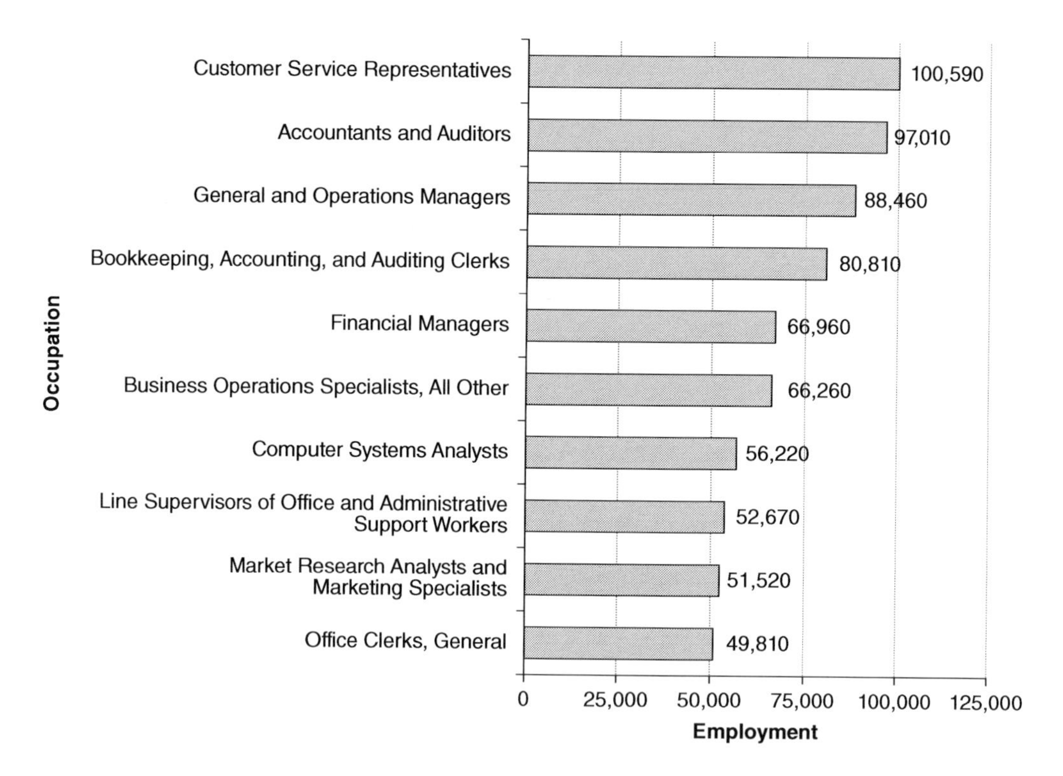

Figure 12.12 Largest Occupations in Sector 55—Management of Companies and Enterprises May 2017
Source: U.S. Bureau of Labor Statistics. https://www.bls.gov/oes/current/ind_emp_chart/ind_emp_chart.htm

for Whites was 6.9 percent, compared with Blacks (3.6 percent), Hispanics (6.4 percent), and Asians (5.6 percent). Incorporated self-employment rates were highest for Asians and Whites (4.0 percent for both groups). High-income occupations for self-employed workers such as dentists and personal financial advisors had a median annual income greater than $60,000 in 2016 (Torpey & Roberts, 2018).

Two key characteristics defining the self-employed are self-direction or *autonomy* in the work and their *private ownership of the means of their own production* (e.g., raw materials, facilities, machinery, and tools used in the production of goods and the provision of services) (Dale, 1986).

People choose to become self-employed for greater independence and autonomy. Evans and Fischer (1992) defined *autonomy* as "the amount of freedom a worker has to schedule their work and determine the procedures to be used in carrying it out." The self-employed exercise greater self-direction and autonomy in their work (Dale, 1986). For example, freelancers may exercise greater control over their work than paid employees. German sociologists Karl Marx and Max Weber regarded the ownership of means of production as crucial for understanding the nature of power and authority relationships between labor and capital (Curran & Burrows, 1986). The earnings of the self-employed represent a return on capital as well as labor, entrepreneurial skill, and risk-taking, whereas the wage an employee receives is a payment for his or her labor (International Labor Organization, 2001).

Over the past years, there has been a rise in unincorporated workers. Researchers in Britain, where self-employment grew remarkably during the 1980s, were among the first to identify the changing nature of self-employment (Cranford *et al.*, 2005). They recorded a *decline* in small business owners who employ other workers and a *rise* in consultants, professionals, and contractors, especially in the service sector (Dale, 1986) or gig work. **Self-service online labor markets** refer to online platforms such as Upwork that make it easier for companies to contract with independent workers for short-term projects. The emergence of self-service online labor markets such as Uber has facilitated the rise of gig work. Nearly 8 percent of Americans have earned money in 2015 from an online "gig work" platform (Smith, 2016). Over the past several decades, there has been growth in the number of nominally self-employed contingent workers (Kalleberg, 2011), including freelancers, home workers, and temporary contractors, who differ from conventionally employed workers only in that they sell their labor power to an employer through a different type of contractual arrangement (Dale, 1986). Working at home is on the rise. In the United States, more people (4.6 percent) age sixteen and over worked at home in 2015 (U.S. Census Bureau, 2017a). A growing number of people find work through such self-service online labor markets. These self-service online labor markets provide a laboratory for researchers to investigate the behavior of employers in the absence of contract enforcement. An audit study of employers reveals that working for a good employer yields 40 percent of higher wages, and employers with good reputations attract work of the same quality (Benson *et al.*, 2015). Increasingly, job seekers are doing online research to learn more about companies before they make employment decisions and thus company reputation matters for employers more than ever before (Reputation Management, 2015).

Self-employed people can receive the traditional benefits of the corporate structure, including limited liability, tax considerations, and enhanced opportunity to raise capital through the sale of stocks and bonds (Hipple & Hammond, 2016), but also consider the downsides, such as the long hours and lack of benefits (Vilorio, 2014). Self-employment is hard work, especially during the first few years. For example, workers may have difficulty finding clients, earning a steady income, securing business loans, and navigating laws. These challenges add up to financial risk and uncertainty (Vilorio, 2014). Empirical studies of self-employment for the Australian, Canadian, Dutch, UK, and U.S. labor markets reveal that self-employment outcomes are significantly affected by factors such as individual abilities, family background, occupational status, liquidity constraints, and ethnic enclaves (Le, 1999). For example, psychological factors such as optimism, confidence, and consciousness positively contribute to the development of self-employment activity (Remeikiene *et al.*, 2011). Research has explored the extent to which children were involved in their parents' businesses. Aldrich and Kim (2007) suggest strong effects from genetic inheritance and parenting practice (during childhood) but little influence from financial support. Parental role modeling is a more important source of the transmission of self-employment than is a child's privileged access to their parent's financial capital (Sørensen, 2007).

Entrepreneurs

Self-made entrepreneurs have taken on financial risks and made contributions to the building of the United States. **Entrepreneurship** involves the process of designing, launching, and running a new business. Entrepreneurship has long been heralded as a key driver of rising living standards (Smith, 1776) and plays a vital role in the growth of the U.S. economy. Famous individual cases such as Bill Gates, Steve Jobs, and Mark Zuckerberg show that people in their early twenties can create eventually world-leading companies (Azoulay *et al.*, 2018). A study of administrative data at the U.S. Census Bureau reveals that successful entrepreneurs are middle-aged and the mean founder age for the 1 in 1,000 fastest growing new ventures is 45.0 (Azoulay *et al.*, 2018). Figure 12.13 shows that among 5.4 million U.S. firms with paid

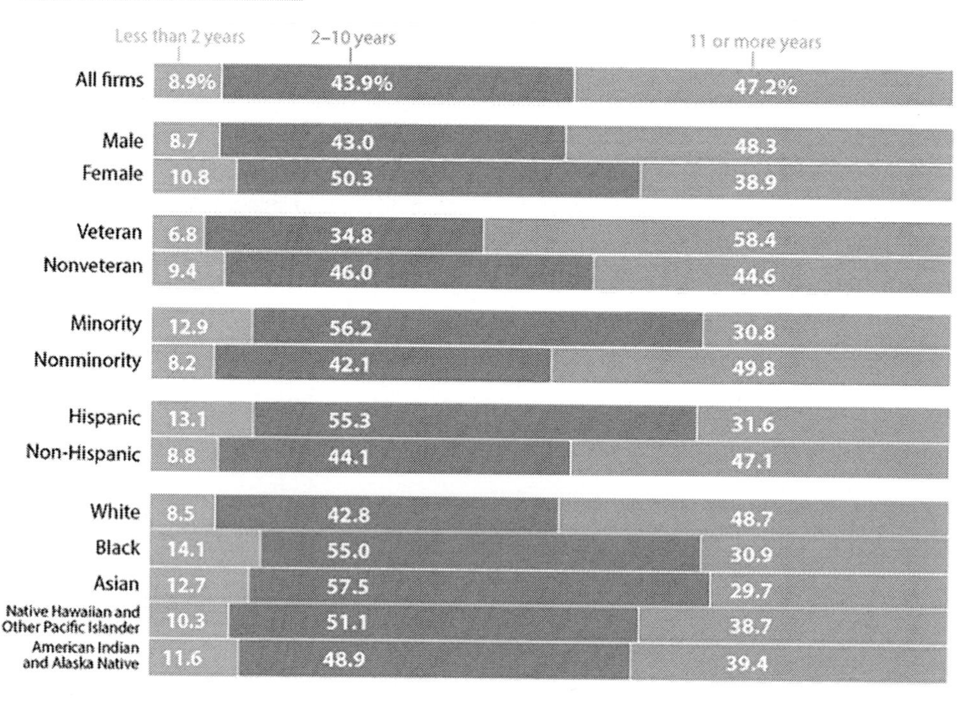

Figure 12.13 America's Entrepreneurs in Business
Source: U.S. Census Bureau, 2016. https://www.census.gov/content/dam/Census/library/visualizations/2016/comm/cb16-148_graphic_ase.pdf

employees, 481,981 (8.9 percent) had been new in business for less than two years in 2014. Research has explored the growth of Latino entrepreneurial activity. Robles and Cordero-Guzman (2007) suggest that low educational attainment may be a driving force in creating the push into self-employment; lack of financial resources for operation and expansion purposes contributes to blocked business stage growth; and Latino entrepreneurial activity is concentrated in the service sector.

Diversity Overview: Military Families

Military sociology is a subfield of sociology that studies the military as a social group. Huntington (1957) and Janowitz (1960) have influenced the concept of military professionalism. Moskos (1977) proposes that the military is shaped by institution and occupation. He argues (1986) that the military is moving from an institutional to occupational format. When men and women serve in the military, so do their families. **Military families** are defined as Active-Duty service members, members of the National Guard and Reserve, and veterans, plus the members of their immediate and extended families, as well as the families of those who lost their lives in service to their country (U.S. Department of Defense, 2011).

Demographics of Military Families

Military families are diverse. The total number of military personnel was over 3.5 million strong during the 2016 fiscal year. Across the Active Duty and Selected Reserve population, there are 2,100,328 military personnel and 2,718,922 family members, including spouses, children, and adult dependents. Of the military personnel, 43.9 percent are single with no children; 34.4 percent are married with children; and 15.6 percent are married without children; 6.2 percent are single with children (Military OneSource, 2018). One-third (31.4 percent) or 404,796 of Active Duty members identify themselves as a racial minority, compared with one-quarter (26.1 percent or 213,664) of the Selected Reserve members. Within the DoD, women make up 15.9 percent of the Active Duty force (204,628) and 19.3 percent of the Selected Reserve (158,173) (Military OneSource, 2018).

Dual-military marriages refer to military members who are married to other military members. In 2016, 6.6 percent of DoD Active Duty members and 2.7 percent of the Selected Reserve members were in dual-military marriages (Military OneSource, 2018). As is the case in civilian families, marriage and divorce are two major life-changing events for military families (National Military Family Association, 2017). In 2016, 3.5 percent of enlisted personnel and 1.6 percent of officers are estimated to have divorced, whereas 2.6 percent of Selected Reserve enlisted personnel and 1.7 percent of Selected Reserve officers are estimated to have divorced (Military OneSource, 2018).

Military Children

Overall, 40.5 percent of military personnel had children in 2016. Of the 1,715,519 total military children, the largest percentage were between birth and five years of age (37.8 percent), followed by six to eleven years of age (31.6 percent), and twelve to eighteen years of age (23.8 percent) (Military OneSource, 2018).

Military children are unique in that they grow up within a military culture, the culture of the home country of their parents, and often the culture of a foreign country in which they reside. When parents are assigned to serve in a number of countries, military children are immersed in military subcultures to the point where American culture can seem foreign. They might move long distances from one non-combat assignment to another during their developmental years. The highly mobile military children are therefore likely to develop a mixed cultural identity due to exposure to numerous national or regional cultures.

Military children may show higher levels of anxiety than their civilian counterparts. Since the war in Afghanistan in 2001 and the war in Iraq in 2002, often referred to as "the war on terror," more than 2 million military children have been separated from their service member parents, both fathers and mothers, because of combat deployments (Cozza & Lerner, 2013). Children of deployed parents are more likely to exhibit anxiety, depression, aggression, attention deficits, and behavioral problems than others, and they are more likely to suffer neglect or maltreatment (Aranda *et al.*, 2011). Teenagers are worried about the absent parent and learn to share family responsibilities at home. Twenty-three percent of children find it easier to reconnect after the deployment if they maintain a strong connection with their deployed parent, and when children experience easier reconnections the spouses of the deployed member (30 percent) are more satisfied with the military way of life and support staying on active duty (MFLP, 2015).

Impact of Spouse Well-being on Military Support

Overall, 66 percent of Active Duty spouses are in the labor force, including 41 percent employed in the civilian labor force, 12 percent seeking work, and 13 percent in the Armed Forces (Military OneSource, 2018).

A better financial status increases support to stay on active duty, while higher levels of stress adversely impact support to stay on active duty. The Military Family Life Project (MFLP) (2015) reports that 38 percent of financially secure spouses are more satisfied with the military way of life and 20 percent of financially secure spouses are more likely to support staying on active duty.

Impact of Military Life on Spouse

Over half (58.0 percent) of the 478,963 spouses who are married to Active Duty enlisted members are 30 years of age or younger. Almost three-quarters (74.6 percent) of the 145,018 spouses who are married to Active Duty officers are 31 years of age or older (Military OneSource, 2018). The military lifestyle impacts families' financial and emotional well-being. Frequent deployments, separations, and relocations are hallmarks of military life and can greatly affect military families (Meadows *et al.*, 2016). Wives of the deployed have higher rates of depression, anxiety disorders, sleep disorders, acute stress reaction, and adjustment disorders than those of non-deployed service members (de Burgh *et al.*, 2011). The MFLP (2015) finds that 14 percent of spouses say that a recent permanent change-of-station move can hurt a family's financial well-being and 35 percent of spouses have difficulty making ends meet. Current deployments can impact a spouse's emotional well-being as 13 percent of spouses have higher stress levels and 5 percent of spouses have higher depressive and anxiety symptoms.

Military Family Life

Military life presents a variety of challenges to military families, including frequent separations and relocations as well as the risks that service members face during deployment (Meadows *et al.*, 2016). For families living in an on-base house, they must follow the rules of the base command and the housing office on how to maintain their property and are afforded little privacy (Baker, 2008). In the context of the wars in Afghanistan and Iraq in the twenty-first century, American military families faced challenges during these operations (Ross, 2014). Many have struggled with combat-related mental health problems, including post-traumatic stress disorder (PTSD), physical injuries, traumatic brain injury (TBI), and death, all of which can affect children and families for years (Cozza *et al.*, 2005). **Family resilience** is defined as the ability of a family to respond positively to an adverse situation and emerge from the situation feeling strengthened, more resourceful, and more confident than its prior state (Simon *et al.*, 2005). This definition focuses on the family as a unit gaining either resources (i.e., acquiring new resources) or competencies (i.e., successfully using existing resources) to "bounce back" from stressors (Meadows *et al.*, 2016).

Family resilience is greatest when protective factors are greatest and risk factors fewest (Hawley, 2000). *The resiliency model of family stress, adjustment, and adaptation* is a prevention-oriented model that explains the behavior of families under stress and the role of the family's strengths, resources, and coping mechanisms (McCubbin & McCubbin, 1988). Practitioners use this model to help families find mechanisms to cope with stressors, i.e., any demand, problem, or loss that may affect the family's functioning or relations (McCubbin & McCubbin, 1993). Families may use protective factors or incorporate recovery factors to help them cope during and/or after a crisis or trauma (Meadows *et al.*, 2016). For families facing prolonged deployment, key recovery factors are self-reliance (the degree to which family members can act independently in the family's best interest), family advocacy (the extent to which the family is part of collective efforts of other families in the same situation), and family meanings (how families define their demands and capabilities and see themselves in relation to the outside world) (McCubbin & McCubbin, 1988). The rapid emergence of family resilience as an area of policy relevance for DoD has led to two groups of policies: (1) policies about programs such as youth programs, which are then modified to address family readiness; and (2) new policies that establish or address programs that specifically target family resilience (Meadows *et al.*, 2016).

Upon completion of this section, students should be able to:
LO5: Describe how the division of paid work and unpaid work led to gender inequality.
LO6: Compare work experience and amount of earnings between women and men.

Women in the Labor Force

The sociological study of gender and work emerged during the 1960s and 1970s, as women's labor force participation rates rose and as the women's movement began calling attention to gender inequality at home and on the job (Wharton, 2018). Industrialization created two new distinctions between men's and women's work roles—*division of paid work and unpaid work, and occupational gender segregation*, leading to **gender inequality** (i.e., a social process by which people are treated differently on the basis of gender) in the workplace and at home. Let's look at historical experiences of women in the labor force.

Division of Paid and Unpaid Work

Milkman (2016) argued that women have acted as *a reserve labor force*, only participating in the labor force when there have been an insufficient number of men to take all available jobs. Before industrialization shifted production to factories, the predominant work was in agriculture and peasant women made time for cottage industry by laboring every available minute (Padavic & Reskin, 2002). Cottage workers worked from home, spinning wool, making lace, weaving cloth, or attaching shirt collars and earning wages on the basis of piece work. **Piece work** is a type of employment in which a worker is paid a fixed piece rate for each unit produced or action performed regardless of time. As cottage industries gave way to small textile factories, many employers continued to hire women and children because they worked for lower wages and were more available for hire than men in some areas (Valenze, 1995).

Throughout the nineteenth century, the labor force became increasingly male (Padavic & Reskin, 2002). One reason was that employment became urbanized as factories were in cities. Secondly, early unions viewed women and children as a threat to men's jobs and wages and thus they set out to drive children and women out of all factory and mining jobs (Pinchbeck, 1930). Pressure from unions and reformers led the nineteenth-century lawmakers in Europe and the United States to pass protective labor laws prohibiting children and women from working at night, and from holding certain jobs (Padavic & Reskin, 2002). As a result, women's labor force participation fell. In 1840, women and children made up about 40 percent of the industrial workforce in the United States; by 1870, three out of four industrial workers were male (Baxandall *et al.*, 1976); and by 1890, only 17 percent of women worked for pay outside the home (Goldin, 1990). In putting many lines of work off-limits to women, protective labor laws contributed to the masculinization of the labor force and reinforced the assumption that it was the male's responsibility to support their families (Padavic & Reskin, 2002).

Ideology of Private Sphere and Public Sphere

After industrialization shifted production to factories, *the first division of labor* assigned men to paid work and women to the unpaid work of running a household (Padavic & Reskin, 2002). A major factor of the low labor participation rate among married women was the **ideology of separate spheres**—holding that a woman's proper place was in the home and a man's place was in the world of commerce or at his job (Davidoff & Hall, 1987). Historians of Britain often call the period between 1780 and 1850 the age of "separate spheres" or "domestic ideology" for men and women (Steinbach, 2012). This ideology of separate spheres encouraged men to work away from home and women to confine productive activities to the household (Padavic & Reskin, 2002). Women were transformed into housewives as a result of men's successful defense of patriarchal privileges under changing economic and social conditions (Jackson, 1991). Established first among the middle classes, this pattern spread widely throughout the working class with increasing momentum from the middle decades of the nineteenth century (Creighton, 1996). At the turn of the twentieth century, only 6 percent of married women were in the labor force, compared to 40 percent of single women (Folbre, 1991). Thus, female work takes place in the private sphere of the home and involves childrearing and homemaking, whereas male work takes place in the public sphere and involves bread-winning.

Female Entry into the Labor Force

Women have entered the labor market for many reasons. The division of paid and unpaid work between spouses is essential for the placement of women within *paid work*, and hence implies several consequences—for the returns, which women receive for their education; for women's employment status in the active age; for women's horizontal and vertical labor segregation; and for their amount of pensions after retirement (Hofäcker *et al.*, 2013).

Many countries experienced a great increase in women's labor market participation and advances in other areas during the 1970s and 1980s and this so-called *gender revolution* (Hochschild & Machung, 1989) affected mainly women who entered male-dominated areas (Nyman *et al.*, 2018). In the United States, at the start of the twentieth century, only a handful of career opportunities were available to women, including domestic service, teaching, and some factory work (U.S. Bureau of Labor Statistics, 2017a). The onset of the Great Depression, which lasted from October 1929 to the 1930s, damaged more heavy industries like steel that were dominated by male workers than light manufacturing industries and the service sector where most women worked. Many husbands were therefore unemployed and were not able to provide for their families. More women entered the labor force also because there was a shortage of labor when men joined the army during World War II. In the second half of the twentieth century, there was further increase in the labor force participation rate of women. Due to the economic growth that increased the demand for labor, baby boomers born between 1946 and 1964 began entering the labor force in large numbers in the early 1960s as they reached working age. In addition to structural forces, individual value systems about personal life might have influenced couples' decisions about maternal employment. For most White married mothers, the decision to work outside the home was best characterized as a personal choice to seek an ideal lifestyle combining family and employment rather than economic necessity (Eggebeen & Hawkins, 1990). The number of women working outside of the household for pay exceeded the amount of housewives on record (Eitzen *et al.*, 2011) and peaked at 60.0 percent in 1999 (Toossi & Morisi, 2017). In June 2018, 76 million females age sixteen and older participated in the civilian labor force, compared with 86 million male counterparts (U.S. Bureau of Labor Statistics, 2018d).

Occupational Gender Segregation

The second division of labor among people who worked in the labor force segregated women and men into different jobs (Padavic & Reskin, 2002). **Occupational gender segregation** is the construction of female and male work based on cultural practices and gender differences. Studies focus on how culture shapes men's and women's work experience. Gender stratification through the division of labor is a cultural universal. All societies classify work by gender (Murdock, 1954). In an examination of the division of labor among 324 societies around the world, Murdock and White (1969) found that in nearly all cases the jobs assigned to men were given greater prestige. Although every society assigns tasks on the basis of people's sex, the kinds of tasks that go to women and men have varied over time and around the world (Padavic & Reskin, 2002). For example, before the initial Industrial Revolution, among the serfs and peasants, men usually plowed and threshed, women weeded, and both sexes harvested. Globally, work that some societies view as female is assigned to the male gender in other societies. For example, men are responsible for growing rice in most countries. In Gambia, however, women have cultivated rice since the fourteenth century. During food shortages in the nineteenth century, the government tried to encourage men to help grow rice but the men refused, insisting that rice was "a woman's crop" (Carney & Watts, 1991). In Muslim societies where religious law requires strict sex segregation, men hold positions as elementary school teachers, secretaries, and nurses, whereas traditionally Westerners think of these as women's work (Papanek, 1973).

Occupational gender segregation is a feature of the labor market. Employers usually delegated tasks on the basis of gender and constructed work. For example, high-paying jobs were offered to men while low-paying and dead-end positions such as secretary were created for women (Milkman, 2016). The rise of the highly feminized service sector prompted an interest in the characteristics of these jobs. To distinguish female-orientated jobs from the blue-collar worker in manual labor, **pink-collar work** is used to refer to non-manual and semiskilled positions primarily held by women in service-oriented work. Pink-collar workers are employed in service occupations such as preschool teacher, dental assistant, secretary, and clerk (Hodson & Sullivan, 2002). **Emotion work** is defined as work in which workers must manage their own feelings in order to create certain feelings in others (Hochschild, 1979). For example, retail clerks try to put a smile on their face and greet customers in a friendly way to impress them and make them feel good about the products they want them to purchase. Employers are likely to hire women for these jobs as they assign tasks based on being female or male. Figure 12.14 shows the 25 most common occupations for full-time female and male workers in 2015.

Today, women make up 47 percent of the U.S. labor force, up from 30 percent in 1950 (Geiger & Parker, 2018). Women now work in a much broader array of fields and although women's earnings have risen compared with men's, differences remain (U.S. Bureau of Labor Statistics, 2017b).

Equal Pay Is a Family Issue

Wealth and income inequality within families is a growing issue. A study of eleven European countries reveals that households that receive gifts and bequests own considerably more wealth than non-receiving households (Korom, 2018). The majority of parents are working as a means of earning income and supporting the family. Gender is embedded in how work is rewarded and women are still struggling to make ends meet.

Occupational gender segregation plays a role in the pay gap. In 1937, the U.S. Supreme Court upheld Washington state's minimum wage laws for women. In 1938, the Fair Labor Standards Act established minimum wage without regard to sex. Women earned 59 cents for every dollar that men earned in 1963. The Equal Pay Act of 1963 requires that men and women

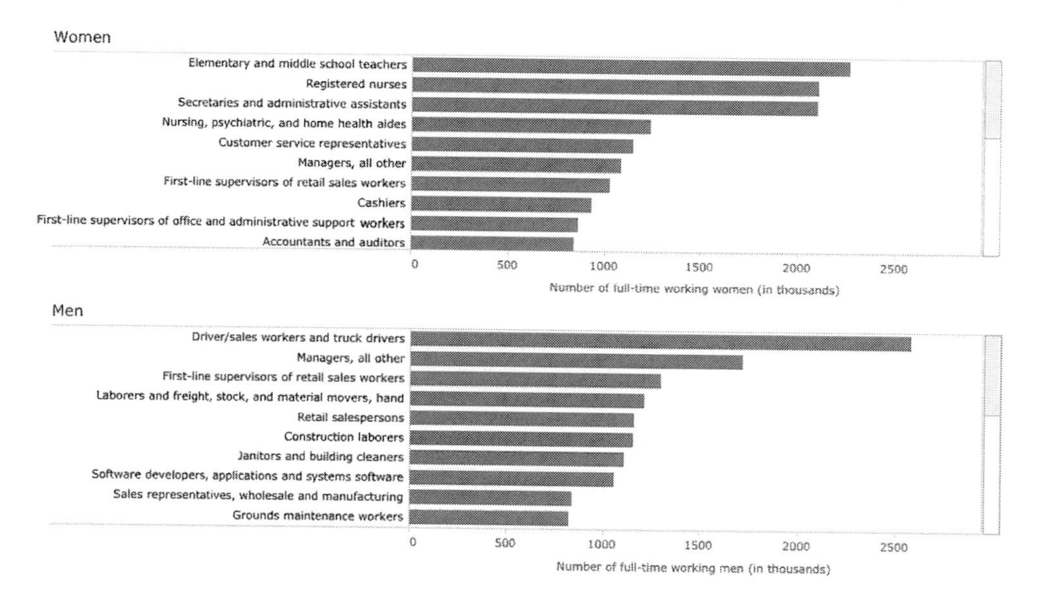

Figure 12.14 The 25 Most Common Occupations for Full-Time Workers, by Sex and Number of Workers, 2015 Annual Averages

Source: 2015 Current Population Survey, U.S. Bureau of Labor Statistics. https://www.dol.gov/wb/stats/most_common_occupations_for_women.htm

Notes: 'Most common occupations' are selected based on the number of full-time workers in the civilian noninstitutional population 16 years of age and older. Full-time workers are persons who usually work 35 hours or more per week.

in the same workplace be given equal pay for equal work. In 2015, women earned 83 percent of what men earned; based on this estimate, it would take an extra 44 days of work for women to earn what men did in 2015 (Brown & Patten, 2017). Women were paid less than males for similar roles (see Figures 12.15 and 12.16).

Between-group and Within-group Income Inequality

Research has explored *between-group inequality*—inequality between groups of individuals or households classified according to various criteria (e.g., gender, race, class) (McKay, 2002)—and *within-group inequality*—inequality occurring among workers who are similar in education, age, and other characteristics (McCall, 2000). Foad (2018) has exploited the variation in income inequality *between groups* to estimate the determinants of income inequality. A study of 32 immigrant groups across a wide range of ethnic enclaves in the United States reveals that most of the inequality between immigrant groups can be explained by education (Foad, 2018). In the second quarter of 2018, full-time workers age 25 and over without a high school diploma had median weekly earnings of $554, compared with $726 for high school graduates and $1,310 for those with a bachelor's degree (U.S. Bureau of Labor Statistics, 2018a). Median weekly earnings of full-time workers vary by gender, age, and race. In the first quarter of 2018, the women's-to-men's earnings ratio (women's earnings as a percent of men's) was highest for younger workers ages 16 to 19 (92.1 percent) and for workers ages 20 to 24 (97.8 percent), compared with workers ages 45 to 54 (76.3 percent) and workers ages 55 to 64 (74.3 percent), and 65 and older (75.3 percent) (see Figure 12.17). In the second quarter of 2018, median weekly earnings for women ($780) were 81.3 percent of those for men ($959) (U.S. Bureau of Labor Statistics, 2018a). Overall, median earnings of Hispanics ($674) and Blacks ($696)

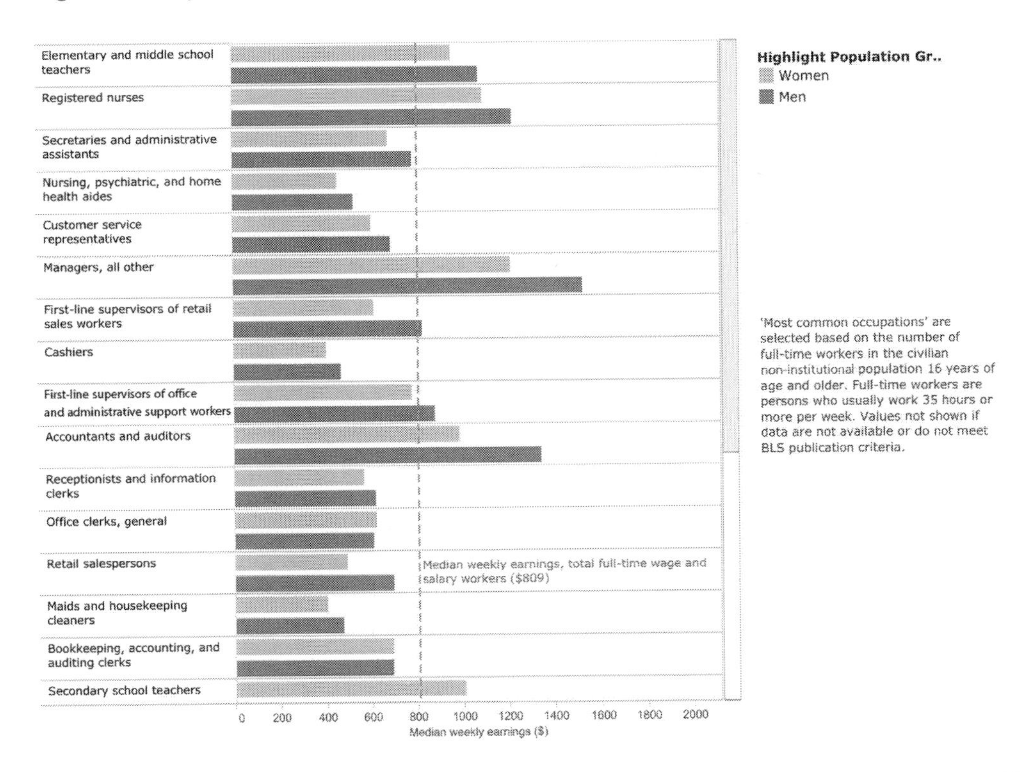

Figure 12.15 The 25 Most Common Occupations for Full-time Working Women, by Median Weekly Earnings and Sex, 2015 Annual Averages
Source: 2015 Current Population Survey, U.S. Bureau of Labor Statistics. https://www.dol.gov/wb/stats/most_common_occupations_for_women.htm

were lower than those of Whites ($911) and Asians ($1,066) (U.S. Bureau of Labor Statistics, 2018a).

In the United States, women have made gains in educational attainment and outpaced men in four-year colleges and postgraduate education (Geiger & Parker, 2018), but women still received less income than men in all occupations. At every level of education, women earned less than men. In the second quarter of 2018, women with a master's or professional degree and above had median weekly earnings of $2,625, compared with their male counterparts ($3,900) (U.S. Bureau of Labor Statistics, 2018a). In 2017, women earned 82 percent of what men earned and, based on this estimate, it would take an extra 47 days of work for women to earn what men did (Graf *et al.*, 2018). As you can see, a pay gap still exists. Milkman (2016) argued that the differences among labor unions (due to historical waves of union organizing), the relationship between union members and their leadership, and the role of women within different types of labor organizations were related to the unequal labor market outcomes between genders.

The *within-group inequality* has accounted for one-quarter to over one-half of the overall increase in inequality (Dinardo *et al.*, 1996) than between-group inequality (Katz & Murphy, 1992). A study of 1.36 million full-time prime-age male and female workers from 28 countries spanning 40 years reveals that in nearly all countries, within-group inequality is the primary driver of levels and trends in inequality (VanHeuvelen, 2018). In the United States, the median weekly earnings of full-time workers vary by gender and age. Median weekly earnings of full-time workers were highest for both men and women ages 35 to 64 but women were paid less.

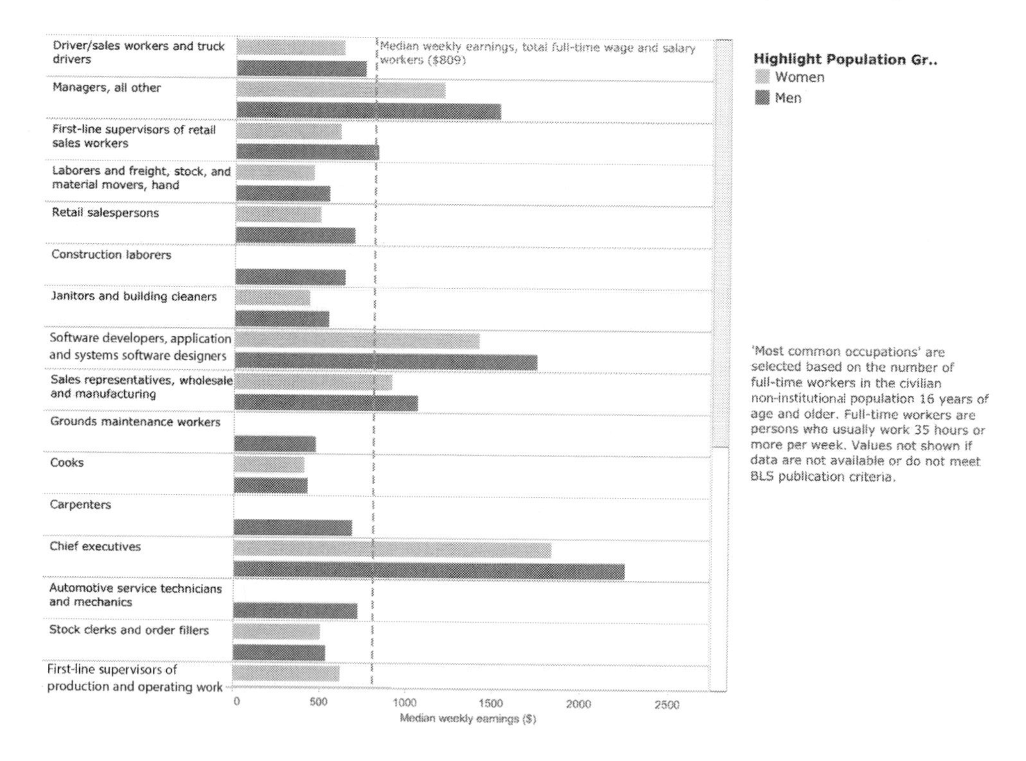

Figure 12.16 The 25 Most Common Occupations for Full-time Working Men, by Median Weekly Earnings and Sex
Source: 2015 Current Population Survey, U.S. Bureau of Labor Statistics. https://www.dol.gov/wb/stats/most_common_occupations_for_women.htm

Weekly earnings were $1,085 for *men* ages 35 to 44, $1,108 for men ages 45 to 54, $1,127 for men ages 55 to 64, and $1,074 for men age 65 and over, compared with $866 for *women* age 35 to 44, $854 for women age 45 to 54, $856 for women age 55 to 64, and $925 for women age 65 and over. Men and women age 16 to 24 had the lowest median weekly earnings, $528 and $511, respectively (U.S. Bureau of Labor Statistics, 2018a). By race, in the third quarter of 2017, median earnings of Hispanics ($698 for men and $597 for women) and Blacks ($744 for men and $658 for women) were lower than those of Whites ($965 for men and $791 for women) and Asians ($1,147 for men and $902 for women) (see Figure 12.18). In the second quarter of 2018, median weekly earnings for Black men ($720) and for Hispanic men ($704) were 73.1 percent of and 71.5 percent of those for White men ($985), respectively (U.S. Bureau of Labor Statistics, 2018a).

Research has examined discriminatory behaviors against obese men and women. For example, overweight women, for the same work, receive less pay than their thin counterparts and overweight men sort themselves into lower-level jobs (Puhl & Brownell, 2001). In Germany, overweight and obese men earn less due to productivity-related variables (e.g. education, work experience, occupation, and physical health), whereas overweight and obese women suffer from **taste-based discrimination**—suggesting that employers discriminate against minority applicants to avoid interacting with them, regardless of their productivity (Bozoyan & Wolbring, 2018).

Research has explored the determinants of within-group inequality. A study of 500 labor markets in the United States in 1990 reveals that flexible and insecure employment conditions (e.g., unemployment, contingent work, and immigration) are associated

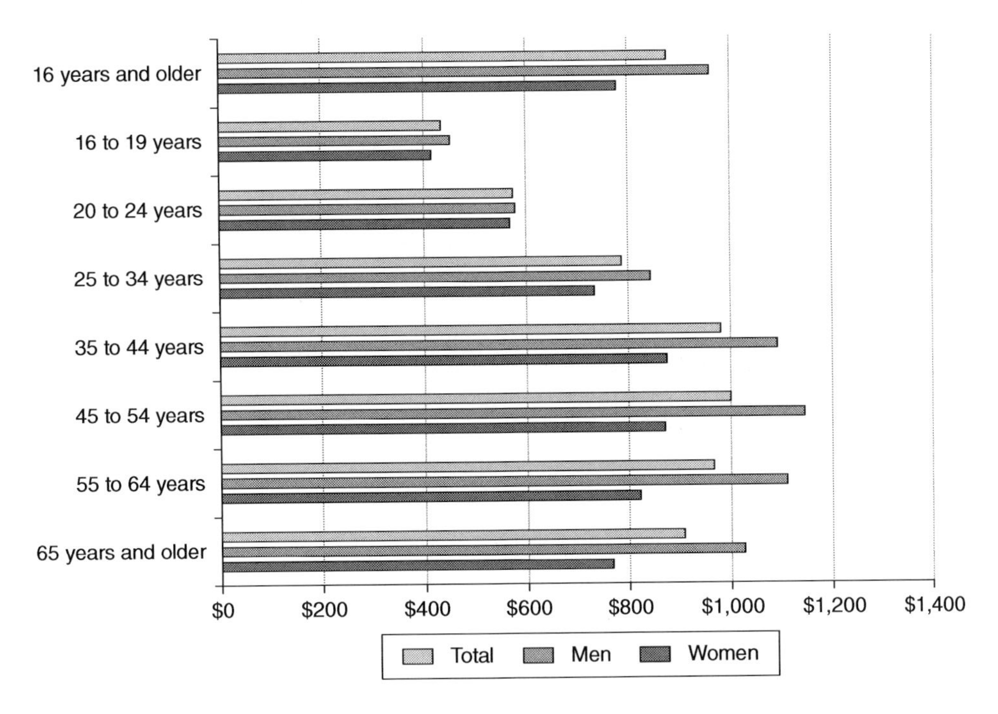

Figure 12.17 Median Usual Weekly Earnings of Full-time Wage and Salary Workers, by Age and Sex, First Quarter, 2018>
Source: U.S. Bureau of Labor Statistics 2018. https://www.bls.gov/opub/ted/2018/median-weekly-earnings-783-for-women-965-for-men-in-first-quarter-2018.htm?view_full

strongly with high local levels of within-group wage inequality, especially among women (McCall, 2000). Within immigrant groups in the Unites States, Foad (2018) finds that for Mexican immigrants, the largest determinant of inequality is gender, whereas for Iranian immigrants, inequality is driven by educational differences. Based on the assumption that individuals are heterogeneous in their innate abilities, Miyake *et al.* (2009) has used the heterogeneity to explain the between-group wage inequality and the within-group wage inequality in the UK and the United States. To reduce inequality, Miyake *et al.* (2009) suggest that each individual should acquire the skill for the new high-tech sector in a competitive labor market.

Division of Unpaid Domestic Work

The distinction between paid work in the labor market and unpaid domestic work (or household labor or domestic labor) has had important consequences for gender inequality (Padavic & Reskin, 2002). Because paid work generates income and is viewed by many in society as "real work," household labor or the work that people do at home does not generate income and

Upon completion of this section, students should be able to:
LO7: Compare the amount of housework and caring activity between women and men.
LO8: Explain factors affecting the unequal division of domestic work between women and men.

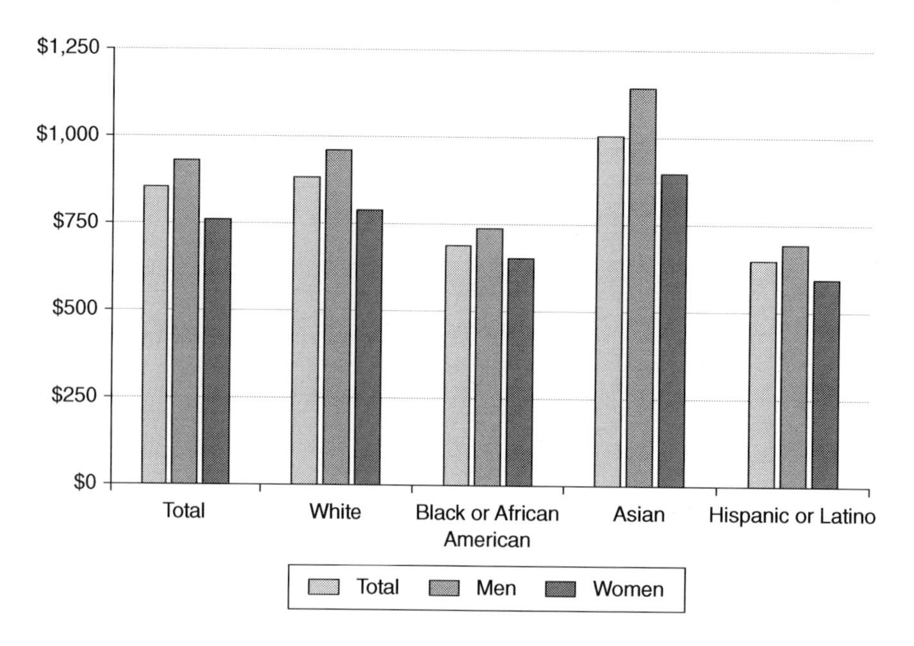

Figure 12.18 Median Weekly Earnings of Full-time Wage and Salary Workers Age 16 and Older by Sex and Race
Source: U.S. Bureau of Labor Statistics 2017. https://www.bls.gov/opub/ted/2017/median-weekly-earnings-767-for-women-937-for-men-in-third-quarter-2017.htm

is therefore devalued. The time spent by a male or female in different activities of their own household is termed as **unpaid domestic work** and unpaid domestic work is often invisible (Sarkar, 2005).

Throughout history, privileged classes have been exempt from productive work and domestic unpaid work. However, a growing group of unemployed workers cooked and cleaned for family members, raised children, cared for sick relatives, and provided social and emotional support to family, friends, and community (Padavic & Reskin, 2002). For women who are full-time homemakers or housewives, household labor is obviously their main "occupational" activity (Shaw, 1988). Although women's presence in the labor force has shown a marked increase, much of the existing research on housework suggests that for heterosexual families, men's and women's housework contributions remain unequal (Moras, 2017). Household labor is time consuming. For employed women, involvement in household work and childcare is generally regarded as a "second job" (Shaw, 1988). Let's look at the *types of unpaid domestic work*: housework and childrearing.

Housework

The first component of domestic labor is to maintain the physical infrastructure of the household (Luxton, 1987). **Housework** refers to all unpaid household activities involving homemaking or the management of a home. It involves day-to-day operations of a home property, domestic chores, etc. Household work has been depicted as a low-prestige activity that is also physically demanding, routine, and isolating (Tao *et al.*, 2010). Robinson and Spitze (1992) explored the links of the performance of family work with psychological well-being. Because family work involves engaging in many intrinsically unlikeable activities, more time spent in such activities will lead to lower levels of psychological well-being and depression or psychological distress (Escribà-Agüir & Tenías-Burillo, 2004).

Researchers have long been interested in how household labor is divided up between parents. Housework is traditionally divided into external or outdoor work (e.g., lawn and garden care, exterior maintenance, and repair) and internal or indoor work (e.g., food preparation and cleanup, interior cleaning, laundry, and household management). Research has examined the nature of housework in terms of *schedule control*—"ability to schedule tasks to reflect one's personal needs" (Barnett & Shen, 1997). **Low-schedule-control tasks** refer to household labor that must be done every day and at certain times with little discretion in the scheduling of tasks. These domestic chores such as grocery shopping, laundry and cooking, cleaning, and childrearing are commonly associated with women and have to be done every day. **High-schedule-control tasks** are often initiated and performed without any time urgency. Men are likely expected to mow the yard, take out the trash, maintain the car, fix broken things, etc. These high-schedule-control tasks can be completed occasionally and delayed. Indeed, the performance of high- and low-schedule-control activities is highly gendered within households, with women typically spending more hours on low-schedule-control tasks and men on high-schedule-control tasks (Marshall, 2006). A study of 233 dual-earner couples revealed that wives were likely to be involved in low-control household tasks and perceptions of marital equity were influenced by time spent in low-control household tasks for both couples (Bartley *et al.* 2005). For husbands and wives, more time spent performing low-schedule-control tasks was associated with greater distress, whereas the amount of time spent on high-schedule-control tasks was unrelated to mental health outcomes (Barnett & Shen, 1997).

Although society now expects married women to participate in the labor force, it continues to define domestic work as women's sphere (Padavic & Reskin, 2002). Women's entry into traditionally male spheres was not matched (at least to the same degree) by men's increased involvement in the domestic sphere (England, 2010), despite a slight increase in men's unpaid labor in the home (Hook, 2006). Research on the division of household labor has typically focused on the division of household tasks or the division of time devoted to such tasks between husbands and wives (Shaw, 1988). In fact, women still do more unpaid housework than men in the United States (Bianchi *et al.*, 2012) and internationally (Sani, 2014). In the United States, the average total hours per day women ages fifteen and over spent in all selected household activities was 2.19 hours, compared with 1.41 hours for men in 2017 (see Figure 12.19).

In the United States, 84 percent of women ages fifteen and over engaged in all household activities, compared with 68 percent of their male counterparts in 2017 (see Figure 12.20).

Childrearing

The second component of domestic labor is childbearing and childrearing (Luxton, 1987). *Childbearing* includes pregnancy, birth, and the early nursing period. Pregnant and nursing mothers experience enhanced narcissism, describe themselves as precious, and occupy an almost mystical position in the cultural perception of our society (Novick, 1988). When it comes to childbearing, the key issues in developing countries include poverty, lack of childcare, and high rates of maternal and infant mortality, whereas industrialized societies are faced with isolation of the nuclear family and economic pressure for mothers to work (Steinberg, 1996). In the United States, some policies may help working parents with a newborn or adopted child.

Parental Leave

The Family and Medical Leave Act (FMLA) of 1993 guaranteed unpaid maternity leave and mandated a minimum of twelve weeks unpaid leave to mothers for the purpose of attending to a newborn or newly adopted child (Curtin & Hofferth, 2006). *Unpaid family leave* refers to an employee benefit (maternity, paternity, and adoption leave) provided to care for

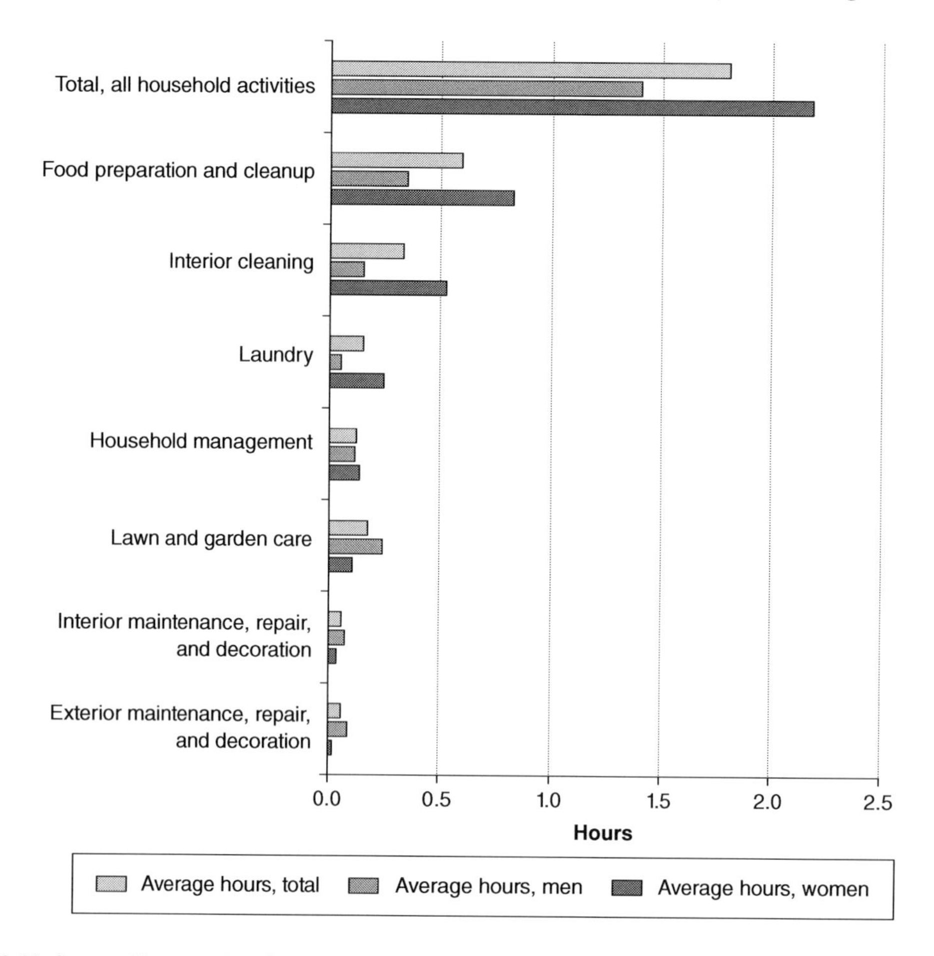

Figure 12.19 Average Hours per Day Spent in Selected Household Activities, 2017 Annual Averages
Source: U.S. Bureau of Labor Statistics. https://www.bls.gov/charts/american-time-use/activity-by-hldh.htm

a newborn baby or ill family members without pay if working in a firm with at least 50 employees and meeting minimum tenure and working hour requirements. *Paternity leave* is when a father takes time off to support his newly born or adopted baby. *Maternity leave* is a temporary period of absence from employment granted to new mothers during the months before and after childbirth. These policies support the mother's full recovery from childbirth, facilitate parent–child attachment, and improve the health and economic situation of women and children. Having a child (including the cost of high-quality childcare) costs families approximately $11,000 in the first year and these financial issues contribute to new mothers in the United States returning to work quicker than new mothers in European countries (Miller *et al.*, 2009). But taking unpaid time off is not a realistic option for families who cannot afford to do so. In addition, because of coverage restrictions based on firm size and number of working hours, only about 60 percent of workers are actually eligible for the protections of FMLA (Child Trends Databank, 2017b).

Paid family or parental leave refers to paid leave granted to an employee to care for a newborn or adopted child, a sick child, a sick adult or short-term disability. Getting paid family leave depends on one's

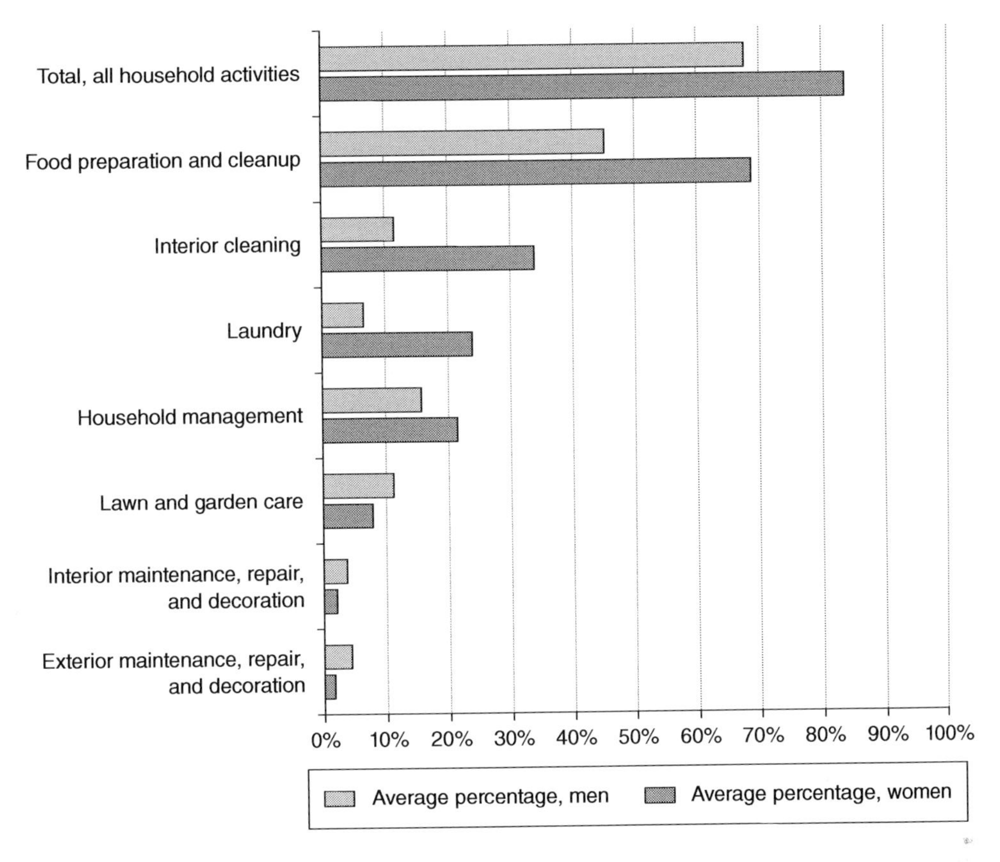

Figure 12.20 Percent of the Population Engaging in Selected Household Activities, Averages per Day by Sex, 2017 Annual Averages
Source: U.S. Bureau of Labor Statistics. https://www.bls.gov/charts/american-time-use/activity-by-hldp.htm

employer and workplace. Many states have enacted policies to provide workers with paid family leave, for example, California, New Jersey, Rhode Island, New York, Washington, DC, and Washington state have paid leave plans in place (Child Trends Databank, 2017b). About eight in ten Americans (82 percent) support paid leave for mothers following the birth or adoption of a child, while 69 percent support paid paternity leave for fathers (Stepler, 2017).

Leave takers with lower incomes are far less likely than those with higher incomes to receive pay when they take time off from work for family or medical reasons (Stepler, 2017). Workers making more than $75,000 per year are more likely than others to have been paid while taking leave (Child Trends Databank, 2017b). In fact, 24 percent of workers in management, professional, and related occupations had more access to paid family leave than did those in other major occupations (U.S. Bureau of Labor Statistics, 2016c). In 2017, 43 percent of civilian workers had access to paid personal leave and sick leave (U.S. Bureau of Labor Statistics, 2018q).

Childrearing refers to the bringing up and taking care of children (Novick, 1988). The parenting of young children is typically among the most challenging phases in a parent's life, particularly when both partners are employed outside the home (Baxter *et al.*, 2008). The average total hours per day mothers spent caring for and helping household children under age eighteen was 1.74 hours, compared with 1.04 hour for

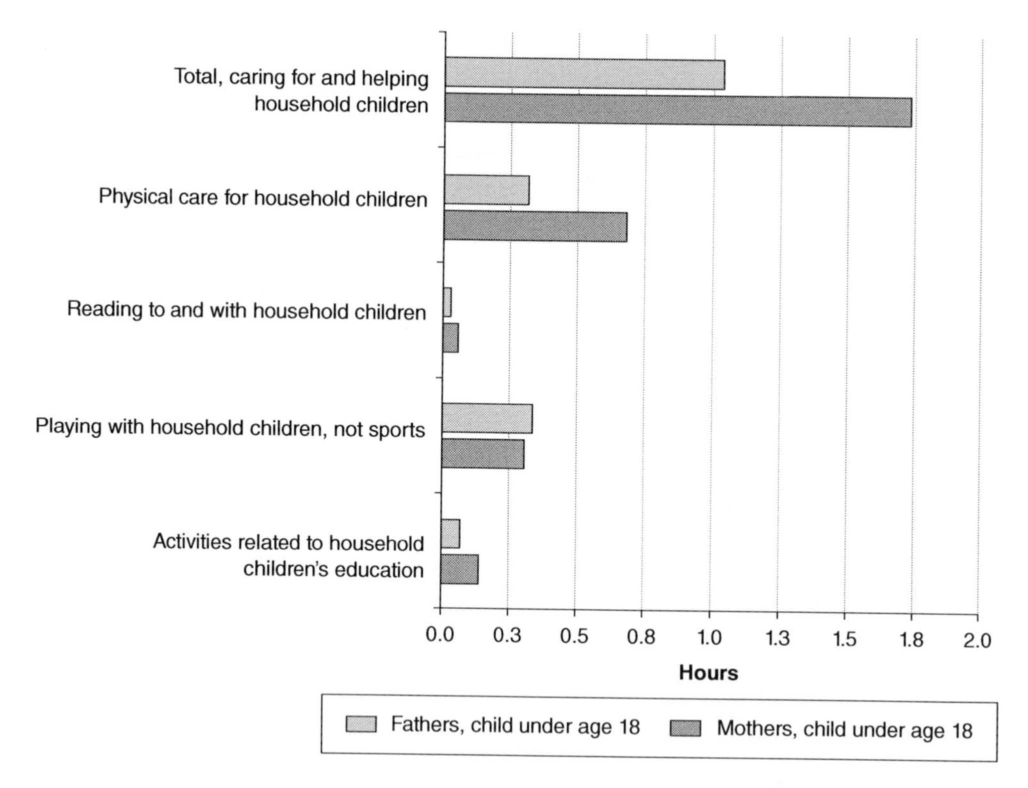

Figure 12.21 Average Hours per Day Parents Spent Caring for and Helping Household Children as Their Main Activity, 2017 Annual Averages

Source: U.S. Bureau of Labor Statistics. https://www.bls.gov/charts/american-time-use/activity-by-parent.htm

fathers in 2017 (see Figure 12.21). Research indicates that an unequal division of childrearing may be more strongly associated with women's psychological distress than an unequal division of housework (Des Rivieres-Pigeon *et al.*, 2002). A study of 293 employed parents (182 mothers and 111 fathers) revealed that for mothers, more perceived time spent in childrearing and high-schedule-control housework tasks (e.g. yard work) relative to one's partner was associated with greater distress (Tao *et al.*, 2010).

In addition to childcare, care work is focused on the responsibility of caregivers to provide for the sick, the elderly, and pets. *Eldercare providers* are individuals who provide *unpaid care* to someone age 65 or older who needs help because of a condition related to aging. Eldercare involves any care activities such as assisting with grooming, preparing meals, and providing transportation and companionship. In 2015–2016, almost 50 percent provided care for two years; 14 percent provided care for ten years or more; most providers ages 35 to 64 cared for a parent (U.S. Bureau of Labor Statistics, 2017e). In 2016, among the 51.0 million eldercare providers aged fifteen and older, most were female (56.4 percent) and those aged 45 to 64 accounted for 43.8 percent (He *et al.*, 2018). Caregivers who report feelings of time pressure or stress related to second- and third-shift caregiving are more likely to have poor health (Pinquart & Sörensen, 2003).

Factors Affecting Unequal Division of Household Labor

When both parents work outside the home, women still do more housework than their male counterparts. What are the social and cultural reasons for this unequal division of household labor? Structural and cultural norms regarding the family and personal autonomy are transferred from the system to the individuals in the couples (Öun, 2013). On a societal level, gendered roles associated with masculinity and femininity play an active role in shaping how housework is shared in couples (Nyman *et al.*, 2013). Traditionally, the household was *patriarchal* (male dominance) in structure. The man was head of the household and to whom all its members and effects were subject: wife, children, and property. Housework recognizes the long history of patriarchal exploitation of women's labor, but demonstrates the historical specificity of modern housework (Jackson, 1991). Using a case study analysis of a three-generation family, McMullin (2005) suggests that gender, class, social context, and family background influence how paid and unpaid work is divided within families. Indeed, the social context of a given time conditions the options women and men have available to them in negotiating the balance of work and family responsibilities. For example, in 1965, fathers' time was heavily concentrated in paid work, while mothers spent more of their time on housework or childcare (Parker & Livingston, 2018). Research also indicates that spouses' relative socioeconomic resources and work hours are linked to their housework allocation. For example, at times when husbands earned more income and worked more hours than wives, wives were responsible for larger portions of household tasks (Lam *et al.*, 2012).

Gender egalitarianism may result in less housework for wives, more for husbands, and more equal sharing of housework by couples. A study of 16,633 men and women across 29 countries reveals that in gender egalitarian countries where women spend more time in the labor market, women and men are more likely to consider doing a larger share of housework to be unfair (Jansen *et al.*, 2016). Even though structural and cultural changes regarding gendered practices have taken place, the convergence and desegregation of domestic labor occurs slowly (England, 2010). For example, most people in Sweden claim to embrace gender equality but the majority of couples still organize their family lives according to traditional gendered patterns (Nyman *et al.*, 2018). Couples in Britain, Norway, and Australia that aspire towards gender equality still have an unequal division of housework (Van Hooff, 2011; Skarpenes & Nilsen, 2015; Walters & Whitehouse, 2012).

Research has explored the relationship between division of unpaid domestic work and marital satisfaction. Both men and women are happier and more satisfied in marriages that equally divide household chores (Frisco & Williams, 2003). Heterosexual couples with equality of domestic labor are less likely to experience marital conflict (Stohs, 2000). When it comes to women's family life satisfaction, researchers have explored two major dimensions of women's fairness comparison—*inter-gender relational comparison* between partners and *intra-gender referential*

Student Accounts of Work and Life

Some people are believers in having career, marriage, and motherhood, three at the same time as they get married while working on their career and growing their family.

Upon completion of the next section, students should be able to:
LO9: Describe the benefits and challenges of dual-earner families and multigenerational families.
LO10: Apply structural functionalist theory to the analysis of stay-at-home parents.

comparison with other women from the same society. An analysis of data across 30 countries reveals that women's family life satisfaction is adversely affected by both a lack of relational fairness and unfavorable referential comparison (Hu & Yucel, 2018).

Work and Life Options

Families are classified as married-couple families or as families maintained by women or men without spouses present. Today people live in dual-earner families and in multigenerational families but some families live with stay-at-home mothers or fathers, or have other forms of living arrangements.

Dual-earner Families

Dual-earner couples represent a large segment of the U.S. workforce. In **dual-earner families** both spouses participate in paid work. As noted earlier in the chapter, this form of family became popular in the 1960s. Among married-couple families in 2017, both the husband and wife were employed in 48.3 percent of families (U.S. Bureau of Labor Statistics, 2018r). In 2016, married-couple families whose youngest child was six to seventeen years old were more likely to have both parents employed (64.7 percent) than families in which the youngest child was under six years old (56.3 percent) (see Figure 12.22).

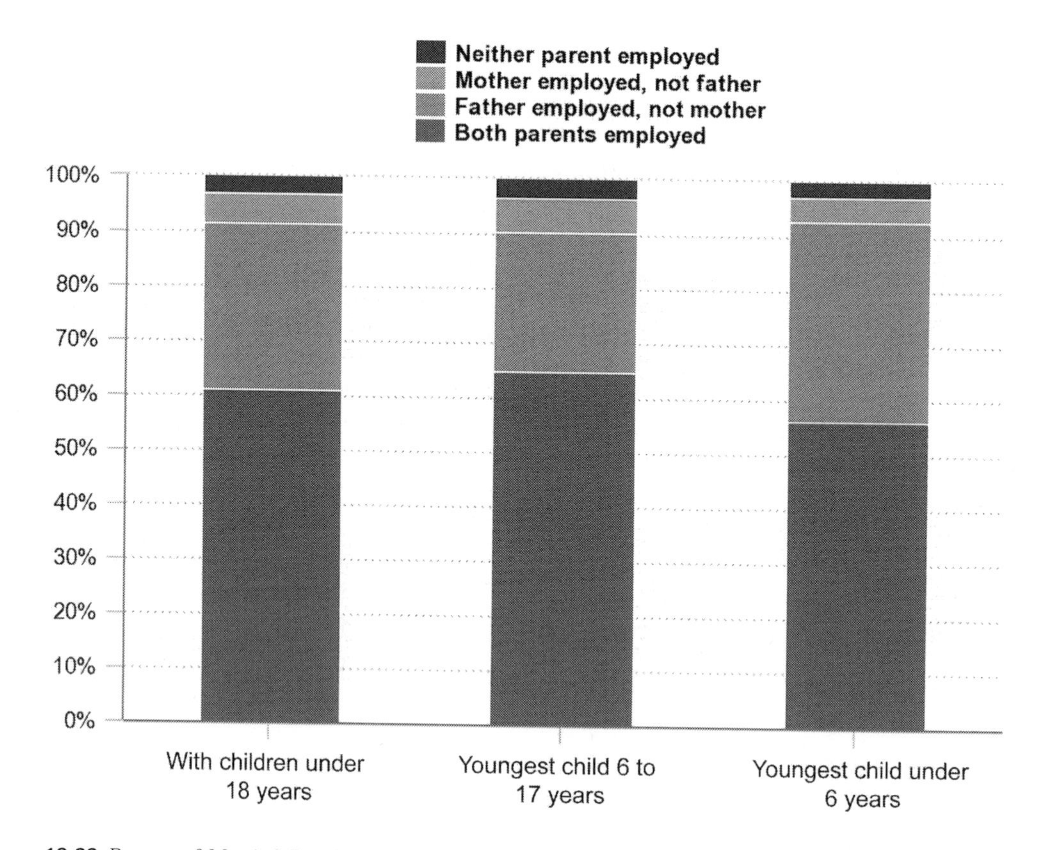

Figure 12.22 Percent of Married-Couple Families with Children, by the Employment Status of the Parents and the Age of the Youngest Child, 2016 Annual Averages

Source: U.S. Bureau of Labor Statistics. https://www.bls.gov/opub/ted/2017/employment-in-families-with-children-in-2016.htm

Benefits and Challenges of Dual-earner Families

Mother's Day or Father's Day is a day to celebrate the contributions that mothers and fathers make to all of our families. One of the contributions any parent makes is earning wages to provide for the family. The benefits of working outside the home include increased self-esteem, independent identity, enhanced social contacts, higher standard of living (Barnett & Baruch, 1985; Eggebeen & Hawkins, 1990; Goldberg & Easterbrooks, 1988), and less depression (Barnett & Hyde, 2001). A family earning two incomes is likely to have more financial resources. In 2015, married-couple households had the highest median income ($84,626), followed by households maintained by men with no wife present ($55,861), and by women only ($37,797) (Proctor *et al.*, 2016).

Another contribution any mother or father makes is performing the household labor it takes to cook meals, clean a house, and raise children. Increases in female participation in the workforce and changes in gender roles have driven up the demand for housework and caring tasks. In the last three decades, many working parents learned to fulfill unpaid work after getting home from the workplace. This has resulted in the second shift for working parents after paid work. The **second shift** refers to the unpaid work performed at home in addition to the paid work performed in the labor force. Kanter (1977) coined the term "myth of separate worlds" of paid work and family work to illustrate that paid work "spilled" over into family life. Many therapists harbor negative and stereotypical assumptions of the quality of dual-earner family life and appear to believe that there are inevitable and significant challenges that are intrinsic to the dual-earner family arrangement (Haddock & Bowling, 2001). Gender roles are now converging as the share of dual-earner households has risen and spouses are expected to share household activities. Artis and Pavalko (2003) used two waves of longitudinal data from a national sample to show that the percentage of wives' household responsibilities declined by about 10 percent over a thirteen-year period. Fathers have more than doubled the time they spend doing household chores since 1965 (see Figure 12.23). Meanwhile, women have increased the time they spend doing paid work and still spend more time on unpaid household work. In 2016, women spent an average of 2.6 hours on housework activities compared with 2 hours for men (U.S. Bureau of Labor Statistics, 2017c).

Many parents turn to daycare centers to help juggle parenthood and paid work. Childcare workers provide care for children when parents and other family members are unavailable. Between 1977 and 1993, the proportion of children in center-based programs increased from 13 percent to 30 percent (Child Trends Databank, 2016). From 2011 to 2015, adults with at least

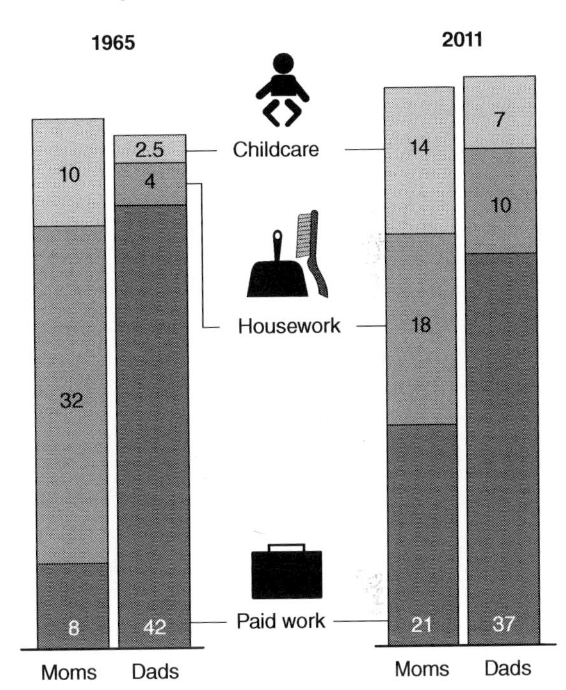

Figure 12.23 Moms and Dads, 1965–2011: Roles Converge but Gaps Remain

Source: Pew Research Center, June 18, 2015. http://www.pewresearch.org/ft_moms-dads-family-roles-2/

Note: Based on adults ages 18–64 with own child(ren) under age 18 living in the household.

one child under age six spent an average of 5.3 hours per day providing secondary childcare (U.S. Bureau of Labor Statistics, 2017c). Dual-earner families have also resulted in an increasing number of children being left to care for themselves. A **latchkey child** is a child who returns from school to an unsupervised home because his or her parents are away at work. In 1995, 15 percent (5.2 million) of children ages five to thirteen were reported to be in self-care regularly during a typical week (Casper & Bianchi, 2001). Among grade-schoolers, self-care increased, with 2 percent of children ages five to eight, 10 percent of children ages nine to eleven, and 33 percent of children ages twelve to fourteen taking care of themselves (see Figure 12.24).

Self-care was much more prevalent among middle-school-aged children than among those in elementary schools (Laughlin, 2013). Because welfare reform in 1996 aimed to encourage recipients to leave welfare and go to work, and move single parents with children off welfare and into jobs, many mothers who could not afford childcare options are forced to leave their children in self-care. Middle school students left home alone for more than three hours a day reported higher levels of behavioral problems, higher rates of depression, and lower levels of self-esteem than other students (Mertens & Flowers, 2003).

Paid Domestic Labor

Highly educated people choose both career and parenthood. **Dual-career couples** refers to both heads of household having "jobs which require a high degree of commitment rather than being simply employed and pursue careers and at the same time maintain a family life together" (Rapoport & Rapoport, 1969). The home environment is a challenge to dual-career couples as they try to meet the demands of careers and build a family life together (Haddock & Rattenborg, 2003). A study of 96 dual-career couples reveals that the amount of labor performed is associated with childcare arrangements, but not the household task

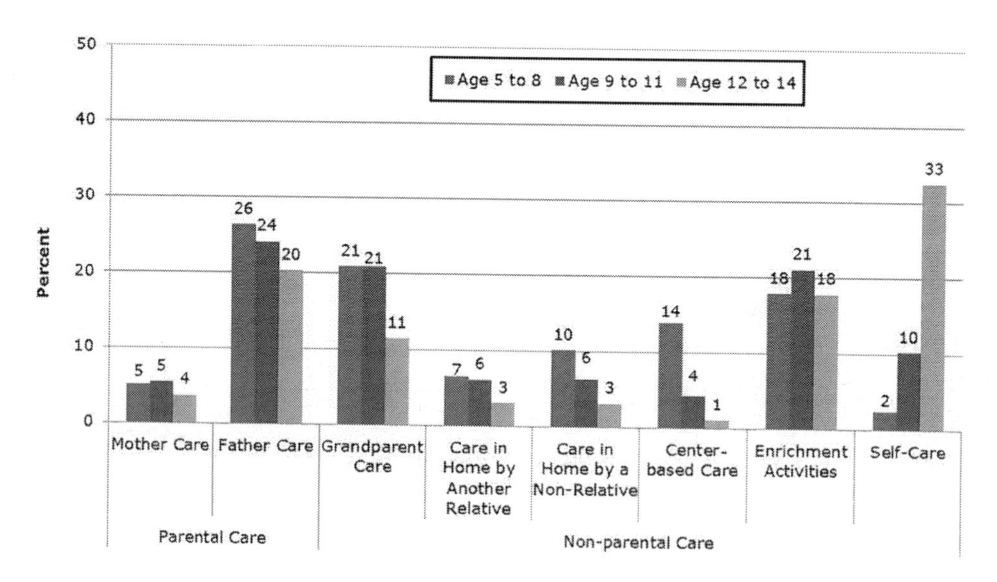

Figure 12.24 Percentage of School-aged Children in Parental and Non-parental Care, by Age and Type of Care, 2011
Source: Federal Interagency Forum on Child and Family Statistics. American Children: Key National Indicators of Well-Being, 2015. Table FAM3.C and Child Trends, May 2016. https://www.childtrends.org/indicators/child-care/
Note: Some children participate in more than one type of non-parental care arrangement. Thus, details do not sum to the total percentage of children in non-parental care.

arrangement (Stevens *et al.*, 2005). Many spouses believe that to be successful in their careers they have to make sacrifices in another domain of their life, usually resulting in less "couple time" and more "work time" (Haddock & Rattenborg, 2003). Most couples try to share the load while others pay for household work. A study of 71 dual-career couples revealed that the husband and wife accepted joint responsibility, but the wife usually and the husband occasionally performed these tasks (Hester & Dickerson, 1984).

Many households regularly outsource unpaid domestic labor by purchasing services to help with cleaning, cooking, and other chores. The daily activities performed by the maid in the respective employer's household may be termed as **paid domestic work** (Sarkar, 2005). Paid domestic work is a feature of households all over the world and it constitutes the single largest female employment sector (Anderson, 2001). In London and other parts of the world, domestic employment has been shown to be of low status and to be denigrated by those involved (Cox, 1997). Paid domestic work is the result of income polarization (Cox, 1997). The earnings of domestic servants are low and many women are looking after more than one household but spend more time and work far more for their employer's than they work in their own household (Sarkar, 2005).

Class divides housework time between women with a college degree and from higher-income families and women of lower socioeconomic status (Schneider & Hastings, 2017). Paid domestic work is usually performed by women, and is managed by privileged women. The influx of privileged women into the labor force, coupled with growing income inequality, has caused an increase in the demand for paid domestic labor (Moras, 2017). In a study of White privileged women who hire domestic workers, Moras (2017) finds that this is not necessarily challenging the existing gendered division of labor but rather displacing housework along raced and classed lines. Indeed, wives in dual-career families who work full-time and who earn more than 50 percent of the family income do less housework than if they earned less than 50 percent of the family income (McFarlane *et al.*, 2000). A study of 3,540 dual-earner households in Spain reveals that women who hire do about 30 minute less housework per day than non-hiring women, but in relation to their partners these women continue to do the same share of housework (Gonalons-Pons, 2015).

Multigenerational Families

Historically, as nations industrialized, it became common for children to strike out on their own after reaching adulthood, often leaving for towns far from their families as they went to college and then work (Muennig *et al.*, 2018). After a long post-war decline in **multigenerational families** (i.e., households with multiple generations of family members living under the same roof), the proportion of multigenerational households in the United States and other wealthy nations is again increasing (Taylor *et al.*, 2010). In the United States, 64 million American people (20 percent of the U.S. population or one in five Americans) lived in multigenerational families in 2016 (Cohn & Passel, 2018), compared with 60.6 million (19 percent) in 2014 and 57 million (18.1 percent) in 2012 (Fry & Passel, 2014). Changes in family structure, as well as greater longevity, have led to an increased reliance on kin to perform family functions (Bengtson, 2001). Families that are more familistic in orientation (e.g. Hispanics, Catholics) are more likely to coreside (Oropesa & Landale, 2004). Growing racial and ethnic diversity in the U.S. population helps explain the rise in multigenerational living. In 2016, people living in multigenerational households included 29 percent of Asians, 27 percent of Hispanics, 26 percent of Blacks, and 16 percent of Whites (Cohn & Passel, 2018).

Multigenerational families have benefits for family members. Multigenerational households share common resources, such as food, childcare, eldercare, heat, electricity, transportation, and rent, thereby reducing the cost of living relative to individual or single family living arrangements (Muennig *et al.*, 2018). In a multigenerational family, grandparents, siblings, or other relatives play the role of childcare provider. "For many Americans multigenerational bonds are becoming more important than nuclear family ties for well-being and support over the course of their lives" (Bengtson, 2001). Kin support through coresidence is an important source of support for families with young children and in particular families that are unwed at the birth of their child (Pilkauskas, 2012). Multigenerational living arrangements might improve financial resources, buffer stress, reduce loneliness, enhance intellectual sharing, and generate structural social capital, thereby elevating the level of one's health (Adler & Kwon, 2002). Healthy people living in two-generation households have longer survival than healthy people living on their own (Muennig *et al.*, 2018).

Living in a multigenerational family has rewards for all generations. In 2016, the most common type of multigenerational household—home to 32.3 million Americans—consisted of two adult generations (parents and their adult children); households with three or more generations (grandparents, parents, and grandchildren) housed 28.4 million Americans; and 3.2 million Americans lived in households consisting of grandparents and grandchildren (Cohn & Passel, 2018). Members of certain immigrant communities may be more likely to live in a three-generation family household (Pilkauskas, 2012). Three-generation households likely play an important role in the lives of children, mothers, and grandparents (Pilkauskas, 2012). *Generational needs*, such as the needs of the parent generation, may influence the decision to coreside (Aquilino, 1990). Assistance generally flows from the grandparent to the parent generation (Fingerman *et al.*, 2009). Grandparents can help with childcare and provide role models in the socialization of grandchildren. Mothers can go to work and go to school. Young mothers or those having their first child may be more likely to live with their own parents (Trent & Harlan, 1994). Mothers in poor health or with a needy baby (low birth weight or disabled) may need to coreside (Pilkauskas, 2012). Equally, the needs of the grandparent generation, such as poor physical or mental health, may also influence the decision to coreside (Choi, 2003).

Living in a multigenerational family can be challenging. For example, intergenerational cultural differences may lead to stress. Multigenerational households may be more crowded than single-generational households, a risk factor for poor health in its own right (Gove *et al.*, 1979). Research indicates that three-generation family households are shortlived and transitions are frequent. Using data from the Fragile Families and Child Well-being Study (n=4,898), Pilkauskas (2012) found that mother's relationship status was associated with three-generation family coresidence.

Stay-at-home Mothers

A **stay-at-home mother** is a mother who was not in the labor force for the past 52 weeks, whose husband was in the labor force for the past 52 weeks, and who cares for the family and serves as the primary caregiver of children younger than age fifteen. Actually, stay-at-home mothers care for their families at home because they are traditionally married with working husbands, or they are unable to find work, are enrolled in school, or desire to raise their children, or are disabled. In 2017, married mothers of children under age eighteen (68.6 percent) remained less likely to participate in the labor force than mothers with other marital statuses (76.5 percent) (see Figure 12.25).

For decades, the number of stay-at-home moms had been declining, but the number of stay-at-home mothers rose in recent years. The share of mothers staying home with their children rose from 2000 to 2004, but dropped in 2005 amid economic uncertainty that foreshadowed the official start of the Great Recession in 2007 (Cohn *et al.*, 2014). The number of stay-at-home mothers in married couple family

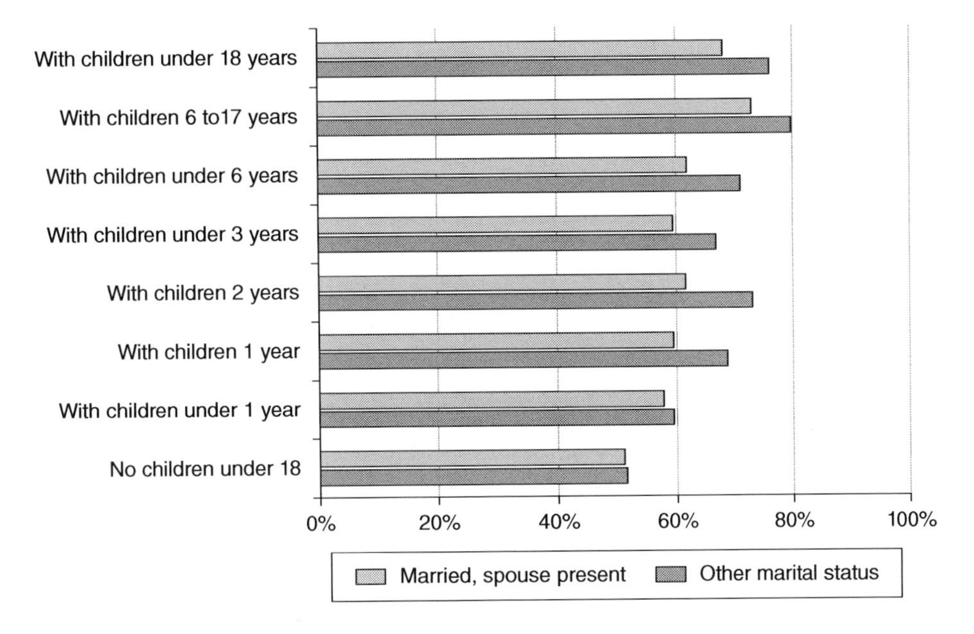

Figure 12.25 Labor Force Participation Rates of Women by Marital Status and Presence and Age of Own Children Under 18, 2017 Annual Averages

Source: https://www.bls.gov/opub/ted/2018/married-mothers-less-likely-to-participate-in-labor-force-in-2017-than-other-moms.htm

groups was 4.9 million in 2017 (U.S. Census Bureau, 2018a), compared with 5.0 million in 2016 (U.S. Census Bureau, 2017b). Figure 12.26 shows an increase in stay-at-home mothers from 20 percent in 1994 to 23 percent in 2017. Factors in the increase of the stay-at-home rate include declines in the labor force participation rate and rising immigration (Galley, 2014).

According to structural functionalist theory, men play an instrumental role in the family as providers, whereas women play an expressive role as homemakers. Sociologist Talcott Parsons (1943) argues that the division of traditional gender roles within the family or the two roles of the husband as breadwinner and wife as homemaker complement each other and contribute to the functioning of families. In estimates of the size and sectoral allocation of the non-market household workforce in the United States between 1800 and 1860, Folbre and Wagman (1993) conclude that women engaged primarily in the provision of domestic services for family members made an important contribution to total output. However, when it comes to the impact of traditional gender roles and the persistence of the traditional male breadwinner mentality, Hajdu and Hajdu (2018) found that the association between the woman's relative income (the woman's share of the couple's total income) and life satisfaction is negative both for men and for women in Hungary.

Stay-at-home Fathers

A **stay-at-home father** is a father who was not in the labor force for the past 52 weeks, whose wife was in the labor force for the past 52 weeks, and who serves as the primary caregiver of children younger than age fifteen. If the father is looking for work or going to school or if the mother changes jobs and is out of work for a week or more, the father does not count as a stay-at-home father. The number of stay-at-home fathers in married-couple family groups was 267,000 in 2017 (U.S. Census Bureau, 2018a), compared with 209,000 of stay-at-home fathers in 2016 who cared for about 392,000 children under age fifteen (U.S. Census Bureau, 2017b/2017c). Figure 12.26 shows an increase in stay-at-home fathers from 1994 to 2017.

High unemployment rates have contributed to the recent increase in stay-at-home fathers, and to the rising number of fathers who are at home to care for their family (Livingston, 2014). Another reason for the increase in stay-at-home dads is that wives may earn more. Between 2000 and 2015, the share of married couples where the wife earned at least $30,000 more than the husband increased from 6 to 9 percent; married couples where the husband earned at least $30,000 more than the wife decreased from 38 to 35 percent (U.S. Census Bureau, 2015).

Studies have examined how breadwinning mothers (BWMs) position their identities and tasks as earners. In an analysis of 44 female breadwinners married to stay-at-home fathers, Medved (2016) presents five aspects of conventional breadwinning that emerged across BWMs' discourse: breadwinning as career-primary, as obligation, as suitability and personality, as relational power, and as ideal workers. In Hong Kong, 0.5 percent of the total male population choose to become stay-at-home fathers. Liong (2017) finds that middle-class men with social and cultural capital are less reflexive to accept their carer identity, whereas working-class fathers, who lack the appropriate capital to resume their provider status, define themselves as family carers. In terms of the time use of stay-at-home fathers in female breadwinner families, Latshaw and Hale (2016) report that some stay-at-home fathers and breadwinning mothers shift or swap their domestic responsibilities. Some mothers continue to perform housework during evenings and weekends and allot husbands more time to pursue other activities.

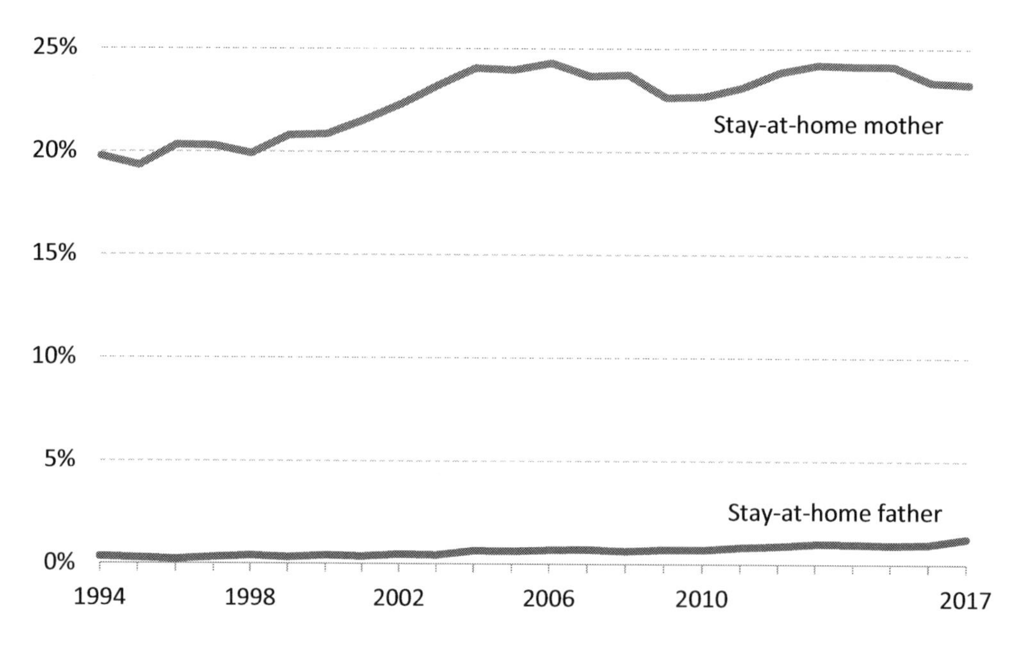

Figure 12.26 Percent of Married Couple Families that Have a Stay-at-home Parent (Couples with Children Under 15)
Source: U.S. Census Bureau, Current Population Survey, Annual Social and Economic Supplements, 1994–2017. https://www.census.gov/content/dam/Census/library/visualizations/time-series/demo/families-and-households/shp-1a.pdf

Upon completion of this section, students should be able to:
LO11: Apply role theory to the analysis of the multiple roles of working parents.
LO12: Explain strategies for balancing paid work, family life, and school work.

Juggling Multiple Roles: Paid Work, Family Work, and School Work

Today, working parents contribute financially to the household in addition to attending to the needs of their children, all the while they may be pursuing a higher education to move up the career ladder. These working parents therefore juggle paid work, unpaid work, and school work. This section discusses how role theory explains how all working parents juggle multiple roles as well as strategies parents use to do the balancing.

Role Theory

As discussed in Chapter 2, **role theory** suggests that most of our everyday activities consist of acting out socially defined roles associated with a given status. Role theory examines the roles that individuals play in the family and explores how people juggle multiple roles. People play different *roles* (i.e., socially defined positions), including marital roles such as husband and wife, or parental roles such as mother and father, student or employee role, etc. For example, mothers and fathers play parental roles in the growth and development of their children, and such parental roles are essential to the family and child's life. An explanation linking family work with mental health is the notion of role strain (Robinson & Spitze, 1992). **Role strain** refers to conflicting demands connected to one status. For example, a single mother is financially, physically, and emotionally overburdened. **Role conflict** refers to a conflict among the roles corresponding to two or more statuses. A commonly experienced role conflict is that among work, family, and school. Role conflict occurs when student parents are required to leave home for paid work or school but have to stay at home and take care of their children or aging parents. **Role overload** is based on the premise that human energy is limited, and the more demands within a role, or the more roles a person plays, the more strain experienced and the greater the likelihood of negative effects on health and well-being. Thus, more time spent in housework and childrearing may create role overload, particularly if combined with paid work, resulting in time pressure and subsequent psychological strain (Tao *et al.*, 2010).

Human behavior is guided by **role expectations**—the expectations required of a role and held both by the individual and by others. The role expectations correspond to different roles individuals perform in their daily lives. For example, the role expectation attached to the status of mother is that she cares and provides for her child. The role expectations of a student include attending class, participating in class discussions, submitting completed school work, and taking tests. **Role performance**—the actual delivery of those expectations—is based on how they accomplish their day-to-day activities. If their role performance meets their role expectations, the student parent is likely to play a good spousal, parental, and student role.

Expansionist Theory

Barnett and Hyde (2001) proposed an **expansionist theory**—holding that multiple roles (e.g., worker, spouse, and parent) are beneficial for both women and men. Hyde (2016) confirms that multiple roles appear to be beneficial for adults' mental, physical, and relationship health, and for children of parents who hold multiple roles. Newer scientific evidence continues to support expansionist theory as women outnumber men in college, and constitute 49.4 percent of the U.S. workforce (Hyde, 2016). Working parents are likely to experience role conflict because they are engaged in both paid work and family work. Role conflicts can become opportunities to increase intimacy and strengthen the relationship. Healthy relationships help foster a positive environment for dealing with work-related issues (Haddock & Rattenborg, 2003). A survey of 404 dual-earner couples with young children living in Torino and Italy revealed that the majority (55 percent) of both wives and husbands perceived fairness and satisfaction, even if most of the chores (about two-thirds) fell on wives' shoulders (Carriero, 2011).

Strategies for Balancing Paid Work, Family Life, and School Work

With more people in the workforce, balancing work and family has become a key issue for working parents. Some 52 percent of working dads say that it is very or somewhat difficult to balance work and family life, while 60 percent of working mothers say the same (Parker & Livingston, 2018). Let's examine some strategies parents can use to balance paid work, family life, and school work.

Student Parents and Childcare

As many jobs require college degrees, working parents pursue higher education and play multiple roles of student, spouse, parent, and employee. Nationwide, of the nearly 1 million low-income working parents in education in 2010–11, most were female (71 percent) and half were White (52 percent), with smaller numbers of African American and Hispanic students (24.9 percent and 18.1 percent). Most low-income student parents (91.6 percent) worked one job, with 8.4 percent working two jobs, and about half of Whites and Hispanics worked full time (49.5 percent and 55.6 percent respectively), compared with about a third (36.9 percent) of African Americans (Spaulding *et al.*, 2016).

Juggling among these roles can be stressful and challenging. For example, student parents must manage complex schedules combining full-time work and full-time school, working irregular schedules, and managing childcare for multiple children. Regardless of income, mothers with non-standard schedules are more likely to use combinations of childcare arrangements (Enchautegui *et al.*, 2015). Childcare for young children is less available than care for older children (Chaudry *et al.*, 2011), who can generally get care (at least during traditional hours) through the public school system. With nearly round-the clock demands on their time for daycare, low-income parents may find it difficult to complete required school work (Johnson *et al.*, 2009). Finances are a major reason students drop out of school (Johnson *et al.*, 2009).

Spaulding *et al.* (2016) suggested strategies that organizations such as state policymakers, workforce boards, and educational institutions can use to help meet student parents' needs. For example, more childcare can be provided during non-traditional hours and for young children. Partnerships between workforce agencies or educational institutions and childcare referral agencies, delivering services on or off campus, may strengthen these supports (Adams & Heller, 2015). The response of policymakers to childcare concern has focused on providing childcare assistance or policies that increase the time off from work for childbirth and parenting (Berg *et al.*, 1999). The Child Care Access Means Parents in School Program (CCAMPIS) provides grants to two- and four-year colleges and universities and can support campus-based childcare programs (Green, 2017). Low-income working families can obtain free or low-cost childcare through the Child Care and Development Fund (CCDF). Some families never apply because the application process is difficult to navigate or they don't know about the subsidies (Child Trends Databank, 2017a).

Family members, educational financial assistance, and social service benefits provide some low-income working student parents with supports to offset expenses and care for their children (Spaulding *et al.*, 2016). About half of low-income student parents with children ages five and younger relied exclusively on family members to care for their children while they were working or in school; more than two-thirds of student parents reported receiving some form of education assistance and 48.1 percent of students received Pell grant aid (the primary source of federal grant money for low-income students). Approximately 47 percent of respondents received social service benefits, with almost half (46 percent) saying that they received Supplemental Nutrition Assistance Program (SNAP, also known as food stamps) benefits. Many colleges and universities offer online courses that provide an alternative way to take working parents into an era of anytime, anywhere and help them obtain degrees.

Sharing Family Work

Family caregiving responsibilities require some time away from full-time work and school work. Mothers were traditionally expected to do housework and still are more likely to do domestic work. Women are aware of their unequal position and they express their dissatisfaction and a desire to change the situation, but they cannot see a way to do so (Nyman *et al.*, 2018). Goldscheider *et al.* (2015) argues that the *gender revolution* has entered a new phase, in which men will significantly increase their participation in childcare and unpaid domestic work. Indeed, fathers today are more involved in childcare and housework.

The purpose of sharing family work is to find a way to balance paid work and family work so that both partners feel happy and appreciated. Lennon and Rosenfield (1994) focus on the proportion of family work done relative to one's partner according to *equity theory*—suggesting that couples evaluate both what they put into a relationship and what they get out of a relationship. Indeed, equity between partners is attained when both contribute and benefit fairly within the relationship (Thompson, 1991). For example, couples can take turns performing time-consuming household tasks so that one partner does not become overburdened. Couples benefit when they work together to maintain a balance at home by dividing up the work, such as preparing dinner, doing laundry, feeding, and bathing the children. Fathers who become involved in childcare find it easier to balance work/family stress than fathers who are less involved in childcare (Berry & Rao, 1997). In 2016, fathers reported spending eight hours a week on childcare, tripling the time they provided in 1965, and fathers put in about ten hours a week on household chores in 2016, up from four hours in 1965 (Parker & Livingston, 2018). Virtually all of the couples indicated that marital equality or partnership was one of the strategies they believed was central to their successful balance of family and work (Haddock & Rattenborg, 2003). Thus, the division of housework could contribute to couples' perceptions of equity in a relationship, and impact psychological well-being.

Global Overview: Future of Work and Work–Life Balance

As noted earlier in this chapter, researchers are aware of the potential benefits and risks of the Fourth Industrial Revolution. Let's look at the future of work, future workforce, and work–life balance in the European Union.

Future of Work

As the Fourth Industrial Revolution occurs in developed countries, artificial intelligence and robotics-enabled technologies are getting increasingly better at the cognitive and physical tasks that comprise much of today's work (Sundararajan, 2017). Many technological breakthroughs associated with the Fourth Industrial Revolution are changing the nature of work globally and are rapidly shifting the frontier between activities performed by humans and ones performed by machines, transforming the world of work (Copley, 2018). Steady technological change has led to large-scale automation of manual and routine tasks (World Economic Forum, 2018a). For example, in agriculture, intelligent robots are reducing applications of inputs by over 90 percent; in manufacturing, advanced robots will eliminate the advantage of cheap labor; and in services, driverless cars will eliminate jobs in transportation (Copley, 2018). A comparative study of 600 occupations' digital scores to their "automation" potential shows a solid negative correlation between a job's increased digital content and the share of its tasks vulnerable to automation. Therefore, workers with superior digital skills tend not only to earn higher wages, but also appear less exposed to automation-driven displacement (Muro & Liu, 2018). As the types of skills needed in the labor market change rapidly, individual workers will have to engage in life-long learning and prepare for *reskilling* (i.e., acquiring the skills and capabilities needed for the future workplace) if they are to achieve fulfilling and rewarding careers (World Economic Forum, 2018b).

Future Workforce

Contrary to fearsome scenarios about robots replacing humans, cognitive computing, robotics, and workforce automation feature prominently in most projections of the fourth industrial workplace (Miscovich, 2017). As noted earlier in the chapter, the trend toward a gig economy has begun. The **gig economy** is an environment in which workplaces contract with independent workers for short-term engagements. New platforms allow economic activity to be organized in ways that shift much of what was traditionally accomplished by full-time workers within an organization to a crowd of individual entrepreneurs and on-demand workers (Sundararajan, 2017). The gig economy has provided two new workforce modules: the "liquid workforce" and the "human cloud." The **liquid workforce** refers to employees who are able to retrain and adapt to their environment in order to stay relevant during the digital revolution. For example, companies can serve as internal labor markets and allow their employees to move fluidly among tasks. The **human cloud** is a sub-segment of the gig economy that includes firms that match workers to work through an online/digital platform (Staffing Industry Analysts, 2017). For example, the Hollywood model brings together autonomous, on-demand workers for project-based work onsite or to perform work remotely. Human cloud freelance workers comprise 35 percent of the workforce, and their numbers are expected to reach 75 percent to 80 percent of the enterprise workforce by 2030 (World Economic Forum, 2017). Work ranges from typing data into a spreadsheet to complex tasks such as writing a piece of code on demand from any location and through online platforms such as Upwork, Freelancer, People per Hour, Hubstaff, and Guru.

Employment and Work–Life Balance in the European Union

The EU's (Member States of the European Union) unemployment rate (7.1 percent) in the first quarter of 2018 has dropped to its lowest point in almost a decade, compared with the first quarter of 2008 (6.8 percent) and the first and second quarters of 2013 (11 percent) (Flores & Krogstad, 2018). Joblessness varies widely among EU countries. In the first quarter of 2018, the Czech Republic, Germany, Hungary, and Poland had the lowest unemployment rates, while Greece, Spain, Italy, Croatia, and Cyprus had the highest (Flores & Krogstad, 2018). For example, in Greece, unemployment climbed to 27.7 percent in the second and third quarters of 2013, the highest unemployment rate over the past two decades. In terms of age, youth unemployment remains high in Europe. As of the first quarter of 2018, a quarter of workers ages 15 to 24 were unemployed in Greece (43.8 percent), Spain (35.5 percent), Italy (32.5 percent), Croatia (23.6 percent), the Netherlands (7.2 percent), the Czech Republic (6.9 percent), and Germany (6.3 percent) (Flores & Krogstad, 2018).

Across the European Union, the reconciliation of work and life responsibilities is an issue that extends over the entire course of one's working life. More men participate in paid work than women. The female employment rate (20–64 years) was lower than that of men (65.3 percent versus 76.9 percent) in 2016 (Eurofound, 2017). Part-time work as a percentage of total employment increased from 17.7 to 20.5 percent between 2005 and 2015. Women still take on the main caring responsibilities over the life course and more frequently work part-time than men. Some 88 percent of women and 64 percent of men report providing care every day (Eurofound, 2017).

Place of work is a new opportunity to promote work–life balance. **Telework** is defined as a work flexibility arrangement under which an employee performs the duties and responsibilities of such an employee's position and other authorized activities from an approved worksite other than the location from which the employee would otherwise work (Office of Personnel Management, 2010). Teleworkers use mobile telecommunications technology such as Wi-Fi-equipped computers to work from coffee shops or at home. In ten EU Member States (Belgium, France, Finland, Germany, Hungary, Italy, the Netherlands, Spain, Sweden, and the United Kingdom), home-based teleworkers and T/ICTM (i.e., Telework/Information and Communication Technology Mobile) workers use new technologies to work outside the employer's premises (Eurofound and International Labor Organization, 2017). Across the European Union countries, an average of 17 percent of employees are engaged in telework/ICT-mobile (T/ICTM) work (Eurofound, 2017). Teleworkers can facilitate a better work–life balance, reduce commuting time, and boost productivity. Disadvantages include lengthening working hours and blurring lines between work and life, leading to stress harmful to health and well-being. The ambiguous and even contradictory effects of T/ICTM on working conditions represent a real-world example of the challenges of the future of work (Eurofound and International Labor Organization, 2017).

Summary

This chapter has examined work as a means of earning income and supporting the family. The chapter has explored the types of paid work that families are doing and the ways through which domestic unpaid work affects family life. The chapter has also explored dual-earner families, multigenerational families, stay-at-home father and stay-at-home mother families. Finally, the chapter has examined strategies for balancing paid work, housework, and school work from sociological perspectives. From the diversity and global approach, the chapter discussed military families and the future of work and work–life balance in the European Union.

Key Terms

Alternative employment arrangements include independent contractors, on-call workers, temporary help agency workers, and workers provided by contract firms.

Artificial intelligence (AI) or machine intelligence refers to intelligence demonstrated by machines.

Care work refers to occupations that provide services that help people develop their abilities to pursue the valuable aspects of their lives.

Childrearing refers to the bringing up and taking care of children.

Coercive work refers to activities that people are forced to do against their will and with little or no pay.

Contingent workers are people who do not expect their jobs to last or who report that their jobs are temporary and they do not have an implicit or explicit contract for ongoing employment.

Discouraged workers are persons not currently looking for work because they believe no jobs are available for them.

Dual-career couples refers to both heads of household having "jobs which require a high degree of commitment rather than being simply employed and pursue careers and at the same time maintain a family life together."

Dual-earner families is a form of family in which both spouses participate in paid work.

Dual-military marriages refer to military members who are married to other military members.

Emotion work is defined as work in which workers must manage their own feelings in order to create certain feelings in others.

Employed persons consist of persons who did any work for pay or profit during the survey reference week; persons who did at least fifteen hours of unpaid work in a family-operated enterprise; and persons who were temporarily absent from their regular jobs because of illness, vacation, bad weather, industrial dispute, or various personal reasons.

Entrepreneurship involves the process of designing, launching, and running a new business.

Family resilience is defined as the ability of a family to respond positively to an adverse situation and emerge from the situation feeling strengthened, more resourceful, and more confident than its prior state.

First Industrial Revolution used water and steam power to mechanize production.

Fourth Industrial Revolution refers to new technologies that are fusing the physical, digital, and biological worlds that alter how we live and work in societies.

Full-time work refers to employment in which an employee works a minimum number of defined hours (35 hours or more per week) and receives benefits provided by an employer.

Gender inequality is a social process by which people are treated differently on the basis of gender.

Gig economy is an environment in which workplaces contract with independent workers for short-term engagements.

Gig work refers to technology-enabled contingent work provided by agencies or contract firms to independent contractors, freelancers, or on-call workers.

Goods refers to tangible creations that people make and use.

High-schedule-control tasks are often initiated and performed without any time urgency.

Housework refers to all unpaid household activities involving homemaking or the management of a home.

Human cloud is a sub-segment of the gig economy that includes firms that match workers to work through an online/digital platform.

Ideology of separate spheres holds that a woman's proper place is in the home and a man's place is in the world of commerce or at his job.

Incorporated workers are small-business owners who have established a legal corporation and employed others.

Jobs are specific positions in specific establishments in which workers perform specific activities.

Labor force is the sum of the employed or labor pool in employment and unemployed persons in a country.

Labor force participation rate is the percentage of the population that is either employed or unemployed (that is, either working or actively seeking work).

Labor market or job market refers to the supply and demand for labor, in which employees provide the supply and employers provide the demand.

Labor market segmentation theory argues for a dual split between primary (or independent) and secondary (or subordinate) segments.

Latchkey child is a child who returns from school to an unsupervised home because his or her parents are out at work.

Liquid workforce refers to employees who are able to retrain and adapt to their environment in order to stay relevant during the digital revolution.

Low-schedule-control tasks refer to household labor that must be done every day and at certain times with little discretion in the scheduling of tasks.

Marginally attached workers are persons who are not in the labor force, wanted and were available for work, and had looked for a job sometime in the prior twelve months.

Military family is defined as active-duty service members, members of the National Guard and Reserve, and veterans, plus the members of their immediate and extended families, as well as the families of those who lost their lives in service to their country.

Military sociology is a subfield of sociology that studies the military as a social group.

Multigenerational family is a form of family in which households have multiple generations of family members living under the same roof.

Occupational gender segregation is the construction of female and male work based on cultural practices and gender differences.

Occupational segregation refers to the concentration of one group to an occupation.

Occupations are categories of jobs that involve similar activities at different work sites.

Paid domestic work refers to the daily activities performed by the maid in the respective employer's household.

Paid work (or market work) generates an income.

Part-time work is a form of employment in which an employee works fewer hours per week than a full-time job and receives no or few benefits provided by an employer.

Persons marginally attached to the labor force are persons who had not searched for work for reasons such as school attendance or family responsibilities.

Piece work is a type of employment in which a worker is paid a fixed piece rate for each unit produced or action performed regardless of time.

Pink-collar work involves non-manual and semiskilled positions primarily held by women in service-oriented work.

Primary labor market is the labor market consisting of high-wage-paying jobs, benefits, and longer lasting careers.

Role conflict refers to a conflict among the roles corresponding to two or more statuses.

Role expectations refer to the expectations required of a role and held both by the individual and by others.

Role overload is based on the premise that human energy is limited.

Role performance is the actual delivery of role expectations.

Role strain refers to conflicting demands connected to one's status.

Role theory suggests that most of our everyday activities consist of acting out socially defined roles associated with a given status.

Second Industrial Revolution or Technological Revolution used electric power to create mass production.

Second shift refers to the unpaid work performed at home in addition to the paid work performed in the labor force.

Secondary labor market is the labor market consisting of high-turnover, low-pay, part-time or temporary work with few benefits and little job security or possibility for future advancement.

Self-employment is the state of working for oneself as a freelancer or the ownership of a business, rather than for an employer.

Self-service online labor markets refer to online platforms such as Upwork that make it easier for companies to contract with independent workers for short-term projects.

Services refers to intangible creations provided by others.

Stay-at-home father is a father who was not in the labor force for the past 52 weeks and whose wife was in the labor force for the past 52 weeks.

Stay-at-home mother is a mother who was not in the labor force for the past 52 weeks and whose husband was in the labor force for the past 52 weeks, and who cares for the family and serves as the primary caregiver of children younger than age fifteen.

Taste-based discrimination suggests that employers discriminate against minority applicants to avoid interacting with them, regardless of their productivity.

Telework is defined as a work flexibility arrangement under which an employee performs the duties and responsibilities of such an employee's position and other authorized activities from an approved worksite other than the location from which the employee would otherwise work.

Third Industrial Revolution or Digital Revolution used electronics and information technology to automate production.

Unemployed persons consist of persons who do not have a job, have actively looked for work in the prior four weeks, and are currently available for work, and persons who were not working and were waiting to be recalled to a job from which they had been temporarily laid off.

Unincorporated workers are people who have not established a corporation and often operate alone.

Unpaid domestic work refers to the time spent by a male or female in different activities of their own household.

Unpaid work refers to non-market work that people voluntarily perform for their families and others.

Volunteer work refers to activities and tasks performed that mainly benefit others without direct reward in money or in kind.

Work refers to any activity involving mental or physical effort done in order to achieve a purpose.

Discussion Questions

- Student parents play multiple roles. Work out a proposal for policymakers to help student parents graduate from school, make income to support the family, and raise healthy children.
- Explore strategies for balancing paid work, family work, and school work.

Suggested Film and Video

Elwes, C. Producer (1990). *Men at Work*. Epic Productions.
Maternity Protection, Parental Leave, Childcare, Working Time: http://www.ilo.org/global/topics/care-economy/parental-leave/lang—en/index.htm
The Future of Work Centenary Initiative: http://www.ilo.org/global/topics/future-of-work/WCMS_448448/lang—en/index.htm
What Types of Work are Contingent or Alternative Employment Arrangements?: https://www.bls.gov/video/?video=DehS8Rnb5SA

Internet Sources

Eurofound: https://www.eurofound.europa.eu/publications/report/2017/occupational-change-and-wage-inequality-european-jobs-monitor-2017
Work–Life Balance: https://hbswk.hbs.edu/Pages/browse.aspx?HBSTopic=Work-Life%20Balance

References

Acemoglu, D. & Restrepo, P. (2017). *Robots and Jobs: Evidence from US Labor Markets*. Retrieved on July 8, 218 from https://economics.mit.edu/files/12763

Adams, G. & Heller, C. (2015). *The Child Care Development Fund and Workforce Development for Low Income Parents: Opportunities and Challenges with Re-authorization*. Washington, DC: Urban Institute.

Adler, P. S. & Kwon, S. W. (2002). Social Capital: Prospects for a New Concept. *Academy of Management Review*, 2(27), 17–40.

Aldrich, H. E. & Kim, P. H. (2007), A Life Course Perspective on Occupational Inheritance: Self-employed Parents and their Children. In M. Ruef & M. Lounsbury (Eds.), *The Sociology of Entrepreneurship* (Research in the Sociology of Organizations, Volume 25) (pp. 33–82). Bingley, UK: Emerald Group Publishing.

Anderson, B. (2001). Just Another Job? Paying for Domestic Work. *Gender & Development*, 9(1), 25–33.

Aquilino, W. S. (1990). The Likelihood of Parent Adult Child Coresidence: Effects of Family Structure and Parental Characteristics. *Journal of Marriage and the Family*, 52, 405–419.

Aranda, M. C., Middleton, L. S., Flake, E. & Davis, B. E. (2011). Psychosocial Screening in Children with Wartime-deployed Parents. *Military Medicine*, 176(4), 402–407.

Artazcoz, L., Benach, J., Borrell, C. *et al.* (2005). Social Inequalities in the Impact of Flexible Employment on Different Domains of Psychosocial Health. *J Epidemiol Community Health*, 59, 761–767.

Artis, J. E. & Pavalko, E. K. (2003). Explaining the Decline in Women's Household Labor: Individual Change and Cohort Differences. *Journal of Marriage and Family*, 65, 746–761.

Azoulay, P., Jones, B. F., Kim, D. & Miranda, J. (2018). *Age and High-Growth Entrepreneurship*, Working Paper No. 24489. Cambridge, MA: The National Bureau of Economic Research (NBER).

Baker, A. (2008). *Life in the U.S. Armed Forces. (Not) Just Another Job*. Westport, CT: Praeger Security International.

Barnett, R. C. & Baruch, G. K. (1985). Women's Involvement in Multiple Roles and Psychological Distress. *Journal of Personality and Social Psychology*, 49, 135–145.

Barnett, R. & Shen, Y. (1997). Gender, High- and Low-Schedule-Control Housework Tasks, and Psychological Distress: A Study of Dual-Earner Couples. *J Fam Issues*, 18(4), 403–428.

Barnett, R. C. & Hyde, J. S. (2001). Women, Men, Work, and Family: An Expansionist Theory. *American Psychologist*, 56(10), 781–796.

Bartley, M. (2005). Job Insecurity and Its Effect on Health. *Journal of Epidemiology & Community Health*, 59, 718–719.

Bartley, S. J., Blanton, P. W. & Gilliard, J. L. (2005). Husbands and Wives in Dual-earner Marriages: Decision-making, Gender Role Attitudes, Division of Household Labor, and Equity. *Marriage & Family Review*, 37(4), 69–94.

Bauder, H. (2001). Culture in the Labor Market: Segmentation Theory and Perspectives of Place. *Progress in Human Geography*, 25(1), 37–52.

Baxandall, R., Ewen, E. & Gordon, L. (1976). The Working Class Has Two Sexes. *Monthly Review*, 28(3), 1–9.

Baxter, J., Hewitt, B. & Haynes, M. (2008). Life Course Transitions and Housework: Marriage, Parenthood and Time on Housework. *J Marriage Fam.*, 70, 259–272.

Bengtson, V. L. (2001). Beyond the Nuclear Family: The Increasing Importance of Multigenerational Bonds (The Burgess Award Lecture). *Journal of Marriage and Family*, 63(1), 1–16.

Benson, A., Sojourner, A. & Umyarov, A. (2015). *Can Reputation Discipline the Gig Economy? Experimental Evidence from an Online Labor*. Retrieved on August 26, 2017 from http://ftp.iza.org/dp9501.pdf

Berg, P., Appelbaum, E. & Kalleberg, A. L. (1999). *The Roles of the Work Environment and Job Characteristics in Balancing Work and Family*. U.S. Department of Labor.

Berry, J. O. & Rao, J. M. (1997). Balancing Employment and Fatherhood: A Systems Perspective. *Journal of Family Issues*, 18, 386–402.

Bewley, T. (May 1995). A Depressed Labor Market as Explained by Participants. *The American Economic Review*, 85(2), 250–254.

Bianchi, S. M., Sayer, L. C., Milkie, M. A. & Robinson, J. P. (2012). Housework: Who Did, Does or Will Do It, and How Much Does It Matter? *Social Forces*, 91(1), 55–63.

Bohle, P., Quinlan, M. & Mayhew, C. (2001). The Health and Safety Effects of Job Insecurity: An Evaluation of the Evidence. *Economic and Labor Relations Review*, 12, 32–60.

Bozoyan, C. & Wolbring, T. (2018). The Weight Wage Penalty: A Mechanism Approach to Discrimination. *European Sociological Review*, 34(3), 254–267.

Bratton, P. C. (2012). *When Is Coercion Successful? And Why Can't We Agree on It.* Washington, DC: BiblioGov.

Brint, S. (1994). *In an Age of Experts: The Changing Roles of Professionals in Politics and Public Life.* Princeton, NJ: Princeton University Press.

Brown, A. & Patten, E. (April 3, 2017). *The Narrowing, but Persistent, Gender Gap in Pay.* Washington, DC: Pew Research Center.

Brynjolfsson, E. & McAfee, A. (2016). *The Second Machine Age: Work, Progress, and Prosperity in a Time of Brilliant Technologies.* New York: W. W. Norton & Company.

Cain, G. (1976). The Challenge of Segmented Labor Market Theories to Orthodox Theory: A Survey. *Journal of Economic Literature*, XIV(4), 1215–1257.

Carney, J. & Watts, M. (1991). Disciplining Women? Rice, Mechanization, and the Evolution of Mandinka Gender Relations. *Signs*, 16(4), 651–681.

Carriero, R. (2011). Perceived Fairness and Satisfaction with the Division of Housework among Dual-Earner Couples in Italy. *Marriage & Family Review*, 47(7), 436–458.

Casper, L. M. & Bianchi, S. M. (2001). *Trends in the American Family.* Thousand Oaks, CA: Sage.

Chaudry, A., Pedroza, J., Sandstrom, H., Danziger, A., Grosz, M., Scott, M. & Ting, S. (2011). *Child Care Choices of Low-Income Working Families.* Washington, DC: Urban Institute.

Child Trends Databank (2016). *Child Care.* Retrieved on November 26, 2016 from http://www.childtrends.org/?indicators=child-care

Child Trends Databank (May 12, 2017a). *Why Millions of Eligible Children Miss Out on Federal Child Care Grants.* Retrieved on August 24, 2017 from https://www.childtrends.org/millions-eligible-children-miss-federal-child-care-grants/

Child Trends Databank (July 20, 2017b). *Paid Parental Leave Is Rare, but Good for Kids.* Retrieved on August 24, 2017 from https://www.childtrends.org/paid-parental-leave-rare-good-kids/

Choi, N. G. (2003). Coresidence between Unmarried Aging Parents and Their Adult Children: Who Moved in with Whom and Why? *Research on Aging*, 25, 384–404.

Clairmont, D., Apostle, R. & Kreckel, R. (1983). The Segmentation Perspective as a Middle Range Conceptualization in Sociology. *Canadian Journal of Sociology*, 8, 245–272.

Cohn, D.'V. & Passel, J.. (April 2018). *A Record 60.4 Million Americans Live in Multigenerational Households.* Washington, DC: Pew Research Center.

Cohn, D.'V., Livingston, G. & Wang, W. (2014). *After Decades of Decline, A Rise in Stay-at-home Mothers.* Washington, DC: Pew Research Center. Retrieved on November 27, 2016 from http://www.pewsocialtrends.org/2014/04/08/after-decades-of-decline-a-rise-in-stay-at-home-mothers/

Conkin, P. (2008). *Revolution Down on the Farm: The Transformation of American Agriculture since 1929.* Lexington, KY: University Press of Kentucky.

Copley, A. (2018). *Figures of the Week: The Future of Work in Africa.* Washington, DC: Brookings Institution.

Cox, R. (1997). Invisible Labour: Perceptions of Paid Domestic Work in London. *Journal of Occupational Science*, 4(2), 62–67.

Cozza, S. J. & Lerner, R. M. (2013). Military Children and Families: Introducing the Issue in Military Children and Families. *The Future of Children*, 23(2).

Cozza, S. J., Ryo, S. C. & Polo, J. A. (2005). Military Families and Children during Operation Iraqi Freedom. *Psychiatric Quarterly*, 76, 371–378.

Cranford, C., Fudge, J. & Tucker, E. (2005). *Self-employed Workers Organize: Law, Policy, and Unions.* Montreal, Quebec: McGill-Queen's University Press.

Creighton, C. (1996). The Rise of the Male Breadwinner Family: A Reappraisal. *Comparative Studies in Society and History*, 38(2), 310–337.

Curran, J. & Burrows, R. (1986). The Sociology of Petit Capitalism: A Trend Report. *Sociology*, 20(2), 265–279.

Curtin, S. & Hofferth, S. (2006). Parental Leave Statutes and Maternal Return to Work after Childbirth in the United States. *Work and Occupations*, 33(1), 73–105.

Dakin, S. & Armstrong, J. S. (1989). Predicting Job Performance: A Comparison of Expert Opinion and Research Findings. *International Journal of Forecasting*, 5, 187–194.

Dale, A. (1986). Social Class and the Self-employed. *Sociology*, 20(3), 430–434.

Davidoff, L. & Hall, C. (1987). *Family Fortunes: Men and Women of the English Middle Class 1780–1850.* Chicago, IL: The University of Chicago Press.

Davis, N. (2016). *What is the Fourth Industrial Revolution?* World Economic Forum. Retrieved on July 8, 2018 from https://www.weforum.org/agenda/2016/01/what-is-the-fourth-industrial-revolution/

de Burgh, H. T., White, C. J., Fear, N. T. & Iversen, A. C. (2011). The Impact of Deployment to Iraq or Afghanistan on Partners and Wives of Military Personnel. *Int Rev Psychiatry,* 23(2), 192–200.

Des Rivieres-Pigeon, C., Saurel-Cubizolles, M. J. & Romito, P. (2002). Division of Domestic Work and Psychological Distress 1 Year after Childbirth: A Comparison between France, Quebec and Italy. *J Commun Appl Soc Psychol.,* 12, 397–409.

Dinardo, J., Fortin, N. M. & Lemieux, T. (1996). Labor Market Institutions and the Distribution of Wages, 1973–1992: A Semiparametric Approach. *Econometrica*, 64(5), 1001–1044.

Dunn, M. (March 2018). Who Chooses Part-Time Work and Why? *Monthly Labor Review*. U.S. Bureau of Labor Statistics.

Eggebeen, D. J. & Hawkins, A. J. (1990). Economic Need and Wives' Employment. *Journal of Family Issues,* 11(1), 48–66.

Eitzen, S. D., Wells, B. & Zinn, M. B. (2011). *Diversity in Families.* Boston, MA: Hanson Karen.

Enchautegui, M. E., Johnson, M. & Gelatt, J. (2015). *Who Minds the Kids When Mom Works a Nonstandard Schedule?* Washington, DC: Urban Institute.

England, P. (2005). Emerging Theories of Care Work. *Annual Review of Sociology,* 31, 381–399.

England, P. (2010). The Gender Revolution: Uneven and Stalled. *Gender and Society,* 24(2), 149–166.

Escribà-Agüir, V. & Tenías-Burillo, J. M. (2004). Psychological Well-being among Hospital Personnel: The Role of Family Demands and Psychosocial Work Environment. *Int Arch Occup Environ Health,* 77(6), 401–408.

Eurofound (2017). *Work–Life Balance and Flexible Working Arrangements in the European Union*. Dublin: Eurofound.

Eurofound and International Labor Organization (2017). *Working Anytime, Anywhere: The Effects on the World of Work.* Luxembourg; Geneva: Publications Office of the European Union and the International Labor Organization.

Evans, B. K. & Fischer, D. G. (1992). A Hierarchical Model of Participatory Decision-Making, Job Autonomy, and Perceived Control. *Human Relations*, 45, 1169–1189.

Fingerman, K., Miller, L., Birditt, K. & Zarit, S. (2009). Giving to the Good and the Needy: Parental Support of Grown Children. *J Marriage Fam.,* 71(5), 1220–1233.

Flores, A. & Krogstad, J. M. (July 18, 2018). *EU Unemployment Rate Falls to Near Pre-Recession Low.* Washington, DC: Pew Research Center.

Foad, H. (2018). *Inequality Between and Within Immigrant Groups in the United States.* Nashville, TN: American Economic Association.

Folbre, N. (1991). The Unproductive Housewife: Her Evolution in Nineteenth-Century Economic Thought. *Signs: Journal of Women in Culture and Society,* 16(3), 463–484.

Folbre, N. & Wagman, B. (1993). Counting Housework: New Estimates of Real Product in the United States, 1800–1860. *The Journal of Economic History,* 53(2), 275–288.

Freeman, C. & Louçã, F. (2002). *As Time Goes By: From the Industrial Revolutions to the Information Revolution.* New York: Oxford University Press.

Frisco, M. L. & Williams, K. (2003). Perceived Housework Equity, Marital Happiness, and Divorce in Dual-Earner Households. *Journal of Family Issues,* 24(1), 51–73.

Fry, R. & Passel, J. (July 2014). *In Post-Recession Era, Young Adults Drive Continuing Rise in Multi-Generational Living.* Washington, DC: Pew Research Center.

Galley, J. (2014). *Stay-at-home Mothers through the Years.* U.S. Bureau of Labor Statistics.

Geiger, A. & Parker, K. (2018). *For Women's History Month, a Look at Gender Gains – and Gaps – in the U.S.* Washington, DC: Pew Research Center.

Gerke, S. & Evers, H.-D. (1993). *Labour Market Segmentation in West Sumatra* (Working Paper/Universität Bielefeld, Fakultät für Soziologie, Forschungsschwerpunkt Entwicklungssoziologie, 197). Bielefeld: Universität Bielefeld, Fak. für Soziologie, Forschungsschwerpunkt Entwicklungssoziologie.

Goldberg, W. A. & Easterbrooks, M. A. (1988). Maternal Employment When Children Are Toddlers and Kindergartners. In A. E. Gottfried & A. W. Gottfried (Eds.), *Maternal Employment and Children's Development.* Springer Studies in Work and Industry. Boston, MA: Springer.

Goldin, C. (1990). *Understanding the Gender Gap: An Economic History of American Women.* New York: Oxford University Press.

Goldscheider, F., Bernhardt, E. & Lappegård, T. (2015). The Gender Revolution: A Framework for Understanding Changing Family and Demographic Behavior. *Population and Development Review*, 41(2), 207–239.

Gonalons-Pons, P. (2015). Gender and Class Housework Inequalities in the Era of Outsourcing Hiring Domestic Work in Spain. *Social Science Research,* 52, 208–218.

Gordon, D. M., Edwards, R. & Reich, M. (1982). *Segmented Work, Divided Workers: The Historical Transformation of Labor in the United States.* Cambridge: Cambridge University Press.

Gove, W. R., Hughes, M. & Galle, O. R. (1979). Overcrowding in the Home: An Empirical Investigation of Its Possible

Pathological Consequences. *American Sociological Review*, 44(1), 59–80.

Graf, N., Brown, A. & Patten, E. (2018). *The Narrowing, but Persistent, Gender Gap in Pay*. Washington, DC: Pew Research Center.

Green, S. (May 25, 2017). *Federal Grants Help Student Parents Succeed, But Trump Budget Threatens Their Elimination*. Washington, DC: Institute for Women's Policy Research.

Haddock, S. A. & Bowling, S. W. (2001). Therapists' Approaches to the Normative Challenges of Dual-Earner Couples: Negotiating Outdated Societal Ideologies. *Journal of Feminist Family Therapy*, 13, 91–120.

Haddock, S. & Rattenborg, K. (2003). Benefits and Challenges of Dual-Earning: Perspectives of Successful Couples. *American Journal of Family Therapy*, 31(5), 325–344.

Hajdu, G. & Hajdu, T. (2018). Intra-Couple Income Distribution and Subjective Well-Being: The Moderating Effect of Gender Norms. *European Sociological Review*, 34(2), 138–156.

Hanawalt, B. (1986). *The Ties that Bound: Peasant Families in Medieval England*. New York: Oxford University Press.

Hawley, D. R. (2000). Clinical Implications of Family Resilience. *American Journal of Family Therapy*, 28, 101–116.

He, W., Weingartner, R. M. & Sayer, L. C. (February 2018). *Subjective Well-Being of Eldercare Providers: 2012–2013*. U.S. Census Bureau.

Hester, S. B. & Dickerson, K. G. (1984). Serving Dual-Career Families: Problem or Opportunity? *Journal of Extension*, 22(4).

Hipple, S. F. & Hammond, L. A. (2016). *Self-employment in the United States*. U.S. Bureau of Labor Statistics. Retrieved on June 29, 2018 from https://www.bls.gov/spotlight/2016/self-employment-in-the-united-states/pdf/self-employment-in-the-united-states.pdf

Hochschild, A. (November 1979). Emotion Work, Feeling Rules, and Social Structure. *American Journal of Sociology*, 85(3), 551–575.

Hochschild, A. R. & Machung, A. (1989). *The Second Shift: Working Parents and the Revolution at Home*. New York: Viking Penguin.

Hodson, R. & Sullivan, T. A. (2002). *The Social Organization of Work*, 5th edn. Belmont, CA: Cengage Learning.

Hofäcker, D., Stoilova, R. & Riebling, J. R. (2013). The Gendered Division of Paid and Unpaid Work in Different Institutional Regimes: Comparing West Germany, East Germany and Bulgaria. *European Sociological Review*, 29(2), 192–209.

Hook, J. L. (2006). Care in Context: Men's Unpaid Work in 20 Countries, 1965–2003. *American Sociological Review*, 71, 639–660.

Hu, Y. & Yucel, D. (2018). What Fairness? Gendered Division of Housework and Family Life Satisfaction across 30 Countries. *European Sociological Review*, 34(1), 92–105.

Human Rights Center (2004). *Hidden Slaves Forced Labor in the United States*. Retrieved on October 16, 2016 from http://www.freetheslaves.net/wp-content/uploads/2015/03/Hidden-Slaves.pdf

Huntington, S. (1957). *The Soldier and the State: The Theory and Politics of Civil-Military Relations*. New York: Belknap Press.

Hyde J. (2016). Women, Men, Work, and Family: Expansionist Theory Updated. In S. McHale, V. King, J. Van Hook & A. Booth (Eds.), Gender and Couple Relationships. *National Symposium on Family Issues*, 6. New York: Palgrave Macmillan.

International Labour Organization (2001). *International Labor Standards: A Global Approach*. Retrieved on June 30, 2018 from http://www.ilo.org/wcmsp5/groups/public/—-ed_norm/—-normes/documents/publication/wcms_087692.pdf

International Labour Organization (2016). *Forced Labor*. Retrieved on August 17, 2017 from http://www.ilo.org/dyn/normlex/en/f?p=NORMLEXPUB:12100:0::NO::P12100_ILO_CODE:C029

Jackson, S. (1991). Towards a Historical Sociology of Housework: A Materialist Feminist Analysis. *Women's Studies International Forum*, 15(2), 153–172.

Janowitz, M. (1960). The *Professional Soldier: A Social and Political Portrait*. New York: The Free Press.

Jansen, L., Weber, T., Kraaykamp, G. & Verbakel, E. (2016). Perceived Fairness of the Division of Household Labor: A Comparative Study in 29 Countries. *International Journal of Comparative Sociology*, 57(1–2), 53–68.

Johnson, J. & Rochkind, J. with Ott, A. N. & DuPont, S. (2009). *With Their Whole Lives Ahead of Them*. New York: The Public Agenda.

Jones, C. & Roemer, M. (1989). Modeling and Measuring Parallel Markets in Developing Countries. *World Development*, 17(12), 1861–1870.

Kalleberg, A. L. (2011). *Good Jobs, Bad Jobs: The Rise of Polarized and Precarious Employment Systems in the United States, 1970s to 2000s*. New York: Russell Sage Foundation.

Kanter, R. M. (1977). *Work and Family in the United States: A Critical Review and Agenda for Research and Policy*. New York: Russell Sage Foundation.

Katz, L. F. & Krueger, A. B. (2016). *The Rise and Nature of Alternative Work Arrangements in the United States, 1995–2015*. Retrieved on November 27, 2016 from https://krueger.princeton.edu/sites/default/files/akrueger/files/katz_krueger_cws_-_march_29_20165.pdf

Katz, L. F. & Murphy, K. M. (1992). Changes in Relative Wages, 1963–1987: Supply and Demand Factors. *The Quarterly Journal of Economics,* 107(1), 35–78.

Keynes, J. M. (1931). *Essays in Persuasion.* London: Macmillan.

Korom, P. (2018). Inherited Advantage: Comparing Households that Receive Gifts and Bequests with Non-receiving Households across the Distribution of Household Wealth in 11 European Countries. *European Sociological Review,* 34(1), 79–91.

Lam, C. B., McHale, S. M. & Crouter, A. C. (2012). The Division of Household Labor: Longitudinal Changes and Within-couple Variation. *Journal of Marriage and the Family,* 74(5), 944–952.

Latshaw, B. A. & Hale, S. I. (2016). 'The Domestic Handoff': Stay-at-home Fathers' Time-use in Female Breadwinner Families. *Journal of Family Studies,* 22(2), 97–120.

Laughlin, L. (2013). Who's Minding the Kids? Child Care Arrangements: Spring 2011. *Current Population Reports,* P70–135. Washington, DC: U.S. Census Bureau.

Le, A. T. (1999). Empirical Studies of Self-Employment. *Journal of Economic Surveys,* 13(4), 381–416.

Leary, K. (2018). AI Can Diagnose Heart Disease and Lung Cancer More Accurately than Doctors. *Futurism.* Retrieved July 5, 2018 from https://futurism.com/ai-diagnose-heart-disease-lung-cancer-more-accurately-doctors/

Lennon, M. & Rosenfield, S. (1994). Relative Fairness and the Division of Housework: The Importance of Options. *Am J Soc.,* 100(2), 506–531.

Leontief, W. (1952). Machines and Man. *Scientific American,* 187, 6, 93–95.

Liong, M. (2017). From Control to Care: Historicizing Family and Fatherhood in Hong Kong. Chinese Fatherhood. *Gender and Family,* 26(4), 39–71.

Livingston, G. (June 5, 2014). *Growing Number of Dads Home with the Kids: Biggest Increase among those Caring for Family.* Washington, DC: Pew Research Center.

Luxton, M. (1987). *More Than a Labor of Love: Three Generations of Women's Work in the Home.* California: Women's Press.

Marshall, K. (2006). Converging Gender Roles. *Perspect Labor Income,* 7(7), 5–17.

McCall, L. (2000). Explaining Levels of Within-group Wage Inequality in U.S. Labor Markets. *Demography,* 37(4), 415–430.

McCubbin, H. I. & McCubbin, M. A. (1988). Typologies of Resilient Families: Emerging Roles of Social Class and Ethnicity. *Family Relations,* 37(3), 247–254.

McCubbin, M. A. & McCubbin, H. I. (1993). Families Coping with Illness: The Resiliency Model of Family Stress, Adjustment, and Adaptation. In C. B. Danielson, B. Hamel-Bissell & P. Winstead-Fry (Eds.), *Families, Health, & Illness: Perspectives on Coping and Intervention* (pp. 21–63). St. Louis, MO: Mosby.

Mcfarlane, S., Beaujot, R. & Haddad, T. (2000). Time Constraints and Relative Resources as Determinants of the Sexual Division of Domestic Work. *Canadian Journal of Sociology,* 25, 61–82.

McKay, A. (2002). Defining and Measuring Inequality. Briefing Paper No. 1 (1 of 3). London: Overseas Development Institute.

McMullin, J. A. (2005). Patterns of Paid and Unpaid Work: The Influence of Power, Social Context, and Family Background. *Can J Aging,* 24(3), 225–236.

Meadows, S. O., Beckett, M. K., Bowling, K., Golinelli, D., Fisher, M. P., Martin, L. T. & Osilla, K. C. (2016). Family Resilience in the Military: Definitions, Models, and Policies. *Rand Health Quarterly,* 5(3), 12.

Medved, C. E. (2016). The New Female Breadwinner: Discursively Doing and Undoing Gender Relations. *Journal of Applied Communication Research,* 44(3), 236–255.

Merson, J. (1990). *The Genius that Was China: East and West in the Making of the Modern World.* Woodstock, NY: The Overlook Press.

Mertens, S. B & Flowers, N. (May 2003). Should Middle Grades Students Be Left Alone After School? *Middle School Journal,* 34(5), 57–61.

Military Family Life Project (MFLP) (2015). *Military Family Life Project: Active Duty Spouse Study Longitudinal Analyses 2010–2012 Project Report.* Retrieved on December 4, 2018 from http://download.militaryonesource.mil/12038/MOS/Reports/MFLP-Longitudinal-Analyses-Report.pdf

Military OneSource (May 2018). *2016 Demographics Report.* Retrieved on August 1, 2018 from http://download.militaryonesource.mil/12038/MOS/Reports/2016-Demographics-Report.pdf

Milkman, R. (2016). *On Gender, Labor, and Inequality.* Champaign, IL: University of Illinois Press.

Miller, K., Suppan Helmuth, A. & Farabee-Siers, R. (2009). The Need for Paid Parental Leave for Federal Employees: Adapting To A Changing Workforce- Executive Summary. Washington, DC: Institute for Women's Policy Research.

Mingay, G. E. (Ed.) (1977). *The Agricultural Revolution: Changes in Agriculture 1650–1880.* London: A & C Black.

Miscovich, P. (2017). The Future Is Automated. Here's How We Can Prepare for It. World Economic Forum.

Miyake, A., Muro, K., Nakamura, T. *et al.* (2009). Between- and Within-group Wage Inequalities, and the Advent of New Technology. *J Econ Inequal,* 7, 387.

Moras, A. (2017). This Should be My Responsibility: Gender, Guilt, Privilege and Paid Domestic Work. *Gender Issues*, 34, 44.

Moskos, C. C. (1977). From Institution to Occupation. Trends in Military Organization. *Armed Forces & Society*, 4(1), 41–50.

Moskos, C. C. (1986). Institutional/Occupational Trends in Armed Forces: An Update. *Armed Forces & Society*, 12(3), 377–382.

Muennig, P., Jiao, B. & Singer, E. (2018). Living with Parents or Grandparents Increases Social Capital and Survival: 2014 General Social Survey-National Death Index. *Population Health*, 4, 71–75.

Murdock, G. P. (1954). *Outline of World Cultures*. New Haven: Human Relations Area Files.

Murdock, G. P. & White, D. R. (1969). Standard Cross-Cultural Sample. *Ethnology*, 9, 329–369.

Muro, M. & Liu, S. (2018). *Highly Digital Jobs Are Less Likely to Be Automated*. Washington, DC: Brookings Institution.

National Military Family Association (2017). *Mission*. Retrieved on August 18, 2017 from http://www.military family.org/?referrer=https://www.google.com/

Noël, R. A. (2018). *Race, Economics, and Social Status*. U.S. Bureau of Labor Statistics.

Novick, K. K. (1988). Childbearing and Child Rearing. *Psychoanalytic Inquiry*, 8(2), 252–260.

Nyman, C., Reinikainen, L. & Stocks, J. (2013). Reflections on a Cross-national Qualitative Study of Within-household Finances. *Journal of Marriage and Family*, 75(3), 640–650.

Nyman, C., Reinikainen, L. & Eriksson, K. (2018). The Tension between Gender Equality and Doing Gender: Swedish Couples Talk about the Division of Housework. *Women's Studies International Forum*, 68, 36–46.

O'Hara, S. (2014). Everything Needs Care: Toward a Context-based Economy. In M. Bjørnholt & A. McKay (Eds.), *Counting on Marilyn Waring: New Advances in Feminist Economics* (pp. 37–56). Bradford: Demeter Press.

Office of Personnel Management (OPM). (2010). The Telework Enhancement Act of 2010. Retrieved on August 23, 2017 from https://www.gpo.gov/fdsys/pkg/BILLS-111hr1722enr/pdf/BILLS-111hr1722enr.pdf

Oropesa, R. S. & Landale, N. S. (2004). The Future of Marriage and Hispanics. *Journal of Marriage and Family*, 66, 901–920.

Öun, I. (2013). Is it Fair to Share? Perceptions of Fairness in the Division of Housework among Couples in 22 Countries. *Soc Justice Res.*, 13(26), 400–421.

Overton, M. (1996). *Agricultural Revolution in England: The Transformation of the Agrarian Economy 1500–1850*. Cambridge, UK: Cambridge University Press.

Padavic, I. & Reskin, B. (2002). *Women and Men at Work*. Thousand Oaks, CA: Sage.

Papanek, H. (1973). Purdah: Separate Worlds and Symbolic Shelter. *Comparative Studies in Society and History*, 15, 289–325.

Parker, K. & Livingston, G. (June 13, 2018). *7 Facts about American Dads*. Washington, DC: Pew Research Center. Retrieved on August 20, 2018 from http://www.pewresearch.org/fact-tank/2017/06/15/fathers-day-facts/

Parsons, T. (1943). The Kinship System of the Contemporary United States. In *Essays in Sociological Theory*. New York: Free Press.

Patel, N. M., Michelini, V. V., Snell, J. M., Balu, S., Hoyle, A. P. *et al.* (2017). Enhancing Next-generation Sequencing–Guided Cancer Care through Cognitive Computing. *The Oncologist*, 23(2), 179–185.

Peck, J. A. (1996). *Work-place: The Social Regulation of Labor Markets*. New York: Guilford Press.

Pilkauskas, N. (2012). Three Generation Family Households: Differences by Family Structure at Birth. *J Marriage Fam.*, 74(5), 931–943.

Pinchbeck, I. (1930). *Women Workers in the Industrial Revolution, 1750–1850*. London: Virago.

Pinquart, M. & Sörensen, S. (2003). Differences between Caregivers and Noncaregivers in Psychological Health and Physical Health: A Meta-Analysis. *Psychology and Aging*, 18(2), 250–267.

Piore, M. J. (1983). Labor Market Segmentation: To What Paradigm Does It Belong? *American Economic Review*, 73, 249–253.

Proctor, B. D., Semega, J. L. & Kollar, M. A. (September 2016). *Poverty in the United States: 2015*. U.S. Census Bureau.

Puhl, R. & Brownell, K. D. (2001). Bias, Discrimination, and Obesity. *Obesity Research*, 9, 788–805.

Qvist, H.-P. Y. & Munk, M. D. (2018). The Individual Economic Returns to Volunteering in Work Life. *European Sociological Review*, 34(2), 198–210.

Rajkomar, A., Oren, E. & Dean, J. (2018). Scalable and Accurate Deep Learning with Electronic Health Records. *NPJ Digital Medicine*, 1(18).

Rapoport, R. & Rapoport, R. N. (1969). The Dual-Career Family. *Human Relations*, XXII(1), 3–30.

Reich, M., Gordon, D. M. & Edwards R. C. (1973). Dual Labor Markets: A Theory of Labor Market Segmentation. *American Economic Review*, 63, 359–365.

Remeikiene, R., Startienė, G. & Vasauskaite, J. (2011). The Influence of Psychological-Sociological Factors on Self-Employment. *Economics and Management*, 16.

Reputation Management (2015). *Employer Reputation Management: More Important Now Than Ever Before*. Retrieved on June 29, 2018 from https://www.reputationman

agement.com/blog/employer-reputation-management-more-important-now-than-ever-before/

Richards, D. & Hunt, J. W. (1983). *An Illustrated History of Modern Britain: 1783–1980*, 3rd edn. Hong Kong: Longman Group.

Robinson, J. & Spitze, G. (1992). Whistle While You Work? The Effect of Household Task Performance on Women's and Men's Well-being. *Soc Sci Quar.*, 73(4), 845–861.

Robles, B. J. & Cordero-Guzman, H. (2007). Latino Self-employment and Entrepreneurship in the United States: An Overview of the Literature and Data Sources. *ANNALS, AAPSS*, 613.

Ross, S. M. (2014). 21st Century American Military Families: A Review in the Context of the Wars in Afghanistan and Iraq. *Social Psychology & Family*, 8(6), 888–902.

Russell, S. J & Norvig, P. (2009). *Artificial Intelligence: A Modern Approach*, 3rd edn. Upper Saddle River, NJ: Prentice Hall.

Ryan, P. (1981). Segmentation, Duality and the Internal Labor Market. In F. Wilkinson (Ed.), *The Dynamics of Labor Market Segmentation* (pp. 3–20). London: Academic Press.

Sani, G. M. D. (2014). Men's Employment Hours and Time on Domestic Chores in European Countries. *Journal of Family Issues*, 35, 1023–1047.

Sarkar, S. (2005). Women as Paid Domestic Workers. *J. Soc. Sci.*, 11(1), 35–41.

Schneider, D. & Hastings, O. P. (2017). Income Inequality and Household Labor. *Social Forces*, 96(2), 481–506.

Schwab, K. (2016). *The Fourth Industrial Revolution, What It Means and How to Respond.* World Economic Forum.

Schwab, K. (2017). *The Fourth Industrial Revolution.* New York: Crown Publishing Group.

Shambaugh, J., Nunn, R., & Bauer, L. (2018). *Independent Workers and the Modern Labor Market.* Washington, DC: Brookings Institution.

Shaw, S. M. (1988). Gender Differences in the Definition and Perception of Household Labor. *Family Relations*, 37(3), 333–337.

Simon, J., Murphy, J. & Smith. S. (2005). Understanding and Fostering Family Resilience. *The Family Journal*, 13, 427–436.

Skarpenes, O. & Nilsen, A. C. E. (2015). Regional Gender Inequality in the Norwegian Culture of Equality. *Gender Issues*, 32, 39–56.

Smith, A. (1776). *An Inquiry into the Nature and Causes of Wealth of Nations.* New York: Modern Library.

Smith, A. (November 2016). *Labor Platforms: Technology-enabled "Gig Work."* Washington, DC: Pew Research Center. Retrieved on November 27, 2016 from http://www.pewinternet.org/2016/11/17/labor-platforms-technology-enabled-gig-work/

Sørensen, J. B. (2007). Closure and Exposure: Mechanisms in the Intergenerational Transmission of Self-employment. In M. Ruef & M. Lounsbury (Eds.), *The Sociology of Entrepreneurship* (Research in the Sociology of Organizations, Volume 25) (pp. 83–124). Bingley, UK: Emerald Group Publishing Limited.

Spaulding, S., Derrick-Mills, T. & Callan, T. (2016). *Supporting Parents Who Work and Go to School.* Washington, DC: Urban Institute.

Staffing Industry Analysts (SIA) (June 2018). *The Gig Economy and Human Cloud Landscape: 2018 Update.* Retrieved on August 23, 2017 from https://www2.staffingindustry.com/row/Research/Research-Reports/APAC/The-Human-Cloud-Landscape-2018-Update

Steinbach, S. (2012). Can We Still Use "Separate Spheres"? British History 25 Years after Family Fortunes. *History Compass*, 10(11), 826–837.

Steinberg, S. (1996). Childbearing Research: A Transcultural Review. *Soc Sci Med.*, 43(12), 1765–1784.

Stepler, R. (March 23, 2017). *Key Takeaways on Americans' Views of and Experiences with Family and Medical Leave.* Washington, DC: Pew Research Center.

Stevens, D. P., Kiger, G. & Mannon, S. E. (2005). Domestic Labor and Marital Satisfaction: How Much or How Satisfied? *Marriage & Family Review*, 37(4), 49–67.

Stohs, J. H. (2000). Multicultural Women's Experiences of Household Labor, Conflicts, and Equity, *Sex Roles.* 42, 105–126.

Sundararajan, A. (June 2017). The Future of Work. *Finance & Development*, June 2017, 54(2).

Sverke, M., Hellgren, J. & Naswall, K. (2002). No Security: A Meta-Analysis and Review of Job Insecurity and Its Consequences. *J Occup Health Psychol.*, 7, 242–264.

Tao, W., Janzen, B. L. & Abonyi, S. (2010). Gender, Division of Unpaid Family Work and Psychological Distress in Dual-earner Families. *Clinical Practice and Epidemiology in Mental Health: CP & EMH*, 6, 36–46.

Taylor, P., Passel, J., Fry, R. *et al.* (2010). *The Return of the Multi-Generational Family Household.* A Social & Demographic Trends Report, March; 18. Washington, DC: Pew Research Center.

Thompson, L. (1991). Family Work: Women's Sense of Equity and of Fairness. *J Fam Issues*, 12, 181–196.

Thompson, F. M. L. (1993). Town and City. In F. M. L. Thompson, *The Cambridge Social History of Britain, 1750–1950: Regions and Communities* (pp. 27–28). Cambridge: Cambridge University Press.

Toossi, M. & Morisi, T. L. (July 2017). *Women in the Workforce Before, During, and After the Great Recession.* U.S. Bureau of Labor Statistics.

Torpey, E. & Roberts, B. (2018). Small-business Options: Occupational Outlook for Self-employed Workers. *Career Outlook*. U.S. Bureau of Labor Statistics.

Torpey, E. & Hogan, A. (May 2016). *Working in a Gig Economy.* Washington, DC: U.S. Bureau of Labor Statistics. Retrieved on July 28, 2017 from https://www.bls.gov/careeroutlook/2016/article/what-is-the-gig-economy.htm

Trent, K. & Harlan, S. L. (1994). Teenage Mothers in Nuclear and Extended Households. Differences by Marital Status and Race/Ethnicity. *J Fam Issues*, 15(2), 309–337.

U.S. Bureau of Labor Statistics (2012). *Charting International Labor Comparisons* (2012 Edition). Retrieved on July 20, 2018 from https://www.bls.gov/fls/chartbook/2012/section2a.htm

U.S. Bureau of Labor Statistics (2015). *Projections of Industry Employment, 2014–24.* Retrieved on June 28, 2018 from https://www.bls.gov/careeroutlook/2015/article/projections-industry.htm

U.S. Bureau of Labor Statistics (March 09, 2016a). *Volunteer Rate Down Slightly for the Year Ending in September 2015.* Retrieved on December 4, 2018 from https://www.bls.gov/opub/ted/2016/volunteer-rate-down-slightly-for-the-year-ending-in-september-2015.htm

U.S. Bureau of Labor Statistics (2016b). *Major Industries with Highest Employment, by State, 1990–2015.* Retrieved on June 27, 2018 from https://www.bls.gov/opub/ted/2016/major-industries-with-highest-employment-by-state.htm

U.S. Bureau of Labor Statistics (November 4, 2016c). *13 Percent of Private Industry Workers Had Access to Paid Family Leave in March 2016.* Retrieved on November 24, 2016 from https://www.bls.gov/opub/ted/2016/13-percent-of-private-industry-workers-had-access-to-paid-family-leave-in-march-2016.htm

U.S. Bureau of Labor Statistics (2017a). *Industries at a Glance. Industries by Supersector and NAICS Code.* Retrieved on August 17, 2017 from https://www.bls.gov/iag/tgs/iag_index_naics.htm

U.S. Bureau of Labor Statistics (March 8, 2017b). *Women's Median Earnings 82 percent of Men's in 2016.* Retrieved on August 17, 2017 from https://www.bls.gov/opub/ted/2017/womens-median-earnings-82-percent-of-mens-in-2016.htm

U.S. Bureau of Labor Statistics (June 27, 2017c). *American Time Use Survey – 2016 Results.* Retrieved on July 28, 2017 from https://www.bls.gov/news.release/pdf/atus.pdf

U.S. Bureau of Labor Statistics (2017d). *Average Hours Employed People Spent Working on Days Worked by Day of Week.* Retrieved on July 28, 2017 from https://www.bls.gov/charts/american-time-use/emp-by-ftpt-job-edu-h.htm

U.S. Bureau of Labor Statistics (September 20, 2017e). *Unpaid Eldercare in the United States 2015–2016. Summary.* Retrieved on December 4, 2018 from http://www.bls.gov/news.release/elcare.nr0.htm

U.S. Bureau of Labor Statistics (July 17, 2018a). *Usual Weekly Earnings Summary.* Retrieved on July 21, 2018 from https://www.bls.gov/news.release/wkyeng.nr0.htm

U.S. Bureau of Labor Statistics (June 22, 2018b). *Labor Force Characteristics.* Retrieved on June 25, 2018 from https://www.bls.gov/cps/lfcharacteristics.htm#unemp

U.S. Bureau of Labor Statistics (June 25, 2018c). *E-16. Unemployment Rates by Age, Sex, Race, and Hispanic or Latino Ethnicity.* Retrieved on June 25, 2018 from https://www.bls.gov/web/empsit/cpsee_e16.htm

U.S. Bureau of Labor Statistics (July 06, 2018d). *The Employment Situation – June 2018.* Retrieved on July 07, 2018 from https://www.bls.gov/news.release/empsit.htm

U.S. Bureau of Labor Statistics (June 1, 2018e). *Frequently Asked Questions about Data on Contingent and Alternative Employment Arrangements.* Retrieved on June 25, 2018 from https://www.bls.gov/cps/contingent-and-alternative-arrangements-faqs.htm

U.S. Bureau of Labor Statistics (June 7, 2018f). *Contingent and Alternative Employment Arrangements News Release.* Retrieved on June 25, 2018 from https://www.bls.gov/news.release/conemp.htm

U.S. Bureau of Labor Statistics (2018g). *Standard Occupational Classification.* Retrieved on August 17, 2017 from https://www.bls.gov/soc/

U.S. Bureau of Labor Statistics (July 6, 2018h). *Goods-providing Industries.* Retrieved on July 7, 2018 https://www.bls.gov/iag/tgs/iag06.htm

U.S. Bureau of Labor Statistics (July 6, 2018i). *Service-providing Industries.* Retrieved on July 7, 2018 from https://www.bls.gov/iag/tgs/iag07.htm

U.S. Bureau of Labor Statistics (July 6, 2018j). *Agriculture, Forestry, Fishing and Hunting: NAICS 11.* Retrieved on July 7, 2018 from https://www.bls.gov/iag/tgs/iag11.htm#about

U.S. Bureau of Labor Statistics (April 13, 2018k). *Agricultural Workers.* Retrieved on June 27, 2018 from https://www.bls.gov/ooh/farming-fishing-and-forestry/agricultural-workers.htm

U.S. Bureau of Labor Statistics (June 11, 2018l). *Home Health Aides and Personal Care Aides.* Retrieved on June 28, 2018 from https://www.bls.gov/ooh/healthcare/home-health-aides-and-personal-care-aides.htm

U.S. Bureau of Labor Statistics (July 06, 2018m). *Professional, Scientific, and Technical Services.* Retrieved on July 07, 2018 from https://www.bls.gov/iag/tgs/iag54.htm

U.S. Bureau of Labor Statistics (June 28, 2018n). *Management of Companies and Enterprises: NAICS 55.* Retrieved on June 29, 2018 from https://www.bls.gov/iag/tgs/iag55.htm

U.S. Bureau of Labor Statistics (March 30, 2018o). *11–1011 Chief Executives*. Retrieved on June 29, 2018 from https://www.bls.gov/oes/current/oes111011.htm

U.S. Bureau of Labor Statistics (June 18, 2018p). *Military Careers*. Retrieved on June 29, 2018 from https://www.bls.gov/ooh/military/military-careers.htm#tab-2

U.S. Bureau of Labor Statistics (March 30, 2018q). *Access to Paid Personal Leave*. Retrieved on August 2, 2018 from https://www.bls.gov/ebs/paid_personal_leave_122017.htm

U.S. Bureau of Labor Statistics (April 19, 2018r). *Employment Characteristics of Families Summary*. Retrieved on July 25, 2018 from https://www.bls.gov/news.release/famee.nr0.htm

U.S. Census Bureau (November 23, 2015). *Wives' Earnings Make Gains Relative to Husbands', Census Bureau Reports*. Retrieved on November 26, 2016 from http://www.census.gov/newsroom/press-releases/2015/cb15-199.html

U.S. Census Bureau (August 9, 2017a). *FFF: Labor Day 2017: Sept. 4*. Retrieved on August 17, 2017 from https://www.census.gov/newsroom/facts-for-features/2017/labor-day.html

U.S. Census Bureau (May 12, 2017b). *Facts for Features: Mother's Day: May 14, 2017*. Retrieved on August 20, 2017 from https://www.census.gov/newsroom/facts-for-features/2017/cb17-ff09-mothers-day.html

U.S. Census Bureau. (June 8, 2017c). *Facts for Features: Father's Day: June 18, 2017*. Retrieved on August 20, 2017 from https://www.census.gov/newsroom/facts-for-features/2017/cb17-ff12-fathers-day.html

U.S. Census Bureau (May 13, 2018a). *Stats for Stories: Mother's Day: May 13, 2018*. Retrieved on July 25, 2018 from https://www.census.gov/newsroom/stories/2018/mothers-day.html

U.S. Census Bureau (October 2018b). *Measuring America: Manufacturing in America 2018*. Retrieved on July 25, 2018 from https://www.census.gov/library/visualizations/2018/comm/manufacturing-america-2018.html

U.S. Department of Defense (2011). *Strengthening Our Military Families: Meeting America's Commitment*. Retrieved on August 1, 2018 from http://warriorcare.dodlive.mil/2011/01/24/the-presidential-study-directive-on-stregthening-our-military-families/

Valenze, D. M. (1995). *The First Industrial Woman*. Oxford, UK: Oxford University Press.

Van Hooff, J. H. (2011). Rationalising Inequality: Heterosexual Couples' Explanations and Justifications for the Division of Housework along Traditionally Gendered Lines. *Journal of Gender Studies*, 20(1), 19–30.

VanHeuvelen, T. (2018). Within-group Earnings Inequality in Cross-National Perspective. *European Sociological Review*, 34(3), 286–303.

Vilorio, D. (2014). Self-employment: What to Know to Be Your Own Boss. *Career Outlook*. U.S. Bureau of Labor Statistics.

Vogel, A. I., Tatchell, A. R., Furnis, B. S., Hannaford, A. J. & Smith, P. W. G. (1996). *Vogel's Textbook of Practical Organic Chemistry*, 5th edn. London: Prentice Hall.

Wadsworth, A. P. & Mann, J. De Lacy (1931). *The Cotton Trade and Industrial Lancashire, 1600–1780*. Manchester: Manchester University Press.

Wagner, D. (2008). *Science and Civilisation in China:* Vol. 5, Part 11: Ferrous Metallurgy. Cambridge, UK: Cambridge University Press.

Walters, P. & Whitehouse, G. (2012). A Limit to Reflexivity: The Challenge for Working Women of Negotiating Sharing of Household Labor, *Journal of Family Issues*, 33(8), 1117–1139.

Wharton, A. (2018). Gender and Work. *Oxford Bibliographies*. Retrieved on July 23, 2018 from http://www.oxfordbibliographies.com/view/document/obo-9780199756384/obo-9780199756384-0127.xml

White, M. (2009). *The Industrial Revolution*. British Library. Retrieved on June 27, 2018 from https://www.bl.uk/georgian-britain/articles/the-industrial-revolution

Wilkinson, F. (Ed.) (1981). *The Dynamics of Labor Market Segmentation*. New York: Academic Press.

Winefield, A. H., Tiggeman, M. & Goldney, R. D. (1988). Psychological Concomitants of Satisfactory Employment and Unemployment in Young People. *Soc Psychiatry Psychiatr Epidemiol.*, 23, 149–157.

Works, R. (2017). The Impact of Technology on Labor Markets. *Monthly Labor Review*. Bureau of Labor Statistics.

World Economic Forum (January 2017). *The Future Is Automated. Here's How We Can Prepare for It*. Retrieved on August 23, 2017 from https://www.weforum.org/agenda/2017/01/the-future-is-automated-here-s-how-we-can-prepare-for-it/

World Economic Forum (January 2018a). *Eight Futures of Work Scenarios and their Implications*. Retrieved on July 28, 2018 from http://www3.weforum.org/docs/WEF_FOW_Eight_Futures.pdf

World Economic Forum (January 2018b). *Towards a Reskilling Revolution. A Future of Jobs for All*. Retrieved on July 28, 2018 from http://www3.weforum.org/docs/WEF_FOW_Eight_Futures.pdf

13
FAMILY CRISES, INTIMATE PARTNER VIOLENCE, AND ALCOHOL AND DRUG ABUSE

Learning Outcomes

Upon completion of the chapter, students should be able to:

1. distinguish the concepts *stress* and *stressors*
2. describe the different types of stressor events in families
3. distinguish the concepts *family crises* and *family resilience*
4. describe different types of crises in families
5. describe ways that families can cope with family crises
6. describe the four forms of family abuse and their effects
7. describe theories explaining abuse in the family and in intimate partner relationships
8. identify constructive responses and strategies in preventing abuse and neglect in families
9. explain family problems related to alcohol and drug abuse and dependency
10. distinguish the roles of genetics and the environment in alcohol abuse and dependency
11. describe other family problems affecting the family and family members
12. explain popular misconceptions regarding family therapy.

Brief Chapter Outline

Pre-test
Stress and Stressor Events in Families
Types of Family Stressor Events
Family Crises in the Family Life Cycle
Young Adults and Family Crises
The Death of a Spouse
The Death of a Child
The Death of a Sibling
Managing Aging Parents

Coping with Family Crises
- Defensive Mechanisms
- Family Resilience

Abuse in the Family System
- Intimate Partner Violence
  - Types of Abuse against an Intimate Partner
    - Physical Abuse; Emotional Abuse; Sexual Abuse
  - Effects of Intimate Partner Violence; Continuation of Intimate Partner Abuse in Family Relationships; Causes for Remaining in an Abusive Relationship
    - Ferraro and Johnson's Typology of Defense Mechanisms; Men in Abusive Intimate Partner Relationships
- Child Abuse and Neglect
  - Types of Abuse against Children
    - Child Physical Abuse; Child Emotional Abuse; Child Sexual Abuse; Child Neglect
  - Effects of Child Abuse and Neglect
    - Battered-child Syndrome
  - Why Child Abuse and Child Neglect Occur: The Risk Factors
    - Age as Risk; Sex as Risk; Caretakers as Risk
- Sibling Abuse
  - Types of Abuse against a Sibling
    - Emotional Sibling Abuse; Sexual Sibling Abuse
  - Effects of Sibling Abuse; Why Sibling Abuse Occurs: The Risk Factors
- Elder Abuse and Neglect
  - Types of Abuse against the Elderly
    - Physical Elder Abuse; Sexual Elder Abuse; Emotional Elder Abuse; Elder Neglect; Elder Financial Abuse
  - Effects of Elder Abuse and Neglect; Why Elder Abuse and Neglect Occur: The Risk Factors
    - Caretakers as Risk; Health and Health Setting as Risk

Theories of Abuse in Families and Intimate Partner Relationships
- Social Learning Perspective
- Resource Perspective
- Feminist Perspective

Preventing Abuse in the Family
- Confronting Intimate Partner Violence
- Confronting Child Abuse and Neglect
- Confronting Child Sexual Abuse
- Confronting Elder Abuse and Neglect

Diversity Overview: Gender, Culture, Religion, and Intimate Partner Violence

Alcohol and Drug Abuse and the Family
- The Case of Alcohol Abuse in the Family System
- The Abuse of Alcohol: A Family-system Problem

Pre-test

Engaged or active learning is a powerful strategy that leads to better learning outcomes. One way to become an active learner is to begin the chapter and try to answer the following true/false statements from the material as you read. You will find that you have ready answers to these questions upon the completion of the chapter.

Stress is a state of tension caused by problems and pressures in our lives.	T	F
Child physical and sexual abuse involves three phases: tension-building, acute battering, and the "honeymoon" phase.	T	F
An early explanation for violence against women suggested that abused women found violence pleasurable.	T	F
Alcohol dependency is also known as alcohol addiction or alcoholism.	T	F
Having access to pornography does not place children at risk for sibling abuse.	T	F
Some children avoid family therapy for fear that therapists will blame them for their parents' marital problems.	T	F
Boomerang children is a term associated with injuries sustained by a child that result from physical abuse inflicted by a parent or adult caregiver.	T	F
Some religious support systems may prolong time in an abusive relationship.	T	F
A maladaptive coping mechanism, such as denying that a family member is alcohol dependent, is an example of family resilience.	T	F
Family stressors occur involuntarily and are never voluntary in nature.	T	F

Caving into their own pressure to please family and friends, Cindy and Sam's 25th wedding anniversary celebration has grown from a small reception of 15 to over 500 people. Now, the Louisiana couple is balancing what seems like a million tasks and thousands of dollars in unplanned expenses. Panic has set in.

Alex in Minneapolis is a top candidate for a Fortune 500 company that she has been following for five years. At first, Alex is ecstatic. However, closer to the interview, her elation begins to wane over the stakes: Could she live down the failure of not getting the job? Or if she gets the position, will moving to a large East Coast city and working long hours bring more problems than hope for her twelve-year-old son, Collin? Anxiety has set in.

Cindy, Sam, Alex, and Collin's stories depict different actors and situations. However, each has something in common that perhaps may sound recognizable to you or someone you know. That is, each family's scenario palpates with a demand or disturbance that has caused some form of stress. To better understand the effects of stress, we examine stressors familiar to many families, including those involving young adults, the death of a family member, and caregiving to aging parents. First, we identify what stress is and distinguish it from family stressors, which, if not handled properly, may transition into far-reaching family crises.

Stress and Stressor Events in Families

Stress is a state of tension and worry caused by problems and pressures in our lives. Sooner or later, we may experience a certain life event whose demands exceed our capabilities—a **stressor**—and whose effects on the family have a potential somehow to change it. Often, the feeling is that our experiences contradict what is normal (Rout & Rout, 2007). Stressors provoke a disquieting sense that our goals or values may be (or become) out of balance with what we want or where we need to be. Yet, as the scenarios above suggest, not all stressful events are necessarily cataclysmic. There are movements within the life cycle that are long-awaited or predictable although inherently stressful. For example, being wedded to a soul mate or landing the ultimate job opportunity are positive, even celebratory, but stressful events.

Types of Family Stressor Events

The stressors we face (an unplanned, exorbitant wedding anniversary reception; a new life, job, and residence) may come from *within* (personal feelings of guilt or obligation to family and friends; a fear of failure)—*an internal stressor*. Traditionally, families have been viewed as sanctuaries from internal stressors, and this is still true for many. However, today's families are increasingly met with external stressors and the needs they present. Often unpredictable or out of our hands, *external stressors* are stressful events that are imposed on a person by forces or circumstances from *outside* the individual. Some examples include experiencing the death of a loved one, physical danger (theft, war, acts of terrorism), major illness, and unemployment. Catastrophic weather (tornadoes, hurricanes, floods) that can destroy property and sentimental mementos is another example of an external stressor (Price *et al*., 2010). In some way, each of these external stressors has the potential for family disruption, because they force family members to realign the boundaries, goals, values, or roles that once sustained or were familiar to them. In Minnesota, for example, Alex's Fortune 500 company interview may lead to that long-dreamt new life in a major corporate town. However, an unforeseen external stressor like a global economic downturn may cause work to disappear in two years. Meanwhile, in

Upon completion of this section, students should be able to:
LO1: Distinguish the concepts *stress* and *stressors*.
LO2: Describe the different types of stressor events in families.

Source: Creativa Images/Shutterstock

Louisiana, Cindy and Sam may find themselves struggling with the turning economic tide, in addition to a recent catastrophic natural disaster, another external stressor. Plans made to assist their children and grandchildren may be overridden or severely derailed, and both spouses may quietly question if the marriage can withstand the strain.

Aside from their external or internal scope, family stressors are typed by the length of time they intervene or present themselves in family members' lives (Adams, 1975). Experts characterize stressors that are inherently transitory and eventually pass from the family cycle as *temporary stressors*. The lists of temporary stressors that press upon the family system are ever-changing and unlimited. Raising an adolescent, finding competent long-term childcare upon short notice, or recovering from an unplanned monthly bill are just a few examples. *Permanent stressors* are events that are lasting, intended to last, or remain unchanged indefinitely. The death of a family member is a sobering example of a permanent family stressor. A surviving relative may learn to adjust to a family member's absence. Still, bouts of grief may resurface throughout the life course. Stressors such as family members dealing with a relative with a long-term illness tend to reside in between a temporary timeframe and permanency.

Family stressors are also identified by their voluntary versus involuntary nature (Adams, 1975). *Involuntary stressors* are events outside the control or conscious efforts of the individual. A couple's frustration over infertility or anger over a child who conducts criminal activities is an example of an involuntary stressor. On the other hand, a couple may experience *voluntary stressors*, which are events related to the volition or the conscious effort of the individual. Tensions from marital separation and divorce are two of many examples. It is important to point out that although voluntary stressors entail more choice than involuntary stressors, this does not necessarily make voluntary

Upon completion of this section, students should be able to:
LO3: Distinguish the concepts *family crises* and *family resilience*.
LO4: Describe different types of crises in families.

stressors easier to manage. Voluntary stressors present some personal responsibility, increasing the potential for blame and other repercussions.

Family Crises in the Family Life Cycle

If change and life adjustment continue to be insurmountable or improperly negotiated, family stressors can turn into family crises. **Family crises** are difficulties that significantly disrupt normal family functioning through a continuous state of disorganization in the family unit. Let's look at some examples of family crises, including surviving the death of a spouse, child, and sibling, and managing aging parents. But first, we will explore family tensions involving adult children who leave and return to their parents' homes.

Young Adults and Family Crises

Some crises may be a consequence of family members coresiding with **boomerang children**—adults between 25 and 34 years of age who return home to live with their parents due to unemployment, divorce, extended education, drug or alcohol problems, or some other temporary transition. Boomerang children may sharply counter our assumptions of autonomy in adulthood. However, boomerang adult children are increasingly common.

Boomerang children are generally frowned upon by the public and media as placing a strain on the family system. Returning adult children may poorly adapt to their parents' codes of authority, especially when parental ground rules are limited to little or no negotiation. Power struggles may compound if the adult child coresident is a "serial home leaver," an adult child who leaves and returns multiple times over the family life course. Yet not all families experience returning adult children as a negative. For example, the pooling of emotional, practical, or financial resources may be a welcomed asset for some parents (Mitchell & Gee, 1996).

The Death of a Spouse

In the inventory of family life events, the death of a spouse may be one of the most distressing. Many spouses not only function as comanagers of their household and families; they also may be parents, sexual partners, exercise buddies, tennis partners, and confidants, who unite the family with other social groups and relationships. Consequently, the trauma following the death and absence of a husband or wife may be worsened by feelings of personal abandonment, even anger. Further, losing a spouse may alter the surviving partner's social role or participation when most needed. Surviving wives, who once had participated in activities as a couple, may feel isolated or marginalized, as may surviving husbands, who may have relied on their wives to manage their leisure time (Osterweis *et al.*, 1984).

On top of normal responsibilities, a surviving partner may be left with essential yet unfamiliar tasks when a spouse dies. The death of a spouse may mean losing the family's chief means of income. Insurance and pension expenditures may augment finances for a while. However, such expenditures are often one-time events, and some widows and widowers may need to learn how to distribute these funds wisely. The death of a spouse may also mean losing the partner who had been responsible for childcare, home management, and sustaining the overall family culture. Many may perceive themselves ill-equipped to do so and may employ housekeepers, share daily tasks with their children, or rely on relatives and their families to compensate (Osterweis *et al.*, 1984).

The Death of a Child

The death of a child shatters our expectations of a long, happy, and healthy life cycle. Its effects on surviving parents may be more pronounced than those not having had the experience might begin to imagine. Psychologist Catherine Rogers and colleagues (2008) conducted a longitudinal study that followed 428 parents of deceased children (from infancy to age 34) and an equal number of non-bereaving parents with similar backgrounds. The bereaved parents were still at risk for depressive symptoms, marital disruption, and poorer well-being for an average of eighteen years after the death of a child. The parents' recovery from grief was not related to the amount of time since death, but rather their reorganizing a sense of purpose and having other children in the family. Whereas disruptions of occupational, social, and family roles are generally associated with grief and grieving, Rogers' bereaving respondents showed little apparent reactions in these areas. Because outward signs of grief are not always obvious, the study also suggests that family and friends continue to support grieving parents as they return to normal life roles.

The Death of a Sibling

The grief of a child whose sibling has died may be overlooked. Often, focus or empathy is innocently placed on grieving parents or, if an adult, the children and/or spouse of the deceased. However, grief reactions may appear throughout a surviving sibling's growth and maturity, making his or her experience with death just as difficult and enduring. For example, some surviving children may become overly anxious about who will care for them if something were to happen to surviving family members. Others acquire a more mature persona and may begin to assume adult responsibilities. If the deceased sibling died of an illness to which the surviving child is predisposed, the surviving sibling may experience worry or fear of impending death (Osterweis *et al.*, 1984).

It is important that surviving siblings learn how to grieve and that adults learn to identify behaviors not always recognized as grief reactions. Bereavement may be more obviously manifested as withdrawal, irritability and sleeplessness, or social isolation. But there are less obvious grief reactions—aches and pains, new or emergent fears, problems in school, regression to earlier behaviors, and risky conduct—that, if identified, should be closely followed (National Child Traumatic Stress Network Child Traumatic Grief Committee, 2009).

Managing Aging Parents

During the early twentieth century, chances were that parents died before their children had reached the age of 50. Currently, better health provisions and a longer life span have positioned more middle-aged children as caregivers to their aging parents. Consequently, findings of depressive symptoms among caregivers of the elderly have been robust. In her review of over 30 studies on adult child caretakers of elderly parents, Rosemary Ziemba (2002) found caregivers at risk for "burnout" and isolation, with a majority reporting feeling demoralized by the experience. In the United States, most elder carework is provided by family members, primarily a spouse or an adult daughter. While managing their own emotional health, daughters also must negotiate a parent's moodiness, depression, or trauma over the loss of independence. For adult children, but especially career-pursuing

Upon completion of the next section, students should be able to:
LO5: Describe ways that families can cope with family crises.

women, elder carework can disrupt work schedules, concentration, and productivity, any of which can risk or limit much-needed earning power. The daughters in Ziemba's research characterized their care experiences as "role reversal": their carework required that they repurpose themselves as providers for those who once gave care to them.

Coping with Family Crises

Losing a loved one or managing a dependent parent or child can be an enduring emotional challenge, which can destabilize many family relationships. Normal family member ties and the family unit's normal environment are often frustrated or besieged by unwanted or unforeseen change. This aside, there may be lingering feelings of guilt, abandonment, and even anger over loss of privacy and independence. Making matters more difficult, integrating personal meaning with new family roles can take a toll. More "reliable" family members may be compelled to adopt new and unfamiliar responsibilities, and unwanted dependency from surviving relatives may set in. At a time when effective communication is critical, some families may poorly express their feelings. Unresolved family problems may resurface as a result.

When a family is met with a crisis, its members generally will try to master, tolerate, or reduce the problem. This collective effort is known as **family coping**, which has several important functions: to reduce the family's vulnerability to the stressor, to strengthen or maintain family system traits, and to reduce or eliminate stressor events and their hardships (McCubbin *et al.*, 1980). How family members reduce or eliminate stress can help or hinder how they adapt to the conflict or recover from the strain. There are coping strategies that effectively reduce tensions; however, they may promote or further family disruption. We examine one such strategy, *defense mechanisms*, in addition to the challenge of approaching crises with resiliency.

Defense Mechanisms

Defense mechanisms are behaviors, thoughts, and other channels of coping that individuals use to distance themselves from an unpleasant experience, often without their knowledge. For example, a distressed family member may refuse to accept the existence or scope of the crises. He or she may overreact to unthreatening or benign situations or act passive aggressively toward others. Some may become fixated with a certain developmental stage, such as developing an oral fixation that leads to overeating or smoking. Essentially, these are maladaptive or negative coping mechanisms in which family members misrepresent, deny, or suppress the crisis to avoid internal stress in order to function. Adaptive, more constructive forms of coping involve deliberate effort, including positively reframing or accepting the problem, injecting humor or religion, and receiving support offered by others (Dyk & Schvaneveldt, 1987).

Family Resilience

When crises encroach upon a family, a constructive response would be to carve out a much-needed but complex balance: to moderate stressors pressing against a family's adaptive resources while promoting family reorganization and the growth of individual members. This ideal, to a great degree, characterizes the challenge of working toward what family researchers call **family resilience**: a family's ability to positively respond to an adverse event to become more resourceful, confident, and stronger. On an optimistic note, research shows that *problem-focused coping*—defining, understanding, and assuming a special action to return to a more orderly daily existence—has a better chance for positive family outcomes. *Emotion-focused coping*—emotional venting and problem avoidance—obscures the source of the problem and complicates family crisis management (Dyk & Schvaneveldt, 1987).

Upon completion of this section, students should be able to:
LO6: Describe the four forms of family abuse and their effects.

Abuse in the Family System

In this section, we examine aggression either by or against a family member through acts of physical, sexual, and/or emotional abuse. In doing so, we highlight the family relationships in which abusive acts often take place, including parent-to-child (child abuse and **neglect**), sibling-to-sibling (sibling abuse), and caregiver-to-elder (elder abuse and neglect). We begin with the types of acts and consequences associated with violence between intimate partners, along with theories explaining its development and continuation.

Intimate Partner Violence

Since first exposed as a broad-based social problem, the perpetration of violence against an intimate partner has undergone multiple terminologies. Early on, *domestic violence* and *battering* had been used to describe abuse between heterosexual intimates, particularly those married and residing in one household. We have since learned that abuse occurs among dating, same-sex, and cohabiting couples. The variance in abusive partner arrangements has prompted a more inclusive, less particular characterization of the problem—**intimate partner violence**. In the

Source: Sinisha Karish/Shutterstock

following section, the term *intimate partner violence* will be used interchangeably with *intimate partner abuse* and *partner abuse,* language current in the literature. But, regardless of how it is phrased, the fundamental definition of intimate partner violence remains: the actual or threatened physical, sexual, psychological, or emotional abuse by a current or former spouse (including a common-law spouse), dating partner, or boyfriend or girlfriend of the same or opposite sex.

Types of Abuse against an Intimate Partner

There are three primary types of intimate partner abuse.

Physical Abuse **Physical abuse**, also called physical violence, is the intentional use of physical force with the potential for causing death, disability, injury, or harm. Acts of physical abuse involve the scratching, shoving, throwing, grabbing, biting, choking, shaking, aggressive hair pulling, slapping, punching, or burning of another individual. The use of a weapon, restraints, body, size, or strength against an intimate also is manifest of physical abuse, as is coercing another person to commit any of the aforementioned acts (Breiding *et al.*, 2015).

Emotional Abuse **Emotional abuse,** also called psychological abuse or psychological aggression, is the use of verbal and non-verbal communication with the intent to mentally and emotionally harm or exert control over another. For example, an emotional abuser may express verbal aggression as a means of harm (name-calling and humiliation). Non-verbal emotional abuse can be just as or more injurious. Examples include the use of coercive control (limiting access to transportation, money, family, and stalking or the excessive monitoring of whereabouts), control of reproductive health (refusal to use birth control, coerced abortions), and threats of violence, in addition to exploiting the victim's vulnerabilities (immigration status, disabilities) or employing "mind games" (providing false information to raise doubt about one's perceptions) (Breiding *et al.*, 2015). Unfortunately, the covert, non-physical nature of emotional abuse can make it difficult for some to perceive it as "real" aggression.

Sexual Abuse **Sexual abuse**, also called sexual violence, involves a number of completed or attempted acts: the rape or penetration of the victim, unwanted sexual contact or touch by the perpetrator, the misuse of authority to force consent to being penetrated, and coercing the victim to sexually penetrate someone else. Offenses in which an individual cannot permit sexual contact due to the voluntary or involuntary use of drugs or alcohol also are considered sexual abuse (Breiding *et al.*, 2015).

A wife's forced sexual intercourse by her husband—marital rape—was generally not considered a crime before the 1970s, a time when advocacy for victims of marital violence and battered women grew more organized. The popular assumption had been that a wife should be available to her husband sexually and that the marriage contract left her with little right to refuse, even in cases of separation (Bergen, 2006). In 1993, marital rape became a crime in all 50 states under at least one section of the sexual offense codes. Nevertheless, marital rape has yet to receive the same analytical attention as other forms of violence against women. Further, societal attitudes questioning if rape can actually occur within the confines of marriage show a need for greater awareness of the problem (Bergen, 2006).

Effects of Intimate Partner Violence

Physical abuse may lead to substantial adverse health effects. A direct outcome of physical violence by an intimate partner may be bruises, knife wounds, broken bones, traumatic brain injury, back or pelvis pain, and headaches. Some victims die. Another consequence of physical abuse involves the state of the survivor's emotional well-being, which tends to worsen as violence becomes more severe. These effects may include antisocial behaviors, suicidal behaviors (in females), a fear of intimacy, inability to trust others (especially in intimate relationships), depression, and low self-esteem. Some report symptoms of post-traumatic distress disorder (PTSD), during which the survivor experiences emotional detachment, sleep disturbances, and mental "replays" of the assault. Compared to their non-abused counterparts, physically abused women undergo higher rates of unintended pregnancies and abortions as well as sexually transmitted infections, including HIV (Intimate Partner Violence: Consequences, 2017).

As a receiver of emotional abuse, a victim's sense of self-worth is diminished and replaced by his or her abuser's verbal and, sometimes, non-verbal dominance. The effects of emotional abuse involve, but are not limited to, anxiety, depression, symptoms of PTSD, antisocial behaviors, suicidal behavior (especially among female victims), and low self-esteem. It is not uncommon for victims to assert that the distress of emotional abuse is more psychologically harmful than that of physical abuse. For example, physical abuse may not occur every day in even the most violent families or relationships. However, emotional abuse may occur often or daily, making its cumulative affect more influential (Intimate Partner Violence: Consequences, 2017).

Like other forms of intimate partner violence, the effects of sexual abuse are considerable and health-related in scope. Physical injuries may include chronic pain, gastrointestinal disorders, gynecological complications (in women), genital injuries, sexually transmitted infections, frequent headaches, and chronic pain. Emotionally, victims of sexual violence are often left to manage both long-term and immediate complications. For example, they may experience shock, denial, fear, confusion, anxiety, shame, guilt, distrust of others, and social withdrawal. Enduring psychological distress may stem from symptoms of depression and PTSD. Completed or attempted suicide, an avoidance or diminished interest in sexual intercourse, and low self-esteem also are of concern (Centers for Disease Control and Prevention, 2011).

Continuation of Intimate Partner Abuse in Family Relationships

First developed to address clinical depression, psychologist Lenore Walker (1979) re-codified the concept *learned helplessness* to explain women's reactions to intimate partner violence. In her **theory of learned helplessness**, Walker argues some abused women may effectively give up hope, resist attempts to escape their abusive relationship, or downplay resources that may change their abusive situation. The theory also describes intimate partner abuse as a cycle that comprises three distinct stages, which only worsen a victim's feelings of disempowerment. Stage one, the *tension-building stage*, is marked by fear, the escalation of verbal abuse, and throwing objects, hitting, shoving, and grabbing. During the *acute battering incident*, the second stage, the tension peaks and the impending assault or threat occurs. Stage three, *the honeymoon phase*, is the respite before the next tension-building and acute battering incidents, if intervention is not successfully achieved. Walker emphasizes that abused women's learned sense of helplessness is an acquired effect that should not be interpreted as inherent or personal weakness. Rather, she argues, learned helplessness is a result of fear, a chronic loss of control, and a cultural belief that "perfect" relationships are attainable, given hard work.

Some have questioned the usefulness of Walker's learned helplessness theory, arguing that a recurring expression of anger and mistreatment on the part of the abuser is not consistent with all abused women's experiences. These critics invariably view each of Walker's phases as only a part of an ongoing spectrum of hostilities, which are intended to exert power and maintain control over a victim. An established, alternate framework to understand reactions to intimate partner violence is the Duluth Model. Rather than identifying violence as a cyclical process, the

Duluth Model demonstrates that an abuser may also employ power and manipulation through male social privileges, victim isolation, economic deprivation, and harm or blocked access to children. Any of these factors can greatly compromise a victim's escape (Domestic Abuse Intervention Programs, 2017).

Causes for Remaining in an Abusive Relationship

Why a partner would remain in an abusive intimate relationship has been a source of unease and inquiry from casual observers to interventionists to academic scholars alike. Regrettably, early investigations did little to advance the realities of how intimate partner violence is experienced. A classic example is a 1964 *Archives of General Psychiatry* article by Snell *et al.*, "The Wifebeater's Wife: A Study of Family Interaction." Following a psychoanalytic model, the authors suggested that married women assaulted by their husbands demonstrated gratification or a masochistic pleasure in the violence against them.

That a woman finds gratification in abuse violation is the personification of victim blaming. Nonetheless, the premise has led to the effect that abuse and suffering emanate from something that the *victim* is failing to do. In one community-based sample of non-victims, the solution to abuse was attributed to the victim (Taylor & Sorenson, 2005), not the abuser. Alongside their accusatory focus, masochistic theories oversimplify the fear, power dynamics, and safety concerns that constrain a victim's options, including escaping from the abusive relationship.

Ferraro and Johnson's Typology of Defense Mechanisms In their study of abused women, sociologists Kathleen Ferraro and Michael Johnson (1983) yielded six maladaptive defense mechanisms used by victims to diffuse or avoid conflict and, in effect, remain in an abusive relationship. Their research uncovered an "appeal to salvation effort," in which battered women perceived their abusers as "sick" or pathological and, thereby, dependent upon them for nurturance. Alternately, some victims perceived abuse as temporary or a response to some external circumstance, like work-related pressures. The researchers identified this pattern as a "denial of the victimizer." Some study respondents rationalized a "denial of victimization." In this defense, the victim believed that she had no right to escape because of a perceived contribution to the abusive episode. Further, some victims were shown to demonstrate a "denial of options" in which practical opportunities toward escape are disregarded because of fear and uncertainty about survival outside the relationship. Another pattern uncovered, "denial of injury," involved the belief that abuse was a part of normal occurrences, particularly if daily routines were met without disruption. In the study's final defense, "appeal to higher loyalties," religious doctrine on the permanence of marriage and reward in the afterlife encouraged some victims to further endure their abusive relationships.

Men in Abusive Intimate Partner Relationships Those puzzled over why a wife would remain with an abusive husband may be more bewildered when the offenders are reversed. The reality is that not only wives abused by husbands but also husbands abused by wives are aware that simply leaving the abusive relationship is not always simple. An abused husband may be afraid to stay for fear that violence inevitably will be directed toward his children but be compelled to leave if violence appears to affect his children's physical and emotional well-being. Yet, legally obtaining child custody may be a challenge for men as a group, and even the most confident fathers may be overwhelmed by the prospect of rearing a child alone. Further, as a group, abused men often find their experiences shameful, dismissed, or minimized by law enforcement or other authorities, a factor in men's underreporting of intimate partner abuse.

Like abused heterosexual men, gay and bisexual men may find it difficult to see themselves as victims for fear of appearing non-masculine or unable to defend themselves (Dunn, 2014). But abused men in same-sex relationships may have other concerns that are no less intimidating. Disclosing abuse may mean disclosing their sexuality to unsuspecting family or friends.

Child Abuse and Neglect

Child abuse, also called child maltreatment, is broadly defined in many states as acts of cruelty inflicted upon a child by a parent or caretaker (e.g., clergy, coach, teacher) that result in serious threat, physical injury, emotional harm, sexual abuse, exploitation, or death. Identifying child abuse can be complex and those most able to end it may be among the last to know. Like abused adults, abused children may be reluctant to disclose maltreatment based on fear of being blamed or disbelieved. The presence of one sign of child abuse does not mean that maltreatment is evident; however, a closer look at the situation may be warranted if signs occur repeatedly.

Types of Abuse against Children

Child abuse or maltreatment is generally codified as physical abuse, emotional abuse, sexual abuse, and child neglect. While each form of abuse may take place separately, they more commonly occur in some combination.

Child Physical Abuse Child physical abuse is the intentional or unintentional physical injury of a child due to punching, beating, kicking, biting, shaking, throwing, stabbing, choking, hitting, burning, scalding, or otherwise harming a child by a caregiver (parent, relative, or another individual responsible for the child). Acts leaving no observable marks, such as smothering a child, also are considered physical abuse. Physical discipline, such as spanking or paddling, can be distinguished from physical abuse as long as actions are within reason and cause no bodily injury to the child. Some bodily evidence of child physical abuse may include unexplained burns, bites, bruises, black eyes, and broken bones, in addition to fading bruises that are noticeable after an absence from school. Physical injuries from sexual contact are defined as child sexual abuse, which we will discuss later (Child Welfare Information Gateway, 2013; Leeb *et al.*, 2008).

Child Emotional Abuse Child emotional abuse, also called *child psychological abuse,* is a pattern of behaviors that impairs a child's emotional development or self-worth through constant criticisms, threats, blaming, degradation, and intimidation. Conduct otherwise harmful or insensitive to a child's developmental needs—such as isolating a child or withholding love, support, or guidance—also are identified as child emotional abuse (Child Welfare Information Gateway, 2013). Caregiver behaviors that are life-threatening or make a child feel unsafe, such as creating situations that could physically hurt, abandon, or kill a child, are forms of child emotional abuse characterized as *terrorizing*. Terrorizing also extends to behaviors or unrealistic expectations that threaten a child or cause loss, harm, or danger if a caregiver's requests are not met. A parent or custodial adult who tells a child that his or her pet will be killed if he or she does not make good grades demonstrates terrorizing behavior (Leeb *et al.*, 2008).

Possible signs of child emotional abuse may entail clear and disturbing psychological and behavioral extremes. A child may be aggressive, overly demanding, and exhibit inappropriate adult behaviors (parenting other children). On the other hand, he or she may be overly compliant, even infantile (frequent rocking or head banging) (Child Welfare Information Gateway, 2013).

Child Sexual Abuse Child sexual abuse is the use, persuasion, or coercion of a child to engage in or to assist another in the engagement of sexually explicit or sexually simulated conduct. Sexual contact can be performed by the caregiver on the child, by the child on the caregiver, or a caregiver can coerce a child to commit a sexual act on another child or adult. Acts involving sexual abuse are related to penetration involving a child's mouth, penis, vulva, or anus by the use of a hand, finger, or other object, although not in all cases. For example, non-contact child sexual abuse excludes acts like touch or physical contact of a sexual nature. Nonetheless, it exposes a child to sexual activity. Examples of non-contact sexual abuse include exploiting a child through prostitution, sexual harassment, the production of pornographic materials, or by filming a child in a sexual manner (Leeb *et al.*, 2008).

Signs suggestive of sexual abuse among children up to three years old may include excessive crying, vomiting, feeding problems, bowel and sleep disturbances, and a general failure to thrive. Children in this age group also may have difficulty walking or sitting, as would victims up to nine years old, who may exhibit feelings of guilt and shame, withdrawal, fear of certain people and contexts, behavioral regression (bed-wetting, stranger anxiety), and excessive masturbation. Older children and adolescents are likely to experience depression, aggression, substance abuse, poor school performance, suicidal gestures, and eating and sleeping disturbances. Older children and adolescents also may run away from home, display pseudo-mature behaviors, and unusual sexual knowledge. Girls risk early pregnancy (Child Welfare Information Gateway, 2013).

Child Neglect Involving acts of omission, *child neglect* is frequently defined as the failure of a parent or caretaker to provide needed food, clothing, shelter, medical care, or supervision to the degree that the child's health, safety, and well-being are threatened with harm. Child neglect can take place in the most familiar contexts and settings. A caregiver's failure to provide adequate hygiene or shelter is an example of physical child neglect as is a child's malnourishment due to being denied meals, or harm committed to a two-year-old left under the sole supervision of a seven-year-old. A caregiver not administering prescribed medications or refusing to take the child for needed medical attention may be identified as child medical neglect. Allowing a child to miss 25 or more school days in one academic year without excuse may meet the criteria of child educational neglect (Leeb *et al.*, 2008).

Effects of Child Abuse and Neglect

Child abuse and child neglect may manifest significantly in personal ways, some of which can be long term. Follow-up studies on child abuse reported three years earlier show that 28 percent of children had chronic health and other conditions. These conditions included poor psychological health (depression, anxiety, and other psychiatric disorders), developmental challenges (grade repetition, low academic achievement), behavioral difficulties (substance abuse, delinquency, pregnancy, and sexual risk-taking), and social obstacles (antisocial traits, aggression, modeling affectionate behaviors with unknown/little-known people) (Child Welfare Information Gateway, 2013). Whether these outcomes eventually dissipate or last a lifetime has a lot to do with a child's age and developmental status at the onset of maltreatment; the relationship between the child and the perpetrator, and the type, frequency, duration, and severity of maltreatment. For example, victims of child sexual abuse already risk a higher rate for rape in adulthood, but that rate of risk increases based on the severity of sexual abuse (Child Welfare Information Gateway, 2013).

Battered-child Syndrome Conduct typifying child neglect and physical, sexual, and emotional abuse have occurred for centuries; however, few medical or legal frameworks existed for interpreting such acts as child abuse or criminal. In nineteenth-century New York, for example, child battering could only be prosecuted with references to policies protecting animals from cruelty.

By the twentieth century, the *Journal of the American Medical Association's* publication "The Battered-Child Syndrome" brought evidence of child abuse to the public sphere. Bearing social, clinical, and diagnostic importance, the article highlighted factors to help physicians assess cases of suspected child abuse. Among these factors were discrepancies between reported injury history and clinical findings, physician-to-caregiver questioning in suspected cases of child abuse, key medical exams regarding abused children, and findings related to poor hygiene and a child's failure to thrive (Leventhal & Krugman, 2012). The term **battered-child syndrome** characterizes injuries sustained by a child resulting from physical abuse usually inflicted by a parent, custodian, or other adult caregiver. Only after this 1962 medical "discovery" was child abuse recognized as a recurring aspect of family life and not just many isolated exceptions.

Why Child Abuse and Child Neglect Occur: The Risk Factors

Child maltreatment is disturbing, and it is understandable for one to ask why it occurs. Despite our need to know, no single, simple answer can explain why a parent or another caretaker would neglect, hurt, or kill a child. Ultimately, making sense of child abuse and neglect and planning solutions toward its end require that we learn the factors that increase children's susceptibility to injury or worse. In the following, we hope to increase your awareness of risks associated with why child abuse and neglect take place.

Age as Risk A child may be vulnerable to child abuse and neglect depending on his or her age and sex. The World Health Organization (WHO) finds that children under four are at the greatest risk for severe injury and death. Notably susceptible are children with special needs: premature or handicapped children, children who cry persistently, and children with non-normative physical features. Fatal cases of physical abuse are largely found among infants (Runyan *et al.*, 2002).

Sex as Risk In a report by the World Health Organization (WHO) (Runyan *et al.*, 2002), boys were at greater risk for harsh physical punishment compared to girls. Researchers suspect that the severe treatment stems from the belief that boys require more physical "discipline" to be better, more resourceful men. Girls, meanwhile, are at a higher risk to be murdered before the end of the first year of life. Educational and nutritional neglect, in addition to forced prostitution, were more apparent for girls in most of the countries sampled. In fact, by international accounts, the rates of sexual abuse are 1.5 to 3 times higher for girls than for boys.

Caretakers as Risk The WHO also states that abusive parents tend to exhibit significant annoyance when responding to their child's moods or behaviors (Runyan *et al.*, 2002). Caregivers who have difficulty nurturing or bonding with their newborn or who have had childhood trauma in their lifetime may show an increased risk for abusing or neglecting a child. Further, misusing alcohol and other substance abuse (including during pregnancy), involvement in criminal activities, and chronic financial difficulties can misdirect time and effort away from effective parenting. Having unrealistic expectations about a child's level of behavioral maturity or emotional development also can place some children at risk. Finally, caretakers who glorify violence against others or who practice cultural traditions on the use of extreme corporal punishment may cultivate an environment that is unsafe for some children.

Sibling Abuse

Sibling abuse is described as the use of physical, emotional, and/or sexual abuse by one sibling upon another, who is related by marriage, biology, adoption, or living arrangement. Sociologists have argued that sibling abuse is the most common form of family violence. Nevertheless, sibling abuse continues to be underreported and incites less broad-based attention than child abuse and child neglect.

Types of Abuse against a Sibling

There are primarily three types of sibling abuse: physical sibling abuse, emotional sibling abuse, and sexual sibling abuse. *Physical sibling abuse* includes, but is not limited to, slapping, hitting, shoving, punching, biting, hair pulling, pinching, and scratching (Wiehe, 1998).

Emotional Sibling Abuse Emotional sibling abuse is considered the most frequently occurring type of sibling abuse. Name-calling, intimidation, ridicule, destruction of property, teasing, torture, or abuse of a pet, in addition to rejection, terrorization, isolation, corruption, and emotional unresponsiveness are examples. Many perceive this form of sibling-to-sibling maltreatment as one of the more psychologically damaging. However, in many cases, younger, more dependent siblings feel that they have little choice but to submit to intimidation and ridicule. Alongside infliction of emotional abuse, sometimes daily, is the threat of retaliation should the victim protest or disclose mistreatment (Wiehe, 1998).

Sexual Sibling Abuse Sibling sexual abuse occurs when a more powerful sibling, who may be older or stronger, induces or threatens another sibling into sexual activity. Coerced sexual activity may occur between a foster, adopted, step-, or biological brother and sister as well as between two sisters or two brothers. As with other types of sexual assault, sibling sexual abuse does not exclusively involve unwanted, inappropriate touching. Exposing one's own genitals, speaking in a sexual manner to a brother or sister, or looking at his or her genitals may be viewed as acts of sexual abuse. In addition to physical sexual assault and rape, more serious acts of sexual sibling abuse involve sexualized touching, coerced sexual acts, and voyeurism (watching the victim dress, shower, or use the toilet when they do not want to be watched). It is not uncommon for the abusing sibling to use force, the threat of force, bribes, or gifts to coerce the victim into keeping violation a secret (Napier-Hemy, 2008).

Effects of Sibling Abuse

The effects of sibling abuse include a personal sense of entrapment, betrayal, guilt, and shame. Unlike bullying that occurs outside the home, for example, an abused brother or sister may be unable to retreat to an alternate haven for safety. Given what seems like little to no recourse, he or she may feel trapped and powerless to stop the aggression. This especially is apparent if the offending sibling is physically larger, older, stronger, and threatening, a combination that may isolate the victim from those who could intervene. Further, a sense of betrayal is likely imminent with sibling abuse. Victims usually at first trust their abusers based on kinship ties; however, an especially clever offender may use a victim's trust to create a sense of complicity. A victim may feel "bad" or "dirty," if guilt or shame is successfully groomed by the offender. Each of these effects is compounded when the victim is disbelieved or led to feel responsible (Napier-Hemy, 2008). Help professionals who work with troubled adults advise that past experiences with sibling abuse may explain some clients' present-day conduct. An abused sibling may suppress pain and humiliation and even socially withdraw yet readily demonstrate aggression in other contexts, including intimate partner relationships.

Why Sibling Abuse Occurs: The Risk Factors

Sibling abuse exceeds "normal" sibling competition or rivalry and is often motivated by the abuser's own prior mistreatment or a perceived crisis. In one study, siblings who inflicted serious injuries on younger siblings had been physically abused themselves; were undergoing family crises that heightened a parent's rejection; were burdened with excessive caretaking of the target sibling, who was perceived as the parental favorite; or had experienced the recent loss of their father or another paternal caretaker (Green, 1984). The researcher also suggests that abuse can function as maladaptive coping: abuse allows retaliation against a higher-regarded sibling, offers a context for attention-getting or rage toward a parent, and cultivates a sense of mastery over abuse trauma.

Additional factors that may contribute to forms of sibling abuse include the following:

- *Child access to pornography*: Sibling aggressors have been shown to coax or force younger siblings into sexual conduct similarly depicted in pornographic material. Moreover, unsupervised internet access may expose children not only to pornographic images but also to adults or older teens who sexually prey on children.
- *Level of caretaker awareness:* Parents may be tempted to ignore fighting and aggression between children or "explain away" unusual sibling behaviors. While denial does not *cause* sibling-to-sibling abuse, it may promote its persistence. Parents also should emphasize to older siblings that supervising a younger brother or sister does not extend to the right to threaten, degrade, or injure (Wiehe, 1998).

Elder Abuse and Neglect

Elder abuse and neglect, also called elder maltreatment, is any intentional action or lack of action that causes or facilitates risk, distress, or harm to an adult age 60 or older. Although a form of family violence, elder abuse and neglect may be perpetrated by an individual outside the family whom the victim trusts. Each year, in the United States, an estimated 5 million Americans may be affected by some type of elder abuse, although only 1 in 24 cases are reported to authorities (Connolly *et al.*, 2014).

Types of Abuse against the Elderly

There are five types of elder abuse: physical elder abuse, sexual elder abuse, emotional elder abuse, elder neglect, and financial elder abuse.

Physical Elder Abuse Physical elder abuse involves the intentional use of physical force that results in chronic illness, injury, pain, functional impairment, distress, or death of an older adult. Examples of physical acts of elder abuse include, but are not limited to, hitting, beating, scratching, biting, choking, suffocation, pushing, shoving, shaking, kicking, stomping, pinching, and burning (Hall *et al.*, 2016). Bruises, abrasions, pressure marks, broken bones, and burns are possible signs of physical abuse. Such punishment tends to be more compromising for older victims, as bones may become brittle with age and relatively minor injuries can lead to serious, even permanent medical conditions (World Health Organization, 2015).

Sexual Elder Abuse Acts of sexual elder abuse include, but are not limited to, forced or unwanted, completed or attempted contact or touching of the genitals, anus, groin, breast, inner thigh, or buttocks by the use of a hand, finger, or other object. Any of these acts are identified as sexual abuse when committed against an elderly person who is incapacitated or not competent to give informed consent. Bruises found around the breasts or genitals may be signs of sexual violation (Hall *et al.*, 2016).

Emotional Elder Abuse Emotional elder abuse or the psychological abuse of an elderly person involves verbal or non-verbal behaviors that inflict mental distress or fear on an older adult. Behavioral tactics include acts intended to humiliate (name-calling, insults), threaten (expressing malicious intent to initiate nursing home placement), isolate (segregation from family and friends), or control (prohibiting or restricting transportation, social interaction, money, and other resources) the victim. Destroying the victim's property also constitutes emotional elder maltreatment. Possible signs may be a victim's unexplained withdrawal from normal activities, a sudden change in alertness, and unusual depression (Hall *et al.*, 2016).

Elder Neglect Elder neglect occurs when a caregiver or other responsible individual fails to protect an elderly person from exposure to unsafe activities and environments. Elder neglect also extends to the omission of provisions that are required to meet essential health needs, such as depriving a victim of food, water, or clothing or demonstrating a denial or oversight of basic activities related to shelter, daily life, and thriving. Bedsores, unattended medical needs, poor hygiene, and unusual weight loss are possible indicators (Hall *et al.*, 2016).

Elder Financial Abuse Elder financial abuse often is the result of a caregiver or another trusted individual who illegally uses the victim's assets without authorization or in ways that benefit someone other than the older adult. Depriving older persons of rightful access to or information about their financial resources is also financially exploitive. Theft of money, coercion, or the misuse of financial or legal systems in inheritance or property transfers (changing a will without permission), unauthorized credit card use, and the improper use of guardianship are equally exploitive although common (Hall *et al.*, 2016).

Effects of Elder Abuse and Neglect

Elderly individuals who are victimized by maltreatment may experience a worsening of pre-existing illnesses, a susceptibility to new illnesses (including sexually transmitted diseases), nutrition and hydration issues, or even death. The immediate physical effects of elder abuse include welts, wounds, and other injuries, along with persistent pain, soreness, and sleep disturbances. Physical wounds may heal in time. Nevertheless, the experience may arouse fearfulness, depression, and self-blame, along with lingering PTSD, anxiety disorders, and a sense of helplessness (Wolf, 1997).

Why Elder Abuse and Neglect Occur: The Risk Factors

There are fewer studies on the risk factors of elder abuse and neglect than those of child abuse and intimate partner abuse. One reason may be inaccessibility to victims, as many are unable to raise awareness of their violation or seek help on their own. Learning the risks of elder abuse may help to identify populations that are vulnerable to maltreatment as well as reduce a victim's need for extensive medical treatment.

Caretakers as Risk It is unclear if spouses or adult children of older people are more likely to abuse. More established is that the quality of the abuser–victim relationship within a shared living situation can increase the likelihood of elder maltreatment. As elderly victims become more dependent, a strained family history or a poor relationship with the caregiver may further descend into financial abuse, neglect, or violence. Moreover, a family caretaker's participation in the workforce may mean less spare time, more stress and burden, and a higher risk of elder abuse or neglect. Caretaker depression and a lack of social support or training in caring for an older adult also may increase the likelihood of abuse (World Health Organization, 2015).

Health and Health Setting as Risk Elders who receive care in hospitals, nursing homes, and other long-term care institutions may be vulnerable to elder abuse. An American survey of nursing home staff suggests rates of elder abuse may be considerable: 36 percent witnessed at least one incident of physical abuse of an elderly patient in the previous year; 10 percent committed at least one act of physical abuse towards an elderly patient, and 40 percent admitted to psychologically abusing patients (World Health Organization, 2015).

Upon completion of this section, students should be able to:
LO7: Describe theories explaining abuse in the family and in intimate partner relationships.

Theories of Abuse in Families and Intimate Partner Relationships

Many theories are used to explain why abuse occurs among intimates and other family members. Each distinguishes specific but sometimes overlapping factors that result in a particular understanding of the same conduct. Below, we explore three influential theoretical frameworks: the social learning perspective, the resource perspective, and the feminist perspective.

Social Learning Perspective

The **social learning theory of abuse**, also referred to as the *intergenerational transmission theory of violence,* provides one of the most consistent and widely accepted explanations of violence within the family. The theory's central premise is that children who have witnessed or experienced violence by primary role models are more likely to use violence compared to children who have been exposed to little or no violence.

The social learning perspective underscores that the transmission of violence is facilitated by a family of origin's mismanagement stressors, which reinforces how some children display anger and aggression later as adults. The approach also emphasizes that the family indirectly or directly cycles violent behavioral patterns insofar that future generations may continue to accept violence as a method of solving problems or changing the behaviors of others. Researchers at Columbia University followed 543 children for twenty years to test the effects of exposure to parental domestic violence, in addition to other disruptive behaviors. Findings show that children's exposure to domestic violence conferred the greatest risk of receiving partner violence. Moreover, for both men and women, early exposure to parental violence was second to persistent disorderly conduct as the strongest risk for partner violence perpetration as adults (Ehrensaft *et al.*, 2003). In a more recent study supporting the theory, results showed that witnessing intimate partner violence as a boy and being physically abused as a boy almost doubled the odds of perpetrating wife abuse. Of the 522 married male respondents, over 36 percent reported having ever perpetrated psychological, physical, or sexual abuse against their current wife (Yount *et al.*, 2015).

We emphasize that neither all abused children nor all children who witness family abuse grow up to become violent offenders. A child may view a parental conflict or beatings as troubling, scary, or "bad" but decidedly reject abusiveness as an adult. One parent may act violently toward the other, but the abused parent may tell the child that violence is wrong and that future intimate partners do not deserve abuse. In light of these competing reactions, one criticism of social learning theory is that it inadequately explains why some abused children do not offend or why some non-abused children do offend (Dobash & Dobash, 1979).

Resource Perspective

The **resource theory of abuse** asserts that the partner who commands the resources (education, money, property, social prestige) in the family or relationship also commands the dominant position in the relationship. As a result, the dominant partner is less likely to exert force against the partner with less relative resources. Alternatively, the theory posits, the individual who aspires to be the dominant partner but has little education, few marketable skills, and a job with low pay, prestige, and promise for advancement is more likely to exert aggression to acquire or recoup power. The theory has been criticized for downplaying the role of gender inequities in explaining the use of aggression to win dominance in a relationship.

When observing its most fundamental premise, the resource perspective depicts abuse as a "weapon" of sorts through which power and status are acquired, if not seized. Findings from a Norwegian study appear to support this explanation. Using a sample of 1,640 male and 1,791 female cohabitants, women of higher income or education than their partners were more likely to experience violent incidents by their intimate partners. Women earning more than 67 percent of the total household income were seven times more likely to suffer psychological and physical abuse compared to women who earned less than 33 percent. Researchers describe the prevalence of abuse in the sample as an attempt to balance the uneven division of power of the relationship (KILDEN, 2014).

Feminist Perspective

The *patriarchal theory of abuse* offers a **feminist theory of abuse** on the experience of physical, emotional, and sexual abuse against women by men, including husbands. As a theoretical lens, feminist approaches to explaining social life invariably premise that institutional and social norms, behaviors, privileges, and expectations shape perceptions of gender that encourage men's control of women. Similarly, the central thesis of the patriarchy theory of abuse is that male-to-female violence and women's historical subordination to men are sourced by norms, belief systems, and policies that have institutionalized and enforced a sexist social and family order. The theory further posits that aggression is used to maintain patriarchal control.

Sociologists Rebecca Dobash and Russell Dobash (1979), who pioneered the patriarchal theory of abuse against women, write that "the seeds of wife beating lie in the subordination of females and in their subjection to male authority and control" (pp. 33–34). Pioneering family violence researchers, sociologists Murray Straus and Richard Gelles (1990) similarly discern a relationship between gender inequity and violence in the family. Calling sexism a "contributing factor" of violence in the home, they add, "We cannot overemphasize the preventative value of promoting sexual equality and eliminating sexism" (p. 203).

The limitations of the patriarchal theory of abuse become especially apparent when explaining abuse in lesbian intimate partnerships. Violence in lesbian relationships and families demonstrates that women can be controlled and victimized by women, something that disrupts the theory's premise that gender inequity and male dominance promotes violence against women. Other critics, including some feminists, point out that not all violent acts used by women are contained to retaliation or self-defense against male assault. They argue that women also can demonstrate aggression towards men as well as the elderly, children, and peers.

Preventing Abuse in the Family

It is not enough to learn the risks, effects, or physical and emotional magnitude of abuse in the family system. It is just as important to learn what individuals can do to prevent and put an end to the problem. In the following section, we offer recommendations on confronting

Upon completion of this section, students should be able to:
LO8: Identify constructive responses and strategies in preventing abuse and neglect in families.

intimate partner violence, child abuse and neglect, child sexual abuse, and elder abuse and neglect.

Confronting Intimate Partner Violence

Public health models that highlight policies, education, and health promotion are a vital step toward identifying family abuse as a persistent social problem. At the same time, awareness and prevention campaigns are no substitute for direct, everyday actions that could undermine the incidence of abuse in families. Experts and interventionists offer these recommendations:

1. Reject norms that legitimate sexism and the use of violence to solve problems.
2. Promote role models who say that abuse against men, women, and children is wrong.
3. Establish support networks that include relatives, friends, and community help services.
4. Work with community efforts to coordinate abuse prevention initiatives, to reduce self-blame, and to promote the disclosure of abuse.
5. Support education and training that provide support for parent-to-child conflicts.

Confronting Child Abuse and Neglect

Among the 315,000 children monitored by the National Children's Alliance's Children's Advocacy Centers (CACs), 211,831 received forensic interviewing and case investigation at a CAC facility following a report by child protective services or law enforcement (National Children's Alliance, 2015). Still, high levels of casework have overburdened many child welfare personnel. Institutional child abuse prevention efforts are better equipped when they collaborate with parents and caretakers. These strategies may help them to establish change and betterment.

1. Teach children constructive, non-violent problem-solving and communication strategies, and practice these skills in personal relationships.
2. Engender a safe, stable, and nurturing relationship between children and their caregivers.
3. Acquire safe, adequate childcare.
4. Promote and participate in caregiver support groups, where individuals work together to strengthen their families and build social networks.
5. Consider respite and crisis care programs that offer temporary relief to caregivers in stressful situations by providing short-term care for children.

Confronting Child Sexual Abuse

It can be a distressing paradox when seeking justice and intervention for an abused child. Family members may not believe the victim or deny that abuse occurred. If believed, friends and other family members may divide loyalties between the victim and the offender or exert pressure to keep abuse a secret. However, child advocates emphasize that the culpability of the adult offender should be neither excused nor misdirected to the victim. Strategies that may help parents and caretakers protect children from child sexual abuse include the following.

1. Stay alert for possible problems by taking an active role in children's lives.
2. Teach young children about the privacy of their own bodies and that no one has the right to touch their bodies in ways that make them feel uncomfortable.
3. Be aware of possible "grooming" behaviors in adults: frequently finding ways to be alone with the child, ignoring the child's privacy, or giving gifts for no particular occasion.
4. Create a home environment in which topics can be discussed comfortably as children age.
5. Take disclosure about sexual abuse seriously and inform the child that he or she is not responsible for abuse.

Confronting Elder Abuse and Neglect

Local, national, and global territories differ in their comprehensive responses to the care and protection of the elderly. In America, most states have penalties for those who victimize older adults. Adult protective services may be deployed to investigate suspected abuse and neglect as well as arrange medical, social, economic, legal, housing, protective, and emergency service provisions. Other nations, particularly developing countries, are only capable of a more limited response. Meanwhile, abused and neglected elderly adults continue to have pressing needs. Below are five recommendations for the caregiving families of the elderly.

1. Arrange for a trustworthy money manager or legal authority to help with bill paying.
2. Investigate home assistance programs that offer help in daily activities and meal delivery.
3. Reach out to older adults through home visitations and routine "check-in" calls.
4. Address the stressors of providing elder care by participating in caregiver support groups.
5. Participate in respite networks that serve to give caregivers a break. The older adult may temporarily enroll with a care agency to provide a few hours of relief to the caregiver or an attendant or volunteer may go to the older adult's home.

Diversity Overview: Gender, Culture, Religion, and Intimate Partner Violence

Some aspects of religious belief systems can sway an abused woman's decision to stay or leave. An abused religious woman with ideological ties to a religious community may look to peers and faith-based traditions to restore a sense of stability, support, divine protection, and an end to violation. At the same time, the result may mean prolonged time spent in an abusive relationship.

Abused Conservative Christian Women

In her review of abused Conservative Christian wives' faith-based challenges, sociologist Shondrah Tarrezz Nash and colleagues found that conceptions of wifely submission, marital permanence, infinite forgiveness, pious suffering, and other religious schemas may obstruct the safety and self-assertion of some abused Christian women (Nash *et al.*, 2013).

Abused American Muslim Women

Dena Saadat Hassouneh-Phillips' (2001) research offers that the importance of marriage, a faith-based entitlement for husbands to chastise their wives, and the insular nature of some Muslim communities may be especially limiting to American Muslim women in need of abuse intervention. Similar to the abused Conservative Christian wives in Nash's study, fear of displeasing God and reactions from religious community peers may weigh on abused American Muslim women as they negotiate whether to leave.

Abused Orthodox Jewish Women

Researchers have exposed an obligation to religious ideals that may deter some abused Orthodox Jewish women from seeking help and finding solutions. The "evil tongue" (*lashon hara*), a custom that dissuades disparaging speech about another, was an influential hindrance among Shoshana Ringel and Rena Bina's (2007) interviews with Orthodox Jewish women survivors of intimate partner violence. Of concern was that the tradition could be used to prevent an abused woman from speaking up and being heard.

Upon completion of this section, students should be able to:
LO9: Explain family problems related to alcohol and drug abuse and dependency.
LO10: Distinguish the roles of genetics and the environment in alcohol abuse and dependency.

Alcohol and Drug Abuse and the Family

The topic of **drugs**—medicines or other substances that have a physiological effect when introduced into the body—has been infused in television shows, movies, internet and, for some, frequent conversations with peers. Because different drugs have a different level of access and legal status, the public has branded some drugs as "good" and others as "bad." For example, there are heroin, cocaine, and other illicit drugs, whose production and use are prohibited or strictly controlled, and whose disuse is encouraged by law enforcement, the legal system, and social stigma. Then, there are pharmaceutical drugs like pain relievers, stimulants, and sedatives that alleviate everything from toothaches to low-grade fevers and which may require a legal medical prescription. However, the reality about the use and effects of drugs is not really so cut-and-dry. Tolerance for drugs differs from person-to-person; therefore, it is difficult to predict the effect of any drug. It should also be recognized that some legal drugs can be as dangerous as illegal drugs, and that chronic use can lead to abuse, addiction, serious health problems, and even death. When family is added to this mixture, the consequences can be destabilizing and well worth analytical attention. To illustrate the role of substance abuse in family dysfunction and maladjustment, we explore the repercussions of alcohol abuse and dependency and how they burden the family system.

The Case of Alcohol Abuse in the Family System

People have consumed beer, wine, and spirits—alcohol—to underscore times of celebration and leisure throughout history. Drinking alcohol, in itself, is not necessarily a problem. Actually, most can consume alcohol in moderate amounts with no harm to others or themselves. The other reality, a stark contrast, is that alcohol use disorder (AUD) is a leading cause of illness and death and a prevalent mental health disorder. AUD is linked to excessive drinking that causes distress or harm. It is composed of two distinct health threats: alcohol dependency and alcohol abuse.

Alcohol abuse is a pattern of drinking resulting in harm to one's health, interpersonal relationships, and/or ability to work. An abuser of alcohol may drink while driving or operating machinery. Arrests and legal problems may occur due to drunk drinking or physically hurting someone under the influence of alcohol. Long-term alcohol abuse can turn into the chronic condition of alcohol dependence. **Alcohol dependency**, also known as *alcohol addiction* or *alcoholism*, is a strong craving for alcohol and an inability to limit drinking, despite repeated physical, psychological, or interpersonal problems due to drinking. In 2015, 15.1 million American adults age eighteen and older had an AUD, including 9.8 million men and 5.3 million women. About 6.7 percent of adults who had AUD in the past year received treatment (National Institute of Alcohol Abuse and Alcoholism, 2017).

The Abuse of Alcohol: A Family-system Problem

Individuals addicted to alcohol or other drugs often will lament their loss of control or feel embarrassed about the substance's effect on their behavior. However, when any addiction touches family members, they are not entirely islands to themselves. Problems with alcohol, drugs, and unsafe conduct tend to ripple incrementally throughout the family system in often negative ways. Some examples are offered below.

Spouses and Alcohol Abuse

Joan Jackson (1954) chronicled wives' experiences with an alcoholic dependent spouse; however, her findings also may be extended to other family members. Jackson found that when incidents of excessive drinking began, the wives' first reaction was **denial**—a state in which one distorts or refuses the problem—followed by attempts to eliminate the issue. When efforts to stop the drinking proved unsuccessful, **family disorganization**—a serious maladjustment in family functioning—became more apparent. For example, the problem was accepted as permanent, attempts to control the alcoholic's behavior were ended, and family members braced for any accompanying stressors. To reorganize the family toward a more workable, constructive state, the sample took on responsibilities normally assigned to their male spouses. They became the child disciplinarians as well as the managers and decision-makers of their households. When hope for change was exhausted, the respondents might remove themselves from their alcoholic spouses to reorganize their families without them. If sobriety was achieved, the family reorganized, but this time around the spouse's sobriety.

Parents and Alcohol Abuse

For parents, having AUD may impair skills that otherwise would serve to guide and protect their children. Alcoholic and drug-dependent parents may poorly monitor a child's curfew or whereabouts, become more permissive about their adolescents' drinking, or inconsistently address a child's failure to perform normal tasks. Parents also may forget to prepare meals for their child, help wash or dress their child, assist with homework, or provide medical attention. Money may be scarce due to supplementing substance abuse, and some children may go without electricity or stable housing as a result. Further, arguing, and verbal and, sometimes, physical abuse may ensue. As a whole, the family's dysfunction can be severe enough to work against a child's sense of order. It can also undermine the security needed to form healthy choices when negotiating friendships, drinking, and other risky behaviors.

Adolescents and Alcohol Abuse

Alcohol misuse poses problems for many youth and their families. An estimated 623,000 adolescents from age twelve to seventeen had an AUD in 2015, including 325,000 girls and 298,000 boys. That same year, about 5.2 percent of youth who had AUD in the past year received treatment. Yet alcohol remains the drug of choice among youth, and drinking continues to be widespread among the group (National Institute of Alcohol Abuse and Alcoholism, 2017).

Puberty and increasing independence have been linked to the use of alcohol. When viewed this way, an adolescent's transitional development may be a factor in starting to drink. However, it should also be recognized that the younger children and adolescents are when they start to drink, the greater the chance of them harming others or themselves. For example, frequent binge drinking—drinking that typically occurs after four drinks for women and five drinks for men within a two-hour time span—can produce many risky or unsafe behaviors: using other drugs, having sex with six or more partners, and earning frequently low grades in school (U.S. Department of Health and Human Services, 2006).

Additional Family Problems Associated with Alcohol and Drug Abuse

Families battling alcohol and drug use by a family member often toil at a distinct set of challenges: denial, impairment of parental skills, family disorganization, and unsafe or risky conduct. In this next section, we explore other family problems related to alcohol and drug abuse.

Negativism Communication among family members generally may take the form of complaints, criticisms, arguments, and other expressions of discontent. Negativity may actually reinforce alcohol or drug abuse.

Anger and Self-medication Children or adults who resent their emotionally deprived home, but who are afraid to express their resentment, may use alcohol or drugs to manage their anger. Family members might also "self-medicate" through alcohol or drug use to cope with intolerable thoughts or medical conditions, such as depression and severe anxiety.

Family Members Enabling Alcohol Abuse A family member may unwittingly enable a relative's addiction when trying to hide the effects of an alcohol disorder. The family member may volunteer to purchase the relative's alcohol, rationalizing that at least he or she will not have to drink and drive. A spouse or sibling may contact a relative's employer with a feigned illness when, in fact, the relative is recovering from a hangover. Although intended to keep harm or complications to a minimum, such efforts remove the substance abuser from full accountability.

Adopting New Family Roles Family members of substance abusers often take on unfamiliar roles to manage everyday family life disruptions. For example, substance abusers often isolate themselves from family members to associate with other abusers. Consequently, their relatives must work especially hard to keep the household financially solvent and family members emotionally intact. Children of single-parent households may have no such defenses and may take on age-inappropriate duties or assume the role of parent to younger siblings.

The Abuse of Alcohol: The Roles of Family Genetics and Family Environment

Alcohol addiction seems to run in families, and genetic background can influence the likelihood for developing alcoholism. Even so, alcohol dependency is neither entirely determined by genetics nor do genetics predestine a family member to alcohol dependency. Genetics is only responsible for about half of the risk for alcoholism. How parents behave and how they treat their children and each other also influence children's risk for alcohol addiction (U.S. Department of Health and Human Services, 2012).

When thinking about alcohol dependency, we should keep in mind that our parents' influence is not limited to the genetic material they give us. A parent's conduct and experiences also have a potential to impress upon a growing child. The child's risk of alcohol dependency increases if both parents severely abuse alcohol and other drugs, if an alcoholic parent is depressed or has other psychological problems, and where there are conflicts leading to family violence and aggression. Even then, more than half of all children of alcoholics do not become alcohol dependent themselves (U.S. Department of Health and Human Services, 2012).

Other Family Problems Affecting the Family and Family Members

There are times in a family member's life in which harmony seems to be the exception and not the rule. While some expect families to be a respite from society's hardships, we now know that some family disturbances are inevitable and, unfortunately, more commonplace than perhaps imagined. In the next section, we examine other events that may interfere with the normal course of family life, including suicide, eating disorders, and mental illness.

Upon completion of this section, students should be able to:
LO11: Describe other family problems affecting the family and family members.

Mental Illness

Mental illness generally entails a range of disorders that affect mild to severe disturbances in mood, thinking, and behavior. There are roughly 200 classified forms of mental health disorders, the most common of which include depression, bipolar disorder, dementia, schizophrenia, and anxiety disorders. There are also mental health issues associated with obsessive thoughts and compulsive behaviors, in addition to eating disorders and substance use. Some mental health factors, like depression, carry the risk of suicide.

Having a spouse or partner with a mental illness can be a challenge for the non-diagnosed partner and the relationship. If not properly diagnosed, what may be criticized as laziness, irritability, emotional distance, or being prone to distraction could be a need for mental health intervention. Still, if diagnosed and treated, it may be hard for the non-diagnosed partner to keep in mind that most serious mental illnesses tend to improve over time. He or she may have to assume additional responsibilities (maintaining the household, taking care of the children), at least for the short term. Meanwhile, others intolerant of a mental illness diagnosis may ask, "Why can't he/she just 'snap out' of it?" That some mental illnesses, unlike physical disabilities, carry little or no physical signs may be partly to blame. In such cases, the non-diagnosed partner may be required to educate friends and family about mental illness and to equip them in providing needed support and encouragement.

The idea that children do not experience mental problems is a common myth. The risk of mental illness can be genetically inherited from parent to child or developed through biochemical imbalances. When in combination with biological factors, environmental contributors could manifest in serious mental health concerns. For example, growing up in chronic poverty exposes a child to scarce resources, greater exposure to physical risks, substandard housing, parenting styles undermined by chronic stress, and other environmental factors that can influence a child's mental well-being. In fact, research suggests that persistent poverty predicts children's internalizing symptoms (withdrawal, anxiety, loneliness, and depression) above and beyond the effect of current poverty (McLeod & Shanahan, 1993).

Suicide

There is evidence that higher suicidal behaviors and attempts can run in families. In a Danish study of over 21,000 suicides over a seventeen-year period, death from suicide among the children, father, mother, and siblings of people who committed suicide was about 3.5 times greater than the first-degree relatives of the control group. The study also found that those with a family history of completed suicide are at a 2.1 times greater risk of committing suicide, even when adjusting for differences in individual socio-economic status and psychiatric history. Findings like these suggest that suicide can cluster in families and, to some extent, may be attributed to the genetic component of depression and depression-related illnesses (Qin, 2003).

Combined with family history, the role of social environment should not be dismissed in addressing the increased risk for suicide. Research shows that among gay, lesbian, and bisexual youth, the risk of suicide attempt is 20 percent greater in unsupportive environments than in supportive ones. In fact, supportive settings are significantly related to fewer suicide attempts, when controlling for demographic variables and risks factors such as depressive symptoms, binge drinking, peer victimization, and physical abuse by an adult (Hatzenbuehler, 2011).

Eating Disorders

An eating disorder is a psychological condition distinguished by abnormal or disturbed eating habits. In the United States, 20 million women and 10 million men suffer from a clinically significant eating disorder at some time in their lives (Wade *et al.*, 2011). The most common forms of eating disorder include anorexia nervosa (inadequate food intake leading to a weight that is clearly too low), bulimia nervosa (frequent episodes of consuming very large amounts of food followed by behaviors intended to prevent weight gain, such as self-induced vomiting), and binge eating disorder (frequent episodes of consuming large amounts of food but without behaviors intended to prevent weight gain).

Although often perceived as less threatening than many other health disorders, eating disorders do have harmful mental and physical health effects and may lead to death. In a study of 1,885 respondents with anorexia nervosa, bulimia nervosa, and an unspecified eating disorder, the mortality rate for anorexia nervosa was 4.0 percent, 3.9 percent for bulimia nervosa, and 5.2 percent for an eating disorder not otherwise specified (Crow *et al.*, 2009). The study also showed a high suicide rate for respondents with bulimia nervosa. Respondents with unspecified, "less severe" eating disorders showed elevated mortality risks that were similar to those found in anorexia nervosa (Crow *et al.*, 2009).

Parents once were viewed as the sole source of their child's eating disorder; however, we now know that parents do not *cause* the condition. Family stress and inadequate social support compounded with body dissatisfaction, thinness ideals, dieting, and other factors may contribute to eating disorders. At the same time, some parent–child dynamics can influence some incidents of eating disorders. For example, a lowered capacity to emotionally or physically separate from a child may unwittingly pose risk. Regularly urging a child to be "perfect," emotionally repressive, or non-expressive may lead to distress and insecurity from which eating disorders may develop.

Restoring Families: Family Therapy

Family therapy—a form of psychological counseling that addresses the resolution of family conflict and the improvement of family member communication—may not magically make an unpleasant family situation disappear. Yet it may be beneficial for some families in the following ways: pinpointing specific challenges, examining the family's ability to solve problems and express thoughts and emotions, exploring existing patterns of family behavior and their links to conflict, identifying the family's strengths, learning new patterns of family interaction, and setting individual and family goals. Regrettably, there are enduring misconceptions about family therapy that hinder its reputation and use (Wedge, 2011), including the following.

- *Length of time:* The impression that family therapy is too long-term or even infinite may come from those who have experienced individual talk therapy or know of someone who has. Family therapy is often short term. Sessions are 50 minutes to an hour and generally continue for less than six months.
- *Parental blame:* Some parents avoid family therapy for fear that therapists will blame them for their children's problems. However, therapists recognize the effectiveness of forming collaborative relationships with parents, although there may be occasions when a therapist will ask parents to reconsider certain aspects of parenting or communication (more

Upon completion of this section, students should be able to:
LO12: Explain popular misconceptions regarding family therapy.

Source: Rido/Shutterstock

consistency in enforcing their rules, refraining from arguing in of front of their children).

- *Ineffectual:* Those with little to no success with individual talk therapy may feel that family therapy will yield similar results. But family therapy works from the premise that family members are also members of many other social groups. This broader scope of analysis helps a therapist to better determine the problem's source or sources. Also, a therapist may ask a family member to maintain his or her prior individualized plan of treatment, which may include prescribed medication or additional one-on-one counseling.

Global Overview: Domestic Violence and Resolution

Violence against women is a serious *and* global problem. Some cultural traditions and expectations regarding women and girls make them more vulnerable to violence by intimate partners and increase their risk for sexual assault. We provide a few examples below (World Health Organization/London School of Hygiene and Tropical Medicine, 2010).

1. A man has a right to assert power over a woman and is considered socially superior in parts of India, Nigeria, and Ghana.
2. A man has a right to physically discipline a woman for "incorrect" behavior in parts of India, Nigeria, and China.
3. Physical violence is accepted as conflict resolution in parts of the United States.
4. Intimate partner violence is a "taboo" subject in parts of South Africa.

5. Divorce is shameful, and sex is a man's right, in Pakistan.
6. Sexual activity (including rape) is a marker of masculinity in parts of South Africa.
7. Girls are responsible for controlling a man's sexual urges in parts of South Africa.

Summary

This chapter began with a discussion of family stress: a state of tension and worry caused by problems and pressures in family life. Stressors, which lead to a state of stress, are life events whose demands exceed our perceived capabilities. There are multiple family stressors, and they come in many forms. A stressor can be internal, external, temporary, and/or permanent. Others are voluntary and involuntary in nature. When we encounter family stressors so significant that they disrupt normal family functioning, they can transition into family crises. Family members often try to cope with the problem by launching defense mechanisms to reduce grief or frustration. However, not all defense mechanisms are constructive. Actions that support more adaptive coping include positively reframing or accepting the problem, receiving support from others, and other strategies that facilitate family resilience.

The forms, effects, and risks related to aggression within the family system also were discussed. They include the experience and offense of intimate partner violence as well as child abuse and neglect, sibling-to-sibling abuse, and caregiver-to-elder abuse and neglect. Each of these offenses is manifested through physical abuse, sexual abuse, and/or emotional (psychological) abuse. The chapter focused on three theories that try to explain why abuse occurs among intimates and other family members. The social learning perspective theorizes that children who have witnessed or experienced violence by a primary role model are more likely to use violence. The resource perspective argues that aggression is more likely used by partners who have little education, skills, or prestige but aspire to dominance in the relationship. The feminist theory of abuse states that male-to-female violence is sourced by norms, belief systems, and policies that have institutionalized a sexist social and family order.

Another family stressor, alcohol and drug abuse, can affect spouses, parents, and children. Its results can influence the family system in several negative ways: denial, impairment of parental skills, family disintegration, unsafe or risky behaviors, negative communication, anger and self-medication, enabling the abuser, and adopting new family roles to manage everyday family disruptions. Genetics may play a role in alcohol abuse. However, it is important to also keep in mind the role of the social environment, as the disruptions of everyday family life may place other family members at risk. Family therapy may help family members learn how to constructively address conflict and improve communication.

Key Terms

Alcohol abuse is a pattern of drinking resulting in harm to one's health, interpersonal relationships, and/or ability to work.

Alcohol dependency is a strong craving for alcohol and an inability to limit drinking, despite repeated physical, psychological, or interpersonal problems; also known as *alcohol addiction* or *alcoholism*.

Battered-child syndrome is a term used to characterize injuries sustained by a child, which result from physical abuse usually inflicted by a parent, custodian, or other adult caregiver.

Boomerang children are adult children who leave and later return to live with their parents.

Child abuse is the neglect, physical, sexual, and/or emotional maltreatment by a parent, caregiver, or another person in a custodial role; also called *child maltreatment*. Emotional maltreatment that is life-threatening or makes a child feel unsafe is called *terrorizing*.

Defense mechanisms are behaviors, thoughts, and other coping strategies individuals use to distance themselves from an unpleasant experience or adverse event.

Denial is a state in which one distorts the severity of a problem or refuses its existence.

Drugs are legal or illicit medications or substances that have a physiological effect when ingested.

Elder abuse is the lack of appropriate action or an intentional action that results in the cause or facilitation of risk, distress, or harm to an adult age 60 or older; also called *elder maltreatment.*

Emotional abuse is to inflict mental or emotional harm on another person; also called *psychological aggression* and *psychological abuse.*

Family coping is the effort of family members to master, tolerate, or reduce a crisis.

Family crises are family difficulties that disrupt the family's normal functioning through a continuous state of disorganization in the family unit.

Family disorganization is severe maladjustment in the family's functioning.

Family resilience is a family's ability to positively respond to an adverse event and become stronger, more resourceful, and more confident.

Family therapy is a form of psychological counseling that addresses the resolution of family conflict and the improvement of family member communication—it may not magically make an unpleasant family situation disappear.

Feminist theory of abuse is the premise that institutional and social norms, behaviors, privileges, and expectations shape perceptions of gender that facilitate men's control of women. The *patriarchal theory of abuse* similarly premises that male-to-female violence is sourced by norms, belief systems, policies, and histories that have institutionalized and enforced a sexist social and family order.

Intimate partner violence is the actual or threatened physical, sexual, or emotional abuse by a current or former spouse, dating partner, or boyfriend or girlfriend of the same or opposite sex; also identified as *intimate partner abuse, partner violence,* and *partner abuse.*

Neglect is a form of child and elder abuse in which a caregiver fails to provide for the person's basic physical, emotional, or educational needs or fails to protect the person from harm or potential injury.

Physical abuse is the intentional use of physical force with the potential for causing death, disability, injury, or harm; also known as *battering* or *physical violence.*

Resource theory of abuse is the premise that the partner with fewer skills, lower prestige, or less education is more likely to use aggression to acquire dominancy in the relationship.

Sexual abuse is the completed or attempted sexual assault against another individual.

Sibling abuse is described as the use of physical, emotional, and/or sexual abuse by one sibling upon another, who is related by marriage, biology, adoption, or living arrangement.

Social learning theory of abuse is the premise that children are taught to use violence by experiencing violence or witnessing primary role models' use of violence; also called the *intergenerational transmission theory of violence.*

Stress is the physical, mental, and emotional response to a particular demand or disturbance, which can be negative or positive.

Stressor is an event that causes a stressful reaction.

Theory of learned helplessness is the premise that some abused women may give up hope, resist attempts to escape the abusive relationship, or downplay resources that may change their abusive situation. Three phases of women's disempowerment are outlined: *the tension-building stage, the acute battering incident,* and *the honeymoon phase.*

Discussion Questions

- Create a scrapbook showing how drugs are presented in the media. The scrapbook should include items from newspapers, magazines, the internet, television, and movies. Some possible examples may include legal drugs (caffeine, medication for headaches, colds, and other illnesses), legal drugs for adults (alcohol and nicotine), and illegal drugs (state-prohibited marijuana, cocaine).
- Answer the following questions when discussing your findings: Was it difficult to find pictures of drugs commonly used? Where did you find cigarette ads? Where did you find alcohol ads? Why do you think these kinds of media advertise these substances? How are the scrapbooks similar? How are they different?

Suggested Film and Video

Rosenberg, K. P. (Director). (2001). *Drinking Apart: Families under the Influence.* (Film). Films for the Humanities Publisher and Home Box Office.

Silent Screams. (2014). (Film). A MAKE production for Channel News Asia.

Wilson, R. (Director). (2006). *Stephen Fry: The Secret Life of the Manic Depressive.* (Film).

Internet Sources

Centers of Disease Control and Prevention: Intimate Partner Violence: http://www.cdc.gov/violenceprevention/intimatepartnerviolence/index.html

Centers of Disease Control and Prevention: Child Abuse and Neglect Prevention: http://www.cdc.gov/ViolencePrevention/childmaltreatment/index.html

Centers of Disease Control and Prevention: Elder Abuse: http://www.cdc.gov/violenceprevention/elderabuse/index.html

National Institute on Alcohol Abuse and Alcoholism: https://www.niaaa.nih.gov/

National Institute on Drug Abuse: https://www.drugabuse.gov/

References

Adams, B. (1975). *The Family: A Sociological Interpretation.* Chicago: Rand McNally.

Bergen, R. (2006). *Marital Rape: New Research and Directions.* The National Online Resource Center on Violence against Women. Retrieved on August 25, 2016 from http://new.vawnet.org/Assoc_Files_VAWnet/AR_MaritalRapeRevised.pdf.

Breiding, M. J., Basile, K. C., Smith, S. G., Black, M. C. & Mahendra, R. (2015). *Intimate Partner Violence Surveillance: Uniform Definitions and Recommended Data Elements.* Centers for Disease Control and Prevention National Center for Injury Prevention and Control. Retrieved on January 3, 2018 from https://stacks.cdc.gov/view/cdc/31292.

Centers for Disease Control and Prevention (2011). *Sexual Violence.* Retrieved on August 30, 2016 from http://www.cdc.gov/violenceprevention/sexualviolence/consequences.html.

Child Welfare Information Gateway (2013). What Is Child Abuse and Neglect? Recognizing the Signs and Symptoms. Washington, DC: U.S. Department of Health and Human Services, Children's Bureau.

Connolly, M., Brandl, B. & Breckman, R. (2014). *The Elder Justice Roadmap: A Stakeholder Initiative to Respond to an Emerging Health, Justice, Financial and Social Crisis.* Washington, DC: U.S. Department of Justice.

Crow, S., Peterson, C., Swanson, S., Raymond, N., Specker, S., Eckert, E. & Mitchell, J. (2009). Increased Mortality in Bulimia Nervosa and Other Eating Disorders. *American Journal of Psychiatry*, 166, 1342–1346.

Dobash, R. & Dobash, R. (1979). *Violence against Wives: A Case of the Patriarchy.* New York: Free Press.

Domestic Abuse Intervention Programs (2017). *What is the Duluth Model?* (2017). Retrieved on January 3, 2018 from https://www.theduluthmodel.org/what-is-the-duluth-model/

Dunn, N. (September 18, 2014). *Domestic Violence Likely More Frequent for Same-sex Couples.* Northwestern University. Retrieved on June 12, 2016 from http://www.northwestern.edu/newscenter/stories/2014/09/domestic-violence-likely-more-frequent-for-same-sex-couples.html

Dyk, P. & Schvaneveldt, J. (1987). Coping as a Concept in Family Theory. *Family Science*, Review, 1, 23–40.

Ehrensaft, M., Cohen, P., Brown, J., Smailes, E., Chen, H. & Johnson, J. (2003). Intergenerational Transmission of Partner Violence: A 20-year Prospective Study. *Journal of Consulting and Clinical Psychology*, 71, 741–753.

Ferraro, K. & Johnson, J. (1983). How Women Experience Battering: The Process of Victimization. *Social Problems*, 30, 325–339.

Green, A. (1984). Child Abuse by Siblings. *Child Abuse and Neglect*, 8, 311–317.

Hall, J., Karch, D. & Crosby, A. (2016). *Uniform Definitions and Recommended Core Data Elements for Use in Elder Abuse Surveillance.* Atlanta, GA: National Center for Injury Prevention and Control, Centers for Disease Control and Prevention.

Hassouneh-Phillips, D. S. (2001). Marriage is Half of Faith and the Rest is Fear Allah: Marriage and Spousal Abuse among American Muslims. *Violence against Women*, 7, 927–946.

Hatzenbuehler, M. (2011). The Social Environment and Suicide Attempts in Lesbian, Gay, and Bisexual Youth. *Pediatrics*, 127, 896–903.

Intimate Partner Violence: Consequences (2017). National Center for Injury Prevention and Control, Division of Violence Prevention. Retrieved on January 3, 2018 from https://www.cdc.gov/violenceprevention/intimatepartnerviolence/consequences.html.

Jackson, J. (1954). The Adjustment of the Family to the Crisis of Alcoholism. *Quarterly Journal of Studies on Alcohol*, 15, 562–586.

KILDEN: Information Centre for Gender Research in Norway (March 5, 2014). Higher Status than One's Partner Makes Both Men, Women Vulnerable to Intimate Partner Violence. *Science Daily.* Retrieved on February 13, 2016 from www.sciencedaily.com/releases/2014/03/140305084753.htm.

Leeb R., Paulozzi L., Melanson C., Simon T. & Arias I. (2008). *Child Maltreatment Surveillance: Uniform Definitions for Public Health and Recommended Data Elements.* Atlanta, GA: Centers for Disease Control and Prevention, National Center for Injury Prevention and Control.

Leventhal, J. & Krugman, R. (2012). The Battered-Child Syndrome 50 Years Later: Much Accomplished, Much Left to Do. *Journal of the American Medical Association*, 308, 35–36.

McCubbin, H. I., Joy, C. B., Cauble, A. E., Comeau, J. K., Patterson, J. M. & Needle, R. H. (1980). Family Stress and Coping: A Decade Review. *Journal of Marriage and the Family.* 42, 855–871.

McLeod, J. & Shanahan, M. (1993). Poverty, Parenting, and Children's Mental Health. *American Sociological Review,* 58, 351–366.

Mitchell, B. & Gee, E. (1996). Boomerang Kids and Midlife Parental Marital Satisfaction. *Family Relations,* 1, 442–448.

Napier-Hemy, J. (2008). *Sibling Sexual Abuse: A Guide for Parents.* Public Health Agency of Canada. National Clearinghouse on Family Violence. Vancouver, British Columbia: Family Services of Greater Vancouver.

Nash, S., Faulkner, C. & Abell, R. R. (2013). Abused Conservative Christian Wives: Treatment Considerations for Practitioners. *Counseling and Values*, 58, 205–220.

National Child Traumatic Stress Network Child Traumatic Grief Committee (2009). *Sibling Loss Fact Sheet: Sibling Death and Childhood Traumatic Grief.* Retrieved on August 20, 2016 from http://www.nctsnet.org/nctsn_assets/pdfs/Sibling_Loss_Final.pdf.

National Children's Alliance (2015). *National Children's Alliance Statistical Fact Sheet.* Retrieved on August 20, 2016 from http://www.nationalchildrensalliance.org/sites/default/files/download-files/2015NationalChildrensAlliance-NationalAbuseStatistics.pdf.

National Institute of Alcohol Abuse and Alcoholism (2017). *Alcohol Facts and Statistics Factsheet.* Retrieved on January 5, 2017 from https://pubs.niaaa.nih.gov/publications/AlcoholFacts&Stats/AlcoholFacts&Stats.pdf.

Osterweis, M., Solomon, F. & Green, M. (1984). Reactions to Particular Types of Bereavement. In M. Osterweis, F. Solomon & M. Green (Eds.), *Bereavement: Reactions, Consequences, and Care* (pp. 71–83). Washington, DC: Institute of Medicine, National Academy Press.

Price, S., Price, C. & McKenry, P. (2010). Families Coping with Change. A Conceptual Overview. In S. J. Price, C. A. Price, & P. C. McKenry (2010). *Families and Change: Coping with Stressful Events and Transitions* (pp. 1–23). Los Angeles, CA: Sage.

Qin, P. (2003). The Relationship of Suicide Risk to Family History of Suicide and Psychiatric Disorders. *Psychiatric Times*, 20, 62.

Ringel, S. & Bina, R. (2007). Understanding Causes of and Responses to Intimate Partner Violence in a Jewish Orthodox Community: Survivors' and Leaders' Perspectives. *Research on Social Work Practice,* 17, 277–286.

Rogers, C., Floyd, F., Seltzer, M., Greenberg, J. & Hong, J. (2008). Long-term Effects of the Death of a Child on Parents' Adjustment in Midlife. *Journal of Family Psychology,* 22, 203–211.

Rout, U. R. & Rout, J. K. (2007). *Stress Management for Primary Health Care Professionals.* New York: Springer Science & Business Media.

Runyan, D., Wattam, C., Ikeda, R., Hassan, F. & Ramiro, L. (2002). Child Abuse and Neglect by Parents and Other Caregivers. In E. Krug *et al., World Report on Violence and Health.* Geneva, Switzerland: World Health Organization.

Snell, J., Rosenwald, R. & Robey, A. (1964). The Wifebeater's Wife: A Study of Family Interaction. *Archives of General Psychiatry,* 11, 107–112.

Straus, M. & Gelles, R. (1990). *Physical Violence in American Families.* New York: Transaction.

Taylor, C. & Sorenson, S. (2005). Community-based Norms about Intimate Partner Violence: Putting Attributions of Fault and Responsibility into Context. *Sex Roles,* 53, 573–589.

U.S. Department of Health and Human Services (2006). Underage Drinking: Why Do Adolescents Drink, What Are the Risks, and How Can Underage Drinking Be Prevented? *Alcohol Alert,* 67, 1–7.

U.S. Department of Health and Human Services, National Institute of Health, National Institute of Alcohol Abuse and Alcoholism (2012). *Family History of Alcohol: Are You at Risk?* Retrieved on May 15, 2016 from http://pubs.niaaa.nih.gov/publications/FamilyHistory/famhist.htm.

Wade, T., Keski-Rahkonen, A. & Hudson, J. (2011). Epidemiology of Eating Disorders. In M. Tsuang & M. Tohen (Eds.), *Textbook in Psychiatric Epidemiology* (pp. 343–360). New York: Wiley.

Walker, L. (1979). *The Battered Woman.* New York: Harper & Row.

Wedge, M. (2011 March 17). Four Misconceptions about Family Therapy. *Psychology Today.* Retrieved August 17, 2016 from https://www.psychologytoday.

com/blog/suffer-the-children/201103/4-misconceptions-about-family-therapy.

Wiehe, V. (1998). *Understanding Family Violence: Treating and Preventing Partner, Child, Sibling, and Elder Abuse*. Thousand Oaks, CA: Sage.

Wolf, R. (1997). Elder Abuse and Neglect: Causes and Consequences. *Journal of Geriatric Psychiatry*, 30, 153–174.

World Health Organization (2015). *Elder Abuse: Fact Sheet*. Geneva, Switzerland: World Health Organization. Retrieved from May 1, 2016 from http://www.who.int/mediacentre/factsheets/fs357/en/

World Health Organization/London School of Hygiene and Tropical Medicine (2010). *Preventing Intimate Partner and Sexual Violence against Women: Taking Action and Generating Evidence*. Geneva, Switzerland: World Health Organization. Retrieved on September 17, 2016 from http://www.who.int/violence_injury_prevention/publications/violence/9789241564007_eng.pdf.

Yount, K., Higgins, E., VanderEnde, K., Krause, K., Minh, T., Schuler, S. & Anh, H. (2015). Men's Perpetration of Intimate Partner Violence in Vietnam: Gendered Social Learning and the Challenges of Masculinity. *Men and Masculinities*, 19, 64–84.

Ziemba, R. (2002). Family Health and Caring for Elderly Parents. *Michigan Family Review*, 7, 35–52.

14
COMMUNICATION, POWER, CONFLICT, AND RESILIENCY

Learning Outcomes

Upon completion of the chapter, students should be able to:

1. compare and contrast verbal and non-verbal communication
2. identify the functions of non-verbal communication
3. explain factors associated with communicating and listening effectively
4. explain Deborah Tannen's concept of *genderlects* and her views on gendered communication
5. recognize patterns of effective communication in marriage
6. describe communicative barriers in multicultural families
7. explain French and Raven's bases of power in marriage and families
8. compare and contrast the resource, cultural resource, and principle of least interest perspectives on the sources of power in marital and intimate partner relationships
9. compare and contrast traditional, egalitarian, and neo-traditional norms of marital power
10. recognize the sources of conflict in marriage
11. explain common couple conflicts
12. explain positive and negative aspects of conflict among couples
13. apply Virginia Satir's four-part model of conflict
14. recognize factors associated with maintaining resilient relationships.

Brief Chapter Outline

Pre-test

Communication in Marriage and Family Relationships

Verbal and Non-verbal Forms of Communication

Body Language

Facial Expressions; Eye Contact; Proxemics; Touch; Gestures

Paralanguage

Functions of Non-verbal Communication

Substitution; Complementing; Contradiction; Accenting; Repetition

Pre-test

Engaged or active learning is a powerful strategy that leads to better learning outcomes. One way to become an active learner is to begin the chapter and try to answer the following true/false statements from the material as you read. You will find that you have ready answers to these questions upon the completion of the chapter.

Body language is a form of verbal communication.	T	F
Active-empathic listening consists of three stages: sensing, processing, and responding.	T	F
Linguist Deborah Tannen argues that men conduct "report" talk.	T	F
A person with expert power has knowledge that is unavailable or unknown to another.	T	F
There is no "positive" side to conflict in marriage and family.	T	F
In their cultural resource theory, sociologists Robert Blood and Donald Wolfe argue that the spouse having the greater resources is apportioned more marital power.	T	F
Support for egalitarian marital norms saw a slower increase and, at times, a decrease around the late 1990s.	T	F
Sociologist Pepper Schwartz argues that peer marriages are currently the cultural norm.	T	F
Psychologist Virginia Satir argues that all twelve areas of conflict must be considered to manage conflict effectively.	T	F
Maintaining resilient relationships is dependent on espousing what others may criticize as "unattainable" expectations or "dream-like" fantasies.	T	F

Upon completion of this section, students should be able to:

LO1: Compare and contrast verbal and non-verbal communication.

LO2: Identify the functions of non-verbal communication.

LO3: Explain factors associated with communicating and listening effectively.

LO4: Explain Deborah Tannen's concept of *genderlects* and her views on gendered communication.

LO5: Recognize patterns of effective communication in marriage.

LO6: Describe communication barriers in multicultural families.

Communication in Marriage and Family Relationships

Andy complains when he sees his wife, Alice, spending so much time talking on the phone and texting in between calls. Annoyed, Alice snaps back that her friend is having problems and that she is trying to help her. Andy ignores the explanation, adding that he always figured that her friend was "too unstable" and overly dependent. Alice, unmoved, ignores Andy's comment.

Later, Andy asks Alice if she wants to go out for dinner; however, Alice states that she has work to finish. Feeling snubbed, Andy complains that Alice does not appreciate him. Alice, pretending not to hear her husband's remark, tries to change the subject; Andy repeats himself. Growing more frustrated by the minute, Alice yells at Andy, telling him that he should spend more time cultivating his own friendships and not worry about her choices. Andy then starts to compare Alice with his mother. The scenario ends with husband and wife angrily screaming at each other

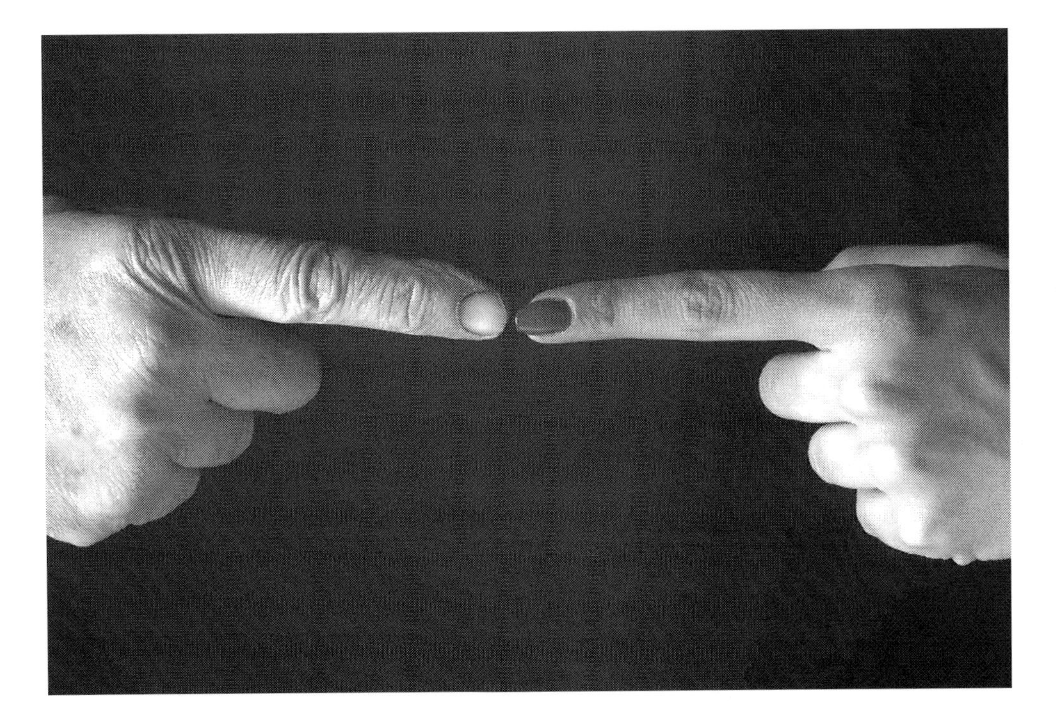

Source: bg_knight/Shutterstock

and, later, feeling troubled over what had happened. If one were to map this communication breakdown, it might look like this (adapted from Khong, 2002):

- Husband directs and demands wife.
- Wife replies to husband in an annoyed way.
- Husband criticizes wife.
- Wife ignores husband's criticism.
- Husband complains to wife.
- Wife deflects husband's complaint.
- Husband complains to wife.
- Wife yells at husband.
- Husband feels depressed and ashamed about their communication.
- Wife feels angry but also ashamed about their communication.

When steeped in conflict, it may be difficult to see how we miscommunicated or what went wrong as we exchanged our ideas. Knowing how to send and receive information effectively is essential to improving and maintaining strong relationships between couples, siblings, and parents and their children. Learning the forms of verbal and non-verbal communication and their roles in creating meaning is a great place to start.

Verbal and Non-verbal Forms of Communication

Verbal communication is the process of sharing information between individuals through sound and speech. However, verbal communication accounts for only a portion of the meanings and messages individuals send and receive. Whether we realize it or not, we reveal ourselves apart from the words we speak. Our social interactions and the meanings others derive from them also are supplemented by **non-verbal communication**: the process of sharing information through nonlinguistic means, which stimulates meaning in others. Through non-verbal communicative acts—eye contact/movements, gestures, facial expressions, touch, vocal tone, pitch, rate, or inflection, proximity, body positioning, and even our silence—others "read" our intentions, authenticity, and how well we listen in any given moment. Below, we examine forms of non-verbal communication and their use in family and other social relationships.

Body Language

Body language is a form of non-verbal communication in which thoughts, intentions, or feelings are displayed by physical movement of the body as well as the physical distance between the communicators. Common examples of body language include facial expressions, physical distance, eye contact, touch, and gestures.

Facial Expressions The human face allows us to convey a number of emotions without a single word. Those specific to happiness, sadness, anger, disgust, fear, and surprise are a persistent, nearly universal form of body language, and their shared interpretation cuts across language and culture. How, when, or if we facially display our attitudes, thoughts, and emotions are socialized early in life.

As we practice facial behavior, we control our expressions through four facial management techniques: intensification, neutralizing, masking, and deintensification (Richmond *et al.*, 2000). Say, for example, that you are teasing your twin sister about her failed prediction that your favorite team would never enter the playoffs. In general, individuals intensify or exaggerate their facial expression to indicate strong emotion (intensification). Likewise, you mischievously arch your eyebrows and smile much more broadly than normal, leaving your sister irritated over your revelry. Your sister, although annoyed, presents you with a perfect poker face. That is, she eliminates any facial behaviors (neutralizing), causing you to display annoyance toward her. A parent who has remained completely indifferent throughout, chooses to intervene but does so by repressing emotions and displaying facial behaviors appropriate to the situation (masking). Feigning a general disinterest and appearing to not side with either child, for example, the parent smiles when expressing enthusiasm over your personal victory but frowns when expressing compassion over your twin sister's misprediction. The other parent, whose favorite team also made the playoffs, controls or subdues his or her facial expression (deintensification), so as not to aggravate the situation. For example, despite great excitement over the victory, he or she smiles less and looks less happy.

Eye Contact Eye contact, a form of non-verbal communication involving the meeting of the eyes between individuals, may be one of the oldest means of interpersonal communication and is best understood in the context of facial expression. Through eye contact, individuals manifest signals that help to synchronize their dialogue and meaning-making during talk. For example, people tend to look away when thinking or hesitating. Acts of laughter, attempted verbal or non-verbal interruptions, and responses to short questions are often underscored by mutual eye contact. Gazing or prolonged eye contact identifies or signals close or intimate expression. For example, if a person steadily gazes at another from across the room, the target is poised to interpret the eye contact as an invitation for interaction. If, by chance, the target averts his or her eyes, the gazer may interpret the signal as either rejection, embarrassment, or some other type of discomfort. If a couple falls in love and enters into an intimate partner relationship, they are statistically more likely to stare into each other's eyes that those of their friends (Furnham, 2014).

Proxemics *Proxemics* is a form of body language that involves the physical distances individuals maintain during social interaction. In studies, close proximity has been shown in the interactions of those in non-marital romantic relationships, whose touch, gaze, and use of silence are more pronounced than relationships involving friends (Guerrero, 1997). However, such observations are culturally socialized and not the universal rule. For example, Americans tend to prefer a greater distance between individuals than do some Latin Americans, Italians, and Middle Easterners (Simmering, 2018).

Touch Also called *haptic communication*, touch is an expansive non-verbal language and an important component of interpersonal interaction. In parts of America, it may be used to convey intimacy (such as kissing), emotional closeness (hugging), platonic relationships (shaking hands, light touching of the arm), comfort (brushing the shoulder or back), immediacy (tugging of the hand of another), or dominance (physical brutality against the body). There are some who might add personal wellness to the list. As one psychotherapist observes, "Being touched and touching someone else are fundamental modes of human interaction, and increasingly, many people are seeking out their own 'professional touchers' ... — chiropractors, physical therapists, Gestalt therapists ... [and] massage therapists" (Farber, 2013).

Gestures Verbal speech often is accompanied by a gesture or body movement, forming altogether an integrated system of communication. When people talk to others, they often punctuate that talk with gestures. For some, gesturing is so intrinsic to their expression that verbal communication seems almost impossible without some accompanying body movement.

Gestures alone may be influential enough to take over the full task of interpersonal communication. Some common gestures you may know include the use of hands to convey "please come here," the use of the "okay" or "thumbs up" sign to convey that all is well, and pointing the index finger to convey a point or a direction. Crossing the fingers commonly conveys luck, but to cross the fingers behind one's back negates whatever statement or assurances you have made or were thinking.

Paralanguage

Paralanguage is another conduit of non-verbal interaction. Paralanguage involves tone, inflection, volume, emphasis, silence, pitch, and other vocal qualities of a spoken message. Through paralanguage, the receiver is able to perceive different messages—some of which may compete with the spoken word—as voice emphases change.

Consider, for example, the powerful effect your tone of voice can have on a statement as fundamental as "I'm doing fine." If you blurt the statement using a cold, isolating tone of voice, the receiver will suspect that you are not altogether truthful and, perhaps, unwilling to discuss your feelings or the situation. A solemn, downtrodden vocal tone can indicate that you are not what your words convey, but that the receiver should inquire further. On the other hand, a bright, cheerful tone of voice is strong indication that you are actually doing quite well.

How we apply and decode paralanguage can affect how we believe others perceive us. A study examining how people perceive others' evaluation of them found that perceived likability was correlated with positive tone of voice, in addition to expressions of similarity and dissimilarity and other factors (Curtis & Miller, 1986). Research exploring the link between body language, paralanguage, and well-being in relationships suggests that errors in decoding facial expressions and voice tone are associated with less relational well-being and greater depression (Carton *et al.*, 1999).

Functions of Non-verbal Communication

Non-verbal actions may play a greater role in how our words are interpreted than our words' intended meaning. For example, a wife may coldly blurt to her husband "I love you" at the same time she is walking out the door. Conceivably, her spouse may find her words hurtful, in poor taste, the "final straw," or grounds for rethinking the stability of the relationship. But what if the wife's coarse "I love you" was because she was late for an undisclosed doctor's appointment to learn if the couple should prepare for their long-anticipated firstborn child? Indeed, how messages are interpreted is subject to imprecision. To increase the quality of words and interactions, it is important to consider how non-verbal communication influences how others decipher what we say and what we mean. To avoid these communicative landmines, we identify five functions of non-verbal communication (Wertheim, 2008).

Substitution

Substitution occurs when a non-verbal act takes the place of a verbal message. It is morning, and Alice, Andy, and their five-year-old daughter are preparing to start the day. It is snowing, so Andy gently goads the kindergartner to put on her coat. She refuses and barks, "No! I don't want to!" Using non-verbal cues in place of words, Andy says nothing but looks directly at her while pointing toward the coat. The kindergartner puts on her coat.

Complementing

Complementing adds to a verbal message by increasing the impact of that message. Running late, for example, Alice miraculously beats the morning traffic, arriving at work twenty minutes before an important meeting. Taken aback by her good luck, she announces to her nearby colleagues, "I'm here—and I'm early!" Alice's boss, also impressed, pats Alice on the back and tells Alice how much she appreciates her efforts. Alice's boss uses a non-verbal cue (patting on the back) to complement her verbal message about Alice's timeliness.

Contradiction

Contradiction is using non-verbal cues to hide, contradict, or negate a verbal message that the individual is trying to convey. Say that Andy's day is not starting out as well as Alice's day. He arrives at work on time but finds that he misplaced his jump drive, which he needs for a meeting in ten minutes. Struggling to hide his panic, he screeches, "I am just fine!" when asked by a passing colleague how he is doing. The colleague, startled by the incongruity of Andy's words and behavior, quietly walks away.

Accenting

Manipulating the intonation of words or regulating the speed of a series of words to underscore a verbal message is *accenting*. Say that Andy shares his jump drive panic with Alice over dinner. Re-enacting the exact moment he realized he had no access to his documents, Andy, with a slow, deliberate rate of speech, says, "I cannot believe this is happening." Andy's slow word delivery highlights his verbal sentiments about the day's catastrophe.

Repetition

A non-verbal act that functions to repeat the message the person is making verbally is *repetition*. Hearing about her husband's day, Alice hugs Andy saying, "It sounds like you need a hug." Here, Alice's non-verbal act of touch reinforces her message of compassion.

Effective Family Communication

We communicate continuously; however, being able to send and receive messages in ways that create *mutual* understanding requires more strategy than we think. Try as we might, we cannot control all of the non-verbal cues we give off about how we really feel or think. Even the best intended can say one thing while non-verbally contradicting the message that he or she wants to convey. Further, tensions can compromise the skills needed to communicate effectively. When stressed, people are more likely to misread verbal and non-verbal messages, send out confusing or negative non-verbal signals, or settle into impulsive behaviors that may worsen a bad situation.

Knowing the dynamics as well as the uncertainties surrounding what facilitates **effective communication**—that is, the sending, receiving, or understanding of verbal and non-verbal information by someone in the way it was intended—has its rewards. There is a strong link between communication patterns and satisfaction in family relationships. The more positively couples rate their communication, the more satisfied they are with their relationship more than five years later (Markman, 1981). Next, we discuss ways through which family members can recognize and, thus, avoid communication debacles and experience effective communication.

Self-disclose

Self-disclosure is the process of granting access to something personal about ourselves, including indiscretions that others do not know. In their interpersonal process model of intimacy, psychologists Harry Reis and Phillip Shaver (1988) suggest that the discloser self-discloses with the hope that his or her expression will draw emotional or behavioral reward from the relationship partner, namely feelings of caring, understanding, and validation. The reaction of and the interpretation by the receiver are important in determining how disclosure will affect the relationship. For example, when initially becoming acquainted, receiving self-disclosure may lead to greater liking, closeness, and other positive impressions (Sprecher *et al.*, 2013). Over time, individuals in relationships generally cycle between whether to be open or selective about what they reveal. Still, the takeaway from many who study relationship maintenance is that self-disclosure may help develop intimacy and satisfaction.

When self-disclosure works well, individuals in relationships earn the trust needed to reciprocate and to remain connected as they grow and change over time. Personal disclosures ("I had a really hard day at school today") and relational disclosures about one's interactions with others ("I can't think of a better way to spend this holiday than with you") keep partners attuned to what each is experiencing and inform the other about how the relationship is moving along. Verbally disclosing a commitment and intention to remain in a relationship ("I am committed to this marriage") helps to reinforce satisfaction with and confidence in the relationship for both partners.

It also should be emphasized that simply increasing self-disclosure may not directly lead to satisfying or improved marital relationships for everyone. Compared with total transparency, selective self-disclosure has been shown to increase marital satisfaction. For example, information that is highly intimate or offensive could make the overall effect of self-disclosure a negative one (Cozby, 1972). At the same time, refraining from disclosure that has the potential to provoke stigma or gridlock has been positively linked to marital satisfaction in some situations. In effect, what is self-disclosed, how positive or negative the disclosed information may be, and the way that disclosure is communicated (superficially, deeply meaningful, intellectually, or emotionally) influence how well self-disclosure is accepted.

Openly Communicate

When it comes to families with children, a free and mutually directed flow of information in which emotions, ideas, plans, and experiences are disclosed on a regular basis—**open communication**—is integral to securing and maintaining parent–children attachment. For parent–adolescent relationships, warm, open communication serves as an opportunity to discuss each other's behaviors, to understand the reasons certain behaviors or concerns exist, to inhibit inaccurate perceptions of the other, to prevent adolescent risk involvement, and for parents to retain influence despite peer pressures. Researchers emphasize that consistent appraisal and alterations are necessary as adolescents' developmental needs change, and that preserving open, positive communication with adolescents is not incompatible with the goal of their achieving autonomy (Leong, 2008). Patterns of open communication in families have been associated with desirable personality traits in young adults. For example, in Huang's (1999) sample of undergraduate students, individuals from conversation-oriented families exhibited not only greater degrees of self-disclosure, but also self-esteem and sociability. Students from less-conversational, more conformity-oriented families were more likely to be self-monitoring, shy, and of lower self-esteem (Huang, 1999).

Listen Actively and Empathically

How we listen has significant bearing on our ability to recognize alternate points of view, to negotiate mutual understanding, to benefit job performance, and to strengthen our social interactions with others. However, time and availability can prevent us from listening well. Many modern families are inundated with work, appointments, extended kin, children's extracurricular demands, or simply getting to their destinations safely and on time. The results may be fragmenting, even if only occasionally. Insufficient time spent in meaningful conversation leaves family members with fewer chances to take stock of how pressures are affecting the family. The practice of clear, direct communication between members may be strained or compromised.

Researchers who study listening skills emphasize that good listening is best realized when listeners or receivers of information are willing to do two things. The first is to empathize or understand what is going on inside the speaker's mind by putting himself or herself in place of the speaker. The second is to listen as well as respond in ways that improve mutual understanding. Below, we further discuss these important aspects of listening well.

The Active-Empathic Listening (AEL) Model is a three-stage listening paradigm that emphasizes empathic engagement and mutual meaning-making.

1. First, the listener absorbs the speaker's communication, how it is said, and shows sensitivity to the emotional state of the speaker (sensing stage). The listener demonstrates AEL by going beyond the literal meaning of the speaker's words to consider the emotions attached to them. The listener also conveys the message that he or she is "taking in" the information being shared.
2. Second, the listener pulls together what is being communicated by asking that points be clarified while integrating parts of the speaker's talk into a coherent narrative (processing stage).
3. Third, paraphrasing and raising further questions, the receiver uses verbal and non-verbal behaviors to demonstrate that he or she is paying attention to the speaker (responding stage).

By these means, the actively empathic listener is not only focused on paying attention to what the speaker is saying but also engages in a purposeful attempt to demonstrate the importance of the speaker and the speaker's communication (Gearhart & Bodie, 2011).

Components of empathic and active listening have been repeatedly identified as contributors to effective family communication. For example, when a young child is faced with a crisis, parents who paraphrase what the child is saying, summarize the child's feelings, and demonstrate interest show commitment and responsiveness from the child's point of view. In Garland's (1981) study of marital enrichment and active listening among couples, the experimental group trained in active listening skills became significantly more accurate in their perceptions of their spouses' attitudes and feelings than did those in groups without training. Meanwhile, Bodie *et al.* (2015) found that listeners who ask open questions, paraphrase content, reflect feelings, and elicit non-verbal active listening behaviors signaled more emotional awareness and promoted a greater degree of emotional improvement (Bodie *et al.*, 2015). In personal relationships, empathic listening behaviors function similarly to expressions of interpersonal affection. Consider, for example, how empathic listening behaviors may be experienced by a receiver: a sense of immediacy or presence is developed, the recipient often feels better understood and validated, and the recipient acknowledges an investment of time and energy that implies their worth to the listener.

Ineffective Family Communication

If we are the speakers, we want others to listen; however, compared to others forms of communication (reading, writing, and speaking), listening may be the skill to which we accord the least preparation or consequence. With effort, inadequate listening does not have to be as common. Recognizing the styles of listening that often hinder good communication can go a long way toward overcoming miscommunication.

How does poor listening "look"? The following scenario observes Alice's use of five common styles of ineffective listening. Alice is, by no means, exceptional. Studies show that a large percentage of individuals listen less effectively than they believe, and many of us are poor listeners (Auxley, 1996). Below, we identify the five common styles of poor listening that Alice displays (Covey, 2014).

Say, for example, that Andy is now home, where he recounts to Alice his panic about being unable to access documents for a meeting. Clues that Alice is listening ineffectively would be:

- Alice really does not want to listen to Andy's story and finds herself thinking about heading to the gym or their daughter's doctor appointment while Andy is speaking. In this example, Alice is *spacing out*. Spacing out occurs when the listener is caught up in his or her own thoughts and focuses on something other than what the speaker is saying.
- Alice, who is more interested in heading to the gym or her daughter's doctor appointment than in Andy's story, fakes attentiveness by saying, "yeah," "uh huh," and "oh ... wow" at intermittent points in the conversation. This is an example of *pretend listening*. Pretend listening is quite common. The listener is not attentive to the speaker, but rather pretends to be by making insightful comments. The speaker may eventually catch the hint and feel that he or she is not important enough to be heard.
- Alice is sort of intrigued by Andy's volatile reaction toward his colleague and, thus, centers her attention on Andy's story at that moment. Here, Alice is displaying *selective listening* or paying attention to only the parts of the conversation that interest her. Since the listener is more considerate of his or her own thoughts and not the speaker's, there is a reasonable chance that a lasting trust may not develop.
- Alice pays attention to Andy's story but primarily zones in on certain aspects of the message, in particular the words, *disaster* and *scared senseless*. This is an example of *word listening*: the listener limits attention to the speaker's words and not the speaker's body language or feelings. By focusing entirely on the speaker's words, it is difficult for the listener to connect with the speaker's emotions or overall message.
- Alice, upon hearing Andy's story, interrupts, "This happened to me just the other week!" and goes on to recount her similar experience. This example reflects *self-centered listening*, which occurs when the listener perceives the speaker's message from the listener's point of view. Although we say we know how the speaker feels, we really are assuming that they feel the same way we do. By listening from only personal points of view, we may respond by judging, advising, or questioning the speaker, any of which may cause the speaker

to "close up" immediately. Self-centered listening also may be experienced as a game of one-upmanship, as if the discussions were competing.

Effective Communication in Marriage

Differences of opinion in marriage may deliver long-term benefits if partners apply a mutual intent to repair any emotional damage (we will discuss this later in the chapter). However, when couples "dig in"—that is, show inflexibility, defensiveness, and withdrawal—their opposition could be destructive to the relationship. To learn what communicative behaviors promote satisfying marital relationships, some family researchers have turned to the interactions, intentions, and outcomes of couples who enjoy happy marriages. Their general findings suggest that certain ways of relating to a partner may increase the chances of maintaining long-term marital satisfaction. Below are some suggestions based on their research (Greenberg, 2015).

Openly Listen

Happy couples make an effort to listen to their partners and seriously consider their needs. Each partner communicates to the other that he or she has an understandable point of view, given personal experience or the current circumstances. Being "present" with each other creates a sense of being on the same side and having each other's backs. For unhappy partners, the chances are that no one is heard or understood. Unhappy partners also tend to feel judged, criticized, or dismissed.

Strive for Closeness and Courtesy

Happy couples make emotional intimacy and positive reinforcement priorities. Affection and appreciation are often expressed through encouraging words and gestures. Basic respect and warmth for each other are shown in many small but potentially lasting ways (bringing each other coffee or offering to help each other). To the contrary, unhappy couples may come across as roommates at first glance. They generally are emotionally distant, lack intimacy, and their communication is focused on mundane life events, like picking up the kids and running errands. Unhappy couples also may not exhibit courtesy and sensitivity in the way they treat each other.

Constructively Engage in Conflict and Conflict Resolution

Conflict is opposition that results when one individual does something to which another individual objects. Happy couples are not immune to conflict; however, their resolve to diffuse tensions may exceed a conflict's negativity or relational harm. For example, happy couples avoid name-calling, refrain from bringing up every unpleasant thing their partner did, and offer conciliatory comments to show their care for each other and the relationship. For unhappy couples, negative interaction (non-constructive, angry comments, contempt, and disgust) is more common, and arguments may turn into ongoing hostility or a silent treatment that can continue for an extended period.

Ineffective Communication in Marriage

As one family researcher wrote, "Misunderstanding is the natural state of the world—and of intimate relationships like marriage" (Cordova, 2011). Consider the following as an example. You are angry with your partner, and you want to constructively express your frustration. You struggle to find the exact words to convey your thoughts, and risk having your meaning or intent get lost in translation. To add to the dilemma, you must precisely hear what your partner is saying, while you accurately translate his or her unique way of saying things into your unique way of hearing things. However, throughout your partner's life, the word *angry* has been used to refer to relatively civil interactions about frustrations and disappointment.

You, on the other hand, were reared in a family where the word *angry* reflected emotionally chaotic, hateful, and traumatizing events. As a result, the simple statement "I'm angry with you" perhaps will imply a different set of meanings to you than to your loved one (Cordova, 2011).

No one can change his or her communication style overnight. Therefore, being aware of how we convey or conceal our emotions or intent has value. Family therapist Virginia Satir (1972) offered four miscommunication styles that surface when individuals are hurt or feel threatened by their partners. We discuss these ineffective styles of communication in relationships below.

The Placater

The Placater is fearful that others will become angry, leave, and not return. Therefore, the Placater communicates in an ingratiating way, attempting to please, apologize, or never show disagreement. Typical Placater speech may be, "Whatever you do is fine with me."

The Blamer

Outwardly, Blamers are tense, verbal, always correct, fault-finding, and unwilling to own up to their own responsibility in matters. Blamers act superior and exert their power. Inside, a Blamer feels that he or she is not given respect, but hides such feelings from everyone, the Blamer included. Typical Blamer speech may be, "The problem is that I was fired because you did not care enough to wake me up and get me out of the house on time."

The Computer

Like their namesake, Computers "process" and are reason-based in that processing. Using a limited set of facial expressions and body positions, Computers create the impression of rigidity and emotional detachment, often to mask feelings of vulnerability. A Computer may dryly blurt, "No logical person would consider this situation to be a problem," leaving the receiver to feel unsupported, not understood, or judged.

The Distracter

If someone wants to discuss a serious matter with a Distracter, expect the Distracter to change the subject. On the inside, the Distracter feels out of place or afraid of the situation. Additionally, the underlying feeling of the Distracter is panic, often resulting in little constructiveness or efforts toward a resolution. A Distracter may express, "Problem? What problem? Let's watch a movie!"

Miscommunication and Gendered Communication Styles

A quick look at such online retailers as Amazon, Barnes & Noble, and other booksellers shows an ongoing interest in closing the gap between men and women's communication styles. Publications tackle what men say, what women hear, and vice versa, as well as how neither, at times, seems to "get it." There also are scholarly treatises shedding light on the frames of reference that regulate how men and women express and exchange ideas, in addition to manuals on creating effective gender communication in the workplace. What is important to think about is that, behind the media blitz, a message is being exposed: the diversity of men's and women's interpretations of messages is a sort of gendered "battlefield." In what follows, we take a closer look at the "front line" that is language.

Linguist Deborah Tannen (1990) argues that miscommunication between men and women often occurs because they fail to realize that they are engaging in cross-cultural communication. Tannen views each sex as socialized to speak and interpret conversations using two cultural dialects: one is masculine; the other is feminine. She calls this gender-typed dialect **genderlects**.

"Report Talk" versus "Rapport Talk" Tannen describes men's conversational language as *report talk*, which seeks to establish status or maintain ascendancy. In "doing" report talk, men may engage in detailed, straightforward, and instrumental speech, as if to win or control the conversation. Tannen describes women's conversational style as *rapport talk*, which seeks to establish a balanced connection and maintain relationships with others. In "doing" rapport talk, women may freely express their emotions, share personal feelings, relate stories, and listen empathically. Women, Tannen also argues, are more conversational in private situations than in the public sphere. Men, in contrast, are likely to engage in the public sphere with talk that makes them look strong, competitive, and independent.

Gendered Patterns of Listening Tannen's research also shows how men's and women's particularities are filtered through the way they listen. Often, the result is confusion. According to Tannen, women are more accommodating in their listening and are likely to demonstrate engagement by asking questions and prompting the speaker to talk more. Men may interrupt the speaker for longer periods than women, which may be difficult for some women to decode. For example, a man's interjecting an experience about the subject matter, although intended to demonstrate consideration, could be misinterpreted as a sign of self-interest. When men are interrupted, they may decode the break in conversational flow as a threat to their status.

Tannen found that men's communicative motive to establish and maintain status in the conversation helps them engage more comfortably in topics that often draw tension. Women, although not necessarily uninterested in conflictual topics, may refrain if they feel that social connection or relationship building is threatened. However, an additional risk may exist for women having a conflict style of talk: employing male patterns of communications when engaging in controversial topics may lead to stigma.

Pegging men's communication as task-focused or aggressive and women's as nurturing or sentimental may not settle well with some, as men and women can do both. Tannen's dichotomy also contradicts the convention that men and women are separate, distinct individuals with behaviors as varied as the backgrounds, experiences, and the contexts informing their actions. She explains that gender greatly distinguishes the way that we communicate and that ignoring the differences in men's and women's communicative patterns is more costly than naming them.

Communication Barriers in Multicultural Families

Bilingual or multilingual families, particularly those with children, often must negotiate how or whether alternate languages should be acquired. In some families, the use of two or more languages is a necessity, as parents may not be fluent in the majority or dominant language. As a result, a child may learn one language at school but practice another at home. On the other hand, bilingualism or multilingualism can be a choice. Parents may wish for their children to acquire more than one language, even if they do not speak a second language themselves.

Advocates for bilingualism in the home argue that bilingual children are better equipped to focus on relevant information, to ignore distractions, and to more creatively plan and effectively solve complex problems when compared to monolingual children. Those who support a monolingual approach to childrearing question if communicative problems may arise in bilingual families with children. They argue, for example, that as bilingual children acquire the dominant language, the language used in the home may become weakened. Bilingual parents also must negotiate how much non-majority language each can use with their child, in addition to how much transitioning between languages should take place (Grosjean, 2015).

Upon completion of this section, students should be able to:
LO7: Explain French and Raven's bases of power in marriage and family.
LO8: Compare and contrast the resource, cultural resource, and principle of least interest perspectives on the sources of power in marital and intimate partner relationships.
LO9: Compare and contrast traditional, egalitarian, and neo-traditional norms of marital power.

Power in Marriage and Families

Once two people come together as a couple, many more decisions follow. For one, who makes the decisions and are there judgments that are "out of bounds" or "best suited" for either partner? There also is the issue of finances and money. Who controls it? Who spends it—and on what? Also, who will provide the skills and labor required for the family's income? Who will provide the skills and labor required to maintain the family household? The ability to exercise one's will or influence over how partners organize their lives as a couple—**power**—has been defined and measured by researchers in several ways. We provide some examples below.

Decision-making as a Function of Power

A number of researchers who investigate marital decision-making underscore the means and extent to which power is positioned in the relationship and family. Their studies try to explain how judgments in marriage and families are negotiated, how household labor and childrearing are apportioned, how finances are allocated, and who has the power to affect, sway, oppose, or determine these and other areas of family preservation (Blood & Wolfe, 1960; Fry & Cohn, 2010; Lee & Petersen, 1983; Schwartz, 1994).

Power as Objective and Subjective

Marital power has been assessed with regard to how influence is perceived and experienced both objectively and subjectively. For example, power may be perceived within the sphere of practical, objectively measurable outcomes and tasks. The number of hours per week a partner contributes to housework or childrearing and the annual earnings a partner contributes to the household are examples. Subjective dimensions of power, such as equity or fairness, address whether a partner thinks that the rewards of an arrangement are equal or proportional to the other's contribution.

Bases of Power in Marriage and Family

Another way to think about how power is positioned in marriage and family is to consider how it is sourced, exacted, and received in a relationship. Defining power as the influence to provoke change in behavior, opinions, attitudes, goals, needs, and values, social psychologists J. R. P. French and Bertran Raven (1959) yielded six bases of power often exerted in social contexts: legitimate, informational, referent, coercive, reward, and expert. French and Raven drew their observances from the standpoint of how individuals establish direction over others' thoughts and actions and how others react to that influence. Below, we explore their typology to explain the power bases within the family system.

Legitimate power is based on social norms requiring that we obey others who are in a superior position. Examples of legitimate power in families involve the rights of parents to influence their children or the rights of adult children to regulate the critical care and housing of their ailing elderly parents.

Informational power is based on one individual having a specific knowledge that is unavailable or unknown to another. For example, if one spouse can assemble pertinent information about the benefits of purchasing a new home in a new location, he or she may be able to convince the other spouse to move to another neighborhood.

Referent power reflects a spouse or family member's power that comes from another spouse or family member's respect, loyalty, or admiration for that person. Those with referent power are perceived by others as role models or as having opinions that carry considerable influence. A parent's referent power may coalesce around actions by family members that demonstrate deference for that parent. For example, a son or daughter may enroll in a grueling weight-loss program to meet his or her parents' body standards or to foster the parents' acceptance.

Coercive power is based on the ability to impose physical, sexual, or psychological harm on others in the family unit. A spouse who incites fear or helplessness in order to prevent his or her spouse from gaining access to their children wields coercive power.

Reward power is based on one family member's ability to influence another by awarding a tangible or intangible gift, favor, or a desired outcome in exchange for compliance. Parents who award their children with money or the latest cell phone as an incentive for earning good grades are exercising reward power.

Expert power stems from a spouse's relevant knowledge or ability to manage a certain issue. For example, one spouse—a tax attorney with an undergraduate degree in finance—may wield expert power when it comes to preparing the family's annual tax returns.

Perspectives on the Sources of Power in Marital and Intimate Partner Relationships

If bases of power exist in marriages and family systems, what factors determine who directs and who is influenced by that power? To enlighten, we offer some perspectives on the balance of power within the family structure, particularly marital relationships. We primarily focus on resource theory, cultural resource theory, and the principle of least interest.

Resource Theory

In their resource hypothesis, sociologists Robert Blood and Donald Wolfe (1960) posit that the spouse having the greater resources is apportioned the most power in the marriage. To test the theory, Blood and Wolfe surveyed over 900 wives to learn which spouse had the "final say" in general family decisions. Economic earnings, participation in the community, and educational attainment were used as indicators of the relative resources said to predict the distribution of power in marriage. Blood and Wolfe concluded that men tended to have more decision-making power, a finding supportive of the researchers' belief that the balance of marital power is determined by the "comparative resourcefulness of the two partners and the life circumstances within which they live" (p. 29). Their study also found that most families had a more egalitarian decision-making arrangement than was then presumed. Of the sample, 72 percent had a "relatively egalitarian" power structure in which husbands and wives held an *almost* equal decision-making influence. Twenty-five percent of the respondents were in marriages in which the husband made the decisions; 3 percent of the families represented were headed by wives. Findings also suggest that wives' relative power increased when they no longer had young children and after they acquired their own wage-earning resources.

Analyses of **resource theory** have been supportive and critical of the perspective. Findings from sociologists Phillip Blumstein and Pepper Schwartz's (1983) study of married, lesbian, gay, and heterosexual cohabitants supported the assumption that a partner's relative power may result from that partner's having greater resources. In particular, men in the sample tended to have more power over their partners when they earn more money than their partners. The effect was similar for married, cohabiting, gay male couples, all of whom were identified as having a committed relationship. On the other hand, in Veronica Tichenor's (1999) study, wives who outearned their husbands did not have greater marital power than their spouses. In fact, both partners ignored or minimized their differences in status and income, positioned the husband as powerful, or repurposed the family-provider role to accommodate the activities of the husband.

Critics also charge that resource theory poorly recognizes the influential sway of non-material resources, including unpaid family work, intelligence, access to certain social networks, attractiveness, likeability, love, and comfort. Women's influence over the socialization of children, generational kin keeping, and the preservation of family culture are similarly offered little recognition. Further, wives' contributions to alternatives that go into their husbands' "final say" may be as powerful as they are unnoticed. David Vogel and colleagues' (2007) survey of 72 married couples showed that, on average, wives appeared to have more power during problem-solving discussions than their husbands. The researchers add that the husbands' agreement or acquiescence was not influenced by their wives' initiation of the subject or their talking more about the subject. Rather, the wives sampled conveyed more powerful messages and behaviors than their husbands.

Resources in a Cultural Context

Other research on power in marriage suggests that dominant cultural norms and behaviors may override the influence of certain resources. Hyman Rodman (1972) analyzed studies across eleven countries to examine the relationship between husbands' resources and power. Rodman's findings were conflicting: in some countries, a husband's greater socioeconomic resources did not increase their power relative to their wives' power; in other countries, wives' employment had no impact on their power in marriage. Rodman explained that cultural norms conferring authority can be so pervasive that they supersede any effect that resources might have on power. His conclusion is characterized as the **cultural resource theory**. Rodman also concluded that the relative power attributed to resources might be greater in cultures transitioning from patriarchal to egalitarian norms.

Principle of Least Interest

Another way to think about sources of power in a relationship is to consider who in the relationship is the most and the least invested. In his perspective, the **principle of least interest**, sociologist Willard Waller (1937) hypothesized that one way to gain power in a relationship is for one partner to withhold love or affection, thus creating a power disadvantage for the partner who is most positive about the relationship. For example, if you only marginally care for your partner but your partner is passionate about you, the power pendulum will likely swing toward you. This being said, when the less interested partner makes requests for things like money, extra time, or affection, the more interested partner may be poised to comply due to fear of risking the relationship.

Susan Sprecher and Diane Felmlee's (1997) study of 101 heterosexual couples supported Waller's principle of least interest. The sociologists found less emotionally involved partners were associated with greater power. Men were more likely to perceive themselves as the less emotionally invested partner. Caldwell and Peplau (1984) found that lesbians who were more involved and committed to their relationships tended to have less power than their partners.

Diversity in Marital Power Norms

In this section, we examine three primary patterns of influence and authority in families—traditional, egalitarian, and neo-traditional—and the social norms that comprise them. We start with traditional norms of power in marriage.

Traditional Norms of Marital Power

The traditional or nuclear family is a group comprising of a husband/father, wife/mother, and their biological or adopted children, all of whom reside in a single household. Historically, traditional norms of power in marriage involve a hierarchal family arrangement. For example, women work to rear children and to secure the family's home life. In historical nuclear family parlance, a woman of the traditional feminine ideal is expected to be cooperative, nurturing, available for emotional support, not too oppositional, not too contentious, not too threatening, and a good listener. As the "keeper" of family culture, she also is socialized to preserve the relationships (extended families, social networks, voluntary organizations, etc.) that connect family to the greater society. Men, because of their physical strength and labor, function to head the family and to procure the needs of dependent others. In view of these expectations, a man who conforms to the traditional model of power historically has been characterized as a good provider. That is, he seeks paid labor outside the home and an occupational success that will ensure a comfortable wage and the material resources required for family preservation. Authority in the traditional family model has stressed that decision-making is vested partly or autonomously in the father and/or husband.

Egalitarian Norms of Marital Power

Cultural and economic shifts, including women's entry into the workplace, their outpacing of men in education and earnings, couples' mutual economical dependency, and the emergence of dual-earner couples, have resulted in an attitudinal shift toward egalitarian marital power norms (Fry & Cohn, 2010). As opposed to the traditional patriarchal marriage model, couples who adhere to egalitarian norms of marital power are generally spouses who both work and divide power in some fair or equal way between them.

Sociologist Pepper Schwartz (1994) uncovered a group of husbands and wives who work to successfully recognize each as having joint status in marriage. Schwartz characterizes these egalitarian marriages as *peer marriages*. Peer marriages are not the cultural norm; however, they suggest a clear effort by some couples to share decision-making and establish equity in marriage. Couples in peer marriages have a tendency to recast gendered marital norms in several ways, including the following:

- *Equity in parenting and housework:* In peer marriages, division of labor is agreeably redistributed around practicality and not traditional male/female roles or expectations.
- *Equity in managing family finances:* Couples in peer marriages feel they have equal control of the family economy and reasonably equal access to discretionary funds. Automatic "veto power" or the question of what one partner will or will not allow another to do with family funds would be opposed.
- *Equity in appreciation for each spouse's labor:* Individuals in peer marriages are given equal consideration in the couple's life plans. Regardless of whether both work outside the home, no one systematically sacrifices the other's work. The partner who earns the least is not the partner who is exclusively given the most childcare or household duties.

Currently, there is no fixed "blueprint" for couples who adhere or want to adhere to peer marriage principles. Therefore, there may be little or no outside validation for peer couples' equal domestic arrangements or their altering career goals to meet family aspirations. Nonetheless, Schwartz found that peer couples produce valuable friendships and deeply collaborative unions, as well as a meaningful psychological connection.

Shifts in Egalitarian Attitudes

Statistics have shown a shift toward egalitarian attitudes in marital roles and power arrangements, especially between 1977 and the mid-1990s (Cotter *et al.*, 2014). Family scientists found the rate and extent of change remarkable. In 1977, 66 percent of Americans agreed, "It is much better for everyone involved if the man is the achiever outside the home and the woman takes care of the home and family." By 1994, just over one-third of respondents thought the husband-achiever model was the ideal family arrangement; 63 percent disagreed. Also in 1977, 52 percent of Americans believed that a working mother could not establish as warm a parent–child relationship as would a full-time homemaker. Only 31 percent of Americans retained this belief in 1994. Around the late 1990s, support for **egalitarian marital norms** saw a slower increase and, at times, a decrease. The portion of Americans who disagreed that a woman's place was in the home had slipped from 63 percent in 1994 to 58 percent in 2000. The portion of Americans who agreed that a working mother could have an equally warm relationship with her child as a full-time homemaker fell from 69 to 60 percent (Cotter *et al.*, 2014).

In their report, "Back on Track?: The Stall and Rebound in Support for Women's New Roles in Work and Politics, 1977–2012," sociologists David A. Cotter, Joan M. Hermsen, and Reeve Vanneman (2014) state that since 2006, "there has once more been upward progress in the values associated with approval of new gender roles and relationships." Between 1994 and 2000, the rate of people who disagreed that it is better for men to earn the money and for women to tend the home slipped from 62 to 58 percent but by 2012 rose to an all-time high of 68 percent. Less than one-third of Americans now say that a male breadwinner family is the ideal arrangement. Further, only about half of Americans in 2000 disagreed that preschool children suffer if their mother works outside the home. By 2012, 65 percent disagreed. The rate of respondents in agreement that a working mother can establish as warm a relationship with her children was 60 percent by 2000. By 2012, agreement increased to 72 percent.

Neo-traditional Norms of Marital Power

Attitudinal shifts historically have gone from a movement toward egalitarian family norms to a stall in egalitarian sensitivities to a regrowth of support for joint-minded couples. Currently, egalitarian tendencies regarding power in marriage and family seem to be undergoing a sort of retooling. The power dynamics of current family life reflect a mixture of conventional expectations and present-day compromises: a **neo-traditional norm of marital power**. Combining elements of the old and the new, neo-traditional families show egalitarian sensitivities insofar that both partners participate in childcare, housework, and formal systems of labor. At the same time, traditional commitments to the institution of marriage remain intact and daily tasks still are largely divided along gendered lines.

Integrating egalitarian principles into everyday marital conduct may be the goal for many couples. However, gender-specific roles and expectations continue in families, making a joint-couple influence or ideal difficult for some to realize. Sociologist Suzanne Bianchi (in Mundy, 2013) explains that, in addition to their paid labor, full-time working wives with children conduct 32 hours a week of housework, childcare, shopping, and other family-related services. Moreover, husbands who work full time put in 21 hours of unpaid labor and have more free time: 31 hours compared with 25 for women. The maternal to paternal ratio of primary childcare among married parents fell from 5.6 in 1975 to 2.2 in the 1990s and has stayed there since: about 69 percent of childcare among married couples is carried out by mothers (Wilcox & Dew, 2013).

Diversity Overview: Power in Same-sex Committed Relationships

Men and women organize traditional heterosexual marriage around two basic principles: a division of labor based on gender and a norm of greater male power and decision-making authority. But how do same-sex couples, who lack biological sex as a basis for assigning tasks and status, organize their lives together? In their review of empirical studies of U.S. same-sex couples, psychologists Letitia Anne Peplau and Adam Fingerhut (2007) explore the division of household labor, power, satisfaction, and other relational aspects. Their work offers a view into how influence is apportioned and experienced in committed gay and lesbian relationships.

Peplau and Fingerhut's investigation found striking contrasts in the division of housework (cleaning, cooking, and shopping) between cohabiting and married heterosexual couples and same-sex couples. Wives of heterosexual couples typically did most of the housework. Conversely, lesbian couples generally shared household tasks more equally. Gay male partners were more likely to have each partner specialize in certain tasks. Stopping short of calling gays' and lesbians' division of household labor "perfectly equal" (Kurdek in Peplau & Fingerhut, 2007), one psychologist explains that same-sex couples tend to negotiate a balance between a fair distribution of labor and accommodating partners' skills, interests, and work schedules.

Some scholarship comparing equity between committed homosexual partners has uncovered competing outcomes. Among the gay couples in Mally Shechory and Riva Ziv's (2007) sample, one partner considered the relationship less equitable than did his partner. Christopher Carrington's (1999) in-depth interviews and observations of gay and lesbian partners resulted in "contradictory accounts of many aspects of domesticity" (p. 14). For example, Carrington's gay male respondents denied having more domestic duties than their partners; however, their claims of egalitarianism were misleading in some cases. It could be that an equal sharing of household activities may reflect couples' ideals but not observable differences with regard to what each partner actually contributes.

Research on inequities in same-sex relationships suggests that biological sex is not always the basis of power in relationships. Earnings, for example, may carry considerable influence. Phillip Blumstein and Pepper Schwartz (1983) found that men in committed opposite-sex relationships tend to have more power over their partners when they earn more money than their partners. This effect was similar for gay male couples, who tended to equate power with monetary resources. Conversely, for lesbian couples, the contested area of power was suggested to lie elsewhere: interacting more with the family's children could be interpreted as a "power move" by some lesbians in committed relationships.

Upon completion of the next section, students should be able to:

LO10: Recognize the sources of conflict in marriage.

LO11: Explain common couple conflicts.

LO12: Explain positive and negative aspects of conflict among couples.

LO13: Apply Virginia Satir's four-part model of conflict.

LO14: Recognize factors associated with maintaining resilient relationships.

Conflict and Relationship Resiliency

You may have heard the saying, "I *love* my husband (wife or partner), but sometimes I do not *like* my husband (wife or partner)." As it turns out, this happens to many couples. It is the feeling that you and your partner cannot get along, or that others cannot possibly argue as much as you and your partner. Despite appearances, there is a good chance that couples in our social circle argue as often and as intensely. In what follows, we discuss conflict in relationships, a reality that is more ongoing than we would like but not always as devastating as many imagine.

Sources of Conflict in Couples

The social contexts influencing why couples experience conflict can be diverse. The absence or presence of those who either stimulate conflict or diffuse provocation may lead to confusion between couples. For example, an in-law's attentiveness or advice may offer consolation to one partner; however, the other may find the relative's presence as nothing less than interference. There also are personal sources of conflict to consider. For example, individual stressors may find their way into couples' daily lives. Problems may be exacerbated by a partner's ineffective communication skills as well as degree of agreeability (likeability or good-naturedness) or disagreeability (prone to mood swings or anxiety). Further, each partner's goals in the discussion (change, harmony, or defense of character), in addition to differing opinions on who determines control or decision-making, may establish whether a conflict will fester or be resolved.

A good deal of couple conflict is interpersonal: conflict created and sustained between the two relationship partners. Because interpersonal conflict involves two individuals who act interdependently, the disagreements that underlie interpersonal conflict tend to be complex. For example, one partner may perceive interference from the other in achieving his or her personal goals or both partners' goals may be viewed as too incompatible or too frictional to be ignored.

Common Couple Conflicts

Even the happiest couples "clash" or engage in conflict. But over what subjects do couples argue? In a year-long study of 75 gay, 51 lesbian, and 108 heterosexual couples, psychologist Lawrence Kurdek (1994) found that couple conflict clustered around six areas: (1) power (being overly critical), (2) social issues (political and cultural matters), (3) personal flaws (drinking or smoking), (4) distrust (doubt, lying), (5) intimacy (sexual matters), and (6) personal distance (work and school commitments). Arguments over power and intimacy and the frequency of these arguments negatively affected satisfaction in the existing relationship to a strong degree. A decrease in each partner's relationship satisfaction over a one-year period was linked to frequent arguments regarding power. Sociological research found that conflict over jealousy, consumption of alcohol, spending money, a lack of communication, moodiness, sexual infidelity, and anger tend to increase the odds of divorce.

Obviously, intimate partners, including married couples, argue over topics that are specific to their relationship and individual circumstances. However, some areas of contention are more anticipated in relationships than others. Below, we expose four additional but common areas of couple conflict: money, housework, sexual intimacy, and children.

Conflict over Money

Family members' desire to spend money can very easily exceed the funds available to the family. But regardless of funds available, blame or unrest are almost certain if a family member feels deprived of the chance to spend family funds or finds inequities between his or her personal access to money and others' ability to spend it. Disagreements over money also may follow conflicts about work, which may generate tensions that are unlikely to go away on their own. Without clear, open dialog about money and controlled spending, couples with different spending practices are likely to experience recurring conflict. Some couples may need to confront the reality that they may never have as much money as they want.

Conflict over Housework

"Routine" housework (meal preparation, housecleaning, grocery shopping, washing dishes, doing laundry) consumes more time than "occasional" housework (household repairs, yard care). Yet the average married woman conducts about three times as much "routine" housework per week as the average married man, who conducts about twice as much "occasional" housework per week as the average married woman (Coltrane, 2000). Couples should keep in mind that resentment may occur, as one partner may be "keeping score" if the load is perceived to be too unequal or unfairly disbursed. In addition, housework disagreements may be interpreted as a lack of appreciation for what each partner does to contribute to the stability of the family and household. If not confronted, a sense of fairness or shared responsibility may dissolve over time.

Conflict over Sexual Intimacy

Partners with or without children can witness to being "too tired," "not in the mood," "not interested right now," in addition to other codes suggestive of dissimilar sexual desires. There are many reasons for dissatisfaction or incompatibility with regard to sexual intimacy in committed relationships. Over time, for example, men and women's libidos may fluctuate due to stress, diet, age, health, quality of sleep, and the level of connection partners feel toward each other. Sexual satisfaction is also affected by miscommunication surrounding the initiation of sexual intercourse, response to initiation, and a poor coordination of sexual episodes. Even so, research on sex and the breakup of relationships suggests that sexual incompatibility and sexual issues are often rated as only a moderately important factor in relationship dissolution (Sprecher & Cate, 2004). At the same time, its impact on relationship quality can be substantive. Sexual satisfaction is associated directly or indirectly with overall relationship stability and, some argue, it is plausible that relationship satisfaction leads to sexual satisfaction. Open communication that is offered in a cooperative, compassionate way may allow both partners to share feelings of embarrassment or shyness and, if needed, to face the presumption that their partner should already know their needs.

Conflict over Children

In a 2014 study by the University of Wisconsin–Madison Couples Lab, 100 married couples kept a diary of their "differences of opinion" for fifteen days. Among the 748 disagreements recorded, the most frequent conflict reported involved children: what the child is doing and why, disagreements about what to do in response to what the child has done, who will take care of the child, and how to discipline the child (Bradbury, 2014).

Parents with incompatible styles of childrearing may experience conflict. For example, there are partners who value a high degree of structure between parents and children and whose style of parenting demands obedience without question or negotiation. Their partner's parenting style may value flexibility, parent-to-child sensitivity, and a willingness to accommodate a child's changing needs or circumstances. The outcomes may be problematic: conflict can degenerate into a win/lose competition between parents, only one parent may participate in decision-making, or a parent may withdraw and abandon responsibility over the issue entirely. Family scientists generally find resolution in parents working toward an agreement on rules and standards that both are willing to enforce with their children. Otherwise, they add, a child may play one parent against the other (Pickhardt, 2014).

The Negative Side of Conflict in Relationships

Couples in distressed relationships are likely enmeshed in cycles of negative conflict behaviors that can be a challenge to break. That is, negative conflict behaviors often are repaid with more negative conflict behaviors, which are inflammatory, even hostile, responses to disagreement. Emotionally, these behaviors can range from tension and whining to coldness and impatience to sarcasm, blame, and anger. Couples also may store old grievances and "dump" them on their partners during moments of conflict (*gunnysacking*). Other conflict styles that generate non-constructive responses or impose negativity on relationships are provided below.

Serial Arguing

Serial arguing is a conflict pattern in which partners argue about the same issue two or more times or focus on a particular problem that has no discernible beginning or end. The pattern begins with an antecedent condition, namely a disagreement on an issue. Two primary processes may follow as the conflict develops: one person decides to attack the other and an actual argument ensues. Two secondary processes also surface: a "heating up" period in which the partners ask themselves if the issue or the relationship is important enough to attempt resolution, and a "simmering down" period that precedes the partners' further attempt to resolve the problem. The secondary processes of serial arguing can happen any number of times, ranging from a few hours to throughout the course of the relationship (Bevan *et al.*, 2004).

Constructively managed disagreements should be the goal of partners who argue serially. If this is properly achieved, past resentments and disappointments that resurface to obscure the original problem are discouraged. Also important is that conflicts are not drawn out longer than they should be or escalated to the point of personal attacks.

Negative Attributions

Viewing one's partner as highly flawed or untrustworthy may lead to assuming the worst or redirecting the blame when a partner behaves badly. For example, dissatisfied spouses in unhappy relationships tend to make negative, distress-maintaining attributions for their partners' undesirable actions, such as perceiving bad behavior as a result of his or her personality or character or downplaying a partner's positive behaviors. In an analysis of 130 couples who were examined at two intervals separated by twelve months, partners who made negative attributions for their spouse's unwanted behaviors at the beginning of one year became less satisfied with their marriages over the course of that year. The researchers emphasized that their findings were not due to depression, self-esteem, initial level of marital satisfaction, or chronic individual or marital disorder (Fincham & Bradbury, 1993)

What occurs, then, with negative attributions is a recurring, socially regressive cycle. The cycle promotes overwhelming feelings of blame or displaced responsibility and decreases satisfaction with the relationship. To make matters worse, couples experiencing less relational satisfaction tend to perceive their partner's problematic behaviors as intentional and selfishly motivated. Indeed, both outcomes have the potential to not only diminish relationship fulfillment, but also increase distance and isolation over time.

Demand–Withdrawal Patterns

In *demand–withdrawal conflict behavior*, the one partner (the "demander") expresses anger and complains while the other (the "withdrawer") retreats from the argument. Sensing dismissal, the demanding partner is often triggered to pursue the "withdrawer" at an increasingly insistent level. The "withdrawer," then, identifies his or her recoil through passivity or defensiveness. In a mutual withdrawal response, both partners either actively (physically walk away) or passively ("tune out" or superficially discuss the issue) remove themselves from the matter. Altogether, demand–withdrawal conflict behavior is a regressive, non-constructive cycle of conflict communication. Partners may prevent a heated disagreement; however, the problem remains unresolved.

It is not unreasonable that conflict withdrawal may be motivated by our own insecurities: the need for order or resolution can be desperate, yet some conflict demands can feel insurmountable. The desire to withdraw can also be motivated by a fear of emotional or physical harm. This is especially the case in conflicts where the discourse becomes overtly intense. But what happens when one partner responds to conflict in a constructive manner and the other responds by passive inaction? A sixteen-year study found that divorce rates are higher in marriages in which one partner withdraws from conflict and the other calmly discusses the situation, listens to the partner's point of view, or tries hard to find out what their partner is feeling. Lower divorce rates were documented when both spouses used constructive strategies to manage conflict. Researchers explain that spouses who manage conflict constructively may view their partner's withdrawal as a lack of relationship investment rather than an attempt to defuse tension (Birditt *et al.*, 2010).

The Positive Side of Conflict in Relationships

Despite the chance that partners will have some differences of opinion, more than a few people believe that all conflict should be avoided (more on this problematic response later). For those who do, here is, perhaps, a novel concept: there may be some benefit in giving conflict a chance. Indeed, it is not always easy to see arguments as opportunities; therefore, we offer observations that may inspire a more optimistic view of relationship conflict.

Well-managed Conflicts

Confrontation involves an open conflict of opposing ideas. At the same time, confrontation invites an opportunity to bring together ideas for comparison or a synthesis of those ideas.

In addition, when in conflict, we may focus on what we have already decided or what we feel is right. However, could it be that the underlying concern is actually the need for security, belonging, recognition, fairness, or emotional well-being? Is there something else involved that has not yet surfaced? Thinking of conflicts as opportunities to learn what disagreements *really* are allows partners to recognize basic needs, to formulate decisions on what should be done, and to stir motivation to satisfy each other.

Managing Developing Disagreements May Defuse Later Conflicts

What we fear about conflict—namely, its potential to damage relationships—is *precisely* what we get from chronically avoiding conflict. For example, say that your partner unpredictably shirks on your mutual agreement to pick up the children from school two days a week. If you avoid conflict and do not express your dislike, your partner may feel that his or her lapse of responsibility is inconsequential to you. The take-away is that not voicing your opinion may suppress conflict today; however, larger conflicts such as lost trust, hurt feelings, and other problems in the relationships may evolve. For example, partners who feel that they cannot risk conflict by expressing their true feelings may find more passive aggressive means to meet their needs.

Gendered Differences in Couple Conflict Responses

The demand–withdrawal pattern conflict response has been the focus of studies on couples' approaches to managing couple disagreement. The pattern appears to transpire around gendered lines: the "demander," usually the wife, exerts emotional requests, desires of change, complaints, and criticism; the "withdrawer," usually the husband, retreats from conflict through defensiveness and withdrawal from the interaction (Christensen & Heavey, 1990).

Long-held stereotypes frame gender differences in demand–withdrawal patterns as essentialist or based on men's and women's inherent personalities. However, Holley *et al.* (2010) suggest an alternate explanation. The researchers uncovered no support for the essentialist argument in their study. Rather, they argue, the persistence of demand–withdraw gender stereotypes may have developed from wives' demand or desire for more change and husbands' insistence on less change, an incentive that may lead to withdrawal and keeping with the status quo. To meet demands, women may perform low-power roles that, over time, have been labeled as "women's behavior"; husbands' withdrawal and avoidance reactions have been labeled as "men's behavior."

In Christensen and Heavey's (1990) study, the wives-as-demanders and husbands-as-withdrawers binary was reconfirmed. However, in a supplemental analysis of specific demand–withdraw behaviors, both sexes were more likely to demonstrate demand when discussing a change that they wanted and more likely to use withdrawal when discussing a change that their partner wanted. So, why the persistent stereotype of women as demanding and men as prone to withdrawal when in conflict? Some question whether male–female differences are overestimated in not only popular culture but also scholarly literature. For example, research outcomes may meet the criteria for statistical significance yet the differences uncovered may not be sizable.

Managing Couple Conflict

Individual peculiarities, dashed expectations, incompatible moods, lack of cooperation, feelings of unfairness, and confrontational "back-and-forths" can try even the most committed intimate partnership. When negotiating conflict resolution, it is important for all involved to stand back, take a breath, and confront each part of the whole problem. In the following section, we examine how couples might maintain a more resilient relational bond, despite the inevitability of conflict. First, we explore psychologist Virginia Satir's four-part model of conflict to better understand the segmented yet interdependent nature of conflict.

Satir's Four-part Model of Conflict

Satir perceived that effective conflict management requires that all parts of conflict receive equal consideration. She identified the four interdependent segments of conflict as *you* (one conflict participant), *me* (the other conflict participant), *context* (the emotional background surrounding the conflict), and *the subject* (the topic of the conflict). According to Satir, a conflict may not be fully resolved if all four segments are not properly recognized and addressed. However, partners generally dislike conflict, and they may try to resolve disagreements too soon by ignoring one or more parts. The following scenario illustrates Satir's model and her views on what could happen if any part of a conflict is disregarded: placating, pouncing, computing, and distracting.

Placating Say that Andy wants to travel to a California beach for the family's vacation. Alice, who wants to do some rock climbing in Kentucky, reluctantly defers to Andy's destination choice. Alice is *placating* in this example, and its effects are directed toward the "me" in the conflict. Although Alice has avoided conflict, her passivity effectively disqualifies her own needs in the matter.

Pouncing Another ineffective response to conflict occurs when one individual disqualifies the other by not acknowledging the other's needs. Satir identifies this problem as *pouncing*, and its aftermath is directed toward the "you" in the conflict. If Andy becomes angry and yells at Alice—or if he quietly but condescendingly explains that rock climbing is a stupid idea—Andy is seizing or pouncing on Alice's views to advance his own.

Computing When a conflict's "context" or emotional aspects are ignored, disqualified, or overridden by the rational aspects of the disagreement, the outcome is what Satir calls *computing*. Say that Alice is visually upset about Andy's choice to go to the beach for the family's vacation. Angrily, she reminds Andy that she halfheartedly went to his vacation spot a year earlier and that, in fairness, Andy should abandon his beachside plans. If Andy responds by saying that Alice should not be upset and that her explanation is groundless, Andy would be neglecting the emotions involved in the conflict, thus hindering resolution.

Distracting It would be *distracting*, on the other hand, if Andy changes the topic or "subject" while Alice tries to explain that weather permits only a few months for rock climbing and that she has been awaiting the opportunity for some time. Actions like crying, laughing, or leaving the premises also are forms of distraction, as any can be used to dismiss the subject of conflict by leading partners away from that subject.

Maintaining Strong, Resilient Relationships

Understanding how to maintain constructive communication during conflict is a more realistic expectation than never having to experience conflict. In this final section, we explore how partners facilitate resilient relational bonds despite inevitable conflict.

Be Realistic

Many intimate partners envision that their relationship will retain the level of agreeability experienced early on: each will love the other always, remain forever mutually attracted, and will retain the closeness both currently enjoy. Indeed, cultivating positive expectations is an admirable place to begin as couples organize their future. However, the reality is that not all relationships ride at a continuous high. If con-

Source: Glaze Image/Shutterstock

fronted with unavoidable discrepancies, hanging on to unattainable expectations or dream-like fantasies about a relationship or a partner may lead to frustration, disillusionment, anger, and disappointment. It may be more realistic to cultivate possibilities that could manifest and strengthen areas of the relationship that prove challenging.

Take Responsibility and Repair Problems

Two people may express two separate views on "how things should be." In the process, many partners never admit that they might be mistaken or at fault. Yet such attitudes only escalate problems, as couples further engage in conflict to prove who is right or who is wrong.

Succinctly, a willingness to make up after an argument is important to every long-term relationship. It is important for partners to take responsibility for their part in interpersonal conflicts. Partners should also try to repair problems *as they arise,* acting with good intentions, empathy, active listening, and a clear intent to avoid hurt feelings, name-calling, and defensiveness.

Know What Affects Personal Behaviors

Our needs and fears may influence how we behave in our relationships or react toward our partners. For example, a person may behave too passively because of a desire to be loved. Another may behave in a controlling manner due to fear of losing his or her independence. When couples are aware of what drives their individual reactions, they are in a more informed position to approach conflict resolution constructively.

Keep in Mind that All Relationships Have Problems

Adoration for a partner may go through ebbs and flows, which are dependent on how partners treat each other. Therefore, partners should mutually strive to interact in ways that stir or strengthen affection and respect. In doing so, remember that most modern relationships no longer rely on roles traditionally dictated by culture and that it is acceptable for couples to negotiate roles that work well in their relationships.

Global Overview: Attitudes on Gender, Marriage, and Work

The Pew Research Center's Global Attitudes Project (2010) conducted a 22-nation survey to examine men and women's views on gender equality. On the topic men, women, and power in marriage and work, the following opinions were uncovered.

Egalitarian Marriage Is Satisfying

In 19 of 22 countries, majorities say that a marriage where both spouses have jobs and conduct childcare and housework is a more satisfying way of life than having the husband singularly provide financial support while the wife cares for the home. This assertion is particularly widespread in western Europe, with 91 percent in France and Spain and 85 percent in Germany showing agreement. Support for both spouses working is mixed across predominantly Muslim countries. About 92 percent in Lebanon and 72 percent in Turkey favor a double-income household and an egalitarian approach to household tasks. Indonesia reported a narrower majority: 56 percent are in favor of dual workers both inside and outside the home; 43 percent say that a marriage where the husband provides for the family and the wife takes care of the house and children is preferable. Egyptians and Jordanians are more divided: 48 percent and 47 percent, respectively, support an egalitarian marriage approach.

Women Should Be Able to Work Outside the Home

In 17 of the 22 countries polled, most say they *completely* agree with the assertion that women should be able to work outside the home. In a number of these countries, women are more likely than men to strongly support this idea. For example, 65 percent of Pakistani women *completely* agree that women should have the option to work outside the home; only 31 percent of Pakistani men hold the same view. South Korea, Kenya, Spain, Lebanon, and Indonesia also showed significant gender differences.

Men Should Receive Preferential Treatment in Tough Economic Times

In Nigeria, in addition to many Asian and predominantly Muslim countries, most respondents say that men should receive preferential treatment when jobs are scarce. To the contrary, majorities in 11 of the 22 countries surveyed disagree, including Spain (87 percent), the UK (85 percent), the United States (85 percent), France (80 percent), and Germany (80 percent). Majorities in Mexico (69 percent), Brazil (63 percent), Argentina (56 percent), Kenya (53 percent), and Poland (51 percent) also reject this notion.

Summary

Whether we realize it or not, we are constantly exchanging messages and meaning with our family members. It is important to keep in mind the dynamics of verbal and non-verbal communication and to learn how to convey both effectively. For example, self-disclosure and open communication can help family members earn the trust needed to reinforce confidence and connection as families change. Listening actively and empathically may help family members to recognize alternate points of view, to negotiate mutual understanding, and to make their interactions stronger. Becoming aware of ineffective listening styles may go a long way toward overcoming them. In marriage, differences of opinion may deliver long-term benefits if partners demonstrate a mutual intent to repair any emotional harm. In fact, research shows that couples who constructively engage in conflict resolution strive for closeness and courtesy.

Once two people decide to become a couple, many other decisions tend to follow. Who makes decisions in the relationship is based on power, the ability to exercise one's will or influence over how both partners will organize their lives together. Subjective views (equity and fairness in household arrangements) have been used by researchers to understand decision-making power in the family, in addition to objective measurements (annual earnings a partner contributes to the household or the number of hours per week contributed to housework or childrearing). Another way to think about power in marriages and families is to consider its bases or sources—legitimate, informational, referent, coercive, reward, and expert—and their sway in how individuals establish, exert, and react to influence.

Three perspectives have been used to explain the existence of power in marital and intimate partner relationships. Resource theory proposes that the partner having the greatest resources generally is apportioned the greatest power in the marriage. However, cultural resource theory argues that authority norms may be so pervasive that they supersede any effect that resources have on power. The principle of least interest perspective proposes that the partner who cares the least about the relationship has the most power in the relationship. That there are traditional, non-traditional, and neo-traditional family forms speaks to the diversity of power distribution in today's families.

Couple conflict may be more pervasive than what we would like; however, it is not always as devastating as many imagine. When seeking resolution, it is important to stand back, take a breath, and, as psychologist Virginia Satir asserts, confront each part of the whole problem. Taking responsibility for one's role in the conflict, being mindful of unattainable expectations, knowing how personal fears affect one's responses, and recognizing that all relationships have problems also may encourage couple resiliency.

Key Terms

Active-empathic listening is a three-stage model of listening in which the listener is attentive to what the speaker is saying while demonstrating sensitivity to the speaker's emotional state.

Body language is the conscious or unconscious use of physical distance (proxemics), gestures, and other forms of body movement to non-verbally communicate thought, intentions, or feelings.

Conflict is a serious disagreement or argument.

Cultural resource theory is the premise that a culture's norms of authority may supersede any effect that resources have on power.

Effective communication is the sending, receiving, or understanding of verbal and non-verbal information by someone in the way it was intended.

Egalitarian marital norm is a marital norm in which power and household labor are fairly or equally distributed between spouses; also characterized as *peer marriages*.

Genderlects is the theory that men and women are socialized to communicate in separate and gender-specific conversational styles; developed by linguist Deborah Tannen.

Neo-traditional norm of marital power is a marital norm in which each spouse works as well as participates in childcare and household tasks, which are unevenly divided on traditional, sex-specific lines.

Non-verbal communication is the process of exchanging information through wordless communicative acts such as eye contact, gestures, facial expressions, touch (haptic communication), body language, and physical distance between interactions, and functions to substitute, complement, contradict, accent, or repeat a verbal message or meaning.

Open communication is a condition of communication in which individuals are allowed to express ideas, emotions, plans, and experiences to each other.

Power is the ability to exercise one's will or influence over how individuals or couples organize their lives.

Principle of least interest is the premise that the person who cares the least about the relationship has the most power in the relationship; developed by sociologist Willard Waller.

Resource theory is the premise that the spouse having the greatest resources is apportioned the greatest power in the marriage; developed by sociologists Robert Blood and Donald Wolfe.

Self-disclosure is a process of communication by which one person reveals information about himself or herself to another.

Serial arguing is a conflict pattern in which partners argue about the same issue two or more times or focus on a particular problem that has no discernible beginning or end.

Verbal communication is the process of exchanging information through the sending and receiving of sound and speech.

Discussion Questions

- Recall the active-empathic listening model of communication. Why is empathy an important component of good listening? How might communication be altered if empathy is not practiced?
- Recall Virginia Satir's four miscommunication styles: Placater, Blamer, Computer, and Distracter. Which miscommunication style best describes your common reaction to managing conflict in family relationships? Has your miscommunication style changed over time?
- According to the text, when having problems with our partner, there may be some benefit in giving *conflict* a chance. In what ways might conflict benefit some relationships? In what ways is conflict negative or damaging to relationships?

Suggested Film and Video

Dinozzi, R. (Director). (2001). *He Said, She Said: Gender, Language and Communication with Deborah Tannen.* (Film). Into the Classroom Media.

Smith, H. (Executive Producer). (2001). *Juggling Work and Family.* (Film). Films Media Group.

Internet Sources

Dealing with Stress: Stress and Families (2016). University of Minnesota Extension: http://www.extension.umn.edu/family/live-healthy-live-well/healthy-minds/dealing-with-stress

Families First – Keys to Successful Family Functioning: Communication (2009). Virginia Cooperative Extension, Virginia State University: https://pubs.ext.vt.edu/350/350-092/350-092.html

References

Auxley, S. (1996). *Communication at Work: Management and the Communication-Intensive Organization.* Westport, CT: Quorum Book.

Bevan, J., Hale, J. & Williams, S. (2004). Identifying and Characterizing Goals of Dating Partners Engaging in Serial Argumentation. *Argumentation and Advocacy*, 41, 28–40.

Birditt, K., Brown, E., Orbuch, T. & McIlvane, J. (2010). Marital Conflict Behaviors and Implications for Divorce over 16 Years. *Journal of Marriage and Family*, 72, 1188–1204.

Blood, R. Jr. & Wolfe, D. (1960). *Husbands and Wives: The Dynamics of Family Living.* Oxford, UK: Free Press Glencoe.

Blumstein, P. & Schwartz, P. (1983). *American Couples: Money, Work, and Sex.* New York: William Morrow & Company.

Bodie, G., Vickery, A., Cannava, K. & Jones, S. (2015). The Role of "Active Listening" in Informal Helping Conversations: Impact on Perceptions of Listener Helpfulness, Sensitivity, and Supportiveness and Discloser Emotional Improvement. *Western Journal of Communication*, 79, 151–173.

Bradbury, T. (2014). *Which Conflicts Consume Couples the Most?* University of Wisconsin Couples Lab. Retrieved from http://relationships.sohe.wisc.edu/2014/which-conflicts-consume-couples-the-most

Caldwell, M. & Peplau, L. (1984). The Balance of Power in Lesbian Relationships. *Sex Roles*, 10, 587–599.

Carrington, C. (1999). *No Place Like Home: Relationships and Family Life among Lesbians and Gay Men.* Chicago, IL: The Chicago Series on Sexuality, Gender and Culture.

Carton, J., Kessler, E. & Page, C. (1999). Nonverbal Decoding Skills and Relationship Well-being in Adults. *Journal of Nonverbal Behavior*, 23, 91–100.

Christensen, A. & Heavey, C. (1990). Gender and Social Structure in the Demand–Withdraw Pattern of Marital Conflict. *Journal of Personality and Social Psychology*, 59, 73–81.

Coltrane, S. (2000). Research on Household Labor: Modeling and Measuring the Social Embeddedness of Routine Family Work. *Journal of Marriage and Family*, 62, 1208–1233.

Cordova, J. (September 9, 2011). Healthy Miscommunication: We Miscommunicate More than We Communicate. *Psychology Today.* Retrieved from https://www.psychologytoday.com/blog/living-intimately/201109/healthy-miscommunication

Cotter, D., Hermsen, J. & Vanneman, R. (2014 July 30). *Back on Track?: The Stall and Rebound in Support for Women's New Roles in Work and Politics, 1977–2012.* The Council on Contemporary Families. Retrieved from https://contemporaryfamilies.org/gender-revolution-rebound-brief-back-on-track/

Covey, S. (2014). *The Seven Habits of Highly Effective Teens: The Ultimate Teenage Success Guide.* New York: Touchstone.

Cozby, P. (1972). Self-disclosure, Reciprocity, and Liking. *Sociometry*, 35, 151–160.

Curtis, R. & Miller, K. (1986). Believing Another Likes or Dislikes You: Behaviors Making the Beliefs Come True. *Journal of Personality and Social Psychology*, 51, 284–290.

Farber, S. (2013 September 11). Why We All Need to Touch and Be Touched. *Psychology Today.* Retrieved from https://www.psychologytoday.com/blog/the-mind-body-connection/201309/why-we-all-need-touch-and-be-touched

Fincham, F. & Bradbury, T. (1993). Marital Satisfaction, Depression, and Attributions: A Longitudinal Analysis. *Journal of Personality and Social Psychology*, 64, 442–452.

French, J. & Raven, B. (1959). The Bases of Social Power. In D. Cartwright (Ed.), *Studies in Social Power* (pp. 150–167). Ann Arbor, MI: Institute for Social Research.

Fry, R. & Cohn, D. (2010). *Women, Men and the New Economics of Marriage.* Washington, DC: Pew Research Center. http://www.pewsocialtrends.org/2010/01/19/women-men-and-the-new-economics-of-marriage/

Furnham, A. (December 10, 2014). The Secrets of Eye Contact Revealed. *Psychology Today.* Retrieved from https://www.psychologytoday.com/blog/sideways-view/201412/the-secrets-eye-contact-revealed

Garland, D. (1981). Training Married Couples in Listening Skills: Effects on Behavior, Perceptual Accuracy and Marital Adjustment. *Family Relations,* 30, 297–306.

Gearhart, C. & Bodie, G. (2011). Active-empathic Listening as a General Social Skill: Evidence from Bivariate and Canonical Correlations. *Communication Reports,* 24, 86–98.

Greenberg, M. (2015, March 30). Seven Secrets of Happy Couples: The Most Crucial Fundamentals for Any Relationship. *Psychology Today.* Retrieved from https://www.psychologytoday.com/blog/the-mindful-self-express/201503/7-secrets-happy-couples?collection=1083340

Grosjean, F. (2015, April 1). One Person—One Language and Bilingual Children: The Amazing Beginning of Modern OPOL. *Psychology Today.* Retrieved from https://www.psychologytoday.com/blog/life-bilingual/201504/one-person-one-language-and-bilingual-children

Guerrero, L. (1997). Nonverbal Involvement across Interactions with Same-sex Friends, Opposite-sex Friends and Romantic Partners: Consistency or Change? *Journal of Social and Personal Relationships,* 14, 31–58.

Holley, S., Sturm, V. & Levenson, R. (2010). Exploring the Basis for Gender Differences in the Demand–Withdraw Pattern. *Journal of Homosexuality,* 57, 666–684.

Huang, L. (1999). Family Communication Patterns and Personality Characteristics. *Communication Quarterly,* 47, 230–243.

Khong, B. (2002). Communication in the Family. *UNIBUDES Annual Magazine.* Kingsford, NSW: UNSW Buddhists' Society.

Kurdek, L. (1994). Areas of Conflict for Gay, Lesbian, and Heterosexual Couples: What Couples Argue about Influences Relationship Satisfaction. *Journal of Marriage and Family,* 56, 923–934.

Lee, G. & Petersen, L. (1983). Conjugal Power and Spousal Resources in Patriarchal Cultures. *Journal of Comparative Family Studies,* 14, 23–38.

Leong, F. (2008). *Encyclopedia of Counseling.* Thousand Oak, CA: Sage Publications.

Markman, H. (1981). Prediction of Marital Distress: A 5-year Follow-up. *Journal of Consulting and Clinical Psychology,* 49, 760–762.

Mundy, L. (June 2013). The Gay Guide to Wedded Bliss. *The Atlantic,* 56–70.

Peplau, L. & Fingerhut, A. (2007). The Close Relationships of Lesbians and Gay Men. *Annual Review of Psychology,* 58, 405–424.

Pew Global Attitudes Project (2010). Gender Equality Universally Embraced, but Inequalities Acknowledged. Retrieved from http://www.pewglobal.org/2010/07/01/gender-equality/

Pickhardt, C. (2014, June 13). When Parents Conflict over Their Adolescent. *Psychology Today.* Retrieved from https://www.psychologytoday.com/blog/surviving-your-childs-adolescence/201406/when-parents-conflict-over-their-adolescent

Reis, H. & Shaver, P. (1988). Intimacy as an Interpersonal Process. In S. Duck (Ed.), *Handbook of Personal Relationships* (pp. 367–389). Chichester, UK: Wiley.

Richmond, V., McCroskey, J. & Hickson, M. (2000). *Nonverbal Behavior in Interpersonal Relations.* Boston, MA: Allyn & Bacon.

Rodman, H. (1972). Marital Power and the Theory of Resources in Cultural Context. *Journal of Comparative Family Studies,* 3, 50–69.

Satir, V. (1972). *Peoplemaking.* Palo Alto, CA: Science and Behavior Books.

Schwartz, P. (1994). *Love between Equals: How Peer Marriage Really Works.* New York: Free Press.

Shechory, M. & Ziv, R. (2007). Relationships between Gender Role Attitudes, Role Division, and Perception of Equity among Heterosexual, Gay and Lesbian Couples. *Sex Roles,* 56, 629–638.

Simmering, J. (2018). Body Language. *Encyclopedia of Management.* Retrieved from www.referenceforbusiness.com/management/A-Bud/Body-Language.html

Sprecher, S. & Cate, R. (2004). Sexual Satisfaction and Sexual Expression as Predictors of Relationship Satisfaction and Stability. In J. Harvey, A. Wenzel & S. Sprecher (Eds.), *The Handbook of Sexuality in Close Relationships.* (pp. 235–256). Philadelphia, PA: Psychology Press.

Sprecher, S. & Felmlee, D. (1997). The Balance of Power in Romantic Heterosexual Couples over Time from "His" and "Her" Perspectives. *Sex Roles,* 37, 361–379.

Sprecher, S., Treger, S. & Wondra, J. (2013). Effects of Self-disclosure Role on Liking, Closeness, and Other Impressions in Get-Acquainted Interactions. *Journal of Social and Personal Relationships,* 30, 497–514.

Tannen, D. (1990). *You Just Don't Understand: Women and Men in Conversation.* New York: Morrow.

Tichenor, V. (1999). Status and Income as Gendered Resources: The Case of Marital Power. *Journal of Marriage and Family,* 61, 638–650.

Vogel, D., Murphy, M., Werner-Wilson, R., Cutrona, C. & Seeman, J. (2007). Sex Differences in the Use of Demand and Withdraw Behavior in Marriage: Examining the Social Structure Hypothesis. *Journal of Counseling Psychology,* 54, 165.

Waller, W. (1937). *The Family: A Dynamic Interpretation.* New York: Gordon.

Wertheim, E. (2008). *The Importance of Effective Communication.* Northeastern University. Retrieved from http://web.cba.neu.edu/~ewertheim/interper/commun.htm

Wilcox, W. & Dew, J. (2013). No One Best Way: Work–Family Strategies, the Gendered Division of Parenting, and the Contemporary Marriages of Mothers and Fathers. In W. Wilcox & K. Kline (Eds.), *Gender and Parenthood: Biological and Social Scientific Perspectives* (pp. 271–303). New York: Columbia University Press.

PART IV

DYNAMICS IN FAMILY LIVING ARRANGEMENTS

15
BREAKING UP, MARITAL SEPARATION, AND DIVORCE

Learning Outcomes

Upon completion of the chapter, students should be able to:

1. identify the factors associated with the dissolution of a dating relationship
2. recognize how partners may react to deteriorating satisfaction in their dating relationships
3. recognize the role of technology in how couples manage breaking up
4. compare and contrast the crude divorce rate and the refined divorce rate
5. describe divorce from an historical perspective
6. define marital separation
7. identify the different forms of marital separation
8. describe the process of uncoupling
9. explain the functions of divorce counseling, divorce mediation, spousal support, child custody, and child support
10. explain sociodemographic and socioeconomic predictors of divorce
11. describe the intergenerational transmission of divorce
12. describe how premarital cohabitation may be a predictor of divorce
13. describe interpersonal models of marital relationship dissolution
14. describe the effects of divorce on children
15. identify the six stations of divorce
16. describe divorce stressors
17. explain the positive effects of divorce.

Brief Chapter Outline

Pre-test

Breaking Up

Factors of Dating Relationship Dissolution

Race; Availability of Other Relationship Partners; Time Spent in the Relationship

Pre-test

Engaged or active learning is a powerful strategy that leads to better learning outcomes. One way to become an active learner is to begin the chapter and try to answer the following true/false statements from the material as you read. You will find that you have ready answers to these questions upon the completion of the chapter.

The rate of divorce per 1,000 people is the refined divorce rate.	T	F
The three forms of marital separation are trial sabbatical, legal sabbatical, and permanent separation.	T	F
Psychologist John Gottman's model of interaction styles that are harmful to relationships is known as "World War III."	T	F
In the historical court decision *Obergefell v. Hodges,* the U.S. Supreme Court legalized same-sex marriage in all 50 states.	T	F
Couples tend to enroll in marriage counseling about three years earlier than preferred by most professionals.	T	F
A psychic divorce is not one of Paul Bohannon's six stations of divorce.	T	F
In her book *The Good Divorce,* E. Mavis Hetherington argues that over 80 percent of children from divorced families show signs of adjustment and coping a year after their parents' breakup.	T	F
Divorce can lead to negative mental and physical health outcomes.	T	F
There are no encouraging outcomes with regard to marital separation or divorce.	T	F
Couples with covenant marriage licenses surrender their rights to a no-fault divorce.	T	F

Upon completion of the next section, students should be able to:
LO1: Identify the factors associated with the dissolution of a dating relationship.
LO2: Recognize how partners may react to deteriorating satisfaction in their dating relationships.
LO3: Recognize the role of technology in how couples manage breaking up.

Breaking Up

Not all dating relationships share the level of commitment as with marital relatonships. Nonetheless, some in committed dating relationships experience benefits that are consistent with couples who enjoy satisfying marriages. Dating relationships may be positively related to young adults' life satisfaction and having a higher subjective well-being than those not dating at all or dating multiple people. A study of 1,621 young college adults found that those in committed romantic relationships had fewer mental health problems and engaged in less risky health and safety behaviors (binge drinking, driving while intoxicated) than their single counterparts. Respondents with more sexual partners were directly associated with more physical and mental health problems. One explanation could be that time with a partner leaves less time for risky behaviors. Or perhaps some risky behaviors (multiple sex partners, substance abuse) are incompatible with the less impulsive lifestyle of a committed relationship (Braithwaite *et al.*, 2010).

But not all relationships last "forever." People emotionally, even physically, cut ties from biological family members. But how often do we hear of someone becoming an ex-mother, ex-father, ex-sibling, or any other "former" or "ex" biological kin? The answer is, perhaps, rarely; however, other close or intimate partner relationships are a different story. Whether inevitably or decidedly, relationships within our most intimate circle may become terminated. As a result, our former intimates become our ex-spouses, ex-boyfriends, ex-girlfriends, ex-romantic partners, or ex-fiancées (Sprecher & Fehr, 1998).

In this chapter, we focus on the problems and processes that come with the decline and termination of marriage. In doing so, we learn that ending

Source: Rawpixel.com/Shutterstock

a marriage can be a stratifying event. Consequently, we also explore how former spouses and children of divorced parents navigate their social, economic, and legal uncertainties toward a new state of family life. The chapter begins with a discussion on the dissolution of dating relationships and how intimate couples may respond.

Factors of Dating Relationship Dissolution

If committed relationships are seemingly more reliable, beneficial, or satisfying, why don't more of them last "forever"? Maintaining a relationship can be complex; therefore, it is not surprising that more than one factor can influence its end or *dissolution*. One factor facilitating dissolution may be simply a function of the partners themselves. A partner may hold a condescending attitude, or he or she may forget your birthday one time too many. Other factors facilitating relationship dissolution have to do with the broad social contexts in which a couple is situated.

In their study of almost 600 undergraduate students, sociologists Diane Felmlee, Susan Sprecher, and Edward Bassin (1990) found that agreeability with social networks close to each partner is important to keeping dating relationships intact. Not perceiving support from a partner's friends and family and disliking a partner's relatives or social network significantly predicted breakups among ongoing relationships. Other predictors of relationship dissolution found in the study are below.

Race

Dissimilarity in race may lead to greater rates of breakup, a conclusion that is similar to a great deal of research on divorce rates among interracial marriages. Researchers speculate that societal disapproval of interracial dating may explain higher dissolution rates. It also is possible that partners of the same race/ethnicity have more or enough in common to maintain a more stable relationship.

Availability of Other Relationship Partners

A couple may be able to endure inequities in a relationship, that is until one partner finds a "better" alternative. Relationship dissatisfaction is less likely to determine whether or how soon a relationship ends than the presence or absence of a favorable alternative to that relationship.

Time Spent in the Relationship

The time spent with a partner and the duration of the relationship may significantly predict a relationship's rate of dissolution. Spending a large amount of time with a partner is one reason why some couples are likely to remain together compared to those who spend less time with each other.

Reactions to Deteriorating Satisfaction in Dating Relationships

Several well-trusted signals may suggest that a relationship is in trouble. Increasing cynicism or distrust in a relationship is an example, in addition to having one's aspirations, expectations, or needs less fulfilled. Using a sample of 128 undergraduate students, psychologist Caryl Rusbult and colleagues (1982) developed a model of how individuals may react to deteriorating satisfaction in romantic relationships:

- *Voice reaction: "Let's talk it over and try to make this work."* In this reaction, one or both partners decide to actively, constructively express dissatisfaction with the intent to maintain the relationship. Partners ask, "What is bothering you? Why are you unhappy?" Either one or both partners will try to change their own or the other's behavior.
- *Loyalty reaction: "Things will get better if we give it some time."* In this reaction, one or both partners wait or hope for the relationship to improve, as they remain passively loyal to the relationship. The loyalty reaction is viewed as less constructive than the voice reaction.

Because reactions tend to be indirect, behaving loyally may be misinterpreted or go unnoticed.

- *Neglect reaction: "Let's just let this 'relationship' fall apart, as it has been."* Neglect can be expressed by spending less time together, refusing to discuss problems, ignoring the partner, treating the partner badly emotionally or physically, or criticizing the partner for issues unrelated to the real problem. The chances of becoming intimately involved with others outside the relationship are increased.
- *Exit reaction: "Let's get this relationship over with now."* The exit reaction is just as it suggests: partners separate, stop seeing each other entirely, or move out of a joint residence. Each may decide to "just be friends" or, if married, begin divorce proceedings. The exit and neglect reactions are relatively destructive to a relationship. Lower levels of investment in the relationship appear to facilitate both reactions.

Taking a Break from a Dating Relationship

If a relationship is in trouble, "taking a break" from it—a temporary or trial separation from a partner—may appear counterproductive. Yet, a trial separation may be a time to contemplate the relationship, develop actions to work through problems, make efforts to improve the relationship, and reflect on how partners should treat each other and themselves. A clear, mutual understanding of what separation means and should accomplish first must be established. Consider the following suggestions.

- *Set ground rules.* Each partner needs to agree that "taking a break" means that both are still in a relationship. Consequently, clear parameters on how the relationship will function during separation must be established. For example, should both date other people? If so, is sexual contact permissible? How frequently should partners communicate? Is there a specific date when the separation period will end?
- *Keep communication open.* Relationship problems need to be verbalized. Severing communication while separated will do little to develop a more stable or satisfying future. If communication is avoided, the same issues may resurface if or when partners reunite.
- *Seek help alternatives.* Couples counseling may be an asset if a problem needs to be addressed. If the relationship is not the actual reason for either partner's stress or unhappiness, time apart and professional intervention may expose complications.

"Taking a break" from a relationship may not be the best decision for every couple or individual. Again, there may be a temptation to avoid constructive communication or to neglect any substantive discussion, for that matter. One or both partners may misguidedly view separation as a convenient "escape" when things become too intolerable. Some also have learned that "taking a break" from their partners may lead to the breakup of the relationship.

Reactions to Breaking Up a Dating Relationship

Breaking up is hard to do. When researchers asked young adults about how a recent breakup affected them, their responses included loneliness, distress, and a lapse of who they were as individuals (Lewandowski *et al.*, 2006). Problematic reactions like these may not come as a surprise. As one might expect, two people's lives become more intertwined the longer they are involved in a relationship. In addition to memories, partners also share intimate knowledge of each other, mutual friends, and possessions. In one study of 1,295 unmarried 18- to 35-year-olds, researchers associated intimate partner breakup to a decline in life satisfaction and emotional distress, which had started before the breakup and continued after dissolution. Larger declines in life satisfaction were noted among those who had been cohabitating or who had plans for marriage. Individuals who had dated their partners for a longer time were more distressed after a breakup than those who dated their partners for less time (Rhoades *et al.*, 2011).

Technology has influenced how many relationships begin. One of the largest matchmaking outlets, Match.com, boasts of websites in 15 languages and service to singles in 24 countries (DatingSites Reviews.com, 2018). A perhaps unintended outcome of online social media and networking services, one study uncovered a tendency for users to develop negative thoughts or mood states while perusing a former partner's Facebook profile (Tran & Joorman, 2015). As if taking a page from the aforementioned study, in late 2015 Facebook announced an initiative to help users end relationships on Facebook "with greater ease, comfort, and sense of control." New features allow users to see less of a former partner's name and profile picture without having to "unfriend" or block the individual. For example, a former partner's posts will not appear among a user's personal news feeds. Nor will his or her name be suggested when friends are tagged in a photograph (Seetharaman, 2015).

Marital Separation and Divorce

Weighing "should I stay" or "should I go" is unlikely to be a casual matter if a marriage is in trouble. Distressing questions often surface. Answers may bring more concerns than clarity.

- Part of me loves, or at least cares for, my spouse. So, isn't this reason enough that there is still hope for us?
- A lot of people—our children, most importantly—will be affected if we decide to divorce. Is this really the best decision? Am I being selfish here?
- Some days I feel confident that things will get better, but more and more I feel less assured about this. What should I do?
- Have I done all I can to repair this marriage?

For some, enduring a dissatisfying marriage and living with a spouse may become too overwhelming. Consequently, one or both spouses may decide to "cool off" and enter into a respite from the relationship and each other. Living apart—**separation**—actually may be a practical means to improve the marriage and appraise the potential of certain aspects of the relationship. Separation also may be a step toward divorce in some marriages. Regardless of intent, a couple may find it helpful to enlist a therapist, mediator, lawyer, or other trusted third party to establish expectations about what separation will entail. For example, what are the mutual goals in separating? Should spouses continue to communicate? Are spouses free to enter into other intimate relationships?

Most who receive couples therapy are positively affected by the experience. The effectiveness of couples' therapy is comparable to individual therapies, in fact, and more successful than control groups not receiving treatment (Lebow *et al.*, 2012). Even then, both partners must work to take advantage of what

Upon completion of this section, students should be able to:
LO4: Compare and contrast the crude divorce rate and the refined divorce rate.
LO5: Describe divorce from an historical perspective.
LO6: Define marital separation.
LO7: Identify the different forms of marital separation.

a good therapist can do to help tilt chances in their favor. Often, too many couples enroll in therapy six years later than is ideal. Marital issues may be longstanding and, some, too ingrained by then. Not only timing but also motivation may pose a problem in stabilizing a troubled marriage. One or both partners may have decided to divorce but may use counseling to announce his or her intent. Other partners simply are not honest or open with their spouses or therapist about their true feelings, which makes it nearly impossible for counseling to be worthwhile (Bodenmann, 1997).

Locating a counselor with whom both partners feel a connection is an absolute must. Experts recommend that couples "shop" for a seasoned professional who has intense marital therapy training, who will explain his or her treatment approach, and who is adept in how the couple's situation fits within treatment. It deserves repeating that unless both partners want to repair the relationship, they may depart from therapy more disappointed than before. Succinctly, partners who do not want to be married should make it known early on in therapy.

Forms of Marital Separation

What if a couple decides that separation is the best way to move forward? There are several arrangements to consider: trial separation, permanent separation, and legal separation.

Trial Separation

A *trial separation* is an informal arrangement in which marriage partners choose to live apart. Its intent is to provide partners with a limited severance—a few days, weeks, or longer—as they decide how to reconcile or whether to end the relationship. The legal obligations under trial separation are the same as if living under the same roof. Property ownership will continue to be shared, if not state-prohibited. Debts and assets earned also are likely to remain jointly owned.

Permanent Separation

A *permanent separation* may follow or immediately start once a couple begins to live apart. Couples who separate permanently are still legally married. However, spouses decide to live in separate residences and to re-establish their lives without divorce or the intent to reconcile. A marriage partner is neither entitled to the other's assets nor responsible for debts incurred by a spouse in most states.

Legal Separation

A *legal separation* can be acquired from most U.S. courts by filing a request in family court. The couple is no longer married; however, they are not divorced nor granted permission to remarry. Rather, similar to divorce, a legal separation establishes agreements pertaining to property division, alimony, child custody, child support, and living arrangements.

A couple may prefer legal separation to divorce for several reasons. If parents, a couple may want to keep their family intact and under the framework of law. Fear of losing health insurance or other benefits due to obtaining a divorce may be a strong inducement as well. Some find legal separation within the bounds of their religious ideals on marital permanency. For others, legal separation is viewed as a practical opportunity to establish separate lives. In fact, some ex-partners comfortably live under this form of marital separation for many years (see Doskow, 2014).

Reactions to Marital Separation

Marital separation may not necessarily transition into divorce; however, the experience can be just as disruptive or demanding emotionally. It is not unordinary to experience a host of grief reactions before denial subsides and a sense of reassurance sets in. As we examine these responses, keep in mind that, as difficult as separation may be, many recount separation as a journey toward personal development.

Denial It is plausible that a person may "settle into" denial in order to soften the emotional impact of separation. However, denial may begin before a couple parts ways. A spouse may believe that the marriage is not as bad as it seems or that problems will diminish or become more tolerable over time.

Anger Feelings of anger may follow, which may be directed toward the situation, the spouse, or oneself. Managing the resentment of others may compound these feelings. For example, children, friends, or family may "pick sides" and cast blame.

Bargaining, Guilt, and Regret One spouse may bargain with the other spouse or beseech a higher power to restore former feelings and fix the marriage. A spouse also may go into cycles of self-blame and seemingly endless questions: "What if I had…?", "Why couldn't I have done…?", or "Why didn't I do more…?" The point of such self-talk is to determine whether something could have been done that would have kept the relationship intact.

Grief and Fear There comes a time when one must confront the fact that the relationship may be close to its end. There may be sadness over the life that will not happen and unease over an uncertain future. For example, thoughts like, "Can I make it own my own?" or "Is there another relationship in the future?" are not uncommon. In light of these feelings, grief and fear may present through depression, sleeplessness, loss of appetite, anger, substance abuse, withdrawal, fatigue, or concentration difficulties. Other family members, including children and in-laws, may be similarly affected.

Acceptance and Reinvention Partners' acceptance of separation should not be confused with being happy with what has happened. Rather, acceptance of separation may be best understood as a "new normal": a time to move on emotionally and to experience the hope of a satisfying future.

Divorce, also called *marital dissolution*, is the legal termination of a marital union. If you observe media and social commentaries on the state of marriage, you may hear or have read that "half of all marriages in the United States end in divorce." There often is the added disclaimer that the rate is "climbing." But could the "50 percent" statistic actually be misleading?

Pretend that 100,000 legal weddings were performed last year, the same year that 50,000 divorces also were granted. At first, it may seem reasonable to think that the scenario is declaring a "50 percent" divorce rate. However, the "50 percent" assumption seems less reasonable once you consider that the scenario measures the number of weddings in one year against all weddings performed over years, even decades. Even then, there may be fluctuations in the timing of marriage and divorce within the sample of cohorts. As you come across statistical interpretations like these, take into account that the exact predicted percentage of divorce has plenty to do with sample composition and the analytical instruments used. Below, we discuss ways family scientists measure divorce.

Measuring Divorce

Measuring the **crude divorce rate** (CDR)—the rate of divorce per 1,000 people in a population—is one of several ways family scientists make sense of the frequency of divorce. In 2017, the Centers for Disease Control and Prevention and the National Center for Health Statistics recorded a 3.2 CDR. These current findings suggests a decline in the nation's crude divorce rate: in 2007, the CDR was 3.6 per 1,000 people; in 2000, the rate was 4.0.

Other family scientists take issue with the CDR's primary method. They argue that to measure the divorce rate by the number of divorces per 1,000 people will include individuals not at risk for divorce, including unmarried or single men and women. Many of these experts find better accuracy in using the

refined divorce rate (RDR)—the annual number of divorces that occur per every 1,000 married women. A study by the National Marriage Project used the RDR formula and found 20.9 divorces per 1,000 married women ages fifteen and older, a rate of about 2 percent a year (Stanton, 2015).

Divorce in an Historical Perspective

Statistically predicting or "capturing" the movement of divorce within a year or an extended period is challenging to pin down. What we do know is that the social, economic, and legislative path to divorce in the United States has almost never been a smooth one. To better understand the history of divorce, we trace divorce from Colonial America to the twenty-first century to learn its emergence and evolution, in addition to those who, even now, seek its reform.

Divorce in Colonial America

Legal prerequisites for granting divorce and alimony—support from the financially stronger spouse—were beginning to take shape in America during the 1600s. In 1639, Mr. and Mrs. James Luxford were documented by a Puritan court, the Court of Assistants in the Massachusetts Bay Colony, as the first married couple to divorce in America. Mr. Luxford, a bigamist, also was levied a fine, ordered to forfeit all property to Mrs. Luxford, and banished to England.

Awarding alimony to the wronged or rejected spouse became law in 1641 and was primarily designed for the protection of divorced wives. Court-recognized, fault-based grounds for divorce also became more standard around the mid-1600s, and generally involved proof of cruelty by a male, adultery by a female, bigamy, desertion, failure to secure provisions, impotence, and fraudulent marital contract (Hawkins, 2013a; Hawkins, 2013b). Given the strict social conformity of the times, the act of ending a marriage may seem somewhat revolutionary. However, Puritan leaders perceived *sound* marriages as helpful in stabilizing the larger social order and a context for sexual activities that kept individual "vices" at bay. Ending *troubled* marriages of apparent irreversible proportions had seemed less harmful to society and family.

Divorce in Eighteenth-century America

During the eighteenth century, the unhappily married increasingly recognized divorce as an actionable concept. For example, only 40 marriages had been dissolved in Connecticut between 1636 and 1698. By the 1700s, Connecticut granted more divorces than Massachusetts, despite the state's smaller population (Hawkins 2013a; Hawkins, 2013b).

Adultery and desertion remained among the most accepted fault-based grounds for marital dissolution during this time; however, there were slight variations across provinces and territories. For example, from the late 1600s to 1786, infidelity, fraudulent marriage, and spousal desertion (for three years), as well as having a partner presumed dead (for three years), could be grounds for dissolution in Connecticut. But New York solely required proof of adultery and, consequently, denied remarriage if proven (Hawkins, 2013a; Hawkins, 2013b). When legal prerequisites were unmet or unconfirmed, couples desperate to end their marriages colluded. One partner might falsely accuse his or her spouse of adultery; the other, committing perjury, would agree to the false accusation.

Divorce in Nineteenth-century America

The divorce rate in America exceeded that of Europe beginning in the 1850s to the turn of the century. After the Civil War, the 1888 U.S. Census confirmed that marital dissolution had peaked nationwide: one out of fourteen to sixteen marriages ended in divorce (Hawkins, 2013a). Among the divorced were wives who filed on the grounds of desertion when their husbands disappeared to the West rather than reunite with their families.

Divorce jurisdiction began to move from an overburdened state legislature system to a court system, as rates of dissolution continued to climb. By the late 1880s, divorce was allowed in all the then 47 U.S. territories and states, except South Carolina. Seven states refused to recognize spousal cruelty as grounds for divorce. To work around this and other legal limits, some couples flocked to Indiana, the Dakotas, and the Utah and Oklahoma territories, where there were more lenient divorce laws and shorter residency requirements (Hawkins, 2013a). Meanwhile, religious and civic leaders began to rally support for a more uniform basis for granting divorce. Motivated by "crimes against chastity … illegitimate births … and a decline in marriage and birthrates in some states," the National Divorce Reform League sought more stringent legal grounds and residency requirements for obtaining divorce (National Divorce Reform League, 1885, p. 3).

Divorce in Early Twentieth-century America

The economic expansion of the 1920s ushered in a higher increase in marriage as well as increased rates of divorce: 201,468 divorces were granted nationwide from 1922 to 1929 (Hawkins, 2013b). That same decade, 71 percent of all American divorces were sought by women; 48 percent of divorces were awarded on the grounds of cruelty (Hawkins, 2013b). Changing attitudes on marital partnership may be one explanation. Men and women sought romantic love and companionship with whom they married, and relationships that could not deliver were less tolerated. Further, these statistics overlap with women's preparedness for life outside of marriage. Women were more educated than in earlier decades, and some were fiscally independent even before marriage.

Maintaining a happy marriage during the Great Depression era of the 1930s was not easy. Massive poverty and economic difficulties were so encumbering that some couples bore fewer children than they wanted. Others simply chose to remain childless. The rate of divorce was 1.3 per 1,000 people in the early 1930s, likely because many couples could not afford to divorce and chose to remain married. The divorce rate climbed to 1.9 per 1,000 people by 1939 (U. S. Bureau of the Census, 1949).

Approaching World War II, young couples who were uncertain of their futures due to wartime rushed to marriage in increasing numbers. After World War II, during the mid-1940s, 3.5 per 1,000 people divorced (Plateris, 1974). The impulse to marry before deployment, warfare stressors, and post-war problems with reunifying relationships are attributed to this temporary surge. Further, families of the period had experienced increasing geographic mobility and urbanization, which tends to weaken traditional family–community ties.

Divorce in Mid Twentieth-century America

The ideal of the American Dream and an increased rate in marriage began to take hold in America by the start of the mid-twentieth century. Husband/breadwinner, wife/homemaker couples and their children settled into an idealized life of material comfort and low rates of divorce. Marital dissolution dropped to 2.1 per 1,000 people by the close of the 1950s (Plateris, 1974).

By the Vietnam era of the mid-1960s and 1970s, many aspects of American family life had shifted. Childbearing had curtailed, marriage rates had declined, and divorce rates increased dramatically. During a span of a decade and a half, divorce rates for married women more than doubled from 10.6 per 1,000 in 1965 to 22.8 per 1,000 in 1979 (Furstenberg, 1994).

Arguably, obtaining divorce was becoming simpler during this time. State legislators defended laws supporting **no-fault divorce**, in which neither partner had to prove spousal wrongdoing to dissolve the marriage. The original intent of no-fault divorce was to expand the legal definition of marital termination to factors other than conventional fault-based grounds (adultery, desertion, spousal cruelty). As a result, unhappily married couples no longer needed to fabricate divorce grounds or migrate to states with more lenient divorce provisions.

Divorce in Late Twentieth-century America

The frequency of divorce began to decline in the 1980s and continued into the 1990s. During the 1980s, the divorce rate peaked early in 1981 with 5.3 divorces per 1,000 people before falling to 4.7 in 1990. In 1999, the year before the new millennium, there was an even lower divorce rate at 4.1 per 1,000 people. Family scientists attribute the drop to the rising number of cohabitating couples, couples marrying with a higher level of education, and couples marrying at an older age at their first marriage (Norton and Miller, 1992; Ruggles, 1997).

The legal grounds for obtaining a divorce appeared to gather further scrutiny. Fault-based divorce advocates viewed no-fault laws as a factor in the distress and impoverishment of divorced women and their children. Others saw the no-fault procedure as one that devalued marriage, undermined the ideal of lifelong marital commitment, and a practice that made divorce too easy (Kohnke, 2013). On the other side were legal organizations that either worked to consolidate no-fault divorce regulations or advocated that fault-based divorce grounds be eradicated. Meanwhile, a growing number of courts and commentators were reassessing whether fault-based factors still served a legitimate function in granting divorce.

The **covenant marriage** was a response to late-twentieth-century divorce rates. Although covenant marriage licenses slightly differ between states, invariably, couples consent to relinquish their rights to a conventional no-fault divorce. If divorce appears inevitable, covenant couples agree to additional counseling, to a pre-divorce waiting period, and to divorce under fault-based grounds, including adultery, separation, felony conviction, physical or sexual abuse, or separation for a lengthy period (Kohnke, 2013). Three state legislatures—Louisiana, Arizona, and Arkansas—have adopted the covenant marriage. The concept generally is viewed as a controversial one, as many twentieth-century family laws have sought not to limit or intervene in a couple's access to divorce.

Proponents of covenant marriage believe that it promotes stronger marriages and recasts marriage as a worthy and more desirable institution. They also argue that its counseling requirements and pre-divorce waiting period allow couples the time to reconcile their relationship. Critics of covenant marriage believe that a required waiting period may have the unintended effect of trapping a spouse in a dysfunctional, even dangerous marriage. Some point out that covenant marriage partners already may be at a lower risk for divorce due to attitudes and other factors each brings to the marriage.

Divorce in Early Twenty-first-century America

So far, men and women married in the 2000s still are divorcing with less frequency. More than 1 million divorces per year were granted throughout the 1990s. In 2000, however, 944,000 marriages—less than a million—ended by divorce or *annulment,* a legal procedure declaring that the marriage never technically existed or was never valid. In 2014, 813,862 divorces or annulments were reported, a decline from the previous year (National Marriage and Divorce Rate Trends, 2015).

Divorce and Same-sex Couples Early twenty-first-century Americans witnessed an unprecedented U.S. Supreme Court decision. The court ruled in *Obergefell v. Hodges* that the Fourteenth Amendment requires all 50 states to permit marriage between gay and lesbian couples. The High Court ruling allowed all same-sex and opposite-sex couples to share the potential to marry and, in effect, to divorce.

The rate of marital dissolution between homosexual men and women appears to be generally lower than that of heterosexual men and women.

For example, data collected before the Supreme Court ruling show that 5.4 percent and 3.6 percent of married same-sex couples in New Hampshire and Vermont, respectively, had divorced in the first four years or so of enacting marriage equality in these specific states. The numbers correspond to an average annual 1.1 percent divorce rate among New Hampshire and Vermont same-sex marriages compared to a 2 percent annual divorce rate among the states' opposite-sex marriages. Before national legalization of same-sex marriage, the average dissolution rate for couples in civil union and domestic partner relationships—a status that once permitted same-sex and cohabiting opposite-sex couples to receive some equivalency of spousal rights without marrying—was 1.7 percent annually across California, Washington DC, New Hampshire, New Jersey, Washington State, and Wisconsin. In comparison, the average dissolution rate for any form of legal heterosexual relationship across the seven states surveyed was 1.6 per year (Badgett & Mallory, 2014).

The Beginning of the End of Marriage: The Processes toward Divorce

Divorce is more than a singular legal event. A marriage's deterioration and end involve often-overlapping processes, whose complications can equal or outpace

Upon completion of this section, students should be able to:

LO8: Describe the process of "uncoupling."

LO9: Explain the functions of divorce counseling, divorce mediation, spousal support, child custody, and child support.

Source: Gajus/Shutterstock

the pressures of mandated court protocols and compromises. Next, we identify what these legal procedures and obligations are and what they entail. We begin with the ending of the marital relationship: "uncoupling," the process of relationship decline.

Uncoupling

A relationship's beginning, deterioration, and finally, permanent breakdown is a painful trajectory to witness, much less experience. In her research on relationship breakups, sociologist Diane Vaughan (1990) captures the decline in detail. According to Vaughn, an observable pattern of uncoupling can be traced in all intimate partner breakups, including marriages. The process of uncoupling looks something like this.

- *Secretive dissatisfaction.* As unhappiness inches into a relationship, there is a tendency for the dissatisfied partner to feel fear or ambivalence toward expressing his or her discontent.
- *Seeking validation elsewhere.* The dissatisfied partner may try to seek validation elsewhere. He or she may spend more time at work, devote energy to a hobby, or consult with family and friends. The dissatisfied individual also may try to make changes in the relationship, without at first intending to leave the relationship.
- *More distancing.* The dissatisfied partner may become preoccupied with his or her partner's perceived flaws and failures, effectively minimizing the individual's more complimentary traits. What Vaughan calls *displays of discontent* may be more apparent around this time: a disgruntled glance, a goodnight kiss omitted, or more infrequent attentive probes ("How was your day?"). These hints of disillusionment may escalate to mundane or intermittent complaints that may bewilder some unsuspecting partners. Declaring, "You laughed so loud that you embarrassed me" or "forgetting" to wear one's wedding band or engagement ring are two examples. In essence, the dissatisfied partner is transitioning out of the relationship.
- *Recognizing the end.* The dissatisfied partner may tell the other that the relationship is over or commit some infraction to hasten its end. Some initiators, at some point, may find a "transitional person" to help bridge the gap between their old and new life. Either way, by the time the non-initiating partner has learned the extent of dissatisfaction, the initiator has tested or established independence from the relationship. Uncoupling ends when both partners acknowledge that the relationship is unsalvageable.

Negotiating Resources: Divorce Counseling, Mediation, and Other Support

Even when one or both spouses decide that the marriage is irreparable, some professional counseling may still be useful. Divorce counselors and divorce mediators possess specialized skills and information that could make divorce less contentious and more agreeable to each party. In what follows, we examine counseling and mediation options for married couples on the verge of divorce. The section follows up with legal actions often required of couples who agree to end their marriage.

Divorce Counseling and Divorce Mediation

Couples negotiating whether to end their marriage require a "safe" communicative space where one or both can air their feelings. If divorce indeed is imminent, a skilled divorce counselor—one credentialed in marital therapy—will be essential. A divorce counselor may need to offset negative attributions and partner-blaming. When necessary, a divorce counselor may need to help a partner or a couple come to terms with the finality of their marriage. If effective, a couple will avoid framing divorce as the fault of the other and will enter divorce or divorce mediation in a cooperative, non-combative state.

In divorce mediation, spouses discuss and try to resolve issues before a neutral third party, usually a family law mediator, prior to legal divorce proceedings. Rather than making decisions for the couple, a divorce mediator helps both parties understand what is best for their situation. A mediator, for example, may offer insight regarding the legal system and how an issue may be viewed by a judge. He or she may alert the couple to potential complications of which they may be unaware. Plans regarding the distribution of marital assets, debts, spousal payments, and the care of children also are drawn. Successful mediation is dependent on both spouses' openness to compromise and understanding each other's points of view.

Once negotiations are completed and agreements are reached, the divorce mediator or the couple's attorneys can help the couple prepare court documents for the dissolution of the marriage. These and other finalizing documents make up part of the terms supporting a partner's divorce judgment, which the court enforces. Obligations surrounding the distribution of marital debt and property, judgments on child custody and support, and, if necessary, spousal support are a critical part of mediated talks (Doskow, 2014).

Spousal Support

Alimony is spousal payment from the financially stronger spouse to the spouse with fewer financial resources. The purpose of this form of spousal support is to correct inequities created by divorce. The amount of payment takes into consideration the periods of time and conditions established by the court as well as certain circumstances that had occurred during the marriage. For example, the court may award spousal support based on a spouse's contributions to the marital home or if he or she subsidized the paying spouse's education. If getting a divorce generates a need or a deficiency in the lifestyle, that spouse may petition the court for more alimony. Women should not overlook the possibility that they may have to pay alimony to their former spouses. The incident is rare among most socioeconomic groups; however, marriages in which wives earn a very high salary may experience this reversal (see Doskow, 2014).

Child Custody

Child custody, a legal term involving the care, control, and maintenance of a child, may be court-awarded to one or both parents following divorce or separation. There are four primary forms of child custody: physical custody, joint custody, legal custody, and sole custody (see Doskow, 2015). Let's examine each.

Physical Child Custody Having *physical custody* of a child means that only one parent and the child have the right to live in the same household. The parent determined by the court to provide the child's daily care and residence is the *custodial parent.* The other parent, known as the *non-custodial parent*, has the right of visitation or parenting time with the child. Court-appropriate visitation privileges generally allow non-custodial parents exclusive parent–child interaction for a number of weeks during the summer months, every other weekend, and alternating major holidays.

Joint Child Custody *Joint child custody*, also called *shared custody*, occurs when both parents share in the decision-making, responsibilities, and physical control of their child. Parents are court-authorized to maintain a predetermined schedule around their children's needs and in accordance with parents' work and housing arrangements. The most common joint-custody arrangement is for the child to divide parental time for a number of weeks per month or to spend weekends and holidays with one parent and weekdays with the other. Joint custody may be arranged whether both parents are divorced, separated, no longer cohabitating, or never had lived together.

Legal Child Custody Parents who have been awarded *legal custody* of their child have court authority to decide their child's upbringing. Parental cooperation is important to agreeable decision-making with regard to the child's education, medical regimen, religious affiliation, and other considerations. Many states regularly award joint legal custody after a divorce.

Sole Child Custody The court may award one parent exclusive physical and legal rights to a child. This form of child custody arrangement, *sole custody*, is commonly ordered when one parent is deemed unfit or incapable of responsible parenting (evidence of substance abuse, child abuse, financial difficulties).

Child Support

In the eyes of the court, a parent has a standing fiscal responsibility to his or her child, despite parental disputes over child visitations or other divorce matters. The custodial parent generally meets his or her child's financial obligations through the custody itself. For a non-custodial parent, child support payments are made to contribute to the custodial parent's expenses toward childrearing. For example, if one parent is awarded sole or physical custody, the other may be required to make child support payments to the custodial parent.

Parents who are awarded joint physical custody have support payments that are based on the earnings of each parent and the percentage of time the child spends with each parent. Additional factors differentiating the amount of joint-physical child support between parents involve a parent's ability to earn a greater income, whether the child has special needs, and which parent pays for the child's care and health insurance. Child support payments can last until the child is eighteen, or older if the child has confirmed special needs (see Doskow, 2015).

Why Do Couples Divorce?

Divorce is no longer a whispered event. The topic has made its way into television screenplays, movies, and theatre productions. Websites on "Hollywood's Shortest Marriages" and "Hollywood's Messiest Divorces" are a keystroke away, and internet links like "Celebrity Marriages That Are Still Going Strong" allude to the irregularity of intact celebrity couplings. For the rest of us, books on how to navigate the emotional, social, economic, and parental aftermath of divorce can be ordered online, mailed to our homes, or purchased from local and national booksellers. Taken as a whole, our culture seems to have recognized divorce as an option to solve marital conflict, unhappiness, and spousal incompatibility.

Sociodemographic Predictors of Divorce

But how can we know that a marriage has "run its course"? Research suggests that variables predicting divorce manifest separately but, more often, in combination. This section addresses sociological and communicative variables, whose effects may forecast the dissolution of some marriages: socioeconomic status, intergenerational divorce in families, pre-marital cohabitation, and styles of interpersonal interaction. We begin our analysis of divorce predictors with a look

Upon completion of this section, students should be able to:
LO10: Explain sociodemographic and socioeconomic predictors of divorce.
LO11: Describe the intergenerational transmission of divorce.
LO12: Describe how premarital cohabitation may be a predictor of divorce.
LO13: Describe interpersonal models of marital relationship dissolution.

at the effects of gender, race, ethnicity, and culture—sociodemographic predictors of divorce.

Gender

People's experiences in marriage are as gendered as men's and women's personal accounts of why divorce occurred. Sociologists Paul Amato and Denise Previti (2003) found that former wives are more likely to self-report problematic personalities or behaviors on the part of their former husbands (infidelity, substance use, mental and physical abuse). Former husbands recall "poor communication" or express uncertainty as to what caused divorce. The researchers made special note of men's self-report of communication issues, suggesting that many married men have grown more sensitive to relationship dynamics. They add, however, that their respondents' frequent reference to "poor communication" is consistent with other studies suggesting that some men still have a tough time articulating problems in the relationship. Results from other studies show that men more often attribute divorce to external factors, such as work or issues with in-laws (Kitson, 1992).

Men and women also may have different views on what is sufficient for a sustainable marriage. Spousal affirmation was found to predict marital stability among husbands but not their wives. Husbands who reported feeling affirmed by their wives also showed a lower risk for divorce than husbands who did not feel affirmed (Orbuch *et al.*, 2002). Meanwhile, wives are more likely to initiate divorce (Hetherington & Kelly, 2002; Kitson, 1992), particularly when young children, physically abusive husbands, or husbands with substance abuse problems are apparent (Amato & Previti, 2003). In a sense, a motivation to protect oneself or one's children may explain some women's departure from marriage.

Race and Ethnicity

The probability of divorce or separation varies by race and ethnicity. For Asian women, first-time marriages dissolve at a rate considerably slower than other racial/ethnic groups: after ten years of marriage, only 20 percent were disrupted by divorce or separation. Thirty-two percent of White women's first marriages dissolved after ten years compared to 47 percent of first marriages by Black women. At 34 percent, the probability of divorce or separation among Hispanic women is close to that of White women (Bramlett & Mosher, 2002).

The decision to divorce or to separate may pattern around race/ethnicity and socioeconomic status (SES). About two-thirds of all Blacks, Whites, and Hispanics say that it is better for the children if their "very" unhappily married parents choose divorce over remaining together. However, Whites with troubled marriages are more likely to complete a legal divorce; Blacks and Hispanics are more likely to separate permanently (Bramlett & Mosher, 2002).

Incidents of marital dissolution also pattern around race/ethnicity, education, and gender. For example, a higher education level tends to lower the risk of divorce among White husbands, White wives, and Black wives but not Black husbands (Orbuch *et al.*, 2002). Hispanics with less than a high school education are less likely than Whites to divorce (National Healthy Marriage Resource Center, 2016). Hispanics also experience longer first-time marriages than other racial/ethnic groups, a possible result from marrying at a younger age (Aughinbaugh *et al.*, 2013).

Effects of Culture

Culture may explain why racial/ethnic groups differ in certain aspects of divorce and separation. Culture-specific observances that prioritize group needs over the personal needs of group members partly may explain lower divorce rates among many Asian couples. Faith-based observances that regulate marital life and support marital permanence are unique to many Hispanics, a factor that may explain some couples' reluctance to transition from separation to divorce. We also have learned that successful unions and intact marriages have some association with a couple's economic prosperity. As a group, Blacks' relatively low median income would suggest that it may be difficult for some couples to meet either bar. Moreover, the stressors that often follow economic disparities may overflow into the relationships of Blacks who marry (Bramlett & Mosher, 2002).

Socioeconomic Predictors of Divorce

Low-socioeconomic-status (SES) individuals are more likely to divorce than high-SES individuals, and they may divorce for different reasons compared to their more affluent counterparts. In recalling their experiences with marital conflict and divorce, couples of low SES tend to contour former grievances around more instrumental factors: financial difficulties, employment problems, and leisure time with friends, in addition to gambling and criminal activities and intimate partner violence (Amato & Previti, 2003). High SES divorced partners, who are often college-educated and less pressured by severe economic adversity, cite different terms of incompatibility. Specifically, high SES divorced couples tend to attribute marital dissolution to expressive, more relationship-centered explanations: poor couple communication, conflict in values or interests, gender role conflict, ex-partner self-centeredness, a lack of love, excessive demands, and marrying too young.

Intergenerational Transmission of Divorce

Children growing up in households in which one or both parents have divorced show a higher than average risk to see their own marriages end in divorce. This increased possibility of divorce among children of divorced parents is known as the **intergenerational transmission of divorce**. Researchers explain that children's prolonged exposure to ineffective communicative behaviors may encourage, as adults, interactions that interfere with sustaining mutually rewarding intimate relationships (Amato, 1996; Wolfinger, 1999).

Premarital Cohabitation and Divorce

The incidence of premarital cohabitation or living together before marriage is viewed as a predictor of divorce, but only under certain circumstances. If partners show less commitment to the relationship or hold a lower standard for those with whom they live than those whom they marry, for example, the likelihood of discord and divorce may increase. Conceivably, there are barriers to ending a cohabiting relationship (having children, sharing possessions, co-owning pets), which may accumulate after a couple lives together and later marries. However, these buffers may facilitate future marital disruption, as they essentially "pull together" two people who may have not married under other circumstances. There also is evidence that married couples who were engaged to be married prior to cohabitation are likely to report fewer relationship problems and greater stability. Researchers speculate that both partners demonstrate a deep mutual commitment not only to each other but also to the relationship (Amato, 2010).

Interpersonal Models of Marital Relationship Dissolution

It is perhaps not surprising that certain interpersonal behaviors may provoke the likelihood of divorce. In one national study, sociologists reported infidelity or "cheating" as the most common self-reported explanation for divorce. Whether infidelity was the direct cause of divorce or a consequence of estrangement was beyond the study's scope. Nonetheless, infidelity was found to influence how some people assess why their marriage unraveled. Spousal incompatibility, growing apart, personality clashes, and lack of communication—negative interpersonal actions that can adversely affect a relationship's quality—also were prominent among personal accounts (Amato & Previti, 2003). Other marriage and family researchers have examined the interpersonal dynamics of marital relationships and their role in marital dissolution. We examine two of them below.

Gottman's Model of Marital Dissolution

Psychologist John Gottman and colleague (1999) identified four destructive interaction styles that often deepen two partners' isolation from each other. Gottman explains that, if done with high frequency, these interactions may signal a strong potential for divorce. We explore Gottman's model, which he characterizes as the **Four Horsemen of the Apocalypse**.

- *Horseman 1: Criticism:* A complaint may focus on a specific behavior; however, a criticism tends to attack a person's character. To tell a partner, "You always talk about you—what you want" or "You are self-absorbed and arrogant" will only escalate the problem.
- *Horseman 2: Defensiveness:* Defensiveness often is a response to a criticism, and it usually manifests as blame. In effect, one partner is saying to the other, "The problem isn't me. The problem is really you."
- *Horseman 3: Contempt:* Contempt is marked by angry statements that are seemingly from a position of superiority. To verbalize, "I'd never sink so low as to do something like that! Exactly, what kind of person are you?" suggests to a partner that he or she is unwanted or devalued. Sarcasm, cynicism, name-calling, eye rolling, sneering, and hostile humor, which often accompany contempt, are often thinly veiled and obvious.
- *Horseman 4: Stonewalling:* Stonewalling occurs when one partner feels overwhelmed and, therefore, shuts down or withdraws during interactions with the other partner.

Levinger's Model of Marital Dissolution

Psychologist George Levinger (1966) studied how partners process their decisions on whether to divorce or remain married. Levinger's approach to understanding how couples weigh and balance the consequences of dissolution lies in three principles: the *attractions of marriage* (rewards including positive social status, wealth accumulation, coparenting, sexual intercourse, and supportiveness), the *barriers to divorce* (the loss of reward, if the marriage ends), and *alternative attractions* which may draw a spouse away from marital commitment (the prospect of sexual freedom and withdrawal from family finances or other pressures).

For example, a spouse may choose to leave his or her marriage if support or other marital rewards are perceived as too low or too infrequent. But not all negotiations are as clear or inevitable, especially if the consequences of divorce are too costly. An individual who is miserable and married may not decide to divorce, as pending singlehood, divorce expenses, or disapproval from family and peers may be even more troubling. On the other hand, another individual may be willing to accept the costs of divorce if dissolution opens up the prospect for more satisfying relationships, greater well-being, or a sense of freedom.

Diversity Overview: Divorce and Families Headed by Same-sex Couples

For divorcing same-sex couples, some areas of legal settlement may prove uniquely arduous. Keep in mind that many same sex-couples may have been in their relationship for years, sometimes decades, before being allowed to legally marry. During that time, a couple may have purchased property together, merged assets or finances, and reared children. Still, most divorce courts divide assets accumulated only from the time a couple legally married.

Another concern for divorced gay and lesbian couples has to do with parental rights and parental classification. For example, one or both parents may not be the biological mother or father of a child who was jointly reared during the marriage. As a result, additional legal concerns and costs may accumulate if a divorcing parent or divorcing couple has not taken appropriate steps to secure their legal parental standing. Obtaining precise answers to some legal questions can lead to a serious delay in progress as well as personal frustration. This is because the recent legality of same-sex marriage and divorce may require legal motions and arguments where none had previously existed (Kim, 2013).

Upon completion of this section, students should be able to:
LO14: Describe the effects of divorce on children.

Effects of Divorce on Children

Some children and adults of divorced parents may come across as minimizing the effects of marital dissolution. They may assert that they "got over it" or that they "turned out ok." This indeed may be true. However, this is not to say that children and adults of divorced parents are in no way poorly affected or that recovery is easy.

Few family scholars would disagree that, as a group, the children of divorced families appear to encounter additional life issues. What those lasting effects may be has led to sometimes polarizing debates among family scientists. The most watched have been from researchers who focus on children's resiliency despite divorce and those who caution the short- and long-term difficulties of being reared in a divorced family. The often-opposing observations of two eminent family scholars, Judith Wallerstein and E. Mavis Hetherington, are an example.

Divorce as a Lasting Harm?

In the 1970s, psychologist Judith Wallerstein, along with colleague Julia Lewis (Wallerstein & Lewis, 2004), began a study of 131 children age three to eighteen whose parents had divorced. The sample of children encountered significant disruptions during their parents' economic, social, and marital transitioning. They had undergone a change of neighborhoods and schools, a loss of friends, concerns over the non-residential parent's well-being, and disturbances in otherwise normal activities. Wallerstein's findings established that, even a decade later, many children of divorced parents experienced not only stress and pain but also promiscuity, fear of betrayal and abandonment, and other maladjustments while appearing to have adjusted to divorce. In summarizing the study's findings, she emphasized, "Out of their experience of the parental breakup, children of all ages reached a conclusion that terrified them: Personal relationships are unreliable, and even the closest family relationships cannot be expected to hold firm" (Wallerstein & Lewis, 2004, p. 359).

For 25 years and during regular intervals, Wallerstein tracked the sample, which, by the end of her study, was age 28 to 48 years. In her follow-up efforts, she again cautioned that marital dissolution was more than an acute stressor from which children recover. Rather, for Wallerstein, the complications of post-divorce family life were influential enough that children experienced distress well into adulthood. (Wallerstein & Lewis, 2004).

Divorce as a Recoverable Harm?

Other researchers suggest less dire prospects for children of divorced families. Psychologist E. Mavis Hetherington sampled close to 1,400 families and over 2,500 children—some of whom were followed for 30 years—to learn the implications of marital breakup for adults, families, and children. Like Wallerstein, Hetherington's sample of children depicted the challenges of post-divorce life. Children demonstrated social, emotional, or psychological issues at a rate 15 percent higher than did children of intact families. Twenty to 25 percent reported depression, impulsiveness, or anti-social behaviors (Hetherington & Kelly, 2002). Even so, Hetherington maintained that divorce is not synonymous with "inflicting a terminal disease on your children" (Peterson, 2002), a stark contrast to the tone of Wallerstein's research.

Hetherington found that 75 to 80 percent of children from divorced marriages were beginning to cope reasonably well within two years after their parent's divorce and were functioning within the normal range. She emphasized that most young adults from divorced families are not unlike their peers: they were choosing careers, developing permanent relationships, and adapting to their new lives as young people. In fact, 70 percent of young adults from divorced families described divorce as acceptable, even if children are present. Forty percent of young adults from intact families feel this way (Hetherington & Kelly, 2002; Peterson, 2002).

Many researchers have found support for Hetherington's claim that the majority of children from divorced parents are resilient. These often-quantitative researchers invariably look to peer relations, socioeconomic status, general distress, or poor parenting skills to explain maladjustment behaviors in children. There also has been disagreement over Wallerstein's research design (no control group); the homogeneity of her sample (largely Caucasian, well educated, and middle class); her findings' lack of generalization to other groups, and her basic conclusion that satisfying adult relationships will likely elude the children of divorce. Contradicting Wallerstein's work by name, one researcher found that young adults of divorced parents can experience agreeable parental ties, particularly when parents were supportive of time spent with the other parent (Fabricius, 2003). Generally, these family scientists believe that evidence accumulated over decades of divorce research suggests that the estimated effects of divorce are not as strong as Wallerstein claims.

Is divorce the *direct* cause of a child's dysfunction as suggested by Wallerstein and even Hetherington? The answer is less precise than one might hope, as scholars progressively have formed a more complex understanding of how divorce impacts children. For example, some findings suggest that the negative effects experienced by some children may not start until after parents begin divorce proceedings (Kim, 2011). Yet, other findings suggest that children's psychological problems may have begun in the years preceding their parents' divorce (Amato *et al.*, 2011). In light of these competing messages, experts have called for a reconsideration of divorce, per se, as the principal basis of maladjustment among children of divorced parents (Kelly, 2000). More certain is that even the most resilient child can develop problems over divorce if risk factors for child maladjustment outweigh otherwise protective factors. Protective factors include a solid parent–child relationship, protection from parental conflict, and having parents of sound psychological well-being.

A "Good Divorce" for Children?

Some experts have asserted another analytical approach to address whether divorce principally causes maladaptive behaviors in children. Rather than ask if divorce affects children, these researchers question how and under what circumstances divorce may affect children either positively or negatively (Amato, 2010). Also important, does the character of divorce—a "good," parentally cooperative divorce versus a "bad," contentious one—have more influence on a child's well-being than divorce itself?

The concept of a "good divorce" was derived from the findings of a two-decade longitudinal study by psychologist Constance Ahrons (1994). Ahrons points out that a "good divorce" is one that reduces two major divorce-related stressors among children: conflict between parents and weakening father–child relationships. Its affect is a result of former spouses developing a parenting partnership that preserves kinship bonds and parents' responsibilities to their children's emotional, economic, and physical needs (Ahrons, 2011). In recent years, divorce researchers have placed the validity of the "good divorce" concept under social science scrutiny. In their conclusions, they argued that Ahrons's premise may be more limited than characterized (Amato *et al.*, 2011).

Amato *et al.* (2011) assessed the "good divorce" premise using a 944 family sample, which they coded into three parenting clusters: cooperative coparenting, parallel parenting, and single parenting. Their study concluded that children in the cooperative coparenting or "good divorce" cluster had the smallest number of behavior problems and the closest ties to their fathers. Nonetheless, adolescents in the "good divorce" cluster were found to be no better off than adolescents in the single parenting cluster with respect to self-esteem, school grades, liking school, substance use, or life satisfaction. Further, young adults in the "good divorce" cluster were shown to be no better off than were young adults in the single parenting cluster in the areas of substance use, early sexual activity, number of sexual partners, cohabiting or marrying as a teenager, and closeness to mothers. Although the researchers added that they had hoped otherwise, their study's results showed only partial support for Ahrons's "good divorce" hypothesis. Ahrons, in return, offered that even though the authors depicted the "good divorce" as a "panacea" for all stressors experienced by all children during and after divorce, it was not what she had claimed and such descriptions are exaggerations (Ahrons, 2011).

A Word of Caution

It is sobering, if not disheartening, to "take in" these data on children, divorce, and the potential for negative behavioral outcomes. But keep in mind that it is not inherent that children are socially, relationally, or otherwise "doomed" by their parents' divorce. Nor should these risk patterns be generalized to every child or adult with divorced parents. Recall that research findings offer only a correlation for the likelihood of problems when compared to the problems of children whose parents are not divorced. True, behavioral risks involving poor school performance, maladaptive psychological well-being, and "acting out" have been cited with regard to children of divorced parents. However, it is also true that none of these concerns is restricted to children of divorced families or that a child may demonstrate behavioral problems from eight to twelve years prior to divorce (Amato & Booth, 1996).

We also emphasize that some children of divorced parents quickly adjust to marital dissolution. Children may develop better relationships with their non-residential parent and extended families when both parents are involved positively in their children's lives (Ahrons & Tanner, 2003). Further, children tend to show improved well-being when divorce removes them from high-conflict households (Amato *et al.*, 2011). In fact, many children of divorced parents who had displayed long-standing problems with transitions during childhood and adolescence have gone on to lead happy, well-adjusted lives. You may know one or several. You, yourself, may be an example.

Effects of Divorce on Adults

When couples say "I do," most enter matrimony with hopes of sustaining a dual respect and a lasting unity. When divorce looms and a marriage ends, some former spouses suffer only temporary setbacks. For others, however, the dissolution of marriage is further

Upon completion of this section, students should be able to:
LO15: Identify the six stations of divorce.
LO16: Describe divorce stressors.
LO17: Explain the positive effects of divorce.

aggravated by a range of stressors and uncertainties (Amato, 2000). In this section, we focus on the challenges wrought by getting a divorce and their effects on adults.

The Six Stations of Divorce

Before cultivating a sense of resolve, many divorced men and women first travail a series of stages that may occur in varying order and different levels of intensity. These stages, or stations, are central to anthropologist Paul Bohannon's (1970) theory, the **six stations of divorce**.

Emotional Divorce

In an *emotional divorce*, partners increasingly are aware of their discontent and dissatisfaction. Mutual trust erodes, criticism accelerates, and charges and countercharges of unfilled promises are leveled. Despite their dislike or suspicion of marital deterioration, some spouses may normalize their experiences; others may disbelieve that the marriage is seriously at risk. Bohannon explains that if marital therapy offers no reconciliation, the grief for some may be comparable to the death of a spouse.

Legal Divorce

Divorce is not acquired through simple consent. It is preceded by lawful actions and Bohannon's second station, a *legal divorce*. In this station, spouses are represented by legal counsel, whose role is to protect spousal support and property interests and, if necessary, child support and clients' custodial or visitation rights. Former spouses are free to remarry during this station. Nonetheless, legal proceedings may become competitive, personal, and embittered at this point, giving further cause for negative feelings.

Economic Divorce

The *economic divorce* points out how divorce can be a stratifying event. The station is characterized by court-ordered decisions regarding monetary settlements, including financial and property divisions, spousal maintenance, and child support payments.

Co-parental Divorce

In the *co-parental divorce*, former spouses decide upon child custody, child visitation arrangements, and other adjustments regarding their parental obligations. The co-parental divorce station can often intersect with the legal and economic stations. For example, the custodial parent may petition the court to terminate visitation and compel the non-custodial parent to increase or to reinstate child support payments.

Community Divorce

A *community divorce* reflects the experience of separating from one's circle of friends and other social relationships. Mutual friends may feel torn or pressured to take sides. Some may remain friendly with only one former spouse. In the meantime, the divorced couple may no longer feel as comfortable with their still-married friends.

Psychic Divorce

In the *psychic divorce*, Bohannon indicates a movement toward autonomy; however, the station still has its challenges. Each former spouse is separated from the other's influence; there is no longer a partner to object to or to rely upon. If unaccustomed to accepting full responsibility for his or her own thoughts or actions, a divorced individual may find this station especially intimidating. Further, the psychic divorce also is marked by deep personal introspection: individuals may ask why they chose their former partners, why they married, and why they ultimately divorced.

Many unfavorable emotional, financial, and legal measures may occur before Bohannon's final station of divorce is reached. But how do social science theories explain the all too apparent link between stress and marital dissolution? To find out, researchers tested three basic theories: that the role of being divorced is more stressful than that of being married (social role theory); that the life changes and role transitions of dissolution produce higher stress (crisis theory); and that married persons with pre-existing mental health problems are often inadequate marital partners and more likely to be selected into divorce (social selection theory). Findings suggest that the higher stress levels of the divorced primarily reflect the effects of social role; the selection and crisis effects made small contributions only (Johnson & Wu, 2000).

Additional Divorce Stressors

Divorce can impose an array of additional stressors. These may unfold over a course of months or, sometimes, years. These affronts include economic pressures as well as individual health-related responses. Each is discussed below.

Post-divorce Economic Stressors

Divorce is a costly process and long-time financial hurdles may follow. Income once allotted for one household is split into two economies of scale, and debts and assets are divided between former spouses. On average, an estimated 30 percent increase in income is required to maintain the same standard of living as prior to divorce (Sayer, 2006).

A woman may find herself financially ill-equipped for solo parenting soon after divorce. While there are recently divorced women who earn a decent income and whose former spouses reliably conduct child-support payments, others may be financially challenged and have greater use of public assistance (Elliot & Simmons, 2011). The proportion of custodial mothers with incomes below poverty level (31.8 percent) is almost twice as high as that for custodial fathers (16.2 percent) (Grall, 2013). Moreover, family responsibilities require about four in ten mothers to take time off from work, to reduce work hours, to care for a child or other family member, or to quit work altogether (Blau & Kahn, 2007).

The attention given to women and children after divorce may imply that men are financially better off after a divorce. The truth is that some research contests this implication. Divorced men's financial loss usually depends on the amount contributed to the family's income. Former husbands who contributed less than 80 percent of a family's income before divorce are likely to suffer financially from marital dissolution. Aside from paying for a residence, men may experience a post-divorce decline in their standard of living due to compulsory or voluntary child support payments and income distributed to former spouses. Former husbands contributing more than 80 percent of a family's income before a divorce do not suffer as much financial loss and may even improve their financial situation (McManus & DiPrete, 2001).

Post-divorce Health Stressors

Men and women may experience decreased mental and physical health when involved in a divorce. Divorced individuals are shown to experience less happiness, more symptoms of depression, more social isolation, more negative life events, and more health problems than their married counterparts (Amato, 2014). Other health outcomes associated with divorce include a lack of sleep, increased tension, and weight gain or loss. Metabolic syndrome, a medical condition that increases the potential for heart disease, stroke, and diabetes, is more likely to develop among divorced women when compared to women in happy marriages (Troxel *et al.*, 2005).

Despite the popular perception that men generally are less vulnerable to trauma than women, research finds that divorce can take a toll on men's health. Non-resident parents, usually fathers, may find the loss of daily contact with their children a new and distressing reality (Amato, 2000). Compared to married men, divorced men are more likely to participate in high-risk activities (abusing alcohol or drugs) and have higher mortality rates from cardiovascular disease, hypertension, and stroke. The rate for suicide among separated and divorced men is 39 percent higher than that of married men. Depression is more common for divorced men as well, who undergo psychiatric care ten times more often than those who are married. And while popular media depict bachelors as having better or more frequent sex, married men report more frequent and satisfying sexual relations than do divorced men (Felix *et al.*, 2013). In sum, divorce can generate significant mental and physical effects on former husbands and former wives. The fact that spouses often encourage each other to lead healthier lives or to reduce potentially harmful behaviors (excessive alcohol use, smoking) suggests that divorce may remove some beneficial effects of marriage as well (Amato, 2014).

Former Spouses Adjusting to Divorce

As with any major life change, the rate of recovering from divorce will differ from individual to individual. Success often is tied to a person's access to social resources (family, friends, new romantic partners, and other systems of emotional support) and economic resources (having an adequate income and a high level of education to absorb the costs of divorce and post-divorce life). Additionally, having or acquiring good coping skills—that is, the ability to "see" divorce as something other than a tragedy or personal failing—has restorative effects. Later in the chapter, we will examine some positive outcomes associated with separation and divorce. First, we revisit psychologist E. Mavis Hetherington (2003) and her classification of former partners' adjustment after divorce.

Hetherington's Typology of Post-divorce Adjustments

In Hetherington's ten-year study of 238 divorced wives and 216 husbands, 20 percent of participants, the *Enhancers*, showed success as parents and grew more fulfilled after divorce. This subcategory, primarily divorced women, progressed in their social and professional interactions and often remarried. The *Competent Loners*, also mostly divorced women, comprise 10 percent of Hetherington's sample. In many respects, Competent Loners and Enhancers are fairly comparable. However, Competent Loners better sustained themselves emotionally and engaged more often in short-lived intimate relationships than Enhancers. Another 10 percent of the sample, the *Defeated*, comprised divorced men and women who were psychologically vulnerable to depression, a lack of direction, feelings of purposelessness, and self-medication via substance abuse. Contrary to the Defeated, Hetherington's subcategory, the *Seekers*, tended to remarry soon after divorce to regain a sense of purpose and structure. The *Swingers*, like the Seekers, were primarily composed of men. Unlike Seekers, Swingers engaged in faster, trendier lives and likely divorced due to moral infractions. By the end of the first year after divorce, Swingers tended to view their lives as empty and sought committed intimate relationships. Seekers and Swingers totaled 20 percent of Hetherington's sample.

Hetherington's last subcategory, the *Good Enoughs*, was the largest with a 40 percent composition. Although Good Enoughs reported that divorce was stressful and discomforting, marital dissolution led to no lasting negative or positive impression. Years later, many in this category exhibited the same pre-divorce problems but with different intimate partners.

Positive Effects of Separation and Divorce

The negative effects of divorce are documented by family scientists and implicit to most others. So, it may be awkward to think of divorce as having some appreciable outcome. That the state and legal entities may have good cause to grant divorce is also worth considering.

A Respite from an Unsafe Marriage

Divorce can be a strenuous, although constructive decision for adults and children in turbulent marriages. Not having to endure fear of personal safety or constant fighting is an appreciable experience for some family members. Separation offers the space to work towards personal healing or possible couple

Source: Sangoiri/Shutterstock

reconciliation, if determined safe and appropriate. In a more practical sense, time apart allows the opportunity to arrange for critical support services, an important step if divorce is pending or if threat to personal safety is a concern.

At the very least, children are removed from daily household tensions when conflict is controlled through their parents' living apart. In their study on marital disruption and child well-being, Morrison and Coiro (1999) asked the question, "How do children fare when their high-conflict parents remain together?" The researchers found that separation and divorce were associated with increased behavioral problems in children regardless of the level of conflict between parents. High levels of parental conflict were linked to even greater increases in behavioral problems among children in marriages that did not break up. In fact, many children who witnessed chronic parental conflict are relieved after separation.

A Time for Personal Development

Divorce removes individuals from a dysfunctional marriage and puts an end to dashed hopes of trying to cultivate a happy, purposeful, or workable family alliance. Divorced adults are permitted to live separately, allowing them the freedom to develop identities and aspirations that are untied to their former marriage. A former wife, for example, may decide to return to the labor force or enhance her education. Most divorced individuals eventually make new friends and establish new romantic relationships (Amato, 2000).

Divorced Parents Cooperating

Divorce or separation may serve as a chance for former spouses to stabilize their relationship by forming an alliance for their children. As stress diminishes and adjustments settle, parents can make more impartial plans for their children's future. Time apart also can encourage some parents to learn to value their time with their children, which may lead to a deeper parent–child connection. Non-resident parents, for example, may put forth greater effort to spend quality time with their children or give more attention to their children's desires.

Global Overview: Associations between Divorce and Murder in the Caribbean

In their 2014 journal article, Paul Andrew Bourne of Socio-Medical Research Institute in Jamaica and colleagues (2014) questioned whether marriage is linked to homicides in Jamaica and if legal separation is a "public health concern." According to researchers, their investigation is a timely one, highlighting gaps in the existing literature on the risk of fatality in relationship dissolution in Jamaica. Accounting for 82.2 percent of the variance in murders in Jamaica, the study found a strong direct correlation between murder and divorce between 1950 and 2013. Separations involving cohabiting couples and couples involved in sexual relationships demonstrated a similar correlation of fatalities.

The researchers characterized the homicide rate as "pandemic," a "separation-health phenomenon," and an outcome of severe dysfunction in relationship decline. In previously documented cases in which husbands killed their wives, risk factors for homicide were attributed to sexual promiscuity, reproductive health issues, unsatisfied sexual needs, having a relationship of short duration, and a lack of stability, in addition to a spouse wanting to leave the relationship and wives wanting to move on with their lives. Findings also showed that these issues not only resulted in husbands committing homicide against their wives but also their resorting to suicide. The current study did not examine the psychological state of husbands following legal separation. However, the researchers deduced that the psychological state of spouses during and after divorce was lower than of spouses in stable, loving marital unions.

That some women will be killed by their spouses is by no means limited to Caribbean nations. The study cites similar homicides in the United States, where married women were murdered by their partners at a rate of 13.11 women per million married women per year. Cohabiting women were murdered at a much higher rate: 116.06 women per million cohabiting women per year. Researchers have called for public health intervention prior to a couple's legal separation with additional prevention services following spousal separation.

Summary

Several signs may suggest that a relationship with an intimate partner is in trouble. "Taking a break" from a dating relationship may seem counterintuitive to some. However, time apart can offer the space to reflect on how partners treat each other and themselves. If married, a couple's decision to separate demands cooperation and some adjustment but should not be viewed as tantamount to getting a divorce. Most couples are positively affected by therapy, although too many couples enroll in therapy six years later than is ideal.

If divorce is imminent, the relationship may experience an observable trajectory from deterioration to permanent breakdown, a process known as "uncoupling." A divorce counselor may help couples to offset negative attributions and partner-blaming. To navigate the legal aspects of divorce, a divorce mediator should be considered. Successful mediation depends on both spouses' openness to compromise, particularly with matters of child and/or spousal support and child custody arrangements.

Few family researchers would disagree that children of divorced families appear to encounter additional life challenges; however, the lasting effects of those challenge have led to sometimes polarizing debates. Some experts focus on children's resiliency despite divorce; others focus on the short- and long-term difficulties from being raised by divorced parents. Meanwhile, a cadre of experts has taken a different analytical approach. These researchers explore how and under what circumstances divorce may influence children either positively or negatively. The question of whether there is a "good divorce" for children has been raised and, subsequently, placed under analytic scrutiny.

The effects of divorce on adults also have undergone analytical attention. Before cultivating a sense of resolve or stability, divorced individuals may experience a series of phases, which may occur in varying order and different levels of intensity. Affronts involving economic pressures or health-related reactions also may occur. Although an unsettling experience for many, divorce or separation can serve as a respite from an unsafe marriage, a time for individual growth, and an opportunity for former spouses to develop an alliance for their children.

Key Terms

Alimony is spousal support payments from the financially stronger spouse to the divorcing spouse with fewer financial resources.

Child custody is a term used to describe the legal relationship between a child and a parent who has been awarded the care, control, and maintenance of that child; it includes *physical, joint, legal,* and *sole child custody* arrangements.

Covenant marriage is a legally designated form of marriage in which marrying spouses consent to pre-marital counseling and more limited grounds in seeking divorce.

Crude divorce rate is the divorce rate of a given population in a given timeframe as measured by the number of divorces per 1,000 people.

Divorce is the legal terminology of a marital union; also called *marital dissolution*.

Four Horsemen of the Apocalypse is a characterization of four negative communicative behaviors—criticism, defensiveness, contempt, and stonewalling—that, if occurring regularly, may signal a failed or terminally unhappy marriage; developed by psychologist John Gottman.

Intergenerational transmission of divorce is an occurrence in which one or more parents' experience with divorce increases the likelihood that their child's marriage will end in divorce.

No-fault divorce is a divorce in which neither spouse is required to prove fault or wrongdoing.

Refined divorce rate is the divorce rate of a given population in a given timeframe as measured by the number of divorces per 1,000 married women.

Separation is the condition of two people who had been married or lived together, living apart; it includes *trial separation, legal separation*, and *permanent separation*.

Six stations of divorce are the six distinct but interrelated features of the divorce process: the *emotional divorce, legal divorce, economic divorce, co-parental divorce, community divorce*, and *psychic divorce*; developed by anthropologist Paul Bohannon.

Discussion Questions

- Interview two groups of heterosexual adults about their views on the U.S. Supreme Court decision to permit same-sex marriage. Your respondents should include six men and women between the ages of 18 and 40 and six men and women ages 41 and over. Compare and contrast the two age groups' responses, paying close attention to how age, gender, and other social demographics (race/ethnicity, religion, education, or contact with gays and lesbians) may have influenced their opinions. When examining the data collected, what similarities and differences were found between the two age groups?
- Consider the following statement: "Fifty percent of all marriages in the United States end in divorce." Explain the problem with this statement. What are the complications of determining the exact percentage of marriages that end in divorce?

Suggested Film and Video

Hawkins, I. (Director). (2013). *A Nuclear Family*. (Film). Hawkinsian Productions.

Real Life Teens: Broken Homes. (2003). (Film). Films on Demand/Film Media Group.

Why Do We Cheat? (2011). (Film). Films Media Group.

Internet Sources

Divorce Magazine. Retrieved on September 28, 2016: http://www.divorcemag.com/home

KidsHealth: A Kid's Guide to Divorce. Retrieved on September 28, 2016: http://kidshealth.org/en/kids/divorce.html

References

Ahrons, C. (1994). *The Good Divorce*. New York: Harper Collins.

Ahrons, C. (2011). Commentary on Reconsidering the "Good Divorce." *Family Relations*, 60, 528–532.

Ahrons, C. & Tanner, J. (2003). Adult Children and Their Fathers: Relationship Changes 20 Years after Parental Divorce. *Family Relations*, 52, 340–351.

Amato, P. (1996). Explaining the Intergenerational Transmission of Divorce. *Journal of Marriage and the Family*, 58, 628–640.

Amato, P. (2000). The Consequences of Divorce for Adults and Children. *Journal of Marriage and Family*, 62, 1269–1287.

Amato, P. (2010). Research on Divorce: Continuing Trends and New Developments. *Journal of Marriage and Family,* 72, 650–666.

Amato, P. (2014). Drustvena Istrazivanja. *Journal for General Social Issues*, 23, 5–24.

Amato, P. & Booth, A. (1996). A Prospective Study of Divorce and Parent–Child Relationships. *Journal of Marriage and the Family*, 58, 356–365.

Amato, P. & Previti, D. (2003). People's Reasons for Divorcing Gender, Social Class, the Life Course, and Adjustment. *Journal of Family Issues,* 24, 602–626.

Amato, P., Kane, J. & James, S. (2011). Reconsidering the "Good Divorce." *Family Relations,* 60, 511–524.

Aughinbaugh, A., Robles, O. & Sun, H. (2013). Marriage and Divorce: Patterns by Gender, Race, and Educational Attainment. *Monthly Labor Review*, 136, 1.

Badgett, M. & Mallory, C. (2014). *Patterns of Relationship Recognition for Same Sex Couples: Divorce and Termination.* Los Angeles, CA: The Williams Institute, UCLA School of Law.

Blau, F. & Kahn, L. (2007). The Gender Pay Gap: Have Women Gone as Far as They Can? *The Academy of Management Perspectives*, 21, 7–23.

Braithwaite, S., Delevi, R. & Fincham, F. (2010). Romantic Relationships and the Physical and Mental Health of College Students. *Personal Relationships,* 17, 1–12.

Bramlett, M. & Mosher, W. (2002). Cohabitation, Marriage, Divorce, and Remarriage in the United States. *Vital and Health Statistics, Series 23.* Washington, DC: U.S. Government Printing Office.

Bodenmann, G. (1997). Can Divorce Be Prevented by Enhancing the Coping Skills of Couples? *Journal of Divorce and Remarriage,* 27, 177–194.

Bohannon, P. (1970). *Divorce and After: An Analysis of the Emotional and Social Problems of Divorce.* Garden City, NY: Anchor.

Bourne, P. A., Hudson-Davis, A., Sharpe-Pryce, C., Clarke, J., Solan, I. *et al.* (2014). Does Marriage Explain Murders in a Society? In What Way Is Divorce a Public Health Concern? *International Journal of Emergency Mental Health and Human Resilience,* 16, 84–92.

DatingSitesReviews.com (2018). Match.com Information, Statistics, Facts and History. Retrieved from https://www.datingsitesreviews.com/staticpages/index.php?page=2010000100-Match

Doskow, E. (2014). *Nolo's Essential Guide to Divorce.* Berkeley, CA: Nolo Publishers.

Doskow, E. (2015). *Nolo's Essential Guide to Child Custody and Support.* Berkeley, CA: Nolo Publishers.

Elliott, D. & Simmons, T. (2011). *Marital Events of Americans: 2009.* Washington, DC: U.S. Department of Commerce, Economics, and Statistics Administration, US Census Bureau.

Fabricius, W. (2003). Listening to Children of Divorce: New Findings that Diverge from Wallerstein, Lewis, and Blakeslee. *Family Relations*, 52, 385–396.

Felix, D., Robinson, W. & Jarzynka, K. (2013). The Influence of Divorce on Men's Health. *Journal of Men's Health,* 10, 3–7.

Felmlee, D., Sprecher, S. & Bassin, E. (1990). The Dissolution of Intimate Relationships: A Hazard Model. *Social Psychology Quarterly*, 53, 13–30.

Furstenberg, F. (1994). History and Status of Divorce in the United States. *The Future of Children,* 4, 29–43.

Gottman, J. & Silver, N. (1999). *The Seven Principles for Making Marriage Work.* New York: Harmony Books.

Grall, T. (2013). Custodial Mothers and Fathers and Their Child Support: 2011. *Current Population Reports.* Washington, DC: U.S. Census Bureau.

Hawkins, C. (2013a). Grounds for Divorce. In R. Emery (Ed.), *Cultural Sociology of Divorce: An Encyclopedia* (pp. 515–518). Thousand Oaks, CA: Sage.

Hawkins, C. (2013b). The Law. In R. Emery (Ed.), *Cultural Sociology of Divorce: An Encyclopedia* (pp. 683–687). Thousand Oaks, CA: Sage.

Hetherington, E. (2003). Intimate Pathways: Changing Patterns in Close Personal Relationships Across Time. *Family Relations,* 52, 318–331.

Hetherington, M. & Kelly, J. (2002). *For Better or For Worse: Divorce Reconsidered.* New York: W.W. Norton and Company.

Johnson, D. & Wu, J. (2002). An Empirical Test of Crisis, Social Selection, and Role Explanations of the Relationship between Marital Disruption and Psychological Distress: A Pooled Time – Series Analysis of Four – Wave Panel Data. *Journal of Marriage and Family,* 64, 211–224.

Kelly, J. (2000). Children's Adjustment in Conflicted Marriage and Divorce: A Decade Review of Research. *Journal of the American Academy of Child and Adolescent Psychiatry*, 39, 963–973.

Kim, E. (2013, August 6). For Gay Couples, Divorce Comes with Extra Costs. NBC Today. Retrieved on July 16, 2016 from http://www.today.com/news/gay-couples-divorce-comes-extra-costs-6C10660976

Kim, H. S. (2011). Consequences of Parental Divorce for Child Development. *American Sociological Review,* 76(3), 487–511.

Kitson, G. C. (1992). *Portrait of Divorce: Adjustment to Marital Breakdown.* New York: Guilford.

Kohnke, K. (2013). Covenant Marriage. In R. Emery (Ed.), *Cultural Sociology of Divorce: An Encyclopedia* (pp. 328–329). Thousand Oaks, CA: Sage.

Lebow, J., Chambers, A., Christensen, A. & Johnson, S. (2012). Research on the Treatment of Couple Distress. *Journal of Marital and Family Therapy*, 38, 145–168.

Levinger, G. (1966). Sources of Marital Dissatisfaction among Applicants for Divorce. *American Journal of Orthopsychiatry*, 36, 803–807.

Lewandowski, G., Aron, A., Bassis, S. & Kunak, J. (2006). Losing a Self-expanding Relationship: Implications for the Self-concept. *Personal Relationships*, 13, 317–331.

McManus, P. & DiPrete, T. (2001). Losers and Winners: The Financial Consequences of Separation and Divorce for Men. *American Sociological Review*, 66, 246–268.

Morrison, D. & Coiro, M. (1999). Parental Conflict and Marital Disruption: Do Children Benefit When High-Conflict Marriages Are Dissolved? *Journal of Marriage and the Family*, 61, 626–637.

National Divorce Reform League (1885). *An Abstract of Its Annual Reports, October, 1885*. Montpelier, VA: Vermont Watchman and State Journal Press.

National Healthy Marriage Resource Center (2016). *Hispanics and Latinos*. Oklahoma City, OK: National Healthy Marriage Resource Center.

National Marriage and Divorce Rate Trends (November 23, 2015). Atlanta, GA: CDC/NCHS National Vital Statistics System.

Norton, A. & Miller, L. (1992). *Marriage, Divorce, and Remarriage in the 1990s*. Washington DC: U.S. Government Printing Office.

Orbuch, T., Veroff, J., Hassan, H. & Horrocks, J. (2002). Who Will Divorce?: A 14-Year Longitudinal Study of Black and White Couples. *Journal of Social and Personal Relationships*, 19, 179–202.

Peterson, K. (2002). Divorce Need Not End in Disaster. *USA Today*. Retrieved on April 18, 2016 from http://usatoday30.usatoday.com/news/nation/2002/01/14/usatcov-divorce.htm

Plateris, A. (1974). *100 Years of Marriage and Divorce Statistics, United States, 1867–1967*. Rockville, MD: National Center for Health Statistics.

Rhoades, G., Kamp Dush, C., Atkins, D., Stanley, S. & Markman, H. (2011). Breaking Up Is Hard to Do: The Impact of Unmarried Relationship Dissolution on Mental Health and Life Satisfaction. *Journal of Family Psychology*, 25, 366–374.

Ruggles, S. (1997). The Rise of Divorce and Separation in the United States, 1880–1990. *Demography*, 34, 455–466.

Rusbult, C., Zembrodt, I. & Gunn, L. (1982). Exit, Voice, Loyalty, and Neglect: Responses to Dissatisfaction in Romantic Involvements. *Journal of Personality and Social Psychology*, 43, 1230–1242.

Sayer, L. (2006). Economic Aspects of Divorce and Relationship Dissolution. In M. A. Fine & J. Harvey (Eds.), *Handbook of Divorce and Relationship Dissolution* (pp. 385–406). Mahwah, NJ: Lawrence Erlbaum.

Seetharaman, D. (November 19, 2015). Facebook Offers to Ease the Pain of Breaking Up. *Wall Street Journal*. Retrieved on June 18, 2016 from http://blogs.wsj.com/digits/2015/11/19/facebook-offers-to-ease-the-pain-of-breaking-up

Sprecher, S. & Fehr, B. (1998). The Dissolution of Close Relationships. In J. Harvey (Ed.), *Perspectives on Loss: A Sourcebook* (pp. 99–112). Philadelphia, PA: Brunnel and Mazel.

Stanton, G. (2015). What Is the Actual US Divorce Rate and Risk? *Public Discourse*. Princeton, NJ: The Witherspoon Institute.

Tran, T. & Joormann, J. (2015). Facebook in Mediating the Relation between Rumination and Adjustment after a Relationship Breakup. *Computers in Human Behavior*, 49, 56–61.

Troxel, W., Matthews, K., Gallo, L. & Kuller, L. (2005). Marital Quality and Occurrence of the Metabolic Syndrome in Women. *Archives of Internal Medicine*, 165, 1022–1027.

U. S. Bureau of the Census (1949). *Historical Statistics of the United States, 1789–1945*. Washington, DC: U.S. Government Printing Office.

Vaughan, D. (1990). *Uncoupling: Turning Points in Intimate Relationships*. New York: Vintage.

Wallerstein, J. & Lewis, J. (2004). The Unexpected Legacy of Divorce: Report of a 25-year Study. *Psychoanalytic Psychology*, 21, 353–370.

Wolfinger, N. (1999). Trends in the Intergenerational Transmission of Divorce. *Demography*, 36, 415–420.

16
REMARRIAGE AND TRENDS IN LIVING ARRANGEMENTS

Learning Outcomes

Upon completion of the chapter, students should be able to:

1. describe the remarriage rate and factors relating to remarriage
2. explain factors relating to the success of remarriage
3. compare the stepfather and stepmother roles among diverse stepfamilies
4. explain the strengths and challenges of stepfamilies
5. explain the increase in cohabitating families
6. describe the effects of serial cohabitation on family and children
7. describe the causes of the increase in single-parent families
8. explain the issues relating to single-parent families
9. describe the family life of late adulthood age groups
10. explain the characteristics of grandparent families
11. describe retirement inequality across diverse social groups
12. explain Living Apart Together (LAT) as a living arrangement on a global basis.

Brief Chapter Outline

Pre-test

Remarriage

- The Remarriage Rate
- Factors Related to Remarriage
  - Remarital Timing after Divorce and Widowhood
- Benefits of Remarriage
- Duration and Factors Relating to Success of Remarriages

Pre-test

Engaged or active learning is a powerful strategy that leads to better learning outcomes. One way to become an active learner is to begin the chapter and try to answer the following true/false statements from the material as you read. You will find that you have ready answers to these questions upon the completion of the chapter.

Americans have higher remarriage rates compared with European nations.	T	F
Remarriage is a marriage that occurs after a previous marriage has ended through divorce or widowhood.	T	F
Stepfamilies are formed through cohabitation, marriage, and births outside of marriage.	T	F
Cohabiting families are the most common living arrangement.	T	F
Multiple partner fertility (MPF) is the pattern of a man or a woman having biological children with more than one partner.	T	F
Many unmarried births now occur to cohabiting parents.	T	F
Serial cohabitation occurs when an individual has cohabited with more than one partner and often has children with different partners.	T	F
Centenarian is an old adult who lives to or beyond the age of 100 years.	T	F
Many grandparents are in the labor force and taking care of grandchildren.	T	F
LAT stands for living apart together.	T	F

Upon completion of this section, students should be able to:
LO1: Describe the remarriage rate and factors relating to remarriage.
LO2: Explain factors relating to the success of remarriage.

Since the 1960s, the family has undergone significant transformation in union formation and marital behavior across Europe and the Americas (OECD, 2012). A declining share of children reside with two married biological parents, and a growing share of children live in an array of other arrangements, including married stepfamilies, cohabiting families, single-parent families (Bumpass & Lu, 2000), and grandparent families. We see a pattern of marriage, divorce, and remarriage, or "marriage go-round" as people are partnering and re-partnering at a dizzying pace (Cherlin, 2010).

Remarriage

Today, men and women have the freedom to decide how, when, and with whom they want to form relationships and in what context they wish to have children. For some people, this means marrying more than once. In the United States, between 2008 and 2012, about half of all men (50 percent) and women (54 percent) aged fifteen and over had married only once (Lewis & Kreider, 2015). In 2015, among people aged fifteen and older who have been married, 19.1 percent of men and women have been married twice and 5.4 percent have been married three or more times (U.S. Census Bureau, 2017a). **Remarriage** is a marriage that occurs after a previous marriage has ended through divorce or widowhood.

The Remarriage Rate

Americans have higher **remarriage rates** (i.e., the number of marriages per 1,000 persons) because marital disruption by divorce is higher in the United States compared with European nations (Andersson, 2003). The difference comes from Americans' embrace of two contradictory cultural ideals of family and personal life: *marriage*—a formal commitment to share one's life with another—and *individualism*—personal choice and self-development (Cherlin,

2010). In the United States, remarriage rates rose during the early 1960s, fell through the 1970s, and have continued to decline into the twenty-first century (see Figure 16.1). Much of the drop is due to the rise of cohabitation and older ages for first marriage (Jayson, 2013).

In the United States, the remarriage rate is consistently higher for men than for women—39 per 1,000 versus 21 per 1,000 in 2016 (see Figure 16.2). In 2013, more divorced men or widowers (64 percent) took a second walk down the aisle, compared with female counterparts (52 percent) (Geiger & Livingston, 2018). Remarriage rates declined among American men and women. In 2016, 27 remarriages occurred per 1,000 men and women, down from 31 in 2008 (Payne, 2018).

Remarriage rates among men and women have changed in different ways according to race/ethnicity. Between 2008 and 2016, Asian men had the highest remarriage rate (66 in 2008 and 58 in 2016 per 1,000); Black and White men tied with the lowest remarriage rate (47 in 2008 and 38 in 2016 per 1,000); the decline was greatest for Hispanic men, down from 54 remarriages to 42 per 1,000. Among women, Hispanic women had the highest rate of remarriage (27 per 1,000) and Black women had the lowest rate of remarriage (16 per 1,000) in 2016 (Payne, 2018).

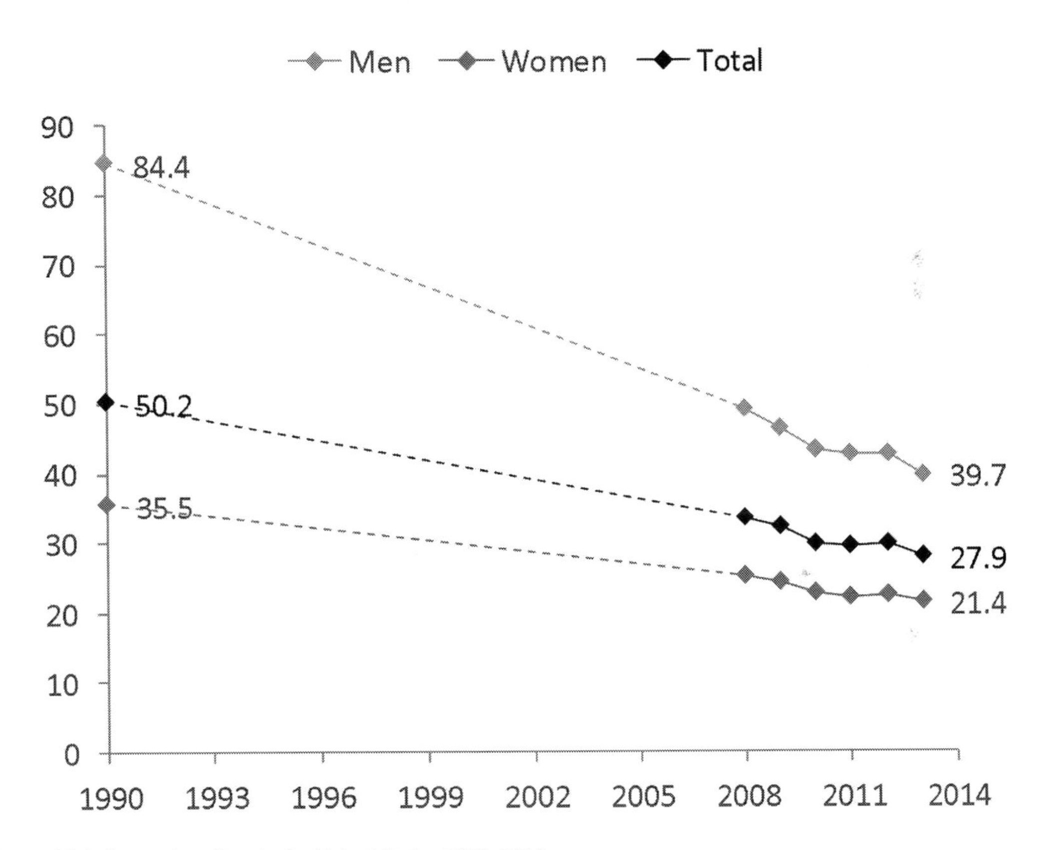

Figure 16.1 Remarriage Rate in the United States, 2008–2013

Source: FP-15–08. National Center for Family & Marriage Research. https://create.piktochart.com/output/6118754-geovar-remarrt_2013_web/

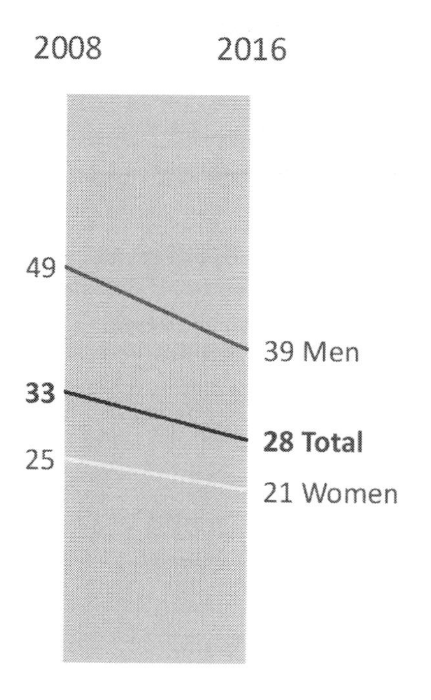

Figure 16.2 U.S. Remarriage Rate by Gender, 2008–2016
Source: FP-18–16. National Center for Family & Marriage Research. https://www.bgsu.edu/ncfmr/resources/data/family-profiles/payne-change-remarriage-rate-fp-18-16.html

Factors Related to Remarriage

The exit status (i.e., how the first relationship ended, widowhood or divorce) could account for differences in remarriage patterns between men and women. *Widowhood* (the marital status following the death of one of the partners) is likely to be an involuntary exit from marriage (Wu & Schimmele, 2005). *A widower* is a man whose spouse has died and *a widow* is a woman whose spouse has died. In preindustrial societies, marital dissolution was likely to occur through *widowhood* as divorce was a rare event. For example, of all marriages 1700–1899 in German villages, 48 percent ended through the death of a wife and 52 percent through the death of a husband (Knodel & Lynch, 1985). The experience of widowhood was gendered in the traditional society. In *patriarchal* societies where the male is dominant and the husband is the sole provider, his death can leave his family destitute or a widow would carry on her spouse's business. The loss of a wife rarely altered a man's status, while the loss of a husband invariably and irrevocably brought about a change in a woman's life (Cowan, 1999).

Today, more marriages end in divorce and remarriage is popular in many countries. *Divorce* (dissolution of marriage) is a voluntary exit for at least one of the two people. Both parties must acknowledge to themselves that the marriage failed (Sweeney, 2002). A *divorcé* is a man who has divorced and a *divorcée* is a woman who has divorced. Educational attainment, race or ethnicity, age, nativity, disability, and personal income are all associated with the probability of experiencing multiple marriages (Siordia, 2014). Remarriage after divorce varies positively with income level for men and negatively with both education and income for women (Glick, 1980). For example, higher income and education were predictors of becoming romantically involved for men (Oh, 1987) but highly educated mothers who separated at older ages and who were divorced were less likely to remarry (Glick & Lin, 1986) because they were occupationally and financially independent (Oh, 1987). The median age at remarriage was 45.5 for men and 42.8 for women in 2013, with the remarriage rate consistently higher for men (40 per 1,000) than for women (21 per 1,000) (Payne, 2014). Young single mothers with more children are more likely to remarry (Schmiege *et al.*, 2001). The needs of children, the difficulties associated with single parenting, and increased financial strain may combine to increase the likelihood of remarriage among divorced women (Stewart *et al.*, 2003). Compared to the U.S. population, service members marry, divorce, and remarry *earlier*. Divorced and remarried service members are overrepresented among the enlisted ranks, joint service couples, and lower education categories (Adler-Baeder *et al.*, 2006).

Interestingly, "quality of life" is important for widows and widowers in Spain who wish to have a partner; "enjoying life more and not feeling alone" is the principal motives for those who have a new partner (Ayuso, 2018).

Remarital Timing after Divorce and Widowhood

Research has shown that the transition to remarriage is quicker for individuals whose first marriage ended in divorce than individuals whose first marriage ended in spousal death, especially for fathers (Watkins & Waldron, 2017). **Marital search theory** suggests that within the limits of a *marriage market* where the search for a second marriageable partner occurs, seekers desire to maximize the quality of the match, but are willing to settle for a minimally acceptable match, called a *reservation quality partner* (England & Li, 2006). Research has drawn from the theory to explain why remarital timing may differ between divorced and widowed persons. In a marriage market, market factors influence the availability of marriage partners—especially those who meet or exceed minimally acceptable standards (James & Shafer, 2012). For example, marital searches can be affected by the demographic and life course factors such as *parity* and *age*. Parity may increase the likelihood of remarriage for divorced men and women because they partner with individuals who are also parents (Goldscheider & Sassler, 2006). Younger age was a predictor of becoming romantically involved for women (Oh, 1987). The median age of divorce was 35 for men and 33 for women, while the median age of widowhood was over 60 for both men and women (U.S. Census Bureau, 2008). A high percentage of divorced women would remarry because they reported wanting an additional child or children because many were in their prime childbearing and childrearing years when their first marriages dissolved (Guzzo & Furstenberg, 2007).

Research has also indicated that remarriage is more common among widowers than widows. For example, a study of 6,500 widows and widowers in the nineteenth century in Breda (south Netherlands) and Gouda (west Netherlands) revealed that the probability of remarriage was much greater for widowers than for widows (Poppel, 1995). In the United States, more widowers (61 percent) than widows (19 percent) were either remarried or involved in a new romance (Schneider *et al.*, 1996). Widowers remarry frequently because they have more opportunity to do so (Kenny, 2006) and maximize the quality of the match. For example, in the Flanders region of Belgium in the nineteenth century, the most common remarriage type was that between a widower and an unmarried woman; the least common was between a widower and a widow (Matthijs, 2003). In some instances, men on their own are needier than women in similar circumstances: emotionally needier, sexually needier, and in greater need of conjugal support in bringing up a family (Kenny, 2006). Men remarry also because they are the beneficiaries of wedlock, whereas women do most of the servicing and structure most of the domestic arrangements and take responsibility for everything from the Christmas card list to remembering in-laws' birthdays (Bernard, 1982). Widowers remarry also because they had a higher incidence of mental and physical illness and higher mortality risk than widows (Skulason *et al.*, 2012) if their wives' deaths were unexpected rather than expected (Sullivan & Fenelon, 2014). The **widowhood effect**—the increased likelihood for a recently widowed person to die—is one of the best documented examples of the effect of social relations on health (Schaefer *et al.*, 1995). Studies have shown that in a robust relationship social and emotional support from others can be protective for health (Reblin & Uchino, 2008). In some way, men remarry because women are the emotional, psychological, and familial net contributors (Bernard, 1982). The major factors preventing widows remarrying or rejecting a new relationship include worries about the future of their children, lack of motivation for remarriage, not finding an appropriate man, loyalty to their previous husbands (Ahmady *et al.*, 2007), and the belief that their lost partner is irreplaceable (Ayuso, 2018). Since widowers remarried faster than widows, their numerical presence in society was smaller and less prominent than widows (Cowan, 1999).

Benefits of Remarriage

Remarriage provides a new opportunity to share economic resources, give/receive emotional support, and experience companionship and sexual intimacy, and thus may offset some of the negative consequences of divorce (Amato, 2010). The physical health benefits of remarriage are well documented. For example, greater psychological well-being was highly correlated with being remarried or in a new romance 25 months after a spouse's death (Schneider *et al.*, 1996). Major symptoms among unmarried widows are somatoform disorder, obsession, depression, anxiety, and paranoid ideation (Ahmady *et al.*, 2007). Adults and children all benefit from the stability of remarried families (Ehrenberg *et al.*, 2012).

If dissolution occurred at a younger age, the benefits of remarrying, such as another income or greater help with parenting, may outweigh the costs of doing so, such as integrating into and dealing with new kinship relationships (Sweeney, 1997). The economic consequences of divorce may differ for men and women, where women often experienced a decrease in socioeconomic well-being whereas formerly married men often remained financially stable (Avellar & Smock, 2005). Indeed, women may benefit more from remarriage than do men (Ozawa & Yoon, 2002), because remarriage improves the financial situations of mothers and their children substantially (Teachman & Paasch, 1994). For this reason, young children with remarried parents may be more likely than those with single parents to obtain education beyond high school and avoid early family transitions (Amato & Kane, 2011). Children also benefit from positive role models as growing up with two parents is strongly associated with more education, work, and income among children (Wilcox & Lerman, 2014).

Duration and Factors Relating to Success of Remarriages

A **remarried couple household**—a household maintained by a married couple, one or both of whom have been previously divorced or widowed—can be thought of as the center of intersecting sets of relationships among parents and children who are linked by the ties of previous marriages (Cherlin & McCarthy, 1985).

One measure of the success of remarriage is the **duration of marriage**—the interval of time between the day, month, and year of the marriage to date. The average duration of marriage increased from about fifteen to twenty years in preindustrial Europe to about thirty-five years in 1900 (Phillips, 1991). Would a second marriage increase a person's likelihood of having a lasting relationship today? In fact, the divorce rate for remarriages is higher than the divorce rate for first-time marrieds (Ehrenberg *et al.*, 2012). Second marriages are shorter than first marriages, with a median length of 14.5 years for second marriages versus 20.8 years for first marriages (Elliott & Simmons, 2011). In fact, those in their first marriages are more likely to report higher levels of marital satisfaction than those in their second marriages (Mirecki *et al.*, 2013b). **Relationship self-regulation** (RSR) is actionable, goal-oriented work aimed at improving relationships in both the short and long term (Halford *et al.*, 2007) and is related to relationship satisfaction. Meyer *et al.* (2012) also found that remarrieds reported significantly lower RSR levels than first marrieds and cohabiting groups. Indeed, remarriage is "difficult and different" from first marriage. In the first marriage, the couple has time alone to set up their own culture, whereas in the second marriage, a parent with his or her children is deeply bonded and already has a culture of their own without the stepparent (Scarf, 2014). Fortunately, as stepparents engage with their pre-adolescent stepchildren in various settings, more opportunities for establishing and sustaining a secure attachment are likely provided (Jensen & Pace, 2016).

Financial security, gender role, and religion are related to the success of remarriages. Longer duration of a first or a remarriage was associated with higher financial security (Van Eeden-Moorefield *et al.*, 2007). When remarriages failed, women spoke of the husband's failure to provide as an important factor in their decision to divorce (Schmiege *et al.*, 2001). Education levels are found to be influential to the level of marital satisfaction in second marriages (Mirecki *et al.*, 2013b). Furthermore, gender role attitudes influence marital satisfaction as they are treated as stable traits in adulthood. Over a twenty-year period, 590 married individuals all demonstrated a shift toward more egalitarian attitudes (Lucier-Greer & Adler-Baeder, 2011). The remarried couples are more likely to minimize the influence of their own actions (Amato & Previti, 2003). Religion may play a role related to duration of remarriage. A study of Christian, Jewish, and Muslim families from seventeen states of the United States revealed that all families consisted of married or remarried parents who had an average of twenty years of marriage and an average of 3.3 children per couple (Dollahite & Marks, 2018).

Couples who remarry often remain optimistic that a new partnership will lead to better results (Bradbury & Karney, 2010). However, like first marriages, second marriages bring on problems of their own for remarrieds (O'Flaherty & Eells, 1994). We would suggest considering major points such as recoupling and adjustment process.

Recoupling Process

One factor that contributes to the success of remarriages is length of courtship. The exit status (divorce or widowhood) could account for length of courtship between divorced and widowed persons. Dating longer gives both partners the chance to get to know one another and their respective children and to learn how they will tackle the problems of everyday life together. As the tasks of the early period of remarriage were completed, each spousal subsystem (i.e., wife and husband roles) gradually experienced a **recoupling process**—the occurrence of an emotional bonding (Kvanli & Jennings, 1987). A study of remarried couples who recounted the developmental tasks of the remarriage process including first marriage, divorce, and courting revealed that the mean length of time of the remarriages was 7.8 years (Kvanli & Jennings, 1987). In cases where a remarriage follows the death of a spouse, the loss can haunt a new marriage and leave it vulnerable. Couples might wait more years after the death of a spouse to remarry. For example, Christian and Jewish traditions have wisely set out a specific period of mourning after a spouse's death, not just to give the bereaved person time to heal, but to prevent the folly of an ill-judged remarriage (Kenny, 2006). In a study of widows and subsequent remarriage, Engblom-Deglmann & Brimhall (2016) suggest that the level of family acceptance includes the length of time between death and courtship, the length of the courtship itself, and the level of family involvement in the courtship. Problem areas might be reduced by adequate counseling prior to the remarriage (O'Flaherty & Eells, 1994).

Because the couple relationship quality serves as an indicator of a growing sense of attachment between two remarried individuals, it is important to build emotional bonds between two people. **Attachment theory** emphasizes the primacy of human emotional bonds (Mikulincer & Shaver, 2007). Research has drawn from the theory to explore how remarried couples' attachment styles influence perspectives on marital commitment and strengthen the quality and duration of remarriages. A study of 145 remarried parents reveals that individuals with a more secure attachment style report higher marital commitment scores than insecure individuals (Ehrenberg *et al.*, 2012). When it comes to weddings, most couples replicated gendered patterns from their first weddings in subsequent weddings, although remarriages tended to involve smaller and less complicated weddings (Humble, 2009).

Adjustment Process in Remarriages

Factors that influence the adjustment of any remarriage include role expectations and role performance, sex, finances, social activities, religion, values, shared goals, gratification of emotional needs, and creative growth and development of individuals, both as individuals and as members of a family (Schlesinger, 1972), relationship with friends and relatives, family cohesion, and attachment with the former spouse (Roberts & Price, 1989). Second marriages might be successful when couples are not too young and have a more profound knowledge of marriage. A study of 450 widows revealed that the age of widows was the only factor correlated with the marital adjustment (Ahmady *et al.*, 2007).

A key issue for remarried couples is interpersonal communication when it comes to finances, child discipline, and personality conflicts in the newly created family, and rivalries between family members (Gaspard, 2016). Studies have shown that uncertainty is a significant predictor of topic avoidance for those in both first marriages and remarriages. People in remarriages were found to have higher uncertainty than those in first marriages, but engage in topic avoidance at similar levels to those in first marriages (Wilder, 2012). **Uncertainty reduction theory** holds that when interacting, people need information about the other party to reduce their uncertainty. In gaining information, people can predict the other's behavior and resulting actions and all of this is crucial in the development of any relationship (Turner & West, 2010). A comparative study of first-married and remarried couples revealed that the remarried people increased *mutual constructive communication* (i.e., mutual expression and discussion of feelings about the problem and problem resolution), decreased *demand–withdraw* (one partner attempts to discuss a problem, while the other avoids the issue), and decreased *avoidance and withholding* (i.e., avoiding discussion of problems and stressors) in their current remarriage relative to their former marriage (Mirecki *et al.*, 2013a). Using uncertainty reduction theory to examine family commitment, Downs (2004) found that the participants were highly committed to their families, perceived moderately high levels of support for their current relationship from family members, and had a high degree of role certainty and role clarity.

Many remarried couples are plagued with anxiety that stems from unresolved issues in their previous marriages. By acknowledging the influence of their previous marriage, couples can work through this anxiety and become less emotionally reactive (Faber, 2004). **Emotion regulation** (ER) refers to attempts to influence which emotions they have, when they have them, and how they experience and express these emotions (Gross, 1998). Effective emotion regulation in remarriage is crucial for different aspects of healthy affective and social adaptation (John & Gross, 2004) and can reduce emotional reactions to stressful, anxiety-provoking situations (Werner *et al.*, 2011). The commonly used emotion regulation strategies are *cognitive reappraisal* (changing the way one thinks about potentially emotion-eliciting events) and *expressive suppression* (changing the way one behaviorally responds to emotion-eliciting events) (Cutuli, 2014), and *mindfulness* (the psychological process of bringing one's attention to experiences occurring in the present moment) (Creswell, 2017). A study of remarried women in Iran reveals that suppression is positively correlated to difficulties in emotion regulation, whereas reappraisal and mindfulness are negatively related to difficulties in emotion regulation (Jelvani *et al.*, 2018).

Couples today are struggling to improve the institution of marriage for both partners, by working out dual careers, shared parenthood, and a combination of personal autonomy and family cooperation (Bernard, 1982). **Resilience factors** enable remarried families to withstand and rebound from the disruptive challenges they face, including supportive family relationships and communication, a sense of control over outcomes in life, activities, and routines that help the family to spend time together, a strong marriage relationship, support from family and friends, redefining stressful events and acquiring social support, and spirituality and religion within the family (Greeff & Toit, 2009).

Child Custody Arrangements

Remarriage can affect child custody arrangements. **Joint custody** is a court order whereby custody of a child is awarded to both parties regarding the care of their children. **Joint legal custody** allows parents to make decisions for their children on education, medical care, etc. **Joint physical custody** means that both parents share physical custody and children spend substantial time living with both custodial parents after a parental divorce. By **sole physical custody**, it means that one parent resides with the child and becomes the custodial parent, while the other parent of the child has visitation time with the child.

Across countries, people encourage courts to consider the option of joint physical custody. For example, the long battle to reform Australia's custody law between 1995 and 2011 resulted in a law that emphasizes the need to protect children from harm (Parkinson, 2018). In the United States, the introduction of the "best interests of the child" standard in family law in the 1970s signaled an important transition away from a maternal preference standard in child custody disputes toward recognition of the importance of both parents in the lives of children after parental separation (Kruk, 2018). An evaluation of Arizona's 2013 revisions to the child custody statutes reveals that the new law is functioning as a rebuttable presumption of equal parenting time (Fabricius *et al.*, 2018). It means that children have increased parenting time with fathers. In most cases, mothers are likely to secure custody when their Order of Protection (OPs) petitions are ordered or withdrawn. Fathers restrained by OPs were more likely to secure visitation orders (64 percent), whereas 80.8 percent of fathers' custody petitions were dismissed when they were restrained by OPs (Rosen & O'Sullivan, 2005). In cases of domestic violence against the mother, Morrill *et al.* (2005) found that more orders gave legal and physical custody to the mother and imposed a structured schedule and restrictive conditions on the father's visits.

Research indicates that joint custody is advantageous for children during child adjustment. Children in joint physical or legal custody were better adjusted; parents with joint custody reported better cooperation with former spouses and greater financial resources than did those with sole custody (Pearson & Thoennes, 1990). Consequentially, joint physical custody gives benefits to children on average (Braver & Votruba, 2018), facilitated ongoing positive involvement with both parents, and was associated with a higher inclination to use both parents as a source of emotional support and better subjective health than a single-parent household in Sweden (Låftman *et al.*, 2014).

The Effects of Remarriage on Children

Consequences of remarriage have impacted children in many ways. Research has explored the effect of parental remarriage on child survival. A study of 7,077 boys and 6,906 girls born between 1720 and 1859 in Krummhörn (Germany) and 31,490 boys and 33,109 girls whose parents married between 1670 and 1750 in Québec (Canada) reveals that the mother's marriage was beneficial in Québec, whereas the father's remarriage reduced the survival chances of the children born from his former marriage in Krummhörn (Willführ & Gagnon, 2013).

Children of remarriage may experience a considerable adjustment challenge. Adjustment to remarriage seems to be easier for boys than for girls, and easier for young children than for adolescents (Jolliff, 1984). Sometimes, adolescents felt excluded and resentful about reduced intimacy with their biological custodial parent (Stoll *et al.*, 2006). Adolescents' well-being was lower if their mothers were in less stable partnership situations or had several relationships since their divorce and when their fathers had remar-

Upon completion of this section, students should be able to:
LO3: Compare the stepfather and stepmother roles among diverse stepfamilies.
LO4: Explain the strengths and challenges of stepfamilies.

ried or begun living with a new partner since their divorce (Bastaits *et al.*, 2018). When it comes to the impact of family structure on school performance, Ham (2004) found that adolescents from intact families outperformed students from other family structures, especially for female students.

Parental remarriages might affect adult children's family formation transitions. A **modeling perspective** assumes that children living in non-traditional households (with single parents or remarried parents) adopt non-traditional views about family life. Parental remarriage increases the likelihood that youth will form cohabiting relationships (Teachman, 2003). **Social learning perspective** holds that people learn from one another through observation, direct instruction, imitation, and modeling. Marital dissolution might be a learned behavior as divorced parents have demonstrated that union dissolution is an acceptable solution to an unsatisfying relationship (Amato & Kane, 2011).

Stepfamilies

Couples who choose to remarry often join families together into a new blended family. The terms "blended families" or "reconstituted families" have been coined to describe the growing phenomenon of second marriages in Canada and the United States (Schlesinger, 1977). A **stepfamily** is a form of family in which at least one marital or cohabiting partner has children who are not biologically related to the other spouse or partner.

In ancient countries, the common pathway to stepparenthood was through individuals marrying a parent whose previous spouse had died (Chapman *et al.*, 2016). Using accounts of guardianship of orphans from Paris and Châlons-sur-Marne between 1630 and 1790, Perrier (1998) explored the relationships between the individuals involved in such *reconstituted* families, in a society where divorce was illegal and patrimonial transmission was organized by customary law. In the early 1970s onwards, more stepfamilies were formed after divorce than after the death of a parent (Strow & Strow, 2006).

Diversity of Stepfamilies

Most stepfamilies are not formed in *first union* (i.e., cohabitation or marriage) as the majority (76 percent) involved two childless individuals (Guzzo, 2016). Stepfamilies are formed in *second and third unions* (i.e., unions referring to remarriage or cohabitation) when an individual brings a child or children from a previous relationship into a new committed relationship (Ganong & Coleman, 2004). In 2016, two-thirds of women in their *second* union (cohabitation or remarriage) and three-fourths of women in their *third* union (children from prior relationships) formed stepfamilies (Guzzo, 2016). Although most women in their third union formed stepfamilies, the levels are higher among women who married directly (84 percent) rather than those who cohabited before marriage (74 percent) (see Figure 16.3).

Berger (1995) identified three types of stepfamilies: *integrated stepfamilies* (families formed to make a new couple and combine old and new perspectives); *invented stepfamilies* (families formed to raise children and in which the past family is erased); and *imported families* (the family functions as if it were a continuation of the old family). Stepfamilies vary by race and ethnicity. Among women ages 15 to 44 in their first union, Black women (57 percent) were more likely to

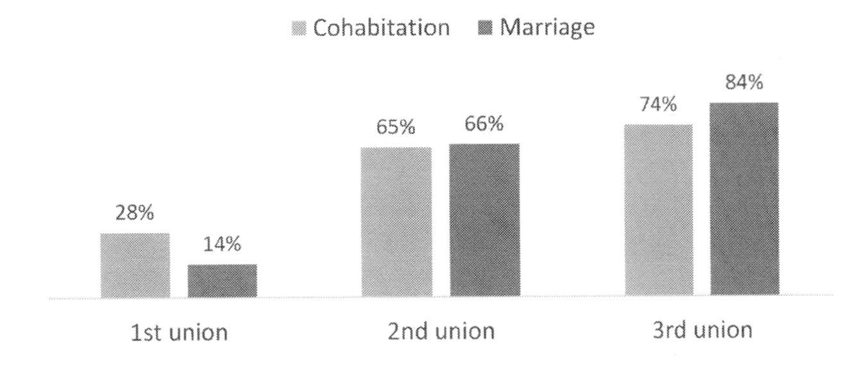

Figure 16.3 Stepfamilies by Union Type
Source: National Survey of Family Growth, 2006/2010 & 2011/2013. http://www.bgsu.edu/ncfmr/resources/data/family-profiles/guzzo-stepfamilies-women-fp-16-09.html

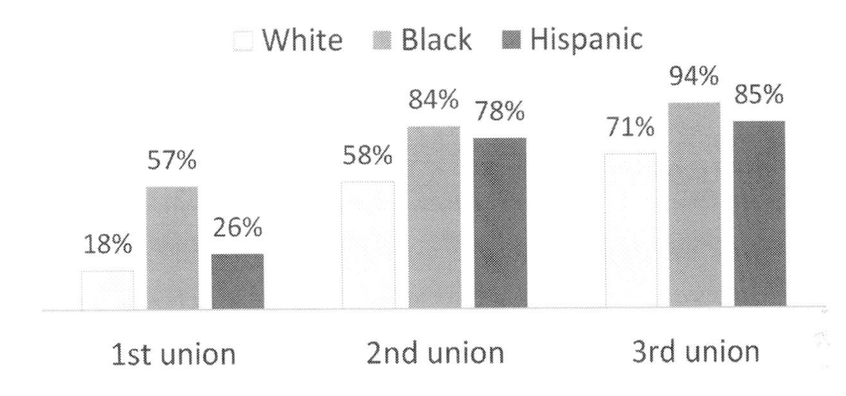

Figure 16.4 Stepfamilies by Union Order and Race-Ethnicity
Source: National Survey of Family Growth, 2006/2010 & 2011/2013. http://www.bgsu.edu/ncfmr/resources/data/family-profiles/guzzo-stepfamilies-women-fp-16-09.html

form stepfamilies than Hispanic (26 percent) or White (18 percent) women (see Figure 16.4).

Stepfamilies include married stepfamilies and cohabiting stepfamilies. A **remarried stepfamily** is a form of family in which at least one partner is remarried and at least one partner has children with someone other than their current spouse. In remarried couple households, stepparents and stepchildren are residing together (Cherlin & McCarthy, 1985). The security of marriage via both a legal tie and a public commitment—what Cherlin calls "enforceable trust" (Cherlin, 2004)—circumscribes the roles of parent and spouse, establishing a cohesive family unit intended to foster joint parental investment in children (England & Farkas, 1986). Remarried stepfamilies include (1) only a stepfather (mother–stepfather family), (2) only a stepmother (father–stepmother family), and (3) both a stepfather and a stepmother. Nearly half of remarried stepfamilies include only a stepfather (46 percent), only a stepmother (one in four, or 23 percent), and both a stepmother and a stepfather (one in three) (Stykes & Guzzo, 2015). Research has explored the effect of the previous marriage on the remarried couples in stepfamilies and found four themes: (1) most couples exhibited caution when considering marrying their spouse; (2) differing loyalties caused stepparents to feel that their needs came last and the biological parents to feel caught in the middle; (3) antagonism toward the ex-spouse brought couples together but also resulted in their feeling powerless and frustrated; and (4) *role confusion created frustration for the stepparents* and guilt for their spouses (Martin-Uzzi & Duval-Tsioles, 2013). Let's look at the parental roles in stepfamilies.

Redefining Roles

One of the major changes that take place is the redefinition of each person's role in the newly formed stepfamily. A **stepparent** is a non-biological parent who is married to or cohabiting with the child's mother or father who had a previous marriage or intimate relationship. There are *four pathways* through which adults become stepparents (Ganong & Coleman, 2004): *marrying* a parent who is divorced or widowed, or whose children were born within non-marital unions, and *sharing* a residence with a partner who has children from a prior partnership.

Stepfamilies are often fraught with uncertainties about "family" norms and unstable alliances between various family members. For example, **role ambiguity**—the absence of normative clarity about obligations and commitments between stepparents, absent parents, and children—constrains resource investments in children (Cherlin, 1978). Similarly, **family boundary ambiguity**—inconsistency in reporting who is in and who is out of the family—is more frequent in stepfamilies, compared to fully biological or adoptive families (Brown & Manning, 2009). Family members often disagree about who is in and who is out of their family (White, 1998). Even in long-term relationships, stepparent–stepchild bonds are more ambiguous than parent–child ties (Ganong & Coleman, 2004). Role ambiguity and family boundary ambiguity likely extend to non-resident parents. Non-resident fathers' child support contributions and contact with children diminish when either person in a former relationship repartners and has a child with a new partner (Tach *et al.*, 2014). Consequentially, family boundary ambiguity is linked to poorer family functioning (Carroll *et al.*, 2007). Such realities can complicate everyday routines and decision-making (Marsiglio, 2004) and increase risk for other problems, and ultimately divorce (Michaels, 2006). Stepfamilies need help in creating roles for stepparents that complement biological parents' roles (Crosbie-Burnett, 1989).

Stepfather's Role

A **stepfather** refers to a non-biological father who is married to or cohabiting with the child's mother who had a previous marriage or intimate relationship. Many young children born to parents from previous relationships or born to unwed parents currently live with their biological mothers and their mothers' new partners or *social fathers* (Bzostek, 2008). *Three types of stepfathers* include (1) married stepfathers living with both step- and biological children, (2) cohabiting stepfathers living with only stepchildren, and (3) unmarried stepfathers living apart from the mother and the child.

In the United States, 91 percent of fathers lived with only their biological children, whereas 4 percent and 5 percent of fathers lived with only stepchildren or both step- and biological children, respectively, in 2016; it was more common for stepfathers with only stepchildren to live in cohabiting unions (13 percent) than for fathers with only biological chil-

Student Accounts of Stepfathers

My parents divorced when I was five years old. Two years passed before my mother met the man I am proud to call Dad today. He was more than just my stepfather; he was my Dad who was willing to guide me through my childhood years and raise me when my own blood would not (personal communication, July 30, 2017).

dren (7 percent) or both step- and biological children (8 percent) (Eckmeyer, 2018).

Married biological fathers invest more in children than social fathers, primarily because they are both legally and socially obligated to do so (Cherlin, 1978), typically engage in higher quality parenting practices than unmarried biological fathers, and generally exhibit higher quality parenting than social fathers (Hofferth *et al.*, 2007). *Evolutionary perspectives* on child-rearing emphasize the importance of genetic links between fathers and children in shaping fathers' parenting practices such as providing material resources and direct caregiving, protecting children from harm, transferring knowledge, maintaining children's homes, preserving kin networks, and economically, socially, or emotionally supporting children's mothers (Hewlett, 2000). These domains represent three crucial dimensions of *fathering*—engagement, availability, and responsibility (Lamb *et al.*, 2000).

Married stepfathers may be more invested in the family unit (children) than those selecting into cohabitation (Nock, 1995) and exhibit equivalent or higher quality parenting behaviors than married and cohabiting biological fathers (Berger *et al.*, 2008).

Since their obligations to children are less fully institutionalized (Cherlin, 1978), *cohabiting stepfathers* lack a legally defined parental role, which may be linked to increased parental role ambiguity and relationship instability (Berger *et al.*, 2008). Stepfathers are often fraught with uncertainties about the role of stepfather as their bonds with stepchildren are constructed outside the normative model of biological fatherhood (Marsiglio, 2004). Stepfathers who had no prior children reported more competition with the biological father than stepfathers who were also fathers to their own biological children (Crosbie-Burnett, 1989). Because many stepfathers have their own biological children, loyalty conflicts sometimes exist between their commitments to their own offspring and to stepchildren (Clingempeel *et al.*, 1994). Stepfamily education courses are believed to improve family relationships. For example, the fourteen European American stepfathers who participated in a stepfamily education course reported the benefits of taking the course including increased family bonding, relationship skill development, and enhanced fathering practices in their roles (Higginbotham *et al.*, 2012).

Marsiglio (2004) analyzed the role of stepfather relevant to *an expanded view of fathering, interpersonal bonding*, and a deeper understanding of a *stepfather's identity* as social father. Most stepfathers described acting "like a parent and a friend," as the position of "the father" was assigned to the biological father (Blyaert *et al.*, 2016). Meriam (1940) emphasized a stepfather's obligations and rights regarding his stepchild. Financial contributions are an important part of the roles of these stepfathers (Shehan, 2016). Residential stepfathers, like fathers in first marriages, are expected to play the role of breadwinner (Marsiglio, 2004) and married social fathers exhibited higher levels of financial investment and participation in caregiving for their children than cohabiting fathers (Landale & Oropesa, 2001). Also, research has emphasized the role of stepfather relevant to interpersonal bonding. A secure stepfather–child attachment is facilitated by **stepfather involvement**—interactive behaviors or shared activities between stepparents and children that take place on a regular basis and involve various facets of family life (e.g., doing chores together, playing sports, watching shows, engaging in everyday talk) (Jensen & Pace, 2016). Stepfather involvement is positively associated with high-quality stepfather–child relationship among racially diverse and low-income stepfamilies with pre-adolescent children (Jensen & Pace, 2016). Indeed, involvement by *resident social fathers* is as beneficial for child well-being as involvement by resident biological fathers (Bzostek, 2008). Social father engagement was associated with declines in young children's behavior problems and better overall health (Bzostek, 2008). Gradually stepfathers "feel more companionship with their stepchildren, experience more intimate stepfather–stepchild interactions, are more involved with their stepchildren's friends, feel fewer negative feelings about stepchildren, and have fewer desires to escape" (Ganong & Coleman, 1994).

Research also has explored whether the stepfathers or social fathers should carry the meaning of legal fatherhood. In fact, remarried stepfathers are likely to legally adopt stepchildren as it lends them legitimate rights to take on the role of father. However, cohabiting stepparents lack legal standing. Based partly on the case of Estonian law which was amended precisely to address a changing society, Joamets (2016) concludes that the concept of the social father protects the interests of the child more effectively and is more suitable for the current social climate as well as for the near future, to be considered as legal fatherhood. When it comes to **paternal claiming**—a stepfather's readiness to nurture, provide for, protect, and claim a stepchild as his own—Marsiglio (2004) suggests that there are ten properties of paternal claiming: *timing*, degree of deliberativeness, degree of identity conviction, paternal role range, solo-shared identity, *mindfulness*, propriety work, naming, seeking public recognition, and biological children as benchmarks. For example, arriving at that mental and emotional space of feeling comfortable with a fatherlike identity is part of a transitional process that unfolds in various ways for different men. Mindfulness captures a stepfather's attentiveness to a child's likes, dislikes, moods, needs, and the times they have shared together (Marsiglio, 2004).

Stepmother's Role

A **stepmother** refers to a non-biological mother who is married to or cohabiting with the child's father who had a previous marriage or relationship. In the United States, there are more than 1 million stepmothers living with their stepchildren (Stykes & Guzzo, 2015). Historically stepmothers are the evil, unkind villains in fairy tales (Cann-Milland & Southcott, 2018). This can impact their own understanding of themselves as women, mothers, and stepmothers, which affects their relationships within their stepfamily (Latrofa *et al.*, 2009).

The *motherhood mandate* may contribute to making stepmothers' roles and relationships challenging (Shehan, 2016). In fact, stepmothers are often expected to function as caregivers, nurturers, housekeepers, mediators, and facilitators of family relationships (Coleman & Weaver, 2005), and thus are more vulnerable to role conflicts than stepfathers (Shehan, 2016). Stepmother anxiety is related to three areas: their relationship with the biological mother, their relationship with their stepchildren, and lack of clarity regarding the stepmother role (Doodson, 2014). Gates (2018) has explored stepmothers' coparenting experiences and reported their struggles with the mother–father–stepmother relationship triangle, their internal struggles with their role, their perception of the mother as powerful, and their view of themselves as powerless. Stepmothers want the fathers to support them and to mediate problems with biological mothers and stepchildren (Craig *et al.*, 2012).

The growing pains of adjusting to a new role can lead stepmothers into positive self-discovery (Cann-Milland & Southcott, 2018). Church (1999) identified five models of kinship (network of people) that informed stepmothers' perceptions of their roles: *nuclear model* (stepmother defines the stepfamily as a nuclear family and wants to be seen as the mother); *extended model* (stepmother defines the family broadly and doesn't try to take a mother role); *couple model* (stepmother's primary focus is on her relationship with her partner); *biological model* (stepmother defines her family along biological lines and focuses on her own children); and *no family model* (stepmother defines herself as alone and not related to anyone in the stepfamily). Interestingly, White, middle-class stepdaughters highlighted five styles of the positive stepmother role: "*my father's wife*" (stepmother perceived as disengaged), "*a peer-like girlfriend*" (stepmother younger than biological parent), "*older close friend*" (stepmother as a friend), "*a type of kin*" (stepmother as a cousin, older sister, or aunt), and "*like another mother*" (stepmother taking a very nurturing role and giving a great deal of support) (Crohn, 2006).

Stepmothers relate to their stepchildren in diverse ways. The primary task facing stepmothers was to undergo a major change in how they thought about their roles. Understanding the stepmother role to be a protector, caregiver, carer, guide, and nurturer was like giving her a job description that need not be affected by the times her stepchildren showed negative behaviors towards her (Cann-Milland & Southcott, 2018). Usually, stepmothers were careful to share that they did not want to either replace or displace the mother (Sanner & Coleman, 2017). For example, "*good enough stepmothering*" entailed stepmothers finding a place doing what they called "second tier mothering" (Hart, 2009). In a study of low-income, unmarried mothers whose partners had children from previous relationships, Burton and Hardaway (2012) described a strategy of *othermothering* (i.e., culturally scripted practices of sharing parenting with biological mothers) as a way to give care to a partner's offspring without usurping the biological mother's status. Indeed, stepchildren's positive experiences include receipt of material and emotional support (Hoenayi & Yendork, 2018). Sometimes, a key role as stepmother was serving as the link between the stepfather (her husband) and her children (Burton & Hardaway, 2012). When it comes to non-residential or visiting stepchildren, stepmothers fulfill the role function of mother when the children are with them but it is "black market mothering," because they have no authority, legal or otherwise, to do it (Weaver & Coleman, 2005). Stepmothers who are also biological mothers are ambivalent about developing close relationships with non-residential stepchildren because they are afraid it might strain the bonds they have with their own children. When conflicts arose between children and stepfathers, the biological mothers engaged in protective behavior manifested in four ways: defender, gatekeeper, mediator, and interpreter as their loyalties lay with their children (Weaver & Coleman, 2010).

Research has explored stepmothers' stress-reduction strategies. Doodson (2014) suggests that interventions designed to clarify the stepmother role and strengthen relationships between stepfamily members would help alleviate anxiety and thus improve stepmother well-being. Cann-Milland and Southcott (2018) reflected upon the specific situations that caused stepmothers to question, alter, and sustain a healthy sense of self and in turn create a secure environment that supports ongoing connections within the stepfamily. For example, stepmothers allow the children to input into the family rules, boundaries, and expectations or give some feedback on ways they can improve the new family situation (Schrodt, 2017). Some stepmothers serve as "family carpenters" who tried to help rebuild relationships between their husbands and their estranged children (Vinick, 1998). Today, many stepmothers reach out to others online (Craig *et al.*, 2012) and seek support from friends and family (Whiting *et al.*, 2007). For example, stepmothers make concerted efforts to get to know the biological mothers, to recognize their concerns, and to appreciate the perspectives of stepchildren.

Stepgrandparent

Stepgrandparent refers to an individual who is not biologically related to children who are two generations below him or her, but serves in a grandparent-like role because of a remarriage within the family (Chapman *et al.*, 2016). There are several pathways to stepgrandparenthood. In some cases, when a grandparent remarries, the grandparent's marital union results in the addition of a stepgrandparent. Also, an adult child with biological parents can marry an individual with children from a prior relationship. The adult child's parents become stepgrandparents as a result of the marriage. In other instances, an adult stepchild may have a child, thus making his/her stepparent a stepgrandparent (Chapman *et al.*, 2016).

Stepgrandparents can serve as important sources of support to younger generations (Chapman *et al.*, 2016) and despite the unique relational challenges facing the stepgrandparent–stepgrandchild, intergenerational relationships are highly valued in families

(Kemp, 2007). Research indicates that the stepgrandparent–stepgrandchild relationship is strongest between *same-gendered pairs* (e.g., stepgrandmothers and stepgranddaughers). Indeed, stepgrandaughters and stepgrandmothers are likely to exhibit companionship, intimacy, and nurturance in their step roles, subsequently promoting the formation of positive step relationships (Monserud, 2008).

Stepchildren

A **stepchild** (stepdaughter or stepson) is the offspring of one's spouse or partner who had a previous marriage or intimate relationship. Both stepmothers and stepfathers can formalize and solidify their relationships with their stepchildren through adoption. Three times as many women and men in second or later marriages had adopted a child compared with those in first marriages (Jones, 2008). Stepchildren are more likely to live with stepfathers (50 percent) than stepmothers (9 percent) in married stepfamilies (Stykes & Guzzo, 2015). Stepfamily formation is marked by unique stressors that place children at risk for adjustment problems (Papernow, 2013), including conflicting expectations, stepcouple disagreements on parenting, coparenting conflict, loyalty binds, shifts in parent–child relationships, stepparenting challenges, and clashing family cultures (Jensen *et al.*, 2015). From the perceptions of stepchildren, Gross (1987) identified four patterns of stepfamily: *augmentation* (including the divorced biological parents and the stepparent); *reduction* (including only the custodial biological parent); *alteration* (including the custodial biological parent and the stepparent, but excluding the non-custodial parent); and *retention* (including the divorced biological parents and excluding the stepparent). As stepparent selection is not the decision of the children, it requires an adjustment period. It takes time for both stepparent and stepchild to get to know each other and to get a feel for what their new life promises.

The ages of the stepchildren affect stepparenting. Pre-adolescence is a developmental stage marked by significant cognitive growth, notable social transitions, and a continued reliance on family relationships to help regulate emotions and provide social structure (Charlesworth, 2015). Stepchildren are more accepting of stepparents at early ages because they are more dependent on their stepparents to meet various needs at these times and stepparents are more likely to spend time with younger children than with adolescents (Ganong *et al.*, 1999). However, tension between children and stepparents is common during adolescence (Hetherington & Clingempeel, 1992) as adolescents are increasingly less reliant on family relationships, more focused on peer relationships, and in pursuit of greater autonomy (Jensen & Pace, 2016). Adolescents differ on measures of adjustment, and these differences persist into the early adult years (Amato *et al.*, 2016). Girls in stepfamilies emitted less positive verbal and more negative problem-solving behavior toward their stepparents than did boys and had distinct adjustment problems that girls in nuclear families do not experience (Suh *et al.*, 1996). Compared with children from low-conflict households, children from high-conflict households tend to have weaker ties to parents and lower levels of psychological well-being (Benson *et al.*, 2008). Sweeney (2010) has explored the intersection of racial identity and stepfamily processes. African American children in both biologically intact families and stepfamilies report similar levels of mental health, self-esteem, and conflict management skills—a similarity not found among White children (Adler-Baeder *et al.*, 2010). Papernow (2013) explains that African American children are more accustomed than White children to receiving discipline from adults who are not biological parents (e.g., grandparents, neighbors, extended kin) and may thus adjust to stepfamily life more readily than their White counterparts. Similarly, Latino stepfamilies from different countries with unique cultures value *familism*—suggesting interdependence, close family ties, mutual support, and cooperation that override personal needs (Coltrane *et al.*, 2008).

Studies have indicated the importance of stepfather–child closeness and parental support with respect to stepchild adjustment. Close ties to married stepfathers are more likely to develop when adolescents are closer to their mothers before stepfather entry (King, 2009). Also, when stepfathers offered practical support (e.g., help with homework or other tasks) and made efforts to be involved in shared family activities, children reported higher quality relationships with their stepfathers (Kinniburgh-White *et al.*, 2010), promoting stepchild adjustment (Papernow, 2013) and improving academic performance among adolescent stepchildren (King, 2006). The *friendship style* (i.e., a stepparent being a friend) is reported to be the role stepchildren prefer (Erera-Weatherly, 1996). Level of open self-disclosure to a stepparent is a positive predictor of stepfamily satisfaction (Metts *et al.*, 2017). In analysis of nationally representative data (n=2,085), King *et al.* (2015) found that qualities of both the stepfather–adolescent relationship and the mother–adolescent relationship were associated with feelings of family belonging in stepfamilies. When stepfathers reported more positive relationships with their stepchildren, they also reported more positive marital quality and a higher frequency of positive marital interactions in terms of intimacy, shared activities, and verbal communication (Bryant *et al.*, 2016).

Predictors of Stepfamily Satisfaction

In a study of 1,646 adolescents residing in married and cohabiting mother–stepfather families, Jensen and Weller (2018) have identified high-quality couple relationship, high-quality parent–child relationships, and low-quality different profiles with respect to family stability.

High-quality Couple Relationship

High-quality stepparent–child relationships are central to stable and satisfying stepfamilies (Papernow, 2013) and are linked to stepfamily stability and children's well-being (Jensen & Pace, 2016). Bray and Kelly (2011) classify stepfamilies into matriarchal, romantic, and neotraditional. *Matriarchal stepfamilies* are headed by independent women who remarried to gain a companion instead of parenting partner and are found to be relatively successful. Their husbands were devoted to these women but had distant relationships with the children (Bray & Kelly, 2011). In *romantic stepfamilies*, couples had unrealistic expectations, wanting instant happiness and wanting the biological parent to disappear for all purposes. Romantic families were the most divorce-prone (Bray & Kelly, 2011). *Neotraditional stepfamilies* are characterized by a committed partnership and fare best as the couples nurture their marriage and raise their children.

Visher (1994) suggests three important lessons that stepfamilies can learn: dealing productively with losses and changes; accepting and appreciating differences; and enhancing relationships by increasing dyadic interactions. Interestingly, the extent to which the father has now fully explained the circumstances of the divorce is a positive predictor of current stepfamily satisfaction, whereas the extent to which the mother has now fully explained the circumstances of the divorce is a negative predictor of current stepfamily satisfaction (Metts *et al.*, 2017). When it comes to the sexual relationship in stepfamilies, Musavi *et al.* (2018) find that remarried women in Iran are still thinking about their previous relationship and are influenced by how their ex-husband evaluated their body image.

Childrearing Practices

Another issue that stepfamilies must confront is potential differences in childrearing practices between parents. Hoenayi and Yendork (2018) find that stepchildren' difficult experiences include (1) subjection to severe disciplinary measures and (2) favoritism. With respect to *disciplines*, Svare *et al.* (2004) suggest that stepparents should develop distinct approaches related to handling discipline, spending time with their children, and defining themselves as a parent figure in response to the needs of the spouse, children, and stepparent, and the involvement of the non-residential parent. Bray and Kelly (2011) suggest a parenting plan with the stepparents playing a *secondary, non-disciplinary role* for the first year or two because even if the stepparents are doing a good job, the children are likely to rebuff them. Both children and parents alike need to learn how to adapt to any new rules initiated by the stepfamily. Consistent family rules about school work and strategies for conflict resolution can help keep the playing field level. When it comes to coparenting in stepfamilies, Repond *et al.* (2018) identified cooperative, complementing, and conflictual coparenting and found no differences for marital satisfaction and child adjustment between the three coparenting types. Also, stepparents must seek to avoid *favoritism* (the practice of giving unfair preferential treatment to one sibling at the expense of another) between biological and stepchildren. Differential treatment may have different implications depending on the domain (e.g., warmth, discipline, chores) in which it occurs (McHale *et al.*, 2000). Adler (1956) advocates the importance of egalitarianism (equal treatment of siblings) as a preventive measure in promoting self-esteem because parental favoritism of one sibling over the other is linked to poorer sibling relationships. Surprisingly, maternal differential treatment was protective in one case. A study of 708 sibling pairs revealed that for adolescents with low self-esteem who were at least three years older than their siblings, maternal differential treatment predicted reduced externalizing behaviors (Rolan & Marceau, 2018). Despite social norms in Western culture that call for parents to treat their children equally (Parsons, 1942), differential treatment of siblings is common across the life span (Tucker *et al.*, 2003). Even in adulthood, differential treatment dynamics persist, with parents making distinctions among their offspring across a variety of domains, including closeness, intimate disclosure and confiding, the provision of emotional as well as instrumental support, and the inheritance process (Fingerman *et al.*, 2009; Suitor & Pillemer, 2000; Suitor *et al.*, 2006; Rolan & Marceau, 2018). With respect to inheritance, Jensen *et al.* (2018) emphasize the key factors of fair inheritance processes, including who participates, what structures are relied upon, the role of information, and ethical interpersonal treatment.

Financial Responsibility

One big contributing factor of stepfamily satisfaction is financial responsibility. Usually, the couples in stepfamilies adopt the *one-pot* or common pot method (pooling all their resources for household expenses) or the *two-pot* method (keeping their money separate). Studies have shown that couples who used the one-pot method reported higher family satisfaction than those who kept their money separate. The common pot or one-pot method lends itself to unifying the stepfamily while the two-pot method encourages biological loyalties and personal autonomy (Fishman, 1983). A comparative study of married and cohabiting couples across four countries (Denmark, Spain, France, and the United States) reveals that the legal status of the union is one of the strongest predictors of money pooling practices in these four countries (Hamplova *et al.*, 2009). Also, stepfamilies must learn how to take on additional financial responsibilities. For example, child support payments are critical to custodial parents following divorce. In 2016, about 22.4 million children (one-fourth or 27 percent of all U.S. children) under 21 years of age had a parent who lived outside their household and about one out of every five custodial parents were fathers (19.6 percent), but about seven in ten custodial parents (69.3 percent) who were supposed to receive child support in 2015 received some payments (Grall, 2018). Partners must work out a plan together as financial obligations to ex-spouses such as child support are likely to cause depletion of savings and place additional financial strains on second marriages.

Multiple Partner Fertility

Another side of stepfamilies is **multiple partner fertility (MPF)**—the pattern of a man or a woman having biological children with more than one partner. In the United States, a sizeable fraction of individuals across demographic groups have MPF (Guzzo & Dorius, 2016), including low-income teenage mothers (Furstenberg & King 1999), people at younger ages with unintended first births among the disadvantaged (Guzzo, 2014), adolescent and early adult women (Guzzo & Furstenberg, 2007), mothers receiving welfare (Meyer *et al.*, 2005), low-income unmarried mothers (Burton & Hardaway, 2012), low-income and less educated parents, parents growing up in single-parent families, and fathers who were incarcerated (Cancian *et al.*, 2016). In 2014, one in ten U.S. adults (10.1 percent) aged fifteen and older experienced MPF; mothers (16 percent or one in six) and fathers (14.6 percent or one in seven) had MPF (Monte, 2017). The prevalence of MPF increases among parents with biological children (15.7 percent) and increases more among parents of two or more biological children (20.6 percent) (see Figure 16.5).

Studies have indicated that the increase in MPF is due to non-marital births, socioeconomic status, welfare policies, and union dissolution. MPF is more common among unmarried couples (ElHage, 2017). The fraction of all mothers who have children with two or more fathers was 23 percent in the United States, compared with 12 percent in Australia, 16 percent in Norway, and 13 percent in Sweden (Thomson *et al.*, 2014). MPF rates are higher in countries where policies allow women to better balance work and family commitments (Rindfuss *et al.*, 2010). For example, the Scandinavian countries offer more generous welfare policies that provide higher levels of support and allow parents to better balance work and family obligations, compared to the United States which offers minimal support (Gornick *et al.*, 1997). MPF is higher in contexts of high union dissolution. Among British women, almost half of those who experienced a marital dissolution subsequently experience a conception within twelve months (Balbo *et al.*, 2013).

MPF can diminish a parent's ability to parent well (ElHage, 2017). Mothers with MPF are more likely to experience depression and parenting stress

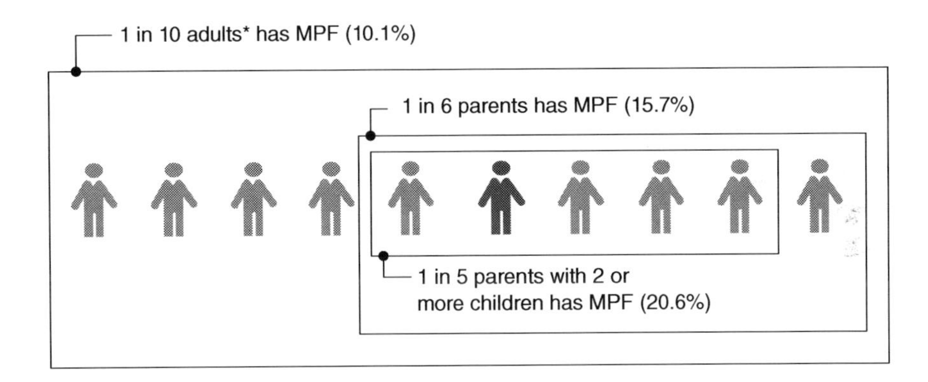

Figure 16.5 Prevalence of Multiple Partner Fertility (MPF), 2014

Source: U.S. Census Bureau, Survey of Income and Program Participation, 2014 Panel, Wave 1. https://www.census.gov/content/dam/Census/library/publications/2017/demo/p70br-146.pdf or https://ifstudies.org/blog/how-multiple-partner-fertility-influences-child-well-being

Note: *People aged 15 or older.

than mothers who have biological children with only one father (Fomby, 2018), because having children with multiple partners divides a mother's loyalty (ElHage, 2017). MPF was "robustly related to self-reported delinquency and teacher-reported behavior problems" (Fomby & Osborne, 2017). However, not every child growing up in an MPF family will experience these adverse outcomes (ElHage, 2017).

Sibling Relationships in Stepfamilies

Baham *et al.* (2008) distinguish three types of **siblings** (two or more individuals having one or both parents in common): **full-siblings** (a sibling of the target child who has the same biological parents as the target child), **step-siblings** (a sibling of the target child who is not biologically related to the child, and has entered the family system via the child's stepparent), and **half-siblings** (a sibling of the target child who shares one biological parent with the child, but the sibling's other biological parent is not biologically related to the child) (Baham *et al.*, 2008). Usually, partners who already have children from previous unions are more likely to have a child together. For example, Swedish couples wanted a *shared biological child*, regardless of how many children they had before their current union (Vikat *et al.*, 1999). In the United States, more than one in six children lives with a half- or step-sibling at age four (Fomby *et al.*, 2016). Step-siblings and full-siblings in stepfamilies are distinct from those in intact families in that they have experienced the loss of a former family system, have gained a new family system, and must adjust to their new family members such as new step- and half-siblings (Baham *et al.*, 2008).

The quality of sibling relationships in stepfamilies is measured in terms of *positivity* (e.g., warmth and affection) and *negativity* (e.g., conflict and rivalry) (Baham *et al.*, 2008). With respect to positivity, step-siblings were found to have more positive interactions than did half-siblings (Anderson, 1999). Step-siblings taught and helped each other less than did full-siblings (Ganong & Coleman, 1993). A study of full-siblings whose parents still live together as a couple or separated or died in Germany revealed that all full-siblings had more contact and felt closer to one another than half-siblings and step-siblings (Steinbach & Hank, 2018).

With respect to negativity, studies have shown that step-siblings exhibited significantly less negativity in their relationships than half- and full-siblings (Anderson, 1999), appeared to engage less frequently in negative behaviors than do full-siblings (Baham *et al.*, 2008), whereas full-siblings had more conflicts than half-siblings and step-siblings (Steinbach & Hank, 2018). Children residing with step- or half-siblings had higher average aggressive behavior (disruptive behavior such as temper tantrums, physical aggression, destruction of others' property, and displays of anger), compared to their peers whose parents had the same union status with no step-/half-siblings (Fomby *et al.*, 2016). The presence of half- or step-siblings in the household is also associated with negative outcomes, including greater depression and delinquency and worse academic performance (Halpern-Meekin & Tach, 2008).

Sulloway (1996) placed sibling rivalry at the core of family relationships and personality development. Feinberg *et al.* (2003) suggest that sibling differentiation may be a strategy for managing sibling conflict and rivalry. Evolutionary perspective suggests that sibling differentiation serves to minimize sibling competition (Sulloway, 1996), is associated with improved sibling relationships (Whiteman & Christiansen, 2008), and serves to mitigate interpersonal rivalry and to ease internal conflict associated with the lateral dimension (Vivona, 2007). Social learning perspective holds that parents can influence sibling relationships when they serve as models in the context of marital interactions and in their dyadic exchanges with their children (Baham *et al.*, 2008).

Research has explored sibling bond and sibling strengths. Attachment (emotional bonds) perspective may provide a solid foundation on sibling relationships across the life span. In addition to their primary caregiver, children can form attachments to a range of familiar others in their social worlds (Whiteman *et al.*, 2011). Siblings serve as companions, confidants, and role models in childhood and adolescence (Dunn, 2007) and as sources of support throughout adulthood (Connidis & Campbell, 1995). Sibling relationships are unique in that they are characterized by both hierarchical and reciprocal elements, which change across place and time (Dunn, 2007). In some cases, older siblings may take on the role of an attachment figure, facilitating exploration (Samuels, 1980) or providing comfort during a distressing situation when the primary caregiver is unavailable (Stewart, 1983). Like parents, older brothers and sisters act as role models and teachers, helping their younger siblings learn about the world (Jambon *et al.*, 2018). Consistent with the potential of siblings to serve as attachment objects, Jenkins (1992) finds that some siblings turn to each other for emotional support in the face of parents' marital conflict. Sibling relationships are often the longest-lasting relationship in individuals' lives (U.S. Bureau of the Census, 2005). In early adulthood, White (2001) suggested a distancing in the sibling relationship, with decreases in contact and proximity. In middle and later adulthood, contact between siblings stabilizes (White, 2001) and most siblings maintain contact with one another throughout the life course (Cicirelli, 1995).

Cohabiting Families

Cohabiting families are the most common living arrangement among people. A **cohabiting family** is a form of family in which a couple, with or without children, lives together without being legally married. Coresidential repartnership occurs through either remarriage or cohabitation. Many stepfamilies have been formalized through marriage, but others are maintained informally through cohabitation (Stewart, 2005). Cohabiting stepfamilies are predicated on informal ties between two adults and their partner's children (Stewart, 2007). Let's start with trends in cohabiting families.

Trends in Cohabiting Families

Cohabitation was often called "living in sin" but that taboo has faded (Jayson, 2013). In Europe, the increase in cohabitation and in childbearing with cohabiting unions has been one of the most striking changes in the family in the past few decades (Perelli-Harris & Gassen, 2012). The Scandinavian countries have led this trend, and most likely have high levels of cohabitation (Noack, 2010). Although once looked down upon, cohabitation is socially accepted in most areas of the United States. More men repartnered than women before 2002 in the United States and in other countries. For example, among Czechs, the odds of repartnering were higher among men than among women in the late twentieth and early twenty-first centuries (Kreidl & Hubatková, 2017). In the United States, both women and men have ever cohabited in the past but now more women cohabited than men as their first union. In 2011 to 2015, more women age 15–44 (14.7 percent) were currently cohabiting compared with men (13.3 percent). Table 16.1 shows the increase in cohabitation.

The increase in cohabitation is a remarkable demographic shift in the United States. The U.S. Census Bureau measured cohabitation by determining the number of *POSSLQ*—partners of opposite sex

Upon completion of this section, students should be able to:
LO5: Explain the increase in cohabitating families.
LO6: Describe the effects of serial cohabitation on family and children.

Table 16.1 Current Cohabitation Status

Percentage of men and women 15–44 years of age who are currently cohabiting with an opposite sex partner

	1982	*1995*	*2002*	*2006–10*	*2011–15*
Women	3.0%	7.0%	9.0%	11.2%	14.7%
Men	n.a.	n.a.	9.2%	12.2%	13.3%

Source: NHSR No. 49, NCHS, and CDC March 1, 2018. https://www.cdc.gov/nchs/nsfg/key_statistics/c.htm#currentcohab

sharing living quarters (Brown & Manning, 2009). However, measurement of children's living arrangements requires including cohabitation as a *family type*. Increases in the number of cohabiting parents raising children have contributed to the overall growth in unmarried parenthood (Livingston, 2018). In 2017, there were 2.9 million heterosexual cohabiting couples with children under age eighteen compared with 1.2 million in 1996. The number of all unmarried couples has increased from 2.9 million to 7.8 million during the same period (see Figure 16.6).

When rates of cohabitation increase, gender disparity might shrink (Kreidl & Hubatková, 2017). Table 16.2 shows the increase in cohabitation for both women and men from 1995 to 2015. Men (56.6 percent) aged 15 to 44 who ever cohabited in 2011–2015 was up from 48.8 percent in 2002 compared with women (52.4 percent), up from 50 percent in 2002.

There is also a growing trend among older adults to cohabit with a partner—2.4 percent of unmarried persons age 60 and older were cohabiting in the United States in 1990, up from virtually 0 percent in 1960 (Chevan, 1996). In 2014, 2 percent of non-institutionalized adults age 65 and older were

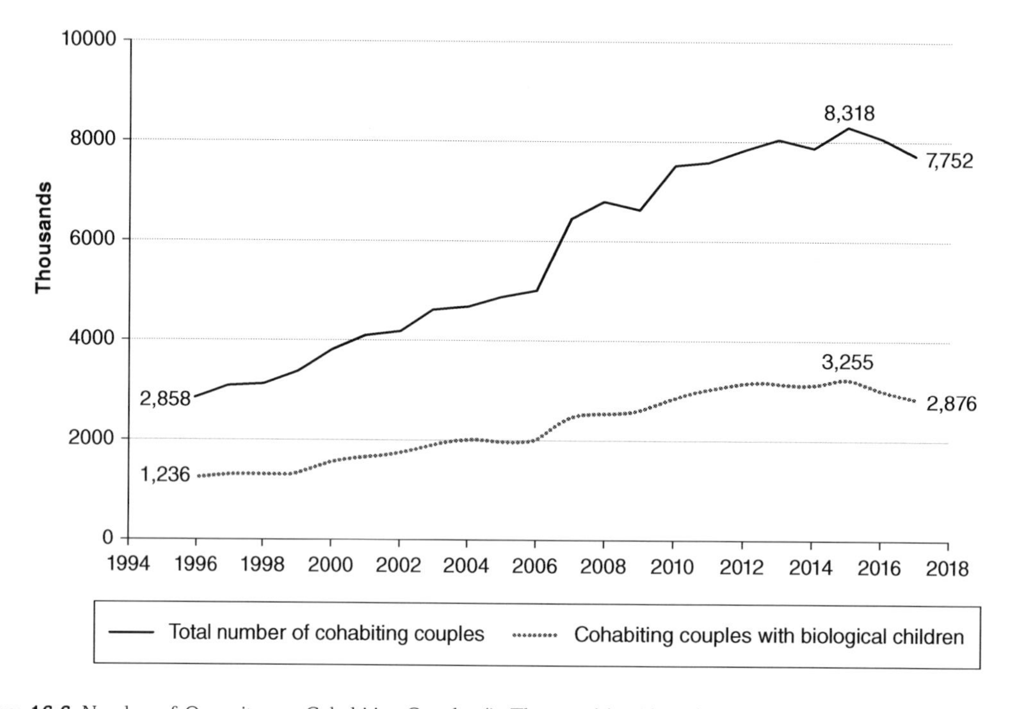

Figure 16.6 Number of Opposite-sex Cohabiting Couples (in Thousands), with and without Children under 18, 1996–2017
Source: Child Trends' calculations using U.S. Census Bureau (2017). https://www.childtrends.org/?indicators=family-structure
Note: Children under 18 include biological children of either partner, who have never been married.

Table 16.2 People Saying They Have Ever Cohabited

	1995	*2002*	*2006–10*	*2011–15*
Percent of men 15–44 years of age who have ever cohabited with a female partner	n.a.	48.8%	49.0%	56.6%
Percent of women 15–44 years of age who have ever cohabited with a male partner	41.2%	50.0%	52.0%	52.4%

Source: NCHS and CDC March 1, 2018. https://www.cdc.gov/nchs/nsfg/key_statistics/c.htm#currentcohab

unmarried and living with a partner (Stepler, 2016a). Older adults have additional reasons for why they consider cohabitation the ideal partnership instead of remarriage. For some, remarriage inspires feelings of disloyalty. Also, adult children may discourage remarriage based on concerns about inheritance.

Factors Related to the Increase in Cohabiting Families

People choose to cohabit for several reasons. It affords the benefits of marriage without the legal constraints (Brown & Booth, 1996) and allows individuals to maintain financial autonomy that would not be possible within the legal confines of marriage (Brown *et al.*, 2016). Cohabiting couples enjoy a close, intimate coresidential relationship akin to marriage, but can retain economic independence, allowing them to continue to receive a former spouse's pension or Social Security benefits, for instance, and ensuring that their assets can be bequeathed to their offspring (Chevan, 1996).

Some view cohabitation as a testing ground for or a prelude to marriage. Young people have consistently been more likely to agree with cohabitation as a testing ground for marriage. For example, high school seniors' attitudes towards cohabitation as a testing ground for marriage have become increasingly favorable over the past 40 years (Anderson, 2016). A cross-national study on the diverse meanings of unmarried cohabitation in Europe reveals that cohabiters differ in their reasons and motives to live together, the way they manage money, and their subsequent relationship decisions (Hiekel, 2014). For example, young adulthood is characterized by uncertain external circumstances when romantic love and independence are highly valued. People in Austria regard cohabitation as a prelude to marriage and regard marriage as preferable later in life when they desire security and commitment (Berghammer *et al.*, 2014).

Demographic characteristics such as age, cohabiting births, education, and race are related to repartnership status. Cohabiting parents are younger on average than single or married parents. The median age of cohabiting parents is 34 years, compared with 38 among single parents and 40 among married parents (Livingston, 2018). In cohabiting couples with children, two-fifths of parents are 25 to 34 years old, compared with around one-quarter of parents in married couples (Child Trends, 2018). The trend toward non-marital or cohabiting births is attributed to an increase in cohabiting unions. One out of four women (25 percent) under age twenty in their first premarital cohabitation experienced a pregnancy (Copen *et al.*, 2013). Due to their relative youth, cohabiting parents (54 percent) have lower levels of educational attainment, compared with single parents (45 percent) and married parents (31 percent) in 2017 (Livingston, 2018). Although between 1990 and 2016, the percentage of non-marital births rose across all levels of education, births to women who did not finish high school were more than six times as likely to be non-marital (62 percent) as births to women with a bachelor's degree or more (10 percent) (see Figure 16.7). And 55 percent of cohabiting parents are White, compared with Black (13 percent), Hispanic (25 percent), and Asian (3 percent) in 2017 (Livingston, 2018).

Other reasons for couples opting to cohabit include the increasing cost of marriages. People may opt to cohabitate if they wish to save money but not commit to a marriage. An **escape from stress perspective** focuses on the level of stress associated with various family arrangements. Marriage and cohabitation both represent opportunities to escape from unhappy family environments and find emotional support and intimacy elsewhere (Amato & Kane, 2011). For example, adolescent females with weak emotional ties to parents or low levels of psychological well-being were especially likely to cohabit, have non-marital births, marry, and have marital births at early ages (Amato *et al.*, 2008).

Outcomes of Cohabiting Relationships

Entry and exit from a series of coresidential intimate relationships is more likely to form and end relationships. **Serial cohabitation** occurs when an individual has cohabited with more than one partner and often has children with different partners. Cohabiting couples with children tend to be younger, less-educated, lower-income, and hold less-secure employment (Child Trends, 2018). Research identified significant differences among cohabiters, remarrieds, and unpartnereds across economic resources, physical health, and social relationships. Women cohabiters were more disadvantaged than remarrieds (Brown *et al.*, 2006). Compared to children born to married parents, children's living arrangements have become increasingly complex and unstable (Raley & Wildsmith, 2004). Children born to cohabiting parents experience higher levels of socioeconomic

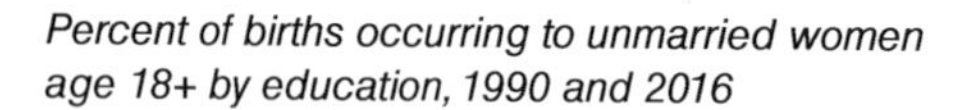

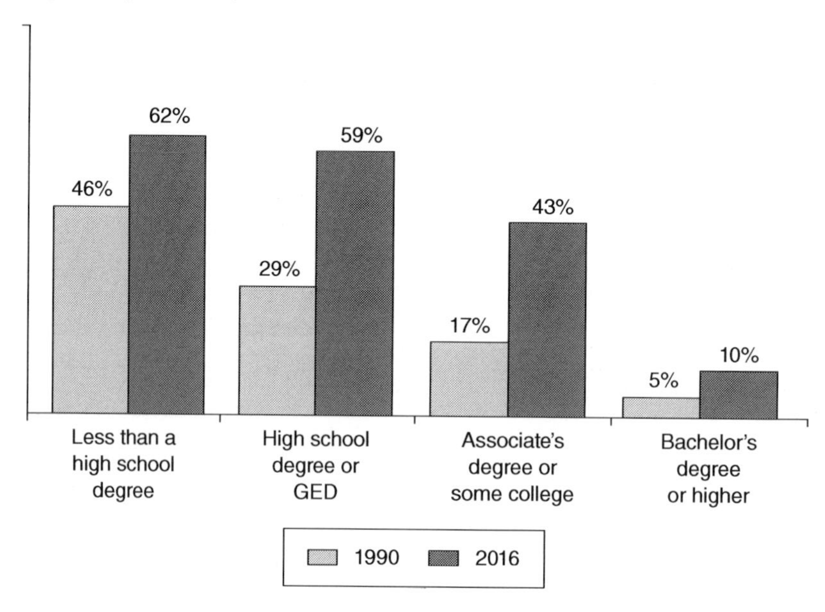

Figure 16.7 Non-marital Childbearing Has Increased Dramatically for Women of All Education Levels
Source: Vital Statistics birth data 1990, 2016 and Child Trends 2018. https://www.childtrends.org/publications/dramatic-increase-in-percentage-of-births-outside-marriage-among-whites-hispanics-and-women-with-higher-education-levels

disadvantage, and fare worse across a range of behavioral and emotional outcomes than those born to married parents (Brown, 2004). Two-thirds experience family instability before age twelve, compared with about one-quarter of children with married parents in 2017 (Child Trends, 2018).

Cohabiting couples are staying together longer than in the past. The length of first premarital cohabitations for women ages 15 to 44 was 22 months in 2006–2010, compared with 20 months in 2002 and 13 months in 1995. More people choose to spend more time living together and premarital cohabitation may contribute to the delay in first marriage for both women and men. In 2006–2010, 40.3 percent of women ages 15 to 44 spent three years living together in their first premarital cohabitations and transitioned to marriage; 32.2 percent remained in cohabitating relationships; and 27.4 percent dissolved (see Table 16.3).

Cohabitation may serve as a step toward marriage. Couples cohabiting often feel that they can assess their compatibility prior to marriage by living together (Bumpass *et al.*, 1991). Between 2010 and 2014, seven in ten cohabited with their spouse prior to the wedding, up from 75 percent over a 30-year period (Hemez & Manning, 2017). Transitions to marriage through cohabitation are more likely for cohabiting women with higher levels of education and income than for cohabiting women of lower socioeconomic status (Lichter *et al.*, 2006). Factors that may influence the progression from cohabitation to marriage include relationship commitment and attitudes toward marriage (Sassler & Miller, 2011). For 679 committed couples, both higher commitment and forgiveness were associated with higher perceptions of their partner's relationship self-regulation (Novak *et al.*, 2018). A study of 8,006 men and women in a variety of romantic unions reveals that both aspects of relationship self-regulation are positively associated with both aspects of relationship quality (Shafer *et al.*, 2016).

The most obvious issue in cohabiting stepfamilies is an elevated level of *family boundary ambiguity* (inconsistency in reporting who is in and who is out of the family) (Boss & Greenberg, 1984). Family members in cohabiting families reported higher levels of family boundary ambiguity than those in married stepfamilies (Brown & Manning, 2009) and experienced greater ambiguity with respect to the addition of the stepparent due to the legal and social uncertainties that coincide with non-marital stepfamily status (Jensen & Pace, 2016). Family ambiguity is linked to lower levels of family connectedness (Brown & Manning, 2009) and is negatively associated with the quality of the couple's relationship and stability of the union (Stewart, 2005). Gaspard (2018) suggests that discussing the role a stepparent will play in raising their new spouse's children, as well as changes in household rules and routines, can help couples build a strong family bond. A study of Norwegian survey data on 4,061 partnered individuals ages 18 to 55 revealed that parents with preschool children met

Table 16.3 Outcome of First Premarital Cohabitation

Probability that the first premarital cohabitation will remain intact, transition to marriage, or disrupt (break up) by duration of cohabitation among women 15–44 years of age, 2006–10

	By Year 1	*By Year 3*
Remain intact	67.0%	32.2%
Transition to marriage	19.4%	40.3%
Disrupt (break up)	13.6%	27.4%

Source: NHSR No. 64, Table 3 and CDC March 1, 2018. https://www.cdc.gov/nchs/nsfg/key_statistics/c.htm#currentcohab

their in-laws more frequently and reported better relations with their partner's parents than childless cohabiters (Wiik & Bernhardt, 2017).

Single-parent Families

In the United States, single-parent families were the second-most common living arrangement of minor children in 2016, with just over 20 million children living with either a single mother or a single father (U.S. Census Bureau, 2016a). A **single-parent family** is a form of family that includes children under age eighteen and that is headed by a parent who is widowed or divorced and not remarried, or never married. Let's examine some aspects of single-parent families, starting with demographic trends.

Trends in Single-parent Families

Single-parent households in most OECD (Organization for Economic Cooperation and Development) member countries from North and South America to Europe and Asia-Pacific are likely to increase by between 22 percent and 29 percent with the bulk of projections to 2025–2030 (OECD, 2012). In Australia, Austria, Japan, and New Zealand, single-parent families' share of all family households with children could reach between 30 percent and 40 percent (up from 28 percent, 26 percent, 22 percent and 31 percent, respectively, in the mid-2000s). The proportion of children living in single-mother families almost tripled from 8 percent in 1960 to 22.7 percent in 2017 in the United States (Figure 16.8).

Upon completion of this section, students should be able to:
LO7: Describe the causes of the increase in single-parent families.
LO8: Explain the issues relating to single-parent families.

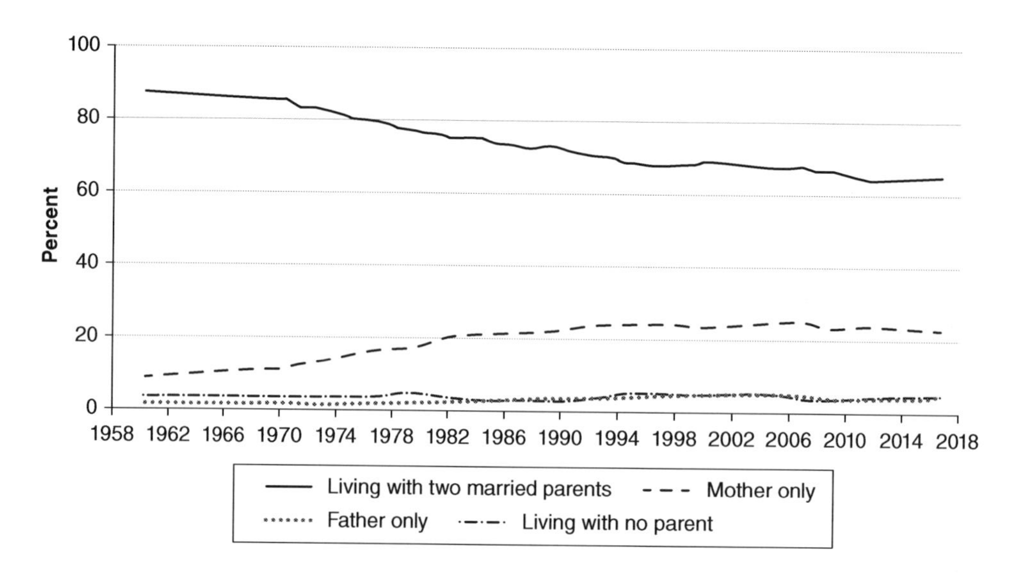

Figure 16.8 Living Arrangements under Age 18, 1970–2017
Source: Child Trends and U.S. Census Bureau. https://www.childtrends.org/indicators/family-structure

Births Outside of Marriage

A reason for the increase in single-parent families is an increase in the number of births taking place outside of marriage. One in four U.S. parents are unmarried due to declines in marriage and increases in births outside of marriage (Livingston, 2018). In 1968, 13 percent of children (or 9 million) were living in this type of arrangement, and by 2017, that share increased to about one-third (32 percent or 24 million) (Livingston, 2018). The percentage of births that occurred outside of marriage increased for Black women from 63 to 69 percent between 1990 and 2016 and grew most rapidly among White and Hispanic women (see Figure 16.9).

Factors Related to Rising Female Family Headship

Since 1970, out-of-wedlock birth rates have soared. In 1965, Black infants (24 percent) and White infants (3.1 percent) were born to single mothers, compared with Black infants (64 percent) and White infants (18 percent) in 1990 (Akerlof & Yellen, 1996). Three causal theories that explain the increase in the number of children born out-of-wedlock are the public policy theory, dislocation theory, and cultural theory (Baker, 1999).

Murray (1984) argues that much of the increase in single-parent families is a direct result of federal welfare policy. The Social Security Act of 1935 was enacted to protect poor single mothers who were likely widows. This changed dramatically in the 1960s and 1970s as the divorce rate soared. Women increasingly decided to raise children on their own and supplement their own income at times with welfare benefits available for single mothers (Murray, 1984). Single-parent families have often been negatively linked with welfare in the public's eye. Reducing the number of out-of-wedlock births was one of the national goals of the Personal Responsibility and Work Opportunity Reconciliation Act of 1996 (PROWRA), signed into law by President Bill Clinton in August 1996. The goal of this welfare reform was to set up welfare eligibility limits and mandatory job

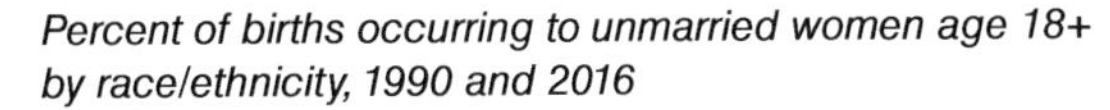

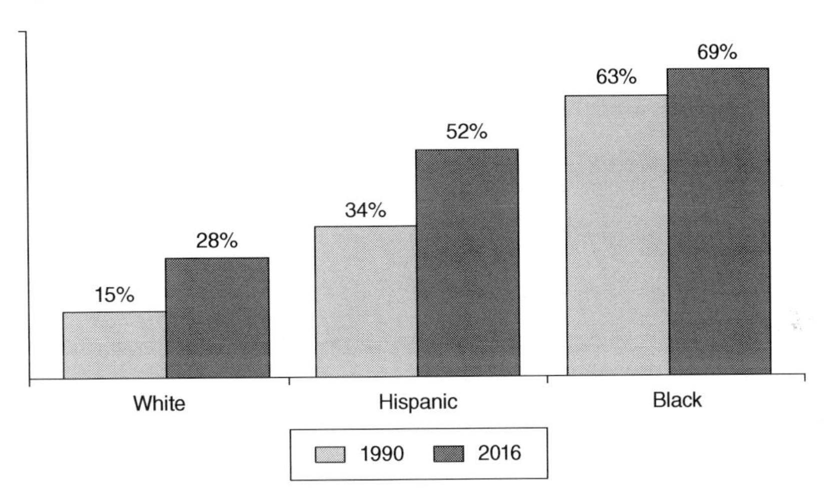

Figure 16.9 The Percentage of Births Occurring to Unmarried Women Has Grown Most Rapidly among White and Hispanic Women

Source: Vital Statistics birth data 1990, 2016 and Child Trends 2018. https://www.childtrends.org/publications/dramatic-increase-in-percentage-of-births-outside-marriage-among-whites-hispanics-and-women-with-higher-education-levels

training requirements to put welfare recipients back to work. Research shows that welfare benefits policy theory could not play a role in the rise of out-of-wedlock births because benefits rose sharply in the 1960s and then fell in the 1970s and 1980s, when out-of-wedlock births increased most (Ellwood & Summers, 1986).

The dislocation theory asserts that the increase in out-of-wedlock births results from the unavailability of single, employed African American males in urban America (Wilson, 1987). This increase was attributed to a decline in the marriageability of Black men due to a shortage of jobs for less educated men. As the U.S. economy experienced a restructuring in the post-industrial society, more jobs in the service industry were available for women while the job market in the manufacturing industry became tightened for men. Men's weakened economic position and women's economic independence led to an erosion of marriage and rising single mother families in the 1970s. In general, men were less likely to be attractive marriage partners as men without a college degree were less likely to earn enough to support a family or did not have steady employment. However, the dislocation theory might not play a role in the rise of out-of-wedlock births either because Wood (1995) estimated that only 34 percent of the decline in Black marriage rates can be explained by the shrinking of the pool of eligible Black men.

The cultural theory posits that the increase in out-of-wedlock births is an outgrowth of the decline in *shotgun weddings* (an enforced wedding because of a pregnant bride) (Baker, 1999). In the late 1960s and early 1970s, many major states such as New York and California liberalized their abortion laws. At about the same time, in July 1970, the Massachusetts law prohibiting the distribution of contraceptives to unmarried people was declared unconstitutional. The increase in the availability of both abortion and contraception (*reproductive technology shock*) is implicated in the increase in out-of-wedlock births; although abortion and contraception are expected to lead to fewer out-of-wedlock births, the opposite happened—the erosion in the custom of shotgun marriages (Akerlof & Yellen, 1996).

Issues Relating to Single-parent Families

Single parents face the daily challenge of "making ends meet," balancing their commitments to work and family, and navigating an increasingly punitive welfare system (Yarber & Sharp, 2010). In 2017, about one-fourth of single parents (27 percent) are living in poverty, compared with married parents (8 percent) (Livingston, 2018). Reliance on kin networks (for example, living with grandparents) can provide social and financial support for some single-parent families (Dunifon & Kowaleski-Jones, 2007). In 2017, single fathers (31 percent) and single mothers (21 percent) are residing with at least one of their parents (Livingston, 2018).

Most children who live with just one parent, regardless of their race, live with their mothers. Figure 16.10 shows that 48.5 percent of Black children live with their mothers only, compared with Hispanic (24.7 percent), White (15.7 percent), and Asian (7.9 percent). The **feminization of poverty** is the phenomenon that women represent disproportionate percentages of the poor. Growing up in a never-married single-mother family can be especially difficult. This is because never-married mothers tend to be younger than divorced mothers and are less likely to have higher education and savings than divorced mothers. Because many daughters from disadvantaged families have low expectations for higher education, they tend to begin their families at relatively early ages but daughters from economically advantaged families tend to delay family formation until they have completed college (Amato *et al.*, 2008). Offspring with single parents and remarried parents had an elevated risk of non-marital births and non-marital cohabitation (Amato & Kane, 2011).

Economic deprivation perspective suggests that most single mothers have a low level of financial resources due to the loss of economies of scale associated with splitting one household into two and the lower earnings of mothers compared with fathers (Amato & Kane, 2011). By contrast, families with greater socioeconomic resources can make greater investments (of both time and money) in their children (Kalil, 2015). These differential investments may be an important factor in growing inequality across generations (Reeves, 2017). For example, single women are less likely to have retirement account savings than single men and couples. Figure 16.11 shows that the typical single woman ($30,000) had lower balances than the single man ($34,000) and the married couple ($78,000) in 2013.

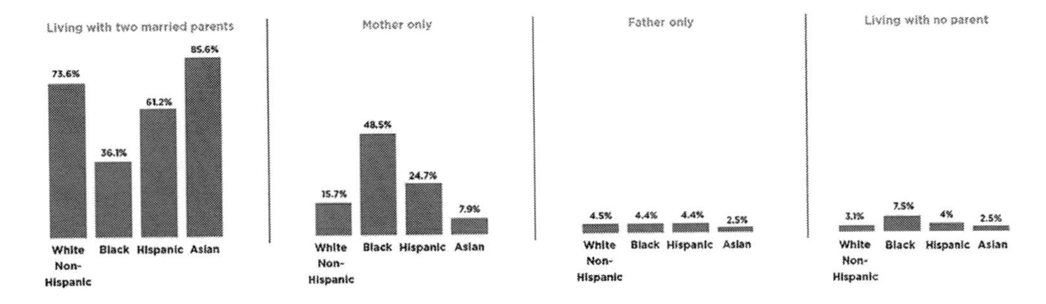

Figure 16.10 Living Arrangements of Children by Race and Hispanic Origin, 2017
Source: Child Trends and U.S. Census Bureau. https://www.childtrends.org/indicators/family-structure

Figure 16.11 Single People Have Less, but Retirement Savings Are Too Low across the Board
Source: EPI analysis of Survey of Consumer Finance data, 2013. http://www.epi.org/publication/retirement-in-america/

Diversity Overview: Retirement Inequality by Race, Class, and Education

Compared to other industrialized OECD (Organization for Economic Cooperation and Development) countries, the United States has very high levels of income inequality. After accounting for taxes and transfers, Gini coefficients (i.e., measure of the income or wealth distribution of a nation's residents) across nineteen countries ranged from 0.24 in Norway to 0.37 in the United States (Gornick & Milanovic, 2015).

Race and ethnicity affect family inequality and retirement inequality. Since the Colonial era, America has been racialized, allocating resources such as income, wealth, neighborhoods, schools, and health insurance unequally along racial lines. Today, Americans rely on savings in 401(k) accounts to supplement Social Security in retirement. Pensions are one form of retirement savings. Only 41 percent of Black families and 26 percent of Hispanic families had retirement account savings in 2013, compared with 65 percent of White families (Morrissey, 2016). Figure 16.12 shows that for families with retirement account savings, the median amount is $22,000 for Black and Hispanic families, compared with $73,000 for White families.

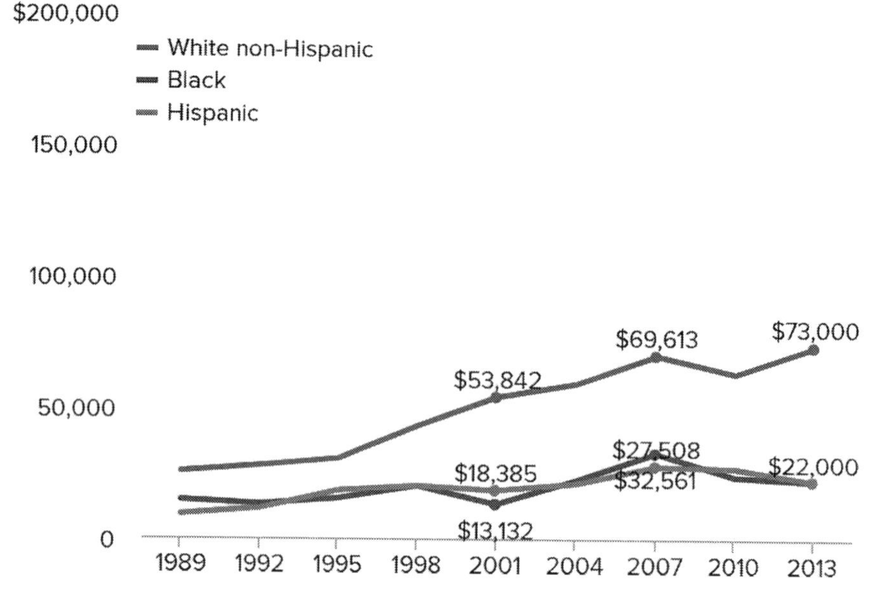

Figure 16.12 Racial and Ethnic Gaps in Retirement Savings

Source: EPI analysis of Survey of Consumer Finance data, 2013. http://www.epi.org/publication/retirement-in-america/

Socioeconomic class also affects retirement inequality. Class inequality affects patterns of partnership and childbearing and, in turn, family change feeds economic inequality (Cahn *et al.*, 2018). Throughout much of North America, well-educated and more affluent families tend to be headed by stably partnered parents who enjoy high levels of relationship quality but working-class and poor families face higher levels of family instability and single parenthood, and lower levels of relationship quality (Cahn *et al.*, 2018). Figure 16.13 shows that the median working-age family had only $5,000 saved in retirement accounts, while the top 1 percent of families had $1,080,000 or more (not shown on chart) (Morrissey, 2016).

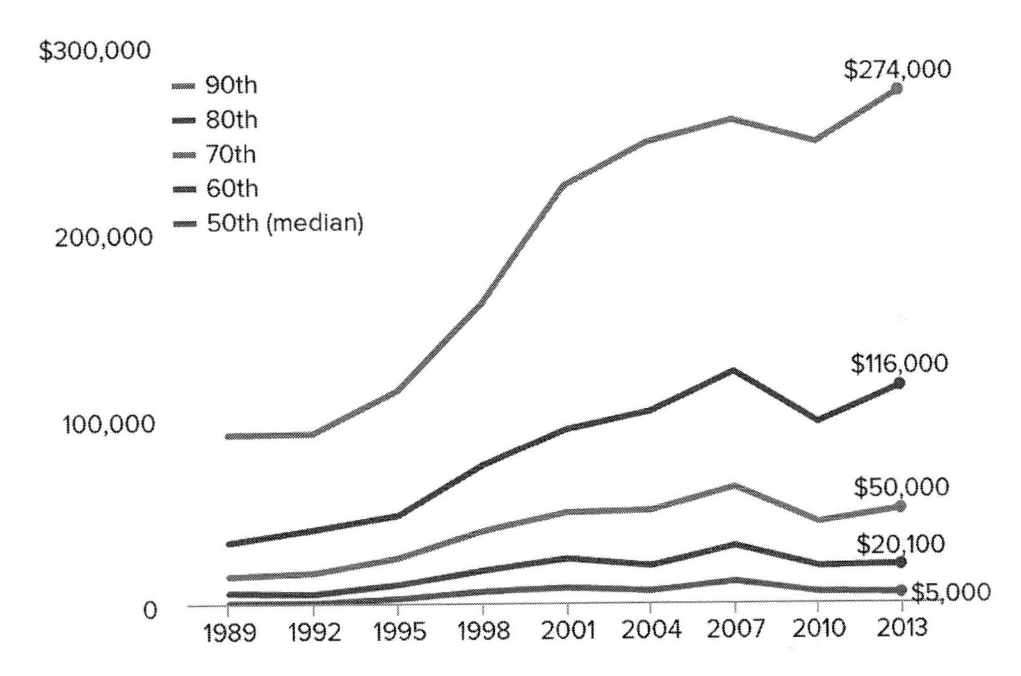

Figure 16.13 The Gaps between the Retirement Haves and Have-nots
Source: EPI analysis of Survey of Consumer Finance data, 2013. http://www.epi.org/publication/retirement-in-america/

Level of education is also linked to retirement inequality. Over three-fourths (76 percent) of families headed by someone with a college degree or more education had savings in retirement accounts in 2013, compared with 43 percent and 18 percent of families headed by someone with and without a high school diploma or GED (see Figure 16.14).

Share of families age 32–61 with retirement account savings by education, 1989–2013

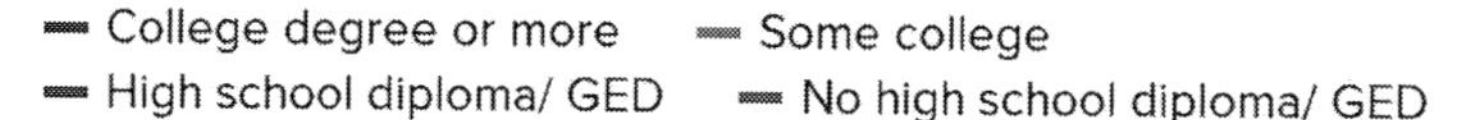

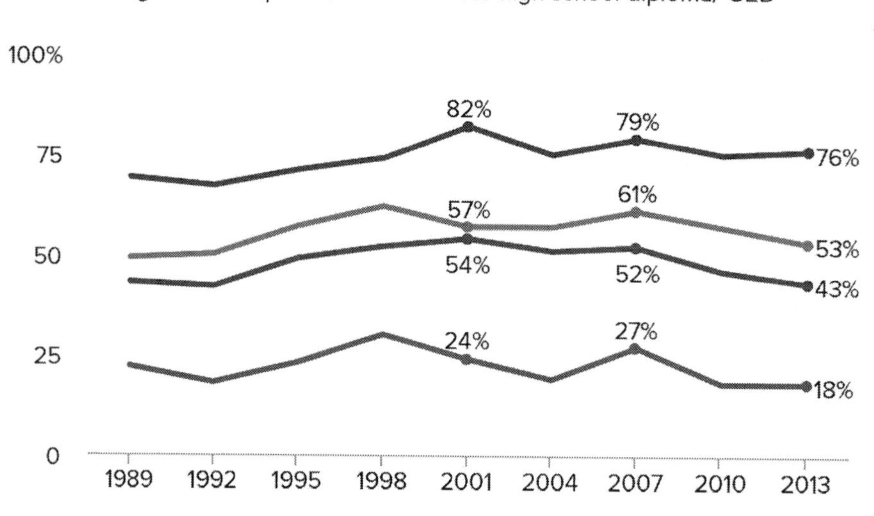

Figure 16.14 College-educated Families Are Much More Likely to Have Retirement Savings

Source: EPI analysis of Survey of Consumer Finance data, 2013. http://www.epi.org/publication/retirement-in-america/

Upon completion of this section, students should be able to:

LO9: Describe the family life of late adulthood age groups.

LO10: Explain the characteristics of grandparent families.

Aging Families

Population aging is a global phenomenon. From 1980 to 2017, the number of persons aged 60 years and over in the world more than doubled from 382 million to 962 million (UNDESA, 2017). An **aging family** is a form of family in which an adult age 65 and older resides in a household alone or with a spouse, adult child, grandchild, or others.

Late Adulthood Age Groups

Older adults are a heterogeneous group in terms of levels of activity and health. Older adults are divided into three age-cohort groups: *the young-old* consist of adults between the ages of 65 and 74, *the middle-old* consist of adults between the ages of 75 and 84, and *the oldest-old* consist of adults age 85 and over. When it comes to the role of the elderly in family and society, it is important to measure the share of young-old, middle-old, and oldest-old in the total population (OECD, 2012). The young-old adults (65–74 years) are likely physically independent and healthy. People born in 2007 in OECD member countries can expect to live in "full health" to their seventies (WHO, 2009). Young-old families are likely to be happy, healthy, and financially sound. Many young-old couples are still working to better prepare for the future and have many years ahead of them to participate in society, to work and provide care (OECD, 2012). However, some young-old families do experience health issues. The rate of disability accounted for 25.3 percent of the young-old population ages 65 to 74 (U.S. Census Bureau, 2018a). The old adults live throughout the United States, though in Sumter, Florida, 54.8 percent of the population reached retirement age in 2015, up from 53.0 percent in 2014 (U.S. Census Bureau, 2016b/2017b).

Many middle-old adults (75–84 years) attend continuing education class and many are technically savvy. In fact, 80 percent reported living in homes with computers and 62.4 percent accessed the internet through a high-speed internet connection in 2016 (U.S. Census Bureau, 2018b). Gray divorce is more common among middle-aged than older adults, non-Whites than Whites, and those with a high school compared with a college degree (Brown & Lin, 2012).

The health gap grows larger through middle and early old age, and then retreats at older ages (House *et al.*, 2005). The rate of disability accounted for 49.5 percent of the middle-old population ages 75 and older (U.S. Census Bureau, 2018a). Lower levels of education, income, and wealth are strongly associated with mortality risk among older middle-aged Americans (Feinglass *et al.*, 2007). From 2009 to 2010, the death rate for middle-old adults decreased by 0.7 percent (Tabloski, 2013).

The number of the oldest-old adults (85 years and over) is increasing. Lowsky *et al.* (2014) find that a significant proportion of older Americans are healthy within every age group ages 51 to 85+. The three convergent environmental factors such as neighborhood problems, access to public services and programs, and community engagement were significantly associated with life satisfaction across all age groups (Park & Lee, 2018). A **centenarian** is an old adult who lives to or beyond the age of 100 years. The world was home to nearly half a million centenarians in 2015—more than four times as many as in 1990. There are 4.8 centenarians for every 10,000 people in Japan, 4.1 in Italy, and 2.2 centenarians per 10,000 people in the United States (Stepler, 2016b). In the United States, over half (62.5% percent) of the 53,364 centenarians were age 100 or 101 in 2010; 43.5 percent of centenarian men lived with others in a household, and for their female counterparts the most common living arrangement was residing in a nursing home (35.2 percent) (U.S. Census Bureau, 2017b). An increasing number of the oldest-old, however, will face disability and a loss of identity, autonomy, and control (e.g., psychological mortality) (Baltes & Smith, 2003), are likely to suffer from bad health, and are often dependent on others. The proportion of three or more types of disabilities reached 41.5 percent for adults age 85 and older (He & Larsen, 2014).

Later-life Marital Dissolution

In 2017, 56.7 percent of adults ages 65 and older were married and 5.6 percent were never married (U.S. Census Bureau, 2018b). Trends in marriage, remarriage, divorce, and living arrangements will change the role of the elderly in their families. Later-life marital dissolution is a stressful life event whether it occurs through spousal death or divorce (Kitson *et al.*, 1989).

Later-life marital dissolution commonly occurs through spousal death and it affects older adults in many ways. In 2017, 22.9 percent of adults ages 65 and older were widowed (U.S. Census Bureau, 2018b). In some instances, widow(er)s have unique institutional protections such as Social Security and spousal retirement benefits that may buffer the disadvantage associated with marital loss (Bair, 2007).

Unfortunately, widowhood is associated with a reduction in one's economic standard of living, especially among women. For example, most widows in Germany and the United States experienced a decline in living standards, and many fell into poverty at widowhood (Hungerford, 2010). Also, spousal loss is related to declines in physical and mental health, particularly for men (Hughes & Waite, 2002). Furthermore, smaller social networks and less contact with children increase emotional and social loneliness in later life (Jong Gierveld *et al.*, 2009). Barriers to remarriage include personal, normative, interactive, and financial factors (Osmani *et al.*, 2018). For example, women expressed more negative feelings about forming new romantic relationships (Schneider *et al.*, 1996). Research has explored the influence of loss on continuing bond expression within an attachment framework. In a study of remarried widows and their husbands in Israel, Bokek-Cohen (2014) found the widows' continuing connection to the deceased husband and husbands' expression of their relationship as triadic. The **theory of continuing bonds** suggests that grief is continuing ties to loved ones and extraordinarily stressful. Klass *et al.* (1996) suggest creating a new relationship with the deceased. For example, clients can use some symbolic objects, either abstract or concrete, to help widows rebuild their self-identity (Bokek-Cohen, 2014).

Later-life marital dissolution increasingly occurs through divorce rather than widowhood (Brown *et al.*, 2018). In 2017, 12.2 percent of older adults were divorced and 1.1 percent were separated (U.S. Census Bureau, 2018b). The **gray divorce** refers to the trend of increasing divorce for older adults or grey-haired couples after year 50. Marriage dissolved more among men ages 50 to 64 (78.9 percent) and men ages 65 to 74 (59 percent) than women ages 50 to 64 (56.2 percent) and women ages 65 to 74 (27.8 percent) (Brown *et al.*, 2016). Divorce may have more benign effects than widowhood because the transition to divorce often involves a calculated decision-making process, at least for the initiator (Bair, 2007). Divorced parents not coresiding with their young children often cannot maintain long-lasting relationships with them and receive less support in old age (Albertini & Saraceno, 2008). Gray divorced and never-married women may face considerable economic insecurity. In a comparative study of divorcées before age 50, divorcées after age 50, and *divorce careerists* (i.e., individuals who experienced a divorce both before 50 and after 50), Spangler *et al.* (2016) find that divorce careerists have the fewest assets such as homeownership.

Exit status (widowhood or divorce), age, education, wealth, health, and race may affect repartnership. Divorced people (aged 50–64 years) are most likely to form a new union (Brown & Lin, 2012) and those whose marriages dissolved after age 55 were much less likely to have formed a new relationship, especially a remarriage (Jong Gierveld, 2004). A study of divorce versus widowhood across four age groups (50–64, 65–74, 75–84, and 85+) reveals that divorcées are more often repartnered through either remarriage or cohabitation than are widows/widowers (Brown *et al.*, 2018). Daters are more likely to be men. For example, 14 percent of unmarried adults age 57 to 85 were in dating relationships and two-thirds of daters were men. Daters were more likely to be college-educated and had more assets, were in better health, and reported more social connectedness (Brown & Shinohara, 2013). Good health is positively associated with remarriage; wealthier older adults, regardless of gender, are more likely to repartner than stay single (Vespa, 2012). Older cohabiters are less often White, compared with both remarrieds and unpartnereds (Brown *et al.*, 2006). Older adult cohabiting unions were quite stable and unlikely to culminate in either marriage or separation (Brown *et al.*, 2012). In later-life remarriage, the emotional bonds between older stepparents and adult stepchildren are less close than parent–child bonds, because the older stepparents have spent little time with their adult stepchildren and neither person may be motivated to develop a relationship (White, 1994). When it comes to eldercare, a study of later-life remarried wives caring for husbands with dementia revealed that caregivers reported receiving support from their own family, but noted an absence of support or active interference from stepfamily members (Sherman, 2012).

Grandparent-headed Families

As older adults are flocking to Florida, Arizona, or similar locations to spend their retirement years, many grandparents become the primary caretakers of their grandchildren. A **grandparent-headed family** is a form of family in which grandparents take care of a grandchild or grandchildren with or without a parent of the child being present. In 2016, 7.3 million grandchildren under age eighteen were living with grandparents (U.S. Census Bureau, 2018c). Let's start with the profile of *coresident grandparents and grandchildren* (i.e., grandchildren and grandparents living in the same household).

Grandparenthood is more common among older adults (aged 65 and older) than those in midlife (aged 40–64) and was most common among widowed adults (78 percent), compared with married (59 percent), divorced/separated (57 percent), and cohabiting grandparents (46 percent) (Wu, 2018b). Whites comprise a higher proportion of older grandparents (79 percent) than midlife grandparents (66 percent); Black (15 percent) and Hispanic (16 percent) comprised a higher proportion of midlife grandparents than Asian (3 percent) older grandparents (Wu, 2018a). In 2015, about three-quarters of children in a grandparent-headed household lived with their parents (three-generation household) and one in four children lived with their grandparent and without a parent (Wu, 2018c).

Grandparent-headed families varied by socioeconomic status (SES). The composition of the households that grandchildren live in is associated with poverty rates, access to health care and other resources, and educational outcomes (Bryson & Casper, 1999). Grandchildren living in households with two grandparents were less often in poverty compared with grandchildren living with only a grandmother (Ellis & Simmons, 2014). Social Security is the foundation of retirement and the most important retirement income source. The median income for families with grandparent householders and/or spouse responsible for grandchildren under age eighteen was $52,610 (U.S. Census Bureau, 2018c). Among these families, where a parent of the grandchildren was not present, the median income was $38,599. Less educated, minority, single, and female seniors are most likely to have low incomes in retirement. Racial and ethnic disparities are significant, with more than half of Hispanic seniors and nearly half of Black seniors living on less than full-time minimum-wage earnings ($15,080 per year) (Morrissey, 2016). Social Security income is lowest for seniors in their late 60s, many of whom are still working. More men age 65 and older (2.8 million) were in the labor force than their female counterparts (2.4 million) in 2016 (U.S. Census Bureau, 2018b).

Recent trends in increased life expectancy, single-parent families, and female employment increase the potential for grandparents to play an important role in the lives of their grandchildren (Ellis & Simmons, 2012). Grandparent relationships play a beneficial role especially after parents separate or divorce (Jappens, 2018) through three perspectives: that of the older persons themselves, that of the divorcing couple, and that of the children of the divorcing couple (Kalish & Visher, 1982). For example, living with a single mother and a grandparent is associated with increased cognitive stimulation and higher reading recognition scores for White children, compared to living with a single mother alone (Dunifon & Kowaleski-Jones, 2007). In Europe, Sheppard and Monden (2018) find that having two highly educated grandfathers shows a stronger association with higher educational outcomes for grandchildren than does having just one highly educated grandfather.

Eldercare

The future role of the elderly in OECD member countries is as both providers and recipients of care (OECD, 2012). Socioeconomic disparities in healthy life expectancy are large and growing (Crimmins & Saito, 2001). Lower income (Deaton & Paxson, 1998) and education (Dupre, 2007) are correlated with poorer mean health outcomes (House *et al.*, 2005). Older Americans live at home and many have functional limitations or ill health (Iriondo, 2018). Eldercare needs are growing as the nation's older population increases and lower fertility has reduced the number of family members available to provide care (He *et al.*, 2018). **Eldercare** providers are individuals who provide unpaid care to someone age 65 and older who needs help because of a condition related to aging. Research attests to the resilience of couples during later-life stages in which health issues may severely inhibit domestic productivity. For example, a study of 3,889 retired couples aged 60–85 reveals that wives' and husbands' housework time declined with health status (Leopold & Schulz, 2018). Indeed, older people often need help with daily tasks as they age. The probability of getting paid help from outside the household increased with declining health, and this increase was more strongly tied to wives' health declines than to husbands' health declines (Leopold & Schulz, 2018). Eldercare services include nursing homes, assisted living centers, adult daycare centers, long-term care facilities, hospice care facilities, and in-home care facilities. There were 4,815 continuing care retirement communities in 2012 and 25,964 business establishments that provide services for the elderly and people with disabilities in 2012 (U.S. Census, 2017b). They provide a range of residential and personal care services with on-site nursing care facilities.

Older people living in a household who have physical, mental, or cognitive functioning limitations are often cared for by unpaid, informal caregivers and rely on home eldercare (He *et al.*, 2018). Spouses and family members take responsibility for eldercare in many societies. For example, in China, the Law of Protection of Rights and Interests of Elderly People lays out the duties and obligations of children and requires children of parents older than 60 to visit their parents "often" and tend to their financial and spiritual needs (Wong, 2013). The United States has no such law. Concern for the well-being of providers is growing as the need for eldercare rises in an aging nation (Iriondo, 2018). In fact, family members provide more than 95 percent of the informal care for older adults who do not live in nursing homes in the United States (Scommegna, 2016). In 2014, 3 percent of older adults lived in nursing homes or other group quarters, down from 6 percent in 1990. This decline may reflect both better health among older adults and increased preference for living within a multigenerational family instead of nursing home care (Stepler, 2016a).

Global Overview: Living Apart Together (LAT)

Living Apart Together (LAT) refers to committed couples who live in different places. LAT offers an appealing living arrangement to couples who wish to engage in intimate ongoing companionship yet maintain their autonomy and independent households. LAT couples have been the focus of much recent research in the United States and around the world.

Research indicates that there has been growing recognition in Europe of LAT relationships. One study found that LAT couples account for around 10 percent of adults in Britain, and similar figures are recorded for other countries in northern Europe, including Belgium, France, Germany, the Netherlands, Norway, and Sweden (Duncan & Phillips, 2010). LAT relationships are more common in western than eastern Europe and among young people enrolled in higher education, people with liberal attitudes, highly educated people, and those who have previously cohabited or been married (Liefbroer *et al.*, 2015). Most people in LAT unions intend to live together but are apart for practical reasons.

In a study of 54 men and women engaged in Living Apart Together (LAT) partnerships in Belgium (Flanders), Lyssens-Danneboom and Mortelmans (2014) find that heterogeneity of LAT partnerships is reflected in couples' monetary behavior, with couples who foresee cohabitation (transitional LATs) showing more marriage-like exchanges of financial support than those who perceive living apart together as a more permanent arrangement (permanent LATs). Traditional notions of gender and coupledom continue to be strongly influential, even in non-traditional types of partnerships.

Based on a survey from ten western and eastern European countries, Liefbroer *et al.* (2015) find that LAT is more often a stage in the union-formation process than an alternative to marriage and cohabitation. Although some groups view LAT as substituting for marriage and cohabitation, these groups differ between the east and west. In eastern Europe, the highly educated elite tend to have been the first to resist traditional marriage norms and embrace LAT and cohabitation as alternative living arrangements. In western Europe, LAT unions are mainly an alternative for people who have been married before or had children in a prior relationship. In Europe, older people and divorced or widowed persons are more likely to choose LAT to maintain independence (Liefbroer *et al.*, 2015). Surprisingly, attitudinal and educational differences are more pronounced in eastern Europe than in western Europe.

In Australia, 7 percent to 9 percent of the adult population has a partner who lives elsewhere (Reimondos *et al.*, 2011). In Canada, the proportion of people in a LAT couple was 7.4 percent in 2011. Young adults, many of whom live with their parents, were much more likely to be in a LAT couple. The duration of such relationships was relatively short—for two in three people in a LAT couple, the relationship was less than three years old (Turcotte, 2013).

In the United States, nearly 40 percent of daters are in LAT relationships (Brown *et al.*, 2013) and they are classified as "single" in conventional studies that focus on coresidential unions (Strohm *et al.*, 2009). They include opposite-sex or same-sex partners who are in a long-term relationship but do not live together. Gay men are somewhat more likely than heterosexual men to be in LAT relationships, while for heterosexuals and lesbians, LAT relationships are more common among younger people. Regardless of sexual orientation, people in LAT relationships perceive similar amounts of emotional support from partners, but less instrumental support than cohabiters perceive (Strohm *et al.*, 2009). Like in Europe, many older U.S. adults are likely to choose LAT as they are interested in companionship but may not want long-term obligations. In the United States, people in LAT relationships are more likely to have been previously married and to own homes and work in different cities (Brown *et al.*, 2013). Adolescents' well-being was lower if their mothers were in a LAT relationship since their divorce. For fathers, adolescents' well-being was lower when their fathers began living with a new partner since their divorce (Bastaits *et al.*, 2018). Many women view dating as a social activity that provides a unique form of companionship not achieved through friendships with other women (Watson & Stelle, 2011). These women desire a close companion but at the same time want to be autonomous and ultimately are not interested in a long-term, conventional commitment (e.g., marriage or cohabitation) (McWilliams & Barrett, 2012).

Other countries practice similar arrangements. In Saudi Arabia, an arrangement called *misyar marriage* involves the husband and wife living separately but meeting regularly (Karam, 2006). Something else similar to LAT is the practice of *walking marriages* in Mosuo, which is close to the border with Tibet, China. A **walking marriage** refers to a sexual relationship in which male partners live elsewhere and make nightly visits to women in their homes. When a woman is interested in a potential male partner, the woman gives him permission to visit her. The male will walk to the house of the woman at night, spend the night with her, and return to his own home before dawn. When a child is born, the father may have little or no responsibility for his offspring. Children of such relationships are raised by their mothers' families. Men may share responsibilities in caring for children born to women within their multigeneration families. All extended family members share the duties of supporting and raising the children. There are no divorce or child custody issues.

Summary

Many people have chosen to remarry due to divorce and death of a spouse or births outside marriage. In stepfamilies, cohabiting or remarried couples and children from previous marriages are blended into a new family. Two types of stepfamily are formed through cohabitation and marriage. Most stepfamilies are formed in second and third unions through marriage and cohabitation. Remarried stepfamilies are diverse; they can include only a stepfather, only a stepmother, or both stepfather and stepmother.

Other people choose to cohabit without being legally married. Cohabiting couples may view their relationship as a step toward marriage or as a testing ground for marriage. Others view cohabitation as an alternative living arrangement or as a substitute for marriage. Another common living arrangement is the single-parent family. Meanwhile, rates of grandchildren living with their grandparents are at an historical high in the United States. From a diversity perspective, the chapter has examined retirement inequality across social groups. From a global perspective, the chapter has explored Living Apart Together (LAT), which has become a new focus in living arrangements globally.

Key Terms

Aging family is a form of family in which an adult age 65 and older resides in a household alone or with a spouse, adult child, grandchild, or others.

Attachment theory emphasizes the primacy of human emotional bonds.

Centenarian is an old adult who lives to or beyond the age of 100 years.

Cohabiting family is a form of family in which a couple, with or without children, lives together without being legally married.

Duration of marriage refers to the interval of time between the day, month, and year of the marriage to date.

Economic deprivation perspective suggests that most single mothers have a low level of financial resources due to the loss of economies of scale associated with splitting one household into two and the lower earnings of mothers compared with fathers.

Eldercare providers are individuals who provide unpaid care to someone age 65 and older who needs help because of a condition related to aging.

Emotion regulation (ER) refers to attempts to influence which emotions we have, when we have them, and how we experience and express these emotions.

Escape from stress perspective focuses on the level of stress associated with various family arrangements.

Family boundary ambiguity refers to inconsistency in reporting who is in and who is out of the family.

Feminization of poverty is the phenomenon that women represent disproportionate percentages of the poor.

Full-sibling is a sibling of the target child who has the same biological parents as the target child.

Grandparent-headed family is a form of family in which grandparents take care of a grandchild or grandchildren with or without a parent of the child being present.

Gray divorce refers to the trend of increasing divorce for older adults or grey-haired couples after year 50.

Half-sibling is a sibling of the target child who shares one biological parent with the child, but the sibling's other biological parent is not biologically related to the child.

Joint custody is a court order whereby custody of a child is awarded to both parties regarding the care of their children.

Joint legal custody allows parents to make decisions for their children on education, medical care, etc.

Joint physical custody means that both parents share physical custody and children spend substantial time living with both custodial parents after a parental divorce.

Living Apart Together (LAT) refers to committed couples who live in different places.

Marital search theory suggests that within the limits of a marriage market where the search for a second marriageable partner occurs, seekers desire to maximize the quality of the match.

Modeling perspective assumes that children living in non-traditional households (with single parents or remarried parents) adopt non-traditional views about family life.

Multiple partner fertility (MPF) is the pattern of a man or a woman having biological children with more than one partner.

Paternal claiming refers to a stepfather's readiness to nurture, provide for, protect, and claim a stepchild as his own.

Recoupling process is the occurrence of an emotional bonding.

Relationship self-regulation (RSR) is actionable, goal-oriented work aimed at improving relationships in both the short and long term.

Remarriage is a marriage that occurs after a previous marriage has ended through divorce or widowhood.

Remarriage rate refers to the number of remarriages per 1,000 persons.

Remarried couple household consists of a household maintained by a married couple, one or both of whom have been previously divorced or widowed.

Remarried stepfamily is a form of family in which at least one partner is remarried and at least one partner has children with someone other than their current spouse.

Resilience factors enable remarried families to withstand and rebound from the disruptive challenges they face.

Role ambiguity refers to the absence of normative clarity about obligations and commitments between stepparents, absent parents, and children.

Serial cohabitation occurs when an individual has cohabited with more than one partner and often has children with different partners.

Siblings are two or more individuals having one or both parents in common.

Single-parent family is a form of family that includes children under age eighteen and that is headed by a parent who is widowed or divorced and not remarried, or never married.

Social learning perspective holds that people learn from one another through observation, direct instruction, imitation, and modeling.

Sole physical custody means that one parent resides with the child and becomes the custodial parent, while the other parent of the child has visitation time with the child.

Stepchildren (stepdaughters or stepsons) are the offspring of one's spouse or partner from a previous marriage or intimate relationship.

Stepfamily is a form of family in which at least one marital or cohabiting partner has children who are not biologically related to the other spouse or partner.

Stepfather refers to a non-biological father who is married to or cohabiting with the child's mother who had a previous marriage or intimate relationship.

Stepfather involvement refers to shared activities between stepfathers and children that take place on a regular basis.

Stepgrandparent refers to an individual who is not biologically related to children who are two generations below him or her, but serves in a grandparent-like role because of a remarriage within the family.

Stepmother refers to a non-biological mother who is married to or cohabiting with the child's father who had a previous marriage or intimate relationship.

Stepparent is a non-biological parent who is married to or cohabiting with the child's mother or father who had a previous marriage or intimate relationship.

Step-sibling is a sibling of the target child who is not biologically related to the child, and has entered the family system via the child's stepparent.

Theory of continuing bonds suggests that grief is continuing ties to loved ones and extraordinarily stressful.

Uncertainty reduction theory holds that when interacting, people need information about the other party to reduce their uncertainty.

Walking marriage refers to a sexual relationship in which male partners live elsewhere and make nightly visits to women in their homes.

Widowhood effect refers to the increased likelihood for a recently widowed person to die.

Discussion Questions

- Traditional married couples with biological children have been on the decline while forms of family such as single-parent families, cohabiting families, and stepfamilies have been on the rise. What are the strengths and challenges of these family forms?

Suggested Film and Video

Most Popular Living Arrangement is Living with Parents: https://www.census.gov/library/stories/2017/08/young-adult-video.html

Ren, W. (Director). 2009. *Parallel Tracks*. [Video file] Video posted to https://www.youtube.com/watch?v=nXyIx4YX6wc

Schwartz, S. (Producer). (1969–1974). *The Brady Bunch* (Television series). The United States.

Stevens, L. (Producer). (1961). *The Marriage-Go-Round* (Film). The United States.

Internet Sources

Families and Living Arrangements: https://www.census.gov/topics/families.html

Social Security: https://www.ssa.gov/ssi/text-child-ussi.htm

References

Adler, A. (1956). Understanding Life. In H. L. Ansbacher, & R. R. Ansbacher (Eds.), *The Individual Psychology of Alfred Adler: A Systematic Presentation in Selections from His Writings*. New York: Harper Torch Books.

Adler-Baeder, F., Pittman, J. F. & Taylor, L. (2006). The Prevalence of Marital Transitions in Military Families. *Journal of Divorce & Remarriage*, 44(1–2), 91–106.

Adler-Baeder, F., Russell, C., Kerpelman, J., Pittman, J., Ketring, S., Smith, T. & Stringer, K. (2010). Thriving in Stepfamilies: Exploring Competence and Well-being among African American Youth. *Journal of Adolescent Health*, 46(4), 396–398.

Ahmady, K., Ashtiani, A. F., Mirzamani, S. M. & Arabnia, A. R. (2007). The Effect of Remarriage on Mental Health of Widows of the Killed-in-Action versus Other Widows. *Journal of Divorce & Remarriage*, 48(1–2), 141–154.

Akerlof, G. A. & Yellen, J. L. (1996). *New Mothers, Not Married: Technology Shock, the Demise of Shotgun Marriage, and the Increase in Out-of-wedlock Births*. Washington, DC: Brookings Institution.

Albertini, M. & Saraceno, C. (2008). Intergeneration Contact and Support: The Long-term Effects of Marital Instability in Italy. In C. Saraceno (Ed.), *Families, Ageing and Social Policy: Intergenerational Solidarity in European Welfare States* (pp. 194–216). Cheltenham, UK & Northampton, MA: Edward Elgar.

Amato, P. R. (2010). Research on Divorce: Continuing Trends and New Developments. *Journal of Marriage and Family*, 72(3), 650–666.

Amato, P. R. & Kane, J. B. (2011). Parents' Marital Distress, Divorce, and Remarriage: Links with Daughters' Early Family Formation Transitions. *J Fam Issues*, 32(8), 1073–1103.

Amato, P. R. & Previti, D. (2003). People's Reasons for Divorcing. *Journal of Family Issues*, 24(5), 602–626.

Amato, P. R., Landale, N. S., Havasevich, T. C., Booth, A., Eggebeen, D. J., Schoen, R. & McHale, S. M. (2008). Precursors of Young Women's Family Formation Pathways. *J Marriage Fam.*, 70(5), 1271–1286.

Amato, P. R., King, V. & Thorsen, M. L. (2016). Parent–Child Relationships in Stepfather Families and Adolescent Adjustment: A Latent Class Analysis. *Journal of Marriage and the Family*, 78(2), 482–497.

Anderson, E. R. (1999). Sibling, Half Sibling, and Stepsibling Relationships in Remarried Families. In E. M. Hetherington, S. H. Henderson & D. Reiss (Eds.), *Adolescent Siblings in Stepfamilies: Family Functioning and Adolescent Adjustment* (pp. 101–126). Monographs of the Society for Research in Child Development, 64(4), Serial No. 259.

Anderson, L. R. (2016). High School Seniors' Attitudes on Cohabitation as a Testing Ground for Marriage. *Family Profiles*, FP-16-13. Bowling Green, OH: National Center for Family and Marriage Research, https://www.bgsu.edu/content/dam/BGSU/college-of-arts-and-sciences/NCFMR/documents/FP/anderson-teen-attitudes-cohab-marriage-fp-16–13.pdf

Andersson, G. (2003). *Dissolution of Unions in Europe: A Comparative Overview, MPIDR Working Paper, WP 2003–004.* Rostock, Germany: Max Planck Institute for Demographic Research.

Avellar, S. & Smock, P. J. (2005). The Economic Consequences of the Dissolution of Cohabiting Unions. *Journal of Marriage and Family,* 67, 315–327.

Ayuso, L. (2018). New Partnerships in Widowhood in Spain: Realities and Desires. *Journal of Women & Aging,* doi: 10.1080/08952841.2018.1463128

Baham, M. E., Weimer, A. A., Braver, S. L. & Fabricius, W. V. (2008). Sibling Relationships in Blended Families. In J. Pryor (Ed.), *The International Handbook of Stepfamilies: Policy and Practice in Legal, Research, and Clinical Environments* (pp. 175–207). Hoboken, NJ: John Wiley & Sons.

Bair, D. (2007). *Calling it Quits: Late-life Divorce and Starting Over.* New York: Random House.

Baker, D. (1999). The Increase of Single Parent Families: An Examination of Causes. *Policy Sciences,* 32(2), 175–188. Retrieved from http://www.jstor.org/stable/4532456

Balbo, N., Billari, F. C. & Mills, M. (2013). Fertility in Advanced Societies: A Review of Research: La fécondité dans les sociétés avancées: un examen des recherches. *European Journal of Population,* 29(1), 1–38.

Baltes, P. B. & Smith, J. (2003). New Frontiers in the Future of Aging: From Successful Aging of the Young Old to the Dilemmas of the Fourth Age. *Gerontology,* 49(1), 123–135.

Bastaits, K., Pasteels, I. & Mortelmans, D. (2018). How Do Post-Divorce Paternal and Maternal Family Trajectories Relate to Adolescents' Subjective Well-being? *J Adolesc.,* 64, 98–108.

Benson, M. J., Buehler, C. & Gerard, J. M. (2008). Interparental Hostility and Early Adolescent Problem Behavior–Spillover via Maternal Acceptance, Harshness, Inconsistency, and Intrusiveness. *Journal of Early Adolescence,* 28, 428–454.

Berger, L. M., Carlson, M. J., Bzostek, S. H. & Osborne, C. (2008). Parenting Practices of Resident Fathers: The Role of Marital and Biological Ties. *Journal of Marriage and the Family,* 70(3), 625–639.

Berger, R. (1995). Three Types of Stepfamilies. *Journal of Divorce & Remarriage,* 24 (1/2), 35–49.

Berghammer, C., Fliegenschnee, K. & Schmidt, E.-M. (2014). Cohabitation and Marriage in Austria: Assessing the Individualization Thesis across the Life Course. *Demographic Research,* 31(1), 1137–1166.

Bernard, J. (1982). *The Future of Marriage.* New Haven, CT: Yale University Press.

Blyaert, L., Parys, H. V., Mol, J. D. & Buysse, A. (2016). Like a Parent and a Friend, but Not the Father: A Qualitative Study of Stepfathers' Experiences in the Stepfamily. *Australian & New Zealand Journal of Family Therapy,* 37(1), 119–132.

Bokek-Cohen, Y. (2014). Remarriage of War and Terror Widows: A Triadic Relationship. *Death Studies,* 38(10), 672–677.

Boss, P. & Greenberg, J. (1984). Family Boundary Ambiguity: A New Variable in Family Stress Theory. *Fam Process,* 23(4), 535–546.

Bradbury, T. N. & Karney, B. R. (2010). *Intimate Relationships.* New York: W. W. Norton & Company.

Braver, S. L. & Votruba, A. M. (2018). Does Joint Physical Custody "Cause" Children's Better Outcomes? *Journal of Divorce & Remarriage,* 59(5), 452–468.

Bray, J. H. & Kelly, J. (2011). *Stepfamilies: Love, Marriage, and Parenting in the First Decade.* New York: Broadway Books.

Brown, S. (2004). Family Structure and Child Well-Being: The Significance of Parental Cohabitation. *Journal of Marriage and the Family,* 66(2), 351–367.

Brown, S. L. & Booth, A. (1996). Cohabitation versus Marriage: A Comparison of Relationship Quality. *Journal of Marriage and Family,* 58(3), 668–678.

Brown, S. L. & Lin, I-F. (2012). The Gray Divorce Revolution: Rising Divorce Among Middle-aged and Older Adults, 1990–2010. *The Journals of Gerontology: Series B,* 67(6), 731–741.

Brown, S. L. & Manning, W. D. (2009). Family Boundary Ambiguity and the Measurement of Family Structure: The Significance of Cohabitation. *Demography,* 46(1), 85–101.

Brown, S. L. & Shinohara, S. K. (2013). Dating Relationships in Older Adulthood: A National Portrait. *Journal of Marriage and the Family,* 75(5), 1194–1202.

Brown, S. L., Lee, G. R. & Bulanda, J. R. (2006). Cohabitation among Older Adults: A National Portrait. *The Journals of Gerontology: Series B,* 61(2), S71–S79.

Brown, S., Bulanda, J. R. & Lee, G. R. (2012). Transitions Into and Out of Cohabitation in Later Life. *Journal of Marriage and Family,* 74(4), 774–793.

Brown, S. L., Manning, W. D., Payne, K. K. & Wu, H. (2013). *Living Apart Together (LAT) Relationships in the U.S.* Bowling Green, OH: National Center for Marriage & Family.

Brown, S. L., Lin, I. F., Hammersmith, A. M. & Wright, M. R. (2018). Later Life Marital Dissolution and Repartnership Status: A National Portrait. *The Journals of Gerontology: Series B,* 73(6), 1032–1042.

Brown, T. N., Romero, A. P. & Gates, G. J. (July 2016). *Food Insecurity and SNAP Participation in the LGBT Community.* Los Angeles: The Williams Institute.

Bryant, C. M., Futris, T. G., Hicks, M. R., Lee, T.-K. & Oshri, A. (2016). African American Stepfathers–Stepchild Relationships, Marital Quality, and Mental Health. *Journal of Divorce & Remarriage*, 57(6), 375–388.

Bryson, K. & Casper, L. (1999). *Coresident Grandparents and Grandchildren, Current Population Reports, P23–198.* Washington, DC: U.S. Census Bureau.

Bumpass, L. & Lu, H. (2000). Trends in Cohabitation and Implications for Children's Family Contexts in the United States. *Population Studies,* 54, 29–41.

Bumpass, L. L., Sweet, J. A. & Cherlin, A. (1991). The Role of Cohabitation in Declining Rates of Marriage. *Journal of Marriage and Family,* 53(4), 913–917.

Burton, L. M. & Hardaway, C. R. (2012). Low-income Mothers as "Othermothers" To Their Romantic Partners' Children: Women's Coparenting in Multiple Partner Fertility Relationships. *Fam Process.*, 51(3), 343–359.

Bzostek, S. H. (2008). Social Fathers and Child Well-being. *Journal of Marriage and Family,* 70(4), 950–961.

Cahn, N., Carbone, J., DeRose, L. & Wilcox, W. (Eds.) (2018). *Unequal Family Lives: Causes and Consequences in Europe and the Americas.* Cambridge: Cambridge University Press.

Cancian, M., Chung, Y. & Meyer, D. R. (2016). Fathers' Imprisonment and Mothers' Multiple-Partner Fertility. *Demography*, 53(6), 2045–2074.

Cann-Milland, S. & Southcott, J. (2018). The Very Perplexed Stepmother: Step Motherhood and Developing a Healthy Self-Identity. *The Qualitative Report,* 23(4), 823–838.

Carroll, J. S., Olson, C. D. & Buckmiller, N. (2007). Family Boundary Ambiguity: A 30-Year Review of Theory, Research, and Measurement. *Family Relations,* 56, 210–230.

Chapman, A., Ganong, L., & Coleman, M. (2016). Step-grandparenting. In C. L. Shehan (Ed.), *Encyclopedia of Family Studies* (pp. 1863–1867). New York: Wiley-Blackwell.

Charlesworth, L. W. (2015). Middle Childhood. In E. Hutchison (Ed.), *Dimensions of Human Behavior: The Changing Life Course.* Thousand Oaks, CA: Sage.

Cherlin, A. (1978). Remarriage as an Incomplete Institution. *American Journal of Sociology,* 84(3), 634–650.

Cherlin, A. J. (2004). The Deinstutionalization of American Marriage. *Journal of Marriage and Family*, 66, 848–861.

Cherlin, A. J. (2010). *The Marriage-Go-Round: The State of Marriage and the Family in America Today.* New York: Vintage.

Cherlin, A. & McCarthy, J. (1985). Remarried Couple Households: Data from the June 1980 Current Population Survey. *Journal of Marriage and Family,* 47(1), 23–30.

Chevan, A. (August 1996). As Cheaply as One: Cohabitation in the Older Population. *Journal of Marriage and Family,* 58(3), 656–667.

Child Trends (August 23, 2018). *Family Structure.* Retrieved on August 9, 2018 from https://www.childtrends.org/?indicators=family-structure

Church, E. (1999). Who Are the People in Your Family? Stepmothers' Diverse Notions of Kinship. *Journal of Divorce & Remarriage*, 31(1/2), 83–105.

Cicirelli, V. G. (1995). *Sibling Relationships across the Life Span.* New York: Plenum Press.

Clingempeel, W. G., Colyar, J. J. & Hetherington, E. M. (1994). Toward a Cognitive Dissonance Conceptualization of Stepchildren and Biological Children Loyalty Conflicts: A Construct Validity Study. In K. Pasley & M. Ihinger-Tallman (Eds.), *Stepparenting: Issues in Theory, Research, and Practice* (pp. 151–173). Westport, CT: Greenwood Press.

Coltrane, S., Gutierrez, E. & Park, R. D. (2008). Stepfathers in Cultural Context: Mexican-American families in the United States. In J. Pryor (Ed.), *The International Handbook of Stepfamilies: Policy and Practice in Legal, Research, and Clinical Environments.* Hoboken, NJ: Wiley.

Connidis, I. A. & Campbell, L. D. (1995). Closeness, Confiding, and Contact among Siblings in Middle and Late Adulthood. *Journal of Family Issues,* 16, 722–745.

Copen, C. E., Daniels, K. & Mosher, W. D. (2013). *First Premarital Cohabitation in the United States: 2006–2010 National Survey of Family Growth.* National health statistics reports; no 64. Hyattsville, MD: National Center for Health Statistics.

Cowan, A. (1999). Review of Widowhood in Medieval and Early Modern Europe (review no. 165). Retrieved on August 17, 2018 from https://www.history.ac.uk/reviews/review/165

Craig, E. A., Harvey-Knowles, J. A. & Johnson, A. J. (2012). Childless Stepmothers: Communicating with Other Stepmothers about Spouses and Stepchildren. *Qualitative Research Reports in Communication,* 13(1), 71–79.

Creswell, J. D. (2017). Mindfulness Interventions. *Annual Review of Psychology,* 68, 491–516.

Crimmins, E. M. & Saito, Y. (2001). Trends in Healthy Life Expectancy in the United States, 1970–1990: Gender, Racial, and Educational Differences. *Soc Sci Med.*, 52, 1629–1641.

Crohn, H. M. (2006). Five Styles of Positive Stepmothering from the Perspective of Young Adult Stepdaughters. *Journal of Divorce & Remarriage*, 46(1–2), 119–134.

Crosbie-Burnett, M. (1989). Impact of Custody Arrangement and Family Structure on Remarriage. *Journal of Divorce*, 13(1), 1–16.

Cutuli, D. (2014). Cognitive Reappraisal and Expressive Suppression Strategies Role in the Emotion Regulation: An Overview on Their Modulatory Effects and Neural Correlates. *Frontiers in Systems Neuroscience*, 8, 175.

Deaton, A. & Paxson, C. (1998). Aging and Inequality in Income and Health. *Am Econ Rev.*, 88, 248–253.

Dollahite, D. C. & Marks, L. D. (2018). Introduction to the Special Issue: Exploring Strengths in American Families of Faith. *Marriage & Family Review*, doi: 10.1080/01494929.2018.1469569

Doodson, L. J. (2014). Understanding the Factors Related to Stepmother Anxiety: A Qualitative Approach. *Journal of Divorce & Remarriage*, 55(8), 645–667.

Downs, K. J. M. (2004). Family Commitment, Role Perceptions, Social Support, and Mutual Children in Remarriage. *Journal of Divorce & Remarriage*, 40(1–2), 35–53.

Duncan, S. & Phillips, M. (2010). People Who Live Apart Together (Lats): How Different Are They? *The Sociological Review*, 58(1), 112–134.

Dunifon, R. & Kowaleski-Jones, L. (2007). The Influence of Grandparents in Single-mother Families. *Journal of Marriage and Family*, 69(2), 465–481.

Dunn, J. (2007). Siblings and Socialization. In J. E. Grusec & P. D. Hastings (Eds.), *Handbook of Socialization: Theory and Research*. New York: Guilford Press.

Dupre, M. E. (2007). Educational Differences in Age-Related Patterns of Disease: Reconsidering the Cumulative Disadvantage and Age-as-leveler Hypotheses. *J Health Soc Behav.*, 48(1), 1–15.

Eckmeyer, K. J. (2018). Fathers with Resident Minor Children, 2016. *Family Profiles*. FP- 18–06. Bowling Green, OH: NCFMR.

Ehrenberg, M. F., Robertson, M. & Pringle, J. (2012). Attachment Style and Marital Commitment in the Context of Remarriage. *Journal of Divorce & Remarriage*, 53(3), 204–219.

ElHage, A. (2017). *How Multiple Partner Fertility Influences Child Well-Being*. Charlottesville, VA: Institute for Family Studies.

Elliott, D. B. and Simmons, T. (August 2011). *Marital Events of Americans: 2009*. Washington, DC: U.S. Census Bureau.

Ellis, R. R. & Simmons, T. (October 2014). *Coresident Grandparents and Their Grandchildren: 2012*. Washington, DC: U.S. Census Bureau.

Ellwood, D. T. & Summers, L. H. (1986). Poverty in America: Is Welfare the Answer or the Problem? In S. Danziger & D. Weinberg (Eds.), *Fighting Poverty: What Works and What Doesn't*. Cambridge, MA: Harvard University Press.

Engblom-Deglmann, M. & Brimhall, A. S. (2016). Not Even Cold in Her Grave: How Postbereavement Remarried Couples Perceive Family Acceptance. *Journal of Divorce & Remarriage*, 57(3), 224–244.

England, P. & Farkas, G. (1986). *Households, Employment and Gender: A Social, Economic, and Demographic View*. New York: Aldine.

England, P. & Li, S. (2006). Desegregation Stalled: The Changing Gender Composition of College Majors, 1971–2002. *Gender and Society*, 20, 657–677.

Erera-Weatherly, P. (1996). On Becoming a Stepparent: Factors Associated with the Adoption of Alternative Stepparenting Styles. *Journal of Divorce & Remarriage*, 25(3/4), 155–174.

Faber, A. J. (2004). Examining Remarried Couples through a Bowenian Family Systems Lens. *Journal of Divorce & Remarriage*, 40(3–4), 121–133.

Fabricius, W. V., Aaron, M., Akins, F. R., Assini, J. J. & McElroy, T. (2018). What Happens When There Is Presumptive 50/50 Parenting Time? An Evaluation of Arizona's New Child Custody Statute. *Journal of Divorce & Remarriage*, 59(5), 414–428.

Feinberg, M. E., McHale, S. M., Crouter, A. C. & Cumsille, P. (2003). Sibling Differentiation: Sibling and Parent Relationship Trajectories in Adolescence. *Child Dev.*, 74(5), 1261–1274.

Feinglass, J., Lin, S., Thompson, J. *et al.* (2007). Baseline Health, Socioeconomic Status, and 10-Year Mortality among Older Middle-Aged Americans: Findings from the Health and Retirement Study, 1992–2002. *J Gerontol B Psychol Sci Soc Sci.*, 62, S209–S217.

Fingerman, K., Miller, L., Birditt, K. & Zarit, S. (2009). Giving to the Good and the Needy: Parental Support of Grown Children. *Journal of Marriage and Family*, 71, 1220–1233.

Fishman, B. (1983). The Economic Behavior of Stepfamilies. *Family Relations*, 32(3), 359–366.

Fomby, P. (2018). Motherhood in Complex Families. *Journal of Family Issues*, 39(1), 245–270.

Fomby, P. & Osborne, C. (2017). Family Instability, Multipartner Fertility, and Behavior in Middle Childhood. *Journal of Marriage and Family*, 79(1), 75–93.

Fomby, P., Goode, J. A. & Mollborn, S. (2016). Family Complexity, Siblings, and Children's Aggressive Behavior at School Entry. *Demography*, 53(1), 1–26.

Furstenberg, F. F. & King, R. B. (1999). Multipartnered Fertility Sequences: Documenting an Alternative Family Form.

Earlier version presented at the 1998 Annual Meetings of the Population Association of America, Chicago, IL.

Ganong, L. H. & Coleman, M. (1993). An Exploratory Study of Stepsibling Subsystems. *Journal of Divorce and Remarriage,* 19, 125–141.

Ganong, L. H. & Coleman, M. (1994). *Remarried Family Relationships.* Thousand Oaks, CA: Sage.

Ganong, L. H. & Coleman, M. (Eds.) (2004). *Stepfamily Relationships: Development, Dynamics, and Interventions.* New York: Kluwer Academic/Plenum Publishers.

Ganong, L., Coleman, M., Fine, M. & Martin, P. (1999). Stepparents' Affinity-Seeking and Affinity-Maintaining Strategies with Stepchildren. *Journal of Family Issues,* 20, 299–327.

Gaspard, T. (2016). *10 Rules for a Successful Second Marriage.* Seattle, WA: The Gottman Institute.

Gaspard, T. (2018). *Navigating the Challenges of Stepfamily Life.* Seattle, WA: The Gottman Institute.

Gates, A. L. (2018). Stepmothers' Coparenting Experience with the Mother in Joint Custody Stepfamilies, *Journal of Divorce & Remarriage,* doi: 10.1080/10502556.2018.1488124

Geiger, A. & Livingston, G. (2018). *8 Facts about Love and Marriage in America.* Pew Research Center.

Glick, P. C. (1980). Remarriage: Some Recent Changes and Variations. *Journal of Family Issues,* 1(4), 455–478.

Glick, P. & Lin, S. (1986). Recent Changes in Divorce and Remarriage. *Journal of Marriage and Family,* 48(4), 737–747.

Goldscheider, F. & Sassler, S. (2006). Creating Stepfamilies: Integrating Children into the Study of Union Formation. *Journal of Marriage and Family,* 68(2), 275–291.

Gornick, J. C. & Milanovic, B. (2015). *Income Inequality in the United States in Cross-National Perspective: Redistribution Revisited.* LIS Research Center Brief (1/2015). New York: Luxembourg Income Study Center at the Graduate Center, City University of New York.

Gornick, J. C., Meyers, M. K. & Ross, K. E. (1997). Supporting the Employment of Mothers: Policy Variation across Fourteen Welfare States. *Journal of European Social Policy,* 7(1), 45–70.

Grall, T. (January 2018). *Custodial Mothers and Fathers and Their Child Support: 2011.* Current Population Reports. Washington, DC: U.S. Census Bureau.

Greeff, A. P. & Toit, C. D. (2009). Resilience in Remarried Families. *The American Journal of Family Therapy,* 37(2), 114–126.

Gross, J. J. (1998). The Emerging Field of Emotion Regulation: An Integrative Review. *Review of General Psychology,* 2, 271–299.

Gross, P. (1987). Defining Post-divorce Remarriage Families: A Typology Based on the Subjective Perceptions of Children. *Journal of Divorce,* 10(1/2), 208–219.

Guzzo, K. B. (2014). New Partners, More Kids: Multiple-Partner Fertility in the United States. *Annals of the American Academy of Political and Social Science,* 654(1), 66–86.

Guzzo, K. B. (2016). Stepfamilies in the U.S. *Family Profiles.* FP-16–09. Bowling Green, OH: NCFMR.

Guzzo, K. B. & Dorius, C. (2016). Challenges in Measuring and Studying Multipartnered Fertility in American Survey Data. *Population Research and Policy Review,* 35(4), 553–579.

Guzzo, K. B. & Furstenberg, F. F. (2007). Multipartnered Fertility among American Men. *Demography,* 44(3), 583–601.

Halford, W. K., Lizzio, A., Wilson, K. L. & Occhipinti, S. (2007). Does Working at Your Marriage Help? Couple Relationship Self-regulation and Satisfaction in the First 4 Years of Marriage. *Journal of Family Psychology,* 21(2), 185–194.

Halpern-Meekin, S. & Tach, L. (2008). Heterogeneity in Two-Parent Families and Adolescent Well-Being. *Journal of Marriage and Family,* 70(2), 435–451.

Ham, B. D. (2004). The Effects of Divorce and Remarriage on the Academic Achievement of High School Seniors. *Journal of Divorce & Remarriage,* 42(1–2), 159–178.

Hamplova, D., Le Bourdais, C. & Le Broudais, C. (2009). One Pot or Two Pot Strategies? Income Pooling in Married and Unmarried Households in Comparative Perspective. *Journal of Comparative Family Studies,* 40(3), 355–385.

Hart, P. (March 17, 2009). On Becoming a Good Enough Stepmother. *Clinical Social Work Journal,* 37, 128–139.

He, W. & Larsen, L. J. (December 2014). *Older Americans with a Disability: 2008–2012 American Community Survey Reports.* Washington, DC: U.S. Census Bureau.

He, W., Weingartner, R. M. & Sayer, L. C. (February 2018). *Subjective Well-being of Eldercare Providers: 2012–2013.* Washington, DC: U.S. Census Bureau.

Hemez, P. & Manning, W. D. (2017). Attitudes towards Marital Infidelity. *Family Profiles.* FP-17–05. Bowling Green, OH: NCFMR.

Hetherington, E. M. & Clingempeel, W. G. (1992). *Coping with Marital Transitions.* Monographs of the Society for Research in Child Development, Vol. 57. Chicago: University of Chicago Press.

Hewlett, B. S. (2000). Culture, History, and Sex: Anthropological Contributions to Conceptualizing Father Involvement. *Marriage & Family Review,* 29(2/3), 59–73.

Hiekel. N. (2014). *The Different Meanings of Cohabitation across Europe. How Cohabiters View Their Unions and Differ in Their Plans and Behaviors.* Amsterdam: Amsterdam University Press.

Higginbotham, B., Davis, P., Smith, L., Dansie, L., Skogrand, L. & Reck, K. (2012). Stepfathers and Stepfamily Education, *Journal of Divorce & Remarriage,* 53(1), 76–90.

Hoenayi, R. K. & Yendork, J. S. (2018). I Wouldn't Say We Were Treated Equally: Experiences of Young Adult Stepchildren in the Ghanaian Context, *Journal of Divorce & Remarriage,* doi: 10.1080/10502556.2018.1466250

Hofferth, S. L., Cabrera, N., Carlson, M., Coley, R. L., Day, R. & Schindler, H. (2007). Resident Father Involvement and Social Fathering. In S. L. Hofferth & L. M. Casper (Eds.), *Handbook of Measurement Issues in Family Research.* Mahwah, NJ: Lawrence Erlbaum.

House, J. S., Lantz, P. M. & Herd, P. (2005). Continuity and Change in the Social Stratification of Aging and Health over the Life Course: Evidence from a Nationally Representative Longitudinal Study from 1986 to 2001/2002 (Americans' Changing Lives Study). *J Gerontol B Psychol Sci Soc Sci.,* Spec No. 2, 15–26.

Hughes, M. & Waite, L. (2002). Health in Household Context: Living Arrangements and Health in Late Middle Age. *Journal of Health and Social Behavior,* 43(1), 1–21. Retrieved August 9, 2018 from http://www.jstor.org/stable/3090242

Humble, A. M. (2009). The Second Time 'Round: Gender Construction in Remarried Couples' Wedding Planning. *Journal of Divorce & Remarriage,* 50(4), 260–281.

Hungerford, T. L. (2010). The Economic Consequences of Widowhood on Elderly Women in the United States and Germany. *The Gerontologist,* 41(1), 103–110.

Iriondo, J. (2018). *Census Research on Well-Being of Eldercare Providers.* Washington, DC: U.S. Census Bureau.

Jambon, M., Madigan, S., Plamondon, A., Daniel, E. & Jenkins, J. (2018). The Development of Empathic Concern in Siblings: A Reciprocal Influence Model. *Child Development,* doi: 10.1111/cdev.13015

James, S. L. & Shafer, K. (2012). Temporal Differences in Remarriage Timing: Comparing Divorce and Widowhood. *Journal of Divorce & Remarriage,* 53(7), 543–558.

Jappens, M. (2018). Children's Relationships with Grandparents in Married and in Shared and Sole Physical Custody Families. *Journal of Divorce & Remarriage,* 59(5), 359–371.

Jayson, S. (2013). Remarriage Rate Declining as More Opt for Cohabitation. *USA Today.* Retrieved on August 14, 2018 from https://www.usatoday.com/story/news/nation/2013/09/12/remarriage-rates-divorce/2783187/

Jelvani, R., Etemadi, O., Jazayeri, R. & Fatehizade, M. (2018). Difficulties in Emotion Regulation among Iranian Remarried Women: The Role of Mindfulness, Thought–Action Fusion, and Emotion Regulation. *Journal of Divorce & Remarriage,* doi: 10.1080/10502556.2018.1488112

Jenkins, J. M. (1992). Sibling Relationships in Disharmonious Homes: Potential Difficulties and Protective Effects. In F. Boer & J. Dunn (Eds.), *Children's Sibling Relationships: Developmental and Clinical Issues.* Hillsdale, NJ: Erlbaum.

Jensen, E. J., Stum, M. & Jackson, M. (2018). What Makes Inheritance Fair in Stepfamilies? Examining Perceptions and Complexities. *Journal of Divorce & Remarriage,* doi: 10.1080/10502556.2018.1488123

Jensen, T. M. & Pace, G. T. (2016). Stepfather Involvement and Stepfather-Child Relationship Quality: Race and Parental Marital Status as Moderators. *Journal of Marital and Family Therapy,* 42(4), 659–672.

Jensen, T. M. & Weller, B. E. (2018). Latent Profiles of Residential Stepfamily Relationship Quality and Family Stability. *Journal of Divorce & Remarriage,* 1–19. doi: 10.1080/10502556.2018.1488111

Jensen, T., Lombardi, B. & Larson, J. (2015). Adult Attachment and Stepparenting Issues: Couple Relationship Quality as a Mediating Factor. *Journal of Divorce & Remarriage,* 56, 80–94.

Joamets, K. (2016). European Dilemmas of the Biological versus Social Father: The Case of Estonia. *Baltic Journal of Law & Politics,* 9(2), 23–42 http://www.degruyter.com/view/j/bjlp DOI: 10.1515/bjlp-2016–0010

John, O. P. & Gross, J. J. (2004). Healthy and Unhealthy Emotion Regulation: Personality Processes, Individual Differences and Life Span Development. *J. Pers.,* 72, 1301–1333.

Jolliff, D. (1984). The Effects of Parental Remarriage on the Development of the Young Child. *Early Child Development and Care,* 13(3–4), 321–333.

Jones, J. (2008). Adoption Experiences of Women and Men and Demand for Children to Adopt by Women 18–44 Years of Age in the United States, 2002. National Center for Health Statistics. *Vital Health Stat.,* 23(27).

Jong Gierveld, J. de (2004). Remarriage, Unmarried Cohabitation, Living Apart Together: Partner Relationships Following Bereavement or Divorce. *Journal of Marriage and the Family,* 66(1), 236–243.

Jong Gierveld, J. de, Broese van Groenou, M., Hoogendoorn, A. W. & Smit, J. H. (2009). Quality of Marriages in Later Life and Emotional and Social Loneliness. *The Journals of Gerontology: Series B,* 64B(4), 497–506.

Kalil, A. (2015). Inequality Begins at Home: The Role of Parenting in the Diverging Destinies of Rich and Poor Children. In P. R. Amato, A. Booth, S. M. McHale & J. Van Hook (Eds.), *Families in an Era of Increasing Inequality, Vol. 5* (pp. 63–82). National Symposium on Family Issues. Berlin: Springer International Publishing.

Kalish, R. A. & Visher, E. (1982). Grandparents of Divorce and Remarriage. *Journal of Divorce,* 5(1–2), 127–140.

Karam, S. (July 21, 2006). Misyar Offers Marriage-lite in Strict Saudi Society. *The Boston Globe.* Reuters.

Kemp, C. L. (2007). Grandparent–Grandchild Ties. *Journal of Family Issues,* 28, 855–881.

Kenny, M. (2006). The World Is Full of Remarried Men. *The Telegraph.* Retrieved on September 6, 2018 from https://www.telegraph.co.uk/culture/3655111/The-world-is-full-of-remarried-men... . html

King, V. (2006). The Antecedents and Consequences of Adolescents' Relationships with Stepfathers and Nonresident Fathers. *Journal of Marriage and the Family,* 68(4), 910–928.

King, V. (2009). Stepfamily Formation: Implications for Adolescent Ties to Mothers, Nonresident Fathers, and Stepfathers. *Journal of Marriage and the Family,* 71(4), 954–968.

King, V., Boyd, L. M. & Thorsen, M. L. (2015). Adolescents' Perceptions of Family Belonging in Stepfamilies. *Journal of Marriage and the Family,* 77(3), 761–774.

Kinniburgh-White, R., Cartwright, C. & Seymour, F. (2010). Young Adults' Narratives of Relational Development with Stepfathers. *Journal of Social and Personal Relationships,* 27(7), 890–907.

Kitson, G. C., Babri, K. B., Roach, M. J. & Placidi, K. S. (1989). Adjustment to Widowhood and Divorce. A Review. *Journal of Family Issues,* 10(1), 5–32.

Klass, D., Silverman, P. R. & Nickman, S. (1996). *Continuing Bonds: New Understandings of Grief (Death Education, Aging and Health Care).* New York: Taylor & Francis.

Knodel, J. & Lynch, K. A. (1985). The Decline of Remarriage: Evidence from German Village Populations in the Eighteenth and Nineteen Centuries. *Journal of Family History,* 10(1), 34–59.

Kreidl, M. & Hubatková, B. (2017). Rising Rates of Cohabitation and the Odds of Repartnering: Does the Gap Between Men and Women Disappear? *Journal of Divorce & Remarriage,* 58(7), 487–506.

Kruk, E. (2018). Arguments against a Presumption of Shared Physical Custody in Family Law. *Journal of Divorce & Remarriage,* 59(5), 388–400.

Kvanli, J. A. & Jennings, G. (1987). Recoupling. *Journal of Divorce,* 10(1–2), 189–203.

Låftman, S. B., Bergström, M., Modin, B. & Östberg, V. (2014). Joint Physical Custody, Turning to Parents for Emotional Support, and Subjective Health: A Study of Adolescents in Stockholm, Sweden. *Scand J Public Health,* 42(5), 456–462.

Lamb, M. E., Pleck, J. H. & Levine, J. A. (2000). The History of Research on Father Involvement: An Overview. *Marriage & Family Review,* 29(2/3), 23–42.

Landale, N. S. & Oropesa, R. S. (2001). Father Involvement in the Lives of Mainland Puerto Rican Children: Contributions of Nonresident, Cohabiting and Married Fathers. *Social Forces,* 79, 945–968.

Latrofa, M., Vaes, J., Pastore, M. & Cadinu, M. (2009). United We Stand, Divided We Fall! The Protective Function of Self-Stereotyping for Stigmatised Members' Psychological Well-being. *Applied Psychology,* 58, 84–104.

Leopold, T. & Schulz, F. (2018). Health and Housework in Later Life: A Longitudinal Study of Retired Couples. *J Gerontol B Psychol Sci Soc Sci.* doi: 10.1093/geronb/gby015.

Lewis, J. M. & Kreider, R. M. (March 2015). *Remarriage in the United States.* American Community Survey Reports. Washington, DC: U.S. Census Bureau.

Lichter, D. T., Qian, Z. & Mellott, L. M. (2006). Marriage or Dissolution? Union Transitions among Poor Cohabiting Women. *Demography,* 43(2), 223–240.

Liefbroer, A. C., Poortman, A.-R. & Seltzer, J. A. (2015). Why do intimate partners live apart? Evidence on LAT relationships across Europe. *Demographic Research,* 32(8), 251–286.

Livingston, G. (April 25, 2018). *The Changing Profile of Unmarried Parents: A Growing Share Are Living with a Partner.* Washington, DC: Pew Research Center.

Lowsky, D. J., Olshansky, S. J., Bhattacharya, J. & Goldman, D. P. (2014). Heterogeneity in Healthy Aging. *The Journals of Gerontology Series A: Biological Sciences and Medical Sciences,* 69(6), 640–649.

Lucier-Greer, M. & Adler-Baeder, F. (2011). An Examination of Gender Role Attitude Change Patterns among Continuously Married, Divorced, and Remarried Individuals. *Journal of Divorce & Remarriage,* 52(4), 225–243.

Lyssens-Danneboom, V. & Mortelmans, D. (2014). Living Apart Together and Money: New Partnerships, Traditional Gender Roles. *Journal of Marriage & Family,* 76(5), 949–966.

Marsiglio, W. (2004). When Stepfathers Claim Stepchildren: A Conceptual Analysis. *Journal of Marriage and Family,* 66(1), 22–39.

Martin-Uzzi, M. & Duval-Tsioles, D. (2013). The Experience of Remarried Couples in Blended Families. *Journal of Divorce & Remarriage,* 54(1), 43–57.

Matthijs, K. (2003). Frequency, Timing and Intensity of Remarriage in 19th Century Flanders. *The History of the Family,* 8(1), 135–162.

McHale, S. M., Updegraff, K. A., Jackson-Newsom, J., Tucker, C. J. & Crouter, A. C. (2000). When Does Parents' Differential Treatment Have Negative Implications for Siblings? *Social Development,* 9, 149–172.

McWilliams, S. & Barrett, A. E. (2012). Online Dating in Middle and Later Life: Gendered Expectations and

Experiences. *Journal of Family Issues.* Advance online publication. doi: 10.1177/0192513X12468437.

Meriam, A. S. (1940). The Stepfather in the Family. *Social Service Review,* 14(4), 655–677.

Metts, S. M., Schrodt, P. & Braithwaite, D. O. (2017). Stepchildren's Communicative and Emotional Journey from Divorce to Remarriage: Predictors of Stepfamily Satisfaction. *Journal of Divorce & Remarriage,* 58(1), 29–43.

Meyer, D. R., Cancian, M. & Cook, S. T. (2005). Multiple-partner Fertility: Incidence and Implications for Child Support Policy. *Social Service Review,* 79(4), 577–601.

Meyer, M. J., Larson, J. H., Busby, D. & Harper, J. (2012). Working Hard or Hardly Working? Comparing Relationship Self-regulation Levels of Cohabiting, Married, and Remarried Individuals. *Journal of Divorce & Remarriage,* 53(2), 142–155.

Michaels, M. L. (2006). Factors that Contribute to Stepfamily Success. *Journal of Divorce & Remarriage,* 44(3–4), 53–66.

Mikulincer, M. & Shaver, P. (2007). *Attachment in Adulthood: Structure, Dynamics, and Change.* New York: The Guilford Press.

Mirecki, R. M., Brimhall, A. S. & Bramesfeld, K. D. (2013a). Communication during Conflict: Differences between Individuals in First and Second Marriages. *Journal of Divorce & Remarriage,* 54(3), 197–213.

Mirecki, R. M., Chou, J. L., Elliott, M. & Schneider, C. M. (2013b). What Factors Influence Marital Satisfaction? Differences between First and Second Marriages. *Journal of Divorce & Remarriage,* 54(1), 78–93.

Monserud, M. A. (2008). Intergenerational Relationships and Affectual Solidarity between Grandparents and Young Adults. *Journal of Marriage and Family,* 70, 182–195.

Monte, L. (2017). *Multiple Partner Fertility Research Brief, Current Population Reports, P70BR-146.* Washington, DC: U.S. Census Bureau.

Morrill, A. C., Dai, J., Dunn, S., Sung, I. & Smith K. (2005). Child Custody and Visitation Decisions When the Father Has Perpetrated Violence against the Mother. *Violence against Women,* 11(8), 1076–1107.

Morrissey, M. (March 3, 2016). *The State of American Retirement.* Washington, DC: Economic Policy Institute.

Murray, C. (1984). *Losing Ground: American Social Policy, 1950–1980.* New York: Basic Books.

Musavi, S., Fatehizade, M. & Jazayeri, R. (2018). Sexual Dynamics of Iranian Remarried Women in Blended Families: A Qualitative Study on Remarried Women's Life. *Journal of Divorce & Remarriage,* doi: 10.1080/10502556.2018.1488122

Noack, T. (2010). En stille revolusjon: Det moderne samboerskapet i Norge. (A Silent Revolution. The Modern Form of Cohabitation in Norway). *Institutt for sosiologi og samfunnsgeografi.* Oslo: University of Oslo.

Nock, S. L. (1995). Comparison of Marriages and Cohabiting Relationships. *Journal of Family Issues,* 16(1), 53–76.

Novak, J. R., Smith, H. M., Larson, J. H. & Crane, D. R. (2018). Commitment, Forgiveness, and Relationship Self-regulation: An Actor Partner Interdependence Model of Relationship Virtues and Relationship Effort in Couple Relationships. *J Marital Fam Ther.,* 44(2), 353–365.

OECD (2012). *The Future of Families to 2030.* Paris: OECD Publishing.

O'Flaherty, K. M. & Eells, L. W. (1994). The Development of Remarriage-Preparation Programs. *Journal of Divorce & Remarriage,* 20, 1–2, 229–245.

Oh, S. (1987). Remarried Men and Remarried Women. *Journal of Divorce,* 9(4), 107–113.

Osmani, N., Matlabi, H. & Rezaei, M. (2018). Barriers to Remarriage among Older People: Viewpoints of Widows and Widowers. *Journal of Divorce & Remarriage,* 59(1), 51–68.

Ozawa, M. N. & Yoon, H.-S. (2002). The Economic Benefit of Remarriage. *Journal of Divorce & Remarriage,* 36(3–4), 21–39.

Papernow, P. (2013). *Surviving and Thriving in Stepfamily Relationships: What Works And What Doesn't.* New York: Routledge.

Park, S. & Lee, S. (2018). Heterogeneous Age-friendly Environments among Age-cohort Groups. *Sustainability,* 10, 1269.

Parkinson, P. (2018). Shared Physical Custody: What Can We Learn From Australian Law Reform? *Journal of Divorce & Remarriage,* 59(5), 401–413.

Parsons, T. (1942). Age and Sex in Social Structure. In R. L. Coser (Ed), *The Family: Its Structures and Functions.* New York: St. Martins.

Payne, K. K. (2014). Median Duration of First Marriage at Divorce and the Great Recession. *Family Profiles* (FP-14–20). Bowling Green, OH: NCFMR.

Payne, K. K. (2018). Change in the U.S. Remarriage Rate, 2008 & 2016. *Family Profiles,* FP-18–16. Bowling Green, OH: NCFMR.

Pearson, J. & Thoennes, N. (1990). Custody after Divorce: Demographic and Attitudinal Patterns. *Am J Orthopsychiatry,* 60(2), 233–249.

Perelli-Harris, B. & Gassen, N. (2012). How Similar Are Cohabitation and Marriage? Legal Approaches to Cohabitation across Western Europe. *Population and Development Review,* 38(3), 435–467. Retrieved from http://www.jstor.org/stable/41857400

Perrier, S. (1998). The Blended Family in Ancien Régime France: A Dynamic Family Form. *The History of the Family,* 3(4), 459–471.

Phillips, R. (1991). *Untying the Knot: Divorce in Western Society*. Cambridge: Cambridge University Press.

Poppel, F. V. (1995). Widows, Widowers and Remarriage in Nineteenth-Century Netherlands. *Population Studies*, 49(3), 421–441. Retrieved September 6, 2018 from http://www.jstor.org/stable/2175256

Raley, R. K. & Wildsmith, E. (2004). Cohabitation and Children's Family Instability. *Journal of Marriage and Family*, 66, 210–219.

Reblin, M. & Uchino, B. N. (2008). Social and Emotional Support and its Implication for Health. *Current Opinion in Psychiatry*, 21(2), 201–205.

Reeves, R. (2017). *Dream Hoarders: How the American Upper Middle Class Is Leaving Everyone Else in the Dust, Why That Is a Problem, and What to Do About It*. Washington, DC: Brookings Institution.

Reimondos, A., Evans, E. & Gray, E. (2011). Living Apart Together (LAT) Relationships in Australia. *Family Matters*, 87, 43–55.

Repond, G., Darwiche, J., Ghaziri, N. E. & Antonietti, J.-P. (2018). Coparenting in Stepfamilies: A Cluster Analysis. *Journal of Divorce & Remarriage*, doi: 10.1080/10502556.2018.1488121

Rindfuss, R. R., Guilkey, D. K., Morgan, S. P. & Kravdal, O. (2010). Child-Care Availability and Fertility in Norway. *Population and Development Review*, 36(4), 725–748.

Roberts, T. W. & Price, S. J. (1989). Adjustment in Remarriage. *Journal of Divorce*, 13(1), 17–43.

Rolan, E. & Marceau, K. (2018). Individual and Sibling Characteristics: Parental Differential Treatment and Adolescent Externalizing Behaviors. *Journal of Youth and Adolescence*, https://doi.org/10.1007/s10964-018-0892-8

Rosen, L. N. & O'Sullivan, C. S. (2005). Outcomes of Custody and Visitation Petitions When Fathers Are Restrained by Protection Orders: The Case of the New York Family Courts. *Violence against Women*, 11(8), 1054–1075.

Samuels, H. R. (1980). The Effect of an Older Sibling on Infant Locomotor Exploration of a New Environment. *Child Development*, 51, 607–609.

Sanner, C. & Coleman, M. (2017). (Re)constructing Family Images: Stepmotherhood before Biological Motherhood. *Journal of Marriage and Family*, 79, 1462–1477.

Sassler, S. & Miller, A. J. (2011). Class Differences in Cohabitation Processes. *Family Relations*, 60(2), 163–177.

Scarf, M. (2014). *The Remarriage Blueprint: How Remarried Couples and Their Families Succeed or Fail*. New York: Scribner.

Schaefer, C., Quesenberry, C. P. Jr. & Wi, S. (1995). Mortality Following Conjugal Bereavement and the Effects of a Shared Environment. *Am J Epidemiol.*, 141, 1142–1152.

Schlesinger, B. (1972). The Adjustment Process in Remarriages: A Canadian Experience. *Australian Social Work*, 25(3), 5–15.

Schlesinger, B. (1977). Husband–Wife Relationships in Reconstituted Families. *Social Science*, 52(3), 152–157. Retrieved from http://www.jstor.org/stable/41886175

Schmiege, C. J., Richards, L. N. & Zvonkovic, A. M. (2001). Remarriage, for Love or Money? *Journal of Divorce & Remarriage*, 36(1–2), 123–140.

Schneider, D. S., Sledge, P. A., Shuchter, S. R. & Zisook, S. (1996). Dating and Remarriage over the First Two Years of Widowhood. *Ann Clin Psychiatry*, 8(2), 51–57.

Schrodt, P. (2017). Relational Frames as Mediators of Everyday Talk: Relational Satisfaction in Stepparent–Stepchild. *Journal of Social and Personal Relationships*, 33(2), 217–236.

Scommegna, P. (February 2016). *Today's Research on Aging*. Population Reference Bureau. No. 33.

Shafer, K., James, S. L. & Larson, J. H. (2016). Relationship Self-regulation and Relationship Quality: The Moderating Influence of Gender. *J Child Fam Stud.*, 25, 1145.

Shehan, C. L. (2016). *The Wiley Blackwell Encyclopedia of Family Studies*, 4-Volume Set (Wiley Blackwell Encyclopedias in Social Sciences). Hoboken, NJ: John Wiley & Sons.

Sheppard, P. & Monden, C. (2018). The Additive Advantage of Having Educated Grandfathers for Children's Education: Evidence from a Cross-National Sample in Europe. *European Sociological Review*, 34(4), 365–380.

Sherman, C. W. (2012). Remarriage as Context for Dementia Caregiving: Implications of Positive Support and Negative Interactions for Caregiver Well-Being. *Research in Human Development*, 9(2), 165–182.

Siordia, C. (2014). Married Once, Twice, and Three or More Times: Data from the American Community Survey. *Journal of Divorce & Remarriage*, 55(3), 206–215.

Skulason, B., Jonsdottir, L. S., Sigurdardottir, V. & Helgason, A. R. (2012). Assessing Survival in Widowers, and Controls – A Nationwide, Six- to Nine-Year Follow-Up. *BMC Public Health*, 12, 96.

Spangler, A., Brown, S. L., Lin, I.-F., Hammersmith, A. & Wright, M. (2016). *Divorce Timing and Economic Well-being (FP-16-01)*. Bowling Green, OH: NCFMR.

Steinbach, A. & Hank, K. (2018). Full-, Half-, and Step-Sibling Relations in Young and Middle Adulthood. *Journal of Family Issues*, 39(9), 2639–2658.

Stepler, R. (February 18, 2016a). *Smaller Share of Women Ages 65 and Older Are Living Alone*. Washington, DC: Pew Research Center. Retrieved on December 7, 2018 from http://www.pewsocialtrends.org/2016/02/18/smaller-share-of-women-ages-65-and-older-are-living-alone/

Stepler, R. (April 21, 2016b). *World's Centenarian Population Projected to Grow Eightfold by 2050.* Washington, DC: Pew Research Center. Retrieved on December 6, 2018 from http://www.pewresearch.org/fact-tank/2016/04/21/worlds-centenarian-population-projected-to-grow-eightfold-by-2050/

Stewart, R. B. (1983). Sibling Attachment Relationships: Child–Infant Interactions in the Strange Situation. *Developmental Psychology,* 19, 192–199.

Stewart, S. D. (2005). Boundary Ambiguity in Stepfamilies. *Journal of Family Issues,* 26(7), 1002–1029.

Stewart, S. D. (2007). *Brave New Stepfamilies: Diverse Paths Toward Stepfamily Living.* Thousand Oaks, CA: Sage.

Stewart, S. D., Manning, W. D. & Smock, P. J. (2003). Union Formation among Men in the U.S.: Does Having Prior Children Matter? *Journal of Marriage and Family,* 65, 90–104.

Stoll, B. M., Genevieve, L. A., Fromme, D. K. & Felker-Thayer, J. A. (2006). Adolescents in Stepfamilies. *Journal of Divorce & Remarriage,* 44(1–2), 177–189.

Strohm, C. Q., Seltzer, J. A., Cochran, S. D. & Mays, V. M. (2009). "Living Apart Together" Relationships in the United States. *Demographic Research,* 21, 177–214.

Strow, C. W. & Strow, B. K. (2006). A History of Divorce and Remarriage in the United States. *Humanomics,* 22(4), 239–257.

Stykes, B. & Guzzo, K. B. (2015). Remarriage and Stepfamilies (FP-15–10). *Family Profiles.* Bowling Green, OH: NCFMR.

Suh, T., Schutz, C. G. & Johanson, C. E. (1996). Family Structure and Initiating Non-medical Drug Use among Adolescents. *Journal of Child & Adolescent Substance Abuse,* 5, 21–36.

Suitor, J. J. & Pillemer, K. (2000). Did Mom Really Love You Best? Exploring the Role of Within-family Differences in Developmental Histories on Parental Favoritism in Later Life Families. *Motivation and Emotion,* 24, 105–120.

Suitor, J. J., Pillemer, K. & Sechrist, J. (2006). Within-family Differences in Mothers' Support to Adult Children. *Journal of Gerontology: Social Science,* 61, S10–S17.

Sullivan, A. R. & Fenelon, A. (2014). Patterns of Widowhood Mortality. *The Journals of Gerontology Series B: Psychological Sciences and Social Sciences,* 69B(1), 53–62.

Sulloway, F. J. (1996). *Born to Rebel: Birth Order, Family Dynamics, and Creative Lives.* New York: Pantheon Books.

Svare, G. M., Jay, S. & Mason, M. A. (2004). Stepparents on Stepparenting. *Journal of Divorce & Remarriage,* 41(3–4), 81–97.

Sweeney, M. M. (1997). Remarriage of Women and Men after Divorce: The Role of Socioeconomic Prospects. *Journal of Family Issues,* 18, 479–502.

Sweeney, M. M. (2002). Remarriage and the Nature of Divorce: Does it Matter which Spouse Chose to Leave? *Journal of Family Issues,* 23(3), 410–440.

Sweeney, M. M. (2010). Remarriage and Stepfamilies: Strategic Sites for Family Scholarship in the 21st Century. *Journal of Marriage and Family,* 72(3), 667–684.

Tabloski, P. A. (2013). *Gerontological Nursing,* 3rd edn. Upper Saddle River, NJ: Prentice Hall.

Tach, L., Edin, K., Harvey, H. & Bryan, B. (2014). The Family-Go-Round: Family Complexity and Father Involvement from a Father's Perspective. *The ANNALS of the American Academy of Political and Social Science,* 654(1), 169–184.

Teachman, J. (2003). Childhood Living Arrangements and the Formation of Coresidential Unions. *Journal of Marriage and Family,* 65, 507–524.

Teachman, J. D. & Paasch, K. M. (1994). Financial Impact of Divorce on Children and their Families. *Future Child,* 4(1), 63–83.

Thomson, E., Lappegard, T., Carlson, M., Evans, A. & Gray, E. (2014). Childbearing across Partnerships in Australia, the United States, Norway, and Sweden. *Demography,* 51(2), 485–508.

Tucker, C. J., McHale, S. M. & Crouter, A. C. (2003). Dimensions of Mothers' and Fathers' Differential Treatment of Siblings: Links with Adolescents' Sex-typed Personal Qualities. *Family Relations,* 52, 82–89.

Turcotte, M. (2013). Living Apart Together. Ottawa, Ontario: Insights on Canadian Society, 75–006-X. Retrieved on December 6, 2018 from https://www150.statcan.gc.ca/n1/pub/75-006-x/2013001/article/11771-eng.htm

Turner, L. H. & West, R. (2010). *Introducing Communication Theory,* 4th edn. New York: McGraw Hill.

UNDESA (UN Department of Economic and Social Affairs) (June 2017). *Population Facts.*

U.S. Bureau of the Census (2005). *Statistical Abstracts of the United States: 2000.* Washington, DC: U.S. Bureau of the Census. Retrieved on December 6, 2018 from https://www.census.gov/library/publications/2000/compendia/statab/120ed.html

U.S. Census Bureau (2008). *Statistical Abstract of the United States, 2007.* Washington, DC: Department of the Treasury.

U.S. Census Bureau (April 20, 2016a). *FFF: Mother's Day: May 8, 2016.* Retrieved on July 22, 2016 from http://www.census.gov/newsroom/facts-for-features/2016/cb16-ff09.html

U.S. Census Bureau (June 23, 2016b). *Sumter County, Fla., is Nation's Oldest, Census Bureau Reports.* Retrieved on December 7, 2018 from https://www.census.gov/newsroom/press-releases/2016/cb16-107.html

U.S. Census Bureau (January 23, 2017a). *Valentine's Day 2017: Feb 14.* Retrieved on November 9, 2017 from https://www.census.gov/newsroom/facts-for-features/2017/cb17-ff02.html

U.S. Census Bureau (April 10, 2017b). *Facts for Features: Older Americans Month: May 2017.* Retrieved on August 15, 2018 from https://www.census.gov/newsroom/facts-for-features/2017/cb17-ff08.html

U.S. Census Bureau (June 6, 2018a). *FFF: Anniversary of Americans with Disabilities Act: July 26.* Retrieved on December 7, 2018 from https://www.census.gov/newsroom/facts-for-features/2018/disabilities.html

U.S. Census Bureau (May 2018b). *Older Americans Month: May 2018. Table A1.* Retrieved on September 2, 2018 from https://www.census.gov/newsroom/stories/2018/older-americans.html

U.S. Census Bureau (September 9, 2018c). *FFF: National Grandparents Day: Sept. 09. 2018.* Retrieved on December 7, 2018 from https://www.census.gov/newsroom/stories/2018/grandparents.html

Van Eeden-Moorefield, B., Pasley, K., Dola, E. M. & Engel, M. (2007). From Divorce to Remarriage. *Journal of Divorce & Remarriage,* 47(3–4), 21–42.

Vespa, J. (2012). Union Formation in Later Life: Economic Determinants of Cohabitation and Remarriage among Older Adults. *Demography,* 49(3), 1103–1125. https://doi.org/10.1007/s13524-012-0102-3

Vikat, A., Thomson, E. & Hoem, J. M. (1999). Stepfamily Fertility in Contemporary Sweden: The Impact of Childbearing before the Current Union. *Population Studies,* 53(2), 211–225.

Vinick, B. H. (1998). *Older Stepfamilies: Views from the Parental Generation.* Washington, DC: AARP-Andrus Foundation.

Visher, E. B. (1994). Lessons from Remarriage Families. *The American Journal of Family Therapy,* 22(4), 327–336.

Vivona, J. M. (2007). Sibling Differentiation, Identity Development, and the Lateral Dimension of Psychic Life. *Journal of the American Psychoanalytic Association,* 55(4), 1191–1215.

Watkins, N. K. & Waldron, M. (2017). Timing of Remarriage among Divorced and Widowed Parents. *Journal of Divorce & Remarriage,* 58(4), 244–262.

Watson, W. K. & Stelle, C. (2011). Dating For Older Women: Experiences and Meanings of Dating in Later Life. *J Women Aging,* 23(3), 263–275.

Weaver, S. E. & Coleman, M. (2005). A Mothering but Not a Mother Role: A Grounded Theory Study of the Nonresidential Stepmother Role. *Journal of Social and Personal Relationships,* 22, 477–497.

Weaver, S. E. & Coleman, M. (2010). Caught in the Middle: Mothers in Stepfamilies. *Journal of Social and Personal Relationships,* 27(3), 305–326.

Werner, K. H., Goldin, P. R., Ball, T. M., Heimberg, R. G. & Gross, J. J. (2011). Assessing Emotion Regulation in Social Anxiety Disorder: The Emotion Regulation Interview. *J Psychopathol Behav Assess.,* 33, 346–354.

White, L. (1994). Growing Up with Single Parents and Stepparents: Long-term Effects on Family Solidarity. *Journal of Marriage and the Family,* 56, 935–948.

White, L. (1998). Who's Counting? Quasi-facts and Stepfamilies in Reports of Number of Siblings. *Journal of Marriage and the Family,* 60, 725–733.

White, L. (2001). Sibling Relationships over the Life Course: A Panel Analysis. *Journal of Marriage and the Family,* 63, 555–568.

Whiteman, S. D. & Christiansen, A. (2008). Processes of Sibling Influence in Adolescence: Individual and Family Correlates. *Family Relations,* 57, 24–34.

Whiteman, S. D., McHale, S. M. & Soli, A. (2011). Theoretical Perspectives on Sibling Relationships. *Journal of Family Theory & Review,* 3(2), 124–139.

Whiting, J. B., Smith, D. R., Bamett, T. & Grafsky, E. L. (2007). Overcoming the Cinderella Myth. *Journal of Divorce & Remarriage,* 47(1–2), 95–109.

WHO (World Health Organization) (2009). *Global Health Observatory.* Retrieved on August 10, 2018 from http://apps.who.int/ghodata, accessed 28 July 2010.

Wiik, K. A. & Bernhardt, E. (2017). Cohabiting and Married Individuals' Relations with Their Partner's Parents. *Journal of Marriage and Family,* 79(4), 1111–1124.

Wilcox, B. W. & Lerman, R. I. (October 28, 2014). *For Richer, For Poorer: How Family Structures Economic Success in America.* Washington, DC: AEI and Charlottesville, Virginia: Institute for Family Studies.

Wilder, S. E. (2012). A Comparative Examination of Reasons for and Uses of Uncertainty and Topic Avoidance in First and Remarriage Relationships. *Journal of Divorce & Remarriage,* 53(4), 292–310.

Willführ, K. P. & Gagnon, A. (2013). Are Stepparents Always Evil? Parental Death, Remarriage, and Child Survival in Demographically Saturated Krummhörn (1720–1859) and Expanding Québec (1670–1750). *Biodemography and Social Biology,* 59(2), 191–211.

Wilson, W. J. (1987). *The Truly Disadvantaged: The Inner City, The Underclass, and Public Policy.* Chicago, IL: University of Chicago Press.

Wong, E. (July 2, 2013). A Chinese Virtue Is Now the Law. *Asian Pacific.* Retrieved on September 3, 2018 from https://www.nytimes.com/2013/07/03/world/

asia/filial-piety-once-a-virtue-in-china-is-now-the-law.html?_r=0

Wood, R. G. (1995). Marriage Rates and Marriageable Men: A Test of the Wilson Hypothesis. *Journal of Human Resources,* XXX, 163–193.

Wu, H. (2018a). Grandparents' Characteristics by Age. *Family Profiles.* FP-18–04. Bowling Green, OH: NCFMR.

Wu, H. (2018b). Prevalence of Grandparenthood in the U.S. *Family Profiles,* FP-18–03. Bowling Green, OH: NCFMR.

Wu, H. (2018c). Grandchildren Living in a Grandparent-Headed Household. *Family Profiles,* FP-18–01. Bowling Green, OH: National Center for Family & Marriage Research. Retrieved on August 12, 2018 from https://www.bgsu.edu/ncfmr/resources/data/family-profiles/wu-grandchildren-living-with-grandparent-hhh-fp-18-01.html

Wu, Z. & Schimmele, C. M. (2005). Repartnering after First Union Disruption. *Journal of Marriage and Family,* 67(1), 27–36.

Yarber, A. & Sharp, P. M. (2010). *Focus on Single-Parent Families: Past, Present, and Future.* Westport, CT: Praeger.

INDEX

Note: Page numbers in *italics* refer to information in figures.